50th Annual Loss Prevention Symposium 2016 (LPS)

Topical Conference at the 2016 AIChE Spring Meeting and 12th Global Congress on Process Safety

Houston, Texas, USA
10-14 April 2016

Volume 1 of 2

ISBN: 978-1-5108-2477-5

Printed from e-media with permission by:

Curran Associates, Inc.
57 Morehouse Lane
Red Hook, NY 12571

Some format issues inherent in the e-media version may also appear in this print version.

Printed by Curran Associates, Inc. (2016)

For permission requests, please contact AIChE
at the address below.

AIChE
120 Wall Street, FL 23
New York, NY 10005-4020

Phone: (800) 242-4363
Fax: (203) 775-5177

www.aiche.org

Additional copies of this publication are available from:

Curran Associates, Inc.
57 Morehouse Lane
Red Hook, NY 12571 USA
Phone: 845-758-0400
Fax: 845-758-2633
Email: curran@proceedings.com
Web: www.proceedings.com

TABLE OF CONTENTS

VOLUME 1

VOLUME 2

A Lesson Learned By Attending Thirty Seven Years of Loss Prevention Symposia

Thomas H. Pratt, PhD, PE
175 Weatherstone Parkway
Marietta, GA 30068

thomaspratt@att.net

Prepared for Presentation at
American Institute of Chemical Engineers
2016 Spring Meeting
12th Global Congress on Process Safety
Houston, Texas
April 11-23, 2016

A Lesson Learned By Attending Thirty Seven Years of Loss Prevention Symposia

Thomas H. Pratt, PhD, PE
175 Weatherstone Parkway
Marietta, GA 30068

thomaspratt@att.net

Keywords: Networking

Abstract

In responding to the request to give a paper at the 50th Loss Prevention Symposium, this paper gives a brief synopsis of my career, and how attending the annual LPS meetings over the last 37 years has tremendously supported my professional work. It relates how a theoretical physical chemist was transformed into a consultant having an expertise in electrostatic ignitions of fires and explosions. Of all the reasons one could have for attending an LPS, networking is singled out, emphasized and supported by examples.

1 Introduction

At last year's GCPS Ron Willey approached me at the break and asked me if I would give a talk for the 50th Loss Prevention Symposium to be held at the 2016 GCPS. He said that I could talk about "anything you want to". I told him that I would before I really thought about it – so here I am writing a paper and preparing a presentation. My first thought was that I am not even a chemical engineer; I spent 23 years in the classroom (interrupted by four years in the Navy) and graduated with a PhD in theoretical physical chemistry. What do I know that would interest a bunch of guys who work with things like pump curves, heat transfer, unit processes? I suppose I could relate some of the interesting or humorous things which have happened over the years, but the attendees do not come to a symposium just to hear a bunch of old jokes; they come to learn something. After reaching this profound conclusion I decided to put on my academic hat and present on one of the benefits I have derived from attending the LPS symposia over the years. This sent me searching for the one thing that I can draw from my experience to pass on to the safety professionals of the future. I selected *NETWORKING.* In order to emphasize how networking can be a tremendous benefit, I will give an overview of my career to show how networking has worked for me. In doing this, I am going to use one of my many projects as an example and use it as a scaffold to tie things together. Hopefully, this will give some idea of how the AIChE LPS helped turn me from being an academic to someone who doesn't mind getting his hands dirty.

2 Beginnings

My career began in 1959 as a rocket scientist with Rohm and Haas Co., Redstone Arsenal which had a contract with the U. S. Army to do basic research in solid propellants, which are themselves the next thing to high explosives. Propellants burn slowly while high explosives give up their energy all at once. A few weeks before I showed up to work they had a 1,400 lb Nike Zeus sustainer detonate on the test stand instead of burning slowly like it should have. Needless to say, the whole of the U.S. Army was interested in the reasons why it detonated instead if burning normally and thus my career started with the explosive properties of solid propellants and propellant ingredients. I soon became active in the Armed Services Explosive Safety Board where I presented some of my results and sat on various committees. During those first few years, working with the ASESB, I did not realize that I was building a network of people who were doing the same kind of work as myself, I attended many meetings, and rather than actively networking at them I would hang out with my buddies, much to the displeasure of our boss. I recall the time he caught some six or eight of us having dinner by ourselves in the midweek of a big meeting.

However, I did manage to tie up with several people who came to know me and what I did. One particular person was R.H. Richardson who was with Hercules Incorporated. Our work at Redstone was primarily with composite propellants while Hercules was big in double base propellants, and thus Richardson and I were on the other side of the industry – more like rivals than competitors. As things evolved we both would attend the same meetings to give presentations, and I enjoyed popping up after he gave a paper to ask questions and bring out some of his soft spots. Likewise he would retaliate, in spades, when I gave papers; so it became an enjoyable sport. We became good friends.

When the government began to cut back on our contracts, the Rohm and Haas Company decided to shut down the Redstone Division. To cut a long story short, I was out of a job and looking for some sort of position in explosives or propellant manufacture. This is perhaps the first time I realized the value of having a network of people to call upon for help. I called my ASESB acquaintances to no avail because everyone was cutting back at that time; but I did hit pay dirt when I contacted Richardson, and I was able to join Hercules in 1971.

3 Some Serendipity

The next, and most significant, turn of my career came about in 1972. On March 27, 1972 the Amcel plant of American Celanese, which was just down the road, had an explosion which killed an operator and injured others. Celanese was looking for help and asked us to help them uncover the particulars of how and why the explosion happened; after all we were the ones who knew all about explosions and could get to the bottom of things. The problem was that none of us had any significant hands-on experience with vapor phase explosions – rockets run on liquids or solids. Of course we did not point this out, and five of us responded by going to the site where we were introduced to George Fisk, the plant manager, Jim Neely, the department manager, and Joe Gay, a safety manager sent in from the home office.

The explosion had occurred near the end of emptying a batch of rayon dope being pushed out with air at 45 psig and 80° F from a 10,000 gal. paddle mixer. The dope was a viscous mix of cellulose acetate dissolved in a 94/6 mix of methylene chloride/methanol. The explosion had blown the mixer apart at the waist by ripping apart a bolt circle of 144 three quarter inch studs, The operator had been standing by the mixer and was killed by the hot gasses escaping as the top parted from the base.

On the first day of the investigation, one of our group postulated that "it certainly was not electrostatics because it was inside of a Faraday cage"; a supposition frequently made by inexperienced investigators. There were many other suppositions made that day while each of us were getting started in this new arena; and as our investigation proceeded, I was somehow designated to ride point.

Early on, I decided that it would be nice to reckon something about the maximum pressure of the explosion. In my naiveté, I found the yield strength of a stud and multiplied it by 144 and got some outrageous number. (As if all the studs had pulled apart simultaneously!) Fortunately, I had been working closely with Joe Gay and when I casually mentioned this calculation, he came unglued and took me aside and gave me the first of many "you dumb ass" tutorials. This one of course is the rule of thumb that the maximum pressure in vapor phase explosions in air it is usually 8 to 10 times the initial pressure. This was what we used in this case to estimate the pressure. (I am still not sure this rule of thumb is good for methylene chloride; e.g., phosgene is one of the products of combustion etc.)

The project progressed in an orderly manner and I wrote a well received report which described in detail how some dope had dried on the mixer blade and was subsequently slung onto the walls of the mixer causing a spark which ignited the vapors. So the cause of the explosion was indeed electrostatics where the charges were generated, accumulated, and discharged entirely within a Faraday cage; i.e., the mixer. Lesson learned: Stand around with your hands in your pockets, ask a lot of questions until you know what is going on, and keep your mouth shut until you are ready to write the report.

4 A Change in Direction

At this point R. Richardson put two and two together; (1) Doing projects for industry could be a source of profit by doing something we already knew something about, and (2) OSHA was just beginning to create requirements for the chemical industry to do things that we were already doing. So decisions were made to expand our efforts into the chemical/petroleum industry and the Hazards Evaluation and Risk Control, HERC®, unit was born. I was to do the marketing and bring in the alligators and the group was to skin them. We had no problem hitting the ground running but there was also a lot of on-the-job training. We had good people, so we were successful.

Since I was doing the marketing and recognized that we had to have visibility in the chemical/petroleum industry, we began to give papers at various selected meetings and symposia. I selected AIChE where I gave a paper [1] in 1974 at the AIChE National Meeting, and in 1978 I gave another paper at the Loss Prevention Symposium [2], where I found a home. I haven't kept track, but I think I have attended all but one of them since 1978.

In the beginning the HERC® group touted all the forms of process hazards analysis, anything the customer wanted. In responding to explosion incidents, we gravitated into the legal arena where we acted as expert witnesses for our industrial clients, which we found to be rather lucrative. All the while, networking was on the front burner because we wanted the telephone to keep ringing. One of our better networking activities was by giving two-day seminars on various topics. Among them were *Explosions, Hazard Identification Techniques,* and *Electrostatics.*

5 Another Change in Direction

As in any such group, there were constant reorganizations and changes of job descriptions, and to cut another long story short I wound up as Division Manager. I did not particularly like this position because I still wanted to get out and crawl through things in the field rather than push paper around the office. The money was good but I could see that I had another gold mine in the network of people who called me directly for my services. In 1989 I left Hercules and joined Burgoyne Incorporated as a consultant, essentially working on my own.

In chasing after fires and explosions over the years I have noticed that most of the projects involve electrostatics, because if it is some mundane ignition source (e.g., a welder) the client can sort it out by himself. (I start my *Electrostatics* seminar by saying "electrostatics is what you blame it on when you can't find anything else".) Writing a book is not exactly networking but it is a good way to get your name and expertise out there, so I took my seminar handouts and compiled them into a book [3]. Thus, I have established myself as an electrostatics guru and as a result have pretty well limited my consultations to electrostatics. Furthermore, Burgoyne Inc. has shut the Atlanta office so now I am truly on my own.

6 Benefits of Attending LPS/GCPS

It is obvious; if you are going to establish a presence in loss prevention in the chemical industry, the best way to do that is to go to where the professionals in loss prevention gather, and the best place to do that is at AIChE meetings, and the best activity to attend is the Global Congress on Process Safety. When listening to the papers I always made a mental note of the speaker and the chair of the session so that I could talk to them during the breaks. The people you meet and talk to at the breaks, bars and dinners are the ones who will remember you.

In the old days the LPS was held in conjunction with the spring national meeting and we did not have the stickums on the badges, but now-a-days things have evolved into the GCPS where the stickums flag the people you need to know, especially the yellow *Fellow* stickums. AIChE has done an excellent job of making the meetings worthwhile by making it easy to find out someone's name and affiliation. (I wish they would put the names on both sides so when they get turned around you could still read them.)

Then, after you establish meaningful acquaintances, don't let them die on the vine – keep coming to GCPS and keep them up to date about what you are doing and where you are headed. You will find that people will tell you things at the GCPS that they don't "talk about" except in a one-on-one conversation. This is particularly true even when talking to competitors; you might have to ask them for a job one of these days.

Also, keep up with new technology as it evolves. As an example, analytical methods come and go; such as fault trees which were all the rage several years ago for analyzing most everything, but experience has shown that they are now applicable only in a few specific cases. The presentations will point out the squirrelly ones which fall by the wayside and the good ones which flourish; and you can then keep up with other people's experience to point you in the right direction.

All of us have dealt with OSHA over the years in one way or another. I have often heard the complaint that OSHA, et.al., were "coming out with regulations faster than we can keep up", but nowadays it seems as if they have ascended up their own learning curve and morphed into a more reasonable agency. They often give papers at GCPS with their latest nuance and direction, and are always willing to discuss issues if you catch them at the breaks. Also, talking with others who have dealt with them is a good way to orient yourself with regulatory problems.

7 Network Example

A typical example of how a network can work is my relationship with Celanese Corporation that began with my first encounter with an industrial explosion, described above. (As one of my bosses used to say "a 'typical example' is the best one you've got.")

- In 1972 I worked with Jim Neely on a follow-on contract. (what would now be called a PHA)
- In 1973 the HERC® group reviewed alternate dressings for plate and frame filters, for Jim Neely. (potential electrostatics)

- In about 1974 Joe Gay sent HERC® a project to determine what occurred when a bucket of toluene ignited. (electrostatics)
- In about 1977 I received a call from Herb Kolodner, who was in corporate safety in Charlotte NC, to go to Rock Hill, SC and sort out a still explosion. (I called Joe Gay and he found that the same thing had happened years ago at another Celanese plant. To quote Trevor Kletz "corporations have no memory")
- In about 1978 I worked with Herb Kolodner on a cool flame problem [4]
- In the '80s I helped Joe Gay, who had moved to Champlin Petroleum, with several problems, some pro bono and some contracted.
- In about 1982 I heard from Jim Neely, who was then with Sun Chemical, to sort out an explosion while drying pigments. (electrostatics in a bag house)
- In 2000 Jim Neely found me on the Internet (In 1989 I left Hercules and joined Burgoyne) and engaged me to attend the gathering of experts responding to the explosion at the Lomac facility in Muskegon MI. (Nitration in herbicide manufacture)

8 Conclusions

- Become active in a committee which aligns with your work, whether it be with AIChE, or some other professional organization which may be a better fit.
- Don't just come to the meetings to listen to the papers, use all the down time to network.
- Make it a point to talk with your contacts when the opportunity arises

9 References

[1] T. H. Pratt, "A Technique for Analysis of Hazards in a Novel Process", AIChE National Meeting, Tulsa, OK, March 10-13, 1974.

[2] L. R. Albaugh, G. M. Williams, J. Katz and T. H. Pratt, "Hazards Analysis, A Practical Loss Prevention Tool", AIChE Loss Prevention Symposium, Atlanta, GA, February 1978

[3] T.H. Pratt, *Electrostatic Ignitions of Fires and Explosions*, Center for Chemical Process Safety of the American Institute of Chemical Engineers, 2000, 182 pages, ISBN 0-8169-9948-1.

Surviving the Black Swan, Strategies for Process Safety Specialists and Companies to Survive Unpredicted Catastrophic Events

John F. Murphy, PE
Process Safety Services
2304 Kenya Lane
Punta Gorda, FL 33983
hamjfm@embarqmail.com

Prepared for Presentation at
American Institute of Chemical Engineers
2016 Spring Meeting
12th Global Congress on Process Safety
Houston, Texas
April 11-23, 2016

Surviving the Black Swan, Strategies for Process Safety Specialists and Companies to Survive Unpredicted Catastrophic Events

John F. Murphy, PE
Process Safety Services
2304 Kenya Lane
Punta Gorda, FL 33983
hamjfm@embarqmail.com

Keywords: black swan, lessons learned, survival, incident investigation

Abstract

This article is a continuation of three articles that I have authored or co-authored on the "black swan", the rare and unpredictable catastrophic event. The first two papers discussed the concept of the black swan as applied to the chemical and petroleum processing industries and the difficulty in predicting such events. The third paper discussed the fact that we have difficulty maintaining the lessons learned from black swan events that have occurred. I have survived and witnessed several black swans. Each has affected me personally and professionally. Black swan events are rare but inevitable. Some individuals and companies survive black swan events and continue on while others do not. I have learned that many black swan events, although initially viewed as catastrophes, can result in opportunities from a process safety improvement, career and personal perspective. This paper presents examples of black swan events that have impacted my career in process safety, often in a positive way, and strategies for surviving these "black swan" events from a personal, professional and a company point of view.

1. Introduction

As I discussed in previous papers, the concept of the black swan is not new. The Greek philosopher Aristotle first used the term white swan as an example of necessary relations and the term black swan as the expression of the improbable.[1] In his 2010 book entitled, *The Black Swan, The Impact of the Highly Improbable*, Nassim Taleb discusses the extreme impact of rare and unpredictable events (black swans) and the human tendency to find simplistic explanations of these events after they occur. According to Taleb, the black swan event has three attributes: "First is an outlier, as it lies outside the realm of regular expectations. Second, it carries an extreme impact (unlike the bird). Third, in spite of its outlier status, human nature leads us to concoct explanations after the fact, making the previously unimaginable event to be both explainable and predictable."[2]

When I use the term "black swan", I agree with Taleb's definition that black swans are rare, catastrophic, and unpredictable events. After the black swan event occurs, explanations of how the event occurred are developed. The fact that the black swan event was not predicted does not mean that the event could not have been prevented. The steps necessary to prevent future similar events become clear after root cause analysis discovers the sequence of events that lead to the black swan event. However, before the event the conventional hazard identification and risk assessment analysis does not predict the sequence of events that led to the catastrophe. Or, the catastrophic event scenario is identified but not thought to be possible. If you have identified a catastrophic scenario during a hazard identification and risk assessment analysis and dismissed the scenario as not credible, you may have let the black swan escape. There is great uncertainty in our ability to predict the likelihood of a catastrophic process safety event.

Those of you who are interested in the black swan concept can view my AIChE webinar entitled *Black Swans in the Chemical Processing Industries* presented March 16, 2016 by logging on to the AIChE continuing education website.

The 50 years of the Loss Preventions Symposium that we are celebrating coincides with my career in process safety. This paper will discuss black swan events that have happened at Dow Chemical and in the chemical industry while I worked there and other black swan events that affected me personally after my career at Dow. I will discuss how we (myself, Dow and the industry) have survived these catastrophic events and even turned many into opportunities. From this experience, strategies for surviving black swan events will be proposed.

2. My Process Safety Career in a Nut Shell

It is important to have a general idea of what has been my career in process safety. I have over 47 years of experience in chemical manufacturing and process safety with jobs in the chemical industry, consulting, and government. I am currently an independent consultant, primarily in a retirement mode, and previously worked as a lead investigator for the U. S. Chemical Safety and Hazard Investigation Board and as a leader in process safety for the Dow Chemical Company. I am a Fellow of the American Institute of Chemical Engineers where I have been a leader in the Safety and Health Division and recipient of the AIChE Safety and Health Division's 2008 Walton/Miller Award for outstanding chemical engineering contributions and achievements in loss prevention, safety, and health. I am presently Co-editor of *Process Safety Progress,* the Division's magazine. I am a Fellow of the Center for Chemical Process Safety (CCPS) and an Emeritus member and Staff Consultant of the CCPS Technical Steering Committee. Among the many CCPS projects I have been involved in, perhaps the most significant is Process Safety Boot Camp, a four-day course on risked based process safety and process safety technology. I am one of the course developers and instructors. I have facilitated courses around the world. I am professional engineer in the State of Texas and graduated with a B. S. in Chemical Engineering from Tufts University in 1968 and an M. B. A. in management from Central Michigan University in 1974.

3. Black Swan number one: The Dow Chemical Company, Freeport, TX, December 12, 1966

The first black swan that affected my career occurred at the Dow Chemical Company prior to my arrival in 1968. I mention this incident because of its impact on the process safety performance of the Dow Chemical Company and my career in process safety.

I have labelled this as a black swan event because it was not predicted by Dow because they did not have a management system in place at the time to ensure that the hazards of new chemistry and modified chemistry were identified before implementation. As story tells us, the event resulted in a needed management system to identify and control reactive chemical hazards, an opportunity to improve future process safety performance at Dow.

On December 12, 1966 at the Dow plant in Freeport, Texas, propargyl bromide and chloropicrin were being blended with a mixture of dichloropropane and dichloropropene in a tank car when one of the tank cars exploded, causing the other tank car to explode. The combined explosions left a large crater, 87 ft. long by 60 ft. wide by 11 ft. deep. The blast caused serious damage to a wooden wall building and a concrete building. Two tank cars, one tank truck and a length of railroad were destroyed. There were three fatalities; two operators and a truck driver.

The energy release of the explosion was estimated as the equivalent of between 6.6 and 14 tons of TNT. Much of the evidence was destroyed, but it is thought that a valve closed causing a gear pump to run dry igniting a mixture of chloropicrin and propargyl bromide. The immediate cause of the accident was the mixing of chemicals in equipment not designed for this process and probably a pump running hot and dry. A contributing cause of the incident was a lack of understanding of the chemistry involved by the people supervising and doing the work.[5]

I asked Kevin Shaughnessy, Dow's Chief Process Safety Engineer to comment on this event on December 22, 2015. He pointed out that the incident occurred 49 years ago from the time he sent his email. Dow has had several reactive chemical events that have led to fatalities since the birth of the reactive chemical program but only one in the past 15 years. Kevin explained: "Dow developed its Reactive Chemical Program in 1966 following an incident just three days before Christmas, where organic chemicals were being mixed at Glycerine I in Freeport using tank cars and a tank truck. The shock sensitivity of these materials had not been appreciated and resulted in an explosion resulting in three fatalities along with three large craters.

Figure 1: Dow Chemical Reactive Chemical Incident, December 22, 1966 (used with permission of the Dow Chemical Company)

Dow's Reactive Chemical Program protocols continue to evolve but it is a foundation of our Risk Management program. We believe strongly in the value of an extensive testing and reactive chemicals evaluation program; to examine the desired reaction and understand potential situations that can cause uncontrolled reactions (such as from process deviations in materials or operating conditions)."[6]

Why did I choose this as an important black swan in my career? It was certainly catastrophic, and at the time not predicted. The story has been told again and again at Dow in an attempt to keep the lessons learned alive. Dow survived this black swan because it determined that a route cause of the incident was a lack of a management system that required that all new chemistry and changes to chemistry be reviewed for reactive hazards and appropriate safeguards be implemented to prevent incidents. Dow's Board of Directors at the time established the Reactive Chemical Program which addressed this issue to prevent future incidents.[5] The program is part of the Dow Chemical's process safety program and continues to be supported by top management demonstrating Dow's commitment to a strong positive process safety culture. Dow's business model at the time required a commitment to research and development. I believe the implementation of the reactive chemical program at Dow was necessary for the survival of Dow's business model that included this commitment to research and development.

How did this event affect my career? Essentially, it helped to structure my career path within Dow as I was assigned to manage reactive hazard's analysis at various levels. After twelve years in chemical manufacturing at Dow, in 1980 I became a loss prevention engineer in Dow's Michigan Division. As a part of my job I became a member of the site reactive chemicals committee that reviewed all new chemistry and changes to chemistry so that reactive chemical incidents can be prevented. In 1988, I was promoted to Process Safety Manager at the Dow Freeport site and became the chair of the site's reactive chemical committee. In 2000 I joined the U. S. Chemical Safety Board as an investigator. My expertise in reactive chemicals resulted in me leading the Reactive Chemical Hazard Investigation. This investigation reviewed major reactive chemical incidents in the United States over a twenty-year period (1980 to 2000) and made recommendations to industry, government and others to prevent future reactive chemical incidents.

How did Dow survive this black swan? The company had a strong positive safety culture that drove Dow management to understand why the event had happened and develop a strong management system to prevent future reactive chemical incidents.

4. Black Swan number two: Nypro (UK) Ltd. chemical plant in Flixborough, UK, June 1, 1974

I was working in manufacturing at the Dow Michigan Division at the time of this incident. This black swan had a major effect on Dow, the chemical industry and my career in process safety.

I labelled this as a black swan because before this event, the chemical industry had not recognized the devastating effects of a vapor cloud explosion and its impact on occupied buildings.

On the afternoon of Saturday, June 1, 1974 a catastrophic vapor cloud explosion occurred at the Nypro (UK) Ltd. chemical plant in Flixborough, UK. Twenty-eight employees were killed and 36 injured, and the plant was totally destroyed. There were an additional 56 injuries in the local community coupled with significant damage to homes and businesses. After a thorough investigation by the government, the technical cause of the incident was determined to be the failure of an inadequately engineered bellows expansion joint and the release of approximately 30 tons of cyclohexene, a flammable vapor that subsequently ignited resulting in a vapor cloud explosion. As with most major incidents there were a number of significant root causes (failure of several management systems including hazard evaluation, management of change, and asset integrity) and many lessons learned.[7]

Of the 28 people that were killed, 18 were in the control room. One of the lesson-learned was that control rooms and other occupied buildings must be kept away from major hazards (such as large quantities of flammable materials) and where this is not feasible, such buildings must be constructed such that if an explosion does occur, the building will withstand the blast and not put people at risk who are inside the building.[4]

Why did I choose this black swan as significant to my career? This incident had a major effect on the Dow Loss Prevention program. Dow's response to Flixborough was to develop engineering guidelines (called Loss Prevention Principles), that offered calculation methods to estimate the consequences of vapor cloud explosions and criteria for designing control rooms and other occupied buildings that could be affected by such explosions. Separation distances were also recommended from process hazards to control rooms and other occupied buildings. Other major chemical companies took similar action. The chemical process industry recognized and learn the lesson of occupied buildings in close proximity to hazardous processes from the Flixborough incident, turning the Black Swan of the hazards of occupied buildings into a White Swan. However, over time, for a variety of reasons, the white swan turned gray for some companies in the chemical process industries and similar incidents occurred.

The plant in Flixborough was never rebuilt, but Dow and other chemicals companies strengthen their Loss Prevention programs to prevent major chemical plant accidents like this from happening again in their facilities. I remember attending a loss prevention training session at Dow where this incident was discussed. The loss prevention training peaked my interest in process safety. In 1980 I became a loss prevention engineer at the Dow Midland Division and I have been involved in process safety ever since.

How did Dow and the chemical industry survive this black swan? Because of a strong process safety culture, Dow and other progressive companies in the chemical and petroleum industry recognized the need to strengthen process safety management systems to prevent similar incidents from occurring.

5. **Black swan number three: Union Carbide SEVIN Manufacturing plant, Bhopal India, December 3, 1984**

I was working in loss prevention at the Dow Michigan Division at the time of this incident. This black swan had a major effect on Union Carbide, Dow Chemical, the chemical industry, the regulatory process worldwide, and my career in process safety.

I labelled this event as a black swan because prior to this incident, a massive release of acutely toxic material that could result in such a catastrophe was also not thought possible. After this event, the chemical industry embraced the concept of inherently safer design and the elimination or minimization of hazardous material became a primary strategy for the design and operation of chemical plants.

On December 3, 1984, in Bhopal, India, the accidental release of 40 metric tons of methyl isocyanate resulted in over 2,000 fatalities, 100,000 injuries, and significant damage to livestock and vegetation.[7] Much has been written about this incident, and many important details are still in dispute and it will never be known for certain exactly what happened, so I leave it to the reader to read the references and research the literature including the proceedings of the Global Congress on Process Safety to learn more. The December 2014 edition of *Process Safety Progress* features the anniversary of the Bhopal event and has an article with interesting insights of CCPS members that visited the Bhopal site ten years ago.

Why did I choose to discuss this black swan? Because the incident had major effects on Union Carbide, the chemical industry, including Dow Chemical, and me personally.

This event resulted in 17 chemical and petroleum companies creating the AIChE Center for Chemical Process Safety (CCPS) in 1985. The purpose of this new organization was to share best practice in process safety throughout the chemical industry for the purpose of preventing catastrophic process safety incidents like Bhopal. This past year marks the 30th anniversary of CCPS. The June 2015 edition of *Process Safety Progress* has an article that discusses the past and future of CCPS written by past and present CCPS leaders.[9] Today CCPS has approximately 200 member companies from around the world. It has produced over 100 books and products and conducts process safety conferences internationally including the Global Congress on Process Safety. CCPS continues to advance process safety technology and knowledge with these efforts.

The Dow Chemical Company was one of the 17 charter members of CCPS. Don Rickard, the Vice President of Manufacturing and Engineering and member of Dow's Board of Directors at the time, was a member of the original CCPS Advisory Board and Bob Smith, Corporate Director of Safety and Services, was a member of the original Technical Steering Committee.[9] My first involvement with CCPS at Dow was in the early 1990's as a subcommittee member on the *Guidelines for Implementing Process Safety Management Systems* book. I continued my involvement in CCPS when I was at the U. S. Chemical Safety Board (CSB) as a member of the TSC. As a CSB investigator, I considered CCPS as major stakeholder when developing recommendations to improve process safety performance. When I retired from the CSB in 2005, I became an emeritus member of the TSC and CCPS Fellow. I continue to be involved in CCPS activities today as discussed above.

In summary, Union Carbide did not survive the black swan of Bhopal but the chemical industry did with the creation of CCPS. Through CCPS, Bhopal has had a very positive impact on process safety from an industry, company and personal perspective.

6. Black Swan number four, Freeport, TX, July, 1996

The next three black swans that impacted my life and career are of a more personal nature, but I thought process safety practitioners need to consider strategies to survive the personal black swans of their lives and careers.

June 12, 1996 was my 50th birthday. Time to fine tune plans for retirement, which I anticipated to be ten years away, for me and my wife. However, these were unsettling times at Dow. Up to this time, Dow had a reputation for keeping good performing employees until retirement, but things were changing. I was aware that downsizing was being implement at all levels of the organization. I was certain my value to the organization was apparent and my job was necessary for the company's future. Others that I knew were terminated but certainly I would be retained. A week after my 50th birthday, my new boss called me into his office and announced that my job was eliminated and I could retire early. The black swan has struck again, this time on a personal basis.

I left Dow on good terms and value my 28 years of employment with the company. I had made significant contributions to the company's process safety program and was allowed to grow professionally and develop a passion for process safety and skills to implement process safety. I was allowed to stay until the end of July and presented a paper I prepared on Dow's new engineering process that required early and continuing input of the environmental, health, safety and process safety functions during engineering for the World Congress of Chemical Engineers in San Diego.

I was of course too young to retire. I deferred my pension and took advantage of a placement service provided by Dow to find employment and through networking, I found a consulting job with RMT, a consulting company in Austin, TX before my accumulated vacation had run out. My new boss was Sam Mannan, now the Director of the Mary K. O'Connor Process Safety Center at Texas A & M University. My process safety career continued.

How did I survive this back swan? First, I had developed a resume that made me marketable. I had been successful at process safety jobs at increasing levels of responsibility. I had led many Dow projects developing process safety management systems. I had participated in CCPS projects. I had attended courses on new process safety topics. I had obtained a PE license in the state of Texas. I would recommend that all process safety practitioners be sure they develop a marketable resume. Second I had developed a network of process safety colleagues to use in my job search. Global Congress participation, Safety and Health Division membership and CCPS participation are all excellent ways to build a network. Finally, downsizing should not be a black swan for those of you practicing process safety today. In 1996 downsizing was a new corporate strategy. Today it is commonplace. New people in process safety today should anticipate working for several different companies in their careers. Downsizing and job change should be anticipated and not be a black swan for those still working.

This black swan also became an opportunity to use to and expand my process safety skills while working as a consultant, investigator for the CSB, and staff consultant for CCPS, and developer and instructor of CPS Process Safety Boot Camp.

7. Black Swan number five, Silver Springs, MD, 2000

I had just started my job at the U. S. Chemical Safety Board in July 2000 as a chemical accident investigator. I had moved from San Antonio, TX to Silver Spring, MD to work down town DC. While living in San Antonio I had noticed a tremor in my left arm at times when I answered the phone. My doctor told me to watch it and see how it progresses.

I went to a new doctor in Maryland after I moved. He sent me to a neurologist. After a thorough examination he diagnosed that I had Parkinson's disease. Another black swan in my life. Parkinson's disease is a neurological disorder that affects one person in a thousand. The major symptoms are tremor, muscle stiffness, and problems with walking and balance. It also affects speech. The disease can be controlled with medication but is progressive.[10]

All black swans result in change and this disease has changed my lifestyle. I take medication periodically during the day to control symptoms. I exercise three times a week on the treadmill, ride a stationary bike, and lift weights. This slows the progression of the disease and controls the symptoms. I perform voice exercises when I know that I will be speaking publically. My neurologist assures me I will die of a heart attack or stroke in old age like everyone else.

Although I have retired two times so far (I may retire again), I have not let the disease interfere with activities I enjoy. A psychologist I heard speak recently had these words of wisdom: "Whatever you can't fix or control, you need to release. Most challenges in life are a combination of making changes and accepting what cannot be changed, and have the wisdom to know the difference."[10] I have made the changes necessary to live my life the way I want to and hope I have the wisdom to continue to adapt as my life goes on.

This black swan was an opportunity to reflect upon and appreciate the life I have and take advance of the skills I have developed over the years. I have learned to live one day at a time as I contribute toward improving process safety and enjoying my life as well.

I mention this black swan because we all have medical issues to deal with as we get older. We need to recognize them so we can adjust our life style, including work activities, to address the new norm.

How did I survive this black swan? First I recognize warning signs, the tremor, and took action. This resulted in my being diagnosed and treated. Second I accepted the new "normal" and adjust my life style. Exercise, taking my medication on schedule, and voice exercises are part of my life routine and have improved my enjoyment of life.

8. **Black Swan number six, Punta Gorda, FL, July 19, 2013**

I had been in retirement since 2005, but still working part time as a consultant and as a CCPS staff consultant when the last black swan arrived.

A friend of mine from high school, a retired pathologist, visited me for a few days. He mentioned that he had done a stress test several years ago and the test indicated a partial blockage in one of the arteries to his heart that required a stent to be inserted in the artery. I had not had a stress test in ten years and he suggested that I should have one now. I had an appointment with my doctor the next day and requested a stress test. He agreed and I scheduled the test. The test indicated a problem in the top area of my heart, but the doctor thought it would not be a major problem and scheduled me to visit a cardiologist after my vacation. After my vacation, the cardiologist conducted a heart catheterization to get a detailed picture of the arteries around the heart. The test showed one artery completely blocked and three arteries severely blocked. On July 19, 2013 I had quadruple heart bypass, another black swan. I was fortunate. I had no heart damage. I had none of the typical warning signs before the surgery; my cholesterol was low, my blood pressure was normal, I was not obese, did not smoke, and exercised at the gym three times a week.

How did I survive this black swan?

The fact that I previously exercised routinely had probably prevented a heart attack and gave me time to have the bypass. I have change my lifestyle again because of this. I continue to exercise at the gym three times a week, take more potent medications that further lower my cholesterol, and a beta blocker to lower my heart rate. Also I try to control my diet and my weight.

I tell people you need to be in good health to endure all the diseases I have.

This black swan was another opportunity to reflect upon and appreciate the life I have and take advance of the skills I have developed over the years. It has reinforced my determination to live one day at a time as I contribute toward improving process safety and enjoying my life as well.

9. Conclusions: Strategies for surviving the black swan

In conclusion, although, individually, black swan events are rare, they have happened with surprising frequency in my personal and professional life as well as in industry and impacted my career in process safety, often in a positive way. I suspect they happen with similar frequency in most people's lives. Because of this, it is wise to have strategies to survive black swans.

From my experience I suggest the following:

1. Companies should continue to use hazard identification and risk assessment analysis to identify most process safety incident scenarios with the caveats not to underestimate the likelihood of catastrophic incidents that are identified and to consider other suggestions in our paper *Beware of the Black Swan: The Limitations of Risk Analysis for Predicting the Extreme Impact of Rare Process Safety Incidents.*[3]
2. Strong, robust risked based process safety management systems will minimize the impact of black swans that do happen to companies.
3. Similarly, individuals that develop strong process safety competency will minimize the likelihood of downsizing by companies and this will also make the individual more marketable if job change becomes a reality.
4. From a personal health standpoint, the negative effects of black swans can be lessened by exercise, appropriate diet, and other good heath practices before disease strikes.
5. Both companies and individuals should pay attention to warning signs so that potential black swans can be avoided or at least diagnosed and dealt with.
6. If the black swan process safety event does happen, a company with a strong process safety culture will require that root cause investigation identifies the underlying management system failures so that they can be fixed to prevent similar incidents from happening again.
7. Black swan events will result in change. If managed correctly, the change can result in improved process safety performance improvement for companies and career opportunities for process safety professionals. Both companies and

individuals should remember the words of the psychologist: "Whatever you can't fix or control, you need to release. Most challenges in life are a combination of making changes and accepting what cannot be changed and have the wisdom to know the difference."[11]

8. Finally, maintain a sense of vulnerability and keep asking what can go wrong? **Beware of the Black Swan!**

10. References

[1] "The Black Swan Theory," Wikipedia, 2011.

[2] N. N. Taleb, *The Black Swan: The Impact of the Highly Improbable,* Random House Digital, New York, 2010.

[3] John F. Murphy, Jim Conner, *Beware of the Black Swan, The Limitations of Risk Analysis for Predicting the Extreme Impact of Rare Process Safety Incidents,* ***Process Safety Progress,*** Volume 31, No. 4, December 2012.

[4] John F. Murphy, Jim Conner, *Black Swans, White Swans and Fifty Shades of Grey: Remembering the Lessons Learned from Catastrophic Process Safety Incidents*, ***Process Safety Progress,*** Volume 33, Number 2, June 2014.

[5] John F. Murphy, *Dow's Reactive Chemicals Evaluation Process*", Process Plant Safety Symposium, October 26-27, 1998, Houston, Texas.

[6] Email communication with Kevin Shaunessey, December 22, 2015.

[7] CCPS, *Incidents that Define Process Safety,* [2008], AIChE, New York.

[8] John F. Murphy, Dennis Hendershot, Scott Berger, Angela E. Summers, Ronald J. Wiley, *Bhopal Revisited,* ***Process Safety Progress,*** Volume 33, Number 4, December 2014.

[9] John F. Murphy, Bob Perry, Louisa A. Nara, Shakeel Kadri, *Thirty Years of Leadership; The Center for Chemical Process Safety,* ***Process Safety Progress,*** Volume 34, Number 2, June 2015.

[10] Marjan Jahanshahi, C. David Marsden, ***Parkinson's Disease, A Self Help Guide,*** Demos Medical Publishing, Inc., New York, NY, 2000.

[11] Jill Scarpellini Huber, *Challenges of Living with Advanced Parkinson's Disease,* hand out at a seminar, Punta Gorda, FL, November 7, 2015.

A Career in Process Safety: 50 Years of LPS

Arthur M. (Art) Dowell, III
A M Dowell III PE PLLC, President
2437 Bay Area Blvd, PMB 260
Houston, TX 77058-1519
Adowell3@comcast.net

Prepared for Presentation at
American Institute of Chemical Engineers
2016 Spring Meeting
12th Global Congress on Process Safety
Houston, Texas
April 11-23, 2016

A Career in Process Safety: 50 Years of LPS

Arthur M. (Art) Dowell, III
A M Dowell III PE PLLC, President
2437 Bay Area Blvd, PMB 260
Houston, TX 77058-1519
Adowell3@comcast.net

Keywords: risk assessment; hazard evaluation; layer of protection analysis; process safety management; process safety culture

Abstract

Drawing from a career from 1963 through 2016 in the chemical and process industries, this article will describe some informal practices of process safety elements observed during the Master-Apprentice era of the 1960s. The article will trace the development of incident investigation, Process Hazard Analysis, and Fault Tree Analysis tools used by the author. It will touch on safety instrumented systems and the development of Layer of Protection Analysis to determine how strong the safety instrumented functions should be.

1 Background

In 1990, when the author asked operators, mechanics, and engineers what they needed to do their jobs safely and to meet operational excellence objectives, they responded with the principles that were codified in the U.S. OSHA PSM (process safety management) rule and later in the U.S. EPA RMP (risk management plan) rule.

Of course, the Loss Prevention Symposium (LPS) was there every year (the 50th is this meeting) with good papers on new techniques, new concepts, and reminders of what we used to know and have forgotten. The author has been to many sessions and has contributed papers to quite a few.

So with the regulations, the good practices, the good papers from LPS and the other symposia in the Global Congress on Process Safety, why isn't our record any better? In fact, our record is better than it was 50 years ago, but shouldn't it be even better today than it actually is?

It is observed: "The message of the budget has been heard throughout the globe, but the process safety message has not been heard to the same extent." Or, as Walt Howard said in an early LPS paper, "We ain't farmin' as good as we know how now." We have the tools, but we have to use them every day on every project. We can identify the hazards and the independent protection layers (IPLs), but we have to install them, maintain them, test them, and repair them.

2 Introduction

I have more than 50 years' experience in the chemical industry and thus equivalent experience in process safety. I was a vacation relief operator for a chemical plant during the summers of 1963 and 1964; an intern chemical engineer for a chemical facility in the summer of 1965; and an intern reservoir engineer for an oil company in the summer of 1966. I started my professional career in 1967 at Rohm and Haas Company as a technical support process engineer to a unit that made hydrogen cyanide. We did not have the policies, procedures, and resources that we think of today when we mention process safety, but both occupational and process safety were very important to the managers, operators, and the engineer for the unit (me).

In this article, I will talk about my experiences learned from both my employer and from my colleagues in other companies. The opinions expressed here are my own and do not necessarily represent the perspective of other individuals.

As it turned out, my subsequent jobs in engineering design, startup, technology management, and research had elements of process safety included. We did want to understand the hazards and risks of our facilities and to prevent incidents. These assignments were a preparation for the process safety job that I entered in 1989—the job I was meant to do when I grew up. Some highlights from my experience might be helpful in understanding where we are in process safety today.

2.1 "Master-Apprentice"

Over the years, I saw informal management of change (MOC) reviews. I also saw the hazardous results of changes made without any review. In that era, a lot of people learned from "Master-Apprentice" relationships. The Masters retired and now everyone tries to do more with less, so we need systems to capture the procedures and learnings that Masters carried in their heads.

In 1965, two chemical engineers stopped by the lab where I was working to consult with two chemists who had knowledge of one of the process technologies. The engineers were planning a change and wanted input on the potential risks and problems that might be encountered with that change. The chemists gave advice and recommended additional personnel to consult. I did not realize it at the time, but I had just observed my first MOC or environmental, health and safety review. "What is the change, why are we making it, what are the risks of the change, how can we manage the risks?" From my memory, that review contained most of the elements that were ultimately incorporated into our formal MOC system in 1991. In addition, that informal MOC review was much more effective than pro forma MOC reviews observed in the early 2000s (at unnamed locations) in which all the attention was focused on getting through the checklist rapidly, and no effort was used to really understand the changes and the potential risks.

Early in my first assignment, the operators showed me a pipeline in the unit called the "rattlesnake" line. They said, "You don't ever want to use this line. If you do, it will bite you big time!" They told me the line had been put in during the middle of the night. I concluded, without any MOC.

2.2 Multiple Causes for Major Incidents

Over time, as my experience grew, I was asked to lead investigations into the causes of explosions, fires, and other process upsets. I observed that there was always more than one cause (typically, at least three) for a serious incident in a reasonably well-designed and well-maintained facility.

2.3 Inherently Safer

In 1974, I was working in the north of England on a startup, when the Flixborough explosion occurred. It particularly caught my attention because of the local media reports. One of the causes was that the process had a large inventory of a liquid phase flammable liquid above the boiling point. When the leak occurred, a large vapor cloud found an ignition source, exploded, and destroyed the facility with 28 fatalities. Subsequently, Trevor Kletz suggested "What You Don't Have, Can't Leak" [1]. Years later, I was cochair of the Center for Chemical Process Safety (CCPS) subcommittee that wrote the "gold book" on inherently safer design for chemical processes. I was a member of the subcommittee that produced the second edition of the book [2, 3].

2.4 When You Do Not Know That You Do Not Know

A second cause of the Flixborough incident was MOC. One of the reactors was removed and replaced with a temporary pipe supported by scaffolding with the existing expansion joints at both ends of the pipe. The replacement was not engineered (a mechanical engineer was not on site) and the staff ***did not know*** that they ***did not know*** about appropriate support for the temporary pipe. Table 1 shows the four quadrants of knowledge.

- If I **know** that I **know** something, I can apply that knowledge appropriately (upper left quadrant).

- If I **know** that I **do not know** something, I can seek expertise from someone who does know (upper right quadrant). For example, as a chemical engineer, I know that I do not know specifics of piping support; I know I should talk to a piping engineer.

- In the lower left quadrant, I **do not know** that I **know** something. Usually, I have to think about it, analyze it, or, possibly, derive it from what I do know.

- The lower right quadrant is the danger quadrant. I **do not know** that I **do not know** something. The Flixborough staff did not know that they did not know whether the temporary bypass needed appropriate support or not.

The Flixborough incident taught us about inherently safer design, and the importance of knowing what we do not know.

Table 1. Four Quadrants of Knowledge

	Information that I KNOW	Information that I do NOT know
I know whether I know the information	Know that I know	Know that I do **NOT** know
I do **NOT** know whether I know the information	Do **NOT** know that I know	Do **NOT** know that I do **NOT** know = DANGER!

2.5 Process Hazard Analysis and Fault Tree Analysis

In 1978, I was introduced to process hazard analysis (PHA), specifically, HAZOP for hazard identification. I was also introduced to fault tree analysis (FTA) as a tool to identify the specific combination of causes for an undesired event to occur and to calculate the frequency of the occurrence. "Break it on paper, fix it on paper, before the incident in the real world." In 1979, we conducted the first HAZOP at the plant where I was working and we continued to do about 2 weeks of HAZOP every year, working our way through one of the more hazardous processes. The unit leader thought that this activity was a good investment of resources to identify hazards and to implement risk mitigation layers. Remember this time period was more than 10 years before the U.S. OSHA Process Safety Management (PSM) rule.

2.6 Safety Instrumented Functions and Layer of Protection Analysis

In the mid-1980s, I became involved in a program to better understand and to establish standards for "interlocks"—what we now call safety instrumented functions (SIFs). In this work, I first became involved in layer of protection analysis (LOPA) to answer the question, "how many layers of protection do we need, and how strong should they be?" LOPA is a simplified process risk assessment tool. We invented LOPA to make risk-based decisions more efficiently than we could with FTA or full quantitative risk assessment (the latter including day/night and seasonal weather consequence modeling). Our goal is to make the appropriate risk based decision with efficient use of resources. CCPS published three books with LOPA guidance [4–6]. I was subcommittee chair and a principal coauthor for the first book, and a peer reviewer for the other two books.

2.7 Incident Investigation

Also in this time frame, we developed an incident investigation tool based on FTA principles. The tool began with the consequence and worked backward step-by-step to determine what had to happen for the consequence to occur. We emphasized sticking to the facts, and working back to system causes. Correction of a system cause should prevent the incident from occurring, not only on that shift, but also on all shifts in that unit, in that plant, and throughout the organization. And, if the incident were shared through the LPS, or a similar forum, the incident would be prevented throughout the industry. I was privileged to be a member of the subcommittee for the first CCPS incident investigation book and a peer reviewer for the second edition [7, 8].

3 First Full-Time Process Safety Job

In 1989, I was asked to take a 6-month temporary assignment for our plant to "get all our process safety systems in order." When I retired from that assignment in 2009, our group had grown from two to seven people, and the workload had transformed as well.

This was the ultimate job that all my previous assignments had prepared me to accept, even though I was the first person to take this job. However, my learning continued in this new assignment.

3.1 Process Integrity Audits

In 1990, we did a series of half-day process integrity audits for our units focusing on safety and operational excellence. We suggested five principles for consideration during the audit:

1. Know what you want to do.
2. Do it well.
3. Have it documented.
4. Manage change.
5. Have the discipline to do 1–4 all the time [9, 10].

Applying these five principles to the areas of operations, maintenance, change, training, and culture, the operators, mechanics, and engineers asked for systems to help them do their jobs safely. The systems were the same elements that were codified in the U.S. PSM rule and later in the EPA RMP. They wanted process documentation and operating instructions that were accurate and correct (process safety information [PSI]).

During the audits, we also learned that managers, operators, mechanics, and engineers could see issues more quickly in a different operating area than the one in which they were working. Having seen an issue in another operating area, they realized that they had the same issue at home.

3.2 MOC Procedure and Action Tracking

By 1991, we had developed a formal MOC procedure and struggled to change the culture of the plant to follow the procedure. Ultimately, we wrote a training package complete with overhead projector slides, a script, exercises, and a short test. The plant (works) manager trained his direct reports in MOC with the training package; his reports trained their reports; and so on until every employee and every contractor on-site had all seen the same information and requirements for MOC. In the first year or so, operators and mechanics frequently called the process safety folks to ask if an MOC review was required for a particular change. We also ensured that each MOC included a person with a qualified occupational or process safety perspective. In the first years, the qualified person was a process safety engineer. In later years, other personnel were trained to fill this role. I knew we had been successful in training when I audited a MOC review and one of the managers was saying to the other participants precisely the same words that I had said in a MOC review the year before. The difference was that, in the prior year, it was essential that I direct the requirements to the manager as well as the other participants. A year later, the manager was insisting on the requirements himself.

We learned that everyone in the organization requires the same understanding of the procedures and needs the commitment to follow the procedures.

We also recognized that we needed an action tracking system to ensure that the actions from HAZOPs, MOC reviews, and incident investigations were completed in a timely way. Without a good tracking system, it was easy for actions to become lost. At the same time, the ISO-9002 quality effort and the ISO-14,000 environmental initiative generated actions that also required tracking. We were able to develop a single action tracking system that met the needs of all of these programs.

As we entered the cycle of PHA renewals, we found that action tracking to closure by itself is not sufficient. Some actions had been closed arbitrarily, some actions had been implemented incorrectly or incompletely, and some actions had been misunderstood. We concluded that an audit of the quality of action closure was required by personnel who were knowledgeable in the process technology and the technology involved in the action.

3.3 Documentation

In response to the 1990 process integrity audit report from operators, mechanics, and engineers about the difficulty of obtaining up-to-date documentation, I articulated a vision that each person in the organization could view an electronic copy of the up-to-date documents needed for his or her job at their workstation in real time. It took several years for the technology (hardware and software) to catch up to the vision. An additional driver was the ISO-9002 quality effort and ISO-14,000 environmental effort.

Remember, if the PSI is not correct, then the PHA has identified hazards for the facility described in the paperwork, but it has not correctly identified the hazards in the actual process facility.

In an unnamed facility, the plant could not be started up after the test of one of the SIFs. The investigation found that the design set point for the SIF was incorrect and prevented the plant from running. The SIF specification was corrected, but the corrected paper document had not been filed. In fact, the correct specification was sitting on top of the file cabinet because the file clerk had been eliminated during the recent downturn. The instrument technician had used the original design set point in the documentation in the SIF folder.

3.4 *Temporary Repair Procedure or "You Do Not Make Enough Money to Make That Decision"*

In the late 1990s, I was involved in implementing a temporary repair procedure for the plant that basically required an engineering design or analysis for any temporary repair. This was a step change from the prior practice of sticking on some fiberglass, starting up, and seeing if it holds. This effort evolved into mechanical integrity inspections and fitness for service systems.

Sometime in the late 1990s, I heard from an unnamed manager at an unnamed facility, "We can't afford to follow these new procedures. We can't run our process profitably." My response was, "We can't afford not to follow the new procedures. Neither of us makes enough money to accept the risk of not following the procedures. It's our responsibility to communicate up the ladder to our management. We need to make sure they understand the cost, the time, and the resources needed to run our processes according to the process safety requirements that the corporation has established. Top management can make the decision, 'yes, we will give you more resources,' or, 'no, we will shut the process down,' or, 'we will sign off on tolerating the additional risk.' We hope that top management does not choose the third option."

3.5 *Human Factors*

Along the way, I took a human factors course from Alan Swain. I was amazed at the number of systems that build in error by the nature of their design. Instead, equipment and procedures should be designed to ensure that the right tasks are done in the right sequence and in the right way. Attention to human factors is critical for good process safety.

In the preliminary design of the second unit for a facility, the engineers wanted to copy the layout of the original unit so that many of the drawings could be reused by simply adding a digit to the equipment numbers. The original unit required the operator to move up and down the stairs several times during a startup to manipulate the valves in the correct sequence. (In practice, the elusive "E" shift probably started the unit by setting all the valves on the second floor, then setting the valves on the first floor even though that was not the correct sequence. Note that the "E" shift makes all the errors in a 4-shift operation [A, B, C, and D shifts].) The operators, supervisors, and I were able to convince the engineers to redesign the valve layout so that all the valves needed for startup were on the same level. The valves were also located about waist high for ease of operation. Yes, the initial capital cost was slightly higher, but the return on investment came from reduced errors over the life of the plant.

3.6 *Do Not Add Layers to Meet Regulations and Requirements, Build a System*

Over the years, I saw the addition of requirements from regulation, competitive pressures, and energy efficiency. New programs with additional requirements came regularly and added burdens to the operating and support personnel. I suggested that we need to integrate all the requirements into a single system—the way that we make chemicals [11].

Recall the discussion under Action Tracking and Documentation.

3.7 Policies Are Not Enough

In 2007, I concluded from the Baker report for the BP Texas City explosion [12] that having policies for process safety in place at the top of the organization does not necessarily mean that the appropriate procedures and practices are in place in the field. The incident reiterated the importance of commitment to process safety by everyone in the organization from the top to the bottom.

4 Why Are Not We Better

From 1966 through 2016, the LPS has been there every year with good papers on new techniques, new concepts, and reminders of what we used to know and have forgotten. While I did not attend the first LPS, I have been to many sessions and have contributed papers to quite a few. I wish I could have gone to every session.

For number of years, I was privileged to hear Walt Howard come to the microphone to ask a question or offer a comment after a paper. He typically said, "Walt Howard, process safety consultant. Thank you for your paper…" As the years have advanced, now that I am retired, I find myself going to the microphone, "Art Dowell, process safety consultant. Thank you for your paper…" So, I have gone from a young man listening to the old-timers at LPS to a gray-haired veteran making comments and asking questions that I hope are of value.

In reviewing the past 50 years, I am very impressed with the amount of effort that has gone into loss prevention and process safety. A large number of people in the industry, insurance, academia, the professional societies, and government have contributed their efforts in designing tools, in building and giving training, and in sharing the tools and the lessons learned across industry. Organizations that are leaders in process safety have freely shared their systems and observations.

So with the regulations, the good practices, the CCPS books, National Fire Protection Association publications, American Petroleum Institute practices, investigation reports from the Chemical Safety Board, and the good papers from LPS and the other symposia in the Global Congress on Process Safety, why isn't our record any better?

I have observed:

"The message of the budget has been heard throughout the globe, but the process safety message has not been heard to the same extent."

Or, as Walt Howard said in an early LPS paper, "We Ain't Farmin' as Good as We Know How Now." Quoting from that paper [13]:

"An enthusiastic young farm agent, recently graduated from agricultural college, was visiting a kindly old farmer. While the farmer patiently listened, the young man spoke at length and glowingly of the many ways by which the farmer could improve his farming practices. When the young farm agent finally finished his long discourse, the farmer smiled a bit and said, 'Wal, thank you young feller. But shucks, I ain't farmin' as good as I know how now.' In the prevention of accidents in the process industries we really ain't farmin' as good as we know how now."

We have the tools, but we have to use them every day on every project. We can identify the hazards and the IPLs, but we have to install the IPLs, maintain them, test them, and repair them. We have to train our people and manage change.

5 Conclusion

Yes, as an industry we are doing a better job than we were 50 years ago when the LPS was started. But, "we still ain't farmin' as well as we know how now", as Walt Howard pointed out 32 years ago. What should we do? Here is what I learned.

1. Keep it simple and safe. (Yes, I know you have heard the last word differently, but this expression is more powerful.)
2. Everyone in the organization must be on the same page for process safety. Everyone must be committed.
3. A management system is needed to manage process safety.
4. The PSM system must include audits.
5. MOC is critical, including management of personnel change.
    a. "No good deed goes unpunished." That is, hazards can be introduced by systems intended to prevent other hazards.
6. The personnel at the bottom and middle of the organization must communicate process safety issues and deficiencies to the top of the organization.
    a. The PSM system should protect the messenger from being "shot."
7. Use the tools we have.
8. Learn from using the tools we have and close the loop to improve the tools.
9. Build systems to identify what we ***do not know*** that we ***do not know***, and then seek the expertise that is needed.
10. When using the tools such as PHA and MOC, apply the sports analogy "Keep our heads in the game!"
    a. Go beyond pro forma checklists to really understand the hazards and the risks and the layers of protection.
11. Always look at inherently safer design, but recognize it is always a trade-off.
12. Apply human factors tools to minimize mistakes and to provide recovery. We must recognize that humans are not perfect. Our job is to build systems that detect and correct mistakes before something serious occurs.
13. Investigate incidents and near misses and seek system level changes.
14. Measure what you want to know.
    a. But be careful what you measure, because unexpected results can occur.
15. Keep the documentation up to date and available to those who need it.

6 References

1. T.A. Kletz, What you don't have, can't leak, **Chem Ind 42** (1978), 287–292.
2. CCPS, Inherently Safer Chemical Processes: A Lifecycle Approach, AIChE, New York, 1996.
3. CCPS, Inherently Safer Chemical Processes: A Lifecycle Approach, 2nd Edition, AIChE, New York, 2009.
4. CCPS, Layer of Protection Analysis: Simplified Process Risk Assessment, AIChE, New York, 2001.
5. CCPS, Guidelines for Enabling Conditions and Conditional Modifiers in Layer of Protection Analysis, AIChE, New York, 2014.
6. CCPS, Guidelines for Initiating Events and Independent Protection Layers in Layer of Protection Analysis, AIChE, New York, 2015.
7. CCPS, Guidelines for Investigating Chemical Process Incidents, AIChE, New York, 1992.
8. CCPS, Guidelines for Investigating Chemical Process Incidents, 2nd Edition, AIChE, New York, 2003.
9. A.M. Dowell III, S.E. Anderson, and D.K. Martin, An audit system for process safety, In: Texas Chemical Council Process Safety Management Seminar, Galveston, TX, USA, June 1990.
10. A.M. Dowell III and S.E. Anderson, An audit system for process safety one year later… Discussion of phase II system", In: Texas Chemical Council Process Safety Management Seminar, Galveston, TX, USA, June 1991.
11. A.M. Dowell III, Regulations: build a system or add layers, AIChE Spring National Meeting, March 21, 1995, **Process Saf Prog 20** (2001), 247–252.
12. J.A. Baker, *The Report of the BP US Refineries Independent Safety Review Panel*, U.S. Chemical Safety Board, www.csb.gov, January 30, 2007.
13. W.B. Howard, Efficient time use to achieve safety of processes or 'we ain't farmin' as good as we know how, **Plant/Oper Prog 3** (1984), 129–132.

Revisions to FM Global Property Loss Prevention Data Sheet 7-14, *Fire Protection for Chemical Plants*

Matthew Daelhousen
FM Global
1151 Boston-Providence Turnpike
P.O. Box 9102
Norwood, MA 02062
matthew.daelhousen@fmglobal.com

Henry L. Febo Jr. PE
Ben Ditch
FM Global
1151 Boston-Providence Turnpike
P.O. Box 9102
Norwood, MA 02062
henry.febo@fmglobal.com
benjamin.ditch@fmglobal.com

Prepared for Presentation at
American Institute of Chemical Engineers
2016 Spring Meeting
12th Global Congress on Process Safety
Houston, Texas
April 11-13, 2016

Revisions to FM Global Property Loss Prevention Data Sheet 7-14, *Fire Protection for Chemical Plants*

Matthew Daelhousen
FM Global
1151 Boston-Providence Turnpike
P.O. Box 9102
Norwood, MA 02062
matthew.daelhousen@fmglobal.com

Henry L. Febo Jr. PE
Ben Ditch
FM Global

Keywords: fire protection, ignitable liquid, flammable gas

Abstract

For nearly two centuries, businesses around the world have relied on FM Global and its property loss prevention data sheets to reduce risk at their existing facilities and ensure their new facilities are built to world-class property protection standards. Data Sheet 7-14, *Fire Protection for Chemical Plants*, addresses fire and explosion hazards at chemical manufacturing plants and similar processing facilities, particularly those associated with ignitable liquid, flammable gas, and liquefied flammable gas where a release could result in a fire of long duration. In 2014, FM Global assigned a team of engineers to update the data sheet's technical guidance and general usability. Although the data sheet had undergone occasional partial revisions, the technical content of the entire document had not been examined in close detail in over 15 years.

A primary focus of the revision was to reduce reliance on prescriptive guidelines when designing fire protection systems at chemical plants. Renewed emphasis was placed on identifying potential ignitable liquid and flammable gas release and fire/explosion scenarios, and designing fire protection systems appropriate for the anticipated loss scenario. Additionally, safeguards to minimize the volume of the potential liquid or gas release were developed, such as accidental release control systems, siting guidance, and other passive methods to mitigate damage in the event of a release. This new guidance is based on a mixture of scientific research, loss history, and engineering experience.

1 Background

What is a chemical plant?

While the answer to this question may seem simple at first, it's actually rather tricky from a fire protection point of view. A typical response might be intuitive in nature; something like "I know one when I see one." While this answer may be sufficient in many situations, it falls a bit short when trying to establish fire protection for facilities processing hazardous chemicals. At what point does the fire and explosion hazard become so severe to necessitate multiple deluge systems, massive water supplies, spill containment systems, and passive fire protection for exposed steel?

Yet this was the issue a team of 12 FM Global employees from across the globe needed to resolve in order to revise Data Sheet 7-14. The team was comprised of chemical engineers, mechanical engineers, and fire protection engineers with a combined 350 years of experience at FM Global. Collectively the group had performed several thousand evaluations at an estimated 1500 chemical plants, including facilities that manufactured paints, agricultural chemicals, dyes and pigments, specialty and commodity chemicals, pharmaceuticals, olefins, and other petrochemicals. Drawing on this experience, the team reexamined the fire protection typically provided at chemical plants, and their work resulted in several significant improvements to Data Sheet 7-14.

2 Scope of Changes

The team's first matter of business was to address the unwieldly title of the data sheet: *Fire and Explosion Protection for Ignitable Liquid, Flammable Gas, and Liquefied Flammable Gas Processing Equipment and Supporting Structures*. The much simpler title of *Fire Protection for Chemical Plants* was proposed, but this would require a clear definition of what, exactly, was meant by "chemical plant." After all, not all chemical plants are created equal; a paint processing facility and an olefins plant may both be categorized as "chemical plants," yet the latter represents a much more severe fire protection challenge.

This is because a direct relationship exists between the quantity of fuel released and the corresponding severity of a fire. If the volume of an ignitable liquid release is small, the fire severity may be limited. Conversely, a large and continuous flow of liquid can result in a growing and sustained fire. FM Global's most fundamental recommendation pertaining to ignitable liquid and flammable gas is to limit the amount of fuel that can become involved in a fire.

At a chemical processing plant, however, prompt and automatic isolation of an ignitable liquid or flammable gas release is often impossible. For example, the consequence of shutting off the heat transfer system providing cooling for a reactor running an exothermic reaction that has a runaway potential could be the violent rupture of the reactor. Similarly, process conditions can result in significant releases. A flow rate of 1000 gpm (3780 L/min) will result in a considerable amount of liquid being discharged even if the system is shut down quickly.

The scope of the previous version of Data Sheet 7-14 applied to chemical plants whose volume of ignitable liquid or flammable gas exceeded the following thresholds:

- 200 gal (0.8 m^3) for any one container or 2000 gal (7.6 m^3) in separate containers over a base area of 2500 ft^2 (232 m^2) for indoor locations.
- 500 gal (1.9 m^3) for any one container or 5000 gal (18.9 m^3) in separate containers over a base area of 2500 ft^2 (232 m^2) for outdoor locations.

The team revising Data Sheet 7-14 quickly realized these thresholds did not sufficiently reflect the fire hazard at a chemical plant and that, rather than basing fire protection recommendations on prescriptive container volume thresholds, they should base them on expected release volumes, including the potential for continuous release. With this in mind, the team identified several chemical plant characteristics that could result in fire scenarios requiring higher levels of passive and active fire protection. These characteristics include the potential release of large volumes of ignitable liquid or flammable gas, high flow rates in piping systems, and the processing of unstable materials. Regardless of the capacity of a vessel, if interlocks are installed to promptly and automatically shut down the release of fuel, the fire hazard can be significantly reduced, and other FM Global data sheets exist to address those reduced hazards.

Thus, having defined the characteristics of the chemical plants Data Sheet 7-14 would address, and with an understanding of the processes that drive the need for enhanced levels of fire protection, the team was able to establish passive and active protection guidance appropriate for the anticipated long-duration fire scenario.

3 Fire Protection Modifications in Data Sheet 7-14

3.1 Passive Protection

The extent of fire or explosion damage to a chemical plant can be minimized by providing separation between process units and other important buildings. The spacing needed is based on many factors, including the following:

- The severity and probability of the potential fires and explosions
- The relative value and importance of facilities
- The susceptibility of buildings and equipment to direct or indirect damage (e.g., from a fire following an explosion)

To gain an understanding of what is needed to minimize potential damage, it is necessary to make a comprehensive analysis of the chemical plant. This analysis will help determine appropriate separation between process units, maintenance and administration buildings, utilities, ignitable liquid storage tanks, and other buildings at the plant. Site-specific conditions must be accounted for, including expected liquid and gas release types, locations, and volumes. Although this analysis is crucial to ensure a layout that is appropriate for the plant, the team established some "rules of thumb" (Table 1) that can be used as a starting point for discussions prior to a more rigorous study. The team also developed guidance for related passive features, such as containment, emergency drainage, and configurations, to limit the potential for three-dimensional spill fires, among other scenarios.

Table 1. Approximate Spacing Guide

The spacing in this table is related to fire and explosion protection only, not for the hazards of toxic or corrosive releases.	*Boiler Houses, Power Generation*	*Electric Substations, Main*	*Service/Administration Buildings*	*Underground High Pressure Flammable4 Gas Lines*	*Fire Protection Water Sources*	*Control Rooms, Central*	*Main Pipe Racks*	*Process Unit, High Hazard*	*Process Unit, Low Hazard*	*Tank Farms, Pressurized Flammable*	*Tank Farms, atmospheric*
Tank Farms, atmospheric											C
Tank Farms, Pressurized Flammable										B	100
Process Unit, Low Hazard									100	200	C
Process Unit, High Hazard								200	200	200	C
Main Pipe Racks							NA	100	50	A	100
Control Rooms, Central						A	100	200	100	200	150
Fire Protection Water Sources					NA	A	200	200	100	200	150(D)
Underground High Pressure Flammable Gas Lines				NA	300	300	A	150	150	150	100
Service/Administration Buildings			NA	300	A	A	100	200	200	200	C
Electric Substations, Main		A	100	300	100	100	100	200	100	200	C
Boiler Houses, Power Generation	A	A	100	300	A	100	100	200	100	200	C
A = Space as needed for operation and maintenance.											
B = Refer to Data Sheet 7-55 for limitations on LPG storage and Data Sheet 7-91 for limitations on hydrogen storage.											
C = Refer to Data Sheet 7-88, Tables 2 and 3. No multi tank quantity limit at chemical plants.											
D = One of multiple sources; single source at least 200 ft.											

3.2 Active Protection

While the passive protection features recommended in Data Sheet 7-14 will serve to mitigate damage if a large, out of control fire or explosion occurs, active fire protection systems also provide a critical layer of protection to minimize the extent of loss. FM Global conducted a series of large-scale fire tests in the 1970s to assess the fire hazard of a multi-level ignitable liquid process structure and evaluate the effectiveness of various forms of fire protection. This test program established the following:

- Three-dimensional (i.e., upper open level) spills create a more severe fire exposure than floor pool fires. Three-dimensional fires produce a much higher heat release rate than pool fires because the liquid is vaporized at a higher rate. The potential for thermal damage from these types of fires is high.
- Sprinklers and foam-water sprinklers can control, and even extinguish, a pool fire but are not effective in reducing exposure from a 3-D spill fire.

The 1970s test series established an initial comprehension of the three-dimensional spill and pool fire hazard at chemical plants, and this was used as a basis for guidance in the previous versions of Data Sheet 7-14. More-recent fire testing had advanced this knowledge, however, and the data sheet's recommendations needed to be revised accordingly.

3.2.1 Area-Level Sprinkler and Deluge Protection

In 2005, FM Global completed a series of tests for the power generation industry in order to establish the necessary sprinkler protection to control and extinguish lube-oil pool fires [1, 2]. Based on this testing, the criteria shown in Table 2 was established.

Table 2. Sprinkler Protection Necessary to Control or Extinguish Oil Pool Fires

Protection Objective	Maximum Roof Height, ft (m)	Density, gpm/ft^2 (mm/min)
Pool fire extinguishment	15 (4.6)	0.3 (12)
	30 (9.1)	0.4 (15)
	40 (12)	0.7 (29)
	45 (14)	0.8 (33)
Fire control only	40 (12)	0.3 (12)

Similarly, a test series completed in 2010 examined common fire protection schemes for multi-level ignitable liquid process structures [3, 4]. This test series reaffirmed the significant hazard that a large ignitable pool fire can pose to structural members and sensitive equipment, but once again demonstrated that fire control can be achieved through proper fire protection system design. The project evaluated the effectiveness of several single-level and multi-level sprinkler systems at reducing the fire hazard, and demonstrated the impact of various sprinkler discharge pressures, discharge densities, and installation heights (Figure 1).

Figure 1. Pool Fire with area protection using a 0.3 gpm/ft² (12 mm/min) discharge density under the lower mezzanine

Based on these two research projects, the recommended design of area-level sprinkler protection was modified in Data Sheet 7-14. Area-level sprinkler protection is a grid of sprinklers, typically installed under a roof, canopy, or mezzanine, designed to provide fire protection to the area below the grid. Industry codes and standards specify a range of sprinkler densities and installation height limits for area-level protection. The previous version on Data Sheet 7-14, for example used a maximum height of 15 ft (4.6 m).

Data Sheet 7-14 now recommends a density of 0.3 gpm/ft² (12 mm/min) up to 30 ft (9.1 m) above a pool fire. This design is based on FM Global's large scale fire testing, but has also proven to be effective fire protection from a practical standpoint for a variety of ignitable liquid exposures. Based on FM Global client experience and loss history, this density has been successful in controlling ignitable liquid pool fires when used in combination with passive features such as containment and emergency drainage.

The design of area protection will vary depending on whether the facility is located indoors or outdoors. For indoor locations, automatic wet sprinkler, deluge, or preaction systems may be installed. When the occupancy to be protected is located outdoors, the installation of deluge systems is recommended for area protection. Automatic sprinklers are not recommended because of wind effects and the potential for delayed actuation.

The design will also depend on the presence of intermediate floors or mezzanines. For indoor locations, a design density of 0.3 gpm/ft^2 (12 mm/min) is recommended below solid levels, such as under roofs and solid mezzanines. A density of 0.3 gpm/ft^2 (12 mm/min) is also recommended above areas where the potential exists for holdup of ignitable liquids, such as solid intermediate mezzanines. Above open-grated mezzanine levels, where the potential for ignitable liquid holdup does not exist, the design density can be reduced to 0.15 gpm/ft^2 (6 mm/min).

For outdoor process structures, a design density of 0.3 gpm/ft^2 (12 mm/min) per protection level is necessary, regardless of the configuration of intermediate levels. The increased density is due to the potential negative impact of the environment, including the effects of wind. Figure 2 represents an example of recommended active fire protection for a large, outdoor chemical processing structure.

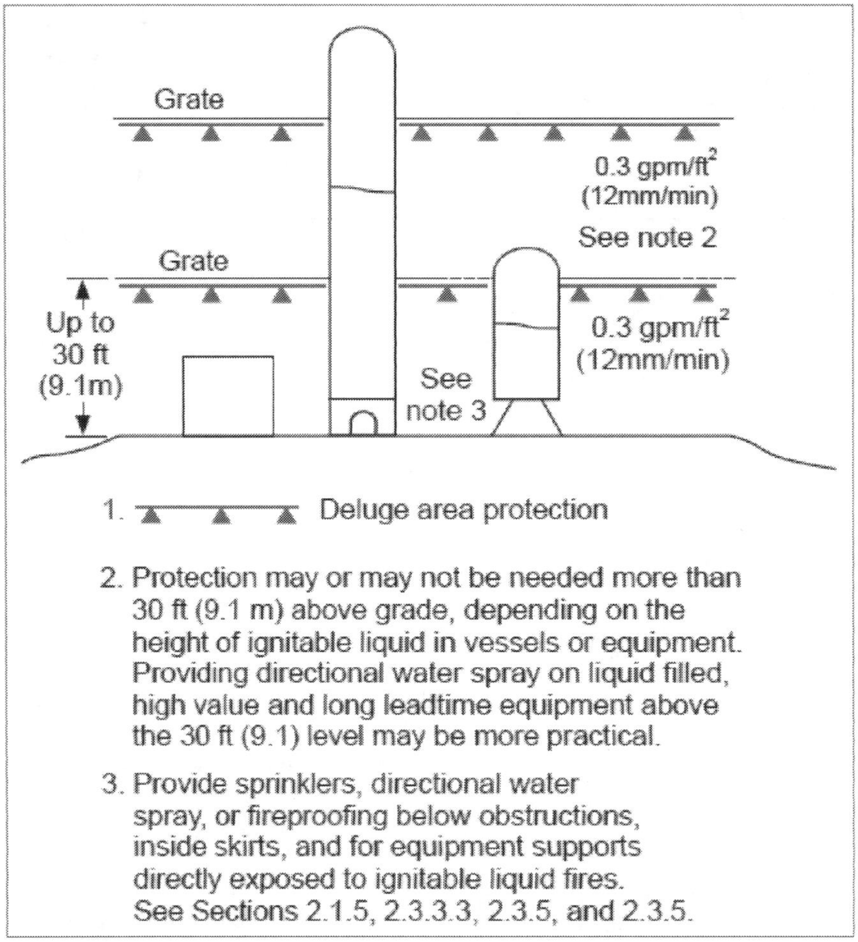

Figure 2. Example of area-level deluge protection for an outdoor processing structure

3.2.2 Supplemental Water Spray Protection

The effectiveness of area protection to provide cooling for equipment and structures is greatly diminished as the clearance between the sprinklers and the fire increases. Therefore, local water spray protection may be necessary in addition to area protection, depending on the configuration of the structure. In cases where area protection is installed more than 30 ft (9.1 m) above a pool fire, additional local protection may be necessary. Chemical plants may also have a significant quantity of conduit, piping, and other obstructions that reduce the effectiveness of area protection. If these items serve to obstruct roof- or mezzanine-level sprinklers, supplemental protection is necessary. This may include placement of additional sprinklers below the obstructions or local water spray protection of equipment supports. In many cases, a combination of area protection and local water spray for exposed equipment and structures will be necessary. The physical arrangement of the processing structure and equipment will determine what type of protection (area-level, directional water spray, or a combination of the two) is most practical.

Supplemental water spray protection may also be necessary where equipment or structures will be directly exposed to a long-duration ignitable liquid spill or pool fire. In both the initial process structure tests conducted in the 1970s and the more recent test series completed in 2010, FM Global has consistently witnessed a significantly increased fire hazard where structures are directly exposed to flames, particularly those associated with an ignitable liquid three-dimensional spill fire. An ignitable liquid spill or pool fire will achieve flame temperatures ranging from 1000°F (538°C) to 2000°F (1093°C), resulting in damaging heat fluxes anywhere near the fire. A steel structure exposed to these conditions can reach the point of failure within minutes.

The 2010 research project examined this type of three-dimensional spill fire scenario and its impact on the distillation column of a multi-level process structure. The test setup included a 2.5 ft (0.76 m) diameter by 32 ft (9.75 m) tall mock distillation column (Figure 3).

Figure 3. Three-dimensional spill and mock distillation column

A spill fire originated immediately above the lower mezzanine level and flowed down a mock distillation column at a rate of about 20 gpm (80 L/min). Temperature was measured at multiple places on the steel structure and along the length of the distillation column with mechanically fastened thermocouples. To match common industry practice, area-level sprinkler protection was provided under this lower mezzanine at a discharge density of 0.3 gpm/ft^2 (12 mm/min).

The test demonstrated that an area-level sprinkler system provides minimal protection for this type of fire scenario; the temperatures on the structural column that were located within the fire spill area exceeded 1000°F (538°C). As shown in Figure 4, while the sprinkler system reduced the severity of the pool fire, there was no observable reduction in the fire intensity along the distillation column. This result highlights the importance of combining area-level sprinkler protection with localized water spray directly onto sensitive equipment to reduce the overall fire hazard. Failure to do so can leave a critical part of the structure vulnerable.

Figure 4. Three-dimensional spill fire before (left) and after (right) sprinkler activation

3.2.3 Steel Column Protection

The previous version of Data Sheet 7-14 recommended passive protection for steel columns where a "greater than average explosion hazard or severe three-dimensional fire hazard" existed. The vagueness of the terms "greater than average" and "severe" caused confusion regarding the need for structural steel protection, and the team worked to clarify this issue.

As discussed in the previous section, a steel element exposed to the high temperatures associated with an ignitable liquid spill or pool fire will quickly reach the point of failure. If the compromised steel element provides critical support for the structure, collapse may occur.

Based on this reasoning, Data Sheet 7-14 was revised to recommend protection of structural steel if all of the following are true:

1. The steel member is load bearing.
2. The steel member supports a structure or equipment that is critical to the facility's operation.
3. The steel member has the potential to be directly exposed to an ignitable liquid three-dimensional spill or pool fire. Typically the structural element must be immersed in the spill or pool fire to necessitate protection.

In addition to the pool and spill fire tests mentioned earlier, the 2010 process structure project included an extensive series of tests examining thermal exposure to structural steel elements [5]. The test series quantified the peak fire severity from a large ignitable liquid fire exposure, as well as the reduction in fire severity provided by several common water spray application schemes. A large fuel pan was partially enclosed on two adjacent sides to reduce radiation loss, producing the hazard scenario present in an actual chemical plant on a reduced-scale.

The hazard evaluation focused on the vertical heat flux profile of a target object located at the centerline of a fire plume. Figure 5 shows the test setup and resulting flame characteristics for a 49 ft^2 (4.6 m^2) pan of heptane. In this test, it was found that exposure to a large unprotected pool fire can expose a steel member to a heat flux over 950 BTU/ft^2/min (180 kW/m^2), resulting from sustained flame temperatures exceeding 1920°F (1050°C). This exposure would be expected to damage an unprotected 1 in. (25 mm) thick steel member within two minutes. Similar testing with a 200 ft^2 (18.9 m^2) ethanol pool fire showed damage to a steel member would be caused within 3 minutes. In either case, the exposure is severe.

Figure 5. Freeburn test representing a large hydrocarbon pool fire

Structural steel exposed to such high temperatures can be protected with fireproofing materials. Fireproofing structural steel provides a thermal barrier to limit temperature rise and maintain structural integrity. Alternatively, water spray protection can be used, but the design of the system is critical; there must be a continuous film of water over the protected area.

The effectiveness of local water spray protection was evaluated for two common system designs (Figure 6). In a direct-impingement system, a uniform water density is applied across all exposed surfaces of an object. By comparison, in a run-down system the same quantity of water is applied only at the top of the object and the delivered density is then averaged over the entire area of the object. This design requires the water to flow down the object's surface, which means it may be evaporated by the fire. The evaluation looked at the system design characteristics of discharge density, nozzle elevation, and number of nozzles.

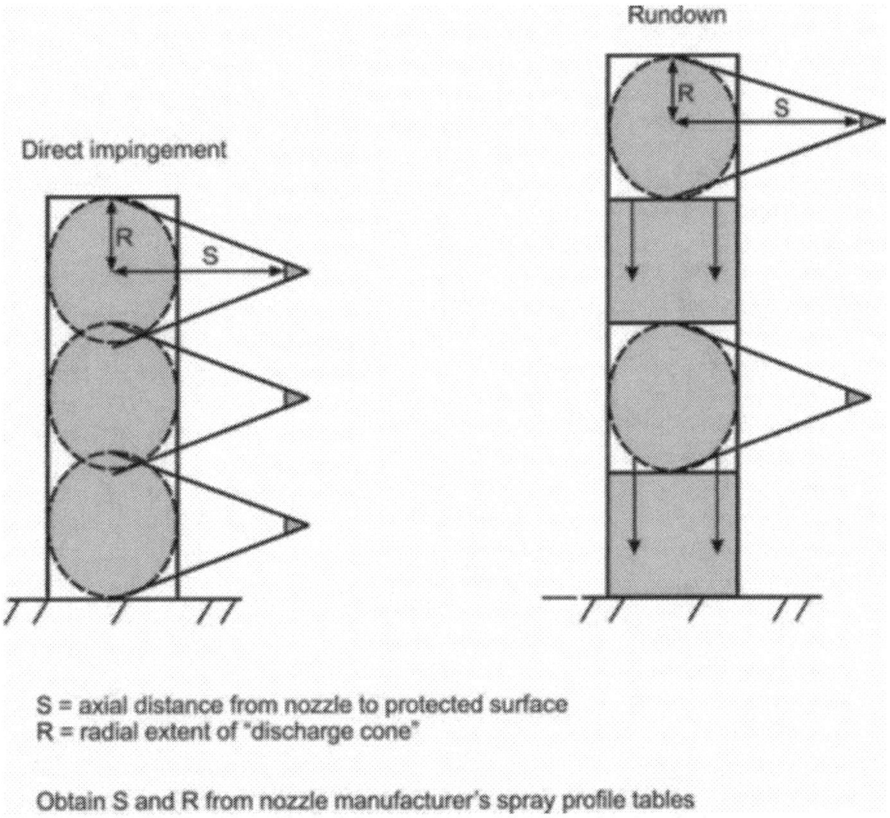

Figure 6. Example of local water spray protection schemes used in fire tests; direct-impingement (left) and run-down protection (right)

An example of the test results for water spray systems using a discharge of 0.3 gpm/ft^2 (12 mm/min) is shown in Figures 7 and 8. Figure 7 shows the huge reduction in heat flux imposed by an ethanol pool fire when either direct-impingement or run-down water spray protection is applied. Heat flux values for an unprotected heptane fire are also included for comparison. A large ignitable liquid pool fire can produce peak heat fluxes ranging from 740 to 950 BTU/ft^2/min (140 to 180 kW/m^2), depending on the fuel that is burning. With a properly designed run-down water spray system, the heat flux experienced by protected equipment can be reduced to below 210 BTU/ft^2/hr (40 kW/m^2). With a direct-impingement system, this can be even further reduced.

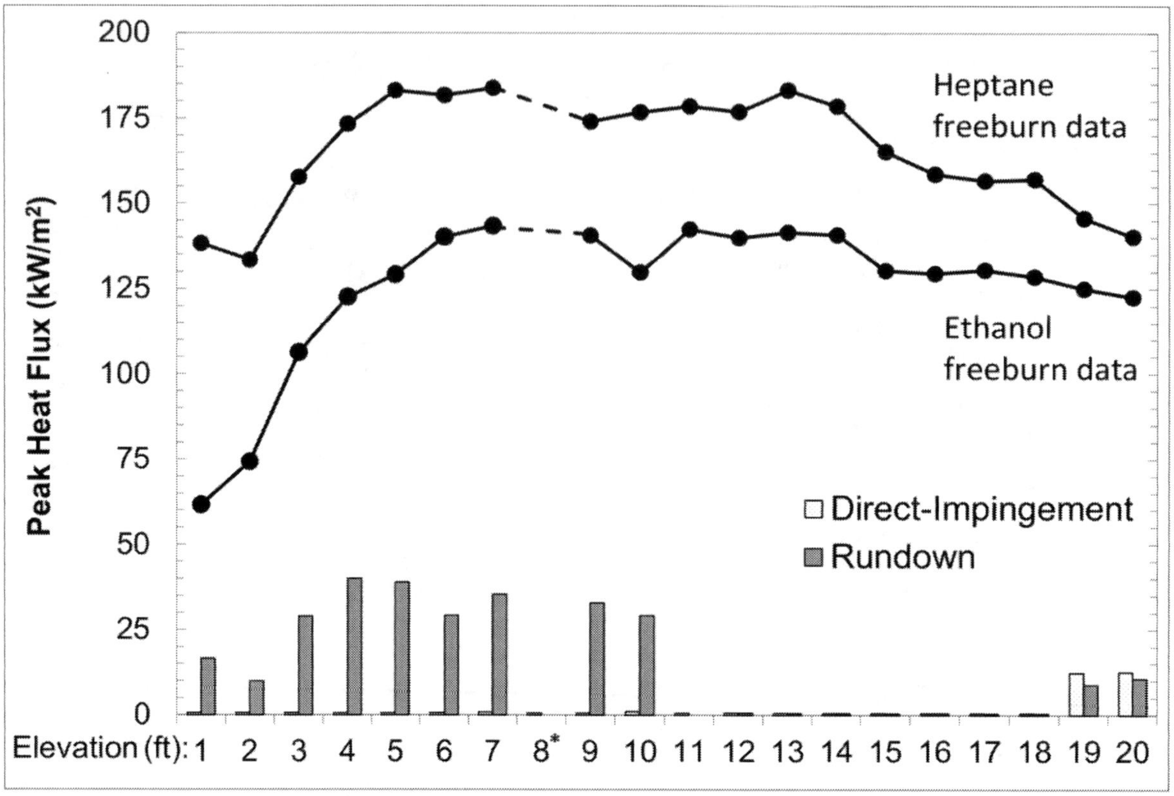

Figure 7. Maximum heat flux imposed on object; effectiveness of spray protection shown for the case of a large ethanol pool fire (note: interpolated values shown for the 8 ft [2.4 m] elevation due to instrumentation damage)

Figure 8 shows how these results relate to temperature using a large ethanol fire as an example. It can be seen that unprotected equipment immersed within the fire will be damaged. Conversely, direct-impingement water spray protection applying a uniform water density exhibited acceptable protection for the same object. The protection became insufficient when a similar volume of water was applied with only a single nozzle located 15 ft (4.6 m) above the pool fire (i.e., run-down protection scheme). However, this run-down protection became effective when the distance below the nozzle was reduced. While a direct-impingement system is clearly more effective, many existing facilities have systems based on the run-down method of protection.

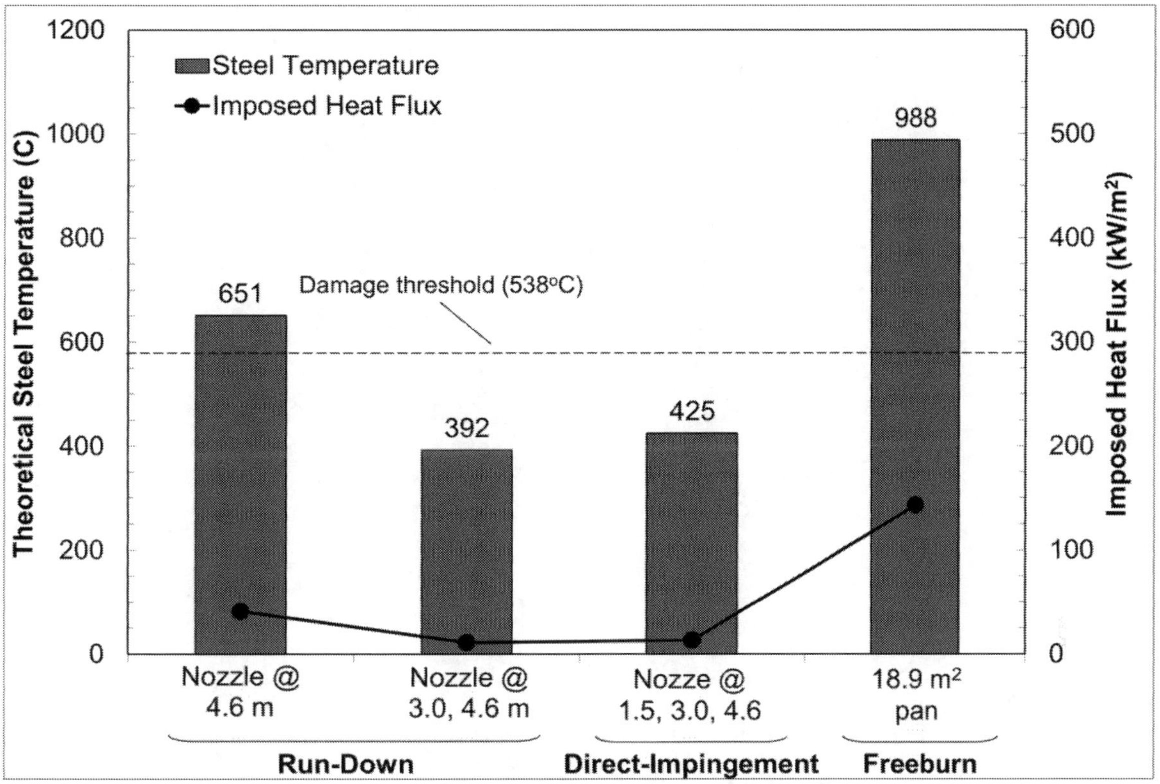

Figure 8. Calculated peak steel temperatures for freeburn and water spray tests using a 0.3 gpm/ft² (12 mm/min) density to protect a large ethanol fire

Based on these results, Data Sheet 7-14 recommends the use of water spray protection using either a direct-impingement or a run-down design, and now includes specific design guidance for each type of system. Spray nozzles must be installed on a maximum of 12 ft (3.7 m) centers, staggered on alternate sides of the column such that a maximum separation of 24 ft (7.3 m) exists between nozzles on one side of the column (see Figure 9). A minimum density of 0.25 gpm/ft² (10 m/min) over the wetted area of one side of the column for the 24 ft (7.3 m) length is recommended ("wetted area" is the surface area on the three sides of the reentrant space formed by the column web and flanges). Guidance is also provided in the data sheet for the protection of horizontal elements.

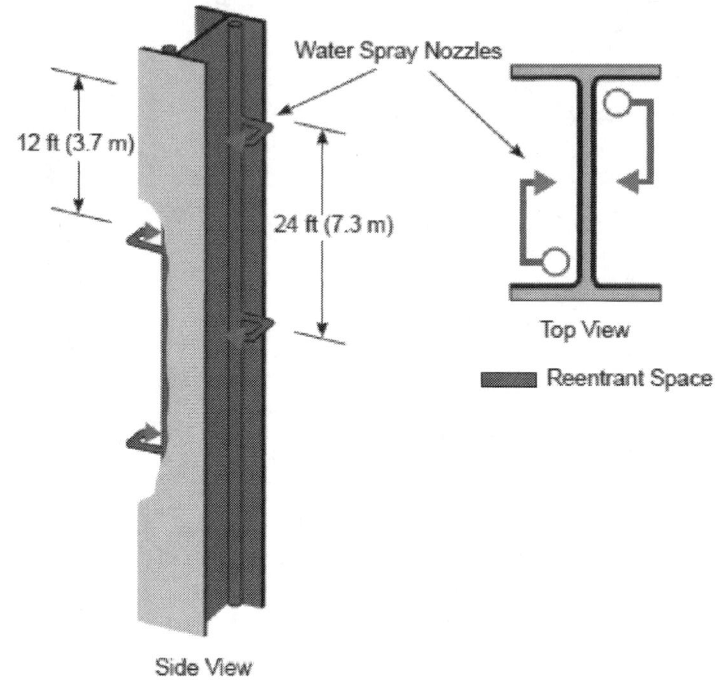

Figure 9. Water spray protection for steel columns

4 Conclusion

Data Sheet 7-14 has been a valuable resource since its initial publication in 1997. Since that time, several significant research projects were completed to better understand the hazards associated with large ignitable liquid releases and how to mitigate these hazards. In 2014, FM Global initiated an effort to improve Data Sheet 7-14 based on the knowledge gained from these research projects, as well as from historical loss data and the work experience of a number of highly skilled FM Global loss prevention engineers.

The results of this effort include an up-to-date standard on the protection of indoor and outdoor process areas, supporting structures, and processing equipment that handles ignitable liquid, flammable gas, and liquefied flammable gas. What is meant by "chemical plant" is now specifically defined in the document, and a number of passive and active fire protection safeguards are recommended to protect against the long-duration fire events that can occur at these facilities.

FM Global Property Loss Prevention Data Sheets are available for download, free of charge, from FM Global's website, www.fmglobal.com.

5 References

[1] T. Cooper, "Turbine Lube-Oil Fires: A Frightening Hazard," *Energy Tech,* February 2005.

[2] C. Wieczorek and J. LeBlanc, "Large-Scale Fire Tests of Oil Systems in Power Generation Facilities." NFPA World Safety Conference and Exposition, 2006.

[3] B. Ditch, "The Fire Hazard of an Ethanol Process Facility." 7th Global Congress on Process Safety, 2011.

[4] B. Ditch and G. Mahnken, "Fire Protection for Ethanol Process Structures." *FS-World,* Spring 2012.

[5] B. Ditch, "Evaluating Pool Fire Severity and the Cooling Effect of Local Water Spray Using a Continuous Plate Thermometer." 10th International Symposium for Fire Safety Science, 2011.

New Model for Predicting Thermal Radiation from Flares and high Pressure Jet Fires for Hydrogen and Syngas

Presenter:
Derek Miller, Air Products and Chemicals, Inc., Allentown, PA
Millerd3@airproducts.com

Prepared for Presentation at
American Institute of Chemical Engineers
2016 Spring Meeting
12th Global Congress on Process Safety
Houston, Texas
April 10-14, 2016

New Model for Predicting Thermal Radiation from Flares and High Pressure Jet Fires for Hydrogen and Syngas

Presenter:
Derek Miller, Air Products and Chemicals, Inc., Allentown, PA
Millerd3@airproducts.com

Keywords: *Hydrogen, syngas, jet fire, flare, model*

Abstract

Current flare and jet fire models used in the process industry are primarily based on hydrocarbon gases and have been found not to be particularly effective for low luminosity gases such as hydrogen and syngas mixtures. This paper presents a new model for such gases, for low to high pressure releases in both vertical and horizontal orientations. The model, which predicts flame geometry and thermal radiation, allowing for jet momentum, buoyancy, wind and flame radiant fraction, is based on extension of established models, with the addition of several new correlations. Validation against existing published data and new previously unpublished test data collected at both pilot and full scale is presented.

The paper provides the complete set of equations for the model.

1.0 Overview. Hydrogen and syngas (mixtures containing hydrogen, carbon monoxide and other components) are routinely used across a wide range of industrial applications. One important hazard to manage is jet fires, both planned (example flares) and unplanned (example process leaks). There is currently no published model that adequately predicts flame geometry and thermal radiation for these gases across a wide range of gas compositions and release conditions (velocity, temperature and orientation), taking adequate account of ambient conditions, such as wind speed - Air Products previously presented a comparison between existing models and published data at GCPS in 2014 [9]. A new model is presented here, built on extension of published models primarily developed for hydrocarbon flames, which has been validated against available published data, as well as new test data specially collected for this purpose. This report provides the equations required for the model with the available experimental data and validation between data and model. The new model is called AP Flame.

2.0 Comparison of models. To introduce the subject, it is worthwhile to briefly compare results of two established models, Chamberlain [1] and API 521 [21] with the new AP Flame model. The Chamberlain model is really only for light hydrocarbons, but is often used more broadly, API 521 is for light hydrocarbons and hydrogen. Figure 1 shows results for a typical hydrogen and syngas flare, across a range of conditions up to choked flow. The example flare is actually located at the Air Products Garyville hydrogen plant, where tests were performed at turndown rates for both hydrogen and syngas (73% H_2, 7% CH_4, 4% CO and 16% CO_2) [20]. Terminology is described in section 3.0 of this paper.

2.1 **API 521:** matches hydrogen flare test data points quite well, but overestimates radiant fraction and flame length at high rates, as choke flow is approached. API 521 model is only valid for sub-sonic flares. API 521 over-predicts flame length and tilt for syngas.

2.2 **Chamberlain:** matches hydrogen test data points quite well, but overestimates radiant fraction at lower rates. Air Products has observed flames getting longer at high tilt angles, (which has caused operational issues) but this is not picked up by Chamberlain. To compound this issue, hydrogen flames are more greatly impacted by wind than natural gas flames, which is why Chamberlain under-predicts tilt for hydrogen. Chamberlain over-predicts radiant fraction and tilt for syngas.

2.3 **AP Flame:** matches test data points well (expected since data is part of overall dataset). General graph trends supported by results from other tests in dataset.

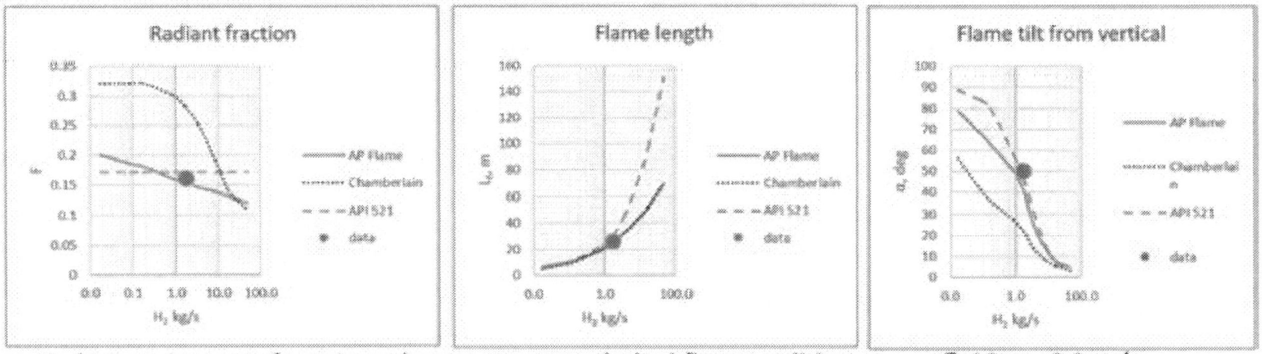

a. Hydrogen. Impact of varying release rate up to choked flow condition, u_{wind} @ 10m = 2.3 m/s

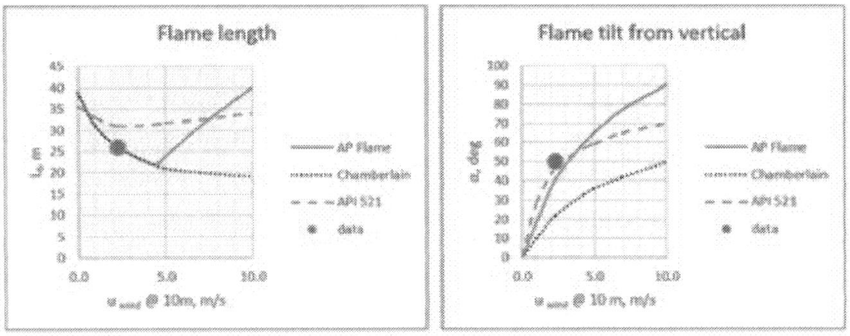

b. Hydrogen. Impact of varying u_{wind} @ 10m, release rate = 1.74 kg/s

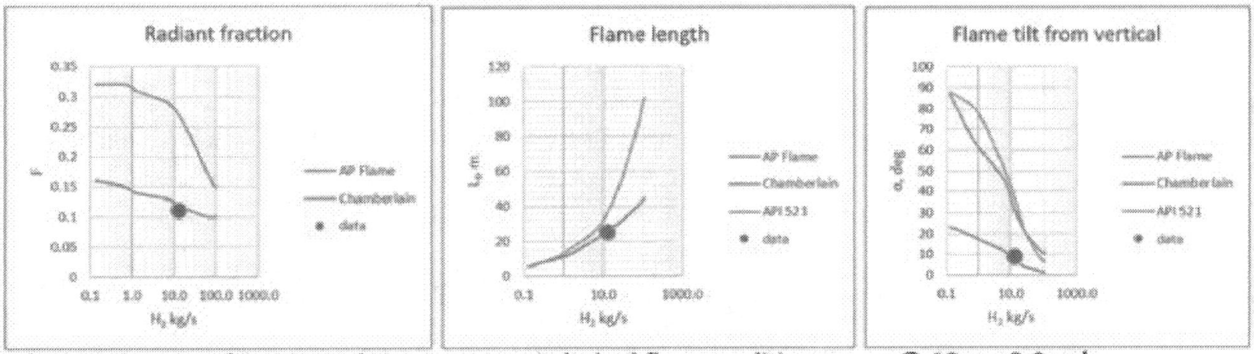

c. Syngas. Impact of varying release rate up to choked flow condition, u_{wind} @ 10m = 2.0 m/s

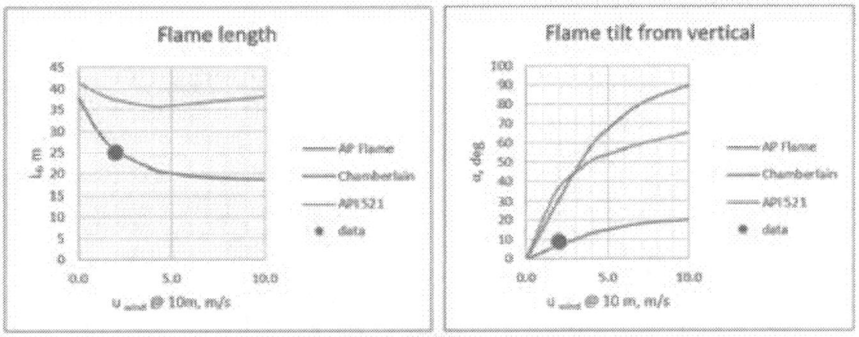

d. Syngas. Impact of varying u_{wind} @ 10m, release rate = 13 kg/s

Figure 1. Comparison of model results: Chamberlain, API 521 and AP Flame for 27" diameter, 53m high vertical hydrogen and syngas (73% hydrogen) flare, rural topography

3.0 Terminology

Figure 2 illustrates key terms used in this paper. AP Flame models vertical, horizontal and 45 degree releases in the wind direction.

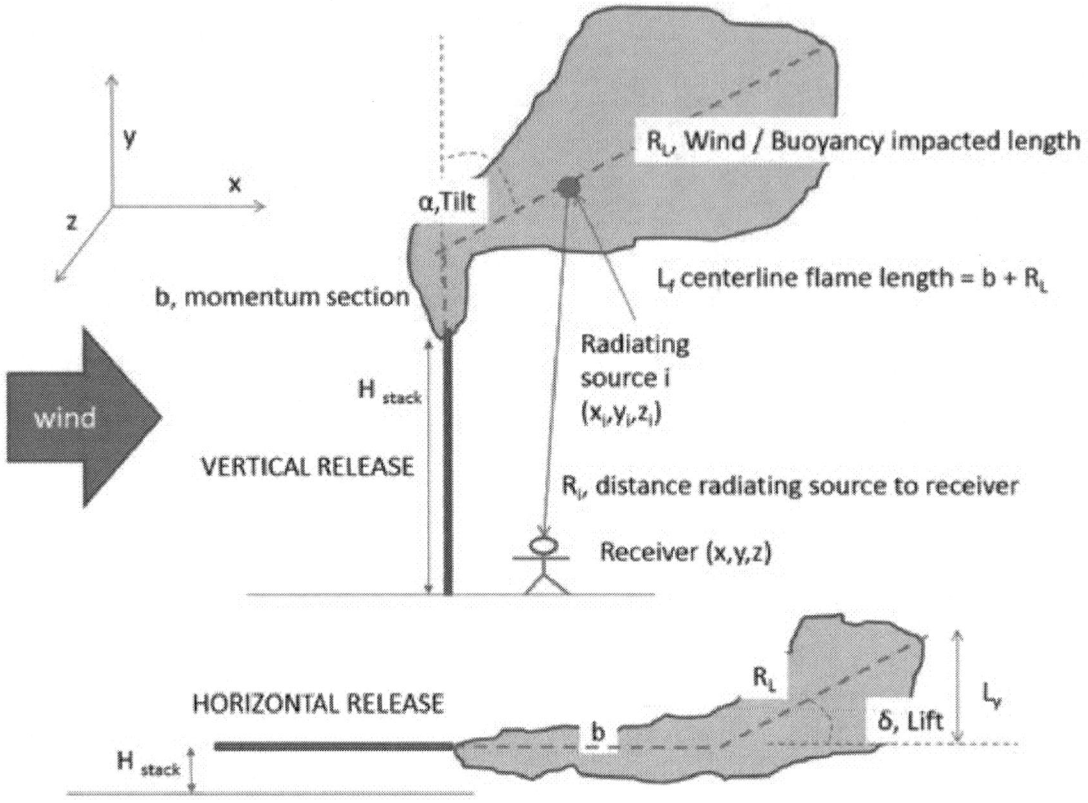

Figure 2. Flame geometry

3.1 **Coordinates.** X is in wind direction. Y is in the vertical direction. Z is in the sideways direction, perpendicular to the wind direction

3.2 **Flame dimensions.** *Initial momentum section length, b*: Flame initially travels straight in release direction due to momentum. For some gases with relatively low burning velocities, the flame may lift off from the flare tip (no combustion at tip). For purposes of this model, b is the total length that is momentum dominated, included any lift off. *Wind - buoyancy section length, R_L*: flame travels in direction determined by buoyancy and wind conditions. Vertical flame direction defined by tilt angle, α. Horizontal flames defined by lift, L_y or lift angle, δ. *Centerline flame length, L_f* is the overall flame length (b + R_L)

3.3 **Flame radiation.** The fraction of the total heat released in the flame that is in the form of thermal radiation is called the radiant fraction, F determined empirically from test data. The total heat radiating from the flame in AP Flame is distributed as individual point sources along the flame centerline, weighted to be a maximum at the widest point in the flame (about 2/3 along the flame, the radiant section), with low contributions from the momentum section and at the tip, where the flame is wispy [11]. The heat flux experienced at the receiver is the sum of the heat received from each point source (function of heat emitted and the distance between the source and the receiver, R_i), with allowance for the air transmissivity, the fraction of thermal radiant passing through a thickness of air (not absorbed). Heat fluxes from other thermal radiation sources, such as solar, should be added to determine total heat flux at any point of interest.

4.0 Summary of data. Table 1 summarizes all data sources used in this analysis. Full set of data used in the development of AP Flame is provided at the end of this paper.

4.1 **Previously published data:** Fishburne, (1979) [3] performed large scale flare testing of ambient and cold hydrogen vertical releases through a variety of vent sizes. Flame length and radiant fraction results provided. Chamberlain (1987) [1] performed large scale vertical natural gas (and similar) flare testing at off-shore and on-shore facilities and presented a flare model derived from this data which is used as the basis for the new model presented in this paper. Shirvill (2005) [9] performed tests on high pressure horizontal hydrogen releases, both ignited and unignited and presented flame length results. Schefer (2007) [16] performed tests with high pressure vertical hydrogen releases and presented flame lengths and radiant fractions, as well as a model for high pressure, vertical jet fires in the absence of wind. DNV GL (previously Advantica) (2008) [7] performed tests on high pressure horizontal natural gas and natural gas / hydrogen mixtures and presented flame lengths and heat flux measurements, from which this author has derived radiant fractions. Willoughby (2011) [5] performed tests on vertical jet fires and provided photos of flames, with weather conditions as well as heat flux values at different locations, from which this author has extracted flame lengths, flame tilt and radiant fractions. Air Products (2014) [9][20] performed tests on full scale hydrogen plant vertical flares with both hydrogen and syngas (73% H_2) and presented results on flame length, heat flux and radiant fraction.

4.2 **Air Products previously unpublished test data**: In 2008, the company commissioned DNV GL (then Advantica) to perform tests on high pressure horizontal hydrogen jet fires, collecting information on flame length, heat flux and radiant fraction. In 2015, Air Products completed testing at its R&D facility on horizontal and 45 degree jet fires at different elevations, of mixtures of natural gas, hydrogen, nitrogen and carbon dioxide and obtained data on flame length, heat flux and radiant fraction to develop methods to model syngas mixtures. Results of these tests are presented in this report.

Table 1. Summary of data sources

data source	gas	release direction	flare diameter "	release rate kg/s	flame length m	Test location	Ref.
Fishburne, 1979	H_2	Vertical	6 - 27	2 - 126	48 - 110	USA Rocketdyne test site	[3]

Chamberlain, 1987	NG	Vertical	6-42	4 - 63	-	UK on & off shore	[1]
Shirvill, 2005	H2	Horizontal	0.12 – 0.24	0.02 - 1	4 - 6	UK HSL site	[8]
Schefer, 2007	H_2	Vertical	0.2 - 0.3	0.06 - 0.4	-	USA SRI test site	[16]
DNV GL, 2008	NG and NG/H2 mixtures	Horizontal	0.8 - 2	3 - 20	20 - 50	UK Spadeadam	[7]
Willoughby, 2011	H_2	Vertical	0.75 -2	0.04 - 0.4	3 - 7	UK HSL test site	[5]
Air Products, 2014	H_2	Vertical	36	1.3	30	Pasadena, TX H2 plant	[9]
Air Products, 2014	H_2 and syngas	Vertical	27	1.7 - 13	25	Garville, LA H2 plant	[20]
Air Products / DNV GL , 2009	H_2	Horizontal	0.7 - 2	1 - 7.5	17 - 49	UK Spadeadam	Previously unpublished
Air Products, 2015	NG, H_2, mixtures with N_2, CO_2	Horizontal	1	0.02 - 0.07	3 - 5	USA R&D lab	Previously unpublished

5.0 Typical flames observed during Air Products tests. Figure 3 shows a selection of flames observed during Air Products tests. Figure 3a and 3b were obtained during the Air Products / DNV GL tests at Spadeadam, for hydrogen releases from ¾" and 2" horizontal pipe. Figure 3c and 3d were obtained during the Air Products Garyville full scale vertical flare tests for hydrogen and syngas using IR camera. Figure 3e and 3f were obtained during Air Products R&D tests for 1" 45 degree natural gas and horizontal hydrogen releases.

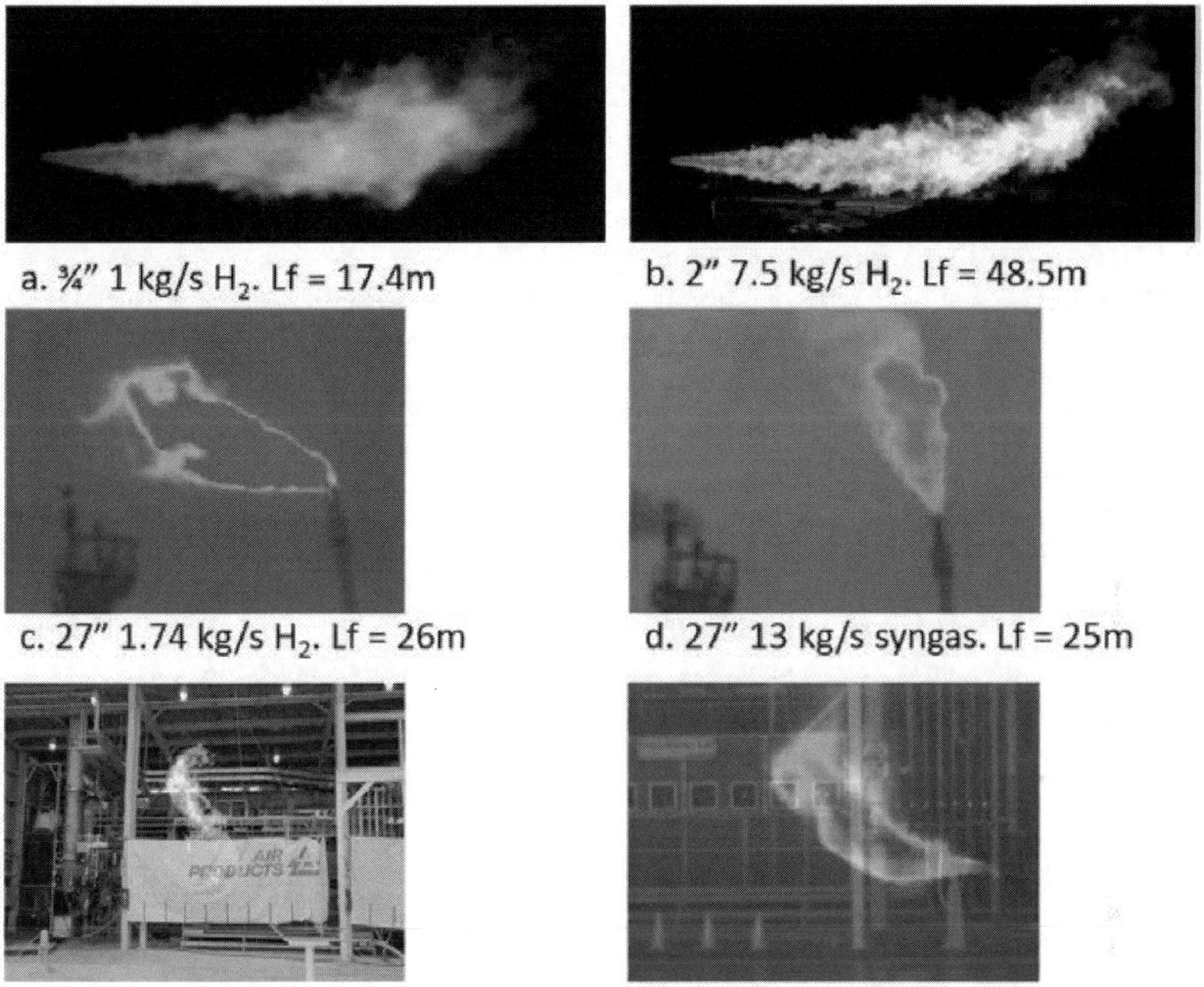

a. ¾" 1 kg/s H_2. Lf = 17.4m
b. 2" 7.5 kg/s H_2. Lf = 48.5m
c. 27" 1.74 kg/s H_2. Lf = 26m
d. 27" 13 kg/s syngas. Lf = 25m
e. 1" 0.03 kg/s natural gas. Lf = 3.5m
f. 1" 0.005 kg/s H_2. Lf = 3m

Figure 3. Typical flames observed during the various Air Products tests

6.0 AP Flame basis. AP Flame is based on the Chamberlain model [1], developed in the 1980's from extensive data collected from natural gas vertical flares. The new model also includes features from the model developed by Sandia National Labs [11] for no wind high pressure flames, primarily hydrogen. AP Flame includes a number of newly developed correlations to adequately model mixtures and horizontal / 45 degree release flames. None of the modifications impact the predictions for natural gas from the original Chamberlain model. All equations from the Chamberlain model are clearly identified in the text with ^{Chamberlain} designation after the equation number. All other equations represent new developments unless otherwise referenced. All symbols used are described at the end of the paper. To help follow the series of equations, this is a simple overview of the calculation steps

6.1 Determine source condition for the flame, expanded source velocity u_{exp}, density ρ_{exp} and diameter d_{exp}.

6.2 Determine Richardson number ζ (L_{BO}) and zero wind flame length L_{BO}, both used extensively as correlation variables.

6.3 Determine flame dimensions b, R_L, L_f and tilt α (for vertical releases) or lift δ (for horizontal releases)

6.4 Determine flame radiant fraction F based on gas mixture and release orientation

6.5 Determine overall heat flux, q at point of interest, by summing heat fluxes received from thermal radiation sources along flame.

7.0 Jet release expanded source condition for choked flow releases it is necessary to determine conditions downstream of the choke, where pressure is atmospheric, called the expanded (or effective) source condition. The expanded source condition defines the source term used in AP Flame. AP Flame uses the Yuceil methodology [21] to determine conditions at the expanded source, using conservation of mass, momentum and energy. Expanded source conditions (u_{exp}, ρ_{exp}, d_{exp}) are calculated from the release throat (pipe exit) condition, where isentropic expansion and choked flow, if applicable, are applied, (ρ_{exp} is determined from H_{exp} and $P_{exp} = P_{atmos}$). Note that the expanded source Mach number > 1 when throat is choked. For non-choked releases, conditions at the throat and at the expanded source are the same, Mach number < 1. The mass release rate, $\dot{m}$ is either specified or calculated based on upstream pressure.

The equations for conservation of momentum, energy, and mass are written as [1] – [3],

$$u_{\exp} = u_{throat} + \frac{P_{throat} - P_{atoms}}{u_{throat}\rho_{throat}} \quad [1]$$

$$H_{\exp} = H_{throat} - \frac{1}{2}\left(u_{\exp}^2 - u_{throat}^2\right) \quad [2]$$

$$\frac{\dot{m}}{\rho_{\exp}u_{\exp}} = \frac{\pi}{4}d_{\exp}^2 \quad [3]$$

8.0 Flame geometry fundamentals

8.1 **Richardson number, ζ (L_{BO}).** This dimensionless number is used extensively by Chamberlain and represents the ratio of buoyancy to momentum flux for the zero wind flame length, L_{BO}.

$$\text{Richardson number, } \xi(L_{BO}) = \left(\frac{g}{D_s^2 u_{\exp}^2}\right)^{0.333} L_{BO} \quad [4]^{\text{Chamberlain}}$$

$$\text{Equivalent air source diameter, } D_s = d_{\exp}\left(\frac{\rho_{\exp}}{\rho_{air}}\right)^{0.5} \quad [5]^{\text{Chamberlain}}$$

Here ρ_{air} is ambient air density, and ρ_{exp} is expanded source density.

8.2 **Zero wind flame length, L_{BO}.** This parameter is used extensively in the model and Chamberlain uses a previously developed correlation, equation [6] by Kalghatgi [20] which is solved iteratively to obtain L_{BO}, using the Richardson number. mw_{air} = mol wt. air, $mw_{products}$ = mol wt. stoichiometric combustion products. T_{ad} = Adiabatic combustion temperature (constant pressure), T_{air} = Ambient air temperature. F_s = stoichiometric mass fraction fuel in fuel + air *(note: simplified equation for Fs provided in original Chamberlain paper only applicable to pure fuels of form $C_n H_{2n+2}$ and should not be used for syngas).* Determination of T_{ad} can be done using the following newly developed correlation, based on T_{ad} values for a range of syngas compositions from pure H_2 through to condensate off-gas (15% H_2, remainder CO_2 and water vapor) determined using the online Chemical Equilibrium with Applications (CEA) from NASA[14]. The correlation provide a reasonable fit, see Figure 4.

$$\left(\frac{D_s\beta}{L_{BO}F_s}\right)^{2/3} = 0.2 + 0.024\xi(L_{BO}) \qquad [6]^{\text{Chamberlain}}$$

Where β is the density correction,

$$\beta = \left(\frac{mw_{air}}{mw_{products}}\frac{T_{ad}}{T_{air}}\right)^{0.5} \qquad [7]^{\text{Chamberlain}}$$

In which,

$$T_{ad} = T_{ad,fuel}\left(-0.7395\,CD_{equiv}^{2} + 0.0366\,CD_{equiv} + 0.9972\right)$$
$$T_{ad,fuel} = 321.08\,H2_{equiv}^{3} - 310.88\,H2_{equiv}^{2} + 144.03\,H2_{equiv} + 2223$$
$$CD_{equiv} = x_{CD} + 0.84x_{WA} + 0.59x_{N2}$$
$$H2_{equiv} = \frac{x_{H2_{equiv}}}{x_{H2_{equiv}} + x_{C1_{equiv}}} \qquad [8]$$
$$x_{H2_{equiv}} = x_{H2} + x_{CM}$$
$$x_{C1_{equiv}} = x_{C1} + x_{C2} + x_{C3} + x_{C4}$$

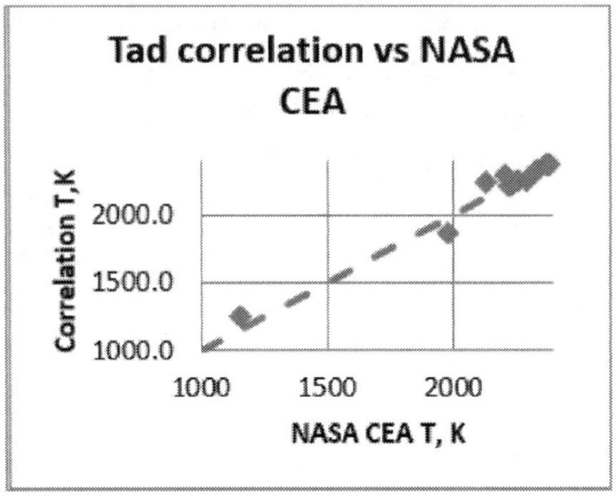

Figure 4. Adiabatic combustion temperature correlation vs. NASA CEA results

8.3 **Wind velocity at flare tip.** Reference wind velocity, u_{wind0} is measured at a reference height, h_0 (10m).Wind velocity at the flare tip, u_{wind} is scaled according to flare stack height, where a = power exponent depends on ground roughness of surrounding area: rural (0.14); built up rural (0.27); suburban (0.34). If release elevation < 10m, the 10m value is used.

$$u_{wind} = u_{wind,0}\left(\frac{h_{stack}}{h_0}\right)^a \quad [9]$$

9.0 Vertical Flames

9.1 **Vertical Flame tilt**. Angle between vertical and flame buoyancy section axis, α. Chamberlain predicts tilt for natural gas flames, α'

$$\alpha'\xi(L_{BO}) = \begin{cases} 8000R, & for \quad R < 0.05 \\ 1726(R-0.026)^{0.5}+134, & for \quad R \geq 0.05 \end{cases} \quad [10]^{\text{Chamberlain}}$$

$$\text{Flame velocity ratio, } R = \frac{u_{wind}}{u_{exp}} \quad [11]^{\text{Chamberlain}}$$

For gases other than natural gas, it is found that a correction term is required since hydrogen tilts more than predicted by Chamberlain, and syngas less, see Figure 5. A correction can be applied based on the ratio of mass release rates (basically a momentum correction) using the concept of $\dot{m}_{NG}$ = mass release rate of natural gas that creates the same L_{BO} length as the gas of interest, for the same flare diameter. A correlation for $\dot{m}_{NG}$ has been developed through multiple runs of the chamberlain model .This method successfully corrects both hydrogen under-prediction and syngas over-prediction.

$$\alpha = \alpha' \frac{\dot{m}_{NG}}{\dot{m}} \qquad [12]$$

Where,

$$\dot{m}_{NG} = 255.42 d_0^2 Y_{NG} \qquad [13]$$

$$Y_{NG} = \begin{cases} 2.0074X_{NG}^2 - 1.0103X_{NG} + 0.0185, & for \quad X_{NG} > 0.665 \\ 0.7873X_{NG}^{2.9013}, & for \quad X_{NG} \le 0.665 \end{cases}$$

$$X_{NG} = \frac{L_{BO}}{174.83 d_0^{0.8794}}$$

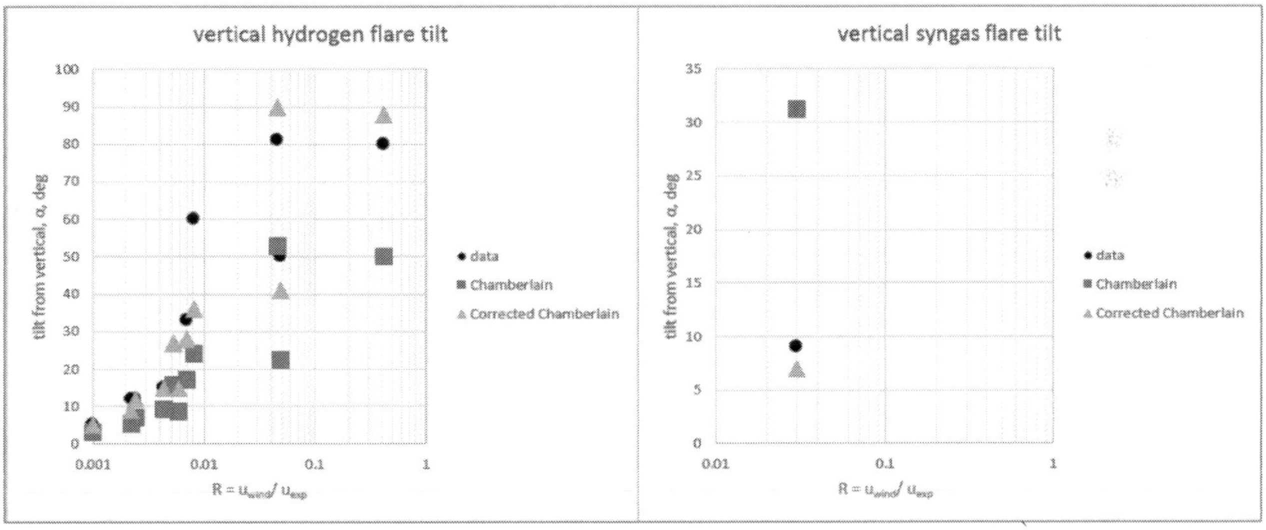

Figure 5 vertical flame tilt vs R for hydrogen and syngas

9.2 **Vertical Flare tip to flame end distance, L_B.** This parameter is used by Chamberlain as an intermediate step in determining flame dimensions L_f and R_L. Flame length decreases with increasing wind, due to air entrainment.

$$\frac{L_B}{L_{BO}} = 0.49 + 0.51 e^{-0.4 u_{wind}} \qquad [14]^{\text{Chamberlain}}$$

For tilt angles $\alpha > \pi/3$ (60 degrees) approximately, it is found that flame length actually increases and obtains a maximum length of approximately L_{BO} at $\alpha = \pi/2$ (90 degrees horizontal), which is not predicted by Chamberlain, see Figure 6. Consequently, a modification is required, which increases L_B from its value at 60 degrees tilt (L_{B60}) to L_{BO} at 90 degrees tilt. This requires a several step process, using R_{60}, the value of R (=u_{wind} / u_{exp}) that yields a tilt of 60 degrees:

$$\frac{L_B}{L_{BO}} = ae^{b\alpha} \text{ when } \alpha < \pi/3 \quad [15]$$

$$a = e^{-90b}$$

$$b = \frac{1}{30}\ln\left(\frac{1}{(L_B/L_{BO})_{60}}\right)$$

$$(L_B/L_{BO})_{60} = 0.49 + 0.51e^{-0.4u_{wind,60}}$$

$$u_{wind,60} = R_{60}u_{exp}$$

$$R_{60} = \begin{cases} 0.026 + \left(\left(60\xi(L_{BO})\frac{\dot{m}}{\dot{m}_{NG}} - 134\right)\Big/1726\right)^2, & for \quad \frac{60\dot{m}}{\dot{m}_{NG}}\frac{\xi(L_{BO})}{8000} > 0.05 \\ \frac{60\dot{m}}{\dot{m}_{NG}}\frac{\xi(L_{BO})}{8000}, & for \quad \frac{60\dot{m}}{\dot{m}_{NG}}\frac{\xi(L_{BO})}{8000} \le 0.05 \end{cases}$$

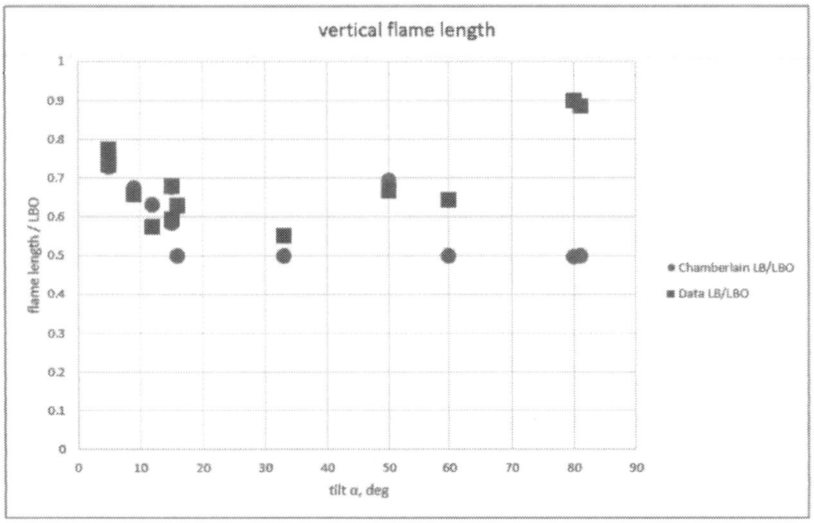

Figure 6. Vertical flame length as a function of tilt

9.3 **Vertical flare momentum section length, b**. The Chamberlain method is used directly in the new model since it found to be adequate for hydrogen and syngas

$$b = \begin{cases} 0.2L_B, & for \quad \alpha = 0 \\ L_B\frac{\sin(K\alpha)}{\sin(\alpha)}, & for \quad \alpha \neq 0, \alpha \neq \pi \\ 0.015L_B, & for \quad \alpha = \pi \end{cases} \quad [16]^{\text{Chamberlain}}$$

$$K = 0.185e^{-20R} + 0.015$$

9.4 **Vertical flare wind – buoyancy section length, R_L**

By geometry,

$$R_L = \left(L_B^2 - (b\sin(\alpha))^2\right)^{0.5} - b\cos(\alpha) \qquad [17]^{\text{Chamberlain}}$$

9.5 **Vertical flare centerline flame length, L_f**

By definition,

$$L_f = b + R_L \qquad [18]$$

10.0 Horizontal Flames

10.1 **Horizontal flame length, L_f.** L_f is found to be independent of wind within range of data obtained and can be approximated to be L_{BO}, see Figure 7a.

$$L_f = L_{BO} \qquad [19]$$

10.2 **Horizontal flame momentum section length, b.** Momentum section length is found to be a function of Richardson number, see Figure 7b. Switching L_{BO} for L_f from [19] gives:

$$\frac{b}{L_{BO}} = -0.125\xi(L_{BO}) + 1.25, \quad 0 \le \frac{b}{L_{BO}} \le 1 \qquad [20]$$

10.3 **Horizontal flame lift, Ly and lift angle, δ.** Horizontal flames lift in the wind – buoyancy section, due to buoyancy. Lift is found to be a function of Richardson number. See Figure 7c. Switching L_{BO} for L_f from [19] gives:

$$\frac{L_y}{L_{BO}} = 0.125\xi(L_{BO}) - 0.25, \quad 0 \le \frac{L_y}{L_{BO}} \le 1 \qquad [21]$$

Equations [20] and [21] are mutually consistent. High jet momentum drives Richardson number down and b/L_{BO} trends towards 1 as L_y / L_{BO} trends towards 0: a flame that stays horizontal. Low jet momentum drives Richardson number up and b/L_{BO} trends towards 0 as L_y / L_{BO} trends towards 1: a flame that rapidly becomes vertical.

The horizontal flame correlation does not include the impact of wind (no available data) and so it is mathematically possible for a horizontal release to become more vertical than the same flame if released vertically (where AP Flame does account for wind), which is clearly wrong. To avoid this, the maximum value for δ is capped so that it cannot exceed (π/2– α) where α is the tilt from vertical, for the same flame released vertically. Lift angle, δ is obtained by simple geometry:

$$\delta = MAX(0, MIN\left\{ ASIN\left(\frac{L_y}{L_f - b}\right), \left(\frac{\pi}{2} - \alpha\right)\right\}) \qquad [22]$$

This means that each time a horizontal release is run in AP Flame, an equivalent vertical release must also be run to determine α. Note that if δ is limited to $(\pi/2 - \alpha)$, then L_f must be adjusted to match, $L_y = (L_f - b) \sin(\delta)$.

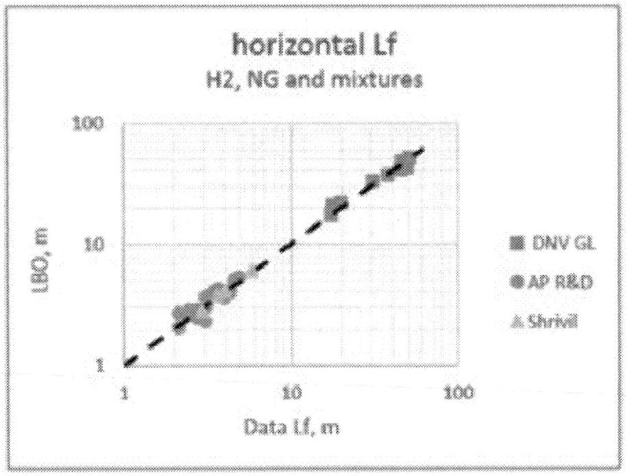

a. Horizontal flame length

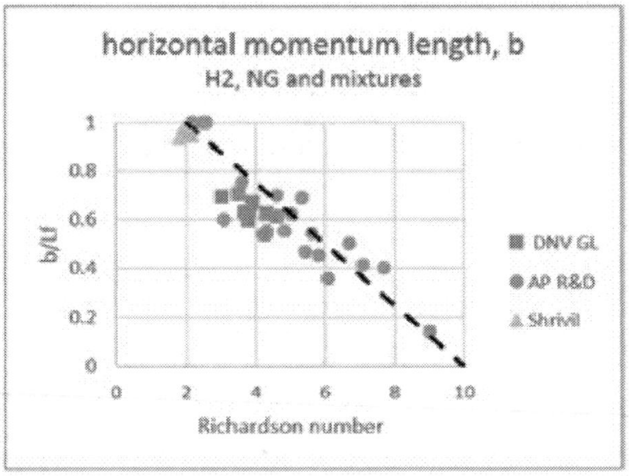

b. Horizontal flame momentum section

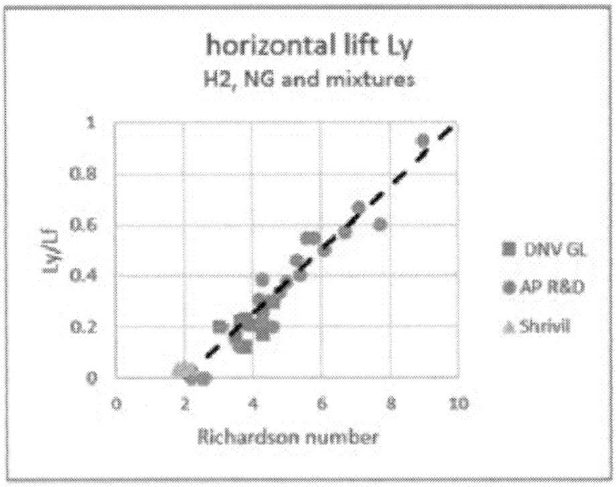

c. Horizontal flame lift

Figure 7. Horizontal flame dimensions for nat. gas, H_2, and mixtures of nat. gas, H_2, N_2 & CO_2

11.0 45 degree release flames

11.1 **Flame length.** Based on Air Products R&D tests, the length of 45 degree release flames is found to be similar to horizontal flames at the same release rate:

$$L_f = L_{BO} \qquad [23]$$

11.2 **Flame tilt.** Based on limited testing, 45 degree release flames can be approximately modelled assuming that they are relatively straight and continue at approximately a 45 degree angle from release point to the end of the flame.

12.0 Flame radiant heat fraction, F. Total heat radiating from flame, Q_{rad} is determined from F = radiant heat fraction, $\dot{m}$ = mass flow rate (kg/s), $\Delta H_{combustion}$ = gas heat of combustion (LHV) using equation [24].

12.1 AP Flame uses the Chamberlain correlation $F = 0.21 \exp(-0.00323\, u_{exp})$ for vertical natural gas flames. Data indicates that this correlation is also reasonable for horizontal natural gas flames, see Figure 8a.

12.2 Figure 8b shows radiant fractions for all available hydrogen data, indicating that vertical release values of F can be well correlated with M_{flux} mass flux at pipe exit (not expanded source). This figure also shows that equivalent horizontal and 45 degree release hydrogen flames have higher F values.

12.3 To predict F values for horizontal and 45 degree release hydrogen flames, a correction factor to the predicted vertical F value (from the M_{flux} correlation) is applied, based on the ratio of flame length, L_f, to elevation of radiant center above ground, h_{RC} see Figure 8c. The correction factor increases as the distance between the flame radiant center and the ground, h_{RC} decreases. Where the flame is very near the ground, it will impinge on the ground and AP Flame is no longer really valid and a cap of 10 is applied to L_f / h_{RC}.

12.4 Radiant fractions are found to be similar for 45 degree and horizontal flames at the same flow rates and example for hydrogen / natural gas mixtures is shown in Figure 8d. Note that for this data F values are similar for natural gas and hydrogen, although this is not generally the case.

12.5 Addition of inert gases reduces radiant fraction and a simple relationship where radiant fraction is proportional to the fuel fraction in the gas does a reasonable job capturing the effect within the range tested. Figure 8e summarizes Air Products R&D test results which were obtained for 45 degree flames, which proved to be more stable than horizontal flames at this scale. The fuel flow was kept constant and inert added as required. It was not possible to obtain readings for higher inert addition due to flame instability / blow-off.

12.6 For purposes of the overall correlation for radiant fraction, the syngas composition is split up into two effective components: $C1_{equiv}$ which is the total mole fraction of "sooting" components C1 through C4 (assumes C2, C3, C4 are only minor components in gas), $H2_{equiv}$, which is the total mole fraction of "non-sooting" components H_2 and CO - no data available on CO flames, so assumed similar to H_2 based on similar stoichiometry and flame temperature. Note that $C1_{equiv} + H2_{equiv}$ add up to less than 1 when inerts are present, reflecting the "dilution" effect of inerts. F_{C1} and F_{H2} are the radiant fractions of "sooting" and "non-sooting" components respectively.

12.7 The minimum values for F will occur for small high pressure release flames, lowest value reported is 0.05 [17], which acts as a lower limit in AP Flame.

$$Q_{rad} = F\dot{m}\Delta H_{combustion} \quad [24]$$

$$F = MAX\{(C1_{equiv} F_{C1} + H2_{equiv} F_{H2}), 0.05\}$$

$$C1_{equiv} = x_{C1} + x_{C2} + x_{C3} + x_{C4}$$

$$H2_{equiv} = x_{H2} + x_{CO}$$

$$F_{C1} = 0.21 e^{-0.00323\, u_{exp}}$$ Chamberlain

$$F_{H2} = F_{corr}\left(0.1691 - 0.01 \ln\left(M_{flux}\right)\right)$$

$$M_{flux} = \frac{4\dot{m}}{\pi d_0^2}$$

$$F_{corr} = \begin{cases} 1, & for \quad vertical \\ 1.36 + 0.076\, MIN\{10, \mathrm{L_f}/\mathrm{h_{RC}}\}, & for \quad horizontal\ /\ 45^{\circ} \end{cases}$$

$$h_{RC} = \begin{cases} h_{stack} + MAX(0, (0.66 L_f - b)\sin(\delta)), & for \quad horizontal \\ h_{stack} + 0.66 L_f \sin(\pi/4), & for \quad 45^{\circ} \end{cases} \qquad [25]$$

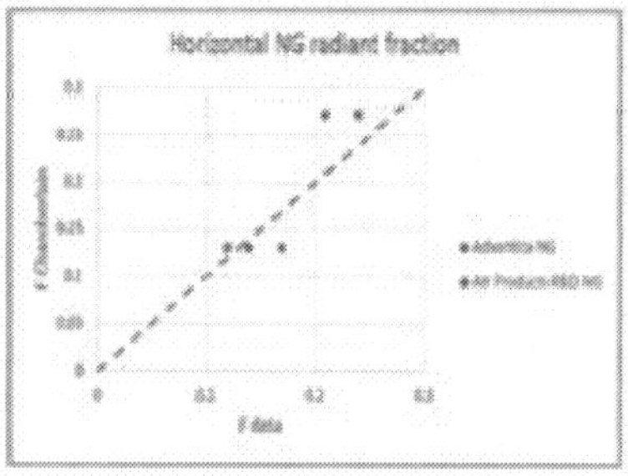

a. Horizontal natural gas

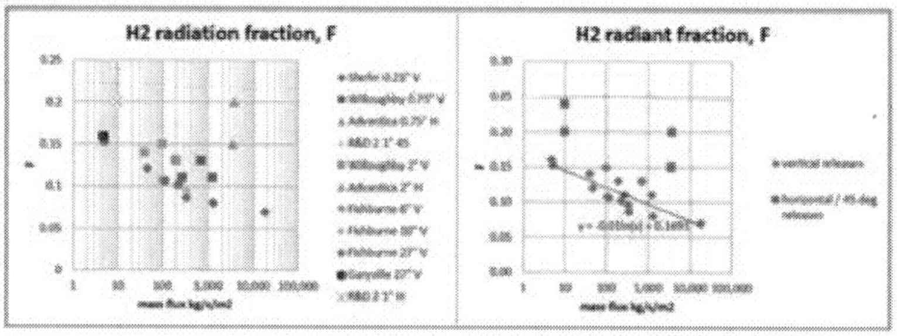

b. Vertical and horizontal / 45 deg H_2

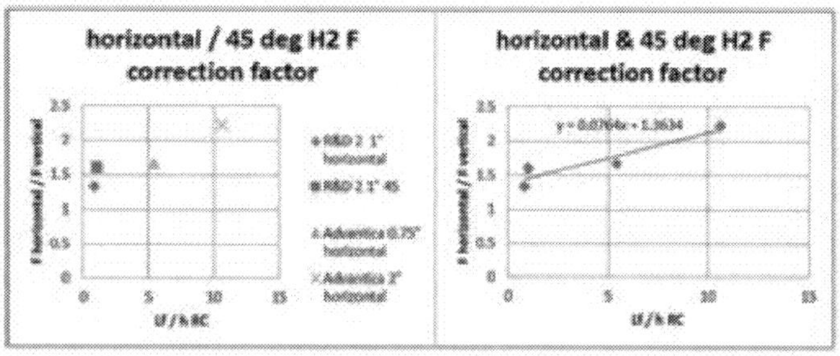

c. Impact of elevation on horizontal / 45 degree H_2

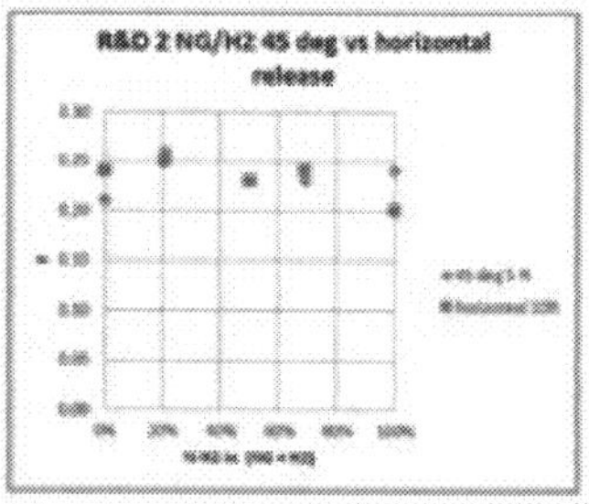

d. horizontal / 45 deg H_2 / NG mixes

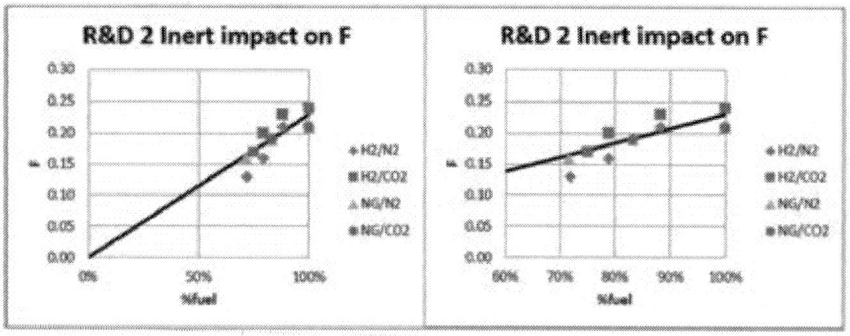

e. Impact of inert addition

Figure 8. Radiant fraction

13.0 Distribution of radiating sources in flame. The model splits the total radiated heat up into 30 point sources located evenly along the flame centerline (length L_f), following method by Ekoto [11]. Each point source, i, has a discrete location (x_i, y_i, z_i) following along the flame centerline path of b and R_L. The amount of heat radiating from each point source has a weighting factor, w_i arranged to be a maximum at the radiant center of the flame, 2/3 along L_f (=b + R_L) and a minimum at each end. Figure 9 shows how the radiating heat is distributed.

The radiating heat weighting factor for ith point in flame, w_i, is defined as,

$$w_i = \begin{cases} 0.003273i, & for \quad 1 \le i \le 20 \\ 0.003273(20 - 1.9(i - 20)), & for \quad 20 \le i \le 30 \end{cases} \qquad [26]$$

Where $Q_{rad,i}$ is the Heat radiation from the ith point source, $Q_{rad,i}$,

$$Q_{rad,i} = Q_{rad} w_i \qquad [27]$$

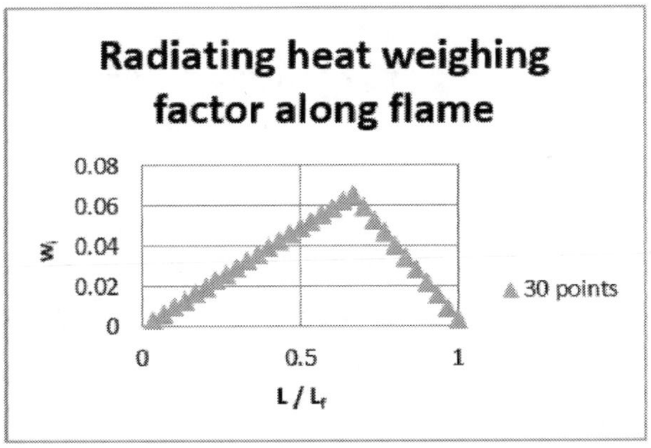

Figure 9. Distribution of radiant heat sources along flame centerline

14.0 Air transmissivity. Water and carbon dioxide in the air adsorb thermal radiation as a function of the concentration of carbon dioxide in the air (fixed), water in the air (function temperature and relative humidity) and distance traveled through air (distance between radiation source and receiver). Method from Wayne [14] τ_i = Air transmissivity = fraction of radiated heat passing through distance R_i of air, R_i = distance from source i to receiver, R_H = Ambient relative humidity (fraction) S_{mm} = saturated water pressure, mm HG at ambient air temperature, T_{air}

$$\begin{aligned} \tau_i = {} & 1.006 - 0.01171\log_{10}(X_{air}(WA)_i) - 0.02368(\log_{10}(X_{air}(WA)_i))^2 \\ & - 0.03188\log_{10}(X_{air}(CD)_i) + 0.001164(\log_{10}(X_{air}(CD)_i))^2 \end{aligned} \qquad [28]$$

$$X_{air}(WA)_i = 288.651\frac{R_h R_i S_{mm}}{T} \quad [29]$$

$$X_{air}(CD)_i = 273\frac{R_i}{T}$$

$$S_{mm} = 0.0075\, e^{\left(23.18986-\left(\frac{3816.42}{T_{air}-46.13}\right)\right)} \quad [30]$$

15.0 Heat flux at point of interest receiver. The heat total heat flux from the flame, at a point of interest receiver, q kw/m2 is the sum of heat fluxes received from each radiating point (i) in the flame. (x,y,z) is location of receiver, (x_i, y_i, z_i) is location of ith radiating source.

$$q = \sum_{i=1}^{30}\frac{Q_{rad,i}\tau_i}{4\pi R_i^2} \quad [31]$$

Where R_i is the distance from i^{th} point in flame to receiver,

$$R_i = \left((x-x_i)^2+(y-y_i)^2+(z-z_i)^2\right)^{0.5} \quad [32]$$

16.0 Overall model predictions for flame length and radiant fraction vs all available data. Figure 10 shows overall comparisons of model vs. data for some key parameters of vertical, horizontal and 45 degree flames. Overall, the model performs well. It is encouraging, for example, to see the good prediction of the full scale Garyville vertical syngas flame radiant fraction (0.12 model vs. 0.11 data) and flame length (25m model vs. 25m data) using the method developed from small scale R&D tests on horizontal syngas mixture flames

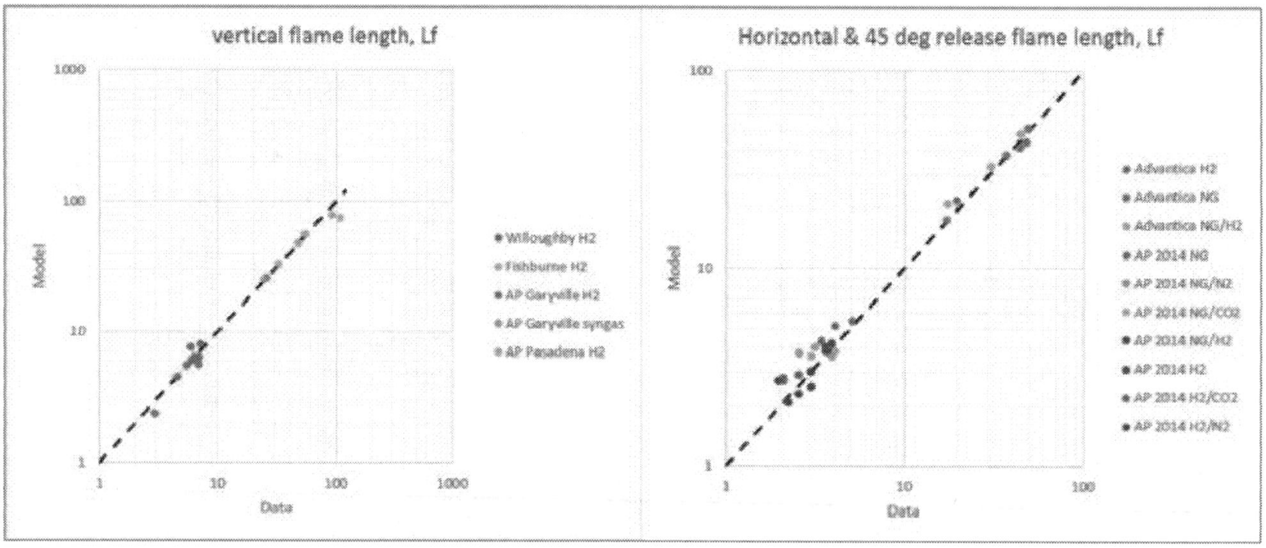

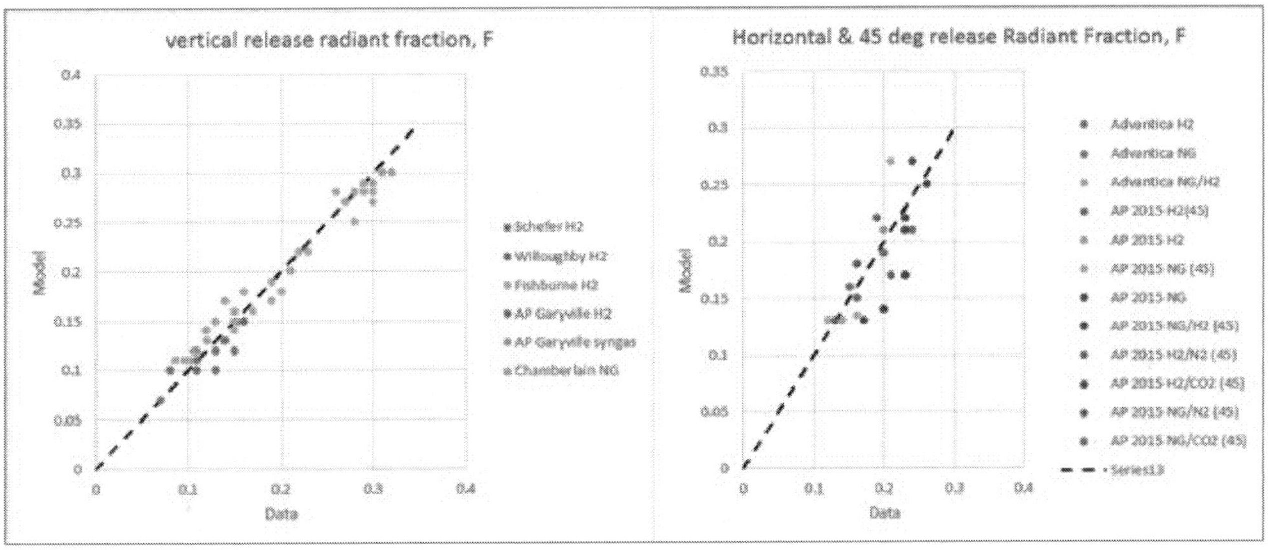

Figure 10. Overall comparison of model predictions vs. data for some key flame parameters

17.0 Summary and Conclusions

17.1 The new AP Flame model is based on the established Chamberlain model which has proven to be a strong foundation. The new model extends the scope to include non-sooting gases such as hydrogen and carbon monoxide as well as inerts and it also covers horizontal and 45 degree release flames.

17.2 The model is based on a reasonably extensive dataset of previously published and new test data on natural gas and hydrogen and more limited data on syngas mixtures. Data covers a wide range of scales from lab and pilot scale up to and including large industrial flares (0.2" to 42" pipe diameters) and low pressure up to high pressure releases (non-choked and choked).

17.3 Given the inherent variability in the natural of flames and the many challenges involved with setting up tests and measuring needed parameters, the new AP Flame performs quite well across the full range of flames studied.

18.0 List of variables

symbol	units	description
a		power exponent for scaling wind at different heights
a		Parameter in vertical flame length correction
b		Parameter in vertical flame length correction
b	m	length of vertical flame momentum dominated
CD_{equiv}		weighted carbon dioxide equivalent for CD, N2, WA components
$C1_{equiv}$		sum C1, C2, C3, C4 mol fractions
d	m	diameter
d_0	m	flare exit diameter
D_s	m	equivalent air source diameter

F		radiant fraction of flame
F_s		stoichiometric mass fraction fuel in fuel + air
g	m/s^2	gravitational constant, 9.81
H	J/kg	enthalpy
ΔH_{comb}	KJ/kg	gas heat of combustion (LHV)
$H2_{equiv}$		sum H2, CO mol fractions
h_{stack}	m	height of flare tip above grade
h_0	m	reference height, 10m
h_{RC}	m	Height of flame radiant center above ground
$\dot{m}$	kg/s	release rate
mw		mol wt. of released gas
K		variable used to determine b
L	m	dimension of interest, such as flame length
L_B	m	distance from flare tip to flame end
L_{BO}	m	flame length in no wind
L_{mb}	m	mean beam length
L_f	m	centerline flame length
L_y	m	Horizontal flame lift
P	Pa	pressure
q	Kw/m2	Heat flux
Q_{rad}	kw	total heat radiated from flame
R		ratio wind velocity / expanded source velocity
R_H		relative humidity
R_{H2}		Ratio H2 to total fuel fraction of gas, used in adiabatic flame temperature correlation
R_i	m	distance from radiating source to receiver
R_L	m	length of flame buoyancy / wind dominated
R_{WA}		ratio water / (water + carbon dioxide)
S_{mm}	mm Hg	ambient air saturated water pressure
T	K	temperature
u	m/s	velocity
X		ratio of LBO to LBO at given Mach number
x		mol fraction
x	m	distance from flare tip in wind direction
X_{air} (CD)		variable representing impact of CD in air on τ
X_{air} (WA)		variable representing impact of WA in air on τ
Y		ratio of release rate / release rate at given Mach number
y	m	vertical distance above grade
Y_L	m	horizontal flame lift due to buoyancy
z	m	cross wind distance
α	radians	vertical flame tilt from vertical due to wind
α'	radians	vertical flame tilt from vertical due to wind for natural gas

β		density correction factor
δ	radians	Horizontal flame lift angle
ρ	kg/m^3	density
ζ (L)		Richardson number
τ		air transmissivity

SUBSCRIPTS		
ad		adiabatic flame condition
ad fuel		adiabatic flame condition for fuel components only (no inerts)
air		ambient air condition
atmos		atmospheric condition
C1		methane
C2		ethane
C3		propane
C4		butane
H2		hydrogen
CM		carbon monoxide
CD		carbon dioxide
choke		choked flow condition
exp		Expanded (effective) source condition
flame		flame condition
i		i^{th} radiating source in flame
NG		natural gas
N2		nitrogen
O2		oxygen
products		combustion products condition
throat		release exit condition
WA		water
wind		wind condition at flare tip
wind0		wind condition at reference height (10m)
0		initial condition
60		condition at 60 degree tilt

19.0 Full dataset

Source	H2	C1	CO	CO2	N2	direction	h m	d "	T K	mdot kg/s	Ta K	wind m/s @ 10m	F	Lf m	b m	α deg	Ly m
Schefer [19]	1	0	0	0	0	V	5	0.2	288	0.36	288	0	0.07			0	
Schefer [19]	1	0	0	0	0	V	5	0.3	288	0.06	288	0	0.08			0	
Willoughby [5]	1	0	0	0	0	V	5.5	0.75	272	0.08	272	1.6	0.11	4.6		5	
Willoughby [5]	1	0	0	0	0	V	5.5	0.75	275	0.2	275	1.9	0.13	6.4		5	
Willoughby [5]	1	0	0	0	0	V	5.5	0.75	271	0.35	271	4.5	0.11	6.4		12	

Willoughby [5]	1	0	0	0	0	V	5.5	2	275	0.08	275	2.5	0.14	5.5		15	
Willoughby [5]	1	0	0	0	0	V	5.5	2	275	0.2	275	4.2	0.15	6.1		15	
Willoughby [5]	1	0	0	0	0	V	5.5	2	273	0.4	273	3.2	0.13	7.3		12	
Willoughby [5]	1	0	0	0	0	V	10	0.75	273	0.35	273	10		7		16	
Willoughby [5]	1	0	0	0	0	V	10	0.75	273	0.04	273	10		3		60	
Willoughby [5]	1	0	0	0	0	V	10	2	273	0.4	273	10		7		33	
Willoughby [5]	1	0	0	0	0	V	10	2	273	0.04	273	10		6		81	
AP / DNV GL	1	0	0	0	0	H	3.25	0.8	279	1	279	3.4	0.15	17.4	12.1		3.5
AP / DNV GL	1	0	0	0	0	H	3.25	2.1	279	7.5	279	2.6	0.20	48.5	30.8		11
AP / DNV GL	1	0	0	0	0	H	3.25	2.1	279	6.6	279	2.6	0.20	44.6	28.4		9.8
DNV GL [7]	0	1	0	0	0	H	3.25	0.8	290	2.9	290	5	0.12	19.8	11.9		2.4
DNV GL [7]	0	1	0	0	0	H	3.25	1.4	290	9.6	290	5	0.14	37.8	23.5		6.4
DNV GL [7]	0	1	0	0	0	H	3.25	2	290	19.5	290	5	0.17	49.9	30.8		15
DNV GL [7]	0.24	0.76	0	0	0	H	3.25	0.8	290	2.7	290	5	0.12	17.6	12.4		2.8
DNV GL [7]	0.24	0.76	0	0	0	H	3.25	1.4	290	7.2	290	5	0.14	30.7	20.6		6.4
DNV GL [7]	0.24	0.76	0	0	0	H	3.25	2	290	16.9	290	5	0.16	45.2	28.5		12
Fishburne [3]	1	0	0	0	0	V	10	6	125	1.9	290	1.5	0.11				
Fishburne [3]	1	0	0	0	0	V	10	10	125	17	290	1.5	0.10				
Fishburne [3]	1	0	0	0	0	V	10	27	125	1.9	290	1.5	0.15				
Fishburne [3]	1	0	0	0	0	V	10	27	125	17	290	1.5	0.12				
Fishburne [3]	1	0	0	0	0	V	10	27	125	43	290	1.5	0.11				
Fishburne [3]	1	0	0	0	0	V	10	27	250	43	290	1.5	0.11				
Fishburne [3]	1	0	0	0	0	V	10	27	125	84	290	1.5	0.10				
Fishburne [3]	1	0	0	0	0	V	10	27	125	126	290	1.5	0.09				
Fishburne [3]	1	0	0	0	0	V	10	27	125	5.5	290	1.5		48			
Fishburne [3]	1	0	0	0	0	V	10	27	125	9.1	290	1.5		55			
Fishburne [3]	1	0	0	0	0	V	10	27	125	22	290	1.5		110			
Fishburne [3]	1	0	0	0	0	V	10	27	125	26	290	1.5		93			
AP Garyville [23]	1	0	0	0	0	V	53	27	311	1.74	290	2.3	0.16	26		50	
AP Garyville [23]	0.73	0.07	0.04	0.16	0	V	53	27	311	13	290	2	0.11	25		9	
AP Pasadena [10]	1	0	0	0	0	V	30	36	310	1.3	290	9		33		80	
AP R&D	1	0	0	0	0	45 deg	1.5	1	290	0.005	290	2	0.24				
AP R&D	1	0	0	0	0	H	3	1	290	0.005	290	2	0.20				
AP R&D	0	1	0	0	0	45 deg	1.5	1	290	0.027	290	2	0.21				
AP R&D	0	1	0	0	0	H	3	1	290	0.027	290	2	0.24				
AP R&D	0.21	0.79	0	0	0	45 deg	1.5	1	290	0.025	290	2	0.26				
AP R&D	0.5	0.5	0	0	0	45 deg	1.5	1	290	0.023	290	2	0.23				
AP R&D	0.69	0.31	0	0	0	45 deg	1.5	1	290	0.019	290	2	0.23				
AP R&D	0.88	0	0	0	0.12	45 deg	1.5	1	290	0.014	290	2	0.21				
AP R&D	0.79	0	0	0	0.21	45 deg	1.5	1	290	0.024	290	2	0.16				
AP R&D	0.72	0	0	0	0.28	45 deg	1.5	1	290	0.032	290	2	0.13				
AP R&D	0.88	0	0	0.12	0	45 deg	1.5	1	290	0.02	290	2	0.23				
AP R&D	0.79	0	0	0.21	0	45 deg	1.5	1	290	0.034	290	2	0.20				

AP R&D	0.75	0	0	0.25	0	45 deg	1.5	1	290	0.042	290	2	0.17				
AP R&D	0	0.83	0	0	0.17	45 deg	1.5	1	290	0.036	290	2	0.19				
AP R&D	0	0.71	0	0	0.29	45 deg	1.5	1	290	0.045	290	2	0.16				
AP R&D	0	0.83	0	0.17	0	45 deg	1.5	1	290	0.041	290	2	0.19				
AP R&D	0	1	0	0	0	H	2.8	1	290	0.027	290	2		3.7	1.5		2.4
AP R&D	0	1	0	0	0	H	2.8	1	290	0.053	290	2		4.6	2.1		1.8
AP R&D	0	1	0	0	0	H	2.8	1	290	0.064	290	2		4.9	3.0		1.8
AP R&D	0	0.83	0	0	0.17	H	2.8	1	290	0.036	290	2		3.4	1.8		1.8
AP R&D	0	0.71	0	0	0.29	H	2.8	1	290	0.045	290	2		3.0	2.1		0.6
AP R&D	0	0.66	0	0	0.34	H	2.8	1	290	0.05	290	2		4.0	2.1		1.2
AP R&D	0	0.83	0	0.17	0	H	2.8	1	290	0.041	290	2		4.0	2.7		1.8
AP R&D	0	0.73	0	0.27	0	H	2.8	1	290	0.053	290	2		4.0	2.1		1.5
AP R&D	0.21	0.79	0	0	0	H	2.8	1	290	0.025	290	2		4.3	2.1		2.4
AP R&D	0.5	0.5	0	0	0	H	2.8	1	290	0.023	290	2		4.3	1.5		2.1
AP R&D	0.69	0.31	0	0	0	H	2.8	1	290	0.019	290	2		3.4	1.5		1.8
AP R&D	1	0	0	0	0	H	2.8	1	290	0.003	290	2		2.1	0.3		2.0
AP R&D	1	0	0	0	0	H	2.8	1	290	0.005	290	2		3.0	1.2		1.8
AP R&D	0.93	0	0	0.07	0	H	2.8	1	290	0.012	290	2		2.7	1.5		0.9
AP R&D	0.88	0	0	0.12	0	H	2.8	1	290	0.02	290	2		2.4	1.8		0.3
AP R&D	0.79	0	0	0.21	0	H	2.8	1	290	0.034	290	2		2.4	2.4		0.0
AP R&D	0.74	0	0	0.26	0	H	2.8	1	290	0.043	290	2		2.4	2.4		0.0
AP R&D	0.88	0	0	0	0.12	H	2.8	1	290	0.014	290	2		2.7	1.5		0.6
AP R&D	0.79	0	0	0	0.21	H	2.8	1	290	0.024	290	2		3.0	1.8		0.6
AP R&D	0.72	0	0	0	0.28	H	2.8	1	290	0.032	290	2		2.1	2.1		0.0
Chamberlain [1]	0	1	0	0	0	V	10	12	290	10.3	290	8	0.23				
Chamberlain [1]	0	1	0	0	0	V	10	12	290	17.4	290	8	0.20				
Chamberlain [1]	0	1	0	0	0	V	10	12	290	22.3	290	8	0.19				
Chamberlain [1]	0	1	0	0	0	V	10	12	290	24.6	290	8	0.15				
Chamberlain [1]	0	1	0	0	0	V	10	8	290	5.9	290	8	0.21				
Chamberlain [1]	0	1	0	0	0	V	10	8	290	8	290	8	0.20				
Chamberlain [1]	0	1	0	0	0	V	10	8	290	11.2	287	8	0.17				
Chamberlain [1]	0	1	0	0	0	V	10	8	290	14.9	290	8	0.15				
Chamberlain [1]	0	1	0	0	0	V	10	8	290	20.3	290	8	0.15				
Chamberlain [1]	0	1	0	0	0	V	10	6	290	4.6	290	8	0.16				
Chamberlain [1]	0	1	0	0	0	V	10	6	290	5.9	290	8	0.14				
Chamberlain [1]	0	1	0	0	0	V	10	6	290	7.6	290	8	0.13				
Chamberlain [1]	0	1	0	0	0	V	10	6	290	19.1	290	8	0.12				
Chamberlain [1]	0	1	0	0	0	V	10	42	290	27.2	290	8	0.29				
Chamberlain [1]	0	1	0	0	0	V	10	42	290	35.8	290	8	0.26				
Chamberlain [1]	0	1	0	0	0	V	10	42	290	43.6	290	8	0.28				
Chamberlain [1]	0	1	0	0	0	V	10	42	290	55.6	286	8	0.27	55.5		36	
Chamberlain [1]	0	1	0	0	0	V	10	6	290	1.54	290	8	0.28				
Chamberlain [1]	0	1	0	0	0	V	10	6	290	2.63	290	8	0.22				

Chamberlain [1]	0	1	0	0	0	V	10	6	290	3.8	290	8	0.19				
Chamberlain [1]	0	1	0	0	0	V	10	24	290	4.6	290	8	0.32				
Chamberlain [1]	0	1	0	0	0	V	10	24	290	5.7	290	8	0.31				
Chamberlain [1]	0	1	0	0	0	V	10	24	290	8.96	290	8	0.30				
Chamberlain [1]	0	1	0	0	0	V	10	24	290	12.2	290	8	0.30				
Chamberlain [1]	0	1	0	0	0	V	10	24	290	12.4	290	8	0.29				
Chamberlain [1]	0	1	0	0	0	V	10	24	290	14.9	290	8	0.30				
shirvil [9]	1	0	0	0	0	H	1.5	0.2	290	0.1	290	2		5.8	5.5		0.2
shirvil [9]	1	0	0	0	0	H	1.5	0.1	290	0.018	290	2		2.9	2.7		0.1
shirvil [9]	1	0	0	0	0	H	1.5	0.1	290	0.035	290	2		3.6	3.5		0.1
shirvil [9]	1	0	0	0	0	H	1.5	0.1	290	0.045	290	2		4	3.8		0.2

Notes:

(1) Chamberlain and Fishburne do not provide flare elevation, but do provide wind speed. For this data, a 10m nominal elevation has been selected.
(2) Willoughby radiant fractions estimated from averaging published heat fluxes at different locations
(3) All flame dimensions determined from photographs. Since flames can move significantly over time due to wind fluctuation, values provided are subject to error. AP Garyville hydrogen and syngas flame dimensions obtained accounting for perspective, according to location of camera vs. flame.
(4) Chamberlain radiant fractions vs. release conditions obtained through interpretation of plots provided in paper

20.0 References

[1] Chamberlain, G.A., "Developments in Design Methods for Predicting Thermal Radiation from Flares," Chemical Engineering Research and Design, Volume 65, 1987.

[2] Advantica. "Report on Hydrogen Jet Fires Carried Out On Behalf of Air Products & Chemicals Inc.". Private Report. 2009.

[3] Fishburne, E. S & Pergament, H. S "The dynamics and radiant intensity of large hydrogen flames", Colloquium on Fire and Explosion.1979.

[4] Schefer, R. W. et al. "Spatial and radiative Properties of an open- flame hydrogen plume", International Journal of Hydrogen Energy, Volume 31, 2006.

[5] Willoughby, D. B., et al. "Hydrogen Venting under Variable Flow Conditions", Health and Safety Laboratory, 2011.

[6] Lowesmith, B.J., Chamberlain,G., et al. "An Overview of the Nature of Hydrocarbon Jet Fire Hazards in the Oil and Gas Industry and a Simplified Approach to Assessing the Hazards". Transcript IChemE, Part B. 2007.

[7] Lowesmith, B.J. "An Experimental Programme to Study High Pressure Jet fires following Releases of Natural Gas / Hydrogen Mixtures". R0051-WP2-R-0 (Part of Deliverable D37). 2008.

[8] Shirvill, L.C., et al. "Characterisation of the Hazards from Jet Releases of Hydrogen", Health and Safety Laboratory. 2005.

[9] Miller, D., Jung, S., Lutostansky, E., "Applicability of Currently Available Flare Radiation Models for Hydrogen and Syngas", Proceedings of 10th Global Congress on Process Safety. 2014.

[10] Ewan, B.C.R., Moodie, K., "Structure and velocity Measurements in Underexpanded Jets", Combustion Science and Technology. 1986.

[11] Ekoto, I.W., Ruggles, A.J., Creitz, L.W., Li, J.X., "Updated Jet Flame Radiation Modeling with Buoyancy Corrections", International Conference on Hydrogen Safety. 2013.

[12] Birch, A.D., et al. "Velocity Decay of High Pressure Jets", Combustion Science and Technology. 1987.

[13] Wayne, F.D., "An Economical Formula for Calculating Atmospheric Infrared Transmissivities". Journal of Loss Prevention in the Process Industries. 1991.

[14] McBride, B.J., Gordon, S., "NASA CEA". http://cearun.grc.nasa.gov/index.html

[15] Birch, A.D., et al. "The Structure and Concentration Decay of High Pressure Jets of Natural Gas". Combustion Science and Technology. 1984.

[16] Schefer, R.W., "Spatial and Radiative Properties of an Open-Flame hydrogen Plume.", Hydrogen Energy. 2006.
[17] Molina, R.W., Schefer, R.W., et al. "Radiative fraction and Optical Thickness in Large – Scale Hydrogen – Jet fires", Proceedings of the Combustion Institute. 2007.
[18] Kalghatgi, G.T., "Flowout Stability of Gaseous Jet Diffusion Flames. Part 1: In Still Air.", Combustion Science and Technology, 1981.
[19] Yuceil, K.B., Otugen, M.V., "Scaling Parameters for Underexpanded Supersonic Jets", Physics of Fluids. (2002)
[20] Miller, D, et al. "Hydrogen and syngas flares – comparison between experimental data and model predictions", AFPRC Industrial Combustion Symposium, 2014.
[21]. "Pressure Relieving and Depressurizing Systems", API 521 6th edition, 2014.

12TH GLOBAL CONGRESS ON PROCESS SAFETY

50th LOSS PREVENTION SYMPOSIUM

CFD Modeling of Jet Fires of Different Configurations

Jin Zhao
Baker Engineering and Risk Consultants, Inc.
11011 Richmond Ave, Suite 700, Houston, TX, 77042
jzhao@bakerrisk.com

Luca Magenes
Baker Engineering and Risk Consultants, Inc.
11011 Richmond Ave, Suite 700, Houston, TX, 77042
lmagenes@bakerrisk.com

Prepared for Presentation at
American Institute of Chemical Engineers
2016 Spring Meeting
12th Global Congress on Process Safety
Houston, Texas
April 11-23, 2016

CFD Modeling of Jet Fires of Different Configurations

Jin Zhao
Baker Engineering and Risk Consultants, Inc.
11011 Richmond Ave, Suite 700, Houston, TX, 77042
jzhao@bakerrisk.com

Luca Magenes
Baker Engineering and Risk Consultants, Inc.
11011 Richmond Ave, Suite 700, Houston, TX, 77042
lmagenes@bakerrisk.com

Keywords: Jet fire, CFD, Impinged flame

Abstract

Jet fire is a major hazard in the chemical processing industry. During a fire incident, one or more obstruction(s) may be present between the flames and the target of interest, which can affect the flame shape and position, and partially block the thermal radiation. As a result, the thermal radiation intensity and resulting level of damage experienced by the target of interest may be significantly affected by the obstruction(s). Thus, Computational Fluid Dynamic (CFD) modeling of the fire is required, in order to accurately capture these effects. Like any other modeling, CFD modeling requires validation against experimental data before it can be used as a defensible, predictive tool for hazards/risk assessment and fire protection design.

In this study, CFD code was used to model jet fires in three different configurations: (1) a horizontal free field jet fire with the wind blowing in the same direction, (2) a jet fire released 45 degrees upward with a cross wind blowing, and (3) a jet fire impinged on a wall with a cross wind blowing. Full-scale testing was performed to validate the calculations, and in all three cases the predicted flame size and shape match closely to actual experiment results. For the impinged case, the wall surface temperature calculated as a function of time agreed reasonably well with the test results obtained.

1 Introduction

Since jet fire presents a major hazard for the chemical processing industry, it is crucial to accurately model the consequences of jet fire when performing risk analyses and fire protection design. Semi-empirical models such as Shell Chamberlain model (1) and Johnson model (2) are widely used to predict free field vertical and horizontal jet fire, respectively. However, actual jet fire events that have happened in the process industry are rarely purely free field fires. In most cases obstructions such as buildings, fire walls, process equipment, or other structures stand between the fire source and the target of interest. This is particularly true for offshore platforms, which are normally highly congested. Thus, the necessity for modeling "non-free field" jet fire makes computational fluid dynamic jet fire modeling inevitable. For such a model to be considered reliable, however, it is necessary for the CFD model to be validated against experimental data before being used as a predictive tool in consequence and risk analysis.

In this study, three jet fire events in different configurations were modeled using a CFD program and the predicted flame shape, dimension and temperature profile were then compared with actual test results.

2 Computational details

In this study, the commercial CFD software PHOENICS developed by CHAM, UK (PHOENICS 2014) was used for the jet fire modeling. The standard k-ε turbulence model (Launder 1982), the Simple Chemically-Reacting System (SCRS) model and the IMMERSOL radiation model as implemented in PHOENICS were used for turbulence, reaction, and radiation modeling, respectively. Details are explained in previous work (3). Steady state calculations were performed to model two free field fires (Case 1 and Case 2), dynamic calculations were performed to model an impinged jet fire case (Case 3). In this study, the jet flame predicted by CFD study is defined as the envelope of combustion surface where 99% fuel is consumed, which is consistent with the parameters established for the fire dynamics simulator (FDS) (4)).

3 Results and Discussion

3.1 Case 1: Free field horizontal jet fire

The fire source for Case 1 is a ½-inch diameter horizontal release of saturated propane, and the wind is blowing in the same direction as the jet release at 4.5m/s.

Figure 1 shows a profile of the predicted flame shape compared to a photo of the actual flame profile from the experimental jet fire test. The photographic image of the jet flame was obtained by processing the video footage of the test using the software ImageJ[1] and taking the average flame shape throughout the test duration.

As shown in Figure 1 and Table 1, the shape and dimension of the jet flame predicted by CFD simulation match closely to the image obtained from the experiment observation.

[1] Public Domain Code – National Institute of Health (NIH), http://imagej.nih.gov/

(a) Predicted Flame (b) Observed Flame

(c) Predicted Flame Overlay on Observed Flame

Figure 1: Case 1 Flame Shape Comparison – Side View

Table 1. Case 1 Flame Dimension Comparison

Source of Results	Flame Length (m)	Flame Height (m)
Test result	8.4	1.3
PHOENICS	9.7	1.2

3.2 Case 2: Free field jet fire pointing 45 degrees upward, with wind blowing 45 degrees toward jet

The source for Case 2 is a ¼-inch diameter release of saturated propane. The release points to the north and is 45 degrees upward from horizontal. Wind is blowing from the northeast at an average speed of 4 m/s. Figure 2 shows a schematic representation of the jet fire set up for Case 2.

As shown in Figure 3, the flame shape predicted by the CFD model for Case 2 correlates well with the actual flame shape that was observed during testing.

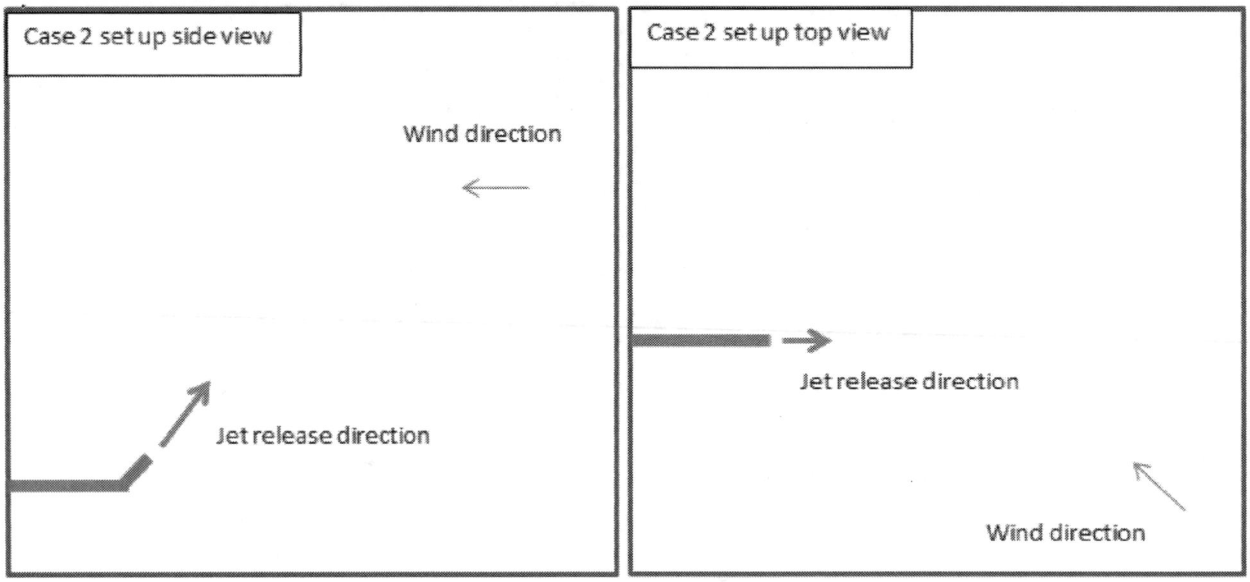

(a) Side View (b) Top View

Figure 2: Case 2 Set-up Layout

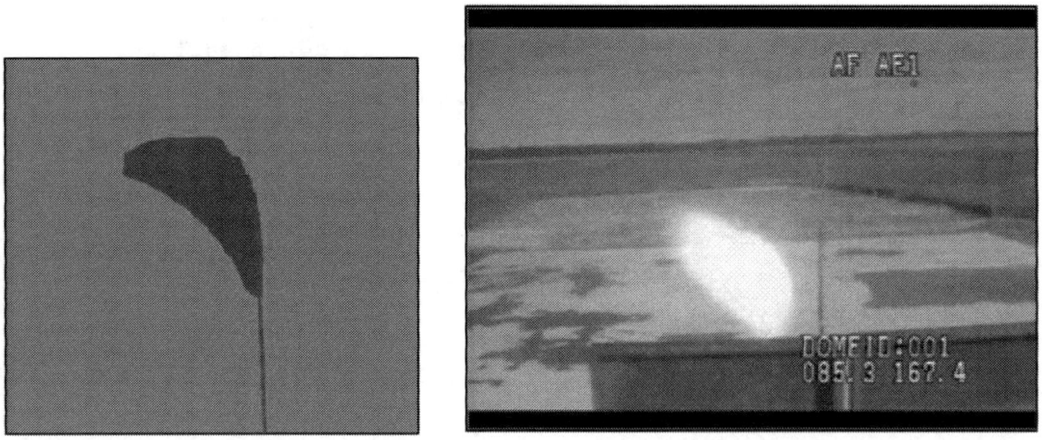

(a) Predicted Flame Top View (b) Observed Flame Top View

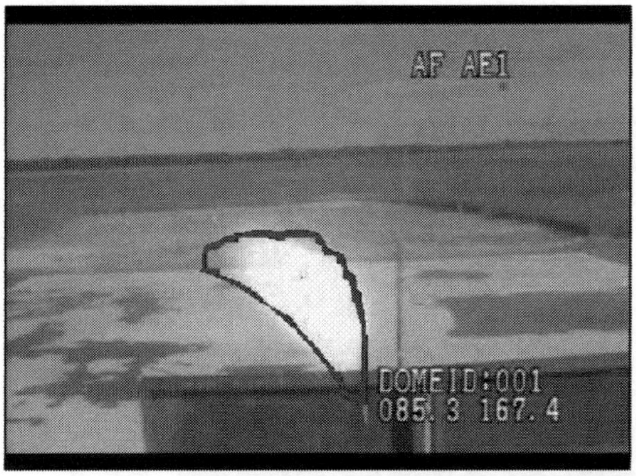

(c) Predicted Flame Top View Overlay on Observed Flame

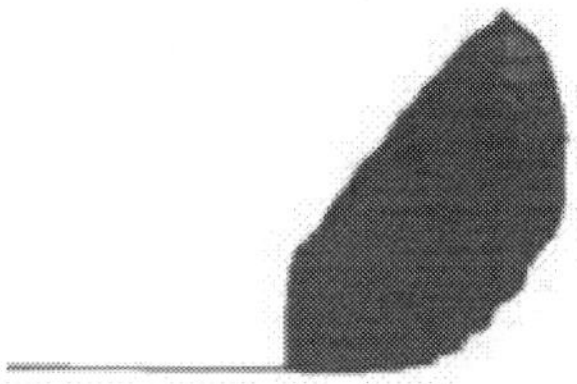

(e) Predicted Flame Side View

(f) Observed Flame Side View

(g) Predicted Flame Side View Overlay on Observed Flame

Figure 3: Case 2 Flame Shape Comparison – Side View

3.3 Case 3: Jet Fire impinging on building

Case 3 is a 3/8-inch saturated propane jet fire with a horizontal release impinging on a building measuring 16 feet long × 8 feet wide × 8 feet tall and located 20 feet away from the release point. The building wall facing the jet impingement is made of 7.6 cm thick Autoclaved Aerated Concrete (AAC). The release is blowing from the north to the south and the wind is blowing from northwest at an average speed of 5 m/s. Figure 4 shows a schematic representation of the jet fire set up for Case 3.

High temperature thermocouples were mounted on the outside wall surface of the building facing the jet fire. Their locations and labels are marked in Figure 7.

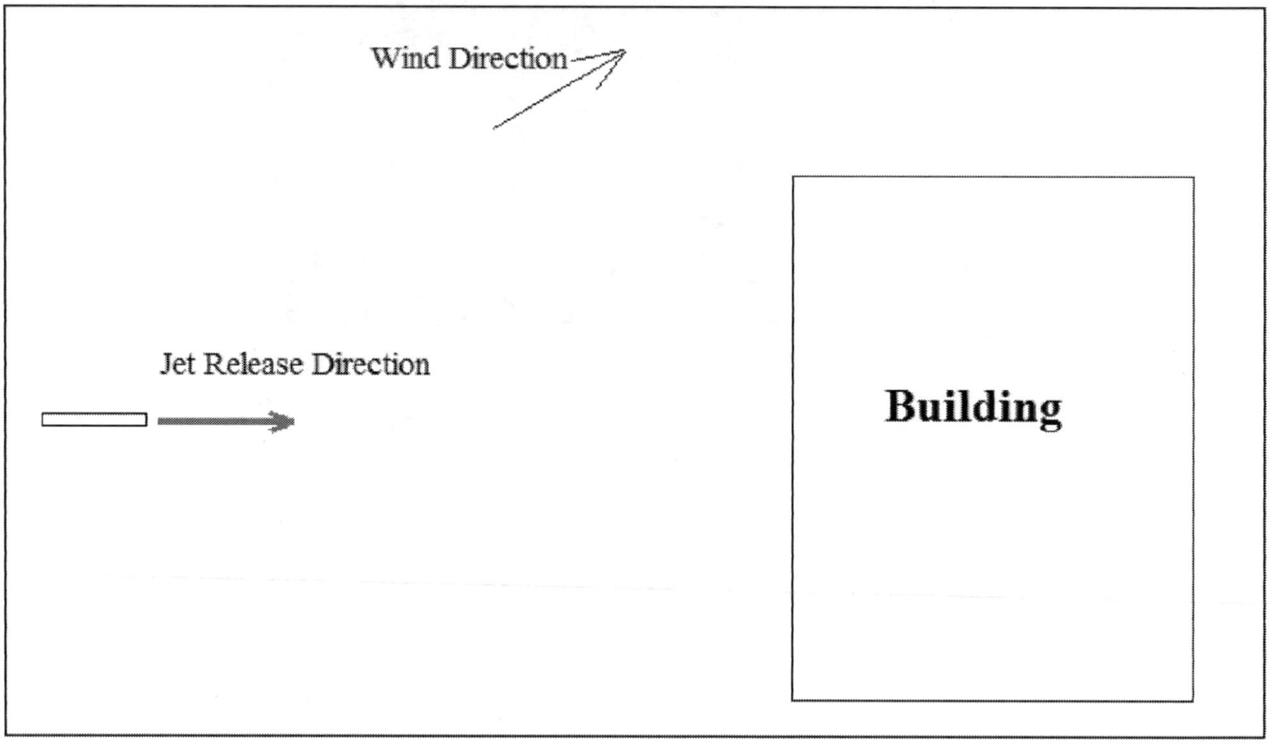

Figure 4: Case 3 Set-up Layout

3.3.1 Flame shape

As shown in Figure 5, both the CFD calculation results and experiment observation identified the following key factors about the flame shape:

- The jet flame impinged on the building wall.
- At ground level, flame is reflected by the wall and burns back along ground level.
- Flame slightly goes past the upper edge of the building wall facing the fire.

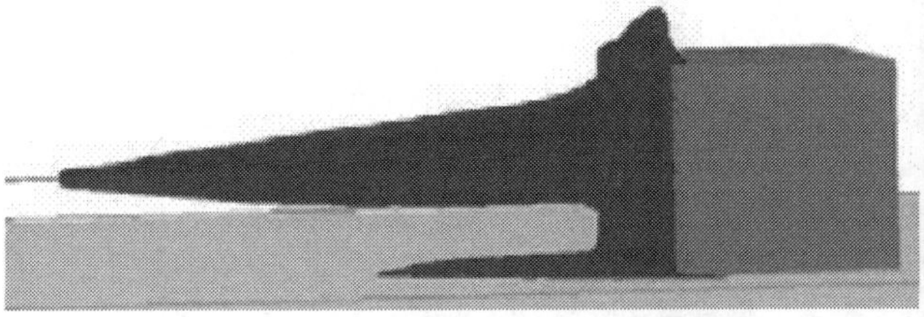

(a) Predicted flame side view (b) Observed flame side view

(C) Predicted flame overlay on observed flame

Figure 5: Case 3 Flame Shape

3.3.2 Building Surface Temperature

Figure 9 is a time-temperature plot reporting the predicted maximum surface temperature of the wall impinged by jet flame and the surface temperature measured during the test at three different locations on the wall (T11, T12, and T13). Locations and labels are marked on Figure 7.

Figure 7 shows the predicted surface temperature of the wall impinged by the jet flame 20 seconds after the fire started. As shown in Figure 9, the maximum surface temperatures were predicted to rise to a steady state temperature of 1,100 °C within 20 seconds after the fire started, while in the experiment, the actual measured temperature at three different locations reached steady state in about 60 seconds. Note that in the experiment, thermocouples had been embedded approximately ¼-inch in the wall so they would not move during testing, which may help explain the 40-second difference in results. Figure 6 shows a photo and a section of the thermocouples installed during testing.

Because thermocouples measure temperatures at their extremity, measurements reported during testing may have been cooler than at the surface of the wall. Also, the thermocouple itself may have blocked some of the radiation and air flow directed to the measurement point.

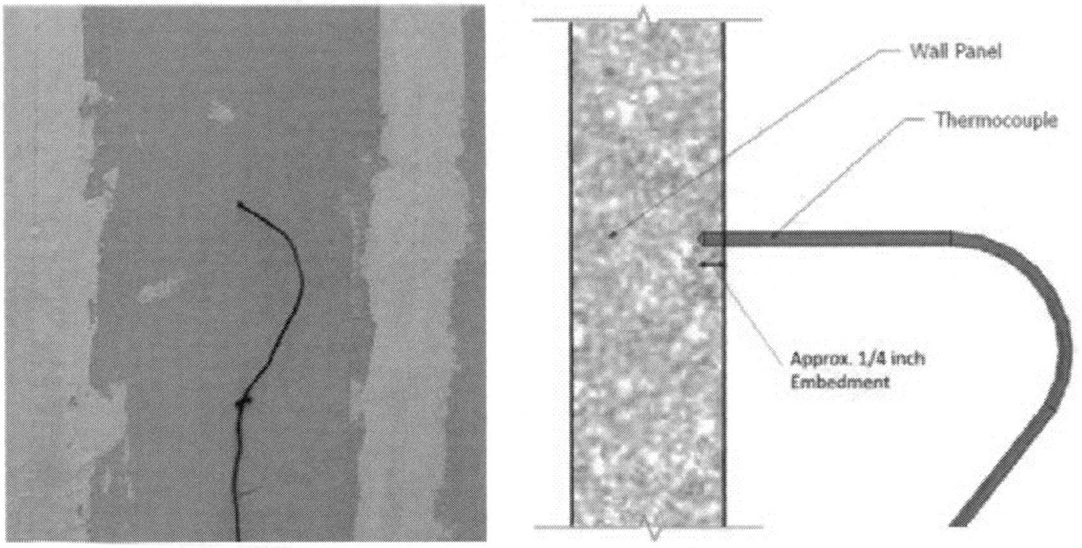

Figure 6. Thermocouple Mounted on Wall during Testing – Photo (left) and Schematic Section (right)

Another difference between the test and the model was observed when comparing the location of the hottest point on the wall. Because the wind pushes the flame towards the south side of the building, the highest temperature is expected to be skewed to the south of the wall center. As shown in Figure 7, the maximum surface temperature is predicted by the model at locations slightly left of the center of the wall surface (between T12 and T11); however, in the experiment, maximum temperatures were recorded at T13, which indicates that the flame was being pushed further to the south. An infra-red (IR) thermo-graphic photo of the building (Figure 8) taken a few minutes after the test provides an additional indication that the maximum surface temperature may have been reached to the left of T12 rather than to the right of it. This could be due to the significant variation of the wind speed and direction during the test, whereas in the CFD model, a constant average wind speed and direction are assumed.

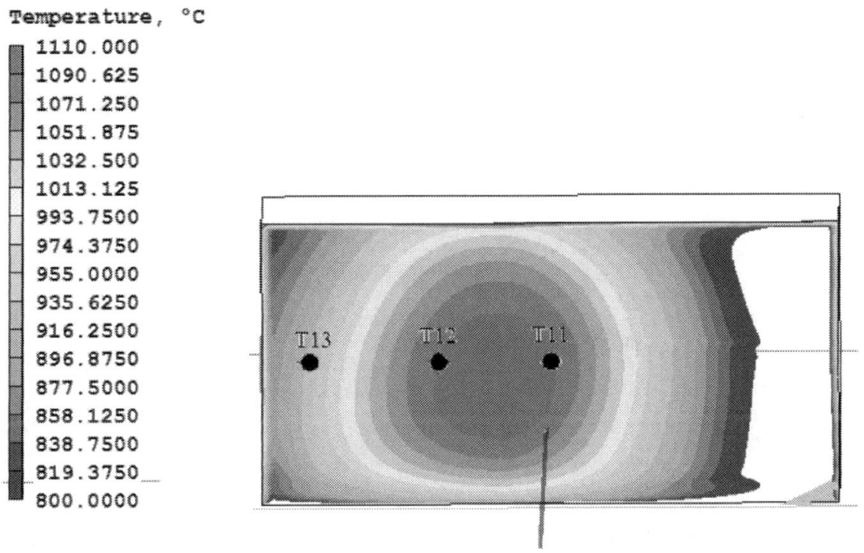

Figure 7: Temperature of Wall Surface Impinged by Jet Flame 20 seconds after Fire Start (Case 3)

Figure 8: IR Photo of the Building Taken Minutes after the Test (Case 3)

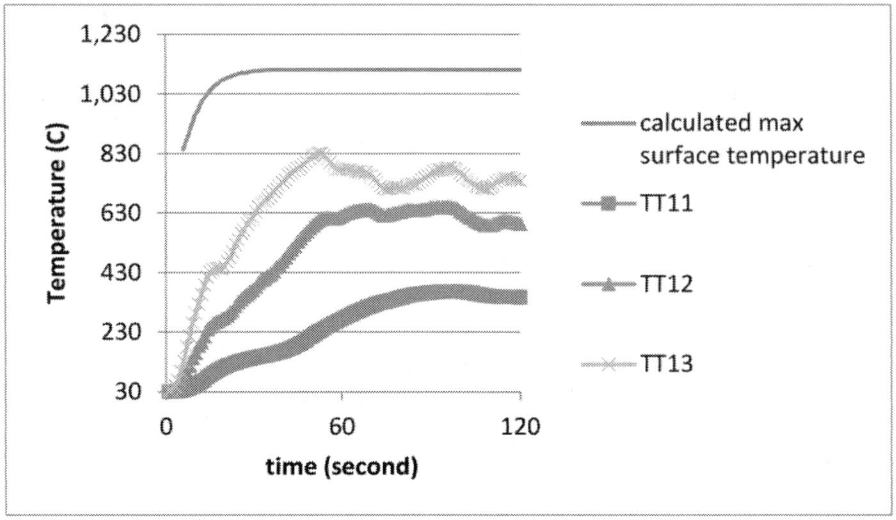

Figure 9: Predicted Max Wall Surface Temperature vs. Measured Wall Surface Temperatures

4 Conclusion

For this study, a CFD code was used to model jet fires in three different configurations: (1) a horizontal free field jet fire with wind blowing in the same direction, (2) a jet fire released 45 degrees upward, with a cross wind, and (3) a jet fire impinged on a wall with a cross wind. Full-scale testing was performed to validate the calculations, and in all three cases the predicted flame size and shape match closely to experiment results. For the impinged case, wall surface temperature as a function of time was calculated and agreed reasonably well with test results.

5 References

1. *Developments in design methods for predicting thermal radiation from flares.* **Chamberlain, G. A.** 1987, Chem. Eng. Res. Des., pp. 299-309.

2. *A model for predicting the thermal radiation hazard from large scale horizontally released natural gas jet fires.* **Johnson, A. D.** 1994, Trans. IChemE., pp. 157-166.

3. *Jet Flame Length and Thermal Radiation: Evaluation with CFD Simulations.* **Aloqaily, A. M.** San Antonio, TX : s.n., 2010. 2010 Global Congress on Process Safety.

4. **Kevin McGrattan, Simo Hostikka, Randall McDermott, Jason Floyd, Craig Weinschenk, Kristopher Overholt.** *Fire Dynamic Simulator Technical Reference Guide.* s.l. : NIST Special Publication.

Why DDT is the Only Way to Explain Some Vapor Cloud Explosions

D.M. Johnson
DNV GL
Spadeadam Testing and Research Centre, MOD R5, Gilsland, Brampton, Cumbria, CA8 7AU, UK
michael.johnson@dnvgl.com

V.H.Y. Tam
DNV GL
Spadeadam Testing and Research Centre, MOD R5, Gilsland, Brampton, Cumbria, CA8 7AU, UK
vincent.tam@dnvgl.com

Prepared for Presentation at
American Institute of Chemical Engineers
2016 Spring Meeting
12th Global Congress on Process Safety
Houston, Texas
April 11-23, 2016

Why DDT is the Only Way to Explain Some Vapor Cloud Explosions

D.M. Johnson
DNV GL
Spadeadam Testing and Research Centre, MOD R5, Gilsland, Brampton, Cumbria, CA8 7AU, UK
michael.johnson@dnvgl.com

V.H.Y. Tam
DNV GL

Keywords: Vapor Cloud Explosion, VCE, Deflagration, Detonation, Incident,

Abstract

The possibility that Deflagration to Detonation Transition (DDT) occurs in vapor cloud explosions (VCEs) in industrial accidents has generally only been recognized for very reactive fuels such as hydrogen and ethylene. Assessment of the explosion hazards and risks associated with less reactive fuels such as propane and other alkanes has generally been based around the assumption that, in practice, only a deflagration can occur. This has the benefit that the magnitude of the maximum hazard to surrounding areas is defined by the physical parameters of any congested process region and not by the extent of the vapor cloud, as would be the case if there was a sustained detonation. However, following the Buncefield incident in the UK in 2005, a considerable amount of effort was expended in explaining the major VCE that occurred. This initially involved gathering and interpreting the evidence from the incident but then extended to experimental and theoretical research over two phases and lasting five years. Further research since showed that DDT can be achieved within dense vegetation and regions of pipework with a relatively small extent (of the order of a few meters). The results from this project will be summarized, indicating why the only explanation that is consistent with all of the evidence is a DDT soon after ignition of the cloud, with a sustained detonation then propagating through the majority of the cloud. Reference will be made to other incidents where similar evidence was observed. Awareness of the significance of such evidence will be important in the investigation of future VCEs. In addition, the possibility of DDT needs to be recognized in the methodologies used to assess VCE hazards.

1 Introduction

The assessment of vapor cloud explosion hazards for less reactive fuels such as propane generally involves the identification of congested process pipework regions, estimation of the pressure generated in these regions and the decay of this pressure with distance from the region. Research dating from more than thirty years ago [1] provided the basis of this approach, showing that a flame propagating through a flammable vapor cloud engulfing such congested regions could accelerate to high speeds and generate potentially damaging pressures. Models have been developed that can be used to assess the hazards, ranging from the simpler screening approaches to detailed Computational Fluid Dynamics.

While such explosions are clearly possible, it is suggested that in at least some (possibly many) of the major vapor cloud explosion incidents, deflagration is only the first stage of an explosion process that then leads to detonation. The evidence to support this contention comes from a combination of large scale experimental research, review of evidence from vapor cloud explosion incidents and consideration of how significant quantities of any vapor cloud can be involved in the generation of damaging overpressures.

Though the experimental evidence for deflagration to detonation transition can be traced back over thirty years and interpretation of incidents as detonations over forty years, it was only the detailed assessment of the vapor cloud explosions at Buncefield, UK in 2005 [2] and Jaipur, India in 2009 [3] that has provided confirmation that detonations are events that can occur in realistic conditions involving common loss of containment scenarios in modern facilities..

Once this step has been taken however, a realization follows that it actually becomes difficult to justify the absence of detonation in any vapor cloud explosion involving a large cloud engulfing some form of congestion.

This suggests that the methods currently used in the design and location of buildings on process sites are based on an incomplete picture of vapor cloud explosions.

This paper first reviews the evidence for DDT and then discusses the implications of the findings.

2 Vapor Cloud Explosion Experiments

2.1 Early Experimental Studies

Experimental studies carried out during the early research into vapour cloud explosions provided evidence that deflagration to detonation transition (DDT) can occur at least within the idealised conditions of the experiments. Given that it is known that the factors that play a role in generating the energetic deflagrations capable of transition to detonation are scale dependent, reference is only made to large scale experimental studies in partially confined or open congested regions in this paper.

There are two aspects of DDT that need to be demonstrated. Firstly, it needs to be shown that a detonation can be initiated by a deflagration in conditions that are representative of conditions that might occur in reality. Secondly, there is a need to show that this detonation is then sustained, not only through the congested region, but also through any open, uncongested part of the vapour cloud.

DDT was observed in stoichiometric propane-air and cyclohexane-air experiments carried out in the 1980s at the DNV GL Spadeadam Testing and Research Centre [1]. The test rig was 3m square in cross section and 45m in length, as shown in Figure 1.

Figure 1: Test Rig at Spadeadam (circa 1980s)

The experiments involved ignition within partial confinement (shown on the left of the picture) with a flame acceleration venting out at a speed of about 200ms^{-1}. Flame acceleration continued over a distance of just 6m in the open congested region to speeds in excess of 600ms^{-1} (~Mach 2) when DDT consistently occurred. Importantly, this detonation was then sustained through the remaining length of the test rig, even where no pipework obstacles were present.

Experimental studies carried out as part of the EU co-funded explosion projects MERGE and EMERGE [4] [5] involved congested regions up to 9m square and 4.5m height. High pressures involving shockwaves were measured at the edges of the congested regions in some ethylene-air and propane-air experiments. These pressures were consistent with the occurrence of DDT, however this could not be demonstrated conclusively as there was no extended vapor cloud for the detonation to propagate through.

A series of large scale experiments involving the interaction of vented confined explosions with an external region of pipework congestion also showed evidence of DDT and the continued propagation of the detonation through unobstructed vapor cloud [6]. This DDT occurred in both ethylene-air and propane-air experiments. Figure 2 shows the experimental arrangement for the propane-air experiments. The region outside the vent was either unobstructed space or occupied by a congestion array.

Figure 2: Experimental Arrangement for Vented Explosion Tests

2.2 More Recent Experimental Studies

More recent experiments conducted with an ethylene-air mixture engulfing a 9.1m long congested area [7] showed that DDT occurred and that the detonation continued to propagate through the remaining 7.3m of unobstructed vapour cloud.

Other recent experiments involving a 30m long ethane-air vapour cloud engulfing a process pipework region [8] also resulted in DDT and continued propagation throughout the unobstructed part of the cloud.

Finally, as will be discussed in more detail later, experiments have been carried out as part of a joint industry project (JIP) studying explosion mechanisms involved in the Buncefield explosion in 2005. This has included experiments examining flame acceleration in propane-air clouds engulfing regions of dense trees and bushes. In some of these experiments, DDT occurred and the detonation continued throughout the unobstructed part of the cloud [9].

In summary, there is ample evidence that in the idealised conditions used in large scale explosion experiments, DDT can occur and that once initiated, the detonation can continue through unobstructed regions of the cloud. In all cases, DDT occurred after flame speeds in excess of the ambient speed of sound had been generated. The interaction of shock waves generated by such flame speeds with the flame is probably a key aspect leading to DDT.

The experimental evidence essentially relates to propane, ethane and ethylene mixtures. There is no direct evidence of DDT in experiments conducted with natural gas-air mixtures, though there is some indirect evidence of limited DDT in full scale experiments conducting using offshore geometries [10].

3 Buncefield

The Buncefield incident in 2005 [2] provided a considerable amount of evidence related to the effects of a vapour cloud explosion in a large pancake vapour cloud. The evidence included damage to items such as cars, oil drums and instrument boxes within the vapour cloud. In addition, directional indicators in the form of bent posts, scoured paintwork, scorched tree trunks and displaced items were present throughout the vapour cloud as illustrated in Figure 3 [11].

Figure 3: Directional Indicators within the Vapor Cloud at Buncefield

An important aspect of this evidence is that there was a very distinct change across the boundary of the cloud, in particular:

- Inside the cloud, pressure damage was severe and, importantly, the degree of damage was very similar, wherever items were found in the cloud.
- The severity of pressure damage reduced quickly once outside the cloud boundary.
- Inside the cloud, directional indicators were everywhere in the direction opposite to that of the flame propagation. In addition, they pointed back towards a single location within the cloud.
- Outside the cloud, the directional indicators all pointed away from the cloud.

The directional indicators are illustrated in Figure 4, which shows red arrows for directional indicators inside the cloud boundary and yellow arrows for those outside.

Figure 4: Directional Indicators at Buncefield (Red Inside the Cloud, Yellow Outside, Taken from [11])

It can be seen from Figure 4 that most of the vapour cloud at Buncefield was over open areas with only occasional cars, lampposts etc. The only significant congestion was provided by trees and bushes along the site boundary.

Phase 1 of a JIP into the Buncefield explosion mechanism [11] conducted analysis of the evidence and modelling studies, indicating that:

- The 'reverse' nature of the directional indicators in the cloud was due to the expansion of the hot combustion products behind the flame. Numerical simulations showed that this induced a net load on items opposite to the direction of the flame propagation.
- This 'reverse' loading on items occurred for both detonations and fast deflagrations. However, the widespread presence of such directional indicators throughout open and uncongested areas was not consistent with a deflagration as there was no means of sustaining the fast deflagration.
- The unvarying degree of pressure damage to items in the cloud was also inconsistent with deflagration through the tree lines (the only real congestion).

This led to the conclusion that the Buncefield incident was caused by an explosion mechanism that could sustain high flame speeds and pressures throughout the vapour cloud. This could not be a deflagration of the form recognised in methods generally used in assessing vapour cloud explosions hazards, as in these the damaging pressures are only generated within the congested areas.

This conclusion in itself is sufficient to say that there is a need to re-examine the potential hazards from vapour cloud explosions. However, it is apparent that the only known explosion mechanism that could explain the evidence is a detonation of the vapour cloud [12]. The evidence indicated that the flame accelerated in the line of trees and bushes and quickly underwent a DDT. The location of the DDT is indicated by the red arrows in Figure 4, pointing to a region at the north end of the site.

4 Other Incidents

A vapour cloud explosion with many similarities to Buncefield occurred in Jaipur, India in 2009 [13] [14]. The evidence of severe pressure damage and directional indicators was again spread throughout the vapour cloud. Examples of the damage and directional indicators are shown in Figure 5 and Figure 6 respectively. Many measurements were made of directional indicators and a summary of the measurements is shown in Figure 7. As with Buncefield, these indicators point back to a single location.

Again, most of the areas where the directional indicators and severe pressure damage was present had little or no congestion that could sustain a deflagration. The conclusion was therefore that there had been a DDT at a location in the North East corner of the site, as indicated by the directional indicators, with most of the cloud (enclosed by the yellow line in Figure 7) undergoing detonation.

Figure 5: Pressure Damage at Jaipur

Figure 6: Directional Indicators at Jaipur

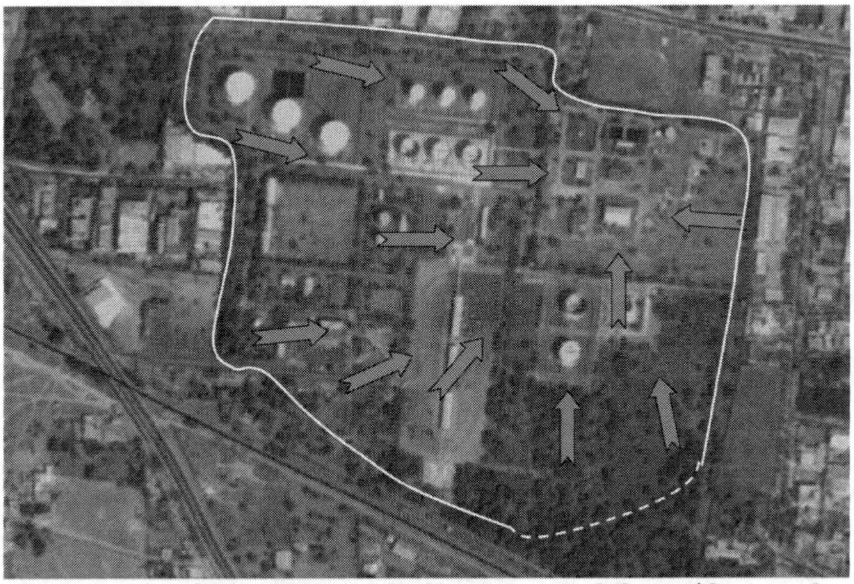

Figure 7: Summary of Directional Indicators at Jaipur (Approximate Cloud Boundary Indicated by Yellow Line)

Looking further back into the incident record, there are two incidents of particular note:

- Port Hudson, Missouri, 1970, in which a propane cloud was ignited in a rural area, generating severe explosion damage [15]. No pipework congestion was present; however, the cloud engulfed buildings and wooded areas.
- Ufa, Russia, June 4, 1989, in which a propane cloud produced by a pipeline failure was ignited by passing trains [16]. There were widespread directional indicators throughout the vapor cloud in the form of fallen trees. These were viewed at the time as being due to the wind generated by a rising fireball, but are entirely consistent with the event being caused by a detonation of the cloud.

The Port Hudson incident was reported to involve a detonation of the propane vapor cloud and shared many characteristics with the Buncefield explosion, including the same pattern of directional indicators within the cloud. In an analysis of the Port Hudson incident [17] it was stated in relation to the damage inside the Port Hudson cloud, "We think that it is significant that the wind direction was everywhere opposite to the postulated direction of the detonation" ('wind direction' in this case is taken to be the implied direction of the gas flow associated with the propagating detonation). These are exactly the same types of directional indicators observed at Buncefield and Jaipur.

It is notable that all of these incidents involved low lying dense vapor clouds spread over a large area with significant parts of the cloud containing no obstacles. This type of cloud will show evidence of detonation much more clearly as:

- High levels of pressure damage throughout the cloud, including the open areas, cannot be explained by deflagrations in the more limited congested areas.
- The mechanism that results in the directional indicators, expansion of combustion products in the opposite direction to the detonation, will be more evident. In fact, if a hemispherical vapor cloud were ignited centrally, the reverse directional indicators would not occur because no reverse flow would be possible.

It is also worth noting that the far-field pressure effects of a detonation when the vapor cloud is limited to the congested area are likely to be similar to those from a severe deflagration in that area. It is the continuation of the high pressure into uncongested cloud that is the issue.

5 Damage Caused to Common Objects

As already described, during the Buncefield investigation and subsequent research [9][11][18][19] objects located in and around the Buncefield site that could help in the explanation of the explosion mechanism were identified. If the response characteristics of such objects can be fully described, they would form an important tool in the diagnosis of a VCE event.

The reliability with which the characteristics and severity of a gas or vapor cloud explosion can be interpreted is dependent on information on the overpressure distribution and propagation direction. In a controlled experiment, there are sensitive instruments that measure the rapid changes in overpressures with time, flame speed and direction and stresses and strains in objects responding to the explosion loads. These characteristics have to be inferred from observations of damage in the post incident survey. However, the vast majority of objects in a typical refinery or oil and gas facility, e.g. structures and equipment, are complex objects with unpredictable response. This makes diagnosis of VCE characteristics difficult.

The majority of objects found in the Buncefield accident are complex. The three types of objects used in the investigation and research projects to interpret the Buncefield explosion mechanism were cars, boxes and drums. For the purpose of this paper, examples are given of boxes and drums, however the evidence for cars is consistent with the overall conclusions.

5.1 Physical Testing

Boxes and drums were exposed to a range of loading: (a) shock load originated from detonations, typically, the duration is short (~ a few milliseconds (ms)) and (b) a longer duration (~10's of ms up to about 100ms) loading from deflagration with a similar rise time and decay time before and after the peak. The tests were performed at the Spadeadam Testing and Research Centre.

Detonation tests were carried out in a gas cloud 30 m long, with a range of widths from 4 m to 10 m and height of between 2 and 4 m. The gas cloud was contained by polythene sheet which was supported by a steel frame and the detonation was initiated by a small explosive charge. Objects were placed inside and outside of the gas cloud and pressures were measured close to the objects. The pressure profiles were always shock waves with the duration of the external overpressure being longer, at about 10ms to 20ms, than that inside (about 5ms).

Deflagration tests were carried out using the 182m^3 bang box illustrated in Figure 2, but without the external congestion. The overpressures characteristics are controlled by the choice of gas, the vent area and the amount of congestion inside. Figure 8 a typical overpressure-time profile used in these experiments. In this case the maximum pressure is about 1.4 bar with a duration of about 110ms.

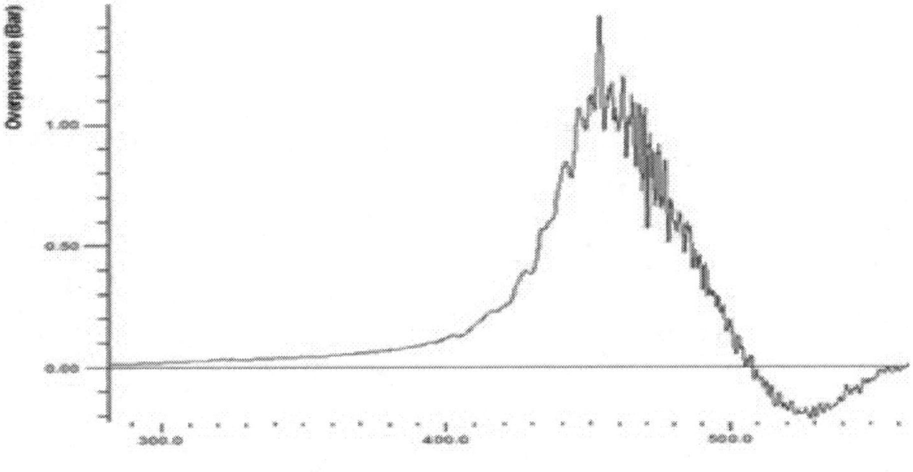

Figure 8: Example Pressure Profile

5.1.1 Instrument Box Tests

Standard 150 cm electrical switch boxes were exposed to shocked load from a detonation outside the detonating gas cloud. Figure 9 shows the response of the boxes to a range of overpressures and durations. Overpressures from detonations typically have very short duration. As the shock wave propagates away from the cloud, dispersion effect lengthens the duration. In the example shown in the figure, this was increased to a maximum of 10 ms in the range of distances that the boxes were located. At less than 3 bar, these boxes had no visible permanent deformation. At higher pressures, permanent damage increases as overpressure increases. The damage was predominant on one side. However, the deformation is different from the box that was located within the detonation; the box was squashed on all sides.

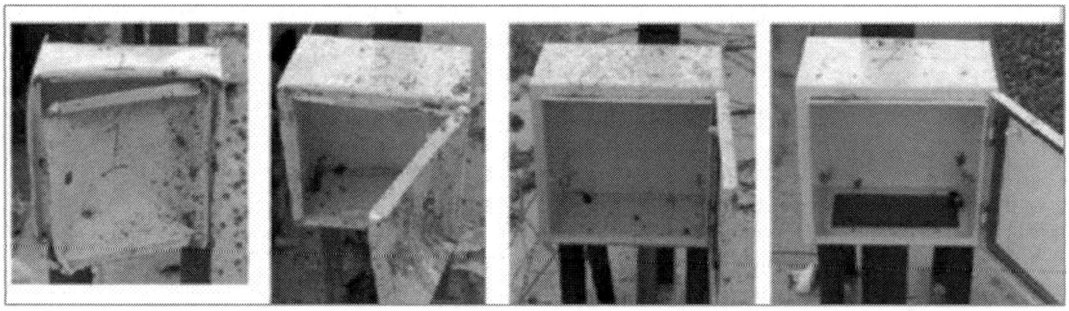

Figure 9: Switch Boxes Exposed to Shock Loading
(Incident overpressures were (left to right) 8.2 bar, 3.4 bar, 2.1 bar and 0.7 bar. Duration ranges from 4ms to 10ms.)

5.1.2 Steel Oil Drums

Standard steel oil drums were located outside a detonation cloud at various distances. Figure 10 shows the final state of the drums after exposure to a range of overpressure between 0.4 bar and just over 4 bar with duration of around 10 ms. For comparison, the drum located within the detonation is also shown. At less than 2 bar, the standard steel oil drum suffered very minor deformation.

Figure 10: Drums Exposed to Shock Loading
(From left to right, the incident overpressures were 4.4 bar, 2 bar, and 0.4 bar and the respective durations were 10 ms, 11 ms and 15 ms. The final drum on far right was inside a detonation cloud.)

Steel drums exposed to a range of deflagration overpressures are shown in Figure 11. At 0.7bar and a duration of 160ms, the drum did not show any sign of damage. At a pressure of over 1bar, permanent damage was observed.

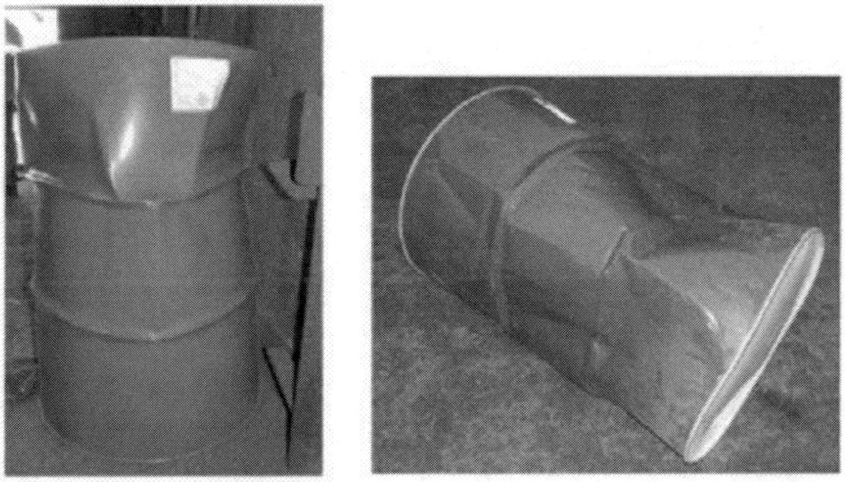

Figure 11: Effect of Deflagration Pressure on Oil Drum
(1.4 bar and duration of 110 ms (left), 1.8 bar with a duration of 43 ms (right))

5.2 Discussions on physical testing

Comparison of the damage observed to instrument boxes in the Buncefield and Jaipur incidents with the experimental results indicates that the incident overpressures must have been well in excess of 2bar, see Figure 12 for an example. As such widespread pressures of these magnitudes could not be generated by a deflagration, they are consistent with a DDT occurring. It is also notable that work to use finite element to calculate the response of the box has shown some success [20].

Figure 12: Pressure Damage to an Instrument Box (Buncefield) and Oil Drums (Jaipur)

Tests show that steel oil drums are more resilient to overpressure loading than switch boxes at short durations of less than 10 ms. A steel drum is a more complex object than a box with many modes of vibration and can also be affected by the fluid contained within the drum.

Work by Horn et al [21] has suggested that creasing of oil drums would be observed at low deflagration overpressures (< 2 bar). However, there was no pictorial information was provided and the short description of damage does not allow comparison with the data presented here. It can be seen that the nature of the damage to the drums shown in Figure 12 is not consistent with that seen when drums are exposed to longer duration deflagration pressures up to 1.8bar (Figure 11). The damage is however consistent with that seen when drums are located within a detonating cloud (Figure 10).

6 Why Would DDT not Occur?

The experimental evidence demonstrates that DDT is possible given flame speeds in excess of the ambient speed of sound and also that flame acceleration to these speeds can be rapid. It should be noted however that the experiments were all conducted with fuel concentrations near to stoichiometric, where the mixture is more detonable. It could therefore be argued that in a realistic case, and concentrations may be well away from stoichiometric, then DDT would not occur because it requires considerably more energy to initiate a detonation [22].

However, this hypothesis is not really sustainable for major vapour cloud explosions because:

- Explaining the level of damage caused by vapour cloud explosions as deflagrations requires that high flame speeds had been generated in the explosion, generally in excess of 200ms^{-1}. It is notable that in a review of pressure estimates for the Flixborough explosion [23] indicates pressures of several bar within the vapour cloud. Such pressures would need flame speeds well in excess of the ambient speed of sound.
- Damage observed to oil drums shown in Figure 5 cannot be experimentally replicated if flame speeds are less than the ambient sound speed.
- Such flame speeds are much more likely to be generated in a deflagration if the fuel concentration is close to stoichiometric (note this is not true for all fuels, e.g. hydrogen).
- The conditions that produce the severe deflagration (required to explain the damage) are therefore exactly those where DDT becomes highly likely.

An additional factor that should also be taken into account is that the fuel concentration within a vapour cloud will have fluctuations. If a deflagration traveling at sub-sonic speeds encounters such fluctuations, the speed of the flame will change. This would produce oscillatory pressure pulses as the flame repeatedly speeds up and slows down. The pressure energy would therefore not be contained within a single pressure pulse and would consequently give low pressure loadings as the pressures decay away from the cloud.

A detonation is not particularly sensitive to concentration variations within the cloud provided they are within the detonable range. Pressure energy will therefore be contained within a single shock wave propagating away from the vapour cloud. Even if a concentration fluctuation outside the detonable range is encountered in the vapour cloud, the detonation is likely to be able to go around or above or below the non-detonable area and continue to propagate.

The same considerations apply to conditions where there are fluctuations in the degree of congestion, particularly with congested regions with relatively high aspect ratios. Oscillatory pressure pulses will be produced as the flame speeds up and slows down.

DDT therefore provides a much simpler means of explaining large vapour cloud explosion incidents in realistic plant.

It should be noted that it is not being proposed that all major vapour cloud explosions involved detonation of almost all of the cloud. There is the potential that ignition of the cloud may occur in an open area and a significant proportion of the cloud may be burnt before the flame enters congestion and accelerates to DDT. In the Buncefield incident for example, the ignition location was close to a line of dense trees and bushes, which quickly led to the DDT. If ignition had occurred in one of the open areas well away from the tree congestion, then significant portions of the cloud could have been consumed prior to DDT.

7 Implications for the Assessment of Vapour Cloud Explosion Hazards

7.1 Existing Guidance

It is a routine part of the design of new facilities, and operation of existing facilities, that vapour cloud explosion hazards are assessed in order to determine where onsite buildings can be located and define the pressure loading that the buildings should be able to withstand. The assessment of the potential for explosions to cause offsite damage is also part of the design process and can be a requirement for land use planning.

The assessment methods are defined within guidance [24][25] and are based around the size of the congested regions and the strength of the deflagration that could be generated within that region. Design and location of buildings can be based on either consequence calculations or they can be risk based. Where a risk based approach is used, the return period used to define the ‘designe vent’i soft enonc e in 10,000 years.

If vapour clouds produced by accidental releases do not extend significantly beyond the congested region, then an approach based on the congested region may well be adequate. However, where a vapour cloud can extend well beyond the congested region, which is more likely for denser than air fuels released in calm conditions, then the possibility of DDT can significantly change the potential consequences.

By way of example, risk based calculations were recently carried out for buildings on a major onshore development of an onshore oil and gas field. The proposed facility involved the processing of significant quantities of higher hydrocarbons. The analysis was carried out on the basis of considering the congested regions only as pressure sources and also repeated to include detonations of the vapour clouds predicted by the dispersion analysis. For the main operations building, the once in 10,000 years incident blast wave was calculated as 130mbar, however when detonations and cloud sizes were included in the analysis, the 10,000 year incident blast wave increased to 200mbar.

It is worthy of note that this assessment did not involve any attempt to model the DDT process. If the deflagration explosion modelling suggested that flame speeds approaching the ambient sound speed were possible, then the possibility of detonation of part or all of the cloud was included within the analysis.

7.2 So Does Something Need to Change?

The fact that the worst vapour cloud explosion in the UK occurred at Buncefield on a site that had been assessed to have no significant explosion hazard, strongly suggests that something needs to change. It was only through good fortune that this incident did not result in any fatalities. Industry cannot continue to make design decisions that are effectively based on an incomplete understanding of vapour cloud explosions.

The exclusion of detonations from most assessments of vapour cloud hazards has been based on the general belief that DDT is not possible in realistic conditions on most facilities. However, if DDT is included, then the assessment process becomes more complicated as the location of the detonation would be dependent on the location of the vapour cloud rather than the fixed position of the congested region. There would therefore be a need to include dispersion calculations within the analysis to determine the potential extent of the detonation effects. Secondly, if DDT is possible on a facility, then it becomes much more difficult to use a consequence based approach given the wide range of possible dispersion scenarios, risk based decisions become almost essential.

The industry and the research community have not been in a position, until now, to be able to properly interpret the forensic evidence present following a vapour cloud explosion. The work that has been carried out following the Buncefield has led to a much better understanding of what certain evidence provides in this respect.

It is suggested that guidance should be updated to highlight the potential for DDT. This update could provide guidance on:

- The types of fuels where DDT is possible.
- The conditions required to produce DDT.
- The types of conditions where DDT may make more difference to design decisions.
- Methods to be used in the definition of design loads for occupied buildings and critical structures.

In addition, the interpretation of evidence present after a vapour cloud explosion needs to be improved. Guidance should be provided based on the research work carried out in the wake of the Buncefield incident.

8 Conclusions

The fact that clear evidence of DDT is limited to a relatively small number of vapor cloud explosion incidents does not mean that DDT is limited to these incidents. Explosions involving more hemispherical vapor clouds that are largely contained within congested process areas will not have such unequivocal evidence. Secondly, it is likely that historically investigators would not have appreciated the types of evidence that might occur within the vapor cloud because it is only following the Buncefield incident that there has been a systematic assessment of the effect of detonations.

DDT provides a relatively simple means of explaining the extensive and wide ranging damage observed following major vapour cloud incidents, removing the sensitivities deflagrations have to concentration and congestion fluctuations. In fact, based on experimental evidence, it is strongly suspected that many major vapour cloud explosions have involved DDT and detonation of part of the vapour cloud.

Guidance and assessment methods need to be improved to assist industry as to when the inclusion of DDT is appropriate and also how design loads can be defined when this is the case.

Guidance also needs to be provided on the interpretation of evidence from incidents.

9 References

[1] Harris R.J. and Wickens M.J., Understanding Vapour Cloud Explosions – An Experimental Study, IGE comm. 1408, 1989.

[2] Buncefield Major Incident Investigation, Initial Report to the Health and Safety Commission and the Environment Agency of the investigation into the explosions and fires at the Buncefield oil storage and transfer depot, Hemel Hempstead, on 11th December 2005. Buncefield Major Incident Investigation Board (BMIIB), 13th July 2006

[3] [1]Independent Inquiry Committee Report on Indian Oil Terminal Fire at Jaipur on 29th October 2009; Committee constituted by MoPNG Govt. of India, completed 29th January 2010.

[4] Mercx, W.P.M., Modelling and Experimental Research into Gas Explosions: Overall Final Report on the MERGE Project. Commission of the European Communities Report, Contract STEP-CT-011 (SSMA), 1993.

[5] Mercx, W.P.M., editor, Extended Modelling and Experimental Research into Gas Explosions, Final Summary Report of the Project EMERGE, CEC contract EV5VCT930274, European Commission, Directorate General XII, Brussels, Belgium (1997).

[6] Tomlin G.B. and Johnson D.M., A Large Scale Study of the Venting of Confined Explosions into Unobstructed and Congested Flammable Vapour Clouds, Seventh International Seminar on Fire and Explosion Hazards, Providence, RI, USA, 5 – 10 May 2013

[7] Duran R.J, Goodrich M.L. and Thomas J.K., Propagation of a Vapor Cloud Detonation from a Congested into an Uncongested Area. Chemical Engineering Transactions, Vol 26, 2012

[8] Pekalski A., DDT in a Vapour Cloud Explosion in Unconfined and Congested Space, Large Scale Test. UKELG 50th Anniversary Discussion Meeting, Cardiff University, 9th to 11th July 2013, http://ukelg.ps.ic.ac.uk/50AP.pdf

[9] Dispersion & Explosion Characteristics of Large Vapour Clouds, Steel Construction Institute, Document ED024, 2015, available at http://www.fabig.com/video-publications/OtherPublications.

[10] Hansen O.R., DDT – Experiments Revisited and New Insight. UKELG 50th Anniversary Discussion Meeting, Cardiff University, 9th to 11th July 2013, http://ukelg.ps.ic.ac.uk/50OH.pdf

[11] Buncefield Explosion Mechanism Phase 1, Joint Industry Project Report, 2009, available at http://www.hse.gov.uk/research/rrpdf/rr718.pdf.

[12] Johnson D.M., Puttock J.S., Richardson S.A. and Betteridge S., "Investigation of Deflagration and Detonation as an Explanation for the Buncefield Vapour Cloud Explosion", Sixth International Seminar on Fire and Explosion Hazards, Leeds, April 2010.

[13] Johnson, D.M. Characteristics of the Vapor Cloud Explosion Incident at the IOC Terminal in Jaipur, 29th October 2009. 8th Global Congress on Process Safety, Houston, TX, April 1-4, 2012.

[14] Johnson D.M Characteristics of the Vapor Cloud Explosion Incident at the IOC Terminal in Jaipur, 29th October 2009, DNV GL Report 11510, August 2011, available at http://www.fabig.com/video-publications/OtherPublications.

[15] US National Transportation Safety Board, Pipeline accident report, Phillips Pipeline Co., Propane gas explosion, Report No. NTSB-PAR-72-1, March 1971.

[16] Makhviladze G.M. and Yakush S.E., Large Scale Unconfined Fires and Explosions. Proceedings of the Combustion Institute, Volume 29, 2002, pp195-210.

[17] Burgess D.S. and Zabetakis M.G., Detonation of a flammable cloud following a propane pipeline break, US Bureau of Mines, Report of investigation 7752, US Department of the Interior, 1973.

[18] Atkinson, G.A., Cusco, L. and Tam, V.H.Y., Interpretation of overpressure marker and directional indicators in fuall scale deflagrations and detonations, Hazards XXI conference, Manchester, UK, Nov, 2009.

[19] Tam V.H.Y., The Buncefield accident, why was the explosion so severe? Loss Prevention Bulletin 222, published by the Inst. Chemical Engineers, December 2011.

[20] Burgan B, Tam V.H.Y, Johnson D.M., Chen A and Allason D. Ordinary Objects come to the aid of Investigation of Gas Explosion Incidents, Hazards XXXI conference, Edinburgh, UK, May 2016.

[21] Horn B.J., Goodrich M.L. and Thomas J.K., Deflagration load generator: repeatability and application to test article blast loading. 11th Global Congress on Process Safety, Texas, U.S.A., 27-29 April 2015

[22] Bull D.C., Elsworth J.E., McLeod M.A., Hughes D., Initiation of Unconfined Gas Detonations in Hydrocarbon-air Mixtures by a Sympathetic Mechanism. Progress in Astronautics and Aeronautics, Volume 75, Gas Dynamics of Detonations and Explosions, Pages 61-72, 1981.

[23] Høiset S, Hjertager B.H., Solberg T. and Malo K.A, Flixborough Revisited – an Explosion Simulation Approach. Journal of Hazardous Materials A77 (2000) 1-9

[24] American Petroleum Institute, Management of Hazards Associated with Location of Process Plant Permanent Buildings. API Recommended Practice 752, 3rd Edition, December 2009.

[25] Chemical Industries Association, Guidance for the location and design of occupied buildings on chemical manufacturing sites. 3rd edition, RC21/10, 2010.

12TH GLOBAL CONGRESS ON PROCESS SAFETY

50th LOSS PREVENTION SYMPOSIUM

Ammonia Vapor Cloud Explosion Testing and Implications for Facility Siting Studies, Consequence Analyses and Risk Assessments

A.M. (Tony) Downes & Sam Rodgers
Honeywell International, Inc.
115 Tabor Road,
Morris Plains, NJ 07950
Anthony.Downes@Honeywell.com
samuel.rodgers@honeywell.com

Darren R. Malik & J. Kelly Thomas
Baker Engineering and Risk Consultants
3330 Oakwell Court, Suite 100
San Antonio, TX 78218
dmalik@BakerRisk.com
kthomas@BakerRisk.com

Prepared for Presentation at
American Institute of Chemical Engineers
2016 Spring Meeting
12th Global Congress on Process Safety
Houston, Texas
April 11-23, 2016

Ammonia Vapor Cloud Explosion Testing and Implications for Facility Siting Studies, Consequence Analyses, and Risk Assessments

A.M. (Tony) Downes & Sam Rodgers
Honeywell International, Inc.
115 Tabor Road,
Morris Plains, NJ 07950
Anthony.Downes@Honeywell.com, samuel.rodgers@honeywell.com

Darren R. Malik & J. Kelly Thomas
Baker Engineering and Risk Consultants
3330 Oakwell Court, Suite 100
San Antonio, TX 78218
dmalik@BakerRisk.com, kthomas@BakerRisk.com

Keywords: Ammonia, low LBV, vapor cloud explosion, VCE

Abstract

An open question with regards to facility siting studies, consequence analyses, and risk assessments for release scenarios involving very low laminar burning velocity (LBV) fuels (i.e., significantly lower than methane's LBV) has been whether such releases pose an unconfined vapor cloud explosion (VCE) hazard. There are no reports of an accidental unconfined VCE with such fuels, and applicable VCE test data has not been published. One approach taken within industry has been to use methane as a reference for such fuels. This is clearly a conservative approach, as the flame speeds achieved with very low LBV fuels would be expected to be much lower than those achieved with methane in a VCE under the same conditions. An alternative approach taken by some has been to discount the potential for a VCE with very low LBV fuels.

In order to better define the flame speeds and VCE blast loads from very low LBV fuels, Honeywell commissioned BakerRisk to perform VCE tests with ammonia and methane. Ammonia was adopted as a conservative representative of very low LBV fuels. Methane was included as a reference benchmark. The rig used for these tests was 72 feet (22 m) long, 12 feet (3.7 m) wide, and 6 feet (1.8 m) high, unconfined, and filled with a high level of congestion made up of a regular array of vertical circular tubes [2.375-inch (6 cm) diameter area and volume blockage ratios of 23% and 5.7%, respectively]. A near-stoichiometric quiescent fuel-air mixture completely filled the test rig, which was ignited against a large wall placed at one end of the rig. Blast pressure histories were recorded using an array of pressure gauges within and external to the test rig. Flame speeds were determined via high speed video recordings of the tests.

This paper discusses the tests performed, results obtained, and the implications with regard to facility siting studies, consequence analyses, and risk assessments for release scenarios involving very low LBV fuels.

1 INTRODUCTION

An open question with regards to facility siting studies, consequence analyses, and risk assessments for release scenarios involving very low laminar burning velocity (LBV) fuels (i.e., significantly lower than methane's LBV) has been whether such releases pose an unconfined vapor cloud explosion (VCE) hazard. There are no reports of an accidental unconfined (external) VCE with such fuels, and applicable VCE test data has not been published. One approach taken within industry has been to use methane as a reference for such fuels. This is clearly a conservative approach, as the flame speeds achieved with very low LBV fuels would be expected to be much lower than those achieved with methane in a VCE under the same conditions. An alternative approach taken by some has been to discount the potential for a VCE with very low LBV fuels.

The current BST methodology considers all pure fuels with a LBV less than 45 cm/s to be in the low reactivity group [1]. This includes the refrigerant HFO-1234yf, which has an LBV of approximately 2 cm/s [2], as well as ammonia, which has an LBV in the range of 7 to 15 cm/s [3]. BakerRisk previously performed VCE tests on behalf of the Explosion Research Cooperative (ERC) with methane-air mixtures (LBV = 40 cm/s), which were used to create the published BST flame speed tables for low reactivity fuels [1]. However, it is recognized that flame speeds based on the methane-air tests are very conservative for fuels with much lower LBV values (e.g., ammonia, HFO-1234yf, etc.).

In order to better define the flame speeds and VCE blast loads from very low LBV fuels, Honeywell commissioned BakerRisk to perform VCE tests with ammonia and methane. Ammonia was adopted as a conservative representative of very low LBV fuels. Methane was included as a reference benchmark. The intent of the test program was to utilize the measured blast loads to develop flame speed data for a new "very low" fuel reactivity group within the Baker-Strehlow-Tang (BST) VCE blast load prediction method. This new "very-low" fuel reactivity group includes ammonia and fuels with lower laminar burning velocities (LBVs), such as HFO-1234yf. The "very low" reactivity group will allow more accurate and less conservative external VCE blast load predictions for such fuels.

2 TEST DESCRIPTION

This section provides a description of the test rig and instrumentation utilized to perform the methane-air and ammonia-air VCE tests. The rig configuration and congestion pattern is described in Section 2.1. The systems used to generate and ignite the fuel-air mixture are described in Sections 2.2 and 2.3, respectively. The instrumentation used to capture the blast loads and flame propagation is discussed in Section 2.4.

2.1 Rig Configuration

The VCE tests were carried out in a relatively long highly congested test rig in order to promote flame acceleration and the generation of measurable blast loads. The test rig was 72 feet (22 m) long, 12 feet (3.7 m) wide and 6 feet (1.8 m) high. A wall [26 ft. (7.9 m) wide by 16 ft. (4.9 m) high] was placed at the end of the rig where the flammable mixture was ignited. An isometric view and photograph of the test rig are provided in Figure 1. With the exception of the wall at the rig end, no confinement was present within the rig.

The same high congestion level used in the creation of the BST flame speed tables [1] was used in these tests. The congestion array was made up of a regular array of vertical circular tubes [2.375-inch (6 cm) diameter area and volume blockage ratios of 23% and 5.7%, respectively]. For each 6 ft (1.8 m) cube, a total of 61 vertical tubes were installed, along with the 4 cube corner supports. A plan view illustrating this congestion pattern is provided in Figure 2. This congestion pattern is intended to be conservative (i.e., highly congested) relative to the congestion in a typical process plant.

2.2 Creation of Fuel-Air Mixture

Prior to the start of the test the entire test rig was covered in a 1.5 mil (0.0015 in., 0.038 mm) thick plastic sheet. The sheet was retained by 12 straps placed along the length of the rig during the generation of the fuel-air mixture. A quiescent wait period was followed once the target concentration was achieved. The straps were released via remote actuators just prior to ignition in order to allow the VCE to expand as freely as possible. The fuel delivery, gas sampling, and mixing systems are described below.

In order to minimize the impact of fuel concentration variations between each test upper and lower acceptable fuel concentration bands were established for each fuel species. The acceptable upper and lower fuel concentrations were based on a 1% decrease in peak LBV for each fuel species. The target fuel concentration for methane tests was 10.0%, with an acceptance range of 9.9% to 10.5%. The target fuel concentration for ammonia tests was 23.2%, with an acceptance range of 22.9% to 23.6%, based on a -1% change from the maximum LBV reported by Duynslaegher [3]. The target concentration and the acceptable upper and lower fuel-air concentrations for each fuel species are summarized in Table 1.

Table 1. Target Fuel Concentrations (vol. %) for each Fuel Species

Acceptable Fuel Concentration Band	Methane	Ammonia
Lowest Acceptable Fuel Concentration (Corresponding to 1% Decrease from Peak LBV)	9.9	22.9
Target Fuel Concentration (Peak LBV)	10.0	23.2
Highest Acceptable Fuel Concentration (Corresponding to 1% Decrease from Peak LBV)	10.5	23.6

Fuel Delivery

The fuel was introduced into the rig through a series of venturis which pull gas mixture in from one end and expel an enriched fuel-air mixture out of the other. The venturis were oriented to expel the enriched fuel-air mixture downward. The rig was divided into 3 fuel delivery zones with four venturis each in order to support the generation of a uniform fuel-air mixture throughout the rig. The locations of the venturis are shown in Figure 3. Each venturi was 3 feet (0.91 m) off the rig centerline at a height of 3 feet (0.91 m).

Fuel Circulation

Fans were used to circulate the fuel-air mixture in order to quickly achieve a uniform mixture throughout the rig. The fans were positioned at the center of each cube at a height of 5.5 ft. and oriented downward. Each fan was rated for a 110 cfm flow rate, which is equivalent to turning the rig volume over every two minutes. The fans were shut off prior to ignition in order to allow the mixture to reach a quiescent state.

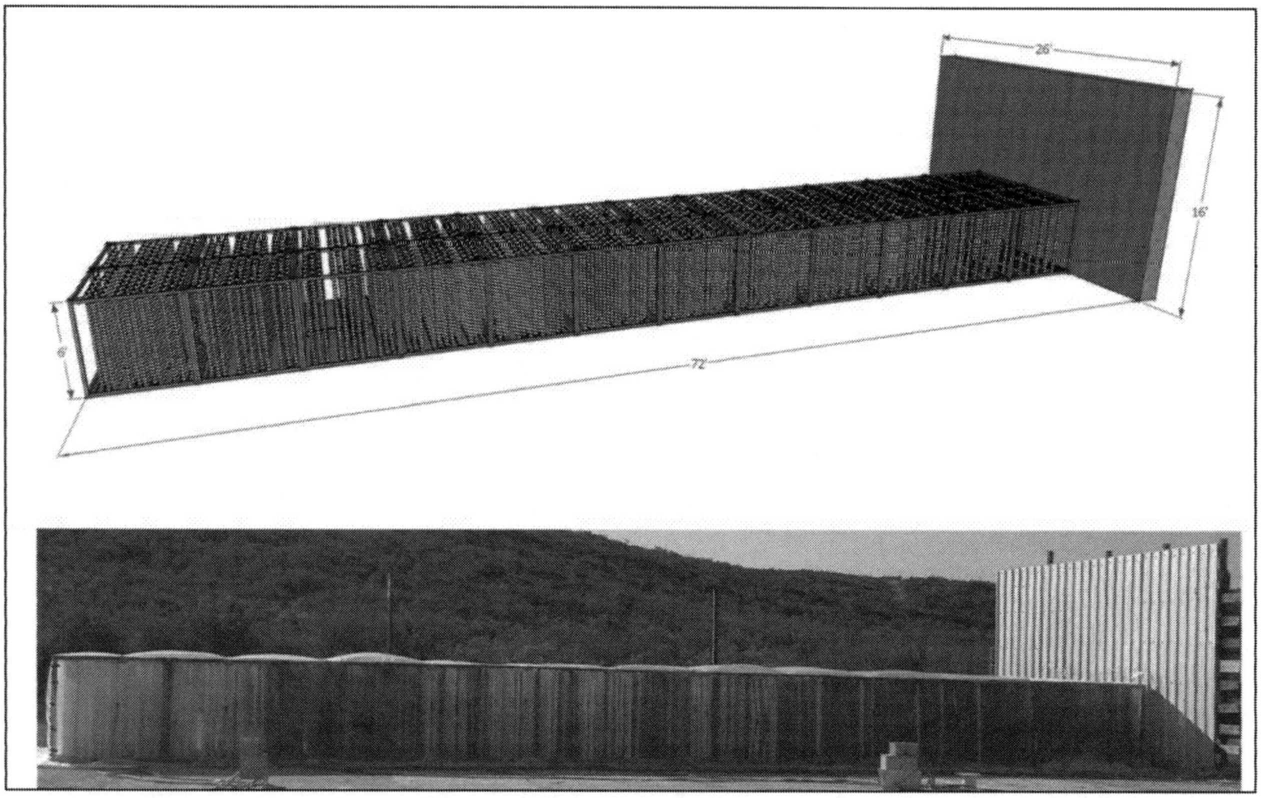

Figure 1. 3D Isometric and Photographic View of the Test Rig Geometry

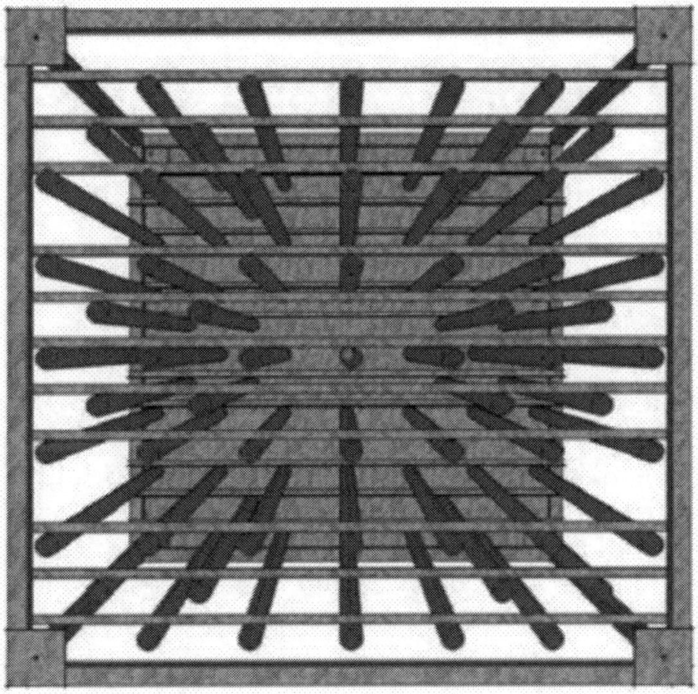

Figure 2. High Congestion Pattern

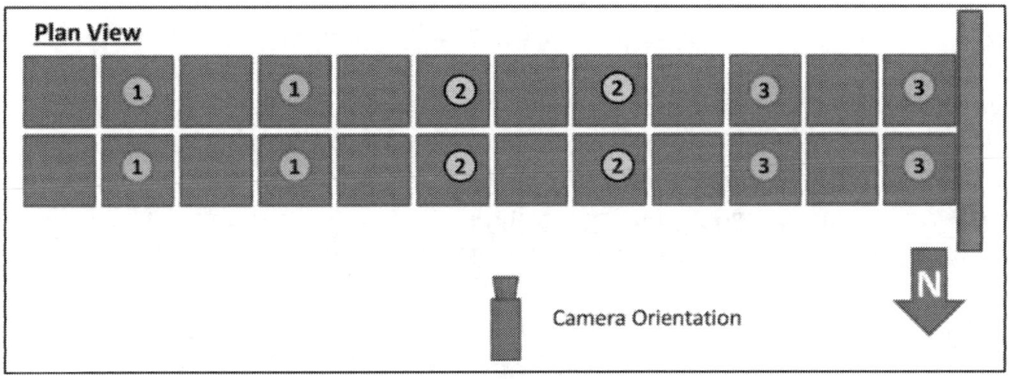

Figure 3. Location of Fuel Delivery Venturis

Gas Sampling

Fuel concentrations were determined indirectly based on oxygen concentration, which was measured using a California Analytical Model 600P Oxygen Analyzer. The analyzer was calibrated prior to each test using a known concentration of a suitable span gas balanced with nitrogen and with nitrogen as the zero gas. The sample points used to monitor concentration during the test were distributed along the long axis of the rig in the pattern, as shown in Figure 4.

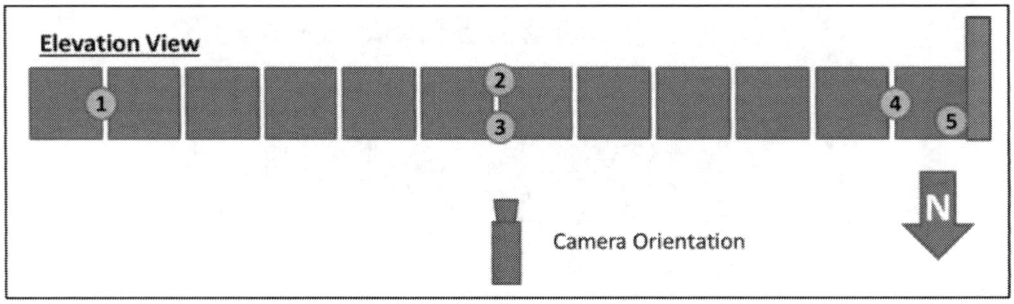

Figure 4. Fuel Sampling Point Locations

Gas samples were pulled through 0.18-in (0.46 cm) inner diameter sample lines, ranging in length from 30 to 80 ft (9 to 24 m), depending on the sample point. The sample pump flow rate was 10 ft^3/hr (17 m^3/min), giving a transit time from the sample point to the analyzer of 2 to 5 seconds. The time required for the analyzer to reach 90% of full scale is 30 seconds. The combined time for a sample to reach the analyzer and for the analyzer to respond was therefore approximately one half minute. This allowed the 5 sample points to be cycled through in approximately 3 minutes.

2.3 Ignition Systems

The location of the ignition source was grade level along the rig centerline near the end wall. Two different ignition systems were used during the course of this test program. The first two methane tests (A01 and A02) were ignited with an exploding fuse wire. A spark gap ignition system was deployed for the last methane test (A03) and all ammonia tests (B01-B03) due to concerns about releasing a toxic cloud of ammonia if the exploding fuse wire system failed to ignite the cloud. The spark gap ignition system was safer in this regard because it could be fired continuously in the event a test needed to be aborted. Conversely, the fuse wire system could only be fired twice (primary and secondary igniters). There was no significant difference in the methane test results due to the ignition system change, as shown in Section 3.1.

Fuse Wire Ignition System

The fuse wire ignition system was composed of equal length wire leads and a Parr exploding fuse wire (part #45C10) approximately 5 cm long. The fuse wire has a heat of combustion of 9.6 J/cm, giving a nominal ignition energy of 50 J.

Spark Gap Ignition System

The spark gap ignition system was composed of a 15 kV transformer, two equal length wire leads and two steel electrodes which are separated by a nominal 0.25 inch (0.64 cm) air gap. The transformer was energized via a remotely activated relay.

2.4 Instrumentation

Pressure Measurement

Blast pressure histories were measured using PCB® Piezotronics general purpose ICP® dynamic pressure sensors (Model CA102B18) connected to a line-powered ICP® sensor signal conditioner using microdot connectors and BNC cables. Each pressure gauge was mounted on a ½-inch steel plate approximately 12 inches (30 cm) in length and width. The pressure gauges internal to the test rig were installed with the pressure gauge mounted in a recess to reduce the effects of thermal radiation from the fireball. The gauges external to the rig were installed with the face of the gauge flush with the top of the mount. The signal from each pressure sensor was recorded using a National Instruments PXIE-1082 Chassis.

A schematic of the pressure transducer array is shown in Figure 5. The distance between the ignition location and each pressure transducer is provided in Table 2.

Flame Speed Measurement

Flame propagation was observed with regular speed high definition and high speed video. The locations of the high speed and high definition cameras are shown in Figure 5. The high-speed camera was set to record between 500 to 1000 frames per second based on the expected flame speed. The high definition video was captured at 30 frames per second.

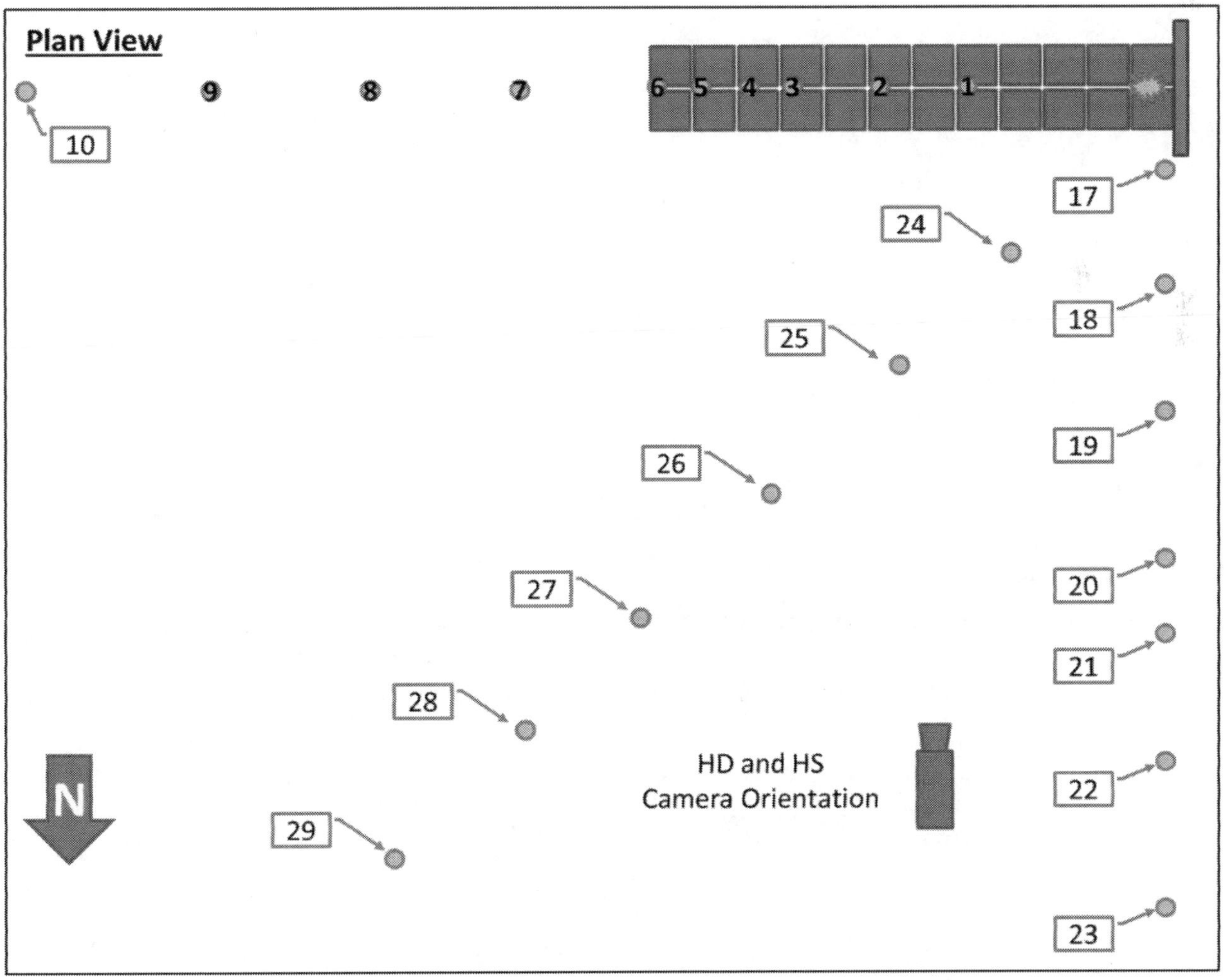

Figure 5. Schematic of the Pressure Transducer Array and Video Cameras

Table 2. Pressure Transducer Locations

Gauge Lane	Gauge Number	Distance from Ignition [ft.]
East Gauge Lane (long axis)	1	29
	2	41
	3	53
	4	59
	5	65
	6	71
	7	97
	8	124
	9	224
	10	324
North Gauge Lane (short axis)	17	14
	18	22
	19	38
	20	45
	21	106
	22	206
	23	306
Northeast Gauge Lane (45 degrees)	24	32
	25	62
	26	100
	27	150
	28	200
	29	300

3 RESULTS

Tests with methane-air and ammonia-air mixtures were performed in the test rig described in the preceding section. The results for each of these tests series are provided in this section. The measured fuel concentration for each test prior to ignition is provided in Table 3.

Table 3. Measured Fuel Concentration Prior to Ignition

Methane		Ammonia	
A01	10.0	B01	23.2
A02	10.4	B02	23.0
A03	10.5	B03	23.1

Table 4 provides a series of still frames which qualitatively shows the relative time of arrival for the flame propagation through the rig for the two fuel species. The HD camera frame rate does not provide sufficient resolution to rigorously match the time of arrival at the same discrete locations for both tests. However, the still frames and approximate times provided in Table 4 provide a qualitative indication of the slow flame speed propagation observed in the ammonia tests in relation to the methane tests. The actual flame speed analysis which is reported in the following paragraphs was determined using the high speed video data.

3.1 Methane Tests

The measured peak pressures for each of the deployed gauge lanes are provided in Figure 6 (east gauge lane), Figure 7 (northeast gauge lane), and Figure 8 (north gauge lane) as a function of the distance from the ignition location. The test data are shown as points in these figures. The line provided in the figures was generated using BakerRisk's SafeSite code [4] with a BST flame speed of Mach 0.19, which was the selected based on a least squares fit analysis of the test data. The location of the end of the test rig is also denoted in these figures.

The observed flame speeds are provided in Figure 9. The maximum and average flame speeds in the rig were approximately 500 ft/s (150 m/s, Mach 0.44) and 350 ft/s (110 m/s, Mach 0.31), respectively. The discrepancy between the observed average flame speed (Mach 0.31) and the least squares fit BST flame speed (Mach 0.19) is due to the test rig aspect ratio, the variance of the flame speed along the length of the rig, and the asymmetric flame propagation (i.e., propagating in a single direction along the rig centerline). The use of the least squares fit BST flame speed is consistent with the derivation of the flame speeds in the published BST flame speed table [1]. As seen in Figure 7, the BST flame speed of Mach 0.19 under predicts the measured peak pressures in the far field of the east gauge lane (along the direction of flame propagation), but over predicts the measured peak pressures in the far field of the north gauge lane, Figure 8.

3.2 Ammonia Tests

The ammonia-air tests did not produce measureable overpressures due to the low flame speed developed in these tests. A plot of the observed ammonia-air flame speeds is provided in Figure 10. The flame speeds observed in Tests B01 and B02 were higher than that observed in Test B03. A possible explanation for this difference is the wind direction. Tests B01 and B02 were performed with a 5-10 mph (7.3-15 ft/s, 2.2-4.5 m/s) wind blowing in the direction of flame propagation, approximately 30 degrees off the long axis of the rig. Test B03 was performed with a 5 mph wind blowing perpendicular to the direction of flame travel. Correcting Test B01 and B02 for an 8 ft/s (2.4 m/s) flow field along the direction of flame propagation results in the wind corrected ammonia flame speeds shown in Figure 11. The wind adjusted flame speeds for Test B01 and Test B02 are in reasonably good agreement with Test B03. The reason for the higher flame speed in Test B01 near the end of the rig is unknown, but it is acknowledged that the derivation of flame speed from video at very low flame speeds is subject to increased uncertainty. Video analysis of all three ammonia test confirmed that the tent remained in place for all three ammonia tests despite the change in wind direction between Test B02 and Test B03. The relative stability of the tent ahead of the flame front can be seen in the still frames provided in Table 4.

Table 4. Selected Still Frames Showing Relative Flame Propagation

Approximate Flame Location	Methane Test A03			Ammonia Test B01		
	Frame #	Time [s]	Still Frame	Frame #	Time [s]	Still Frame
Ignition	0	0.00		0	0.0	
2 cubes	2	0.07		30	1.0	
3 cubes	4	0.13		48	1.6	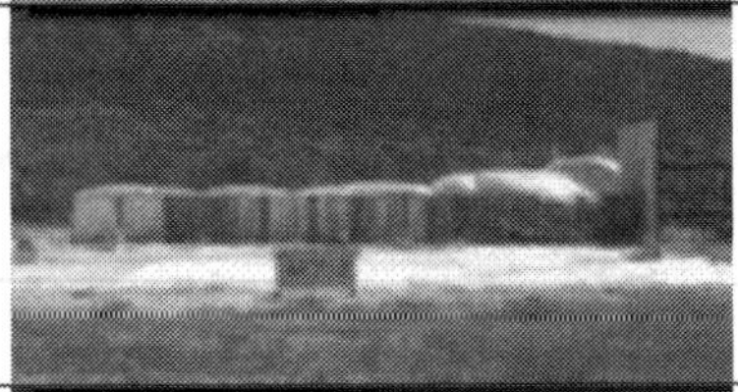
6 cubes	5	0.17		73	2.4	
8 cubes	6	0.20	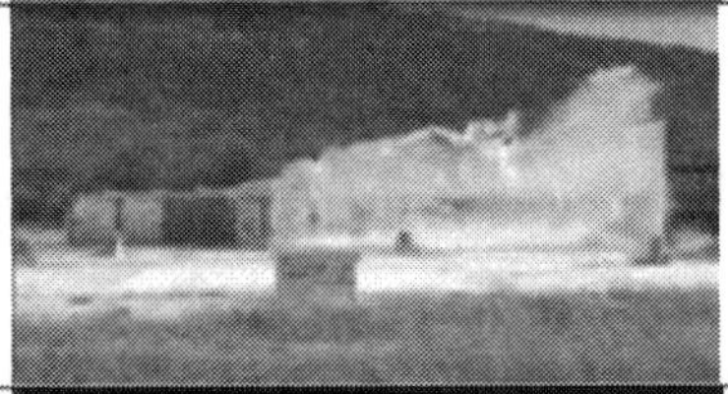	107	3.6	
10 Cubes	7	0.23		117	3.9	
12 Cubes	8	0.27		132	4.4	
End of Camera Frame	9	0.30		142	4.7	

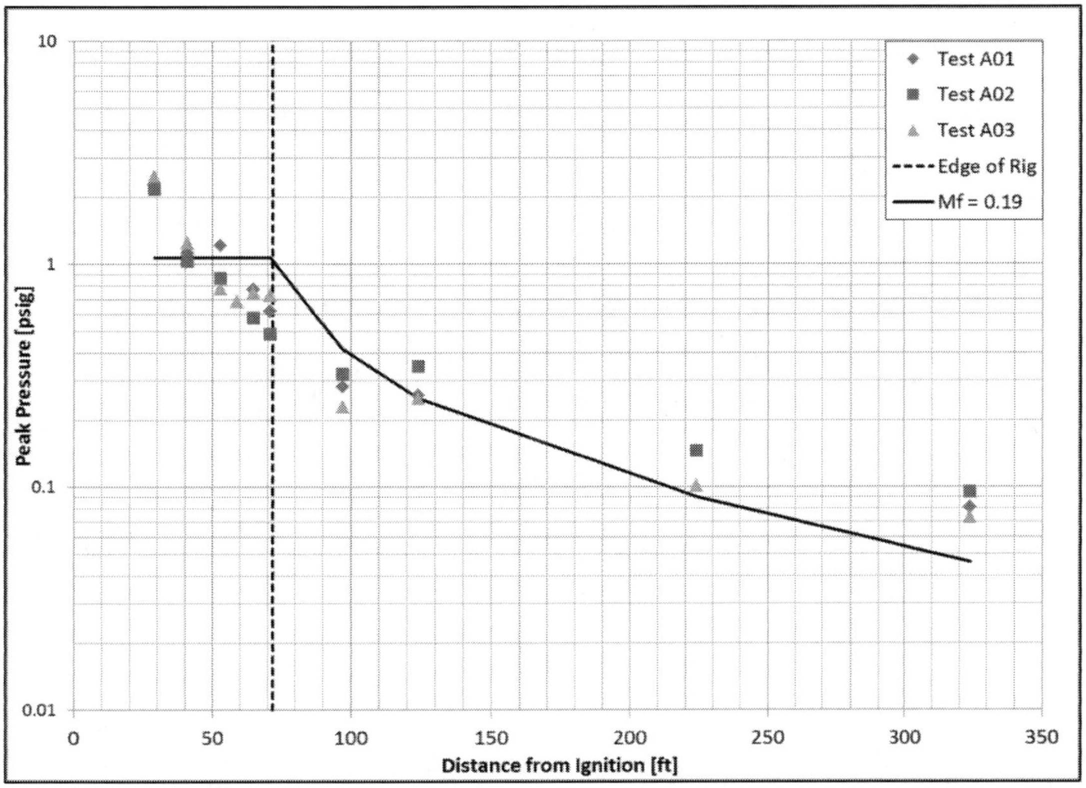

Figure 6. Measured Peak Pressures along the East (long axis) Gauge Lane (methane)

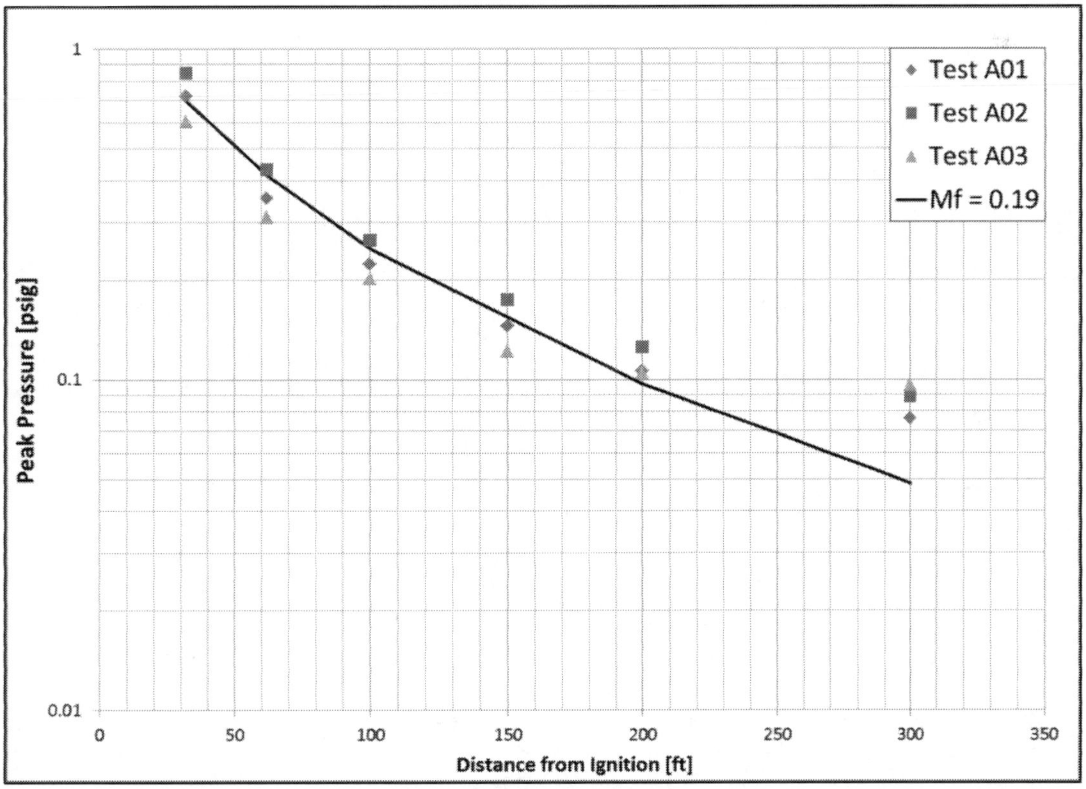

Figure 7. Measured Peak Pressures along the NE (45 degree) Gauge Lane (methane)

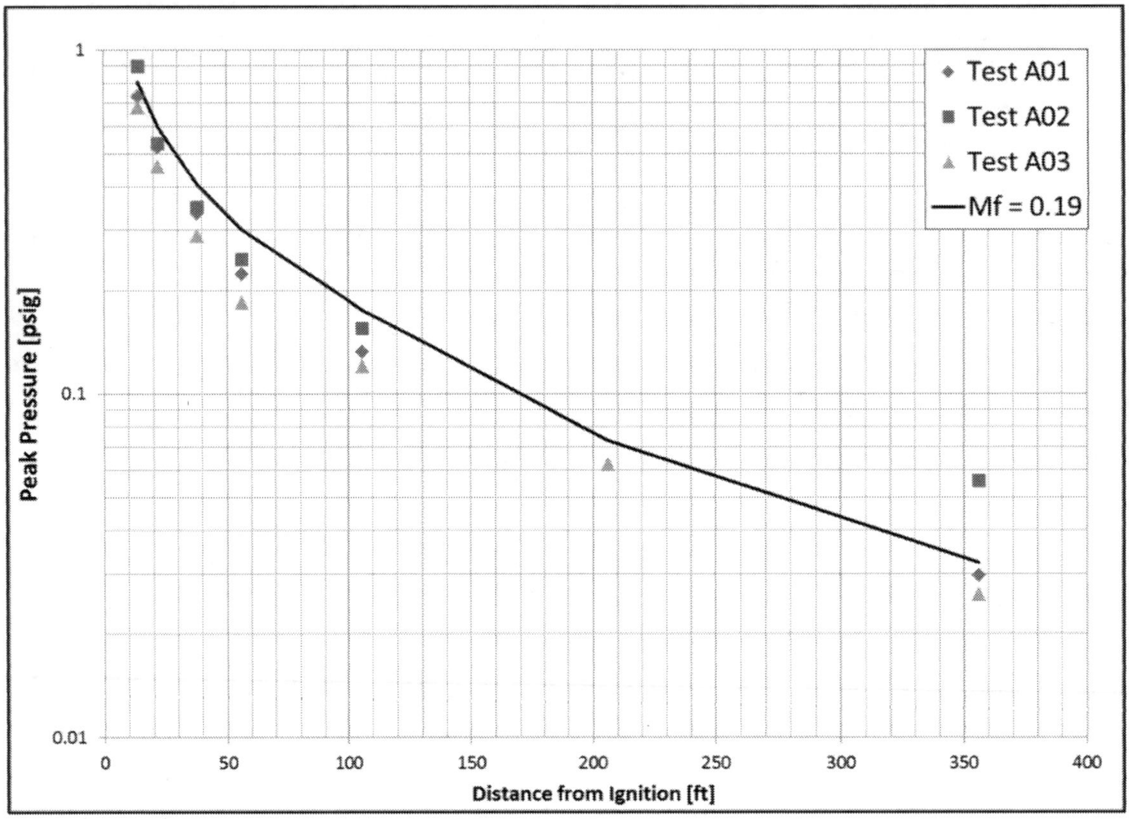

Figure 8. Measured Peak Pressures along the North (short axis) Gauge Lane (methane)

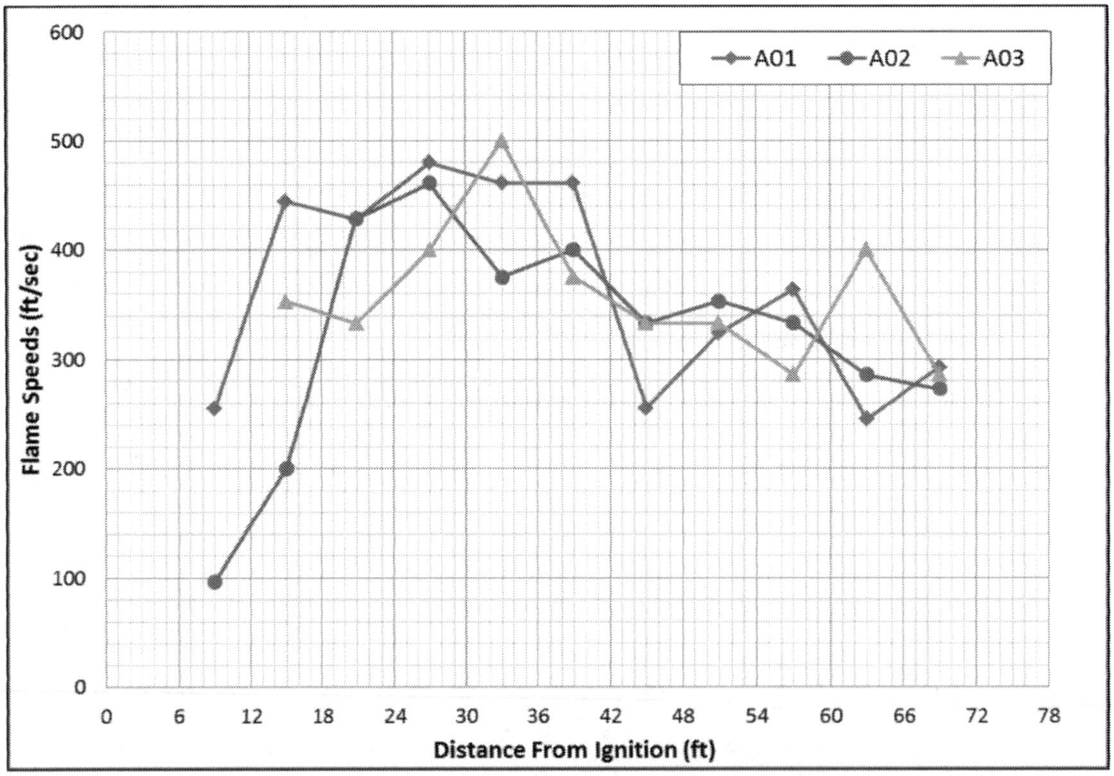

Figure 9. Observed Methane Flame Speeds

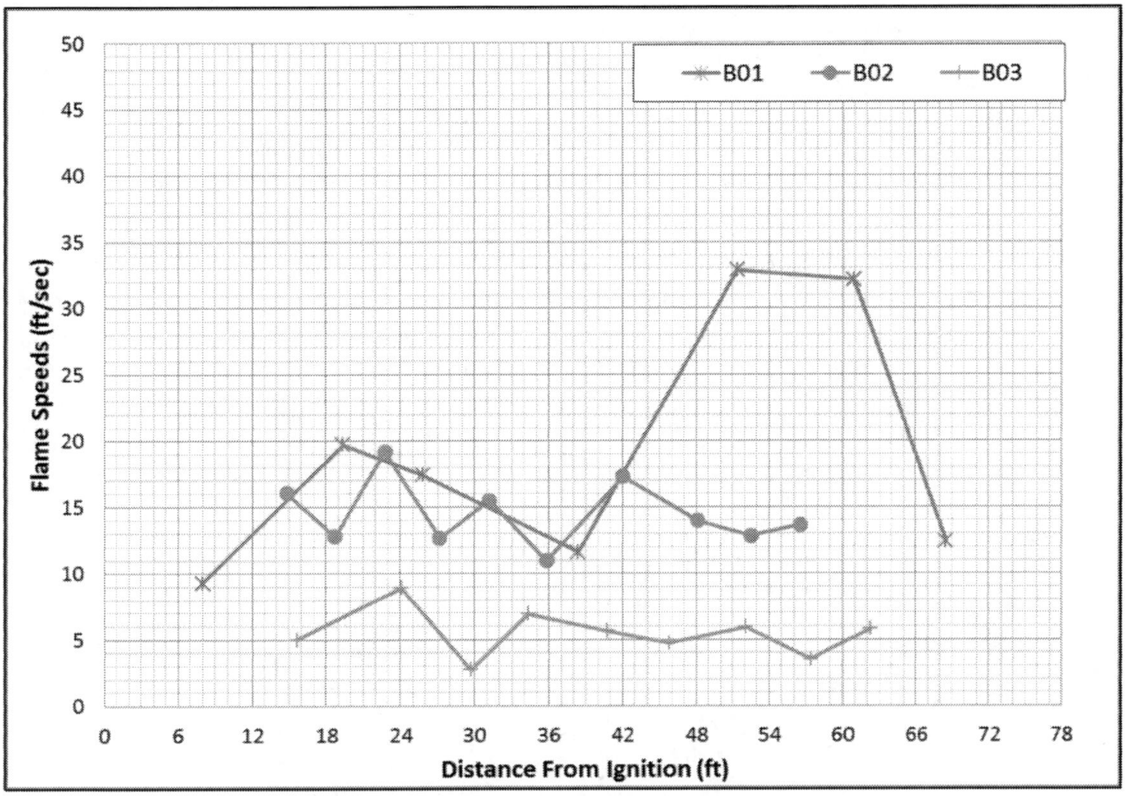

Figure 10. Observed Ammonia Flame Speeds

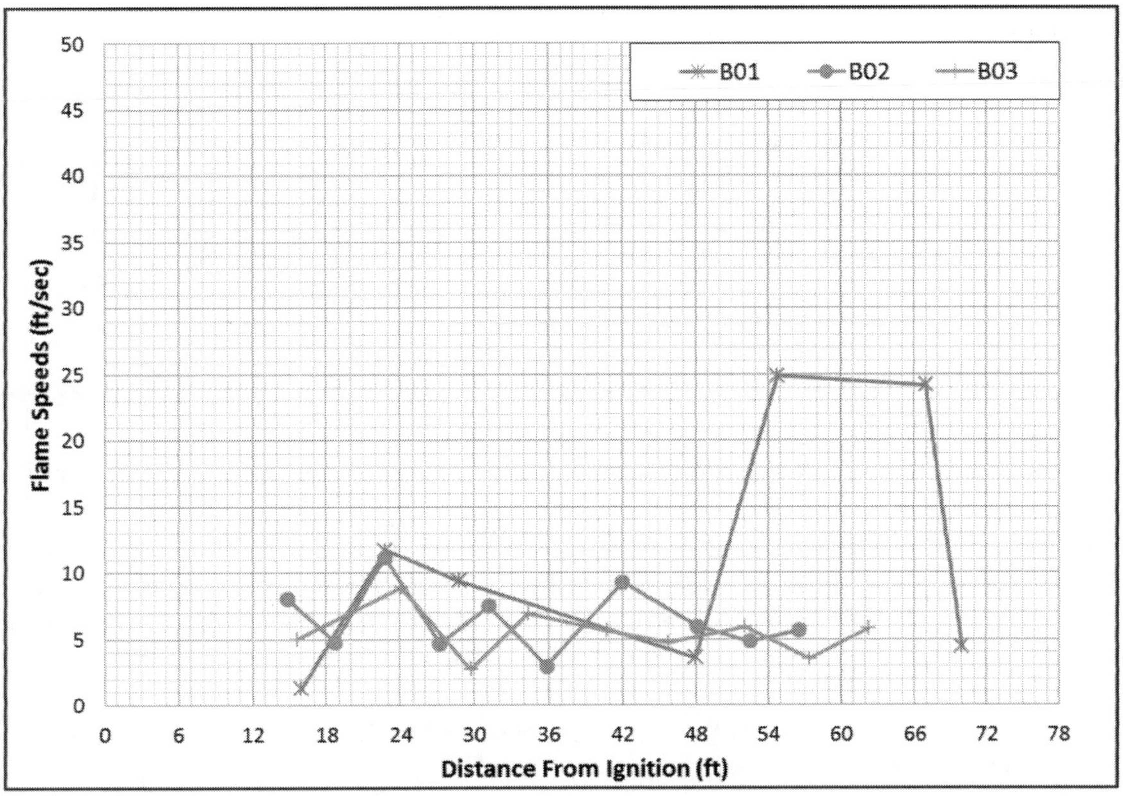

Figure 11. Wind Corrected Ammonia Flame Speeds

4 SUMMARY AND CONCLUSIONS

The methane-air VCE tests produced maximum blast pressures of approximately 2 psig (0.1 kPa) and maximum flame speeds of approximately 500 ft/s (150 m/s, Mach 0.44). The ammonia VCE tests did not produce measureable overpressures and the observed maximum flame speeds were approximately 25 ft/s (7.6 m/s, Mach 0.02), a factor of 20 lower than in the methane air tests.

An updated BST flame speed table with a new "very low" reactivity class is provided in Table 5 in order to scale the results of this test program for use in a consequence or risk-based facility siting ~~study (FSS) or quantitative risk assessment (QRA)~~. The values given for the "very low" reactivity fuel class in Table 5 were calculated by multiplying the low reactivity flame speeds by the ratio observed ammonia-air flame speed (Mach 0.02) and the flame speed based on the least squares fit analysis of the methane blast overpressure data (Mach 0.19); even lower values would result if the maximum observed methane flame speed (Mach 0.44) were employed.

Table 5. Updated BST Flame Speed Table with Very Low Reactivity Class

Confinement	Reactivity	Congestion		
		High	Medium	Low
2-D	High	DDT	DDT	0.59
	Medium	1.6	0.66	0.47
	Low	0.66	0.47	0.079
	Very-Low	0.069	0.049	0.008
2.5-D	High	DDT	DDT	0.47
	Medium	1	0.55	0.29
	Low	0.5	0.35	0.053
	Very-Low	0.053	0.037	0.006
3-D	High	DDT	DDT	0.36
	Medium	0.5	0.44	0.11
	Low	0.34	0.23	0.026
	Very-Low	0.036	0.024	0.003

To illustrate the VCE blast loads expected for the "very low" fuel reactivity class, consider a congested volume with a footprint of 200 ft (61 m) by 200 ft (61 m) and a height of 25 ft (7.6 m), which gives a total volume of 1.0×10^6 ft^3 (28,000 m^3). Even if the entire congested volume is treated as high congestion with 2-D confinement ($M_f = 0.069$), which would be a very extreme case, the blast overpressure at a standoff from the congested volume center of 150 feet based on the BST VCE blast load prediction method would only be 0.1 psig. Hence, unconfined VCEs of fuels in the "very low" reactivity class (e.g., ammonia, HFO-1234yf, etc.) do not appear capable of causing damaging blast loads, even under relatively severe conditions.

It is noted that there is support in the literature and industry guidance for assuming that ammonia and slower burning fuels will not pose a significant external VCE hazard. Baker et al. [5] noted that they were unaware of any cases of ammonia releases in outdoor process areas resulting in a

VCE, and considered only internal release scenarios in their example consequence assessment. The Factory Mutual guideline for VCE evaluation expressly excludes NH_3 outdoor releases as credible VCE scenarios [6].

Further support for the conclusion that external VCEs fueled by flammable mixtures with LBVs less than 10 cm/s should be expected to produce negligible blast loads is provided by the TNO GAMES correlation [7]. Calculating the 3D TNO GAMES correlation with the methane test parameters, (LBV of 0.4 m/s, VBR of 0.057, L_f of 4.1 m, and obstacle diameters of 0.06 meters) result in a calculated P_0 of 0.42 bar (6 psi) which falls below the TNO Severity Level 6 curve. This calculation and the resulting TNO severity curve bounds the observed methane test data. Repeating the 3D TNO GAMES correlation calculation with the LBV for ammonia (0.1 m/s) gives in a calculated P_0 of 0.01 bar (0.14 psi), which corresponds to the TNO 1 curve. The TNO 1 curve would suggest that the corresponding very low fuel reactivity VCE would not produce damaging overpressures.

It is important to note that the results of this test program are only applicable to unconfined VCE scenarios. Confined scenarios (i.e., indoor explosions) would behave differently. Testing of confined VCEs with very low reactivity fuels (e.g., ammonia) would be required to define the corresponding explosion hazard, as well as the level of conservatism in the current blast load prediction methodologies for such scenarios. Given the potential for internal vapor cloud explosions to create damaging overpressures, a logical next step would be to perform such tests to better define the explosion hazard associated with such very low reactivity fuels.

5 ACKNOWLEDGEMENTS

The tests described in this paper were performed under the sponsorship of Honeywell International, Inc., and their support is gratefully acknowledged. Brad Horn, Jef Rowley, Martin Goodrich, Austin Burch and Emeka Ugwu were instrumental in the completion of this test program and their contribution to this work is also gratefully acknowledged.

6 REFERENCES

[1] Pierorazio, A.J., J.K. Thomas, Q.A. Baker and D.E. Ketchum "An Update to the Baker-Strehlow-Tang Vapor Cloud Explosion Prediction Methodology Flame Speed Table," Process Safety Progress, 24(1): 59-65, 2005

[2] Minor, B. and M. Spatz "HFO-1234yf Low GWP Refrigerant Update," Intl. Refrigeration and Air Conditioning Conference, Purdue, 14-17 July 2008, Paper 937.

[3] Duynslaegher, Catherine." Experimental and Numerical Study of Ammonia Combustion" Université catholique de Louvain. Sept. 2011

[4] Version 2012.1.31 Release Date: Aug. 30 2014, DLL Aug. 30 2014

[5] Baker, Q.A., K.A. Dejmek and X. Qin "Facility Siting in Ammonia Facilities," AIChE 53rd Annual Safety in Ammonia Plants and Related Facilities Symposium, Paper 3e, San Antonio, TX, September 9, 2008.

[6] Factory Mutual Global, Property Loss Prevention Data Sheets, 7–42, "Guidelines for Evaluating the Effect of Vapor Cloud Explosions Using a Flame Acceleration Method," Factory Mutual Insurance Company, Johnston, RI, Oct. 2008.

[7] Mercx, W.P.M., A.C. van den Berg. And D. van Leeuwen "Application of Correlations to Quantify the Source Strength of Vapour Cloud Explosions in Realistic Situations, Final Report for the project: 'GAMES'" TNO Prins Maurtis Laboratory, Oct. 1998.

CATASTROPHIC DOMINO EFFECT FROM PROJECTION OF FRAGMENTS ASSOCIATED WITH SPHERICAL VESSEL EXPLOSIONS

Alejandro Castellanos Vargas
Department of Chemical Engineering, Universidad de Los Andes
Cra. 1 Este N° 19A – 40, Bogotá, Colombia
a.castellanos1630@uniandes.edu.co

Felipe Muñoz Giraldo
f.munoz@uniandes.edu.co
Fabio Ocampo
fa.ocampo2005@uniandes.edu.co
Nicolás Villalba
na.villalba340@uniandes.edu.co

Department of Chemical Engineering, Universidad de Los Andes
Cra. 1 Este N° 19A – 40, Bogotá, Colombia

Prepared for Presentation at
American Institute of Chemical Engineers
2016 Spring Meeting
12th Global Congress on Process Safety
Houston, Texas
April 11-23, 2016

CATASTROPHIC DOMINO EFFECT FROM PROJECTION OF FRAGMENTS ASSOCIATED WITH SPHERICAL VESSEL EXPLOSIONS

Alejandro Castellanos Vargas
Department of Chemical Engineering, Universidad de Los Andes
Cra. 1 Este N° 19A – 40, Bogotá, Colombia
a.castellanos1630@uniandes.edu.co

Felipe Muñoz Giraldo, Fabio Ocampo, Nicolás Villalba
Department of Chemical Engineering, Universidad de Los Andes

Key words: Explosion, Domino Effect, Numerical Simulations, Monte-Carlo Simulations, Fragments.

Abstract

The domino effect related to the projection of fragments in an industrial environment occurs when an element of a system, such as a chemical vessel, undergoes catastrophic fragmentation as the result of an explosion. The high-velocity fragments then impact and damage neighboring systems, which then leads to other dangerous scenarios such as loss of containment, fires or even other explosions. Of all industrial accident scenarios, those involving throwing of fragments have the greatest range from the accident point, well over 1 km. This process has been seen in several major accidents worldwide. In this work, this domino effect scenario was modeled in three steps: fragment generation, trajectory calculation and impact damage. In the first step, fragmentation patterns were studied using numerical simulations. For this, the inner explosions of 100 stainless steel 304 vessels, whit spherical geometry and diameters between 1 m and 20 m and thickness in a range of 10 mm to 90 mm, were simulated with the software of material dynamics modeling ANSYS-AUTODYN. In this step, the burst pressure of each vessel was estimated with Svenneson's equations, likewise, the explosion energy was calculated using Baum's equations; with this information, the explosion was modeled with an equivalent quantity of TNT. The detonation was simulated using Smoothed-Particles Hydrodynamics (SPH) method and the tank were meshed with a 3D Lagrangian grid. In addition, Mott's stochastic fragmentation model was implemented in order to simulate a material with heterogeneous microstructure. The probability density functions of mass, velocity and number of fragments were obtained as a result of this step, these data were used as input in the second step, trajectory calculation. In this step, 3D kinematics equations were employed to calculate the fragments paths during their flight and the feasible impact regions taking into account aerodynamic characteristics such as drag and lift coefficients. Additionally, the impact probability was estimated for different targets (cylindrical and spherical tanks of several sizes) located at multiple points from the explosion. Finally, in the step of impact damage, Monte Carlo simulations were implemented using an analytic penetration model in order to estimate the damage probability due to one or more fragments impact a target. The results of this work could be used in industrial risk assessment methods so that plant construction or plant modification can be designed in order to minimize the effects of this kind of scenarios inside the plant, in surrounding communities and in the environment.

1 Introduction

A domino effect involves a primary event that affects a corresponding primary installation, which induces one or more secondary accidents that affect other installations, leading to a major accident. There are mainly three kinds of domino effect scenarios: heat radiation, overpressure and impact of fragments. The first scenario is associated with primary events involved with fire and the last scenarios are related with VCE, mechanical explosion and BLEVE (*Boiling liquid expanding vapor explosion*). Based on the above, Cozzani et al [1] states four elements to characterize a domino effect: (i) A primary accidental scenario, which triggers the domino effect; (ii) A propagation effect following the primary event, due to the effect of escalation vectors caused by the primary event on secondary targets; (iii) One or more than one secondary accidental scenarios, involving the same or different plant units, causing the propagation of the primary event; and (iv) An escalation of the consequences of the primary event, due to the effect of the secondary scenarios.

Some of the major accidents involving domino effect are reviewed in different researches [2], [3], highlighting accidents such as: Feyzin (France, 1966) where 11 storage tanks were destroyed, it caused 18 deaths and injured 84 persons; Mexico City (Mexico, 1984), it caused 550 deaths, 7200 were injured, and almost 200000 were evacuated; Visakhapatnam (India, 1997), a fire and further explosions devastated the terminals and storage tank at the refinery of HPCL causing more than 55 deaths and dozens were injured; and Buncefield (United Kingdom, 2005), a major fire caused by a series of explosions at the Hertfordshire Oil Storage Terminal, it were 43 injured and more than 2000 were evacuated.

This work is focused in a domino effect scenario due to projection of structural fragments, in which a vessel or equipment related with the explosion is divided into many fragments which are projected to different direction with high velocities; these fragments could be impact other equipment or systems causing their failure [4]. Domino effect by impact of fragments is the most interesting escalation scenario taking into account that its affectation range can exceed 1 km; therefore, it corresponding escalation degree is the highest. Also, in this kind of scenarios, the escalation vector is discrete, which for this case is defined as the impact of the fragments generate in the primary event, while in the domino scenarios associated with heat radiation and overpressure the escalation vectors are continuous, which can be defined in units of heat flux density (i.e. W/m^2) and pressure (i.e. Pa), respectively. [3]. In general, available researches evaluate this event in three steps: fragmentation process, fragment trajectories and damage due to the impact of one or more than one fragments.

There are analytical, empirical and numerical approaches to address this domino effect scenario; nevertheless, analytical methods are associated with simplified physical models which do not allow representing this phenomenon in a complete real way; on the other hand, experimental approaches required expensive investments for safety protocols, explosive materials and measuring equipment. Thus, numerical approaches represent a good option to study domino events associated with fragments, because it is possible to model different scenarios considering most variables related with it. Also, it is possible to change easily the characteristics and conditions of the event in order to have a more complete and detailed information of it. Given this, the objective of this study is to develop a numerical simulation of the explosion and fragmentation of a spherical vessel and determine the probabilities of impact and damage of different targets due to the projection of fragments generated after inner explosions of a spherical tank.

2 Methodology

Domino scenarios caused by projection and impact of structural fragments is usually studied in three steps (i) Fragmentation process, (ii) Trajectory calculation and (iii) Impact damage. These steps represent the event chain necessary so that a domino event takes place. This methodology of study of domino scenarios allows an individual assessments of each of the steps. Nevertheless, in this work a complete assessment is performed starting with the information from fragmentation process, which is used as input data for trajectory calculation; likewise, the results from the second step are considered for the impact damage step.

The frequency of occurrence $(\boldsymbol{f_d})$ of a domino event due to projection of structural fragments generated by an inner explosion of a tank or vessel may be expressed as [3]:

$$f_d = f_p \cdot P_d \tag{1}$$

Where f_p is the expected frequency of the primary event, in other words, it is the frequency that an inner explosion of a tank or vessel happens. On the other hand, P_d represent the probability of subsequent events chain established as: (i) Generation of fragments, (ii) Projection and impact of one or more than one of the fragments with a specific target, (iii) Irreversible damage of the target [3].

Each of the fragments generated after the tank explosion can potentially trigger a domino event with the following expected frequency:

$$f_{d,F} = f_p \cdot P_{d,F} \tag{2}$$

Where $P_{d,F}$ is the probability of subsequent events chain for each fragment. This probability is defined as:

$$P_{d,F} = P_{gen,F} \cdot P_{imp,F} \cdot P_{dam,F} \tag{3}$$

Where $P_{gen,F}$ is the probability of *F-th* fragment to be generated, $P_{imp,F}$ is its probability of impact with a given target and $P_{dam,F}$ is the probability that an irreversible damage occurs in the target due to the previous impact of the *F-th* fragment.

Finally, total expected frequency of this kind of domino events can be calculated as sum of the domino probabilities for single fragments as follows:

$$f_d = \sum_F f_{d,F} = f_p \cdot \sum_F P_{d,F} = f_p \cdot \sum_F (P_{gen,F} \cdot P_{imp,F} \cdot P_{dam,F}) \tag{4}$$

The International Association of Oil & Gas Producers (OGP) report the storage incidents frequencies from its risk assessment data directory [5], in Table 1 the expected frequencies for explosions of storage tanks are presented.

Table 1. Storage incidents frequencies

Storage	Incident	f_p [per year]
Atmospheric tank	Internal explosion & full surface fire	9.00E-05
	Internal explosion without fire	2.50E-05
Pressurized vessel	BLEVE	1.00E-06

In the following sections the assessment methodology for the three steps in a domino event triggered by projection of fragments (fragmentation process, trajectory calculation and impact damage) is described.

2.1 Fragments generation step

Fragments originate when a tank or vessel is divided in many parts that are detached from it. Fragmentation process involves the formation and propagation of cracks through the material of the tank [3]. In this work the fragmentation of spherical metallic vessels due to inner explosions is studied using numerical simulations based in hydrocode methods.

2.1.1 Source tank characteristics

In order to determine the probability density functions of the characteristics of the fragments (mass, velocity, number and size) the numerical simulations of the inner explosion of 100 spherical tanks were performed. The diameters of these tanks were defined between 1 m and 20 m, while thickness were established in a range from 10 mm to 90 mm. Austenitic stainless steel 304 (SUS304) was used as material for all tanks, Nara and Miyazaki [6] presented the statistical data for the mechanical properties of this material (Table 2). That information was used in normal distributions with the purpose of generate random values for the material properties of each of the 100 tanks.

Table 2. Statistical data of SUS304

Property	Unit	Mean (μ)	Standard deviation (σ)
Yield stress (S_Y)	MPa	284.7	22.94
Ultimate strength (S_U)	MPa	641.58	40.713
Plastic strain at fracture (δ)	-	0.597	0.064

2.1.2 Burst pressure

When the internal pressure of a vessel increase caused by the loss of control of a chemical reaction, unwanted combustion of gases, vapors, dust or anything else, the required pressure to fragment the spherical vessel is known as burst pressure. Svennson [7] developed a model (Equation 5) to calculate the burst pressure of spherical vessels (P_S) based in their mechanical properties and geometrical characteristics.

$$P_S = S_Y \psi B \tag{5}$$

Where ψ is a geometric factor and B is a burst factor that depends upon the strength ratio β defined as follows:

$$\beta = \frac{S_Y}{S_U} \tag{6}$$

When the spherical tank can be classed as thin walled $\left(\frac{R_i}{T} \geq 10\right)$, Svensson's theory leads to the following equation:

$$P_S = \frac{S_Y}{m} B \tag{7}$$

Where m is defined as the inner radius/thickness ratio:

$$m = \frac{R_i}{T} \tag{8}$$

A relation between the strength ratio (β) and the burst factor B is presented in the book of Huston and Josephs [8], this relation is defined as:

$$B = 4.18\beta^2 - 9.57\beta + 7.38 \tag{9}$$

Although there are other models to calculate the burst pressure of spherical vessel, Svensson's equation is one of the formulas that present the lowest error range (-9.6, 10.1%) in the estimation of burst pressure as Yihui et al [9] report on their work.

2.1.3 TNT equivalent

The burst pressure is used to calculate the internal energy related with the explosion, for this, the explosion is modeled as the expansion of an ideal gas. There are several equations that allow to model this phenomenon such as Brode's equation [10] and Baker's equation [11]; however, these produce a large overestimation of the energy. For that reason, in this work Baum's model [12] was used, because this equation incorporates the work of air pushed away by expanding gas, this equation is expressed as follows:

$$E = \left[1 - \left(\frac{P_0}{P_S}\right)^{\frac{\gamma-1}{\gamma}} + (\gamma - 1) \cdot \frac{P_0}{P_S}\right] \cdot \frac{P_S}{\gamma - 1} \cdot f \cdot V \tag{10}$$

Where P_0 is the atmospheric pressure, γ is the ratio of specific heats, f is the degree of filling of the vessel and V is its volume. In this case the worst case was evaluated, which corresponds to $f = 1$.

In order to achieve modeling the inner explosion of the vessel, the energy calculate with Baum's equation is converted to the equivalent mass of TNT using the following expression:

$$m_{TNT} = \frac{E}{E_{TNT}} \tag{11}$$

A typical value for the energy of explosion of TNT (E_{TNT}) is 4686 kJ/kg [13].

2.1.4 Numerical simulation

The numerical simulations of the inner explosions of the spherical vessels were developed using the software ANSYS AUTODYN. This is a versatile explicit analysis tool for modeling the nonlinear dynamics of solids, fluids, gases and their interactions [14]. Numerical methodology used in this software known as hydrocode. This methodology uses equations of state, constitutive relations and mass, momentum and energy balances to represent dynamic behavior of materials [15]. After develop the numerical simulations of the inner explosions of the spherical vessels the fragment's data such as number, mass, size and velocity are available.

- *Consideration:* Numerical simulation proposed in this work is only focused in the fragmentation process due to formation and propagation of cracks through the material of the tank; in this case, the mechanical and welded joints, valves, pipes, internal corrosion, dust and other characteristics of the tank were not taken into account.

- *Discretization methods:* AUTODYN has mainly three discretization methods: Eulerian mesh, Lagrangian mesh and Smoothed Particles Hydrodynamics. For this work, a coupling between the last two methods was used. In Lagrange method the continuous material is divided into little volumes known as finite elements. In a 3D model, each finite element has 8 nodes, these are connected with nodes of other elements. For each node, the hydrocode equations system is solved considering the information of neighboring nodes [16].On the other hand, Smoothed Particles Hydrodynamics (SPH) is a mesh free technique in which the material is divided into a lot of particles. Hydrocode solution is calculated simultaneously for all particles, for that reason this method generates a high computational cost. However, with SPH method is possible to represent large distortion; thereby, it can model blast, high explosions and gas expansions [17].

- *Simulation model:* Each of the 100 spherical vessels of SUS304 are modeled in a 3D environment and a Lagrangian mesh with 60000 elements was used as discretization method. On the other hand, as mentioned above, the explosive load is modeled with TNT, the quantity of explosive was calculated using Baum's model (Equation 10) and TNT equivalent (Equation 11). Because the energy model depends of the material properties and geometric characteristics of the vessel, an equivalent quantity of TNT was calculated for each tank. The equivalent mass of TNT was represented as a sphere with a volume defined as:

$$V_{TNT} = \frac{m_{TNT}}{\rho_{TNT}} \tag{12}$$

The explosive load was modeled with SPH method, for all simulations 80000 particles were used to discretize the volume of TNT. Additionally, the detonation point was located at the coordinates (0,0,0) and the detonation time was defined at 0ms. The previous description is illustrated in Figure 1.

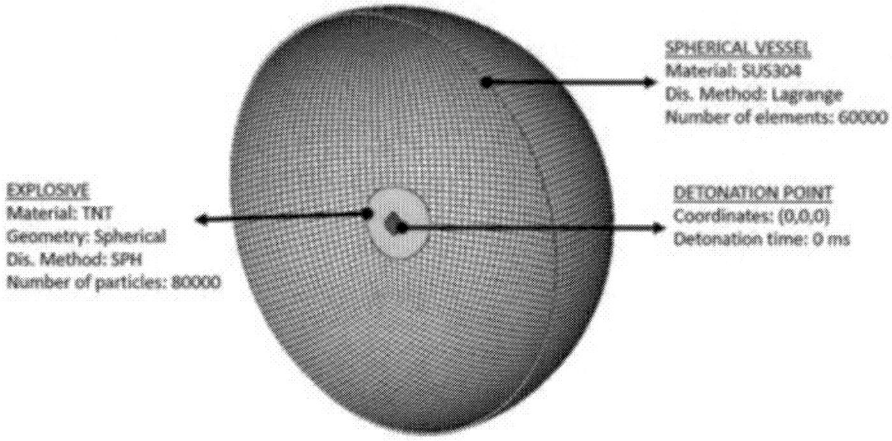

Figure 1. Simulation model

2.2 Trajectory calculation step

After the vessel explosion, fragments are projected in different directions, their paths depend of 3 forces acting on these. The first one, is the weight $(\vec{W})$ which is related with gravitational acceleration. The second one, is the inertia force $(\vec{I})$ due to the fragment acceleration related with its velocity. The last one, is the aerodynamic force which involves 2 components: the drag force $(\vec{D})$ resulting from friction between the projectile and the air, and the lift force $(\vec{L})$ caused by the pressure difference between the projectile intrados and the extrados [18], [19]. From the balance of the forces acting on the projectile is possible to define the general movement equations of a fragment in a three-dimensional system (Figure 2) as follows:

$$(-k_D \cdot \cos(\varphi) \cdot \cos(\theta) - (-1)^q \cdot k_L \cdot \sin(\varphi) \cdot \cos(\theta)) \cdot (\dot{x}^2 + \dot{y}^2 + \dot{z}^2) - \ddot{x} = 0 \qquad (13)$$

$$(-k_D \cdot \cos(\varphi) \cdot \sin(\theta) - (-1)^q \cdot k_L \cdot \sin(\varphi) \cdot \sin(\theta)) \cdot (\dot{x}^2 + \dot{y}^2 + \dot{z}^2) - \ddot{y} = 0 \qquad (14)$$

$$(-(-1)^q k_D \cdot \sin(\varphi) + k_L \cdot \cos(\varphi)) \cdot (\dot{x}^2 + \dot{y}^2 + \dot{z}^2) - \ddot{x} -= 0 \qquad (15)$$

Where $q = 1$ at the descending part and $q = 2$ at the ascending part. k_D and k_L (Equations 16 and 17) are terms related with the drag and lift force, respectively.

$$k_D = \frac{1}{2}\frac{\rho_{air} \cdot C_D \cdot A_D}{m_f} \qquad (16)$$

$$k_L = \frac{1}{2}\frac{\rho_{air} \cdot C_L \cdot A_L}{m_f} \qquad (17)$$

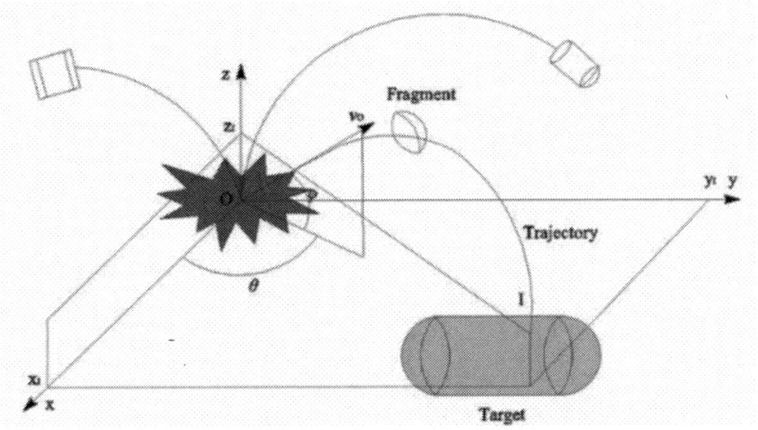

Figure 2. Description fragment trajectory [20]

When the drag and lift coefficients do not change their values with the time, Equations 13, 14 and 15 lead to the following expressions:

$$(k_D + (-1)^q \cdot k_L) \cdot \dot{x}^2 + \ddot{x} = 0 \quad (18)$$

$$(k_D + (-1)^q \cdot k_L) \cdot \dot{y}^2 + \ddot{y} = 0 \quad (19)$$

$$((-1)^q \cdot k_D - k_L) \cdot \dot{z}^2 + \ddot{z} + g = 0 \quad (20)$$

The complete solution for above equations are developed by Mébarki et al [18].

2.3 *Impact damage step*

Damage model is defined from the penetration model proposed by Mébarki et al [21], who related the penetration depth (h_p) with the mass (m_f), the velocity (v_p) and the impact angle (α) of the fragments and the mechanical properties of the target.

Penetration depth for the case $\alpha \neq 0$:

$$h_p = \frac{\left(-d_p \cdot \cos(\alpha) + \sqrt{(d_p \cdot \cos(\alpha))^2 + \frac{4}{\pi} \cdot \tan(\alpha) \cdot \left(\frac{E_c}{f_u \cdot \varepsilon_u}\right)^{\frac{2}{3}}}\right)}{2 \cdot \tan(\alpha)} \quad (21)$$

Penetration depth for the case $\alpha = 0$:

$$h_p = \left(\frac{E_c}{f_u \cdot \varepsilon_u}\right)^{\frac{2}{3}} \cdot \frac{1}{\pi \cdot d_p} \quad (22)$$

Where the kinetic energy is defined as $E_c = (m_f \cdot v_p^2)/2$, f_u and ε_u are the ultimate strength and ultimate strain of the target's constitutive material, respectively.

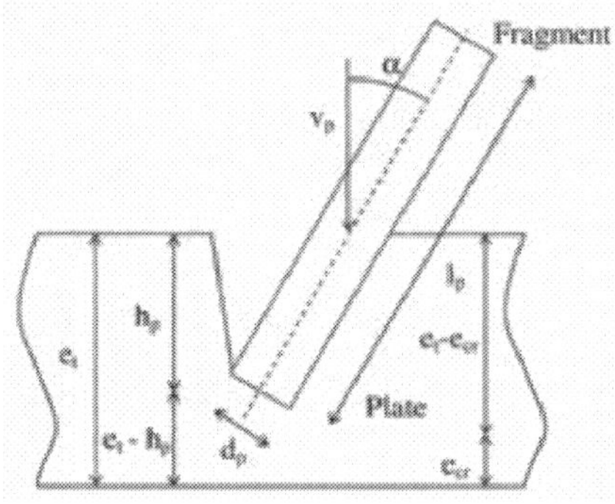

Figure 3. Fragment penetration scheme [22]

In the Figure 3 the penetration scheme of a fragment is showed, where e_t is the target's thickness and e_{cr} is its critical thickness. e_{cr} is calculated as the minimum thickness according the code *API 620* for atmospheric tanks and *ASME boiler and pressure vessel code* for pressurized tanks. Based in Figure 3, the target resistance against critical damage after impact corresponds to the following limit sate function:

$$E = e_t - h_p - e_{cr} \tag{23}$$

In this way, a failure of a target tank takes place when $E \leq 0$.

2.4 Case study

A storage area with 9 tanks was defined as case study, in Table 3 the tank information are showed and in Figure 4 the storage area distribution is illustrated.

Table 3. Tanks information

ID	Type	Diameter [m]	Thickness [mm]	Length/Height [m]	Location X [m]	Location Y [m]	Operating Pressure [Pa]	Material
S	Sphere	10	25	-	0	0	1.5E+06	SUS304
1	Sphere	10	25	-	25	0	1.5E+06	SUS304
2	Sphere	10	25	-	0	-25	1.5E+06	SUS304
3	Sphere	10	25	-	25	-25	1.5E+06	SUS304
4	Vertical cylinder	15	10	20	100	35	Atmospheric	SUS316
5	Vertical cylinder	15	10	20	100	0	Atmospheric	SUS316
6	Vertical cylinder	15	10	20	100	-35	Atmospheric	SUS316
7	Horizontal cylinder	2.5	30	10	-50	0	1E+06	SUS316
8	Horizontal cylinder	2.5	30	10	-60	0	1E+06	SUS316
9	Horizontal cylinder	2.5	30	10	-70	0	1E+06	SUS316

S: Source tank

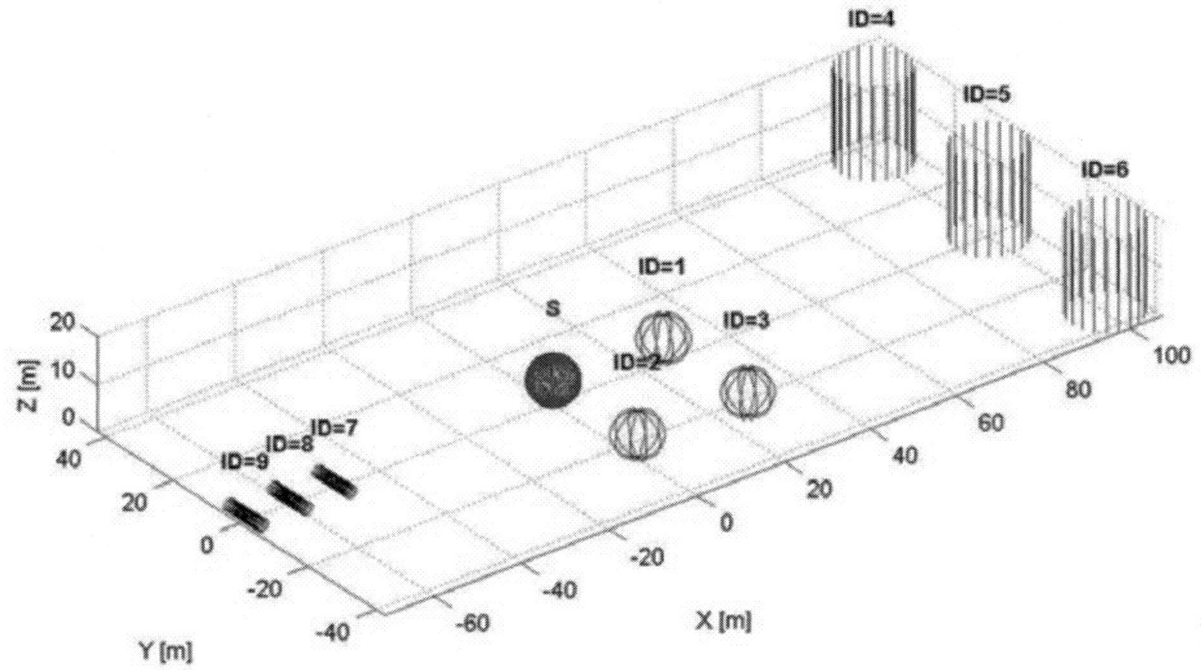

Figure 4. Storage area distribution

For the frequency of a domino scenario takes place the frequency of the primary (f_p) event was defined as 1E-06.

2.5 Monte-Carlo simulations

Monte-Carlo simulations were developed in order to calculate the probability of a domino scenario takes place. For this work, 2500 simulations were performed, this means that 2500 explosions of the source tank were assessed. The methodology for calculating the probability of generation, impact and damage is shown below.

2.5.1 Probability of generation

The probability of generation is defined as the probability at least one fragment is generated after the source tank explosion. In this case, in each of the 2500 simulations more than one fragment was generated, because in each simulation the failure and fragmentation process of the source tank takes place. On that basis, the probability of generation is defined as $P_{gen} = 1$.

2.5.2 Probability of impact

An impact occurs when the volume of one or more fragments intersect the volume of the target, this is defined in a probabilistic way as:

$$P_{imp} = \frac{1}{N} \cdot \sum_{i=1}^{N} \frac{1}{n_i} \cdot \sum_{j=1}^{n_i} k_{i,j} \quad and \quad k_{i,j} = \begin{cases} 1, & if \ (V_{i,j} \cap V_{target}) \neq \emptyset \\ 0, & otherwise \end{cases} \tag{24}$$

Where N is the number of simulations, n_i is the quantity of fragments of the *i-th* simulation, V_{target} is the volume of the tank and $V_{i,j}$ is the volume of the *j-th* fragment of the *i-th* simulation.

2.5.3 Probability of damage

Damage is calculated taking into account the limit sate function presented in Equation 23 as follows:

$$P_{dam} = \frac{1}{N} \cdot \sum_{i=1}^{N} \frac{1}{(n_{imp})_i} \cdot \sum_{j=1}^{(n_{imp})_i} q_{i,j} \quad and \quad q_{i,j} = \begin{cases} 1, & if\ E_{i,j} \leq 0 \\ 0, & otherwise \end{cases} \qquad (25)$$

Where $(n_{imp})_i$ is the number of fragments that impacted the target in the *i-th* simulation.

3 Results and discussion

In the fragmentation step, when the blast wave reaches the inner wall of the source tank, the expansion of the vessel begins. In the first instants, an elastic deformations occurs causing a slight increase in diameter; however, the blast continues applying pressure and plastic deformation takes place. Finally, the failure of the tank occurs when the strain is equal or greater than the strain at fracture defined for each tank. At this moment, the fracture process starts due to formation and propagation of cracks and the fragments are generated. The above is shown in Figure 5.

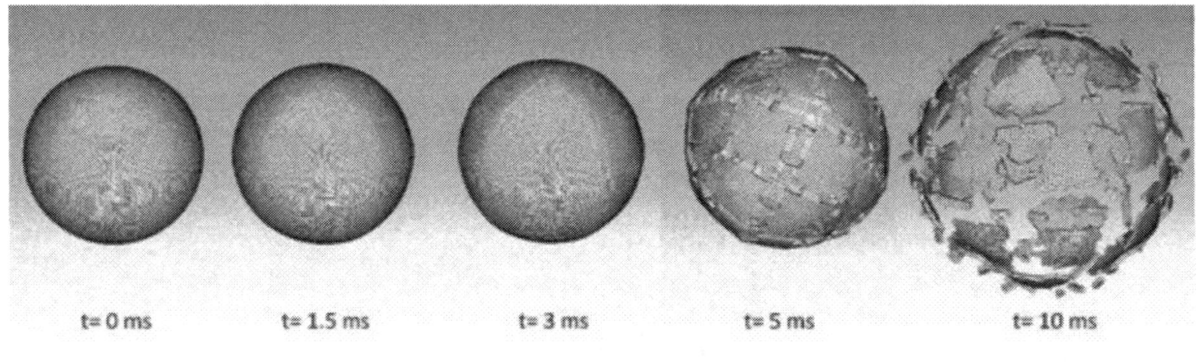

Figure 5. Fragmentation process of a spherical vessel

With the information obtained with the numerical simulations (number of fragments, mass, velocity and size data), the trajectory calculation and impact damage steps were developed. In Figure 6 the coordinates of the centers of mass of the fragments that impact each tank are shown (green points), as was to be expected, tanks closer to the source had more impacts than others. On the other hand, in Figure 7, the fragments that caused damage on the different targets due to the impact of them are represented with red points. In this case, it is possible to observe that the amount of red points is less than the green points, because not all fragments that impacted cause damage on the targets. This could be due to the mass, the velocity and the angle of impact of the fragments or the material properties of the target and its critical thickness, which are the variables that determine whether or not the tank fails.

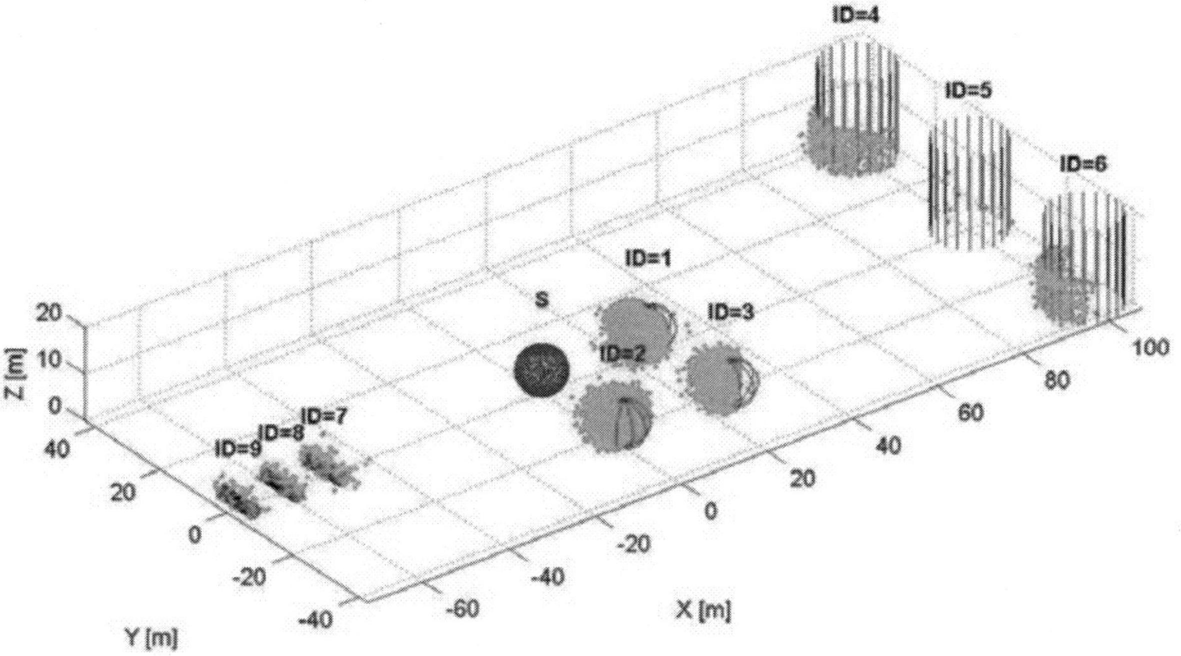

Figure 6. Impact distribution

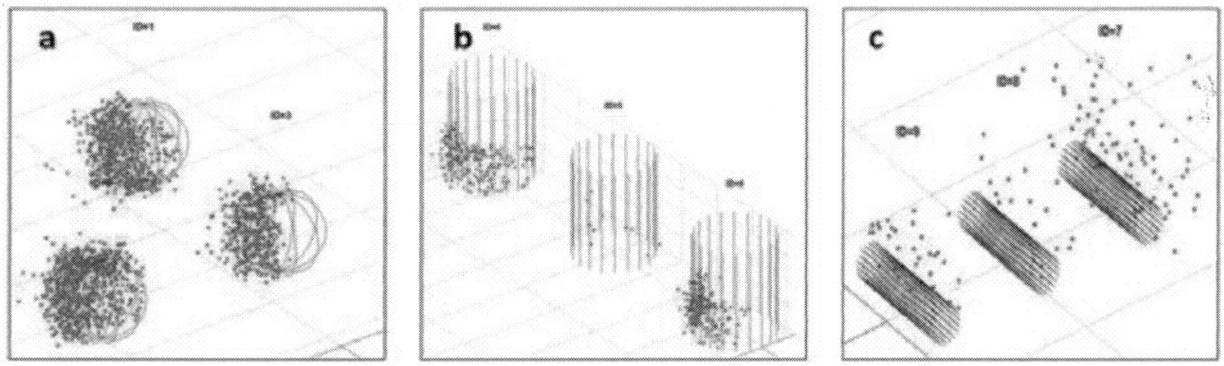

Figure 7. Fragments that caused damage, a. Spherical tanks, b. vertical cylindrical tanks, c. horizontal cylindrical tanks

The probabilities of impact and damage related with Monte-Carlo simulations are shown in Figure 8. The probability of generation not was plotted because it takes the value of 1 for any number of simulations.

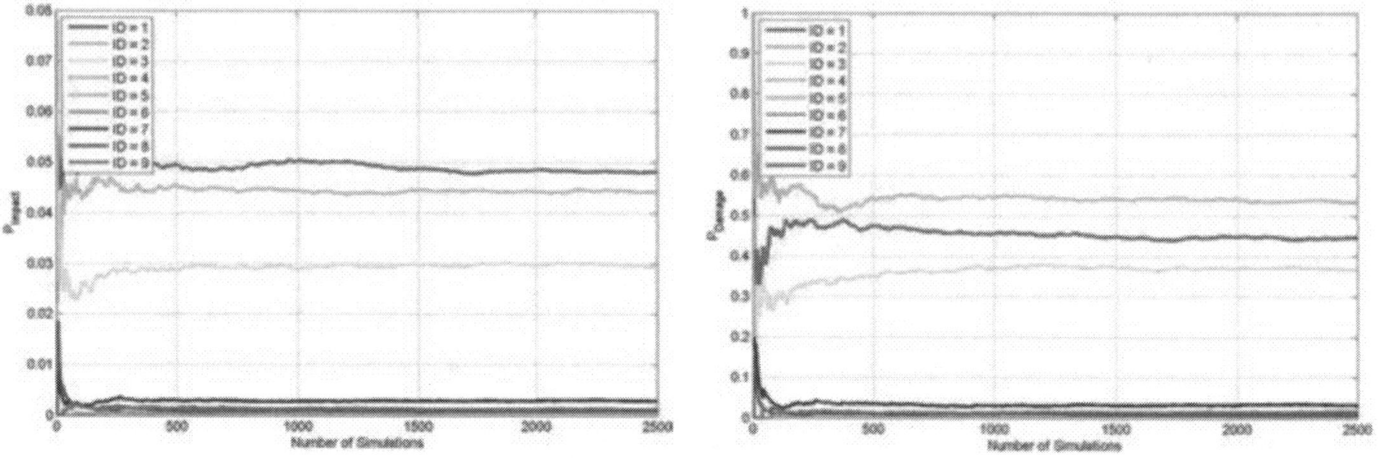

Figure 8. Probability of impact and damage calculated with Monte-Carlo simulations

A clear tendency is observed in the above figures, in all cases the probabilities converge to a constant value, this indicates that the model proposed in this work is stable. In the Table 4 the probabilities calculated for each target are shown. The couple of tanks 1-2 and 4-6 have the same or almost the same impact probability, because these couple of tanks are located at the same distance from the source tank. In the case of the probability of damage the values are also very close; however, in this case the probability associated to each target depends of the material properties of them and the mass, velocity and angle of the fragments that impact them. The tank 5 has the lowest frequency of domino effect scenario, this is due to its value of impact probability; in this case, the tank 1 is located between the source and the tank 5, for that reason, many of the fragments impact the tank 1 instead of the tank 5; in simple words, the tank 5 uses the tank 1 as a shield. A similar case occurs with the horizontal tanks (ID: 7,8 and 9).

Table 4. Probabilities for the case study

ID	P_{gen}	P_{imp}	P_{dam}	P_d	f_d [per year]
1	1	4.81E-02	4.45E-01	2.14E-02	2.14E-08
2	1	4.41E-02	5.34E-01	2.35E-02	2.35E-08
3	1	2.95E-02	3.66E-01	1.08E-02	1.08E-08
4	1	7.00E-04	1.80E-02	1.26E-05	1.26E-11
5	1	1.78E-04	4.40E-03	7.83E-07	7.83E-13
6	1	5.66E-04	1.64E-02	9.27E-06	9.27E-12
7	1	2.85E-03	3.18E-02	9.05E-05	9.05E-11
8	1	1.17E-03	7.20E-03	8.39E-06	8.39E-12
9	1	1.07E-03	1.18E-02	1.27E-05	1.27E-11

4 Conclusions

Numerical simulations of the inner explosions of 100 spherical SUS 304 tanks were developed using a Lagrangian discretization for the tank material and SPH discretization for the explosive material. The fragment data such as number, mass, length and velocity obtained from the numerical simulations were used to calculate the impact and damage probabilities of the fragments using kinematic model which takes into account the drag and lift forces, also a penetration model and probabilistic equations defined from Monte-Carlo simulations were used. With this information, the probability of a domino scenario takes place due to the projection of fragments was calculated for a storage area with 9 target tanks with spherical and cylindrical geometries. In general, the calculated probabilities depend of the distances of the targets from the source tank as well as the distribution of them in the storage area.

The methodology proposed in this work could be used in industrial risk assessment methods and decision making process so that plant construction or plant modification can be designed in order to minimize the effects of fragments impact inside the plant, in surrounding communities and in the environment.

5 Nomenclature

A_D: Drag surface [m^2]
A_L: Lift surface [m^2]
C_D: Drag coefficient [-]
C_L: Lift coefficient [-]
$\vec{g}$: Gravitational acceleration [m/s^2]
m_f: Fragment mass [kg]
ρ_{air}: Density of air [kg/m^3]
ρ_{TNT}: Density of TNT [kg/m^3]
φ: Vertical departure angle [°]
θ: Horizontal departure angle [°]

6 References

[1] V. Cozzani, G. Antonioni, and G. Spadoni, "Quantitative assessment of domino scenarios by a GIS-based software tool," *J. Loss Prev. Process Ind.*, vol. 19, no. 5, pp. 463–477, 2006.
[2] F. Kadri and E. Châtelet, "Domino effect analysis and assessment of industrial sites: A review of methodologies and software tools," *Int. J. Comput. Distrib. Syst.*, vol. 2, no. III, pp. 1–10, 2013.
[3] G. Reniers and V. Cozzani, *Domino Effects in the Process Industries: Modelling, Prevention and Managing*. Newnes, 2013.
[4] J. Casal, *Evaluation of the effects and consequences of major accidents in industrial plants*, vol. 8. Elsevier, 2007.
[5] B. Frequencies, "Risk Assessment Data Directory, OGP (International Association of Oil and Gas Producers)," Report, 2010.
[6] S. Nara and Y. Miyazaki, "Influence of Variation in Material Strength on Ultimate Strength of Stainless Steel Plates under In-Plane Bending and Compression."
[7] N. L. Svensson, "The bursting pressure of cylindrical and spherical vessels," *J. Appl. Mech.*, vol. 25, no. 1, pp. 89–96, 1958.
[8] R. Huston and H. Josephs, *Practical stress analysis in engineering design*. CRC Press, 2008.

[9] Y. Yihui, L. Xinfu, S. Yin, T. YUN, M. JUN, L. YUANDONG, and Y. SHAORONG, “Amendment and reconstruction of empirical formulas calculating bursting pressure of special spherical vessels (I),” in *Proceedings of the 10th asia-pacific conference on enginnering plasticity and its applications, Wuhan, China*, 2010.
[10] H. L. Brode, “Numerical solutions of spherical blast waves,” *J. Appl. Phys.*, vol. 26, no. 6, pp. 766–775, 1955.
[11] W. E. Baker, *Explosions in air*. University of Texas Press, 1973.
[12] M. R. Baum, “The velocity of missiles generated by the disintegration of gas-pressurized vessels and pipes,” *J. Press. Vessel Technol.*, vol. 106, no. 4, pp. 362–368, 1984.
[13] D. A. Crowl and J. F. Louvar, *Chemical process safety: fundamentals with applications*. Pearson Education, 2001.
[14] Engineering Eye, “AUTODYN.” [Online]. Available: http://www.engineering-eye.com/AUTODYN/.
[15] C. Y. Tham, “Numerical and empirical approach in predicting the penetration of a concrete target by an ogive-nosed projectile,” *Finite Elem. Anal. Des.*, vol. 42, no. 14, pp. 1258–1268, 2006.
[16] J. Zukas, *Introduction to hydrocodes*. Elsevier, 2004.
[17] G. R. Liu and M. B. Liu, *Smoothed Particle Hydrodynamics: A Meshfree Particle Method*. World Scientific, 2003.
[18] A. Mébarki, Q. B. Nguyen, and F. Mercier, “Structural fragments and explosions in industrial facilities: Part II – Projectile trajectory and probability of impact,” *J. Loss Prev. Process Ind.*, vol. 22, no. 4, pp. 417–425, Jul. 2009.
[19] L. M. Milne-Thomson, *Theoretical aerodynamic*. Macmillan, 1966.
[20] D. Sun, J. Jiang, M. Zhang, Z. Wang, G. Huang, and J. Qiao, “Parametric approach of the domino effect for structural fragments,” *J. Loss Prev. Process Ind.*, vol. 25, no. 1, pp. 114–126, 2012.
[21] A. Mebarki, R. A. Saada, Q. B. Nguyen, F. Mercier, F. Meftah, and M. Reimeringer, “Mechanical behavior of metallic plates under metallic rods impact and industrial risks,” in *International Conference on Computational & Experimental Engineering and Sciences (ICCES'05)*, 2005.
[22] Q. B. Nguyen, A. Mebarki, R. A. Saada, F. Mercier, and M. Reimeringer, “Integrated probabilistic framework for domino effect and risk analysis,” *Adv. Eng. Softw.*, vol. 40, no. 9, pp. 892–901, 2009.

Modeling of LNG Spreading and Vaporization Using Smoothed Particle Hydrodynamics Method

Anh Viet Bui
AnhBui@flowsim3d.com

Morgan Reed, Tony Liu, Eric Peterson
MMI Engineering Inc.
11490 Westheimer Road, Suite 150, Houston, Texas 77077
MReed@MMIEngineering.com
TLiu@MMIEngineering.com
EPeterson@MMIEngineering.com

Prepared for Presentation at
American Institute of Chemical Engineers
2016Spring Meeting
12thGlobal Congress on Process Safety
Houston, Texas
April 11-23, 2016

Modeling of LNG Spreading and Vaporization Using Smoothed Particle Hydrodynamics Method

Anh Viet Bui
AnhBui@flowsim3d.com

Morgan Reed, Tony Liu, Eric Peterson
MMI Engineering Inc.
11490 Westheimer Road, Suite 150, Houston, Texas 77077

Keywords: LNG, vaporization, CFD, Smoothed Particle Hydrodynamic (SPH)

Abstract

Smoothed Particle Hydrodynamics (SPH) has been used for numerical simulation of the release of LNG and the subsequent spreading of the liquid. Although the SPH method has been used for years in the astrophysics industry, its application to the LNG industry is quite novel. The spreading and vaporization of LNG creates the potential for a large vapor cloud with the risk of an explosion or flash fire. The SPH method offers excellent mass conservation, especially when modeling small scale changing fluid interfaces (i.e. thin film or small vapor bubbles). Based on LNG consequence hazards, many regulations (i.e. NFPA 59A) restrict vapor cloud dispersion extent to the facility property perimeter; less expensive (time and money) calculation methods are sought.

Vaporization of LNG during its spreading over water or solid substrate involves either evaporation on the air-LNG interface or nucleate/film boiling on the contact of cold LNG with relatively high-temperature substrate surface. For the latter vaporization mechanism to be correctly modeled, surface temperature and its subsequent change due to contact with LNG needs to be described. Consequently, a conjugate heat transfer model which includes wall heat conduction is necessary.

This paper discusses the continued development of a multiphase flow model with heat transfer and phase change based on the SPH model available in the LAMMPS Molecular Dynamics Simulator software. A mechanistic film boiling model is implemented and incorporated into our SPH flow model. Limited validation of the SPH prediction of the LNG spreading will be conducted and examples showing the enhanced boiling and heat transfer capabilities of the proposed LNG spreading model are outlined and discussed.

1 Introduction

LNG release, spreading, and vaporization and the formation of a flammable vapor cloud present a significant fire and explosion hazard to oil & gas facilities. NFPA 59A Standard for the Production, Storage and Handling of Liquefied Natural Gas, requires that the periphery of a vapor cloud (where fuel gas concentration becomes 50% of LFL) formed as a result of LNG release be within the property limits of LNG facilities.

Gravity-driven spread of LNG is a non-trivial process due to the intensive heat exchange of cryogenic LNG with the surrounding environment (soil, water, wind) which involves a rapid thermal exchange and evaporation of LNG. Various boiling regimes, such as nucleate, film, and/or explosive, may happen depending on the thermal and mechanical interaction of LNG and the substrate. In a scenario of LNG release to water, for instance, mechanical mixing of very cold LNG and relatively "hot" water may occur leading to a chain reaction of explosive increase of LNG-water interphase area and fiery LNG boiling dynamics (i.e. rapid phase transition).

The presence of a large free surface of spreading LNG and possible thermal and mechanical interactions of LNG with complex terrain and structural geometries pose significant difficulty to the modeling of LNG spreading using traditional grid-based CFD approaches. Such CFD approaches usually require a refined grid to resolve the physics happening inside an increasingly thinning layer of LNG when it spreads further away from the release origin, which is computationally expensive. Complex geometry of and abrupt change of material properties across the gas-liquid interfaces may also pose a significant problem to numerical methods used to solve the system of transport equations.

The Smoothed Particle Hydrodynamics (SPH) method has been proposed to model the spreading of LNG and the mechanical and thermal interactions with the surrounding environment and structures [1]. In comparison to the traditional grid-based CFD method, SPH can naturally and efficiently deal with complex (moving) fluid interfaces and terrains. This paper discusses the continued development of the SPH LNG spreading model with heat transfer and phase change [1] which is based on a SPH model available in the LAMMPS Molecular Dynamics Simulator software [2]. The model has been revised to enable the simulation of LNG spreading over complex solid boundaries represented by the STL (STereoLithography) geometries. The model of thermal interaction of LNG with such complex boundaries has been implemented which may include a mechanistic LNG boiling model. Limited testing and validation of the model will be conducted and examples showing the enhanced capabilities of the proposed LNG spreading model will be outlined and discussed.

2 Methodology

2.1 A brief description of the SPH method

Smoothed Particle Hydrodynamics method is based on a Lagrangian description of a fluid domain in which a limited number of particles with their own individual masses, momenta, and energies are employed for the modeling of fluid motion. The governing equations of fluid motion, i.e. Navier-Stokes equations, are solved with use of discrete (particle) approximations of field variables, i.e. density, velocity and energy, as well as their derivatives and integrals.

In the SPH method, the system of Navier-Stokes equations is expressed using particle approximations of the field variables and their derivatives as follows

$$\frac{D\rho_i}{Dt} = -\sum_{j=1}^{N} m_j W_{ij},$$

$$m_i\frac{D\mathbf{v}_i}{Dt} = -\sum_{j=1}^{N} m_i m_j\left(\frac{P_i}{\rho_i^2}+\frac{P_j}{\rho_j^2}\right)\nabla_j W_{ij} + \sum_{j=1}^{N} \frac{m_i m_j}{\rho_i\rho_j}(\mu_i+\mu_j)\mathbf{v}_{ij}\left(\frac{1}{r_{ij}}\frac{\partial W_{ij}}{\partial r_{ij}}\right) + \boldsymbol{F}_i,$$

[Eq.1]

$$m_i\frac{De}{Dt} = -\frac{1}{2}\sum_{j=1}^{N} m_i m_j\left(\frac{P_i}{\rho_i^2}+\frac{P_j}{\rho_j^2}\right):\mathbf{v}_{ij}\nabla_j W_{ij} - \sum_{j=1}^{N} \frac{m_i m_j}{\rho_i\rho_j}\frac{(k_i+k_j)(T_i-T_j)}{r_{ij}^2}\mathbf{r}_{ij}\cdot\nabla_j W_{ij}.$$

where $\boldsymbol{r}_{ij}= \boldsymbol{x}_i - \boldsymbol{x}_j$, $r_{ij} = |\boldsymbol{r}_{ij}|$, $\mathbf{v}_{ij}=\mathbf{v}_i - \mathbf{v}_j$; e, P, μ and k are the particle energy, pressure, dynamic viscosity and thermal conductivity, respectively; and m_i, ρ_I and N denote the particle i mass, density, and the total number of particles located within the influence area. W_{ij} or $W(\boldsymbol{x}_i - \boldsymbol{x}_j, h)$ is a smoothing function which defines the influence or support area around each particle. This function determines the accuracy of the SPH approximations and efficiency of the computation. A *cubic spline smoothing function* is used in this work.

A weakly-compressible approach to simulate incompressible fluids is applied which is based on the following equation of state (EOS):

$$p(\rho) = \frac{\rho_0 c^2}{\gamma}\left[\left(\frac{\rho}{\rho_0}\right)^{\gamma} - 1\right]$$ [Eq.2]

where c is the sound speed, ρ_0 is the reference density, and γ is a factor defining fluid stiffness. The time integration of Eqn.2 is conducted using a Velocity-Verlet scheme [3].

2.2 Treatment of complex wall boundary conditions

Most SPH models in the literature employ the so-called ghost (or dummy) particles technique to represent solid wall boundaries. In this approach, two or more layers of ghost particles are placed around a boundary (Figure 1). These ghost particles interact with the fluid particles confined inside the boundary, but are stationary during simulation. For complex boundary geometry with curved walls, appropriate generation and placement of ghost particles can be a challenging task.

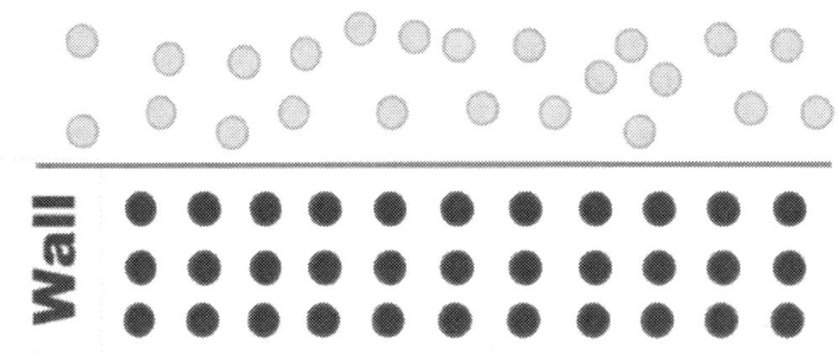

Figure 1: SPH particles – fluid (yellow) and boundary (purple) particles.

In this work, complex wall geometry is treated with use of a repulsive force boundary condition which prevents the penetration of interior fluid particles through the wall [4]. The wall is represented by STL geometries which approximate complex surfaces with triangular meshes. The repulsive force is calculated for every particle which is in the proximity of a boundary triangle. This force acts normal to the wall triangle and is given by

$$F_{rep,i}(d) = \begin{cases} D\left[\left(\frac{r_0}{d}\right)^a - \left(\frac{r_0}{d}\right)^b\right], & d < r_0 \\ 0, & d \geq r_0 \end{cases} \quad \text{[Eq.3]}$$

where d is the distance from a particle to a wall triangle and D, r_0, a, and b ($a > b$) are some regulation parameters.

The viscous forces acting on the directions tangential to the wall surface are accounted for by a modified viscous term which has the following form [4]:

$$\boldsymbol{F}_{vis,i}(d) = \frac{2m_i\mu_i\mathbf{v}_{i,tan}}{\rho_i^2}\left(\frac{1}{d}\frac{\partial W_d}{\partial d}\right) \quad \text{[Eq.4]}$$

with $\mathbf{v}_{i,tan}$ being the projection of the velocity vector of particle $i$on the wall surface plane.

2.3 Models of surface tension and cohesion forces

The surface tension force is defined simply by [5]

$$\boldsymbol{F}_{s,i} = -\sigma_s \sum_j m_j \boldsymbol{r}_{ij} W_{ij} \quad \text{[Eq.5]}$$

with σ_s being the surface tension coefficient.

A cohesion force can also be added which attracts nearby small bubbles (vapor particles) to form a large one. This force is given by [5]

$$\boldsymbol{F}_{cohesion,i} = -k_c m_i \sum_j \rho_j \boldsymbol{r}_{ij} \quad \text{[Eq.6]}$$

with k_c being a parameter that controls the strength of the cohesion force.

2.4 Modeling of wall heat transfer

Constant temperature or heat flux can be specified on the solid wall boundaries. When a constant wall temperature is set, the heat flux from the wall to the fluid particle $i$is defined by

$$Q_i = h(T_w - T_i) \quad \text{[Eq.7]}$$

where T_w and T_i are the wall and particle temperatures, respectively, and h is the convective heat transfer coefficient.

This wall heat flux can be added to the energy equation using the following term:

$$Q_i = \frac{2m_i^2 h(T_w - T_i)}{\rho_i^2}\left(\frac{\partial W_d}{\partial d}\right). \quad \text{[Eq.8]}$$

Upon a transition from convective heat transfer to nucleate or film boiling heat transfer, an appropriate boiling heat transfer coefficient should be utilized for h.

2.5 Modeling of LNG-vapor coexistence and evaporation

The flow of LNG and its vapor can be modeled with use of a *multiphase* or *multi-fluid* SPH model where separate transport equations and equations of state (EOS) are utilized for liquid and vapor LNG [1,6,7]. The mechanical and thermal interactions between phases including phase transition (LNG evaporation) are described by the interphase exchange terms. Different particle types are used to represent LNG and its vapor and the mass source or sink terms define the phase change rates.

Alternatively, the coexisted LNG liquid and vapor phases can be described by a single set of transport equations and a common EOS. The van der Waals equation of state is normally used in such a model, although it may impose some stability constraint on the numerical solutions.

Modeling of the near-wall boiling is complicated due to small scale physics involved. The physics of boiling and the transition from nucleate to film boiling are still not well understood and the split of wall heat flux (to convective and boiling heat transfers) and the wall evaporation rate are still mostly defined based on experimental data.

Phase change, i.e. evaporation or condensation, happens as a result of heat transfer between heating wall and flow or thermal exchange across the fluid-vapor interface. In this work, only evaporation near a heating wall is considered with the wall heat flux (Eq.7) assumed to be decomposed into a fluid heating and an evaporation components [6,7]. The evaporation heat flux Q_{evap} results in an evaporation mass flow rate of $\dot{\boldsymbol{m}}_{\boldsymbol{evap}} = \boldsymbol{Q}_{\boldsymbol{evap}}/h_{evap}$ (h_{evap}– the latent heat of evaporation). A corresponding number of vapor particles needs to be generated and inserted into the computational domain. In the current model, a simple evaporation model based on simple phase/particle type switching is employed.

3 Simulations of Fluid Spreading and Phase Change

Limited testing of the SPH method in modeling the spread of water over dry and wet flat surfaces has been conducted as described in a previous work [1]. In this study, new capabilities of the further developed SPH model are tested.

3.1 Simulation of fluid spreading over STL boundaries

The latest approach tracks the liquid running pool as it transverses a terrain other than a flat surface. Figure 2 shows an example of what can be considered a trench system common at many LNG terminals throughout the United States. The fluid temperature is set at 20 °C, whereas the surface temperature is at 100 °C. While the problem is essentially 2D, the simulation is 3D due to the presence of 2D STL boundary. The initial water column has the dimensions of 0.1m × 0.1m × 0.2m, and is described by approximately 3000 fluid particles.

Figure 2: A simple STL surface.

The simulation results are show in Figure 3 which indicates that the SPH model can successfully capture the dynamics of fluid spreading over a relatively complex boundary described by a STL surface meshe. The SPH method is seen to be able to capture the void space developed under the fluid jet in the beginning stage of the spreading (which can hardly be described by a shallow water model relying on a fluid height equation). Heat transfer between the wall and the fluid is also found to be reasonably simulated. It is worth noting that only constant temperature or constant heat flux boundary condition can be specified for the STL walls. If heat conduction inside the wall needs to be considered, the traditional approach of using stationary ghost particles needs to be applied.

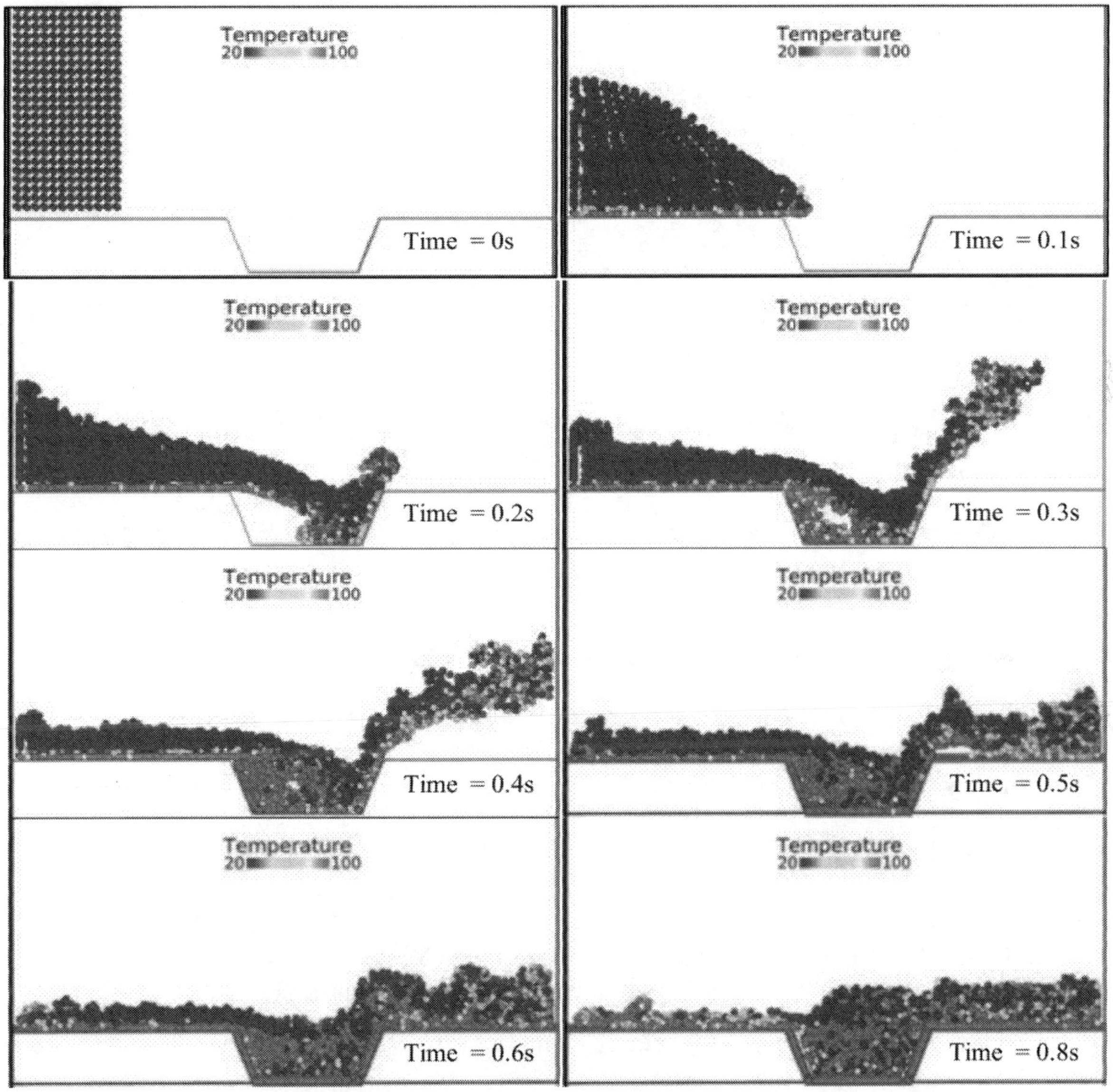

Figure 3: Spreading and heating of fluid over a STL surface.

3.2 Simulation of fluid evaporation during spreading

The SPH has been tested for its capability to correctly simulate the evaporation of fluid during spreading and the dynamics of the vapor phase. The fluid spreading is initiated by the collapse of a fluid column, which had the dimensions of 0.5m width by 1m height. The fluid and vapor densities are set at 450 kg/m^3 and 2 kg/m^3, respectively. The temperature difference between the spreading surface and the fluid is set at 900°C to facilitate faster heat transfer and evaporation of fluid.

The simulation results are presented in Figure 4, which shows the creation of an unstable vapor layer along the hot spreading surface. In comparison to previous simulations with similar settings and conditions [1], predictions of the wall evaporation dynamics and the rising of vapor phase across the fluid layer are more accurately predicted in Figure 4. The progressive thinning of the fluid layer is also seen.

To simulate the atmospheric transport and possible combustion of the vapor phase, the open-sourced CFD software OpenFOAM will be used. A coupling of OpenFOAM and this SPH spreading model is currently under development.

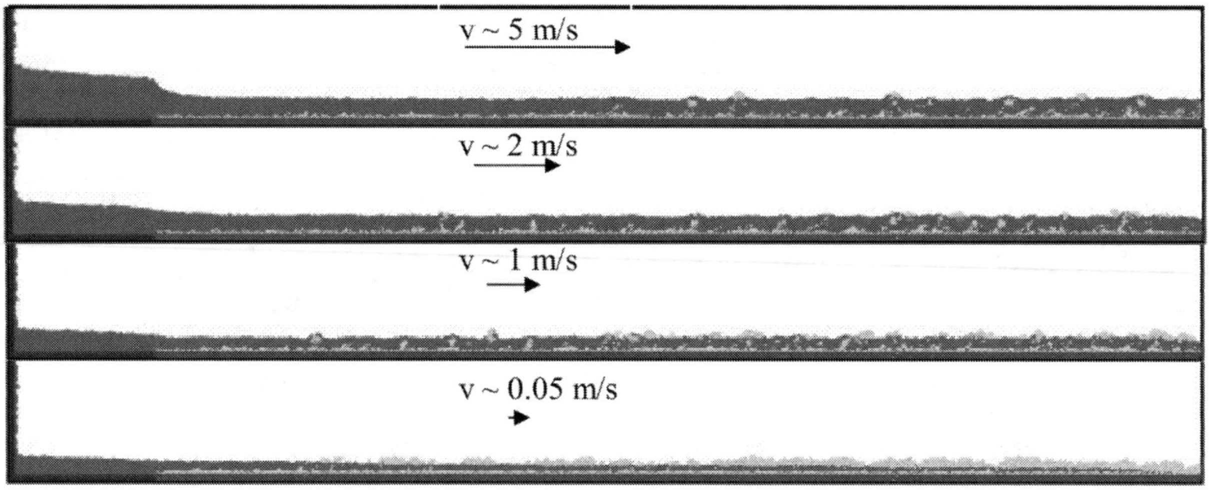

Figure 4: Fluid spreading, evaporation and dryout.

4 Conclusion

A SPH model of LNG spreading and evaporation [1] has been further developed in this work to include the capabilities of

- Modeling the mechanical and thermal interactions of the spreading fluid with terrain and objects of arbitrarily complex 3D geometries represented by STL surface meshes,
- Taking into consideration the effect of interactions between phases, i.e. surface tension and cohesion forces.

The wall evaporation model has also been improved to account for the effect of subcooled boiling.

Limited testing of the model has been conducted, but further model verification and validation are needed before it can be applied to the simulation of realistic industrial scenarios of LNG spreading.

Using CFDEM®coupling simulation engine [9], the proposed LNG spreading model can easily be coupled with the OpenFOAM software to enable 3D CFD modeling of LNG vapor dispersion and fire.

5 References

[1] A.V. Bui, T. Liu, M. Reed and E. Peterson. "CFD Modeling of LNG Spreading and Vaporization," Proceedings of the 11thAIChEGlobal Congress on Process Safety, Austin, Texas, April 2015.

[2] LAMMPS Molecular Dynamics Simulator, Available at http://lammps.sandia.gov, Assessed on 02/15/2016.

[3] G.C. Ganzenmuller, M.O. Steinhauser, and P.V. Liedekerke, The implementation of SPH in LAMMPS – A guide to the SPH-USER package, Germany, 2011.

[4] A. Eitzlmayr, G. Koscher, and J. Khinast, "A novel method for modeling of complex wall geometries in smoothed particle hydrodynamics," Comput. Phys. Comm., 185, pp.2436-2448, 2014.

[5] M. Ihmsen, J. Bader, G. Akinci, and M. Teschner, "Animation of air bubbles with SPH," Proceedings of International Conference on Computer Graphics Theory and Application, 2011.

[6] N. Shirakawa, Y. Yamamoto, H. Horie, and S. Tsunoyama, "Analysis of subcooled boiling with the two-fluid particle interaction method," J NuclSci Tech, 40(3), pp.125-135, 2003.

[7] A.K. Das and P.K. Das, "Modeling of liquid-vapor phase change using smoothed particle hydrodynamics," J Comput Phys, 303, pp.125-145, 2015.

[8] J.L. Woodward and R.M. Pitblado, "LNG Risk Based Safety – Modeling and Consequence Analysis," John Wiley & Sons, Inc., Hoboken, New Jersey, 2010.

[9] CFDEM®project, Available at http://www.cfdem.com, Assessed on 02/15/2016.

LNG Facility Siting Regulations – Home (USA) and Away (International)

Chaitanya Gottimukkala
RISKbytes Inc
1318 Nance St Houston TX 77002
Chaitanya@riskbytes.com

Prepared for Presentation at
American Institute of Chemical Engineers
2016 Spring Meeting
12th Global Congress on Process Safety
Houston, Texas
April 11-23, 2016

LNG Facility Siting Regulations –

Home (USA) and Away (International)

Chaitanya Gottimukkala
RISKbytes Inc
1318 Nance St Houston TX-77002
Chaitanya@riskbytes.com

Keywords: LNG, Facility Siting, Regulations, USA, International, Risk-based, Prescriptive

Abstract

Several LNG facilities are being proposed, designed and built around the world. These facilities will have to be designed, built and operated according to their local and national codes and standards. This paper provides an overview of the various codes, regulations and standards that govern the siting of the onshore LNG Facilities in the U.S.A and around the world. This paper also highlights some of the important siting requirements including risk criteria, vapor dispersion and overpressure criteria, acceptable thermal radiation levels found in these regulations and standards.

The primary siting regulations for U.S. onshore LNG facilities are 49 Code of Federal Regulations (CFR), Part 193 and NFPA 59A, "Standard for the Production, Storage, and Handling of Liquefied Natural Gas (LNG)". The design and siting standard for Europe and some facilities internationally is the European Standard, EN 1473, "Installation and equipment for liquefied natural gas - Design of onshore installations".

1 Introduction

With the demand for cleaner fuels rapidly growing around the world, Liquefied Natural Gas (LNG) is playing a vital role in this market. In order to meet the growing demand, LNG has to be transported from places of availability to places of demand. Therefore, a number of LNG Liquefaction and Regasification facilities are in proposal, design or construction around the world.

These LNG facilities must follow the industry codes, local and national regulations, and standards applicable to their location. The next section of this paper provides an overview of these required codes, regulations and standards for siting of LNG facilities both in the USA (home) and internationally (away). This paper's major focus is on land-based large scale LNG facilities; however, some of the standards for offshore based LNG facilities are also listed.

The major difference between the U.S. and international (for example - Europe & Australia) LNG siting requirements is that the United States (U.S) generally follows a "Prescriptive" approach and majority of the rest of the world follows a "Performance or Risk-based" approach. These approaches are consistent with the general approaches the onshore industries follow in these countries.

Prescriptive approaches in the U.S must follow a defined set of spacing criteria as outlined in NFPA 59A [3] and 49 CFR Part 193 [4]. In the risk-based approach, the proposed facility's spacing needs to demonstrate that it meets the tolerable risk criteria established by the host Nation, company or as outlined in codes such as EN 1473 [5], whichever is most stringent.

The subsequent sections of this paper highlight some of the important requirements with respect to the facility siting process including determination of design spills, risk criteria, allowable vapor dispersion and overpressure criteria, allowable thermal radiation levels.

2 Regulations, Codes, and Standards

The regulations, codes and standards applicable to facility siting in the US and internationally are summarized in the subsections below.

Local regulations, codes, and standards exist for countries such as Australia, United Kingdom, Canada, China, Europe, India, Japan and Mexico. Both the NFPA 59A (Prescriptive) and EN 1473 (Risk-based) have been adopted globally either individually or as a combination of both. Table 1 below summarizes the approaches followed by some of the countries.

Table 1. Summary of Some of the Countries with LNG Facilities and their Approaches

Country	Approach
U.S.A - NFPA 59A 2001 edition	Prescriptive
Canada - CSA Z276	Prescriptive
Japan - JGA-102	Prescriptive
Mexico - NOM-013-SECRE-2004	Combination of Prescriptive and Risk-based
India - OISD-STD-194	Combination of Prescriptive and Risk-based
Europe (U.K, Netherlands, France) - EN 1473	Risk-based
Australia - AS3961	Risk-based
Singapore	Risk-based

For example, the Canadian standard, CSA Z276, adopted NFPA 59A's prescriptive approach and the Australian standard, AS3961, adopted a risk-based approach similar to EN1473.

The primary focus of this paper is to compare the Prescriptive approach in the U.S (home) versus the Risk-based approach followed in Europe and other parts of the world (away). For purposes of this paper, when "U.S" or "Home" is mentioned it refers to the countries following Prescriptive approach and when "International", "EN1473" or "Away" is mentioned it refers to countries following Risk-based approach.

Offshore LNG facility siting process is outside the scope of this paper. However, some of the industry standards applicable to the offshore LNG facilities are listed below:

- DNV-OSS-103, Rules for Classification of Floating LNG/LPG Production, Storage and Loading Units
- ABS Guide for Building and Classing Floating Offshore Liquefied Gas Terminals

2.1 U.S (Home) Regulations, Codes and Standards – Onshore LNG Facility Siting

The following are the facility siting regulations and standards applicable to the U.S land based LNG facilities.

- 49 CFR Part 193, Liquefied Natural Gas Facilities: Federal Safety Standards
- 33 CFR 127 Subpart B, U.S Coast Guard - LNG Facility Inspection
- PHMSA, U.S Department of Transportation Pipeline and Hazardous Materials Administration
- FERC, Federal Energy Regulatory Commission
- NFPA 59A (2001/2009) – Standard for the Production, Storage, and Handling of Liquefied Natural Gas (LNG)
- API RP 752, Management of Hazards Associated with Location of Process Plant Permanent Buildings
- API RP 753, Management of Hazards Associated with Location of Process Plant Portable Buildings
- API RP 500, Recommended Practice for Classification of Locations for Electrical Installations at Petroleum Facilities
- API RP 505, Recommended Practice for Classification of Locations for Electrical Installations at Petroleum Facilities Classified as Class I, Zone 0, Zone 1, and Zone 2
- API STD 2510, Design and Construction of LPG Installations

2.2 International (Away - European Norm) Regulations, Codes and Standards – Onshore LNG Facility Siting

The following are the facility siting regulations and standards applicable to the land based LNG facilities following Risk-based approach, specifically, countries adopting European Standard - EN 1473:

- EN 1473 (2007), Installation and equipment for liquefied natural gas - Design of onshore installations

- EN 1160, Installations and equipment for liquefied natural gas - General characteristics of liquefied natural gas
- GAP 2.5.2, Oil and chemical plant layout and spacing
- GAP 2.5.2 A : Hazard Classification of Process operations for spacing requirements
- IP15, Area Classification code for installation handling flammable fluids Part 15, 2d edition (2002), by the Institute of Petroleum
- Site selection and design for LNG Ports and jetties: SIGTTO

The Host Nations have specific standards and regulations for their LNG facilities in addition to the above standards, which have not been included in this paper.

3 Facility Siting Process

U.S (Home) LNG facilities follow a prescriptive approach to siting as described in US 49CFR193 and NFPA 59A (2001) requiring strict compliance to the prescribed thermal radiation levels, vapor dispersion levels and overpressure levels at the facility's property line that can be built upon. The primary focus of the NFPA 59A and US 49CFR193 regulations is offsite risk to public. Other standards and regulations such as API, Environmental Protection Agency (EPA), Occupational Safety and Health Administration (OSHA) provide guidance for onsite risk criteria.

Some of the International (Away) LNG facilities follow a risk-based approach as described in EN 1473. Three categories of are risk listed in EN 1473 [5]:

- Level 3: Situation which is undesirable and cannot be tolerated. Remedial action required (Not Acceptable);
- Level 2: Situation which shall be improved. A level at which it shall be demonstrated that the risk is made As Low As Reasonably Practical (ALARP);
- Level 1: Normal situation (Acceptable).

These facilities should meet the risk acceptance criteria i.e risks in the "Acceptable" and "As Low As Reasonably Practicable (ALARP)" region for both public and on-site personnel either set by the Host Nation, Company or the EN 1473 standard, whichever is more stringent. If the quantified risk estimate lies in the Acceptable and ALARP regions, then the risk is "tolerable". The acceptability criteria are more stringent for hazards to the public.

The first step for facility siting in both the Home and Away approaches is hazard analysis to determine the maximum design spillage. The criteria for calculation of design spills are different in both approaches. Figure 1 provides a flow chart for facility siting process for both the Prescriptive (US) and Risk-based (International) approaches.

Home (US) Regulations and codes require the determination of design spill, referred to as Single Accidental Leakage Source (SALS) in NFPA 59A based on a maximum 10 minute release. Recent Nominal Failure Rate data published by FERC [2] and PHMSA [1] should be utilized to determine the failure rate criteria of equipment, piping and other appurtenances. Approved modeling tools should be used for evaluating the prescribed criteria for vapor dispersion, thermal radiation and overpressure at the property line that can be built upon. In case the prescribed siting criteria is not satisfied, passive mitigation measures can be specified. Some of the passive mitigation measures are rerouting of piping and relocation of hazardous equipment / tanks; modifying design conditions such as changing flow rates, pipe sizes, and capacities; constructing vapor barriers; or buying more land to extend the property line.

Away (EN 1473) requires the determination of the design spill either following a deterministic approach and/or probabilistic approach as described in Section 4.4.2 of the EN 1473 standard and Figure 1 below. Deterministic approach requires identification of hazards, establishment of credible hazards, determination of consequences, and mitigation measures to reduce consequences if risk is not tolerable. Probabilistic approach requires identification of hazards, determination of consequences, application of failure rate data, determination of frequency of hazards, and quantifying risk level from these data. If risk level is not tolerable, it can be reduced by active or passive mitigation measures including water deluge, fire and gas detection, vapor barriers and thermal shelters. The facility layout can be optimized and finalized following careful evaluation of various "tolerable" options.

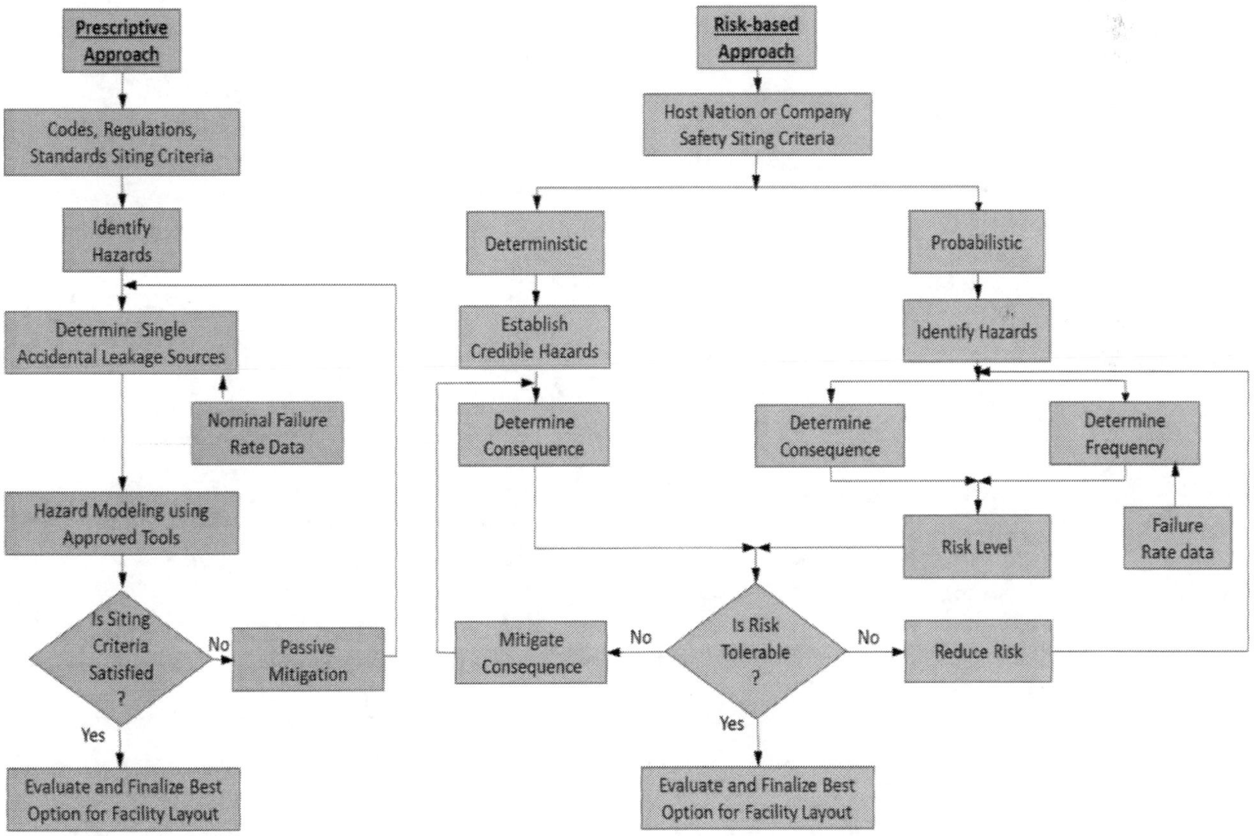

Figure 1. Facility Siting Process – Prescriptive (U.S) vs. Risk-based (International – EN1473) Approaches

The following subsections describe and compare the ambient conditions, required thermal radiation levels, vapor dispersion and overpressure criteria for both the Home (U.S) and Away (European) facilities.

3.1 Ambient Conditions

Table 2 describes the ambient conditions used for both the Prescriptive (U.S) and Risk-based (International) LNG facilities.

Table 2. Ambient Conditions for Home (49CFR193 / NFPA 59A, 2001) and Away (EN 1473, 2007)

Parameter	Home (49CFR193 / NFPA 59A, 2001)		Away (EN 1473, 2007)	
	Thermal Radiation	Vapor Dispersion	Thermal Radiation	Vapor Dispersion
Wind Speed	Wind speed producing the maximum exclusion distances shall be used. [1]	Wind speed producing longer predicted downwind dispersion distances than other weather conditions at the site at least 90 percent of the time. Alternatively, 4.5 miles per hour (2.01 meters/sec)	Combination of wind speed and atmospheric stability that can occur simultaneously and result in the longest predictable downwind dispersion distance that is exceeded less than 10 % of the time. Alternatively, 10 meters/sec	Combination of wind speed and atmospheric stability that can occur simultaneously and result in the longest predictable downwind dispersion distance that is exceeded less than 10 % of the time. Alternatively, 2 meters/sec
Temperature	Ambient temperature that produces the maximum exclusion distances shall be used.[2]	Average ambient temperature in the region	Ambient temperature in the region	Average ambient temperature in the region
Relative Humidity	The ambient relative humidity that produces the maximum exclusion distances shall be used.[2]	50%	50%	50%
Pasquill Stability Class	N/A	F	N/A	Combination of wind speed and atmospheric

Parameter	Home (49CFR193 / NFPA 59A, 2001)		Away (EN 1473, 2007)	
	Thermal Radiation	Vapor Dispersion	Thermal Radiation	Vapor Dispersion
				stability that can occur simultaneously and result in the longest predictable downwind dispersion distance that is exceeded less than 10 % of the time. Alternatively, Stability Class "F" can be used.
Surface roughness	N/A	0.03 m, higher values may be used based on the nature of terrain	N/A	Not specified. Industry standard values can be used.
Average gas concentration in air	N/A	½ of Lower Flammability Level (LFL) or 2.5 %	N/A	Not specified. Industry standard values can be used.

1. Per 49 CFR 193.2057, 'The wind speed producing the maximum exclusion distances shall be used except for wind speeds that occur less than 5 percent of the time based on recorded data for the area."
2. Per 49 CFR 193.2059, "in calculating exclusion distances, the ambient temperature and relative humidity that produce the maximum exclusion distances shall be used except for values that occur less than five percent of the time based on recorded data for the area."

3.2 Allowable Thermal Radiation Levels

The allowable thermal radiation levels are described in both the standards, U.S and International.

3.2.1 U.S (Prescriptive) Thermal Radiation Criteria

U.S 49CFR193 and NFPA 59A specify the thermal radiation flux levels and the exclusion zones for impounding areas containing LNG spills as listed in Table 3. U.S standards do not account for mitigation measures such as water deluge, shelters in reducing the heat flux level to the required limit. This often results in conservative estimates for thermal radiation hazards leading to larger exclusion zones. LNG spills from loading/unloading arms and LNG carriers should also be included when calculating the incident thermal radiation flux. Details on thermal radiation and impounding area design and siting requirements can be found in Section 2.2 of NFPA 59A and 49CFR193.2057 Subpart B. Thermal Radiation calculations should be calculated using the approved computer model LNGFIRE3.

Table 3. Allowable Thermal Radiation Flux Levels per Section 2.2.3 NFPA 59A

Exposure at Exclusion Zone	Thermal Radiation Flux	
	BTU/hr/ft^2	KW/m^2
At a property line that can be built upon upon ignition of a design spill	1600	5
At the nearest point located outside the owner's property line that, at the time of plant siting, used for outdoor assembly by groups of 50 or more persons for a fire over an impounding area containing a volume, V, of LNG	1600	5
At the nearest point of the building or structure outside the owner's property line that is in existence at the time of plant siting and used for occupancies	3000	9
At a property line that can be built upon for a fire over an impounding area containing a volume, V, of LNG	10,000	30

3.2.2 International (Risk-based: EN 1473) Thermal Radiation Criteria

EN 1473 specifies the thermal radiation flux levels excluding solar radiation for equipment inside the boundary and for areas outside the plant boundary as listed in Table 4. There is a provision for reducing the heat flux to the required limit by means of spacing, water sprays, fireproofing and radiation barriers. The main advantage with this approach over prescriptive standards is that overly conservative thermal exclusion zones will not be required after application of mitigation measures to reduce the risk to be in the "Tolerable" risk region.

LNG spills from loading/unloading trucks or LNG carriers should also be included when calculating the incident thermal radiation flux. It also specifies the allowable thermal radiation flux from flares or ignited vent stacks on equipment inside the plant boundary and on areas outside the plant boundary. This standard does not specify a required computer model to estimate thermal radiation and specifies that the radiation flux from an LNG fire shall be calculated using appropriate and validated models.

Table 4. Allowable Thermal Radiation Flux Levels per Table A.1 & A.2 of EN 1473

Exposure at Equipment Inside Boundary or Area Outside Boundary	Maximum Thermal Radiation Flux KW/m^2
Equipment Inside Boundary	
Concrete outer surface of adjacent storage tanks [1]	32
Metal outer surface of adjacent storage tanks	15
The outer surfaces of adjacent pressure storage vessels and process facilities	15
Control rooms, Maintenance workshops, laboratories, warehouses	8
Administrative buildings	5

Exposure at Equipment Inside Boundary or Area Outside Boundary	**Maximum Thermal Radiation Flux** KW/m²
Area Outside Boundary	
Remote area [2]	8
Critical area [3]	1.5
Other areas [4]	5

1. For pre-stressed concrete tanks, maximum radiation fluxes may be determined by the requirements given in A.1.1 of EN 1473.
2. An area only infrequently occupied by small numbers of persons, e.g. moor land, farmland and desert.
3. This is either an unshielded area of critical importance where people without protective clothing can be required at all times including during emergencies or urban area (defined as an area with more than 20 persons per square kilometre) or a place difficult or dangerous to evacuate at short notice (e.g. hospital, retirement house, sports stadium, school, outdoor theatre).
4. Other areas typically include industrial areas not under control of the operator/occupier of the LNG facilities.

3.3 Vapor Dispersion Criteria

The vapor dispersion criteria have been specified in both the standards, U.S and International.

3.3.1 U.S (Prescriptive) Vapor Dispersion Criteria

In the U.S, the regulation 49CFR193, and code NFPA59A, prescribe the criteria. Clarifications and additional details for these criteria can be found on the PHMSA website [1]. These regulations require assessment of vapor dispersion hazards resulting from design spills of LNG and other flammable substances. Flammable vapor dispersion distances should be calculated at ½ Lower Flammability Limit (LFL) concentrations for both LNG and other flammable fluids.

FERC and PHMSA provide guidance on determining the single accidental leakage source to assess the flammable vapor dispersion hazards. Process and ambient conditions, among other information is used to determine the design spills. The potential release source is determined using nominal failure rate criteria recently published by FERC and PHMSA. However, these failure rates do not account for the consequence part of risk in terms of injury or fatality onsite and/or offsite.

The approved computer models for vapor dispersion exclusion zones calculations are DEGADIS 2.3, FEM3A, FLACS 9.1 Release 2 (Oct. 7, 2011) and PHAST-UDM Version 6.6 and 6. 7 (Oct. 7, 2011).

In addition to the flammable vapor hazard analysis, U.S regulations also require the toxic gas (for example - benzene) hazard analysis, as applicable.

3.3.2 International (Risk-based: EN 1473) Vapor Dispersion Criteria

EN1473 specifies a risk-based approach for vapor hazard analysis. The determination of consequence, probability and associated risk level of a hazardous event is specified in the standard. If the local, country or company criteria are more stringent, then that criteria should be utilized. The criteria for estimation of consequences of LNG and other hazardous fluids is either detailed in EN 1473 or a reference is cited. More information is available in Section 4.4 and Annexures J, K and L of EN 1473.

Risk to personnel and public should be calculated and it should be within the "Tolerable" range of the risk criteria.

3.4 Overpressure Criteria

3.4.1 U.S (Prescriptive) Overpressure Criteria

There is no prescriptive requirement for overpressure and explosion in U.S 49CFR193 but Chapter 2.1.1(d) of NFPA59A requires that all hazards having an impact on public or plant personnel should be evaluated. FERC and PHMSA require the use of the 1.0 psi to determine the offsite impact, and are consistent with the EPA's "Risk Management Program for Chemical Accidental Release Prevention" (RMP, 40 CFR 68).

FLACS 9.1 Release 2 (Oct. 7, 2011) and PHAST-UDM Version 6.6 and 6. 7 (Oct. 7, 2011) are accepted models for use in determination of the overpressure hazards.

3.4.2 International (Risk-based: EN 1473) Overpressure Criteria

EN 1473 describes the methodology for overpressure analysis in Section 4.4.2.5 d. It also specifies a risk-based approach for overpressure analysis. The determination of consequence, probability and associated risk level of a hazardous event is specified in the standard. If the local, country or company criteria are more stringent, then that criteria should be utilized.

EN 1473 refers to EN 1160 for flammability range of mixtures of gas and air. The standard indicates that some of the recognized methods and models such as the multi energy method and/or deflagration at constant speed method can be used to calculate the over-pressure.

Risk to personnel and public should be calculated and it should be within the "Tolerable" range of the risk criteria.

3.5 Other Siting Criteria

3.5.1 U.S (Prescriptive) – Other Siting Criteria

U.S Federal regulations prescribe additional siting criteria for impoundments, equipment, buildings and structures. Spacing criteria within the facility as well as with respect to the property line is described in detail. Details of these prescriptive requirements are in U.S 49 CFR 193 and NFPA 59A.

Consequences from Boiling Liquid Expanding Vapor Explosions (BLEVE) should be considered during siting and design.

Permanent and portable buildings siting shall be in accordance with API 752 and 753.

Escape routes, SIMOPS and construction activities should also be considered during siting.

3.5.2 International (Risk-based: EN 1473) – Other Siting Criteria

EN 1473 specifies a risk-based approach for facility siting based on hazard assessment. Separation distances should include radiation flux levels, LFL contours, noise, blast effects. It also describes the siting of buildings, equipment and tanks. It refers to industry standards such as GAP 2.5.2 and GAP 2.5.2.a for additional guidance on facility siting.

Escape routes, SIMOPS and construction activities should also be considered during facility siting.

4 Prescriptive (US) vs Risk-based (International) Approaches – Pros and Cons

Based on the defined criteria, the siting of facilities using Prescriptive and Risk-based approaches have their share of pros and cons. Table 4 below describes a few pros and cons for both these approaches.

Prescriptive approach warrants the LNG facilities to follow the facility siting requirements without specific consideration for their location with respect to population density in that area, while the Risk-based approach takes into consideration the overall risk to public outside the fence based on the quantitative risk estimates included during modeling. The prescriptive approach does not take into account active mitigation measures for risk reduction while in the risk-based approach risk levels may be reduced using appropriate mitigation measures.

For example, U.S 49 CFR 193 and NFPA 59A require that the LNG facilities should adhere to the thermal radiation flux of 1600 BTU/hr-ft^2 and vapor dispersion exclusion criteria of ½ LFL at the property line that can be built upon. These criteria have to be followed without consideration of the population density in that area which may result in the LNG facility needing to own more acreage than required to satisfy the facility siting requirements. International standard, EN1473, the facility siting requirements are satisfied as long as the thermal radiation, vapor dispersion and other hazards associated with the LNG facility comply with the Company or Country required "Tolerable" risk levels.

Table 5. Prescriptive (Home) vs Risk-based (Away) Approaches – Pros & Cons

Prescriptive (Home - US)	Risk-based (Away - International)
Pros	
Application of Defined criteria is easy to follow	Flexibility in facility siting and layout
Defined criteria provide consistency across LNG industry based on industry knowledge and best practices.	Takes Location's population density into account
Less time consuming during design phases	Active and passive mitigation measure accepted to demonstrate tolerability
Cons	
Strict compliance irrespective of Location's population density	Definition of Tolerability Criteria
Limited mitigation measures (passive)	Requires complex calculations and assumptions
Special cases require longer processing time	Longer Project duration for finalizing Facility layout to demonstrate tolerable risk.

5 Conclusion

The growing demand for LNG requires proposal, design and construction of LNG facilities around the world, U.S (home) and International (away). This paper highlighted some of the regulations, codes and standards distinguishing between U.S (Prescriptive approach) and International (Risk-based approach: EN 1473) requirements.

U.S facilities follow a "Prescriptive" approach (U.S 49CFR Part 193 and NFPA 59A, 2001) and some of the International facilities follow a "Risk-based" approach (EN 1473). This paper highlighted the various requirements between these two approaches with respect to facility siting.

Criteria for ambient conditions, allowable thermal radiation levels, vapor dispersion, overpressure and additional spacing requirements for both approaches has been described.

The pros and cons of the Prescriptive and Risk-based approaches have been discussed.

In the U.S, the newer editions of NFPA 59A such as the 2013 edition include QRA (Quantitative Risk Assessment) methodology for facility siting. However, US 49 CFR 193 and FERC still refer to the 2001 edition and have not adopted the latest edition. Future adoption of the risk-based approach into the U.S regulatory ecosystem such as the U.S 49 CFR 193 and FERC will have its own challenges from public, judicial and regulatory systems.

Acknowledgment:

The author would like to thank his supervisor, Robert McDonnell, for his valuable guidance and review of the manuscript.

6 References

[1] U.S Department Of Transportation Pipeline Transportation Pipeline and Hazardous Materials Administration, Available at http://primis.phmsa.dot.gov/ptr.htm, Accessed on March 15, 2016

[2] Federal Energy Regulatory Commission, Available at http://www.ferc.gov/, Accessed on March 15, 2016

[3] National Fire Protection Agency 59A, "Standard for the Production, Storage, and Handling of Liquefied Natural Gas (LNG)", 2001

[4] Code of Federal Regulations, 49 CFR Part 193, Subpart B, Available at http://www.ecfr.gov/, Accessed on March 15, 2016

[5] European Standard, EN 1473, "Installation and Equipment for Liquefied Natural Gas - Design of Onshore Installations", 2007

[6] Australian Standard, AS 3961, "The Storage and Handling of Liquefied Natural Gas", 2005

[7] Canadian Standard, CSA Z276, "Liquefied Natural Gas (LNG) - Production, Storage, and Handling", 2011

[8] International Association of Oil and Gas Producers, OGP Draft 116901, "Guidance on Performing Risk Assessment in the Design of Onshore LNG Installations including the Ship/Shore Interface", 2013

[9] TNO, "Guidelines for Quantitative Risk Assessment, 'Purple Book,'" CPR 18E, Publication Series on Dangerous Substances (PGS 3), 2005

[10] T.A. Williams and N. Assing, " Risk-Based LNG Facility Siting and Safety Analysis in the U.S.: Recent Developments", LNG 17 Proceedings, Available at http://www.gastechnology.org/, Accessed on March 15, 2016

[11] J.L. Woodward and R.M. Pitblado, "LNG Risk based Safety Modeling and Consequence Analysis", John Wiley & Sons, Hoboken, New Jersey, 2010.

Flare Reduction Strategies for LNG Plant Design

Meghana Cherukuri
Process Engineering Professional
Houston, TX - 77054
Meghana.Cherukuri@yahoo.com

Sumit Sharma
Senior Technical Professional – Process
KBR
601 Jefferson St., Houston, TX - 77002
Sumit.Sharma@kbr.com

Kristina Martinez
Principal Technical Professional – Process
KBR
601 Jefferson St., Houston, TX - 77002
Kristina.Martinez@kbr.com

Prepared for Presentation at
American Institute of Chemical Engineers
2016 Spring Meeting
12th Global Congress on Process Safety
Houston, Texas
April 11-23, 2016

Flare Reduction Strategies for LNG Plant Design

Meghana Cherukuri
Process Engineering Professional
Houston, TX - 77054
Meghana.Cherukuri@yahoo.com

Sumit Sharma
Senior Technical Professional – Process
KBR, 601 Jefferson St., Houston, TX - 77002
Sumit.Sharma@kbr.com

Kristina Martinez
Principal Technical Professional – Process
KBR, 601 Jefferson St., Houston, TX - 77002
Kristina.Martinez@kbr.com

Keywords: Flare system, flare optimization, flare minimization, LNG, Plant design.

Abstract

In liquefied natural gas (LNG) plants, the flare system provides a reliable and safe method for vapor and liquid hydrocarbon disposal. The flare system enables controlled handling of off-spec streams during plant upsets, emergencies, and short term operations such as start-up, shutdown, or equipment maintenance activities. Although the flare system is essential for the safe operation of the plant, proper design and optimization is necessary to control capital cost.

A typical LNG flare system includes Warm (or Wet), Cold (or Dry), and Low Pressure (or LP/Marine) flares. The disposal of streams with potentially freezable components originating from plant inlet and gas treating sections is achieved through the Warm flare. The Cold flare is utilized to dispose of streams with non-freezable components and very low operating temperatures. Low pressure streams from the storage or loading area of the LNG plant are routed through the LP flare for safe disposal. Based on project design criteria, the flare system can require an elevated, ground, enclosed or smokeless flare design.

Addressing flare optimization issues provides opportunities for controlling material loss during plant operation. In this paper, an overview of factors that influence flare type selection and determine flare size are discussed. Significance of flare type and size for overall plot space requirements is also addressed. Based on typical major relieving scenarios, flare load reduction strategies for each of the LNG plant flares are presented.

LNG plant design configurations for multiple trains, refrigerant compressors, and process recycle streams can significantly affect design flare loads. High design flare loads increase flare size and require large plot space due to flare radiation limits. Proper selection of design options for major LNG plant sections, guided by flare reduction strategies, can help in reducing flare design flow rate, flare size, and plot space requirements.

1. Introduction

The flare system in Liquefied Natural Gas (LNG) plants provides a reliable and safe disposal method for hydrocarbon vapor and liquid streams. These streams may result from a deviation from normal plant operation or from intermittent operations such as start-up, shutdown, venting, draining, or purging. Typical plant safety design manages any potential material loss during plant operation by routing the material streams to the flare system for disposal.

Although the flare system is essential for the safe operation of the plant, proper design and optimization is necessary to control capital cost. Flaring design flow rates affect the size of the flare and determine the plot space requirements. [1] High flare loads increase flare stack and header size and require large plot space due to flare radiation limits which altogether can increase the overall project capital cost.

LNG plant design, train, and equipment configurations can significantly affect flare design loads. Addressing flare optimization at the design stage provides opportunities for controlling material loss during plant operation. Proper selection of design options for major LNG plant sections, guided by flare reduction strategies, can help in reducing flare design flow rate, size, and plot space requirements.

2. The Flare System

Disposal of gas and liquid streams in a process plant is achieved through flaring which involves controlled combustion of hydrocarbon streams. Flared hydrocarbons are routed to a remote location and are typically burned in an open flame during which stream components are ideally converted to carbon dioxide and water.

Proper design of the flare system creates an efficient mechanism so that complete combustion of hydrocarbon can be achieved. Without flaring, the alternate options of venting or draining of the process streams can create health, safety, and environmental problems. Local regulations may also make it mandatory to include a flare system for safe disposal of process streams from an operating plant.

A typical flare system (Figure 1) in a process plant consists of three primary components: the flare structure, a plant wide flare header piping system, and plot area for radiation zone. The process streams for disposal are sent to appropriate flares where the streams undergo combustion. Routing of the process streams to the flare is achieved through a plant wide piping system that also includes knockout drums for phase separation of any two phase streams. Liquid content to the flare must be avoided for safe operation.

At the flare tip, the continuous flare flame is maintained using a pilot system, and purge gas is used to prevent oxygen ingress into the flare system. In order to delineate safe work zone conditions, a sterile radiation zone is maintained around the flares.

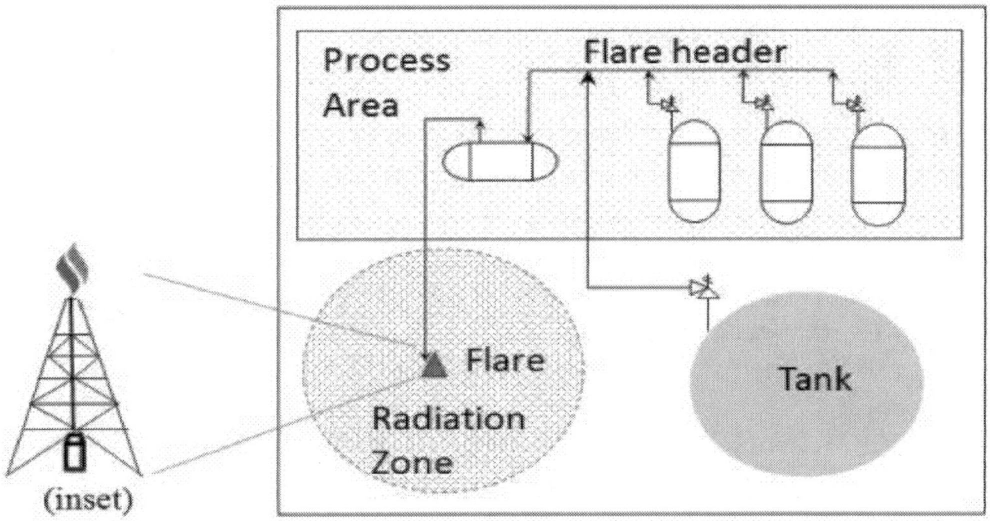

Figure 1. Simplified flare system sketch

2.1. Flare type:

The main component of the flare system is the flare structure that can be classified as elevated or ground flare based on the location of the open flame relative to the flare structure. Selection of the flare type is based on project and process requirements.

- Elevated flares: Elevated flares include a vertical stack with a flare tip that is located at the highest point of the stack. A process stream routed to the flare stack for disposal undergoes combustion at the top of the stack. An elevated combustion point helps in the dispersion of the flare products into the atmosphere.

  The support structure for the elevated flares can be self (Figure 2a), guy wire (Figure 2b), or derrick supported (Figure 2c) based on the height of the flare stack. High elevation of the flare also creates visual and noise impact from the flaring process. Efficient mixing of flaring stream with combustion air is important in achieving smokeless flare design.

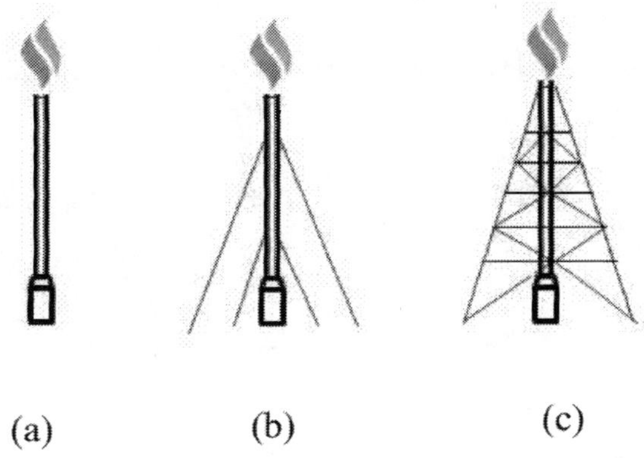

Elevated flares:

(a) Self supported
(b) Guy wire supported
(c) Derrick supported

Figure 2. Elevated flares

While there are no regulatory restrictions for the stack height, Federal Aviation Administration (FAA) may require height limits for flare placements in the vicinity of an airport. Flare dispersion modeling during the design phase is essential to ensure that concentrations of toxic emissions do not exceed safe levels at any location where personnel can be present. Flare dispersion is affected by the type of flare tip. In the US, 40 CFR 60.18 (4)(iii) limits flare tip velocity to 400 ft/s with a flared gas lower heating value of at least 1000 Btu/SCF. [3]

- Ground Flares: Open multi-point ground flares (Figure 3a) or enclosed ground flares (Figure 3b) are the preferred option when project requirements prohibit visual and noise disturbances from the flaring process. Ground flares are designed smokeless but dispersion of flare combustion products is typically inefficient. Stream combustion in ground flares is achieved at grade level.

  In an enclosed ground flare the burners are located within a refractory lined physical structure that encloses the flame and contains thermal radiation. Open ground flares are not enclosed but typically are surrounded by a radiation fence which similarly blocks radiation.

  Open ground flares can be designed to handle full flare capacity similar to the elevated flares whereas enclosed ground flares are used for routine small scale flaring. Compared to the elevated flares, the ground flares are costlier and require larger plot area (open ground flares).

(a) Open ground flare

(b) Enclosed ground flare

Figure 3. Ground flares [2]

2.2. Flare header piping system

Plant equipment are protected from overpressure scenarios using relief valves. The outlet flow from the relief valve is collected using a plant wide flare header system and transported to the flare for disposal. Process streams from intermittent plant operations such as start-up, shutdown, and emergency blowdown are also routed to the flare through the flare header piping to dispose of hydrocarbon inventory in a controlled manner to prevent or mitigate escalation of any unsafe event. Flare and flare header piping can be classified based on the operating conditions of the flaring stream:

Table 1. Flare and flare header classification:

• Based on flaring stream operating pressure:	
High Pressure (HP) Flare:	Sonic flare. Handles large high pressure flaring streams.
Low Pressure (LP) Flare:	Sub-sonic flare. Typically used for equipment with low design pressure specifications such as storage tanks.

• Based on flaring stream composition and operating temperature:	
Warm (Wet) Flare:	High temperature flaring streams with components that can potentially freeze at low temperatures
Cold (Dry) Flare:	Low temperature flaring streams with no freezable stream components

Separation of the flares in terms of pressure, temperature, and stream composition is necessary for safe and reliable flare operation. Depending on the operating conditions of the relieving stream, the flare header material can range from carbon steel to stainless steel.

Typically, for the Warm flare, carbon steel piping can be used to handle the flaring stream whereas the Cold flare will require stainless steel piping because of the possibility of very low temperatures that may exist in the flaring streams during relieving scenarios. High design flow flaring rate for the flare results in large flare header size and increases the cost of the system. Stainless steel metallurgy makes the Cold flare relatively more size and cost sensitive compared to the Warm flare.

2.3. Radiation zone

Flaring large quantities of hydrocarbon can result in high intensity thermal radiation and noise around the flare during plant operation. In order to maintain safe working conditions, the area around the flare is demarcated as sterile zone. The permissible levels for flare radiation are summarized in Table-3 below.

Table 3. Radiation and noise limits

• **Radiation: Permissible Radiation Level** [4]	
KW/m^2	
1.58	maximum radiant heat intensity at any location where personnel with appropriate clothing can be continuously exposed
4.73	maximum radiant heat intensity in areas where emergency actions lasting 2 min to 3 min can be required by personnel without shielding but appropriate clothing
6.31	maximum radiant heat intensity in areas where emergency actions lasting 30 seconds can be required by personnel without shielding but appropriate clothing
9.46	maximum radiant heat intensity at any location where urgent emergency action by personnel is required. When personnel enter or work in an area with the potential for radiant heat intensity greater than 6.31 KW/m^2, radiation shielding and/or special protective apparel (e.g. a fire approach suit) should be considered. Safety precaution- It is important to recognize that personnel with appropriate clothing cannot tolerate thermal radiation at 9.46 KW/m^2 for more than few seconds

• **Flare noise limits**		
85 dB (A)	max	continuous exposure
115 dB (A)	max	for short term exposure (up to 15 mins)
115 dB (A)	max	with hearing protection

Heat radiation calculations for flare typically take into consideration the wind velocity, relative humidity, and solar radiation at the flaring location to assess the heat radiation impact of flaring. The impact of flare radiation and flare front can be identified early in the project to determine locations for flare stack and plot area for radiation zone.

3. Flares in LNG Plant

In a typical LNG plant the feed gas to LNG conversion involves four primary process units that include feed gas pre-treatment, refrigeration, liquefaction, and product storage. The feed gas pre-treatment section removes water, acid gas, and mercury from the feed gas stream. The treatment process occurs at high pressure and warm temperatures. Refrigeration and liquefaction units operate at high to low pressure and temperature ranges. The LNG product storage unit operates at close to atmospheric pressure and very low temperature range.

The flare system in an LNG plant is comprised of different flares that are designed to handle streams originating from various units for safe disposal. The hydrocarbon flare streams mainly contain methane, ethane, propane and other heavy components from the natural gas streams or refrigerant streams:

- Warm (Wet) Flare: A high pressure carbon steel flare for warm and wet hydrocarbons, designed to handle streams containing heavy hydrocarbons, contaminated water streams and streams with any freezable components. These streams mainly originate from the pre-treatment units.

- Cold (Dry) Flare: A high pressure stainless steel flare for cold and dry hydrocarbon vapors. These light hydrocarbon streams originate from the refrigeration and liquefaction units of the plant.

- Low Pressure (LP) or Marine Flare: A low pressure stainless steel cold flare for emergency vapor (low pressure) releases from the LNG storage and loading system.

Apart from the Warm, Cold, and LP flares, the LNG plant may also include a Sour Gas Flare, and Thermal Oxidizer or Incinerator for flaring hydrocarbon streams containing a high percentage of acidic gases such as H_2S, CO_2, and volatile hydrocarbons (VOC) originating from pre-treatment units. Some facilities also include dedicated warm and cold liquid collection header systems to separately handle liquid disposal.

4. Flare Design Loads

The design flare loads in LNG plants are determined by identifying major relieving scenarios during process upset, intermittent or normal operations. The Warm and Cold flares are designed to handle the largest gas flows in the LNG plant while the LP or Marine flare is designed based on major flaring incidents from the tank or product storage area.

In the following sections, all discussions related to controlling cases for flare design are based on the Propane Pre-Cooled Mixed Refrigerant Process (C3MR) by Air Products and Chemicals Inc. (APCI).

4.1. Warm Flare Design Load:

The largest flow at ambient or warm temperatures (above 0 °C) in the LNG plant is the feed gas flow into the facility. The inlet pressure to the process units is typically maintained by single or multiple (parallel) pressure letdown valves. Any overpressure scenario in the feed gas line in the inlet or pre-treatment facilities can result in large feed gas flaring to the Warm flare. Another large flow that can determine the Warm flare capacity is the blowdown loads from the inlet and pre-treatment units of the LNG plant. The blowdown operation removes hydrocarbon inventory from the units and routes the hydrocarbon streams to the flare system.

4.2. Cold Flare Design Load:

The Cold flare design flow is typically based on refrigerant flows in the refrigeration and/or liquefaction units of the LNG plant. The primary refrigerants for the APCI process are Propane which is used for pre-cooling the feed gas and a Mixed Refrigerant (typically consists of nitrogen, methane, ethane, and propane) is used for liquefaction process. The Propane (PR) and Mixed Refrigerant (MR) flows are circulated using Propane and MR Compressors respectively. Process upset cases such as PR or MR compressor blocked outlet or local power failure may require routing PR or MR circuit flow to the Cold flare. Blowdown load from the cryogenic heat exchanger and emergency depressuring events are also sent to the Cold flare.

4.3. LP /Marine Flare Design Load:

Low Pressure or Marine flare handles any relieving streams from the LNG storage tanks or product loading operations. The capacity of the LP/Marine flare is typically set by the Boil-Off Gas (BOG) compressor blocked outlets case. The largest flow to the Boil-Off Gas compressor occurs during LNG product loading operations which may utilize multiple BOG compressors operating in parallel.

5. Flare System Optimization

Collectively, a significant quantity of gas is estimated to be flared annually from LNG export terminals. [5] Optimization of the flare system and minimization of flaring incidents can potentially reduce LNG plant's carbon foot print and overall plant cost. Flare design optimization and flare minimization will require consideration of design options that can minimize flaring incidents or reduce the amount of flare load during a flaring incident.

5.1. Warm Flare:

For the Warm flare, the largest flare load is typically from the feed gas pre-treatment section located upstream of the liquefaction unit. An overpressure scenario in which the feed gas forward flow is blocked in the pre-treatment section can result in routing all feed gas flow to the Warm flare through the activated relief valves.

In another scenario the failure of plant inlet control valve can significantly increase flow to the plant inlet. This control valve failure will activate relief valves downstream from the valve and route the relief valve outlet flow to the Warm flare (assuming that there is no additional heating provided and that the upstream pressure is not high enough to cause the downstream gas temperature to be too cold for the Warm flare metallurgy). The flow resulting from the control valve failure is a function of the pressure drop across the control valve. The higher the difference between the inlet and outlet pressures across the valve, the higher the potential flow to the flare subsequent to the control valve failure. High flare load will increase the flare height, the radiation zone, and associated costs.

5.1.1. High Integrity Pressure Protection System (HIPPS) for the feed gas line

An preventative safety system such as High Integrity Pressure Protection System (HIPPS) can be used to eliminate high Warm flare load from any blocked outlet overpressure scenario in the feed gas flow at plant inlet. For the pressure letdown control valve failure case, consideration of a HIPPS downstream of the plant inlet control valve will prevent a flaring incident by shutting down the forward flow to the plant. Implementation of a HIPPS at the plant inlet will trip any dedicated emergency shutdown valves at the plant inlet before reaching a scenario in which the operating pressure exceeds the design conditions. This scheme will significantly reduce the flow to the flare by eliminating full feed gas forward flow as a consideration for the Warm flare sizing.

HIPPS can provide a completely independent safety layer. Inclusion of redundant features such as additional sensors and instrument logic control increases the HIPPS reliability. ASME Sec VIII, Division 1, UG-140 accepts the instrumented protective device in consultation with and on the advice of the owners for certain services.

A HIPPS can provide sufficient reliability to replace a mechanical device such as a pressure relief valve. HIPPS is a particularly attractive option for revamps where limited capacity of an existing flare can prohibit any flare load increase from revamp changes. Implementation of HIPPS for such cases will reduce the occurrence of any flaring scenarios that can impact flare design loads.

HIPPS application requires frequent testing and maintenance to keep the device operative and available when needed. This can be a significant burden for the operator. The use of a HIPPS requires a hazard and risk analysis (per ISA S84.00.01/2/3) to determine the required risk reduction that must be achieved by the overpressure protection. This then sets the required safety integrity level for the instrumented protection. Any HIPPS based design should be thoroughly reviewed in HAZOP, and clearly understood by both the owner management and the plant operation personnel.

5.2. Cold Flare:

The design flare load for the Cold flare usually constitutes the highest flaring rate in LNG plants. Refrigerant compressor blocked outlet cases result in routing the refrigerant system flow to the Cold flare. The refrigerant system flow and inventory capacity affect the size and capacity of the Cold flare and the overall flare system.

5.2.1 Implementation of HIPPS for refrigerant compressor system

Similar to the HIPPS for feed gas inlet control valve, HIPPS can be considered for refrigerant compressor system. A two out of three pressure transmitter voting can be implemented on the refrigerant compressor outlet to trip the machine during a high pressure case. Using HIPPS, the overpressure scenario and high flare load from compressor blocked outlet can be eliminated for the Cold flare.

HIPPS application requires frequent testing and maintenance to keep the device operative and available when needed. This can be a significant burden for the operator. The use of a HIPPS requires a hazard and risk analysis (per ISA S84.00.01/2/3) to determine the required risk reduction that must be achieved by the overpressure protection. This then sets the required safety integrity level for the instrumented protection.. Any HIPPS based design should be thoroughly reviewed in HAZOP, and clearly understood by both the owner management and the plant operation personnel.

5.2.2 Partial condensation at propane condenser by natural convection.

Failure of the propane condenser results in overpressure in the propane refrigerant system. For air cooled LNG plant design, the resulting calculated flare load for the Cold flare sizing can be reduced by considering partial condensation at the propane condenser air coolers.

Power failure affects the propane condenser air cooler fans however partial condensation at elevated pressure will still occur due to natural convection cooling at the air coolers which are usually located high up on the top of pipe racks. Consideration of partial condensation will reduce the relieving load and the flaring requirement.

5.2.3 Recycle relief valve flow to compressor suction [6]

The relief valve downstream of the refrigerant compressor discharge sends the outlet flow to the Cold flare in case of overpressure scenarios in the compressor system. Recycling the relief valve outlet flow to suction of the compressor can significantly reduce the Cold flare load (Figure 4).

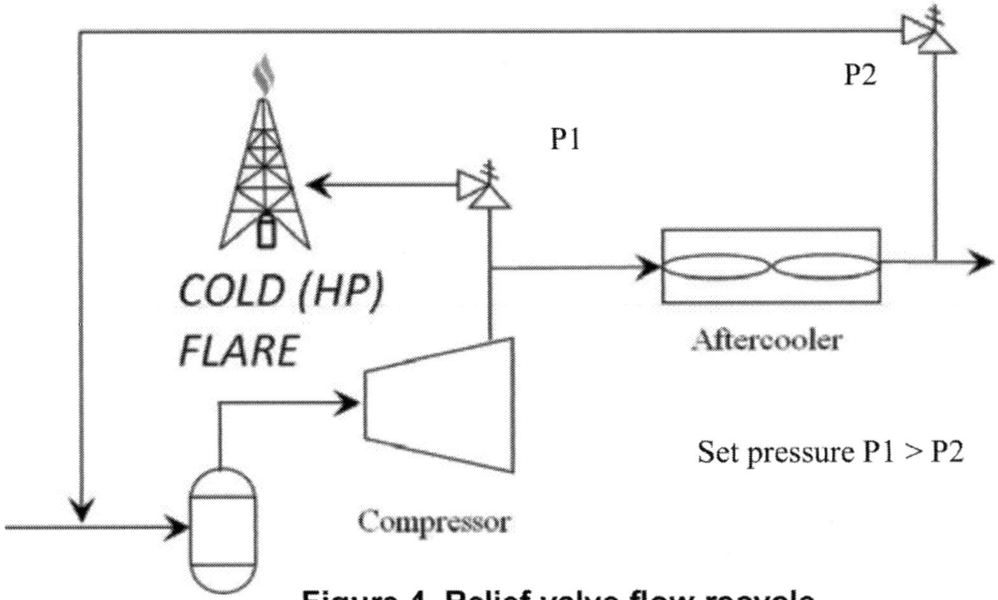

Figure 4. Relief valve flow recycle

The relief valve outlet flow recycle scheme eliminates the flare load from compressor blocked outlet case. The recycle route is parallel to the anti-surge valves that protect the compressors. This scheme can be implemented for both propane and mixed refrigerant compressors.

The proposed relief valve recycle scheme works based on differential pressure which makes it susceptible to unintended relief valve opening. Using pilot operated relief valves can potentially solve this problem. This scheme may require multiple relief valves within the compressor system. Long term operation and heat built up effects may have to be completely analyzed with inputs from compressor supplier and dynamic simulation analysis.

5.2.4 Split scheme for the flare load from the refrigerant compressors

The relief valve flow from refrigerant compressors blocked outlet or power failure cases is conventionally sent to the Cold flare. However the outlet flow from the relief valves may be routed to the warm flare if the stream temperature is at a higher temperature during the compressor system overpressure scenario. A high stream temperature can result in warmer letdown temperature downstream of the relief valves, and may not reach an unacceptable limit.

This scheme (Figure 5) will substantially reduce the size of the cold flare which is all stainless steel material and lower the cost by routing the relieving load to the warm flare which is typically based on carbon steel material.

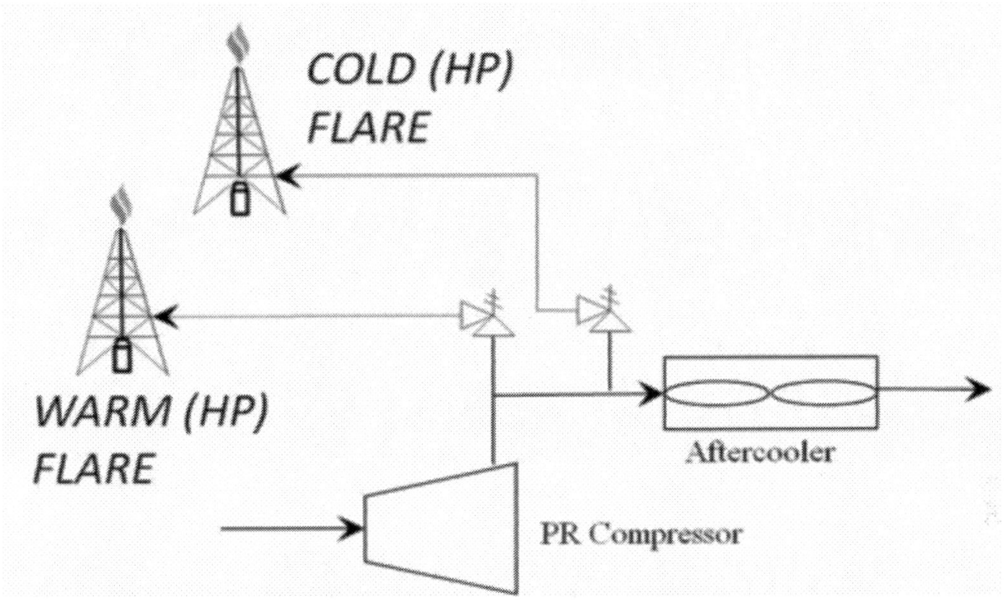

Figure 5: Flare load split scheme

If all of the flare stacks (Warm, Cold, and LP flares) are located at a single location, the heights of the flare stacks are usually the same height so that there is no flame impingement on any flare stack from an adjacent active stack. The choice to route the refrigerant flow to the Warm flare may not provide any benefit if it leads to increasing the Warm flare size beyond the Cold flare size and if other controlling scenarios for the Cold flare maintain the same initial Cold flare size.

The total radiation level from this scheme may stay the same because the total propane system load to flare remains the same based on the overpressure scenario for the compressor system.

5.2.5 Increasing the propane system design pressure

Increasing the propane system design pressure can potentially reduce the flare load to a level so that it no longer controls the Cold flare design. The propane refrigeration system is generally designed in carbon steel materials. The design pressure of the propane system can be increased to a level that is within flange limits of the piping material class. A higher set pressure for the relief valve will reduce flare flow rate from the compressor during overpressure scenarios. Higher relief valve set pressure may also reduce the frequency of overpressure scenarios for the compressor system.

This option will lead to higher design pressure for the air coolers, compressor casings, and other equipment in the propane compressor system. These design modifications can lead to the compressor surging at elevated pressure (if the anti-surge control system fails) which may damage the compressor. Many owners also view the propane compressor discharge relief valve as providing a back-up means of machinery protection in addition to its overpressure protection function.

5.2.6 Dynamic simulation for compressor overpressure scenarios

Dynamic simulation can provide realistic insight into the impact of overpressure scenario on the overall compressor system. Using information on inventory, driver power, compressor curves and other details, Dynamic simulation can predict realistic reliving rates during overpressure incidents. Based on dynamic simulations, the relief flow can be significantly lower compared to a conservative estimate of relieving all compressor refrigerant forward flow and system inventory capacity to the flare during overpressure scenarios. [7]

5.3. Flare system for multiple train design:

In multiple train design each train can have a common or independent pre-treatment, refrigeration, and liquefaction sections. For the independent train design, both Warm and Cold flare systems can be common for the overall plant with each flare sized for their single train controlling cases plus the respective start-up load for one additional train. For independent trains, lacking any common cause failure, double jeopardy considerations can eliminate multiple train failures and limit overall flare loads. Provision of common spare stack can be determined by the project philosophy and safety review of the flare design.

For a single train design with multiple parallel compressor strings (2x50%), the cold flare load may not decrease compared to a same capacity single compressor string (1x100%). In multiple parallel compressor string design, the blocked outlet for the compressor strings can occur at a location where compressor outlet flows from both the strings combine, sending the full combined 100% compressor outlet flow to the Cold flare.

6. Flaring Minimization

Various plant operations can result in flaring in an LNG plant. With increased emphasis on containing plant emissions and operating costs, design modifications can be considered to reduce flaring incidents.

Intermittent planned or unplanned operations such as start-up, shutdown, or plant trips can result in significant flaring as a percentage of feed gas [8]. A full pressure re-start/or partial depressurization for faster re-start of the liquefaction plant after a trip, based on available power, can reduce the amount of gas flared. Using aeroderivative gas turbine driven compressors for refrigeration service can also reduce flaring during start-up activities. Consideration of isolation and maintenance activities during the design phase can provide easy access and means of equipment isolation with minimum flaring.

Design options should also consider recycling or re-routing any gas stream from flare to either front-end of the plant for processing, or to the fuel gas system for consumption.

7. Conclusion

The flare system design in an LNG plant is affected by equipment and system configurations. With increased emphasis on containing plant emissions and costs, design modifications can be considered to reduce flaring incidents during normal or intermittent plant operations. An optimized flare system design will require design considerations for major LNG plant units including the feed gas inlet, pre-treatment, and refrigeration sections.

High flaring rates increase the plant cost by increasing the size of the flares, the flare headers and the required plot area for radiation zone. Flare system optimization for single or multiple train configurations reduces the flare loads by recycling, re-routing, or eliminating the flaring streams from various units of the LNG plant. Implementation of flare load optimization and minimization options during the design phase provides opportunities to limit the overall plant flaring rates, incidents, and related costs.

8. References

[1] L. B. Evans, W. M Vatavuk, D. K. Stone, S. K. Lynch, R. F. Pandullo, W. Koucky, Flares, Available at http://www3.epa.gov/ttncatc1/dir1/cs3-2ch1.pdf. Accessed on February 01, 2016.

[2] TxOGA, Flare Operation Discussion, Available at http://www.tceq.state.tx.us/ assets/public /implementation/air/rules/Flare/Resources_TXOGAFlareOperationDiscussion.pdf, Accessed on February 01, 2016

[3] Electronic code of federal regulations, General control device and work practice requirements, Available at http://www.ecfr.gov/cgi-bin/text-idx?SID=7ae903b94d41527119c 08be5a939da ed&mc= true&node=se40.7.60_118&rgn=div8, Accessed on February 01, 2016.

[4] API Standard 521. Pressure - relieving and Depressuring Systems. American Petroleum Institute, Sixth Edition, January 2014.

[5] C. D. Elvidge, M. Zhizhin, K. Baugh, F. Hsu, and T. Ghosh, Methods for Global Survey of Natural Gas Flaring from Visible Infrared Imaging Radiometer Suite Data, Energies, 9, (2016), 14, Available at http://www.mdpi.com/1996-1073/9/1/14, Accessed on February 01, 2016.

[6] C. E. Lucas, P. A. Brochu, W. C. Rooney, Gas compressor, WO 2005010375 A1, Available at https://patentscope.wipo.int/search/en/detail.jsf?docId=WO2005010375&redirectedID=true, Accessed on February 01, 2016.

[7] Y. N. Liu, C. E. Lucas, R. W. Bower, and J.C. Bronfenbrenner, Reducing LNG Costs by Better Capital Utilization, Available at http://www.airproducts.com/industries/Energy/LNG/ resource-center/whitepapers/reducing-lng-costs-better-capital-utilization.aspx, Accessed on February 01, 2016.

[8] Qatar Gas, Flare reduction programme, Available at https://www.qatargas.com/English /SafetyAndEnvironment/Documents/Qatargas%20Flaring_Fact%20Sheet_20%20Nov%202012. pdf, Accessed on February 01, 2016.

441507 Using Data Analytics to Improve Process Safety

Abstract:

Using Data Analytics to Improve Process Safety

Increasingly, data from multiple sources are compiled into central storage systems for entire facilities and in some cases entire corporations. The low cost of digital memory and the increasing process speed of modern microprocessors and computers continue to make it easier to analyze, manipulate and merge very large databases. Big Data and Data Analytics have been touted as allowing staff at all levels to make real time, data-driven business decisions. These concepts can be used to improve process safety at facilities by proactively analyzing sources of data to identify opportunities for improvement in operating practices, identify and track near misses, and in real time detect incipient instrument and equipment failures that could lead to catastrophic incidents. The aviation industry is one example that is voluntarily implementing similar analytics. Over the last decade the aviation industry has implemented flight operations quality assurance (FOQA) or flight data monitoring (FDM) programs which collect, store, and analyze recorded flight data with the goal of increasing overall safety.

In this presentation we show examples of analytics Exponent performed on data from facilities. These analyses were used to analyze alarm frequencies during specific types of operations, identify near misses, and proactively detect failures of instrumentation and equipment in order to prevent or mitigate catastrophic incidents.

442485 Safety Culture and Leadership- Independent of Language

Abstract:

The existence of safety in the workplace is mandatory for a sustainable business. Strong safety cultures are characterized by human factors, workforce engagement and fully committed safety leadership. With that being said, increased diversity in work environments and international recruitment can oftentimes complicate the process of implementing a common safety culture.

More specifically, those who come from different types of work environments than their current environment may have trouble interpreting the procedures and guidelines that are enforced at their new place of employment. Differences in languages and dialects may also lead to misinterpretations of intended safety guidelines. Lastly, there may be some instances where the workplace follows a non-flexible and hierarchical structure that is not very open to change.

In order to best develop safety practices in the workforce, the senior leaders need to first develop an environment based on the understanding of the different rationales and expectations of a multi-cultural workforce. Once that has been established, communicating with the workforce on safety expectations of the organization and optimizing interpretation of this message is critical.

This paper will continue the discussion from a previous publication; how safety leadership can consider linguistic differences when implementing a safety plan for a diverse workforce. The discussions will emphasize the learnings of the Bradley Curve, showing how important it is to follow safe practices to reach the proximity of an almost 100% incident free area, and will provide real examples from international work assignments and diverse work teams, highlighting this need and offering potential approaches

439897 Indulging in Invincibility (III)- Twelve Steps to Sobriety

Abstract:

<>8/3/2015 <>Indulgence in Invincibility (III)- Twelve Steps to Sobriety <>by Brian P. Kolodji, PE

Could Bhopal, Macondo, or Texas City have been prevented? Directly related examples are presented where PS cultural indulgences contribute to a false sense of invincibility. Light is shed on where the perennial dilemmas of misplaced loyalty, unachievable goals, and belief in corporate higher power contribute to and enable denial of process safety culture failure. Successes and a program are presented where these failed practices are brought out of the shadows and recovery begins.

Case Study of Relief Analysis of Compressor Stations

Kumud Chaurasia
Process Engineer
Lloyd's Register Energy Americas, Inc.
1330 Enclave Parkway, Suite 200, Houston, TX 77077, USA
Kumud.chaurasia@lr.org

Aniket Patankar
Senior Process Engineer
Lloyd's Register Energy Americas, Inc.
1330 Enclave Parkway, Suite 200, Houston, TX 77077, USA
Aniket.patankar@lr.org

Prepared for Presentation at
American Institute of Chemical Engineers
2016 Spring Meeting
12th Global Congress on Process Safety
Houston, Texas
April 11-23, 2016

Case Study of Relief Analysis of Compressor Stations

Kumud Chaurasia, Process Engineer

Aniket Patankar, Senior Process Engineer

Lloyd's Register Energy Americas, Inc.

1330 Enclave Parkway, Suite 200, Houston, TX 77077, USA

kumud.chaurasia@lr.org, aniket.patankar@lr.org

Keywords: Relief analysis, Compressor Station, case study, concern identification

Abstract

A compressor station is a facility which helps the transportation process of natural gas from one location to another. Compressor stations include several key component parts, the primary being the actual compressor unit, either centrifugal or reciprocating. They also typically include scrubbers, strainers or filter separators which remove liquids, dirt, particles, and other impurities from the natural gas. This study will give an overview of the relief analysis conducted for multiple compressor stations. The relief analysis was system based as opposed to relief device based so as to identify any unprotected systems that may need overpressure protection. It will discuss the overpressure scenarios identified for these systems. It will provide some general statistics on the total number of applicable overpressure scenarios. It will give a detailed breakdown of the various concerns found including, but not limited to, undersized devices, unprotected systems, installation concerns, data needs, and documentation discrepancies. The study will discuss possible mitigation solutions for these concerns.

1. Introduction

Compressor stations are typically located along a natural gas pipeline to facilitate the compression of the gas to a specified pressure, which allows the gas to flow along the pipeline to a target location. Since these pipelines typically extend for several hundreds of miles, the gas tends to lose pressure frequently. This necessitates the requirement to compress it frequently during the transportation process.

The size of the station and the number of compressors (pumps) varies, based on the diameter of the pipe, and the volume of gas to be moved. Compressor stations include several key components, the primary being the actual compressor unit, either centrifugal or reciprocating, scrubbers, and strainers or filter separators which remove liquids, dirt, particles, and other impurities from the natural gas.

It is critical to ensure that the natural gas transported to its intended target is free from any contaminants. This is accomplished through four mains steps. The first step is contaminant removal, which occurs upstream of a compressor station near the well head. Here, the raw natural gas, at very high temperature and pressure, is passed through separators in order to remove oil, water, sand etc. It is then passed through a slug catcher, typically located within the compressor station, to remove any leftover condensate. The third contaminant removal step occurs downstream of the compressor unit (after the gas has been pressurized) via a contactor unit that predominantly removes liquid water and carbon dioxide using triethyleneglycol (TEG) as the absorbent. The TEG that separates out at the bottom of the scrubber unit, carrying the water and carbon dioxide with it, is typically regenerated in a glycol regenerator unit before it is recirculated back into the scrubber unit. The final step prior to delivery of the gas to its intended target is to remove any leftover TEG in the gas using a scrubber unit. There are also several tanks within the compressor station for storage of the condensates and TEG. Additionally, the compressor stations contain several components that are part of the glycol regenerator unit including filters, flash tank separators, pumps, heat exchangers, etc. A general process flow diagram of the compression station is shown in Figure 1.

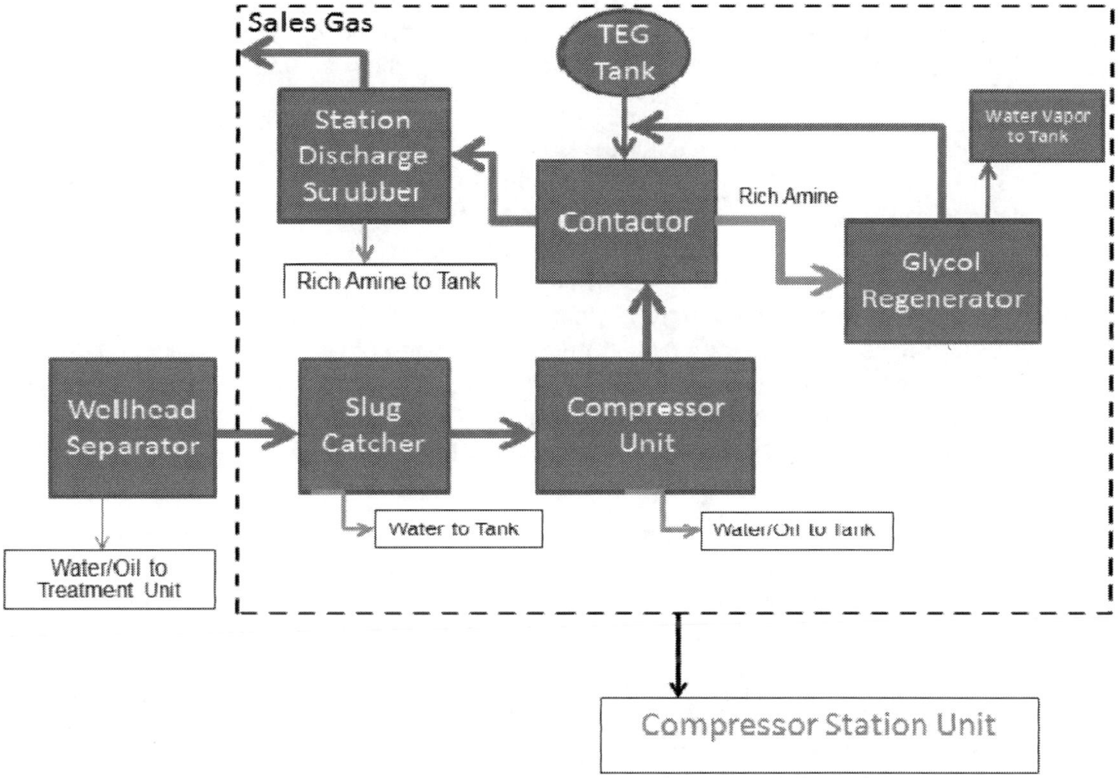

Figure 1: Process flow diagram of Compressor Station

2. Relief Analysis

For this case study, twenty separate compressor stations were analyzed. Specifically, the steps required for performing the analyses, different applicable scenarios, and all the concerns identified with mitigations options are presented. The analysis was conducted using API 520 Part 1 and Part 2 standards, while considering the client's design philosophy as well.

2.1. Steps involved in Relief Analysis

The relief analysis, which is a system based analysis performed for each equipment in the unit, is a methodical four step approach including data and design documentation collection, contingency analysis, pressure relief calculations, concern identification and mitigation. These steps are discussed in detail below.

Data and Design Documentation Collection: Since the client for whom the relief analysis is being performed may not always provide all the information necessary to conduct the analysis, it is imperative that all necessary data and documentation required is collected at the client site prior to commencing the project. Examples of such data collected include isometrics for PSV inlet and outlet piping, PSV and control valve information, equipment and relief device name plate information, U-1A forms, equipment drawings etc. Usually, the piping and instrumentation diagrams (P&IDs) and heat and material (H&M) balance are provided by the client in order to complement the other data collected. All the data collected is then entered into a client project database.

Contingency Analysis: Prior to commencing the contingency analysis, the compressor station is divided into various systems, with each system consisting of several equipment. The interface of systems is defined based on where block valves are situated within the P&ID. Once the systems have been defined, a contingency analysis is performed where all scenarios that could possibly result in an overpressure condition are identified. For the 20 compressor stations that were analyzed in this case study, the following eighteen contingencies were identified: Power Failure, Cooling or Reflux Failure, Steam Failure, Instrument Air Failure, External Fire, Blocked Outlet, Isolated with Heat Input, Equipment Failure, Inadvertent Valve Operations, Failure of Automatic Controls, Gas Breakthrough, Internal Boundary Failures, Reverse Flow, Run Away Reaction, Accidental Mixing, Change in Conditions, Vacuum, Other (applicable scenarios other than listed above).

Calculations: Once the contingency analysis is finished, the required flow rate of the applicable overpressure scenario and the capacity of the relief device (based on the relief device ASME orifice area) are calculated. The inlet and outlet pressure drops, Mach number and sound power level of the piping is also calculated based on the dimensions of the relief device.

Concerns Identification and Mitigation: After the calculations are completed for a system, all the concerns are identified and documented in the database. This is as per the guidelines provided in API 520 Part 1, 2 and the design philosophy provided by client. The concerns are identified based on the calculation results and discrepancies and assumptions made during the analysis. Once all the concerns are identified, recommendations for implementation of mitigation strategies are provided to the client.

3. Results of Case Study

3.1. Relief Device Identification

For the twenty compression stations that were analyzed as part of this case study, the relief devices included all the relief valves, pressure and vacuum vents (PVRVs), goosenecks, and rupture disks. Usually, the relief valves and rupture disks were situated on the pressure vessels while the PVRVs and goosenecks were situated on the low pressure tank. A total of 1099 relief devices were identified for the 20 compressor stations. Depending on the type of relief device, the method of calculation may vary. For example, in the case of PVRVs and goosenecks, it is not necessary to calculate the pressure losses.

The breakdown of the percentage and number of the relief devices, by type, that were analyzed in this study are shown in Table 1 and Figure 2 respectively.

Table 1: Percentage of different relief devices by type

Relief device Type	Percentage
Relief valve	78%
Gooseneck/PVRV's	21%
Rupture Disk	1%

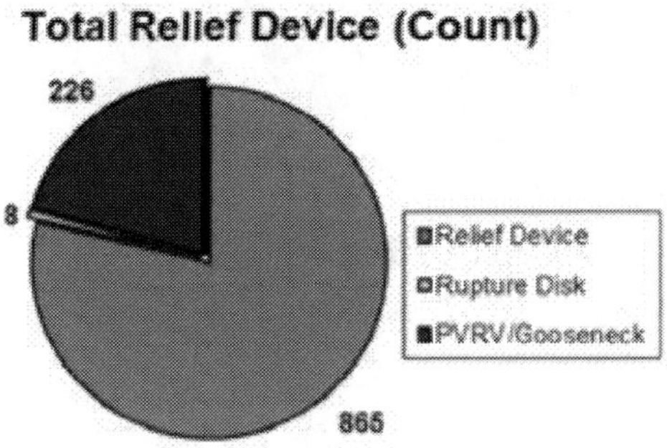

Figure 2: Actual number of different relief devices, by type, for the twenty compressor stations

3.2. Relief Valve Type

There are three types of relief valves that are typically used in the oil and gas industry: conventional, bellows, and pilot valve. Based on the type of valve, the outlet pressure drop limits vary; hence it is important to identify the relief valve type. As shown in Table 2, out of a total of 865 relief valves identified, the majority of the valves in the compression stations analyzed were conventional relief valves,. The breakdown of the number and percentage of the relief valve that were analyzed in this study are shown in Table 2 and Figure 3 respectively.

Table 2: Percentage of different relief valves, by type

Relief Valve Type	Percentage
Conventional	68%
Pilot	23%
Bellows	9%

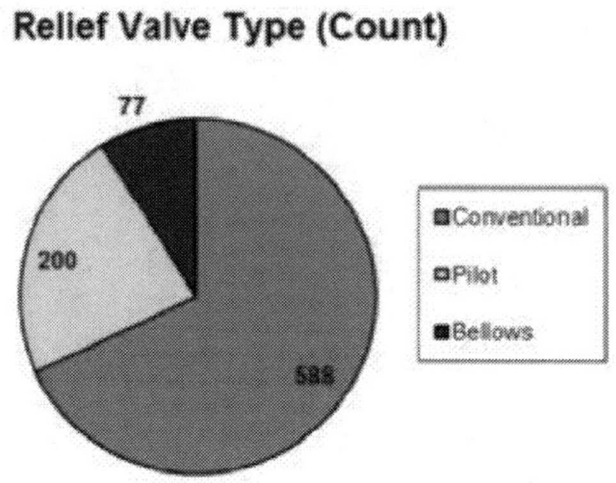

Figure 3: Actual number of relief valves, by type for the twenty compressor stations

3.3. Equipment Type

Based on the equipment type, the overpressure scenario may be applicable or not and the data required to do the analysis will be different. Hence, it is very important to identify the type of the equipment. A total of 2014 equipment were identified in all the compressors stations, of which majority were pressure vessels.

The breakdown of the percentage and number of the equipment type that were analyzed in this study are shown in Table 3 and Figure 4 and respectively. Note that the "Other" type of equipment includes compressors, glycol regenerators etc.

Table 3: Percentage of different equipment, by type

Equipment Type	Percentage
Pressure Vessel	48.5%
Tower	8.1%
Pumps	8.9%
Exchangers	16.6%
Fired Heater	0.1%
Low Pressure Tanks	6.1%
Piping Component	6.4%
Others	5.3%

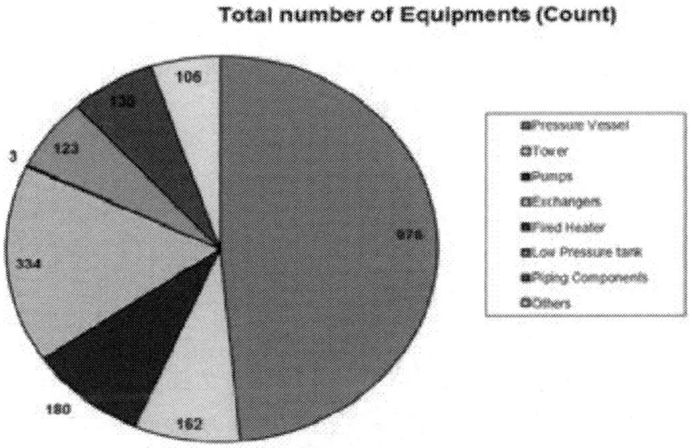

Figure 4: Actual number of equipment, by type, for the twenty compressor stations

3.4. Applicable Overpressure contingencies:

All the applicable scenarios, for each system, were identified after reviewing the P&IDs and other available data. This section shows all the applicable scenarios identified. The breakdown of the percentage and number of the applicable contingencies that were analyzed in this study are shown in Table 4 and Figure 5 respectively. Of a total of 3026 applicable contingencies, the majority of the contingencies in the compression stations analyzed were external or pool fire and blocked outlet.

Table 4: Percentage applicable contingencies by type

Contingency Type	Percentage
External fire	22.8%
Cooling or reflux failure	0.1%
Blocked Outlet	22.6%
Failure of automatic controls	13.3%
Gas breakthrough	8.5%
Inadvertent valve operation	5.6%
Instrument Air failure	2.9%
Internal boundary failure	1.6%
isolated with Heat Input	6.4%
Power failure	0.9%
Reverse flow	3.2%
Vacuum	7.1%
Other	5.0%

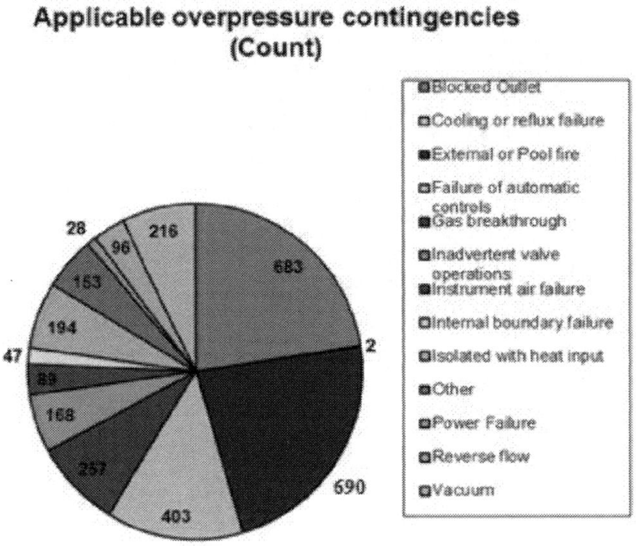

Figure 5: Actual number of applicable overpressure contingencies, by type, for the twenty compressor stations

3.5. Concerns Identification:

This was one of the most important steps in the relief analysis. Based the calculation results, assumptions, discrepancies etc., all the concerns were identified and documented based on the guidelines. The concerns were divided into five major categories: Undersized Concerns, Installation Concerns, Unprotected Systems, Data Requirement, and Documentation Updates. It is important that assumptions made during this stage be documented, verified and approved by the client. The undersized concerns, installation concerns and unprotected system concerns were assigned a higher priority since they could have a direct impact on the operation and failure of the compressor station. Hence, it was recommended that these concerns be mitigated immediately.

The breakdown of the percentage and number of the concerns that were analyzed in this study are shown in Table 5 and Figure 6 respectively. Of a total of 3278 applicable concerns that were identified in this case study, the majority of the concerns in the compression stations were related to Data Requirements.

Table 5: Percentage of concerns identified, by type

Concern Type	Percentage
Undersized	10.4%
Installation	24.6%
Unprotected Systems	9.5%

Data Requirement	50%
Documentation Updates	5.5%

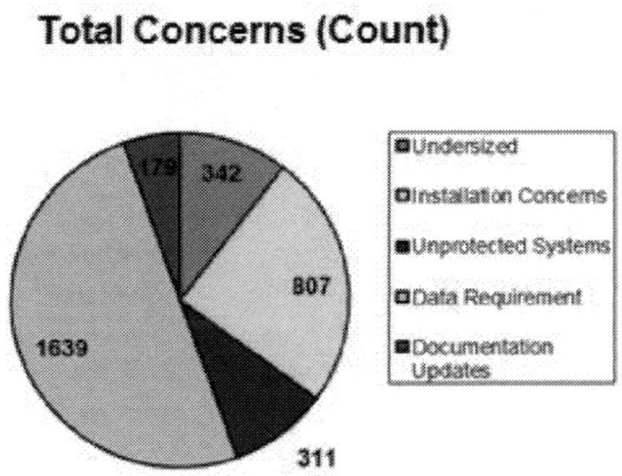

Figure 6: Actual number of different concerns, by type, for the twenty compressor stations

Undersized Concerns:

When the required relief flow rate is more than the capacity of the relief device, the relief device is defined to be undersized. This section shows the overpressure scenarios which were undersized. This includes all the contingencies for which the relief devices were undersized. The most common reasons for relief devices to be undersized were identified as being because of incorrect upstream pressure, incorrect composition of the relief stream, or that the contingency was not identified at all in the previous study.

The breakdown of the number and percentage of the undersized concerns that were analyzed in this study are shown in Table 6 and Figure 7 respectively. Of a total of 342 undersized concerns that were identified in this case study, the majority of the contingencies in the compression stations were Vacuum, Gas Breakthrough, External or Pool Fire and Failure of Automatic Controls.

Table 6: Percentage of type of contingency within undersized concerns

Contingency Type within Undersized Concern	Percentage
Blocked outlet	11.1
External fire	15.2
Failure of Automatic Controls	15.2
Gas breakthrough	15.5
Inadvertent valve operation	1.1
Internal Boundary failure	8.2
Reverse flow	8.5
Vacuum	24

Other	1.2

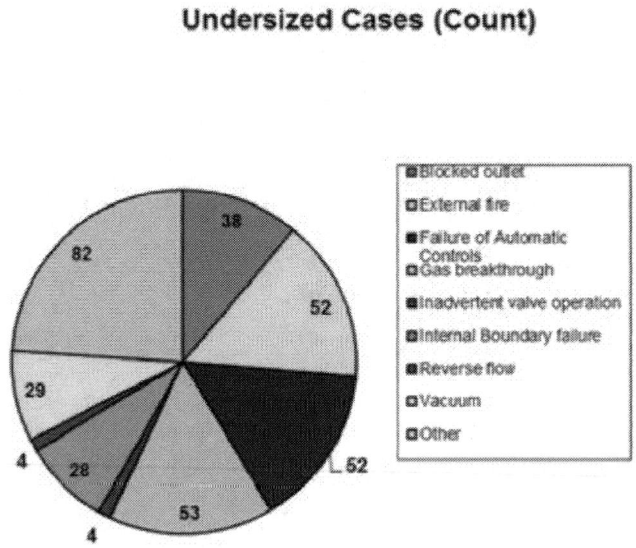

Figure 7: Actual number of different contingency types within the undersized concerns for the twenty compressor stations

Installation Concerns:

The installation concerns (total of 807 were identified) are divided into eleven sub categories as shown below. This is based on the guidelines used to perform the analysis.

Inlet Pressure Drop: The inlet pressure drop is calculated based on the relief device capacity. This includes all the scenarios for which the inlet pressure drop is greater than 3% of set pressure. As per API 520, if the inlet pressure drop exceeds 3%, the potential for capacity reduction and/or mechanical damage may exist for the relief valve. Note that inlet pressure drop concern was not identified for the relief devices which were undersized for one or more overpressure scenarios.

Outlet Pressure Drop: The outlet pressure drop is calculated based on the capacity of the relief device. This includes all the scenarios for which the outlet pressure drop is greater than 10% of set pressure for single conventional valves, or greater than 16% of set pressure for multiple conventional valves except for fire case, greater than 21% of set pressure for conventional valves for external fire, and greater than 50% of set pressure for bellows valves. As per API 520, if the outlet pressure drop exceeds the limit, the potential for capacity reduction and/or mechanical damage may exist for the relief valve. Note that outlet pressure drop concern was not identified for the relief devices which were undersized for one or more overpressure scenarios.

High Exit Velocity: The exit velocity and sound power level of the relief valve piping can affect its stability. All overpressure contingency with Mach number greater than 0.8 and sound power level greater than 155 was documented. Note that the high mach number concern was not identified for the relief devices which were undersized for one or more overpressure scenarios.

High set pressure: The relief device with set pressure greater than maximum allowable working pressure (MAWP) of the equipment was identified since the relief valve can no longer protect the system if this condition exists.

Isolation valve: This includes the block valves that are not chain locked open as required (the most common ones were the block valve on relief device inlet and outlet piping).

Pocketed Piping: In the event of liquid release and if the relief device is pocketed, the liquid can accumulate in the piping and hence the stability of the piping can get affected; hence it is very important to correct this concern.

Restricted line: In the event the inlet segment of the relief device piping is greater than the inlet nozzle as per API 521, these segments needs to be identified and documented.

The breakdown of the number and percentage of the installation concerns that were analyzed in this study are shown in Table 7 and Figure 8 respectively. Of a total of 807 installation concerns that were identified in this case study, the majority was the high outlet pressure drop concern.

Table 7: Percentage of installation concerns by type

Installation Concern	Percentage
High inlet Pressure drop	17.2%
High Outlet pressure drop	24.8%
High Exit Velocity	5.9%
Pocketed Piping	5.3%
Restricted Line	1.7%
Isolation Valves	11.8%
High Set Pressure	0.9%
No Weep Hole	20.8%
Liquid to Atmosphere	1.5%
Valve Stem Rotation	4.0%
Other	6.1%

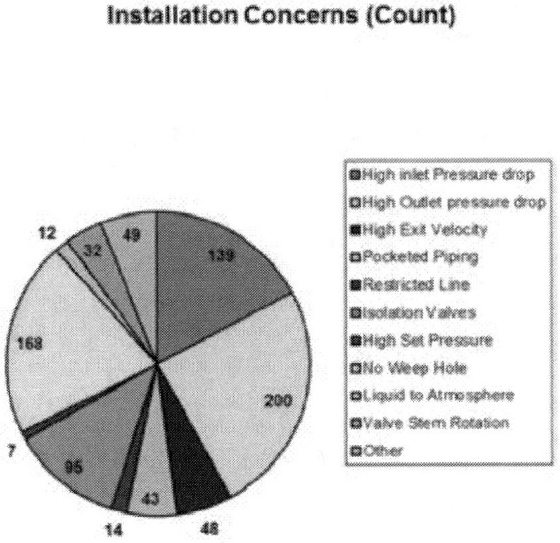

Figure 8: Actual number of installation by type for the twenty compressor stations

Unprotected Systems

These concerns were identified only for those systems that had applicable scenarios without any relief device protecting them. Contingency analysis for all the systems which didn't have any relief device protecting them was performed. As the case study was conducted on system based analysis, it was easier to identify these systems. The systems which had an internal relief device were considered to be unprotected. There were few cases in which there was a block valve present between the system and the relief device; hence these systems were considered unprotected. This list also includes the overpressure contingencies which resulted in over temperature of vessel in the event of external fire. Of a total of 3278 concerns identified, 311 were categorized as unprotected concerns.

Data Requirement

All the assumptions which were made during the analysis were documented under this category. During the analysis, a data need log was created if a proper supporting document was not provided and any follow on assumptions made due to the lack of documentation was approved by the client. Some general assumptions which were made were liquid levels, MAWP of the system (usually was assumed to be same as relief device set pressure), and relief valve information. Of a total of 3278 concerns identified, 1639 was data requirement concerns.

Documentation Updates

During the analysis, some discrepancies were found between different documents such as the U1A-form and the P&ID and these discrepancies were documented, with the correct information needing to be updated on the relevant documents. Documentation updates are usually a low priority concern as they may not impact the operation and safety of plant significantly. Of a total of 3278 concerns identified, 179 were documentation updated.

3.6. Mitigation of Concerns

Once the concerns were identified and documented, different recommendations were provided in order to mitigate the identified concerns. Usually, the clients choose to implement mitigation strategies predominantly for the higher priority concerns such as the unprotected concerns, installation and undersized concerns. An example list (not comprehensive) of the recommended mitigation strategies that were provided to the client for undersized, installation concerns and unprotected systems is detailed below.

Mitigation Strategies for Undersized Concerns

The provided recommendation to resolve undersized concerns depends on the overpressure contingency for which the relief device was undersized.

External Fire: In the event the relief device is undersized for external fire, it was checked if the vessel is insulated and whether credit of insulation could be taken. Usually, in order to overcome this concern, it was recommended that a bigger relief device be installed.

Blocked Outlet: In the event the relief device is undersized for blocked outlet, it was recommended that a bigger relief device be installed.

Isolated with Heat input: In the event the relief device is undersized for "isolated with heat input", it was recommended that a bigger relief device be installed. An alternate recommendation was to see if the undersized concern could be mitigated using administrative controls.

Tank Draining (Vacuum): This was a major concern identified in low pressure tanks. This overpressure scenario was missed in previous studies and hence, a lot of undersized concerns were found. It was recommended that a restriction orifice be installed on the drain line. Another option recommended was to install a bigger PVRV.

Failure of automatic control: For this undersized scenario, it was recommended that a restriction orifice be installed on the line, while still ensuring that it can allow the normal flow through it; however it can regulate the maximum flow such that the relief device is adequately sized. Another option recommended was that a restriction orifice be installed so that the maximum C_v of the control valve can be reduced, which ensures that the relief valve is adequately sized. It was rarely suggested that a bigger relief device be installed for this case.

Inadvertent valve operation: These cases usually included the inadvertent opening of bypass valves around the control valve. For these cases, usually the pipe capacity calculation was reperformed based on the equivalent length of piping from source to destination to check if the concern still existed. If the concern still exists, it was recommended that the bypass valve be chain lock closed or a restriction orifice be installed.

Gas breakthrough: Majority of the time, the failure of automatic controls and inadvertent valve operations resulted in gas breakthrough. For these cases, usually the pipe capacity calculation was reperformed based on the equivalent length of piping from source to destination to ensure if the concern still existed. If the concern still exists, it was recommended that the bypass valve be chain lock closed or a restriction orifice be installed.

Reverse flow: This is one of the most common overpressure contingencies that can be missed while doing the analysis. That is one the reason why the relief device was undersized for this case. In order to overcome this concern, it was recommended that multiple check valves be installed in the direction of flow. An alternate option recommended was the installation of a bigger relief device.

Mitigation Strategies for Installation Concerns

Inlet & Outlet pressure drop: For pressure drop concerns, it was recommended that the concerned piping be modified. The equivalent length of the piping was provided such that it was within the prescribed limit. Alternately, if the relief device was oversized, it was recommended that a smaller relief valve be installed. In this scenario, it is important to still ensure that the relief device is still adequately sized. For inlet pressure drop, it was recommended that a pilot valve with a remote sensing line be installed. For outlet pressure drop, it was recommended that bellows or a pilot valve be installed.

Mitigation Strategies for Unprotected System

If the system is unprotected, the most common recommendation was to install a new relief device. An alternate option recommended was to chain lock open any open block valve to take credit for other relief device. The third option recommended was to mitigate the scenario with administrative controls. This option was common for heat exchangers which have only thermal expansion as the applicable overpressure contingency.

4. Summary

This case study provides an overview of the relief analysis conducted for multiple compressor stations. As stated previously, the relief analysis was system based as opposed to relief device based so as to identify any unprotected systems that may need overpressure protection. It is very important to use the correct operating conditions and fluid composition to perform the relief analysis. The study provided an overview of the various types of overpressure scenarios identified for these systems, with a statistical breakdown of the scenarios by type. It was found that external fire and blocked outlet were the most commonly identified overpressure scenarios. In addition, this case study provided a breakdown of the types of relief devices, a further breakdown of types of relief valves, and the different types of equipment in the compressor stations. A breakdown of the type of concerns identified during the compressor station revalidation was provided along with a general idea of the mitigations that can be implemented to resolve the concerns. When performing relief analysis for compressor stations, it is very important to look out for any contingencies previously not considered, possibly due to changes in facility design or changes in the guidelines.

440451 An Approach to Reducing Risk during Online Isolation, Removal and Maintenance of Relief Devices Using Engineering Assessment

Abstract:

Overpressure protection must be provided for pressurized equipment by means of pressure relief devices, which sometimes require isolation and removal for inspection or repair. It would be advantageous to have a means of isolation without shutting down the process, especially in light of the additional risks associated with shutdown and startup; however, doing so may put the pressurized system at risk of being unprotected in the event of overpressure. Current recognized practices of providing dedicated personnel for monitoring and responding to an overpressure event, or installation of an independent spare, are non-ideal in many instances. An engineering assessment can identify specific ways for providing temporary overpressure protection, including elimination of root causes of overpressure events, provision of alternative pathways to overpressure protection, modification of process operations, and/or establishment of a means of equivalent or better risk reduction for the duration of the isolation. Guidelines and examples are provided to illustrate the analysis, along with recommendations for documentation.

441392 Common "Design Flaws" Identified during Pressure Relief System Analysis

Abstract:

Common "design flaws" identified during a pressure relief system analysis.

It is the goal of a pressure relief system analysis to identify possible gaps in a pressure relief system. In many oil and gas refining and petrochemical plants, it's common to encounter certain "design flaws" which can lead to overpressure scenarios where the system is not adequately protected.

Within these design flaws there are a few which frequently occur, regardless of the process. These can be just simple mistakes or related to consideration not being given to the whole system as part of the design. This paper will focus on five common design flaws:" unusually large bypass valves around a (smaller) control valve"; "liquid head between relief device and equipment being protected"; "how does the suction of the feed pump behaves during the scenario", "tube rupture can affect an entire system and not just the exchanger"; and "common API 650 tank concerns". For each "design flaw" a few common mitigations will also be discussed.

The goal of this presentation is not to discuss the issues in detail, but rather just bring to the industry's attention a few concerns being frequently identified.

Considering Overfilling to be a Non-Applicable Overpressure Scenario

Natalie Doe
Smith & Burgess
190 S. LaSalle St., Ste 550, Chicago, IL 60657
Natalie.doe@smithburgess.com

Dustin Smith, P.E.
Smith & Burgess
7600 W. Tidwell, Ste 600, Houston, TX 77040
Dustin.smith@smithburgess.com

Prepared for Presentation at
American Institute of Chemical Engineers
2016 Spring Meeting
12th Global Congress on Process Safety
Houston, Texas
April 11-23, 2016

Considering Overfilling to be a Non-Applicable Overpressure Scenario

Natalie Doe
Smith & Burgess
190 S. LaSalle St., Ste 550, Chicago, IL 60657
Natalie.doe@smithburgess.com

Dustin Smith, P.E.
Smith & Burgess
7600 W. Tidwell, Ste 600, Houston, TX 77040
Dustin.smith@smithburgess.com

Keywords: PSV, Risk mitigation, liquid overfilling, Safety Instrumented Systems

Abstract

The consequences associated with liquid overfilling are serious for plant safety, as illustrated by many recent incidents (2009, Bayamón, Puerto Rico; 2008, Chesapeake, VA; 2005 Hertfordshire, England; 2005 Texas City, Texas). In many cases, providing overpressure protection solely with pressure relief devices may not protect the facility. Preventing the uncontrolled release of liquid may be the only viable safe option. This paper will present an abbreviated analysis of several incidents that could have been avoided with proper preventive measures. Engineering methods will be presented to eliminate the need to release liquids from a closed system. For instance, the system can be designed such that upstream pressure sources are limited below the allowable downstream pressure. A designer can provide a safety instrumented system with an appropriate Safety Integrity Level (SIL) rating to reduce or eliminate the credibility of overfilling. Alternatively, consideration may be taken for alarms and operator response, with consideration for RAGAGEP, to reduce the credibility of overfilling to an acceptable risk. Systems can be redesigned to be inherently safer by minimizing contained liquids. This paper then gives a few examples of the use of system design, shutdown systems, alarms/operator response combination and inventory controls to eliminate or reduce the potential for liquid release scenarios.

1 Introduction

The consequences of an overfilling overpressure scenario can result in increased risk to the facility when mitigated with pressure relief devices. This could occur due to many different root causes. In order for liquid overfilling to require overpressure protection from a relief device, the following must all be true:

1. The source pressure must be capable of exceeding the system MAWP. Since the relief device is supposed to be set at or below the system MAWP, if the relief device is set lower than the MAWP, a liquid release may occur at a lower pressure.
2. The flow of liquid through the system can be stopped by closing a valve on the system outlet or the loss of a discharge pump (etc). In general, system designers ensure that liquid flows can be stopped, so this is usually a given.
3. There are no acceptable administrative or instrumented controls in place to prevent the scenario.

When reviewing the consequences of an overpressure scenario, the designer must check to ensure that the overfilling event may also not occur during a larger plant upset (power failure, instrument air failure, or other global utility disruption). This would add a simultaneous load to the disposal system, which should be included in the disposal system design.

Overpressure protection

The system designer should consider the root cause of the overfilling scenario to determine how to protect the system or eliminate the scenario. Relief devices are often used to prevent excessive vessel overpressure due to liquid overfilling. For many systems, overfilling is not expected to be the controlling case for relief device sizing; however, a review of incident history shows that many different systems have been overpressured by overfilling with liquid, resulting in relief. Therefore, when relief devices are used to provide protection for overfilling, it is important that proper consideration is given to the effluent handling system.

1. For a system with several pieces of protected equipment, consideration must be taken for the static head throughout the column to ensure equipment at grade is adequately protected. This is particularly an issue for low pressure tanks.
2. When systems overfill, there is the potential for a large amount of liquid to be released to the effluent handling systems. Unless liquid holding drums are sized for the potential liquid rate, there could potentially be an unplanned liquid release with significant consequences.

This paper will outline recent incidents and their consequences due to overfilling, and will overview methods to protect the vessels from overpressure and the system from the consequences of large liquid releases.

2 Overview of recent incidents

The following incidents from the past decade illustrate the consequences of overpressure due to overfilling and the handling of liquid releases.

2009, Bayamón, Puerto Rico;

Per the CSB report, "Tank 409 overflowed with gasoline, resulting in a vapor cloud that encompassed 107 acres of the CAPECO tank farm." [1] The report also states that "Tank 409 overflowed for an estimated 26 minutes before the vapor cloud ignited. The CSB determined nearly 200,000 gallons of gasoline… rushed out of six vents in the tank." These six vents responded to the upset conditions to relieve the excess pressure in the tank; thus, the overpressure protection provided was sufficient to prevent a catastrophic vessel failure. The action of the "engineered" pressure relief devices for this scenario worked and then became the initiating event for this incident. The final result was that the company went bankrupt and 17 of 48 tanks were burned.

2008, Chesapeake, VA;

Per the CSB report, "At about 2:20 pm, as the tank reached a level of 26.72 feet, a vertical split started midway up the shell and rapidly extended to the floor and roof of the tank on the side opposite the workers." [2] In this case overpressure protection to prevent catastrophic failure of the tank would not have worked because the tank was not "full". This incident could not have been protected for by pressure relief devices - only instrumentation, procedures or the like.

2005 Hertfordshire, England;

Per the aforementioned CSB report, "Similar to the CAPECO incident, the vapor cloud explosion and multiple tank fires occurred after a tank was overfilled with gasoline." [1] Thus another incident of overpressure protection for the overfilling scenario working; as such, it operated as designed. The "engineered" overpressure mitigating system was the initiating event for an incident that caused more than a billion dollars' worth of damage to the surrounding commercial and residential property and injured 43 people.

2005 Texas City, Texas

Per the CSB report, "The incident occurred during the startup of an isomerization (ISOM) unit when a raffinate splitter tower was overfilled; pressure relief devices opened, resulting in a flammable liquid geyser from a blowdown stack that was not equipped with a flare. The release of flammables led to an explosion and fire." [3] The pressure relief devices opened, and prevented excessive overpressure of the ISOM tower. As such, the engineered pressure relief devices in this case also worked, preventing a catastrophic overpressure-related loss of containment, but the design for the disposal of the fluid created an additional hazard with significant consequences.

Many of these incidents involved tanks, but that is inconsequential as in each and every case either a relief device was insufficient to prevent overpressure, or the engineered design of the pressure relief device and disposal was the root cause for these scenarios. Not only do the pressure relief devices have to be designed to prevent loss of containment from overpressure, the disposal system has to be designed to ensure the effluent has a safe location to be stored or is adequately destroyed. This can be particularly difficult for liquid releases.

3 Means of overpressure protection

Relief Devices and Disposal

While relief devices are an acceptable method to protect from overpressure due to liquid overfilling, complete risk mitigation requires proper consideration of the ability of the disposal system to handle the liquid effluent. As mentioned previously, additional pressure due to static head should be considered when protecting multiple pieces of equipment. For column systems with relief devices located on the top of the column, this can have significant impact on associated equipment located at grade. It is common practice to perform overfilling relief device calculations at 10% overpressure rather than 16% overpressure when multiple devices are installed, in an attempt to account for the static head. For tall column systems, or for systems with a heavy relief fluid, this may not be sufficient to ensure the allowable accumulation on the equipment located at grade is not exceeded. The relief device set pressure may need to be lowered below the system limiting MAWP to ensure adequate protection for all equipment.

Additionally, the disposal system must be capable of handling the liquid flow from the relief devices. The disposal piping should be evaluated to determine if it can handle the weight of the liquid. A knockout drum must be incorporated as part of the disposal system to ensure liquid is not sent to the flare or vent stack. For large liquid releases, it may not be practical to size one knockout drum to obtain the required particle separation. Multiple knockout drums in parallel can be used to ensure adequate particle separation for these cases. Failure to size the knockout drum(s) correctly can lead to liquid being sent to a system which cannot handle it, which can have serious consequences. The CSB report which reviewed the Texas City incident [3] unduly criticized atmospheric blowdown drums; many of their preferred disposal systems, i.e. flare systems, have had similar incidents [4] with the authors aware of countless near misses from flare systems.

Overpressure protection by system design

The need to design a disposal system for the liquid effluent from relief devices that protect a system of equipment from overpressure by overfilling can be eliminated by designing the system such that relief devices are not needed. The acceptable way to do this is to ensure that the overfilling scenario is not credible. For vessels designed to ASME VIII, UG-140 (formerly known as code case 2211) states "A pressure vessel does not require a pressure relief device if the pressure is self-limiting… and this pressure is less than or equal to the MAWP of the vessel at the coincident temperature…" [5]. This can be achieved in many ways.

Equipment Design

A designer can utilize equipment design to ensure that the disposal system does not need to accommodate effluent from relief devices installed for liquid overpressure protection.

<u>Example 1 – Pump Selection</u>

Instead of feeding a system with a positive displacement pump, a centrifugal pump may be used that is not capable of exceeding the downstream system MAWPs. This is illustrated in Figure 1 below, as P-1 cannot exceed the MAWP of V-1. The defining quality of centrifugal pumps is that their flow will decrease with increased head. Therefore, as the pressure in the downstream system increases, the capacity of a centrifugal pump will decrease. Thus, pump selection can be an effective method of limiting the overfilling rate.

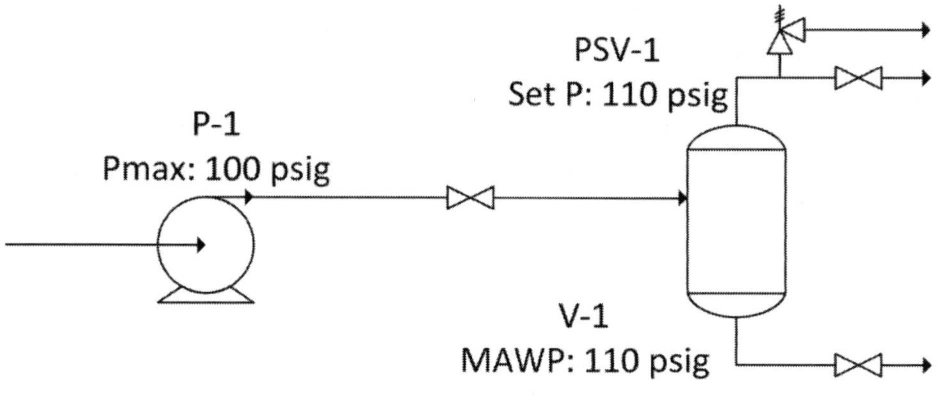

Figure 1: Upstream sources incapable of overpressuring the system

In contrast, positive displacement pumps are constant volume machines that are capable of maintaining high volumetric flow even at increased pressures. Therefore, alternative methods can be applied to protect from overpressure due to positive displacement pumps.

<u>Example 2 – PRV Discharge to Process</u>

If a relief device is installed on the pump instead of the vessel, the discharge piping can be routed to the pump suction. This allows for the liquid relief to be sent to process, rather than to a disposal system. However, the designer is cautioned that this will stop the forward flow of liquid so equipment upstream must be checked for this upset condition. An example is shown in Figure 2 below, as PSV-2A reduces the maximum feed pressure and eliminates the overfilling scenario on V-2.

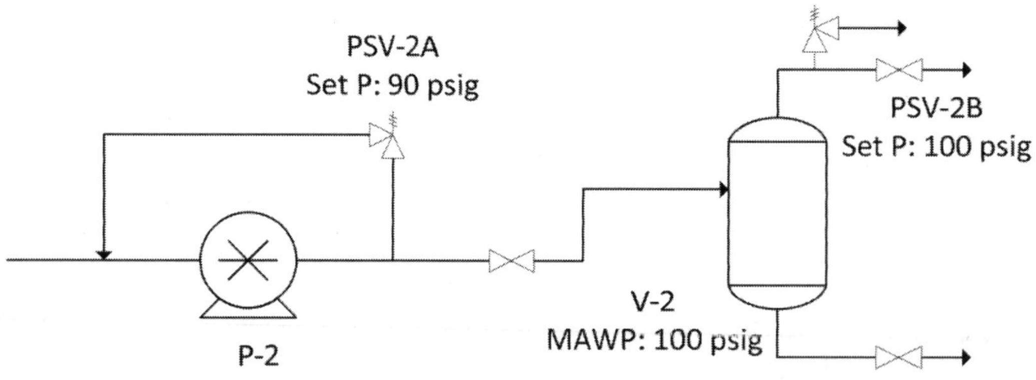

Figure 2: Protecting upstream pressure sources

In other cases, protecting for overfilling by pump selection or upstream protection may not be possible. Inventory controls may be utilized in these situations to ensure overfilling is not credible.

Example 3 - Inventory Controls

If possible, a designer can minimize the liquid inventory in a system. This is also a tactic in inherently safer process design. If the total liquid inventory of a system is insufficient to fill the vessel(s) vapor space then overfilling cannot occur. In this case overpressure does not occur irrespective of the maximum upstream pressure. An example of this is illustrated in Figure 3. The positive displacement pump (P-3) cannot overpressure the vessel (V-3) because the liquid inventory in the upstream tank (T-3) is less than the vessels vapor space.

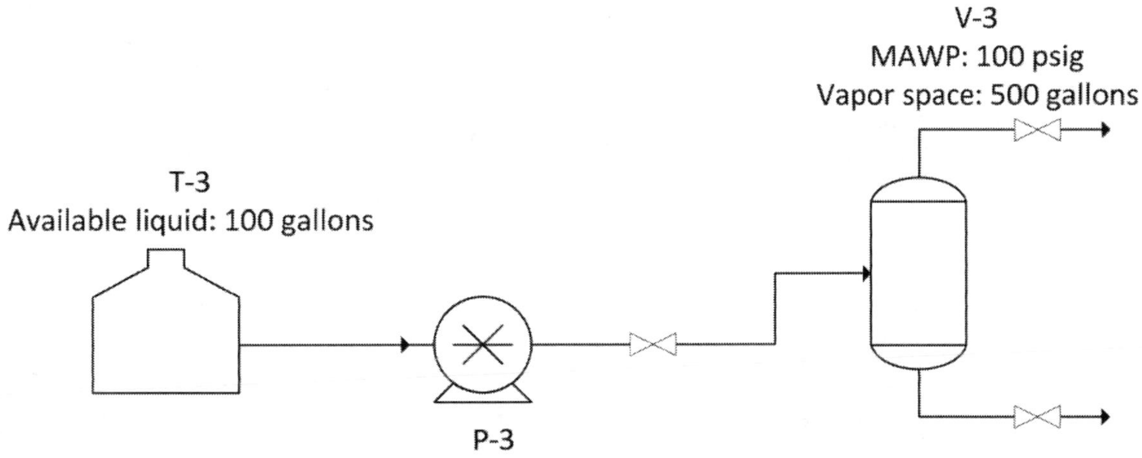

Figure 3: Inventory Controls

Instrumented Solutions

For installations where it is not practical to remove the credibility of the overfilling scenario by equipment design, safety instrumented systems (SIS) can be implemented. There are two ways to protect a facility from the effects of liquid effluent: remove the relief device entirely, or mitigate the consequences of the liquid effluent escaping from the disposal system.

Example 4 – Replace Relief Devices with instrumentation

The system can be designed such that instrumentation is used instead of overpressure protection. A designer can find guidance in API STD 521 which states in Annex E that "High Integrity Protection Systems (HIPS) are designed to avoid overpressure incidents by removing the source of overpressure or by reducing the probability of an overpressure contingency to such a low level that it is no longer considered to be a credible case."[6] ASME VIII UG-140 allows for the use of these systems, provided that detailed description of the safety system is documented, and includes "the identification of all truly independent redundancies and a reliability evaluation... of the overall safety system" [5] The ASME requirements for these systems are further explored elsewhere [7]. Figure 4 below shows an example of a system protected by instrumentation. Three separate level taps are used to determine the liquid level in V-4. Two independent emergency shutdown devices (ESD), each with an independent controller, are used to shut off the feed on high liquid level. The redundancies expressed in this example would be integral to determining the safety integrity level (SIL) of this system.

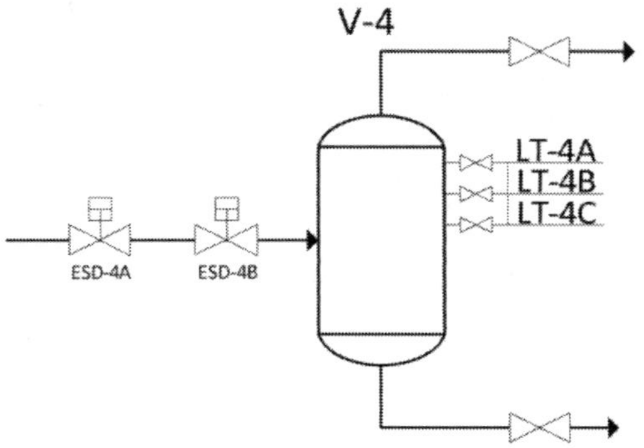

Figure 4: Protection through instrumentation

Example 5 – Add a layer of Protection

Alternatively, the design of a SIS does not require that overfilling is mitigated to a point at which it cannot occur; rather, that the probability is low enough that the total system risk is at an acceptable level. Depending on the existing system, the most cost effective solution may be to mitigate the effluent system release with a SIS but rely on a the relief device to protect the equipment. Due to the uniqueness of each system a risk assessment (such as a LOPA) will be required to determine the reliability of the SIS. The designer may choose to put in a less reliable safety instrumented system in combination with an existing relief device, or the safety instrumented system may replace the relief device. As mentioned above, these redundancies are used to determine the safety integrity level (SIL) of the system. For most cases, a SIL-3 System is needed to remove the credibility of the overfilling scenario. However, a SIL-1 or SIL-2 system may be acceptable when reviewed in a risk analysis [such as Layers of Protection Analysis (LOPA)] to mitigate an incident caused by a release of liquid to the disposal system. Either way, a detailed risk assessment should be performed to determine the SIL rating required to mitigate the hazard [8][1]. An example of a SIS that could provide overfill protection would be level indicators independent of the process control instrumentation from a separate vessel tap feeding a dedicated PLC that triggers a shutdown valve to shut off the feed to the system on a high level alarm.

Operator Response to mitigate overfilling

For many facilities, a combination of instrumentation with an operator response is a reasonable mitigation of risk from overpressure. Research has shown the following factors tend to determine the reliability of operators to take the intended action [9][10]:

1. Trigger – there is a clear indication to the operator that something is wrong triggering the need to take action. It is unreasonable to assume that the operators would have mitigated the overfill scenario at Bayamón PR as they had no indication of level or procedures to know when the tank overfilled [1].
2. Time – Per the work performed by the NRC [9], an individual in a stressful situation with limited time to perform a moderately complex task has a 100 times likelihood of failing compared to the exact same task with sufficient extra time allotted for execution. Others, [10] have put the reliability of an operator completing a clearly understood task on which he/she is trained, with adequate time under stressful conditions, is equivalent to a SIL-III system.
3. Consequences – if the mitigation action causes (or is perceived to cause) a chain reaction that is worse than the failure being mitigated there may be reluctance to perform the action. Additionally, if the operator assumes (or is told) that they will be punished for stopping production, then the likelihood of following the procedure decreases greatly.
4. Training – without the knowledge and training to act, the probability that an action will be taken diminishes greatly, especially if the action is not obvious to the individual operating the facility.

As safety instrumentation systems can be complex and expensive, many facilities choose instead to take credit for operator response to high level alarms. There are several important factors to consider when taking these credits. API 521 4.4.7.3 [6] outlines several factors to consider when relying on level alarms:

a) Whether the level instruments used for safeguards against overfilling are on separate process taps from the process control system
b) Whether level instruments used for safeguards against overfilling are susceptible to the same common mode device failures as those used for the basic process control system.
c) Whether the programmable level transmitter is set to show a low or zero level when the level exceeds the instrument range
d) Whether an open system differential pressure cell is used that can show a false low level due to accumulation of liquid in the upper impulse line if it is overfilled or collects condensed liquid
e) Whether density changes due to temperature effects or composition changes impact readings from differential pressure transmitters
f) Whether the instruments are proven in use for the specific process applications
g) Whether the range of at least one of the level measurement(s) can indicate a valid level reading of the full range between the high critical alarm point and any shutdown or interlock point
h) Whether operating characteristics of the level measurement during off-design, start-up, and shutdown operations are considered in display of level, physical properties, setting alarms, trip points, operator training, and operating procedures
i) Maintenance and testing frequency required for the instrumentation

Even if all of these factors are considered, and the reliability of the operator is evaluated, it is important to analyze the risk associated with human error, and the operator not responding in the allotted time. As with instrumentation systems, relying on level alarms and operator response is not a guarantee overfilling will not occur; rather, it lowers the risk to an acceptable level.

Overfilling of low pressure tanks present a unique challenge as compared to pressure vessels. API has created a specific document to provide further guidance to protect for the overfilling of low pressure storage tanks [11].

Example 6 – Alarms with subsequent operator response

As with Example 4 above, Figure 5 shows a vessel (V-6) with three independent level taps. These level taps are connected to two independent high level alarms (HLA). The designer performed a LOPA, and determined the operator can reliably address the alarm in 30 minutes. This is less than the fill time of V-6 from the time when the alarm sounds (1 hour), so credit may be taken for the operator actuating HV-6A and HV-6B to shut off the feed and prevent overfilling.

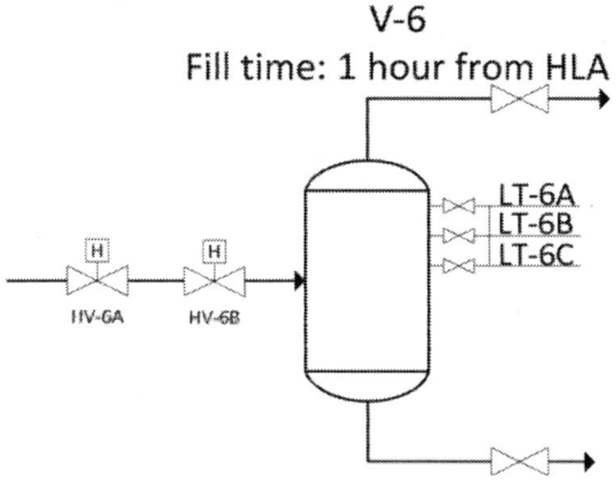

Figure 5: Operator Response

4 Conclusion

Liquid overfill can have serious safety and financial consequences, as seen in the recent incidents outlined in this paper. It is important for the designer to consider both the means to prevent overfilling, as well as methods to respond to an overfilling event. Appropriately designed relief devices and disposal systems are an acceptable method to react to an overfilling event. However, if this is to be the sole means of protection, careful consideration must be given to the large liquid effluent rate sent to the disposal system. Alternatively, the event may be prevented through the proper application of equipment design, instrumented solutions, alarms and operator response, or some combination thereof.

There are a large number of incidents involving overfilling that occurred despite the fact that the installed overfill protection reacted as designed. The protecting relief devices opened and were adequate to prevent overpressure of the overfilled vessel, but the disposal system was not designed for that case. The incidents mentioned in this paper resulted in billions of dollars' worth of damage, and even the bankruptcy of one company. When protecting for overfilling, it is imperative to consider the ramifications of allowing large liquid releases to occur. Overfilling protection design that encompasses consideration of all of the options presented above can prevent an incident with catastrophic consequences.

5 References

[1] U.S. Chemical Safety and Hazard Investigation Board. "Caribbean Petroleum Refining Tank Explosion and Fire" Available at http://www.csb.gov/caribbean-petroleum-refining-tank-explosion-and-fire/. Accessed on 2/5/2016.

[2] U.S. Chemical Safety and Hazard Investigation Board. "Allied Terminals Fertilizer Tank Collapse" Available at http://www.csb.gov/allied-terminals-fertilizer-tank-collapse/. Accessed on 2/5/2016.

[3] U.S. Chemical Safety and Hazard Investigation Board. "BP America Refinery Explosion" Available at http://www.csb.gov/bp-america-refinery-explosion/. Accessed on 2/5/2016.

[4] HSE (Health and Safety Executive) (Eds.), 1997. The explosion and fires at the Texaco Refinery, Milford Haven, 24 July 1994. A report of the investigation by the Health and Safety Executive into the explosion and fires on the Pembroke Cracking Company Plant at the Texaco Refinery, Milford Haven on 24 July 1994, HSE Books, London.

[5] American Society of Mechanical Engineers, 2013 ASME Boiler & Pressure Vessel Code: Section VIII, Division I. s.l.

[6] API Standard 521, "Pressure-Relieving and Depressuring Systems," American Petroleum Institute, 6th ed., 2014.

[7] Smith, Dustin J. "Safety instrumented systems in lieu of pressure relief valves." *Process Safety Progress* 33.4 (2014): 345-349.

[8] A.E. Summers. "Overfill Protective Systems – Complex Problem, Simple Solution," 12th Annual Symposium, Mary Kay O'Connor Process Safety Center, October 2009.

[9] Blackman, Harold S., David I. Gertman, and Ronald L. Boring. "Human error quantification using performance shaping factors in the SPAR-H method." *Proceedings of the Human Factors and Ergonomics Society Annual Meeting*. Vol. 52. No. 21. SAGE Publications, 2008.

[10] Mannan, Sam, ed. Lees' Loss Prevention in the Process Industries: Hazard Identification, Assessment and Control. Vol. 1. Butterworth-Heinemann, 2004.

[11] API Standard 2350, "Overfill Protection for Storage Tanks in Petroleum Facilities," American Petroleum Institute, 4th ed., 2012.

How Reliable Are Your Process Safety Barriers?

Mike Neill
Petrotechnics USA Inc.
10497 Town & Country Way, Suite 700, Houston, TX 77024
mike.neill@petrotechnics.com

Prepared for Presentation at
American Institute of Chemical Engineers
2016 Spring Meeting
12th Global Congress on Process Safety
Houston, Texas
April 11-23, 2016

How Reliable Are Your Process Safety Barriers?

Mike Neill
Petrotechnics USA Inc.
10497 Town & Country Way, Suite 700, Houston, TX 77024
mike.neill@petrotechnics.com

Keywords: Process Safety Management; Process Safety Barriers; Operational Risk Management

Abstract

Using Process Safety Management techniques, we design and implement multiple layers of protection for our hazardous plants. For these barriers to remain effective over the life of the asset is a challenge, particularly with aging assets. We know barriers deteriorate over time without proper, preventative maintenance and inspection. Often a lack of focus on critical barriers can result in activities being deferred or cancelled without recognizing the potential consequence.

There is generally a lack of understanding in our organizations of process safety techniques and approaches. While the criticality of these barriers maybe well-understood in the Process Safety department, this is unlikely true through other organizations, including Operations and Maintenance.

With so much equipment being tagged "safety critical" on our facilities; it is often difficult to ensure safety barriers receive the focus they need to ensure their effectiveness at protecting against high consequences events when needed. Operators need to manage the health of their barriers in near real-time and mitigate the impact arising from any impairments.

By understanding dynamic aspects of cumulative risk, all members of an organization can easily identify barrier impairments. This level of visibility helps make better operational activity management decisions on account of operational activities' potential impact on Major Accident Hazard barriers.

This paper will examine the traditional approach to barrier design and assumptions on reliability, the approach to maintenance, and the reality of operational dynamics. It will introduce the concept of cumulative risk impacts on safety barriers and examine operational dynamics which impact barrier performance.

1 Introduction

Confucius, the Chinese Philosopher from around 500BC is quoted as saying that "a man who lacks reliability is utterly useless." Well the same can be said for our Process Safety barriers, whose sole purpose is to prevent major accidents or mitigate the consequences of such events. These barriers enable us to operate our hazardous facilities, exposing people to potential danger, since we satisfy ourselves that the calculated risk of incident is acceptable, assuming reliable performance. Unfortunately, when we investigate incidents, we see again and again examples where multiple barriers fail to perform. Occasionally we see evidence where the design was at fault and the assumptions made were incorrect, and we see plenty of evidence where Management Systems designed to prevent barrier failure, fail themselves.

All too often the barriers do not work as intended for a variety of reasons ranging from corroded pipe, seized valves, inoperable switches, blocked flowlines to a simple case of responsible people failing to follow operating procedures.

We focus a lot of attention on making sure we design our barriers properly in the first place but do we do enough during the life of the facility to make sure the barriers continually perform per design? Are our organizations set up to optimally manage these barriers and are our frontline workers engaged sufficiently and armed with the right knowledge to play their role in identifying barrier impairments and to make appropriate operating decisions? These are questions this paper will address and will conclude with examining new solutions which are becoming available to help the industry with this challenge.

2 Evidence of Barrier Failure

The failure of barriers designed to protect or mitigate the consequences of Major Accident Hazards (MAH) often results in an incident such as a significant gas release or an over pressure situation. The failure of multiple barriers can escalate into a major accident such as an ignition barrier failure coincident with a gas release which results in an explosion or fire. The term "barrier", refers not just to equipment hardware such as piping, vessels, instruments, actuated valves, etc., but also includes procedural systems, where plant staff follow agreed practices to operate the plant and respond to emergencies.

The latest Marsh and McLennan report (ref1) provides a current update to property damage losses in the hydrocarbon industry. This report, in its 23rd edition, outlines the 100 largest losses across our industry over the past 40 years. A direct quote from that report

> *"Accidents that result in major losses, such as those reported in this document, generally occur because of the failure of a number of barriers within the process-safety management system — all occurring at the same time; typically, none of these losses are the result of the failure of a single barrier or protection measure."*

Furthermore, in the same report, other quotes reflecting on the incidents,.

"N*one of the losses listed in this document should be considered 'black swan' events.*"

("Black Swan" events are those regarded as inconceivable and impossible to consider as a credible threat.)

The implication of this statement is that the scenarios and failure modes were anticipated so either the design of the barriers was inadequate to meet conditions or the design of the barrier was adequate, but its performance did not match expectations. It could also be a combination of both.

In another report, (ref2) Ron Jarvis from Swiss Re looked at a sample of 45 incidents in the Oil, Gas, and Petrochemical fields over the prior 15 years. Despite this being a random sample and not necessarily a statistical analysis, it is worthy of consideration, not least because its aim was to prioritize those risk control elements whose failure resulted in a majority of incidents.

The study concluded 52% of primary incident cause was failure due to corrosion, 27% failure of operating practices and procedures, and 14% was attributed to unsafe maintenance, equipment isolation and permit to work management. Considering secondary and tertiary causes, poor design and/or inadequate hazard identification is another cause which becomes more prominent. The author goes on to identify inadequate piping and equipment inspection programs resulted in 68% of corrosion failures, the remainder being due to incorrect material selection or defective materials when first installed. Regarding the failure of operating practices and procedures, Jarvis highlights poor management of deliberate by-passing of Safety Critical Equipment (SCE) and the management of abnormal situation on the plant. 57% of all incidents occurred during abnormal conditions on the plant, including start-up, shutdown, or maintenance.

The evidence from these two studies is conclusive that Major Accidents occur as the result of multiple barrier failures. You may ask why it is important to recognize this seemingly obvious fact! This Author believes it is having attended many Process Safety Conferences over the years where the emphasis is on barrier design and static risk scenarios with little focus on what happens to barriers over the life of the asset. This, and working in an industry where there seems only to be a tacit recognition of a dynamic component to MAH risk in which a 5 year review of safety management systems seems adequate. The fact is, barriers fail and over time most, if not all, will fail without periodic inspection, testing, and preventative interventions.

The challenge we face is the data on Major Hazard Barriers, which can indicate an impaired status or diminished integrity, is not easily shared with the people who need to know and who routinely make decisions on daily operations based on assumptions that everything is in good health.

3 Barrier Design Versus Barrier Performance

3.1 Design Principles

When designing Oil, Gas, and Petrochemical plants, many of the principles used are determined by regulatory requirements in addition to generally accepted engineering design standards. Wall thicknesses and material specifications for pipes and vessels will be based on the expected fluid properties and operating conditions as will the specifications for valves and rotating equipment selection. HAZOP and HAZID studies look at potential hazard scenarios and failure mechanisms and identify methods for minimizing risk and/or mitigating the outcome. Risk-based studies look at ways to reduce MAH risk to a level which meets the criteria of acceptance. When the risk levels are excessive, additional barriers are designed.

Unfortunately, barriers themselves are not 100% reliable. This can be the result of a material defect in piping or vessel walls, or corrosion of those same walls, a stuck relief valve, a deluge system plugged with scale, or a standard operating procedure which is not followed. A single safety barrier cannot eliminate risk but it can reduce the likelihood of an event happening or mitigate the magnitude of the consequences if it does occur. As such, the traditional design approach is to apply multiple barriers, some preventative and some mitigative. Together they reduce the overall consequential risk.

Modern techniques of Inherently Safe Design, pioneered by Trevor Kletz, seek to build plants where hazards are removed rather than controlled and where volumes of hazardous fluids are minimized to reduce the magnitude of consequences. This is paving a way towards safer plants. At present, the fundamental needs of Oil, Gas, and Petrochemical plants continue to demand processing flammable, explosive, and toxic fluids often at high pressures and high temperatures and in large volumes. As such, the approach of multiple barriers to reduce and mitigate risk is still essential for plants - old and new.

When a hazardous scenario has been identified, the risk of the consequences occurring is calculated using statistical and Quantitative Risk Analysis (QRA) techniques. Risk acceptance criteria is often prescribed or advised by the regulator or set by internal corporate policy for an Operator. For example, the threshold may be a broadly acceptable risk of death of 1×10^{-6}. If a scenario results in the calculated risk of this outcome being higher than the threshold, additional barriers need to be designed to reduce the risk level to one which is acceptable.

Each barrier is credited with a level of risk reduction. This is calculated using performance characteristics of the barrier which, for hardware, is provided by the manufacturer based on empirical data such as "mean time between failure," etc. The assurance from the manufacturers of this performance is founded on a manufacturing quality system designed to provide reproducible results.

For vessels and piping the wall thickness and material specification are the principle drivers of strength and the ability to contain the process fluids. Once again, manufacturers' quality assurances assume manufacturing standards are followed. Mill certificates demonstrate material specification compliance. Hydrostatic testing of vessels and fabricated flowlines / pipelines, at pressures below the material yield points, are part of commissioning new plant and verifying a high level of integrity prior to start-up.

This process of risk reduction through designing additional barriers where necessary is carried out across the facility until the overall level of risk is reduced to an acceptable level.

Unlike hardware barriers, Procedural Barriers typically are less reliable since they require human interventions to be consistent and repetitive and with a discipline to follow standard procedures. Since most Operators are not robotic by nature, reliability is less than for engineered systems. On balance, this is a good thing since, unlike robots, Operators are free-thinking, unconfined by a narrow scope, and are able to identify a wide spectrum of developing situations and respond appropriately to regain control.

Reliability for procedural barriers is improved if the procedures themselves are well-written and workable so have been repeatedly tested and improved over time. The use of checklists provides a simple go/no-go approach, removing ambiguity in the step-by-step procedure. Training, and of course, experience are the other strong determinants of the reliability of procedural barriers.

3.2 Barrier Categories

The diagram below shows an example of fundamental barrier groupings. These are categories of barriers based on their function. While shown as the familiar "Swiss Cheese" model, the layout is similar to a "bow-tie" diagram where the left hand barrier groups are "preventative" and the right hand are "mitigative." It is useful to understand barriers in the context of multiple layers of protection and to appreciate that the term "multiple layers" does not necessarily indicate multiple redundancy.

As an example, if there is a gas leak on a plant than the failure in itself is a serious incident and has the potential for fire and explosion. Another group of barriers would be ignition control, preventing the escaping gas from igniting. Gas detection is another barrier and fire suppression systems a further barrier categorized under Protection. So each of the multiple layers maybe performing different roles to either prevent an event happening or to mitigate the resulting consequences or preventing escalation.

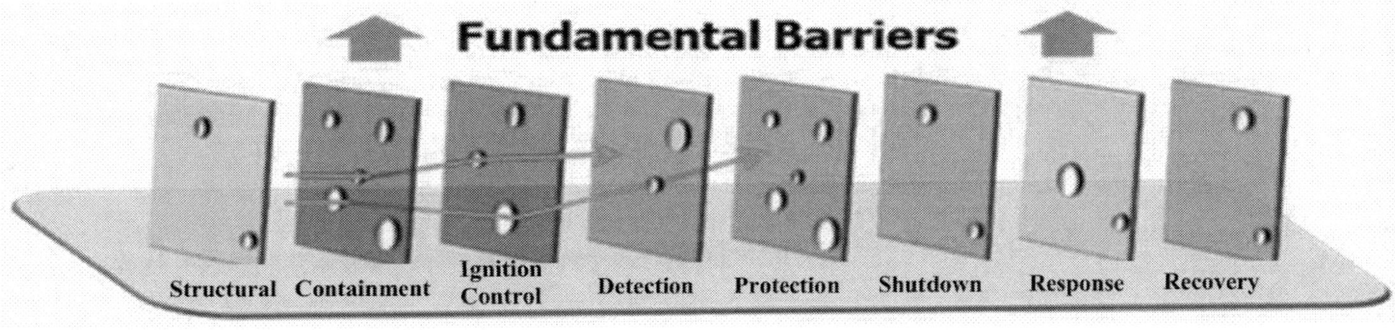

Figure 1. Fundamental Process Safety Barriers

Another example of categorizing is shown in the table below was proposed by Blacklaw, et al. (ref 3), detailing some of the specific barriers for offshore installations:

Table 1. Cumulative Risk Assessment Barrier Model

Barrier System	Example Components
Structural Integrity - (offshore structure)	Jacket, Topsides, Cranes
Process Containment	Vessels, Pipelines, Rotating Equipment, Pressure Systems, Relief Valves
Ignition Control	Ignition Prevention Systems, Flare Systems, Ventilation, Positive Pressure Systems, Gas Seals, etc.
Detection Systems	Fire & Gas Detection and Alarm Systems
Protection Systems	Fire Water Pumps, Deluge & Sprinkler Systems, Safe Refuge, Foam Systems, Blast Walls, etc.
Shutdown Systems	Pressure Trips and Shutdown Valves, ESD Systems, etc.
Emergency Response	Escape Routes, Emergency Lighting, Fire Water Mains & Portable Equipment, etc.
Life-Saving	Life Rafts, Personal Survival Equipment, etc.

3.3 Barrier Failure and Maintaining Barrier Performance

The reliability of engineered barriers will deteriorate over time, eventually to a level where they will just not work. This is simply a factor of material changes due to environmental actions such as corrosion or the rigors of service, erosion, wear, etc. To prevent or slow this deterioration, regular inspection, testing, and preventative maintenance is necessary.

As was explained in section 3.1, the reliability of engineered barrier performance determines the level of risk reduction for which credit is taken resulting in a residual risk which meets the acceptance criteria.

The data used for barrier reliability is based on manufacturer's information but assumes an ongoing schedule of inspection, maintenance, calibration and function testing to maintain that performance. What this means in practice is to take risk reduction credit for a shutdown valve, all the components of the system, from the pressure switch sending the signal to the controller, the actuator, valve stem, valve body and valve seat all have to be periodically inspected, tested, and undergo preventative maintenance. If these activities fail to occur within a specified timeframe, the_only consequence is the risk of major accident increases. Without these sustaining activities, dependability of the system declines, reducing or eliminating the ability to claim a risk reduction credit- thus the consequential risk has to increase.

This simple logic is important because there are often temptations to defer inspection and maintenance of such systems due to other operational priorities. Such deferrals are justified by arguments that it is preventative work and that there have not been reports of any failures…..so there is no reason to assume they will not still work!

The single most important set of barriers on a plant are those classed as "primary containment." Primary Containment includes pipework and vessels, plus a myriad of valves and flanges which connect them all together. It is not only the main flowlines but also includes by-pass lines, impulse lines for pressure switches, temperature gauges, sight glasses, etc. plus small bore pipework for instrument connections, bleed lines, chemical injection, etc.

The performance of these barriers is essential to keep the process fluids contained and under control. They are simpler than instrumented systems, not relying on moveable parts but instead on wall thickness and material specification plus the devices used to couple such as flanges, gaskets, bolt tensions, etc. Corrosion, both internal and external, are the most common failure mechanisms. Erosion, scale deposits, chemical damage and accidental damage will play their part. Inspection programs monitoring for losses in wall thickness are typically in place across the facility. These are supplemented with preventative programs such as protection coatings, corrosion inhibitors, anodic and cathodic protection, etc. Periodic hydraulic testing and recertification of vessels and pipework is also a standard requirement from most regulatory authorities and / or insurance companies.

3.4 Management Systems

Whether specified by the regulator or by internal policy, Management Systems are typically in place to manage the processes described in the sections above. As part of an organization's governance, the suitability of these Management Systems will be periodically reviewed, and their effectiveness tested by audit to ensure they are properly applied. Performance indicators provide a measure of the effectiveness of such systems with API RP 754 (ref. 5) being a standard most commonly used in the O&G & Petrochemical industry for process safety KPI's. In this standard, Tier 1 and 2 measures are known as "lagging indicators" based on historical events such as loss of containment. Tiers 3 and 4 are referred to as "leading indicators" since they are typically based on auditing of management systems in place to prevent or mitigate process safety events. In the UK, the Health & Safety Executive issued guidance, HSG 254, on developing process safety indicators (ref 6) a few years before API published their standard. Here they promoted the identification of MAH scenarios, barrier identification and identifying the "Risk Control System" or RCS which is managing the performance of that barrier. Leading and lagging indicators, similar to API 754 are also recommended in this guide.

While it is essential to the overall safe management of a facility that barrier / risk management systems are in place, both API 754 and HSG254, generate indicators which inform on the system alone and not on specific pieces of equipment or areas of the plant. Furthermore, what are called "leading indicators" are based on data collected months if not years in advance so may not represent what is current. What is clear is that they do not help day to day decisions on risk levels in specific areas of the plant or working on specific equipment.

4 The Reality of Operations and Dynamic Risk

4.1 Deferred Maintenance of Safety Critical Equipment

Maintenance Management Systems are programmed with time-based schedules for inspection and preventative maintenance of all equipment on a facility. Where this equipment forms one of our major accident hazard barriers, it is flagged as "safety critical." A rolling schedule is managed based on different time horizons. As equipment breaks down, repair work is added in based on priority. At two weeks or seven days, the schedule is typically "frozen" and handed over to Operations.

Executing the plan as designed is often a difficult challenge for a variety of reasons and maintenance backlogs, or maintenance activities beyond their planned completion dates, become common as activities slip on the schedule. This becomes more of a concern when the backlog contains safety critical equipment. In such instances, Operators need to justify why the maintenance is not being carried out and a "deferment" review process involving technical and engineering authorities should be conducted.

As we have seen earlier, when inspection and preventative maintenance of safety critical barriers is delayed, the inevitable outcome is that MAH risk levels increase.

Deferred safety critical maintenance is an unfortunate reality in the O&G & Petrochemical industry, especially with old facilities and infrastructure. The magnitude of this issue is difficult to judge since data on the topic is not easy to access as few Operators want to share data which shows them in a bad light; however, there is one exception.

The UK sector of the North Sea is a region where Operators have, for several years, voluntarily submitted data on the maintenance backlog for Safety Critical Equipment (SCE). Oil and Gas UK, the cross-industry body recently published this data in their annual Health & Safety Report in 2015 (ref 4). The diagram below shows how the growth in SCE backlog over the last 4-5 years. The report mentions the regulator has raised concerns the industry has increasingly deferred maintenance backlogs rather than addressing them.

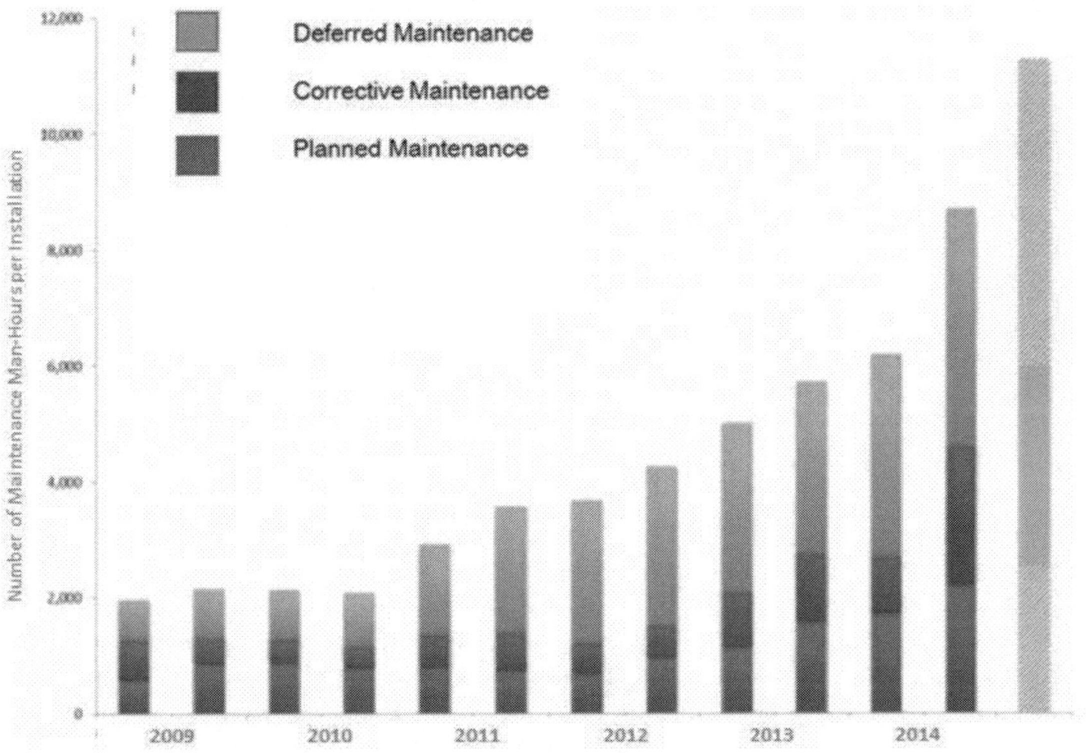

Figure 2. Oil & Gas Health & Safety Report 2015 (ref 4) – Growing Backlog of Safety Critical Maintenance

What this picture says is there are a number of barrier systems on the facility which, with the deferment of inspection and preventative maintenance, are knowingly "impaired. In other words, they are not providing the level of risk reduction that was intended by design. Add to this, the barrier systems which have been inspected but have failed to perform adequately and are therefore also in need of corrective action.

4.2 Deliberate defeat of barrier systems

Another category of compromised barriers must also be considered. Every day, many barriers on a facility are deliberately defeated. This is a necessary on a plant where ongoing maintenance and repair activities have to take place with the plant in a live state or condition. Detection systems, such as gas and fire detectors, have to be deliberately defeated to carry out work in a vicinity which could accidently trip them. Alarms and process trips have to be bypassed for work to be carried out. Ignition sources have to be brought onto the live plant in order for work to be carried out.

The terminology used to describe these situations include: "temporary defeats (of safety system), "instrument override", "interlock by-pass" but they amount to the same thing. A formal, managed process should be followed during this action, often accompanied with a risk assessment. This will often focus on the following critical parameters: **time** – the longer a defeat is in place the greater the risk; **reinstatement** – ensuring the temporary defeat does not get forgotten and becomes permanent; **communication** – ensuring all necessary parties are aware of the status.

Even our most important barrier, primary containment, has to be compromised fairly regularly in order to replace and repair equipment such as valves, pumps, filters, etc. Work involving "breaking containment" is, again, normally managed with a formal process involving energy isolation, draining and purging and gas testing.

All of these situations where safety barriers are deliberately defeated involve an additional level of risk. While the defeats maybe managed with formal process and additional barriers or controls maybe put in place, there will be a residual risk which needs to be considered along with the other sources of risk on the facility.

4.3 Operational Decisions and Corporate Silos

On a daily basis, plant Supervisors make the decision, typically informally and often unconsciously, on whether the plant is safe to operate. They also decide whether it is safe to carry out simultaneously activities - such as maintenance and repair work, project construction work and normal Operational activities. These decisions are often based on incomplete information, specifically with regard to the status of barriers on the plant.

Unfortunately it is quite common within our organizations, that data and information is stored in areas often difficult to access by anyone other than the people who created the data. While this may not be intentional, the result of such "discipline silos" is that updated reports from these discipline groups tend to be too generalized to support specific decision-making.

Given the wide distribution of data, it is not surprising that daily decisions are often made on the basis of exception, i.e. unless told otherwise and in the absence of any visible concerns the assumption is that the barriers are in good health. Visual evidence of barrier impairment cannot be relied upon other than very obvious external corrosion or physical damage to equipment and "local memory" of ongoing status rely on individuals being on shift or very good handover notes and Operations logs. A good supervisor will make great efforts to find the most current information before making decisions but as always when trying to prove a negative….how do you know when to stop looking?

The table below shows some random examples of the types of discrete risk management processes and disconnected data systems which are common in our organizations. This is a short list to demonstrate the principle. The point is that our organizations and discipline groups tend to be very insular and the data they collect is difficult to access from outside the group. This is partly because the work carried out to identify deviations or verify conformance is fragmented and generally covers the whole facility. The output reports tend to be summarized around performance indicators (examples of which are included in the table). These KPI's are useful to judge the effectiveness of the Risk Management Process (RMP) averaged over sizeable time steps (per month , per quarter or per year). They are not necessarily helpful for supporting real time decisions in specific areas of the plant. For that, decision makers need to know the status of specific barriers in particular areas of the plant to identify whether planned activity or plant changes are safe to instigate.

Table 2. Examples of Corporate "Silos" of data and information

Lead Discipline	Risk Management Process (RMP)	Data "silo" system	Performance Indicator
Maintenance	Identified barrier impairment	MMS (Maintenance Management System)	% impaired vs not impaired
Engineering	Corrosion Inspection	Inspection Systems	% systems inspected % failed
Engineering	Safe Operating Limit excursions	DCS / Data Historian	Number of SOL excursions
Instrumentation	Relief Valve testing	Testing Systems	% tests carried out
Instrumentation	Instrument Malfunction	DCS / Data Historian	% malfunctioning instruments
Engineering	Management Of Change	MOC system	% MOC's outstanding
HSE	Incident Reporting	Accident / Incident Database	Tier 1 – 4 reporting statistics
HR	Competency Management	Competency Management / Training database	% workers trained
Process Safety	HAZOP / HAZID	Risk Analysis Database	% Hazops complete
Management	Management Audits	Audit / Action Tracking System	Audit Actions complete

4.4 Cumulative Risk & Dynamic Risk

Two factors of risk management which are gaining some attention in our industry are cumulative risk and dynamic risk. Dynamic risk is perhaps the most simple concept since it is just a recognition that the risk situation is not static in any of our facilities. It changes with a variety of time dependent mechanisms. We have covered extensively the way in which barrier systems fail over time which results in a change in the consequential risk of the considered scenarios. The degree of risk elevation will depend on the length of time the barrier(s) is impaired. Other time dependent components to risk include environmental conditions particularly those at the limit of design such as hurricanes, tornadoes, floods, earthquakes, etc. Ongoing activities on the plant will potentially have an impact on risk either through introduced hazards or introducing potential casualties if an event occurred. Community expansion over time outside the perimeter fence is also a dynamic which changes the risk profile of a facility.

Built into some of our standards and regulatory practice is a 5 year review of hazard and risk scenarios across the facility. This is a suitable time frame for reviewing major equipment items and to ensure that appropriate barriers and risk control systems are in place as well as an opportunity to implement new attitudes towards risk and risk reviews. It does not, however, address the reality of day to day changes in our barriers, temporary situations and the changing hazards of a typical work day and as a result, the changes in overall risk. Instead we rely on informal processes and diligent front line supervisors to recognize when risk levels are reaching unacceptable levels.

Cumulative risk is another component of this dynamic picture which is gaining some recognition in areas, notably, offshore in the North Sea. Oil and Gas UK have recently issued in draft form guidance on the management of Cumulative risk (ref 7). In this document they define Cumulative Risk as follows:

The Cumulative risk on an installation is considered to be the total risk at a particular point in time due to a combination of:

- The risk described in the Safety Case
- Changes to the above caused by one-off operations or other operational changes
- Additional risk from deviations (from a normal state), the interactions between them and the influence of other factors on these risks

The North Sea is a mature basin area with many of the operating facilities having been first installed in the 1970's. Maintaining these facilities is a challenge and, as we saw in section 4.1, a backlog of safety critical maintenance now exists on many facilities. The regulator has a concern on how operators manage the cumulative effect of multiple "deviations", i.e. impaired barriers and how decisions are made.

There is nothing special about the North Sea and it probably represents what Deepwater Gulf of Mexico will look like in 10 to 15 years. Similarly, onshore refineries, chemical plants, storage and distribution facilities suffering the effects of ageing equipment and infrastructure will need to understand and manage the cumulative and dynamic impact of multiple impaired barriers and simultaneous operations. To achieve this successfully, we need to have a better set of tools to help decision makers, in real time, assess overall risk from a number of disparate sources and enable them to more safely and efficiently manage activities and the operation of their plant.

5 A New way to Manage Risk, Barrier Health, and Activity

5.1 New Category of Enterprise Operational Excellence Software

New software platforms are emerging under the category of Enterprise Operational Excellence Management enabling companies to make better decisions in all areas of activity and risk management. These systems are designed to manage activity with the benefit of understanding risk from a number of sources. Barrier impairment, in such systems, is easily tracked and very visible when activities are planned and executed.

These systems help manage some of the following challenges:

- They provide a unified approach to assessing operational risk and governing the status of Process Safety barriers
- They provide opportunities to produce meaningful real-time and predictive risk indicators, to manage the sequence and execution of activities.
- Real time and predictive cumulative risk modeling in time and space dimensions
- Safe execution of frontline activities across the whole planning-to-execution business process.
- Data exchange with other enterprise platforms such as Maintenance Management Systems (CMMS), Inspection databases, Process Data Historians, etc. integrating across many of the data silos in our organizations
- Visualize risk and activity in space and time
- Prioritize work scheduled based on optimizing future reduction of risk
- Easy identification of barrier impairments and management of deviations in risk control systems
- Ability to manage, in real time, all risk and activity in the one place using easy simple graphical representation of the facility.
- Management of all activities including permitted work, energy isolations (LOTO), task risk assessments / Job Safety Analysis, confined space entries, routine activities, shift handover, Operator Duties, etc.
- Efficient and safe execution of work schedule
- Real time and predictive KPI dashboards shwing risk, productivity, plan attainment, time on tools, audit compliance, etc.

The screen capture below shows an example of how graphical views of the plant can indicate where MAH barriers are impaired. Icons show the different categories of impaired barriers. Other icons show ongoing or planned activities. The view can be real-time or predictive depending on the shift date selected. The future position is based on planned activities and estimates for the duration of impacted barriers. The system predicts the overall level of risk for that time period and displays with a simple traffic signal red, orange, green pattern to indicate high, medium, or low level risks. It can also highlight the specific area on the plant where the highest risks are predicted.

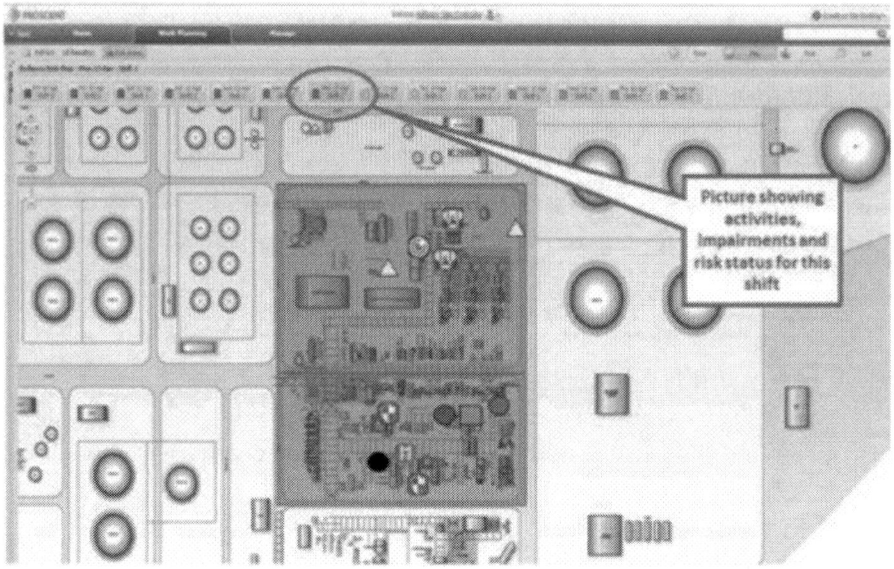

Figure 3. Graphical View of the Plant to Display Operational Risk Impairments and Deviations

Specific risk view screens for each area of the plant allow drill-down into the data to understand which activities or conditions are driving higher risks. This view is also available for current real-time or for future shift periods, with the traffic signals indicating cumulative risk for the facility during that period. This allows an operator to see the current or future state of fundamental safety barriers in each area of the plant.

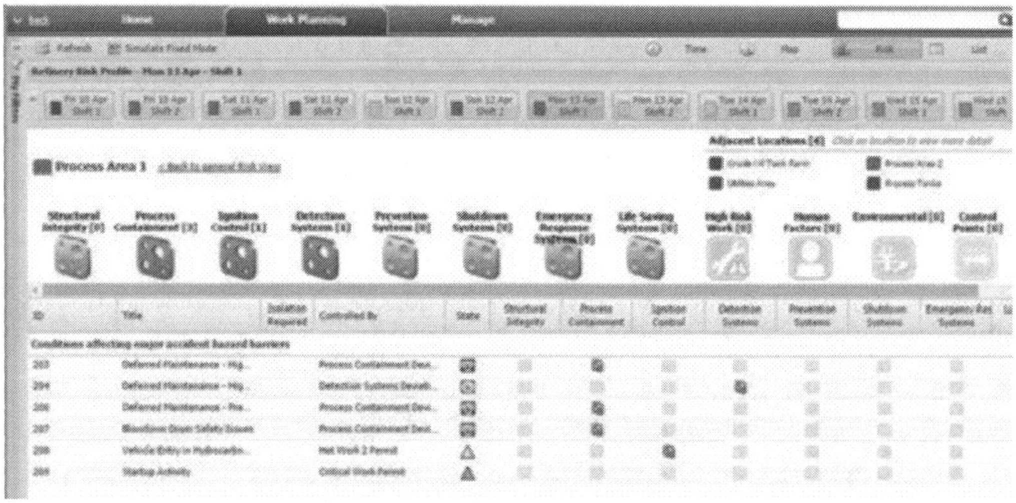

Figure 4. Risk Views Against Specific Plant Areas for Current or Future Shifts

The impact on risk is driven by specific work activities, plant statuses, or barrier conditions. Each of these is captured as a "line item" and displayed in a Gantt chart style. Much of this data is "sucked in" from the disparate silo stores and is displayed in a single place showing all future work and conditions. This screen automatically shows conflicts depending on configuration rules established. For example, if there was a containment issue in an area and an operator was trying to schedule hot work in the same area, the system would prompt a conflict and shows the activities with a red boundary. Even if it was not a conflict rule, the high risk status for the shift would alert the situation.

Simple drag and drop features allow Frontline Leadership and Planning to simply move conflicts or compounding risk events so that they do not coincide and in doing so, reduce the levels of predicted peak risks.

Digital displays help Operators identify prolonged barrier impairments which are elevating risk levels in specific areas of the plant to unacceptable levels. Simple simulate/fix functionality empowers operators to see the risk reduction impact of fixing an impaired barrier earlier than scheduled.

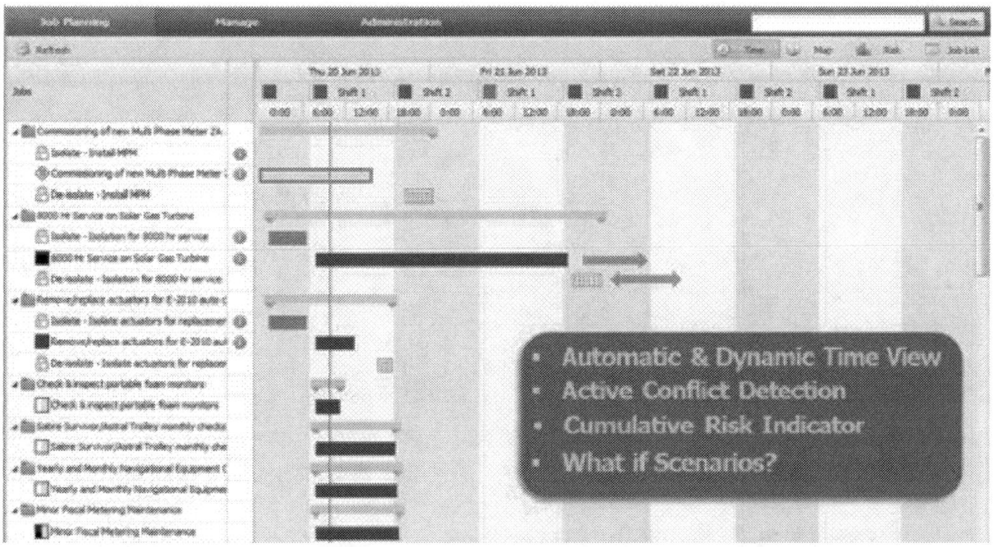

Figure 5. Time View showing Durations of all Planned Activities

A digital risk assessment engine enables Operators to evaluate the risk involved in their planned activities. At the same time, everyone across the plant can see the status of impaired barriers and assess the cumulative risk impact of their activities and those of other departments on all barriers, in all areas, at all times. This means medium-term planning and scheduling (90+ days) can be prioritized using MAH risk criteria. Rather than just blanket labeling of work priorities with the term "safety critical," the true impact on MAH risk can be assessed and activities prioritized accordingly.

With this data being collected centrally and used to assess the overall impact on risks, the quality of analysis and reporting surpasses that which has been achievable to date. The screen below show some of the types of reporting possible such as tracking overdue deviations (i.e. deviations or impairments which are not resolved within an expected timeframe), risk control systems impacts and trends and even trending of MAH risks over time.

These systems are designed for enterprise deployment so reporting can cover a whole organization of assets and identify where barrier impairments and resulting MAH risks are highest.

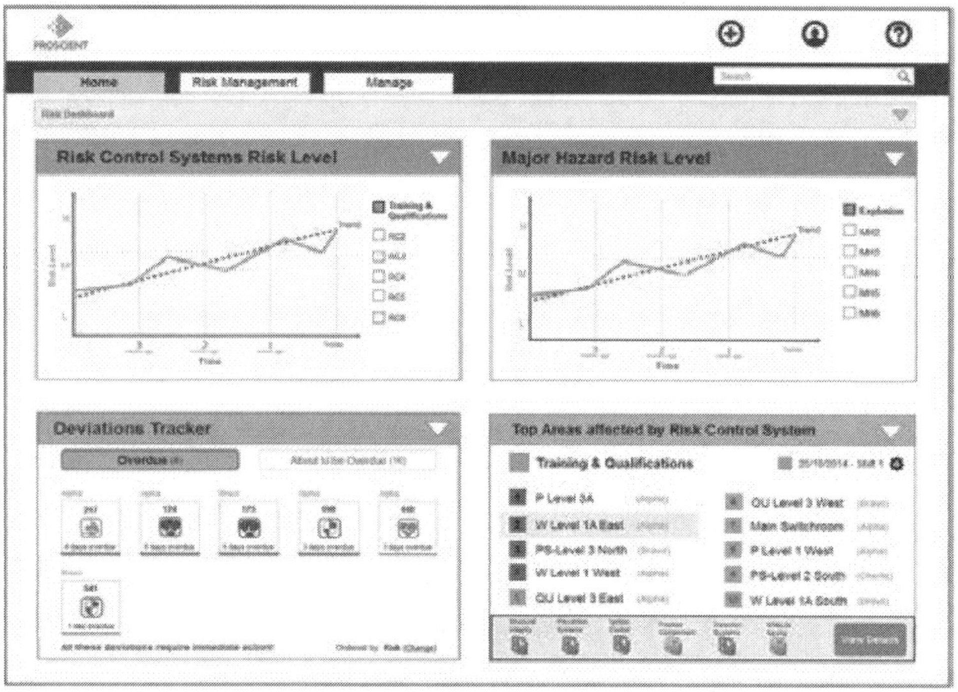

Figure 6. Examples of Dashboard Type Reports

6 Conclusions

We have seen how major accidents occur due to the failure of several of our protective / mitigative barriers. Simply put, the barriers did not perform as designed. Their reliability is dependent on continued inspection and preventative maintenance which, as we have seen, can often be deferred resulting in an increase of MAH risk. This is because the risk reduction credit for the barrier, taken during design, is no longer applicable. Another problem we face is that the data and information regarding these impaired barriers is often not easily accessible, being stored within corporate silos. Frontline Supervisors are making day to day decisions on the basis of assumed barrier status and cannot easily access information which could contradict some of these assumptions. They also need to take into account the dynamic nature of risk which can change from day to day, hour to hour and the cumulative impact of multiple impaired barriers.

Our frontline supervisors and decision makers have the responsibility to manage a safe plant so they need this information in order to make fundamental decisions such as whether the plant is safe to operate and whether the planned program of work is safe to execute. Currently they do not have easy access the relevant data in order to make these decisions.

Fortunately new technology platforms are emerging where the data on impaired barriers, temporary defeats, equipment isolations, work activities, etc. can be sourced through a single place. In this same place, activities can be planned and executed. Planned maintenance schedules can be imported and work carried out safely. Cumulative risk indicators show the aggregate impact of multiple activities and impaired barriers on risk in specific areas.

These new platforms represent an opportunity for a significant change in the way we manage risk on our facilities and should lead to safer and more efficient operations.

7 References.

[1] "The 100 Largest Losses 1974-2013 – Large Property Damage in the Hydrocarbon Industry", 23rd Edition, Marsh Energy Practice 2014.

[2] "An analysis of common causes of losses in the Oil, Gas & Petrochemical Industries" Presented at IChemE Safety and Loss Prevention Special Interest Group, Ron Garvis Swiss Re Risk Engineering Services, July 7th 2015

[3] The Cumulative Risk Assessment Barrier Model, Alan Blacklaw, Al Ward, Kevin Cassity, BG Group, SPE, 146255, 2011

[4] Health & Safety Report 2015 – Oil & Gas UK

[5] API Recommended Practice 754 - Process Safety Performance Indicators for the Refining and Petrochemical Industries, API, April 2010

[6] HSG 254: Developing Process Safety Indicators, UK Health & Safety Directorate, 2006

[7] Cumulative Risk Guidance (Issue Draft for Comment), Oil & Gas UK, February 2016

445034 Donde, Cuando y Que Enseñar y la nueva certification de seguridad de procesos del CCPS

Abstract:

El primer objetivo de este paper es el de tratar de desmistificar que la Seguridad de Procesos debe enseñarse solamente o fundamentalmente en la carrera de Ingenieria Quimica. Basandonos en el concepto de los 20 elementos propuestos por el libro del Center for Chemical Process Safety "Seguridad de Procesos Basada en el Riesgo " (RBPS por sus siglas en ingles) se puede demostrar que ciertos elementos, como integridad mecanica y planificacion de emergencias, cruza transversalmente a otras disciplinas del conocimento academico como Ingenieria Mecanica, Ingenieria Electrica, etc. y ademas, otras no pertinentes al ambito de la ingeniera, como es el caso de divulgacion a la comunidad. Se presentaran ejemplos de ciertas universidades de los Estados Unidos, Canada, y Colombia. Como segundo objetivo, y conscientes del hecho de que no existe en el mundo actual, un modelo unico y probadamente eficaz para implementar la ensenanza de la Seguridad de Procesos, los autores buscan presentar diferentes modelos y llamar a la accion a las universidades de países hispanoparlantes para desacelerar la brecha entre lo que provee la Universidad y lo que necesita la industria.

Ademas, este paper buscara explicar los requerimientos del nuevo proceso de certificación de seguridad de procesos implementado por el CCPS, que es una nueva herramienta para demostrar competencia en seguridad de procesos, que permitirá a los profesionales agregar a sus credenciales "Certified Process Safety Profesional" or CPSP.

440342 PSM Legal Update

Abstract:

Over the last year, there have been many changes to the PSM Standard. There have been two court cases explaining key terms in the PSM Standard such as "change," "process," and "process equipment" and discussing key issues such as the time to implement PHA and audit findings. In addition, OSHA has issued four memoranda expanding the boundaries of PSM by defining RAGAGEP, creating a new mixture rule for Appendix A substances, completely revising the retail exemption and defining "proximate" for the purpose of establishing the boundaries of a process. OSHA issued a Request for Information to gather information for 14 potential changes to the PSM Standard to modernize it. After reviewing all of the responses to the RFI, OSHA has just initiated the review process under the Small Business Regulatory Enforcement Fairness Act (SBREFA) in which a draft rule with the proposed changes to the PSM Standard will be published and small business will have an opportunity to comment on it. The SBRFA process will provide a framework for additional changes being developed for the PSM Standard. This presentation will provide an overview of the key changes in the PSM Standard that OSHA has implemented and a summary of the changes being considered by OSHA.

440951 Application of Computational Fluid Dynamics (CFD) for Liquefied Natural Gas (LNG) Pool Spreading and Vaporization on Water

Abstract:

A marked increase in LNG utilization is forecast with an increase in LNG marine operations. A ship-ship collision of an LNG marine tanker or a rupture of the LNG loading and unloading lines can result in accidental spillage of LNG on water. Upon release, a spreading liquid can form a pool with rapid vaporization, leading to the formation of a flammable vapor cloud. Safety analyses for the protection of the public and property involve the determination of consequences of such accidental releases. The evaluation of consequences resulting from an accidental spill of LNG on water involves the determination of the rate (vaporization rate/source term) at which flammable hydrocarbon vapor is produced and the dynamics of the spreading pool. A specific offshore scenario of rupture of an offloading line is considered resulting in an LNG spill on water, between the space enclosed by the offloading deck and the LNG ship. This scenario was studied experimentally by releasing LNG in a narrow concrete trench filled with water. The experimental study involves the determination of the vaporization rate for a continuous release of LNG in a trench. Along with the vaporization rate, the spreading rate, pool area and pool height were measured. The computational work involves development of a methodology to model the pool spreading and vaporization behavior in Computational Fluid Dynamics (CFD) for the defined scenario. The overall aim of this research is to improve the understanding of pool spreading and vaporization of LNG releases on water.

Modeling an Unconfined Liquid Spill on a Sloped Surface

Robert Deshotels
Retired
21492 Pedroso, Mission Viejo, California, USA
R.L.Deshotels@cox.net

Prepared for Presentation at
American Institute of Chemical Engineers
2016 Spring Meeting
12th Global Congress on Process Safety
Houston, Texas
April 11-23, 2016

Modeling an Unconfined Liquid Spill on a Sloped Surface

Robert Deshotels, PE
Retired
21492 Pedroso, Mission Viejo, California, USA
R.L.Deshotels@cox.net

Keywords: liquid spreading, hydraulic radius, pool evaporation, friction factor, liquid spill modeling, simulation of liquid spreading

Abstract

When determining the consequences of a vapor cloud or fire from a liquid spill, the area of the liquid pool is a key variable. If the pool is contained by walls or berms, dispersion or flame parameters can be evaluated directly from the pool area using correlations that are well known to consequence analysts. For a spill onto an unconfined paved area or sloping deck, there is no simple formula for the pool area. For an unconfined pool, area must be calculated from flow modeling. The progressive movement and size of a liquid spill onto a sloped surface can be accurately modeled with well-known fluid equations using an empirical friction factor. The fluid flow formulas for piping can be applied to a sloping flat spill surface by including a correction for the equivalent hydraulic radius, the ratio of flow area to wetted area of the frictional surface. For uncomplicated flow surfaces, spreadsheet calculation is sufficient. Example calculations point out the need for improved knowledge of friction factors in terms of quantitative surface characteristics, and an improved formula for heat transfer during film boiling of cryogenic liquids.

1 Overview

The surface area of a spill of a toxic or flammable liquid spill is crucial to the analysis of hazardous consequences. The evaporation rate from a pool of volatile liquid, and hence the dispersion rate, depends on liquid properties, meteorological conditions, and the surface area of the pool. In the case of a pool fire, the heat flux radiating to nearby structures or people depends on the height and width of the flame, which depend on the size and shape of the pool. Usually, a liquid spill is confined by berms or curbs, so that the worst-case spill area is known from confinement dimensions. For an unconfined spill, there isn't any obvious maximum pool area, even if the source of the spill has a limited volume, because the pool will tend to continue to spread as time progresses. Therefore, unconfined spills require careful analysis to determine whether the spill can affect sensitive areas or lead to accident escalation. As well as environmentally sensitive areas, sensitive areas include escape paths, structures used for the shelter of employees or the public, and areas where fire or explosion could lead to escalation of the consequences.

Equations for the rate of spread of a liquid pool have been published in standard references on hazard evaluation [Ref. 1, 2], but friction loss was not considered in these reference guides. Friction loss is a major factor in the speed of liquid flow, especially for a thin pool depth. For a fixed-volume release, frictional loss can be the most important factor in whether a pool spreads into sensitive regions.

Much of the recent research on liquid spreading across a surface involves the effects of capillary action and surface tension on the motion of a thin film [Ref. 3 - 5]. There is much to be learned about the interaction between spilled liquid and the surface below in terms of capillary action and the hydrophilic/hydrophobic nature of the surface. However, the complexity surrounding surface tension applies only for very thin films of liquid. It will be shown that very thin films are not usually a significant part of industrial-sized releases.

For liquid spills that present a design challenge to industrial-size facilities, equations that are commonly used for design of piping and open channels, such as the Darcy equation, can be adapted to the purpose of modeling the time-dependent flow of an unconfined spill [Ref. 6 - 8].

For volatile liquids, the open channel flow equations give relatively accurate estimates for dispersion consequences. However, the concern may be that the spill will flow into a sensitive area. For non-volatile pools, the following question will be answered: "At what point is the spill thickness too low to produce a significant hazard?"

With a methodology for evaluating the minimum significant pool thickness in place, the basic steps for modeling an unconfined spill are:

1. Determine an appropriate release scenario.
2. Evaluate flow parameters for a generalized frictional flow formula.
3. Estimate the endpoint minimum significant pool thickness.
4. For sequential time increments, calculate pool dimensions, taking evaporation into account.

At each stage of modeling, uncertainties in the models and parameters should be considered to ensure that conservative choices are made. Uncertainty is especially significant when postulating a release scenario, estimating temperature for the liquid and surroundings, and appropriate modeling for heat transfer and evaporation. The largest pool area and evaporation rate produce the worst toxic or flammable release to the atmosphere. When the concern is distance of pool travel, minimum evaporation will create a worse case. There may be several potential worst case scenarios that must be quantified to determine which one is actually the worst.

Two example cases of unconfined liquid spills are provided in this article:
Example A. A leaking tanker truck located on a large paved area;
Example B. A piping leak onto the sloped deck of a floating oil production platform.

2 Release scenario

Generally, the source of the liquid spill will be a vessel or piping system with many potential leak points and spill trajectories. One possible way to address the multitude of possible release scenarios is to choose a worst credible case. If the worst case results in tolerable risk and acceptable mitigation expense, there is no reason to evaluate any less conservative cases. Sometimes, a release scenario may be so conservative as to be speculative. For example, if a release is conjectured to be the total contents of a tank due to accidental impact, there may not be any good basis for determining the speed at which the tank breaks, the size of the break, or the splashing that would occur upon impingement against nearby objects. In this case, simple assumptions should be made, and there should be a discussion of the sensitivity to uncertainties.

Any individual spill case may require a different approach to conservatism, but in general, the most conservative release scenario involves the greatest volume and flowrate. Concerning scenario parameters, pool spread distance is worse when the slope is high, and evaporation is low. Higher liquid temperature decreases viscosity and increases evaporation. When the spill poses multiple hazards or affects multiple possible targets, several potential worst-cases may have to be run to find the actual worst. When risk acceptability or mitigation cost are sensitive to the choice of release scenario, quantitative risk assessment tools such as event tree analysis should be used to pick the appropriate set of release cases. Data from offshore platform leaks [Ref. 9] can be used to estimate the probability of specific leak sizes from common process piping and equipment items.

Standard fluid dynamics methods [Ref. 7, 10] can be used to calculate the flow for a release of a given opening size and specified liquid conditions. The volumetric flow rate through an orifice, q, can be expressed in terms of the following parameters (in order of mention) flow coefficient, flow area of the orifice, gravitational acceleration, and differential pressure across the orifice, expressed in height of liquid:

$$q = C\,A\,(2g\,h)^{1/2}$$

The value of the flow coefficient, C, is a function of the Reynolds number (discussed in the section "Formulas for Flow of a Liquid Pool") and the ratio of the orifice diameter to the diameter of the pipe or vessel that is leaking. Plots of C are readily available [Ref. 7].

For liquids that flash upon release to atmospheric pressure, classical thermodynamics theory [Ref. 10] will accurately predict the vapor-liquid split and liquid properties after flashing. The mass fraction of the leaking fluid that vaporizes after release, f_v, can be expressed in terms of the exponential of the following parameters (in order of mention) liquid specific heat capacity, initial temperature of the liquid, atmospheric boiling point, and latent heat of vaporization. The heat capacity and latent heat should be the mean values over the temperature range T_i to T_b.

$$f_v = 1 - exp\{-C_p(T_i - T_b) / \Delta H_v\}$$

Depending on release conditions and liquid physical properties, the liquid release may reach the pool surface as a steady stream or a spray of droplets. If the release is in the form of an undispersed stream, the trajectory can be evaluated using classical Newtonian physics, including the effect of wind pressure on the stream. In some cases, the release will immediately begin breaking into droplets, and in all cases a liquid stream will disperse into droplets if it falls enough distance.

Empirical formulas have been developed to predict the distance that a liquid release will travel before breaking-up into droplets. Droplet formation is characterized by the dimensionless Weber number, which is the ratio of shear force on the liquid surface, $\frac{1}{2}\rho_g v_d^2$, to surface tension pressure on the drop, σ/d. The Weber number is defined in terms of the following parameters (in order of mention) gas density, velocity of the drop relative to the air, drop diameter, and surface tension :

$$W_e = (\frac{1}{2}\rho_g v_d^2 d) / \sigma$$

For subcooled liquids, the length between jet release and breakup into droplets, L_b, can be expressed in terms of the following parameters (in order of mention) jet diameter at the narrowest point, Weber number, and Reynolds number [Ref. 11]. This formula applies to leaks large enough to cause a serious hazard, but not to a 2 mm diameter hole or less. The units of measure must be consistent with the dimensionless Weber number.

$$L_b = 540\ D\ W_e^{1/2}\ R_e^{-5/8}$$

For superheated liquids [Ref. 11],

$$L_b = 2.75 \times 10^{10}\ D\ W_e^{1/2}\ R_e^{-2}$$

Based on experimental results for liquids without superheat, the typical droplet size, d, can be expressed in terms of the following parameters (in order of mention) surface tension, gas density, and jet velocity relative to the wind. The drop size depends on the liquid, and could be double the typical value [Ref. 1].

$$d = 24\ \sigma / (\rho_g v_d^2)$$

For superheated liquids, the droplet size depends on the difference between the liquid temperature and the atmospheric boiling point [Ref. 1],

$$d = 3 \times 10^4 / (10 + 4.0\ \Delta T_{sh})$$

The terminal velocity of a droplet in air is a complicated function of the liquid density and drop diameter, which can be read from a plot in a standard reference [Ref. 12].

Comparison of the terminal velocity to the wind speed will show whether droplets will rain out onto the pool or drift away like a cloud. The direction of release and Newtonian physics determine the trajectory that the droplets will follow from the liquid breakup point to the point where the pool begins growing. Different wind conditions may have to be considered. The probability of the combination of accident scenario, wind and temperature should be consistent with a tolerable risk for the accident.

3 Liquid pool growth and shrinkage

The pool volume will increase as long as the liquid release rate into the pool exceeds the evaporation rate. The release rate changes with time, depending on source inventory, pressure and effectiveness of process shutdown controls. The evaporation rate depends on pool area and changing temperatures.

It is possible to determine a worst-case maximum pool area using the initial leak rate and assuming a minimum evaporation rate. In this case, the maximum pool area equals the maximum pool volume created over the interval of concern (e.g., maximum escape and rescue time) divided by the minimum significant pool thickness. If the hazards evaluated from this bounding case give a tolerable risk, there may be no need to actually model the growth and configuration of the pool. However, even if the worst case results in tolerable risk, dynamic modeling of the pool may provide insight that would be valuable for planning and training.

Dynamic modeling consists of stepwise calculation of pool characteristics at small increments of time. At each increment, the following variables are computed:

1. Based on process conditions of pressure, temperature and composition, determine the flow properties and vapor pressure of the liquid just upstream of the release point. Using classical thermodynamics, determine the fraction of vapor flash after release into the atmosphere.

2. For the release scenario chosen, calculate the liquid flow rate into the pool using the orifice flow equation that is appropriate to the scenario. Calculate the liquid trajectory and dispersion from the jet breakup formula that applies to the orifice size, Weber number, and flashing superheat. It can be assumed that the liquid volume will be in a conical shape, with the target area being circular on a flat substrate, or elliptical on a sloped substrate. While a liquid release is in the momentum-driven region of flow (about 6 to 100 orifice diameters from the release point), a liquid jet will diverge at an angle of 10° from the centerline [Ref. 6]. From the flowrate, trajectory and divergence angle of the release, the target area of the liquid added to the pool is determined, as well as the volume increase in the target area during the time interval.

3. From a predicted Reynolds number, or the Reynolds number from the previous time increment, the applicable flow equation for pool spread is chosen. The volume flow and velocity at the edge of the pool can then be calculated, from which pool edge growth can be calculated. The equations for flow of the pool are given in the section "Formulas for Flow of a Liquid Pool".

4. A conservative calculation of pool evaporation is done, based on the new pool area calculated in the previous step. Depending on the purpose of the modeling, the evaporation calculation should be either a low estimate or a high estimate. An exact calculation is generally not possible because of uncertainty about rates of diffusion and heat transfer. An approach to providing a conservative estimate of evaporation is provided in the section "Evaporation". Evaporation may result in a cooling effect on the pool. The evaporation calculation is expressed as a uniform decrease in pool thickness, which will result in an adjustment in the pool surface area.

5. Knowing the inventory volume and process conditions, new process conditions are calculated from the liquid loss through leakage flowrate calculated in step 2 above. As the process pressure decreases, any evaporative cooling inside the process containment should be considered in the revised process conditions and composition.

Considering the accuracy of the postulated release scenario and air dispersion models, a simple model for pool thickness should suffice. It may be reasonable to model the pool volume as a cone with an elliptical base (because of the slope of the substrate), with the peak of the cone located at the target point of the liquid release trajectory.

4 Minimum significant pool thickness

The driving force for pool flow is the difference in height between different parts of the pool. For a pool moving down a sloped surface, the uphill part will become thinner as the pool spreads downhill. Furthermore, it is common for the leak rate from the process to stop or diminish after a while, causing the rate of evaporation rate from the pool to exceed the fill rate. At some point, part or all of the pool may become too thin to pose a significant hazard. To see whether this is possible in a specific case, it helps to categorize the different hazards posed by a pool, as follows.

Pool liquid that has not evaporated can pose a hazard of toxic exposure by direct contact, or a fire hazard.

For liquids that are not flammable, but are combustible upon ignition by flame, the pool may be too thin to maintain combustion. This happens when heat transfer to the substrate will quench any fire that starts by cooling the liquid below the flash point. This thickness could be calculated, but instead of solving unsteady-state heat transfer calculations using uncertain heat transfer coefficients and uncertain burning rates, the minimum fire hazard thickness for a non-volatile liquid should be determined by a field test. For one rough example, in a backyard test in a one-foot square confined pool of kerosene on concrete sidewalk at 20°C in breezy air, it was found that the minimum liquid thickness that would maintain a flame was between 2 and 3mm.

Where a liquid is highly toxic to people or the environment (e.g., aqueous ammonia) OSHA Standard 1910.119 has established a threshold quantity. A pool below this quantity no longer poses a catastrophic level of hazard, but mitigation is still imperative. Consequently, there is no insignificant pool quantity or thickness in the case of toxic liquids.

Vapor that is evaporating from a pool can create a toxic or flammable cloud leading to toxic exposure, flash fire, or explosion. The severity of exposure of a person to a toxic cloud is measured as the integrated dose of concentration over time. Therefore, for toxics, the quantity of material in the vapor cloud is more important than the peak rate of evaporation. As a result, the thickness of a pool does not affect the severity of a toxic release much, especially for a thin film that will evaporate soon. For a flash fire or explosion, severity is measured as heat flux or blast pressure, which depend on the size and concentration of the flammable cloud. Thus, for a spill of volatile liquid, the pool thickness just prior to complete evaporation does not have much impact on accident consequence.

When there is a concern that a combustible spilled liquid will flow into a sensitive area, the pool can be considered non-hazardous, provided there is no continuous source of ignition, if the entire pool depth is less than the minimum fire hazard thickness, determined from field testing of flame propagation.

5 Formulas for flow of a liquid pool

All practical formulas for the flow of liquids are derived from Bernoulli's equation, an energy balance that includes all of the involved forms of energy: potential energy (elevation difference), pressure, velocity, and frictional loss. The Bernoulli equation for the energy change for a fluid flowing from point 1 to point 2 can be expressed in terms of the following parameters (in order of mention) elevation, pressure, density, velocity, gravitational acceleration, and frictional loss expressed as height of fluid [Ref. 6, 7]. Consistent units of measure must be used. With U.S. customary units of measure, the gravitational acceleration factor of 32.2 ft/s^2 is needed to convert between pounds force and pounds mass.

$$Z_1 + P_1/\rho_1 + v_1^2/2g = Z_2 + P_2/\rho_2 + v_2^2/2g + h_L$$

Because the entire pool is at atmospheric pressure, there is no pressure energy term, P/ρ, to consider for a pool. The change in the elevation term Z is the only driving energy for pool flow, unless the release into the pool has a significant velocity due to the release trajectory at the target point. Further explanation follows for the velocity head term and friction head term.

Flow of a liquid pool on a sloped surface has two components. The first component of pool flow is the downhill motion of the entire pool. The driving force for downhill motion is the gravitational force multiplied by the slope. To determine downhill motion, the downward velocity due to the slope must be added to any velocity head imparted by the trajectory of the release flow into the pool, and the friction head must be subtracted. The downhill momentum increases as time continues. The second component of pool flow is the spreading of the pool outward from the center of mass, which can be modeled from the difference in height between the liquid and the surrounding surface.

Unlike flow through a pipe or in an open channel, liquid spreading on a sloped surface does not have a constant flow depth or cross-sectional flow area. Consequently, velocity varies through the pool. Therefore the value of the velocity used in the velocity head term, $v^2/2g$, and friction head term, h_L, must be considered carefully. For a pool modeled as a cone with a round or elliptical base, spreading begins at the high point as gravity-induced downward flow, then turns outward to be parallel with the base at the outer edge. To determine pool growth, the velocity head due to the elevation difference must be decreased by the friction head to yield the effective elevation change from the high point to the pool edge. From the effective elevation change, a new pool radius can be calculated for a chosen time differential. The new height can be calculated from the mass balance of the pool.

When the substrate is sloped or there is a velocity head imparted by the release stream, the growth in the pool can be evaluated by considering the cone to have four quadrants: uphill, downhill, and two equal sideways quadrants. Each quadrant is broken into differential elements at intervals of radial length along the distance from center of the ellipse outward to a point on the edge in order to evaluate a velocity that applies to each differential element of the quadrant. This allows a calculation of the friction head contribution of each element, from which the radius increase for each quadrant is calculated, and the new pool height determined as discussed in the previous paragraph.

There are different liquid flow formulas (e.g., Darcy, Fanning, Manning) that can be used to estimate the empirical term for frictional loss, h_L.

Regardless of the formula used for friction loss, there are four regimes of flow that can apply to an unconfined liquid spill:

1. Frictionless flow
2. Turbulent flow
3. Laminar flow
4. Thin film flow

5.1 frictionless flow

When the flow path has a low elevation gradient and a high pool depth, frictional loss may be insignificant. Also, it is possible for frictional loss to be very low in the case where the pool is heated by film boiling at the substrate interface [Ref. 4]. Otherwise, in actual spill cases that pose a serious risk to a facility, frictional loss is significant. When significant frictional loss is ignored, the simplified formulas for liquid flow rate [Ref. 1, 2] will overestimate the speed of liquid travel. Because friction loss is not difficult to evaluate, there seems to be little incentive to ignore friction.

5.2 turbulent flow

Flow turbulence is characterized by the Reynolds number. The Reynolds number is dimensionless, and can be expressed in any consistent units of measure, in terms of the following parameters (in order of mention) pipe diameter, velocity, density, and absolute viscosity.

$$\text{For pipe, } R_e = D\, v\, \rho / \mu$$

Commonly used formulas for liquid flow through a pipe or channel [Ref. 6 - 8] express frictional loss in terms of a friction factor, f, which is determined from experiments. However, formulas for pipe and channels cannot be used directly; the formulas must be modified to account for the different geometry of flow.

In an unconfined liquid spill, the flow often transitions from turbulent flow to laminar flow as the pool spreads thinner. Although the Reynolds number can be calculated exactly, the transition between the turbulent and laminar flow regimes is not consistent from one case to another. The transition is usually gradual, and can depend on flow disturbances and flow history. Turbulent flow is ordinarily exhibited when the Reynolds number is above 4000, with laminar flow exhibited below 2100. Laminar flow has less friction loss than turbulent, resulting in a higher velocity for a given slope. However, laminar flow results in a lower heat transfer rate. In the transition range between flow regimes, the flow should be modeled conservatively as either fully turbulent or laminar. The conservative choice will depend on the situation.

Unconfined liquid flow on a horizontal or sloped surface is analogous to gravity-induced flow in a pipe, except that wall friction only occurs along the bottom of the spill. The concept of hydraulic radius, R_H, [Ref. 6, 7] must be applied.

$$R_H = \text{cross sectional flow area /wetted perimeter}$$

For flow through a round pipe full of fluid, the flow diameter equals $4R_H$. For unconfined pool flow, the hydraulic radius is the pool height, z, at the point where velocity is measured, and the equivalent flow diameter equals $4z$. The Reynolds number for a round pipe full of liquid is expressed in terms of the inside pipe diameter. For full flow through a narrow slot, the Reynolds number is expressed in terms of the height of the slot. For open flow across a flat surface, we typically assume that there is no friction at the top surface of the liquid or the edges, although this is not strictly true due to relative motion between the air and liquid surface. With negligible air friction, the zero-shear level is at the top of the liquid pool, rather than at the centerline as in a round or rectangular pipe. Therefore, the equivalent hydraulic radius used to evaluate the Reynolds number for a round liquid pool, with a flow area (circumference multiplied by height) of $2\pi\ r\ z$ and a perimeter of $2\pi\ r$, should be corrected to twice the usual value, or $2z$. In the Reynolds number and friction head equations, the D used for full-flow round pipe should be replaced by $8z$ for a pool.

$$\text{For a pool, } R_e = 8z\ v\ \rho\ /\mu$$

Wind friction can affect pool motion, but the effect is expected to be minor. Weather stations report wind speed at a reference height of 10 m above grade. If the spill is at grade, the wind is reduced to about 8% of the reference value in the daytime, or 5.5% at night (the difference is due to air stability) [Ref. 4]. It should be noted that wind wave height cannot exceed the pool thickness; therefore, the surface of the liquid will be smooth, even under the influence of wind. Wind drag can be evaluated from low-speed aerodynamics. The drag pressure caused by air friction is the drag coefficient multiplied by the wind velocity head. For wind blowing parallel to a flat surface the drag coefficient, C_D, is 0.04 [Ref. 13]. To convert into liquid head, the wind head is multiplied by the ratio of gas density to liquid density. If the wind is in the opposite direction of flow, the wind head is added to the friction head. If the wind is in the same direction as flow, the wind head is subtracted from the friction head. The wind head, h_{drag}, can be written in terms of the following parameters (in order of mention) drag coefficient, air density, air velocity, and liquid density.

$$h_{drag} = C_D\ \rho_G\ v_G^2\ /2\rho_L$$

Table 1

Formulas for Turbulent Flow Friction Loss

Darcy (Moody)	$h_L = f\ L\ v^2 /2D$
Fanning	$h_L = 2f\ L\ v^2 /D$
Manning	$h_L = f\ L\ v^2 /2R_H$

Note that the Moody friction factor is four times larger than the Fanning friction factor. This can be confusing because plots of friction factor do not always mention which friction factor is given. For the Darcy and Fanning equations, the friction factor is a function of the Reynolds number and relative roughness of the pipe wall, ϵ/D. For a pool having a height of z, D should be replaced by $4R_H$, or 8z. Roughness is determined by experimentation, rather than surface measurement. Based on the limited data available and the difficulty in defining roughness, the friction factor is the major cause of uncertainty in flow rate calculations. The Manning equation is an empirical equation developed from data for water flow in deep open channels. For the Manning equation, a roughness parameter n is used to correlate the friction factor to the hydraulic radius [Ref. 6, 8].

$$\text{Manning } f = 29n^2 /R_H^{1/3}$$

Table 2

Roughness Factors for Turbulent Flow

Surface	**ϵ** ft	**n** $ft^{1/6}$
Clean commercial steel	0.00015	
Galvanized iron	0.0004	0.012
Cast iron	0.00085	0.011
Concrete pipe	0.001 - 0.01	0.015
Riveted steel	0.003 - 0.03	0.014
Earth, straight and uniform		0.019
Brick open channel		0.017
Concrete open channel		0.022
Gravel open channel		0.033

5.3 laminar flow

Unlike turbulent flow, fully laminar flow along an unobstructed path exhibits no eddies in the axial or radial direction. Friction loss is due to viscous effects alone, and the friction factor is a function of Reynolds number alone, not surface roughness.

Darcy laminar $f = 64/R_e$

Fanning laminar $f = 16/R_e$

The value of f for laminar flow is substituted into the Darcy or Fanning equation for friction head. However, the laminar friction factor was developed for round pipe containing full flow, and will be somewhat in error for open flow on a flat surface, due to the parabolic velocity profile that exists between the pipe wall and zero-shear point in laminar flow [Ref. 6]. Although this paper proposes a correction in the hydraulic radius for the location of the zero-shear point, there is no published data to verify that this proposed correction is accurate or conservative.

Like turbulent flow, as long as there is a height difference between the top of the liquid pool and the surface of the substrate, the Bernoulli equation and derived Darcy formula tell us that the pool will continue to spread without limit, although at an ever decreasing rate. For turbulent or laminar flow, the Darcy formula does not consider the effects of surface tension, capillary action, or hydrophilic/hydrophobic nature of the surface under the spill.

5.4 thin film flow

Recently, investigators have been able to explain why liquid pools do not actually continue to spread endlessly [Ref. 5]. When the liquid height is less than around 100 nm (0.01 cm), long-range intermolecular forces (like van der Waals forces) begin affecting the ability of the liquid to wet the surface of the substrate. There is a minimum thickness of liquid film, beyond which the film no longer wets the substrate; instead, the film breaks into individual droplets. In the case of a level substrate, flow stops. The minimum film thickness is affected by substrate surface properties such as roughness, cleanliness, porosity, and surface tension between the solid and liquid. There are many cases for which the proposed model has not been tested (including a sloped surface), but two conclusions can be drawn:

1. For thin pools, the laminar flow model gives a higher spreading rate than any model that considers the effect of surface tension.

The thin film flow model does not apply for liquid pools thicker than about 0.01 cm.

6 Evaporation

Evaporation will occur if the liquid has a significant vapor pressure at the pool temperature, and the atmosphere above the pool is not saturated with the liquid vapor. The temperature of the pool can change because of heat transfer with the substrate or air, which would affect the vapor pressure.

Evaporation rate from the pool is described in terms of classical thermodynamics [Ref. 10]. The driving force for evaporation is the difference between the vapor pressure of the liquid and the partial pressure of the liquid in the air. The mass evaporation rate, W, can be written in terms of the following parameters (in order of mention) molecular weight, mass transfer coefficient, flow area, vapor pressure of the pure liquid, partial pressure of the vapor in the bulk stagnant air above the liquid, gas law constant, and temperature of the liquid. Pressure and temperature are in absolute units.

$$W = M\,K\,A\,(P_{sat} - P_v)\,/(R\,T)$$

At atmospheric pressure and practical operating temperatures, ideal gas can be assumed. Except in the case of a boiling liquid, the rate of evaporation is restricted by the rate of diffusion from the surface of the liquid. Diffusion coefficients have been measured or estimated for common chemicals [Ref. 14]. The resistance to mass transfer from a homogenous liquid to air is mainly due to the resistance of the air. The resistance of the air depends on the transport properties of the air (expressed as the Schmidt number) and flow regime (expressed as the Reynolds number). There are many theoretical and empirical correlations for evaporation, depending on the geometry of the surroundings, flow regime, concentration of the evaporating component in each phase. A great deal of judgement is required in choosing an appropriate correlation. Key issues involve friction modeling for a boiling pool, wind effects, and heat transfer from the substrate [Ref. 5]. It seems that nearly all of the correlations were developed for equipment design and were tested in laboratories, not outdoors in industrial-sized facilities where nearby objects can obstruct or funnel the wind. Therefore, correlation accuracies claimed in literature may not apply to evaporation from a real pool [Ref. 5]. An empirical correlation, rather than theoretical-based, should be applied when the flow is parallel to the interface through which mass transfer occurs (such as pool evaporation) [Ref. 14]. For a calculation that overestimates evaporation, one can assume that there is no resistance to mass flow within the liquid or across the liquid-air interface, which results in an air concentration just above the interface that is equal to the partial pressure of the liquid relative to atmospheric. When the resistance to diffusion is essentially due to air convection alone, the diffusivity is in the range of 10^{-4} to 10^{-6} m^2/s. A simple but crude estimate of diffusivity, K, based on molecular weight, M, may be sufficient [Ref. 10], by assuming the ratio between the diffusivity and cube root of molecular weight is constant for the liquid of interest and some reference liquid with known diffusivity, K_0.

$$K = K_0\,(M_0/M)^{1/3}$$

An evaporation correlation specifically for a pool of volatile liquid has been developed, accounting for the wind speed and stability class of the air [Ref. 2]. This formula does not consider evaporative cooling of the pool. The evaporation rate, W, involves two parameters, λ and n, which depend on the Pasquill-Gifford air stability classification. The equation is written in SI units in terms of the following parameters (in order of mention) λ, wind velocity, vapor density, vapor pressure, ambient pressure, wind velocity squared, pool diameter, and n.

$$W = \lambda\,v_w\,\rho_v\,P_v\,/P\,\{2/(v_w^2\,D\}^n$$

Table 3

Pool Evaporation Parameters

Pasquill Stability Class	λ	n
A	1.0×10^{-3}	0.078
B	1.2×10^{-3}	0.091
C	1.5×10^{-3}	0.111
D	1.7×10^{-3}	0.130
E	1.8×10^{-3}	0.149
F	1.8×10^{-3}	0.180

To calculate a high estimate of evaporation for non-cryogenic spills, assume the spill occurs on a windy, hot summer day; for a low estimate assume a still, cold winter night.

A separate formula applies to a spill of a cryogenic liquid on land, explicit as a function of time [Ref. 2]. This equation is written in SI units in terms of the following parameters (in order of mention) a surface correction factor, thermal conductivity of the substrate, heat of vaporization, thermal diffusivity of the substrate, substrate temperature, liquid temperature, and time. The surface correction factor, *X*, is assigned a value of 1 for a smooth substrate, or 3 for a rough substrate such as gravel. Note that this equation crudely assumes that heat transfer is proportional to the temperature difference between the substrate and liquid; this assumption does not apply at all beyond the point of maximum nucleate boiling, when partial or full film boiling occurs.

$$W_{cryo} = 2X\, k\, /\Delta H_v\, (\pi\, \alpha)^{-1/2}\, (T_S - T_L)\, t^{1/2}$$

If the pool is boiling, the evaporation rate is usually limited by heat input. If the liquid flashes upon release from the process, the liquid and vapor reach equilibrium soon after release, usually before the liquid touches the ground. The liquid will not continue to boil unless heat is added from the air or substrate. The main uncertainty involves heat transfer from the substrate under the pool. Different modes of heat transfer are possible, nucleate boiling, film boiling or transition [Ref. 15, 16]. The transition from nucleate boiling to film boiling occurs at a critical temperature difference between the substrate and liquid, which depends on liquid properties. Heat transfer rates for boiling LNG and NPG have been documented from field experiments with liquefied gas inside level confinement [Ref. 16]. In the case of film boiling, which often occurs in cryogens (especially LNG), the friction factor at the substrate interface is markedly decreased. Equations for the friction factor during film boiling of LNG spilled over a water surface have been published [Ref. 4]. If the pool is supported by the ground, transfer through the substrate depends on several factors including moisture content. One bounding option that overestimates evaporation is to assume that the substrate temperature remains constant. A bounding assumption that underestimates evaporation is that for fully developed film boiling, the heat flux equals the minimum for film boiling for that liquid.

Several references show an "S" curve for typical boiling liquid heat flux as a function of temperature difference between the substrate and liquid. The typical curve shows a low flux at low temperature difference, rising like the front end of a sine wave to reach a maximum flux at the point of fully developed nucleate boiling. When the temperature difference increases beyond this critical point, the heat flux decreases like the back end of a sine wave, reaching a minimum flux point caused by a layer of vapor that separates the substrate and liquid. Beyond the minimum point, the "typical" flux curve curves up again at a slightly lower slope than the initial curve for nucleate boiling. This increase is attributed to the increase in radiative heat transfer to the liquid from the hotter substrate. The data that supports this "S" curve is mainly drawn from data from experiments with silicone oil, which boils at a high temperature. In the typical experiment, the substrate is heated beyond temperatures where radiant heat transfer is significant. This data may apply to film boiling of cryogenic liquids inside pressure vessels, but an atmospheric spill of LNG is different. If the increase in film boiling heat transfer is due to radiative transfer, it should not be expressed as a linear relationship with temperature. Radiative heat transfer is a function of absolute temperature to the fourth power. The equation for transfer between parallel plates, or a flat substrate to a liquid suspended by a flat vapor layer is [Ref. 15]:

$$Q/A = \epsilon \sigma T_1^4 - \alpha \sigma T_2^4$$

where Q/A is heat flux, ϵ is emissivity, σ is the Stefan-Boltzmann constant, T_1 is the absolute temperature of the substrate, α is the absorptivity of the liquid, and T_2 is the absolute temperature of the boiling liquid. Units of measure must be consistent with the Stefan-Boltzmann constant, which is 1.713 $\times 10^{-9}$ Btu/(ft^2 h °R^4) or 5.67 $\times 10^{-8}$ W/(m^2 °K^4). The emissivity and absorptivity of common materials is given in standard references [Ref. 15]. It would seem that the radiative effect on film boiling heat flux could be calculated, but there has been no experimental data to validate this conjecture.

7 Conclusions

Use of the Bernoulli equation for liquid flow provides a method for calculating the spread of an unconfined liquid pool on a sloping surface. The Bernoulli equation is an expression of the fluid flow energy balance, so there is no need for the user to characterize the flow as gravity-driven or momentum-driven in order to choose an appropriate model. Friction loss can be calculated the same as for piping or open channel flow after flow geometry is accounted for. Although flow can be calculated exactly if all parameters are known, the surface roughness parameter can be uncertain by an order of magnitude. The most important variable, release flow rate into the pool is generally determined by accident conditions that are difficult to predict. Furthermore, the evaporation rate from the pool can be affected by heat transfer rates and mass diffusion coefficients that are somewhat difficult to predict outside of a laboratory. Therefore, care should be taken to assure conservative assumptions, methods and data.

The calculation accuracy of liquid flow across a sloping surface is limited by roughness data and heat transfer knowledge. Within this current limitation, one-dimensional spreadsheet calculations can provide useful bounding results. However, when accurate data has been developed, computational fluid dynamics will provide more accurate results.

The non-skid checkered deck plating used on the decks of land-based processing facilities and floating production platforms has well known surface dimensions, but no documented friction factor for flow calculations. Rather than relying on field testing of qualitatively defined surfaces, isn't it time that someone develop a method for estimating friction factors based on dimensions and some measurable index of roughness?

Spills of cryogenic fuels such as LNG and LPG pose a high risk of catastrophic accident worldwide. The heat flux during film boiling as a function of differential temperature could be more accurately calculated as function of the difference of the fourth power of the absolute temperatures, like the Stefan-Boltzmann equation, rather than a first-order difference. Experiments with cryogenic liquids such as liquid nitrogen could test this possibility and be used to develop a better correlation for film boiling, especially for liquids with low boiling points. Furthermore, experiments with cryogenic liquids flowing on sloped surfaces would be useful for accident analysis.

8 References

1. J.L. Woodward, *Estimating the Flammable Mass of a Vapor Cloud*, "Discharge Characteristics", p 49-59, Center for Chemical Process Safety of the American Institute of Chemical Engineers, 1999.
2. A. Cox, F. Lees, and M. Ang, "Classification of Hazardous Locations," pp 46-47, Inter-Institutional Group Classification of Hazardous Locations of the Institution of Chemical Engineers, Rugby, UK, 1990.
3. M. Apel-Paz and A. Marmur, "Spreading of liquids on rough surfaces," J Hazard Mater, 146, pp 273-279, 1999.
4. S. Brambilla and D. Manca, "Accidents involving liquids: A step ahead in modeling pool spreading, evaporation and burning," J Hazard Mater, 161, pp 1265-1280, 2009.
5. A. A. Pahlavan, et al., "Thin Films in Partial Wetting: Internal Selection of Contact-Line Dynamics," Phys. Rev. Lett. 115, 034502, July, 2015.
6. R. H. Perry and D. W. Green, *Perry's Chemical Engineers' Handbook,* 7th ed., "Incompressible Flow in Pipes and Channels," pp 6-9 through 6-16, 6-20, 6-21 McGraw-Hill, New York, 1997.
7. Crane Co., *Flow of Fluids, Technical Paper 410*, Ver. 15, pp 1-1, 1-4 through 1-7, www.flowoffluids.com, 2009.
8. S.D. Lin, *Water and Wastewater Calculations Manual,* pp 242-252, McGraw-Hill, 2001.
9. UK Health and Safety Executive, "Hydrocarbon Releases Database," https://www.hse.gov.uk/hcr3/, 2005.

10. D.A. Crowl and J.F. Louvar, *Chemical Process Safety Fundamentals with Applications,* 3rd ed., pp 122-140, Prentice Hall, 2011.

11. J.H. Lienhard and J.B. Day, "The Breakup of Superheated Liquid Jets," J Basic Eng. Trans., ASME, pp 515-521, 1970

12. R. H. Perry and D. W. Green, *Perry's Chemical Engineers' Handbook,* 7th ed., "Particles Dynamics," pp. 6-50 through 6-54, McGraw-Hill, 1997.

13. B. Jones, *Elements of Practical Aerodynamics,* 2nd ed., p 17, Wiley & Sons, 1939.

14. R. H. Perry and D. W. Green, *Perry's Chemical Engineers' Handbook,* 7th ed., "Mass Transfer," pp 5-42 through 5-79, McGraw-Hill, 1997.

15. R. H. Perry and D. W. Green, *Perry's Chemical Engineers' Handbook,* 7th ed., "Heat Transfer," pp 5-8 through 5-29, McGraw-Hill, 1997.

16. J.L. Woodward and R.M. Pitblado, *LNG Risk Based Safety: Modeling and Consequence Analysis,* pp 142-152, Wiley, 2010.

17. Western Regional Climate Center, "Cooperative Climatological Data Summaries," http://www.wrcc.dri.edu/climatedata/climsum/ Feb. 2016.

18. ASHRAE, *Fundamentals Handbook,* ASHRAE, Ch. 22, 1977.

Example A

Leaking LNG tanker truck located on a large paved area

A-1 Accident scenario

The concern is that a tanker truck at a LNG terminal could spill LNG into a storm sewer, with the potential that the natural gas could ignite in a sewer line in a public area. A calculation must be performed to see whether the contents of the truck would evaporate before reaching the nearest storm sewer drain.

At a terminal used to fill city vehicles powered by compressed natural gas, the cryogenic storage tanks are filled by a 10,000 g tanker truck. The tanker is double-walled for insulation and impact protection. The tanker contains LNG at about 70 psig and -210°F. In this example, the LNG is assumed to consist of 100% methane. At an actual facility, the LNG composition would be known. At atmospheric pressure, methane boils at -259°F. The tanker is unloaded through a nominal 3" flexible hose.

The terminal equipment and storage tanks are located inside a protective building. The tanker unloading hose is connected to a double-walled pipe that goes through the building wall. To reach the unloading station, the truck must drive along a concrete driveway that is 59 ft wide. The driveway slopes at a 1% grade toward a storm drain that is located a little more than 400 ft down the driveway.

A-2 Release modeling

The intent is to model a bounding case, to see if the worst-case spill of LNG can reach the storm drain. Leaks from tanker trucks seldom occur, and when they do, it is usually due to accidental impact or failure of the unloading hose or coupling. Failures involve some degree of human error, and therefore are difficult to assign a probability. The failure chosen for modeling is a sudden loss of the total contents of the tanker, which may not be credible, and certainly is a bounding case.

The LNG spill quickly flashes upon release to the atmosphere. The flash can be assumed to be adiabatic. After the initial flash, the LNG continues boiling during the life of the pool. The longest spill travel occurs when there is minimum LNG evaporation or lowest surface friction. Evaporation is minimized by cool air, cool pavement, and poor heat transfer. The lowest evaporation would occur on a cold winter night. However, friction is minimized during film boiling, which is favored by a hot summer day. Calculations that follow show that the summer temperature case is worse for pool travel. At the terminal site, the lowest annual temperature is 39°F (4°C), and the highest annual temperature is 96°F (36°) [Ref. 17]. The maximum temperature typically occurs on a clear, breezy afternoon, consistent with a pavement surface temperature of about 180°F (82°) due to solar gain. Frictionless flow is assumed when the temperature difference between the substrate and liquid indicates the possibility of film boiling. The summer case is average for about 60 days a year, or 16% of the year. Therefore, the summer temperatures could be exceeded 16% of the time. It turns out that the ability of the LNG pool to travel 400 ft is not sensitive to surrounding temperature within the range that occurs at the site.

The modeling covers two phases of pool spreading: unconfined and partially confined. For simplicity, the first phase begins with the volume of the tanker in the shape of an upright cylinder, the same height as the center of the tanker, located at the tanker unloading station. After the cylinder expands to the width of the driveway, which takes about 3 seconds, the pool shape changes due to partial confinement imposed by the curbs on each side of the driveway.

A-3 Pool spread methodology

The boiling regime and heat transfer rate depend on the temperature of the pavement, which will vary with time due to contact with the LNG. A preliminary step of analysis is to correlate pavement surface temperature as a function of contact time with boiling LNG. The heat transfer from the pavement to the LNG depends on the temperature profile of the pavement and sub-pavement as a function of depth. The starting point profile was created from typical day and night temperatures as a function of time, using thermal properties of concrete and aggregate and a plot of air-concrete heat transfer coefficient as a function of wind speed [Ref. 18]. The summer daytime case requires data on solar radiation, which varies with location [Ref. 17]. The profile of temperature versus pavement depth depends on the recent history of air temperature and solar radiation. For the winter night case, there is radiative heat loss on a clear night from the pavement to the sky which is at a temperature of nearly absolute zero. However, the winter night radiation effect is insignificant. For the summer day case, there is no historical data of pavement surface temperatures, so a pavement profile was calculated by solving the one-dimensional differential equations for conductive heat transfer, starting with the temperature profile at sunrise at an average daily minimum summer temperature, allowing for night solar radiation. The air temperature and solar gain varied until six hours later, when air temperature and solar gain reached the average summer maximum. Several trial calculations showed that the daily temperature cycles did not have a significant effect on the profile below a depth of about 12". The summer temperature at 12" depth is assumed to be the average of the summer minimum and summer maximum air temperatures for the month. The equations for one-dimensional heat transfer in to a thin section of substrate at the surface of the pavement are:

$$Q_{radiation}/A = F_R \, \Delta t$$
$$Q_{convection}/A = h \, (T_{air} - T_1) \, \Delta t$$
$$Q_{conduction}/A = -k \, x \, (T_1 - T_2) \, \Delta t$$
$$\Delta T = (Q/A)/m \, C_P$$
$$m = \rho \, A \, x$$

where Q is the heat transfer rate
A is the transverse heat transfer area
F_R is the solar heat flux
Δt is the differential time step
T_1 is the temperature in the 1st section of pavement
k is the thermal conductivity of the pavement
x is the thickness of the section of pavement
ΔT is the temperature change in a section during time dt
m is the mass of the section
C_P is the specific heat capacity of the pavement

ρ is the density of the pavement

By taking a small enough time step, the heat transfer equations can be solved accurately enough by using the temperatures from the previous step. The exception occurs at the first section, where the temperature changes rapidly with time. This section was solved by obtaining the difference between the first and second sections by trial and error. Trial and error is easily done by manually adjusting the temperature difference in the spreadsheet for each time step. This procedure was used to calculate the temperature profiles for the summer case influenced by solar gain, and for the profile when pavement is exposed to LNG.

A summary of the summer temperature profile at different pavement depths is tabulated below. The bottom row shows the profile at the hottest part of a typical summer day:

Table A-1
Pavement Temperature Profile before an LNG Spill
Pavement sections are 1" thick. Section numbering begins with the section on top.
The profile for the hottest hour (hour 6) is used for spill calculations.

Time h	T_{air} °F	Solar Btu/h-ft^2	Section 1	Section 2	Section 3	Section 4	Section 5	Section 6	Section 7
0	60.4	0	59.1	61.1	63.1	65.1	67.1	69.1	71.1
1	66.4	50	69.9	68.2	68.1	68.5	69.3	70.3	71.4
2	72.5	100	85.4	80.2	77.6	75.9	74.9	74.4	74.4
3	78.5	150	105.1	96.4	91.1	87.0	83.9	81.7	80.1
4	84.5	200	128.6	116.3	108.0	101.2	95.8	91.4	88.0
5	90.6	250	155.3	139.3	127.9	118.3	110.2	103.4	97.7
6	96.6	250	178.6	161.2	147.7	135.9	125.5	116.4	108.4

When LNG is spilled onto pavement in the summer or winter, the pool stays in the film boiling regime until the surface of the pavement cools to within 20°C (36°F) of the LNG boiling point. In the film boiling regime, there is virtually no friction loss. As the tanker spill spreads to fill the area between the curbs of the driveway, the liquid is modeled as an upright cylinder in film boiling, non-frictional flow. For lack of better information, it is assumed that the vapor film under the liquid is of uniform thickness. Energy loss due to internal turbulence is ignored, which overestimates velocity and distance travelled. In the case of frictionless, gravity-induced open flow, the change in elevation equals the resulting velocity head: horizontal velocity squared divided by twice the acceleration of gravity, where the velocity is the average value across the cross-sectional area of flow.

$$z = v^2/2g$$
$$v = (2g\ z)^{1/2}$$

The velocity head equation applies to both components of pool motion – downhill motion of the entire pool, and radial spreading. In the case of downhill motion, z_d is the horizontal distance travelled, Δy, (including the downhill radius of the cylinder) multiplied by the slope of the surface.

$$v\,\Delta t = \Delta y = (2g\ z_d)^{1/2}\,\Delta t$$

In the case of radial spreading, z_r is the height of the pool at the center

$$\Delta r = (2g\ z_r)^{1/2}\,\Delta t$$

Neglecting viscous forces, the initial radial increase equals $(2g\ z)^{1/2}\ \Delta t$, where Δt is the time interval. Prior to considering evaporation loss, the cylinder volume is the same as before, but the cylinder is shorter by Δz. Therefore, the new pool height is the volume of the pool divided by the area calculated from the new radius.

In the beginning, the pool is deep, and the time required to reach the curbs is short, so it is not yet important to model the effect of the driveway slope. However, when the elevation difference between the uphill and downhill portions of the pool becomes significant relative to the pool thickness, the slope should be considered. The cylinder can be partitioned into four quadrants: uphill, downhill, and two sides. Without surface friction, the top of the cylinder is level, so each point on the circumference has a different height above the pavement. The sloped z values for the uphill and downhill quadrants are used to calculate new radii of the uphill and downhill quadrants. The net effect of the longer downhill radius is that the cylinder will spread farther in the downhill direction and become slightly elliptical.

The evaporation loss is the combination of evaporation from the surface of the liquid plus the vapor generated from boiling. The boiling heat transfer rate for LNG can be conservatively calculated by using the minimum heat flux value of 20 kW/m^2 when the temperature difference between the substrate and LNG exceeds 20°C [Ref. 16]. Evaporation from the surface is due to the combined effects of heat input from solar heat and convective heat transfer with the air. Evaporation is assumed to be limited by heat flux, not diffusion into the air. The heat flux terms for solar heat and air convection are the same as used for calculating the pavement temperature profile for a summer day.

As the pool spreads thinner, there is a time when Δz exceeds the pool thickness on the uphill edge of the pool. When this happens, the uphill edge is moved so that the thickness becomes zero at the uphill edge.

A-4 Results

Upon release to atmospheric pressure, 17.3% of the LNG flashes, based on an adiabatic flash, taking data from the methane pressure-enthalpy diagram.

The one-dimensional, time-dependent heat transfer calculations showed that on an example hot summer day, concrete that is in contact with boiling LNG can take almost 10 minutes before the liquid is in frictional contact with the pavement. This is because a film of natural gas vapor insulates the concrete, maintaining a high temperature difference between the boiling LNG and the pavement surface. Film boiling occurs for about 8.7 minutes in the summer example. For the next minute, boiling becomes more nucleate, finally reaching the point of maximum heat flux. The same spill on a cold winter night showed that the duration of film boiling was about half as long at this site, but still greatly exceeding the time required for the pool to evaporate. Other factors, such as difference in evaporation rate between summer and winter are not important. Therefore, because flow will be essentially frictionless during film boiling, the hot summer case is the worse case for distance travelled by the moving pool of LNG.

Example B

Piping Leak onto the Sloped Deck of a Floating Oil Production Platform

B-1 Accident scenario

There is a potential for a large fire on a lower deck of a floating oil production platform. To assess the potential for this fire to cause escalating damage to the platform and possibly interfere with escape, the size of the pool must be estimated. The area of the oil spill is the major factor affecting the height and heat flux of the potential pool fire. This particular spill case is one possible event of an event tree performed for a quantitative risk assessment. The length of the spill path is known, but the width of the pool as it flows down the deck must be estimated by flow calculation.

The spill source is a manifold of 12-inch pipe on a deck made of metal checker plate of the kind typically used in industrial applications. The piping is supported every few meters, so that the piping does not interfere with the flow of oil across the deck. The deck plates are galvanized metal with tread ridges to provide a non-skid surface. The deck has a slope of 2.5% from the keel line down to the edge of the deck, 31.5 m away. The lower deck is shielded from sunlight, and there is little difference in temperature between the air, deck, piping, and contained oil.

The potential leak consists of light crude oil that has been separated from produced water and gas. The crude oil contains enough dissolved gas to be flammable, but the oil loses an insignificant volume of light components when flashed to the atmosphere. The crude oil is contained at a maximum pressure of 8 barg. The transport properties of the oil at flowing temperature are:

Density	889 kg/m^3
Viscosity	11 cp, or 0.00112 kg s/m^2
Mol Wt	250 kg/mol
Surf tension	27 dyn/cm, or 0.027 N/m

Based on the number of potential leak points and the probability of leakage from each type of potential leakage, the probability of a serious leak is on the order of 10^{-3} per year for a leak equivalent to a 50-mm diameter hole. The most important contributor to serious leakage is gasket failure at a flange or valve. Nearly all flanges are arranged vertically, with the opening aimed downhill. Ordinarily, any leak onto the deck will be quickly detected by flammable gas detectors, resulting in a manual shutdown of the piping system. In the accident case for the fault tree branch of interest, the detector alarm system or operator response fails, which is estimated to happen less than 10% of the time, which lowers the probability to 10^{-4} per year. The probability of ignition is about 5% to 10%, which places the accident scenario in the barely credible range. The average wind speed is 4.3 m/s, directed crosswise to the slope of the deck. Extreme weather conditions at the time of the accident would make the scenario beyond credible, and consequently need not be considered.

B-2 Release modeling

The maximum credible leak, equivalent to a 50-mm hole is likely to occur as a narrow gap rather than a round opening, either as a tear in piping due to impact or a gap between flanges due to improper bolt tightening or gasket failure. Round holes often result from corrosion, which is not a problem with crude oil in this case. The gasket thickness is 2 mm, so a 2 mm gap is assumed. A full blow-out of a 2 mm gasket from a 12-inch line results in an opening with an area of 1920 mm^2, which is about equal to the area of a 50-mm hole.

The flow rate through the leak opening can be calculated from the flow coefficient [Ref. 7], which is a function of the Reynolds number for the pipe. The Reynolds number for the pipe is 17,400 and the ratio of orifice diameter to pipe diameter is less than 0.2, which gives a square-edge flow coefficient, *C*, of 0.60. The standard equation for flow through an orfice gives the leak flow rate. Based on the pressure drop, density, and equivalent orifice size, the leakage flow is 0.0500 m^3/s. The liquid velocity at the release point is the volumetric flow rate divided by the flow area, or 26.0 m/s.

For holes less than 2 mm wide, the drop diameter is about 1.9 times the narrow dimension (2 mm in this case) [Ref. 1, 11]. This correlation gives a drop diameter of 3.8 mm. A second method for calculating drop size is based on the Weber number being between 12 and 22 [Ref. 1]. For a Weber number of 12,

$$d = 24\,\sigma/(\rho_g\, v_d^2).$$

For the site air density of 1.204 kg/m^3, surface tension of 0.027 kg/s^2, and drop velocity of 26.0 m/s, the estimated drop diameter is between 0.80 mm and 1.46 mm. The free fall velocity for a drop between 0.80 mm and 3.8 mm is between 3 m/s and 10 m/s [Ref. 12]. The chosen values are 2.3 mm and 2 m/s.

The release will break up into droplets at a distance of L_b, which is a function of the Weber number and Reynolds number. For a drop diameter of 2.3 mm, and an air density of 1.204 kg/m^3, the Weber number is 34.7. The Reynolds number at the release point is 47,500. The distance from release to break-up into drops is

$$L_b = 540\, D\, W_e^{1/2}\, R_e^{-5/8}$$

In this case, L_b = 0.0874 m.

The trajectory of the release begins as a circular sheet 2 mm thick sprayed from a pair of vertical flanges. Before the sheet contacts the deck below, it breaks into droplets and the velocity slows to 2 m/s, relative to the air. For the portion of the spray directed straight down, that liquid lands directly under the leak point. For the portion of the spray directed up, the droplets will be in the air for about 0.36 s, calculated from the free fall drop velocity. The wind force on a droplet F_d, is a function of the drag coefficient C_d, projected drop area facing the wind A_d, air density ρ_a, and relative velocity v between the drop and air [Ref. 12]:

$$F_d = \tfrac{1}{2}\, C_d\, A_d\, \rho_a\, v^2$$

From Newtonian physics, acceleration of the droplet is force divided by mass:

$$a = F_d/(V_d\,\rho_d)$$

Distance travelled by a droplet is:

$$y = \tfrac{1}{2}\,a\,t^2$$

At the free-fall velocity, the droplet Reynolds number is 12,900. For droplet Reynolds numbers exceeding 1,000, C_d has a value of 0.445. The volume of a spherical droplet is:

$$V_d = \pi\,d^3/6$$

Solving the above, the wind will move liquid droplets downwind around 0.2 m, crosswise to the slope of the deck. At the same time, the trajectory of the leak will move droplets up to 2 m downhill, depending on the position on the vertical flange from which the droplet was released.

B-3 Pool spread methodology

The exact direction of leakage (and splashing against objects) cannot be predicted, but the model should be plausible. The release of crude oil is assumed to be continuous and at a constant flow of 0.0500 m^3/s. The starting point of the pool is directly below the leak point, located 31.5 m from the downhill edge of the deck. The initial pool area is 2 m in radius, based on droplet dispersal radius from the assumed leak due to gasket failure. After the release lands on the deck, the effect of wind is insignificant compared to gravity. Evaporation is not significant. The fluid properties of the oil remain constant. Friction loss is calculated from the Darcy equation for laminar flow. The pool flow is in the laminar regime at the edges of the pool, and is a conservative bound on areas that are in the turbulent or transition flow regime.

The pool is modeled as a conical volume with the peak located directly under the leak point. In the case of gravity-induced open flow, the change in elevation equals the resulting velocity head plus head loss from friction. The velocity head is the horizontal velocity squared divided by twice the acceleration of gravity. The friction loss is a linear function of the area of contact and the Reynolds number.

$$z = v^2/2g + h_L$$
$$v = (2g\,(z - h_L))^{1/2}$$

To include the effect of the sloping deck, the conical pool is partitioned into four quadrants: uphill, downhill, and two sides. The height of the peak is the same for all quadrants, but in each quadrant the vertical distance between the peak and the edge can be different. The radius for the uphill quadrant is adjusted as needed so that the uphill edge is never higher than the peak. As pool spreads faster in the downhill direction, the pool model changes from a right cone with a circular base to an angled cone with an elliptical base. The contact area for a quadrant of a cone is $\tfrac{1}{4}\,\pi\,r^2$. In this example, the quadrants are calculated as sections of cones of different radii, which do not match at the edges. The result is a pool area that looks like a wind rose. This mismatch results in an error in pool area calculated as an ellipse. It would have been possible to calculate the area more correctly or increase the number of sections from four if it had seemed that this error was important.

Because the friction factor is a linear function of Reynolds number in laminar flow, the average friction factor occurs at the radius associated with half of the area, r divided by the square root of

2, or approximately $\underline{r} = r/1.414$. The Darcy equation for friction loss, with the hydraulic diameter, $8z$, substituted for pipe diameter, is

$$h_L = f\,L\,v^2/(16\underline{z}\,g)$$

where $\underline{z}$ and v are evaluated at $r/1.414$, and the quadrant radius, r, is substituted for L.
For laminar flow, the Darcy equation for friction factor is

$$f = 64/Re$$

where

$$Re = 8\,\underline{z}\,v\,\rho/\mu$$

giving the following for laminar flow friction head

$$h_L = \mu\,r\,v/(2\underline{z}^2\,\rho\,g)$$

For turbulent flow ($Re > 4{,}000$), the Churchill equation for the Fanning friction factor can be used [Ref. 6]

$$1/f^{0.5} = -4\log\{0.27\epsilon/D + (7/Re)^{0.9}$$

where ϵ is the surface roughness. The roughness for galvanized steel is 0.152 mm. The variable D is replaced with the equivalent value of $8z$. The Fanning f must be multiplied by 4 for use in the Darcy equation.

After the average friction loss is calculated at $\underline{r}$, the velocity at that point can be calculated. Then the outward spread at that point can be calculated by

$$.\ \Delta\underline{r} = v\,\Delta t$$

The outward spread at the radial edge of the pool, Δr, is calculated by assuming that the ratio between $\underline{r}$ and r does not change:

$$\Delta r = 1.414\,\Delta\underline{r}$$

The equation for Δr and even the proposal that pool spreading can be modeled with a conical shape has not been validated by computational fluid dynamics or actual experiments.

The deck material is assumed to be flat after installation, and has typical checker plate ridges. The ridges rise from the base at an angle of 45° to a height of 1.5 mm. The ridges occupy about 25% of the surface area. Each ridge has about 1.4 times the surface area of the flat base, which increases the overall area of the plate by 8%. Therefore, the ridges increase the hydraulic radius by a factor of 8% above that of flat plate. The roughness value, ϵ, is estimated to be the same as for galvanized steel pipe.

B-4 Results

Table B-1

Oil Spill Spread on a Sloping Ship Deck from a Constant Continuous Leak

The radii for the side, downhill, and uphill quadrants are labelled ***a***, ***b***, and ***c***, respectively.
The height of the conical peak, **z**, is relative to the deck directly below.
The velocities for the quadrants are calculated at the centroid.

Time s	V m^3	a m	b m	c m	z m	v(a) m/s	v(b) m/s	v(c) m/s
0	0.005	2.00	2.00	2.00	0.001	0.004	0.790	0.088
1	0.055	2.28	3.65	0.33	0.012	0.271	1.136	0.175
2	0.105	2.67	5.80	0.44	0.012	0.269	1.409	0.123
3	0.155	3.03	8.40	0.44	0.011	0.238	1.675	0.000
4	0.205	3.35	11.46	0.44	0.010	0.204	1.944	0.000
5	0.255	3.62	14.98	0.44	0.009	0.175	2.207	0.000
6	0.305	3.85	18.96	0.44	0.008	0.150	2.486	0.000
7	0.355	4.05	23.40	0.44	0.007	0.130	2.759	0.000
8	0.405	4.26	28.30	0.44	0.006	0.141	3.034	0.000
8.6	0.435	4.43	31.46	0.44	0.006	0.152	3.199	0.000

As an overall characterization of a spill on a sloped surface, the spill can be viewed as having a roughly triangular footprint. In this case, with a surface slope of 2.5%, the spill could be viewed as a triangle 31.5 m long and 8.9 m wide at the base. This corresponds to a divergence angle of about 8° from the centerline of flow.

Uncertainties in the results include the question of whether the conical model is representative of the action of an actual liquid spill at an industrial site. There is no published data that could be found containing a friction factor for checker plate decking. The checker plate ridges may have an effect on flow pattern, such as deflecting the flow. In this case, the flow pattern that has the largest area (greatest width before stopping at the edge of the deck) gives the worst consequences. There was no study done for this report to investigate the effect of various conditions affecting the width of the flowing pool. However, results are based on recognized flow equations and should be viewed as useful in the absence of experimental or computational fluid dynamics results.

440778 Apparent Discrepancies in Methods for Predicting the Explosion Energies of Bleves

Abstract:

The authors have used calculation methods for BLEVE blasts as described in the CCPS book *Guidelines for Vapor Cloud Explosion, Pressure Vessel Burst, BLEVE and Flash Fire Hazards*, and most recently similar methods in the PHAST process industry hazard analysis/consequence analysis software tool, developed by DNV-GL. Predictions made using PHAST appeared to be overly conservative, which instigated a comprehensive comparison of PHAST to other methods in the literature. There is not universal agreement concerning the dominant mechanisms driving the air blast pressure wave from a BLEVE, and there are major differences in methods used to predict the explosion energy driving the blast wave. The CCPS guidelines book and PHAST consider the explosion energy driving the blast wave to be from both the vapor and liquid phases present at the time of loss of containment. For each phase the contribution to explosion energy is the difference in internal energy between the thermodynamic states of the liquid and vapor at the time of loss of containment and the final states of the materials when isentropically expanded to atmospheric pressure. Recently, Birk compared various prediction methods to experimental data for propane BLEVEs and concluded that the flash vaporization of the liquid may be too slow to contribute significantly to the blast, therefore the blast energy is primarily from the expansion of the vapor in the vessel at the time of loss of containment. Other authors also base the explosion energy on the gas expansion and others on the liquid superheat. Ogle and coworkers have discussed the available energy of a BLEVE (exergy) and compared this to the BLEVE explosion energy predicted using several methods in the literature. These various approaches can lead to drastically different predictions of explosion energy and the resultant blast pressures and impulses. BLEVE theories and models are discussed in this paper and predictions are compared to limited experimental data. This paper is meant to provide comparisons between the many various approaches, including many more that are not listed above, and summarize their differences.

441089 Automated Optimization of Gas Detector Placement

Abstract:

Chemical process facilities are moving toward the use of performance-based design techniques for fire & gas detector placement. These approaches are displacing methods for layout of gas detection based on uniform "grids" that date back more than 25 years. First generation performance-based design techniques utilized 'gas volume detection' principals that are valid in terms of detecting and mitigating a large explosion event, but can result can result in an excessively large number of detectors and a costly design. This paper presents a superior alternative to the first generation technique of volume detection that uses sophisticated modeling of gas release scenarios with computational fluid dynamics (CFD). The techniques described in this paper align with ISA TR 84.00.07 guidelines for scenario-by-scenario basis. Furthermore, the paper breaks the paradigm that CFD can only be used sparingly on complicated gas dispersion problems. With the proper definition of the process, optimal selection of boundry criteria, and automated scripting of multiple scenarios and data collection from those scenarios, the thousands of CFD runs required to develop results can be efficiently and cost effectively performed. This paper will demonstrate practical application of CFD modeling and probabilistic risk analysis in order to improve on the performance of gas detection systems while at the same time, reducing unnecessary conservatism in the number of detectors. The paper will present methodology for determining scenario-based gas detector coverage and include a worked application example for an offshore gas processing platform.

441414 Quantitative Risk Assessment in the Process Industries: Beyond Numerical Estimates of Individual and Societal Risk

Abstract:

The Chemical Process and Oil and Gas industries (both on and offshore) have largely adopted Quantitative Risk Assessment (QRA) with open arms following its application in the nuclear industry. Currently, the methodology not only forms a core aspect of the risk management programs of numerous organizations who design, build, own, manage or operate high hazard facilities (refineries, LNG terminals etc.). It also occupies a central role in the legislative and regulatory framework in many jurisdictions across the world. It has become a tool of choice for both the regulators and the regulated parties.

As its application has become more widespread, it has been the subject of extensive critical inquiry, much of which has revolved around the primary product of the process, namely numerical estimates of risks to exposed people. Advocates highlight the utility and rationality such numbers offer in the often challenging decision making process associated with risk management. Critics make reference to the wide variability and uncertainty associated with such predictions as well as the implicit value judgments they embody. Less emphasis has been placed on the wealth of other insights (often latent) afforded by a well scoped out, executed and documented QRA.

The aim of the paper is to examine and explore the value of QRA over and beyond the well-known orthodox output of individual or societal risk estimates. It identifies and outlines what these added benefits are and elaborates on their significance - particularly in the context of achieving risk management objectives. It argues that such latent benefits should be considered to be equally or more important to the typical output and that more attention should be paid to these whilst conducting a QRA. A popular adage goes thus "sometimes, it's the journey that teaches you a lot about the destination"; so it is with QRA.

12TH GLOBAL CONGRESS ON PROCESS SAFETY

Quantifying the Importance of Accurate Physical Property Data in Dispersion Modeling

Devin Averett
Baker Engineering and Risk Consultants, Inc.
11011 Richmond Avenue, Suite 700
Houston, TX 77042
daverett@BakerRisk.com

Jef Rowley
Baker Engineering and Risk Consultants, Inc.
11011 Richmond Avenue, Suite 700
Houston, TX 77042
jrowley@BakerRisk.com

Prepared for Presentation at
American Institute of Chemical Engineers
2016 Spring Meeting
12th Global Congress on Process Safety
Houston, Texas
April 11-23, 2016

Quantifying the Importance of Accurate Physical Property Data in Dispersion Modeling

Devin Averett
Baker Engineering and Risk Consultants, Inc.
11011 Richmond Avenue, Suite 700
Houston, TX 77042
daverett@BakerRisk.com

Jef Rowley
Baker Engineering and Risk Consultants, Inc.

Keywords: Dispersion Modeling, Physical Properties, DIPPR

Abstract

Modeling discharge and dispersion of materials from processing vessels is an important part of a facility siting study. In this paper, we investigate the dependence of dispersion modeling on physical property data accuracy, including the impact of uncertainty in the lower flammability limit, heat of vaporization, liquid density, and vapor pressure data. Vapor pressure is further analyzed by comparing correlations obtained from the Design Institute for Physical Properties (DIPPR) to the Clausius-Clapeyron equation using the normal boiling point as the reference.

Overall, the predicted dispersion distances are relatively resilient to uncertainty in physical properties. However, errors in dispersion distances can be large under certain conditions. It is important to have accurate data for these cases. To ensure data accuracy, we recommend using a modern, critically reviewed database such as the latest DIPPR database, which contains thermophysical data for a large number of compounds and routinely employs rigorous analysis and consistency checks to assure the reliability of the data.

1 Introduction

Dispersion modeling can be a main driver for many major economic and safety decisions. Modeling of a toxic release may help dictate where and what type of personal protective equipment (PPE) is needed, where to place high-occupancy buildings and work areas, and whether risk to personnel or the surrounding community warrants investment of capital into mitigating the effects or decreasing the likelihood of such a release. Modeling of a flammable vapor cloud may provide similar information regarding safety from the risks posed by flash fires and explosions, but it can also aid in estimating the economic impact from explosion events by determining which buildings and equipment are likely to be damaged. This impact is not just in the cost of replacing equipment, but also in the loss of production during the repair downtime. This information may in turn determine insurance rates and proper amounts of coverage.

With such important decisions hinging on the results of a model, it is vital that the model is reliable and provides as accurately as possible a representation of what will happen during an event, especially when there is already so much inherent uncertainty due in large part to the natural variability in the weather. So selecting a good model that will make the best of a difficult situation is vital. When modeling a dispersion scenario, there are many models available. Typically the most accurate models are based on computational fluid dynamics (CFD), as most other models do not explicitly account for the presence of obstacles; however, use of CFD is computationally expensive and impractical if it is necessary to model numerous release scenarios subject to various weather conditions. Other models have been developed that sacrifice some accuracy in the presence of obstacles to gain computational speed.

Whatever the model, similar inputs are typically required, many of which are scenario-dependent (e.g., atmospheric conditions such as wind direction and speed, weather stability class, etc.). The inputs that remain constant across all different scenarios are the physical properties of the material being released. Unlike the weather, a material's chemical properties do not exhibit large and unpredictable fluctuations. So if there is any input to the model in which one can place a reasonable level of confidence, then surely it is the physical property data.

Of course there is disagreement in the literature pertaining to physical properties. Aside from the more obvious sources of error (human error, bad equipment calibration, etc.), different experimental techniques may yield differing results, and there is even variability in experiments employing the same or similar techniques. The trouble, then, is determining which literature value is likely to be the most correct, which may not be an easy task.

In this paper, we explore whether the effort of finding accurate material properties for consequence modeling is worth the effort and cost. If the physical properties are not exact, does it make a significant difference in dispersion model results? We take a quantitative approach to answer this question by utilizing the dispersion model currently used in SafeSite$_{3G}$®, BakerRisk's in-house software for consequence modeling. By varying a number of different properties, we can determine the impact they have on the final dispersion results. Conclusions can then be drawn on the importance of obtaining the best physical property data in dispersion modeling.

2 Methodology

Dispersion model results can't be summarized in a single number. There are many points of comparison that could have been used in our analysis. In this study, we use for our basis of comparison the horizontal distance at which the centerline of our release reaches a concentration equal to the lower flammability limit of the material.

The dispersion model used is detailed by Rowley [1]. In short, it solves the conservation of mass, momentum, and energy in one dimension, and then concentration profiles in the vertical and cross-wind directions are obtained based on similarity concepts. The model itself actually draws from several other ancillary models for handling phenomena such as droplet trajectory, pool formation, and droplet and pool evaporation. Physical properties for materials used in the model were obtained from the most recent version of the DIPPR® 801 database [2].

Several different physical properties were varied to monitor the effect any variance in the values would have on dispersion distances. These were chosen based on which properties intuitively would seem to have the greatest impact on dispersion results; namely lower flammability limit (LFL), heat of vaporization (HVP), liquid density (LDN), and vapor pressure (VP).

Several chemicals were chosen for the analysis, selected to give a wide range of normal boiling points (NBP) to ensure materials of different volatilities were adequately explored. The chemicals, in order of increasing normal boiling points, are methane, ethylene, propylene, ethylene oxide, methanol, *n*-octane, phenol, 1-decanol, glycerol, and *n*-eicosane. The intent was to select fairly common chemicals that differed in normal boiling point by approximately 50 K.

Dispersions were run at two different weather conditions, both at atmospheric pressure of 1 atm and an ambient temperature of 70 °F. The first weather condition used a wind speed of 1.5 m/s and a Pasquill stability class of F, and the second used a wind speed of 3 m/s with a D stability class. In addition, different dispersion runs were performed for different storage temperatures and pressures. Pressures of 10, 100 and 1000 psig were run at temperatures ranging from -275 to 675 °F in 50 °F intervals. Releases where the material exists as a solid at those storage conditions were ignored.

To establish a baseline set of results, dispersions were run for the different materials with physical properties set equal to the accepted values from DIPPR. Additional dispersion runs were then gathered where one physical property was altered slightly. Each property was altered by 5% in either direction. (Note that if altering the property resulted in a phase change at the storage conditions, these cases were omitted from the analysis, the presumption being that even if a physical property may not be accurately known, it is known in which phase the material is being stored.) We were then able to obtain an estimate as to how much the dispersion distances change in relation to a standard step change in a specific physical property. The vapor pressure analysis was further extended by looking at how dispersion results differed when using the Clausius-Clapeyron equation using the normal boiling point from DIPPR as the reference point. This was done to see how important it is to obtain accurate data across a broad range of temperatures rather than simply at a single point and extrapolating from there numerically with a basis on theory.

3 Results and Discussion

Taking one property at a time, results from the dispersion runs are summarized below, accompanied by some discussion on implications that can be drawn from the data. All results are plotted as:

$$\frac{\%\Delta(Distance\ to\ LFL)}{\%\Delta(Physical\ Property\ of\ Interest)}\ vs.\ \frac{Process\ Temperature\ (K)}{NBP(K)}$$

where $\%\Delta$ refers to a percent change in the pertinent value.

3.1 Lower Flammability Limit (LFL)

The flammability limit has the most straightforward effect on dispersion distances. This is because it has no effect on the dispersion model whatsoever. It merely changes the target concentration we are looking for when calculating the extent of the plume to LFL. Figure 1 shows the results for all runs.

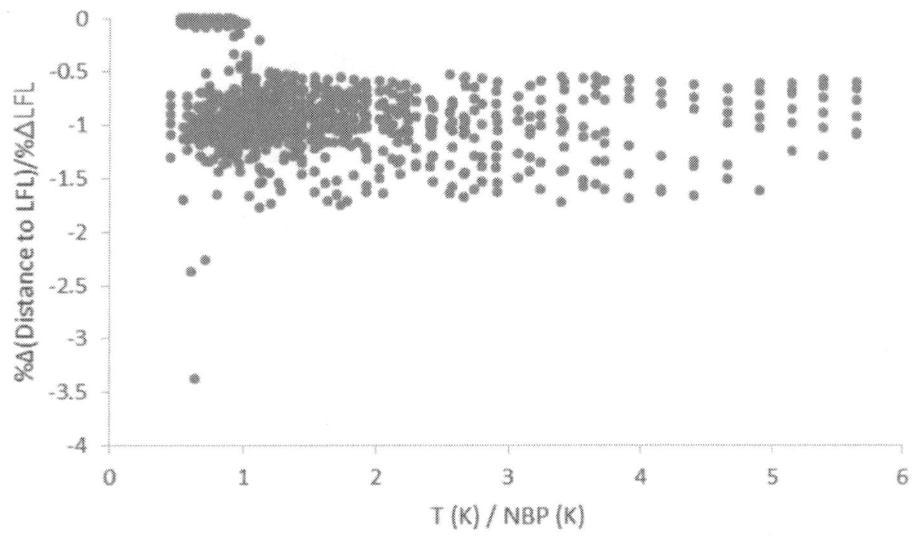

Figure 1. Ratio of change in dispersion distance and change in LFL vs. ratio of process temperature and NBP.

The uppermost cluster showing virtually no change in dispersion distance represents those scenarios that form highly non-volatile pools. Otherwise, all scenarios show a decrease in dispersion distance with an increase in LFL, which intuitively makes sense. The largest deviations occur for materials near saturation (T/NBP = 1). On average, dispersion distances changed only slightly, with an average change of -0.81 % in dispersion distance for a 1 % increase in LFL and a standard deviation of 0.45 %. While this may not appear to be very significant, the trouble with LFL is the inaccuracy in measuring LFL. Literature values may vary widely, and these large discrepancies will result in large differences in dispersion distance. For example, within the DIPPR database, looking at only methane, the accepted value of 5 % vol [3] differs by as much as 12 % from other experimental values (a value of 4.4 % vol [4]). A previous study has shown that large discrepancies can occur simply due to the method of measurement. Data obtained using the US Bureau of Mines tube method can give dispersion distances to LFL that differ by as much as 200 % from data obtained using the EN 1839 tube method [5]. Indeed, the same study discusses the reality that rather than a discrete value, the LFL exists as a range of concentrations that may yield varying degrees of flame propagation, and different methods of measurement may give measurements biased towards one end of this range or the other. Fortunately, in this study we found that small differences in LFL generally didn't have a large impact on dispersion distances, but certainly caution should be taken in cases where the LFL is low or wide ranges of LFL data are available in the literature.

3.2 Heat of Vaporization (HVP)

For heat of vaporization, we examine a much narrower range of conditions. This is because for HVP to play any significant role at all, an appreciable amount of both liquid and vapor phases must be present. Thus high-temperature vapor releases and low-temperature liquid releases would give virtually the same results for large changes in HVP because a negligible amount, if any, of vaporization is occurring. Figure 2 shows the results from the analysis.

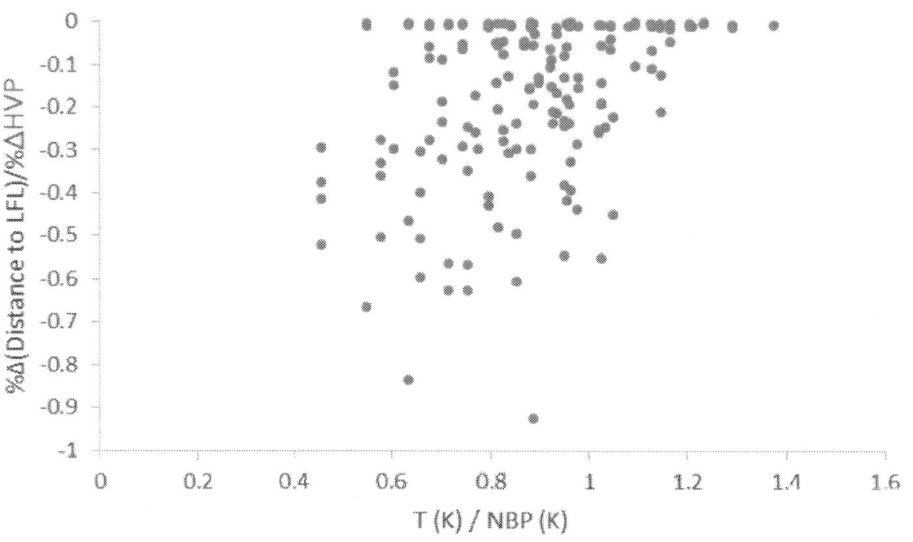

Figure 2. Ratio of change in dispersion distance and change in HVP vs. ratio of process temperature and NBP.

Similar to LFL, we see that small changes in HVP result in approximately the same order of magnitude of change in dispersion distances. For every one percent increase in HVP, the distance to ½ LFL will on average change by -0.17 %, with a standard deviation of 0.19 % and a minimum of -0.93 % occurring in the cases we analyzed. And this, remember, is only for cases where any appreciable phase change is taking place. If at least somewhat reliable HVP data can be obtained, then we can be confident that it will not affect dispersion distances to any significant degree, and especially not if the scenario does not include a phase change.

3.3 Liquid Density (LDN)

The effect of liquid density on dispersion models is much more complex than the effect of LFL or HVP. First of all, liquid density will affect the discharge rate. It would seem logical that a higher density would lead to a material that would discharge at a greater rate from a specific orifice, suggesting greater dispersion distances. However, this would affect other characteristics of the flow as well such as the choke pressure. The change in choke pressure can cause additional issues, particularly in cases where the liquid flashes upon release. Any change in pressure will then have a large effect on the vapor fraction of the discharge, and hence will cause a large swing in the overall density of the initial discharge.

The second issue to consider is of plume buoyancy. For plumes of high liquid fractions, an increase in liquid density will cause a decrease in plume buoyancy, resulting in the liquid raining out at a nearer distance. This would likely result in shorter dispersion distances, which could either supplement or compete with any effects from the change in discharge.

Although there are other issues affected by LDN, the last issue we will mention here is that momentum must be conserved. Given two identical droplet-filled plumes (both plumes start with the same velocity and have the same number of droplets with equal diameter) with the only difference being the density of the material in the droplets, and assuming an equal air entrainment rate between both plumes, conservation of momentum dictates that the denser plume will decelerate at a slower rate than the less dense plume. So with these simplistic plumes, the liquid in the denser plume will rain out at a farther distance, exactly the opposite conclusion that was reached when considering plume buoyancy. We see that with LDN, trying to reason out what effect changing LDN might have on a single scenario without running the model can become very complicated.

We turn to the model, then, for some answers. Figure 3 shows results for LDN dispersion runs. Again, some data was omitted because it was not relevant to the analysis, namely the cases where no liquid was present throughout the scenario. Note that unlike previous properties, both negative and positive deviations in dispersion distances are seen, and the differences become more pronounced the more the temperature is decreased. The point that gives the largest deviation (a surprising 13.1 % increase in dispersion distance per one percent increase in density) is phenol at P = 1000 psig and T = 275 °F for the F-classified weather. Under these storage conditions, phenol exists as a liquid. Looking at this specific case in more detail, we see that the droplets in the denser plume rain out more quickly (less than a second compared to about 11 seconds), but the velocity in the less dense plume is knocked out very quickly, causing rainout to occur at about 28 ft compared to the denser plume at about 60 ft.

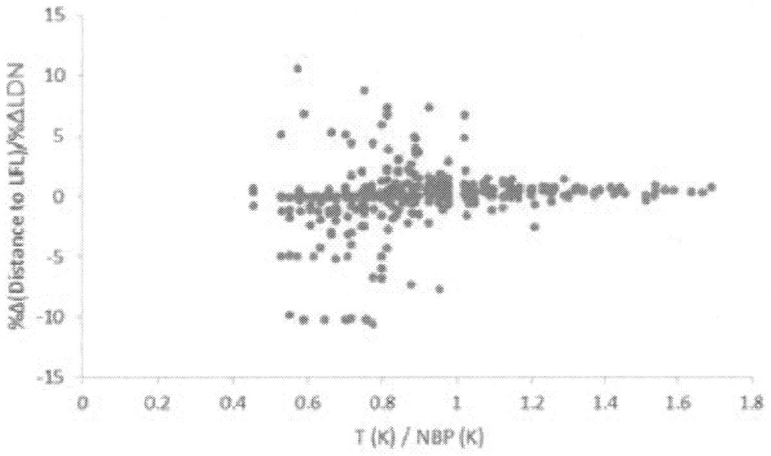

Figure 3. Ratio of change in dispersion distance and change in LDN vs. ratio of process temperature and NBP.

On average, a one percent increase in LDN results in only a 0.08 % increase in dispersion distances, but that has a standard deviation of 2.21 %. Thus, having inaccurate LDN data can cause large differences in dispersion results, and those differences can be difficult to predict. Fortunately LDN is a relatively simple property to measure, but if the scenario deals with a wide range of temperatures it is imperative to have accurate temperature-dependent data for LDN.

3.4 Vapor Pressure (VP)

3.4.1 Deviation in Vapor Pressure

Like HVP, vapor pressure can really only affect releases where two phases are possible. Hence, pure vapor releases are ignored, as well as releases low enough in temperature to form pools with no significant amount of evaporation. Also ignored are cases where the change in VP would result in a phase change at process conditions, the assumption being that the initial phase is known.

Like LDN, VP may have competing effects on dispersion modeling. An increase in VP would cause a flashing liquid upon discharge to have an increased vapor fraction and thus a decreased mass flow, likely suggesting shorter dispersion distances. However, it also means droplets and pools evaporate at higher rates, serving to lengthen dispersion distances.

Figure 4 shows the results from the VP runs. As would be expected, most of the variability occurs when the temperature of the fluid is near the normal boiling point where a two-phase release is much more probable. The most extreme result (an 11 % increase in dispersion distance per 1 % increase in VP) come from *n*-eicosane at P = 100 psig and T = 625 °F for the D-classified weather. In this case, the difference is largely buoyancy driven. For higher VP, the resulting excess vapor is sufficient to increase the buoyancy of the plume and significantly delay rainout, keeping the concentration and mass of material in the plume at higher levels for a longer time.

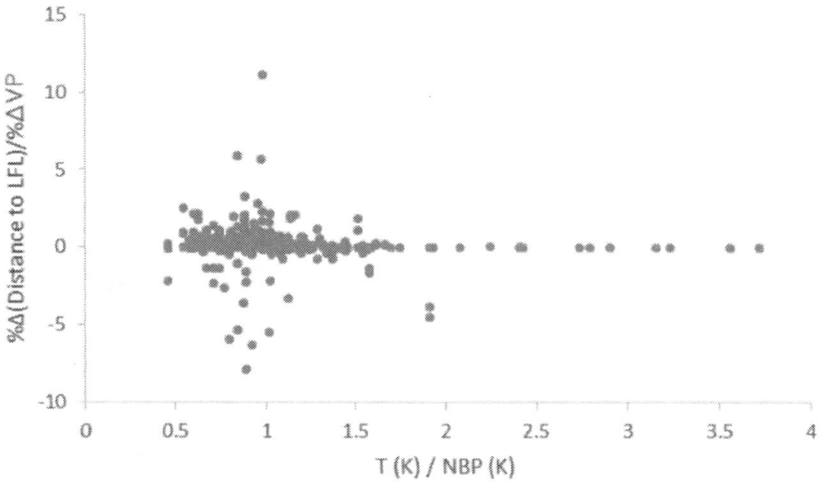

Figure 4. Ratio of change in dispersion distance and change in VP vs. ratio of process temperature and NBP.

The average result is only a 0.18 % increase in dispersion distance per 1 % increase in VP, but with a standard deviation of 1.19 %. While it does not tend to affect dispersion distances to the same degree as LDN, VP does have the potential to significantly alter dispersion results.

3.4.2 Clausius-Clapeyron Equation

It was also of interest to see if one could still get reliable results using the Clausius-Clapeyron equation, or at least the well-known approximation of it, shown here:

$$\ln\left(\frac{P_1}{P_2}\right) = \Delta H_{vap}\left(\frac{1}{T_2} - \frac{1}{T_1}\right) \tag{1}$$

where P_1, T_1 and ΔH_{vap} are the vapor pressure, temperature, and heat of vaporization, respectively, at a reference state, and P_2 and T_2 are the vapor pressure and temperature, respectively, at some new state. Hence, at the reference state, the equation is exact, but errors will develop and worsen as the desired temperature moves farther from the reference temperature. Often, heat of vaporization data is available at the normal boiling point of a material. To establish a basis for this study, then, we select the normal boiling point of the material to be our reference point, using the HVP value obtained from the DIPPR database at this temperature. The DIPPR database base correlations for HVP and VP are based on rigorously evaluated experimental data, so we can be confident that the DIPPR correlation will always result in a more accurate answer than the Clausius-Clapeyron equation, regardless of the temperature. Hence, in the following discussion we assume any deviation from results derived from DIPPR data are unfavorable. Again, pure vapor releases, releases low enough in temperature to form pools with no significant amount of evaporation, and cases where the change in VP would result in a phase change at process conditions are ignored.

Earlier results relating to vapor pressure would suggest that using the Clausius-Clapeyron equation would likely have little effect on dispersion distances since significant deviations seen in Figure 4 are concentrated around the NBP, the area in which the Clausius-Clapeyron (as we have defined it) should be the most accurate. Following the same procedure as the other analyses (testing several materials at several temperatures and pressures), there surprisingly were some significant deviations in dispersion distances. Figure 6 and Figure 6 show the results. Note that because we are not dealing with a strict percentage deviation in vapor pressure, the y-axes in these figures are not presented as ratios, but rather Figure 5 shows straight percent differences in dispersion distance and Figure 6 shows absolute differences in dispersion distances. This makes it difficult to draw a comparison with the results from earlier in this study. Nevertheless, it is clear that particularly at temperatures just below the NBP, significant deviations can occur from the use of Clausius-Clapeyron. Over all runs, there was an average change of 10.0 %, with a standard deviation of 40.4 %.

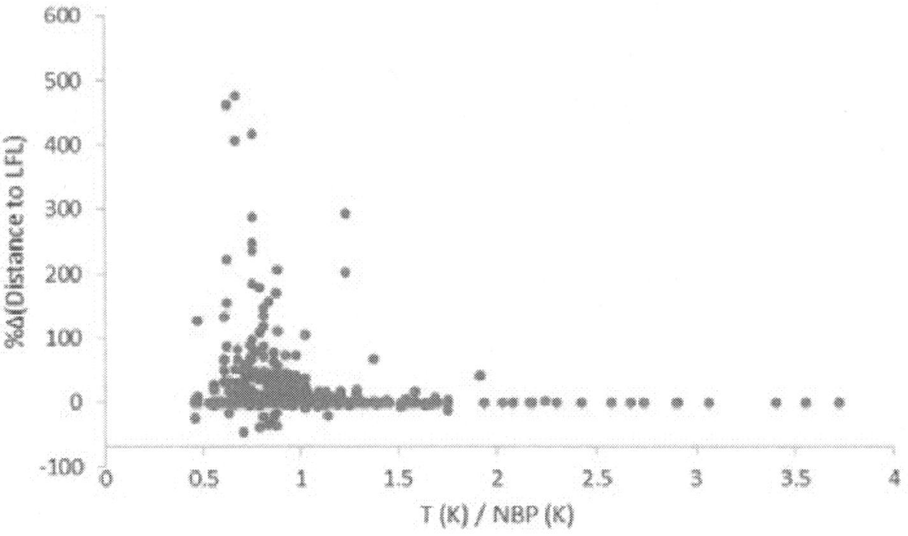

Figure 5. Percent change of dispersion distance vs. ratio of process temperature and NBP.

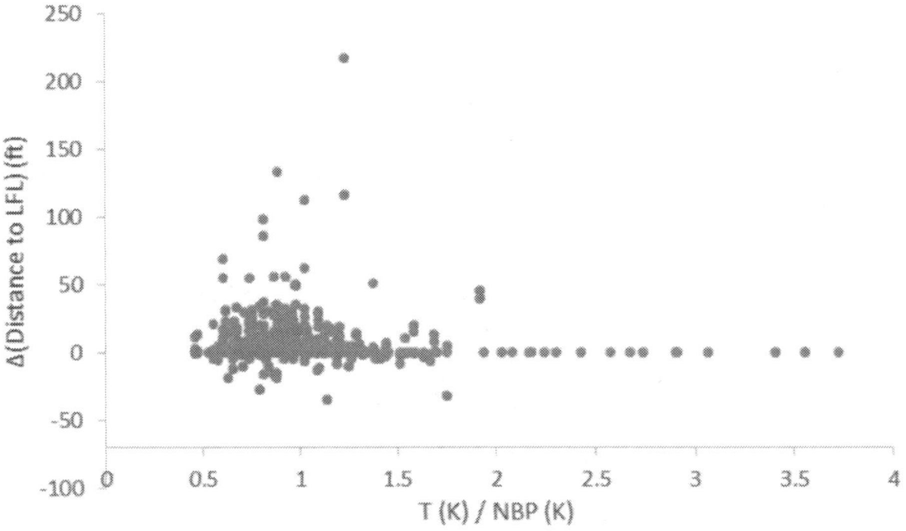

Figure 6. Absolute change in feet of dispersion distance vs. ratio of process temperature and NBP.

Still, the magnitude of these deviations are at times so extreme that it warrants further inspection to determine the reasons behind it. Figure 7 helps make results from Figure 5 and Figure 6 more relatable to results from Figure 4 by showing exactly how much VP as calculated from Clausius Clapeyron differs from VP calculated from the DIPPR correlation, once again plotted against T / NBP. Figure 4 shows deviations per percent change in VP, and Figure 7 shows that using Clausius Clapeyron will frequently result in a deviation much higher. The general trend seen in Figure 7 is that higher-boiling components reach greater deviations at lower T/NBP ratios. (Note that if plotted against T - NBP, the data would look much more uniform across all compounds.) It is the high-boiling compounds that make up the majority of the higher deviations in Figure 5. So not only do we find that Clausius Clapeyron can give very different results from correlations based on temperature-dependent experimental data, but we also see that the magnitude of those differences have a positive correlation with the NBP of the material. For heavier compounds, then, it becomes even more important to use accurate VP data in dispersion modeling.

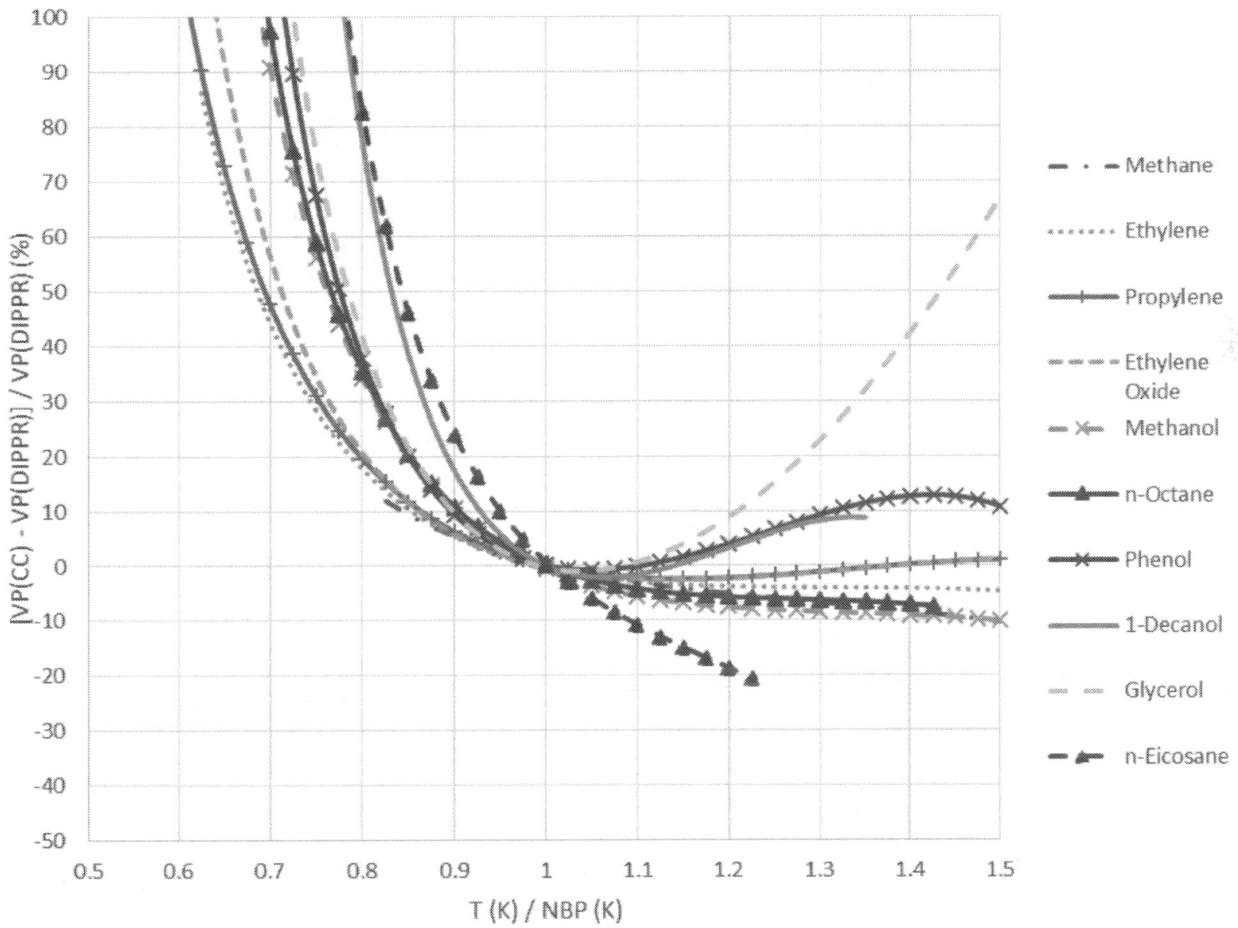

Figure 7. Percent difference between VP as obtained from the DIPPR correlation and the VP as given by Equation 1 plotted against T / NBP.

4 Conclusions

Overall, dispersion distances turned out to be quite resilient to deviations in physical properties. Minor changes in HVP or LFL resulted in only slight deviations in dispersion distances. In many cases, minor deviations in LDN and VP resulted in only minor deviations of dispersion distances. However, under the right conditions (far into the liquid regime for LDN or near the NBP for VP) large deviations in predicted dispersion distances can occur with only minor changes in physical property data. Consequently, great care should be taken as to where one obtains the physical property data used in dispersion modeling.

Furthermore, the Clausius-Clapeyron equation may not work well across the wide range of conditions that may be encountered during even a single dispersion. The processing temperature may be different from the temperature upon discharge, the temperature of the droplets in the plume, or the temperature of any pool formed, and we see in this study that it is vital to have reliable temperature-dependent vapor pressure data rather than rely on an extrapolation from a single point.

Obtaining accurate physical properties may be difficult and time-consuming, and we can often be tempted to settle for any values we can find, regardless of the source. Particularly in the case of LDN and VP, this study has shown that all physical property data should be met with some level of scrutiny when attempting to model dispersions.

We recommend using a critically reviewed database such as the current version of the DIPPR database [2]. The DIPPR database contains thermophysical data for a large number of compounds, and it routinely employs rigorous analysis and consistency checks to assure the reliability of the data. The database is continually updated and reviewed by experts in thermodynamics to ensure the recommended data reflect the latest advancements in thermophysical property measurement and evaluation. For properties where literature values are not available, state-of-the-art prediction methods are utilized to offer a best-guess value for situations where an estimate is needed.

5 References

[1] J. R. Rowley, "A new integral dispersion model based on one-dimensional turbulence theory," in *IChemE's 24th Hazard Symposium*, Edinburgh, UK, 2014.

[2] R. R. Rowley, W. V. Wilding, J. L. Oscarson and N. F. Giles, DIPPR® Data Compilation of Pure Chemical Properties, New York, NY: Design Institute for Physical Properties, AIChE, 2015.

[3] M. G. Zabetakis, "Flammability Characteristics of Combustible Gases and Vapors," in *Bulletin No. 627*, U.S. Bureau of Mines, 1965.

[4] K. Nabert, G. Schön and T. Redeker, Sicherheitstechnische Kenngrößen brennbarer Gase und Dämple, Deutscher Eichverlag, 2004.

[5] J. R. Rowley and J. E. Bruce-Black, Proper application of flammability limit data in consequence studies, Southport, UK: IChemE's 23rd Hazard Symposium, 2012.

[6] G. H. Tryon, Fire Protection Handbook, 12th ed., Boston: National Fire Protection Associations, 1962.

441423 Sustainable Risk Reduction with Dynamic Execution of Procedures for Any Operating Condition

Abstract:

Managing the volume of new information created daily is a challenge. Add to that the complexity of new equipment, employee turnover, and third party contractors working on-site, documenting clearly and concisely is critical to the operation. To truly mitigate risk we need look to operating companies focused on leading indicators of risk management where improved work conditions and profitability of the operation are the by product.

- Reduce human error
- Improve user adoption
- Improve operator efficiency
- Optimize SME time
- Quick resolution incident investigation
- Improve time "on the job"
- Extend equipment life

These objectives, met through the implementation of procedure best-practices to significantly reduce risk and improve the reliability of the operation.

During this presentation we will share the results of our procedure best-practices benchmark report and review the impact of asset level operating procedures for our clients.

As operating companies evolve through the knowledge management maturity model, they not only standardize best-practices but leverage that knowledge to optimize performance with a collaboration hub and dynamically generated procedures to fit any situation.

We will demonstrate how asset level operating procedures, designed for routine operator and facility maintenance, directly relates to improvements in asset life and performance, site-safe incident reduction based on human factors, and improve equipment start-up predictability when returning from planned maintenance.

This is so much more than simply documenting procedures; this is about sustainable adoption to improve operating discipline. Innovatia will introduce a complete procedure management solution to facilitate knowledge transfer in real-time to mitigate risk and improve reliability of the operation.

441155 Process Hazard Analysis of River-Crossing Natural Gas Pipeline

Abstract:

In order to control the process hazard of the river-crossing natural gas pipeline, HAZOP analysis is performed to qualitatively analyze its potential risk. Based on the results of HAZOP analysis, high risk scenarios are identified. So as to calculate the risk caused by pipeline's catastrophic rupture, FLACS software which based on CFD data simulation is used to quantitatively calculate the flame propagation and explosion overpressure of the explosion of natural gas. With the results of HAZOP analysis and quantitative calculation, safety measures are put forward to ensure the long-term safe operation of river-crossing natural gas pipeline.

Safe Depressurization of Large Thickness Vessels Under Pool Fire

Subhash S. Kale and George Maliakkal [(1)], Bechtel Corporation

(1) To whom all correspondence should be addressed

Abstract:

Vessels fail at pressures lower than the Maximum Allowable Working Pressure (MAWP) during fire due to degradation of strength at high temperatures. Analytical solution of the vessel wall temperature profiles as a function of fire exposure time available in API 521 Annex A for thicknesses of 3.2 mm (0.125 in.), 12.7 mm (0.5 in.) and 25.4 mm (1 in.) was extended for vessel thicknesses up to 152.4 mm (6 in.) to include high pressure/large diameter vessels encountered in gas processing. For fire depressuring, plots of ratio of the maximum vessel pressures to avoid failure to MAWP were developed for various wall thicknesses and compared to a plot of the ratio of vessel internal pressure to MAWP, both as a function of time. If the vessel internal pressures throughout fire depressuring were less than the reduced vessel failure pressures, then vessel rupture is avoided. The methodology is illustrated for the case of a high pressure vessel of wall thickness 57.15 mm (2.25 in). in natural gas service.

Introduction:

Large vessels, for e.g. vertical compressor suction KO drums, have a substantial flammable hydrocarbon liquid inventory, which if leaked can form a source of pool fire. The vessels usually have a large diameter 3 - 6 m which requires vessels with thicknesses higher than 25.4 mm (1 in.). API 521 [1] Section 4.4.13.2.4.1 gives information on wall temperatures for plates 3.2 mm (1/8 in.), 12.7 mm (1/2 in.) and 25.4 mm (1 in.). Also, Appendix A of API 521 gives information on maximum pressure as % of Maximum Allowable Working Pressure (MAWP) for plates of thickness up to 25.4 mm (1 in.). No data is available in API 521 for plates of thicknesses greater than 1 in. This work uses the analytical method given in Appendix A of API 521 to extend the calculations for vessels of thickness higher than 25.4 mm (1 in.) - for e.g. 50.8 mm (2 in.) to 152.4 mm (6 in.). thicknesses. The analytical equation given in Appendix A of API 521 is solved to develop wall temperature profiles for various wall thicknesses including those beyond 25.4 mm (1 in.). The temperature versus allowable stress data given in Table A-7 of the Appendix A can then be used to develop temporal profiles of maximum allowable pressure to avoid failure as a function of MAWP. The vessel internal pressure experienced during fire case depressurization is calculated using HYSYS v 7.3 depressurization utility. The internal pressures during depressurization, expressed as a ratio of the internal pressure to MAWP, is overlaid on the failure pressure temporal profiles to ascertain whether the vessel can avoid rupture during a pool fire scenario. The following describes the assumptions used in the calculations, the methodology to compute the temporal profiles of wall temperature and the maximum pressure to avoid failure.

Assumptions:

1. A carbon steel vertical vessel with hemispherical heads is assumed.
2. The surfaces are assumed to fully engulfed in flames hence both convection and radiation terms in equation A-1 of API 521 Appendix A apply.
3. The parameters given in Appendix A are slightly modified to better fit the temperature profiles given in Figure C-9 of Appendix C of API 521.
4. Since the vessels have thicknesses above 25.4 mm (1in.), the depressurization is from the design pressure, which is assumed to be same as MAWP, to half the design pressure in 15 minutes per API 521 guidelines. For these calculations, a design pressure of 80 barg is assumed.
5. The calculations are illustrated for a vessel of controlling thickness 57.2 mm (2.25 in.) for the hemispherical heads. The methodology can be extended to larger thicknesses from 25.4 mm (1 in.) to 152.4 mm (6 in.).

Wall Temperature Calculation by Analytical Method:

The analytical equation given in A-1 of Appendix A of API 521 is stated as:

$$q_{absorbed} = \sigma\ (\ \alpha_{surface}{}^{*}\varepsilon_{fire}{}^{*}T_{fire}{}^{4} - \varepsilon_{surface}{}^{*}\ T_{surface}{}^{4}\ ) + h\ (T_{gas} - T_{surface})\quad (1)$$

Also, by a heat balance,

$$q_{absorbed} = m\ c_p\ (\text{Delta } T_{surface} / \text{ Delta time})\qquad (2)$$

The following gives a description and the values used for the various parameters in this equation per API 521 Table C-3:

$q_{absorbed}$, Heat absorbed by the wall

σ, Stefan Boltzmann Constant = 5.67 E-8 W/m^2.K^4 (1.713E-09 BTU/H-ft^2-°R$^{4)}$

$\alpha_{surface}$ is the equipment absorptivity = 0.8

ε_{fire} is the fire emissivity = 0.85

$\varepsilon_{surface}$ is equipment emissivity = 0.75

T_{fire} is the fire temperature = 904 °C (1652 °F)

$T_{surface}$ = equipment temperature initially assumed to be 59 °C (122 °F or 582 °R)

The surface temperature increases as the surface heats up due to radiation from the fire

h is the convective heat transfer coefficient 20 W/m^2-K (3.52 BTU/hr-ft^2-R)

Plate area is assumed to be 2.3 m^2 (24 ft^2) (per API Section 4.4.413, Figure 1)

Density of carbon steel is 7850 kg/m^3 (490.1 lb/ft^3)(from internet)

Plate thickness varies from 3.2 mm (1/8 in.), 12.7 mm (1/2 in.) and 25.4 mm (1 in.).

Heat capacity of carbon steel = 502 J/kg°-K (0.12 BTU/lb-°F)

Additionally, larger thicknesses 50.8 mm (2 in) and 76.2 mm (3 in) are also used.

From equation (1), $q_{absorbed}$ is calculated.

The plate volume is calculated for each thickness from the given area.

Using the given carbon steel density, the plate mass is calculated.

From equation (2) Delta $T_{surface}$ is calculated.

Starting with the initial surface temperature, the change in the surface temperature is calculated from equations (1) and (2) for different times in 0.5 second increments. The resulting wall temperatures are plotted and compared to values given in API 521 Appendix C (Figure C-9) as shown in Figure 1.

Figure 1 Calculated wall temperature profiles as a function of time

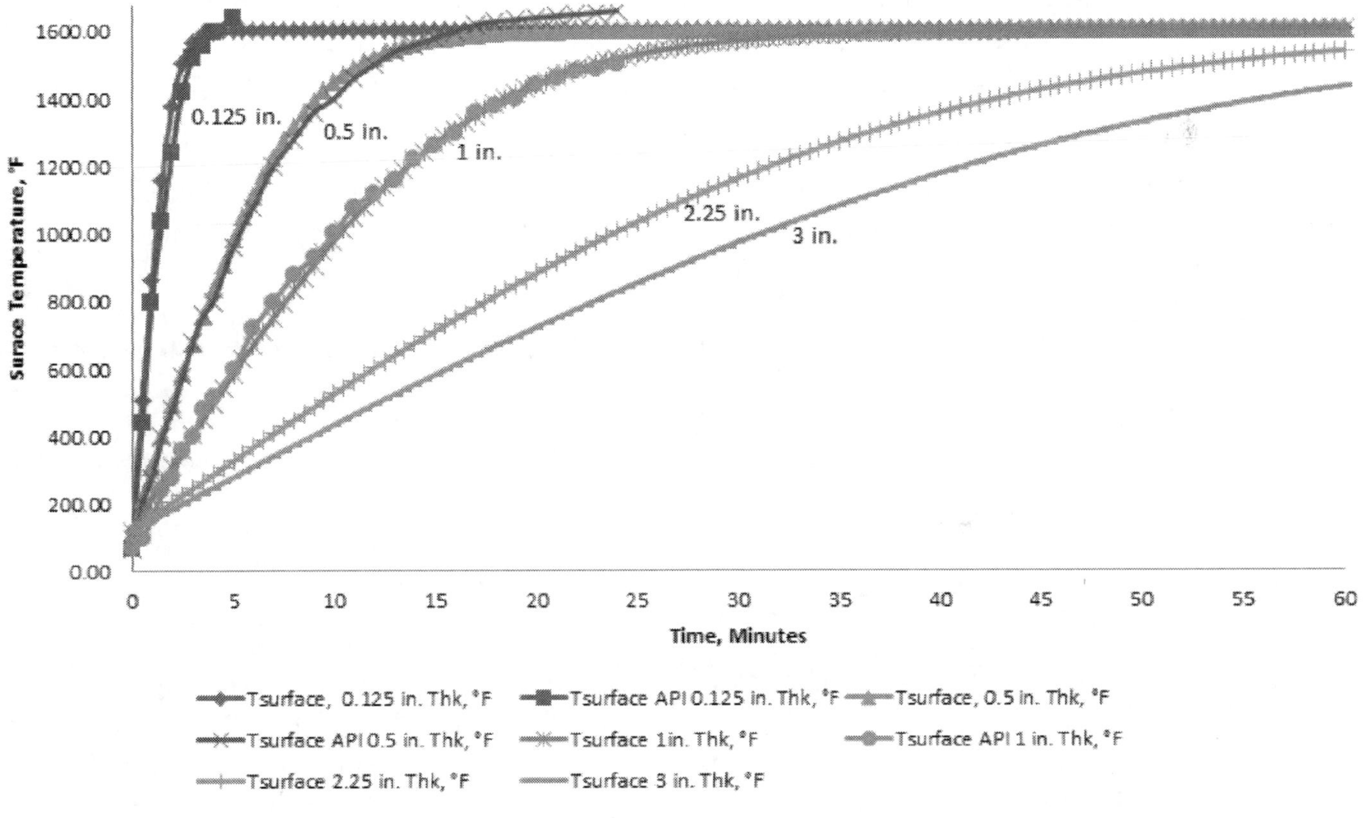

Calculation of the maximum pressure to avoid failure for various wall thicknesses:

As described in Appendix A Section A.3.5.4.6, the wall temperatures can be combined with tensile strength versus temperature data for carbon steel to determine the tensile strength, or the maximum pressure to avoid failure of a vessel, as a function of time in a pool fire scenario.

The tensile strength versus temperature data for carbon steel from Table A-7 in API 521 Appendix A is given in Table 1 for carbon steel.

Table 1 Tensile strength versus temperature for Carbon Steel (SA-515, SA-516)

Temperature (°C)	Temperature (°F)	Tensile Strength (MPa)	Tensile Strength (psi)
482	900	313.7	45,500
538	1000	251.7	36,500
593	1100	187.5	27,200
649	1200	137.9	20,000
704	1300	93.1	13,500
760	1400	62.2	9025

As the wall temperature heats up to 482°C (900°F), the tensile strength or the maximum pressure to avoid failure, decreases. These values are normalized with respect the tensile strength of 313.7 MPa (45,500 psi) and a safety factor of 0.75 is applied to generate Figure A-5 in Appendix A of API 521.

These values are extrapolated to larger thickness greater than 1 in using regression to determine the times to reach failure pressures expressed as a ratio of maximum pressure to avoid failure to MAWP for various thicknesses.

Table 2 Vessel rupture strength as % of MAWP versus tire exposure times

% of MAWP	100%	80%	60%	40%	34%	30%	26%	24%	
Thickness									
in	Time (mins)	Time (mins)	Time (mins)	Time (mins)	Time (mins)	Time (mins)	Time (mins)	Time (mins)	Comment
0.125	1.2	1.5	1.9	2.4	3.1	6.0	12.0	17.5	API 521
0.5	3.6	3.9	4.5	6.1	7.6	10.5	17.0	28.0	API 521
1	12.6	13.1	14.1	17.0	20.0	23.1	31.0	37.0	API 521
2	25.19	25.91	27.58	33.24	38.83	42.15	52.16	59.66	Extrapolated
2.25	28.52	29.28	31.14	37.48	43.74	47.11	57.68	65.18	Extrapolated
3	38.49	39.44	41.81	50.20	58.44	62.00	74.22	81.72	Extrapolated

Figure 2 shows a plot of the maximum allowable pressures to avoid failure for plates of different thicknesses from 25.4 mm (1 in). to 76.2 mm (3 in.)..

Figure 2 Failure pressures for various thicknesses as a function of time

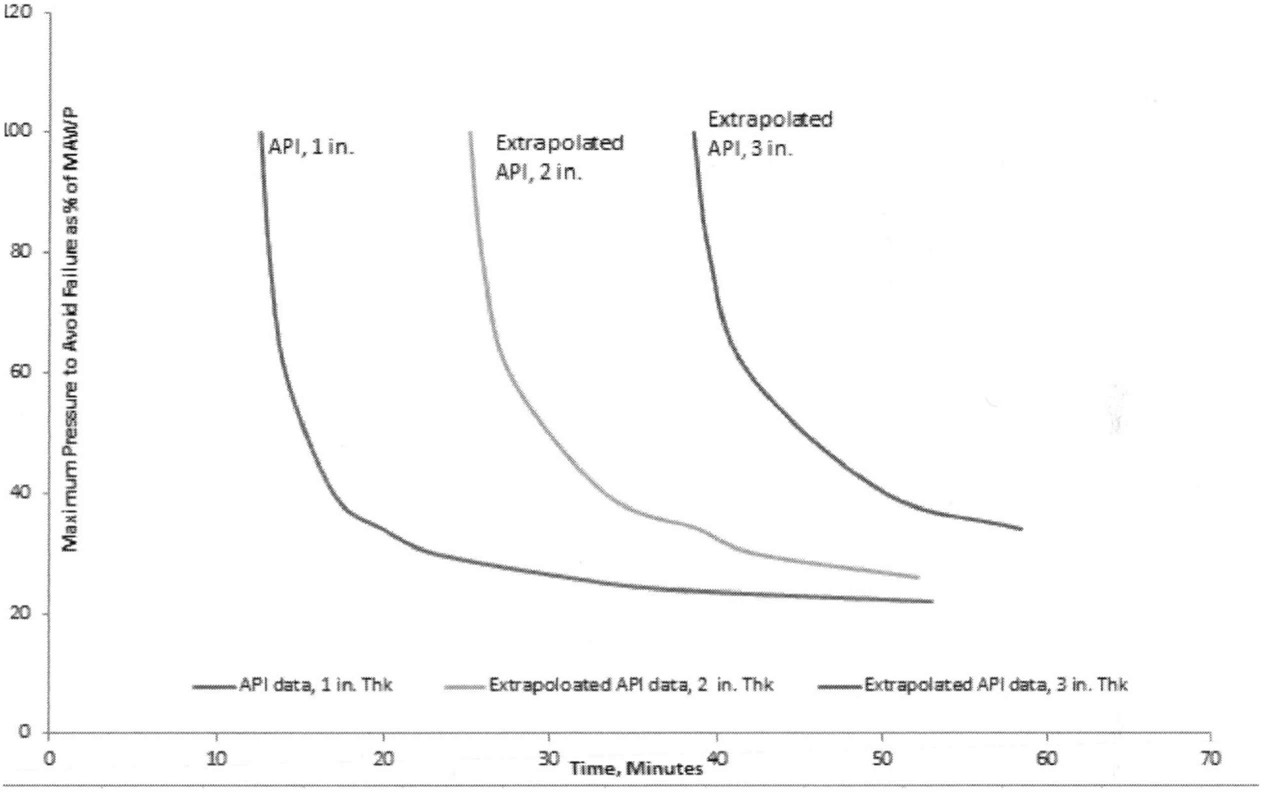

A hypothetical vessel of 4 m diameter and 7 m height with a vessel thickness of 101.6 mm (4 in.) having a hemispherical heads of thickness 57.2 mm (2.25 in.) is depressuerd from a design pressure of 80 barg to a pressure of 40 barg in 15 minutes using HYSYS v 7.3 depressuring utility. The resulting profile of the vessel internal pressure as a function of MAWP is overlaid on the temporal profiles of the maximum allowable pressures to avoid failure for various thicknesses. Data for about 60 minutes is plotted for comparison with the profiles of internal pressures as shown in Figure 3.

Figure 3 Plot of failure pressures for various thicknesses and the vessel internal pressures as a function of time

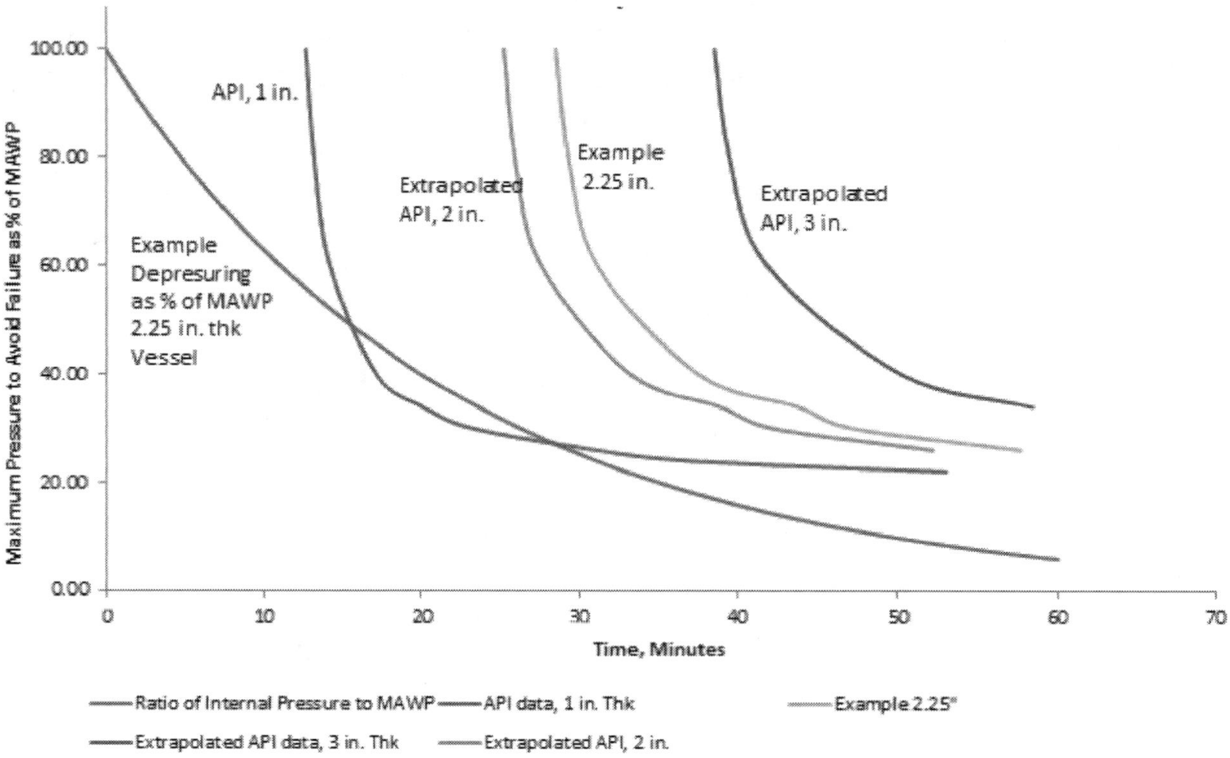

Discussion:

The kinks in the plot are due to round off errors and errors in the statistical regression. However, as seen from Figure 3, the vessel internal pressure for depressuring in a pool fire scenario is less than the maximum allowable pressure to avoid failure for thicknesses greater than 1". Thus, a methodology is developed for calculating wall temperature profiles during pool fire for vessels of various thicknesses and to check if the vessel internal pressures during depressurization exceed the failure pressures for vessel thicknesses greater than 1in.

References:

1. Pressure-relieving and Depressuring Systems, API Standard 521, Sixth Edition, January 2014.

Quantifying the Impacts of Human Factors on Functional Safety

Julia V. Bukowski, Ph.D.
Dept. of Electrical & Computer Engineering
Villanova University, Villanova, PA 19085
Julia.Bukowski@villanova.edu

Loren Stewart, CFSP
exida
64 N. Main St., Sellersville, Pa 18960
lstewart@exida.com

Prepared for Presentation at
American Institute of Chemical Engineers
2016 Spring Meeting
12th Global Congress on Process Safety
Houston, Texas
April 11-23, 2016

Quantifying the Impacts of Human Factors on Functional Safety

Julia V. Bukowski, Ph.D.
Dept. of Electrical & Computer Engineering
Villanova University, Villanova, PA 19085
Julia.Bukowski@villanova.edu

Loren Stewart, CFSP
exida

Keywords: systematic failure rate, product failure rate, site-generated failure rate, FMEDA, PFDavg

Abstract

It is not difficult to think of numerous ways in which human factors might impact functional safety. For example, imperfect repair, improper calibration, faulty installation, etc. can all contribute to decreases in functional safety performance. However, if actions are to be taken to mitigate these impacts then it is essential to have some quantitative measure of the impacts so that the actions' effectiveness can be assessed. This paper describes how the Site Safety Index™ (SSI) is used to adjust safety metrics, computed under the assumptions that human factors play no part in safety system performance, to reflect the effects of human factors on safety system performance on a site by site basis. Examples of the application of the method showing the safety performance changes attributable to human factors are provided.

1. Introduction

A safety instrumented function (SIF) has an inherent product failure rate based on its design and implementation and the assumption that the end user will take all necessary actions (i.e., meet all IEC 61508 [1] and/or IEC 61511 [2] requirements) to insure that the inherent product failure rate will be achieved. In practice, end-user activities and practices often fall short of this ideal and, consequently, the safety performance achieved by the SIF is often less than the ideal [3]. The SSI measures, on a scale of 0 to 4, the extent to which end-user actions support the achievement and retention of ideal SIF safety performance. The SSI is described in detail in [4]. This paper explains how end-user actions as measured by SSI affect various parameters that are used to calculate SIF safety performance and show how different levels of SSI impact various safety metrics.

2. Notation

FFD	field failure data
FMEDA	failure modes, effects and diagnostic analysis

IEC	International Electrotechnical Commission
MTTFS	mean time to fail safe
OREDA	Offshore and Onshore Reliability Data
PFDavg	average probability of dangerous failure on demand
RRF	risk reduction factor
SIF	safety instrumented function
SIL	safety integrity level
SRS	Savannah River Site
SSI	Site Safety Index™
1oo1	1-out-of-1 architecture

3. Using SSI to Measure Impacts of End-User Activities and Practices on Safety Performance of SIFs

3.1 Why a Site Safety Index™?

Studies [5] have shown that the failure rates estimated from field failure data (FFD) for identical modules in the same industrial application exhibit a difference in estimated failure rates for the same product from site to site. Typically, when compared to inherent predicted failure rates for the same products, the FFD estimates are approximately 1.2 to 3 times higher depending on product type. Site audits have attributed these differences to dissimilarities in end-user activities and practices which vary from site to site. This variation in end-user activities and practices can be captured by the SSI [4]. Further, these variations in activities and practices directly impact many variables used in models which assess the safety performance of SIFs.

3.2 The Key Variables Impacting PFDavg and Other Metrics

A key metric for measuring the safety performance of SIFs is average probability of failure on demand (PFDavg). Other closely related metrics are risk reduction factor (RRF), safety integrity level (SIL), and mean time to fail safe (MTTFS). Based on studies of FFD and proof test reports, nine key variables impacting PFDavg have been identified [6]. These nine variables are

1. Failure rates of each device including failure modes and any diagnostic coverage from automatic diagnostics (an attribute of devices chosen and of end-user practices);
2. Proof Test Effectiveness (an attribute of the proof test method and, to the extent that there is more than one possible proof test method, an attribute of end-user practices);
3. Proof Test Intervals (assignable by end-user practices);
4. Mean Time to Restore (an attribute of end-user practices);
5. Probability of Initial Failure (PIF) (an attribute of end-user practices);
6. Mission Time (an attribute of end-user practices);
7. Proof Test Duration (an attribute of end-user practices);
8. Site Safety Index™ (an attribute of end-user practices); and
9. Redundancy of Devices including common cause failures (an attribute of SIF design).

3.3 How Key Variables Are Modified to Quantify the Impacts on PFDavg and Other Metrics

Clearly, redundancy of devices is not an attribute of end-user practices or activities. The SSI itself is not modified in safety performance calculations; rather it is used to modify many of the other variables which impact PFDavg and other metrics. Specifically, whenever SSI is less than 4, the following five key variables are modified:

- failure rates of each device
- proof test effectiveness
- proof test intervals
- mean time to restore
- PIF

Failure rates of each individual device in the SIF are increased by a specific multiplier which is determined by the SSI value and the device itself. It is known that final elements are more likely to be negatively impacted by less than ideal end-user practices than are sensors or logic solvers. By increasing device failure rates on an individual device basis, it is possible to more accurately account for the effects of site practices on safety performance. Proof test effectiveness, i.e., the likelihood that all failures are discovered and completely repaired, is also decreased with decreasing SSI.

Proof test intervals, mean time to restore, and PIF are increased by specific multipliers determined by SSI value. It is common to find that less than ideal end-user practices include delaying proof testing, reducing proof test coverage, requiring additional restoration time, and introducing initial failures through imperfect pre- or post-installation testing. Final elements that may develop failures simply due to lack of motion of the moving parts over an extended time (a few months) are especially vulnerable to initial failures in excess of 1% [7, 8, 9].

The two variables not modified are mission time and proof test duration. Mission time is not modified because its impact is generally small unless the mission time is sufficiently beyond the SIF useful life so as to cause the SIF to see failures due to ageing, i.e., failures no longer modeled by a constant failure rate, without renewal during the mission time. Proof test duration has no impact if proof testing is performed with the process off-line. For proof testing performed on-line, proof test duration generally has only a very small effect unless proof testing frequency is very frequent, i.e., greater than or equal to once per month – an impractical situation for most industrial processes.

3.4 Calibrating the Multipliers

As with the development of the Failure Modes, Effect and Diagnostic Analysis (FMEDA) failure rate predictive technique [10] and its associated validated databases [11, 12, 13], it is necessary to calibrate the database of multipliers for the various modifications and to provide on-going validation against actual FFD. The initial calibration phase is still in progress and uses the extensive FFD of the Offshore and Onshore Reliability Data Project (OREDA) [14, 15] as well as the detailed proof testing and root cause failure records of the Savannah River Site (SRS) [8].

4. Examples of Quantifying the Impacts of SSI on Various Metrics

The same simple 1oo1 SIF was analyzed under five different levels of SSI. Figure 1 shows the impacts on PFDavg and SIL of these variations in SSI. For the same 1oo1 system, Figure 2 shows the impacts on RRF and SIL of variations in SSI. Finally, Figure 3 shows the impact on MTTFS of variations in SSI.

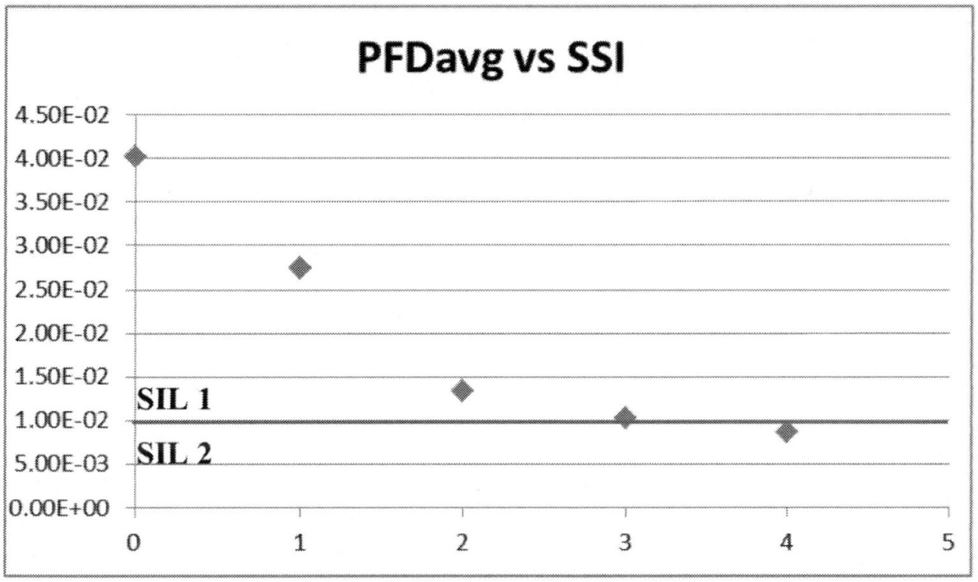

Figure 1. Impacts on PFDavg and SIL of Variations in SSI.

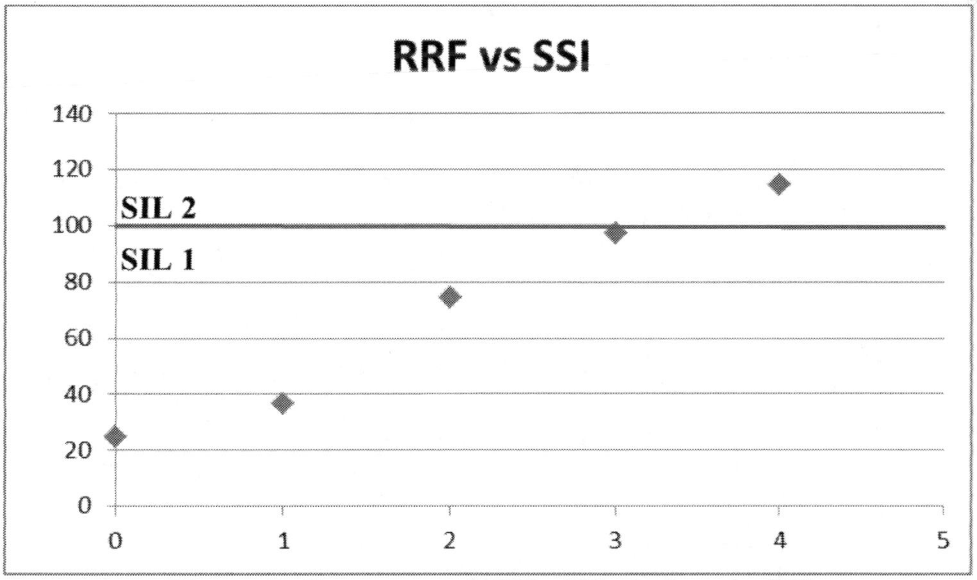

Figure 2. Impacts on RRF of Variations in SSI.

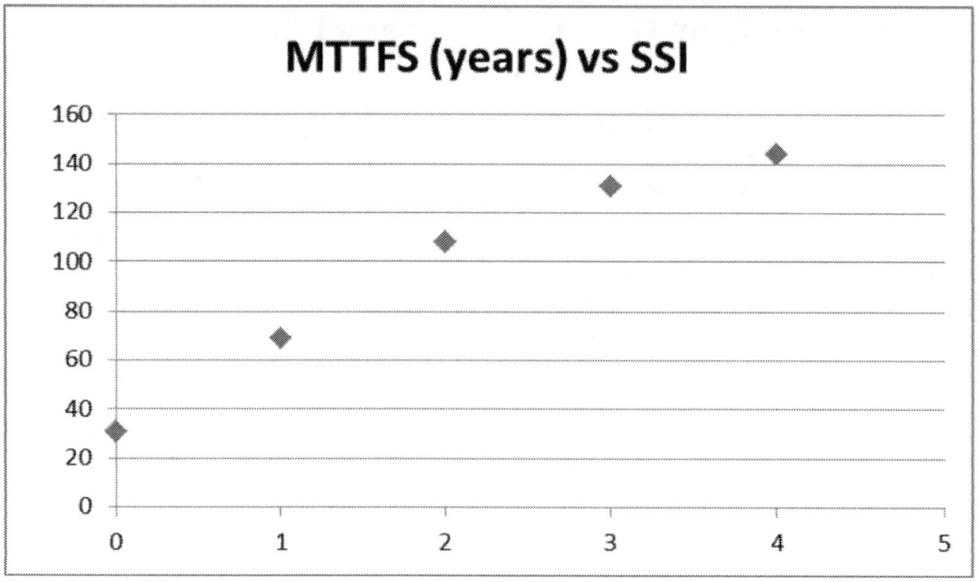

Figure 3. Impacts on MTTFS of Variations in SSI.

5. Summary

This paper clearly shows that the impacts of end-user activities and practices on safety performance of a SIF can be quantified through a simple model like the SSI. It also explains how the SSI modification multipliers are being calibrated and will be updated as increasing amounts of FFD become available. The methodologies presented here and in [4] allow end-users to measure quantitatively their own activities and practices on a site-by-site basis and to measure the impacts of those activities and practices on SIF safety performance.

6. References

[1] IEC 61508, *Functional Safety of electrical / electronic / programmable electronic safety-related systems*, Geneva, Switzerland, 2010.

[2] IEC 61511, *Application of Safety Instrumented Systems for the Process Industries*, Geneva, Switzerland, 2nd edition, 2004.

[3] J. V. Bukowski, W. M. Goble, and I van Beurden, "Product Failure Rates vs Total Failure Rates at Specific Sites: Implications for Safety," *Proceeding AIChE 11th Global Congress on Process Safety*, Austin, TX, April 2015.

[4] J. V. Bukowski and D. Chastain-Knight, "Assessing Safety Culture via the Site Safety Index™," *Proceeding AIChE 12th Global Congress on Process Safety*, Houston, TX, April 2016.

[5] W. M. Goble, "Maintenance capability: Does it matter in the long term?" *Hydrocarbon Processing,* June 1, 2013.

[6] "The Key Variables Needed for PFDavg Calculation," White Paper, exida 2015, http://www.exida.com/Resources/Whitepapers/the-key-variables-needed-for-pfdavg-calculation

[7] J. V. Bukowski, R. E. Gross and W. M. Goble, "Probability of Initial Failure for Spring Operated Relief Valves," *Proceedings of the ASME 2011 Pressure Vessels and Piping Division Conference*, Baltimore, MD, July 2011.

[8] J. V. Bukowski, R. E. Gross and W. M. Goble, "The Adhesion Failure Mode in Stainless Steel Trim Spring Operated Pressure Relief Valves," *ASME Journal of Pressure Vessel Technology*, Vol 135 No 6, December 2013.

[9] L. Stewart, J. V. Bukowski and W. M. Goble, "Improving Reliability & Safety Performance of Solenoid Valves by Stroke Testing," *Proceedings AIChE 9th Annual Global Conf on Process Safety – Process Plant Safety Symposium*, San Antonio, TX, April-May 2013.

[10] W. M. Goble and A. C. Brombacher, "Using a Failure Modes, Effects and Diagnostic Analysis (FMEDA) to Measure Diagnostic Coverage in Programmable Electronic Systems," *Reliability Engineering and System Safety*, Vol. 66, No. 2, Nov 1999, pp. 145-148.

[11] W. M. Goble and J. V. Bukowski, "Development of a Mechanical Component Failure Database," *Proceedings of the 2007 Annual Reliability and Maintainability Symposium*, Orlando, FL, January 2007, pp. 451-455.

[12] J. V. Bukowski and, W. M. Goble, "Validation of a Mechanical Component Constant Failure Rate Database," 2009 *Proceedings Annual Reliability and Maintainability Symposium*, Fort Worth, TX, January 2009, 338-343.

[13] J. V. Bukowski and L. Stewart, "Comparing Failure Rates for Safety Devices: FMEDA Prediction vs OREDA Estimation," exida White Paper available at www.exida.com

[14] SINTEF, *OREDA Offshore and Onshore Reliability Data Handbook, Vol 1. – Topside Equipment and Vol. 2 – Subsea Equipment*, 6th Ed, OREDA Participants, 2015.

[15] SINTEF, *OREDA Offshore Reliability Data Handbook, Vol 1. – Topside Equipment and Vol. 2 – Subsea Equipment*, 5th Ed, OREDA Participants, 2009.

441468 Using HAZOP/LOPA to Create an Effective Mechanical Integrity Program

Abstract:

Many people view the conduct of a HAZOP/LOPA to address regulatory requirements as a chore, and stop there. However, the implementation of a quality HAZOP/LOPA has the potential to provide a framework for addressing numerous safety and operational optimization issues at plants, including the formulation/refinement of the Mechanical Integrity Program. The purpose of this paper is to focus on the mechanical integrity program, illustrate how a quality HAZOP/LOPA can support the effective implementation of some of the new Damage Mechanism Review requirements for California Refineries (e.g., 5189.1(k)), and optimize key elements of an effective Mechanical Integrity Program, e.g.:
- inspection/testing methods
- testing intervals
- maintenance outage periods
- repair prioritization and allowable outage
- identification of low priority equipment

441486 Reactive Chemical Hazards of Azide Compounds: a Case Study and Lessons Learned

Abstract:

While a researcher was servicing a vacuum pump an unintended detonation occurred. This pump had been used in a sulfonyl azide synthesis ($R\text{-}SO_2\text{-}N_3$). The detonation occurred when he was applying effort to the hose barb from the outlet of the pump to connect it with a rubber hose. Differential Scanning Calorimetry (DSC) and headspace Gas Chromatography/Mass Spectrometry (GC/MS) were employed for the root cause investigation. The DSC showed a large exothermic decomposition (-1388J/g) for the residual sample from the pump fitting. Headspace GC/MS identified the presence of hydrazoic acid in the pump oil. The investigation concluded that either the azide precursor (sodium azide) or hydrolysis of the product (sulfonyl azide) was responsible for generating hydrazoic acid, which escaped from the dry ice trap and appeared in the downstream equipment, and probably reacted with metal parts (brass) to form a highly shock sensitive material (copper azide).

This paper reviews the reactive chemical hazards for azide compounds involved in this case study. Sodium azide is not particularly sensitive to mechanical impact, but its propensity to form hydrazoic acid in water is well-known, even at neutral pH. Likewise, the test data on sulfonyl azide product showed that it is not particularly sensitive to mechanical impact, but it can generate hydrazoic acid upon hydrolysis with water. Metal azides which are formed by the reaction between hydrazoic acid and metal surfaces of equipment parts are highly hazardous due to their extreme impact and friction sensitivity.

Based on lessons learned, this paper also describes further steps for synthesizing organic azides using aqueous sodium azide. These steps include, but are not limited to: (1) Utilizing a dedicated hood and equipments to limit potential HN_3 vapor streams from contacting any reactive metal surfaces. (2) Using a dual caustic trap system designed to scrub the outlet gases swept from the azide reaction step, instead of dry ice trap. (3) Using a 5% aqueous solution of ceric ammonium nitrate as a decontaminating agent for metal surfaces inside the fume hood at the completion of the campaign. (4) Eliminating all brass or copper fittings from equipment used in the synthesis, due to the potential formation of the unstable explosive copper azide that can form from exposure of brass or copper parts to HN_3.

441495 The Application of Frequency Data in Quantitative Risk Assessment

Abstract:

Quantitative Risk Assessment (QRA) is a vital resource for estimating risk in the process industry. There are four primary steps involved in conducting a QRA: hazard identification, frequency analysis, consequence analysis, and risk calculation (the combination of frequency and consequence to determine the risk). Frequency analysis is a key component of the QRA and can have a major impact on the risk profile.

The primary method of estimating event frequency for QRA utilizes historical leak frequency data, of which the Hydrocarbon Release Database (HCRD) is the most widely used leak frequency database. One main issue with applying historical frequency is not having transparency of the causes of the events. Using the historical data approach, a single value is generated for the frequency of an event occurring, with no association to the cause (e.g. corrosion, dropped object, etc.). It has also become widely accepted to apply these offshore databases to onshore processes. There are many concerns in applying these databases to onshore processes, as there are differences in the types of equipment, the hazards the equipment is exposed to, and the safety management systems in place.

When QRAs were first conducted during the 1980s, they were an exhaustive and detailed process. In today's fast track industry, often frequency analysis is conducted without undertaking an evaluation to determine if the frequency being applied is appropriate for the hazard at hand. Leak frequency analysis has become an isolated activity, rather than relying on frequencies established during HAZOP and reliability assessments such as fault tree analysis.

An alternative method to employing historical databases to estimate event frequency is conducting fault tree analysis. Fault tree analysis involves dissecting a hazard into its causes, using a "top down" approach, beginning with hazard identification and working down to a level in which applicable frequency data is available. Fault trees provide a logical relationship between the hazard and its causes.

If historical leak frequency data is employed, which is independent of the cause, the resulting risk is also independent of the cause. Therefore, mitigation measures and the decision making process tend to focus on the back-end consequence of the event, rather than the cause of the event.

Fault tree analysis, although it has had limited application in onshore and offshore risk assessment, provides a major advantage of providing a link between the cause of an event and the associated frequency. This transparency between the event frequency and cause enables us to identify major contributors to the risk and provide risk reduction measures based on what is actually causing the event to occur.

The objective of this paper is to provide an alternative to employing offshore historical frequency data to QRA and demonstrate that the alternative method can provide major advantages in mitigations by linking the estimated risk to the cause and focusing the decision making process on the front-end of the problem rather than solely on the back-end.

441528 Uncertainty in Quantitative Risk Assessment and Decision Making

Abstract:

Including uncertainty in Quantitative Risk Assessment (QRA) is recommended in most published Best Practice Recommendations for QRA, including those published by TNO, CCPS, and Norwegian authorities. Unfortunately typical commercial QRA techniques and models used in Process Risk Analysis do not include the capabilities of uncertainty within the model formulations. Effects on uncertainty may be based on qualitative estimates or ignored, or at most accounted for by sensitivity analysis by "swinging" variables to estimate individual variable sensitivity. By not presenting the uncertainty, the risk assessment results may be misrepresented as possessing an accuracy greater than is warranted.

Failure to account for uncertainty also tends to support centrist thinking and poor quality decision making. The natural tendency of users of risk analysis is to believe in their results, especially if much effort has been put into developing the results. If the result is a single number, that number may be assumed by the analyzer and users as "right". In addition, there is a tendency in people to assume that by picking the most likely values for the inputs into a model, the end result is also the most likely result. This may not be true if the model is complex and non-linear. The most likely results must be determined by including variable distributions and uncertainty modeling in the analysis.

Decision analysis is also deceptively simple when single point criteria or simple curves are used for decision making purposes. A challenge in risk assessment is utilizing results and presenting them in a straight forward way that capture the uncertainty and complexity while allowing clear, high quality decisions by end-users.

This paper presents sources of uncertainty, analysis methods, examples of potential errors introduced by ignoring uncertainty, and illustrates how one can include uncertainties in your quantitative risk assessment (QRA) and benefit from it in your risk assessment decisions.

441529 Analysis of Process Systems during Accidental Fire

Abstract:

In the past 40 years, there have been over 60 major fire accidents reported on offshore platforms. The thermal effects imposed on process equipment and systems (such as pressure vessels, flare scrubbers, etc.) has the likely potential to result in catastrophic consequences such as explosions given that these process systems typically operate at very high pressures. In the event of an accidental fire, blowdown valves are engaged to relieve the excess pressure of the process system to acceptable values in a specific time period. During that time, however, the increased temperature of the blowdown system due to the impinging fire may result in the degradation of its structural capacity. The combination of a degraded vessel shell and a high internal pressure typically leads to sudden ruptures and explosions which could result in the escalation of the initial fire or injuries to personnel. Therefore, it is critical to accurately capture the pressure and temperature of the process system in a fire blowdown event.

A Blowdown analysis is typically carried out using simple thermodynamic software. While these software provide cost-effective approaches to studying process systems, they are unable to accurately compute the local effects of heat transfer and fluid behavior. Moreover, stress concentrations on the pressure vessel due to nozzle effects are not captured correctly using classical approaches and advanced analysis is required. Therefore, to accurately capture the rupture time of pressure vessels, scrubbers, etc., during a fire scenario, advanced approaches such as Computational Fluid Dynamics (CFD) are required.

In this paper, we will demonstrate a new approach in analyzing onshore and offshore process systems in an extreme fire scenario. Heat transfer through the vessel shell as well as the internal fluid will be accounted for by computing the rates of conduction, convection, and radiation. The time history of temperature and pressure as well as the change of phase of the fluid inside the pressure vessel will be captured in the analysis. A degradation study of the structural steel will be performed to predict the time of rupture of the process system to internal pressure using advanced analysis. Moreover, the effect of jet fire momentum on the vessels and their supports will be computed according to industry guidelines.

The methodology is validated by comparison with experimental data. The rupture time and rupture location will be computed and design recommendations will be provided to increase the strength of the process system at rupture locations. The current approach could provide accurate results to compute the response of process systems such as high-pressure vessels, scrubbers, etc., during offshore or onshore fire scenarios.

Assessing Safety Culture via the Site Safety Index™

Julia V. Bukowski, Ph.D.
Dept. of Electrical & Computer Engineering
Villanova University, Villanova, PA 19085
julia.bukowski@villanova.edu

Denise Chastain-Knight, P.E., CFSE
exida
64 N. Main St., Sellersville, Pa 18960
dchastainknight@exida.com

Prepared for Presentation at
American Institute of Chemical Engineers
2016 Spring Meeting
12th Global Congress on Process Safety
Houston, Texas
April 11-23, 2016

Assessing Safety Culture via the Site Safety Index™

Julia V. Bukowski, Ph.D.
Dept. of Electrical & Computer Engineering
Villanova University, Villanova, PA 19085
Julia.Bukowski@villanova.edu

Denise Chastain-Knight, P.E., CFSE
exida

Keywords: Safety Lifecycle, safety culture, systematic failure rate, site-specific safety performance, PFDavg

Abstract

How can a company establish a baseline measurement of its safety culture against which to gauge improvement? The Site Safety Index™ (SSI) quantifies in part (on a scale of 0 – 4) the degree to which a company's end-user practices support the attainment and retention of an appropriate safety culture for operations and maintenance. Further, the SSI can be used to appropriately adjust parameters that directly impact measures of safety such as probability of failure on demand (PFDavg). Thus, the impacts of safety culture can be further quantified and the effects of changes to safety culture can be assessed.

1. Introduction

Consider the following hypothetical scenario. A company assesses the risk intrinsic to one of its plant processes (given its current level of mitigation) to be greater than the tolerable risk dictated by company policy or regulatory standards. Further mitigation to reduce the process risk to an overall risk at or below tolerable risk level will require a safety instrumented function (SIF) which will result in a risk reduction factor (RRF) of at least 110, i.e., a SIF of safety integrity level (SIL) 2. The company may choose to design and implement the required SIF itself or to deploy an existing SIF design and implementation.

In either case, the company will need to ascertain that the SIF meets its risk reduction requirements. There are several ways to do this. For example, it may rely on proven-in-use data to claim the company has met its risk reduction goal. This proven-in-use data may come from another company or the database of a consortium of companies. Alternately, the company may predict the risk reduction and SIL level by using manufacturers' published data or other available databases perhaps in conjunction with a commercial tool. If by either or both means of risk reduction/SIL verification the company determines the risk reduction goal is met by the SIF, is this alone sufficient for the company to state that, when implemented in its plant, its process risk will be at or below tolerable levels? **No!**

When the company implements the SIF in its process, the actual risk reduction achieved by the SIF will be determined not only by the SIF attributes taken into account by the verification methods but also by the company's safety culture, i.e., by the overall impact of end-user practices that take place within the company in general and especially within the plant/site itself. Further, these practices may differ from site to site even in the same company.

For example, if the risk reduction/SIL was estimated from proven-in-use data, unless "systematic" failures were eliminated, these data inherently contain the impacts of the end-user actions/omissions which took place at the plant or plants during the time period in which the proven-in-use data were collected. When the company implements the SIF, it may actually experience the same, higher or lower risk reduction than that estimated from proven-in-use data depending on whether the company's end-user current practices are the same, better or worse than those practices in the plants *when* and where the proven-in-use data were collected. Especially for consortium-based data, it is not generally known exactly from where the proven-in-use data were gathered much less what the specific end-user practices were at those plants. Thus, estimated risk reduction/SIL based on proven-in-use data is difficult to adjust to more accurately represent the actual risk reduction/SIL the company will experience.

If the risk reduction/SIL was predicted based on the manufacturers' published failure rates or other failure rate data, these rates are likely predicated on the assumptions that the end-user faithfully performs all required actions to sustain the safety integrity of the equipment. Such actions include, but are not limited to, insuring that

- the equipment implementing the SIF is appropriate to the task and environment,
- the equipment is properly installed and calibrated,
- the equipment is correctly functioning when installed,
- all in-service maintenance is correctly and completely performed on schedule,
- the equipment is maintained so that no ageing occurs prior to proof testing,
- proof testing is completely and correctly performed,
- the equipment is replaced when its useful life can no longer be extended by maintenance and refurbishment.

Note that the list extends beyond the usual concepts of maintenance alone. Any required action that is omitted, delayed, or performed incompletely or imperfectly will reduce the risk reduction (and perhaps SIL) achieved in the SIF's actual use in the plant. However, in the case of predicted risk reduction/SIL, which is based on manufacturers' data or Failure Modes, Effects and Diagnostics Analysis (FMEDA) [1], it is possible to adjust the initially calculated risk reduction/SIL to account for a company's end-user practices. This adjustment occurs through the use of a SSI.

This paper explains the way in which the SSI is measured and illustrates how the SSI can impact typical safety parameters such as PFDavg. A second paper [2] explains how the SSI is used to adjust the initially calculated risk reduction/SIL.

2. Notation

FMEDA	failure modes, effects and diagnostic analysis
IEC	International Electrotechnical Commission

PFDavg	average probability of dangerous failure on demand
RRF	risk reduction factor
SIF	safety instrumented function
SIL	safety integrity level
SSI	Site Safety Index^{TM}

3. Measuring a Site's Safety Culture

3.1 Why a "Site" Safety Index?

The reader may wonder why the terminology "Site" Safety Index is used as opposed, perhaps, to "Company" Safety Index. It is well documented that a safety product may exhibit a safety performance less than that specified by the manufacturer or by a FMEDA [3, 4] and that even within the same company, failure rate estimates derived from field failure data can vary significantly from site to site [5]. Thus, while company-wide policies and procedures will impact the safety achieved, the possibly different ways company-wide directives are implemented on a site-by-site basis requires that a safety index be measured at the site level.

3.2 How is the Site Safety Index^{TM} Evaluated?

There are currently two ways that SSI can be evaluated: via an external audit, which provides the most accurate measure of SSI, and via a self-administered questionnaire, which relies on less information than the audit and provides an approximate measure of SSI. These two evaluation methods are detailed next.

3.2.1 Via an External Audit

The international standard IEC 61511 [6] lists the requirements that must be met for the entire Safety Lifecycle process. In an on-site audit, it can be verified that the site meets these requirements in combination with additional detailed requisites for the operations and maintenance activities. Additionally, all safety and lifecycle procedures are audited to be sure: a) that the procedures are adequate to support the requirements and b) that they are being correctly followed at the site. The outcome of the audit is a report which includes the SSI score obtained as well as recommendations for changes and improvements. The interpretation of the measured SSI level itself is discussed below.

3.2.2 Via a Self-Administered Questionnaire

A site that is interested in acquiring an approximate measure of its current state of safety culture may choose to complete a questionnaire of about 40 items which represents a subset of the information gathered in a complete audit. These questions cover activities and practices spanning a portion of the lifecycle of the SIF and include installation and commissioning, safety validation, operations and maintenance, SIF modification and decommissioning. Table 1 lists the questions for the operations and maintenance phase of the lifecycle. The complete set of questions is available at [7]. The questions, which are answered Always, Often, Never, or Not

Applicable, are subject to weighted scoring. The total score received determines an approximate SSI level for the site.

Table 1 Questions Regarding Site Activities and Practices during the Operations and Maintenance Phase of the SIF Lifecycle

Do your operation and maintenance procedures include competency requirements for those performing tasks to maintain the "as designed" functional safety of the SIF?
Do your operation and maintenance procedures include the information which needs to be maintained on device failures, proof test results, false trips, demand rates and restore times on the SIF?
Do your operation and maintenance procedures include procedures for fault diagnostics and repair after a failure in the SIF?
Do your operation and maintenance procedures require instruments to be rebuilt or replaced at the end of their useful life?
Do your operation and maintenance procedures require the use of a computer tool like SILStat™ to track demands, false trips, failures and maintenance performance?
Do your operation and maintenance work tasks proceed only in accordance with the relevant procedures?
Are operators that work with the SIF in their area trained to ensure they understand how the SIF functions and what hazards are being protected against?
Do you perform periodic proof-tests using the written procedure?
Is your frequency of proof tests decided using the PFD_{avg} calculation?
Are scheduled proof test performed on schedule?
Do you repair any deficiencies found during the proof testing in a safe and timely manner?
Do you maintain records of proof tests including descriptions of tests and inspections performed, equipment tested, dates, and names of personnel performing work?
Do you maintain records of proof tests including "as found" results of the tests and inspections?

The basis of the validity of using a questionnaire for determining SSI is the accepted practice per IEC61508 of using a questionnaire to determine the β factor that describes the extent of common cause failure to be modelled when evaluating the safety performance of redundant SIF architectures [8].

3.3 Interpreting the SSI Level

The SSI is a number from 0 – 4 which is an indication of the level of site activities and practices that contribute to the safety performance of SIF's on the site. Table 2 details the interpretation of each SSI level. Note that the levels mirror the levels of SIL assignment and that SSI 4 implies that all requirements of IEC 61508 and IEC 61511 are met at the site and therefore there is no degradation in safety performance due to any end-user activities or practices, i.e., that the product inherent safety performance is achieved.

Table 2 Details of the Interpretation of Each SSI Level

SSI Level	Description
SSI 4	Perfect - Repairs are always correctly performed, Testing is always done correctly and on schedule, equipment is always replaced before end of useful life, etc.
SSI 3	Superior - Repairs are usually correctly performed, Testing is usually done correctly and on schedule, equipment is usually replaced before end of useful life, etc.
SSI 2	Typical - Repairs are mostly correctly performed, Testing is mostly done correctly and on schedule, most equipment is replaced before end of useful life, etc.
SSI 1	Good - Repairs are often correctly performed, Testing is often done and often on schedule, some equipment is replaced before end of useful life, etc.
SSI 0	None - Repairs are not performed, Testing is not done, equipment is not replaced until failure, etc.

4. An Example of the Impacts of SSI

The specific ways in which SSI modifies key variables (e.g., failure rate, proof test effectiveness, etc.) which impact PFDavg, SIL and RRF, as well as other safety metrics, is explained in detail in [2] where additional examples are presented. For the purposes of this paper, Figures 1 and 2 provide a simple example of how SSI can impact safety performance. In Figure 1, computations were performed under the assumption that SSI level is 4 and thus end-user activities and practices do not degrade the safety performance of the SIF. Note that in this case the SIF SIL level is computed as SIL 2 based on PFDavg, and the RRF is computed as 113 – slightly greater than the requirement of the opening hypothetical example. In Figure 2, the analysis is repeated but this time the SSI level is assumed to be 3. Now, based on PFDavg, the SIL performance is a SIL level 1 and the RRF is computed as 98 – insufficient for the requirement of the hypothetical example. This shows how the impacts of site activities and practices can be quantified and how effects of changes to these activities and practices can be measured.

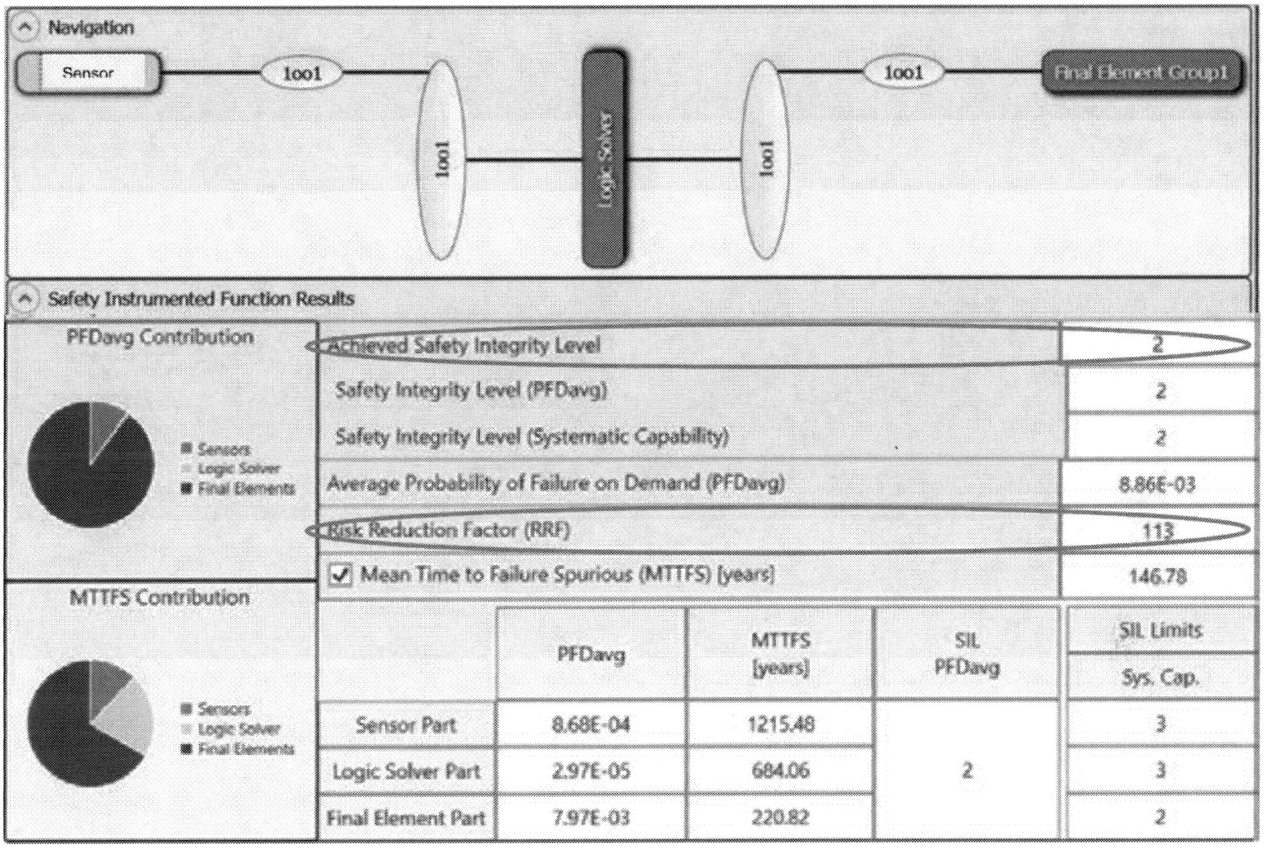

Figure 1. SIL verification/RRF computations performed under assumption SSI = 4.

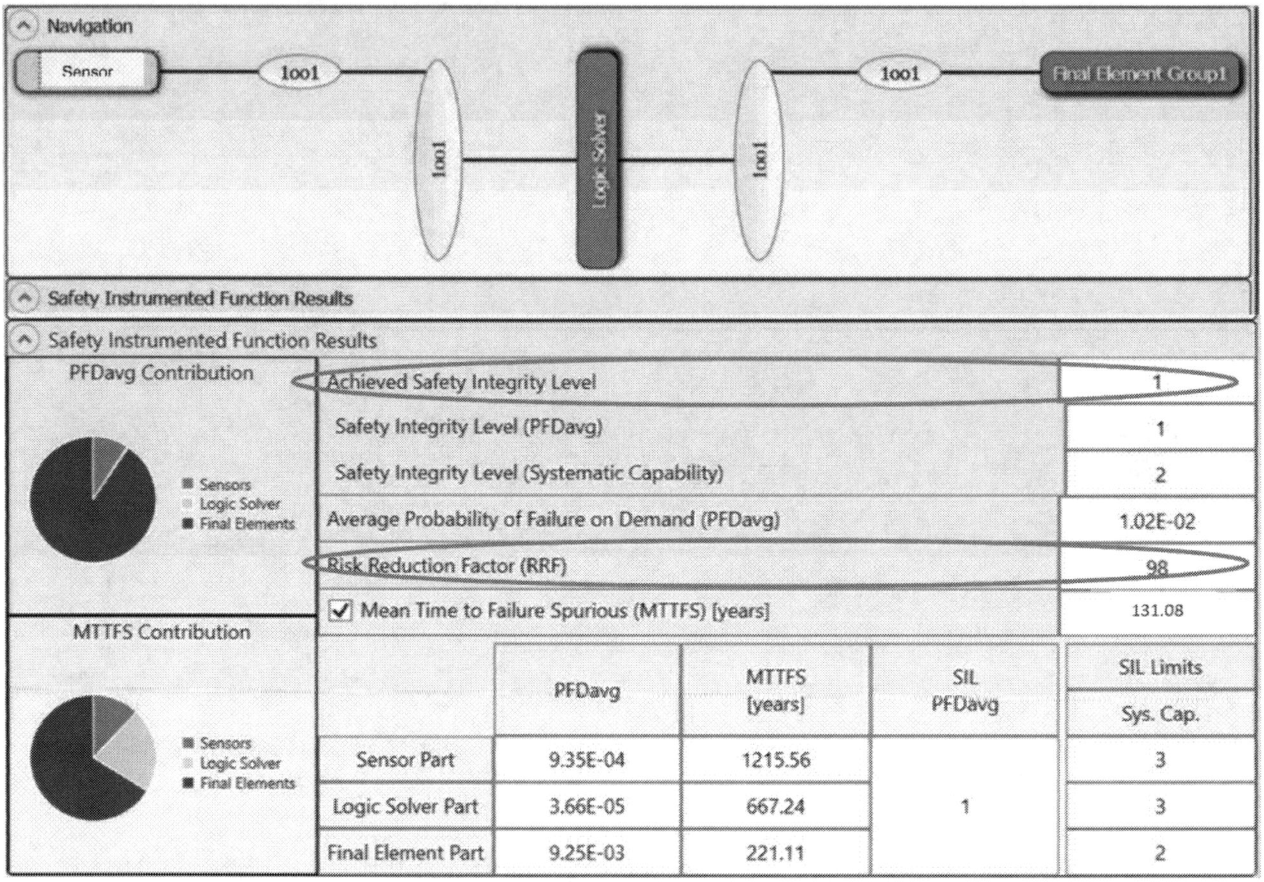

Figure 2. SIL verification/RRF computations performed under assumption SSI = 3.

5. Conclusions

The SSI provides a simple means to measure how well site-specific end-user activities and practices support the achievement and retention of SIF maximal safety performance. Further they can be used to adjust metrics of SIF safety performance from the ideal achievable to a more realistic measure of what is actually attained at a specific site given the site's activities and practices. This last concept is explored more fully in a separate paper [2].

6. References

[1] W. M. Goble and A. C. Brombacher, "Using a Failure Modes, Effects and Diagnostic Analysis (FMEDA) to Measure Diagnostic Coverage in Programmable Electronic Systems," *Reliability Engineering and System Safety*, Vol. 66, No. 2, Nov 1999, pp. 145-148.

[2] J. V. Bukowski and L. Stewart, "Quantifying the Impacts of Human Factors on Functional Safety Failure Rates," *Proceeding AIChE 12th Global Congress on Process Safety*, Houston, TX, April 2016.

[3] J. V. Bukowski, W. M. Goble, and I. van Beurden, "Product Failure Rates vs Total Failure Rates at Specific Sites: Implications for Safety," *Proceeding AIChE 11th Global Congress on Process Safety*, Austin, TX, April 2015.

[4] J. V. Bukowski, W. M. Goble, R. E. Gross and S. P. Harris, "The Effects of Maintenance Actions on the PFDavg of Spring Operated Pressure Relief Valves," *Proceedings of the ASME 2014 Pressure Vessels And Piping Division Conference*, Anaheim, CA, July 2014.

[5] W. M. Goble, "Maintenance capability: Does it matter in the long term?" Hydrocarbon Processing, June 1, 2013.

[6] IEC 61511, *Application of Safety Instrumented Systems for the Process Industries*, Geneva, Switzerland, 2nd edition, 2004.

[7] www.exida.com/SSI

[8] IEC 61508, *Functional Safety of electrical / electronic / programmable electronic safety-related systems*, Geneva, Switzerland, 2010.

441381 Fire Hazard Associated with Polymers and Application of Polymer Nanocomposites for Flame Retardancy

Abstract:

Due to advances in polymer science, their widespread application is found in daily products including vehicles, electronics, clothing, furniture *etc*. However, most of the items that first ignite in any fire scenario includes mostly polymeric materials according to NFPA reports which clearly elucidates the importance into looking for flame retardancy approach for polymers. Polymer nanocomposites is a relatively new approach undertaken for flame retardancy and has the potential to bring multi-functionality including improved thermal behavior, physical barrier effect, catalytic charring effect and better material performance. The objective of this paper is to initially provide the current state-of-the-art review on polymer nanocomposites in terms of their fire retardancy. The paper describes the hazards associated with burning of polymers, flame retardant mechanism, synthesis and characterization of polymer nanocomposites with focus on thermal analysis. Different kinetic model has been studied and applied for analysis of thermogravimetric analysis data. The kinetic parameters of the polymer degradation have been identified. Primarily polystyrene-silica nanocomposite has been prepared by in-situ method and were characterized for their morphology and thermal behavior using characterization methods, such as Scanning Electron Microscopy (SEM), X-Ray Diffraction (XRD), Thermogravimetric Analysis (TGA), and Differential Scanning Calorimetry (DSC). Initial results show nanocomposites with promising fire-retardancy properties.

441530 Applying QRA Study to Select and Design Effective Release Prevention Safeguards

Abstract:

In the past two decades, Quantitative Risk Analysis (QRA) has become a staple safety analysis in offshore oil and gas industry. A comprehensive QRA study could be a powerful tool if the results are understood and changes are implemented to improve safety in the engineering design. Conventional QRA study relies on generic leak frequency and parts count method. One major limitation of conventional QRA study is that it is focused primarily on the loss of containment of hazardous fluids and what happens when they are released. To reduce the risk to acceptable level, usually post-release mitigation measures are addressed. It is a challenge to tackle the effectiveness of release prevention safeguards. This paper presents an approach to address the dominant risk contributors identified in conventional QRA study. The aim is to increase confidence and facilitate decision making in the design selection of loss of containment prevention safeguards.

Sensitivity analysis is usually conducted as part of a QRA study to evaluate the effect of input parameters. However, in most cases, sensitivity analysis focuses on the mitigation measures, such as safety system locations or ratings, personnel distribution or modification to the QRA model itself. Means of release scenario prevention is rarely investigated to the point that it is beneficial to the operation or to the design project. This paper proposes that in order to maximize the use of QRA results, decision makers should investigate means of scenario prevention by performing reliability analysis (e.g., fault tree analysis, human reliability analysis) on the top contributing scenarios. This paper presents some examples of reliability analysis performed on top risk contributors. This paper also presents suggestions on how the results can potentially be applied for the benefit of the operation or design project. The outcome from the QRA study in combination with reliability analysis will increase decision makers' understanding of the top risk contributing scenarios. This approach can be utilized to identify opportunities for improvement in the prevention safeguards.

441536 Managing the Uncertainty in Blast Load Calculations with Enhanced Technical Safety Analysis

Abstract:

The safety of personnel working within onshore and offshore oil and gas processing facilities with respect to vapor cloud explosion (VCE) scenarios revolves around the risk associated with the event. Conducting a risk assessment is quite typical within the offshore and onshore loss prevention communities. The consequence of a VCE event is a very complex phenomenon however. It varies based on several blast load parameters. With improved understanding of blast loading parameters such as blast overpressure, impulse, time lag, net blast loading, etc. structural designers have gained increased confidence in their design and have reduced conservatism in their blast designs. However, there is a significant gap in the risk assessment of onshore and offshore facilities between the process safety and the structural design groups. Using structural insight in combination with rigorous process safety risk evaluation techniques we can reduce the gap between the risk and consequence of a blast design, providing the stakeholders with a more accurate understanding of the risk and consequence of a blast event.

This paper will attempt to address the impact of uncertainty in pressure exceedance curves and;

- how the confidence level impacts the design blast loads;
- how the slope of the exceedance curves impacts the blast design loads;
- two different blast events can result in similar design blast loads for a given risk;
- how a small change is frequency can lead to a significant change in design blast loads;
- how the negative phase of a blast scenario could lead to failure of some specific type of structures;
- how the response of long wall or piping structure is impacted by time-lag of the design blast load;
- how the design blast pressure and design blast impulse impacts a structure;
- how non-structural members, typically not considered in blast design, impact the blast design;
- how simultaneous application of blast load on different parts of structure, taking into account time-lag, affects the blast design.

Currently for a blast event, both blast pressure and blast impulse are defined as independent variables whereas in reality they are coupled. The correlation between the impulse and pressure will also be discussed in the paper. Defining appropriate pressure time-history is critical for structural design. Rise time, positive-phase and negative-phase duration are unknowns in a blast design. How an appropriate time-history is established will also be addressed.

Prior knowledge of implications of a blast event for an associated risk is helpful during structural blast design. Early involvement of structural engineers within the process safety procedure is a recommended solution to such questions. Enhanced loss prevention of offshore and onshore structures can only be established by always keeping the consequence for a blast event for a given risk in mind.

441537 Occupant Vulnerability for Process Plant Building Design Due to Secondary or Domino Events

Abstract:

Statistics indicate that it is in fact not uncommon to have an accidental explosion on an offshore platform, refinery or petrochemical installation. The focus for several years has been on occupied buildings and their resistance to blast loads from accidental releases of hydrocarbons. Common practice dictates that domino effects may not need to be considered given that a large catastrophic release is equivalent in consequence. However, the catastrophic events are filtered out in a risk assessment due to their unlikelihood and as a result, domino effects are not captured. For example, a small release of highly reactive fuel, can lead to a small explosion which can easily damage nearby piping and vessels which can in turn become catastrophic events. The design and assessment of piping systems and vessels against blast loads are of paramount importance and require advanced analysis capabilities. In order to have a safe, i.e. blast-resistant process piping system design, an advanced nonlinear analysis in time domain is required so that we may properly capture the dynamic response of the piping and vessels subjected to blast loading. This type of analysis, in general, should take into account (1) Blast drag load and blast overpressure; (2) Nonlinear material properties, i.e. thermal and strain-rate dependence; (3) Effects of non-structural masses and adjacent piping systems; (4) Pipe supports; (5) Failure of flanges and piping detail; (6) Effects of Operating Temperature on Material Properties; (7) Blast direction/Ignition location/Attenuation sensitivity; (8) Shielding effect, e.g. pipes behind large objects such as vessels; and (9) Effects of pipe insulation, i.e. increased pipe diameters. This paper first discusses the methodology/techniques to account for such effects in finite element analysis for a safe design. The paper then uses case studies to demonstrate the methodology we develop to show how advanced simulation techniques are applied as tools for piping blast response to minimize the risk of secondary failure due to initiating events. These case studies involve a piping system that is simulated from global to detail modeling. The goal of the paper is to discuss how to achieve a safer design for process piping systems by using advanced analysis.

441629 Assessment of the Mechanical Integrity of a Pipeline with Metal Loss, Using Finite Element Method

Abstract:

Following the guidelines proposed by the ASME B31G and API579 standards, for the assessment of systems that compromise its mechanical integrity, given the existence of defects associated with the presence of corrosion damage, the probability of failure of a pipeline, which transports hazardous materials associated with the Oil & Gas industry, and is exposed to pitting corrosión damage was calculated when it overcomes its yield stress. For this, a methodology that combines the corresponding structural analysis of the system and subsequent reliability analysis was proposed. Through the first, the calculation of the von Mises stress present in areas where the pitting defects was made; for this, the corrosion growth model to be used in the present work was defined by implementing the correlation proposed by the Southwest Research Institute (SwRI) to find out the corrosion rate growth in annual terms. Subsequently the pipeline section where the study would be carried out was defined, based on the results obtained in a previous study, which analyzed information related to corrosion damage present on the pipeline, collected through the In Line Inspection (ILI) technique. This allowed to restrict the test area to a total length of 80 cm and 1/8 section of pipe, lastly simulating the system and obtaining the loads associated with the von Mises stress affecting corrosion defined areas. For reliability analysis it was necessary to find the probability distribution function associated with the von Mises stress on the system studied; defining an experimental design which sought to evaluate different values of the random variables defined on the model proposed by SwRI, thus simulating the system a sufficient number of times in order to find the distribution of von Mises stress for this case study. With the distribution function defined, system reliability analysis was carried out through Monte Carlo simulations; thus, it was established that the limit state function would have a conventional structure comparing resistance with loads. Thus, the resistance was set as the yield stress of the material, while the load would be the von Mises stress, and thus the random variable from which random numbers governed by the probability distribution found previously were generated. As mentioned previously, it was defined that the present system fails when the von Mises stress exceeds the yield stress of the material; finding that for the time range evaluated (year one, five, ten and fifteen), the probability of failure is behaved as expected, increasing year by year if no maintenance is not performed to the pipeline section evaluated.

441539 An Improved Risk-Based Approach for Accidental Loadings

Abstract:

Accidental loading such as fire, blast, ship impact, and dropped objects constitute the great majority of potential and actual fatalities in both onshore and offshore operations. In order to help prevent human loss and for increased safe design or assessment of an onshore/offshore asset, the risks of these accidental loadings should be quantified, in terms of both probability/frequency and consequence aspects. In this paper we propose a unified risk-based approach for the assessment of all accidental loadings mentioned above.

This approach will address both human and asset risk assessment against accidental loadings. Technical safety and process safety inputs such as hazard curves, frequency exceedance curves, and return periods are where the process all begins. Based on these inputs, an event frequency analysis is carried out to determine how often one should expect the event to happen, i.e., the frequency aspect of the assessment. Depending on the type of the accidental loadings and the subject of the risk assessment (human or asset), the output can include a fragility curve, damage frequency matrix, and individual risk matrix. Additionally, a consequence analysis is carried out to determine the severity levels of the event. In a risk-based consequence analysis, the severity levels are fully determined based on the risk associated with the event. Based on both event frequency and consequence, the risk ranking is identified using a risk matrix. If the risk is not acceptable according to the risk matrix, mitigation/repair/strengthening options are then developed.

As an example, a risk-based structural fire analysis is compared with the conventional approach which is widely used in the industry. In a conventional Passive Fire Protection (PFP) analysis using Ductility Limit Analysis (DLA) according to API RP 2FB, fire loads are deterministically applied to a structure whose response is then analyzed. The initial PFP scheme is developed based on the analysis and then optimized based on input from all disciplines (e.g. technical safety, operations, structures, etc.). This approach is sometimes misinterpreted as a "risk-based" approach; however, it does not take into account the frequency aspect of the risk assessment. In a risk-based PFP analysis using DLA, fire scenarios are developed in a particular target zone based on exceedance curves (i.e., input from technical safety discipline). The ductility level analysis is performed to determine the structural consequence. If personnel safety is of interest, the consequence of the structure is then linked to individual risk to determine fatalities. The amount of PFP to be applied on the structure is fully based on the risk that is produced by the fire scenarios in the target zones or areas.

Followed by the approach, a new perspective on safe design of onshore/offshore structures for accidental loadings is outlined to estimate the associated risk to potential targets such as human personnel as well as asset components. The proposed assessment methodology will contribute towards identifying the mitigation measures and safety-critical procedures and equipment, and toward a safer design.

441540 Occupant Vulnerability for Temporary Blast Resistant Modular Buildings

Abstract:

Human fatality and injury in the process facility due to overpressure effects can be categorized as (a) direct injury to body due to pressure change, (b) secondary injury resulting from fragment or debris impact, and (c) tertiary injury as a result of body being thrown by explosion drag on stationary objects. Secondary and tertiary effects are typically the dominant causes of fatalities in the process facilities. Direct injuries may result from change in internal pressure as a result of wall/roof response or penetration of blast overpressure into the building. Existing studies allow for an estimation of occupant vulnerability to blast on typical default building construction types. There exists no such building occupant vulnerability calculation method for modular buildings. This paper seeks to develop an improved understanding of the impact of blast loads on modular building and a more comprehensive understanding of the vulnerability to personnel who are within this type of building if they are subjected to a blast load. The study will be carried out in analytical basis which will include basic and advanced numerical methods along with engineering judgment and accident investigation for estimation of vulnerability. A numerical model can include deformable crash test dummies as well as many types of non-structural components in order to estimate occupant vulnerability and maximum individual risk. The outcome of this study will include new response criterion limits related occupant vulnerability in modular buildings as well as design and retrofit guidelines for critical non-structural components.

441541 Occupied Building Blast Retrofit and Upgrade for Personnel Safety

Abstract:

Accidental blast events at petrochemical facilities such as BP Texas City have shown that the occupied buildings near the process areas can be vulnerable to blast and collapse even under low blast overpressures. Additionally, existing permanent and portable buildings may exhibit insufficient capacity to resist blast loading during facility siting studies per API RP 752 and 753. Although it is not possible to completely eliminate risk from blast loading, it is possible to minimize the individual risk, which is a probability an occupant will experience fatality due to an explosion event, by applying retrofit and upgrade strategies to increase the blast-resisting capacity of an occupied building and meet the safety performance criteria.

There are several ways to decrease individual risk due to explosion with respect to occupants in a building:

1. Moving the building from a high blast region near the process area to a lower blast region, and design a new building;
2. Retrofitting existing occupied building;
3. Replacing existing occupied building with a blast resistant modular building; and
4. Replace the existing occupied building with a new blast resistant building.

The first and forth options can be inefficient from construction cost and time perspective. Additionally, the first option may require purchasing an additional land and obtaining permit. The second and third options are significantly cost and time efficient, and causes less interruptions for the operations. However, these options require analysis and engineering judgment from a highly skilled and experienced engineer. In this paper, case study will be presented where a variety of retrofit and upgrade options will be discussed from perspectives such as design, constructability, cost and individual risk.

Implementing Engineered Controls to Reduce the Likelihood of High Severity Scenarios

Alek Hamparian
Kazarians & Associates, Inc.
100 West Broadway, Suite 970, Glendale, California
Alek.Hamparian@Kazarians.com

Mardy Kazarians
Kazarians & Associates, Inc.
100 West Broadway, Suite 970, Glendale, California
MKazarians@Kazarians.com

Prepared for Presentation at
American Institute of Chemical Engineers
2016 Spring Meeting
12th Global Congress on Process Safety
Houston, Texas
April 11-13, 2016

Implementing Engineered Controls to Reduce the Likelihood of High Severity Scenarios

Alek Hamparian
Kazarians & Associates, Inc.
100 West Broadway, Suite 970, Glendale, California
Alek.Hamparian@Kazarians.com

Mardy Kazarians
Kazarians & Associates, Inc.

Keywords: Hazard Identification and Risk Analysis, Process Safety

Abstract

The Center for Chemical Process Safety (CCPS) *Guidelines on Inherently Safer Chemical Process A Life Cycle Approach* (2009) defines a hierarchy of controls that can be used to decrease the risk of a system. These guidelines include a hierarchy of layers of protection that can be used in Process Hazard Analyses (PHAs) to determine adequacy of the existing safeguards. This approach can be particularly useful for companies that have potentially high severity scenarios for which only a procedural layer of protection is available.

Releases of hazardous materials have occurred that could have been prevented by implementation of engineered safeguards. From a review of these incidents, it had become evident that the owners and operators of the affected facilities relied on procedural safeguards to prevent occurrence of those events. Examples of such cases have involved gasoline, chlorine, and anhydrous ammonia, which are three commonly used hazardous materials at facilities in the United States.

Process Hazard Analysis (PHA) provides a thorough and efficient method for systemically reviewing complex systems for safety concerns. A commonly applied PHA method is the Guide-Word Style Hazard and Operability (HAZOP) approach. HAZOPs are a team effort and risk determination varies based on the experiences and safety culture of team members. Some teams may determine that procedural safeguards are adequate to prevent against a scenario that could lead to a severe event (e.g., liquid release of a volatile toxic material such as chlorine). However, accidental releases have been experienced demonstrating inadequacy of procedural safeguards to minimize the risk of a severe event.

The above noted CCPS guidelines consider procedures as the least effective protection layer. An active engineered safeguard is considered a more reliable protection layer. The purpose of this paper is to point out the potential for a pitfall in risk acceptance where a high severity consequence scenario is concluded to be of acceptable risk based solely on operator action. It is recommended for owner and operators of such facilities to require mitigation feature other than procedural controls for high severity scenario regardless of the risk ranking.

1 Introduction

Releases of hazardous materials have occurred in a wide range of systems with safeguards in place. A number of these events have occurred in systems that relied on procedures and operator intervention as the primary safeguard. The Center for Chemical Process Safety (CCPS) [1] and the American Industrial Hygiene Association (AIHA) [2] have each published guidelines on the topic of hierarchy of controls which can be used as a qualitative measure to establish the effectiveness of risk reduction measures. Both CCPS and AIHA have ranked engineered controls as a higher level safeguard than human intervention. The October 23, 2009 explosion at the Caribbean Petroleum Corporation's facility in Puerto Rico shows the importance of implementing engineered controls in addition to human intervention. The Chemical Safety Board (CSB) [3] investigated that event and recommended for similar facilities to add engineered controls such as "automatic overfill protection systems" to prevent future accidents.

Facilities that contain certain hazardous materials above a certain quantity must comply with the Process Safety Management (PSM) [4] and Risk Management Plan (RMP) [5] process safety regulations that require a Process Hazard Analysis (PHA). The purpose of a PHA is to identify potential hazards posed by a system, the safeguards that are in place to minimize the likelihood or severity of the event and risk rank the scenarios that can lead to adverse consequences. In some instances, the probability qualitatively assigned to an adverse scenario with only procedural safeguards may be underestimated. This underestimation of probability can lead to a lack of urgency to implement engineered safeguards to prevent an adverse consequence.

In addition to PHAs, some companies voluntarily perform Layer of Protection Analysis (LOPA), which is a more detailed analysis than PHA. LOPA is used to semi-quantitatively assess whether sufficient safeguards exist to meet a target risk level. In some instances, teams may use LOPA in such a way as to show that adverse consequences are unlikely by crediting conditional modifiers that may not be applicable to the scenario. This can lead to a determination that additional safeguards are not necessary as risk tolerance has been met.

It may be prudent to establish rules to implement engineered safeguards in potentially severe scenarios that rely only on procedural safeguards. The purpose of this paper is to describe the establishment of a criterion in which potentially high severity scenarios are identified that rely primarily on procedural safeguards and how to add an engineered safeguards to reduce the likelihood of the scenarios occurrence.

2 Industry Incidents

Many incidents have occurred in the chemical process industry where a procedural safeguard was the main layer of protection.

The CSB has performed many incident investigations and provides recommendations to different organizations with the main purpose of learning from past events to enhance safety. A recent investigation focused on the CAPECO incident of October 23, 2009 when a gasoline storage tank was overfilled leading to the formation of a vapor cloud and explosion [3]. In Appendix B of that investigation report [3], CSB lists twenty-two major accidents that have occurred in the process industry since 1962 that involved hydrocarbon storage tank overfill. A majority of these accidents involved processes that did not have an automatic overfill protection system and only relied on human intervention to prevent an overfill.

Similar events have occurred at facilities that handle anhydrous ammonia, a highly hazardous material. Ammonia is a toxic material with a low boiling point that vaporizes upon release into the atmosphere. Inhalation of ammonia vapors can result in irritation of the mucous membranes of the nose, throat, and lungs. If a plume of ammonia vapors with a concentration above 5,000 ppm is inhaled, sudden death may occur [6]. Many refrigeration systems use anhydrous ammonia as the refrigerant in the United States. In ammonia refrigeration systems, compressors provide the motive force to circulate the ammonia. Compressors require oil for lubrication, which by design leaks into ammonia side of the system. Entrained oil is removed from the system from the oil pots where oil separates from ammonia and settles in the pot. In order to ensure proper ammonia refrigeration system cooling efficiency, these oil pots must be drained periodically to remove the oil on the ammonia side of the system. This is achieved by opening a drain valve into an open pale. System pressure moves the oil. The operators are required to be vigilant and close the valve as soon as oil composition starts to change indicating that the pot is drained of oil and ammonia may follow.

Releases of ammonia have occurred from ammonia refrigeration systems due to oil drain operations. An engineered safeguard that could prevent this type of release is a self-closing emergency stop valve. It is designed such that an operator must hold the valve open by holding on to it and when operator's hand is removed the spring loaded self-closing emergency stop valve immediately closes. If an emergency stop valve is installed on an oil drain valve, it can limit the amount of ammonia released to atmosphere in the event that the drain operation is performed too long and ammonia starts to release once the oil has been drained from the pot.

The accident at CAI, Inc. and Arnel Inc., [7] is another example of an accident involving a system without an engineered safeguard. The CSB performed an incident investigation on this accident and issued recommendations [7]. According to the CSB, an initiating cause of the explosion was determined to be that a steam valve was left open to a tank containing flammable liquid. This steam valve was periodically used to heat the contents of a tank that stored ink and then manually shutoff by an operator once the desired temperature of the material was reached. On November 22, 2006 the steam valve was forgotten open and the ink in the storage tank overheated and vaporized. The vapor cloud found an ignition source and a large explosion occurred. The CSB recommended that CAI, Inc. install safety controls that would prevent overheating of flammable or combustible liquids.

3 Hierarchy of Controls and its Application

3.1 Hierarchy of Controls

CCPS has published "Inherently Safer Chemical Processes, A Life Cycle Approach", which describes the importance of making chemical processes inherently safer [1]. The CCPS states the following, "A chemical manufacturing process is inherently safer if it reduces or eliminates the hazards associated with materials and operations used in the process and this reduction or elimination is permanent and inseparable" [1]. The CCPS has described methods of making chemical processes safer and an important concept of the guidelines is levels of inherent safety. The CCPS [1] describes the levels of inherent safety as follows from most effective to least effective:

First Order Inherent Safety Measures –
  Avoid Hazards
Second Order Inherent Safety Measures –
  Reduce Severity;
  Reduce Likelihood
Layers of Protection –
  Apply Passive Safeguards;
  Apply Active (Engineered) Safeguards;
  Apply Procedural Safeguards

It is important to note that the least robust layer of protection listed by the CCPS is the application of procedural safeguards. This means that the application of active safeguards is a more robust layer of protection than procedural safeguards. The highest or first order of inherent safety listed is avoiding hazards. This is defined as elimination of a hazard. The two second orders of inherent safety listed are reducing severity and reducing likelihood. The severity of a hazard can be reduced by a number of methods including reducing the quantity of the material in a system, reducing the pressure or temperature within a process, substituting the material for a less hazardous material, etc. The likelihood of a hazard's occurrence can be reduced by a number of methods including using more corrosion resistant material, redesigning a process from four vessels to requiring two vessels, etc. The next order of inherent safety after the first and second orders are layers of protection. Layers of protection are ranked in order as passive safeguards, active safeguards, and procedural safeguards with procedural safeguards being the least robust and passive safeguards being the most robust. An example of a passive safeguard is designing the vessel within a system with a maximum allowable working pressure that is higher than the highest pressure that can potentially be reached within the system. An example of an active safeguard is an automatic overfill protection system that shuts off flow to a tank when high level is reached in the tank. An example of a procedural safeguard is relying on operators to shut off the flow to a tank when a certain level is reached.

The AIHA also addresses hierarchy of control from the standpoint of occupational safety. The "Z10-2012: Occupational Health and Safety Management Systems" Standard, defines hierarchy of controls which can be summarized as follows:

A. Elimination;
B. Substitution of less hazardous materials, processes, operations, or equipment;
C. Engineering controls;
D. Warnings;
E. Administrative controls; and
F. Personal protective equipment

There are many similarities between the CCPS's inherent levels of safety and the AIHA hierarchy of controls. Similar to the CCPS, the AIHA hierarchy of controls considers engineering controls as more robust than human intervention, warnings and administrative controls. The AIHA hierarchy of controls is written for occupational safety whereas CCPS's hierarchy of controls is written specifically for process safety applications. In both hierarchy of controls, the first item listed is eliminating a hazard. The second order of inherent safety listed by the CCPS is "reduce severity, or reduce likelihood", which is very similar to the second hierarchy of control listed by AIHA. Substituting a less hazardous material translates to reduced severity. Substituting less hazardous operations can be considered as a measure that reduces likelihood. The third hierarchy of control listed by AIHA is "engineering controls", which can be considered the same as the first two layers of protection listed by the CCPS, "Apply passive safeguards, Apply active safeguards". The third layer of protection listed by the CCPS, "Apply procedural safeguards" can be considered a combination of the fourth and fifth hierarchy of controls listed by AIHA, "Warnings; Administrative controls". The one major difference is that the CCPS does not call out alarms specifically as the AIHA does with the *warnings* hierarchy of control. The CCPS's inherent safety measures refer to alarms in the procedural controls as all alarms require personnel intervention.

An example of an engineered control referred to by AIHA in regards to occupational safety is a shutoff system for an automated laser welding machine that shuts off immediately upon detection of limbs within the welding area. This is an example of an engineered control designed to prevent personnel exposure to laser burns. An automatic overfill protection system that is considered an engineered safeguard by the CCPS for process safety applications is also considered an engineering controls according to the AIHA hierarchy of controls.

Regulatory bodies involved with PSM and RMP have taken notice of the importance of hierarchy of controls. Current drafts of California's Process Safety Management [8] and Risk Management Plan [9] regulations for refineries will specifically require hierarchy of hazard controls to be formally addressed. The draft regulations state that an analysis must be performed on all existing processes at a refinery that pose a process safety hazard. The analysis team must document the possible inherent safety measures for each hazard that can be implemented to reduce risk. The analysis team must review the feasibility of each level of inherent safety and must document why an inherent safety measure cannot be implemented for the process if that decision is made.

3.2 Application of Hierarchy of Controls

The most preferable time period to apply the hierarchy of controls to a system is during its design phase when changes can be made without severe economic impact. In addition to the design phase of a system, the hierarchy of controls can be applied throughout a system's life cycle when changes are considered. However, the upcoming regulations in California will be requiring application of the concept to built and operated systems.

To be able to apply hierarchy of controls to a system, it is only rational to establish what hazardous scenarios are possible and what safeguards are in place as protection. Process Hazard Analysis (PHA) provides a systematic method to identify and rank potential hazard scenarios and to identify the layers of protection in place to minimize the risk. By conducting the PHA, one can claim that all potential hazard scenarios that a team of experts could identify have been addressed systematically.

The hierarchy of controls can be used during a Process Hazard Analysis (PHA) to determine the adequacy of available safeguards to prevent each release scenario identified. This can be completed by identifying the available safeguards for scenarios that are postulated to have severe consequences and questioning whether or not inherently safer measures can be implemented. In addition, the hierarchy of controls can be applied when a recommendation is made during a PHA. The hierarchy of controls can act as guidance to choosing a higher level control measure to reduce the risk of a scenario. The hierarchy of controls can be reviewed while making any major changes to a system to assess if there are inherently safer options than the change being proposed. The hierarchy of controls can also be reviewed during an incident investigation to determine if a higher order of control should be implemented to prevent the recurrence of an incident.

The thought process of applying hierarchy of controls for a gasoline producer may be the following:

1. Currently there is a need for gasoline as vehicle fuel since a majority of automobiles sold in the United States require gasoline as the energy source. Therefore, this is a viable business venture that society at large needs. Eliminating gasoline from the line of products produced and sold by the company is not a prudent business decision because its safety risk is relatively well understood and has historically been accepted by society. Therefore, eliminating the hazard (i.e., moving out of gasoline business) as the very first item of the hierarchy of control is not a prudent option.
2. The next hierarchy of controls addresses reducing severity and likelihood. In order to reduce the severity of a gasoline storage tank related mishap (e.g., overfill and vapor cloud explosion), the hazard radius of a vapor cloud would need to be reduced. This can be done by reducing the pool surface of released material, as an example. Pool area is a function of the berm walls around the storage tanks and height of the walls. As another example, the owner may decrease the flow rate into a tank allowing the operators more time to discover and shutoff the inflow. These changes may not be feasible or the risk reduction gained from them may be smaller than other options (e.g., automatic shut-off on high level).

3. In order to reduce the likelihood of a gasoline storage tank overfill, the number of times the storage tank filling operation is performed can be reduced. This will require larger storage tanks which may increase the quantity of gasoline release in case of tank failure. Thus, the risk reduction gained from decreasing frequency of fills may be offset by the increase in risk due to larger tank size.
4. Engineered controls are the next available hierarchy of control. The first type of layer of protection based on CCPS's hierarchy of controls is a passive safeguard. One commonly used passive safeguard for large hydrocarbon storage tanks is secondary containment (a berm). The purpose of a berm is to contain a spill. However, in the case of gasoline, the light ends within the mixture would vaporize resulting in a vapor cloud formation with the potential for an explosion. As discussed above, reducing the pool size may reduce the evaporation rate but better risk reduction can be obtained by other means.
5. The next hierarchy of control is an active safeguard such as an automatic overfill protection system that will shut off the pump filling the storage tank when high level is detected. For a built and operated system, this option may provide the best risk reduction.
6. Procedural safeguards are the next hierarchy of control after active safeguards. The procedural safeguard for a gasoline storage tank filling operation would be operator monitoring of the level on the storage tank, installation of a high level alarm, and shutting off the feed pump manually at a certain level.

4 Process Hazard Analysis

4.1 Use of Process Hazard Analysis

A Process Hazard Analysis (PHA) is a hazard analysis tool that must be performed for systems covered under the PSM and RMP regulations [4 and 5]. PHA is a systemic tool for identifying the hazards of a system. PHA also provides a means to establish the potential severity of each hazard scenario, the available safeguards to prevent the hazard and likelihood of the scenarios occurrence. By applying risk ranking protocols, the risk gap of each scenario is identified in the PHA process.

The most common PHA method is HAZOP, which is typically performed in a brainstorming session with the participation of team members with experience in different fields necessary to properly analyze a system.

An important part of a PHA is establishing the risk of a hazard scenario. Risk is defined as "the combination of the expected frequency (events/year) and severity (effects/event) of a single incident or a group of incidents" [10]. Before a PHA study commences, the management responsible for the system must decide what risks they are willing to accept and what risks they will consider unacceptable. The acceptable and unacceptable risks are then translated into a risk tolerance matrix. The risk tolerance matrix contains severity and likelihood pairs that are acceptable and severity and likelihood pairs that are considered unacceptable and require risk reduction. This risk matrix is then used by PHA teams in all the PHA studies owned and operated by the management. The risk matrix provides a basis to identify risk gaps and whether or not system upgrades should be considered to improve safety.

4.2 Augmenting PHAs with Hierarchy of Controls

PHA results can vary from team to team. People working in the chemical process industry have different career experiences, which can result in different risk estimations during a PHA. It is possible and the authors' experience in leading PHA studies corroborates this observation that the likelihood of a severe consequence may be underestimated by PHA Teams. For example, operators who work with a system on a daily basis and have not experienced any incidents may be under the impression that the probability of a specific high severity scenario occurring is very low. Operators may feel that a system has been performing well for many years and if an issue did arise, the operators are knowledgeable enough to stop the progression to the bad event. Also, some PHA team members may believe that certain initiating causes of hazardous consequences are very unlikely based on their experience. Many of these issues can lead to underestimation of the risk of a severe consequence. Therefore, it is possible that the risk of a high severity scenario is estimated within the acceptable range allowing reliance on procedures based safeguards with no other safeguard. Due to the fact that risk ranking issues such as the one described above exist, it may be prudent for companies to use hierarchy of controls as the bases to require that all high severity scenarios to include at least one independent layer of engineered safeguard protection.

5 Layer of Protection Analysis

Layer of Protection Analysis (LOPA) is a tool that can be used to semi-quantitatively estimate the likelihood of a high severity scenario. Typically, LOPA is applied to the high severity scenarios identified during a PHA in order to enhance the accuracy of scenario likelihood. As part of the LOPA process, typically the LOPA Team estimates probabilities on an order of magnitude basis. The values that are estimated are the initiating cause likelihood, failure probabilities of the safeguards, occupancy factor, ignition probability, time at risk, and other conditional modifiers [11]. The initiating event likelihood is an estimate of how likely it is that the initiating cause occurs that can initiate the postulated chain of events. For example, according to the CCPS [12] the probability of a basic process control system (BPCS) failure is 0.1 per year. Also, if the safeguard has special features (e.g., it is designed to be of high reliability) the team may assign 0.01 probability of failure.

Occupancy factor refers to the amount of time personnel are within the hazard zone. For example, the CCPS [13] states that the probability of one person being present in a hazard zone is 0.1 for an area where operators spend 17 hours on average per week. Ignition probability refers to the probability that a released flammable material finds an ignition source resulting in a fire or explosion. Ignition probabilities will vary based on many factors such as the properties of the released material in terms of the released temperature, hot equipment near the release point, etc. Time at risk factor applies to non-normally operating modes of a system. Other conditional modifiers may also be used depending on the scenario. For example, some companies may take a conditional modifier stating that a catastrophic compressor seal failure due to liquid being sent to the suction side may occur once every ten times.

LOPA teams may credit additional conditional modifiers and enabling conditions resulting in a low risk level within the acceptable range. One example is crediting a time at risk factor for scenarios that can only occur during startup. Another likelihood reduction can be achieved through the application of occupancy factor for scenarios where generally personnel are not present within the hazard zone. Application of these factors raises an important question about the possibility of reliance on procedural safeguard as the only safeguard against a high severity consequence. Similar to the PHA, management may mandate that for all high severity scenarios, the LOPA team should consider an engineered safeguard regardless of the likelihood and risk ranking.

Startup is an interesting case as a safety risk contributor. The CCPS states that time at risk is not advised to be used during startup, "Use of a time-at-risk enabling condition in such a situation may lead to a conclusion that the average risk meets a facility's risk criteria; however, the peak risk imposed on persons exposed to the hazard may be inappropriately high" [13].

6 Limitations and Benefits

6.1 Increased Complacency Potentially Leading to Adverse Consequences

A potential adverse outcome of adding engineered safeguards to a system is an increase in operator complacency, which may lead to an adverse consequence. Operators involved in tasks such as storage tank filling may become complacent if an engineered safeguard is available as a backup in case they do not shut off the filling operation in a timely manner. Many engineered safeguards are meant to serve as a final barrier to prevent the occurrence of a release. The knowledge that a final barrier such as an engineered safeguard is available may lead to personnel relying on that safeguard and not being as attentive to system operation as before the addition of the engineered safeguard. Increased complacency may unintentionally lead to an adverse consequence as typical engineered safeguards have a reliability of approximately 90%. This means that approximately 10% of the time that they are required to prevent an adverse consequence, they may not work. If operators become complacent and rely on engineered safeguards to prevent incidents, an incident may occur if the engineered safeguard does not function as required. For this reason, it must be stressed to personnel prior to installation of engineered safeguards that they are not meant to be a replacement of a safeguard (i.e., engineered safeguard replacing a procedural safeguard). Adding engineered safeguards to a system must be understood to be a measure to reduce risk significantly, which it cannot accomplish if it only replaces a procedural safeguard.

6.2 Engineered Safeguards Are Not Perfect

Engineered safeguards should not be considered as means of eliminating the potential for an adverse consequence. Engineered systems are not guaranteed to work every time they are needed. Incidents have occurred in the chemical industry in which an engineered safeguard did not work. One such incident occurred on August 14, 2002 in which a chlorine hose ruptured during a chlorine transfer and resulted in a release of over 48,000 pounds of chlorine [14]. The chlorine transfer system did have automatic shutoff valves that were supposed to activate if chlorine was detected at nearby chlorine detectors. The chlorine detectors did sense chlorine and sent a signal to the automatic shutoff valves to close. However, all automatic shutoff valves did not close as they had experienced some internal corrosion. The automatic shutoff system was not tested properly as part of the Mechanical Integrity Program. Management at the site was not aware of the corrosion issue within the system and did not take proactive measures to ensure the automatic shutoff system activates properly in the event that it was needed.

Another example of an engineered safeguard failing to work is the explosion at the Buncefield Oil Storage Depot on December 10, 2005 [15]. A gasoline storage tank was overfilled and the automatic overfill protection system did not activate as it was not placed into service properly after its previous test. Maintenance errors such as this can happen and it is important for personnel working with engineered safeguards to understand its functionality, and test and maintenance procedures in order for it to activate when needed.

The above two examples of an engineered safeguard failing to activate should not lead to a conclusion that adding engineered safeguards to a system does not reduce risk. Engineered safeguards do indeed reduce risk if the Maintenance Program of the system to which it is added is updated to include the proper procedures and testing frequency recommended by the manufacturer of each piece of equipment associated with the engineered safeguard. It should also be noted that personnel must be aware of the functional details of the engineered safeguard system to ensure that it is operable so that it can prevent an accident when needed as it is designed to.

6.3 Resources

Adding engineered safeguards to a system requires resources from many departments. Engineered safeguards added to an existing system may need to be designed according to the Instrumentation, Systems, and Automation Society (ISA) Standard 84.00.01-2004 [16]. The ISA 84 Standard has many detailed requirements for installing Safety Instrumented Systems (SIS). In order for these requirements to be properly met, input is needed from many different groups within a company.

6.4 Financial Aspect

Adding engineered safeguards may prevent financial losses incurred due to major accidents. Some may argue against adding engineered safeguards to a system that has been found to be of low risk but high consequence severity due to the immediate financial strain it may have on a company. It is true that adding an engineered safeguard requires some financial investment and the system must be included in the periodic test and maintenance program. However, a major accident can cost millions of dollars in monetary losses and other non-tangible impacts (e.g., personnel injuries and reputation impact). For example, in 1999 an explosion occurred due to a gasoline storage tank overfill at the Thai Oil Company that cost $22.3 million and resulted in seven fatalities [3]. The storage tank only had a high level alarm, which was not heard by onsite personnel. The automatic overfill prevention system would have cost a small fraction of that monetary loss, not to mention the loss of life that was experienced.

An opinion that is increasingly being accepted by industry management and safety professionals is that safety is an investment, not a cost. According to OSHA's Business Case for Safety Webpage [17], "Employers that implement effective safety and health management systems may expect to significantly reduce injuries and illnesses and reduce the costs associated with these injuries and illnesses, including workers' compensation payments, medical expenses, and lost productivity." We can extrapolate this statement to process safety. "Z10-2012: Occupational Health and Safety Management Systems" Standard [2] is generally accepted as being a very good guidance document to implementing an effective safety and health management system. A very important section within that standard is the one on hierarchy of controls. Therefore, implementing a system in which engineered safeguards are added for high severity scenarios where only procedural safeguards exist would be a step towards improving process safety. This can result in a decrease in incidents at a chemical processing facility, which can have a direct positive impact on facility profitability.

Adding engineered safeguards to a system does not necessarily mean that future accidents will be prevented. However, it will reduce the likelihood of a major accident occurring at a facility. The investment required to add and maintain an engineered safeguard can be significantly less than the costs associated with an accident.

7 Conclusion

In the chemical process industry, many potentially high severity tasks exist for which procedural safeguards are the main layer of protection. The hierarchy of controls guidance developed by the process industry ranks engineered safeguards as a more robust layer of protection than procedural safeguards. Multiple chemical process industry accidents have occurred in which the procedural safeguard layer of protection failed and an engineered safeguard did not exist. These accidents could have been prevented if the associated system had an engineered safeguard as a layer of protection in addition to the procedural safeguard.

The PHA and LOPA methodologies are tools that can be used to identify high severity scenarios and the existing safeguards. It is possible, to apply these methods properly and arrive at a low risk conclusion for high severity scenarios that solely rely on a procedural safeguard. It is recommended in this article that these methods be augmented by owners and operators responsible for highly hazardous operations with a mandate that all high severity scenarios to include at least one engineered safeguard. This should ensure that regardless of the estimated likelihood of the scenarios, the operators are not the only safeguard preventing the high severity release.

References

1. *Inherently Safer Chemical Process: A Life Cycle Approach.* Center for Chemical Process Safety, American Institute of Chemical Engineers, New York, NY, 2nd Edition, December 2008.

2. *Z10-2012: Occupational Health and Safety Management Systems.* American Industrial Hygiene Association, Falls Church, Virginia, June 2012.

3. *Final Investigation Report – Caribbean Petroleum Tank Terminal Explosion and Multiple Tank Fires, Bayamon, Puerto Rico, October 23, 2009.* U.S. Chemical Safety and Hazard Investigation Board, Report 2010.02.I.PR, October 2015.

4. *Process Safety Management of Highly Hazardous Materials.* Occupational Safety and Health Administration, Federal Code of Regulations, 29 CFR, §119.119.

5. *Risk Management Programs for Chemical Accidental Release Prevention.* U.S. Environmental Protection Administration, 40 CFR Part 68.

6. *Safety Data Sheet (SDS) for Anhydrous Ammonia.* Hill Brothers Chemical Company, HBCC SDS No. CA10000, Revision Issued: 5/14/2015.

7. *Investigation Report – Confined Vapor Cloud Explosion – CAI, Inc. and Arnel Company, Inc. – Danvers, Massachusetts.* U.S. Chemical Safety and Hazard Investigation Board, Report 2007-03-I-MA, May 2008.

8. *Process Safety Management for Petroleum Refineries.* California Occupational Safety and Health Administration, California Code of Regulations, Title 8, §5189.1, Draft Version 5.0, September 14, 2015.

9. *California Accidental Release Prevention (CalARP) Program.* California Governor's Office of Emergency Services. Title 19 of CCR Division 2, Chapter 4.5, Draft, September 24, 2015.

10. *Guidelines for Hazard Evaluation Procedures.* Center for Chemical Process Safety, American Institute of Chemical Engineers, New York, NY, 3rd Edition, April 2008.

11. *Layer of Protection Analysis – Simplified Process Risk Assessment* Center for Chemical Process Safety, American Institute of Chemical Engineers, New York, NY, October 2011.

12. *Guidelines for Initiating Events and Independent Protection Layers in Layer of Protection Analysis.* Center for Chemical Process Safety, American Institute of Chemical Engineers, New York, NY, February 2015.

13. *Guidelines for Enabling Conditions and Conditional Modifiers in Layer of Protection Analysis*. Center for Chemical Process Safety, American Institute of Chemical Engineers, New York, NY, November 2013.

14. *Investigation Report – Chlorine Release – DPC Enterprises, L.P.* U.S. Chemical Safety and Hazard Investigation Board, Report No. 2002-04-I-MO, May 2003.

15. *Buncefield: Why Did It Happen?*. United Kingdom Health & Safety Executive, February 2011.

16. *ANSI/ISA-84.00.01-2004: Functional Safety: Safety Instrumented Systems for the Process Industry Sector*. Instrumentation, Systems, and Automation Society (ISA), September 2004.

17. https://www.osha.gov/dcsp/products/topics/businesscase/. Occupational Safety and Health Administration, Business Case for Safety, Accessed on February 8, 2016.

12TH GLOBAL CONGRESS ON PROCESS SAFETY

Maintaining Consistency in the Determination of Consequence Severity Rankings for Process Hazard Analyses

Kirk Busby
Kazarians & Associates, Inc.
100 West Broadway
Glendale, CA 91210
Kirk.Busby@kazarians.com

Prepared for Presentation at
American Institute of Chemical Engineers
2016 Spring Meeting
12th Global Congress on Process Safety
Houston, Texas
April 11-13, 2016

Maintaining Consistency in the Determination of Consequence Severity Rankings for Process Hazard Analyses

Kirk Busby
Kazarians & Associates, Inc.
100 West Broadway
Glendale, CA 91210
Kirk.Busby@kazarians.com

Keywords: Hazard Identification and Risk Analysis, Process Safety, Consequence Modeling

Abstract

A Process Hazard Analysis (PHA) is an important tool for identifying and managing the hazards associated with a chemical process. One of the most critical aspects of the analysis is the risk ranking of scenarios, which is used to determine which events pose a high risk and may require further mitigation. The risk ranking of a scenario is built on two factors – severity and likelihood. The severity of a scenario is a measure of the potential worst-case impact on facility operation, facility personnel, the public, or the environment, while the likelihood is an estimate of the probability with which an event may occur, given the presence of the safeguards acting to prevent or limit the severity of the event.

Severity of a scenario is generally expressed in qualitative terms. While a typical approach to assigning consequence severity rankings involves categorizing events by impact (e.g., recordable injury, permanent disability, or fatality), this approach relies on the subjective judgment of the analysis team in determining the exact impact of a specific event. Experience has shown that, because of the subjective nature of this step, different PHA study teams have had varying opinions on the potential severity level of a given scenario. It is the aim of this study to provide guidelines to improve consistency of severity rankings both within a single analysis, and across multiple analyses with separate PHA study teams. It is possible to increase the consistency in consequence severity rankings through application of standardized guidelines. A method has been developed to aid PHA facilitators based on dispersion modelling of various conditions typically encountered in a PHA. The underlying data is based on groups of chemicals with similar properties and other conditions (e.g., break size, break location, etc.).

1. Introduction and Background

Process Hazard Analysis (PHA) is an important tool for identifying and managing the hazards associated with a chemical process. The goal of a PHA is to use a systematic methodology to identify each potential hazard source within the process. A typical PHA methodology is the guideword-style Hazard and Operability (HAZOP) study. To ensure completeness, the HAZOP methodology breaks down a complex system into individual nodes, making it easier to thoroughly analyze each section of the process. Further, within each node, hazards are postulated according to a standard combination of guidewords and process deviations. For example, when the guideword 'No' is paired with the deviation "Flow," all scenarios within the node which lead to a loss of flow are considered.

Once a scenario is identified, the risk posed to the facility must then be determined. This is normally achieved through the risk ranking process. To determine the risk ranking, each scenario is assigned a severity and a likelihood level. This pair is then used in a risk ranking matrix, which defines the risk targets developed by company management prior to the start of the PHA. Assigning a risk ranking is one of the most important facets of the PHA process, as it is used to determine which risks are acceptable given the safeguards currently in place, and which require further actions to ensure that the risk is maintained at a tolerable level.

Currently, there is a large potential for variance in risk ranking determination, specifically with regard to consequence severity. While severity determination is meant to be a qualitative ranking, there are a number of benefits to improving the consistency of severity rankings among similar release scenarios. Kazarians & Associates, Inc. has developed a method that has proven effective in reducing variance among consequence severity rankings through the use of pre-determined dispersion modeling results.

2. Current Methods for Consequence Severity Determination

It is common practice to estimate the worst case consequence severity for the identified potential release scenarios. For this, it is assumed that all existing active safeguards are unavailable. Typically, the severity ranking is a qualitative determination made by the PHA study team based on their understanding of the potential impact a given scenario may have on the safety of facility personnel and the public, as well as the environment or business operations. The text provided in defining each severity level is generally used to determine the consequence severity. Table 1 provides an example of severity level definitions available to the PHA study team.

Definitions such as those in Table 1 can provide a good starting point for consequence severity selection. However, use of these limited definitions can result in inconsistent severity rankings, both within an individual PHA, as well as among PHAs. This variation stems from the fact that the PHA study team generally carries the chain of events to the release condition (i.e., flange leak, pipe rupture, etc.) whereas the definitions of consequence severity level is often in terms of level of harm (i.e., first aid injury, fatality, etc.). This then requires the study team to envision how the release will ultimately result in harm, which means that they have to take into consideration factors such as source conditions, dispersion, and ignition, which are often complex phenomena by themselves. Therefore, this leads different PHA study teams to hold varying opinions of what the impact of a certain event may be. Typically, these opinions are based on the experience of the personnel present during the PHA.

Table 1. Typical Consequence Severity Guidance

Consequence Severity	Impact on Personnel	Impact on the Environment	Impact on business
1	Minor first aid injury	Slight spill onto soil, release to flare	< $10,000
2	Recordable injury	Small spill onto water, moderate spill onto soil	< $100,000
3	Permanent disability	Moderate spill onto water, large spill onto soil	< $500,000
4	Fatality	Moderate spill into an environmentally sensitive area	< $1,000,000
5	Multiple fatalities	Large spill into an environmentally sensitive area	< $5,000,000

The following example is taken from PHAs conducted for chlorination systems at two facilities operated by the same organization. The example demonstrates the difference operational experience can make in the severity ranking of a given scenario. During PHA sessions for the two systems, the PHA team encountered the same scenario in which a storage tank is exposed to an external fire, resulting in a pressure increase within the tank, ultimately resulting in tank rupture. In this scenario, the two teams have arrived at an identical endpoint (i.e., release scenario). While the release conditions (i.e., temperature and pressure) will be the same for both cases, the first team concludes that a single fatality may result, the second believes that multiple fatalities may result. The two teams may record the scenarios according to Figure 1.

In each case, the PHA study team has arrived at the same chain of events leading up to an integrity failure of the storage tank, and release of chlorine into the atmosphere. However, the final consequence statement and therefore risk rankings vary based on the estimated severity of the release.

Inconsistency in severity rankings can cause a number of problems following a PHA. Most importantly, they can lead to an over- or under-estimation of the risk a scenario poses. A consequence with the potential to cause severe injury that has only been identified as a first aid issue could result in insufficient safeguarding to protect against the event. This could ultimately lead to major safety incidents during the operating lifetime of the process. Safeguards are typically added when the risk ranking of a scenario does not meet the pre-determined corporate risk target. A lower severity ranking may indicate that the appropriate safeguards are in place, when in reality an additional protection may be necessary.

Team 1:

Causes	Consequences	Safeguards	S	L	RR	HAZOP Recommendations
Greater than 140°F in the Chlorination Room caused by a fire	▪ Pressure buildup in the Chlorine Storage Tanks. ▪ Potential for loss of containment of a Chlorine Storage Tank if the relief valve fails to lift. Release of chlorine into the Chlorination Room. Potential for personnel exposure.	▪ Relief Valves will lift to prevent overpressure. ▪ Detectors alarm locally and on SCADA. ▪ Scrubber activated by detectors at 3 ppm. ▪ Plant housekeeping practices prohibits storage of any combustible materials in the Chlorination Room.	4	D	M	

Team 2:

Causes	Consequences	Safeguards	S	L	RR	HAZOP Recommendations
Fire inside the Chlorine Building.	Potential for increased pressure in the storage tanks. Potential to rupture the storage tank. Potential for release of chlorine inside the building. Potential for personnel and public exposure.	1. PSVs on the storage tanks set at 225 psig. 2. Chlorine detectors alarm at 1 ppm. Scrubber system will start at 3ppm on any one detector. 3. Good housekeeping practices.	5	E	M	No recommendations necessary. Safeguards are considered adequate.

Figure 1. Identical PHA Scenarios with Differing Consequene Statements and Risk Ranking

A second and more subtle consequence of inconsistent severity ranking is the potential discrepancy between processes or facilities. If two PHA study teams assign different severity rankings to similar events on two similar systems, there is a potential for one system to require an extra safeguard to be installed which is not in place on the second system. This discrepancy between units could potentially lead to confusion later in the life of the plant. If a group of operators is responsible for both systems, there could be confusion as to which setup is correct, as well as the incorrect assumption that both systems are configured in the same manner. Additionally if an incident occurs in the system with fewer safeguards, there is a potential for increased regulatory scrutiny and fines due to the fact that the second system is not as well protected as the first.

3. Improving Consistency through Dispersion Modeling

To reduce variation among consequence severity rankings, it is possible to provide PHA teams with improved guidance on the potential impacts of a release. Dispersion modelling is a tool which is commonly used to determine the hazardous area resulting from a specific chemical release scenario. There are a number of sophisticated dispersion models and software packages currently on the market which can provide a picture of the hazard distance presented by various chemical release events. Many of the commercially available packages allow the user to input specific data such as the temperature, pressure, quantity, and release orifice size, and calculate the released quantity, pool size, vaporization rate, and resulting hazard distance. Depending on the model chosen, the user may be able to customize the scenarios to a greater degree, including such factors as terrain type, degree of confinement, explosive characteristics, etc.

Currently, some PHA teams already use dispersion modeling to assist in performing a more in-depth analysis for select scenarios with complex release characteristics. These studies are useful, but require a large amount of input data to be determined by the PHA team in order for them to generate accurate results. Therefore, it is much too time consuming to use such an analysis for every release scenario postulated during a PHA.

To combine the time and cost efficiency of PHA with the increased accuracy afforded by the use of dispersion modeling, the dispersion calculations can be performed generically prior to beginning the PHA discussions. To utilize this method, a database is developed for each hazardous chemical which may be released, and dispersion calculations are performed for the range of temperatures, pressures, phases (liquid or gas), and break sizes which may be encountered during the PHA. Every possible combination of parameters does not need to be considered, as the PHA team may quickly interpolate between results when needed. With a pre-developed matrix in front of them during the PHA sessions, participants can quickly relate a set of known release conditions to an approximate hazard zone, which will allow them to choose a more refined consequence severity. Some large chemical processing and manufacturing corporations have already begun to adopt such methods to improve the consistency of their PHAs.

The method developed by Kazarians utilizes PHAST modeling software by Det Norske Veritas [Reference 1] to create databases for both flammable and toxic chemicals. Specifically, the common alkanes (i.e., methane through hexane), hydrogen, hydrogen sulfide, anhydrous ammonia, and chlorine have been analyzed, in both liquid and gaseous form. In each case, a generic outdoor release event is postulated for varying temperatures, pressures, spill quantities, and release sizes. From this basic input data, PHAST is capable of generating information on the maximum dispersion distance, the flammable range, the range of explosive overpressure, and toxic effects. For each release event, the maximum distance of each hazard type is compared, and the greatest distance is selected. For example, for a specific set of release conditions, the maximum distance at which an injury may occur due to thermal radiation is 200 feet, while the maximum injury distance due to explosive overpressure is 250 feet. For these release conditions, 250 feet will be selected and recorded as the worst-case injury hazard distance. The final result is a set of eight tables for each chemical, with four liquid phase tables and four vapor phase tables. Tables 2 and 3 below show the four hazard distance tables for methane in the liquid phase.

Table 2A. Methane Liquid Injury Hazard Distance (ft)

	Source Temperature (F)			
Release (lbs)	**-250**	**-200**	**-150**	**-117**
10	110	100	140	160
100	350	230	320	350
1,000	610	510	710	780
10,000	1,520	1,130	1,570	750
100,000	4,060	2,480	3,350	3,350
500,000	7,690	4,260	5,360	5,360

Table 2B. Methane Liquid Injury Hazard Distance (ft)

	Source Pressure (psi)				
Break Diameter (in)	**50**	**150**	**300**	**400**	**450**
0.25	20	30	30	30	30
0.5	110	110	110	110	110
0.75	180	210	210	210	210
1.0	210	280	310	310	310
1.5	280	410	480	510	510
2.0	380	540	640	670	700
3.0	510	770	970	1,070	1,100
4.0	610	1,000	1,330	1,430	1,460
6.0	840	1,430	2,020	2,180	2,250

Table 3A. Methane Liquid Fatality Hazard Distance (ft)

	Temperature (F)			
Release (lbs)	**-250**	**-200**	**-150**	**-117**
10	40	40	60	70
100	240	90	130	150
1,000	530	210	310	340
10,000	1,390	470	690	1,690
100,000	2,990	1,070	1,500	1,500
500,000	5,290	1,860	2,400	2,400

Table 3B. Methane Liquid Fatality Hazard Distance (ft)

	Pressure (psi)				
Break Diameter (in)	**50**	**150**	**300**	**400**	**450**
0.25	20	20	20	20	20
0.5	40	50	50	50	50
0.75	110	140	140	140	140
1.0	140	210	240	240	240
1.5	210	340	400	440	440
2.0	300	470	570	600	630
3.0	440	700	890	990	1,030
4.0	530	930	1,260	1,350	1,390
6.0	760	1,350	1,940	2,110	2,170

Tables 2 and 3 present two sets of hazard distances. One represents the distance inside which an injury is expected to occur, and one inside which it is possible for fatalities to occur. Additionally, the sets have been subdivided into Tables A and B, which provide dispersion information for Release Quantity vs Temperature, or Break Diameter vs Pressure. The PHA team may elect to use either of the tables based on what information is available for a given scenario. For example, Break Diameter vs Pressure tables may be more useful for a pipe break scenario, while Release Quantity vs Temperature may be easier to use in the event of a seal failure or liquid overfill scenario.

There are a few notable trends which can be identified from Tables 2 and 3. Specifically, it can be seen that in the tables relating Break Diameter and Pressure, the maximum hazard distances reach a constant value as pressure increases. This is due to the release reaching what is known as "sonic velocity." [Reference 2] When a material under pressure is released through an orifice, the velocity of the release is initially dependent on the pressure within the vessel. Higher release pressure will typically result in a higher release velocity. However, once the release velocity has reached sonic velocity, further increases in pressure will no longer cause the velocity to increase. Put simply, no more material can fit through the release orifice, no matter how much force is behind it.

4. Defining Exposure Thresholds

The final dispersion distances reported in Tables 2 and 3 can vary greatly depending on what thresholds values are chosen for injuries and fatalities. This analysis used three potential sources of impact to the population: overpressure due to explosion, thermal radiation due to fire, and exposure to toxic vapors. Each threat has a unique hazard zone that is determined by its effects on both the human body and surrounding structures.

There is a great deal of research performed on the various effects on the human body caused by exposure to overpressure, thermal radiation, or toxic chemicals [References 3 through 10]. From this research, threshold values can be estimated to determine the exposure limits for injuries and fatalities. Table 4 presents the threshold values used in this article to differentiate between effects.

Table 4. Exposure Threshold Definitions

Explosion Overpressure	
Effect	**Threshold Value**
Injury	0.5 psi
Fatality	2 psi
Thermal Radiation	
Effect	**Threshold Value**
Injury	5 (kW/m^2)
Fatality	10 (kW/m^2)
Toxic Exposure	
Effect	**Threshold Value**
Injury	ERPG-2
Fatality	ERPG-3

It is important to note that these values represent conservative estimates of the potential impact an event may have on a person. For example, the threshold values for thermal radiation exposure are time dependent. Exposure to 10 kW/m^2 will not result in an immediate fatality, but prolonged exposure of approximately 60 seconds could result in death. This value is chosen as it represents the possibility of a fatality since it requires rapid evacuation from the hazard zone.

For each scenario modeled in PHAST, the distances to each threshold value described in Table 4 are compared, and that which results in the largest hazard distance is selected as the worst-case hazard zone. Toxic effects are only calculated for those chemicals which pose a significant risk when inhaled or contacted.

5. Using Hazard Distance Tables During PHA Sessions

Once developed, hazard distance tables can be used during PHA discussions to quickly determine an appropriate severity ranking based on the potential impact to facility personnel and the public. To convert the hazard distance tables into a form which may be more easily used during the PHA, the PHA facilitator may review the company risk matrix, as well as the facility layout prior to starting the PHA sessions. During this time, the facilitator should convert the dispersion distances in Tables 2 and 3 into charts providing guidance on the potential worst-case severity of different release events.

For example, a hypothetical company has developed a risk matrix with the consequence severity rankings described in Table 1. The system under analysis is an LNG storage tank and associated piping, which are located in an isolated part of the facility, but approximately 700 feet from the closest residences. Using this information, the PHA facilitator may translate the worst-case hazard zones in Tables 2 and 3 into a set of tables with suggested consequence severity rankings which are tailored to the system under review. Table 5 provides an example of these consequence severity tables for the specific LNG system described above.

Table 5A. Methane Liquid Consequence Severity Rankings

	Source Temperature (F)			
Release (lbs)	**-250**	**-200**	**-150**	**-117**
10	1	1	2	2
100	3	2	2	3
1,000	4	3	3	4
10,000	4	4	4	5
100,000	5	5	5	5
500,000	5	5	5	5

Table 5B. Methane Liquid Consequence Severity Rankings

	Source Pressure (psi)				
Break Diameter (in)	**50**	**150**	**300**	**400**	**450**
0.25	1	1	1	1	1
0.5	1	1	1	1	1
0.75	2	2	2	2	2
1.0	2	2	3	3	3
1.5	2	3	4	4	4
2.0	3	4	4	4	4
3.0	4	4	4	5	5
4.0	4	5	5	5	5
6.0	5	5	5	5	5

When translating the calculated hazard zone to consequence severity, it is important to note that the dispersion modeling calculations are performed generically, and their results may need to be adjusted based on the specifics of the system under review. The relationship between hazard distance and similar consequence categories (i.e., first aid injury, reportable injury, and permanent disability) should be developed with the primary decision maker for the facility (i.e., company management). To accomplish this, the facilitator should record the decision making process used to relate hazard distances to consequence severity levels, and present these decisions to company management prior to beginning the PHA sessions, and adjust the severity tables based on comments received. .

Additionally, the model calculations make a number of assumptions which may or may not apply to the system being analyzed. Most importantly, the calculations assume that releases occur outdoors, and outside of any secondary containment areas. Indoor releases can have a significantly reduced hazard zone outside of the building. This may depend on the ventilation capabilities of the building, or doors normally kept open. Toxic material may build to higher concentrations within the building than those reported in the hazard distance tables. Similarly, the presence of secondary containment limits the surface area of a liquid pool, which may result in a smaller vapor cloud than an uncontained liquid spill. To adequately assess the severity of a chain of events, these deviations from the assumed model must be accounted for. In cases such as these, the hazard distance tables should be used as a starting point, but adjust their consequence severity ranking according to the physical systems present in the field.

During the PHA discussions, once a scenario has been identified and the release conditions are established, the consequence severity tables can be consulted to assist in selecting the most appropriate consequence severity ranking. To use the tables, a process engineer or other person familiar with the design of the system under consideration should determine the physical parameters of the release. It must first be determined whether the released material is in the liquid or the vapor state. Subsequently, the release conditions must be specified in terms of temperature, pressure, quantity spilled, or breach diameter, depending on what information is most readily available to the team. This determination is not intended to be a rigorous calculation, but rather a quick estimation of the conditions of the release.

Once the release parameters have been agreed upon, the team may then use the consequence severity tables to determine the worst-case severity ranking. For example, in the LNG system described above, it is postulated that a full break occurs in a one inch diameter pipe containing LNG at 170 psig. Given the information at hand, the team consults Table 5B, and identifies 150 psig as the closest pressure rating found on the tables. From the Table, the team selects a severity ranking of 2, indicating that a recordable injury is the anticipated worst-case consequence.

6. Conclusion

Kazarians & Associates, Inc. has developed a method which uses pre-developed consequence severity tables to refine the judgement of PHA teams with regard to the impact that a scenario may have on personnel. This method has been applied, resulting in a more accurate determination of consequence severity. When applied across multiple teams and studies, this increase in accuracy will lead to consistency in severity ranking among similar PHA scenarios. In turn, this will lead to consistent application of safeguards across multiple systems, resulting improved plant operability and avoiding the potential for confusion among operating staff.

7. References

[1] Phast Modeling Software, version 7.11.33.0. DNV GL AS.

[2] *Guidelines for Hazard Evaluation Procedures*, published by the Center for Chemical Process Safety of the American Institute of Chemical Engineers, 3rd Edition, April 2008.

[3] *OGP Risk Assessment Data Directory: Vulnerability of Humans*. The International Association of Oil and Gas Producers. Report No. 434-14.1. March 2010.

[4] Toups, Harry J., *Fires and Explosions - Fundamentals and Design Considerations* [PowerPoint Slides]. Retrieved February 26, 2015 from www.sache.org / links / Pike21Jul2004 / Fires%20and%20Explosions.ppt.

[5] Clancey V J, 1972. Diagnostic features of explosion damage, 6th Intl. Meeting on Forensic Sciences, Edinburgh, Scotland.

[6] ALOHA 5.4.3 Modelling Software, Office of Emergency Management – U.S. EPA and Emergency Response Division – National Oceanic and Atmospheric Administration.

[7] *Handbook of Chemical Hazard Analysis Procedures*. Federal Emergency Management Agency, U.S. Department of Transportation, U.S. Environmental Protection Agency.

[8] Hockey, S M, and P J Rew. *Review of Human Response to Thermal Radiation*. HSE Contract Research Report No. 97/1996. Health and Safety Executive. July 2, 1996.

[9] Raj, Phani K. *A Review of the Criteria for People Exposure to Radiant Heat Flux from Fires*. Technology & Management Systems, Inc. October 25, 2006.

[10] National Fire Protection Association. (2013). *NFPA 59A: Standard for the Production, Storage, and Handling of Liquefied Natural Gas (LNG).*

441564 Application of Open Source Software Tools in the Process Safety Education

Abstract:

Process safety professionals have used, developed, and contributed to open source software tools. They have fascinated more and more notice from the society. The free software will enhance the collaboration and motivate the scientific and technical development. The software users can use and even modify the software modules. A lot of applications can be found in the real industrial practice so the engineering learning process should follow this trend. Before starting the career, the students should get to know how to take the full advantage of these tools in their future jobs because the future engineers should know how to find the free software packages in order to solve process safety problems.

This presentation will tell the audience how to integrate the open source software into process safety education or learning practice. The examples of modeling projects which can be practiced by the students in the classroom will also be introduced. The modeling will give the students more opportunities to recall their knowledge of mass and energy balance, thermodynamics, heat transfer, reaction kinetics, transport phenomena, process safety, statistics and numerical methods. These projects will also help them to learn how to conduct the risk assessment and process safety management through different kinds of model development. These chemical engineering education activities will allow them to explore computer aided tools such as computing and process simulators throughout the curriculum.

441688 Beyond Quantitative Risk Analysis Results - Part I: Explosions and Blast Phenomena Characterization

Abstract:

The present paper illustrates the results of a Quantitative Risk Analysis (QRA) case study of a refinery. The main purpose of a QRA is to evaluate the risk levels (both individual and societal risk) of a process due to Loss Of Containment scenarios (LOCs). A QRA can be used to optimize the risk reduction measures to be implemented to comply with a given tolerable risk level. Moreover, the development of a detailed QRA is the best opportunity to perform a specific analysis addressing structural response due to overpressure impacts caused by explosions and also estimating the Time To Failure (TTF) of structures and equipment due to thermal radiation of fires.

Part I of this paper is focused on illustrating how a detailed analysis of QRA results can provide valuable information for characterizing the blast phenomena parameters for a given location. The analysis gathers three key sources of information: (1) peak overpressure, maximum and minimum positive phase durations and impulses as a function of the cumulative frequency of occurrence; (2) total number of LOC scenarios which contribute to the impact of explosions (including vapor cloud explosions due to both immediate and delayed sources of ignition); and (3) the identification of each LOC scenario and associated individual frequency of occurrence. This information is valuable for ensuring the blast loading function for a subsequent structural response analysis (e.g., equivalent elasto-plastic Single Degree Of Freedom analysis), and also provides specific information regarding which LOCs contribute most on overpressure impact.

441690 Beyond Quantitative Risk Analysis Results - Part II: Fires and Domino Effect Characterization

Abstract:

The present paper illustrates the results of a Quantitative Risk Analysis (QRA) case study of a refinery. The main purpose of a QRA is to evaluate the risk levels (both individual and societal risk) of a process due to Loss Of Containment scenarios (LOCs). A QRA can be used to optimize the risk reduction measures to be implemented to comply with a given tolerable risk level. Moreover, the development of a detailed QRA is the best opportunity to perform a specific analysis addressing structural response due to overpressure impacts caused by explosions and also estimating the Time To Failure (TTF) of structures and equipment due to thermal radiation of fires.

Part II of this paper is focused on illustrating how a detailed analysis of QRA results can provide valuable information for characterizing potential domino-effects due to thermal radiation of fires at a given location. The analysis gathers three key sources of information: (1) thermal radiation as a function of the cumulative frequency of occurrence; (2) total number of LOCs which contribute to the impact of fires; and (3) the identification and characterization of each LOC scenario and associated individual frequency of occurrence. This information not only is valuable for ensuring the heat radiation for a subsequent analysis intended to estimate the TTF and associated conditions remaining in the system (e.g., Wall Dynamics Analysis), but also provides specific information of which LOCs contribute most to the heat radiation impacts. This information ensures which LOC scenarios need to be addressed for optimizing the risk reduction decision-making process. Additionally, the accurate estimation of potential equipment failure due to thermal radiation provides added value for predicting when and how impacts such as BLEVEs (Boiling Liquid Expanding Vapor Explosions) are expected to occur.

441700 A Leadership Vision for Accelerating Learning from Past Lessons to Create a Future Free of Incidents

Abstract:

The practice of process safety management (PSM), involves the analysis and understanding of what can go wrong, what the effects could be, and what should be done to prevent or mitigate an incident. That is: (1) what are the causes (human errors and equipment malfunctions)? (2) What are the potential consequences and impacts (to people, property, process, reputation)? and (3) what are the risk controls needed to prevent, detect, control, mitigate, or respond to such events (safeguards, layers of protection)?

Lessons learned from industrial disasters, as well as successes by industry leaders in preventing such incidents, form the basis for a practical framework for a PSM system as a holistic management tool for managing hazards and their risks. Quality design/development and effective implementation of a robust risk management system driven by a proactive culture of leadership to reduce operational risks and improve performance, enables organizations to avoid future incidents.

However, a conventional approach of longer-term continual improvement of operational risk performance could easily be outpaced by the rapid changes in technology advances, organizational and human factors, and the dynamics of the market. Operating companies, service providers and regulators alike are challenged by such changes and their potential impacts. Lessons of disaster investigations from the process industry have been full of learning opportunities. In fact, many of the lessons and their root causes are common across most process safety incidents and near misses ... yet, 15-20 years has proven to be an unacceptable period for gradual learning.

This presentation will present some common lessons as a priority for establishing improvement strategies as organizations strive for operational excellence. Available proven risk management methodologies and best practices for operational risk reduction will be summarized. Given that the global industry has made remarkable progress toward developing proven solutions aimed at prevention, organizations must now commit to accelerating the adoption of these solutions before at-risk operations experience near future incidents.

The focus of the paper will be the challenge to accept a leadership position to resolve existing but perhaps unidentified risks, then to commit to action to proactively identify hazards and their associated risks, and implement risk reduction measures before losses occur. A framework of a robust risk-based management system, incorporating all of the necessary elements based on lessons learned and including supporting leadership and management processes to achieve risk reduction performance objectives, will be presented.

Attendees will benefit from practical takeaways of how each management system element can be developed with "leadership designed in" to control the process safety risks at three opportune times: before the incident (prevention), during the incident (emergency management) and after the incident (incident investigation, root cause analysis and lessons learned).

The presentation will pose a challenging question: "Knowing what we already know, why would we want to take years to gradually reduce our risks, when we could accelerate the learning from past lessons and prevent incidents using available best industry practices"?

This paper will draw upon the authors' extensive experience in designing, developing and implementing PSM systems.

441698 Analytics to Reveal Meaurable Factors to Process Safety Performance

Abstract:

Analytics on a Large Industry Data Set Reveal the Leadership and Organizational Factors that Drive Down Risk and Reduce Incidents

With access to a multi-year data set of over a million incidents from 1000's of operating facilities worldwide in the Energy and Chemicals industry, the research team mined the data set to explore the differences in risk management practices and behaviors which most effectively explain the variation in process safety outcome performance. By modeling and examining predictive relationships, the study identifies four attributes - worker engagement, process discipline, risk sensitivity, and learning-mindedness – as critical for reducing incidents.

Findings from this study support a progressive and integrated approach and suggest that companies that are proficient in driving process safety performance are companies that, to a greater degree, have successfully engaged their workforce, strengthened their operating discipline, sharpened their risk sensitivity, and become learning-minded. A learning-minded organization manages each incident and responds to each failure deliberately to identify and fully leverage improvement and learning opportunities.

While this study provides statistical evidence to support the strong correlation between leadership, organizational behavior attributes and safety performance, it also emphasizes the organizational factors that are necessary to enable and sustain effective risk mitigation efforts. By deriving these insights from a data set of routine field-level activities, operational measurements may be created to measure progress and drive focus on engagement, discipline, risks, and learning.

This paper will conclude by sharing examples of how companies have applied these findings to establish quarterly and monthly measurements to drive leadership and organizational focus in these critical areas and ultimately to improve outcome performance.

441960 Coupling Dynamic Pressure Relief System Design with Experimental Vent Flow Regime of Dicumyl Peroxide / Toluene System

Abstract:

Abstract

A safe process design requires knowledge of the chemical reactivity of the desired reactions as well as the undesired reactions due to process upset conditions. Techniques used for sizing pressure relief systems for reactive systems under thermal runaway conditions include direct empirical scaling of experimental data obtained from a vessel with very low thermal inertia, analytical methods, and computer simulations using reaction kinetic models developed from adiabatic calorimetry data.

Pressure relief size estimation is based on the required flow capacity through the relief device at a specific set pressure, in order to keep the pressure in the vessel below the maximum allowable accumulated pressure. The flow capacity of the relieving device is highly dependent on the quality of the vent flow regime, whether it is liquid, all-vapor/gas, or two-phase. This requires knowledge of the physical properties of the contents inside the vessel at the time of venting. For a reactive system, these properties can be difficult to determine.

In this paper, the pressure relief size of a reactive system containing 50% dicumyl peroxide in toluene will be estimated using the dynamic simulation method. The vent flow regime of the reactive system will be determined from measured data, according to the methods described in the DIERS (Design Institute of Emergency Relief Systems) project manual. An automatic pressure tracking adiabatic calorimeter (APTAC^{TM}), which is a low thermal inertia calorimeter, will be used to conduct vented tests. The pressure relief size estimates of the above dynamic simulation method will be presented and compared with the DIERS simplified method.

441701 We Did a Quantitative Risk Assessment Study. Now What Do We Do with It?

Abstract:

Quantitative Risk Assessment studies have been extensively used in risk based regulatory regimes and they are becoming more common as a replacement for facility siting requirements in Process Safety Management (PSM) covered regulatory regimes and as a corporate risk management tool. Besides assessing risks to society and personnel, it can also assist in practical risk reduction implementation in design projects and operating assets. This paper intends to discuss these practical risk reduction implementation to show value to projects and operating assets, and how implementation can be achieved to using specific examples. The following examples will be discussed in the paper and will outline a broader set of QRA implementation improvements:

- Operating modes input to isolatable inventories and failure criteria
- Failure mechanisms from Hazards and Operability studies, and corrosion or mechanical integrity surveys
- Survivability criteria for manned and unmanned buildings for upgrades or siting challenges
- Specialized consequence analysis such as uncommon release material, indoor releases, and elevated receptors.
- Spill containment application
- Thermal protection and passive fire protection application
- Personnel and structural cryogenic protection application
- Risk based fire and gas detection mapping
- Escape, evacuation and rescue impacts
- Operational risk reduction using performance standards

The objective of this paper would be to better understand how to utilize QRA studies by improving their content and by practical implementation (as necessary) of mitigating risk reduction measures in design projects and to existing assets.

442063 Effective Implementation of PSM to Reduce Operational Risks

Abstract:

Ultimate success of process safety management (PSM) is measured through effective operational risk reduction. Usually, there is a lot of emphasis on putting the various elements of PSM in place through adoption of established international standards and best practices. Many of these standards and best practices are very well documented and element leaders/champions are appointed. Unfortunately, many of these elements and element champions are in silos and thereby not contributing to the overall reduction in operational risk. The risk is not reduced until the documentation is converted into reality on the ground.

Also, there is a lot of emphasis on doing high quality technical studies and quantitative process hazards analysis but very little is done to follow up on implementing the recommendations and results of these studies. The operational risks are not reduced until the recommendations are actually implemented. So, despite significant amount of effort, the improvement in the effectiveness of PSM implementation is limited.

Learning from this experience, SCG Chemicals has adopted a practical, bi-focal approach that is driven with a specific purpose in mind — operational risk reduction. The short term focus was to address deviations from design intent and established procedures in the field. If these items are not addressed today, they can contribute to a potential catastrophic incident tomorrow. This enhanced operational discipline and prevented 'normalization of deviations'. The longer term approach focuses on redoing structured PHAs and using the risk matrix to establish the levels of acceptable/unacceptable risks. Management decisions were made to address priority of the unacceptable risks with a rigorous follow up on implementing recommendations. This process has contributed to better outcomes in increased effectiveness of PSM while optimizing the money, time, and effort required. This paper will share the experience of SCG, along with DuPont as a partner, and the journey in implementing PSM effectively to reduce operational risks.

442151 Probit Models for the Determination of Damage Probability of Storage Tanks Due to Flooding Impact

Abstract:

Global climate changes may significantly influence temperatures and precipitation patterns, increasing the frequency of severe meteorological events which may lead to flooding. Flooding may trigger severe accidents on population among urban as it may do on chemical process industries too, causing the damage of equipment and the loss of containment of hazardous materials which might transform into toxic dispersions, fires, explosions or severe environmental contamination. This type of events are known as Natech events (Natural-Technological). This type of events may lead to extremely severe consequences bringing up the need to develop approaches for the risk assessment of Natech scenarios. Recently important efforts have been dedicated to develop tools that include Natech events in the quantitative risk assessment framework. In order to allow a quantitative risk assessment of a Natech event, a key point is the definition of tools for the estimation of equipment damage probability on the basis of intensity parameters of flooding. Past accident data analysis evidenced that horizontal and vertical storage tanks were often damaged in Natech events triggered by floods. Because of this the present study presents specific probit models for the damage probability of horizontal and vertical storage tanks due to flooding intensity parameters (flooding water height and flooding water).

Probit models in this study were derived on the basis of a mechanical model used for the assessment of the vulnerability of storage tanks due to the impact of flooding. For the determination of damage probability Montecarlo simulations were applied over the mechanical model in order to obtain a more accurate value and a probabilistic distribution of this parameter in dependence of the flood impact vector. Once the damage probability distribution is determined the probit method was applied using a logarithm approach with the impact flood vector (water height and water velocity) as the dose variables and the probability of damage as the response variable. A logarithm regression using the minimum square method was utilized to determine the probit constants for the model. Since the mechanical model used as basis has a huge dependence on the geometry of the storage tank being studied, it was find the damage probability will be affected and its value will change. To make sure this affections will be taken into account in the probit modeling a relationship between the geometry of the tanks and the probit constants was established using a polynomic regression. Once again the minimum square method was used and these two variables were related using a second order polynomic approach. Probit constants were determined for two substances (gasoline and LPG) and two types of storages tanks (vertical and horizontal).

The results of this study are a useful tool for the development of the quantitative risk assessment of Natech events due to flood impacts in storage parks. Also contributes to the reduction of the uncertainty in the framework for the QRA of Natech events triggered by floods.

442178 Analysis of Fire Extinguishing Methods in Engine Control Rooms

Abstract:

The study of electrical fires is a major concern for fire safety in the industry and more particularly for oil industry. In petrochemical facilities, the fire hazard may cause severe consequences on safety functions inside the fire compartment and furthermore may lead to the loss of one or several compartments, as just as structural damage on the facilities. From the OECD Fire Database a majority of the fire ignition is due to electrical sources (close to 50%, with a significant contribution attributed to electrical or electronic cabinets, with an occurrence of more than 10% of all fire events). Because of the above mentioned, it is necessary to implement tools to study these phenomena without using an experimental method. For example, the simulation of the dynamics of combustion and fire spread, allows an estimation of the effects on the structure of the case study preemptively. This project addresses the comparison of different fire protection philosophies in engine control rooms, using Fire Dynamics Simulator as support tool. The project studies three different cases, one of them without a protection system, another a clean agent as extinguishing tool and a final one, which implements an inherently safe design. For the first case study, the obtained data for the room temperature and the heat release ratio indicate a potential risk for the personnel and the equipment of the facility. It is expected that the study of the remaining cases show the significant advantages of each fire extinguishing philosophy.

12TH GLOBAL CONGRESS ON PROCESS SAFETY

The Study of HAZOP Revalidation Approach

Jing Xu
SINOPEC Research Institute of Safety Engineering
No.218, Yan'an san Road,Qingdao,China
xuj.qday@sinopec.com

Chuntao Ge
SINOPEC Research Institute of Safety Engineering
No.218, Yan'an san Road,Qingdao,China

Yi Zhang
SINOPEC Research Institute of Safety Engineering
No.218, Yan'an san Road,Qingdao,China
Zhangyi.qday@sinopec.com

The Study of HAZOP Revalidation Approach

Jing Xu
SINOPEC Research Institute of Safety Engineering
No.218, Yan'an san Road,Qingdao,China
xuj.qday@sinopec.com

Chuntao Ge
SINOPEC Research Institute of Safety Engineering

Yi Zhang
SINOPEC Research Institute of Safety Engineering

Keywords: HAZOP, HAZOP revalidation, update HAZOP,PHA, MOC

Abstract

Abstract: HAZOP evaluation method is used to analyze the risk of the process by the worldwide chemical enterprises and design units. However, Standard Specification for many countries stipulate the enterprise needs to carry on the process hazard analysis once again (revalidating) in several years later. The goal of periodic revalidation is to identify the cumulative changes of risk over time and to improve understanding of risk. The ultimate goal is to avoid accidents. But there is no systematic approach to guide the HAZOP revalidation at present. If we redo HAZOP blindly (performing a new HAZOP similar to the initial one) will only be time-consuming. This paper introduces the concept of HAZOP revalidation which is based on the analysis of the initial HAZOP and discusses many of the factors that need to be considered for the revalidation such as the change management, accident analysis and HAZOP report quality. At the same time, the specific procedure and approach of the HAZOP revalidation are proposed. The panel of experts scoring method is used to determine the revalidating approach and then decides whether to redo or update the initial HAZOP. If the scoring results suggest redo the HAZOP, and then conduct the Hazard and Operability Analysis in accordance with the complete HAZOP procedures; if the scoring results suggest update the initial HAZOP and then modify and supplement previous HAZOP analysis results to meet enterprise current status. This approach is not only time-saving and more targeted which can guide enterprises to conduct better and faster HAZOP revalidation.

1 Introduction

Hazard and Operability Study (HAZOP) is a kind of process safety analysis methods which is invented by Britain's imperial chemical industries (ICI) in the 1960s[1]. It is carried out by a suitably experienced multi-disciplinary team (HAZOP team) during a set of meetings to identify potential risk and operational problems in the system.

However, the safety status of enterprise is not invariable after the HAZOP.Many changes may occur in the later production process inevitably. The safety status of enterprise will change if the technological process, equipment, safety equipment, operating procedures, personnel, organization structure, and environment changes in the entire life cycle. So getting through the HAZOP analysis is not once for all. HAZOP Revalidation is to conduct a comprehensive evaluating for process hazards situation of the system during the next few years after the first time HAZOP analysis. To eva`luate the validity of the analysis results and determine whether meet the requirements of laws and regulations,and improve the process safety management level of enterprises at the same time. The goal of regular revalidation is to identify the change of the risk as time goes on and to improve the understanding of the risk. The ultimate goal is to avoid the accident. However, HAZOP analysis of a typical chemical process takes a few weeks to a few months. When in the HAZOP revalidation, whether can take advantage of the existing HAZOP analysis report to make the HAZOP revalidation more simplified and efficient and guarantee the results conform to the enterprise current situation. More and more companies try to simplify the method of revalidation, rather than redo HAZOP blindly[2].

2 Status analysis of HAZOP revalidation

HAZOP technology has developed more than 50 years getting a lot of project application, and become one of the general evaluation method on process safety domain at present. Petrochemical companies and design companies worldwide use HAZOP method to analyze the risk of petrochemical process. Such as Dow chemical, Rhodia company, BP, Bayer, Royal DSM, SINOPEC, CNPC, CNOOC and so on[3].

OSHA(the Occupational Safety & Health Administration) put forward this revalidation concept at Process Safety Management in 1992. Claim that "After the first process hazard analysis, revalidation should be done at least every five years, to ensure consistency of process hazard analysis and the current process" [4]. In the risk management procedures of EPA(Environmental Protection Agency) and aEuropean SEVESO II have corresponding requirements in the instruction. CCPS(Center of Chemical Process Safety) published a book Revalidating process hazard analyses Defined that the revalidation process attempts to protect this investment by identifying and building upon the still pertinent portions of the prior PHA .Corrections are made and new content is added to the PHA as required,and the result are documented to serve as the new PHA[5]. In the United States, the large-scale petrochemical enterprises have done 2 ~ 3 times HAZOP, identifying the potential risk in the system, found a lot of potential safety problems. The study of HAZOP technology started late in China. Now a lot of enterprises done HAZOP before will conduct HAZOP revalidation. Guidelines for process safety management of petrochemical corporations (AQ/T3034-2010) standard requires: Guidelines for process safety management of petrochemical corporations (AQ/T3034-2010) standard requires: Companies should conduct a process hazard analysis in the process during construction and should to finish the process hazard analysis to confirm again and updated every three year[6]. The difference between Revalidation and first HAZOP is the former is conduct on the basis of the practical operating experience. The key for the first time HAZOP is to perfect the design and the retrieval permissions[7]. HAZOP Revalidation is to compare the results of the first HAZOP whether still applicable to the current environment, more dependent on site operation actual situation and record filing, etc. There is no systematic method to guide how to HAZOP review again. However, blindly redo HAZOP is not only time-consuming, also need a lot of resources. More and more companies expect to revalidate on the basis of the first HAZOP analysis.It can be more simple and efficient, and at the same time guarantee the quality of process hazards analysis.

3 The Study of HAZOP Revalidation Approach

3.1 The key factor of HAZOP revalidation

HAZOP revalidation using redo or update way is decided by many factors, such as whether there is a big defect of the previous process hazard analysis, whether change management system is perfect, whether there is a major accident happened, whether the information is complete, new standards, etc.

The way for HAZOP revalidation depends on the following three aspects:

a. The design change number, the size of the problem, process complexity, change process hazard analysis quality and management of change since the last HAZOP analysis;
b. The quality of Previous HAZOP analysis, including process safety information comprehensiveness, the comprehensiveness of HAZOP analysis, HAZOP record content integrity, HAZOP results application effect, etc;
c. Previous accidents and device status: the type of process safety accidents, statistical accident frequency, Consequences severity and the reason for the accident , etc.

3.2 The process of HAZOP revalidation

a. Establish revalidation team, the composition of team members are as follows[8]:

- the group leader: trained, rich experience in organization HAZOP analysis; Responsible for communication between HAZOP team and project management personnel; Review plan
- Process equipment technician: familiar with the past project, past and present design/operational problems and change the content
- Site operator: familiar with the situation, the operation experience
- owner (user) : familiar with the operating environment, the consequences of deviation and the degree of danger of deviation
- Experts: providing related professional knowledge. Technology, automatic control experts shall be invited to assist analysis group for part analysis
- Recorder: review record, record the related changes, and update or supplement HAZOP records, to assist the analysis team leader for planning
- Maintenance personnel (if required)

b.Collect information, including the HAZOP report, change management (MOC) data, the previous accident report, HAZOP recommendations, process engineering and safety information, operation procedures, etc

c.Expert assessment team scoring method. Details see Table 1 - the marking table for determining HAZOP revalidation methods and Table 2 - HAZOP revalidation method
d.According to the examination way to implement. Redo HAZOP or update HAZOP.

The flow to determine the specific revalidation way is shown in figure 1

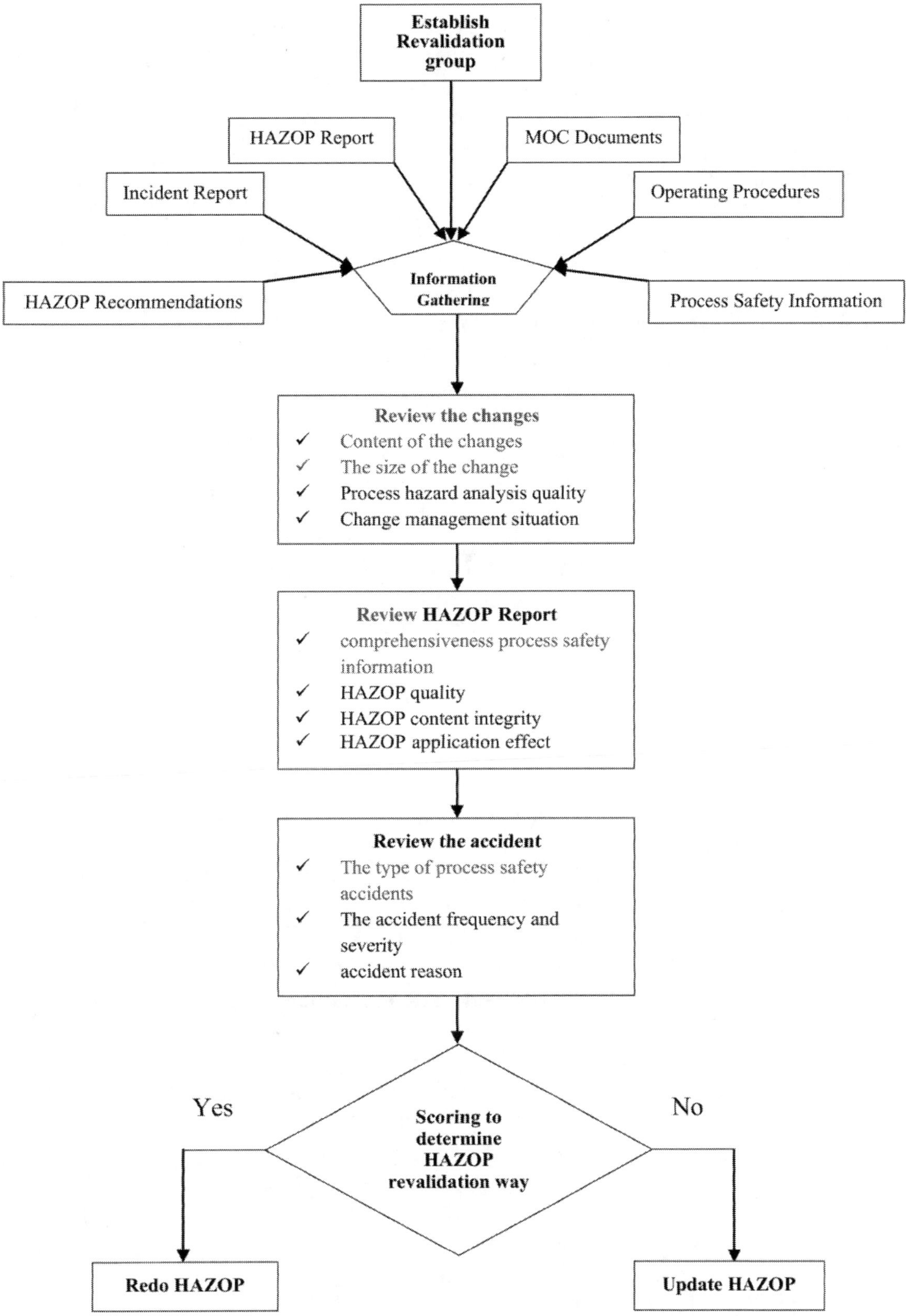

Fig.1 HAZOP revalidation approach decision diagram

Redo HAZOP is not introduced in the paper, the most effective way to update the HAZOP is to collect all data and accident information related to the change, to review the preliminary HAZOP Suggestions, to view the process pipeline flow chart meters (PID), understanding the changes and the defects of the system[9]. Such as process flow, equipment, safety equipment, operating procedures, operating conditions, location and environmental changes, and keep record by the recorder. According to the situation of change size, change management and change of process hazards analysis quality lead revalidation group to carry out different levels HAZOP for change unit by the team leader. Update the HAZOP results that conform to the enterprise the status quo. Update the HAZOP is time saving and maximum use of existing resources compared to redo HAZOP.

Table 1. HAZOP revalidation reference score table

Review items		**Index description**	**Reference scores**
Change Review	Change content (max)	Change of Process/tank type/operating procedure	15*
		Change of safety facilities	10
		Change of pipeline, pump, valve	5
		No change	0*
	Scale of change	Change of working hours more than one month/ 5 types of job post	10*
		Change of working hours about 2 weeks/3 types of job post	5
		Change of working hours about one week/2 types of job post	2
		No change	0*
	Quality of hazard analysis of change	No hazard analysis of change	10*
		Bad hazard analysis of change(lack of pertinence, record, etc)	5
		Good hazard analysis of change(More targeted, detailed record)	2
		Eexcellent hazard analysis of change(effective method, comprehensive analysis)	0*
	Management of change	Records and files are deficient and not combined with operating procedure or job training partly after change	5*
		Records and files are complete and combined with operating procedure or job training partly after change	2
		Complete and electronic records and files,which combined with operating procedure or job training after change	0*
HAZOP quality review	Comprehensiveness of process safety information	Deficient data or big change of site layout	5*
		Partly complete data or small change of site layout	2
		Complete data and less or no change of site layout	0*

	Integrity of HAZOP record	Deficient node description, design intent, drawing name, date, participant and the content of the standard form, bad reference	5*
		Partly complete node description, design intent, drawing name, date, participant and the content of the standard form, good reference	2
		Complete node description, design intent, drawing name, date, participant and the content of the standard form, wonderful reference	0*
	Rationality of HAZOP	Deficient deviation and guide word, unreasonable nodes, bad suggestions and measures	15*
		Partly complete deviations and guide words, reasonable nodes, common suggestions and measures	10
		Complete deviations and guide words, more reasonable nodes, good suggestions and measures	5
		Perfect deviations and guide words, most reasonable nodes, appropriate suggestions and measures	0*
	Application effect	Weak effect, less use of risk identification	5*
		Good effect, general use of risk identification	2
		Perfect effect, effectively use of risk identification	0*
Review of accidents		Vast or serious process safety accidents	30*
		Some specific process safety accidents	15
		Some accidents which has nothing to do with the process	5
		No accidents	0*
* reference point，according to the actual situation within the range of its rated arbitrary integer (interval)			

Table 2. HAZOP revalidation approach determination table

Scores	Review way
⩾60	Redo HAZOP
<60	Update HAZOP

4 Application examples

A spherical tank area of PetroChina has four 3000 m3 spherical tank, two 5000 m3 spherical tank, storage for butene 1, butadiene. The company has completed the HAZOP analysis in 2010. Device in service conduct HAZOP analysis every four years according to the company policy. HAZOP revalidation use the proposed method, completed HAZOP analysis efficiently.

Table 3. Application of HAZOP revalidation approach determination table

Review items		practical situation	Scores
Change Review	Change content (max)	No change	0
	Scale of change	No change	0
	Quality of hazard analysis of change	Good hazard analysis of change(More targeted, detailed record)	2
	Management of change	No change	0
HAZOP quality review	Comprehensiveness of process safety information	Partly complete data or small change of site layout	2
	Integrity of HAZOP record	Perfect deviations and guide words, most reasonable nodes, appropriate suggestions and measures	0
	Rationality of HAZOP	Complete deviations and guide words, more reasonable nodes, good suggestions and measures	5
	Application effect	Good effect, general use of risk identification	2
Review of accidents		No accidents	0
Scores			11

5 Conclusion

This paper introduces the concept of HAZOP revalidation which is based on the analysis of the initial HAZOP and discusses many of the factors that need to be considered for the revalidation such as the change management, accident analysis and HAZOP report quality. At the same time, the specific procedure and approach of the HAZOP revalidation are proposed. The panel of experts scoring method is used to determine the revalidating approach and then decides whether to redo or update the initial HAZOP. If the scoring results suggest redo the HAZOP, and then conduct the Hazard and Operability Analysis in accordance with the complete HAZOP procedures; if the scoring results suggest update the initial HAZOP and then modify and supplement previous HAZOP analysis results to meet enterprise current status. The method has been applied for a spherical tank area of PetroChina. Conclusion is spherical tank area further implement the last countermeasures of HAZOP in view of the actual operation situation, do not need the modified or added. To satisfy the needs of the enterprise for better and faster process hazards analysis.

6 References

[1]Dunjó J, Fthenakis V, Vílchez J A, et al. Hazard and operability (HAZOP) analysis. A literature review[J]. Journal of Hazardous Materials, 2010,173(1):19-32.

[2]Smith K E, Whittle D K. Six steps to effectively update and revalidate PHAs[J]. Chemical engineering progress, 2001,97(1):70-88.

[3]WAN Gujun, LI Qi, ZHANG Guangwen, et al. Implementing HAZOP to Control Major Accidents (Ⅰ)[J]. Safety Health & Environment, 2009,9(011):2-4.

[4]ZHAO Jianmin. Discussion About HAZOP Revalidation [J]. Modern Occupational Safety, 2012(134):83-85.

[5]L. F W, K. W D. Revalidating process hazard analyses[M]. Center for Chemical Process Safety, 2001.

[6]State Administration of Work Safety. AQ/T 3034-2010 Guidelines for process safety management of Petrochemical corporations[S]. 2010.

[7]FU Jianmin, ZHAO Dongfeng, CHEN Guoming, et al. Quantitative study over the probability of HAZOP analyzing technique in petro-chemical plants[J].Journal of Safety and Environment,2008,8(6):130-134.

[8]Herbert I. Learning the lessons-retrospective HAZOPs[J]. Offshore Europe, 2011.

[9]Wagner T, Champion J. A work process for revalidating LOPAs and other risk analyses[J]. Process Safety Progress, 2012,31(2):122-129.

The Importance of Process Safety:

A Young Engineer's Perspective

Waheed Wakil
Smith & Burgess
Waheed.Wakil@smithburgess.com

Travis Bruner
Smith & Burgess
Travis.Bruner@smithburgess.com

Prepared for Presentation at
American Institute of Chemical Engineers
2016 Spring Meeting
12th Global Congress on Process Safety
Houston, Texas
April 11-23, 2016

The Importance of Process Safety: A Young Engineer's Perspective

Waheed Wakil
Smith & Burgess
Waheed.Wakil@smithburgess.com

Travis Bruner
Smith & Burgess
Travis.Bruner@smithburgess.com

Keywords: Process Safety Management, Young Engineers, Overpressure, OSHA, LOPA

Abstract

Most young engineers experience a defining moment or moments in their career that shape their attitude towards importance of Process Safety Management (PSM). Experiencing a plant upset first-hand knowing that there were actual capital or human life impacts or both from an incident at the site, one soon realizes early in their engineering career why process safety is so important. It becomes evident that the most successful corporations are those who are willing to invest large capital costs into PSM documentation projects or projects that have come as a result of findings identified during a study, allowing them to prevent incidents from occurring. The findings generated during the course of a study may result in simple documentation update action items, or they may require a multi-million dollar investment. For young engineers, this exposure puts into perspective the value of their work, and emphasizes the importance of a thorough and adequate study.

1 Introduction

Many young engineers have a point early in his or her career when process safety hits home. Different companies develop their own unique strategies towards safety that they are comfortable with, which results in considerably different safety cultures within industry. Unfortunately, it is all too common for engineers to treat process safety as an afterthought, with priority given to efficiency and profitability. However, as we will see, treating safety with the utmost importance can prevent costly and life-threatening incidents.

2 First Hand Experiences

2.1 Organic Peroxides Plant Background

Two major upsets at a chemical plant within a year may very well serve as a defining "moment" in a young engineer's career that forces him to reflect on and realize the importance of dedication to process safety. A fellow Process Safety Consultant at Smith & Burgess who was a process engineer at a plant that produces organic peroxides was fortunate enough to live through two such accidents unscathed. His story is a lesson and reminder that even the most disastrous accidents can be prevented with due diligence, a safety-focused company culture, and proper attention to basic Process Safety Management.

Based on the nature of the chemicals produced, a Process Hazard Analysis for this particular plant would have revealed scenarios with severe consequences. However, the safety culture at this plant was not adequately focused in preventing and mitigating upsets. Not only were poor preventative maintenance practices used, but lack of written emergency procedures and poor training prevented adequate mitigating protection layers after the initiating events. A few relevant OSHA requirements are worth noting here as failure to meet these requirements was likely the reason for loss of containment:

- Per OSHA 29 CFR 1910.119(f)(1), "The employer shall develop and implement written operating procedures that provide clear instructions for safely conducting activities involved in each covered process consistent with the process safety information...".
- Furthermore, per OSHA 29 CFR 1910.119(j)(2), "The employer shall establish and implement written procedures to maintain the on-going integrity of process equipment."
- Lastly, per OSHA 29 CFR 1910.119(n), "The employer shall establish and implement an emergency action plan for the entire plant in accordance with the provisions of 29 CFR 1910.38. In addition, the emergency action plan shall include procedures for handling small releases."

If these basic elements of PSM would have been implemented at this plant, both accidents may have been prevented or mitigated.

Organic peroxides are highly unstable – their O-O bond breaks very easily, and that is why their decomposition properties are so useful in initiating polymerization reactions. This useful property can also be very dangerous. When an organic peroxide is in its pure form, the rate at which its exothermic decomposition generates heat is likely to be greater than the rate at which the heat can be dissipated to its surrounding environment if proper heat transfer equipment are disabled, resulting in a situation called self-accelerating decomposition in which the reaction continues to accelerate due to its own generated heat.

The organic peroxides produced in this particular plant are characterized by very low Self Accelerating Decomposition Temperatures (SADT), defined as the minimum temperature at which an organic peroxide would experience self-accelerating decomposition after one week of storage. As such, glycol cooling jackets on vessels and piping in this plant maintained operating temperatures in the ballpark of -30 °C (-22 °F) to prevent the possibility of self-accelerating decomposition. For this heat transfer to work properly, the organic peroxides must be continuously flowing in the process piping and equipment and cannot become stagnant. At the same time, the glycol that cools the organic peroxide must also continuously flow. Stoppage of flow for either component may be disastrous. As we will see in the following incident descriptions, loss of this cooling results in rapid increase of temperature and decomposition reaction rate, resulting in a deflagration that cannot be mitigated by a typical relief device such as a safety valve or a rupture disk. To prevent loss of containment and subsequent damage to surrounding structures, this scenario must instead be prevented or mitigated using written operating procedures, regular inspection and maintenance of equipment, and employee training.

2.2 Organic Peroxides Incident #1

What went wrong?

Organic peroxides are so thermally unstable that precautions must be taken when packaging and storing the material. Packaging must occur continuously to prevent the material from becoming stagnant in piping and the temperature must be monitored and controlled at or below the recommended storage temperature. If material remains stagnant without sufficient cooling, organic peroxides can reach their SADT temperature and undergo a violent reaction, which usually results in rupturing the packaging, dispersing peroxide, as well as discharging liquid and gaseous decomposition products considerable distances. The heat generated during this reaction has the potential to auto ignite flammable vapors.

An operator was transferring organic peroxides to 55 gallon drums, which are immediately transported by forklift to an icehouse for storage. An incident occurred which caused the operator to abandon the task in the packaging building to provide assistance elsewhere. While the operator was away, the product remained stagnant in the filter where glycol cooling was inefficient. This caused the material to exceed its SADT. The material began to decompose as temperature continued to increase. When the operator returned the material suddenly exploded propelling the head of the drum through the air. The remains of the drum were embedded in the ground and lights in the facility were shattered. The operator was behind a blast shield, but still suffered severely burnt hands and damage around the ear. Although no life threatening injuries were inflicted, the plant suffered considerable downtime and costly repairs to the filter, packaging equipment, and process piping.

How could it have been prevented?

Proper implementation of PSM standards likely could have prevented this incident from occurring - in particular, OSHA 29 CFR 1910.119(f)(1), mentioned previously. Documented operating procedures that clearly state not to leave the packaging unit unattended during operation to avoid stagnant organic peroxides, inefficient cooling, and the potential for explosion could have prevented (or, at the very least, limited) the disastrous effects of this incident.

2.3 Organic Peroxides Incident #2

What went wrong?

This plant utilizes large plug flow reactors that are jacket-cooled. On one particular day, the site was hit by a thunderstorm with wind speeds high enough to break a tree branch that would damage a power line, resulting in total loss of electricity to the plant. Electric power was vital to this plant as it kept peroxide flowing through the reactors and piping and it kept glycol flowing through the jacketing to keep the peroxide temperatures down to safe levels. The plant had a backup generator, but only a single operator was aware of the location of its switch. It just so happened that it was this operator's day off.

In the event of an emergency such as this, interlocks were in place to purge out residing organic peroxides in the equipment using control valves with actuators regulated by instrument air. Once glycol and peroxide flow was lost in the event of a power failure, the peroxides would rapidly exceed their SADT. The purging action would dump the peroxides into waste water drains that are gravity fed to open tanks, where it is much safer for the peroxides to reach their SADT. However, the instrument air that controlled the valves was supplied from air storage, and during the purging, the instrument air ran out. Nitrogen was used as backup for air during emergency situations, but due to poor inspection and maintenance for cost savings, there was damage to the backup nitrogen header that had gone unnoticed. This damage caused a leak in the header when it was exposed to the nitrogen pressure, rendering the header useless as there was not enough pressure to actuate the purge control valves properly. At this point, purging was halted. The catastrophic failure was now inevitable. With stagnant peroxides in piping and equipment, and no jacket cooling, fluid temperatures rapidly sored past the SADTs and eventually resulted in a deflagration. The reactor top heads failed first, and the building in which the reactors were located was designed to fail at the ceiling and the outer wall. The building and its contents were destroyed, and the plant suffered 6 months of downtime. Fortunately, operators were protected on the other side of a blast wall, and no one suffered major injuries.

How could it have been prevented?

Looking at the number of Independent Protection Layers (IPLs) that were in place at the time of failure, namely the operator's response to activate the backup generator and interlocks that functioned to purge out residing peroxides, there were not enough layers of protection in place to prevent a deflagration. Table 1 summarizes the highest possible credits that could be taken for each IPL based on CCPS's LOPA book [1].

Table 1: IPL Credit Summary

Initiating Event	Initiating Event Frequency	Number of IPL Credits Required for Category IV Consequence	Maximum Number of IPL Credits Available
Loss of Cooling	Frequency ≥ $1X10^{-2}$	**2.0**	Operator Response for Backup Generator 0.5
			SIL 1 SIF 1.0
			Total **1.5**

Like many such accidents, it is not uncommon for multiple independent failures to cascade into a catastrophic event, especially when a key to a successful emergency response from operators is proper training and a key to proper function of mechanical equipment is proper maintenance and inspection. To pinpoint PSM measures that would have likely prevented this accident, we can start by calling on OSHA 29 CFR 1910.119(n) referenced previously in this paper. The first broken link in the series of events leading up to the accident was the lack of training amongst operators with respect to activating the backup generator in the event of loss of power. This should have been a basic emergency response action that every operator at the plant should have been trained on and been aware of, but unfortunately, no one at the plant at the time of the failure was trained on how to activate the generator. The failures cascaded into the second phase of emergency response when the instrument air was exhausted at exactly the wrong time. To address the issue of the backup nitrogen header leak, we call upon OSHA 29 CFR 1910.119(j)(2), "The employer shall establish and implement written procedures to maintain the on-going integrity of process equipment." To cut costs, plant management had reduced regular inspection and maintenance of the backup nitrogen header, and it proved disastrous on this day when they needed it the most.

3 Conclusion

As illustrated in the incident descriptions, a poor safety culture is likely to result in poor preventative inspection and maintenance cycles and lack of clear emergency operating procedures and training. In most process plants, inadequacies in PSM elements can result in disastrous consequences. For a young engineer, it sometimes takes a real-life experience to change their perspective on PSM from one of passive compliance to one of active responsibility. Each element of PSM is important and critical, from the most obvious and simple, like written procedures on activating a backup generator, to the most expensive and extensive, like redesigning a flare system. It makes every detail in every process safety analysis that much more important and that much more rewarding when the benefits are apparent.

References

[1] CCPS. *Layer of Protection Analysis: Simplified Process Risk Assessment.* Center for Chemical Process Safety, American Institute of Chemical Engineers, New York, NY, 2001.

Scaling Up Safely: Making Smarter Decisions about Rate Dependency

Zubin Kumana, P.E.
Smith & Burgess LLC
7600 W. Tidwell Rd, Ste 600
Houston TX 77040
Zubin.Kumana@smithburgess.com

Venkata Badam, P.E.
Smith & Burgess LLC
Venkata.Badam@smithburgess.com

Achilles Arnaez
Smith & Burgess LLC
Achilles.Arnaez@smithburgess.com

Prepared for Presentation at
American Institute of Chemical Engineers
2016 Spring Meeting
12th Global Congress on Process Safety
Houston, Texas
April 11-23, 2016

Scaling Up Safely: Making Smarter Decisions about Rate Dependency

Zubin Kumana, P.E.
Smith & Burgess LLC
7600 W. Tidwell Rd, Ste 600
Houston TX 77040
Zubin.Kumana@smithburgess.com

Venkata Badam, P.E.
Smith & Burgess LLC
Venkata.Badam@smithburgess.com

Achilles Arnaez
Smith & Burgess LLC
Achilles.Arnaez@smithburgess.com

Keywords: Process Safety, Safe Design, and Venting and Emergency Relief

Abstract

Over the lifetime of a facility, it is almost a certainty that the throughput of a process will be increased, often repeatedly. Rate-dependency is frequently used as a way to identify pressure relief system sizing that would be affected by the scaling up of process throughput. This paper will define rate-dependency as it relates to pressure relief systems and describe the various ways in which rate-dependent classifications may be affected by the mechanical and thermodynamic constraints of the process and equipment. In addition, examples of when the "rate-dependent" designation may no longer apply due to common assumptions and methodological choices will be provided.

1 Introduction

In the refining and chemical processing industries, increasing throughput through a processing unit is often considered as a means of maximizing value with existing capital or minimal additional investment. Often, the throughput will be increased many times over the lifetime of the facility. The pressure relief system for a process is designed in part based on the throughput of the unit; therefore, the change in the throughput may affect the adequacy of the pressure relief system design.

Verifying the design of the process unit's pressure relief system for these increased rates is typically done by identifying overpressure scenarios as either rate-dependent or rate-independent. A rate-dependent overpressure scenario is one whose required relief rate varies with changes in process throughput. A common example is a blocked outlet scenario on a separator. With higher the process throughput, the quantity of mass that must be relieved to limit the pressure to the allowable accumulation is expected to increase.

Conversely, rate-independence refers to overpressure scenarios whose required relief rate is independent of process throughput. A common example is exterior fire, where the required relief rate depends on the heat of vaporization of the liquid mixture inside the vessel, the density of the vapor and liquid phases, and the wetted area of the vessel the fluid is residing in. Changing the feed rate to the vessel is not expected to result in a change to the required relief area for the exterior fire scenario. In some circles, these scenarios may also be termed "equipment-dependent". While the equipment and piping remains constant, the required relief rate remains constant. However, as rate scale-up increases, the equipment and piping may be changed to accommodate the additional capacity.

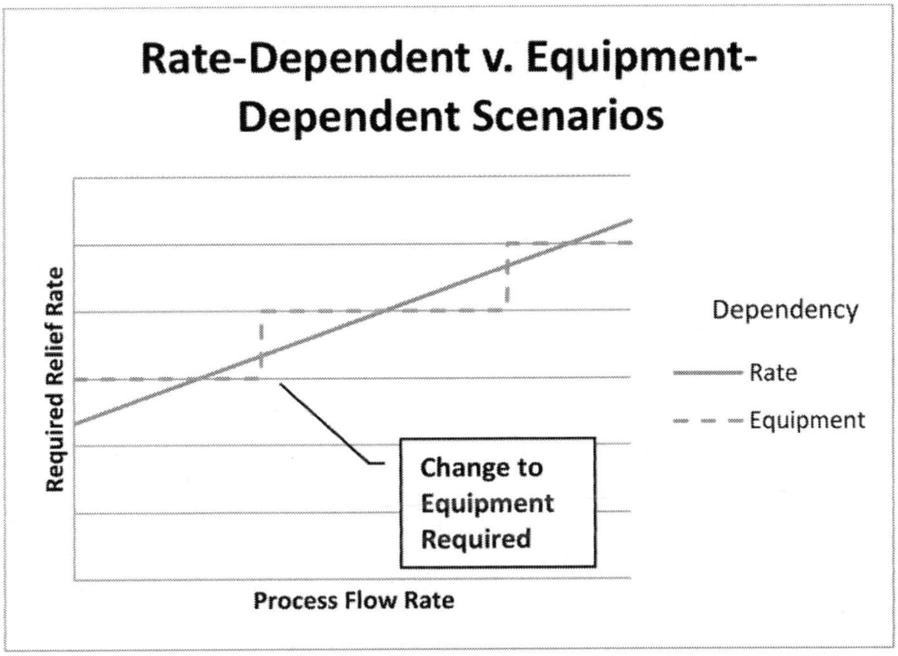

Figure 1: Rate-dependent scenarios increase in required relief rate (and therefore area) with process flow rate increase, while equipment-dependent scenarios are effectively constant until a change to equipment is required.

1.1 How Rate Dependency is used in Pressure Relief System Design

The common practice in pressure relief system design is to select a unit process basis that the user believes represents both the actual operation of the process as well as the worst case for the relief system design – specifically, the case that will result in the largest required areas and piping diameters. Since process units can have a range of operating modes and rates, selection of these cases tends to favor the following:

- Modes of operation with lighter feed, as this can result in more vaporization and therefore larger vapor and two-phase relief loads
- Higher process throughput rates; often the peak rate achieved, or some percentage above the peak rate

The conscientious user might separately specify a worst-case heavy liquid relief load basis in order to properly design the liquid handling capabilities of the disposal system.

As the pressure relief system design is completed, the engineer performing the design will typically identify overpressure scenarios as rate-dependent as the analysis proceeds. This designation is then used when the throughput rate is modified in the future. This allows engineers to minimize the amount of recalculation that needs to be performed, by only performing new calculations for the overpressure scenarios that are dependent on rates; the scenarios that had been identified as rate-independent are believed to be static and are left unchanged.

A generalized expression for pressure relief device sizing is given in API Std 520 Part I (9th ed.) Annex B, equation B.5, rearranged to solve for required area [1]:

$$A = \frac{W}{G \prod[K]} \qquad \text{(Eq. 1)}$$

Where

G	is the theoretical mass flux through the nozzle
W	is the mass flow through the PRV
A	is the effective discharge area
Π[K]	is the product of all applicable correction factors

The scale-up of the pressure relief system design is typically done using the assumption that the only change is to the flow rates through the unit; variables such as fluid composition, operating pressures and temperatures are assumed to remain constant. This effectively limits the relationship between the throughput rate and the relief system required areas to a linear one; as the fluid mass flux is considered constant, and the applicable correction factors are generally constant (exceptions are always possible based on Reynolds' number and actual backpressure resulting from flow), the values of A and W are directly proportional.

For the evaluation of the pressure relief system design, this has led to the following widespread practice:

1. The user takes the heat and material balance at the previous throughput rate
2. The user obtains a list of calculations that had been identified as rate-dependent
3. The user takes the percentage increase in flow and factors this value into the required relief rates
4. The user re-evaluates the pressure relief device requirements vs. capacity (along with other associated concerns where relevant, e.g. pressure losses if calculated at required relief rates)

This process rarely involves revalidating the process unit simulation to confirm the heat and material balance. Typically this practice has been followed for small changes to throughput rates. However, once the procedure has been implemented, the potential exists for repeated applications to increase the degree of deviation from a validated baseline even further. This poses a danger, as excessive extrapolation and repeated application of the scale-up process without carefully considered heat and material balance revalidation can quickly lead to larger and larger deviations, and the potential for undersizing the pressure relief system.

This paper focuses on rate scale-ups that do not involve any change to the as-built system. The rate scale-ups considered are therefore limited to the maximum throughput that the existing equipment can support. Debottlenecking the equipment in the process unit (rotating equipment, exchanger heat transfer area, and pipe sizing) may be required to accommodate further rate scale-up. As a capital project, this effort typically includes development of a new project baseline heat and material balance along with a review of the pressure relief system design.

2 Proper Identification of Rate-Dependency

2.1 How to Identify Rate-Dependency

Smarter decision-making with regard to rate dependency begins with understanding the ways that the traditional definition of rate-dependency may be invalidated by subsequent modeling choices in the determination of the required relief rate.

A general rule of thumb often passed around among novices in pressure relief system design is that overpressure scenario required relief rates that result from equations are rate-independent, and required relief rates that come from a heat and material balance (HMB) are rate-dependent. However, while it may be generally true for many facilities, there are aspects that are commonly overlooked in this approach, and some engineers may take subsequent steps in the required relief rate calculation that can change the classification. These additional steps are often taken as a means to "sharpen the pencil" when the initial determination of the required relief rate results in an undersized pressure relief device. These may include such actions as:

- Taking credit for an outlet that would not be blocked as a result of the scenario (including secondary causes like controller action and shut off systems)
- Reducing heat transfer into the system due to elevated bubble point temperatures
- Limiting flow rates based on the hydraulic limits of the piping, including valves and fittings
- Limiting flow rates to the maximum capacity of rotating equipment that can deliver the stated pressure at the point of relief

For example, if determining the required relief rate in the event of a blocked liquid outlet on an exchanger by taking the normal feed rate as the required relief rate is found to result in a required relief area greater than the currently installed relief area, the engineer may review the inputs and determine that when discharging against the relief pressure, the capacity of the feed pump is reduced below the normal feed rate, potentially making the existing installed relief area adequate. In this example, the overpressure scenario has changed from rate-dependent to equipment-dependent.

Table 1 shows common sources for determining the required relief rate for an overpressure scenario, and potential additional modeling choices that can change the classification.

Table 1. Examples of Dependency and Factors That May Change Classification†

Overpressure Scenario	Rate Commonly Derived From	Dependency for Common Derivation	Factors that may change Classification
Closed outlets	HMB/Simulation, Exchanger duty	Rate-Dependent	Limited by pump/compressor, valve, or pipe capacity
Reflux failure	HMB/Simulation, Exchanger duty	Rate-Dependent	
Cooling failure	HMB/Simulation, Exchanger duty	Rate-Dependent	
Overfilling	HMB	Rate-Dependent	Limited by pump or valve capacity
Inadvertent opening of manual valve	Line capacity	Equipment-Dependent	Credit taken for volume outflow
Failure open of inlet control valve	Valve capacity	Equipment-Dependent	Credit taken for normal minimum flow or outflow
Check valve failure	Valve or line capacity	Equipment-Dependent	
Abnormal heat input	Valve or exchanger capacity, HMB/Simulation	Rate-Dependent	
Exterior fire	Fluid properties, Vessel area	Equipment-Dependent	
Hydraulic expansion (pipe)	Fluid properties, Pipe area	Equipment-Dependent	
Hydraulic expansion (exchanger)	Fluid properties, Exchanger duty	Equipment-Dependent	Exchanger duty driven by hot side flow rate
Tube rupture	Exchanger properties, fluid properties	Equipment-Dependent	
Tube leak	Exchanger properties, fluid properties	Equipment-Dependent	
Plate weld failure	Exchanger properties, fluid properties	Equipment-Dependent	
Column power failure (Complex)	HMB/Simulation, Exchanger duty	Rate-Dependent	
Column power failure (Simple*)	Fluid properties, reboiler duty	Rate-Dependent	
Loss of Absorbent	HMB/Simulation	Rate-Dependent	

*Simple column failure includes overpressure scenarios where the required relief rate can be determined as a function of the reboiler duty and heat of vaporization of the residing fluid.
†Not a comprehensive list

3 Rate-Dependent Factors Influencing Pressure Relief System Design

Although the common practice is to size the pressure relief system as a linear function of process flow rate, the reality is more complex, as other factors that are themselves influenced by process flow rate can also impact the pressure relief system design, primarily through their impact on pressure relief device sizing. These factors can impact the pressure relief device sizing by causing additional deviations in relief rate calculations for rate-dependent overpressure scenarios (i.e., nonlinear relations), as well as creating changes in required relief rates for overpressure scenarios previously categorized as equipment-dependent.

A quick review of three factors influencing pressure relief device sizing will aid in understanding the limits of scaling up the pressure relief system design basis: composition, operating conditions, and heat exchanger duties. Note that this list is intended for illustrative purposes of the effects and is not an all-inclusive list; there are other factors that can influence the design of the pressure relief system that may change during a process rate scale-up.

3.1 Composition

3.1.1 Composition and Pressure Relief System Design

Fluid composition impacts the pressure relief device sizing in two significant ways.

1. Firstly, it can directly impact the determination of the required relief rate; for example, fire calculations use the latent heat of vaporization for a mixture, and depending on the particular modeling choices selected (e.g. inclusion of sensible heat, vaporization interval selected) may include further dependencies on heat capacity, bubble point temperature, vapor to liquid volume ratio, and other variables. Orifice flow calculations (such as are found in tube rupture calculations) depend on the density or molecular weight, compressibility, and specific heat ratio for the fluid.

2. Secondly, the composition of the fluid mixture directly impacts the relationship between the pressure of the fluid and the specific volume of the fluid, which governs the determination of the critical mass flux across the relief device nozzle (G in equation 1).

3.1.2 How Scaling Up the Process Can Affect Composition

It has been commonly observed that the compositional data in a heat and material balance is rarely adjusted when a scale-up of the unit throughput is performed. This assumption can be invalidated by any of the following effects:

1. Changes in the operating pressure and temperature – Changes in operating temperature and pressure can have an impact on the composition by affecting the separation of the vapor and liquid phases within the process equipment. Increasing the operating pressure can raise the bubble point temperature of the fluid mixture, requiring additional heat to achieve the same separation point. Decreasing the operating pressure can reduce the bubble point temperature of the fluid mixture, possibly resulting in more vapor flashing off, with a higher molecular weight.

2. Limitations on feed sources – In order to maximize the throughput for the specific process unit, feeds from multiple sources may be required. If the process unit has excess capacity but insufficient feed is available from on-site units, feed may be supplemented from offsite or from tankage. These feeds may have variations in compositions, and as the new feed is added, the blend ratio can change. For example, a gas oil hydrotreating unit (GOHT) may have additional capacity, but due to changes in product demand the crude column system may be producing a higher proportion of lighter products. To take advantage of the slack

in the GOHT, additional gas oil may be imported, resulting in a change in composition of the feed.

3.2 Operating Conditions

3.2.1 Operating Conditions and Pressure Relief System Design

The operating conditions within equipment in the process unit can directly impact the pressure relief device sizing primarily through the temperature's effect on the fluid conditions (for overpressure scenarios that model the required relief rate using its operating temperature). Temperature also influences the maximum mass flux across the pressure relief device. Since the relief device is sized at the relief pressure, there is no similar direct effect of operating pressure.

More significantly, changes in the normal operating temperature and pressure can affect the required relief rate calculations for overpressure scenarios that require these values as inputs, such as modeling flow across an orifice, flow through piping or valves, or where separation processes are occurring. For example, inadvertent opening of a valve into a separator may involve flow of the fluid mixture across a valve and series of pipe fittings, followed by a separation where the vapor is expected to overpressure the downstream vessel, while the liquid will slowly fill the vessel. The change in overall vapor quality can both affect the required rate to be relieved while the vessel is filling as well as affect the available time for operator intervention – in some cases making the required relief rate better, and in other cases making it worse. Another example is the potential for changes in composition, temperature, or pressure to change the net positive suction head available (NPSHA) for pumps, requiring adjustments to liquid levels in order to maintain proper pump operation, thereby affecting equipment-dependent overpressure scenarios.

It is even conceivable that such a change may alter the applicability of an overpressure scenario, where the existing operating conditions are already near the threshold of applicability – from either direction. Changing the operating pressure may affect the suction pressure of a pump or compressor, thereby altering the deliverable pressure to a downstream system, whose determination of applicability for a control valve failure or inadvertent opening of a manual valve overpressure scenario may depend on the comparison of this pressure to the downstream system maximum allowable working pressure.

Changes to operating conditions can also affect the specification of the pressure relief device. Operating pressures that were previously at the limit of the acceptable range that may now exceed the operating limits of the pressure relief device may require alternative relief device specification (e.g., if the operating pressure exceeds 90% of the PRV set pressure, this may require replacing a conventional PRV with a pilot operated PRV). Significant changes to operating temperature may also affect the set pressure of the pressure relief device. For a spring loaded valve, a new cold differential set pressure may be required.

3.2.2 How Scaling Up the Process Can Affect Operating Conditions

The operating conditions can be impacted in the following ways during a scale-up:

1. The mechanical characteristics of rotary equipment – For example, for a given installed impeller diameter, as the flow rate is increased, the operating point of the rotating equipment must move out on the pump performance curve, resulting in a lower delivered pressure. (This can be countered by changing out the pump impeller, but that is outside the scope of this paper).
2. Piping hydraulic limitations – As flowrate is increased, the pressure drop is expected to increase. With changes in the pressure drop between equipment, the operating pressure profile may be affected. For flow in the turbulent region, for the same diameter of piping, the pressure drop normally follows a trend of

$$\frac{F_1}{F_2}\frac{v_1}{v_2} = \frac{Q_1}{Q_2} = \sqrt{\frac{\Delta P_1}{\Delta P_2}} \qquad \text{(Eq. 2)}$$

Where

F	is the mass flow rate
v	is the specific volume of the fluid
Q	is the volumetric flow rate of the fluid
ΔP	is the pressure drop through the pipe

3. Heat transfer – As flowrates change, when running up against the limits of the heat transfer equipment, changes in temperature and mass quality may result. For example, applying the same duty to a fluid entering at a higher feed rate will soak up more sensible heat and result in a smaller vapor fraction exiting the equipment.

3.3 Heat Exchanger Duties

3.3.1 Heat Exchanger Duties and Pressure Relief System Design

Heat exchanger duties can potentially cause deviations in the relief device sizing basis if the process duties are used as inputs in determining the required relief rates. Note that some users choose to use the design duties of the exchangers in lieu of the process duties. This may mitigate the impact of scale-up (up to the point where new heat transfer equipment is required). Common examples of overpressure scenarios impacted by this factor are blocked outlets on heat exchangers, thermal expansion, and simplified column calculations (i.e. boil-up).

The other route through which heat exchanger duties can impact the pressure relief system sizing is by impacting the operating conditions of the process (see section 3.2.1).

3.3.2 How Scaling Up the Process Can Affect Heat Exchanger Duties

Scaling up the process flow rate through an exchanger can significantly impact the exchanger heat duty if it is limited by flow instead of limited by heat transfer area and overall heat transfer coefficient ("UA-limited"). For the same heat exchanger, the flow pattern correction factor F_T is expected to be constant. In these cases an increase in process fluid flow can result in increased duty, up to the point where the exchanger UA becomes limiting. Note that for forced convection, since the Nusselt number is a function of the Reynolds number, the value for the overall heat transfer coefficient U is not constant; rather, it is itself a weak function of flow rate (as well as composition and operating conditions). If the exchanger duty is limited by UA, then the exchanger heat transfer duty can be still changed by differences in log mean temperature difference (ΔT_{lm}). The temperatures changes may come about as described in section 3.2.2.

3.4 Extent of Deviations and Overall Impact on Pressure Relief System Design

The above discussions have focused on how rate scale-ups can impact the pressure relief system design in ways beyond the typically assumed linear relationship between mass flow rate and required relief area, as well as impacting scenarios that might otherwise not be classified as rate-dependent. While it can be expected that minor changes in flowrate would have a correspondingly minor impact on composition, temperature and pressure, and heat exchanger duties, and therefore a minor impact on the pressure relief system design, it is important to recognize that the extent of deviation from linear behavior can increase with the extent of the process throughput scale-up. Additionally, repeatedly performing scale-ups on a previously scaled-up process, without revalidating the heat and material balance, can lead to the same result in a less obvious manner.

4 Examples

4.1.1 Examples of Compositional Impact on Pressure Relief System Design

For equipment and systems where relief rate is determined by calculating a rate of vaporization, the relief rate typically increases as the proportion of lighter components within the feed increases. Even where the relief rate is not determined by calculating a rate of vaporization, a lighter feed may not change the required relief rate mass flowrate but might instead decrease the relief device capacity, on a mass basis. For example, crude units can be operated with significant variations in the crude slate. These changes in feed composition could cause different effects for relief device design within the crude unit, even with no change in overall unit throughput.

The sizing scenarios for crude column systems commonly include partial power failures that result in continuing crude feed and heat input from the fired heater, with loss of all overhead and pumparound cooling, among other effects (such as backflow and desalter water carryover). In these scenarios, the relief stream is based on the lighter portion of the crude feed, which typically consists of the offgas, liquefied petroleum gas (LPG), naphthas, kerosene, and/or diesel, etc., depending on the heating capability of the feed heater relative to the feed. Therefore, a change in crude slate that increases the proportion of naphtha cuts in the crude while decreasing the proportion of atmospheric gas oil (AGO) cuts can result in a direct increase on the relief rate of the crude column, whereas a change that only changes the heavier portions of the crude feed may not have a significant effect on the crude column. Note that composition changes that do not affect the crude column may affect another system, such as the vacuum column.

4.1.2 Examples of Changes in Operating Conditions on Pressure Relief System Design

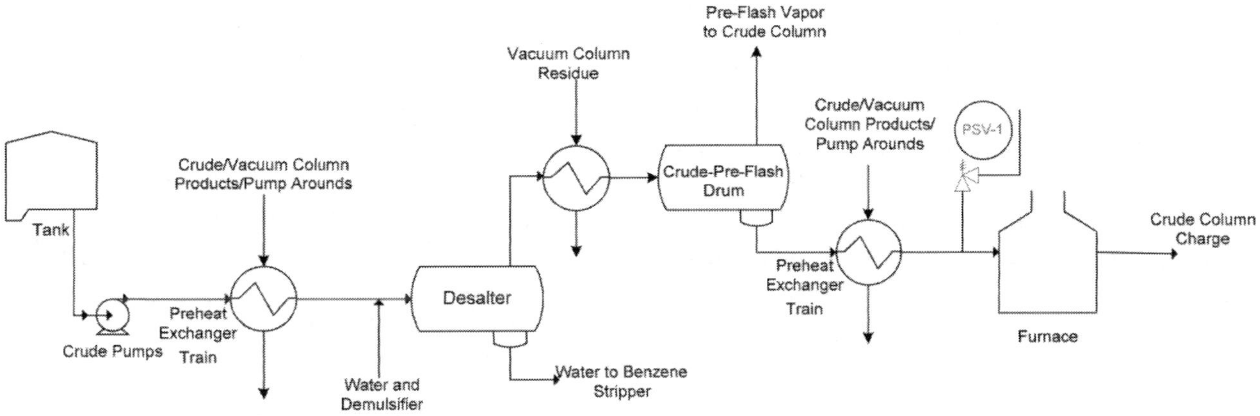

Figure 2: Crude Feed Pre-heating System Process Flow Diagram

The process flow diagram in Figure 2 represents the feed pre-heating section of a Crude Column system. Feed from tankage is pre-heated in a series of exchangers against hot fluid from Crude/Vacuum Column products and pump arounds; the crude feed is then treated for salt content in the Desalter before sending to the Crude Pre-Flash Drum. The Crude Pre-Flash Drum operates significantly lower than the Desalter, allowing the crude to flash. The flashed vapor is directly sent to Crude Column, while the liquid from the drum is sent through another train of exchangers, the furnace, and then to the Crude Column.

a. Effect of Operating Pressure Change on Composition

PSV-1 on the Flashed Crude pre-exchangers is sized for Blocked Outlet, and this case is usually selected as rate-dependent. However, a change in operating pressure of the Crude Pre-Flash Drum, with or without the throughput change, affects the composition of flashed liquid and vapor at the Crude Pre-Flash Drum, because the bubble point of the crude stream components is directly related to the operating pressure. In this case, as the fluid composition is changed, an assumption of a linear relationship between changes in throughput and required area may not be accurate.

b. Effect of Operating Temperature Change on Composition

Similarly, a change in the throughput may result in change in product rates, Crude Column product/pumparound rates and Vacuum Column product/pumparound rates, which could result in change in the heat input at the Crude Preheat exchangers. Similar to pressure, a change in Pre-Flash Drum inlet temperature, may affect the composition of Crude Pre-Flash Drum liquid and vapors, and the rate dependent assumption may no longer apply for Blocked Outlet case for PSV-1.

5 Conclusion

The economic driver for maximizing the production from process units is ever-present. However, it is vitally important to ensure that the techniques used to evaluate the pressure relief system design are appropriate and suitable to the task. Awareness of the assumptions inherent in rate-dependent classifications, and the potential to invalidate this classification with further "pencil-sharpening", is a starting point. Understanding the ways in which process throughput scale-ups can impact other factors in determining the design of the pressure relief system, such as affecting the fluid composition, operating conditions, and heat exchanger duties, and their impact on required relief rate determination or even scenario applicability, helps to illuminate potential problem areas when evaluating a process throughput scale-up, and informs the decision on whether a revalidation of the heat and material balance is appropriate. Altogether, these considerations should enable engineers to make smarter decisions about rate-dependency and ensure safer operation during process throughput scale-ups.

6 References

[1] American Petroleum Institute, *API Standard 520: Sizing, Selection, and Installation of Pressure-relieving Devices Part I – Sizing and Selection*, 9th edition, July 2014.

442298 Learning Lessons from a Big Chemical Disaster in Korea

Abstract:

Learning lessons from a big chemical disaster in Korea

Kwnaghee Lee and Il moon*

Department of Chemical and Biomolecular Engineering, Yonsei University, 50 Yonsei-ro, Seodaemun-gu, Seoul 120-749, Republic of Korea

On September 27, 2012 a massive chemical dispersion disaster occurred at the Gumi National Industrial complex in the southern city of Korea. About 8 tons of hydrogen fluoride gas leaked as the valve of tank lorry was opened due to a worker' mistake. The operators at the factory had never experienced such a big chemical disaster, and the fire station in the area did not have enough equipment at the time of chemical accident, so that an initial counter measurement could not be conducted smoothly. Due to this accident, 5 workers were killed and about 3,600 local residents had to receive a medical treatment at hospitals due to the accident. And the damage proceeded to the second and third stage, causing 17.7 billion KRW of damage to enterprises and spoiling the crops on 323.8ha of land. After the Gu-mi disaster, law and regulations, government organization, and industrial safety system were completely revised and culture have leaded positive effects towards managing and responding to chemical disasters. The governmental action has been led to make more practical plan and manual to control and respond to chemical accident. The governmental reorganization against chemical disaster such as Joint inter-agency Chemical Emergency Preparedness Center which is a specialized agency focused on responding to chemical disasters has also made to increase technicalities of chemical accidents response. In addition, the relevant laws and regulations has clarified the different acts and programs on control, management and assessment of chemicals along as well as its accident. Moreover, private and public enterprises have been also making efforts to systematically manage safety issues and build high-level safety culture. In this paper, we discuss the changes attitudes against chemical accidents from various viewpoints and the successful case in which the relevant agencies responded to major chemical accidents more efficiently after those changes. In order to improve the initial response system against chemical accidents continuously, it is required to strengthen the role of on-site coordinator, establish the Chemical Accident Investigation Committee on a national level. This study covers the big systematic changes in laws, regulations, business culture for the chemical safety management after a big chemical release accidents.

442301 Study on Thermal Radiation of Enclosed Ground Flare

Abstract:

Flare system is an important part of the safety systems in refineries or petrochemical facilities, and on oil and gas platforms. There are generally two types of flares: the elevated flare and the enclosed ground flare. Enclosed ground flares get more and more applications because of its high efficiency, flame invisibility and low nose. During an emergency flaring situation, a portion of the heat produced by the flame will radiate to its immediate surroundings.The amount of radiation heat emitted by flares is important in the design and location of flare systems. But there are no theoretical or empirical methods for predicting the radiation of enclosed ground flares in current literatures. Predicting the thermal radiation becomes one of the greatest challenges in designing a groud flare system. To investigate the radiation of the enclosed ground flare, the flaring process of a series of flare gases were simulated using CFD method.The temperature and radiation field was calculated.Results shows that the radiation flux at ground level were below 1.58kW/m2; at the height above 9m, the 1.58kW/m2 radiation isopleth can reach 60-70 meters radially from the enclosed ground flare axis,so the design of towers or other elevated structures containing operating platforms should consider radiat.

442313 Simple Objects As Indicator of the Severity of Vapour Cloud Explosions

Abstract:

Simple objects such as instrument boxes and oil drums are often found on petro-chemical plant sites. In the event of an accidental vapour cloud explosion (VCE), they serve as reliable indicators of the explosion overpressure at their positions provided there is a good understanding of their behaviour in a VCE.

Such objects were used to great effect in estimating the overpressure magnitude in the Buncefield incident in 2005. In large scale tests conducted since the Buncefield incident (both as part of a joint industry project and subsequently), such objects were located at varying distances from the cloud, as well as some within the cloud. Overpressure measurements were made at or adjacent to the location of these objects and many have been analysed using advanced finite element analysis. The objectives were to (i) gain a better insight into the response of these objects when subjected to different types and magnitudes of pressure loading, (ii) validate finite element models; (iii) assess the suitability of different finite element modelling techniques for predicting the response of these structures and (iv) develop pressure impulse (PI) diagrams which can be used to provide a rapid assessment of explosion magnitude through an examination of the damage level.

This paper presents an overview of the tests performed and the analysis results with comparisons with the actual behaviour of the objects.

Best Practices for Writing Operating Procedures and Trouble-Shooting Guides

Lauren Madden, Process Safety Manager
REC Silicon
Moses Lake, WA 98837
e-mail: Lauren.Madden@recsilicon.com

William G. Bridges, President
PROCESS IMPROVEMENT INSTITUTE, INC. (PII)
1321 Waterside Lane, Knoxville, TN 37922
Phone: (865) 675-3458
Fax: (865) 622-6800
e-mail: wbridges@piii.com

Prepared for Presentation at
12th Global Congress on Process Safety
Houston, TX
April 10-13, 2016

UNPUBLISHED

Best Practices for Writing Operating Procedures and Trouble-Shooting Guides

Lauren Madden, PSM Manager
REC Silicon

RECSiLICON

William G. Bridges, President
PROCESS IMPROVEMENT INSTITUTE, INC. (PII)

Keywords: Procedures, Human Factors, Risk control, process safety management, PSM

Abstract

There is no complete, best practice guideline or textbook for writing operating procedures and trouble-shooting guides. This paper presents the proven, best approach for developing accurate operating procedures and for ensure the format of the page and steps are optimized to reduce human error rates. This approach and the 35 rules established in this paper are based on the foundations set by Swain and others in the 1970 for control of human error rates, but uses experiences from more than 60 sites where this approach has been successfully followed. The approach and rules for developing procedures for responding to process deviations (such as those needed for Human IPL) are again the best approaches found and have been applied successfully since the early 1990s; these are many times called operational trouble-shooting guides. The paper also discusses how and when to integrate trouble-shooting guides into distributed control systems. Several case studies are provided that show the gains from following this approach. The guidelines in this paper build upon ones presented in 1999 at CCPS.

Background

Operating procedures have always been crucial to the safety, quality, and productivity of process systems. And with the advent of new safety and quality standards such as OSHA's process safety management (PSM) regulation, OSHA's personal protective equipment (PPE) regulation, EPA's risk management program (RMP) regulations, and the ISO 9000 quality standard, many companies are facing the daunting task of developing or upgrading their procedures to satisfy varied and sometimes complex or conflicting requirements.

Although there are overlapping characteristics that the various regulations and standards share, each approaches the procedure-writing process from a somewhat different perspective (e.g., quality or safety, protection of workers or protection of the public/environment), and the required level of detail for documenting the procedures differs greatly between these regulations and standards. For instance, the ISO 9000 quality standard advocates documenting procedures and work instructions that impact quality and suggests keeping the procedures as simple as possible, while the OSHA PSM regulation requires detailed procedures that address all operating modes (startup, shutdown, etc.) and that address operating limits, consequences of deviations, means to avoid hazards, safety and health considerations, and safety systems and their functions. How can companies develop and maintain procedures that ensure productivity and simultaneously satisfy the different regulations and standards? The key is to remember that the ultimate goal of the regulations and standards is consistent: reduce human errors that can impact quality, productivity, and/or safety.

Procedure-related errors are errors that occur because some characteristic of the procedure caused task performance to fail. ***This is currently the most critical human factor at most sites since 90% of accidents have at least one root cause related to mistakes within procedures.*** Reducing these procedure deficiencies can reduce human error rates by a factor of 2 to 10, normally. [1, 2]

The concepts and rules presented in this paper are applicable to all industrial settings. Most importantly, this approach **is effective in satisfying both safety and quality concerns.** With this approach, procedures can and have been developed that provide a sound basis for training and that provide the users and supervisors with an understanding of what can go wrong and how it can go wrong. This approach also provides management with an understanding of the inherent risk (to quality, productivity, safety, etc.) of an activity (i.e., the activity addressed by the procedure) and an appreciation of the safeguards that are being relied upon to mitigate this risk.

An extensive case study is presented by REC Silicon, which illustrates this approach, along with supporting the benefits of the accurate and clear procedures. But the same is true at more than 60 sites where the same approach has been followed.

Fundamentals of Writing Effective Procedures

When developing procedures, or any written material, always remember that the burden of written communication is on the writer, not on the reader. It should never be up to the reader to decipher the meaning of a message. Engineers who write procedure manuals may "tend to include information that is more relevant to them than it is to the operators actually running the unit" [Sutton, 1992].[3] When writing procedures, your guiding principle should always be: if the written procedures are difficult to read, difficult to locate (e.g., if titles are do not have a clear meaning), or inconvenient to use, they will seldom be used [Swain, 1983].[1]

Deficient Procedures are the most prevalent problem in process industries since procedures have not traditionally been developed from the perspective of optimizing human factors; instead, procedures have been traditionally developed to meet a compliance requirement to have written procedures. Examples of procedure deficiencies (inaccuracies) include:

- Incorrect/incomplete/nonexistent (most procedures we have audited have been only 70-75% accurate – the inaccuracies include missing critical steps, steps as written are not what needs to be done, or the steps are out of sequence)
- No/misplaced warnings (for a example, a warning should Never **contain** the action to take; it should instead **emphasize** the action to take)
- Poor format and presentation rules

Figure courtesy of Process Improvement Institute, Inc. (PII), copyright 2003-2016[4]

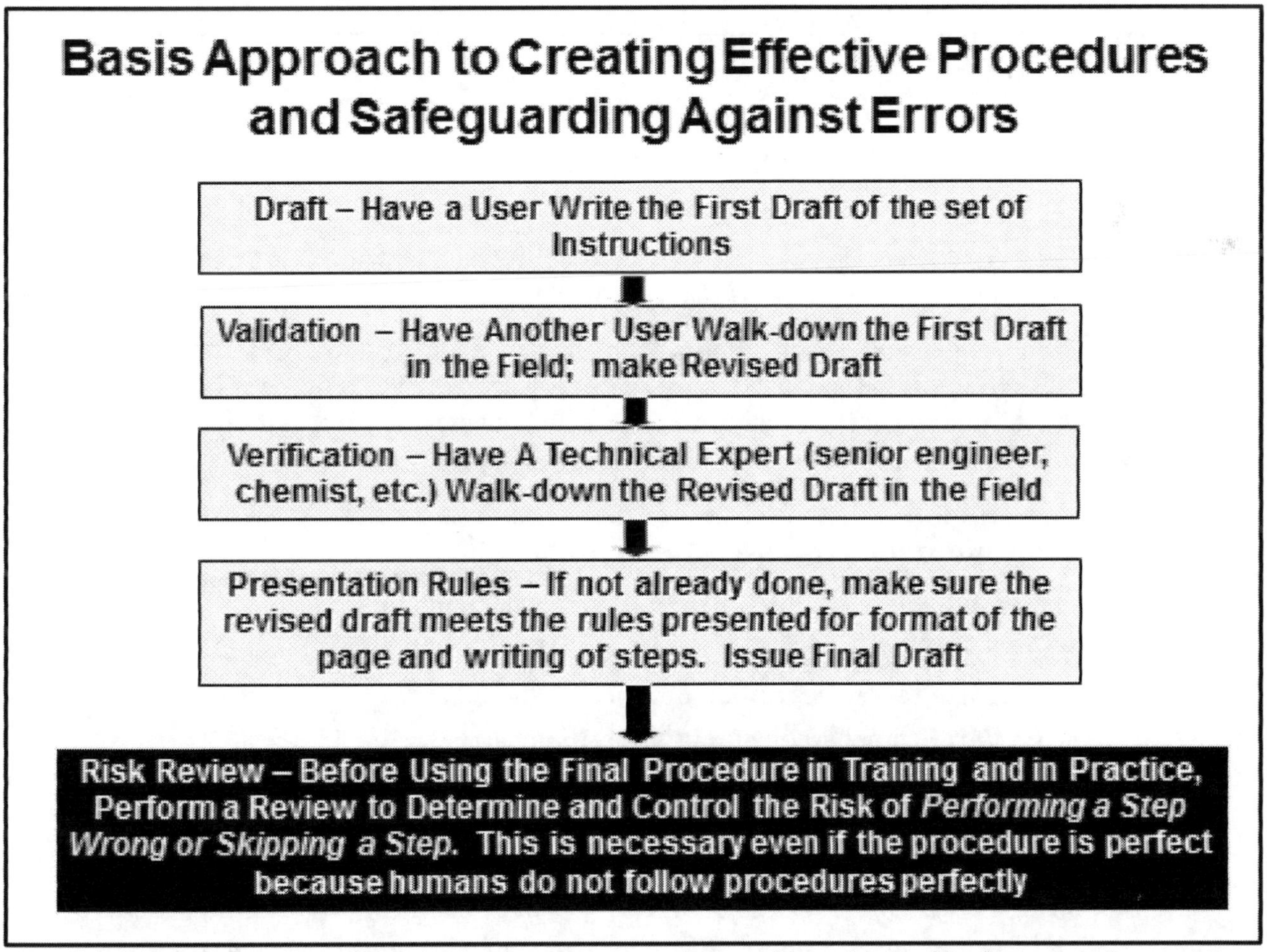

The flowchart above describes the basic sequence of steps for writing effective procedures. Note that the some basic rules are contained in the flowchart, such as "Operators writing operating procedures" and "Maintenance craft-persons write maintenance procedures" and "Lab technicians write lab procedures." Procedures are Not written by engineers or department superintendents. This is a critical rule because it ensures the procedures are written in the common language of the users and is written at the right level of detail.

Universal Purposes of a Procedure[4]

1. A written guide to accomplish initial training of employees
2. A quick reference manual for experienced (trained) employees

The Right Information to Write – Content is KEY[4]

The Content of the procedure controls the human error related to "procedure-based" error – if and content is wrong, the trainee is very likely to learn the step wrong (if the trainer is using the procedure as a basis or when the trainee refers the to the procedure step that is wrong). So:

If the content of the step is wrong it does not matter if the presentation is clear; but if the content is perfect, you can reduce error rates a further 50 to 70% by presenting the information properly.

Ask experienced mechanics or operators to describe a simple task that they perform. You will quickly see that even simple tasks can involve many steps.

You should consider the following questions when writing each procedure step:

- How is the step performed? Does the action involve interacting with a computer terminal, an automatic controller, or simple devices (gauges, valves, micrometers)?
- Can the actions be performed as written and in the sequence written? Can the equipment be operated as specified? Can the steps be physically performed? Does the user have the training or experience to understand and carry out the action using the information available, or is additional information needed?
- Does the user need to be alerted of any potential hazards (Cautions or Warnings) or need any supporting information (Notes) before performing the action?
- Does the user need to know specific operating ranges or limits to:
    - Perform this action
    - Recognize the successful completion of the action
    - Recognize an actual or potential problem
    - Make a decision

For example, does the user have to verify temperature or pressure visually? Is the user waiting for a computer command or verification? How should the process react? Will exceeding a limit result in an accidental chemical release, personal injury, or quality concern? What is the chance of not operating within safe operating limits, and what would be the consequences? If operation of the component is critical and a deviation will probably cause a failure, state the corrective actions. If failure is unlikely, keep going.

- Is needed information found on an instrument, panel, or monitor or is it in the procedure or another source such as a graph, table, drawing, or specification sheet? Should this information be included in the procedure or be referenced?
- What is the next logical step? How is the next step affected by what is performed in the current step?
- What are the results of improper task performance? If an action is critical, spell it out in detail. For example, when starting a special pump that may bind or cavitate if not vented and primed properly, the procedure may require more information than simply stating "start the pump."
- Is the action frequently performed? Is it an action that is easily overlooked? Is this a complex piece of critical equipment that has a low frequency of use? These questions can help determine the level of detail in your procedure.
- Is the action performed so infrequently or is it so complicated that the user is not sure how to do it? Is the action so complicated that nobody is ever certain it's done right the first time?
- Is the decision point clearly defined if a decision is required? Unclear decision points can cause arguments and delays in performing actions.

To address the concerns listed above, a few simple rules are needed for developing content:

- The procedures should be written at the level of someone who has just completed the basic training for that task. Do not write the procedure for someone just hired or for the 10+ veteran.
- Clearly identify ahead of time what activities need procedures to help ensure error rates or controlled low enough, and identify which activities are considered common "skills" of all of the staff.

  Example: Starting a pump is typically considered a "skill." This means a procedure step can simply be "Start the benzene recycle pump (P-119)." The procedure does not need to explain how to start the pump. But, starting a pump is a skill that nonetheless must be learned; it requires doing several sub-steps related to positioning of intake and discharge valves, how the discharge pump is throttled at startup, checking local pressure gage, etc. So, a training module is needed for starting pumps of various types, and the operator needs to learn this necessary skill. Once the skill is learned, it will be applied to so many various cases and so often, that a procedure is no longer needed as a "refresher" on how to do "start the pump."

- The first draft of the procedure must be walked-down in the field by another user. (Simply reading through the procedure does not catch the enough of the mistakes.)
- The revised draft should be reviewed in the field by a technical staff person, such as a process engineer (for operating procedures).

Next:

- The procedure needs to follow the page format and step writing rules described in the next section. This ensures the accurate steps are clearly presented.

The target is to reach an accuracy of 95% or better (so, no more than one wrong or missing step out of 20 steps). The following have been observed in the field by PII staff by direct data collection at more than 110 sites/plants around the world, based on walk-down review of several to dozens of procedures at each unit/plant:

- At 95% accuracy or better and when the same procedures follow 80% of the rules for procedure clarity (presented next), then most users will follow the written procedures and will try to keep the procedures up-to-date.
- At about 85% accuracy or less, about half of the users stop using the procedures
- At about 70% accuracy or less, less than 10% of the users will refer to the procedure or will try to keep it up-to-date. ***So, the written procedures are useless.***
- Unfortunately, the typical Operating procedure walked down by PII staff (accompanied by senior operators, a process engineer, and a shift supervisor) is about 75% accurate (so one step in four is missing or wrong). Usually these inaccurate procedures also follow less than half of the best practices for procedure format, presented next.

PHA/HAZOP of procedures

Just as critical as getting the content accurate and presenting the steps clearly, is performing a PHA/HAZOP of the written steps to determine if you have enough independent protection layers (IPLs) against errors that the "imperfect" operators will eventually make when using the "near perfect" written procedures. This analysis is described in detail in Chapter 9 of the *Guidelines for Hazard Evaluation Procedures*, 3rd edition, 2008[5]. **For an update on the methods for such analysis and for a discussion of the extra emphasis on PHA of procedures by US OSHA, refer to the more recent paper (Bridges, Marshall; 2015).**[6 and 7, 8]

Procedure Clarity

The importance of the accuracy of the procedure steps were discussed above. The clarity of the procedure steps (how they are written and how the page is formatted) is also important. Following best practices for step and page format reduce human errors by a multiplying factor of 3 to 5.[1, 9]

The best practice rules for writing and validating procedures have been published for many years *(see Bridges & Williams, 1997).*[11] These rules include best practices for formatting of the pages and steps; these have been gradually improved over the past decades and now are incorporated into PII's training materials.[4]

The Table on the next is a checklist (in auditing format) based on the current set of best practice rules for developing accurate and clear operating, maintenance, and other work-instructions (procedures).

PROCEDURE QUALITY CHECKLIST *(courtesy PII, copyright 2003-2016)*[4]

#	Issue	Response
	Procedure Content Checklist	
1	Is the procedure drafted by a future user of the written procedure? (Engineers should not author procedures to be used by operators or maintenance staff.)	
2	Is the procedure validated by a walk-down in the field by another future user of the procedures?	
3	Is the procedure reviewed and commented on by technical staff (engineers or vendors)?	
4	Is the procedure checked versus the Page and Step format rules below?	
5	Is a hazard review of step-by-step procedures performed to make sure there are sufficient safeguards (IPLs) against the errors that will occur eventually (when a step is skipped or performed wrong)?	
6	Is the content measured using "newly trained operators" to judge the % of steps that are missing, steps that are confusing or wrong, and steps that are out-of-sequence? (A score of 95% accuracy of content is good; 98% should be the targeted average.)	
	Page Format Checklist	
1	Is the title of the procedure the largest item on the page?	
2	Is the procedure title clear and consistent with other titles, and does it uniquely describes the topic?	
3	Are the document control features the smallest items on the page?	
4	Are temporary procedures clearly identified?	
5	Is white space used effectively? • Is there one blank line between each step? • Does indentation help the user keep their place? • Are the margins large enough to reduce page congestion?	
6	Is type size is 12 pt font or larger?	
7	Is mixed case used for words of steps, with ALL CAPS used only for special cases (such as IF, THEN, AUTO, and WARNING)?	
8	Is the step number very simple (single level of number used)? Only an integer?	
9	Have sections or information not necessary to performing the steps been moved to the back or to another part of the manual or training guide?	
10	Are section titles bold or larger than the text font? Do sections have clear endings?	
11	Is the decision on electronic presentation versus hard copy correct? Are electronic linkages to procedures clear and accurate and easy to use? If paper is chosen, is the color of the paper appropriate?	
12	Is the overall page format (such as Outline format or Two-Column [T-Bar] format) appropriate to the use of the procedure?	
13	Are play script features added for tasks that must be coordinated between two or more users? • Play script is normally used when there are two or more hand-offs of responsibility for steps.	
14	Are rules followed for formatting of Warnings, Cautions, and Notes? (See annotated rules, such as Warnings are for worker safety and Warnings must clearly standout from rest of page.)	
	Step Writing Checklist	
1	Is each step written as a command?	
2	Is the proper level of detail used throughout? This is judged based on: • *Who* will use the procedures • Same level of detail used in similar procedure steps	
3	On average, is there only one implied action per instruction? Best practice is to average 1.2.	
4	Does the procedure indicate when sequence is important? • If sequence matters, each step should be numbered (with an integer or letter) • If sequence does not matter, bullet lists should be used	
5	Are only common words used? Apply "education" level test (5 grade reading level is best)	
6	Do all acronyms, abbreviations, and jargon aid understanding? • Develop a list of such terms for use in procedures *and* communication. • Use terms that users use (within reason)	

7	Is each step specific enough? No room left to guess/interpret: • The meaning of a word or phrase (Check vs. Make sure) • The intent of a step or series of steps • A desired quantity or value • To what equipment the step applies	
8	Is the procedure free of steps that require in-your-head calculations? • Values expressed as ranges rather than targets with error bands • Conversion tables, worksheets, or graphs provided where needed	
9	Are graphics to the user's advantage? • No explanatory paragraphs or lengthy instructions that could be replaced by a picture • No impressive graphics that provide no real advantage	
10	Are references to the user's advantage? • No lengthy explanations or instructions that could be replaced by branching to a reference • No references to a procedure that references still another • No gaps or overlaps between this procedure and a referenced document • If branching, must branch to a procedure, not to a specific step in a procedure	
11	Are rules followed for writing warnings, cautions, and conditional steps? • No more than 2 per page • No actions within a warning or caution (actions must always be numbered steps) • Warnings and Cautions contain descriptions of potential consequences	

Each of these rules is explained in detail in the course *Writing Effective Operating and Maintenance Procedures*, by PII.[4] (Visit www.piii.com for more details.) Each rule is backed by experimental or observed data that indicates there is a clear advantage in error reduction by using the rule. Below are a couple of examples from others on procedure writing rules.

> ***Proper Page Format:*** *In one study, Haney took a procedure presented in narrative format and converted it to column format. He then asked experienced technicians who had been completing the procedure using the narrative format to complete it as presented in the column format. He found that when the technicians used the column format, their errors were reduced by a factor of 3 compared with the narrative format [Haney, 1969].*[10]

> ***Including Supplemental Checklists:*** *Checklists can be effective aids to performance and often should be used to supplement procedures presented in any format. Checklists can be particularly effective in terms of preventing errors of omission. Some of the literature supports using checklists only for simple procedures that do not need to be performed under time or other constraints [Sutton, 1992].*[3] *However, there is also support for using checklists for emergency procedures in which a core set of steps is used to respond to emergency situations, particularly when the actions follow a simple sequence of steps for shutting down a process [Connelly, 1992].*[12] *Overall, experimental data have shown that the probability of errors of omission and commission can be reduced by a factor of 3 to 10 when checklists are used as reminders of key steps in a work instruction [CMA's A Managers Guide to Reducing Human Error, 1990].*[9]

For more data on the history and development of these rules, see Bridges & Williams (1997)[11] and especially refer to originating data and discussion for most of the rules (Swain, 1983).[1] **The procedure on the next page follows the rules**; this is a T-Bar format. Several other page formats are also valid.

PII CHEMICAL COMPANY Page 1 of 2
Unloading Monomer from Tanker to Storage Revised 11/4/02, Printed 2/21/16

Unloading Monomer from Tanker to Storage

STEPS	DETAILS
1. Wear standard PPE, plus rubber gloves and full-face organic respirator.	Standard PPE includes hard hat, safety glasses with side shields, and steel-toe shoes.
2. Weigh in the tanker.	Record weight on GROSS line of form.
3. Check bill of lading.	... to verify correct type of material is in tanker.
4. Sign in tanker driver.	Driver must sign in as a visitor and be escorted at all times.
5. Take Certificate of Analysis to QC.	***CAUTION: To avoid contamination, DO NOT unload until you receive approval from QC.***
6. Spot tanker.	A) Direct driver to location. B) Verify brake is set. C) Chock the wheels on at least one side between the two rear axles. D) Ground the tanker (attach grounding strap to an unpainted metal surface).
7. Have the storage area operator make sure the storage tank can hold contents of tanker.	***WARNING:*** *Failure to perform this step can result in an overflow and spill of hazardous Monomer.* Perform **Tank Gauging Procedure (SOP-01-804)** or check the storage tank load cell readout. Storage tank operating limits are stenciled on the control panel.
8. Place 5-psig nitrogen pad on tanker.	A) Connect from local 5-psig nitrogen drop to aft bottom nitrogen connection on tanker. B) Open nitrogen supply valve. C) Then open valve at tanker.
9. Connect Monomer unloading hoses.	A) Verify hoses and gaskets are in good condition. (Hoses are normally stored on hose rack at unloading spot.) B) Connect from tanker to air pump. (Use aft bottom connection on tanker. Remove cap.) C) Connect from air pump to storage tank. (Connect to storage tank line labeled "from TW.")

Trouble-Shooting Guides (Procedures for Responses to Critical Deviation Alarms) [13]

Trouble-shooting guides are a special form of operating procedure; they are written for the actions that we want the operators to take to recover from a process deviation, *before an emergency situation occurs.* They are called guides since rarely can we predict the process conditions at the time the action is required. Trouble-shooting guides (and training and drills of the action) are required for any action that is considered a Human IPL. The Action Limit is what we show as the Min or Max in a Trouble-Shooting Guide. The action limit triggers the demand to use the trouble-shooting guide.

- The response is typically still possible, but it is time dependent. The time available is called the process safety time (PST). The operator must complete the diagnosis, make the necessary change(s), and make sure they are out of harm's way by the end of the Maximum Allowable Response Time (MART). [13]
- We usually set an alarm or a pre-alarm to trigger this action.
- This is usually before the shutdown triggers (ESD occur automatically) or release points (PSV set points) are reached
- The Min and Max shown in the Trouble-shooting Guide table is not the absolute safety limit for a system, but is instead some value that leaves us some time to take action to prevent from reaching the absolute limits.
- There is still time to prevent or avoid the final consequence that could occur if we reach the ultimate limits of the process. Usually, we want the MART to be ½ or less of the PST, and we want MART > 10 minutes for trouble shooting in the field/plant and MART > 5 min for trouble-shooting only from the control room. [13]

If the unit has a good PHA/HAZOP, then it is best to extract information from the HAZOP (or What-If) analysis tables to start the development of each guide. The guide is then finished with input from the process experts.[4]

Examples of Creating a Trouble-Shooting Guide from a HAZOP Table

HAZOP Table Entry	Trouble-Shooting Guide Entry
Cause: *Bypass valve is open or passing*	Make sure the bypass is tightly closed
Safeguard: *Isolation valves for the vessel*	Isolate the vessel, if necessary
Safeguard: *Relief valve*	Make sure the relief valve block valves/relief path are open

The key categories of information needed in a trouble-shooting guide are:[4]

- IMMEDIATE ACTION (by system or by operator)
- DECIDE IF ALARM is REAL
- FINDING and FIXING the CAUSE
- FIX or BYPASS PROBLEM

The example below is of a trouble shooting guide for one critical alarm/action.

Trouble-Shooting Guide

Alarm or Indicator:	PAL 4446 – Low Pressure Alarm for Suction of Organic Feed Pump 40-PM-18.445

Action Limit:	5 kPa		
Consequence:	Possible pump seal failure, releasing or spraying organic waste into the berm.		
Process Area:	FB&D Incinerator; Liquid Organic Liquid Feed	Oper. Mode:	Normal
Drawing #s:	D-400-PI-013		

IMMEDIATE ACTION (by system or by operator)

- DCS should shut down the organic feed pump (40-PM-18.445).
- From the DCS display, MAKE SURE the organic feed pump is shutdown.
- HAVE the field operator check for leaks near the organic feed pump.
- IF there is a large leak/release, THEN use the ESD switch to shutdown the unit and then follow/complete the shutdown and isolation procedure, OPS-ESD-117.
- IF there is a minor leak or no leak, THEN:
    - COMPLETE the rest of the trouble-shooting,
    - and DECIDE how to contain the leak for now,

DECIDE IF ALARM is REAL

- From the DCS, CHECK the pressure and feed tank level trends. IF the trends indicate the alarm if valid, THEN continue with finding the cause or fixing or bypassing the problem.

FINDING and FIXING the CAUSE

- CHECK valves upstream of the organic feed pump to see if any are closed too far, including checking ESD valves.
- CHECK, by feel with hand, if the heat tracing is on; IF Not, then TURN ON or open heat trace valves
- MAKE SURE nitrogen to the pump seal is at the normal operating pressure.
- CHECK if the line if plugged or frozen (skill)

ABC Chemical Company Prepared by: *Printed copy of this procedure is good for one job task duration.*	OPS-76-TSG-233	Incinerator Unit Revised 8/24/2015 Printed 2/21/16 Page 1 of 2

Trouble-Shooting Guide

Alarm or Indicator:	**PAL 4446 – Low Pressure Alarm for Suction of Organic Feed Pump 40-PM-18.445**

- CHECK if the line if plugged or frozen (skill)
- CHECK if the level is actually low, use the Organic Feed Tank.
- IF the cause is low level in the feed tank, THEN resolve the problem if necessary based on cause that is found (skill)

FIX or BYPASS PROBLEM

- IF necessary, SHUT DOWN the Unit to allow fixing of the problem.
- FOLLOW the proper procedure to resolve problems (repair procedure, line clearing procedure, etc.)
- IF the decision is made to continuing operation without all equipment in normal condition, THEN:
  - FOLLOW Temporary Operating mode, if there is a temporary procedure already written for this possible problem/condition
  - FOLLOW MOC procedures to obtain approval for any non-standard temporary operating procedure or mode

END

Optimal presentation of trouble-shooting information.

Ideally (or ultimately) such trouble-shooting information should be imbedded in the basic process control system (DCS) so that the operators can access the information on demand, with the click of mouse or key.

Comments on Current Industry Consensus Guides for Writing Procedures

None of the consensus textbooks available for procedure writing cover all of the major points of this paper. The best "guidelines" for writing procedures currently exist in the course notebooks used by PII (and likely in the course notebook used by ABS Consulting, as they had similar origins).

Notably, the AIChE/CCPS textbook, *Guidelines for Writing Effective Operating and Maintenance Procedures*, 1996[14]:

- has NO description at all on the development and use and format of Trouble-Shooting Guides. This should have been a major chapter in the textbook.
- mentions only about half of the rules listed in this paper (and the related PII course notebook) for formatting of pages and steps
- emphasizes technical staff or professional procedure writers writing procedures, rather than the Users themselves writing the procedures (it violates the first rule of procedure writing)
- Fails to cover the competency needed by procedure writers.
- recommends and emphasizes having subsections in each work instruction/procedure (between the Title of the task and Step 1) rather than relegating such information either to other sections of the operating manual or rather than turning this information into steps.
- fails to mention the need for a PHA of procedures that focuses on hazard review of deviations from the steps; this is required by best practices and by US OSHA for PHAs under PSM.

This guideline textbook is not worth the investment.

As a better example, but one that nonetheless still falls short of what is needed, the ASM Consortium *Guidelines for Effective Procedural Practices,* 2010[15]:

- has good discussion on the need and thought processes related to handling abnormal situations (trouble-shooting or response to critical deviations), but there is little description on the format and detailed content of the trouble-shooting guides; in fact, the guideline fails to differentiate between routine procedures for startup, shutdown, and online maintenance, and those used for response to critical alarms (trouble-shooting)
- mentions only about half of the rules listed in this paper (and the related PII course notebook) for formatting of pages and steps
- mentions that experience users themselves should write their own procedures, but it does not emphasize this rule and leaves the authorship open to engineers or a committee. And it specifically mentions use of retired senior operators for writing operating procedures. While this is one good way to write the procedures (while also capturing the knowledge from this experienced group is prior staff), we have found that the current users do a better job overall, once they have achieved competency in how to write procedures.
- mentions the competency needed for procedure writers in general.
- mentions the need for a PHA of procedures, but does not reference the method for this in Chapter 9 of *Guidelines for Hazard Evaluation Procedures*, CCPS/AIChE, 3rd edition, 2008,[5] or other related papers on the topic. Also, the description of the PHA is misleading because it focuses on how PHAs should be used in writing procedure content (which is fine for trouble-shooting guides) but fails to emphasize that a step-by-step analysis of deviations from the steps are needed to ensure that the process has enough IPLs against mistakes that can occur when imperfect humans use the procedure.

Though this guideline has deficiencies, it is worth the investment and provides useful insight not found elsewhere. Just be sure to address the deficiencies noted above.

Implementation of this approach

PII staff has helped more than 60 sites implement the approach above. The typical approach for successful implementation is:

1. Perform a gap analysis to determine the average accuracy of procedure content, average clarity (as judged versus the procedure format rules, and identify any examples of excellent procedure writing.
2. Convince management of the problem and get their buy-in on the solution.
3. Train (1.5 to 2 days for operators; 1 day for lab or maintenance staff) and coach the users of procedures (normally for 3 to 8 days, to have them achieve full competency) so they can be effective authors. Normally, 20 to 25% of the users (hands-on operators, hands-on maintenance and inspection staff, lab staff) need to be trained as authors. Not all of those trained will make good writers; the organization needs to judge who does the best jobs. About 15 to 20% of the staff should end up being authors.
4. Get organized, like a project, for tracking and scheduling of the effort. Use master spreadsheets such as those used at REC Silicon (shown later).
5. Use a Traveler form (described later) to make sure each new procedure gets a proper amount of review by other users and by technical staff (such as engineers)
6. Reward procedure writers for their efforts. Especially ensure there are no disincentives for taking the extra effort to write procedures.
7. Track the progress on the effort; fix problems as they arise; enforce schedules (to show management cares).
8. Perhaps use to specialized software make it easy to output the content of procedures into different user-friendly formats such as training tools, training competency records, and field checklists. WordPerfect (formerly from Corel) made this relatively easy with the use of Primary and Secondary merge files; MS Word makes this more difficult. As a result, some specialized software have been developed to help keep formats consistent and to help reformat the content, on demand.
9. Issue procedures through a document control system that does not add extra workload.
10. Of course, ensure there is JSA performed for procedures that have "personal injury" potential and perhaps a PHA of the steps in the procedures whenever there is "process safety" potential.
11. Train all affected workers in the new procedures.
12. Identify the best writers to become the trainers of new authors, over the long run.

Overview of REC Silicon

REC Silicon is one of the world's largest manufacturer and suppliers of

- Silane Gas and specialty silicon gases.
- Granular Polysilicon from Fluid Bed Reactor technology for solar photovoltaic applications
- Ultra-pure float zone Polysilicon for the electronic industry

REC Silicon's Product and Service Offerings

Solar Grade
Polysilicon with various applications for the solar industry

Silane Gas
Product used in production of semiconductors, flat panel displays & thin film technologies

Electronic Grade
Polysilicon delivered to the semiconductor industry as well as other electronic market segments, such as optics & MEMS

Float Zone
High-purity polysilicon delivered to the power electronics industry

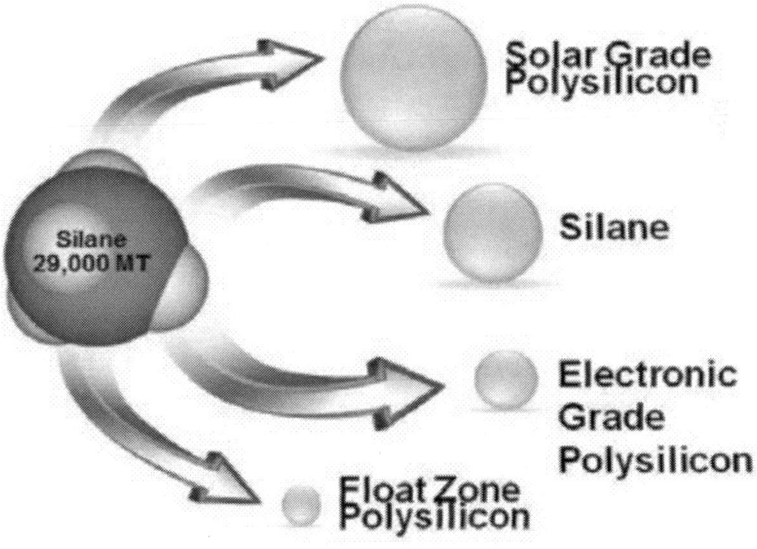

Moses Lake, Washington, USA 396 Employees	Silicon III, IV FBR	Silane gas: 22,000 MT Granular silicon: 16,300 MT
Butte, Montana, USA 255 Employees	Silicon II	Silane gas: 7,000 MT Siemens grade: 3,600 MT

Implementation at REC Silicon

REC production facilities in the USA have been actively implementing process safety for many years, with a renewed interest in 2013 in meeting industry best practices (which goes well beyond US OSHA PSM requirements). To that end, REC commissioned gap analyses versus best practices at both the Butte and Moses Lake production sites. One common finding was that the operating procedures and maintenance procedure for both sites needed improvement with respect to accuracy of the steps as well as clarity (page format and step format). The following were some of the issues with the procedures and related management systems, per the audit results:

- Procedure Inaccuracy (procedure content was less than 80% accurate)
- Technical Staff (such as process engineers) were viewed as owners, not shop floor
- Annual review process ineffective
- Procedure formats were poor; did not follow best practices for reducing human error
- Inconsistent use of Warning, Caution, and Note
- Procedure Naming – not consistent
- Tried to take on all modes of operation in one procedure title
- Procedure accessibility was difficult in the legacy database; multiple searches needed to find tool for job
- Procedure Safety Reviews – no formal methods were applied for procedure safety reviews; i.e., there were no PHA of procedures with process hazards or JSA of tasks with personal hazards

<u>Procedure Author Training – *the first step in any procedure writing effort*</u>

In early 2014, REC Silicon adopted Process Improvement Institute's (PII's) procedure author training module (*Writing Effective Operating and Maintenance Procedures*[4]) with minor customizations. The training consists of 2 days of classroom lecture and about 6 exercises, followed by practical procedure development and field validation steps in the coaching phase. The classroom course covered all aspects of procedure writing with the focus on optimizing the aspects of the procedures that will lower the likelihood of human error when using the procedures. Topics included:

- human factors overview
- getting the content right
- page format rules and selection
- step format rules
- addressing operating limits (including creating trouble-shooting guides
- nuances of emergency shutdown and emergency operating procedures

- implementation guides and high level procedures for the task of getting all of the work instructions written
- examples/exercises on each topic

The coaching consisted of the instructors from PII, who are expert in organizing procedure writing efforts, helping the procedure writers organize and write operating and maintenance procedures, and trouble-shooting guides. Coaching was for 2+ days. About 20 staff from the Moses Lake and Butte sites attended the initial author training and coaching session.

REC Silicon has revised its management systems to require this competency for all personnel developing or changing procedures and to reach full operator certification. Operators must demonstrate an active role in the procedure MOC process to qualify for recertification every 3 years, otherwise they must repeat the classroom and field training. The unit trainers have taken over leading the procedure author training.

The benefits we have seen from these changes include transfer of ownership of procedures from the technical day staff to the operator since these are the people performing the work in the field. This is necessary to have accurate procedures. **The emphasis on intentional competency development for what is required to write effective operating procedures has greatly improved the quality of our procedures and our operator training and certification program.** The quality of the annual reviews and audits has improved tremendously with the rewritten procedure format and content changes.

Safe Operating Limits and Consequence of Deviation

REC Silicon has developed an engineering standard that governs the content and format for this type of special safety troubleshooting procedure. *CCPS Guidelines for Engineering Design for Process Safety*[16] is the basis for determining the safe operating limits. Production engineers are the owners of these procedures. Engineering competency is required to develop the safe operating limits table based on the available process safety information. A consequence of deviation is then developed by the Process Hazard Analysis team. Finally involvement of certified operators is required to develop steps to correct or avoid exceeding the safe limit. This combined effort produces the safety troubleshooting guide, which complies with the approach listed earlier in this paper.

REC silicon has risk ranked its process areas and is focusing on developing safety troubleshooting guides for high risk process first. To date, for the Moses Lake Facility, thirty six (36) guides have been developed. These have been developed in conjunction with PHA updates to the safe operating limits. We have not yet implemented the formal training and competency testing of these safety procedures. This is planned for the next phase of the initiative.

How REC Silicon approached the re-write effort:

To begin this initiative, each operating department listed all of the existing operating procedures into the procedure development tracking form. Each department also developed a list of “skills” that are used across multiple procedures and that operators or maintenance staff are required to attain before using the written procedures that were developed for more complex tasks (such as starting a pump, or hookup of a chemical transfer hose).

The workers (users) then screened each procedure for safety, equipment damage, or quality risk. This assisted in determining whether 2, 3 or 4 peer walk-downs were required on the validation traveler workflow. Since most of the REC Silicon procedures covered more than one mode of operation, the term "parent" was developed for the original procedure and each procedure was rewritten and broken out into a separate "child" procedures. The procedure re-write assignments were divided up between all four rotating shifts and the superintendent assigned certified authors (e.g., operators trained/coached as procedure writers) from their pool. Each department is managing their pool of qualified authors. For example, our fluidized bed reactor (FBR) unit has 11 qualified authors. Silane, which is a much larger department with 3 production units, has 56 qualified authors. Cross shift validation was utilized to ensure all operating approaches were considered. The assigned author was responsible for consolidating red line comments.

The traveler process introduced by PII was customized to consider information that is marked for removal from the parent procedure. **A Traveler form shown to the right accompanies each new or revised procedure through the writing and editing/review process.** When complete, the traveler form is then kept in the QA file for the procedure to validate long-term how the initial procedure was developed.

Procedure Development Traveler - Moses Lake			
New Document Number:		Area:	
Procedure Title:			
Procedure Type:			
Author:			
Initial Draft Completion Date:			
Author Walkdown Date:			
Author Revision 1 Completion Date:			
Peer 1:		Shift:	
Peer 1 Walkdown Date:			
Optional Peer 2:		Shift:	
Peer 2 Walkdown Date:			
Optional Peer 3:		Shift:	
Peer 3 Walkdown Date:			
Optional Peer 4:		Shift:	
Peer 4 Walkdown Date:			
Author Revision 2 Date:			
Engineer/Verifier:			
Verification Completion Date:			
Author Revision 3 Completion Date:			
QA Person:			
Author Revision 4 Completion Date:			

Images and layouts copyrighted by PII, 2008-2016[4]

A color coding process has been developed to indicate if the content can be permanently deleted, or must be moved into the unit training manual. The unit trainers have been responsible for tracking and reporting weekly progress on the initiative. **See an example of the overall tracking form below.** A total of 184 "parent document" procedures for Moses Lake were tracked through the development, review, and revision process (along with about 200 child documents).

Images and layouts copyrighted by PII, 2008-2016[4]

Procedure Development Tracking Form - Moses Lake

							Initial Development			Validation				
Item	New Doc #	Old Doc #	Area	Procedure Title	Proc Type	Author(s)	Initial Draft	Author Walkdown	Author Rev 1	Peer 1 Walkdown	Opt. Peer 2 Walkdown	Opt. Peer 3 Walkdown	Opt. Peer 4 Walkdown	Author Rev 2
1	FBP.3.xxx	FBP.3.160	FBR	Hydrogen Booster Compressor Cleanup	Cleanup	LT/FG	3/19/14	3/19/14	3/20/14	3/21/14	3/21/14			
2	FBP.3.xxx	FBP.3.160	FBR	Hydrogen Booster Compressor Operation and Op Restrictions	Operation	LT/FG	3/20/14	3/20/14	3/20/14	3/21/14	3/21/14			
3	FBP.3.xxx	FBP.3.160	FBR	Hydrogen Booster Compressor LOTO	LOTO	BCA/OL	3/19/14	3/19/14	3/21/14					
4	FBP.3.xxx	FBP.3.160	FBR	Hydrogen Booster Compressor Standby	Standby	BCA/OL	3/20/14	3/20/14	3/21/14					
5			4.0	STC/TCS from Receiving Tank to STC/TCS Storage Transfer Through Metering Line	Transfer	ME/RR	3/19/14	3/19/14	3/20/14	3/20/14	3/21/14	3/21/14	3/21/14	
6			4.0	STC/TCS from Receiving Tank to STC/TCS Storage Transfer Through Bulk Line	Transfer	ME/RR	3/19/14	3/19/14	3/20/14	3/20/14	3/21/14	3/21/14	3/21/14	
7			4.0	H2 Recycle Compressor Switching to Standby	Standby	ME/RR	3/20/14	3/20/14	3/20/14	3/20/14	3/21/14	3/21/14	3/21/14	
8			4.0	H2 Recycle Compressor Cleanup	Cleanup	ME/RR	3/20/14	3/20/14	3/20/14	3/20/14	3/21/14	3/21/14	3/21/14	

From PII's procedure author training,[4] the Quality and Content checklist was modified and used as quick reference guide to make sure that all the content development and formatting rules are being followed during the re-write process. We used Microsoft Word and made templates for the Tbar, Flat and Consequence of Deviations.

Procedure Management of Change

Through this project REC Silicon has adopted formal safety review methods for procedures. These methods standardize the approach to determine if a procedure is safe to use as written or if revisions are required. These details are outlined in a completely new REC Silicon policy - Procedure Management of Change PSM.30.004.01. The REC Silicon IT department developed a custom tool to execute these requirements in an electronic workflow. The benefits from these changes are formally transferring competency to lead and document job safety analysis to the shop floor supervisors. Prior to this effort, JSA was traditionally a review lead by our safety specialists.

Overall Summary of the procedure writing initiative:

For the Moses Lake facility, 398 procedures (of which 184 were unique and 214 were similar to the 184 unique ones) for normal and routine operations, various maintenance tasks, and trouble-shooting guides have been developed through this project (new and rewrites combined.) Similarly, for the Butte facility, 116 procedures (of which 24 were unique and 92 were similar to the 24 unique ones) for normal and routine operations, various maintenance tasks, and trouble-shooting guides have been developed through this project (new and rewrites combined.) In addition, about 56 existing procedures (that were in "okay" state) were updated and the format improved.

We do not have any data yet for reduction in error rates or increase in production, quality or safety; however we have seen great improvements with the new formats with the ability to conduct job observations in the field. Our job observations assess accuracy of the work being performed, accuracy of the written procedures, and an opportunity for on the job operator training.

This type of project can seem overwhelming for a manufacturing site. It is essential to form a project team with key decision makers from all the correct disciplines and have a project sponsor that can remove barriers. The way REC Silicon was able to accomplish this type of effort was dedicating time to meet each week, setting short term goals, and tracking KPIs to meet those goals. The procedure writing initiative at REC Silicon is still very much "a work in progress". Leaders of the project have to make time to support the staff to keep the momentum going on this initiative and also on the related culture change, until it becomes "the way the work always gets done."

Quotes from workers involved in initiative:

- Operators are receiving better work instructions
- Sequence of steps is well understood
- Much easier to audit operator compliance to procedure
- Standardizes how we do work – protects our equipment from damage, protects our people, protects the environment
- No longer need 20 year operator tribal knowledge and interpretation to complete procedure
- Procedures are not cluttered anymore, they just have the necessary information
- Procedure element now requires involvement of many shop floor workers
- Procedures are really safe to use
- JSA technique is alive on the shop floor
- Procedure can be trusted
- Technical staff no longer doing all procedure work. Freed up to optimize processes and realize other cost saving benefits.

Conclusions

Having clear and accurate procedures is a must. Poor procedures show up as one of the root causes in 90% of the accidents. The approach described in this paper has worked for more than 35 years for one of the authors, and continues to work today. Controlling risk is mostly about controlling human error rates, and having accurate and clear procedures are one key to this control. One other gain from the approach described in this paper that the users become the owners, and therefore they have a vested interested to keep the procedures up to date and clear.

No doubt, writing all of the procedures for a facility is a large task. But, it can be accomplish by site staff, with support from process safety or quality control leadership, after enough users become competent authors; attaining this competency takes less than 2 weeks of work, on average.

Acronyms Used

AIChE– American Institute of Chemical Engineers
ASM – Abnormal Situation Management consortium
CCPS – Center for Chemical Process Safety (a division of AIChE)
CMA – Chemical Manufacturer 's Association, now American Chemical Council (ACC)
FBR – Fluidized Bed Reactor
HAZOP – Hazard and Operability Analysis
IPL - Independent Protection Layer
JSA – Job Safety Analysis
LOPA – Layer of Protection Analysis
MART – Maximum Allowable Response Time
MOC – Management of Change
NUREG – Nuclear Regulation, US Nuclear Regulatory Commission
OSHA – Occupational Safety and Health Administration, US Department of Labor
PHA – Process Hazard Analysis
PII – Process Improvement Institute, Inc.
P&ID – Piping & Instrumentation Diagram
PSI – Process Safety Information
PSM – Process Safety Management
PST – Process Safety Time
SOP – Standard Operating Procedure

References

1. Swain, A. D., Guttmann, H. E., *The Human Reliability Handbook - with Emphasis on Nuclear Power Plant Applications*; NUREG/CR-1278, U.S. Nuclear Regulatory Commission, Division of Facility Operations, 1983.
2. Gertman, D., Blackman, H., Marble, J., Byers, J. and Smith, C., *The SPAR-H Human Reliability Analysis Method,* NUREG/CR-6883, U.S. Nuclear Regulatory Commission, Office of Nuclear Regulatory Research, 2005.
3. Sutton, Ian S., "Writing Operating Procedures," *Proceedings of the International Conference on Hazard Identification and Risk Analysis, Human Factors and Human Reliability in Process Safety*, AIChE, Center for Chemical Process Safety, New York, NY, 1992.
4. *Writing Effective Operating and Maintenance Procedures*, Course Notebook, Process Improvement Institute, Inc., 2003-2016.
5. *Guidelines for Hazard Evaluation Procedures*, 3rd Edition, 2008, CCPS/AIChE.
6. "Necessity of Performing Hazard Evaluations (PHAs) of Non-normal Modes of Operation (Startup, Shutdown, & Online Maintenance)", W. Bridges and Mike Marshall (US OSHA), *18th Annual International Symposium, Mary Kay-O-Connor Process Safety Center, College Station, TX, October 2015.*

7. "How to Efficiently Perform the Hazard Evaluation (PHA) Required for Non-Routine Modes of Operation (Startup, Shutdown, Online Maintenance)," W. Bridges and T. Clark, *7th Global Congress on Process Safety*, Chicago, AIChE, March 2011.

8. Bridges, W.G., et. al., "Addressing Human Error During Process Hazard Analyses," *Chemical Engineering Progress*, May 1994

9. *A Manager's Guide to Reducing Human Error*, 1990, Chemical Manufacturers Association (CMA)

10. Haney, R. W., "The Effects of Instructional Format on Functional Testing Performance," *Human Factors*, 11, 1969, pp. 181-187.

11. "Effective Procedure Writing – A Two-Part Series (Part 1: Reducing Risk of Worker Errors; Part 2: Addressing Process Safety Management Requirements for Operating Procedures)," *Chemical Engineering Progress*, 1997, W. Bridges with T. R. Williams (both of JBFA).

12. Connelly, C. S., "Emergency Operating Procedure Writing for Oil Refinery Process Units," *Proceedings of the International Conference on Hazard Identification and Risk Analysis, Human Factors and Human Reliability in Process Safety*, AIChE, Center for Chemical Process Safety, New York, NY, 1992.

13. *Guidelines for Initiating Events and Independent Protection Layers*, 2015, CCPS/AIChE.

14. *Guidelines for Writing Effective Operating and Maintenance Procedures*, 1996, CCPS/AIChE.

15. *ASM Consortium Guidelines – Effective Procedural Practices*, 2010, ASM Joint R&D Consortium.

16. *Guidelines for Engineering Design and Process Safety*, 2nd Edition, 2012, CCPS/AIChE.

How Process Safety Information Affects the Quality of a Process Hazard Analysis

Michael Mayo
Smith & Burgess LLC
7600 W. Tidwell Rd, Ste 600
Houston TX 77040
mike.mayo@smithburgess.com

Dick Baum
dick.baum@smithburgess.com

Prepared for Presentation at
American Institute of Chemical Engineers
2016 Spring Meeting
12th Global Congress on Process Safety
Houston, Texas
April 11-23, 2016

How Process Safety Information Affects the Quality of a Process Hazard Analysis

Michael Mayo
Smith & Burgess LLC
7600 W. Tidwell Rd, Ste 600
Houston TX 77040
mike.mayo@smithburgess.com

Dick Baum
dick.baum@smithburgess.com

Keywords: PHA, PSI, MI

Abstract

This paper is a tutorial that shows the importance of how Process Safety Information affects the quality of a Process Hazard Analysis. The Process Hazard Analysis (PHA) is the one requirement of the PSM standard that has the facility stop and review a process in its entirety to ensure safe operation. However, a high quality PHA is dependent on accurate process safety information (PSI) that is consistent with the current equipment state (Inspection History, Changes from original design, operating conditions etc.). A PSI sub-element, relief system design basis documentation, requires significant amounts of other PSI and if organized with forethought, will greatly improve the quality, effectiveness and, potentially, speed of a PHA. If the relief systems documentation were performed in a system that stored all of the other relevant PSI and was cross referenced with current inspection records (MI systems), then this subset of PSI would contain much of the critical information needed for a PHA. The system would have to be designed such that it was accessible to the PHA team and organized to match the PHA workflow. This paper will present a proposed organization of PSI and then show how with a trained PHA facilitator and organized and accessible PSI/MI data, specific items found in PSM audits of PHAs would have been avoided.

1 Introduction

From defining the Causes, developing the Consequence scenarios and ending with Safeguards, accurate PSI information is required to provide the most credible assessment of risk. The best practice of minimizing the assumptions used during a PHA study has two significant benefits;

- Management can be assured (feel good) that the PHA process utilized all relevant and up-to-date PSI information thereby reducing potential inaccurate assumptions and,
- It will reduce time spent for discussions between participants to come to common conclusions and speed up the PHA process while maintaining the integrity of the study.

As learned from many PSM audits, PSI information is typically spread throughout many locations and maintained by numerous groups;

PSV files - Engineering

Mechanical Equipment Data/Inspection reports - Engineering, Maintenance, Inspection Departments

Critical Instrument Logs - Engineering or Controls

In a perfect world, maintaining a common database to store all up-to-date PSI information would be ideal, but the delegation of maintaining and access would be cumbersome. There are many people involved with PSI information and each has their own responsibilities and actions for validating the data. One way to reduce the roadmap for locating PSI data is to maintain relief systems documentation in such a way that it provides most all relevant data in a simplified format and links to other sources of PSI information.

This poster will walk through the PHA process and how readily available PSI information can be located with robust relief systems documentation.

2 Causes

Typically, the detailed cause of a PHA would not reference back to any PSI data. The failure or misalignment of valves are always credible occurrences that are documented. Where PSI information becomes relevant is in the cause category. The example shown in figure 1 indicates the following;

- Inlet PCV-001 on methane supply to V-001 reactor fails open

Cross-referencing to the P&ID indicated a figure 2, it is noted there is a PSV-002 upstream of PCV-001, a vessel, V-001 with its own relief device, PSV-001. A quick check during the PHA study of the relief systems documentation for PSV-002 will show that the set pressure of 400 psig is indeed above the MAWP of V-001, 285psig.

This validates the cause category of high pressure. Had the set pressure of PSV-002 been documented at 250 psig, this cause probably belongs in another category, i.e. more flow.

Processes: 1. V-001 Reactor

Nodes: 1. Feed streams

Deviation	Causes	Consequence	Risk			Safeguard
			S	L	RR	
1. High Pressure	1. Inlet PCV-001 on methane supply to V-001 reactor fails open	1. Potential for high pressure in V-001 reactor, potential **flange leak**, loss of containment, explosion, jet fire, employee injury, damage to equipment and loss of production				1. PSV-001 on V-001 reactor
		2. Potential for high pressure in V-001 reactor, potential **rupture**, loss of containment, explosion, jet fire, employee injury, damage to equipment and loss of production				

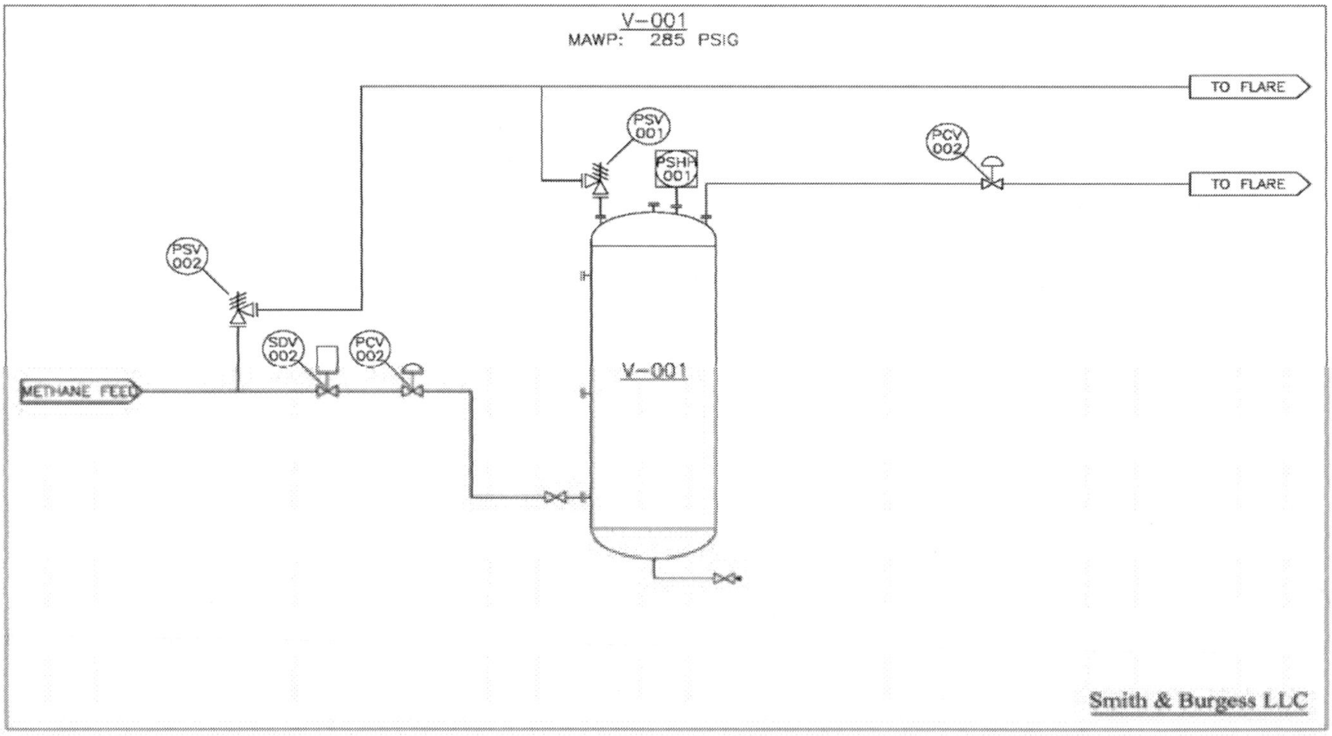

3 Consequences

The consequences developed during the PHA can benefit greatly from accurate and readily accessible PSI information. In the high pressure deviation we can use the PSI information in establishing the overpressure ratio to assist in pinpointing the consequence severity levels. Using the PSI information for PSV-002, a 400 psig supply into vessel V-001 with a MAWP of 285 psig, results in an overpressure ratio of 1.404. Based on your client's guidelines for overpressure and probabilities, this could result in either a flange leak or vessel rupture.

The consequence severity of these two scenarios is quite different but with the PSI data well documented and easily obtained, accurate severities can be obtained during the PHA.

Processes: 1. V-001 Reactor

Nodes: 1. Feed streams

Deviation	Causes	Consequence	Risk			Safeguard
			S	L	RR	
1. High Pressure	1. Inlet PCV-001 on methane supply to V-001 reactor fails open	1 Potential for high pressure in V-001 reactor, potential **flange leak**, loss of containment, explosion, jet fire, employee injury, damage to equipment and loss of production				1. PSV-001 on V-001 reactor
		2 Potential for high pressure in V-001 reactor, potential **rupture**, loss of containment, explosion, jet fire, employee injury, damage to equipment and loss of production				

PSV-002 set pressure can help determine consequence severity

PSI data for V-001 can help determine the consequence severity

4 Safeguards

The safeguards listed during a PHA have the most critical requirement for validation and accuracy during a PHA. They are the protection layers that mitigate or prevent major accidents from occurring and thus leave no room for assumptions or quantitative assessments. PSI information in relief system documentation should be current and accurate for the current operating parameters. A well-documented and arranged system will allow for easy access during the PHA. Information from other PSI data such as P&ID's does not always reflect the purpose of the safeguard. For example, many P&ID's show the governing case of a pressure relief device, which may or may not satisfy the scenario release and could lead to inaccurate assumptions being made during the PHA. (Fire case vs overpressure)

Processes: 1. V-001 Reactor

Nodes: 1. Feed streams

Deviation	Causes	Consequence	Risk			Safeguard
			S	L	RR	
1. High Pressure	1. Inlet PCV-001 on methane supply to V-001 reactor fails open	1 Potential for high pressure in V-001 reactor, potential **flange leak**, loss of containment, explosion, jet fire, employee injury, damage to equipment and loss of production				1 PSV-001 on V-001 reactor set at XXX PSV-001 set pressure can determine the likelihood
		2 Potential for high pressure in V-001 reactor, potential **rupture**, loss of containment, explosion, jet fire, employee injury, damage to equipment and loss of production				

5 Conclusion

Well documented PSI information is critical to the PHA process and provides the validation needed to ensure that the process is being operated and maintained in the best possible safe state. The examples shown indicate how comprehensive pressure relief files can provide much of the needed PSI data or arranged to provide the roadmap for data locations through links to other files. The additional benefit for having accessible PSI information is a smooth transition during the PHA and reduced costs and time for the PHA process.

442025 Experimental Measurements of Flammability at Elevated Conditions Using Thermal Criterion

Abstract:

Accurate measurements of flammability limits for combustible gases and vapors are necessary for the prevention of fire and explosion. The flammability limit, which is the maximum or minimum fuel concentration at which a gas mixture is flammable in a given atmosphere, has been studied for decades. However, due to the measurement conditions and apparatus set difference, the flammability values provided by different experimental methods are inconsistent, especially at elevated conditions. In this study, the flammability limits of methane-air were measured in an 8-L closed cylindrical vessel with upward flame propagation at maximum initial temperature of 300°C and maximum initial pressure of 2 bar. During the experimental measurement, a new thermal criterion was used to define if the gas is flammable and the experimentally determined flammability limits were compared with existing data reported in the literature. Finally, the temperature and pressure effect on flammability was investigated and compared with the modified Burgess-Wheeler Law.

Use of Process Incident Data for Accident Sequence Precursor Analysis

Erin P. Collins
JENSEN HUGHES
333 E. 46th Street, #11L
New York, NY 10017
ecollins@jensenhughes.com

Bijan Najafi
JENSEN HUGHES
2105 S Bascom Ave, Suite 350
Campbell, CA 95008-3278
bnajafi@jensenhughes.com

Prepared for Presentation at
American Institute of Chemical Engineers
2016 Spring Meeting
12th Global Congress on Process Safety
Houston, Texas
April 11-23, 2016

Use of Process Incident Data for Accident Sequence Precursor Analysis

Erin P. Collins
JENSEN HUGHES
333 E. 46th Street, #11L
New York, NY 10017
ecollins@jensenhughes.com

Bijan Najafi
JENSEN HUGHES

Keywords: accident sequence precursor, data, risk analysis

Abstract

Investigations of recent process incidents have identified multiple process safety management deficiencies that led to the accident. Furthermore, audits of the Safety and Environmental Management System (SEMS) program under the U.S. Department of the Interior's Bureau of Safety and Environmental Enforcement (BSEE) have shown that the "management of risk" elements such as Hazards Analysis and Management of Change (MOC) are not consistently implemented as tools to manage risk. In addition, the requirement to perform hazards analysis is frequently documented and often referenced as a corrective action, rather than as a "system driver". There are several pieces to risk assessment and risk management that allow facility operators to evaluate and measure what is actually occurring in their facilities, including the review of historical event data to identify what went wrong and for near misses, what almost happened. This paper describes techniques that have been used in the commercial nuclear power and other industries to collect and apply historical experience data to improve safety and operability. An approach using accident sequence precursor analysis of incident data and risk analysis techniques is recommended to identify potential accidents before they occur as part of a comprehensive treatment of historical data.

1 Introduction

There are several pieces to risk assessment and risk management that allow facility operators to evaluate and measure what is actually occurring in their facilities, including the review of historical event data to identify what went wrong and for near misses, what almost happened.

For example, the U.S. Chemical Safety Board (CSB) investigation of the February 18, 2015, explosion at the ExxonMobil Refinery in Torrance, California, identified multiple process safety management deficiencies that led to the accident and a serious near miss. In the case of the near miss, "...large pieces of debris from the explosion were thrown into other units of the refinery…One of these pieces of debris hit scaffolding in the refinery's alkylation unit, narrowly missing a tank containing tens of thousands of pounds of modified hydrofluoric acid, or HF. The CSB determined that had the debris struck the tank, a rupture could have been possible, resulting in a potentially catastrophic release of extremely toxic modified HF into the neighboring community." [Ref. 1]

Such near misses are not unique to the chemical process industry, but have been documented in the offshore oil industry as well. As described by EnergyWire [Ref. 2], "Partial failures happen regularly on oil rigs, with a worker catching the errors and correcting them before a major blowout. These are not always reported to regulators, but they are logged carefully and stored as raw data on rigs and onshore in Houston."

The Discussion Paper on Precursor Analysis by the research organization Resources for the Future [Ref. 3] identifies the specific challenge faced by the oil and gas industries of "how to reduce the risk of low-frequency oil spill events, and especially high-consequence events like the Deepwater Horizon accident, when historical experience contains few oil spills of material scale and none approaching the significance of the Deepwater Horizon." Based on a review of successful programs in the nuclear power, aerospace and other industries, the paper recommends the consideration of accident precursor analysis as a means for providing "a risk-informed oversight capability focused on preventing catastrophic oil spills and based on data from ongoing operating experience on the Outer Continental Shelf (OCS)."

The ASP approach is similar to Layer of Protection Analysis (LOPA) [Ref. 4] or Bowtie Analysis in that it postulates accident scenarios and the paths that could lead to a potential accident, but differs in that it uses insights from actual historical near-miss event data to identify the branch points which, if different, could have led to a severe accident.

This paper discusses the collection of historical event data, various programs that undertake such data collection, and proposes steps for use of this data toward risk assessment and management using ASP methods.

2 Collection, Evaluation and Application of Historical Events Data

While HAZOP and FMEA are useful tools for postulating incidents and their consequences, historical event data provides important input to the risk assessment process beyond the development of accident frequencies. Incident data collection and analysis can provide multiple qualitative insights, such as information for developing and refining risk models and a better understanding of system operation and failure (including the human-machine interface and efficacy of procedures and practices).

The CSB investigates incidents at chemical process facilities, produces investigation reports such as the one for the West Fertilizer Company Fire and Explosion in April 2013 [Ref. 5] and conducts public meetings to present and discuss their findings, as with the February 2015 ExxonMobil Refinery explosion [Ref. 6].

Following the Deepwater Horizon accident, the Deepwater Presidential commission report [Ref. 7] found at the time that "Although the raw incident data are online, MMS last produced an analysis of offshore incidents—critical data for promoting the safety of offshore operations—for calendar year 2000."

Probably for this reason, incident analysis was highlighted in the discussion of the Workplace Safety Rule in 2009 [Ref. 8] and one of the key areas where it was expected that the U.S. Department of the Interior's Bureau of Safety and Environmental Enforcement (BSEE) would be focusing their attention during the second Safety and Environmental Management System (SEMS) audit cycle in 2015 was on "Developing representative leading indicators for SEMS performance, while continuing to track industry accepted lagging indicators for evidence that SEMS programs are driving an improvement in safety performance." [Ref. 9] The development of these representative indicators by its nature must come from the review of historical evidence.

The narratives and timelines of previous accidents and near-misses can indicate what triggered the event, the automatic system response and failure, subsequent system failures and their causes, and worker involvement either to cause, exacerbate or mitigate the event. These insights in turn feed back into the data collection process by indicating what trends are of interest to monitor and, therefore, what categories of data should be collected.

The Workplace Safety Rule announcement by MMS in 2009 [Ref. 8] indicated that "As a result of MMS research conducted on accident panel investigations and reports, incident analysis, and [Incidents of Non-Compliance] INCs, it appears that equipment failure is rarely the primary cause of the incident or accident. This is due to technological advances which have provided industry with very efficient and reliable equipment for finding, producing, and transporting offshore oil and gas. However, in most cases, accidents and oil spills can be traced to human error and/or organizational failures." This is consistent with the Deepwater Presidential commission report [Ref. 7], which identified several human error-related issues, including the following:

- Numerous changes to the temporary abandonment procedures for securing the well were made in the two weeks leading up to the operations plan for the worker activities, but "[t]here is no evidence that these changes went through any sort of formal risk assessment or management of change process"

- "The drilling crew and other individuals on the rig also missed critical signs that a kick [unplanned influx of gas or fluids] was occurring. The crew could have prevented the blowout—or at least significantly reduced its impact—if they had reacted in a timely and appropriate manner. What is not now clear is precisely why the crew missed these signals."

- "The crew should have diverted the flow overboard when mud started spewing from the rig floor. While that ultimately may not have prevented an explosion, diverting overboard would have reduced the risk of ignition of the rising gas. Considering the circumstances, the crew also should have activated the blind shear ram to close in the well. Diverting the flow overboard and/or activating the blind shear ram may not have prevented the explosion, but likely could have given the crew more time and perhaps limited the impact of the explosion."

While the human involvement in incidents cannot be denied, the ATHEANA method of human reliability analysis (HRA) developed by the US Nuclear Regulatory Commission [Ref. 10] discusses the situation of "error-forcing contexts" or EFCs in which the plant conditions and so-called "performance shaping factors" (context-specific influences on reliable human performance) conspire to make human error nearly unavoidable. The ATHEANA document provides examples of these EFCs that have occurred in past US nuclear plant accidents, such as mismatches between the actual plant conditions and the operator job aids (e.g., training, experience) during the Three Mile Island event. EFC elements were also identified by ATHEANA for each of the four stages of information processing (i.e., detection, situation assessment, response planning, and response implementation) that describe the range of human activities required to respond to abnormal or emergency conditions. The review of past events and near-misses considering these EFC elements may indicate that organizational factors and gaps in procedures or training may in fact be the root cause.

An independent investigation of the Deepwater Horizon by the UC Berkeley Center for Catastrophic Risk Management stated that "[T]he Macondo system's good record of lost time accidents could not provide effective precursors for the major impending system accident. Efforts to prevent system accidents must address a broader set of organizational issues and hazards associated with all of the elements that comprise a given system. The good record of lost time accidents plausibly led to complacency that contributed to the disaster." [Ref. 11]

A program such as ASP looks at historical events from the perspective of commonality of potential outcomes and identifies the organizational issues that can cause near misses to become severe accidents.

3 Chemical Process and Nuclear Power Industry Experience

Examples of the projects that exemplify the type of incident reviews that factor industry experience into lessons learned are the Accident Sequence Precursor (ASP) program [Refs. 12, 13 and 14], the AIChE CCPS Process Safety Incident Database (PSID) [Ref. 15] and Process Equipment Reliability Database (PERD) [Ref.16], the EPRI Fire Events Database (FEDB) project [Refs. 17 and 18], the EPRI Guidelines for PRA Data Analysis [Ref. 19], the AIChE CCPS Guidelines for Process Equipment Reliability Data [Ref. 20], and the EPRI/NRC-RES Fire HRA Guidelines [Ref. 21]. Jensen-Hughes has been an active participant (lead in some cases) in the development of all of these projects except PERD and PSID.

3.1 Accident Sequence Precursor (ASP) Program

The ASP Program began in the early 1980s under the auspices of the U.S. Nuclear Regulatory Commission in response to the Three Mile Island accident. ASP is concerned with the identification and documentation of operational events that have involved portions of accident sequences and with the estimation of associated frequencies and probabilities. Identification of precursors requires the review of operational events for instances in which plant functions that provide protection against severe accidents (core damage, in the case of nuclear plants) have been challenged or compromised. The intention is to understand how these 'near misses' could have propagated into a severe accident had the circumstances been slightly different and thereby inform the risk modeling, data collection and operational insights.

[NOTE: The following explanation of the US NRC's ASP program is extracted from Reference 11.]

Reports of plant incidents are reviewed to identify those potentially involving failures in systems that provide protective functions against severe accident-related initiating events or the initiating events themselves. Those events selected undergo one- or two-engineer review(s) to determine if the reported event should be examined in greater detail. Events determined to be potentially significant as a result of this initial review are then subjected to a thorough, detailed analysis. This extensive analysis is intended to identify those events considered to be precursors to potential severe accidents, either because of an initiating event or because of failures that could have affected the course of postulated off-normal events or accidents. A variety of other data sources about the same event are also reviewed and compared to thoroughly understand the contributors to each event.

Quantification of accident sequence precursor significance involves the determination of a conditional probability of a subsequent severe accident given the failures observed during an operational event. This probability is estimated by mapping failures observed during the event onto the ASP accident sequence models (event trees and linked fault trees modified to reflect the event), which depict potential paths to the severe accident, and by calculating a conditional probability of the severe accident.

The information gained from the ASP process not only provides quantitative data, but provides insights into the types of events that should be included in risk models. In addition, events that provide insights into unusual failure modes with the potential to compromise continued safety functions but are not considered to be precursors are documented by ASP as "interesting" events.

Building upon this effort, the National Academy of Engineering Program Office undertook the Accident Precursors Project in 2003 to examine the complex issue of accident precursor analysis and management. This seven-month project was designed to document and promote industrial and academic approaches to detecting, analyzing, and benefiting from accident precursors, as well as to understand public-sector and private-sector roles in using precursor information. [Ref. 10]

3.2 AIChE CCPS Process Safety Incident Database (PSID)

The purpose of the CCPS PSID is to pool process safety incident experience among participating companies so they can learn from the experiences of others without suffering the consequences of failures, while minimizing corporate liability. PSID includes process safety incidents with a potentially important lesson to be learned from incidents that did or could have resulted in a fire, explosion, fatality, multiple injuries, significant release of hazardous materials, or any other unique process safety incident (including near-misses).

3.3 AIChE CCPS Process Equipment Reliability Database (PERD)

The equipment reliability database is intended to provide high quality, valid, and useful data to the Hydrocarbon Processing Industries (HPI) and Chemical Process Industries (CPI). This type of data provides input to analyses of availability, reliability and equipment design improvements, maintenance strategies and life cycle cost determination. The database includes Lessons Learned and Contributing Factors, which are data input requirements.

3.4 Fire Events Database (FEDB) Project

The FEDB is a collection of fire events that occurred in nuclear power plants since 1967. These events were reported to the U.S. Nuclear Regulatory Commission or the industry groups such as the Institute for Nuclear Power Operations (INPO) and the Nuclear Energy Institute (NEI). These events were reviewed to extract information on the equipment involved in the fire initiation, response of fire prevention and mitigation systems, response of the fire brigade and the consequences of the fire (degraded incipient or catastrophic). Guidance was also provided for the interpretation and categorization of fire event data. Since its publications the event history has been used, at time in combination with tests and a process called Phenomena Identification and Ranking Table or PIRT, to help define and characterize accident initiation, progression, mitigation and consequences, including:

- Types and characteristics, e.g., timing, of accident initiators,
- Sequence of events, including prevention/mitigation and human interface that helps define accident scenarios (or sequences) with risk implications,
- Severity of the event, e.g., heat release rate of an electrical panel fire or zone of influence from a high energy arc fault in a high voltage switchgear,
- Mitigating systems response, e.g., reliability and effectiveness of the fire brigade response to various types of fires,
- Cable damage and quantification of effects from fire, such as spurious indication of instrumentation and control systems leading to failure of accident prevention/mitigation systems or inadequate/erroneous operator response, and
- Qualitative insights on the types of fires that are most likely to occur, their duration, severity and how the plants address them.

3.5 EPRI Guidelines for Data Analysis for PRA

The EPRI Data Guidelines discusses the nuclear power industry approach to collecting, analyzing and providing summary statistics for several major categories of data types: initiating event data, component reliability data, unavailability (downtime) data and common cause failures. Initiating events are quantified using a combination of historical evidence of event occurrence, both from the specific plant and across the nuclear industry.

3.6 AIChE CCPS Guidelines for Process Equipment Reliability Data

Many of the aspects of data analysis discussed in the EPRI Data Guidelines were identified thirteen years earlier for the chemical process industry in the AIChE CCPS Process Equipment Reliability Data Guidelines. This data book built on the Chemical Process Quantitative Risk Assessment (CPQRA) Guidelines to show the process engineer how to make quantitative estimates of the risk of hazards identified, which can in turn identify the major contributors to risk. This project helped demonstrate the need for process industry data collection efforts like PERD and PSID.

3.7 EPRI/NRC-RES Fire Human Reliability Analysis (HRA) Guidelines

Although its focus is on human reliability, the Fire HRA Guidelines discusses the need to review past incidents experienced at a facility to understand whether the event was human-caused or how the workers' responded to the event as it evolved. These insights not only assist the HRA, but provide an understanding of how workers perceive their role during an accident and how they can influence accident progression, for good or for ill. These findings can then be fed back into the improvement of organizational and safety practices for accident prevention.

4 Considerations for Historical Data Analysis Application in the Chemical Process Industry

The Discussion Paper on Precursor Analysis for Offshore Oil and Gas Drilling [Ref. 3] identifies incident reports required under the Incident Reporting Rule (30 CFR 250.188) for oil, gas and sulphur operations in the outer continental shelf as an example of industry precursor data that would allow a system failure of concern [such as "significant uncontrolled escape of hydrocarbons" (SUEH)] to be evaluated through event trees based on generic commonality of events (such as those used in the US NRC ASP program).

Further, the Discussion Paper recommends an approach for the design of an ASP-enabled incident reporting system, which has been made more general here for application to a wider variety of chemical processes:

a) Define the class of events (such as SUEH) related to the system failure of concern.

b) On the basis of PRA models, design the set of generic event trees that have these events as possible endpoints.

c) Ensure that all nodes in these generic event trees are captured in an incident reporting system, such as PSID or under 30 CFR 250.188.

d) Design a control loop, possibly based on the Potential Incident of Non-Compliance (PINC) checklist inspection system, to ensure coverage of the incident reporting system, with appropriate sanctions for noncompliance.

e) Set up an analysis team to map the key even-related incidents onto generic event trees.

f) Set up a review board comprising industry and government agencies.

g) Design feedback mechanisms for involving industry in efforts to optimize industry-wide performance as measured by the ASP indicators.

h) Design an automation plan to streamline the ASP analysis.

The following additional tasks should be considered for translating these steps into industry guidance:

a) Documentation of insights and development of a draft guidelines for a selected process facility (to be defined) reflecting items a) through h) above.

b) Testing, validation and review of the guidelines. This may be done using one or more well documented industry events.

c) Final guideline for the selected process facility.

d) Adaptation of the guideline to other processes.

5 Conclusion

A recent article in CEP Magazine [Ref. 22] profiles five chemical process industry accidents that, while not as severe as Bhopal or Deepwater Horizon, still had serious results. The article states that "Accidents such as those discussed in this article are more likely to occur. These incidents demonstrate that even small mistakes can have disastrous consequences. Therefore it is imperative that engineers learn from previous incidents to reduce their likelihood of recurring." Near misses, like smaller scale events, are also instances that can provide valuable insights on lapses in process safety management.

As stated in EnergyWire [Ref. 2]. "ASP would use the data from these partial failures, including the failure of rig workers to respond appropriately, to compute the likelihood of a catastrophe. It would identify problems when they occur so regulators can stop them from happening again."

6 References

[1] "U.S. Chemical Safety Board Finds Multiple Safety Deficiencies Led to February 2015 Explosion and Serious Near Miss at the Exxon Mobil Refinery in Torrance, California," News Release, U.S. Chemical Safety and Hazard Investigation Board, Washington, DC, 16 January 2016.

[2] Vaidyanathan, G., "GULF SPILL: Could risk analysis prevent future deepwater disasters?" EnergyWire, Environment & Energy Publishing (E&E), http://www.eenews.net/stories/1059963200, 20 April 2012.

[3] Cooke, R.M., Ross, H.L. and Stern, A., "Discussion Paper - Precursor Analysis for Offshore Oil and Gas Drilling, From Prescriptive to Risk-Informed Regulation", RFF DP 10-61, Resources for the Future, Washington, DC, January 2011.

[4] Guidelines for Initiating Events and Independent Protection Layers in Layer of Protection Analysis, American Institute of Chemical Engineers (AIChE), Center for Chemical Process Safety (CCPS), New York, NY, 2015.

[5] "West Fertilizer Company Fire and Explosion", Final Investigation Report 2013-02-I-TX, U.S. Chemical Safety and Hazard Investigation Board, Washington, DC, January 2016.

[6] Presentation slides from CSB Public Meeting on ExxonMobil Refinery Explosion, U.S. Chemical Safety and Hazard Investigation Board, Washington, DC, 13 January 2016.

[7] Deep Water: The Gulf Oil Disaster and the Future of Offshore Drilling, Report to the President, National Commission on the BP Deepwater Horizon Oil Spill and Offshore Drilling, January 2011.

[8] The Bureau of Ocean Energy Management, Regulation and Enforcement, Office of Public Affairs, Fact Sheet, The Workplace Safety Rule on Safety and Environmental Management Systems (SEMS).

[9] Memorandum from Doug Morris, Chief, Office of Offshore Regulatory Programs to Brian Salerno, Director BSEE, SEMS Program Summary—First Audit Cycle (2011-2013), July 23, 2014.
http://www.bsee.gov/uploadedFiles/BSEE/Regulations_and_Guidance/Safety_and_Environmental_Management_Systems_-_SEMS/SEMS%20Program%20Summary%208132014.pdf

[10] Technical Basis and Implementation Guidelines for A Technique for Human Event Analysis (ATHEANA), NUREG-1624, Rev. 1, U.S. Nuclear Regulatory Commission, Washington, DC, May 2000.

[11] Final Report on the Investigation of the Macondo Well Blowout, Deepwater Horizon Study Group (DHSG), Center for Catastrophic Risk Management (CCRM), University of California – Berkeley, Berkeley, CA, March 2011.

[12] Minarick, J. W. and Kukielka, C. A., Precursors to Potential Severe Core Damage Accidents: 1969-1979, A Status Report, NUREG/CR-2497, U.S. Nuclear Regulatory Commission, Washington, DC, June 1982.

[13] Belles, R.J. et al., Precursors to Potential Severe Core Damage Accidents: 1998, A Status Report, NUREG/CR-4674, US Nuclear Regulatory Commission, Washington, DC, July 2000.

[14] Phimister, J.R., Bier, V.M., and Kunreuther, H.C. (Eds.), Accident Precursor Analysis and Management: Reducing Technological Risk Through Diligence, The National Academies Press, Washington, DC, 2004.

[15] Presentation slides, Overview: Process Safety Incident Database (PSID), American Institute of Chemical Engineers (AIChE), Center for Chemical Process Safety (CCPS)

http://www.aiche.org/ccps/resources/psid-process-safety-incident-database

[16] American Institute of Chemical Engineers (AIChE), website "About PERD"

http://www.aiche.org/process-equipment-reliability-database-perd/about-perd

[17] EPRI 1000894, Fire Events Database for U.S. Nuclear Power Plants: Update Through 1999, Electric Power Research Institute (EPRI), Palo Alto, CA, October 2000.

[18] EPRI 1025284, The Updated Fire Events Database: Description of Content and Fire Event Classification Guidance, Electric Power Research Institute (EPRI), Palo Alto, CA, July 2013.

[19] EPRI 3002000774, EPRI Guidelines for PRA Data Analysis, Electric Power Research Institute (EPRI), Palo Alto, CA, December 2013.

[20] Guidelines for Process Equipment Reliability Data with Data Tables, American Institute of Chemical Engineers (AIChE), Center for Chemical Process Safety (CCPS), New York, NY, 1989.

[21] EPRI 1023001/ NUREG-1921, EPRI/NRC-RES Fire HRA Guidelines, Electric Power Research Institute, Palo Alto, CA and U.S. Nuclear Regulatory Commission, Washington, DC, July 2012.

[22] A. Ness, "Lessons Learned from Recent Process Safety Incidents," CEP Magazine, American Institute of Chemical Engineers (AIChE), New York, NY, March 2015.

442384 Quantitative Risk Analysis of the Expansion of the Chlorine Room of the Industrial Water Plant of the Petrochemical Complex Jose Antonio Anzoategui of Petroquimica Venezuela Pequiven S.a

Abstract:

Quantitative risk analysis of the expansion of the chlorine room of the industrial water plant of the petrochemical complex Jose Antonio Anzoategui of Petroquimica venezuela PEQUIVEN S.A.

Venezuela Petrochemical PEQUIVEN SA has a management responsible for risk control, which makes this quantitative risk analysis to estimate the risk to employees and third parties for process safety events that may occur as a result of the loss of containment in the facilities and thus indicate recommendations aimed at reducing the level of risk associated with such events. Quantitative risk analyzes are made from the standard rules of Petroleos de Venezuela , PDVSA as IR-S-02: criteria for quantitative risk analysis and other international standards such as Center of Chemical Process Safety Guidelines for Chemical Process Quantitative Risk Analysis and the netherlands organization Yellow book: Methods for calculation of physical effects, through the computational tool PHAST RISK 6.7 . Developed by DNV Company.

PEQUIVEN has three complexes nationwide, in which fertilizers for the food industry, olefins and plastics for residential construction, among other products are produced. The Jose Antonio Anzoategui complex will make an expansion of the chlorine room of the industrial water plant and because of the dangers associated with chlorination process, a quantitative risk analysis has requested to engineering management risks.

The process of a quantitative risk analysis (QRA) started setting goals, objectives and depth of the study, once these are established, and in conjunction with materials, inventory, environmental conditions and the description of the process, one proceeds to identify scenarios. In this case, the HAZOP tool was applied, which gives the place a list of possible scenarios, of which those that could cause the worst consequences. With the previously selected scenarios, the model building on which the estimated frequency and consequence will be held is started. The frequency is estimated based on a review of probability theory study according to Chemical Process Safety Fundamentals with Applications Standard, through a theoretical data extracted from the norm Guidelines for Process Equipment Reliability Data, a probability of occurrence of the scenarios is established.

The estimation of Consequence answer a series of complex mathematical models that simulate the behavior of these events based on the data of the others that occurred already trough PHAST RISK software. Once the simulation is made, the software generates a series of results of contours risks in function of exposure to chlorine gas to the personnel working in this complex and surrounding facilities, allowing emit a series of observations than are aimed to decrease level of Risk associated with the expansion of the chlorine room.

Assessing Cognitive Human Errors for the Process Industries

Erin P. Collins
JENSEN HUGHES
333 E. 46th Street, #11L
New York, NY 10017
ecollins@jensenhughes.com

Bijan Najafi
JENSEN HUGHES
2105 S Bascom Ave, Suite 350
Campbell, CA 95008-3278
bnajafi@jensenhughes.com

Prepared for Presentation at
American Institute of Chemical Engineers
2016 Spring Meeting
12th Global Congress on Process Safety
Houston, Texas
April 11-23, 2016

Assessing Cognitive Human Errors for the Process Industries

Erin P. Collins
JENSEN HUGHES
333 E. 46th Street, #11L
New York, NY 10017
ecollins@jensenhughes.com

Bijan Najafi
JENSEN HUGHES

Keywords: cognitive, human error, human reliability analysis

Abstract

Through the analysis of actual incidents and the understanding of the underlying human errors that have contributed to them, it is apparent that 'first generation' HRA methods are appropriate for some instances but have their limitations. For example, failure to recognize the significance of high fluid returns during the negative pressure test and failure to perceive the importance of drill pipe pressure during well monitoring activities were key contributors to the Deepwater Horizon incident. These diagnosis and decision-making errors are not well addressed by first generation HRA methods and require particular timing, error mode and performance shaping factor considerations. This paper provides an introduction to HRA methods and then discusses those available for evaluating cognitive errors and their applicability to the chemical process industry.

1 Introduction

Layer of Protection Analysis (LOPA) provides a simplified risk analysis technique for a select set of postulated incident scenarios, evaluating an initiating event (IE) and subsequent independent protection layers (IPLs) as a series of probabilities of failure on demand (PFDs). Human error is assessed in LOPA as either an IE or IPL. Techniques that have been recommended for human reliability assessment (HRA) by the CCPS Guidelines for IEs and IPLs in LOPA [Ref. 1] are generally considered in other industries as 'first generation' HRA methods, meaning that they are appropriate for some instances but have their limitations. These limitations have become more apparent through the analysis of actual incidents and the understanding of the underlying human errors that have contributed to them. For example, failure to recognize the significance of high fluid returns during the negative pressure test and failure to perceive the importance of drill pipe pressure during well monitoring activities were key contributors to the Deepwater Horizon incident. These diagnosis and decision-making errors are not well addressed by first generation HRA methods and require particular timing, error mode and performance shaping factor (PSF) considerations. Several HRA quantification methods are available to assess the human error probability associated with these so-called "cognitive" aspects of a task, as well as the equipment manipulation.

While many HRA approaches had their genesis in the nuclear power industry, they have been successfully applied to other industries, such as chemical weapons destruction plants and nuclear fuel transport and storage, and their results have been accepted during peer and regulatory agency reviews. In each instance where HRA is applied to a new industry, however, an evaluation must be made regarding the degree of applicability of the standard error types, PSFs and quantification methods. Since the error types and PSFs have an underpinning in human factors, psychology and cognitive science, they generally apply regardless of the industry. However, it is important to evaluate the applicability of HRA tools and techniques to ensure that industry-specific issues are being appropriately addressed.

This paper provides an overview of HRA principles and methods, then discusses those available for evaluating cognitive errors and their applicability to the chemical process industry.

2 HRA Principles and Methods

HRA provides a framework to evaluate, prioritize, quantify and improve the human interface with complex systems. HRA tools and techniques were pioneered by the commercial nuclear power industry, but are based on fundamental precepts of human reliability and data collection and analysis that are not technology specific. They have been applied to an increasing number of facilities across various different industries, such as aerospace [Ref. 2], chemical weapons destruction [Ref. 3], HVAC system optimization, nuclear fuel disposal facility design risk evaluation [Ref. 4] and detection and suppression activities in response to fire events in the military [Ref. 5]. This experience and repetition have allowed lessons learned to be identified and fed back into safe practices, the development of a safety culture, and the ability to meet standards [Ref. 6] and regulatory goals (which are quite stringent in the nuclear power industry).

This history and associated lessons learned indicates that wider application of HRA in the chemical process industry could provide greater understanding of risks (both accident and economic), how to control them (by improving procedures, focusing maintenance and surveillance efforts on the actual errors), and how to develop a safety culture that focuses on human error prevention.

The steps of the HRA process discussed below are adapted from NUREG-1921, the EPRI/NRC-RES Fire Human Reliability Analysis Guidelines [Ref. 7]. The process begins with an identification and definition task to identify actions performed by various workers and to define the actions in terms of human failure events or HFEs. These actions may include both failure to perform a required task or undesired actions with deleterious impact on system performance and safety. For the chemical process industry, this identification and definition step could be expected to be very closely tied with the subsequent step of qualitative analysis because the same information gathering that helps define the events also feeds into the development of an "operational story" or HFE narrative that assists in further understanding the events.

A sound qualitative analysis allows the HRA to provide feedback to the facility operator on the factors contributing to both the success and failure of a worker action. This involves the evaluation of performance shaping factors (PSFs), which are situation-specific influences on human error probabilities (HEPs) and include:

- Cues and indications that indicate a system problem
- Timing of the task
- Procedures and training
- Complexity of the task
- Workload, pressure and stress
- Human-Machine Interface (HMI)
- Environmental factors, such as heat or confined spaces
- Special equipment, such as tools and protective gear

These PSFs are used to flesh out the HFE narrative and identify the key elements of the HFEs that require further understanding through the performance of subject matter expert interviews.

Finally, quantification or quantitative analysis is performed by applying the information gained from identification, definition and qualitative analysis of HFEs toward the estimation of HEPs. Quantification can be done at varying degrees of detail and rigor. The three levels of analysis performed in HRA are: Screening, Scoping and Detailed.

Screening analysis involves the assignment of a preliminary and conservative expert judgment value as a means for "screening" out the risk-significant HFEs from the non-contributing ones.

The Scoping approach, discussed in NUREG-1921, was developed to provide additional rigor to a semi-screening process, but still results in conservative values for initial quantification.

Detailed analysis is performed by evaluating the action steps, timing and PSFs in detail and assessing the probability of errors in the cognitive aspects, or diagnosis and decision-making process for the correct action to take, as well as the execution aspects, such as the equipment manipulation steps.

Several HRA quantification methods are available and it has been found to be most effective if the analyst evaluates the strengths and weaknesses of each method against the nature of the human error being evaluated to make the best match. Having a large HRA method toolkit to draw from allows the analyst the flexibility to assess the driving failure mechanism behind the error rather than simply filling out a checklist.

The following HRA quantification methods are most widely used in the nuclear power industry, and are contained in Electric Power Research Institute's (EPRI's) HRA Calculator software [Ref. 8]:

- Human Cognitive Reliability/Operator Reliability Experiment (HCR/ORE) [Ref. 9], which focuses on the cognitive or decision making errors that are constrained by time;

- Cause-Based Decision Tree Method (CBDTM) [Ref. 9] that addresses cognitive errors as well but from the standpoint of the error-causing mechanisms, such as unclear instrumentation or lack of procedural direction;

- The Technique for Human Error Rate Prediction (THERP) [Ref. 10], which has been used for many years for the evaluation of task performance failures (also called execution errors) that can occur either before or after an accident;

- The Accident Sequence Evaluation Program (ASEP) [Ref. 11] that provides a simplified version of THERP for human errors that contribute to but occur prior to the accident; and

- The SPAR-H method [Ref. 12], which only permits highly simplified assessments of cognitive and action errors (and has not been used in the latest generations of nuclear plant risk assessments since it does not allow the depth of analysis to support safety improvements).

For HRA applications other than nuclear power, such as chemical weapons destruction plants and nuclear fuel transport and storage, other quantification methods have been successfully applied and their results accepted during peer and regulatory agency reviews. These include the Cognitive Reliability Error Analysis Method (CREAM) [Ref. 13], and the Human Error Assessment and Reduction Technique (HEART) [Ref. 14]. As the name implies, CREAM focuses on the decision-making errors by providing HEPs for a standard set of task types, but also includes adjustment factors to evaluate facility and scenario-specific influences. HEART contains a list of generic task types and their HEP estimates that are modified using factors associated with 'error producing conditions'.

While the application of a method and generation of HEP values is important, the primary output of the HRA is the identification and prioritization of the human error issues of greatest risk significance and the development of risk insights. The benefit of conducting more detailed HRA is in gaining a better perspective on what influences human failure and therefore what can be done to improve human performance.

The decomposition of human error into cognitive vs. execution tasks and a detailed evaluation of these tasks has proven to be particularly important since the approaches to their improvement are different. For example, more instrumented processes attempt to provide information to the operator that allow him/her to diagnose the situation and take the appropriate action. In other cases, however, the process may not include such instrumentation or it may be not be available due to failure, so diagnosis becomes a critical issue. Even when there is no instrumentation or early warning system provided for an action, the HRA can provide an understanding of the cues required for diagnosis and allow these to be included in training and procedures/manuals. These cues may even be sensory in nature, such as the sound, smell or sight of a process that is going awry. This does not mean that the HRA will recommend instrumenting everything. If correctly done, the HRA instead allows the most critical information used by the worker to be identified and suggests how it should be presented or emphasized to optimize performance.

The reliability of execution tasks is generally improved by providing a formal set of procedures and training that are continually reviewed and refined. However there may be instances where procedures and training are not yet formalized, such as when the facility is a new design or is in the pre-operational testing phase, or because of historical precedent or preference. In these instances, the worker action is evaluated in HRA as a so-called 'recovery' action, where the worker utilizes previous experience with tasks they consider to be similar and applies them to a new situation. These insights are important to evaluating the reliability of task performance and can even be subsequently captured in procedure modifications and training enhancements. The risk benefit assessment might even recommend conducting a design change to the human-machine interface to improve the execution task reliability.

In summary, HRA is

- Widely used
- Applied across a variety of industries, which demonstrates that it provides value
- Allows the user to quantify and prioritize the human influence on safety
- Provides the means for improving process safety and reliability in the short term and a safety culture to reduce future errors over the longer term.

3 Cognitive Errors in the Process Industries

The Deepwater Presidential commission report [Ref. 15] identified several human error-related issues that led to the severity of the incident, including the following:

- Numerous changes to the temporary abandonment procedures for securing the well were made in the two weeks leading up to the operations plan for the worker activities, but "[t]here is no evidence that these changes went through any sort of formal risk assessment or management of change process"
- "The drilling crew and other individuals on the rig also missed critical signs that a kick [unplanned influx of gas or fluids] was occurring. The crew could have prevented the blowout—or at least significantly reduced its impact—if they had reacted in a timely and appropriate manner. What is not now clear is precisely why the crew missed these signals."
- "The crew should have diverted the flow overboard when mud started spewing from the rig floor. While that ultimately may not have prevented an explosion, diverting overboard would have reduced the risk of ignition of the rising gas. Considering the circumstances, the crew also should have activated the blind shear ram to close in the well. Diverting

the flow overboard and/or activating the blind shear ram may not have prevented the explosion, but likely could have given the crew more time and perhaps limited the impact of the explosion."

These issues can be considered as relating to cognitive human errors since they concern the lack of procedural guidance that assists with diagnosis and the decision-making on the part of the crew that appears to have been ill advised or ill informed. Hubbard and Embrey [Ref. 16] assessed these issues from a human factors standpoint and they raised the following concerns related to the cognitive errors during the incident:

- Why were procedures not used – is this typical?
- What was the effect on decision making and on practice of the pressure to get the well tested and capped?
- Why did the crew use the mud gas separator rather than pipework that would have discharged the mud more quickly?
- What emergency training did the team have? Did they have training or job support to help them identify and respond to escalating situations?

Questions such as these are key to the evaluation and assessment of cognitive errors.

A recent article by Ness in the CEP Magazine [Ref. 17] outlined lessons learned from several process accidents, many of which involved cognitive human errors:

- ARCO Channelview wastewater tank explosion, 1990 – "The wastewater tank was not considered part of the operating plant. Hence, the management and workers did not understand that a chemical reaction was taking place in the tank, generating oxygen. The lack of understanding enabled a series of poor decisions, such as discontinuing the nitrogen purge…no management of change (MOC) review of these decisions, and no pre-startup safety review."
- Terra Industries ammonium nitrate explosion, 1994 – "…the lack of procedures for shutdown and monitoring the equipment during shutdown led operators to perform actions that sensitized the AN solution and provided energy to the decomposition reaction."
- Formosa Plastics vinyl chloride release, 2004 - "The operator overrode an interlock, which led to a release of hot, pressurized VCM. Formosa Plastics did not have comprehensive written standards…for managing interlocks on the vessels. Employees were unprepared for a major accident at the facility."

These incidents demonstrate faulty understanding of the process and lack of procedural direction, which in turn led to improper decision-making and catastrophic consequences.

4 Treatment of Cognitive Human Error in Process Guidelines and Studies

The CCPS *Guidelines for Preventing Human Error in Process Safety* [Ref. 18] mentions cognitive failure assessment methods such as CBDTM, CREAM and HEART but does not provide much guidance to the user for implementing them. Furthermore, in the ten years since that guidance was issued, the Deepwater Horizon incident has occurred and much has been learned in various industries on the analysis and mitigation of human error using HRA.

The recently issued CCPS *Guidelines for Initiating Events and Independent Protection Layers in Layer of Protection Analysis* [Ref. 1] states that "developing PFD values for human IPLs is not straightforward." Note that PFDs as defined in these guidelines include not only the HEP, but the PFD of the cue for response, namely the "instrumentation or equipment used to detect the process deviation and/or take action to return the process to a safe state."

The IPLs for human actions that are cited in these guidelines are:

- Human response to an abnormal condition
- Human response to an abnormal condition with multiple sensors and indications and the operator has >24 hours to accomplish the required response action
- Adjustable movement-limiting device – this is actually equipment that is intended to prevent operator misoperation of equipment beyond a limit that would initiate a scenario of concern
- Personal Protective Equipment (PPE) – safety gear that acts as a barrier between a person and a hazard.

In an HRA context, the first two IPLs would be considered Human Failure Events (HFEs), but the specification of sensors/indications (cues) and time available for response would be evaluated in detail for both cases, in addition to other scenario-specific PSFs, such as procedures and training. Most importantly, the time available would be balanced against the time *required* for the action to assess the time margin and the HEP.

Note that the *nature* of the human response is not specified for these IPLs, so there is no differentiation between cognitive and execution errors, which have different aspects and methods for evaluation.

The adjustable movement-limiting device IPL would not itself considered to be a human failure, but would be considered as a mechanical form of interlock that would need to be defeated (either through another human action to leave it out of position after maintenance or through equipment failure) to allow the human error to occur. A Quantitative Risk Assessment (QRA) model allows such equipment and human failures to be functionally linked.

PPE is addressed in HRA not as an HFE but as a performance shaping factor on an existing human response action since it is considered to cause sight, sound and manipulation limitations on the human action, and can even lead to increased stress.

Papers presented in recent years discuss application of HRA to the petroleum industry through the use of the SPAR-H quantification method [Refs. 19 & 20]. Gould, Ringstad and Van De Merwe state that "The main advantages of SPAR-H is that it provides clear guidance on assigning numerical weights to the performance-shaping factors, is not very resource-demanding, and includes worksheets that promote analyst consistency." While SPAR-H provides a useful screening analysis of factors related to cognitive error, such as Available Time, Procedures and Ergonomics, it would seem to be best considered as the first step toward identifying risk-significant cognitive human errors, which would then need to be evaluated in greater detail. The benefit of detailed HRA lies in identifying the driving mechanisms behind the error that can then be applied to steps for error likelihood reduction.

Another current study [Ref. 21] has successfully applied the CREAM cognitive error evaluation technique to the cargo loading process of LPG tankers "to systematically predict human error potentials for designated tasks and to determine the required safety control levels on-board LPG ships." The human cognition and action context are determined in accordance with Common Performance Conditions (CPCs) which provide a well-structured and comprehensive basis for characterizing the circumstances and to define sets of possible error modes and error causes.

5 Information Processing Model for Cognitive Activities

Assessment of HFEs can be guided by a model of higher-level cognitive activities, such as an information processing model. The model that has been used by the authors is based on the discussion in Chapter 4 of the ATHEANA method [Ref. 22] and consists of the following elements:

- Monitoring and detection - Both of these activities are involved with extracting information from the environment. Also, both are influenced by the characteristics of the environment and the person's knowledge and expectations. Monitoring that is driven by the characteristics of the environment is called data-driven monitoring. Monitoring initiated by a person's knowledge or expectations is called knowledge-driven monitoring. Detection can be defined as the onset of realization by operators that an abnormal event is happening.

- Situation awareness - This term is defined as the process by which operators construct an explanation to account for their observations. The result of this process is a mental model, called a situation model that represents the operator's understanding of the present situation and their expectations for future conditions and consequences.

- Response planning - This term is defined as the process by which operators decide on a course of action, given their awareness of a particular situation. Often (but not always) these actions are specified in procedures.

- Response implementation - This term is defined as the activities involved with physically carrying out the actions identified in response planning.

The final bullet concerns the execution of the response, but is predicated upon the successful cognitive assessment of the course of action.

By assessing these features of a human action, the analyst can then select from the available HRA models to evaluate the HEP and moreover, identify the driving issues leading to error that can be factored into error prevention.

6 Conclusions

Recent studies are showing that existing HRA methods can be applied to assess cognitive human errors in the process industries, but these insights have yet to be incorporated into industry guidelines. While initial assessments can be made with simplified tools, the key to addressing and preventing major incidents is the identification of underlying human error mechanisms so that risk-mitigating solutions that truly address the diagnosis and decision-making process can be developed.

7 References

[1] *Guidelines for Initiating Events and Independent Protection Layers in Layer of Protection Analysis*, Center for Chemical Process Safety, American Institute of Chemical Engineers, New York, NY, 2015.

[2] Collins, E.P. and Fragola, J.R., "Advances in Human Error Evaluation", 1995 Reliability, Availability, Maintainability Symposium (RAMS), Washington, DC, January 1995.

[3] Umatilla Chemical Agent Disposal Facility Quantitative Risk Assessment, Report No. SAIC-00/2641. Volume I, Science Applications International Corporation (SAIC), Abingdon, MD, 2002.

[4] Yucca Mountain Project Canister Receipt and Closure Facility Reliability and Event Sequence Categorization Analysis, 060-PSA-CR00-00200-000-00B, Bechtel SAIC Company, LLC (BSC), November 2008.

[5] Garvey, M. (Dept. of the Navy), Joglar, F. and Collins, E.P. (JENSEN HUGHES, Inc.), "HRA for Detection and Suppression Activities in Response to Fire Events", 2014 Annual Reliability and Maintainability Symposium (RAMS), Colorado Springs, CO, January 2014.

[6] ASME/ANS RA-Sa-2009, Addenda to ASME/ANS RA-S-2008, "Standard for Level 1/Large Early Release Frequency Probabilistic Risk Assessment for Nuclear Power Plant Applications", The American Society of Mechanical Engineers (ASME), New York, New York, 2009.

[7] Electric Power Research Institute and U.S. Nuclear Regulatory Commission, EPRI/NRC-RES Fire HRA Guidelines, EPRI 1023001/ NUREG-1921, July 2012.

[8] Electric Power Research Institute, The EPRI HRA Calculator® Software Manual, EPRI 3002004030. Palo Alto, CA, May 2014.

[9] EPRI TR-100259, An Approach to the Analysis of Operator Actions in Probabilistic Risk Assessment, Electric Power Research Institute (EPRI), Palo Alto, CA, June 1992.

[10] Swain, A. D. and Guttman, H., Handbook of Human Reliability Analysis with Emphasis on Nuclear Power Plant Applications, NUREG/CR-1278, Washington, DC: U.S. Nuclear Regulatory Commission, August 1983.

[11] Swain, A., Accident Sequence Evaluation Program Human Reliability Analysis Procedure, NUREG/CR-4772, U.S. Nuclear Regulatory Commission, Washington, DC, February 1987.

[12] The SPAR-H Human Reliability Analysis Method, NUREG/CR-6883, Idaho National Engineering Laboratory for U.S. Nuclear Regulatory Commission, Washington, DC, August 2005.

[13] Hollnagel, E., *Cognitive Reliability and Error Analysis Method – CREAM*, Elsevier, New York, NY, 1998.

[14] Williams, J., HEART - A Proposed Method for Assessing and Reducing Human Error, 9th Advances in Reliability Technology Symposium, National Centre of Systems Reliability, University of Bradford, Bradford, UK, 1986.

[15] Deep Water: The Gulf Oil Disaster and the Future of Offshore Drilling, Report to the President, National Commission on the BP Deepwater Horizon Oil Spill and Offshore Drilling, January 2011.

[16] Hubbard, A. and Embrey, D., "Deepwater Horizon – Summary of Critical Events, Human Factors Issues and Implications," Human Reliability Associates, Ltd., 23 September 2010.

[17] Ness, A., "Lessons Learned from Recent Process Safety Incidents," CEP Magazine, American Institute of Chemical Engineers (AIChE), New York, NY, March 2015.

[18] *Guidelines for Preventing Human Error in Process Safety*, American Institute of Chemical Engineers (AIChE), Center for Chemical Process Safety (CCPS), August 2004.

[19] Gould, K.S. and Ringstad, A.J (StatOil) and Van De Merwe, K. (Det Norske Veritas), "Human Reliability Analysis in Major Accident Risk Analyses in the Norwegian Petroleum Industry", Proceedings of the Human Factors and Ergonomics Society 56th Annual Meeting, San Diego, CA, October 2012.

[20] Boring, R.L (Idaho National Laboratory) and Øien, K. (SINTEF), "Defining Human Failure Events for Petroleum Risk Analysis", Probabilistic Safety Assessment and Management PSAM 12, Honolulu, HI, June 2014.

[21] Akyuz, E. and Celik, M., Application of CREAM human reliability model to cargo loading process of LPG tankers, Journal of Loss Prevention in the Process Industries, Vol. 34, pgs. 39-48, Elsevier Ltd., 2015.

[22] Technical Basis and Implementation Guidelines for A Technique for Human Event Analysis (ATHEANA), NUREG-1624, Rev. 1, U.S. Nuclear Regulatory Commission, Washington, DC, May 2000.

442486 Critical Considerations in the Vacuum Protection of Butane Spheres

Abstract:

Vacuum protection of pressure vessels, including spheres, is not well defined in the ASME Boiler and Pressure Vessel Code (BPVC) Section VIII, Division I or API Std. 520. Vacuum protection of pressure vessels usually follows API Std. 2000 on low pressure tanks for guidance.

Vacuum can occur on butane storage spheres at temperatures of 25 to 30 F due to butane condensing. This paper addresses the challenges in vacuum protection of butane storage spheres, including both the positive pressure rating of vacuum devices and sizing of vacuum vents for a butane condensing scenairo.

442493 Assessing Risk in the Supply Chain Using the Bowtie Method

Abstract:

After a number of recent incidents involving the transportation and storage of hazardous materials, many industry end-users wish to understand their level of risk exposure during the supply chain. This exposure can either be via actual ownership or when the product can be associated with the concerned party. Ultimately, this may mean that an end-user's procurement group will want to see that their suppliers of products and services are performing as needed in the control of these risks.

A qualitative risk assessment is a technique by which these risks can be identified, understood, and evaluated. This paper will describe the application of one such method, where the bowtie barrier analysis is central to the means by which this assessment is completed. The barriers identified are those necessary to prevent and mitigate undesired events that could result in a fatality or a catastrophic incident to the environment, public, or a company's reputation.

The overall method employed in this process includes the following steps:

1. Hazard identification
2. Bowtie diagram development and barrier identification
3. Barrier screening to identify safety critical elements - those that provide a credible reduction in the risk of fatalities or other catastrophe
4. Performance standard development for the safety critical elements

A quantitative risk assessment can also be completed to provide additional risk knowledge to aid in prioritization and decision-making. The performance standards are then enforced appropriately throughout the supply chain. This will ensure that the transportation and handling of hazardous materials meet the purchasing company's expectations with regards to proper risk management.

442386 Reference Framework for the Application of Quantitative Risk Analysis for Hydrocarbon Pipelines, Coupled with Uncertainty Treatment Methods: Techniques for Analysis and Modelling Methods

Abstract:

Risk assessment is imperative to reduce consequences of accidents generated by leaks on pipelines. After a failure in an urban pipeline, the contents can be released rapidly forming a spreading pool in the surface or a vapor cloud that can lead to different events such as fires and explosions, affecting the population in the area. Generally, deterministic assessments are carried out where certain assumptions are made in the models used to calculate failure and event probabilities, event consequences and risk values. This causes the existence of uncertainty that is not considered and therefore it generates results only for "mean" or "worst-case" scenarios. Not taking uncertainty into account can lead to decisions with an unknown degree of conservatism: overestimate results could increase cost of decisions and underestimate results can lead to poor risk management. There is a lack of a clear directive to use methods of uncertainty assessment in risk analysis and for that reason, a framework is needed for carrying out quantitative risk assessment coupled with uncertainty treatment methods in order to handle, reduce, and properly inform uncertainty to the decision makers.

Uncertainty appears when we try to represent a system using different analysis techniques and therefore, depends on how accurately this process can be done. This in turn relies on how complex is the system under analysis and which techniques can be used. As part of the framework mentioned above, a classification is suggested to adequately select methods in probabilistic risk analysis to take into account different types of uncertainty in different kinds of representation and system analysis techniques. The classification is first based on the complexity of the system to be analyzed and then in the complexity of the modeling technique that is used to deal with system complexity. System complexity depends mainly on three variables: 1) understanding about the system (variables and its interactions) 2) the number of elements in the system 3) complexity of system interactions. Depending on system complexity, different techniques to represent it and analyze these systems in risk analysis are proposed and then, methods to propagate uncertainty are mentioned. A comparison of methodologies is presented to provide the engineer with tools to use adequately these methods in different situations and provide guidelines to deal with uncertainty in risk assessment.

Dynamics Operational Risk Management in Organizational Design, the Challenge for Sustainability

Salvador Ávila Filho
Universidade Federal da Bahia, UFBA
Rua Aristides Novis, Federação 02, Salvador Bahia
avilasalva@gmail.com

José Filipe Michel Gagliano Ferreira
Universidade Federal da Bahia, UFBA
Rua Aristides Novis, Federação 02, Salvador Bahia
gagliano.fm@gmail.com

Carlos Roberto de Oliveira Sousa
Universidade Federal da Bahia, UFBA
Rua Aristides Novis, Federação 02, Salvador Bahia
carlosososousa.cs@gmail.com

Ricardo Araújo Kalid
Universidade Federal do Sul da Bahia, Itabuna
Ricardo.kalid@gmail.com

Prepared for Presentation at
American Institute of Chemical Engineers
2016 Spring Meeting
12th Global Congress on Process Safety
Houston, Texas
April 11-23, 2016

Dynamics Operational Risk Management in Organizational Design, the Challenge for Sustainability

Salvador Ávila Filho
Universidade Federal da Bahia, UFBA
Rua Aristides Novis, Federação 02, Salvador Bahia
avilasalva@gmail.com

José Filipe Michel Gagliano Ferreira
Universidade Federal da Bahia, UFBA
Rua Aristides Novis, Federação 02, Salvador Bahia
gagliano.fm@gmail.com

Carlos Roberto de Oliveira Sousa
Universidade Federal da Bahia, UFBA
Rua Aristides Novis, Federação 02, Salvador Bahia
carlososousa.cs@gmail.com

Ricardo Araujo Kalid
Universidade Federal do Sul da Bahia
Itabuna
ricardo.kalid@gmail.com

Keywords: business risk and corporative, human factors and organizational, tools, sustainability

Abstract

The maintenance of the organizational sustainability requires acknowledgment that business risks are dynamic and, therefore, depend on the balanced analysis of issues relating to product and process with people and culture. The analysis of the dynamic risk occurs in three steps as described as follows. The first analyzes the balance between the human and social factors with process and product factors, generating preliminary recommendations from probable scenarios through techniques such as LOPA and PRA with recommendations to avoid accidents and technical failures. Risks related to the environmental and human aspects are analyzed with PRA-SH to anticipate future events that may cause technical failure or evoke a stressor environment in the workplace. The second step requires knowledge of the organization in its aspects of management, technical and non-technical, inputs and productivity benchmark to define the assumptions and criteria used to project performance of operations, including the skills required by the business. The result of this study is the definition of technical, human, organizational and natural factors for the sustainable operation of production system. The techniques applied in this second stage are: HAZOP and HAZOP-Social; requirements for the task (GMTA, PADOP); events cause and consequence analysis; and analysis of barriers for events (BOWTIE). In the third stage of dynamic risk analysis rescues the criteria and assumptions established in the previous step by defining the management model and detailing the technology used. In this phase, it will be handled operational routines, asset control, control of failures and possible crisis. Thus, it will designed the systems of technical area (equipment, processes, products and control modes), interface (task, management, and human-machine interface), and social area (culture and biases, cognitive decision, stress analysis and behavior). The factors that make up the control of operations are analyzed in the standard procedure way and in the failure analysis using the following techniques: equipment (FMEA, FMEAH); product and process (monitoring and statistical control); task in complex mode (GMTA, CARMAN, PADOP); man-machine interface and control room (PRA-SH and Regional Criteria for control room design); culture (human reliability committee for organizational change and archetypes analysis); cognitive decision (study of cognitive failure); stress and decision (LOPA); and, stress and occupational health (LESHA). These techniques for analysis of operations in a dynamic way have metrics based on the operator's speech from the results of routine by reflecting the occupational consequences (accident), environmental (pollution and waste), climate (CO_2 emissions) and economic (competitiveness losses based in the rates of consumption and productivity). The result of these measurements allows the routine adjustments in the design of operations, the organization's project settings and business project. A deployment exercise will be proposed for the chemical industry. Diagram will be presented reflecting the dynamic in the management of operational risks.

Glossary

ANP - National Petroleum Agency
FMEA – Failure Mode and Effect Analysis
FMEAH – Human Failure Mode and Effect Analysis
FPSO – Floating Production Storage and Offloading
FTA – Fault Tree Analysis
GMTA – Goals-Means Task Analysis
HAZOP – Hazard and Operability Study
HAZOP-Social – Social Hazard and Operability Study
HSE – Health, Safety and Environment
LESHA – MultiLayer Stress Effect on Health & behavior Assessment
LODA – MultiLayer Operation Decision Assessment on Progressive Stress
LOPA – MultiLayer Operation Protective Assessment
MDA – Methyl Diphenyl Amine
PADOP – Patterns and Procedures in the Operation
PHR – Preliminary Hazard Analysis
PRA – Preliminary Risk Analysis
PRA-SH – Socio-Human Preliminary Hazard Analysis
THERP – Technique for Human Error Rate Prediction

1 A discussion on reliability and risk – an introduction

1.1 Industrial risk analysis

The initial thought on the issue of risk and subsequently his management was primarily driven by the financial sector, specifically post-World War II [1][2]. However, in parallel with accelerated development of industry and processes increasingly complex - which implies exposure to workers, society and the environment to hazards ever growing - this issue gained notoriety in the industrial environment.

The occurrence of major industrial accidents in Brazil, and around the world, became the subject Risk Analysis and Management more popular to meet the demands of society and governments about the safety of industrial processes. It is in this context [3] that is needed better management and mitigates the impacts of accidents. One reason for the importance of developing risk management was the change in constraints for operating license after the Seveso accident in Italy (1976).

Schneider and Martins [4] define risk as a result of the frequency of occurrence and the consequences generated by an unwanted event, so that risk reduction is by reducing the frequency of undesirable events and the mitigation of the consequences. The risks that can be quantified are risks of injury to the person, damage to property and environment, and risk of loss of reputation. Risk management depends on the level that is being analyzed: equipment, process unit, facilities, and other levels.

According to Bharatiya [5], an industrial hazard example is the handling of flammable products that can be released causing fire and explosion and resulting impacts on workers and third parties, as well as economic losses such as: profit loss caused by shutdown plant; damage to equipment; penalties for release of toxic material, etc. In this regard, it is necessary application of qualitative and quantitative techniques of risk analysis and its review to ensure a safe working environment for employees and neighborhood.

Qualitative techniques of risk analysis such as PHA/PRA (Preliminary Hazard Analysis / Risk) and HAZOP (Hazard and Operability Studies) allow identifying and assessing risks and hazards associated with industrial processes, and identify operational problems through scenarios identification. According to Fleming and Garcia [6] PHA/PRA technique performs a risk assessment for each identified scenario, develops mitigation measures and specific studies to calculate the probability and analyzes the effects through the vulnerability of these scenarios. The HAZOP allows analyzing risk scenarios using a procedure that generates questions and answers in a structured way by guiding words that enable the identification of possible deviations of the operation. The review team examines the links between the causes and consequences of these deviations.

Regarding the semi-quantitative techniques for risk assessment, there is a tool for layers of protective analysis (LOPA) that is used in the safety integrity level setting against unwanted events in safety instrumented system [7]. His concept was first treated by Center for Chemical Process Safety [8].

In quantitative analysis, the various methods are distributed from the analysis of consequences, which is measured the extent of damage caused by adverse events, to quantify the frequencies and probabilities of each event causing accidents [4]. Among the most widely used techniques are Vulnerability, Fault Tree Analysis (FTA) and LOPA. The techniques can be combined in a study of quantitative risk analysis.

The development of quantitative criteria can follow internationally recognized parameters. In general, the aim should be to develop criteria also based on the existing company culture and the most effective process in this direction would be a hybrid of internal and external internationally accepted criteria [9].

1.2 Risk analysis services

Brazil has been placing on a world stage major events that generate therefore great responsibilities. Mega-events such as road races, international shows, World Cup soccer (2014) and the Olympics (2016) demand new ways to manage the risks. Experiences with carnivals indicate the need for organization in public management to meet the security needs of the population, quality in the health area, mobility, and social life.

For each event has a positive impact on society and economy, it is important to have a control of risks in the event of scenarios. According to Correia [10], the factors that pose major difficulties in quality control and risks in the event derive from the fact of being subjective experiences and difficult to measure.

Each event is unique and presents its own security risks. Major events bring large crowds with factors to be analyzed for risk management [11]: demand complex planning with different stages in the design and operation; have different participating organizations that serving people of different customs; It works with many volunteers and inexperienced staff; It has national and international media projection; demands less than the time required for the complexity of the services; It has regional, state and national economic impact; and requires careful consideration for risk management.

For large events are ephemeral, not admit failures, delays, and its facilities and spaces are in most provisional, at the same time these experiences are characterized by subjectivity, It is more efficient to analyze the risk qualitatively [10]. In this way the analysis generates a more tangible and simple approach, easy to study and implementation by the organizing companies.

1.3 Integrating the reliabilities: equipment, operation and human

The first steps in the study of reliability were been conducted in the early twentieth century (the end of the World War I), but the concept of reliability as it is treated today only began to take shape during the World War II in an attempt to improve the reliability of missiles [12]. For this classic model of reliability, Kardec and Nascif [13] define reliability as "the probability of the system or device to perform the function successfully required, under specified conditions of operation and a certain time".

The development process of reliability engineering was important in product development and process improvements. In the latter case your application has a strong relationship with the evolution of maintenance management acting not only in the transition between corrective, preventive and predictive models, but above all, the definition of maintenance strategies where each model has a degree of importance that depends on the results of investigations into the system failure patterns. That is, even if the corrective, preventive and predictive models represent, respectively, the 1st, 2nd and 3rd phase of maintenance management, any one of them may be applicable depending on the operational context of the organization and its failure patterns. The concept of reliability then goes to lead the efforts for efficient maintenance.

To meet the demands of analyzing the nature of the failure and its consequences several tools were developed to improve the reliability of equipment. The most used are: Failure Mode and Effect Analysis (FMEA), and Root Cause Failure Analysis, Method of Analysis and Troubleshooting and Reliability-Centered Maintenance.

However, while many companies design the best systems based on this reliability model, taking into account the great difficulty in maintaining the availability and reliability of equipment and machinery at satisfactory levels and the cases of accidents with environmental impacts and fatalities, it is clear that such practices are not sufficient to maximize system reliability and/or guarantee the effectiveness of crash prevention barriers. Therefore, in order to fill this gap, many studies have been conducted to include the role of human in this process, what we call human reliability analysis. These studies began to gain strength from the sixth annual meeting of the Human Factors Society which had as its theme the quantification of human error. So was born the first formal method for human reliability analysis – the THERP technique [14].

According to Pallerosi *et al.* [15], human reliability can be defined as "the probability that a person does not fail to comply with a task in a given period of time, in appropriate environmental conditions and available resources to do it". The author also classified two types of procedures to quantify the human failures. One is called the prediction methods and includes tools as TESEO, HEART, SPAR-H, THERP, ASEP, MPPS and others. The other method is classified as evaluation methods and use theories and probability distributions as Gamma, Weibull, Exponential, Binomial and other distributions applied to each type of task analyzed.

The elements or factors that contribute to human error happen are related to the cognitive and operational processes. These elements of the cognitive process of decision-making are: perception, memory, abstraction, emotions, stress, fatigue, knowledge, skills, communication, tasks, procedures, organizational culture etc. Ávila [16] discusses these factors in the operator's performance in the control of socio-technical processes in task execution and learning stage. These factors can be measured by individual results in the task accomplishing, from the quantitative approach presented in verification methodology of learning parameterized (RGA - Reliability Growth Analyses).

The human factors analysis allows the projection and the calculation of risk and also the calculation of human reliability. Among the analytical techniques are: HAZOP SOCIAL, PRA-SH, FMEAH, LODA, etc. In the organizational dimension, the API 770 standard [17] aims to assess the company culture and the impact on human and operational reliability diagnosing aspects of policy, procedures, man-machine interface, and issues relating to the roles and task settings. The latter issues are present in David Embrey (2000), Frank Lees (1996), Rasmussen (1997) and Avila studies (2009).

The understanding of cognitive processes is essential to increase the human and operational reliability in organizations. Study the social, economic and emotional utility [16], cognitive function [18][19] and models for decision-making [20] are required to analyze the commitment and quality of the decision.

Therefore, approaches to socio-technical areas of reliability should not be an isolated study for each functional part of the industry [21]. The reliability of each part is interconnected with each other and their level of integration depends on the level of automation, the type of process, the manager's management style and the environment promoted by the organization. By understanding about the relationship between the functions of these systems makes it possible to study the integrated reliability. This knowledge enables the formatting of factors that increase the performance of staff and equipment leading to greater organizational efficiency.

1.4 Routine management and crisis management

The failure analysis and reliability project should be carried out at various stages of the project lifecycle - from conceptual design (detailing and assembly), through operation (control of operations and routine) to crisis management. The study of fault begins with the perception of signals from the process, people, equipment and tasks and advances in the analysis of incidents and / or accidents that cause financial loss (or image). These events, if not treated properly, can lead to loss of control in a chain reaction and It requires contingency actions not always known by the leaders, characterizing a crisis situation.

According Constantinides [22], crisis management is a systematic attempt to identify and detect potential crisis scenarios and take action to prevent them, contain its effects or interruption, and finally the recovery of the system normality. Muller [23] highlights the need for early warning systems projects of undesirable events and resolve them before they turn into a real crisis. He added that in case of failure of preventive measures is essential to take an immediate action.

2 Performance of GRODin and publications

The Group of Dynamics and Operational Risks (Grodin) located at the Federal University of Bahia intends to conduct diagnostic systems from applied research and works to develop solutions to complex problems in industrial sectors and services over the integration of academy, industry, and Public sector. Its constitution allows multidisciplinary making the investigative process dynamic starting in the subjective transformations of human and social behavior until the objective accomplishing of the task in industrial facilities with the search of metrics for decision making in organizations.

Currently (2015) the group consists of 25 masters, 5 doctoral and 10 graduate students with graduate completion forecast for masters and doctors around 10 per year. Large areas of concentration of Grodin's publications are: (a) Process Safety; (b) Human Factors; (c) Reliability and Risk analysis in socio-technical systems; (d) Energy Efficiency and Sustainability.

(a) Process safety

Publications performed in process safety seek always discuss the inclusion of human and organizational factors among the aspects considered in the accident process. Thus, several articles were discussed and presented at conferences as described: (a1) risk analysis of the factors that define human performance for the tasks in equipment and processes classified as critical [24]; (a2) the analysis of the performing task environment, the relationship between man and technology that result in failures and the discussion about leadership and the moment of decision [25]; (a3) analysis of underlying causes (human error) which are involved in the accident process (economic bias, risk management, conflicts between policy and practice, cultural traits, task planning, contracted management) [26]; (a4) analysis of decisions of the operation under progressive stress on the sinister situation [27]; (a5) analysis of high frequency and lower impact accidents considering the inclusion of human factors and complexity of the task [28]; (a6) analysis of social relations in oil platform and the implications for an enabling environment to an accident [29]; (a7) analysis of organizational change, drafting checklist, impact analysis of possible process safety events and the need for a more detailed discussion of the human factors [30]; and (a8) risk analysis of public services in cultural mega events (carnival) [31] and sporting events (World Cup) [32].

(b) Human factors

The publications performed on Human Factors always tries to discuss the analysis of the intrinsic variability of the man in an organizational and cultural environment and that brings consequences as gains or losses for the society and industry. Thus, several articles were discussed and presented at conferences as described: (b1) usability of procedures based in operators opinion of the fertilizer industry [33]; (b2) complex task evaluation taking into account the requirements of criticity [34]; (b3) cognitive deficit analysis during the performance of executive function in performing tasks [35]; (b4) social assumptions that change cognitive decision for the review of ergonomics standards in the control room [36]; (b5) analysis of archetypes and their derivatives to avoid improper behavior on safety of industrial plants [37]; (b6) Operator behavior analysis and leadership in emergency situations, high stress caused by chain reactions [27]; (b7) trace analysis of the Brazilian and global culture, biases, and bad habits that hinder the establishment of operational discipline [38]; (b8) stress analysis in industry, health care and diving and its consequences on the task, behavior and health [39]; (b9) changing operator behavior based on social dynamic culture, difficult to detect changes before impacting events happening like accidents with economic, human and environmental losses [40].

(c) Reliability and risk in socio-technical systems – complex processes

Publications performed in Reliability and Risk Analysis in complex processes discusses how to detect and calculate the risks and failures always considering the presence of human performance factors. Thus, several articles were discussed and presented at conferences as described: (c1) relationship between loss of industrial processes and connection to the cognitive processes with impacts measurement [24]; (c2) the operational failure exergy relating the energy loss in cognitive failure with the loss of energy in the industrial process heat systems [41]; (c3) evaluation of failure in complex processes, complex tasks and complex social relations in the chemical industry [34]; (c4) dynamic risk analysis for industrial sustainability [42]; (c5) human factors and equipment failures analysis using tools such as SPAR-H and Weibull ++ in the nuclear and chicken industries [43][44]; (c6) Analysis of socio-technical reliability in the refining industry [21]; (c7) risk analysis of the pre-salt considering: phenomenology, reliability model and risk factors [45][46].

(d) Energy efficiency and sustainability

Publications performed in energy efficiency and sustainability seek to discuss what are the opportunities to reduce consumption of energy and water and ensuring the sustainability of business. Thus, several articles were discussed and presented at conferences as described: (d1) mass balance and energy in cooling towers [24]; (d2) audit in organic effluent and rainwater to prevent process losses; (d3) audit of environmental deviations in industrial routine [47]; (d4) contamination control program at source by mapping effluent [48]; (d5) application of mass balance techniques and energy to identify opportunities to reduce energy and water losses in the industry; (d6) a discussion of systems and cooling towers, challenges to avoid energy losses and water; (d7) identification and calculation of thermal performance and energy efficiency in equipment and industrial processes [49]; (d8) Analysis of loss of condensate and steam heating systems and energy recovery [50].

3 A vision of sustainable organization based on risk

Maintaining organizational sustainability requires an understanding that the business risks are dynamic and the operational risks depend on a balanced analysis of issues related to product and process with the people and culture issues. So, to start the discussion on sustainability in high-risk industry a management model will be presented.

An efficient risk management of business develops premises to organizational project considering social-technical aspects. This risk management draws an industrial operation format including multidisciplinary analyzes of different levels of normality (default to failure and the crisis). It is intended with this model to measure the effects and impacts on society and then review the criteria and procedures adopted by project operations.

3.1 Inputs in the Risk Management and Organizational Design

Important inputs for the analysis of dynamic risk are: entrepreneur demand (vision and mission), technological conceptual design (product and process), analysis of the availability of resources and production factors (energy, materials and people), and external demands represented by the law and behavior of the stakeholders in the existence of this business. The risk analysis stage to address the dynamic risk intent to insert a discussion about the process and product taking into account the type of task (or the procedure), the cognitive models for decision (local or multicultural) and the social behavior of internal and external community of the organization.

3.2 Step one: Risk Management - socio-technical risk analysis

The first step of socio-technical risk analysis balances human-social factors with process and product factors generating preliminary recommendations. In the case of technical risks are evaluated the product, process and the likely scenarios through LOPA and PRA and getting as result recommendations to prevent accidents and technical failures. For non-technical risks such as environmental and human aspects is performed an PRA-SH to anticipate future events related to human errors that can cause technical failure or generate an stressor environment at the workplace. At this time are considered the risks of events that involving cognitive faults and failure on the task resulting from social processes.

3.3 Step two: organizational project - detailing of technical and non-technical project

The second step requires knowledge of the detailed organization design in the technical and non-technical aspects to define the assumptions and criteria used for the design of operations. Thus, the organizational project aims to open the product technology, process, service, and control to discuss which the necessary demands on leadership, behavior, attitude, and skills.

The intrinsic variability policies, humanities, social and variability of nature (climate, biota, etc.) are scored as important for maintaining the environment in which the industry is installed and intends to stay in a sustainable manner. The organizational design analysis requires detailed information of technology, knowledge in integrated management system, consumption analysis and energy benchmark, water, materials and people (productivity) to establish performance indicators. The result of these studies is the definition of technical, human, organizational, natural factors for the sustainable operation of the production system. The techniques applied in this second phase are: HAZOP, HAZOP-Social; requirements for Task (GMTA, PADOP); cause and effect analysis of the events; barriers analysis of events (BOWTIE).

PADOP is a method that starts with the analysis of the requirements for feeding the following phases: environment, planning and execution of the task. Others steps such as task control and conformity analysis under review are performed in operational routine (see item 3.4).

3.4 Step three: design and control of operations

In the third stage of dynamic risk assessment is intended to rescue the criteria and assumptions set out in the previous step, organizational design, where technology is detailed and an integrated management mode is defined. In the project of operations, it will be handled operational routines, asset control, and control of failures and possible crisis. So, it will be designed systems for operation in the technical area (equipment, processes, products and control modes), the interface (task management, and human-machine interface) up to non-technical area (culture and biases, cognitive decision and stress and behavior analysis). The results of these measurements allow adjustments on the routine of the design of operations and business and organization projects.

These factors and elements that make up the control of the operations are analyzed on standard procedure mode and failure analysis in the following areas and using the following techniques: equipment (FMEA, FMEAH); product and process (monitoring and statistical control); task in a complex way (GMTA, CARMAN, PADOP); man-machine interface and control room (PRA-SH and regional criteria for control room projects); culture (human reliability committee for organizational change and archetypes analysis); cognitive decision (study of cognitive failure); stress and decision (LOPA); and stress and worker health (LESHA). These techniques for analysis of dynamically operations present measurements and indicators that are based on operator's speech. This speech is announced from the routine results and reflects the occupational consequences (accident), environmental (pollution and waste), climate ($CO2$ emissions) and economic (loss of competitiveness based on consumption rates and productivity).

In this case, the PAPOD technique intent to compare the standard designed for task control rates with measurement of the process after the task performed. It is worth noting that these standards include variables of integrated management (safety, quality, environment, occupational health, etc) as well as socio-humans aspects (quality of communications). In case of deviations it will be realized preventive corrections with necessary revision to the task or written procedure.

4 Application of exercise

An implantation exercise is proposed based on the application of different technical and management systems in research in various chemical process industries. Activities and analyzed processes involve energy products (oil and LPG), chemical (sulfuric acid, methyl diphenyl amine – MDA, and explosives), petrochemicals (basic petrochemicals and fertilizers), polymers (polycarbonates) and plastic (plastic bottles).

These various research opportunities involve aspects of routine at the work place and characteristics of management that promote the company's results close to normal or fault situations. The experiences are gathered in an integrated way toward sustainability considering that risks are dynamic and that the systems are socio-technical.

Some of these techniques are applied with quantitative results indicating the intensity of the problem and the volume of investment to be taken and others indicate the existence of the failures and suggest the construction of safeguards, but in a qualitative way.

This exercise will only treat of techniques relating to the socio-environmental-human risk and it will quote only within the management system proposed and which evaluations (of process and product) should be interconnected.

4.1 Inputs in the risk management and organizational design

For this model that considers the dynamic risk, the entrepreneur discusses the social variability in the possibility of failure and practices a vision and mission more sustainable, which admits the existence of a multidisciplinary approach to the definition of the operating pattern. This discussion is part of the control of the operational routine. So, inputs to this management system from experienced cases it is suggested.

Entrepreneur demand

Case 1: the entrepreneur demand to identify the routine deviation is a requirement to draw up barriers that prevent systemic failure. To discuss the causes of failures that lead the operation to a mode of low productivity, an entrepreneur of the LPG industry recognizes the need to analyze the routine. The operator's speech feeds the construction of the fault tree for routine events and together with the task analysis identify the GAPs of competence that lead to failure [51].

Case 2: the perception of the manager must be attentive to behavioral deviations that can lead to accidents. In this case, the corporate HSE manager of the explosive industry believes that the study of human typology is necessary to change the criteria in risk management. This belief is based on the possible change of behavior, which caused two consecutive accidents not expected in different countries, after a history of 20 years without accident. This entrepreneur corroborates the following statement: "cultural changes alter the behavior of the workers and there is a need to study new criteria for dynamic risk management".

Analysis of the availability of resources and production factors (water, energy and people)

Production factor: energy. The water crisis in the state of São Paulo [52] and the reduction of energy quotas for petrochemicals due to its unavailability caused by climate change indicates that the analysis of availability is critical input in the organizational design to address the variability of nature.

Production factor: water. CEOs of Brazilian refineries are being instructed by National Agency of Water (for now voluntarily) for an actual 20% reduction in water consumption. This highlights the urgent need to change behavior and decisions in this organizational design.

Production factor: people. The opening speech of the Brazilian Association of Risk and Reliability (ABRISCO) indicates that the Brazilian engineer does not know and does not quantify industrial risks indicating the unavailability of appropriate skills to quantify investments in safeguards to prevent accidents.

External demands legislation for new behaviors of stakeholders

Stakeholder – National Petroleum Agency (ANP). After the FPSO production ship accident in the city of São Mateus, an investigation realized by the ANP [53] indicates the need to discuss the issues of contract management, omissions, commitment, crisis management and decision in stress. These issues are part of social- human variabilities.

Stakeholder – University which reflects demand for workers. In Risk Management is important to define the requirements necessary from technology and local culture for planning the task (PADOP). From the master's degree studies on the task analysis it was found that most of the 450 operators of a fertilizer industry accepted the importance of establishing social and human premises to carry out the critical task.

Stakeholder – Multinational oil companies. From international meetings with reliability consulting groups that serve the oil industry, it was found out the need to discuss regionalities to change the ergonomic criteria in control room. This indicates that the CEOs of major companies (Chevron, Statoil, BG) are reflecting on the question of risk review to include in their analysis the dynamic and social aspects (cultures and people).

4.2 Step one: risk and opportunity management - socio-technical risk analysis

After the hazard analysis and technical risks by conventional application of the techniques APP and APR (to business, processes and products) it is natural to use the vulnerability analysis in the most critical scenarios to indicate the level of impact on human and nature. Similarly to analyze the social and human risk seeks to detail the various performance shaping factors [54] that indicate the level of consequence in case of human errors.

Fault analysis (in 20 cases) in the chemical process industry indicates that elements such as organizational climate, leadership, skills, knowledge, and decision-making processes are related to human error in performing the task.

Several cases involving deviations, failures, incidents and accidents were analyzed and allowed the construction of the anatomy of failure in complex mode relating human and organizational factors [26]. Among the cases are: director's behavior in the plant starting management (polycarbonates); deviations in mitigation of waste generation reaction (MDA); operational control to avoid stoichiometric reaction stop (MDA); skill in the perception of vacuum loss in sulfuric acid unit; failures caused by improper turn change (MDA); investigation of effluent contamination source (MDA, sulfuric acid, lab, fertilizers); diagnosis to map the gaps in the process (MDA, polycarbonates, LPG); negotiation of new operating parameters without causing downtime (MDA); improper use of individual protection equipment and omission by fear of guilt (MDA); oring diameter change at startup leading to failures with solvent losses (polycarbonate); and others.

Based on the investigation of cases the initiators factors to human error are [26]: design criteria, risk management, economic bias, cultural traits, task planning, and conflict between policy and practice. It is also important to identify the social dangers to avoid accidents before assembling the installation and make the hires to operate the industrial site.

A specific case was applied at the stage of design of an oil exploration platform in northeastern Brazil during the graduate student research in the area of security and human factors analysis in 2013 [55]. After seven risk analysis sessions for human and social factors, the group of six experts (socio-technical) listed the following factors that can cause human errors in this installation that come into operation in 2015: difficulties in routine leadership; non-commitment to the safety result; low perception to failure; and improper internal communication.

It is worth listing the technique steps: identification of social and human dangers, social and human risk analysis, risk classification, interaction between risks (interaction matrix to correct risks), the final risk map, and treatment of social and human risk.

An item that is repeated in the workplace discussion refers to the effect of stress on the effectiveness of the task [56] and occupational health [57]. Thus, it interconnects the worker performance factors to the result of task with economic results for the industry. The causal relationship between work environment, cognitive fault and failure in the task are motivating ongoing research that result in techniques and applications.

Technique 1: analysis of social relations in the workplace. Application: oil platform where the rituals of loading and unloading, the distance and isolation are analyzed to list critical topics that can cause stress and human error [58] [29]. Technique 2: protocol for the analysis of stressor environment. Application: industry [56], nursing services [39] and diving services in deep waters. Technique 3: task analysis [33]. Application: chemical process industry. Technique 4: analysis of cognitive impairment [35]. Application: chemical industry.

4.3 Step two: organizational project - detailing of technical and non-technical project

Step 1 of this management model for sustainability is related to the fit project of product, process, and business, in a discussion of technology and economic viability. The analysis of social and human risk in the task is performed along with the APR technique of this future operation.

Step 2 is the organizational design that includes the adaptation of the technology for the future operation that includes product and process aspects, and non-technical issues related to teamwork in performing the task (culture and people). At this stage (detailing phase) are carried out studies of the conceptual design (made in step 1) and details the risks of failure to cause economic loss and / or the company's image.

Application of HAZOP-Social

The process deviation analysis during operation indicates the possibility of danger containment loss related to the industrial activities. The Social HAZOP tool works similarly to the conventional technique [16]. In its application the cognitive decision-making is tested through the deviations and recommendations are announced to prevent failure of attention, memory, competence, managerial decisions and others. Social HAZOP was applied to a draining operation of a decanter vessel in the refining industry as having maximum result the accident with fatality and with the critical lack of attention, centralized management profile and commitment to the task.

Diagnostic of skills gap

Identification of fault root cause in the chemical industry that has characteristics of processes and complex tasks [59] [34] requires the study of process variables and production beyond the qualitative and quantitative analysis of operator's speech [60][61]. This discourse analysis allows the organization to verify the main failure modes in socio-technical systems by locating the critical variables in the tasks, equipment, skills and operator behavior. When comparing the type of fault carried out (based on competence offered) with the demand for knowledge and skills in technology (competence demanded) have been related what are the gaps in competence.

This technique was applied in PhD research [51] of the gas compression unit (LPG) in northeastern Brazil. In this case, the database was collected from one year of statistical analysis of process and production variables, and four months of shift book analysis in low performance periods. Diagnosis of gaps indicates the lack of skills in the following areas [25][62]: Task control (abstraction and logical); concepts (thermal and material exchange); failure (rotary, motorized valve and burner); skills (thermal cycles, pressurized system and logistics); process control and safety (supervisory and risk in the routine).

The gap of skills already identified in the LPG plant may result from lack of commitment of the people, the team cooperation and inappropriate profile of leadership. A classification model of commitment, the leader's behavior and aggregation of the team is suggested by Avila [51][25][62].

Research on behavioral intrinsic variability

Demand for leaders in chemical industry and oil organizations during the GCPSs events indicates the importance of analyzing the changing worker behavior as a result of Dynamic Culture. A methodology is proposed to find the biases of global and regional culture that promote behavioral changes towards the accident [40]. This method was structured to identify potential changes in teams of explosives chemical company with unusual situations accident in facilities overseas. The suggested steps with the respective models are: (1) analysis of social culture; (2) analysis of the social and human types to work; (3) analysis of the dynamic factors of culture or affective socio-economic utility; and (4) evaluation of the result in the worker's behavior performing the critical task.

Archetypes and regionalities analysis

Entrepreneurs from different areas (chemical and oil) have discussed the ergonomic criteria for control rooms and raise the possibility that tacit aspects influenced by behaviors are not being considered. This hypothesis is due to continuity failures in critical operations such as plant startup and shutdown. These questions led the researchers to analyze the control room construction criteria based on the ergonomics requirements in the EEMUA and ISO standards [36]. The result of the research is the mapping of critical situations that can result in accidents.

In the case of a LNG industry control room were related aspects of local and global culture that impact on the cognitive processing leading to failure in decision making. This tool depends on the study of product risks and social risks (APR and APRSH), the analysis of the major failures and relations with human error in similar facilities, and brings results in a cross-matrix, which relates the social influences, cognitive functions with interfaces and control room devices.

The regional or global behavior that can turn into social risk (eg, default and inadequate leadership) may be due to cultural biases from primary or secondary archetypes. A case of industrial gases plant is analyzed and actions are suggested in the form of ritual to neutralize or deflect improper behavior [37].

Variability of climate change

The urgency of actions to reduce GHG emissions, fuel consumption and water motivate researchers in techniques seek to adjust the demand for water and energy in relation to confirmation of the offer. The demand adjustment actions involving routine management, knowledge of techniques for thermal performance analysis and new technologies such as process intensification or modularization of projects. The research conducted (LAB, Olefins, Urea) indicate that the equipment involved in research and routine adjustment are: cooling towers, boilers, furnaces, condensate systems and unit operations with their reflux [49][50].

4.4 Step three: design and control of operations

The Control of Operational Routine depends on the knowledge of appropriate standards for the maintenance of a stable process with the appropriate cost. It is increasingly announced by managers of high-risk companies that the economic result is guaranteed if the image is positive - that is, the non-occurrence of events related to accidents.

The definition of operating standard to routine management is part of a policy initiated by the management of technical and non-technical risks. So, the foreknowledge keeps the assets in normal mode. The occurrence of deviations, failures and incidents in the industrial operation can cause a crisis situation that requires a new leadership profile (crisis and contingency management).

Cases of task analysis

In LPG industry there are high-risk activities that can cause fire or explosion. Purification of LPG avoids failure through moisture removal. The moisture-removing columns have a saturation point and are removed from operation to be regenerated. This, in turn, requires the moisture expulsion using heated LPG itself resulting from passage through a furnace.

During the investigation concerning the study of root cause analysis in LPG process [51] was made the analysis of the operation in the task, startup and breakdown in regeneration furnaces using techniques such as GMTA[20], CARMAN[63] and PADOP[26][33]. These techniques used for fault analysis in the task involves the logic and the nature of the failure, the chronology of events, connectivity between factors beside the point where failure goes out of the social mode (eg. communication) to the physical mode in GLP [51] and Polycarbonate [60].

The task control actions include the procedure review (requirements, targets, or steps) with training. When the risk, complexity of processes, the complexity of social relationships and stress levels are high (in part or in full), the training actions has no effectiveness demanding action on culture [56][33] or to the redesign of equipment and processes.

Cases of failure in equipment and process

The benzene hydrogenation is an intermediate step in the production of caprolactam. The recycle gas is maintained at constant pressure to control the reaction through a centrifugal turbo-compressor of the brand Demag. Repeated interruptions on Monday (after the weekend) raised the possibility of having human error in the operation and/or maintenance which did not allow the effectiveness of technical actions.

A FMEAH study was performed to identify the failure mode which problems of perception, attention and expertise were occurring in routine production. The method resembles the traditional FMEA, but includes analysis of human error in the equipment failure mode [64].

The degradation systems, changes in technology, organizational changes [30] and climate change may cause variations in the process which results in the modification of the possible event scenario, which in turn increases the risk of accidents. In this regard, through the signals emitted in routine, seeks to review the product risk, process and operations through the application of new techniques such as APR, APRSH, HAZOP, Social HAZOP.

Human performance case in state of stress

A startup case of sulfuric acid purification discussed the influence of stressful environment. The elements that make up the work environment to achieve a state of stress that affects human performance are: task risk level; health impacts, the cost of business and social environment; function of the task in the process; and analysis of the psychological and organizational types. The protocol indicated that the level of stress was 4.8 exceeding the stress level manageable (under 4).

An accident analysis in progressive stress

An accident involving LPG furnace, as discussed in this paper, was used to build a fault tree of technical and non-technical events in different levels of stress.

The application of LODA technique [27] from gradually more dangerous events indicates that the decision of the shift varies according to the level of stress. The progressive stress is promoted by the need for decision-making by building possible increasingly serious consequences scenarios at the following levels: deviation from the routine, lack of process, lack of equipment and task, accident and sinister.

Unfortunately, the psychological unpreparedness of staff in dealing with contingencies, especially the leaderships, makes the human error probability goes from 10%, in steps of process deviations and lack of control of process, to 50% in sinister steps.

Cognitive failure cases and vices in the industry operators behavior

In a startup of a chemical industry reactor [35] are analyzed psychological functions in the executive work of the task. From social relationship models and the executive function analysis it was identified the following types of failure on the team with priority level 1: lack of knowledge and difficulties in memory recovery.

In the sulfuric acid industry [40] were analyzed behavioral addictions through fault trees that indicated the need for vaccines for cognitive diseases, among which: "frozen" perception, low motivation to cooperate, not enabled memory, commitment corrupted, communication and leadership inadequate, and very simple mental map. This finding allows the construction of behavior adjustment programs.

In industrial gases and chemical industries were found two types of behaviors in the control room which allowed the analysis of two different groups archetypes. Some archetypes favor a robust safety culture and others hinder their establishment [37].

Root-cause analysis in complex systems

The operator speech analysis is a tool that links events occurred in the shift such as: problems with equipment, product quality, double-shifts, missing and other facts. The form of communication, the statistics of events, the possibility of omission and team cooperation are some of the aspects included in this speech.

On several occasions in the chemical process industries [60] were applied root cause analysis techniques in complex systems. In the production of alkyl amines it was noted that the delays were the result of errors in the setup of the campaign. In the production of methyl diphenylamine it was observed that the interruptions were strongly influenced by the formation of groups that are not communicated with each other and by the lack of knowledge of the importance of recycle line for the quality in the process. In polycarbonates production sought to reduce loss of methylene chloride. In this case the operator's speech has enabled to review: procedures (from the task analysis) and manager behavior in the plant startup in control room.

In LPG industry in which the concern is the leakage events, fire or explosion, the operator's speech helped to identifying critical areas, gap of competence for decision-making, and team behavior in emergency situations [51].

5 Conclusion

The model of operational risk management requires technical and social aspects are correlated at different levels of normality / abnormality of the processes with the consequences and impacts that produce in their own business, on environment and society. By multidisciplinary involved in the analysis of these risks, the definition of management strategies should take into account the dynamics of the risks in business.

In some cases analyzed were used the operator' speech as a tool for building the event fault tree in the routine that jointly with task analysis identified the gaps of competence that lead to failure where the criteria and procedures adopted can be reviewed in the design of operations.

In another case (explosives factory) was found to be necessary to investigate the human type to change the criteria in risk management. In this case, the entrepreneur recorded accidents with serious injuries to the worker after a history of 20 years without accident (plants located in different countries) and admitted that cultural changes alter behavior which highlights the need to consider new criteria for dynamic risk management.

From the perspective of production factors such as water and energy, is also recorded vulnerability in the sustainability of the business in major change situations in the availability of these resources requiring that teams know how to prioritize investments to overcome such difficulties.

The applications of various techniques mentioned in this work are structured in a sustainable model of risk management. This model requires generation of multidisciplinary expertise (socio-technical aspects) to properly handle the dynamics of risk.

6 Annex

Figure 1 – Risk Management Model for Sustainable Organization

7 References

[1] G. Dionne Risk Management: history, definition, and critique. Risk Management and Insurance Review, 16 (2013), 147-166.
[2] J. L. Vesper. *Risk Assessment and Risk Management in the Pharmaceutical Industry: Clear and Simple*. Parenteral Drug Association, New Vernon, 2006.
[3] M. R. Martins *et al.* Undesirable events Consequences analysis during the transport, loading and unloading of Liquefied Natural Gas (LNG) on ships. Proceedings of Brazilian Association of Risk - ABRISCO. Rio de Janeino, 2013.
[4] A. M. Schleder and M. R Martins. Methodology for quantitative analysis of the consequence of flammable leaks and / or toxic environments with barriers or confined. Proceedings of Brazilian Association of Risk - ABRISCO. Rio de Janeino, 2013.
[5] B. K. Bharatiya. Projet Hazard Analysis In Hydrocarbon Processing Industry. Proceedings of 4th Latin American Conference on Process Safety - CCPS. Rio de Janeino, 2012.
[6] P. V. Fleming and C. B. Garcia. Industrial and Environmental Risk Assessment With Preliminary Hazard Analysis (PHA) and Fuzzy Logic. Proceedings of Brazilian Association of Production Engineering - ENEGEP. 1999.
[7] R. Freeman. Quantifying Lopa Uncertainty. Process Saf Prog 31(2012), 240-247.
[8] CCPS. *Guidelines for Safe Automation of Chemical Processes*. Center for Chemical Process Safety, American Institute of Chemical Engineers, New York, NY, 1993.
[9] J. N. Shan. Understanding and Developing Quantitative Risk Criteria. Proceedings of 8th Global Congress on Process Safety, Houston, 2012.
[10] A. H. L. Correia. Strategic Marketing of Sporting Event. International Seminar of Sports Event Management. Ministério da Juventude e do Desporto. Centro de estudos e formação desportiva. Lisboa, 2001.
[11] Worksafe Victoria. Advice for Managing Major Events Safely. 1st Ed. Austrália. 2006. Available at: https://www.worksafe.vic.gov.au/__data/assets/pdf_file/0010/12502/major_events_organisers_advice.pdf. Accessed on January 3, 2015.
[12] F. S. Flogiatto and J. L. D. Ribeiro. *Reliability and Industrial Maintenance*. Elsevier: ABEPRO, Rio de Janeiro, 2009.
[13] A. Kardec and J. Nascif. *Maintenance – Strategic Function*. 4th Ed. Qualitymark Editora, Rio de Janeiro, 2013.
[14] R. L. Boring. Fifty Years of THERP and Human Reliability Analysis. U.S. Department of Energy National Laboratory. U.S.A., 2012.
[15] C. A. Pallerosi; B. P. M. Mazzolini and L. R. Mazzolini. *Human reliability - Concepts, Analysis, Evaluation and Challenges*. AllPrint Editora, São Paulo, 2011.
[16] ÁVILA FILHO S.; PESSOA F.L.P. and ANDRADE J.C.S. Social HAZOP at an Oil Refinery. Process Saf Prog - 32 (2013), 17-21.
[17] D. K. Lorenzo. API770 – A manager's guide to reducing human errors, improving human in the process industries. API Publishing Services, Washington, 2001.
[18] J. Reason. *Human Error*. Cambridge University Press, New York, 2003.
[19] R. Stenberg Junior. *Cognitive Psychology*. 3rd Ed. Artmed, Porto Alegre, 2008.
[20] E. Hollnagel. *Human Reliability Analysis Context And Control - Computers And People Series*. Academic Press Inc, San Diego, 1993.
[21] S. Ávila Filho. Reliability analysis for socio-technical system, case propene pumping. Engineering Failure Analysis, 56 (2015), 177-184.

[22] P. Constatinides. The Failure of Foresight In Crisis Management: a Secondary Analysis of the Mari Disaster. Technological Forecasting & Social Change, 80 (2013), 1657–1673
[23] R. Muller. Corporate Crisis Management. Long Range Planning, 18 (1985), 38-48.
[24] S. Ávila Filho *et al*. Risk Management and Prevention Program to Avoid Loss of Process (LP3) based on Human Reliability, Practical Application and Measurement. Proceedings of 7th Global Congress on Process Safety – GCPS. Chicago, 2011.
[25] S. Ávila Filho and A. C. Carvalho. Human Factors Assessment to Avoid Accident at LPG Industry. Proceedings of 6th Latin American Conference on Process Safety – CCPS. Rio de Janeiro, 2012.
[26] S. Ávila Filho. Assessment of Human Elements to Avoid Accidents and Failures in Task Perform, Cognitive And Intuitive Schemes. Proceedings of 7th Global Congress on Process Safety – GCPS. Chicago, 2011.
[27] S. Ávila Filho. Dependent Layer of Operation decision analyzes (LODA) to calculate Human Factor, a simulated case with PLG event. Proceedings of 7th Global Congress on Process Safety – GCPS. Chicago, 2011.
[28] S. Ávila Filho. Review of Risk Analysis and Accident on the Routine Operations in the Oil Industry. Proceedings of 5th Latin American Conference on Process Safety – CCPS. Cartagena de Indias, 2013a.
[29] S. Ávila Filho. Risks in the Platform: Social and Human Guidelines Avoid Accidents and Alter Standards in Oil Production. Revista Proteção, 2013b.
[30] S. Ávila Filho; E. T. Dantas and J. Duarte. The Requirements & Tools to Treat Process Safety Risk as Result of Organizational Change. Proceedings of 9th Global Congress on Process Safety – GCPS. Austin, 2015a.
[31] S. Ávila Filho, *et al*. Analysis of risk scenarios and mapping of communication processes to avoid incidents in Salvador Carnival in 2015/16. Proceedings of Brazilian Association of Risk – ABRISCO/PSAM. Rio de Janeiro, 2015a.
[32] S. Ávila Filho, *et al*. Risk Analysis in Sports Mega-events, UFBA Project in the World Cup. Proceedings of 34th Brazilian Association of Production Engineering - ENEGEP. Curitiba, 2014.
[33] S. Ávila Filho; C. R. Sousa and A. C. Carvalho. Assessment of Complexity in the Task to Define Safeguards Against Dynamic Risks. Proceedings of 5th International Conference on Applied Human Factors and Ergonomics – AHFE. Las Vegas, 2015.
[34] S. Ávila Filho. Failure Analysis in Complex Processes. Proceedings of 19th Brazilian Chemical Engineering Congress – COBEQ. Búzios, 2012.
[35] S. Ávila Filho and C. Costa. Analysis of Cognitive Deficit in Routine Task: as a Strategy to Reduce Accidents and Increase Industrial Production. Proceedings of European Safety and Reliability Conference – ESREL. Zurich, 2015.
[36] S. Ávila Filho and F, L. P. Pessoa. Proposition of review in EEMUA 201 & ISO Standard 11064 Based on Cultural Aspects in Labor Team, LNG Case. JofPM, 3 (2015), 6101 – 6108.
[37] S. Ávila Filho, M.L.A. Menezes. Influence of Local Archetypes on the Operability & Usability of Instruments in Control Rooms. Industrial Production. Proceedings of European Safety and Reliability Conference – ESREL. Zurich, 2015.
[38] S. Ávila Filho, *et al*. Influence of Brazilian Culture in the Workers' Behavior and Operational Discipline. Proceedings of Brazilian Association of Risk – ABRISCO/PSAM. Rio de Janeiro, 2015b.

[39] K. R. V. Nunes, *et al.* Working Hours Excessive and its Influence on Accidents, Productivity and Decision Making of Nursing Technicians in an ICU. Proceedings of Brazilian Association of Risk – ABRISCO/PSAM. Rio de Janeiro, 2015b.
[40] S. Ávila Filho; C. C. Silva and C. R. O. Sousa. Worker Behavior Changing by Dynamic Culture. Proceedings of 9th Global Congress on Process Safety – GCPS. San Antonio, 2013.
[41] S. Ávila Filho. The Exergy of Operational Failure, a Philosophical Discussion with Basis Practices. Proceedings of 19th Brazilian Chemical Engineering Congress – COBEQ. Búzios, 2012.
[42] S. Ávila Filho, J. R. N. Lopes and C. R. O. Sousa. Strategies to Industry Sustainability. Annuals of 20th World Renewable Energy Forum – WREF. Denver, 2012.
[43] A. C. Alcoforado and S. Ávila Filho. SPAR-H Application in the Crushing Operation of Uranium Ore. Proceedings of Brazilian Association of Risk – ABRISCO/PSAM. Rio de Janeiro, 2015.
[44] S. S. Dórea and S. Ávila Filho. Failure Investigation in Operations of Chicken Industry with Based Human and Technical Factors. Proceedings of Brazilian Association of Risk – ABRISCO/PSAM. Rio de Janeino, 2015.
[45] M. N. G Abreu and S. Ávila Filho. Assessment of Risk in the Exploration and Production of Oil at Pre-Salt Layer – Brazil, a Discussion about Social-Technical Factors. Proceedings of Rio Pipeline Conference & exposition. Rio de Janeiro, 2015.
[46] M. N. G Abreu and S. Ávila Filho. Assessment of Reliability and Risk at Dynamic Systems: a Discussion of Social-Technical Factors in the Exploration and Production of Oil at Pre-Salt Layer – Brazil. Proceedings of Global Congress on Process Safety – GCPS. Austin, 2015.
[47] S. Ávila Filho, *et al.* Environmental Deviation Assessment, Record, Failure, Probability and Investment. Chemical Eng Trans, 26 (2012a), 441-446.
[48] S. Ávila Filho, *et al.* Program to Avoid Product Leakage at Source, Fertilizer Case. Chemical Eng Trans, 26 (2012b), 447-452.
[49] S. Ávila Filho, *et al.* Iniciatives of Energy Efficiency Toward Sustainability, Opportunities & Challenges In: Rio Oil Gas Conference & exposition. Rio de Janeiro, 2014.
[50] S. Ávila Filho, *et al.* Avoiding Loss of Energy in a Petrochemical Industry, Operation and Design. Proceedings of World Renewable Energy Congress – WREC. Linköping, 2011.
[51] S. Ávila Filho. Etiology of Operational Abnormalities in Industry: a Model for Learning. Doctoral thesis - Chemistry School. Federal University of Rio de Janeiro. Rio de Janeiro, 2010.
[52] ANP. Investigation Report of an Explosion Incident in FPSO Ship in São Mateus Town on November 02, 2015. National Agency of Petroleum, Natural Gas and Biofuels. Available at: http://anp.gov.br/?pg=78836&m=&t1=&t2=&t3=&t4=&ar=&ps=&1452081674560. Accessed on January 15, 2016.
[53] R. Bonalume Neto. 30% Loss of Water During Distribution Contributes to Crisis in São Paulo. Sustainability Forum. Folha de São Paulo. São Paulo, 2014.
[54] D. K Lorenzo. API770 – A manager's guide to reducing human errors, improving human in the process industries. API Publishing Services, Washington, 2001.
[55] M. P. Barroso. The influence of human factors in APP and APR risk analysis techniques: a study case of an oil drilling platform in the northeast. Master's Dissertation – Polytechnic School. Federal University of Bahia. Salvador, 2013.
[56] G. G. Gonçalves, *et al.* Protocol to evaluate the influence of stress on the task involving human types, behavior and health: An exercise proposal. ESREL 2015. Zurick. 2015.

[57] S. Ávila Filho. LESHA – Multi-Layer progressive stress & impact assessment on health & behavior. Proceedings of 9th Global Congress on Process Safety. San Antonio, 2013.
[58] S. Ávila Filho. Human Errors Assessment, Social Environment at Oil Platforms. Proceedings of 4th Latin America Conference on Process Safety. Rio de Janeiro. 2012.
[59] C. Perrow. *Normal accidents: living with high-risk technologies.* NY: Basic Books, 1984. 453 p.
[60] S. Ávila Filho. Methodology to Minimize Generation of Waste at Source from the Investigation of Operational Abnormalities. Master's Dissertation – Polytechnic School. Federal University of Bahia. Salvador, 2004.
[61] E.S. Drigo, S. Ávila Filho and C. A. O Sousa. Operator discourse analysis as a tool for risk management. Proceedings of European Safety and Reliability Conference – ESREL. Zurich, 2015.
[62] S. Ávila Filho. Diagnosis in Socio-Technical Systems Aiming to Increase Human Reliability, a LPG Case. Proceedings of International Applied Reliability Symposium – ARS. Fortaleza, 2011.
[63] D. Embrey. Preventing human error: developing a consensus led safety culture based on best practice. London: Human Reliability Associates Ltd. 2000.
[64] S. Ávila Filho, *et al.* Human Factors Analysis in Turbocharger Equipment Failure FMEAH in Chemical Plant. Proceedings of 27th Brazilian Maintenance Congress. Rio de Janeiro, 2012.

Water and Energy availability analysis, a proposal for management for chemical industry

Salvador Ávila Filho
Federal University of Bahia – UFBA
Salvador – Brazil
avilasalva@gmail.com

José Rafael Nascimento Lopes
National Service of Industrial Learning – SENAI
Salvador – Brazil
jrafael.lopes@gmail.com

Jean marcel Prazeres Silva
Federal University of Bahia – UFBA
Salvador – Brazil
jmarcel_7@yahoo.com.br

Prepared for Presentation at
American Institute of Chemical Engineers
2016 Spring Meeting
12th Global Congress on Process Safety
Houston, Texas
April 11-23, 2016

Water and Energy availability analysis, a proposal for management for chemical industry

Salvador Ávila Filho
Federal University of Bahia – UFBA
Salvador – Brazil
avilasalva@gmail.com

José Rafael Nascimento Lopes
National Service of Industrial Learning – SENAI

Jean Marcel Prazeres Silva
Federal University of Bahia – UFBA

Keywords: water availability, energy analysis, sustainability

Abstract

The evidences that the chemical industry is not ready to deal with interruption of operations resulting from climate change and the complexity of decisions is increasingly apparent. In the state of São Paulo / Brazil, petrochemical industries are using served water from sewage treatment plants to supply the cooling tower makeup water. However, industries are not striving to analyze the availability of water and energy to suit the demand to supply.

The setting of premises to build scenarios of water availability, energy and skills is required. This will allow to make strategic plans, control programs, techniques and methods of investigation that aspects of products and processes are considered as well as the characteristics of the culture and behavior of managers and employees.

The water supply scenario and its use in the agriculture, livestock and cities indicate that there will be reduced availability of water for industry. As these consumer sectors are disorganized and there are losses in the distribution of water in the order of 30%, it is up to industry to worry about meeting their demand while reducing your water consumption.

The aim of this paper is to present an analysis model of availability of water and energy in the chemical industry, along with a behavioral adjustment model to enhance a large reduction in water consumption and energy supply to meet the expectations of these. The reduction in water consumption and energy depends on: management actions; switching procedures; behavior changes in the short term; application reuse techniques and reprocessing wastewater and energy in the medium term; and development and installation of technologies to increase the supply of water and energy and to reduce demand for water and energy in the long run.

Initially normal trend of events without intervention in the provision of management or demand management is considered. If occur climate change altering the supply of water and energy, on the other is the population and industrial growth by increasing demand for water and energy. In this case the industry will become hostage to a situation of lack of water and energy when looking at demand in relation to supply. In our case this will occur in 2022.

The analysis of availability compensates for the reduced supply with even greater reduction in demand. The analysis method of availability is divided into the following steps considering the comparison between the supply of water and energy with climate change with the demand for water and energy with preventive actions. Should perform prediction of what should happen in the future to adjust the present with the routine programs. This methodology is divided into the following steps: (1) Definition of current and future supply scenarios based on IPCC; (2) Identification of current demand scenario and future to compare Offer; (3) Analysis of the availability of water, energy and competence; (4) Culture Analysis and People for behavioral change; (5) Analysis of Routine and Integrated Management; and (6) Compliance Analysis of consumption of water and energy. A case of the chemical fertilizer industry are tested.

1 Introduction: Context and Background

Water is essential for life and it is the base of almost all human activities. In this context, water is used in power generation processes, domestic use, agriculture, cattle raising, and industrial sector among others. Provide access to clear water in quantity and quality represent one of the major contemporary challenges [1].

According to the UN World Report, the demand for water is expected to grow by 55% worldwide by 2050 due to the fast urbanization mainly in developing countries. The report also states that the world's water supply will suffer a deficit by 40% until 2030 if the water management do not improve. The agricultural sector will produce about 60% more food, and the industry sector will increase water demand up to 400% until 2050. Developing countries such as Brazil, Russia, India, China and South Africa will drive this [2].

The concern about water scarcity increased in Brazil after the water crisis in the state of São Paulo the most populated area located in the southeast of the country. This region from Brazil instead of the Northwest was an area that there is not this kind of problems. The water crisis didn't harm only the life of the local population, but also the industrial sector which account for 32% of the Brazilian GDP [3].

The industry needs to be alert and prepared to deal with this kind of event, because this issue affect the running and growth of production [4]. Moreover, scarcity and water pollution have amplified some planetary boundaries known by scientists as safe operating space for humanity. Be in the planetary boundaries is a precondition for sustainable development. According to Johan Rockstrom *et al.* freshwater problems is influencing and affecting some planetary boundaries as biodiversity, food security, population health, and the ecological functioning, such as climate regulation and carbon sequestration [5].

According to ANEEL (National Electric Energy Agency), approximately 75% of Brazilian power generation come from the use of renewable sources mainly hydropower with 62% [6]. Water is also a driving force in the generation of electricity in thermal power generation (steam generation), and necessary to chemical and petrochemical industries for heating (thermal energy) or cooling, among others services [7]. Thus, the availability of water is associated to the availability of energy in socio-economic activities in Brazil and the world.

The lack of structure industrial sector to deal with the interruption of operations resulting from non-availability of water and energy (effects of climate change) is increasingly evident. The structure is related to technological aspects (e.g., automation) and of complex decisions. Many decisions involve multiple factors such as climatic and behavioral aspects.

Although industries are trying to develop projects to reuse of water from sewage treatment plants to supply cooling tower makeup water as was done in the state of São Paulo during the water crises, it is not a tendency. The reuse of wastewater in the industrial sector has been considered synonymous with economic gains, in addition to the positive image transmitted. Nevertheless, the concern is the concentration of pollutants and the possibility of increased losses by corrosion. Although reuse of water from sewage treatment demonstrate that management procedures are related to the basic principles of preventing water shortages, energy and environmental impacts, it isn't clear a commitment to analyze the availability of water and energy in order to adequate demand over supply considering scenarios for 2035 [4].

2 Method

The method intend to compare the supply of water and energy to their demand in the present time and a projection for the future - forecasting. This future scenario permit to take place the necessary actions to maintain the water and energy availability over time – backcasting [8]. Furthermore, prediction scenarios of greenhouse will help adjust the current emissions through routine management programs and new technologies.

Thus, short and long-term actions can be preventive as reducing the emission of greenhouse gases (replacement of energy by renewable or decrease their use or reduction of fuel burning) or corrective as the reduction of water consumption.

The analysis of availability method (shown in Figure 1) is divided into the following steps. (1) Definition of current and future supply scenarios based on the IPCC. (2) Identification of the current demand scenario and future to compare with the supply. (3) Analysis of the availability of water, energy and competence. (4) Culture Analysis and People for behavioral change. (5) Analysis of Routine and Integrated Management. (6) Compliance Analysis of the consumption of water and energy.

2.1 Definition of current supply scenario and future based on IPCC

The scenarios of use and supply of water indicate further reductions in the availability of water for industry. Water users (agriculture, livestock and population) are messy and work with distribution losses that can reach up to 34% [9]. Thus, the industrial sector need to care about increasing the supply of water and reduce their consumption at the same time through demand management and local source control programs.

The discussion about the energy consumption is linked to the greenhouse emission. The rise of GHG into the atmosphere depends on the burning of nonrenewable fuels or the lower conversion efficiency producing CO. The environmental impacts result in climate change and consequently alterations in water availability for society and for the economy as well as impact on biodiversity.

According the annual report about the inventory of CO_2 in the atmosphere, the amount of emission should be minimal to reverse climate change. Thus, the planet earth must send less and less CO_2 into the atmosphere, gradually, even with the growth of economies. The value target is 14 GTY in 2050 [10] and [11]. The forecast on the growth of the economy, with no GHG emission reduction action is to increase the current 50 GTY to above 62 GTY in 2050 (value stated in 2009) [10] - indicating the irreversibility of the greenhouse effect.

This paper aims to demonstrate how an analysis model on the availability of water and energy in the chemical industry and a behavior adjustment model may cause a large reduction in water consumption and energy in order to meet the expectations mapped by future supply of these inputs. The reduction in water consumption and energy depends on: management decisions; switching procedures; behavior changes in the short term; application reuse techniques and reprocessing of waste water and energy in the medium term; and development and deployment of technologies to increase the supply of water and energy and to reduce demand for water and energy in the long run.

2.2 Identification of the actual and future demand scenario to compare with the offer

A analysis of production and shutdown reports allow to build, in a preliminary way, the knowledge about the cause of losses and the behavior type to deal with these losses mainly of time, materials water, and energy. A rapid assessment of water and energy restrictions can give an idea of the scenario in the company. The production growth rate indicates if the demand will increase with the view to forecast the water consumption in the future, without including the preventive and corrective actions to demand management. When the supply and demand are compared, we can estimate the gap and which information is required to improve the analyses.

Water is classified in industry in various levels to meet the demand of the production plant and the administration. Large volumes of water are used in operations such as washing, transportation, ion exchange and others. One of the largest industrial water uses are in the cooling towers that meets the warm currents of the separation unit operations in the chemical process industry. Clarified water is the type of water used in cooling towers and represent 70% of the raw water inlet to the industry [12].

The measurement of water consumption is a legal obligation in order to inform the government environmental agency who granted a license with a utilization limit. This grant lets to use a quantity of surface or underground raw water and returns it (in effluent format) a quality better than the shot. Thus, it is necessary knowing the use of water, qualities (restrictions of equipment and processes), disposal and the behavior of the team and the manager regarding this precious resource to reduce the current water consumption and future need.

Regarding to energy consumption, classifying the type of energy used by industry is the first step to establish a demand profile. The electrical or thermal energy are primary energy and they can be converted into chemical energy and work. Thus, the main inputs in chemical industry are electric power and fuel. The measurement of consumption allows calculating the use of energy per unit of product giving an idea of the efficiency of its use. When a baseline and target (benchmark) is established and these are compared with the requested reduction of emission of GHG it is possible to establish the programs and investments in medium and long term.

2.3 Analysis of the availability of water, energy and competence

The starting point of the availability analysis is building the graph indicating water consumption and how much in terms of grants is authorized consume (immediately). In the case of supply, it is necessary to build an algorithm to increase the supply due to new sources or the reduction of losses. Another algorithm must be built when the source of water decreases due to climate change, increasing urban population, rise of irrigation area to agriculture, and so on. However, demand for water could consider two possibilities: (1) any action to reduce demand during the time (2016-2035) indicating only increase in the forecast increase in production or (2) actions and investments in process design, control routine and team behavior to reduce water consumption.

When the annual supply is projected for the future (including economic changes in society and the climate) and compared to future demand without considering corrective action, there is an indication of interruption of industrial operations for lack of water. Thus, actions to reduce water demand is indicated.

The availability of energy analysis actually compares the possible amount of CO_2 emission (which changes with time due to more restrictive situations in the future) from estimates by the International Energy Agency, with the current and future amount of CO_2 emission in different situations as the action taken or not to prevent irreversible climate change.

The initial approach of water/energy supply and demand needs to be confirm by mapping the thermal systems, equipment controls, water and energy saving programs, and analysis of sources and water and energy uses. The demand mapping is confirmed and compared with the preliminary data, they are made of the demand adjustment exercises in the timeline considering new best-case scenarios or more critical every five years.

Thus, the master plan for water and energy is builds up with the demand adjustment regarding the supply by technologies, programs, and technologies to be installed gradually. It is considered to predict supply by 2035 criteria related to industry growth, population, issues related to agriculture and irrigation, rainfall forecast, and the probable faults keeping the same demand.

This initial analysis should be reviewed after a comparison with external benchmark and improve the quality of information from direct and not secondary measurements.

Among the technologies are: reverse osmosis, geothermal, flexible cooling towers for reuse, process intensification, and others. Among the techniques are: water balance, energy balance, pinch, reuse of wastewater, Exergy analysis, analysis of material losses and energy, and others.

2.4 Analysis of Culture and People for behavioral change

A series of procedures and attitudes in the levels of implementation, management and strategy are based on increasingly complex decisions [13] and [14]. It is necessary to know how these decisions occur based on local cultural characteristics [15] and likely rules that influence not reduce water and energy consumption. It is important to analyze the influence of equipment failure, processes and operations of the direct losses and indirect materials to the reprocessing [16]. The decision to adjust the quality of the product can impact on the increase in the consumption of water, energy and the rework of workers.

An analysis of people and cultures will be able to identify indicators that evidence the inappropriate behavior that enhance water loss and energy. This allows building a program to control losses from these resources. With this identification, it is possible setting a program of training or behavioral adjustment to be implemented in the company.

The cultural influence is discussed for building a robust safety culture and is based on the dynamic behavior of the staff. This behavior is the result of a field of cultural forces: organizational desire, local culture and features of global culture. Some trends should be strengthened and other inhibited and this action are based on archetypes analysis [17].

The human factors analysis allows to establish programs to increase human reliability. Human factors can be [18]: conflict between policies and practices, the conflict between production and safety priorities, control facilities, managerial aspects, skills analysis, and preparation procedures are part of human factors that should be analyzed and monitored.

The moment of decision is then influenced by biases of culture and human factors causing cognitive impairment [19] occurs following human errors mechanisms as the default [20]. In addition to the analysis of cognitive impairment seeks to reduce the risk of accident through the joint analysis of technical factors and not technical. This strategy allows us to discuss what are the aspects that can impact the security processes in an organizational change [21].

2.5 Analysis of Routine and Integrated Management

After identifying the behavioral constraints or characteristics that lead the employee and the manager to start faults that cause loss of energy and water, it is important to analyze the routine. Check the risks of system stop on a daily basis allows you to work in a review of equipment and critical tasks. The knowledge of water systems, heat exchangers, energy integration, wastewater allows for maximum use of energy by increasing the energy efficiency. The water and waste energy reuse will rationalize the consumption reducing dependence on bulk supply facilitating sustainability of the company.

2.6 *Analysis of the availability of water, energy and competence*

The establishment of comparison criteria for use of water, energy and behavior that avoids the technical and human error adjusts techniques, technologies, decisions, behaviors and knowledge. So this step enables the continuous improvement of the risk management program in Energy and Water. (ISO 50001) [22].

3 Demand management

Some global factors have to force the industry to make efforts to increase productivity and reduce the consumption of basic materials as raw materials, water and energy. The water is already being alert reason for society and for the industrial sector, where specific events of lack of water can turn into long periods of shortage impacting on the economy and the population.

Energy is a key input to ensure the economic and social development and the rationalization of its use is an essential strategy to avoid losses during periods of restrictions resulting from the emission of CO_2 and the scarcity of natural resources. Energy rationalization might be to change procedures, behavior change, control techniques and technologies to enhance the use of energy and thus improve efficiency and improved thermal performance of the equipment and systems.

The water consumption in the industry is directly linked with energy demand, for the use and distribution of water in a production process, the energy has a fundamental role to keep the equipment operating. Similarly, the electrical and thermal energy coming from a unit of basic inputs need the aid of water for its existence.

Industrial activities that involve the consumption of basic inputs has suffered deployments due to significant consumption and existing waste. Cycles of concentration in cooling towers, water use outside of specified boilers, mass loss and energy in steam distribution systems, heat losses in electric power generators, discarded wastewater without prior analysis, are some examples that show opportunities for deployment techniques or technologies that contribute to the reduction of utilities, waste and therefore environmental impact.

However, to reduce the consumption of water and energy are proposed a number of solutions management, various techniques used by the STAFF during routine operations and technologies.

3.1 *Management*

The management plays a key role in the implementation of programs to control routine related to the behavior of the staff and control of deviations. Each technique has suggested goal, steps and impacts are described below.

The control program at the source [23] is to identify equipment and pathways of contamination to the final disposal, liquid (effluent), solid (waste inventory) and gas (emissions or losses fugitive emissions). Understanding on which the carrier fluid and how it is possible to establish measures to avoid losses. The additional technical material balance loss is the mass calculation of losses and the impact on fixed costs, variable and billing.

Control of critical task, named as POS, Procedures and Operational Standards [24], aims to identify, in this case of water and energy availability, which tasks are that they can produce losses and analyze the environment, planning, execution, control and the failed task.

The analysis of environmental deviations [25] want based on the fault tree method and field observation, identify deviations that lead to contamination of everyday life. When this technique is well understood in the organization it enables a behavior change operation.

This research is in development and is about identifying what the proper conduct of the operation team to prevent energy loss. Follows the line of CCPS in Operational Discipline [26] and includes aspects of the company culture. It takes into account that energy is intangible and less noticeable to control losses.

The calculation of the reliability of socio-technical systems [27] depends not only on engineering and includes the social and managerial area of the company. For this reason it will be led by the manager. When the reliability of each functional part of the system (process, equipment, operation and human) is identified, after the process of the complexity of the classification, the calculus of the system reliability level is possible. The system reliability is then corrected through social attractiveness that indicates the strength of culture, the level of cooperation (communication) and adaptive leadership to new realities. Through this analysis it becomes possible to more policy action on production controls.

By calculating the reliability of systems and identified the impact of human reliability begins research on each performance factor indicated by SPARH [28] and discussed by API770 [18]. Thus, it becomes possible to define critical areas for action by the manager. These analyzes made by the management has focused on reducing energy losses!

3.2 Engineering

The engineering staff has the role to develop adaptations to current processes and equipment and provide new equipment to avoid or mitigate energy losses. These solutions require equipment yield calculation, energy efficiency and thermal performance.

Initially the operational control equipment and thermal systems adjusts the best operating point through procedures and control systems. It is therefore important that the process engineering and equipment to track and control the income and performance of furnaces, boilers, cooling towers, electric motors, power contracts, distribution of steam, condensate distribution, reflux in the distillation, and balance energy in the process equipment and utility in general.

This control can identify the need to revise the performance calculation methodology to be open systems, such as cooling towers. In furnaces and boilers may be a need to review the maintenance procedures (installation of refractory) or operating as purging. In collecting and recovering the condensate, on the other hand, several factors may impact on live steam loss to the atmosphere, it depends on the thermal load of the plant, steam quality, operation of steam traps, and quality of thermal insulation.

The effluent mapping to identify the loads of contaminants requires the establishment of networks of sources and uses of water by continuous or intermittent processes procedures. It is needed when discussing energy sources and uses, in this case, less visible, and thus requiring additional technical field, such as using pyrometer surface of hot equipment.

The mass balance and energy are more detailed technical and require methodologies for utilities and process equipment. simplifications may be needed, where process streams with low heat capacity variation are classified to 80% balance of the impact, reducing the number of streams.

For combustion systems are used to heat the lower and higher power. For open cooling tower systems we use the useful heat exchanged. In the case of energy balance that includes the axis of work it is important to check the income and losses of energy through heat. The energy balance is important to compare the fuel consumption with the issue of greenhouse gases.

3.3 Technology

In terms of technology we will discuss those involving new criteria for design of industrial plants and new equipment that reduce energy losses and water or increase its offer. Among the technologies involving new standards for process design and its control: modularity, intensification, integration and exergy. Among the technologies involving new equipment to reduce losses and increase in supply: desalination, heat pump, cogeneration, trigeneration and combustion bomb.

4 Application in the Chemical Industry

Applying this analysis methodology on the availability of water and energy it is based on the experiences in research projects to rationalize water use in the chemical industry of linear alkyl benzene (LAB) and the fertilizer industry (UREA) [29]. Other research has enabled the discussion of opportunities to reduce energy losses in the petrochemical industry from energy balances in processes and utilities [30] AND [31].

The experiences in the LPG industry and own fertilizer industry [24] allowed the present discussion of cultures and people. And finally, the argument based on ISO31000 and ISO50001 enabled compliance analysis.

The implementation of these steps during the period of two years will have as a result the establishment of control programs in source and strategic plans for water, energy and skills. Every five years is necessary to readjust the supply condition and analyze the indicators that demand occurred in accordance with the master plan. Organizational changes can alter the demand for water, energy and time.

The intended application for two years will be addressed through the following topics: contextualization, supply management, demand management, strategic analysis of the availability, culture and people analysis, routine management and compliance analysis.

4.1 Contextualization

Audits for use and water sources that used the effluent mapping technique and performance analysis in cooling towers indicated the high water consumption in the industry LAB mainly to feed the cooling towers [32]. It was noted also that occurs excessive contamination of waste water with several points of discharge of effluents and products. The LAB production process is complex, with the possibility of reprocessing with handling high toxicity products such as hydrofluoric acid, but a well-structured staff to decisions simply to achieve better control, access to appropriate techniques for controlling consumption energy and water.

The energy and water balance techniques, due to the complexity of petrochemical processes, indicates that more rigorous methods require mathematics reconciliation data. In other cases seeking the simplification based on the impact of the amount of energy involved, where the reduction of line processes is necessary, within the volume control for mass balance.

In the petrochemical industry of basic inputs where the type of naphtha defines differences in production costs, the equipment, the process parameters and the quantity to be produced of each product (olefins or BTX), it is important to review the energy balance. Flow rates in these processes are very high, so errors in the way of operating equipment can cause the plant to have high energy losses, indicating a need for high fuel consumption and high or make-up water for cooling towers.

The demand by up to 30% water cuts and agreements of pressure climates, and briefly change the intensifying industries in energy use. The lack of assertive decisions or appropriate behaviors have removed the goals in relation to the consumption of energy and water in the routine. This indicates that the transformation requires greater care in the industrial segment!

4.2 Supply management: control and planning for 2035

Petrochemical industry of cases and fertilizer supply of real data and water demand and actual data supply and energy demand are discussed.

4.2.1 Water supply management: premises for the analysis of supply management in the fertilizer industry in northeastern Brazil

Assumptions: (a) average climate change scenarios A1B with rainfall reduction impacts in Bahia (2070-2100) in about 60% on the coast and 35% in semi-arid with 2-5ºC temperature increase [33]. (b) increase in supply due to improvements in the operation of river basins and due to climate change by 2018; (c) reduction due to climate change estimated at 28% to 60% of the current offer in surface and groundwater respectively, for 2035; (d) reduction of reliability in service of the Paraguaçu River Basin [34] for 2011 to 2040, mainly for irrigation, industry and power generation; (e) Increase in the number of months during which there were failures in care:

the Savior supply (1% to 5%) and industrial demand (1% to 9%) with reduced power generation by 66% in Dam of Horse Stone [35].

Supply: the water supply to the industrial sector in the basin of Paraguacu and Salvador city between 2012 and 2034. **Offer 2012:** Surface water = 20.9 m^3/s, groundwater = 3.9 m^3/s, total = 24,8 m^3/s. **Offer 2018:** Climate change with increased rainfall: 24,8 to 78,8 m^3/s (3.2 x). **Offer 2034:** Reduced rainfall = 25% in 2018, 19.7 m^3/s.

Fertilizer unit: in the specific case of fertilizers unit [29] is considered that the initial flow supplied (2012) is equal to the demanded flow rate, that is, 700 m^3/h. **The Offer in 2018:** since part of the overall offering for the industry shifts to supply the city (an increase of 2 X instead of 3.2 X), that is, 1400 m^3 / h. **Offer 2034:** decrease due to climate change by 33%, that is, 462 m^3/h. The demand for water in the fertilizer unit should be less than this flow.

4.2.2 Power supply management: premises for the analysis of supply management in the petrochemical industry in northeastern Brazil

Considers that energy supply is limited by the capacity of the atmosphere of the planet receive GHG to prevent climate change. Thus, the supply limit is based on climate studies by the IPCC. On the other hand, the demand for energy in industry depends on the activities carried out by thermal and electrical systems. Thus, the balance between environmental restriction is sought (supply) and the actual greenhouse gas emissions by industry (demand).

The power availability analysis will be handled by the comparison between supply and demand. Supply management will have an upper limit maximum environmental capacity to receive GHG emissions to prevent climate change (14 GTY in 2050), this is the blue scenario. In this scenario emissions are monitored since 2005 with 28 GTY emissions and tend to grow up to the value of 62 GTY when the climate change process becomes irreversible. This emissions growth occurs if no action is taken towards sustainable economic growth of BRICS and developed countries [10].

Global targets need to be resized to establish regional or industrial sector or by enhancing industry in energy consumption targets. This schedule allows you to define local action strategy seeking to guarantee the reduction of GHG emissions.

In the case of a petrochemical in Camaçari (Brazil): since 1000 there are on planet earth petrochemical same capacity and the global industry emits 100 times the amount of CO_2 (based on fuel consumption) of the global petrochemical segment. The proportion of the global emission reduction for local equivalent to a reduction of 10-5 Industrial overall goal [10] for this local industrial. In the case of switching to renewable matrix, the overall goal is 9 GTY after the local design amounts to 0.09 MTY for this petrochemical industry.

Offer in 2050: (a) replacement by renewable energy → 10 gigatons year of global emissions or 0.1 mega ton year local issue; (b) reduction in CO_2 consumption by effective use of thermal energy resulting from the burning → 11,5 gigatons year of global emissions or 0.115 mega ton year local issue [10].

4.3 *Demand Management: control and planning for 2035*

Petrochemical industry of cases and fertilizer supply of real data and water demand and actual data supply and energy demand are analyzed.

4.3.1 Water demand management: assumptions for demand management analysis on the fertilizer industry in northeastern Brazil

Assumptions: demand management analysis in a fertilizer industry in the Northeast of Brazil: (a) average growth of 2,5% per year of urban demands of the cities of the RMS (increased population projection and reduction of per capita consumption) [36] and (b) growth in consumption by industrial expansion and irrigation.

Urban and irrigation demands will reflect the reduced availability of water for industry.

Demand: the demand for water for industrial thread in the basin of Paraguaçu and nearby cities including Salvador city between 2012 and 2034 considering that there is no demand tuning activities. **Demand in 2012:** 15,5 m^3/s city and 6,2 m^3/s for industry, total = 21,7. **Demand in 2034:** 40% of consumption by the population increase and reduction of losses of approximately 25% due to investments in the operation of water supply equivalent to 5% increase in the city, 16,3 m^3/h. New industry and new units with 15% increase indicating 7,2 m^3/s for the industry. Total 23,5 m^3/s. Indicating that the offer will be lower (19,7) than the demand (23,5 m^3/s). That is, if no action is taken to adjust to demand, the industry will be at high risk of interruption.

Fertilizer unit: in the specific case of fertilizers unit [29] is considered that the initial flow supplied (2012) is equal to the demanded flow rate, that is, 700 m^3/h. **Demand by 2034 without setting activities,** will increase by 15% in water consumption = 805 m^3/h. This unit of fertilizer also have problems which need to consume less than 462 and will be consuming 805 m^3/h.

4.3.2 Energy demand management: assumptions for demand management analysis in the petrochemical industry in northeastern Brazil

The suggestion by the International Energy Agency (IEA) is that there are different ways to reduce GHG emissions: energy efficiency (electrical and thermal), renewable energy (wind, solar, biomass), nuclear energy and carbon capture. The IEA also shows that the main actors for this reduction are: processing industry, power industry, transport and buildings.

Avila *et al.*, 2011, showed that a petrochemical site:

(a) replacing 20% of energy consumption based on fuel oil with biomass energy. Following the assumption that this biomass, which compose a forest previously cut off from the atmosphere by photosynthesis, the same amount of CO^2 emitted in their use. This petrochemical be taking 20% of CO^2 emissions initially issued.
(b) 50% reduces the loss of live steam (estimated at 30%) in the condensing system, steam systems, boilers and furnaces.
(c) recover 50% of losses due to non-use of cooling water recirculation flow in heat exchange in cooling towers (ideal thermal performance (dti) = 95%; dti real = 75%; 20% difference). Thus, an original temperature difference of 10°C, with higher

energy efficiency reaches 11°C, a reduction of 1°C in cold water → bringing as positive impact improving the profile of the separation columns and reduction of fuel demand thereof magnitude and CO^2 emissions as a result, 10%.

Note that:

1. The amount of biomass used (petrochemical-based local → global petrochemical → global industry) inhibits the emission of CO^2 with the change of 20% for renewable matrix 8,45 GTY, compatible with the demands of the International Energy Agency.

2. Increased energy efficiency in thermal systems (petrochemical-based local → global petrochemical → global industry) reduces burn and CO_2 emissions by 12,5 GTY, compatible with the demands of the International Energy Agency.

4.4 Analysis of availability: strategic plan

Exercises for analysis of the availability of water, energy and skills

The program to control wastewater at the source and master plan provided operational activities and new projects to achieve the demand reduction of 750 to 310 m3/h, so below the 462 m3/h.

Scheduled actions are [29]: (2013-2018) (1) pattern of change in the final product plant shutdown; (2) stability in the operation of purification and reuse; (3) reprocessing the waste water treatment, reuse in cooling tower; (4) increase in operational reliability; (5) reuse of disposal with a high level of contaminant; (5.1) purifier contamination reduction - reduction of product load; (6) condensate polishing removal of organics. (2019-2020) (7) energy conservation project, condensate (I); (8) control work in the steam generation (II); (9) condensate return for demineralized water; (10) the entry of new plants with improved design pattern. (2020-2024) (11) of rainwater use project; (12) improvements in the use of cooling towers; (13) Water reuse for fire and utilities; (14) improvements in water treatment plant; (15) rationalize water use in the area and the administrative use. (2025 and 2028) (18) effluent reuse in cooling towers; (19) advanced technologies to reduce vapor loss - cogeneration and heat pumps. (2029 and 2034) (20) new techniques for heating and cooling: intensification; (21) new cooling technology projects: geothermal energy and application of the products.

The availability of energy and competence [8] are discussed from Figures 1 and 2. Figure 1 is considered the emission of CO2 in a temporal analysis which compares the issue (with and without adjustment of shares) with the offer is the maximum capacity of the atmosphere to process GHGs without climate change. In Figure 2 we analyze the defendant competence for technology with the expertise offered by people with and without adjustments regarding the offer.

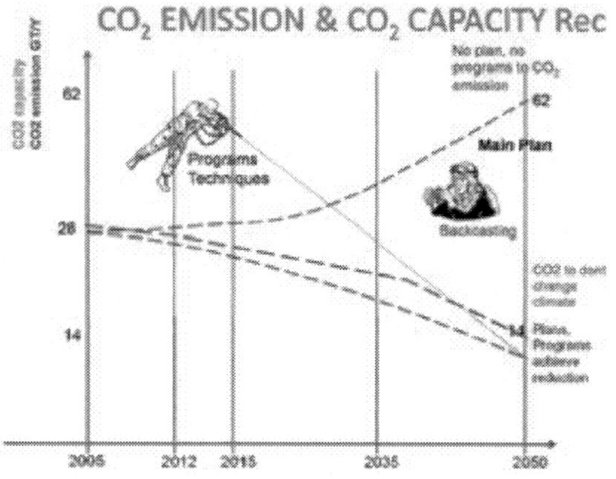

Figure 1 - Supply and demand GHG [8] Figure 2 - Supply and demand Competence [8]

<u>Master Plan for Water and Energy [8]</u>

The Master Plan is a strategic tool that enables meet the challenges of climate change and its consequences for the industry. Planning for adjustment and availability of water & energy for the industry is complemented with technical and management tools for the implementation of the Master Plan.

The Master Plan discusses the legal and institutional aspects of the laws dealing with water, energy and environmental impacts. supply management depends on legal issues that allow water and energy management in industry. The current and future availability is performed by comparing supply of water resources in local basins with the demand of these resources to the industry seeking the necessary adjustments. This discussion should also is made to address the constraints of the Brazilian energy matrix with the conditions for water use and CO2 emissions.

Analysis of water and energy classes and specifications of equipment and processes enables us to diagnose the possibility of water and energy reuse. The identification of classes used in the industry indicates their effective use to have qualities in different sources. The water treatment and effluent depends on the water balance enabling reuse projects.

Concepts and applications are detailed for adaptive management of supply and demand and which proposals for long-term strategic adjustments as credit demand reduction, industrial symbiosis concepts and opportunities, the importance of monitoring indicators, and the risk of lack of water for 10 and 40 years.

The ways to manage supply take into account: availability of water sources, climate change, hydro-geological perspective, costs and energy content. A calculation procedure is suggested for the adjustment of supply relative to the projected demand. In case of power must calculate the CO2 emission target for the specific case of the installation. The management of future demand includes the control program in the source material balance losses and others.

In the Annex of this plan are presented technical and managerial and operational methods. The intention is to keep demand and controlled supply. The main technologies to maintain demand and controlled supply for Industry are classified and detailed.

Control Equipment and Thermal Systems

The thermal control equipment and systems, process control and reliability of the systems are extremely important to regulate the demand for water and energy. The demand adjustment requires lower value to water supply (based on state water policy) and energy (based on local and global constraints of GHG emissions).

The equipment that has high impact on fuel consumption and the emission of greenhouse gases are: (a) cooling towers; (B) heat exchangers that utilize utilities; (C) generating steam boiler; (D) heating stoves or pyrolysis reaction; (E) steam distribution and condensate energy recovery (including traps); (F) current of high enthalpy process; (G) operating point and balance of separations such as distillation; and (h) electrical systems in general.

At this point will be discussed energy efficiency in some of these systems: (a) Deten-water research project in 2003, Ávila noted the importance of working with clean cooling water circuit to avoid blockages in the upper pools and highlighted the potential losses if maintaining the cooling tower structure (and water quality) is poorly made, with losses that may cause heating of water up to 1 ° C. By maintaining one should prevent the cooling water flow paths follow without heat and mass transfer with the dry air. (b) This project was highlighted the importance of working with cycle highest concentration in the towers to reduce water consumption. This control increases the risk of corrosion in order to be header exchangers and hinders water reuse in cooling towers [37] [38].

(c) There are several opportunities to recover waste heat in the boilers that generate steam in enhancing industry in energy consumption [30] [31] [38]. Since preheating of the water, control of stoichiometric combustion, the specification and maintenance of refractories, the utilization of heat of flue gas (new boiler, flue pump). (e) the design assumptions for steam distribution, steam traps installation and condensate energy recovery (including traps) should be based on the expansion possibilities of production and operating limits of the systems [30] [31] [38] . Thus, due to the limitations of operating with traps and thermal systems, search is the best point and with new needs, install new production modules.

Water balances, effluent, fuel, GHG and energy

In research projects to rationalize water and energy were used mathematical techniques mass balance (FAFEN-WATER, DETEN-WATER) and enthalpy (Ecobraskem) with multiplier weights varying according to quality information and then closing balance by data reconciliation. This technique was used for LAB production plants, Olefins, urea and petroleum [30] [31] [38]. As for the balance of utilities in equipment such as cooling towers, boiler and furnace energy balance it was based on ICPs and the formulas simplified the treaters water that is based on the wet bulb temperature, the flow and the temperature difference [38].

Sustainable projects: new techniques and technologies

The need to reduce energy consumption requires changes in equipment design and process. The main reason for these changes is to reduce the gaps, allowing a reduction in the consumption of utilities through direct heat transfer without the need for thermal fluid design adjustments considering the external environmental changes and changes in thermal load.

The spare capacity of operations and equipment is adjusted to increase the capacity of industrial plants. In industrial processes it is possible to carry out this adjustment due conservative design factors that enhance the capacity of the equipment, but in the case of utilities is not always possible. Thus, you can use the spare capacity of reactors, pumps, but it is not possible to trap and electric motors. With restrictions on the use of energy and knowledge in human reliability, it is intended to modularize the projects through more accurate system design [38].

Another option in the project area is to reduce the use of utilities and size of equipment by intensifying operations and processes where the heat necessary to the achievement of operations is transferred directly by the equipment wall. Where the heat is released from certain operation and received in another operation without the need for use of thermal fluids. If energy balance in petrochemical is noted that the use of exchangers with hot thermal fluids (steam) and cold (cooling water) consumes 4000 GJ / h compared effectively with the necessary heat (80 GJ / h) to unit operations work without utilities. With intensive use of energy can significantly reduce power consumption and size of equipment [38].

The rationalization of energy and water can be conducted using techniques which integrate the process with the utility and waste. The identification of water users with the material restrictions for equipment allows the water and energy savings through its reuse in equipment. This technique is applied in sewage networks and networks of heat exchangers.

In the case of wastewater networks (pinch), the contamination levels of waste water streams are analyzed to compose a quality goal that meets the requirements of equipment that can use this water. If exchanger network (pinch), the residual heat through the cooling water reuse allows a better use to reduce energy losses.

The main equipment that consumes water and has high potential for use of waste water is the cooling tower [32]. Thus, it is the integration of processes (sources) with the effluent (waste water standard), and returns to the utilities (water cooling).

Energy recovery from the best thermal performance in cooling tower and application of process integration (pinch) has as a result the reduction of water loss and the reduction of CO_2 emissions. To account for this loss reduction in CO_2 emission area is seeks to study the integration of the cooling tower, heat exchangers, separation operation (reflux), the heat demand (steam oven), burning fuel and CO_2 emissions.

The decrease in water supply from climate change requires research on the feasibility of industrial desalination including comparing other solutions involving energy supply and water (eg lifting deep brackish water). The current technology of reverse osmosis membrane desalination is still not profitable for water treatment industries in the private sector.

4.5 *Culture analysis and people*

Private and public institutions recognize the importance of investigating the matter culture and people for sustainability, but do not yet know what priority to this issue. To establish an organizational and strong safety culture is required: to know the tasks and risks; analyze the behavior through the routine signals; and establish metrics to identify the stage of safety culture.

An onshore oil industry indicated that the image directly affects the profitability of the company, so often the actions prioritize avoid independent accident effect on production and on revenues. A proposed action was suggested for intervention in three years: (1) Task: To establish indicators of efficiency, formalizing forms of communication in organizational levels to review the inadequate standards establishing best practices; (2) Behavior: analyze routine deviations, establish a causal link with the process losses, see how minor emergencies are handled (decisions) and indicate gaps in relation to skills, knowledge and standard of acceptance; (3) Culture: Establish indicators for technical culture (bottom-up) and measure the difference in perception between the entrepreneurial vision and practices. Set organizational practices (rituals) to neutralize or mitigate biases culture.

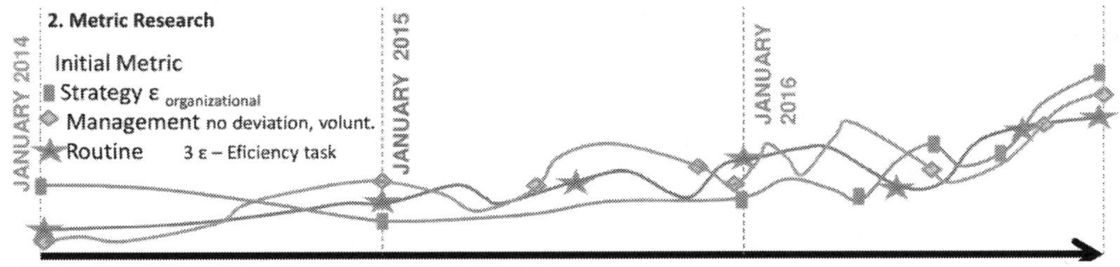

This initiative for the onshore site has been replicated through intervention in the fertilizer industry to analyze events in the routine that involves the effluent, where old vices of operation contribute to the process losses. The suggestion to change behavior and neutralization of decisions that led to contamination of effluent was through the educational program "Friends of emergency pond." The tools used in this program were: lectures in the control room and field, event analysis and hypothesis of the root cause, exchange of information through informative texts and making changes in the decisions of supervision and management.

The process that involves loss materials are more easily perceived as contamination of wastewater, fugitive emissions or the generation of solid waste. Large losses that affect the company's image as an accident or leaks has immediate impact on the stock market. Since energy losses are less visibility difficult behavior change to achieve targets in reducing GHG emissions. So it is important to consider cultural aspects and behavior to take more assertive decisions.

<u>Analysis of elements or human factors [24]</u>

In this GCPS organizational factors related to regional-global culture that are entered in 20 events in the chemical industry were discussed. These factors can initiate failure mode with final impact on energy losses. Thus, conflicts between policies and practices, economic difficulties, poor planning of the task, cultural traits that induce inappropriate decisions, risk management that does not include human factors are important factors to analyze the operational routine which leads to energy losses. The routine in deviant behavior difficult to establish operational discipline to avoid accidents and loss of energy [26].

Cultural analysis based on Archetypes [17]

Cultural phenomena do not exactly follow a logical order and are based on ancient customs rooted through tribal and new rituals introduced by the media. In an attempt to develop cultural barriers to the accident and GHG emissions seeks to analyze the phenomena resulting from the action of primary and derivative archetypes in the organization's tasks. This kind of technique indicates which new rituals should be deployed to vaccinate cognitive impairment.

Analysis of Cognitive Failure [19]

The cultural environment induces undue behaviors in individual or group during the task. Understanding the type of cognitive failure depends on the analysis of executive function in the various stages of cognitive processing: memory operations, environmental awareness for the construction of knowledge and action, attention calibration, job planning, job control and socioeconomic utility affective. It is important to identify the type of human error in the energy losses to prevent the ineffectiveness of action taken.

Reliability Analysis Sociotechnical [27]

The integration of production technology (product, process and control) with its operations on critical equipment indicates the importance of joint analysis of reliabilities. The operation achieves success if the understanding of the task objectives and knowledge of the process conditions (standard and fault). Thus, reliability of operations, processes and equipment depends on human reliability that receives the direct influence of cultural environments.

A propylene pumping operation in the refining industry was analyzed and it was concluded that the redundancy pumps is not enough when cognitive functions are weakened in conflict between policies and practices environment.

4.6 Routine management

The quality brought about the revolution of the administration [39] but did not ensure the sustainability of their results. The complex decisions and the investigation of systems failure mode indicate that to avoid losses should measure the efficiency and effectiveness of the task should be controlled thermal systems, critical procedures and also analyze the level of material losses and energy.

In the area of utilities and processes which circulates high amount of energy it is important to perform measurements and balances to prevent high losses of energy and water industry. The control of thermal systems involves critical steps such as maintaining water quality, purge tower and boiler, the level of control and the use of cells in cooling towers [31].

The production control by type of raw material defines the economic quantities of schedules based on variable cost of energy. In the petrochemical industry this issue has great importance in the definition of production programs knowing input naphtha type (weights and average).

Depending on the input naphtha occur demand for changes in processes, for example, the recycle gas will have another quality and will need to review the burners in the process gas boilers reuse [38]. To carry out the control of these tasks with high impact for the loss of energy is important to apply task analysis techniques in complex processes and have control over the realization schedule these tasks [40].

The mass balance and energy (water, materials and supplies, steam, fuel) enables the detection of losses and control at source [35] [23]. The balance of goods and inputs is closed with the quantification of losses: solid waste; liquid contaminants to the effluent; and fixed and fugitive emissions to the atmosphere.

4.7 Compliance Analysis of water and energy consumption

ISO 50001 [22] are described important procedures for the management and control of energy. It is important to identify sources (equipment and systems) lose water and energy in the production process. It is important to define if the losses are isolated or continuous, which the physical state and properties, the transport fluid and the impact receiving body. The output of thermal systems is measured by measuring the consumed energy and the comparison with expectations. Part of the energy loss and water is recoverable and the cost outweighs the benefit. compliance analysis indicates where the main losses and what the distance from the expected default behavior.

The fixtures and fittings that consume fuel or indirectly affect this consumption are extremely important: boiler, furnaces, heat exchangers, steam traps, cooling towers, water and steam distribution pipes, flash vessels, safety valves, and distillation columns .

It is important to define the consumption profile (baseline), the expected pattern of consumption, benchmark-based target, the difference between the measurement and the expected pattern (loss) and the percentage of impairment. It is also important to describe the operational procedures that can impact on the loss of water and energy consumption, expected standard and goal. These losses may have been caused by misconduct or lack of knowledge.

The criteria for equipment performance analysis are based on rules established by institutions such as ASME (boilers and furnaces) and CTI (cooling tower). Other criteria related to behavior and procedures may be based on API770 and operational discipline guidelines CCPS.

5 Conclusion

Initially considered the normal trend of events without intervention in the provision of management or demand management. On the one hand occur climate change altering the supply of water and energy, on the other hand is the population and industrial growth by increasing demand for water and energy. In this case the industry will be trapped in a situation of lack of water and energy across the demand relative to supply. In our case this will occur in 2022 (for water).

According to the problems already experienced shortages, so there is a balance between supply and water demand and energy are already demonstrated the means and methods that need to be taken into consideration and applied in the productive sectors and society. A case of the fertilizer and petrochemical industry have been discussed.

Already have the knowledge that technologies and techniques have been implemented in the industrial sector considerably reducing the waste of raw materials and fuel and thus environmental damage, but their implementations are not prioritized the way it should, as a general in contrast, there is a cultural influence, behavioral, economic and management that becomes increasingly slow the process of improvement.

With the projection of a future scenario and possible interventions mainly developing countries GHG emissions planned for the year 2050 is approximately 14GTano, very difficult task to achieve.

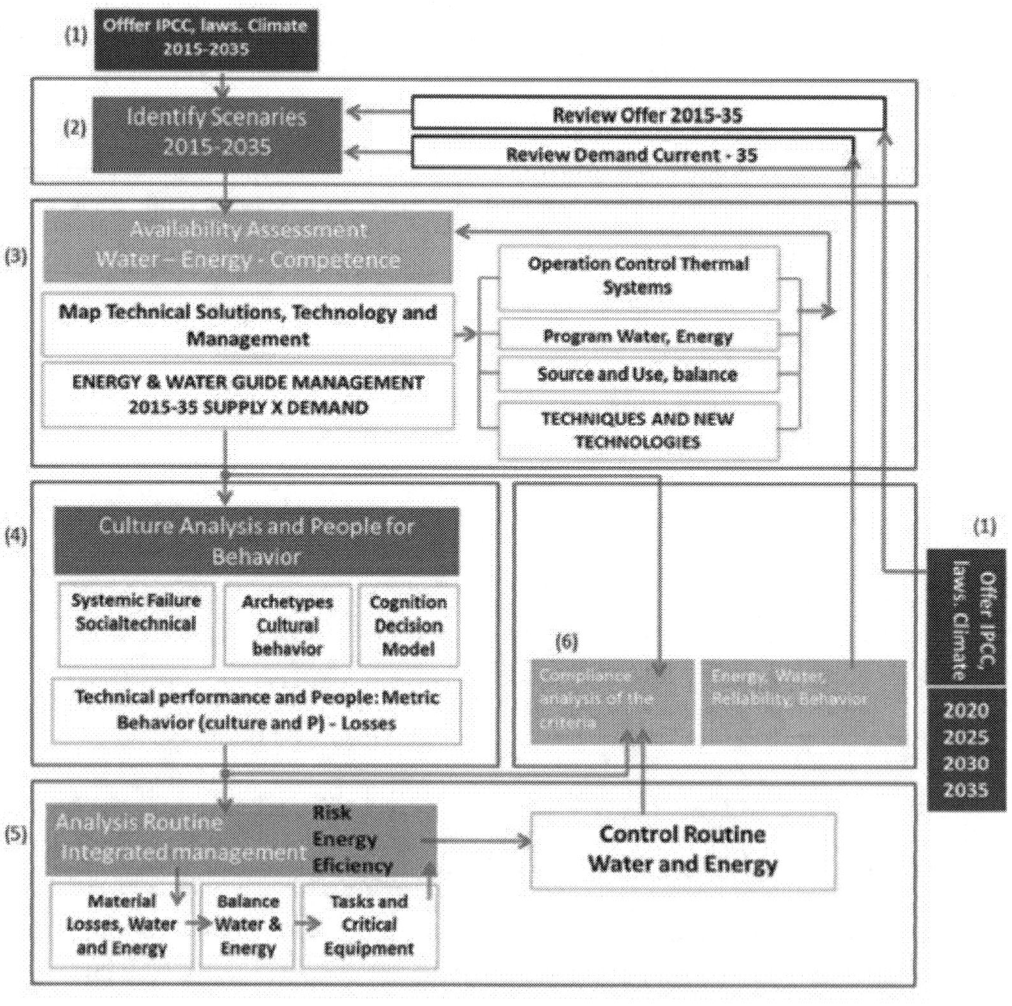

Figure 1. Analysis of Water Availability, Energy and behavior.

6 Bibliography

[1] CARMO, R. L.; DAGNINO, R. S.; FEITOSA, F. F.; JOHANSEN, I. C.; CRAICE, C. População e consumo urbano de água no Brasil: interfaces e desafios. In: SIMPÓSIO BRASILEIRO DE RECURSOS HÍDRICOS, **Anais...** Bento Gonçalves, 2013.

[2] WWDR – The United Nations World Water Development Report 2015: Water for a Sustainable World. Paris, UNESCO, 2015.

[3] FUNDAÇÃO SISTEMA ESTADUAL DE ANÁLISE DE DADOS – SEADE. Disponível em: < http://www.seade.gov.br/pibtrimestral/analise/>. Acessado em: 06 fev. 2016.

[4] PQU. SABESP. Sabesp e Polo Petroquímico lançam o maior projeto de Água de Reuso do País. 2008. Disponível em: <http://www.agenciagabc.com.br/noticias.asp?id=1045>. Acesso em: 06 fev. 2016.

[5] ROCKSTRÖM, J. *et al*. Planetary Boundaries: Exploring the Safe Operating Space for Humanity. Ecology and Society 14(2): 32. 2009. Disponível em: <http://www.ecologyandsociety.org/vol14/iss2/art32/>. Acessado em: 06 fev. 2016.

[6] ANEEL – Agência Nacional de Energia Elétrica. Matriz de Energia Elétrica. 2016. Disponível em: <http://www.aneel.gov.br/aplicacoes/capacidadebrasil/OperacaoCapacidadeBrasil.cfm>. Acesso em: 06 fev. 2016.

[7] MIERZWA, J. C.; HESPANHOL, I. Água na indústria: uso racional e reuso. São Paulo: Oficina de Textos. 2005.

[8] ÁVILA, S. F.; LOPES, J. R. N.; SOUZA, C. R. O. Strategies for industry sustainability. In: WREF 2012 - WORLD RENEWABLE ENERGY FORUM. 2012, Denver.

[9] BONALUME NETO, R. Perda de 30% da água durante distribuição contribui para crise em SP. **Folha de São Paulo**, São Paulo, 05 de jun. 2014. Fórum de Sustentabilidade.

[10] DOORNBOSH, R.; GIELEN, D.; KOUTSTAAL, P. 2008. Mobilising Investments in Low-Emission Energy Technologies on the Scale Needed to reduce the Risks of Climate Change, OECD SG/SD/RT (2008) 1, Paris, France.

[11] IPCC (2007), Metz, B.; Davidson, O. R.; Bosch, P. R.; Dave, R.; Meyer, L. A. Technical Summary in: Climate Change 2007: Mitigation. Contribution of Working Group III to the Fourth Assessment Report of the Intergovernmental Panel on Climate Change, Cambridge University Press, Cambridge, United Kingdom and New York, NY, USA.

[12] FURTADO, M. Água – Clientes exigem resfriamento de baixo custo. Revista Química e Derivados. São Paulo, pp. 32 – 34., 2009.

[13] PARRY, M. L. et al. (Ed.). Climate change 2007: impacts, adaptation and vulnerability; contribution of Working Group II to the Fourth Assessment Report of the Intergovernmental Panel on Climate Change. Cambridge (UK): Cambridge University Press, 2007f. 976p.

[14] ÁVILA, S. F. Análise de falha em processos complexos. In: CONGRESSO BRASILEIRO DE ENGENHARIA QUÍMICA, 19,, 2012. **Anais...** Búzios, Rio de Janeiro.

[15] ÁVILA, S. F.; PESSOA, F. L. P. **Proposition of review in EEMUA 201 & ISO Standard 11064 based on cultural aspects in labor team, LNG case.** Procedia Manufacturing, vol. 3, pp. 6101 – 6108, 2015.

[16] ÁVILA, S. F. **Reliability analysis for socio-technical system, case propene pumping**. Engineering Failure Analysis, vol. 56, pp. 177 – 184, 2015.

[17] ÁVILA, S. F.; MENEZES, M. L. A. Influence of local archetypes on the operability & usability of instruments in control rooms. In: EUROPEAN SAFETY AND RELIABILITY CONFERENCE – ESREL 2015. 25., Zürich, Switzerland.

[18] AMERICAN PETROLEUM INSTITUTE. API 770.2001: A manager´s guide to reducing human errors. Improving Human Performance in the Process Industries. Washington, DC., 2001.

[19] ÁVILA, S. F.; COSTA, C. Cognitive deficit analysis in routine tasks, as a strategy to reduce accidents and increase of industrial production. In: EUROPEAN SAFETY AND RELIABILITY CONFERENCE – ESREL 2015. 25., Zürich, Switzerland.

[20] REASON. J. Human error. New York: Cambridge University Press, 2003. 302p.

[21] ÁVILA, S. F.; DANTAS, E. T.; DUARTE, J. The Requirements & tools to treat process safety risk as result of Organizational Change. In: GLOBAL CONGRESS ON PROCESS SAFETY. 11., 2015. **Anais…** Austin, Texas.

[22] ABNT, Associação Brasileira de Normas Técnicas. **NBR ISO 50001:2011: Sistemas de Gestão de Energia: Requisitos com orientações de uso.** Rio de Janeiro: ABNT, 2011.

[23] ÁVILA, S. F.; KALID, R. A.; KIPERSTOK, A.; CARVALHO, V. S. Program to Avoid Product Leakage at Source, Fertilizer Case. Chemical Engineering Transactions. vol. 26, 2012.

[24] ÁVILA, S. F. Assessment of Human elements to avoid accidents and failures in task perform, cognitive and intuitive schemes. In: GLOBAL CONGRESS ON PROCESS SAFETY. 7., 2011. **Anais…**Chicago, Illinois.

[25] ÁVILA, S. F.; KIPERSTOK, A.; KALID, R. A.; SANTOS, A. C.; CARVALHO, V. S. Environmental Deviation Assessment, Record, Failure, Probability and Investment. Chemical Engineering Transactions. vol. 26, 2012.

[26] CCPS (Center for Chemical Process Safety), Conduct of Operations and Operational Discipline: For Improving Process Safety in Industry, American Institute of Chemical Engineers, New York, NY, 1985.

[27] ÁVILA, S. F. Reliability analysis for socio-technical system, case propene pumping. Engineering Failure Analysis, vol. 56, pp. 177 – 184, 2015.

[28] GERTMAN, D.; BLACKMAN, H.; MARBLE, J.; SMITH, B. C. The SPAR-H Human Reliability Analysis Method. Idaho National Laboratory, Washington DC, 2005.

[29] ÁVILA, S. F.; LOPES, J. R.N.; KALID, R. A.; KIPERSTOK, A. F. Gestão da disponibilidade de água e energia através do Plano Diretor. In: RIO OIL & GAS EXPO AND CONFERENCE 2014. Rio de Janeiro.

[30] ÁVILA, S.; KIPERSTOK, A.; BRAGA, B.; KALID, R. Avoiding loss of energy in a petrochemical industry, operation and design Congress 2011. In: WORLD RENEWABLE ENERGY CONGRESS 2011. Linköping, Sweden.

[31] ÁVILA, S. F.; COIMBRA, H.; SOUZA, C.; GOLÇALVES, A. Iniciativas de eficiência energética para sustentabilidade, oportunidades & desafios. In: RIO OIL & GAS EXPO AND CONFERENCE 2014. Rio de Janeiro.

[32] ÁVILA, S. F.; JESUS, Z. M. R.; KALID, R. A.; NASCIMENTO, V. Good practices in Cooling Towers: Comparison between industrial audits. In: Cooling Technology Institute Annual Conference. Corpus Christi, Texas, 2003.

[33] GENZ, F., TANAJURA, C. A. S., DE ARAÚJO, H. A. Impacto das Mudanças Climáticas nas Vazões do Rio Pojuca sob o Cenário A2 do IPCC - 2070 a 2100. In.: I CONGRESSO BAIANO DE ENGENHARIA SANITÁRIA E AMBIENTAL. Salvador. 2010.

[34] SILVA, S. F. Análise da disponibilidade e demanda para o sistema de abastecimento de água de salvador frente a cenário de mudanças climáticas. Dissertação (mestrado). Programa de pós-graduação em Meio Ambiente, Águas e Saneamento. UFBA. Salvador BA. 2012.

[35] KIPERSTOK, A.; BRAGA, E. A.; CESAR, M.; ÁVILA, S. F.; SAMARA.; KALID, R. A. Eds: 2012. Base de Informações para Plano Diretor de Água Fafen-BA, TECLIM RESEARCH FAFEN – UFBA. Camaçari, Brasil.

[36] SEDUR, SECRETARIA DE DESENVOLVIMENTO URBANO DA BAHIA. Estudos de Aproveitamento dos Mananciais da RMS – Revisão e Atualização das Proposições do Plano Diretor de Abastecimento de Água, Incluindo o rio Pojuca, e Visando o Uso Racional dos Recursos Hídricos Regionais - Estudo de Demandas. Relatório 1100.11-REL. EST. DEMANDAS-00. 73p. Salvador. 2010.

[37] SANTIAGO, P. S., & SANTOS, R. P. Major Improvements in cooling system performance at an ammonia/urea plant. IWC – International Water Conference, Proceedings: Pensylvania: 2000.

[38] ÁVILA FILHO, S., NASCIMENTO LOPES, J.R. Cooling Water Systems: An Overview, in: Z. Amjad, K.D. Demadis (Eds.), Mineral Scales and Deposits: Scientific and Technological Approaches, 2015, pp. 499–531.

[39] DEMING, W. E. Qualidade: A revolução da Administração. Rio de Janeiro: Editora Marques-Saraiva. 1990. 367 páginas.

[40] ÁVILA FILHO S., SOUSA C. R., CARVALHO A. C. Assessment of complexity in the task to define safeguards against dynamic risks. 6th International Conference on Applied Human Factors and Ergonomics (AHFE 2015) and the Affiliated Conferences, AHFE 2015. Procedia Manufacturing 3 (2015) 1772 – 1779.

442403 A Novel Open Source CFD Code for Explosion Modelling

Abstract:

The porosity parametrization of complex geometry is very important in explosion simulation. This project proposes a novel approach to calculate the mesh porosity (volume and area) based on the Gilbert–Johnson–Keerthi (GJK) distance algorithm. The GJK algorithm can be used to check the collision between convex objects. This concept is used in the current work to check the collision between an element of the mesh and an object of the geometrical model. The parametrized geometry is combined with the combustion model based on flamelet formulation in accordance with BML (Bray-Moss-Libby) model. The typical chemical libraries are replaced by the integral length scale of wrinkling based on semi-empirical formulation. One simple step reaction mechanism is considered. The turbulence closure problem is addressed via Boussinesq formulation using two equations model for turbulent kinetic energy and the rate of dissipation of turbulent kinetic energy.

The CFD (Computational Fluid Dynamics) solver relies on an explicit finite-volume code based on a structured Cartesian grid. The classical fourth order Runge-Kutta time marching approach is considered. The PDR (porosity distributed resistance) approach is used to model all geometry under analysis (including the large-scale objects and the small ones). The central differencing scheme is applied to discretize the governing equations and artificial viscosity is used to smooth out numerical wiggling. Additional resistance terms are considered in the momentum equation. Additional generation source term for turbulence is considered in the turbulent kinetic energy equation. The simulation of standard flow across different geometries presented consistency with the literature. Good agreement is found for similar simulations using a commercial CFD software. The developed open CFD code seems to be promising as an alternative to solve complex explosion problems from the engineering view point.

442442 Compliance with Process Safety Management Elements

Abstract:

OSHA promulgated 29 C.F.R. § 1910.119 regulation in 1992 and the fourteen Process Safety Management (PSM) elements. The main objective of defining these elements is to prevent unwanted discharge of highly hazardous chemicals exposing personnel to serious hazards. The elements are linked to each other and have intricate interdependencies. Every element provides information to other element(s) or uses information from other element(s). Given the interdependencies of these elements, an effective Process Safety Management Program (PSM) has to be in place which considers process design, technology, procedures (both normal and emergency), emergency response plans, training programs etc. The presentation talks about all these fourteen elements in details with main emphasis on the relief systems design.

442454 A Fuzzy Logic and Probabilistic Hybrid Approach to Quantify the Uncertainty in Layer of Protection Analysis

Abstract:

Layer of Protection Analysis (LOPA) is a widely used semi-quantitative risk assessment method. It provides a simplified but less precise method to assess the effectiveness of protection layers and the risk of an incident scenario. The outcome frequency and consequence are intended to be conservative by prudently selecting input data, given that design specification and component manufacturer's data are often overly optimistic. There are many influences, including design deficiencies, lack of layer independence, availability, human factors and wear in testing and maintenance, which are not quantified and are dependent on type of process and location. This makes the risk usually overestimated. So, there are different sources and types of uncertainty in the LOPA model that need to be identified and quantified. In this study, a fuzzy logic and probabilistic hybrid approach was developed to quantify the uncertainty of frequency of initiating event and the probabilities of failure on demand (PDF) of Independent Protection Layers (IPLs) based on the available data and expert experience. The method was applied to a distillation system with a capacity to distill 40 tons of flammable n-hexane. The outcome risk of the new method was proven to be more precise comparing to the results of the conventional LOPA approach.

442474 Quantitative Risk Analysis of Nitrogen Fertilizer Plant, of Moron Petrochemical Complex

Abstract:

Quantitative Risk Analysis of Nitrogen Fertilizer Plant, of Moron Petrochemical Complex

Considering the possible risk scenarios associated with new Nitrogen Fertilizer Plants (Ammonia, Urea and Utilities) of the Moron Petrochemical Complex, in general, are common to most petrochemical complexes in the world and are a result of materials hazardous that are manufactured, processed, handled and stored at the facility; the various processing systems and conditions; the different procedures used for the operation and maintenance of the facility; as well as the response of the detection and mitigation of hazards. Previous experiences in similar plants indicate that the most likely hazards are those who identify with the physical-chemical properties of the materials used and the process conditions.

For petrochemical facilities, the most common consequences are:

- Toxic Gas Effects
- Jet Fire
- Flash Fire
- Pool Fire
- Bleve
- Vapor Cloud Explosions.

Based on the premise of "Only what is known can be controlled," This paper aims to develop the Quantitative Risk Analysis (QRA) for new Nitrogen Fertilizer plants (Ammonia, Urea and Utilities) of Moron Petrochemical Complex, which it is located in the jurisdiction of the Municipality Juan Jose Mora, Carabobo State from Bolivarian Republic of Venezuela, considering for it the most severe scenarios identified previously by HAZOP analysis, SIL studies and previous plannings studies for the control and emergency response and contingency, then determining the levels of social and individual risk based on the probability of occurrence of the events of greater impact and its consequences and impact on people, equipment and buildings, comparing the risk levels obtained with tolerance criteria established by the corporation.

The Quantitative Risk Analysis (QRA) are developed from the standard and guidelines of Petroleos de Venezuela, (PDVSA); Center of Chemical Process Safety (CCPS), and The Netherlands Organization (TNO), through the computational tool PHAST RISK 6.7 . from Det Norske Veritas (DNV).

Management of Plant IPL's

Richard Stanley
Tehama Control Systems Consulting Services
Consultant to Mangan Software Solutions
26741 E Clifton Drive
Aurora, CO 80016
rick@tehamacontrolservices.com

Prepared for Presentation at
American Institute of Chemical Engineers
2016 Spring Meeting
12th Global Congress on Process Safety
Houston, Texas
April 11-23, 2016

Management of Plant IPL's

Richard Stanley
Tehama Control Systems Consulting Services
Consultant to Mangan Software Solutions
26741 E Clifton Drive
Aurora, CO 80016
rick@tehamacontrolservices.com

Keywords: IPL, LOPA, Independent Protection Layer, Safety Life Cycle Management, Safety Instrumented System, Safety Instrumented Function

Abstract

Independent Protection Layers (IPL) are usually identified in a Layers of Protection Analysis (LOPA) during a Process Hazards Analysis. The IPL portion of a LOPA identifies Functions or Design Features within a processing facility that serve to prevent or minimize the consequences of a deviation from the expected processing conditions. Many times a LOPA may identify various IPL's, but the facility then has difficulty in clearly associating the functions identified by an IPL with real plant assets. They then have difficulty in assuring that those assets have the necessary operational and maintenance visibility, and are operated and maintained to assure their availability. This paper discusses these issues and describes the characteristics of an effective Management Program for IPL's.

1 Introduction

During Process Hazard Analysis (PHA) for processing facilities, a number of Independent Protection Layers (IPL) may be identified as having the capability to reduce the risk of a given consequence occurring. A given scenario may require that one or more of these IPL's be in place to reduce risk to tolerable levels. Typically, these IPL's are identified during a Layers of Protection Analysis (LOPA), but other methods can be used.

Once these IPL's have been identified, the process of managing them has just started. Designation of an IPL adds Operational and Maintenance requirements that often exceed typical practices. This presents an operating facility with the challenge of adapting Operation and Maintenance practices to assure that the IPL's that are credited are well understood, that their protective functions remain available and meet the performance requirements that were defined for them.

This paper discusses the various type of IPL's and the requirements and methods for assuring their availability. The basic ownership requirements are described along with observations relative to the difficulty that real facilities have had in addressing these requirements. This particularly applies to the less visible IPL types such as those IPL's that are not associated with Safety Instrumented Functions (SIF).

After discussion of the ownership requirements and obstacles to meeting those requirements, the use of a Safety Life Cycle database to overcome these obstacles is discussed. The effective application of a Safety Life Cycle database can enable an operating organization to manage those requirements with maximum effectiveness and value with minimum impact of existing organizational structures and accountabilities.

2 Independent Protection Layers

A key concept in managing Process Hazards is that of an Independent Protection Layer. The formalized concept of an IPL was developed along with the LOPA process. During a PHA the potential consequences of various deviations in plant operation are evaluated, and the potential consequences of those deviations if no preventative measures are taken are defined in qualitative terms.

When an unmitigated risk identified by the PHA is higher than an organization's tolerance, the probability of the event occurring may be further evaluated to better understand the risk and the effectiveness of the various potential safeguards that could prevent consequence from occurring.

The most common technique for performing this analysis is the Layers of Protection Analysis (LOPA). In a LOPA, functions or design characteristics that are credited as being able to prevent the consequence are called Independent Protection Layers (IPL).

2.1 IPL Identification

The measure of effectiveness of identified IPL's is generally assessed against an organization's risk policies using a Target Mitigated Event Likelihood (TMEL). TMEL is usually based upon the severity of the LOPA scenario unmitigated consequences. For example, a severity level may have a required TMEL of 1e-04, or one occurrence of the unmitigated consequence in every 10,000 years.

During the LOPA, various Initiating Causes for a Deviation scenario are identified and each is assigned an expected frequency of occurrence. Some organizations may choose to apply additional frequency mitigation factors, such as the probability of personnel being in the area at the time of the Initiating Cause. Some organizations will choose to sum Initiating Causes, while others do not.

The product of expected frequency of the initiating causes, any additional frequency mitigation and Probability of Failure Upon Demands (PFD's) of the credited IPL's is used to calculate the value of the Total Mitigated Event Likelihood ($TMEL_{act}$). If the $TMEL_{act}$ is below the threshold specified for the consequence level associated with the event, the number of IPL's is considered to be sufficient. If the $TMEL_{act}$ exceeds the threshold the LOPA team recommends corrective actions, often recommending that additional IPL's be added. However, other recommendations such as process modifications to reduce the consequences or reduce the initiating causes may also be recommended. An example of a LOPA TMEL analysis is shown in Figure 1.

Figure 1 – Sample LOPA TMEL Analysis

LOPA TMEL Calculation Example

Consequence TMEL = 1.0E-05 (1 consequence Event per 100,000 years)

IC1 - 0.1 (once per 10 years)
IC2 - 0.1 (once per 10 years)
MF - 0.1 (Personnel present 10% of the time)

IPL1 PFD = 0.1
IPL2 PFD = 0.01

TMELact = (IC1+IC2) * MF * IPL1 * IPL2
TMELact = (0.1+0.1) * 0.1 * 0.1 * 0.01 = 2.0E-06

Actual TMEL (TMELact) is less than TMEL, so the IPL protection is acceptable. No additional protection is required

2.2 What is an IPL?

An Independent Protection Layer is an instrumented or physical function which, if effective, would prevent the unmitigated consequences of a scenario identified in a PHA or LOPA from occurring.

The primary characteristics of an IPL are:

- The proper functioning of the IPL would prevent the consequence associated with a scenario from occurring
- The IPL is independent of any of the scenario initiating causes and is independent of other IPL's cited for the scenario.
- The IPL is reliable enough to perform the expected function with the probability of failure being with the range with which the IPL is credited.

- The physical and functional design of the IPL meets the required independence and availability requirements.
- The IPL is Operated and Maintained in a manner that assures the credited reliability is achieved during plant operation.

CCPS [1] provides guidelines as to what functional, design or physical aspects of a process installation can qualify for being credited as an IPL, and describes the limitations for where multiple IPL credits may be taken. However, it is up to the facility owner or operator to determine specific local requirements. It should be noted however, that a 2015 US OSHA interpretation [2] appears to extend the concept of Recognized and Accepted Good Engineering Practices (REGAGEP) to publications that are not necessarily consensus documents. CCPS publications, such as the LOPA publication cited above, fall into this category. Furthermore, the OSHA interpretation notes that internal standards must represent REGAGEP and will reject standards that do not comply.

2.3 IPL types

Some of the more common IPL types are:

- Operator response to an Alarm
- A Basic Process Control System Automatic Action
- A Safety Instrumented Function (SIF) implemented in a Safety Instrumented System (SIS)
- A non-SIF interlock
- Pressure relief devices
- A physical barrier that would contain process fluids after a loss of primary containment. An example is a tank dike or a drainage system
- Physical equipment that would prevent a scenario from causing the unmitigated consequence associated with it. For example, this could be a valve limit stop, a restriction orifice or other flow restriction that prevents excessive inflow.
- Inherently Safe Design

2.4 IPL's and the Safety Life Cycle

The concept of the Safety Life Cycle is often considered to be associated with Safety Instrumented Systems (SIS) because its prominence in IEC 61508 and IEC 61511. However, the concepts are applicable to any protective functions or systems and are used in other contexts.

An important factor to recognize is that the risk reduction provided by IPL's is the product of their RRF's, and that if any IPL fails to meet the performance credited by a LOPA, the associated risk reduction of that IPL is lost. CCPS recognizes this by requiring that any IPL be Operated and Maintained in manner that assures that the credited Risk Reduction is effective.

The IEC 61511 and ISA 84.00.01-2004 Standards provide a framework for managing of Safety Instrumented Systems (SIS) and the Safety Instrumented Functions (SIF) that were identified by the Hazard Analysis process. However, these Standards are silent on managing of other non-SIF IPL types.

This has led some operating organizations to conclude that there are no corresponding requirements relative to the non-SIF IPL's. This is not the case: non SIF IPL's require sufficient care to assure their availability and cannot just be run to failure or be treated like non-IPL functions. The sections in the remainder of this paper discuss the practicalities of IPL implementation and ownership, the risk gaps that have been observed, and suggested management practices to prevent these gaps from developing.

3 What Should an IPL Management System Look Like

One of the fundamental gaps observed in practice is that there is often a perception that once an IPL is credited in a LOPA, and something exists in the facility that appears to perform the function defined by the LOPA, no additional action is required. This results in a system in which the effectiveness of an IPL is left to chance and the assumed risk reduction becomes an illusion.

An effective IPL Management System must be able to assure that IPL's are visible, that their design meets the functional and independence requirements for an IPL, and that the IPL's are Operated and Maintained to assure the required availability.

IPL's types and their implementation can vary considerably. While SIF IPL's may be paid more attention because of standards such as IEC 61511 and ISA 84.00.01-2004, non-SIF IPL's are usually shared with other basic plant systems such as the Basic Process Control System (BPCS) or blend in with other mechanical systems. Without an effective IPL Management System, non-SIF IPL functions often are either not effective to begin with, or atrophy to the point were they do not deliver the risk reduction with which they were credited.

Table 1 below identifies some of the gaps in IPL Management that have been observed in real operations. Unfortunately, these gaps are more common than not.

Table 1 – IPL Management Gaps

IPL Management Gap	Cause	IPL Management Requirements
Improper IPL credits are taken	Personnel are not properly trained on what the requirements and limitations for IPL qualification are, or local standards do not comply with REGAGEP	LOPA facilitators must be well trained on IPL requirements and advise participants on IPL requirements including IPL Types allowed and restrictions taking multiple credits. LOPA's should be reviewed by qualified and empowered individuals to assure validity.
IPL's as implemented do not meet the effectiveness, independence and reliability requirements defined in the LOPA's	Plant assets are identified of being IPL's, but are not evaluated to verify that they meet the IPL requirements.	Checklists for assessment of IPL qualifications are available and all IPL's are assessed against those requirements.
IPL functions are not understood by Operations and Maintenance personnel	IPL's are identified in LOPA documents and are not clearly identified in documents that are used by Operations and Maintenance personnel. LOPA documents are not readily available to non-PSM personnel and then are difficult to correlate to actual plant installations.	A system is needed that clearly identifies plant assets that are used to implement IPL's. The data should identify real plant assets that perform IPL functions, the Operational and Maintenance expectations for the IPL's, including operator actions, restrictions on disabling or bypassing the IP and Maintenance and Testing Requirements.
IPL's are bypassed or disabled	Assets credited as IPL's are not clearly identified. Operating and Maintenance procedures for IPL's are not prepared, available or enforced	Operating procedures need to exist that define criteria for bypassing or disabling of IPL's and requirements for management or operations while bypassed or disabled.
IPL's are modified or removed without proper MOC	The IPL list is not available, or is out of date. Projects making modifications do not recognize devices which are in IPL service and treat their modification like modification of non-IPL services	A continuously up to date IPL list is provided and linked to all LOPA, IPL design and operations and maintenance data. Projects evaluating proposed modifications have all the data needed to perform a proper MOC available in one place.
IPL's are not tested at the intervals required to maintain performance	IPL's are not identified in maintenance systems or do not have test procedures available. When procedures are available, management is ineffective because of many documents in many places, many of which may not be consistent or up to date	Test intervals, maintenance procedures and testing procedures are integrated in one database. Test records are captured in the Operations and Maintenance Module to allow for identification of next required test interval.

3.1 *Why Is Compliance So Hard?*

Many organizations have demonstrated difficulty in incorporating the requirements of IPL ownership and operation into their day to day practices. The basic reasons for this are that the importance of IPL's is not always recognized, and that effective management of IPL's generally does not fit within traditional plant management structures.

Effective IPL management requires that information cross traditional organizational boundaries. In traditional operating facilities, these boundaries usually result in many parallel information and documentation systems that have limited accessibility or provide only the functionality that a particular group requires and does not necessarily support the needs of other groups. Some examples of this are:

- Process Safety Groups have proprietary software that often has highly restrictive incensing terms that effectively lock up HAZOP and LOPA data access to just a few people.
- Maintenance Management Systems have historically been developed for management of major rotating equipment by and for rotating equipment personnel. Other functions are focused upon parts inventory management and work order tracking. These systems perform poorly when other systems are involved. They are particularly bad at handling Instrumented Systems. Resistance to making them effective for Instrumented Systems, or handling safety related non-rotating equipment has typically been quite high.
- Much of other traditional information management is done with discrete paper files or electronic files that are often not coordinated or readily available. These records are held by various organizations and are difficult coordinate and keep up to date. Often these records do not survive changes in personnel.

Management of IPL related data across these multiple platforms and formats is extremely inefficient and easily broken. Most facilities do not have the personnel or skill sets required to manage these problems. The Systems, even when well intentioned, tend to fail under the burden the highly labor intensive requirements of maintaining this data.

3.2 *Safety Life Cycle Management*

Effective management of IPL's across the entire Safety Life Cycle requires Safety Life Cycle Management systems that address the cross discipline nature of the data required and supports the following types of information:

- Defining relationships between Hazard Identification, Functional Requirements and Design of Protective Functions and Equipment, and Operation and Maintenance requirements.
- Capturing the Operational and Maintenance Events that are related to the IPL functions and equipment and using that data to provide Operation and Maintenance Management a clear picture of compliance and performance.

- Allowing users from various organizations to readily access IPL related data including the Hazard associated with IPL Assets, IPL functional requirements and equipment specifications and IPL Asset events and performance

Figure 2 illustrates a high level view of the relationships that should be defined and maintained and their general relationship to the Safety Life Cycle.

- HAZOP scenarios that have a high consequence are related to LOPA's
- LOPA's and the IPL's that were credited by the LOPA are identified
- Plant Assets that are used to perform the functions identified by a LOPA IPL are identified and their relationships to to the LOPA IPL's are shown. Note that in some cases, multiple LOPA IPL's may refer the the same Plant Asset.
- The Plant Asset data contains definition of the functions required, performance requirements, testing requirements, expected operator responses and physical equipment used to implement the functions.
- Operation and Maintenance Events are recorded and associated with the Plant Assets or Plant Equipment.
- KPI's are derived from the Operation and Maintenance Events and used to manage such things as period testing and assess performance such as demands, failures, bypasses.

Figure 2 – IPL Asset Data and Relationships

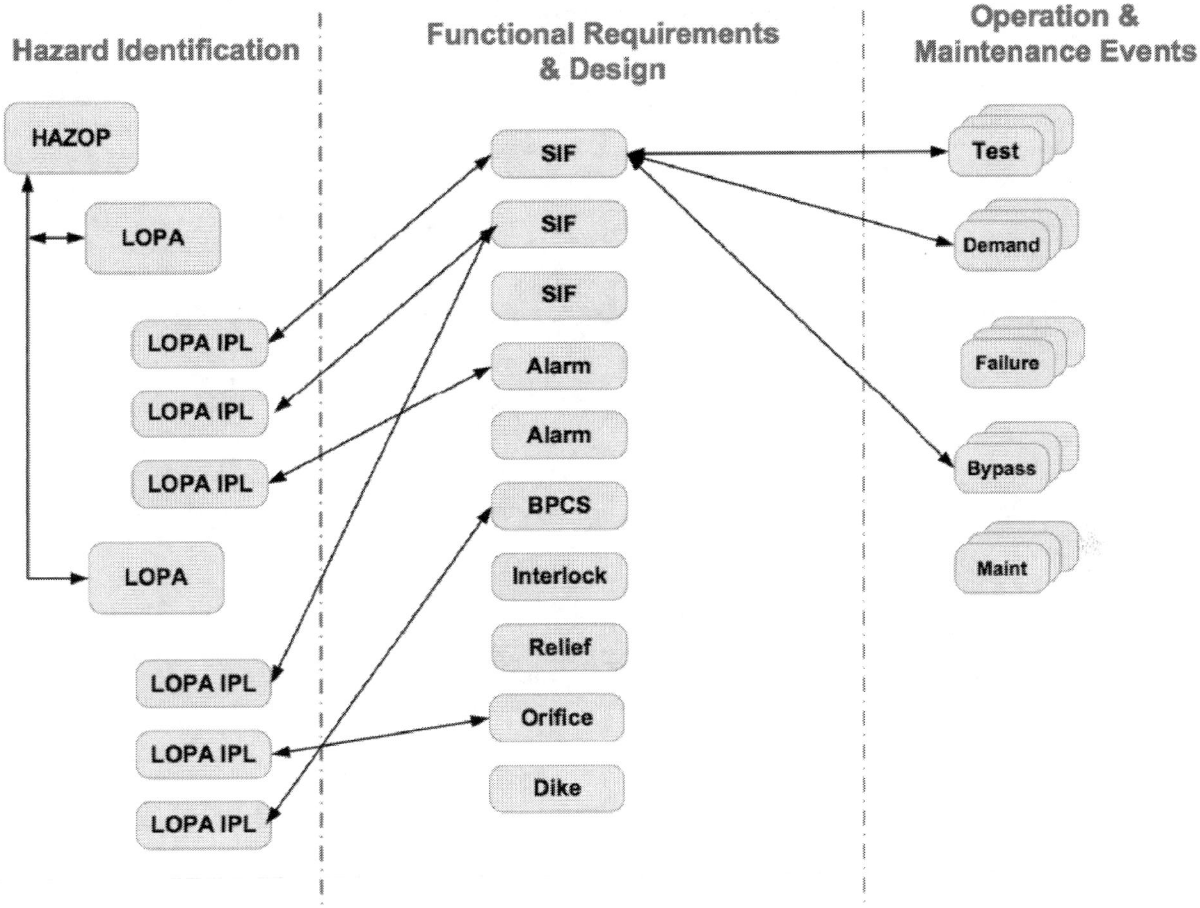

Figure 3 shows a more detailed set of data and relationships that focuses on one Plant Asset that is used to perform IPL functions.

- The Asset is related to one or more LOPA IPL's and to the equipment used to perform the Asset functions
- Plant Equipment that is used to implement an IPL Asset, such as instrumentation or other physical equipment is associated with the IPL Asset
- Assessment checklists and results are associated with the LOPA IPL's. Design Assessment checklists are associated with each Plant Asset and its associated equipment.
- Important documents such as Operating and Maintenance procedures or design documents are associated with the Asset for quick access.
- Operating and Maintenance Events are recorded and associated with their related equipment or Plant Assets

Figure 3 – IPL Asset Data and Relationship Details

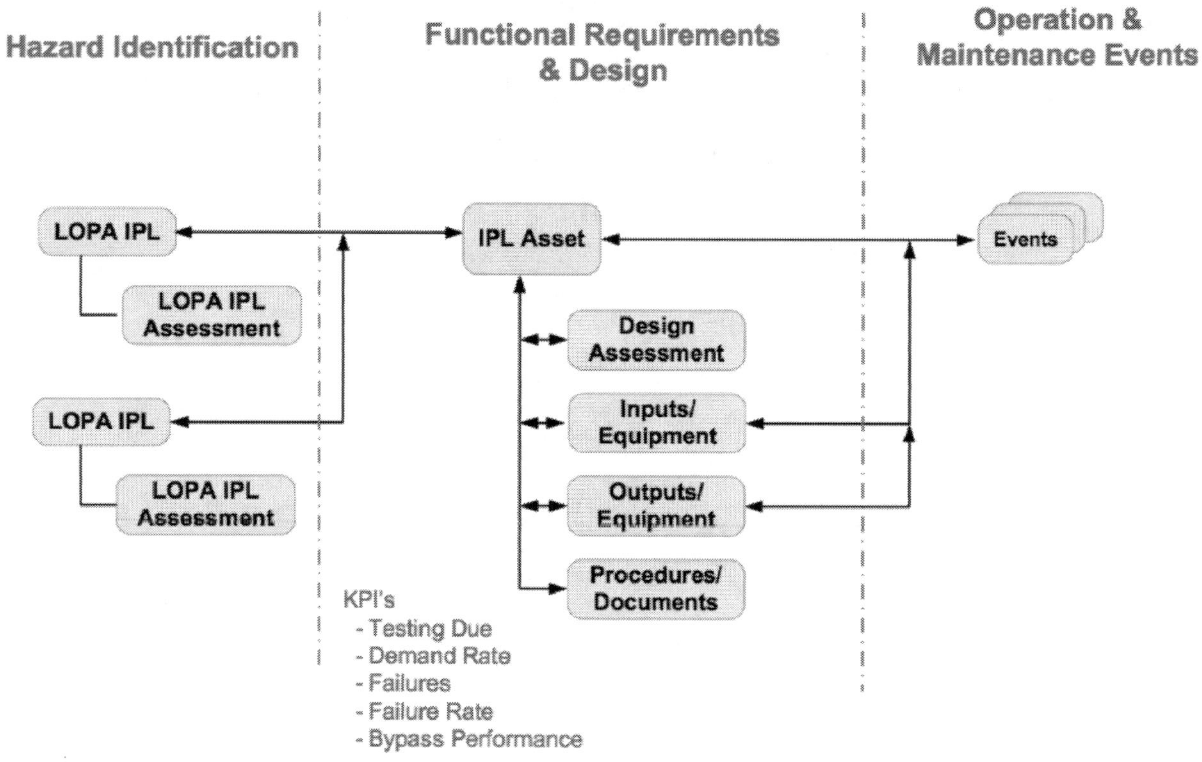

4 Using a Safety Life Cycle Management Data Base

A Safety Life Cycle Management System is best implemented in a database that is constructed specifically to manage the data and events that apply to the Safety Life Cycle. The data should be highly linked, so that a User can readily access related data. For example, an operator would be able to view the functional requirements for an IPL Asset, using a tag or other ID that is used within the Unit, and then directly access the Hazard data associated with the IPL. Appropriate Operations, Maintenance or Engineering personnel should then be able to associate Events with the IPL Assets and view KPI's associated with those Assets.

4.1 Data Relationships

A Safety Management Life Cycle Management Data Base should contain data objects that provide relationships among the various items shown in Figure 3. Among these are

- LOPA IPL's that define basic definitions of the IPL function and performance requirements and relate the LOPA IPL to the appropriate LOPA and HAZOP scenarios

- IPL Assets that define the actual plant assets that are used to implement the IPL functions. These should be identified using the same identifiers used in other plant documentation and by which plant operators, engineers and maintenance personnel know the asset. The IPL Asset data should define the design requirements such as function, expected response, accuracy and test and maintenance requirements
- Assessment Check Lists that define the items relative to performance, design, documentation or testing that is required of the IPL
- Input and Output Asset and Device objects that define the functional requirements for the Asset, Inputs or Outputs that are used to implement the IPL function and the specification data for the Devices that are installed in those Asset locations
- If IPL Performance is to be tracked, Event data objects that allow for tracking of events such as Demands, Failures, Testing, Bypasses, etc.
- Reference Documents such as procedures or design or specification data

4.2 Data Content

The data for each of the objects shown in Figures 2 and 3 should capture information necessary to support IPL Management. Some examples are:

- LOPA IPL's define the type of IPL, the performance requirements - typically in terms of a Risk Reduction Factor (RRF), and a functional description of the IPL. If operator response to an alarm is involved, the expected operator action and response time should be specified
- SIF Asset data should define the functional requirements, performance expectations and testing and maintenance requirements. In some implementations, data for a SIF could include all data necessary for definition of the Safety Requirements Specification (SRS) as defined in IEC 61511 and ISA 84.00.01-2004.
- Non-SIF IPL Asset data should define define the functional requirements, performance expectations and testing and maintenance requirements appropriate to the type of IPL.
- A LOPA IPL Assessment is a check list of items and their assessment results necessary to validate that the IPL will be effective and independent of other IPL's credited for the LOPA scenario
- An IPL Design Assessment is a check list of items and assessment results necessary to verify that the IPL Asset as designed meets the functional, performance and independence requirements for the IPL.
- Input/Equipment and Outputs/Equipment should define the functional requirements for Inputs and Outputs used by the IPL. The Equipment sections should contain or be linked to specification data for the specific equipment performing the functions.
- Procedure/Documents data should identify the specific Operating Procedures, Maintenance and Test Procedures or other Design Documents that apply to the IPL.

- Events – Events are used to describe the date, time and results of various types of Operation and Maintenance Events that occur during the IPL Life Cycle. The data for each event should be appropriate to the Event Type, such as as-found and as-left data for Functions and Equipment, start and end of bypasses, etc.

5 AN EXAMPLE IPL

The Figures below illustrate how a Safety Life Cycle data base application can be used to manage an IPL. An Alarm IPL is shown as an example.

5.1 IPL Functional Requirements

Figure 4 shows an example of an IPL Functional Specification for an Alarm with Operator Action IPL. This allows a User to define the functional requirements for the Alarm including such data as

- The Alarm ID – Typically this is the Tag Number of the Alarm as it appears on P&ID's, Operator Interfaces and other documentation
- Alarm functions, set points and basic implementation requirements
- Expected operator response and the time within which the response is expected
- Alarm operating restrictions such as when the alarm may be disabled or restrictions on changing set points or other configuration data
- Alarm Testing Requirements – type of testing and frequency of testing.
- LOPA Associations including the LOPA scenario for which the Alarm was cited as an IPL, the LOPA IPL and other IPL's that were cited by same LOPA scenario
- Associated Documentation such as Test Procedures, Operating Procedures

Figure 4 – Alarm IPL Functional Requirements Data

Alarm O-M Data | Alarm Functional | Alarm Devices | Revisions

Print | 1 Document ▾ Attach Tasklist | Subscription ▾ Bookmark ▾ Admin Tools ▾

Alarm Functional Requirements - FCC-FAL-138, Feed to East Riser

Alarm Requirements

Alarm ID: FCC-FAL-138
Alarm Description: Feed to East Riser Low Flow
Functional Description: Feed flow to the East Riser is indicated on the BPCS by the analog input poing FCC-FI-138. This point has a low flow alarm confirgued to alert the operator of low riser feed flow
Asset Protection Category: IPL
Operator Repsonse Time: 10 Min
Operator Repsonse Time Notes: Nominal time to responsd to a high priorty alarm
Defined Corrective Action: The operator shall place the East Riser Assist Steam flow controller FCC-FC-282 into automatic with a set point of 1100 PPH. Verify that the controller is properly proving steam to the riser

Protective Category:

Health and Safety	Environmental	Commercial	Reputation
Yes	No	No	No

Performance Requirements: RRF 10 PFD 1.0 E-01
Alarm Operating Requirements: Alarm must be enabled at all times when feed is in the unit. If the alarm must be bypassed or disabled prepare Safety Bypass Authorization. Set point is not to be changed without an MOC being performed

Alarm Response Time: 5 sec
Alarm Respsonse Time Notes: BPCS must activate alarm within 5 seconds of when the transmitter senses low flow

Alarm Location: FCC Operator's Console in Main Control Room
Alarm Implementation: Soft alarm on analog input tag
Alarm Priority: High
Alarm Priority Rationale: The risler low feed flow alarm indicates that conditios may exist the contribute to a riser flow upset but the situation is no yet crtical. Additional SIF layers of protection exist
Alarm Testing Requirements: Test transmitter, calibrate and test alarm activation and verify set points at turnarounds
On-Line Testing Interval: N/A
Off-Line Testing Interval: 60 Months

Alarm Set Point (Software Based)

Set Point ID	Value	Type	Tolerance	Time Delay	Basis
FCC-138-SPL	115	Decreasing	5%	1 sec	Minimum HC flow w/o assist steam

Alarm Set Point (Device Settings)

Set Point ID	Value	Type	Tolerance	Time Delay	Basis
N/A	N/A	N/A	N/A	N/A	N/A

LOPA References

LOPA IPL ID	LOPA ID	LOPA IPL Description	LOPA Hazard	LOPA Consequnces	Required RRF
FCC-LOPA-15-1	FCC-LOPA-15	Low Feed Flow Alarm	Increase in riser catalyst density and loss of flow stability. Potential loss of flow and flow reversal	Hydrocarbon in regenerator. Very high regenerator temperataures and potenial hydrocarbon in preciptator. Potential explosion	10

Associated IPL's

ID	Type	Description
FCC-SIF-02	SIF	East Riser Emergency Steam
FCC-SIF-05	SIF	FCC Catalyst Reversal Protection System

Alarm Testing Procedures

Document ID	Title	Description
CE-TEST-7704	Alarm Testing Procedure	General Refinery Alarm Testing Procedures

5.2 IPL Operating and Maintenance History

Figure 5 shows an example of a view of the Alarm IPL's Operating and Maintenance history for the same Alarm as illustrated above. This view repeats the basic Alarm Functional Requirements data and the provides data specific to the Alarm operating status and the history of the events which have been entered into the database for the Alarm. This view provides the following types of data.

- Basic Alarm Function Requirements
- Alarm Testing and Performance Data – this section lists things such as the testing status for the alarm, the last test date and next test data, commissioning data, hours in service and a summary of performance statistics such as the number of demands and failures and the in-service failure rate
- Lists of the events that have been entered into the database for the alarm listed by type of event. In the example, this includes the Alarm Service Status history, Test Events, Demand Events, Bypass Events, etc.

In this view, many data fields are linked so the User can click on things such as Event ID's to go directly to the event detail view.

Figure 5 – Alarm Operation and Maintenance Data

Alarm O-M Data | Alarm Functional | Alarm Devices | Revisions

Print | 1 Document | Attach Tasklist | Subscription | Bookmark | Admin Tools

Alarm O-M View - FCC-FAL-138, Feed to East Riser

Alarm Data

Alarm ID: FCC-FAL-138

Alarm Description: Feed to East Riser Low Flow

Functional Description: Feed flow to the East Riser is indicated on the BPCS by the analog input poing FCC-FI-138. This point has a low flow alarm confirgued to alert the operator of low riser feed flow

Operator Repsonse Time: 10 Min

Defined Corrective Action: The operater shall place the East Riser Assist Steam flow controller FCC-FC-282 into automatic with a set point of 1100 PPH. Verify that the controller is properly praving steam to the riser

Testing and Performance Data

Alarm Testing Requirements: Test transmitter, calibrate and test alarm activation and verify set points at turnarounds

On-Line Testing Interval: N/A

Off-Line Testing Interval: 60 months

Test Status: Current

Last Test Date: 3/26/12

Next Test Due Date: 3/26/17

Due Date Based Upon: Turnaround Schedule

Determining Devices	ID	Type	Description
	FCC-FT-138	Flow Transmitter	Feed to East Riser

Alarm Service Status: Commissioned

Service Hours: 108,624 hrs

Alarm Commission Date: 4/28/08

Alarm Commission Notes: None

Alarm Decommission Date: N/A

Alarm Decommission Notes: N/A

Alarm Performance	Life of Plant	5 Years	1 Year	Year to Date
No. of Failures	1	1	0	0
No. of Demands	6	6	0	0

In Service Failure Rate: 7.0 E-02

Alarm Service Status History

Event ID	Date	From	To
FCC-FAL-138-Serv-001	4/28/2008	Created	Commissed

Test Groups

Test Group ID	Test Group Description	Date Last Run	Test Event ID
FCC-Alarms-001	FCC Safety Related Alarm Testing at Turnaround	4/4/2008	FCC-Alarms-001-Test-002
FCC-Alarms-002	FCC Safety Related Alarm Testing at Turnaround	3/24/2012	FCC-Alarms-001-Test-001

Functional Test Events

Event Date	Event ID	Parent Event ID	As Found	As Left	Event Notes	Event Status
4/4/2008	FCC-FAL-138-Test-001	FCC-Alarms-001	Failed	Passed	Pre-Commissioning Test	Approved
3/4/2012	FCC-FAL-138-Test-002	FCC-Alarms-002	Passed	Passed	Routine Turnaournd Tst	Approved

Demand Events

Demand Date	Demand Event ID	As Found	As Left	Event Notes	Event Status
4/17/2008	FCC-FAL-138-Demand-001	Passed	Passed	Startup	Approved
9/8/2008	FCC-FAL-138-Demand-002	Passed	Passed	Feed Pump Trip and Spare Start	Approved
12/17/2008	FCC-FAL-138-Demand-003	Passed	Passed	Feed Pump Trip and Spare Start	Approved
7/22/2011	FCC-FAL-138-Demand-004	Passed	Passed	Feed control valve sticking	Approved
10/9/2012	FCC-FAL-138-Demand-005	Passed	Passed	Feed Pump Trip and Spare Start	Approved
11/30/2013	FCC-FAL-138-Demand-006	Passed	Passed	Feed Pump Trip and Spare Start	Approved

5.3 IPL Assessment

As mentioned above, a critical part of the Safety Life Cycle is verification of an IPL's qualification to perform as an IPL. Check Lists can be created and tracked to document this process. Examples for an IPL Independence Check List and an Alarm Design IPL Check List are shown below.

In Figure 6 illustrates a Check List that would be completed near the time an IPL is cited in a LOPA is shown. This task list is intended to verify that the basic requirements to qualify the function as an IPL have been satisfied.

Figure 6 – LOPA IPL Check List

Item No.	Description	Finding	Comments
1.1	Verify that the IPL function, if effective, will prevent the consequence of the initiating cause from occurring	Passed	
1.2	Verify that the IPL is independent of the LOPA scenario imitating cause	Failed	The Alarm appears to use the same transmitter whose failure was the initiating cause.
1.3	Verify that the IPL is independent of any other IPL's that have been credited for the LOPA scenario	Passed	

Figure 7 shows a Check List summary that appears in the LOPA IPL view. This provides the user with a quick overview of the verification status and typically appears on the IPL Functional Requirements view.

Figure 7 – LOPA IPL Check List Status

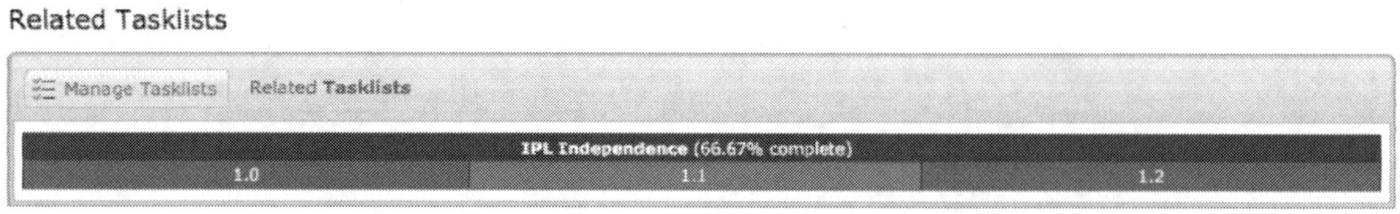

Figure 8 illustrates an IPL Design Check List for an Alarm. This check list addresses design requirements and operational preparedness items such as operation and maintenance preparedness items.

Figure 8 – Alarm IPL Checklist

Item No.	Description	Finding	Comments
3.1	Verify that the Alarm and its associated operator actions are capable of preventing the consequence for which the alarm has been cited as an IPL	Passed	
3.2	Verify that the alarm location is continuously attended	Passed	
3.3	Verify that the initiating or sensing device is independent of the initiating cause of the LOPA scenario	Failed	This alarm is derived from FCC-FT-138 whose failure is the LOPA scenario IC. Need an independent transmitter
3.4	Verify that the initiating device or sensor is independent of other protection layers claimed as IPL credits for the same LOPA scenario	Passed	
3.5	Verify that an operator action has been defined for this alarm, that the action can be accomplished and the result effective within the time between the initiating cause and the consequence occurring	Passed	
3.6	Verify that a written procedure for the operator response to this alarm has been prepared, that operators have been trained to locate the procedure and that the procedure is readily available to the operator	Failed	Operating procedures require updating.
3.7	Verify that this alarm has a suitable priority and it is identified as an IPL Alarm on Operator Graphics and Alarm views	Failed	Alarm needs to be added to operating graphics and IPL Alarm graphic

Item No.	Description	Finding	Comments
3.8	Verify that the settings for this alarm and disabling, inhibiting or bypassing of the alarm is under control of a Bypass Control Procedure and suitable approvals are required for the above actions and that return to operation is managed by this procedure.	Failed	The alarm needs to be added to the list of critical items that require Safety Bypass Authorization.
3.9	Verify that modifications to this alarm are under Management of Change Controls	Passed	Site Standard MOC Procedure
3.10	Verify that Maintenance and Testing Procedures have been prepared which are applicable to this alarm, and that the Procedures include capture of all testing records	Passed	Site Standard Maintenance procedure for testing of alarms
3.11	Verify that the P&ID; correctly shows this alarm	Failed	P&ID has not yet been updated to show this alarm as an IPL
3.12	Verify that the Alarm has been included in the Unit IPL List	Passed	

5.4 KPI's from Data Base

Once IPL data is captured in a data base, KPI's associated with the IPL Operation and Maintenance can be readily created and displayed. Some of the KPI's and summary data that can be developed are

- Identification of IPL Testing that is upcoming or overdue
- Identification on IPL's that have had high Demand or Failure rates
- Calculation of In-Service IPL Failure rates
- Reporting on IPL Assessment status and and completion of open items
- IPL quantities for Units, Sites or Enterprise-wide
- Identification of LOPA IPL's that are not linked to a Plant IPL Asset
- Identification of Plant Assets that are associated with multiple LOPA IPL's

Figure 9 shows a few examples of the types of IPL KPI data that can be provided to Users

Figure 9 – Typical IPL KPI's

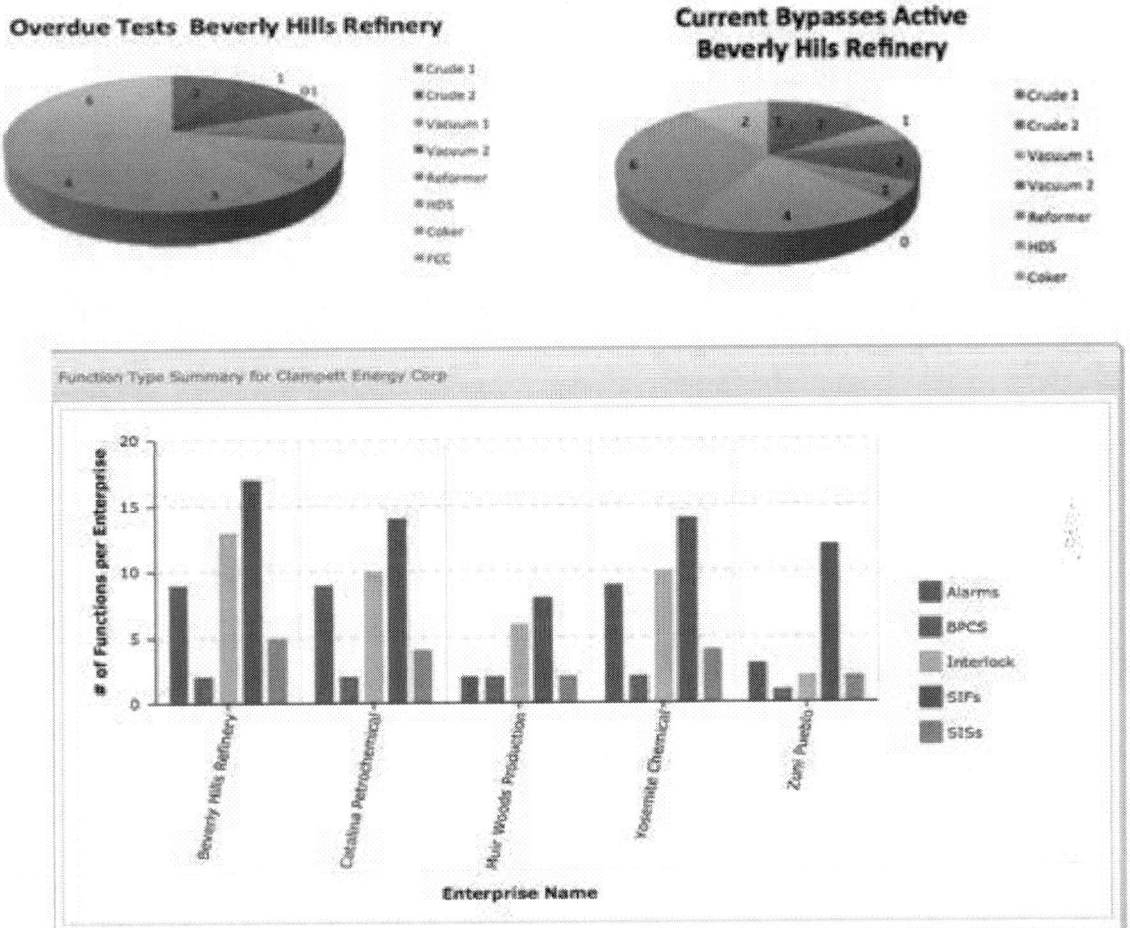

Enterprise Name	Alarms	BPCS	Interlock	SIFs	SISs
Beverly Hills Refinery	9	2	13	17	5
Catalina Petrochemical	9	2	10	14	4
Muir Woods Production	2	2	6	8	2
Yosemite Chemical	9	2	10	14	4
Zuni Pueblo	3	1	2	12	2

6 Conclusions

The Management of IPL's within an Operating Facility presents distinct challenges to Facility Management and trickles down to many organizations within the Facility. Many organizations are having signification problems in managing these requirements due to lack of the clarity of requirements and the cross discipline aspects that do not fit within the practical restrictions of management tools traditionally used. Force fitting IPL Safety Life Cycle management requirements as an overlay to these tools has generally been highly inefficient and generally ineffective.

The compliance gaps that have been observed as operating facilities struggle with management of IPL's and the Safety Life Cycle are a result of these problems. A practical and effective means of compliance is to apply a Management Process that is designed specifically for support of the Safety Life Cycle. This Management Process can maximize effectiveness of compliance and assure that the risk reduction that has been assumed for a facility is actually realized.

7 References

[1] – Layer of Protection Analysis, Center for Chemical Process Safety, ISBN 0-8169-0811-1, 2001

[2] – OSHA Memorandum – June 5, 2015 – Thomas Galassi, RAGAGEP in Process Safety Management Enforcement.

Rick Stanley has over 40 years' experience in Process Control Systems and Process Safety Systems with 32 years spent at ARCO and BP in execution of major projects, corporate standards and plant operation and maintenance. Since retiring from BP in 2010, Rick has consulted with Mangan Software Solutions (MSS) on the development and use of MSS' ProSys SLM Safety Lifecycle Management software and has performed numerous Functional Safety Assessments for both existing and new SISs.

Rick has a BS in Chemical Engineering from the University of California, Santa Barbara and is a registered Professional Control Systems Engineer in California and Colorado. Rick has served as a member and chairman of both the API Subcommittee for Pressure Relieving Systems and the API Subcommittee on Instrumentation and Control Systems.

442704 Jack Rabbit II: Experimental Large-Scale Chlorine Release Trials

Abstract:

In August and September of 2015, the U.S. Department of Homeland Security (DHS) Science & Technology (S&T) Chemical Security Analysis Center (CSAC) conducted the Phase 1 field trials of the Jack Rabbit II (JR II) program, a series of 5 large-scale outdoor chlorine release experiments with a collaborative team of partners from government, industry, and academia. Sponsored by the CSAC, the Defense Threat Reduction Agency (DTRA), and Transport Canada, these field trials were performed to fill critical knowledge, data, and capability gaps to support and improve chemical release modeling, industrial safety, and emergency response procedures. This paper summarizes the execution of the 5 successful chlorine release trials ranging from 5 to 9 tons, the instrumentation deployed, the data collected, and preliminary analyses of the high-definition video footage, cloud movement, and behavior and concentration/time profiles as a function of distance from the release site. This paper will also present plans for the JR II Phase 2 field trials scheduled for late 2016 which will include additional chlorine releases up to 16 tons.

445083 Novel Methodology for Inherent Safety Assessment in the Process Design Stage

Abstract:

Various physical and chemical hazards are intrinsic to the material itself or its manner of storage and use. Inherently safer design of the chemical process and equipment has been increasingly used by the design engineers to minimize or eliminate these hazards. This approach aims to mitigate hazards by design instead of add-on measures. This presentation describes a new design approach, introducing comprehensive inherent safety index (CISI), for use in the early stages of process development. Chemical, process and connectivity scores are calculated for individual units in the process, based on which the CISI assigns safety scores for the equipment. Each chemical in the unit has a weighted severity score. Reactivity score is calculated for mixture of chemicals in that unit. Connectivity score in introduced because the units can be highly interconnected, leading to compounding of hazards. Biodiesel and methyl methacrylate processes were studied to demonstrate the new safety assessment methodology. The results of the assessment are used to clearly identify key areas for improvement in a root-cause analysis.

12TH GLOBAL CONGRESS ON PROCESS SAFETY

50th LOSS PREVENTION SYMPOSIUM

Siting Bulk Hydrogen Storage Systems: Advances in Risk-Informed Siting Procedures in NFPA Codes

Carl Rivkin, P.E.
National Renewable Energy Laboratory
15013 Denver West Parkway
Golden, CO 80401
Carl.rivkin@nrel.gov

Martin Gresho, P.E.
FP2 Fire, Inc.
282 Ronnie Road
Golden, CO 80403
marty@fp2fire.com

Dr. Christine LaFleur, P.E.
Sandia National Laboratories
PO Box 5800 MS0748
Albuquerque, NM 87185
aclafle@sandia.gov

Alice Muna
Sandia National Laboratories
PO Box 5800 MS0748
Albuquerque, NM 87185
amuna@sandia.gov

Prepared for Presentation at
American Institute of Chemical Engineers
2016 Spring Meeting
12th Global Congress on Process Safety
Houston, Texas
April 11-23, 2016

Siting Bulk Hydrogen Storage Systems: Advances in Risk-Informed Siting Procedures in NFPA Codes

Carl Rivkin, P.E.
National Renewable Energy Laboratory
15013 Denver West Parkway
Golden, CO 80401
Carl.rivkin@nrel.gov

Martin Gresho, P.E.
FP2 Fire, Inc.
282 Ronnie Road
Golden, CO 80403
marty@fp2fire.com

Dr. Christine LaFleur, P.E.
Sandia National Laboratories
PO Box 5800 MS0748
Albuquerque, NM 87185
aclafle@sandia.gov

Keywords: hydrogen, compressed gas, safety distance, fuel cell vehicle, storage

Abstract

From 2007-2009, the NFPA 55 Industrial and Medical Gas Technical Committee and the NFPA 2 Hydrogen Technology Technical Committee developed revised separation distances for the bulk gaseous hydrogen storage systems based on an analytical process that integrated both Computational Fluid Dynamic (CFD) modeling to evaluate the consequences of hydrogen releases and probabilistic risk analysis to include the frequency of hydrogen releases, the harm criteria associated with these releases, and the acceptable risk level for the separation distances calculated as a function of these inputs. An NFPA 55 and 2 joint Hydrogen Task Group was formed in 2014 to accomplish the following tasks:

-Develop revised separation distances for bulk liquefied hydrogen storage based on a risk-informed methodology parallel to the process used in the previous update of the gaseous requirements.

-Evaluate hydrogen system safety measures for both gaseous and liquefied hydrogen storage and develop a system for separation distance reduction based on additional layers of safety or new safety measures.

-Evaluate the risk criteria used to develop the separation distances for bulk gaseous hydrogen storage and make recommendations for changes in these distances based on the revised criteria.

This paper summarizes this work including the risk-informed process for developing separation distances for bulk liquefied hydrogen storage which has not previously been addressed in NFPA codes.

1 Introduction

Hydrogen-utilizing technologies have been a part of the industrial work place for decades. [1] These technologies require bulk hydrogen storage systems. In the last twenty years there has been an effort to develop Fuel Cell Electric Vehicles (FCEVs) that require retail Hydrogen Fueling Stations (HFS). Moving hydrogen storage from the controlled industrial environment, where personnel exposure is typically limited, to the retail environment, where the general public is exposed, has triggered a review of the NFPA hydrogen storage requirements. The objective of this review is to establish risk-informed safety requirements for hydrogen storage which are based on scientific analysis of the physical behavior of unintended hydrogen releases as well as the probabilistic analysis of release frequencies. The risk threshold used to inform this analysis is based on risks incurred at conventional gasoline fueling stations which are deemed acceptable to the general public.

The process for developing risk-informed requirements for bulk hydrogen systems in NFPA codes began in 2007 and produced a set of risk-informed gaseous separation distances in the 2010 edition of *NFPA 55 Compressed Gases and Cryogenic Fluids Code* [2] that were also extracted into the 2011 edition of *NFPA 2 Hydrogen Technologies Code[3]*. The basis for these distances was extensively documented in Sandia Technical Report 2009-0874, *Analysis to Support Risk-Informed Separation Distances for Hydrogen Codes and Standards*. [4] The separation distances are the minimum distance between any point on the bulk hydrogen storage system and a defined set of exposures. Typically, the most restrictive exposures are the lot line and openings in nearby buildings. For the 2016 versions of NFPA 55 and NFPA 2, the exposures were sorted into three groups based on the risk criteria used to develop the separation distance. The significance of separation distances is that most HFS will be located in urban areas where there is limited space for siting storage systems. Many HFS will be located at existing gasoline retail facilities where space is also limited.

The separation distances for bulk liquefied hydrogen storage systems were not evaluated during the gaseous requirements revisions. The current requirements are based on distances developed by NASA in the 1960's and incorporated into Occupational Safety and Health Administration (OSHA) requirements. The technical bases for the existing requirements were not documented. In 2014, NFPA created a new NFPA Hydrogen Storage Task Group (2014 NFPA Task Group) to develop bulk liquefied hydrogen storage system separation distances based on scientific analysis and revisit the separation distances for bulk gaseous systems. This analysis was to include a methodology for converting redundant or extra safety measures into reductions in separation distances.

This paper documents progress to date on the efforts to develop science based separation distances for bulk liquefied hydrogen storage and revise the bulk gaseous separation distances based on information derived from project deployment. There is ongoing work that will likely produce further revised requirements and flexibility in bulk hydrogen storage beyond those described in this paper.

2 Historical Separation Distances

A risk-informed approach for selecting the leak diameters was utilized to establish the separation distances in the 2011 edition of NFPA 2 and the 2010 edition of NFPA 55. The risk-informed approach included three considerations: the frequency of leakage for typical hydrogen facilities, the cumulative frequency of system leakage, and the risk from leakage events for the example facilities. All three are dependent upon the establishment of example facilities and the generation of hydrogen-specific component leakage frequencies. Facility descriptions were provided for four typical facilities operating at different pressures.

Hydrogen-specific component leak frequencies were generated as a function of leak size for several hydrogen components using Bayesian analysis. Hydrogen leakage data was provided by members of NFPA Task Group for use in this analysis. The hydrogen-specific data was utilized to generate system leakage estimates for a 20.7 MPa and 103.4 MPa facility. For a 0.1% piping leak size, the system leakage frequency is 3 x 10^{-2}/yr. and 6 x 10^{-2}/yr. for the 20.7 MPa and 103.4 MPa systems, respectively. The higher leakage frequency for the 103.4 MPa system is due to the contribution from the additional compressor and high pressure storage modules. These values are comparative to the fire frequency in gasoline stations (7.4 x 10^{-2}/yr. [7]) and suggest that a 0.1% leak would be expected during the lifetime of these facilities. To reduce the potential for significant consequences to a person at the site boundary due to expected accidents, larger and less frequent leak sizes of at least 1% should be used as the basis for separation distances.

The cumulative probability for different leak sizes was then calculated to determine what range of leaks represents the most likely leak sizes. When evaluated on a cumulative distribution basis, leaks equal or less than 0.1% of the component flow area was estimated to represent 95% of the system leakage frequency. Basing separation distances on this size of leak would ensure that they cover the majority of possible leakage events. However, the use of a larger leak size was recommended to the NFPA Task Group based on consideration of the actual system leakage frequencies discussed above and uncertainty in the component and system leak frequency analysis.

The risk resulting from different size leaks was also evaluated for four standard gas storage configurations. The risk evaluation indicates that the use of 0.1% of the component flow area as the basis for determining separation distances results in risk estimates that significantly exceed the 2 x 10^{-5} fatality/yr. risk guideline selected by the NFPA Task Group for both the 20.7 MPa (3000 psig) and 103.4 MPa (15000 psig) example systems.

On the other hand, use of a leak size equal to 1% to 10% of the component flow area results in risk estimates that are slightly above, but reasonably close to the risk guideline. Based on this input, the NFPA Task Group decided that the separation distances specified in NFPA 2 and NFPA 55 would be based on a leak size equal to 3% of the largest flow area downstream of a gaseous storage system greater than 11.3 m^3 (400 scf).

More details on this entire process can be found in [4].

3 Bulk Liquefied Hydrogen Storage

An NFPA Hydrogen Storage Task Group was formed in 2014 to develop a scientific basis for the separation distances for bulk liquefied hydrogen storage systems. The Task Group had the following objectives:

- Define a representative bulk liquefied hydrogen storage system.
- Perform a risk analysis on the system to determine the release scenarios with the highest risk.
- Evaluate the risk criteria for each of the highest risk release scenarios by using a scientific model that predicts the extent of specific hydrogen concentrations.
- Acquire leak frequency data from the discharge points from the highest risk release scenarios
- Perform an integrated analysis of releases and leak frequencies to estimate the system separation distances using the risk threshold of 2 x 10^{-5} fatalities /yr.

The Task Group decided to use the liquefied hydrogen system described in *CGA P-28 OSHA Process Safety Management and EPA Risk Management Plan Guidance Document for Bulk Liquid Hydrogen Systems* 2009 (CGA P-28) [5] as the representative system. The representative system contains a liquid hydrogen storage tank, a pressure buildup circuit and economizer, vaporizer, and pump. CGA P-28 also contains a representative hazard and operability study (HAZOP) that was performed on the representative bulk liquefied hydrogen storage system. The process was divided into seven segments, called nodes:

1. Delivery trailer, hose, and fill line to the storage vessel
2. Storage tank
3. Pressure buildup circuit and economizer
4. Hydrogen line—liquid through the vaporizer to the final line
5. Hydrogen pump
6. Vent stacks
7. General items

In CGA P-28, each node was examined based on deviations from normal operating parameters that could have potentially hazardous consequences. The Task Group started with this division of the system and analyzed each node deviation to come up with event likelihood and hazard severity. Event likelihood was the annual probability that the deviation may occur for each node, with the highest level expecting the deviation to occur once per year or more and the lowest level expected to occur once per 10,000 years, documented in Table 1. Hazard severity classified the consequence from the node deviation. The hazard severity ranged from negligible to catastrophic, documented in Table 2. The Task Group estimated the event likelihood and severity based on industry experience and subject matter expertise and had to come to an agreement on the levels given to each node deviation.

Table 1. Event Likelihood Classification Table

Level	Annual Probability	Probability Description
1	Frequent > 1.0	Expected to occur once per year or more frequently.
2	Reasonably probable 1.0 to 0.1	Expected to occur once per 10 years.
3	Occasional 0.01 to 0.1	Expected to occur once per 100 years.
4	Remote 0.001 to 0.01	Expected to occur once per 1000 years.
5	Extremely remote 0.0001 to 0.001	Expected to occur once per 10,000 years.
6	Improbable < 0.0001	Expected to occur less than once per 10,000 years. Extremely unlikely to occur.

Table 2. Hazard Severity Classification Table

Level	Description	Potential Consequences
1	Catastrophic	May cause fatality to non-associated members of the public.
2	Critical	May cause severe injury to non-associated members of the public, fatality or serious injury to works of the public, fatality or serious injury to workers of persons conducting business at a refueling site or significant damage to equipment/facilities.
3	Marginal	May cause minor injury, or minor system damage.
4	Negligible	Will not result in injury or system damage.

A few assumptions were made for this analysis. First, the task group assumed an NFPA 55 and CGA H-5 code complaint system. The scenarios would only have a single failure, and cascading failures would not be analyzed. The current setback distances did not provide any credit during this analysis and tornadoes were not included in the natural disaster considerations. Based on the level of likelihood and severity, the node deviation was given a risk ranking, documented in Table 3.

Table 3. Risk Ranking Table

		Likelihood					
		1	2	3	4	5	6
Severity	1	1	1	1	2	3	4
	2	1	1	2	3	3	4
	3	2	2	3	3	4	4
	4	4	4	4	4	4	4
1: High Risk 2: Moderate Risk 3: Low Risk 4: Routine Risk							

After the Task Group had prioritized all 102 node deviations, the risk ranked results were reviewed and discussed. Two scenarios were added, based on Task Group feedback, which analyzed high and normal flow from a trailer vent stack. In total, the Task Group chose nine scenarios with the highest risk to analyze further to define separation distances to exposures. The liquid hydrogen storage exposures are similar to gaseous hydrogen storage exposures and are grouped in Table 4.

Table 4. Bulk Liquid Hydrogen Storage System Exposures

Group 1 Exposures
(a) Lot lines
(b) Air intakes (HVAC, compressors, other)
(c) Operable openings in buildings and structures
(d) Ignition sources such as open flames and welding
Group 2 Exposures
(a) Exposed persons other than those servicing the system
(b) Parked cars
Group 3 Exposures
(a) Buildings of noncombustible non-fire-rated construction
(b) Buildings of combustible construction
(c) Flammable gas storage systems above or below ground
(d) Hazardous materials storage systems above or below ground
(f) Ordinary combustibles
(g) Unopenable openings in buildings and structures
(h) Encroachment by overhead utilities
(i) Piping containing other hazardous materials

(j) Flammable gas metering and regulating stations
(k) Inlet to underground sewers[1]
1. Exposure not included in gaseous hydrogen storage separation distances

The Task Group's high risk release scenarios are shown in Table 5. Note that the first six scenarios listed occur during liquid hydrogen (LH2) transfer operations from a tanker truck to the bulk LH2 storage tank. The other three scenarios occur during normal system operations. The two scenarios that the Task Group added are included in this list. For each of the scenarios, the Task Group decided upon the inputs to liquid hydrogen models and the risk criteria for analysis. Sandia developed a cryogenic hydrogen integral jet and plume model, called COLDPLUME, which is intended to be used to predict release characteristics from various real-world hydrogen storage states. [7] Sandia is also developing a network flow model which accounts for internal system flows from the storage vessel to the release plane, called NETFLOW, which will be suitable for liquid and multiphase cryogen flows. [8] These models are currently under development and experimental validation. Once modeling results are available, the hydrogen exposure risk criteria used for the bulk gaseous separation distances (discussed in the following section) will be used to develop the new liquid separation distance table.

Table 5. High Risk Liquid Hydrogen Release Scenarios

	HAZOP Number and Description		**Modeling Notes**	**Separation Distance Driver**
Release scenarios during liquid transfer to bulk storage tank	1.18	High flow of gaseous hydrogen from trailer vent stack due to venting excess pressure after LH2 transfer	Sandia's NETFLOW and COLDPLUME models will characterize the temperature and concentration from the trailer vent stack. The task group would like to model an estimate of up to 50 kg of cold, gaseous hydrogen.	
	1.19	Normal flow from trailer vent stack due to venting excess pressure after LH2 transfer	Sandia's NETFLOW and COLDPLUME models will characterize the temperature and concentration from the trailer vent stack. The task group would like to model an estimate of up to 50 kg of cold, gaseous hydrogen.	Modeling results will be used to calculate separation distance from air intakes and overhead utilities because this is highest frequency release and release point is closest proximity to these exposures.
	1.6	High flow from line rupture, valve or component failure during transfer process	Sandia's NETFLOW and COLDPLUME models will characterize the temperature and concentration from the release to the air. A model is needed to characterize pooling and evaporation effects.	
	1.4	High temperature due to external fire causes high flow venting through tank vent stack	Sandia's NETFLOW and COLDPLUME models will characterize the temperature and concentration from the trailer vent stack. This is the worst case venting scenario from the pressure relief device to a high ambient temperature.	
	1.8	Reverse flow during transfer process caused by human error and pressure mismanagement	Sandia's NETFLOW and COLDPLUME models will characterize the temperature and concentration from the release to the air. A model is needed to	

			characterize pooling and evaporation effects.	
	1.16	Loss of containment from external impacts, consider all causes	The task group postulated a vehicle crash into the tanker during transfer. This causes a transfer hose rupture with NETFLOW and COLDPLUME models characterizing the release to the air. A model is needed to characterize pooling and evaporation effects.	
Release scenarios during normal system operation	4.15	Loss of containment from pipe leading from tank to vaporizer or vaporizer itself caused by thermal cycles or ice falling from vaporizers	Sandia's NETFLOW and COLDPLUME models will characterize the temperature and concentration from the release to the air. A model is needed to characterize pooling and evaporation effects.	Modeling results of hydrogen concentration plume and heat flux from a subsequent fire will be used for all other separation distance exposures because this is the highest risk priority during normal operations.
	6.15	Misdirected flow caused by operator error resulting in large low level release of cold gaseous hydrogen through bottom drain valve of vent stack during normal tank venting process	Sandia's NETFLOW model will capture how much gas is going out of the drain vent and vent stack. Sandia's COLDPLUME model can characterize the temperature and concentrations from the releases to the air. A model is needed to characterize pooling and evaporation effects.	
	2.1	High pressure because of a leak in inner vessel allowing hydrogen into the vacuum area	Sandia's NETFLOW and COLDPLUME models can characterize the flow out of the casing vent (a large hole instead of a pipe orifice).	

4 Bulk Gaseous Hydrogen Storage

As described in Section 2, the initial gaseous hydrogen setback distances were based on a few key risk criteria. The 2014 NFPA Task Group decided to revisit three of the risk criteria: the leak area as a percent of the pipe diameter, the incident heat flux threshold, and the hydrogen concentration threshold.

The cumulative probability for different leak sizes was calculated to determine what range of leaks represents the most likely leak sizes. The system leakage frequency corresponds to the largest internal pipe downstream of the highest pressure source in the system. The results of this analysis indicated that leaks less than 0.1 percent of the component flow areas represent 95 percent of the leakage frequency for the example systems, however the risk resulting from this small leak size significantly exceeded the 2×10^{-5}/yr. risk guideline set by the Task Group. At the same time, the use of a leak size between 1 percent and 10 percent of the component flow area results in risk estimates that are reasonably close to the risk guideline. [2]

Based on the results of both the system leakage frequency evaluation and the associated risk assessment, the Task Group decided that adjusting to a diameter of 1 percent value, instead of a 3 percent, would remove excess conservatism from the input assumption to the model. The 1 percent value still accounts for 95 percent of the leakage frequency from the example systems and does not exceed the 2×10^{-5}/year risk guideline established in the previous analysis. This results in more permissive separation distance requirements with no change in risk.

The Task Group also reviewed the heat flux values and determined that the use of a "no harm" criterion (1.6 kW/m^2) was overly conservative. This heat flux assumes exposed persons will not take protective actions, such as moving away from the fire scene. The task group deemed it reasonable to assume that exposed personnel will relocate away from a fire scene within a few minutes and therefore the "no harm" criteria is not appropriate for establishing separation distances. Exposures that were analyzed based on this heat flux value were updated to reflect the harm distance for a radiation heat flux level of 4.7 kW/m^2. The Task Group decided to not change the three other heat flux values used in the previous revision of the separation distances in 2009.

The Task Group reviewed the hydrogen concentration threshold. Based on work done at Sandia National Laboratories Combustion Research Facility, the Group concluded that there would not be sustained ignition at hydrogen concentrations of 8% or less. [6] There could be localized hydrogen ignition that would not develop into sustained combustion. Table 6 shows the effect on separation distances of the different leak areas, heat flux harm criteria at different system pressures, without the changes to the risk criteria.

Table 6. Harm Distances for Leak Areas, Harm Criteria, and Pressures [2]

Harm Criteria	Harm Distance (Leak Area)			
	>0.10 to 1.72 MPa (>15 to 250 psig)	**>1.72 to 20.68 MPa (>250 to 3000 psig)**	**>20.68 to 51.71 MPa (>3000 to 7500 psig)**	**>51.71 to 103.43 MPa (>7500 to 15000 psig)**
Un-ignited jet concentration - 4% mole fraction of hydrogen	31.2 m (20% Area) 22.1 m (10% Area) 15.7 m (5% Area) 12.1 m (3% Area) 7.0 m (1% Area)	36.1 m (20% Area) 25.6 m (10% Area) 18.1 m (5% Area) 14.0 m (3% Area) 8.1 m (1% Area)	22.6 m (20% Area) 16.0 m (10% Area) 11.3 m (5% Area) 8.8 m (3% Area) 5.0 m (1% Area)	26.8 m (20% Area) 19.0 m (10% Area) 13.4 m (5% Area) 10.4 m (3% Area) 6.0 m (1% Area)
Radiation heat flux level of 1.6 kW/m^2 (500 Btu/hr-ft^2)	23.4 m (20% Area) 15.9 m (10% Area) 10.7 m (5% Area) 7.9 m (3% Area) 4.1 m (1% Area)	28.1 m (20% Area) 19.0 m (10% Area) 12.8 m (5% Area) 9.5 m (3% Area) 4.8 m (1% Area)	16.6 m (20% Area) 11.2 m (10% Area) 7.8 m (5% Area) 5.5 m (3% Area) 2.6 m (1% Area)	20.5 m (20% Area) 13.8 m (10% Area) 9.6 m (5% Area) 6.8 m (3% Area) 3.3 m (1% Area)
Radiation heat flux level of 4.7 kW/m^2 (1500 Btu/hr-ft^2)	17.0 m (20% Area) 11.6 m (10% Area) 7.9 m (5% Area) 5.9 m (3% Area) 3.1 m (1% Area)	20.2 m (20% Area) 13.8 m (10% Area) 9.4 m (5% Area) 7.0 m (3% Area) 3.7 m (1% Area)	12.2 m (20% Area) 8.2 m (10% Area) 5.5 m (5% Area) 4.1 m (3% Area) 2.1 m (1% Area)	14.9 m (20% Area) 10.0 m (10% Area) 6.7 m (5% Area) 5.1 m (3% Area) 2.6 m (1% Area)
Greater of radiation heat flux level of 25237 W/m^2 or visible flame length[1]	13.0 m (20% Area) 9.2 m (10% Area) 6.5 m (5% Area) 5.0 m (3% Area) 2.9 m (1% Area)	15.0 m (20% Area) 10.6 m (10% Area) 7.5 m (5% Area) 5.8 m (3% Area) 3.4 m (1% Area)	9.4 m (20% Area) 6.7 m (10% Area) 4.7 m (5% Area) 3.6 m (3% Area) 2.1 m (1% Area)	11.1 m (20% Area) 7.9 m (10% Area) 5.6 m (5% Area) 4.3 m (3% Area) 2.5 m (1% Area)

Greater of radiation heat flux level of 20000 W/m^2 or visible flame length[1]	13.0 m (20% Area) 9.2 m (10% Area) 6.5 m (5% Area) 5.0 m (3% Area) 2.9 m (1% Area)	15.0 m (20% Area) 10.6 m (10% Area) 7.5 m (5% Area) 5.8 m (3% Area) 3.4 m (1% Area)	9.4 m (20% Area) 6.7 m (10% Area) 4.7 m (5% Area) 3.6 m (3% Area) 2.1 m (1% Area)	11.1 m (20% Area) 7.9 m (10% Area) 5.6 m (5% Area) 4.3 m (3% Area) 2.5 m (1% Area)

[1]The largest harm distances are predicted for the visible flame length.

Table 7 shows the revised separation distances for the four storage pressure ranges based on the task group's recommended changes in risk criteria. The safety distances in the table are also rounded to the nearest whole number and multiplied by a 1.5 safety factor. The significance of these potential changes can be shown in the high pressure column for the Group 1 Exposures. The reduction in separation distances from 34 feet to 16 feet for bulk gaseous systems would allow these systems to be placed at existing gasoline fueling stations and other space constrained locations. This expansion in the population of locations that could house hydrogen storage systems and therefore HFS would further the deployment of hydrogen fueling infrastructure to support FCEVs.

Table 7. Draft Updated Values to 2016 NFPA 2 and NFPA 55 Tables with 1.5 Safety Factor

Exposures		Separation Distance			
		>0.10 to 1.7 MPa (>15 to 250 psig)	>1.7 to 20.7 MPa (>250 to 3000 psig)	>20.7 to 51.7 MPa (>3000 to 7500 psig)	51.7 to 103.4 MPa (7500 to 15000 psig)
Group 1	Existing	12 m (40 ft)	14 m (46 ft)	9 m (29 ft)	10 m (34 ft)
	Proposed New	**5 m (16 ft)**	**6 m (20 ft)**	**4 m (13 ft)**	**5 m (16 ft)**
Group 2	Existing	6 m (20 ft)	7 m (24 ft)	4 m (13 ft)	5 m (16 ft)
	Proposed New	**5 m (16 ft)**	**6 m (20 ft)**	**3 m (10 ft)**	**4 m (13 ft)**
Group 3	Existing	5 m (17 ft)	6 m (19 ft)	4 m (12 ft)	4 m (14 ft)
	Proposed New	**4 m (13 ft)**	**5 m (16 ft)**	**3 m (10 ft)**	**4 m (13 ft)**

Group 1 Exposures include: lot lines, air intakes, operable openings in buildings and structures, and ignition sources. Group 1 separation distances are based on the higher value of radiation heat flux of 4.7kW/m^2 or the unignited jet concentration decay distance of 8% hydrogen volume fraction concentration. In this instance, the separation distance is higher for the concentration value than the heat flux value.

Group 2 Exposures include: parked cars, exposed persons other than those servicing the system. Group 2 separation distances are based on the higher value of the incident radiation heat flux of 4.7kW/m^2 exposure to employees for a maximum of 3 minutes or the visible flame length.

Group 3 Exposures includes everything else (ex: buildings of combustible construction, ordinary combustibles, openings in buildings and structures, etc.). Group 3 separation distances are based on the higher value of the radiant heat flux for non-combustible equipment of 25.2 kW/m^2 or the visible flame length.

5 Future Work

The NFPA Hydrogen Storage Task Group has several more tasks to complete its objective of developing a scientific basis for hydrogen storage separation distances and a methodology for converting safety measures into reductions in separation distances.

These tasks include:

- Completion of the liquefied hydrogen release analysis.
- Evaluation of liquefied hydrogen leak frequencies.
- Integration of leak frequencies and release analysis into a science-based set of bulk liquefied hydrogen separation distances.
- Creating a risk based methodology to convert both liquefied hydrogen and gaseous hydrogen storage safety measures into separation distance reductions.

6 Conclusion

There has been significant progress made in developing science-based justifications for bulk hydrogen storage separation distances. These changes in bulk hydrogen storage will enable a much broader deployment of HFS in support of HFCEVs.

The initial effort in the 2007-2009 timeframe produced bulk gaseous separation distances based on a risk-informed analytical approach that was well documented, peer-reviewed, and repeatable. The effort started in 2014 has led to a proposed a revised set of separation distances for bulk gaseous storage systems. In the 2016-2017 timeframe, a revised set of separation distances for bulk liquefied hydrogen storage systems will be developed.

7 References

[1] Austin, G.T., Shreve's Chemical Process Industries Fifth Edition, 1984 McGraw Hill, New York.

[2] *NFPA 55 Compressed Gases and Cryogenic Fluids Code*, 2010 NFPA, Quincy, MA.

[3] *NFPA 2 Hydrogen Technologies Code*, 2011 NFPA, Quincy, MA.

[4] J. LaChance, W. Houf, B. Middleton, and L. Fluer. Sandia Technical Report 2009-0874 *Analysis to Support Risk-Informed Separation Distances for Hydrogen Codes and Standards.* Sandia National Laboratories 2009.

[5] *CGA P-28 OSHA Process Safety Management and EPA Risk Management Plan Guidance Document for Bulk Liquid Hydrogen Systems*, CGA 2009, Chantilly, VA.

[6] R.W. Schefer, G.H. Evans, J. Zhang, A.J. Ruggles, and R.Greif. "Ignitability limits for combustion of unintended hydrogen releases: Experimental and theoretical result", International Journal of Hydrogen Energy 36 2011.

[7] W. Houf, R. Schefer, "Analytical experimental investigation of small-scale unintended releases of hydrogen," International Journal of Hydrogen Energy, Vol. 33, 2008, pp. 1435-44

[8] Winters WS, a New Approach to Modeling Fluid/Gas Flows in Networks. SAND2001-8422, Sandia National Laboratories, July, 2001.

Is a Two-Inch Hole Adequate for a Siting Study?

Jeffrey D. Marx
Quest Consultants Inc.
Norman, OK
jdm@questconsult.com

Antonino Nicotra
Bechtel Oil, Gas & Chemicals, Inc.
Houston, TX
anicotra@bechtel.com

Prepared for Presentation at
American Institute of Chemical Engineers
2016 Spring Meeting
12th Global Congress on Process Safety
Houston, Texas
April 11-13, 2016

Is a Two-Inch Hole Adequate for a Siting Study?

Jeffrey D. Marx
Quest Consultants Inc.
Norman, OK
jdm@questconsult.com

Antonino Nicotra
Bechtel Oil, Gas & Chemicals, Inc.
Houston, TX
anicotra@bechtel.com

Keywords: Building siting, API RP 752, two-inch hole, 2-inch hole, consequence analysis, QRA

Abstract

For several decades now, many consequence-based siting assessments (e.g., API RP 752) have relied upon a two-inch hole to represent what is termed a "credible release size" or a "maximum credible event" (MCE). Alternatively, a study may evaluate siting by use of a quantitative risk analysis (QRA), which would apply a full range of hole sizes. Due to QRA complexity and expense, the MCE analysis has often been the preferred approach for siting studies in the US, and the two-inch hole has been one of the more common choices for the MCE. Hole size choices range from small leaks (1/4-inch hole) through one-, two-, four-, and even six-inch holes. Why a two-inch hole? Does a two-inch hole provide a sufficiently accurate qualitative evaluation of risk? When a single hole size is chosen for a given purpose in a consequence analysis, that hole size should not be chosen based solely on engineering judgement, but should be based on more defensible information. This paper seeks to explain the basis for selecting the two-inch hole, including a literature review, concepts from consequence modeling, data from failure frequencies, and engineering judgement. Examples comparing the use of a single hole size (with emphasis on the 2-inch hole) to QRA results will also demonstrate some of the potential benefits and problems of an MCE analysis.

1 Introduction

Siting studies have always been a means to understand the potential impacts that could be imposed by hazardous materials on plant workers, the public, process equipment, or the environment. The most common siting study in the process safety industry is the building siting study, meant to ensure that the potential impacts to persons in buildings result in an acceptably low risk, or are mitigated to the point that the risk to those persons is acceptably low.

While hazard modeling technology and siting study methodology have progressed over the years, it is still a significant resource investment to evaluate siting issues using fully risk-based techniques such as quantitative risk analysis (QRA). Because of this, siting studies are often accomplished with simplifying approaches, such as a consequence-only study. Within this realm of simplifying methodologies, the choice of a given hole size stands out as the one with the most influence on the study's results. Other modeling parameters are also important, but the hole size's qualitative link to probability (providing an insight into risk) makes it the singular choice for defining the magnitude of hazards which are to be considered.

A general list of acute hazards at or near petrochemical facilities includes the following:

- Exposure to a flash fire (an ignited flammable gas cloud)
- Exposure to thermal radiation from a jet (torch) fire, pool fire, or fireball
- Exposure to overpressure following a vapor cloud explosion (VCE) or physical explosion
- Exposure to a toxic gas

These hazards have been evaluated in many consequence and risk analysis studies for persons who are outdoors, for persons indoors (e.g., workers in an occupied building at a petrochemical facility), and for equipment or property within the facility. The effects on persons or equipment is a matter of choosing the hazard impact levels, or endpoints, that are applied in consequence modeling.

1.1 Building Siting

The American Petroleum Institute (API) has published a Recommended Practice (API RP 752) titled *Management of Hazards Associated with the Location of Process Plant Buildings* [1]. It is the recognized standard for building siting assessments, and outlines a methodology for conducting occupied building siting assessments. API RP 752 provides guidance for identifying hazards that could affect the process plant buildings and/or occupants of those buildings, and for managing risks related to those hazards. API RP 752 addresses three categories of hazards:

- Explosion
- Fire
- Toxic material

Evaluation of these hazards is as varied as the number of analysts. Explosions are traditionally the focus of building siting; fires have often been addressed through spacing alone; toxic gas exposure has often been simplified to be a function of the concentration outside a building. While most process plant buildings will provide occupant protection in the event of an external flash fire and provide shielding against thermal radiation, the evaluation of hazards should account for what levels of protection are afforded by the type of building and any mitigation measures in place. For example, where flammable or toxic gases are present in the facility, dispersion analyses should check for infiltration to determine occupant impacts, taking into account the building construction and any air exchange system.

As the potential hazards are better understood, other aspects of the analysis are given consideration. The subject matter of this paper – the selection of release hole size – is arguably the most important variable when the analysis is not a full QRA. Hole size selection can define the general level of risk (larger, high consequence releases are significantly less frequent than small, low consequence leaks) and it can define an event magnitude that has an assumed probability, leading to a qualitative measure of risk. Selection of a maximum credible event (MCE), based on a specific release hole size, then forms an important basis for the building siting evaluation.

2 Methodology

The methodology that forms the background for this paper is consistent with API RP 752 and provides a four stage process that allows the study to stop at any stage by proceeding only as far as necessary to demonstrate that occupants of the subject buildings will be adequately protected. Similar methodologies can be applied for siting studies that involve offsite populations or onsite equipment.

Stage 1 is a screening analysis, serving to identify potential hazard sources and determine which buildings should be included in the analysis. This involves a review of the process to determine the potential hazards that are inherent to the materials being handled. A decision concerning the inclusion of buildings is then based on criteria found in API RP 752. The guidance provided in API RP 752 regarding building inclusion can be summarized by the principle that all buildings intended for occupancy are to be included in the analysis.

Stage 2 is a worst-case consequence analysis, and serves to evaluate the potential consequences of the hazards identified in Stage 1. The siting assessment can be stopped after Stage 2 if it has been shown that adequate protection against worst-case scenarios is provided for building occupants through building spacing, building construction, process controls, or other engineering methods.

Stage 3 is a qualitative risk analysis. This combines consequence analysis and estimates of accident frequencies to arrive at a qualitative measure of risk posed to each building's occupants. This is accomplished by modeling the MCE, often defined by a hole size that is deemed the maximum credible hole size. The intent is to determine if the risk to building occupants is below a qualitative risk acceptability criterion. Depending on the results of this stage, determination of acceptability may be qualitative or semi-quantitative. If the risk is found to be acceptable, the analysis can end after Stage 3.

Stage 4 is a quantitative risk analysis (QRA), a much more rigorous process than previous stages. In this stage, hazard footprints are first generated for the full range of release sizes, locations, and weather conditions. Next, the frequency of occurrence of each hazard footprint is determined using historical equipment failure rates, conditional probabilities, and site-specific weather data. All calculated hazard zones, with their respective probabilities and originating locations, are combined to produce contours which demonstrate the frequency with which an event originating in the facility could adversely impact an occupied building and its occupants due to vapor cloud explosions, fires, and toxic or flammable vapor clouds. If the level of risk is acceptable, the analysis is complete. If not, mitigation measures that reduce event consequences or frequencies (or both) are implemented. Stage 4 can then be repeated, with modifications of the event consequences and/or probabilities based on mitigation measures to determine the risk.

3 The Maximum Credible Event

API RP 752 defines the maximum credible event (MCE) as:

> A hypothetical explosion, fire, or toxic material release event that has the potential maximum consequence to the occupants of the building under consideration from among the major scenarios evaluated. The major scenarios are realistic and have a reasonable probability of occurrence considering the chemicals, inventories, equipment and piping design, operating conditions, fuel reactivity, process unit geometry, industry incident history, and other factors. Each building may have its own set of MCEs for potential explosion, fire, or toxic material release impacts.

The use of the word 'credible' is always problematic because it is inherently subjective. Based on the guidance given in RP 752 regarding the MCE, the selection of MCEs can be done in several ways.

- Those events that have a 'reasonable' probability of occurrence
- Those events that have the largest consequences on building occupants
- Those events that have historical precedence
- Some combination of the above (or other) factors

An investigation into the types of MCE selections (as available in open literature) shows that there is a great deal of variance in how MCEs are defined and applied in siting studies. Some notable examples are:

> In "*Application of Qualitative and Quantitative Risk Analysis Techniques to Building Siting Studies*" [2], the authors present a methodology that integrates a corporate standard of a two-inch hole as the largest credible event. The methodology subsequently applies a quantitative risk analysis (QRA) approach if occupied building impacts due to the two-inch holes are found.
>
> In "*Evaluation of Permanent Buildings at an Existing Upstream Oil and Gas Business*" [3], the authors apply a methodology found in a corporate engineering technical practice that specifies two-inch holes and full-bore pipe ruptures as the events to be evaluated in a building siting study. The case study used these consequence analyses to evaluate varying magnitudes of hazards and then transition to a risk-based evaluation which used the probability of each of the hole sizes in the analysis.
>
> The Center for Chemical Process Safety (CCPS) has published a book titled "*Guidelines for Fire Protection in Chemical, Petrochemical, and Hydrocarbon Processing Facilities*" [4]. Four release categories for fire hazard analysis studies are presented. Small, medium, large, and rupture categories are defined by hole sizes of 3-10 mm, 10-50 mm, 50-150 mm, and full-bore ruptures, respectively. Discussion around these hole sizes also points out that the large and rupture

releases are not always the worst-case events due to event durations and dosage effects that should be accounted for.

The paper "*Methodology for selection hole sizes for consequence analysis studies*" [5] further explores the issue of event duration relative to hole size. Their case studies show that for fires "an intermediate hole size provides the combination of intensity and duration that results in the maximum thermal effect." While the worst case hole size was found to vary based on the effect being evaluated, the process system properties, and the distance from the release point, a two-inch (50 mm) hole could be taken to be representative of the worst case for many of the cases evaluated. A hole size around the two-inch size was also found to be the worst case for toxic effects in their analysis.

"*A Frequency Based Approach to Hole Size Selection for Consequence Analysis*" [6] presents a methodology which selects the MCE based on a likelihood of occurrence approach for specific equipment groups. This results in a variable hole size for the MCE, based on the extent of the equipment group that is selected. While this method does provide a "leveling" effect for release probability, it is dependent on the subjective definition and delineation of equipment groups and results in larger MCE hole sizes for larger (more complex) equipment groups.

In the CCPS book "*Guidelines for Evaluating Process Plant Buildings for External Explosions, Fires, and Toxic Releases*" [7], general guidelines and technical details are presented for complying with API RP 752. The MCE approach offered by this text leads an analyst to consider many possible events, screening out those that are not deemed credible, and selecting an MCE for each building as the event that has the maximum consequences at that building.

The Canadian Standards Association (CSA) standard Z276, *Liquefied natural gas (LNG) – Production, storage, and handling* [8], specifies design spills which define the potential extent of hazard zones that are not allowed to impact the public. Z276 defines, among other (smaller) release scenarios, that the design spill should be "The maximum flow of LNG that could be spilled from … the largest branch connection, 60 mm (2 in) or smaller…." This effectively defines a 2-inch hole as the siting criterion for Canadian LNG plants.

These various approaches to MCE selection demonstrate the difficulty in determining the most technically sound approach. Due to the history of using the two-inch hole and its frequent occurrence in siting studies, further investigation into why it has been, and might continue to be used is warranted.

4 A Frequency Argument

While equipment failure frequencies are typically brought into a siting analysis only if the analysis has moved to a risk-based approach, they are sometimes integrated into a study as a comparative evaluation of various hole sizes. Whether for QRA or comparative purposes failure frequencies must be appropriate for the system being evaluated, and must be statistically valid. There are only a few failure frequency databases that meet these criteria. The most notable, and most popular for use by risk analysts, is the Hydrocarbon Releases Database (HCRD) [9]. This database provides 20+ years of incident data, which are all releases of hydrocarbon, as well as the data for numbers of equipment-years of service. Because it provides detailed information for both failures and the equipment numbers from which failures can originate (the numerator and denominator for calculating a failure rate) it is the most consistent database from which to take data.

Use of the HCRD is generally straightforward, although some manipulation is necessary in order to apply the data to accidental release scenarios [10]. For the purposes of this paper, an evaluation was made, using the HCRD, to determine the overall frequency distribution for the two-inch hole. The analysis took on the following steps:

1. For each equipment type, gather the data for number of equipment-years in service.
2. For each equipment type, gather the data for all hydrocarbon release incidents associated with that equipment type.
3. Remove all incidents which occurred under very low pressure (as these are typically maintenance events with little consequence).
4. Calculate a total release frequency for all incidents involving that equipment type.
5. Sort the remaining release incident data by release hole size.
6. Create a cumulative distribution curve of hole size versus fraction of the total release frequency. An example cumulative distribution curve is shown in Figure 1.

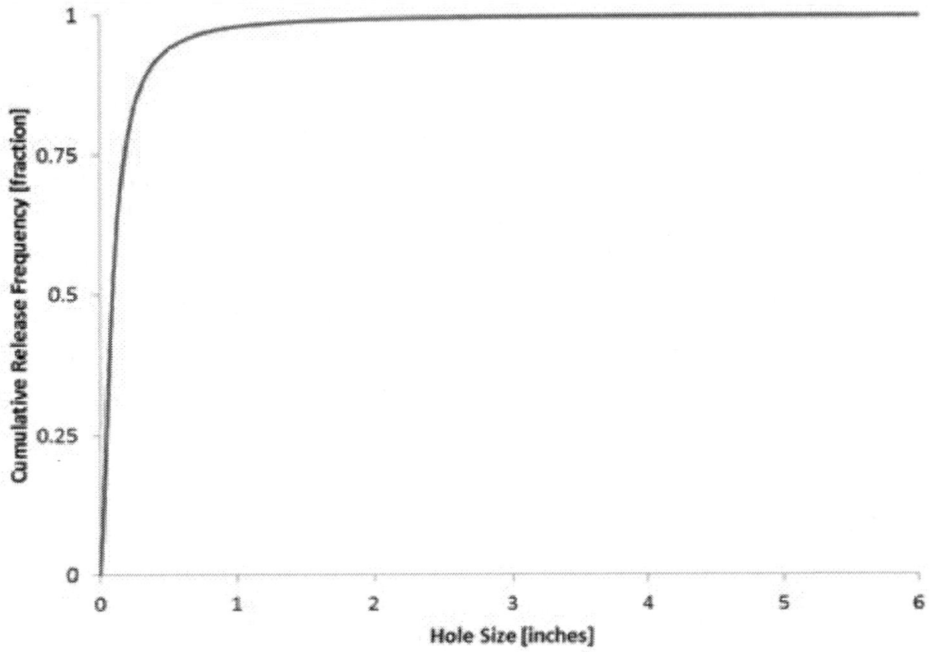

Figure 1. Example Cumulative Distribution Curve

With this approach, any hole size or hole size range can be chosen to evaluate its contribution to the overall failure rate. Taking this approach for the two-inch hole, the following conclusions were reached:

- For small equipment such as flange connections (gaskets), instrument connections, and small diameter pipe ($\leq$ 2 inches), hole sizes up to two inches represent 100% of the failure rate.
- For all other equipment (piping, compressors, pumps, heat exchangers, valves, filters, pressure vessels, storage tanks, and pig launchers/receivers), hole sizes up to two inches represent 95.46% of the total failure rate. This value is an aggregate average over all equipment types, not including the small equipment mentioned above.

Thus, it can be seen that the two-inch hole, if taken to represent all release hole diameters up to and including two inches, accounts for over 95% of the probability of hydrocarbon release events. This conclusion is reached from the evaluation of an offshore process systems database; events that are not well represented in this database (e.g., onshore storage tank overfill) may need to be included in a set of credible events.

5 Example Cases

To further illustrate how the two-inch hole may be an acceptable choice for an MCE, consider two example hydrocarbon facilities.

1) A small gas plant – This facility processes sour natural gas from a gathering system, removing contaminants such as water, carbon dioxide, and hydrogen sulfide (H_2S). It also separates out a stream of heavier hydrocarbons, or natural gas liquids (NGLs), which are separated, stabilized, and stored on site to be exported via truck. The remaining natural gas is compressed and sent out by pipeline.
2) A liquefied petroleum gas (LPG) transfer depot – This facility receives LPG (primarily propane) by pipeline and rail car. Propane is stored in pressure vessels on site and exported via tank truck to serve local markets. Besides propane and a mercaptan addition system for odorization (as needed), the facility does not handle, transport, or store any other hazardous materials.

The layout in Figure 2 represents this example small gas plant. Figure 3 shows the LPG facility used in this study. Both of these facilities are small, relative to many other hydrocarbon processing facilities. However, the principles applied to, and results obtained from, these smaller plants will apply to larger hydrocarbon processing facilities.

5.1 Screening Analysis

In the example gas plant facility, the control room, warehouse, and office are the three occupied permanent buildings on site. All three buildings are small pre-engineered metal buildings with similar sizes. As stated in the facility description, toxic and flammable hazards exist; therefore, the analysis will need to evaluate fires, explosions, flammable gas infiltration, and toxic gas infiltration.

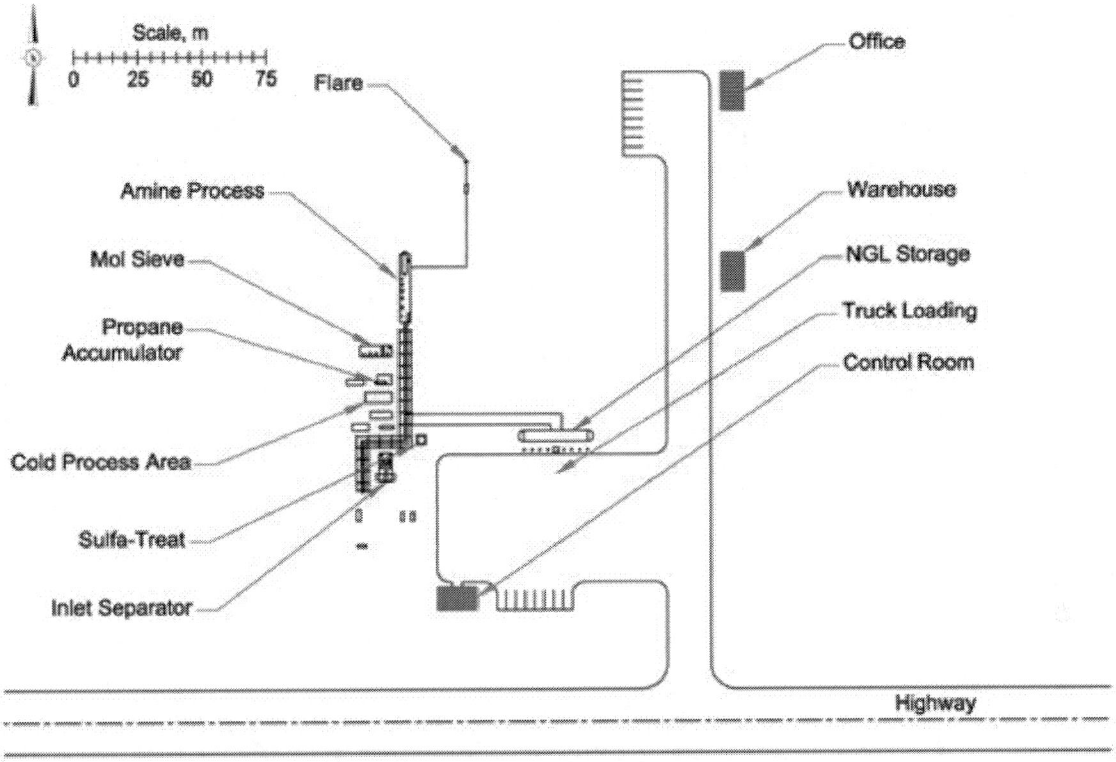

Figure 2. Example Gas Plant Facility Layout

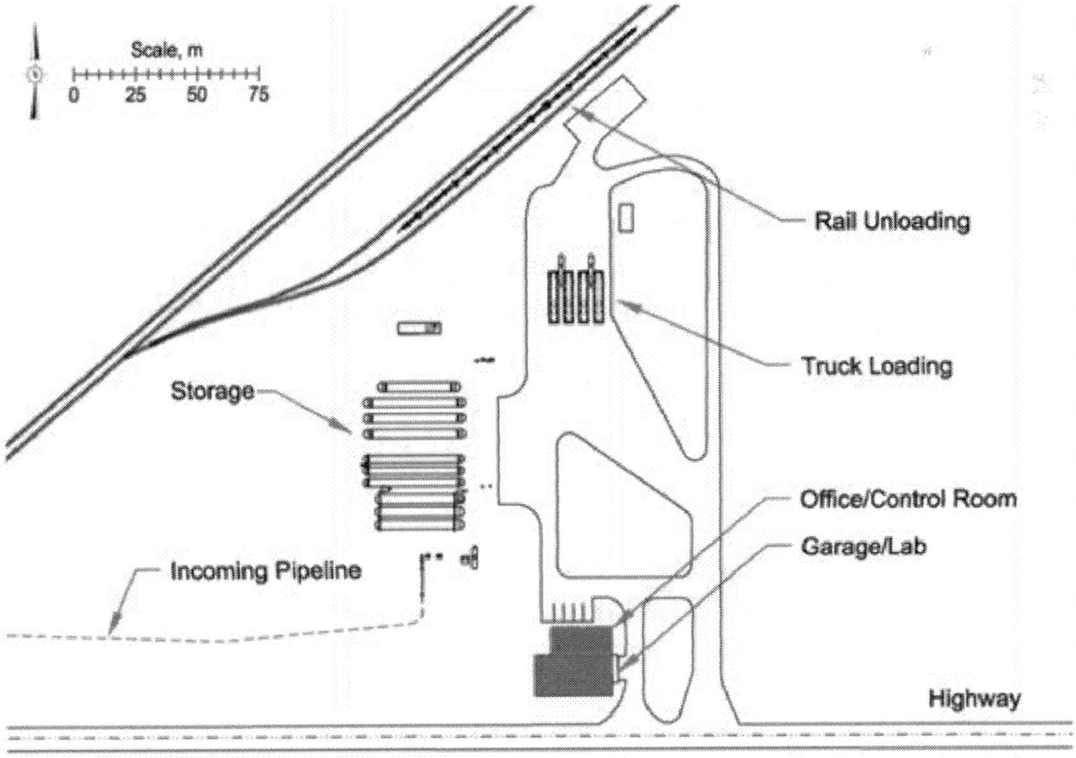

Figure 3. Example LPG Facility Layout

The LPG facility includes only two occupied buildings: a control room/office and garage/warehouse. While these two buildings are physically connected, they will be treated separately for the purposes of a building siting assessment. The site's hazards include fires, explosions, and flammable gas infiltration.

5.2 Consequence Analysis

The inherent hazards associated with these example facilities are well understood. A review of the process shows that there are multiple failure case scenarios that could result in fire, flammable cloud infiltration, toxic cloud infiltration, or explosion hazards that might impact the occupied buildings. A set of failure case scenarios was selected to represent each example facility – eight scenarios were chosen for the gas plant, and ten for the LPG plant (some additional scenarios were modeled to account for emergency shutdown systems and portions of the piping that are buried).

As part of the consequence analysis, locations for potential VCEs are selected to model representative explosion events. Locations within and around the facility that are confined or include congestion are referred to as potential explosion sites, or PESs. As thc amount of confinement or degree of obstruction increases, so does the potential strength of the blast wave that could be created by a vapor cloud explosion within that area. This type of analysis is covered in more detail in other papers [11,12]. The explosion model used in the assessment is the QMEFS model [13]. All other consequence calculations were accomplished with CANARY by Quest® consequence analysis software.

The endpoints chosen for use in this study are listed in Table 1. The endpoints represent the threshold of fatality to building occupants when the building is a pre-engineered small metal building.

Table 1. Hazard Endpoints for Building Occupants

Hazard	Endpoint for Threshold of Fatality	Reference Number
Overpressure	1.13 psi	[14]
Radiation	35 kW/m^2	[15]
H_2S Infiltration	100 ppm†	[16]
Flammable Infiltration	Lower Flammable Limit (LFL)	[1]

† This is the emergency response planning guidelines level 3 (ERPG-3) value, as defined by the American Industrial Hygenists Association

At this point in the siting study, the assessment typically evaluates whether the potential impacts to the occupied building(s) are acceptable by comparison of worst-case and MCE impacts. For the purposes of this paper, both example facilities were taken through a QRA. Within that process, the release hole sizes that would have been assigned as worst-case (usually pipe ruptures) and MCE (two-inch hole) were included, so that a comparison could be made between the consequence-based results and QRA results.

As an example of consequence-based results, consider a range of hole sizes (1/4 inch through 6-inch pipe rupture) for the liquid discharge line of the propane storage tank at the LPG facility. Figure 4 demonstrates the hazards extents for these release scenarios, as a function of hole size, for fire (immediate jet fire) and flammable infiltration. The hazard magnitude values are the maximum found at that hole size, considering all weather conditions and release orientations.

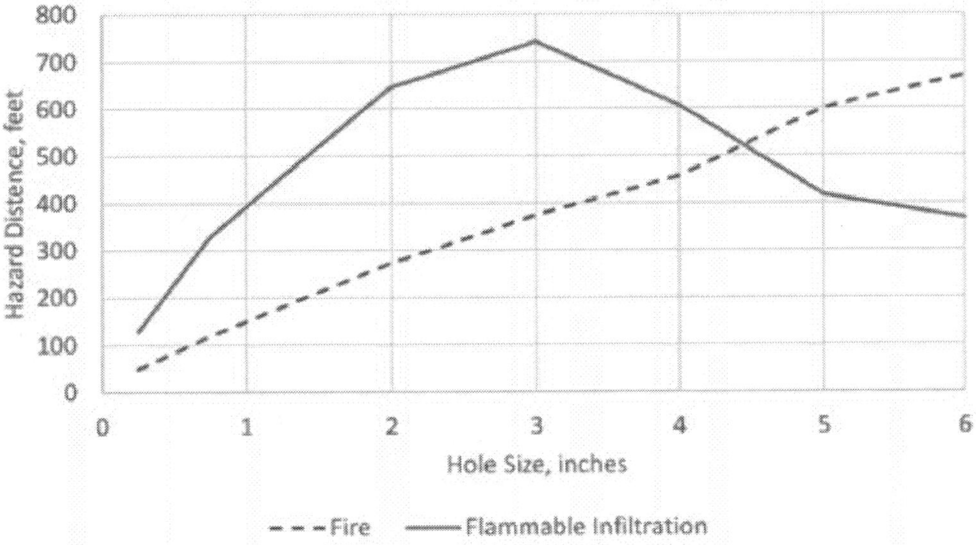

Figure 4. Effect of Hole Size on Maximum Hazard Magnitude

For this analysis the MCE was defined as a release up to a two-inch diameter hole in process equipment or piping. Due to dosage effects (in flammable and toxic infiltration, and occasionally radiation hazards), the worst case is not always the largest hole size of a selection set [5]. By setting the two-inch hole as a maximum, the MCE incorporates some element of frequency, but allowing smaller holes that may produce a larger hazard incorporates a maximum impact effect. Vulnerability zones for the various hazards, as defined by the MCE scenarios, are shown in Figure 5 for the gas plant and Figure 6 for the LPG facility.

5.3 *Quantitative Risk Analysis*

The risk that an individual inside a building is potentially exposed to a fatal hazard, as a result of events that could originate in the facility, can be represented by a numerical measure. This numerical measure represents the chance, or probability, that an individual will be exposed to a fatal hazard during a year-long period. For example, within the context of this example siting study, a value of 1.0×10^{-4} per year (or 10^{-4} in shorthand notation) represents one chance in 10,000 per year of being fatally affected inside a building by a release originating in the facility.

The risks due to all possible unique accidents in each example facility were combined to produce measures of risk for building occupants. The expansion of the failure cases to a full set of unique accidents involved the following parameters:

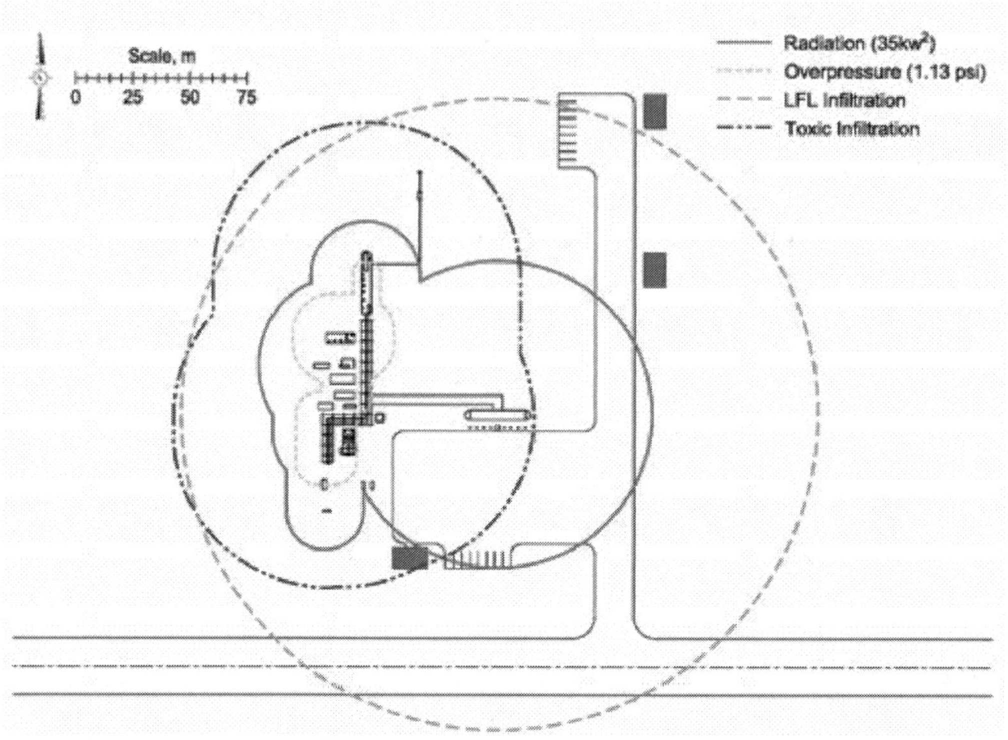

Figure 5. Gas Plant - Composite Vulnerability Zones for MCE Hazards

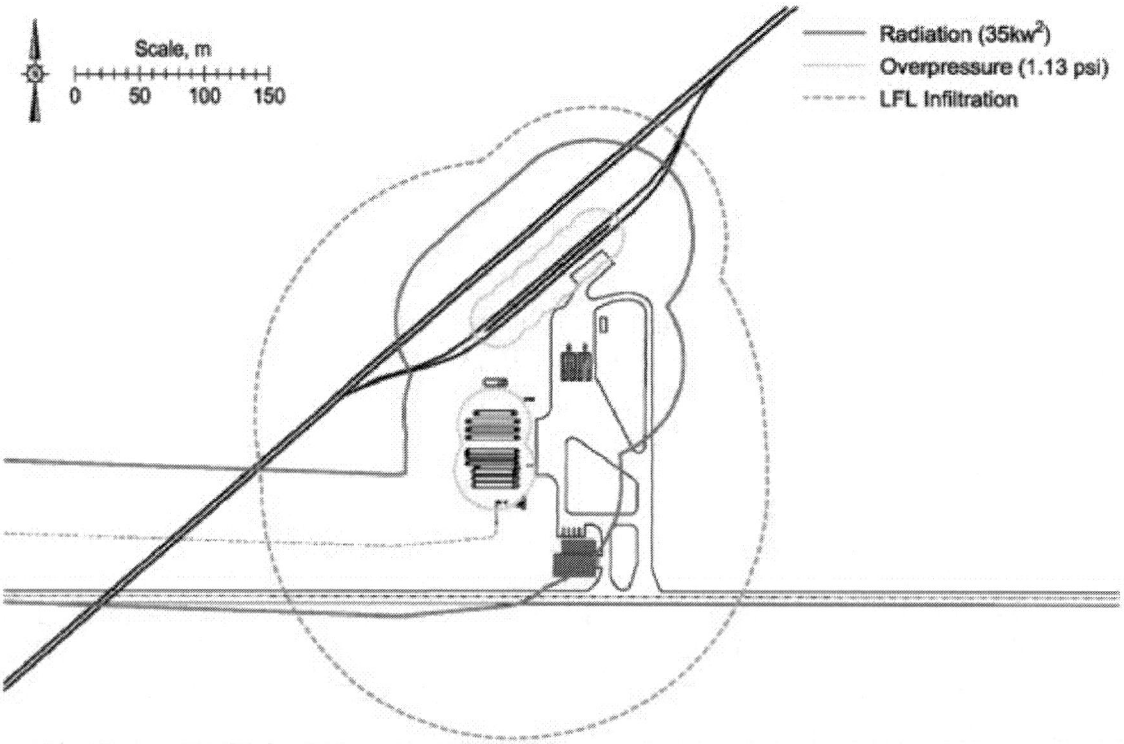

Figure 6. LPG Facility - Composite Vulnerability Zones for MCE Hazards

- 5 release hole sizes for each failure case scenario
- 2 release orientations (horizontal, upward)
- 21 wind speed and atmospheric stability combinations; 64 wind directions
- Fire radiation, vapor dispersion, and vapor cloud explosion hazards

The hazard impacts were computed and an event frequency was established for most of the above combinations (some combinations do not exist; for example, fire radiation models are not sensitive to atmospheric stability, so it is not a variable, and 6 wind speeds were applied instead of the 21 combinations). The event frequencies were calculated based on the Hydrocarbon Releases Database (HCRD), a comprehensive failure rate database [9], and the application of conditional probabilities. For any one unique scenario, the frequency of a release is obtained by following these steps:

- For the set of equipment described by the failure case (an equipment group which can be described by similar event consequences following an accidental release), the various pieces of equipment from which releases could occur are counted.
- The number of each type of equipment is multiplied by its failure frequency from the HCRD. The failure frequencies are summed for all equipment types involved in the failure case. Failure frequencies are developed for each release hole size, so the end result is a release frequency for each of the five hole sizes.
- The conditional probability for release orientation (horizontal=0.5, upward=0.5) is applied.
- The conditional probability for ignition timing is applied. Ignition is broken down into three possibilities: immediate ignition, delayed ignition, or no ignition.
    - Immediate ignition (jet fire, pool fire) is determined by the release rate and material type.
    - Delayed ignition (flash fire, vapor cloud explosion, jet fire, pool fire) is determined by the potential ignition sources encountered by the flammable vapor cloud.
    - The probability for non-ignition (toxic gas exposure, infiltration hazards) is the remainder after immediate and delayed ignition are accounted for.
- The conditional probabilities associated with weather conditions are applied. These include wind speed, atmospheric stability, and wind direction.
- Any conditional probabilities associated with a fractional time-of use for specific failure case scenarios, as applicable, are applied.

The total risk to building occupants is the summation of the potential impacts, at their respective event frequencies, for all combinations of failure case scenarios, hole sizes, release orientations, ignition timing, and weather conditions. The predicted risk represents the potential fatality for building occupants due to radiation from a fire, the effects of a vapor cloud explosion, flammable gas infiltration to the flammable limit (with ignition assumed), and toxic gas infiltration to the ERPG-3 concentration endpoint. The total risk is graphically presented in the form of location specific individual risk (LSIR) contours. The overall risk results are shown in Figure 7 for the gas plant and Figure 8 for the LPG facility. (Note that the LSIR contours are only valid where buildings exist or may be placed.)

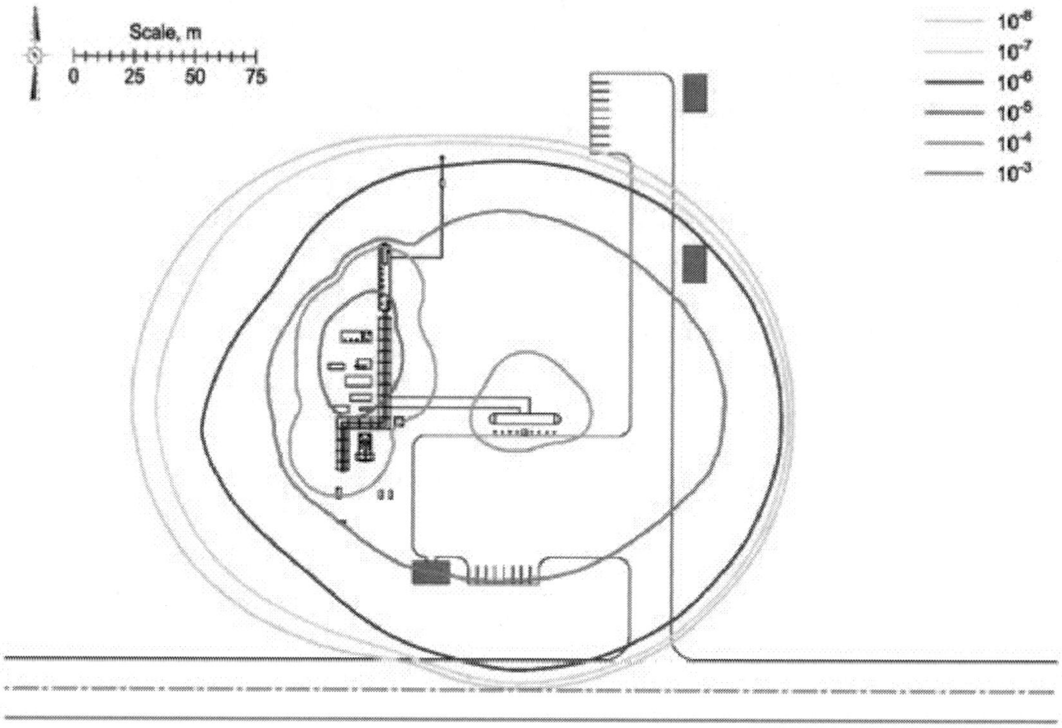

Figure 7. Gas Plant – Overall Risk Contours for Building Occupants

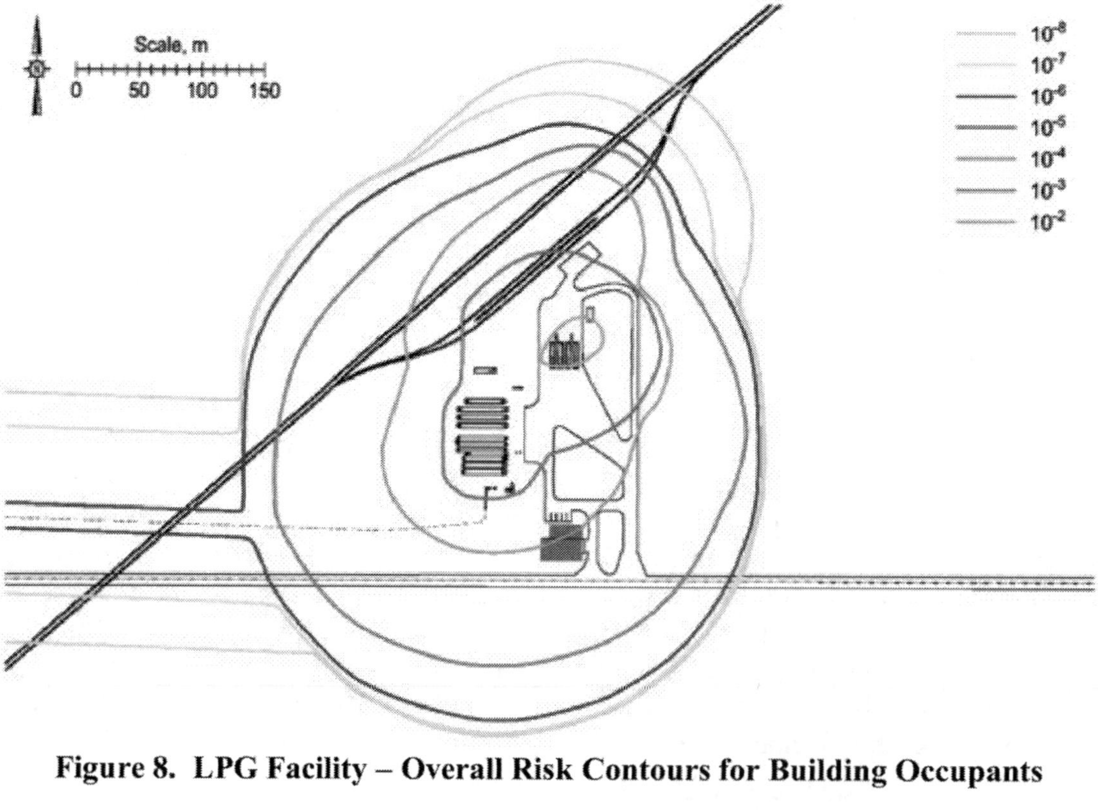

Figure 8. LPG Facility – Overall Risk Contours for Building Occupants

Because the risk contours shown in Figures 7 and 8 are defined based on impacts to occupied buildings, they are only valid at a location where a building exists (or could exist). For the control room in the example gas plant facility, the annual risk of fatality is predicted to be approximately 1.0 x 10^{-5} per year. For the control room in the example LPG facility, the annual risk of fatality is predicted to be approximately 1.0 x 10^{-4} per year. These risk levels assume continuous occupancy of the building(s), meaning that they are occupied 24 hours per day, 365 days per year.

6 Comparison

When the vulnerability zone results presented in Figures 5 and 6 are compared to the risk analysis results in Figures 7 and 8, several key findings can be extracted.

- For both facilities, the explosion overpressure vulnerability zones are within the overall 1.0 x 10^{-4} per year LSIR contour (those associated with all hazards). For these examples, the MCE vulnerability zones for explosions are small, but not expected to be any larger for less probable events due to the limited size of the PESs. This is generally true for most facilities.
- The fire radiation vulnerability zones following two-inch releases (as defined by the 35 kW/m^2 level) are approximately equal to the 1.0 x 10^{-5} per year risk contour at the gas plant. For the LPG facility, the MCE fire radiation vulnerability zones are approximately equal to the 1.0 x 10^{-4} per year risk contour, with the exception of the area surrounding the incoming pipeline.
- For the gas plant, the MCE flammable infiltration vulnerability zones extend well beyond the 1.0 x 10^{-6} per year risk contour. For the LPG facility, the MCE flammable infiltration vulnerability zones are approximately equal to the 1.0 x 10^{-6} per year risk contour.

7 Conclusions

The analysis presented in this paper sought to investigate the application of the two-inch hole to siting studies, with emphasis on building siting.

A review of published work on this subject found that the selection of hole size has been inconsistent, including some methods which select a variable hole size, some which use a fixed size, and some which leave the selection to events deemed credible. However, a more common choice was found to be the two-inch hole.

When evaluated according to release frequency, the two-inch hole was found to represent about 95% of all hydrocarbon release scenarios. This suggests that the two-inch hole is a reasonable selection as an upper bound of release hole sizes, and should be a reasonable choice for the MCE. This applies to general process system accidental releases and is not meant to exclude other events at a particular site that may be deemed credible.

In two example studies, the MCE vulnerability zones were compared to LSIR contours. Using an acceptable risk tolerance value for process plant personnel of 1.0×10^{-4} per year [17], vulnerability zones can be evaluated for suitability in siting studies. Based on this comparison for the example facilities, it was found that:

- When using an MCE defined by releases up to a two-inch hole, the fire radiation vulnerability zone compares well to the overall 1.0×10^{-4} per year LSIR contour for persons in buildings.
- MCE vulnerability zones for explosions are expected to be larger in extent than the 1.0×10^{-4} per year risk of fatality due to explosions, largely following the general observation that explosion impacts are limited by the volume of confined and congested space. Thus, the MCE explosion vulnerability zones can be taken to be representative of explosion risk.
- The flammable vapor dispersion and infiltration risk may not be well represented by MCE vulnerability zones – the vulnerability zones are aligned more with the 1.0×10^{-6} per year LSIR contour than the 1.0×10^{-4} per year contours.
- Toxic vapor dispersion and infiltration hazards may not be well represented; however, this will be a site-specific results that is highly dependent on the quantity and concentrations of toxic chemicals at the site.

As a generalization, the two-inch hole as an MCE selection appears to be a reasonable choice for representation of risk to occupied buildings. If a building is located outside of the MCE fire and explosion vulnerability zones, its risk will likely be acceptable. The exception to this is locations where there may be a high probability of flammable or toxic infiltration impact. This finding is certainly dependent on the size of the facility, the types of hazardous materials processed or stored, and the parameters applied to a siting analysis.

In conclusion, the general trend indicates that the two-inch hole can be used as a reasonable approximation of the risk to process plant building occupants: if a building is outside of the MCE vulnerability zones (as defined by releases up to two inches in diameter), the risk to occupants is generally found to be acceptable.

8 References

[1] API (2009), *Management of Hazards Associated with Location of Process Plant Permanent Buildings*. API Recommended Practice 752, American Petroleum Institute, 1220 L Street, Northwest, Washington, D.C. 20005, Third Edition, December 2009.

[2] Cornwell, J.B., Marx, J.D., and Lee, W.W. (1998), "Application of Qualitative and Quantitative Risk Analysis Techniques to Building Siting Studies". Process Plant Safety Symposium, Houston, TX, October 26-27, 1998.

[3] Aspengren, M., Whipple, T., Vralsted, D., Raghunathan, V., and Pitblado, R. (2009), "Evaluation of Permanent Buildings at an Existing Upstream Oil and Gas Business". 5th Global Congress on Process Safety, Tampa, FL, April 26-30, 2009.

[4] CCPS (2003), *Guidelines for Fire Protection in Chemical, Petrochemical, and Hydrocarbon Processing Facilities*. Center for Chemical Process Safety of the American Institute of Chemical Engineers, 3 Park Avenue, New York, New York, 10016-5991. (ISBN 0-8169-0898-2)

[5] Sanguino, O. and Hissong, D.W. (2013), "Methodology for Selecting Hole Sizes for Consequence Studies". Journal of Loss Prevention in the Process Industries 26 (2013) 483-488.

[6] Stahl, M.W. and Kenady, K. (2011), "A Frequency Based Approach to Hole Size Selection for Consequence Analysis". AIChE 7th Global Congress on Process Safety, Chicago, IL, March 13-16, 2011.

[7] CCPS (2012), *Guidelines for Evaluating Process Plant Buildings for External Explosions, Fires, and Toxic Releases, Second Edition*. Center for Chemical Process Safety of the American Institute of Chemical Engineers, Published by John Wiley & Sons, Inc., 111 River Street, Hoboken, New Jersey, 07030. (ISBN 978-0-470-64367-9)

[8] CSA Z276 (2015), Liquefied natural gas (LNG) — Production, storage, and handling. Canadian Standards Association, 178 Rexdale Boulevard, Toronto, Ontario, M9W 1R3, January 2015.

[9] HSE HCRD (2012), Hydrocarbon Releases System, Health and Safety Executive. https://www.hse.gov.uk/hcr3

[10] Bolsover A., (2013), "A Public Leak Frequency Dataset for Upstream and Downstream Quantitative Risk Assessment". AIChE 9th Global Congress on Process Safety, San Antonio, TX, April 28 – May 2, 2013.

[11] Marx, J.D., and K.M. Werts (2013), *The Application of Pressure-Impulse Curves in a Blast Exceedance Analysis*, Journal of Loss Prevention in the Process Industries, Volume 26, May 2013, pp 478-482.

[12] Marx, J.D., and Ishii, B. (2015), "A Comprehensive Approach to API RP 752 and 753 Building Siting Studies." Mary Kay O'Connor Process Safety Center International Symposium, College Station, TX, October 27-29, 2015.

[13] Melton, T.A. and J.D. Marx (2009), "Estimating Flame Speeds for Use with the BST Blast Curves." *Process Safety Progress*, Vol. 28, No. 1, March 2009, pp 5-10

[14] DOD (2009), *"Approved Methods and Algorithms for DOD Risk-Based Explosives Siting"*. Technical Paper No. 14, Revision 4. Department of Defense Explosives Safety Board, Alexandria, VA, July 21, 2009.

[15] TNO (1999), *Guidelines for Quantitative Risk Assessment* (First Edition), the Purple Book. CPR 18E, the Netherlands Organization for Applied Scientific Research, Committee for the Prevention of Disasters, the Hague, Netherlands, 1999.

[16] AIHA (2013), American Industrial Hygienists Association, 2013 Emergency Response Planning Guidelines, https://www.aiha.org/get-involved/AIHAGuidelineFoundation/EmergencyResponsePlanningGuidelines/Documents/2013ERPGValues.pdf , 2013.

[17] CCPS (2009), *Guidelines for Developing Quantitative Safety Risk Criteria*. Center for Chemical Process Safety of the American Institute of Chemical Engineers. Published by John Wiley & Sons, Inc., Hoboken, New Jersey, 2009. (ISBN 978-0-470-26140-8)

Retrofit of Non-Structural Components on Buildings Subjected to Blast Loading

David A. Holgado, M.S., P.E.
ABS Consulting
Advanced Engineering Division
140 Heimer Rd., Suite 300
San Antonio, TX 78232
dholgado@absconsulting.com

Jerry Collinsworth, M.S.
ABS Consulting
Advanced Engineering Division
jcollinsworth@absconsulting.com

Jeff Goodman
ABS Consulting
Advanced Engineering Division
jgoodman@absconsulting.com

UNPUBLISHED

Prepared for Presentation at
American Institute of Chemical Engineers
2016 Spring Meeting
12th Global Congress on Process Safety
Houston, Texas
April 10-14, 2016

Retrofit of Non-Structural Components on Buildings Subjected to Blast Loading

David A. Holgado, M.S., P.E.
ABS Consulting
Advanced Engineering Division
140 Heimer Rd., Suite 300
San Antonio, TX 78232
dholgado@absconsulting.com

Jerry Collinsworth, M.S.
ABS Consulting
jcollinsworth@absconsulting.com

Jeff Goodman
ABS Consulting
jgoodman@absconsulting.com

Keywords: Hazard Identification and Risk Analysis, Safe Design

Abstract

Structures in chemical and petrochemical facilities are often located in areas that may be subjected to blast loading. Occupied buildings typically have non-structural components located along the interior of the exterior walls and roof such as windows, doors, wall mounted AC units, lights, furniture, storage racks, hanging equipment, and loose articles.

Occupants of buildings subjected to accidental explosions may be injured from glass fragments and interior non-structural items becoming projectiles and impacting building occupants. As a pressure wave from a blast impacts the exterior of a building, the wall and roof components are rapidly accelerated inward. Equipment or contents mounted on or in contact with the exterior façade are also accelerated and may be dislodged and projected as debris. Items anchored to the ceiling structure can be thrown vertically from the initial forward deflection of the supporting member or break free from their supports and become falling debris hazards.

Therefore, evaluation and mitigation of non-structural debris for buildings subjected to blast load is important to further mitigate the potential hazards to personnel occupying these buildings. This paper provides design retrofit recommendations based on accident investigation experience at chemical and refining facilities and engineered solutions for typical hazards commonly observed at these facilities.

1. Introduction

In blast related events, injuries may result from glazing fragment hazards and interior articles that become projectiles and impact building occupants. Hazards associated with non-structural components occur when the pressure wave from an explosion impacts the exterior surfaces of a building. The wall and roof structural components and the attached interior components are rapidly accelerated inward. After reaching peak inward displacement the structural members spring back, creating a dynamically applied reaction force on the connection between the structural components (SC) and non-structural components (NSC). Connections that are too weak to resist these reactions break free from the loaded surfaces and the NSC are propelled into the occupied spaces of the building.

Non-structural hazards result from items of notable weight attached to or in contact with exterior walls or ceiling. Common examples of these items are bookcases, filing cabinets, storage racks, wall mounted fixtures, overhead lights and air conditioning diffusers. As a general rule most items mounted overhead, or located 48-inches or higher above the floor, can pose a hazard to occupants in the event of an accidental explosion.

Figure 1 depicts a typical vertical cross-section of a building with non-structural components secured to exterior structural components. An example of the non-structural hazards that can result even at lower building damage levels is provided below in Figure 2 [1].

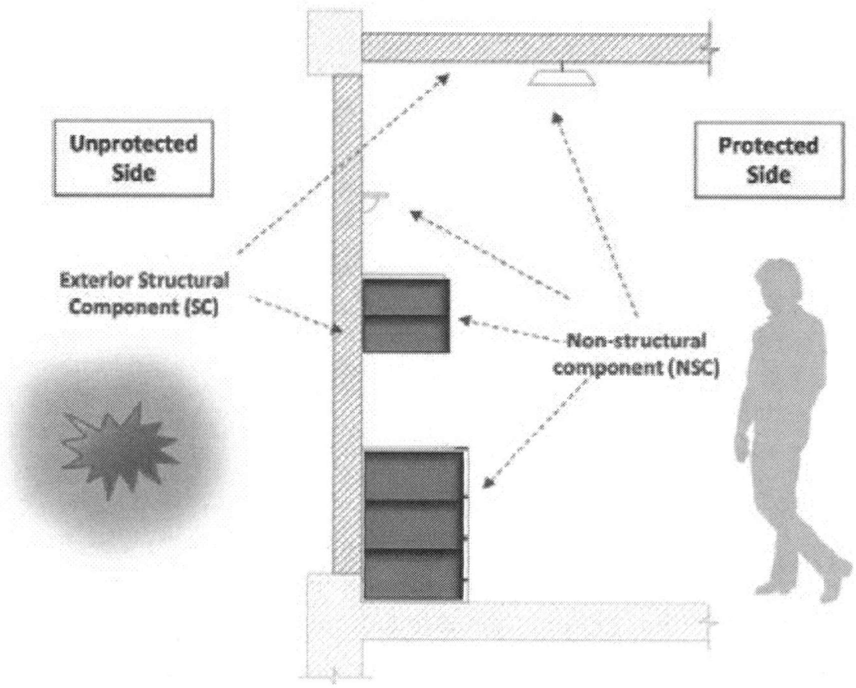

Figure 1. Non-Structural Hazards Adjacent to Exterior Building Surfaces

Figure 2. Lightly Damaged Building and Associated Non-Structural Hazards [1]

The response of non-structural components (NSC) to blast loading is associated with the response of the attached structural component (SC) in order to determine the design basis acceleration for supporting the NSC weight. This paper summarizes the basis of the methodology developed for this purpose.

Note that the upgrade solutions developed in this paper are not intended to prevent damage to the fixture itself (i.e. internal/external mechanical or electronic damage) but rather to mitigate the hazards to occupants. Also, window glazing and door response to blast loads are not included but can also present non-structural hazards. They should be addressed on a case by case basis for each building.

2. Design Methodology

2.1 Structural Response Prediction

Structural components subjected to blast loading are commonly analyzed utilizing the Single Degree of Freedom (SDOF) method [2]. The SDOF solution is obtained using simple numerical integration methods and the Single-Degree-of-Freedom Blast Effects Design Spreadsheet SBEDS [3] is one of the publically available tools.

Closed form solutions of the SDOF response are difficult to obtain because of material non-linearity and variation of loading. However, an analytical solution of the SDOF response has been achieved considering certain assumptions and simplifications as described in the following sections. The typical representation of SDOF system is depicted in Figure 3.

Although the peak deflection of SDOF elasto-plastic system could be predicted using simplified charts, such as the one shown in Figure 4, [2], the charts predicting the acceleration at the moment of maximum deflection, which is the design basis acceleration for restraining the non-structural fixture, are not available.

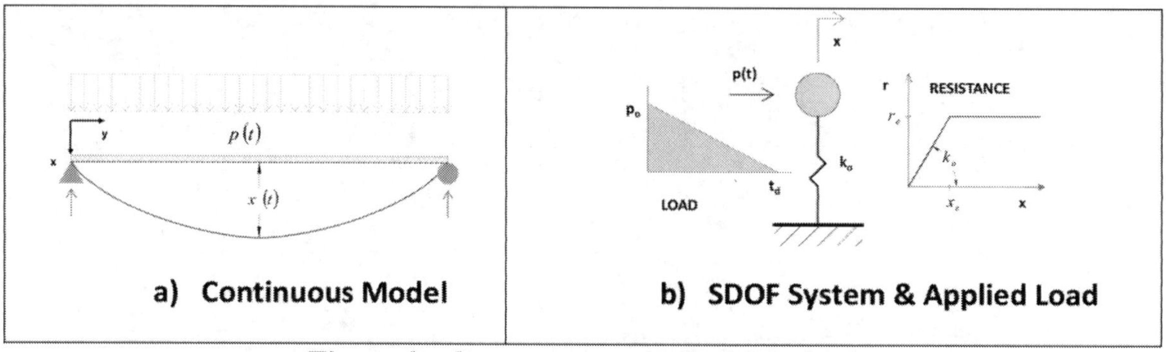

Figure 3. Component and SDOF Model

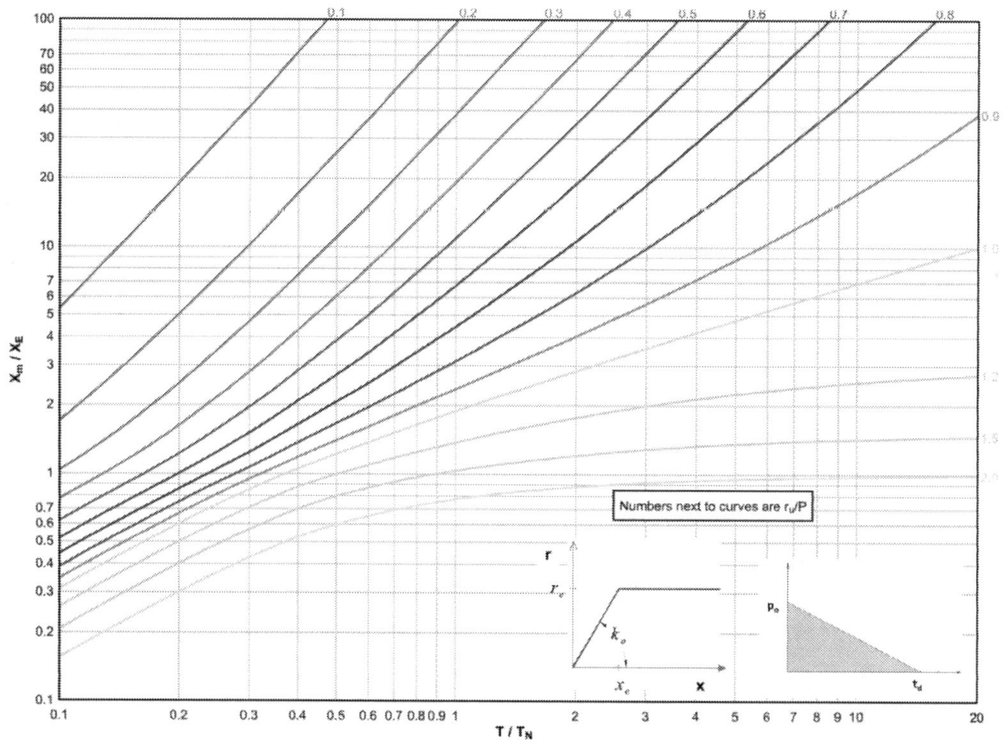

Figure 4. Non-Dimensional Maximum Deflection Response of SDOF [1]

2.2 *Assumptions and Approximations*

- Mass is uniformly distributed along the span of the component.
- The assumed shape function, $\phi(y)$, is applicable for deflection, velocity and acceleration.
- The applied blast load is assumed to be right triangular with an instantaneous rise to peak pressure and a linear decay to ambient as illustrated Figure 3 b).
- The resistance function follows a bi-linear configuration as shown in Figure 3 b).

- The component has adequate shear strength to transfer the reaction forces during the component response.
- For components having resistance other than bilinear due to support constraints (e.g. Fixed-Simple, Fixed-Fixed, etc.), the approach described herein is still applicable. The equivalent elastic stiffness, k_{EQ}, should be computed that results in the resistance deflection function having equal strain energy.

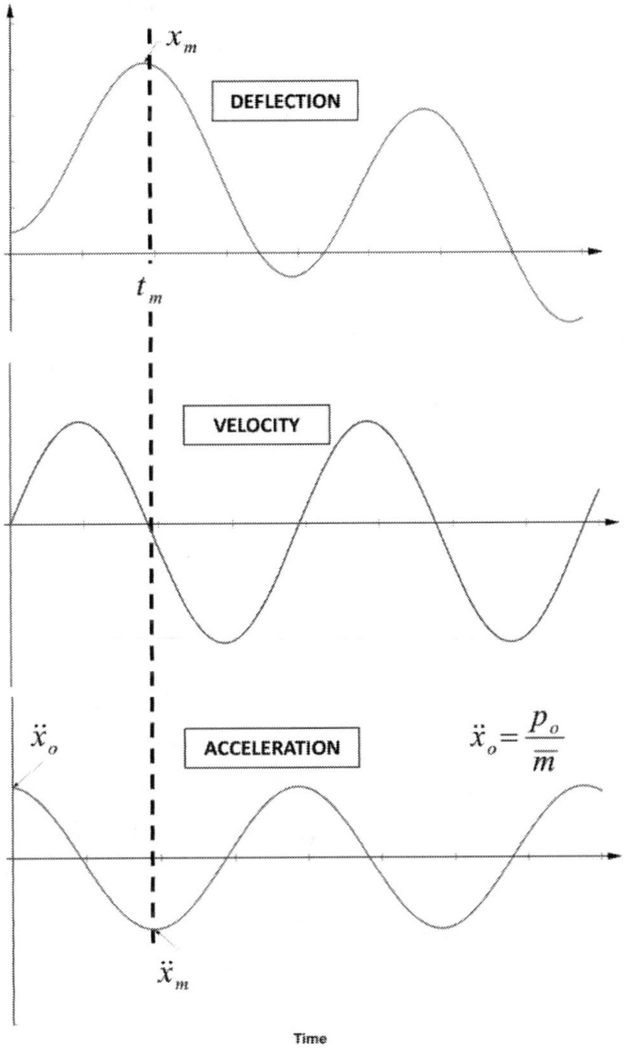

Figure 5. SDOF Dynamic Response

2.3 Component Response Prediction

2.3.1 Deflection Response

The analytical solution is implemented in Visual Basic code and the SC ductility is plotted for the SDOF system as a function of the ratio of the load duration to the natural period (t_d/T_n) and the dynamic load factor (r_e/p_o) = [0.1, 0.2.. 1.5, 2.0] and then, the non-dimensional plot is generated as shown in Figure 6. The computed solution in Figure 6 is identical to plots provided for simple SDOF solutions as shown previously in Figure 4.

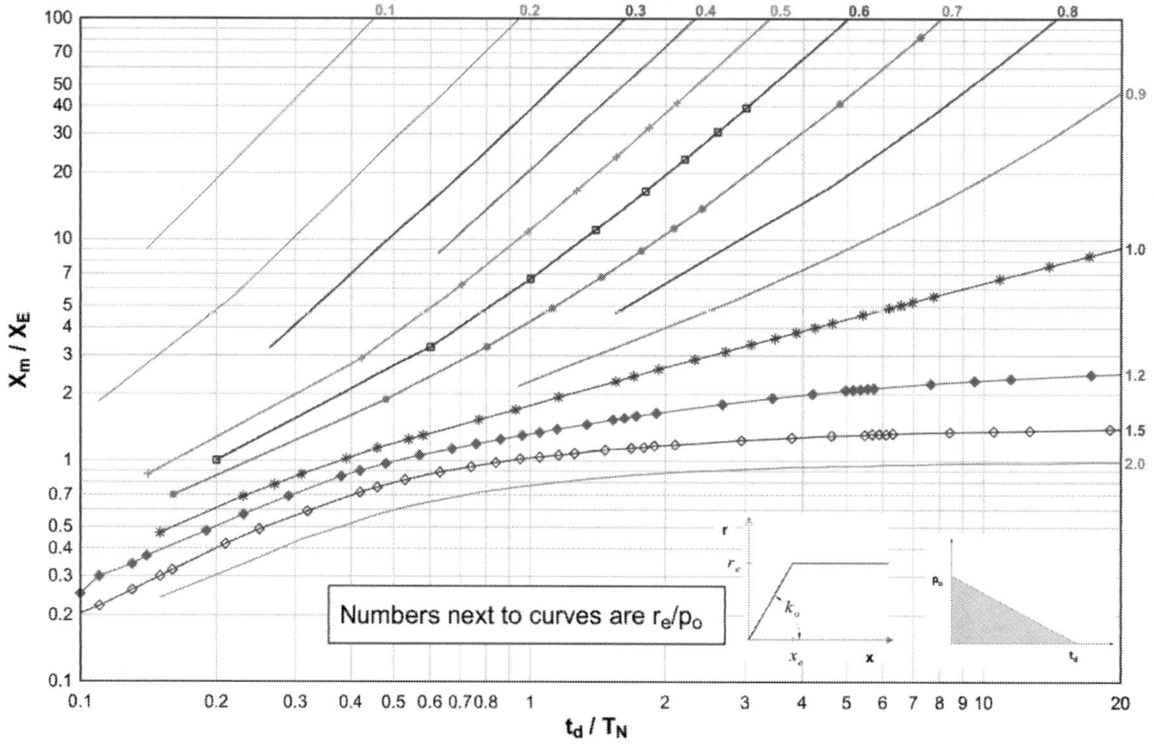

Figure 6. Non-Dimensional Maximum Deflection Response of SDOF

2.3.2 Acceleration Response

In a manner similar to the prediction of peak deflection described in previous section, the design basis acceleration (occurring at the time when the maximum deflection of the SC is achieved) is plotted also as a function of ratio of the load duration to the natural period (t_d/T_n) and the dynamic load factor (r_e/p_o) for $r_u \le p_o$ and $r_u \ge p_o$ as shown in Figure 7 & Figure 8, respectively. Figure 7 & Figure 8 may be utilized to determine the design basis acceleration for retrofitting the non-structural fixture attachment to the supporting SC.

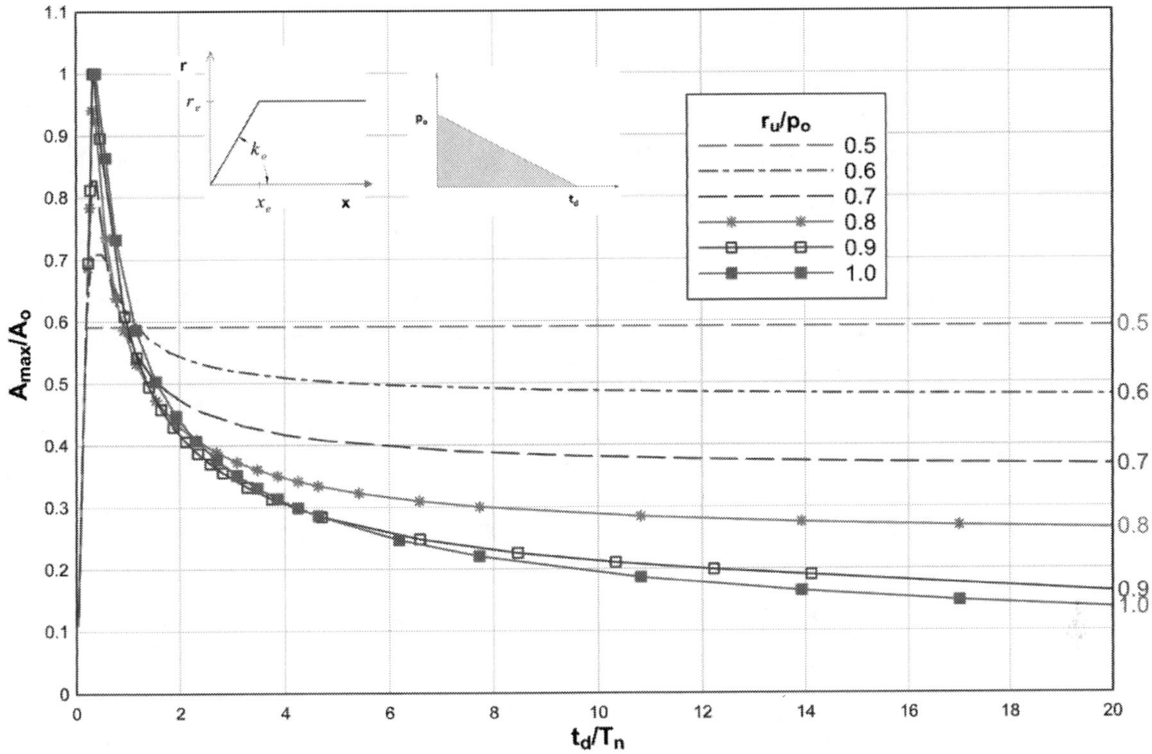

Figure 7. Non-Dimensional Acceleration Response for $r_u \leq p_o$ of SDOF

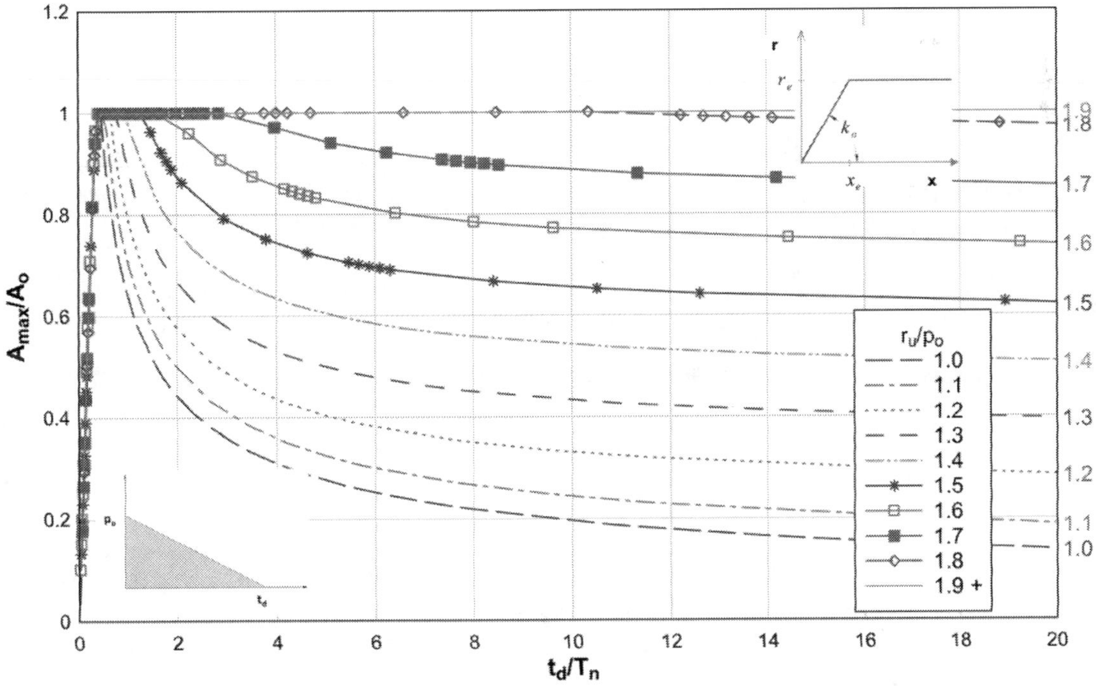

Figure 8. Non-Dimensional Acceleration Response for $r_u \geq p_o$ of SDOF

3. Retrofit Design Techniques

3.1 Assumptions and Limitations

- Structural components should be limited to Medium Response or lower as defined in Appendix 5.B of ASCE [5]; this component response allows certain permanent deformation during blast loading but it is still repairable.
- The building must be designed and constructed according to modern conventional construction practices, within the last 50 years using wood, steel, masonry or concrete. This excludes temporary buildings or trailers.
- The mass of the fixture is insignificant compared with the mass of the structural component and the fixture behaves as a rigid body during dynamic response of the structural component. In other words, there is no interaction effect during the response of the structural component and non-structural fixture as computed using the corresponding SDOF approach.
- The fixture is located near the location of the maximum response of the structural component (i.e. midspan).
- The fixture is in contact with the structural component prior blast loading (i.e. no impact effect).
- The fixture is not attached to cladding but is attached to a secondary framing component (e.g. purlin, girt, joist, etc.) or primary framing component (e.g. column, beam, Concrete Masonry (CMU) wall, etc.).
- The connection between the fixture and structural component is assumed to be rigid.
- The upgrade solutions developed herein are not intended to prevent damage to the fixture itself (i.e. internal/external mechanical or electronic damage) but rather to mitigate the hazards to occupants.
- Window glazing and door response to blast loads are not included but can also present non-structural hazards. They should be addressed on a case by case basis for each building

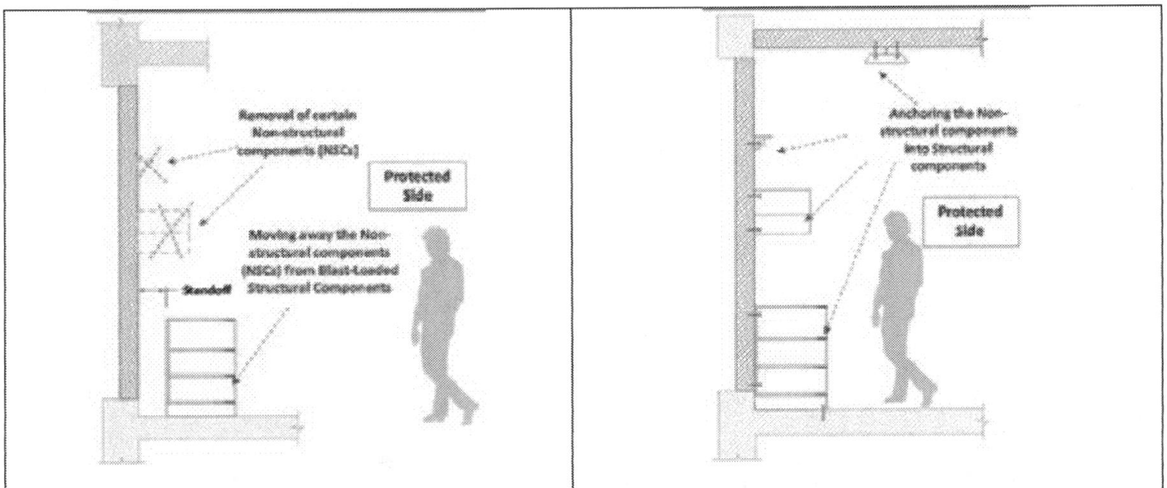

a) Decoupling Retrofit Strategy **b) Coupling Retrofit Strategy**

Figure 9 Non-Structural (NSC) Hazards Retrofit Strategies

3.2 Decoupling Retrofit Strategy

Of the two retrofit options, decoupling is the simpler one to implement. This technique is based on the concept of providing sufficient distance for all heavy fixtures from the exterior building surfaces (i.e. walls and roofs), as shown in Figure 9 a). A safe offset is computed based on the maximum inward displacement of the structural component along the exterior surface as shown in Figure 10. Thus, the maximum displacement of the primary or secondary framing system is computed for the applied blast load. Figure 11 depicts an example of maximum deflection of component under a set of applied blast loads. The non-structural fixture is then separated from the structure by sufficient distance to accommodate the total deflection of the supporting SC.

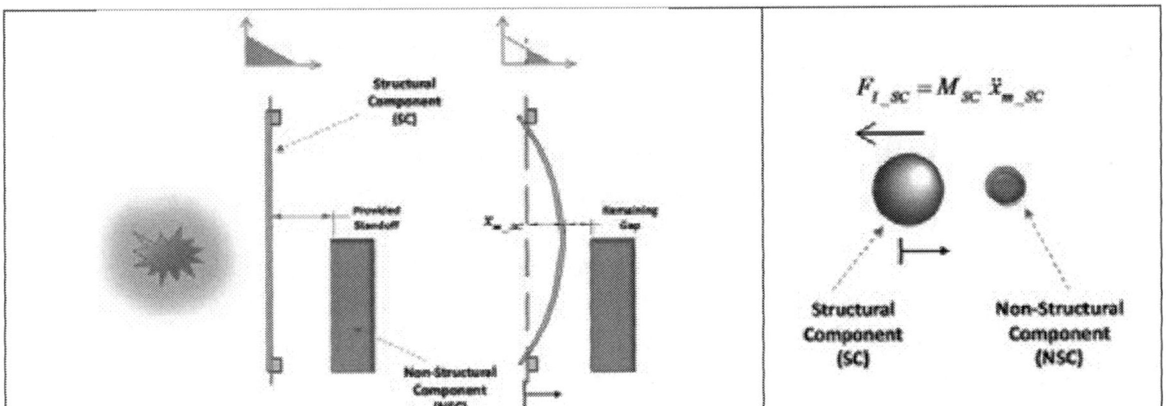

Position at Maximum Displacement **Free Body Diagram of SC & NSC**

Figure 10. Dynamic Response of Decoupled Components

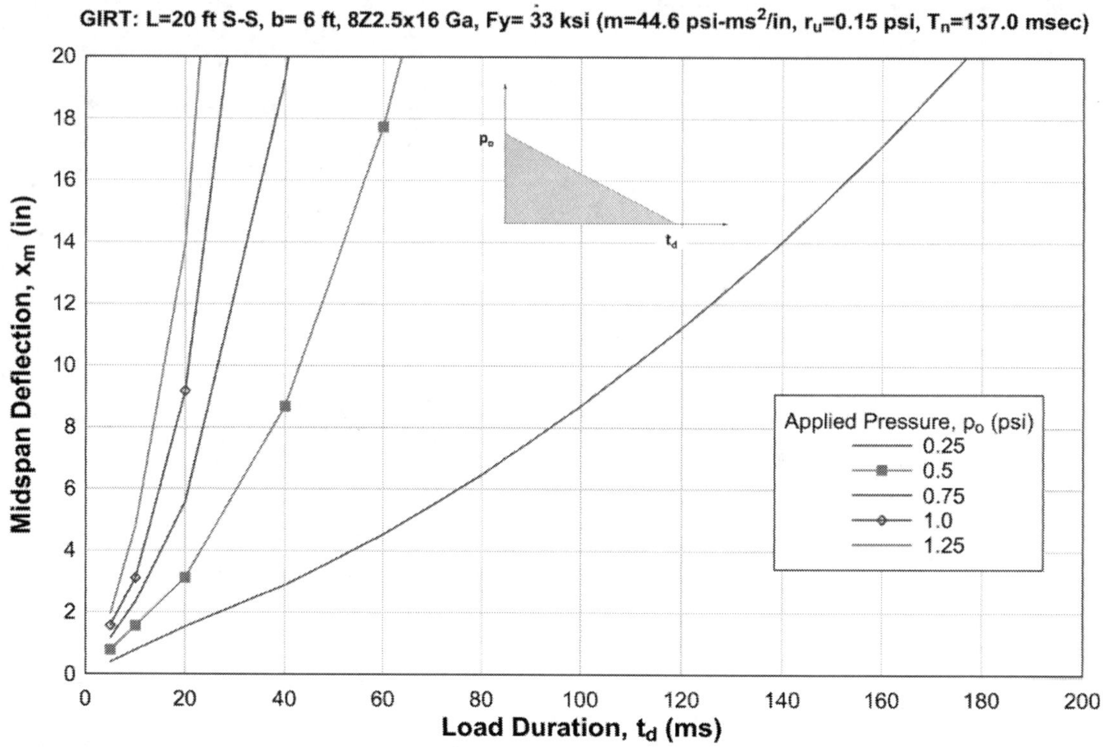

Figure 11. Maximum Deflection Response of Structural Component (SC), x_{m_SC} (example)

3.3 Coupling Retrofit Option

Since the decoupling retrofit technique is based on providing sufficient standoff distance between fixtures and loaded exterior surfaces, the availability of sufficient room is critical and in many cases that may not be a feasible retrofit option. In those instances, the coupling retrofit technique will need to be evaluated for implementation. Figure 9 b) depicts the general concept of the coupling option retrofit.

As the coupling retrofit relies on the fasteners between the SC and the non-structural fixture it is imperative the connection be sufficient to withstand the imparted forces at the time of peak displacement and the onset of structural rebound. The forces used to develop the connection requirements are based on the mass of the fixture and the acceleration of the wall at the point of maximum inward displacement, as depicted in Figure 12. Several assumptions are made in order to compute the acceleration of the fixture which are described in Section 2.3.2.

Figure 13 depicts an example of design basis acceleration of component under a set of applied blast loads

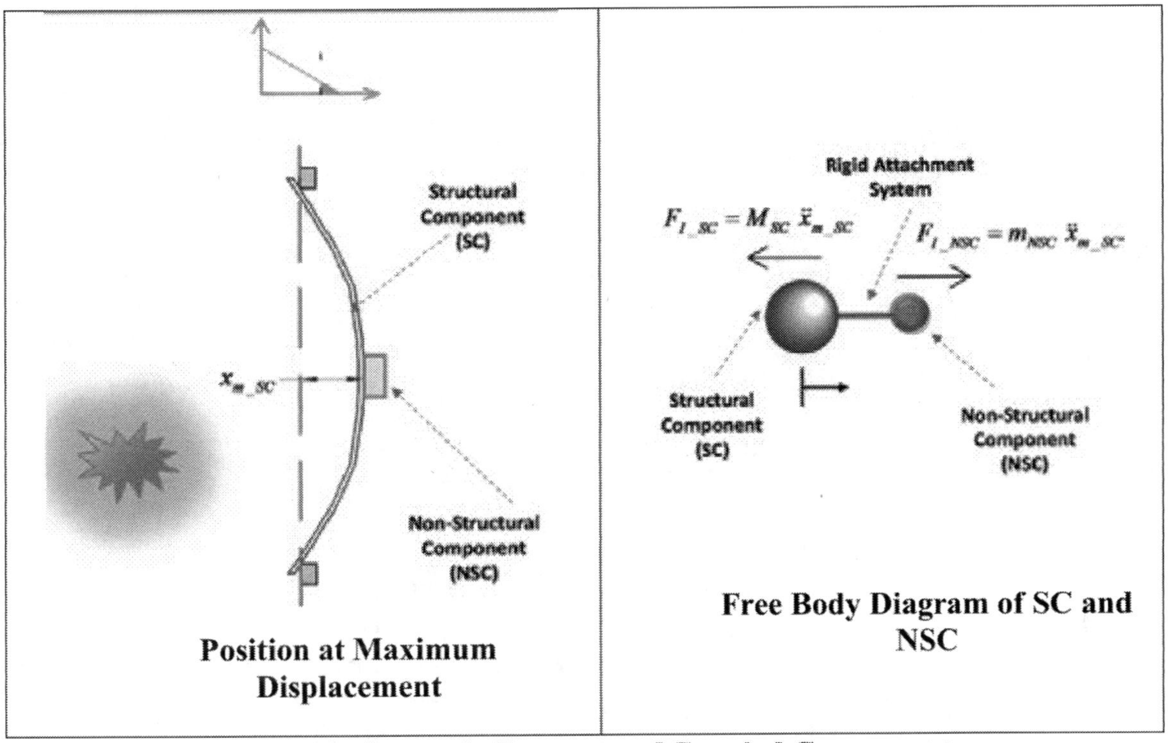

Figure 12. Dynamic Response of Coupled Components

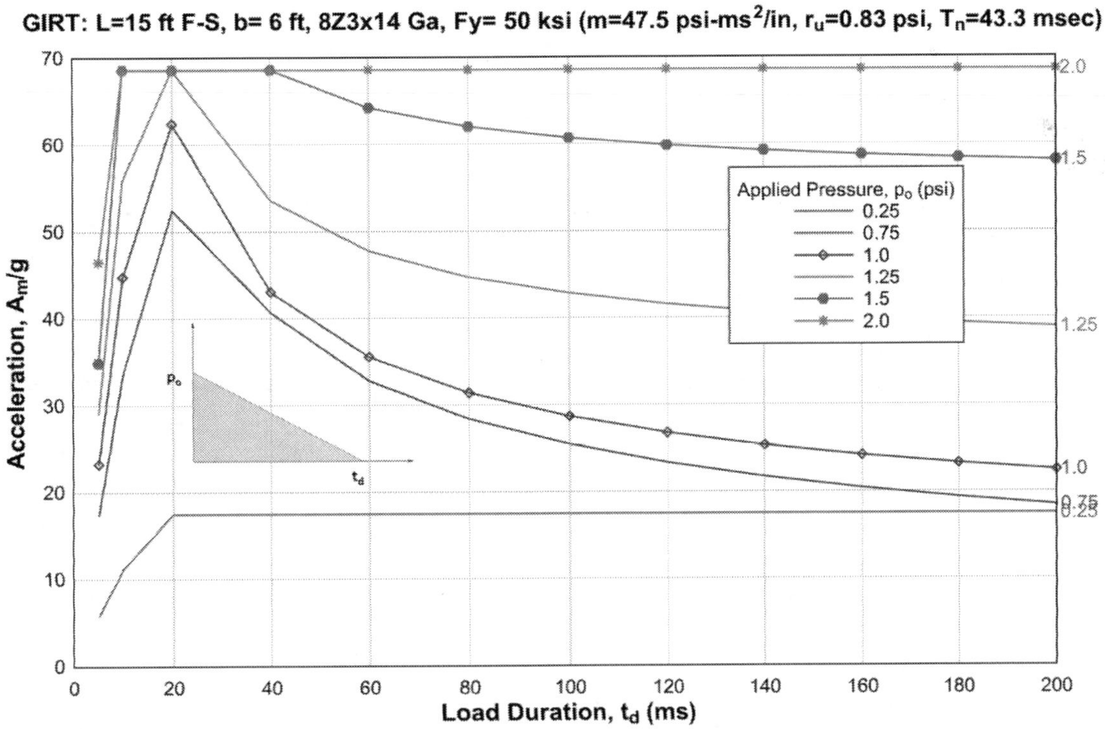

Figure 13. Acceleration of Structural Component (SC), $\ddot{x}_{m\_SC}$ (example)

3.3.1 Computation of Required Anchorage Forces

The required anchorage forces on NSC are based on the predicted acceleration of the structural response (at maximum displacement), as described in Section 2.2, and the mass of the attached fixture. Table 1 shows in example of peak acceleration of components per type of building construction.

Some common NSC to be retrofitted are identified including their corresponding weight as shown in Table 2.

Table 1. Maximum Acceleration of Structural Components (example)

Surface	Substrate	Amax (g)
Wall	R CMU	3.2
Wall	PEM	139.2
Roof	PEM	129.9
Wall	Steel	43.2
Roof	Steel	40.0
Roof	OWSJ	14.1
Wall	Wood	4.6
Roof	Wood	10.3

Table 2. Required Anchorage Forces per NSC (example)

			Required Anchorage Force, Fu (lb)		
Fixture	Weight (lb)	Surface	On Steel	On CMU	On Wood
Multi-Point Hanging Fluorescent Light	4.5	Roof	221		46
Fluorescent Light Fixture	24	Roof	1178		247
Light Fixture	35	Roof	1719		361
Garage light	10	Wall	1392	18	46
Wood cabinet	50	Wall	6960	89	230
Steel Lockers	110	Wall	15312	195	506
Peg board	5	Wall	696	9	23
Ceiling Diffuser	9	Roof	442		93
Bookcase	80	Wall	5568	142	368
Cabinet / Locker	60	Wall	4176	106	276
Metal Lateral File Cabinet (04 drawers)	710	Wall	98832	1257	3266
Whiteboard	53	Wall	7378	94	244
Emergency Respirator Wall Case	14	Wall	1949	25	64
15,000 BTU Window AC	108	Wall	15034	191	497
Metal Conduit, Electric	7	Wall	974	12	32
Fan Wall Mounted, 24 in	50	Wall	6960	89	230
Fan Wall Mounted, 30 in	54	Wall	7517	96	248
Water Dispenser, 5 Gal.	84	Wall	11693	149	386
Fiber Glass Ladder, 6 ft	18	Wall	2506	32	83
Fiber Glass Ladder, 8 ft	26	Wall	3619	46	120
Fiber Glass Ladder, 12 ft	48	Wall	6682	85	221
Hose Reel Retractable	40	Roof	1964		412
Oxygen Tank	120	Wall	16704	212	552
Electrical Box	45	Wall	6264	80	207
Storage Rack	1050	Wall	146160	1859	4830
HVAC Duct	138	Roof	6776		1421
Heater	55	Wall	7656	97	253
Base Cabinet	67	Wall	9326	119	308
Kitchen Cabinet, Wall Mounted	125	Wall	17400	221	575
Book Shelve, Wall Mounted	44	Wall	6125	78	202
TV. 40" Plasma	20	Wall	2784	35	92

In addition to the loading from the blast, gravity loads are imposed on non-structural components affixed to the roof. However, this refinement may be meaningful only for certain components. Using as example the values from Table 1, the gravity effect on wood components may increase the anchorage force by 20%. For the other roof components (i.e. steel purlins, Open Web Steel Joists, (OWSJ)), the gravity effect could be disregarded. Figure 14 and Figure 15 depict the free body diagrams (FBD) of anchorage forces along wall and roof surfaces, respectively.

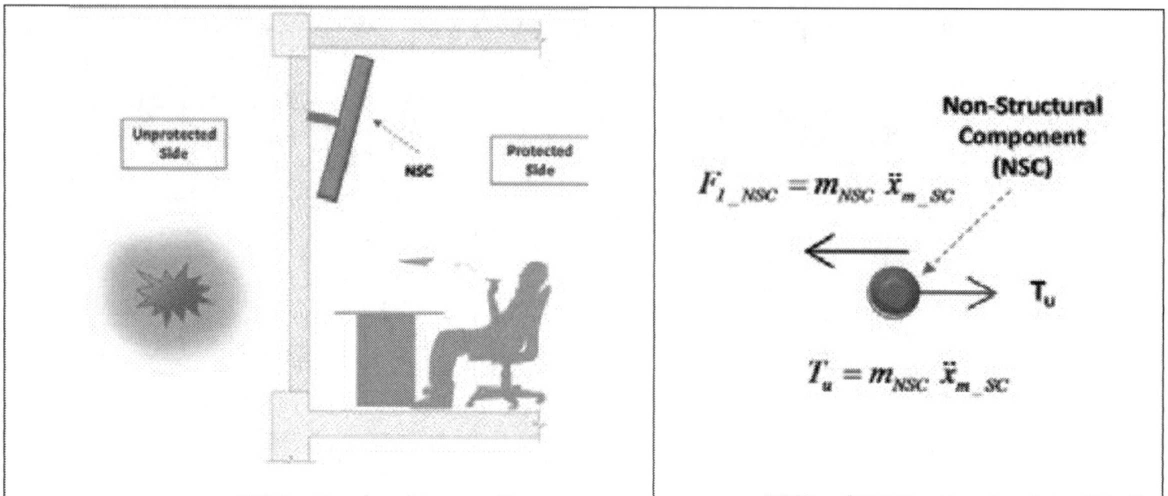

NSC attached to wall | **FBD of NSC attached to Wall**

Figure 14. NSC along Exterior Wall

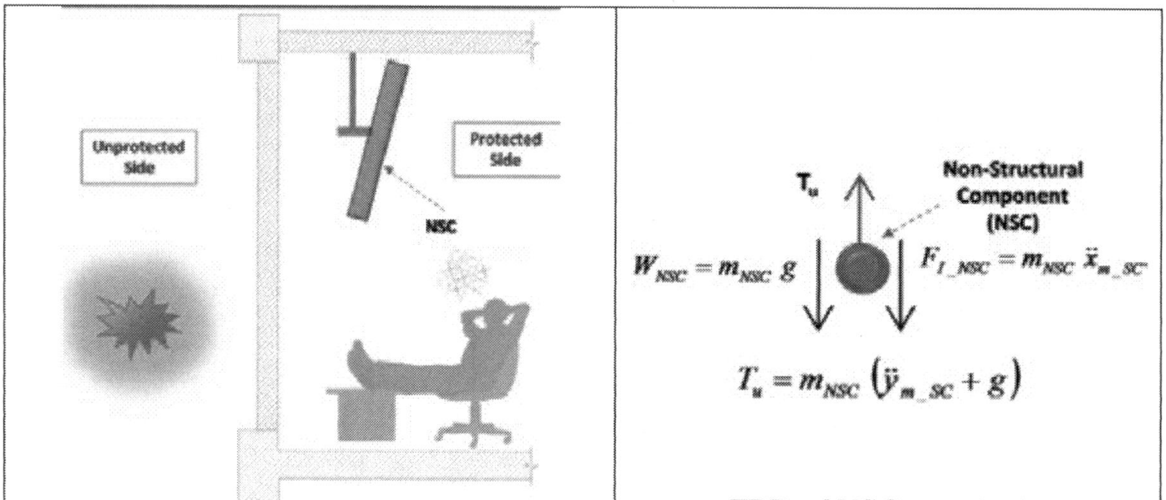

FBD of NSC attached to Roof

NSC attached to roof

Figure 15. NSC along Roof

Finally, the strength of the provided anchorage used to restrain the NSC must be greater than the required force as shown in Table 2. Also, this retrofit technique may require the connection between the fixture and structural component formed by several pieces of hardware rather than single fasteners (i.e. hooks, chains, straps, etc.). Therefore, all hardware involved in the connection must comply with the required strength through the load path. As an example, Figure 16 depicts the retrofit of wall fans mounted on exterior wall.

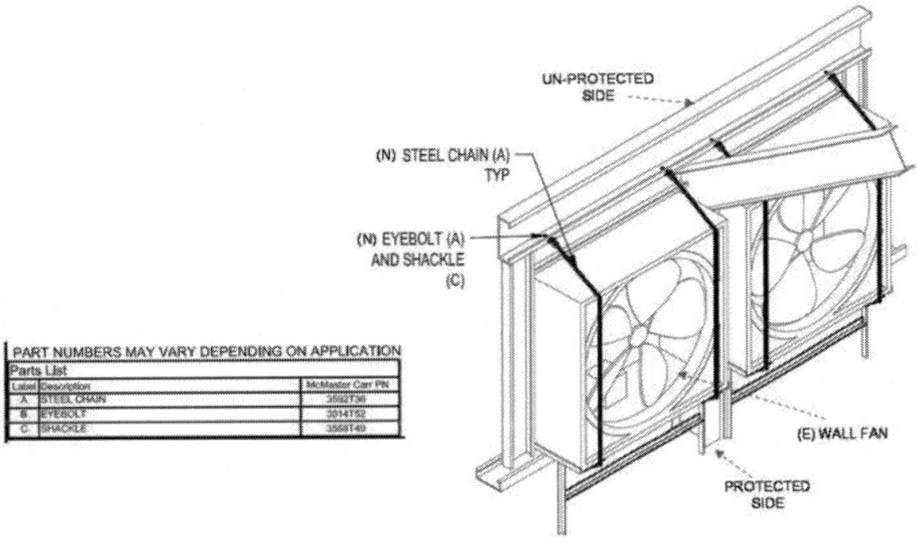

Parts List		
Label	Description	McMaster Carr PN
A	STEEL CHAIN	[illegible]
B	EYEBOLT	[illegible]
C	SHACKLE	[illegible]

Figure 16. NSC Retrofit Sketch (example)

4. Recommendations

In addition to the anchorage requirements for the coupling retrofit technique described in the Section3.3, some wall/roof conditions may require special attention as follows:

- Placement of anchorage in location that does not induce prying in fasteners; see Figure 17.
- Placing the NSC near the support of the structural component and away from the point of peak deflection may significantly reduce the anchorage requirements; see Figure 18.
- NSCs that are relatively large compared with the structural component length may reduce the anchorage requirements because acceleration of fixture varies along its length; see Figure 19.
- Anchorage for retrofit must be installed into secondary or main frame members capable of supporting the design basis forces.

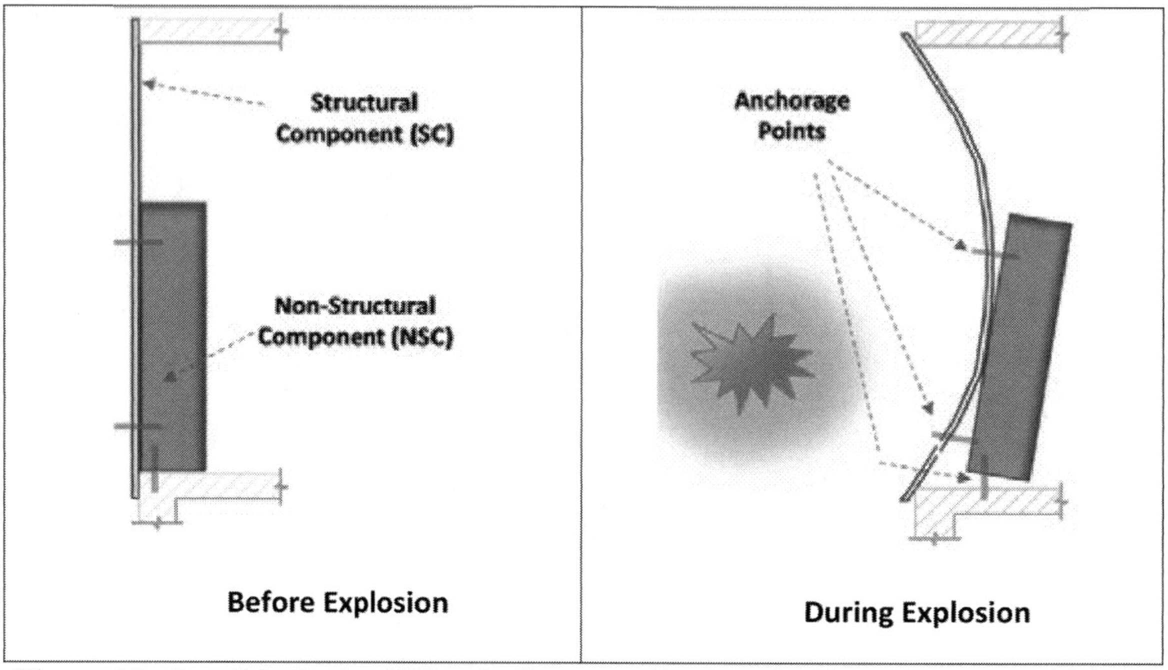

Figure 17. Possible Prying Effect in Fasteners During the Blast Response of SCs & NSCs

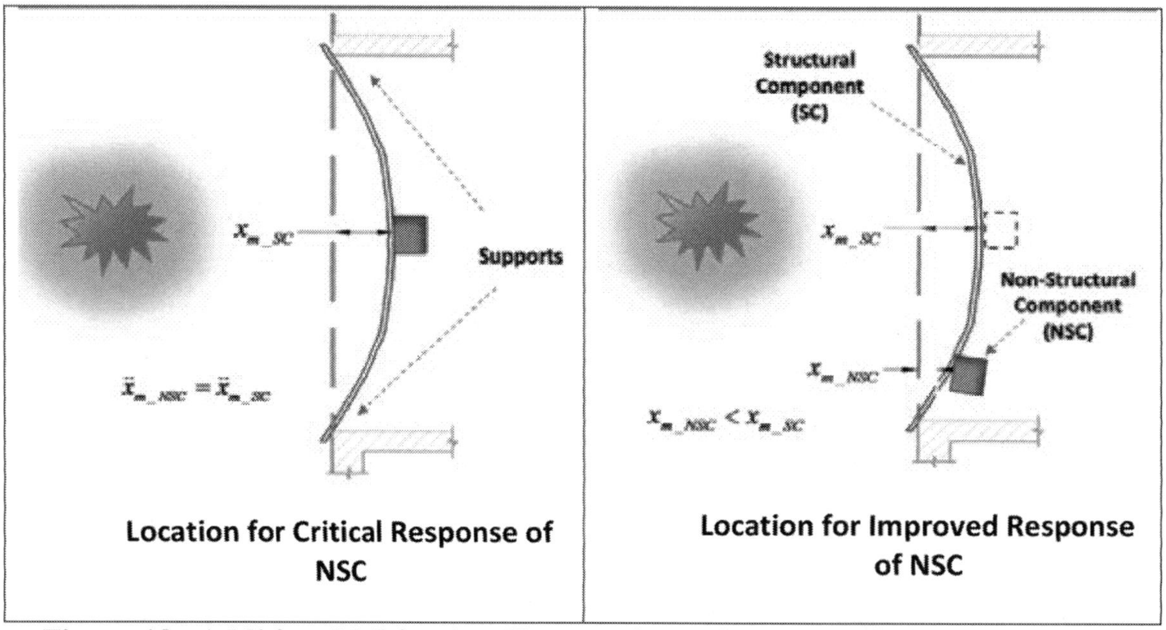

Figure 18. Additional Mitigation Hazard of Non-Structural Component (NSC)

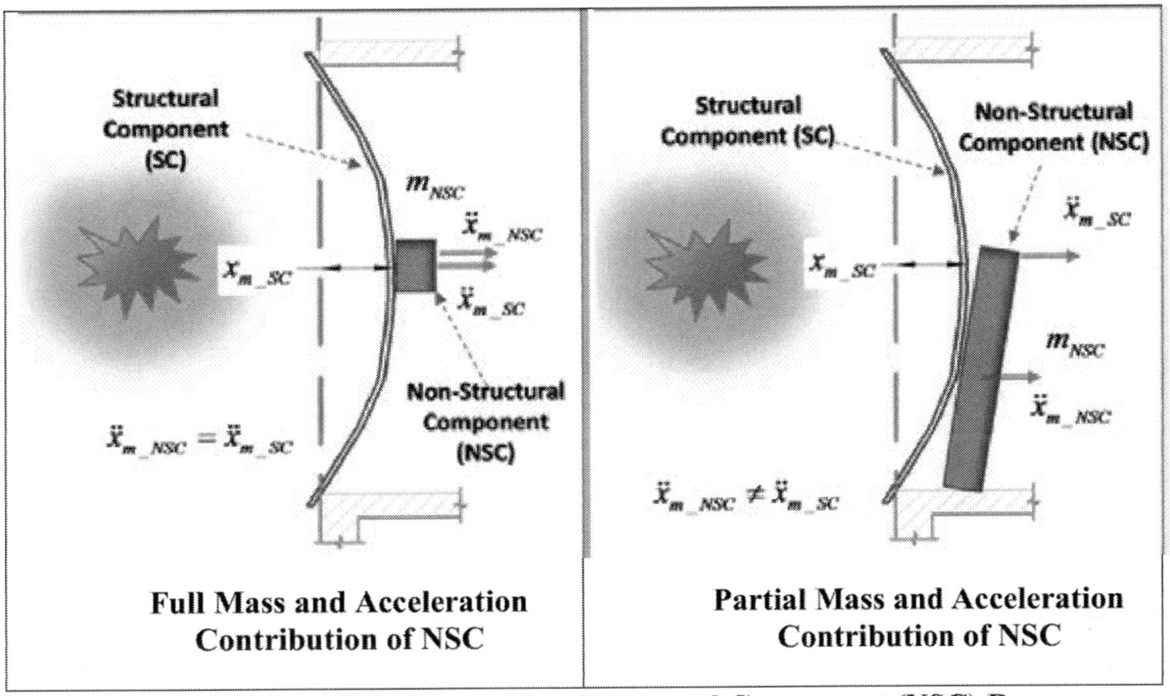

Figure 19. Effect of Location on Non-Structural Component (NSC) Response

5. Conclusions

Evaluation of non-structural building components for the potential to become hazardous debris and mitigation of predicted hazards for buildings subjected to blast load is important to mitigate the hazards to personnel occupying these buildings. A methodology for creating simplified design tool for addressing these hazards was presented and two approaches for retrofitting non-structural components by separating the component from the building structure or by determining the required forces to attach the fixture to the structure were also provided and illustrated.

6. References

[1] Ben Harrison, Sanaa Alaoui, "T2 Laboratories Explosion Damage Assessment", 34th DDESB Explosives Safety Seminar (2010).

[2] Biggs, J.D., "Introduction to Structural Dynamics," McGraw-Hill Publishing Company, New York, NY, (1964)

[3] SBEDS, "Single-degree-of-freedom Blast Effects Design Spreadsheets", Ver 5.0, U.S. Army Corps of Engineers Protective Design Center, December 2012

[4] Department of Defense (DoD), Unified Facilities Criteria (UFC) UFC 3-340-02 "Structures to Resist the Effects of Accidental Explosions" (former TM 5-1300), September 2014

[5] ASCE Task Committee on Blast-Resistant Design. "Design of Blast-Resistant Buildings in Petrochemical Facilities" 2nd. Edition (2010)

AUTHOR INDEX

AUTHOR INDEX

AUTHOR INDEX

50th Annual Loss Prevention Symposium 2016 (LPS)

Topical Conference at the 2016 AIChE Spring Meeting and 12th Global Congress on Process Safety

Houston, Texas, USA
10-14 April 2016

Volume 2 of 2

ISBN: 978-1-5108-2477-5

Printed from e-media with permission by:

Curran Associates, Inc.
57 Morehouse Lane
Red Hook, NY 12571

Some format issues inherent in the e-media version may also appear in this print version.

Printed by Curran Associates, Inc. (2016)

For permission requests, please contact AIChE
at the address below.

AIChE
120 Wall Street, FL 23
New York, NY 10005-4020

Phone: (800) 242-4363
Fax: (203) 775-5177

www.aiche.org

Additional copies of this publication are available from:

Curran Associates, Inc.
57 Morehouse Lane
Red Hook, NY 12571 USA
Phone: 845-758-0400
Fax: 845-758-2633
Email: curran@proceedings.com
Web: www.proceedings.com

TABLE OF CONTENTS

VOLUME 1

VOLUME 2

12TH GLOBAL CONGRESS ON PROCESS SAFETY

Cost and Benefit Considerations of Designing for Higher Pressure

Juan P. Lopez
Smith & Burgess LLC
7600 W. Tidwell Rd., Ste 600
Houston, TX 77040
juan.lopez@smithburgess.com

Guomin Wang
Smith & Burgess LLC
7600 W. Tidwell Rd., Ste 600
Houston, TX 77040
guomin.wang@smithburgess.com

Prepared for Presentation at
American Institute of Chemical Engineers
2016 Spring Meeting
12th Global Congress on Process Safety
Houston, Texas
April 11-23, 2016

Cost and Benefit Considerations of Designing for Higher Pressure

Juan P. Lopez
Smith & Burgess Process Safety Consulting
7600 W. Tidwell Rd., Ste 600, 77040
juan.lopez@smithburgess.com

Guomin Wang
Smith & Burgess Process Safety Consulting

Keywords: inherently safer design, overpressure protection by system design, UG-140, maximum achievable pressure, process design.

Abstract

There are multiple changes within a processing plant that can trigger the need to change the design of equipment, flare, or availability of Safety Instrumented Systems (SIS). For example, a rate increase in the facility may lead to a specific relief valve being undersized, at the same time it may also impact the adequacy of the flare, and lastly in order to adequately manage risk a new SIS may be required. These three action items are due to the necessity of relieving fluid in order to prevent overpressure from a vessel that is not rated for the maximum possible achievable pressure. The proposed approach of designing the equipment for a higher pressure in the design phases, so it is able to contain the fluid without need for relief, could result in a decrease in future expenses related to upgrades of relief systems as well as lower risk. This paper will look into the disadvantages of designing equipment for higher pressure versus the benefit that may come from doing so, in accordance with ASME Section VIII – Division 1 UG-140: Overpressure Protection by System Design.

1 Introduction

The safety of a plant has become one of the foremost priorities due to the increasing pressure from government regulatory bodies and communities. As a result of several high profile incidents, the focus of the chemical process industry and regulatory bodies has shifted towards *Inherently Safer Design* (ISD). Incidents like Bhopal, India, in 1984 emphasized the need for eliminating hazards whenever possible and reducing or minimizing the risk when the hazard could not be completely eliminated through ISD.

Designing process equipment that is strong enough to contain the maximum achievable pressure is just one example of ISD. The option of designing process equipment for higher pressure can bring about many advantages when other ISD strategies may not be feasible. For example, refineries carry out many processes at high pressure (such as hydrotreating reactors). The ideal solution of ISD would be to eliminate the hazard of high pressure inventories by utilizing different processes carried out at lower pressures; however, based on the current technology, the proven effectiveness of these reactions combined with the availability of other forms of mitigating the risk such as process controls, instrumentation, or Safety Instrumented Systems (SIS) present a compelling case for managing the risk associated with having a large inventory of high pressure material. The risk of loss of containment due to overpressure in a hydrotreating reactor could also be lowered by designing the process equipment for the largest achievable pressure produced by a runaway reaction, without relying on SIS, operator intervention, or process controls and cutting the cost of maintenance of SIS or the need for a large emergency relief system.

ISD provides a different approach to process safety: instead of accepting hazards and then controlling the risk by adding layers of protection, the designer is challenged to provide a method to eliminate the hazard altogether. If the hazard cannot be eliminated, then the challenge becomes finding a method to make the process inherently safer. This paper proposes to design process equipment for higher pressure as a viable method to make the plant inherently safer. It also challenges the notion that designing for higher pressures is not cost effective; through careful design it may possible to make processes safer and economically feasible if ISD is considered throughout the early stages of the lifecycle of the facility. The paper will also provide some case studies of operating facilities where designing for the maximum achievable pressure could have helped in lowering the future expenses related to inflexible design.

2 Overpressure Protection by System Design within ISD

Inherently safer design is only one method for managing risk. The strategies for managing risk, according to the book *Inherently Safer Chemical Processes*[1], are classified into four categories:

"Inherent — Eliminating the hazard by using materials and process conditions that are non-hazardous; i.e., substituting water for a flammable solvent;

Passive — Minimizing the hazard through process and equipment design features that reduce either the frequency or consequence of the hazard without the active functioning of any device; .e., providing a diked wall around a storage tank of flammable liquids;

Active — Using controls, alarms, safety instrumented systems, and mitigation systems to detect and respond to process deviation from normal operation; i.e., a pump which is shut off by a high level switch in the downstream tank when the tank is 90% full. These systems are commonly referred to as engineering controls, although human intervention is also an active layer;

Procedural — Using policies, operating procedures, training, administrative checks, emergency response, and other management approaches to prevent incidents, or to minimize the effects of an incident; i.e. hot work procedures and permits. These approaches are commonly referred to as administrative controls."

All four strategies of risk management (or a combination of them) can present an acceptable and safe design. Inherent safety attempts to avoid the hazard completely. For situations in which removing the hazard may not be feasible, accepting the hazard and attempting to manage the risk with *passive*, *active*, and *procedural* strategies may produce a more desirable result. A good example is transportation: travel by airplane is more hazardous than traveling by car; however, there are a lot of safeguards that allow air travel to be ultimately safer than ground transportation and therefore, the benefit of fast air transportation outweighs the hazards associated with it because the risk can be controlled. In the same manner, a more hazardous procedure may be more desirable if the risk is managed appropriately through the utilization of *passive*, *active*, and *procedural* risk management strategies.

Overpressure protection by system design falls under the category of *Passive* risk management as it requires no active functioning of any device. It does not fall under *Inherent* category because it does not eliminate the hazard altogether, but it simplifies the process enough to make it *inherently safer*.

Among inherently safer design there are multiple strategies to eliminate or reduce hazards. According to *Inherently Safer Chemical Processes*[1], the four major strategies of ISD are explained as follows:

"*Minimize*:	Use smaller quantities of hazardous substances (also called *Intensification*).
Substitute:	Replace a material with a less hazardous substance.
Moderate:	Use less hazardous conditions, a less hazardous form of a material, or facilities that minimize the impact of a release of hazardous material or energy (also called *Attenuation* and *Limitation of Effects*).
Simplify:	Design facilities which eliminate unnecessary complexity and make operating errors less likely, and which are forgiving of errors that are made (also called *Error Tolerance*)".

The magnitude of the Bhopal incident occurred as a result of storing a large amount of an extremely toxic material. All four strategies could have helped in preventing the incident or reducing its consequence. Some inherently safer methods to reduce this hazard would have been to *minimize* the inventory of toxic material, *to substitute* it with a less hazardous substance, to *moderate* the impact of the hazard with better emergency response, or *simplify* the process to avoid the opportunity of misoperation. After the incident occurred, it was clear that ISD was required.

ASME Section VIII Div. 1 UG-140 "Overpressure Protection by System Design" states that "A pressure vessel does not require a pressure relief device if the pressure is self-limiting (e.g., the maximum discharge pressure of a pump or compressor), and this pressure is less than or equal to the MAWP of the vessel at the coincident temperature[2]".

Overpressure protection by system design falls under the *Simplify* strategy of ISD. The idea is to reduce the potential for overpressure and reduce the probability of loss of containment by simplifying the process and allowing for larger margin of error. Note that designing process equipment for the maximum achievable pressure does not completely eliminate the hazard, since the possibility of loss of containment of high pressure material still exists by means other than overpressure. However, the hazard is greatly lowered since the probability of loss of containment due to overpressure caused by process upsets, human error, or instrumentation failure is inherently reduced.

This paper focuses on designing for the maximum achievable pressure as a means to make processes inherently safer. There are applications today where this strategy could feasibly be implemented. The following sections discuss some examples on how potential hazards can be avoided if the systems are designed for higher pressure ratings. The examples will demonstrate the potential benefits in terms of relief valve sizing, disposal systems (flare) sizing, and cost reduction, as well as the challenges faced by overpressure protection by system design.

3 Designing for the Maximum Achievable Pressure

Inherently safer design is valuable for process risk management. Regulatory bodies, process safety experts, and policy makers have already embraced the proven concepts related to ISD over the last few years. However, there are still challenges to be faced in the industry.

In the book *Process Plants: A Hand Book For Inherently Safer Design*, the following is stated about equipment and operator failure: "In all industries, equipment failures and errors by operators and maintenance workers are recognized as major causes of accidents, and much thought has been given to ways of reducing them or minimizing their consequences.[2]" A procedure known as PRIM (Process Related Incidents Measure) is used to analyze incident data annually by the Canadian Chemical Producers' Association (CCPA) member companies. The 2014 PRIM analysis indicates the six Process Safety Management (PSM) elements shown in the figure below contribute to 85% of the total incidents.

PSM Element No.	*PSM Element*	*% of Incidents*
6	Process and equipment integrity	23.8
2	Process knowledge and documentation	21.2
4	Process-risk management	16.8
7	Human factors	8.9
5	Management of change	7.3
3	Capital project review and design procedures	6.5

Figure 1. Statistic of Incidents related to PSM Elements

According to PRIM, the largest percentage of incidents fall into the category of process and equipment integrity. Some of this may come from corrosion failure, but it is likely that some of these events may come from overpressure and loss of containment. In fact, out of the six elements of PSM listed in the figure, all of them could be resolved by inherently safer design and potentially through overpressure protection by system design. If the process vessels are designed for the maximum achievable pressure in accordance with ASME VIII Div.1 UG-140, the risk associated with equipment and operator failures could be greatly reduced, since the vessels would be able to contain the pressure without compromising equipment integrity.

Carefully establishing a design pressure with a combination of procedural or active layers of protection may provide the same risk reduction and may be cheaper than designing for the minimum pressure and then installing expensive SIS or complicated layers of protection to manage the risk. Designing for higher pressure does not necessarily mean that every system should be designed for the highest achievable pressure. It is easily recognizable that there are situations in which designing for the maximum achievable system pressure is not economically feasible; however, there are processes in which marginally increasing design pressures in combination with other cheaper active layers of protection (for example relief devices instead of SIS), may provide the same risk reduction level while still being a cheaper option.

If the concepts of Overpressure Protection by Design are applied to an entire facility, the need for large relief systems could potentially be eliminated. If every system within the plant was designed to be able to contain the inventory of material with no need to relieve it, the result would be that the relief systems required to handle the relief loads would be a lot smaller than the actual. Smaller flare systems could be easily achieved if enough process knowledge and research is spent in the early stages of the life cycle of the facility.

Additionally, designing vessels for higher pressure would reduce the risk associated with human or instrumentation error, and the conditions necessary for vessel failure would also be less likely to occur. The cost associated with maintenance of SIS's and training would go down, as well as the expenses resulting from testing and inspection requirements. The loss of product associated with venting requirements also needs to be included in the estimation; designing for the maximum achievable pressure would greatly reduce unnecessary product loss.

Along with designing for the maximum achievable pressure comes the flexibility of being able to increase the throughput of the facility without having to modify the active or procedural layers of protection. This is a common problem for operating facilities, where the change in throughput is increased over the years and this results in having to upgrade, modify or maintain additional layers of protection over the years.

In summary, a passive design is not only safer but it may also prove to be cheaper throughout the lifetime of the facility.

4 Why Not?

The ultimate goal of inherently safer design is to eliminate hazards. However, it has been established that a safe design can be achieved with a combination of passive, active and procedural layers of protection. It is necessary to find a balance between acceptable risk and economic factors using a combination of the risk management techniques. Even though the probability for loss of containment is greatly reduced, the hazard of maintaining large inventories of high pressure hazardous material through overpressure protection by system design is not eliminated.

There is a perceived notion that designing process vessels for higher pressure results in unnecessary cost or that the economic burden would drive capital projects to not be feasible. The reality is that some engineers design process equipment based solely on the operating conditions of process vessels just to have the cheapest and most "cost-effective" design possible. This approach results in design that is not flexible to process upsets caused by human error, controls or utilities failure. Because the risk associated with this design is commonly high, relief systems or SIS are necessary to adequately manage the risk. Designing for the maximum achievable pressure is an inherently safer strategy since there is no need to rely on additional layers of protection in order to reduce the risk. The upfront cost of design would be higher if every process vessel was designed for the highest achievable pressure, but this methodology would not be a fair cost and risk evaluation.

Instead, when considering the cost associated to different process safety alternatives, a truly fair analysis must also consider the cost and risk over the entire life cycle of the facility. It is likely that for some situations simplifying the design by careful consideration of the process and designing for the highest possible pressure would result in a more cost effective option in comparison with the upfront cost of active layers of protection (SIS, relief systems) plus the cost of operation and maintenance of the necessary active and procedural layers of protection. The perception of higher expenses is still an obstacle to the development of inherently safer techniques.

Another challenge is the fact that old designs are already established and there is very little room for innovation. It is easier to just recycle the last design because it has already been proven to work; there is no need to re-think and carefully analyze the process and search for inherently safer strategies for design. Knowledge of the process outside of the design conditions is required in order to fully utilize the concepts of overpressure protection by system design, which adds another challenge when competing with old and proven designs.

5 Case Study (PSV Sizing)

A system with higher design ratings can help minimizing the need for installing additional active layers of protection. Often times, the required relief rates due to certain overpressure scenarios are so large that they cannot be easily mitigated. The figure below represents one example to of the difficulties to mitigate the risk when the chosen design relies solely on overpressure protection.

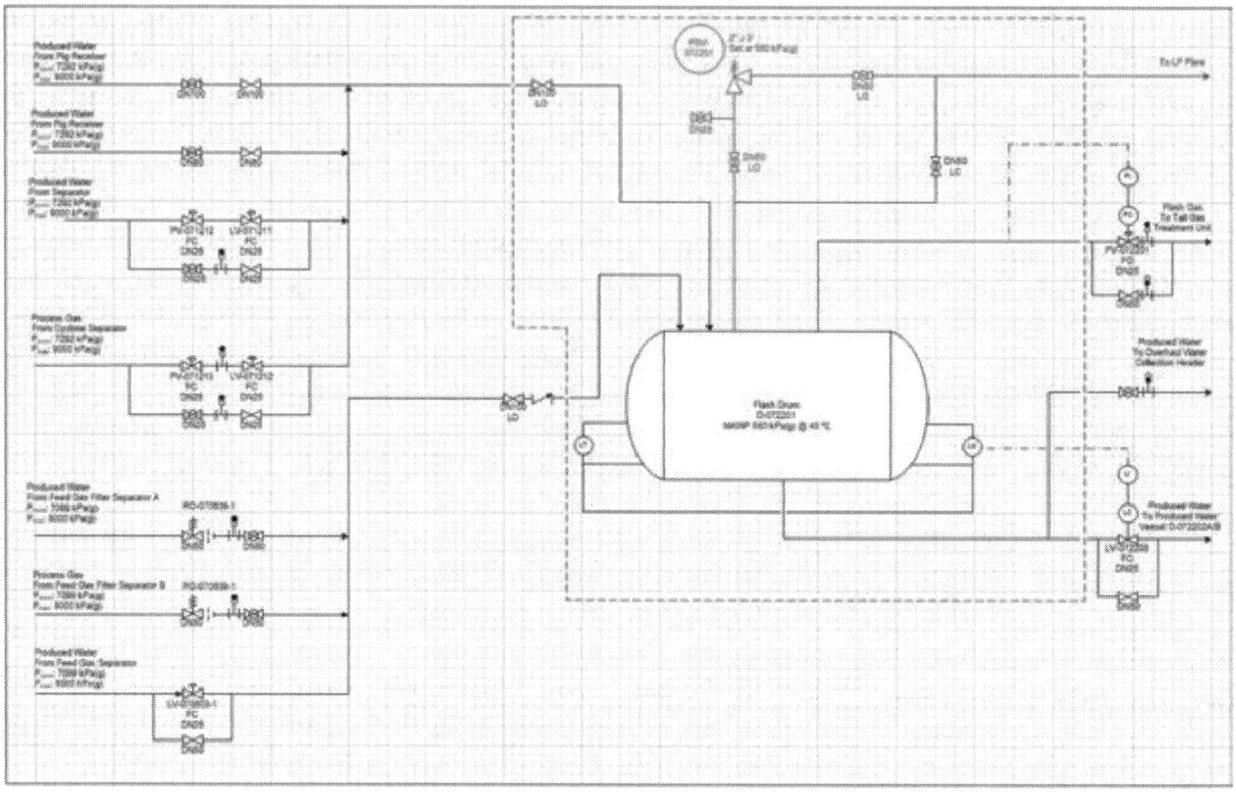

Figure 2. Flash Drum Example

The protected system is the Flash Drum. The controlling scenarios for the system shown in the above are instrumentation failure and the inadvertent valve opening of the normally closed valves. The normal operating pressure upstream of the control valves are around 7200 kPag, which exceeds the MAWP of the system (580 kPag). This risk of loss of containment could not be mitigated by the addition of relief devices (active layer) since addition of relief devices large enough to handle the area was unfeasible. Another possible solution was to install restriction orifices (procedural layer). Due to the large pressure differential between the normal upstream pressure and the PSV set pressure, the required restriction would need to be less than 7mm. An orifice area of 7 mm can potentially be plugged; therefore, this was not a feasible solution either. The only option was to install high integrity SIS in order to shut the feed in the event of low liquid level in the upstream system in order to prevent the possibility of gas-blowby. This approach can mitigate the risk to an acceptable level for one scenario; however, the SIL rated system has to be regularly maintained and tested to ensure its integrity. Note that the relief device still required an increase in size in order to mitigate the liquid overfill scenario. Due to this size increase, the inlet and outlet discharge piping had to be increased in size.

Note that the mitigation could have been simplified if the design pressure of the vessel was higher. The reason for having to install so many additional layers of protection is that there was a real risk of loss of containment since the upstream pressure for this system was approximately 10 times as large as the vessel hydrotest pressure. If the vessel had a higher design rating from the beginning of the life cycle, the entire procedure could have been avoided. It is not necessary in this case to design the vessel up to the maximum achievable pressure, there are situations in which this is economically not feasible; however, designing the vessel for a high enough feasible design pressure in combination with a relief device could produce the same risk reduction with lower maintenance cost when compared to the actual design in combination with SIS systems.

6 Case Study (Flare Design)

With additional process knowledge and carefully analyzed process safety information, designers have the ability to create processes that are safer and more cost-effective over the lifetime of a facility. Due to environmental regulations, a processing plant had action items from the authorities which included routing several relief valves currently relieving to the atmosphere to a closed system. The analysis included considerations of global scenarios (such as Refinery Power Failure) resulting in multiple systems simultaneously relieving to the flare.

Based on the original design pressures for each of the systems, the resulting global loads were larger than any combination of flare headers could handle. The calculated backpressure based on the generally accepted methodology for flare system analysis was so large that the simulation calculated accumulation levels high enough to lead to loss of containment if all the atmospheric relief valves were routed to the closed systems.

The end result of this analysis led to a combination of increased risk (which was accepted by the facility) and the installation of multiple additional active layers of protection to maintain the risk to an acceptable level. In addition, these modifications imposed a large upfront cost as well as future maintenance costs that could have been initially prevented if additional risk considerations were taken into account in the early stages of the lifecycle of the plant.

An advantage of carefully specifying design pressures and overpressure protection by system design is that relief loads can be reduced or eliminated altogether. Through careful analysis the, the designer could specify the maximum temperature of the heating medium to not exceed the bubble point of the working fluid at relieving pressure, this is called reboiler pinch-out. If a similar methodology of reducing or eliminating loads through overpressure protection by system design is followed throughout the entire facility, the disposal system requirements could be reduced.

7 Conclusion

The process safety industry is constantly learning from past incidents. Inherently safer design alternatives are needed in order to reduce or eliminate the hazards resulting from necessary processes. Designing for the highest achievable pressure is one strategy that results in inherently safer designs. There are numerous benefits of designing for the highest achievable pressure in accordance with ASME Section VIII – Division 1 UG-140: Overpressure Protection by System Design:

- Designing for the highest achievable pressure will make any process inherently safer.
- Initial cost of implementation of active and procedural layers of protection as well as the cost of operation and maintenance of active and procedural layers of protection will not be necessary.

- Disposal systems would be smaller if the strategy is implemented in the early stages of the life cycle of the facility.
- Rate increases, debottlenecking, regulatory practices, and changes in technology would be less likely to result in changes to the necessary existing layers of protection due to the added flexibility of designing for the maximum achievable pressure.

In order to more carefully assess the viability of overpressure protection by system design, the assessment of risk and economic variables should be done during the early stages of the life cycle of the facility. Researching the process and carefully analyzing all inherently safer design strategies as options will ultimately lead to a safer design; however, overpressure protection by system design has real and practical applications today, when other inherently safer design strategies may not be feasible.

8 References

1. CCPS (Center for Chemical Process Safety). "Inherently Safer Chemical Processes: A Life Cycle Approach, 2nd Edition". December 2008.
2. Trevor Kletz and Paul Amyotte. "Process plants: a handbook for inherently safer design. 2nd Ed". May 17, 2010.
3. BPVC Section VIII-Rules for Construction of Pressure Vessels Division 1

441692 Practical Risk Based Approach to Pressure Relief and Effluent Handling System Design

Abstract:

Recent updates to pressure relief system design RAGAGEP include provision for risk based overpressure protection. Although risk based approaches are not new, the recent RAGAGEP provides additional guidance and criteria. This study presents a practical risk based approach to pressure relief and effluent handling system design and a supporting case study that results in safe and cost effective design.

Historically, pressure relief system design focused on qualitatively identifying a controlling scenario or design basis, the situation that places the highest demand on the pressure relief system, and calculated that the relief system could adequately provide overpressure protection. The associated design basis documentation was inadequate and was often only a relief device specification sheet.

The promulgation of OSHA 29 CFR 1910.119, the PSM standard, has significantly improved design basis documentation. Since the implementation of the PSM standard, typical steps in relief system design and design basis evaluation and documentation include identification of overpressure scenarios; quantification of the associated relief requirements and relief system performance; and evaluation of the hydraulic performance of the effluent handling system. This approach has improved relief system design and design basis documentation, but has also significantly increased the cost without a corresponding benefit in plant safety.

Risk based pressure relief and effluent handling system design also begins with identification of overpressure scenarios. However, the foundation of the risk based approach leverages existing PHAs and risk tolerability criteria already implemented at PSM covered facilities. When PHAs use LOPA, the consequences and likelihood of the scenarios are determined. The consequence of an overpressure scenario can be further evaluated with computational techniques, which include consideration for the pressure rise, potential release of toxic and flammable fluids, and resulting vapor cloud. Next, using risk tolerability criteria the risk of the identified overpressure scenarios is compared to risk tolerability guidelines. Scenarios with an intolerable risk are then mitigated through enhancements and alternatives including traditional pressure relief systems. The risk tolerability and cost of the mitigation measures are then balanced to ensure a safe and cost effective relief system.

Integrating the pressure relief system design into existing PSM practices reduces the effort and cost of developing and maintaining relief system design basis documentation and focuses the designers' intent on improving plant safety. This risk based approach to pressure relief system design ensures adequate overpressure protection systems and sound investment in human and capital resources focused on risk reduction.

441563 Overpressure Protection By System Design - Use of High Integrity Pressure Protection Systems

Abstract:

Overpressure Protection by System Design - Use of High Integrity Pressure Protection Systems

Use of a safety instrumented system (SIS) in lieu of a traditional pressure relief device is commonly referred to as a high integrity pressure protection system (HIPPS). With the approval of ASME code case 2211 in 1996, ASME Section VIII has been updated to allow for overpressure protection by system design in the BPV Code Part UG-140. A detailed analysis and thorough documentation is required to ensure that the vessel is protected by "system design", or in other words, the pressure is self-limiting, or there are no credible overpressure scenarios that result in the vessel pressure exceeding 116% of the maximum allowable working pressure (MAWP). API 521 Pressure-relieving and Depressuring Systems Appendix E allows replacement of a pressure relief device with an SIS that has a reliability consistent with the risk of an overpressure event. The API approach is similar to the approach outlined by CCPS, i.e. they are risk based accounting for both the probability of the overpressure event and the consequences of such an event. In contrast, the ASME guidance is based only on reducing the likelihood of an overpressure event.

The use of HIPPS to replace conventional pressure relief protection is growing. Some of the main drivers include the need to reduce flare system loads on large green field projects, limit and control the release of green house gases and also to allow for increased capacity in brownfield expansions.

This paper will review the differences in the ASME, API and CCPS approaches for overpressure protection using safety instrumented systems, the "credibility" criteria in ASME and the design of high integrity instrumented systems required to meet the code requirements. The paper will also include a discussion on how different users have implemented HIPPS in the offshore and onshore processes.

442373 RAGAGEP Considerations for Relief and Flare Systems

Abstract:

Abstract

Recent emphasis on RAGAGEP increased the scope of relief systems risk factors that require evaluation in order to develop complete and compliant Pressure Relief and Flare Systems (PRFS) documentation.

This paper discusses how RAGAGEP considerations now require evaluation and proper documentation of risk factors that are often overlooked including but not limited to: dispersion analysis, thermal radiation, noise, vibration risk, reaction forces and structural supports, metal cold temperatures due to expansion cooling and two phase flow, hot temperatures due to fire exposure and/or runaway reactions, PRV stability, and chemical reaction systems.

Important RAGAGEP considerations for these additional risk factors are highlighted and illustrated using several examples and case studies. Recommendations are provided on how to best address these factor in the evaluation and documentation of design basis.

Considering External Fire to be a Non-Applicable Overpressure Scenario

Jessye Palladino
Smith & Burgess: Process Safety Consulting
190 S. LaSalle St.
Ste. 550
Chicago, IL 60603
Jessye.Palladino@smithburgess.com

John Burgess, Ian McPike
Smith & Burgess: Process Safety Consulting
7600 W. Tidwell Rd.
Ste. 600
Houston, TX 77040

Prepared for Presentation at
American Institute of Chemical Engineers
2016 Spring Meeting
12th Global Congress on Process Safety
Houston, Texas
April 11-23, 2016

Considering External Fire to be a Non-Applicable Overpressure Scenario

Jessye Palladino
Smith & Burgess: Process Safety Consulting
190 S. LaSalle St.
Ste. 550
Chicago, IL 60603
Jessye.Palladino@smithburgess.com

John Burgess, Ian McPike
Smith & Burgess: Process Safety Consulting

Keywords: External Fire, Overpressure, Relief Valve, Pool Fire

Abstract

An external fire is usually considered an applicable overpressure scenario as the necessary elements to fuel the ignition of a fire are almost always present within a refinery, gas plant, or petro-chemical plant. This article will demonstrate the situations and conditions when it is acceptable to not consider external fire as an applicable scenario. The primary circumstances that will be covered include: surrounding chemical inventory, ignition source distance, and equipment elevation. In the event that these conditions are such that fire is considered an applicable scenario, the required flammable liquid pool conditions should be evaluated to ensure that an external fire can both be created and sustained to result in an overpressure occurrence.

1. Introduction

In the recent past, there have been multiple catastrophes in the chemical industry as a result of fire. In 1988, explosions devastated a plant owned by the Pacific Engineering and Production Company of Nevada (PEPCON) and again in 2002 a manufacturing facility for Third Coast Industries was completely destroyed. Fire is a dangerous possibility in the chemical industry, and as such, should be considered when discussing and identifying hazardous scenarios for plant operations. After initial consideration, deeper investigation into the mitigating factors and potential root causes of a hypothetical pool fire event may prove beneficial in determining an external fire to be a non-applicable scenario for certain cases. Conservatively accounting for exterior pool fire for every piece of equipment can result in issues such as large disposal requirements or unnecessary modifications to existing relief systems and flare headers. The site layout, the availability of flammable liquid to pool, and facility history can be used to determine if the conditions exist such that a pool of flammable liquid can be created. Society of Fire Protection Engineers (SFPE) literature and American Petroleum Institute (API) standards are useful in quantitatively analyzing if sufficient chemical inventory is present to sustain an exterior fire event. Guidance from these sources can be used to determine the size and burn time of a given pool of flammable liquid. By considering the above criteria, the possibility of overpressure to equipment as a result of pool fire can be determined.

2. Why is Exterior Fire Important to Consider?

The PEPCON facility in Henderson, Nevada manufactured Ammonium Perchlorate, a powerful oxidizing material utilized in solid propellant rocket boosters. On May 4, 1988, workers were performing repairs on a building that had been damaged in a recent sandstorm. Welding led to ignition of the structure as it was covered in Ammonium Perchlorate residue from a nearby drying process. The conflagration eventually spread to the storage building directly adjacent, where several thousand 55-gallon drums of Ammonium Perchlorate were being stored, resulting in two major explosions. The second explosion ignited a high pressure natural gas line located under the facility which burned for approximately 1200 hours until authorities could stop the flow of fuel [1].

More recently, on May 1, 2002, a petroleum products facility owned by Third Coast Industries in Friendswood, Texas was completely destroyed by a fire that consumed approximately 1.2 million gallons of flammable liquid and burned for over 24 hours [4]. The Third Coast Industries facility blended and packaged various petroleum lubricants such as motor and hydraulic oil. A small fire outside one of the packaging warehouses caused nearby small storage drums to fail, contributing additional fuel and growing the fire to affect other larger containers in the area. Eventually, the pool fire consumed the entire facility as more storage tanks failed and released their flammable contents. The U.S. Chemical Safety and Hazard Investigation Board concluded that one of the major contributing factors to the severity of this disaster was that no containment or drainage was available to control the spread of flammable liquids at the site of the initial fire [4]. Additionally, onsite fire water was insufficient to quell the flames once discovered. This allowed sufficient time for the fire to cause numerous storage tanks to fail and release enough flammable liquid to fuel the fire for over 24 hours [4].

These examples highlight the importance of considering factors such as liquid drainage and burn time when determining risk associated with an external pool fire event. In each case, investigation determined that the effect of the fire could have been mitigated by elements that affect the formation and behavior of the flammable liquid pool [1] [4]. For facilities where these elements already exist, it may be possible for credit to be taken when determining the appropriate protection for equipment. Accounting for fire loads for every vessel and equipment can result in very large relief loads to the flare or disposal header. Designing for this basis may be problematic and unnecessary, especially if the flare requirements for other site-wide relief scenarios such power or utility failure are much smaller comparatively. Examining the conditions necessary for a pool fire to occur is useful in analyzing the possibility of overpressure.

3. Determining Applicability of Pool Fire

When determining the applicability of an overpressure event as a result of fire, conditions necessary to create a pool of flammable material may be considered. The formation of a pool of flammable liquid is contingent upon the availability of liquid fuel from nearby process equipment. These flammable pools typically result from a liquid release from tanks, vessels, and processes due to inadvertent valve opening or failures in vessel and piping integrity [5]. Operating conditions for certain processes and areas of a plant may be such that little to no liquid is available in the process equipment. Similarly, the chemical inventory in some units may be of a non-flammable nature. For each of these cases, a loss of containment due to one of the aforementioned initiating events may not pose a risk of exterior pool fire because no liquid would be available to pool during a release or the pool that would be formed would not be flammable.

The applicability of an external pool fire as an overpressure scenario depends on the amount of the flammable liquid inventory and the burn rates of flammable material. This quantitative analysis begins by evaluating the dimensions of a flammable liquid pool that could result from a loss of containment. The size and burn time of a fire is correlated to the area of the spill and the volume of the spill. A larger spill area will result in a larger fire circle and a larger spill volume results in a longer burn time for a fire. Furthermore, the main dimension for determining if a liquid-controlled flame spread pool fire can be sustained is the depth of the liquid pool. Experiments have shown that in fuel spills with depths less than 1.5 mm, the flames do not spread outwardly from the ignition source and thus the pool is not fully ignited [2]. A possible reason for this lack of flame propagation is due to the dampening of liquid recirculation at these shallower depths [2]. By losing this liquid recirculation, the flammable liquid material burns out before the circulation can aid in igniting the surrounding fuel. Further experiments concluded that the minimum depth of a pool fire necessary to support flame spread is two millimeters; however, different molecular weights may result in a deeper depth of liquid needed for a pool fire [2]. For example, 2.0 millimeters is needed for decane versus 2.3 millimeter for dodecane [2]. Since 2.0 millimeters is considered the minimum depth to sustain liquid-controlled flame propagation it is used as the required depth in this analysis. Additionally, minimum fuel depths are only applicable for fires that are in the liquid-controlled spread regime. By using this minimum depth and the spill area, the minimum required volume of flammable liquid can be determined from equation 1 [2].

$$V_{min} = A_{spill}\delta_{min} \tag{1}$$

$$V_{min}(m^3) = A_{spill}(m) \cdot (0.002m) \tag{2}$$

$$V_{min}(ft^3) = A_{spill}(ft) \cdot (0.00656ft) \tag{3}$$

Where V_{min} is the minimum volume of flammable liquid required, A_{spill} is the area of the liquid pool, and δ_{min} is the minimum depth of the pool. However, to calculate the minimum volume, an area of the spill has to be defined. The area of a spill can vary depending on if the spill is confined or unconfined. Confined spills have defined areas as there are physical barriers (open containers with fuel in them, walls, dikes, etc.) that prevent the liquid from spreading. However, free flowing spills are not constrained and can spread laterally depending on several factors such as the spill momentum, surface material, and surface tension of the spreading fluid. The greater the initial spill momentum a larger area is expected to be covered. The surface a spill is occurring on has many factors that influence the spreading of a liquid. A smooth flat concrete flow has less surface friction and thus allows for the liquid to spread easier. Whereas, a more porous rougher surface, such as dirt, limits the lateral expansion of a spill resulting in a smaller spill area with a greater liquid depth. Similarly, a fluid with less surface tension is expected to spread more easily than one with a higher surface tension. These multiple factors make the area of an unconfined spill difficult to define. If there is limited data about the area of a spill, then general assumptions can be made. Conservatively, API STD 521 §5.3.2 states "in the absence of any other governing factors, consideration of a fire-impact area is frequently limited to a ground area of 230 m^2 to 460 m^2 (2,500 ft^2 to 5,000 ft^2)" [3]. Using this range of area for a pool fire, the minimum volume equations are simplified to equation 4 and 5:

$$V_{min}(m^3) = 230m^2 * 0.002m = 0.46\ m^3 \tag{4}$$

$$V_{min}(m^3) = 460m^2 * 0.002m = 0.92\ m^3 \tag{5}$$

$$V_{min}(ft^3) = 2500ft^2 * 0.00656ft = 16.40\ ft^3 = 122.7 \text{ US Gallons} \tag{6}$$

$$V_{min}(ft^3) = 5000ft^2 * 0.00656ft = 32.81\ ft^3 = 245.4 \text{ US Gallons} \tag{7}$$

resulting in a range of liquid volume of 122.7 to 245.4 US gallons. By calculating the minimum volume needed to sustain a pool fire for the API suggested area (2,500 ft^2 to 5,000 ft^2), the flammable liquid of a plant can be inventoried to verify that there is a sufficient volume present to sustain a pool fire. If the fire area does not contain at least 122.7 gallons of flammable liquid then an external pool fire due to an uncontained spill may not be considered an applicable source for overpressure. Similarly, if the area of a spill can be confined then the minimum volume of fuel needed for the known area can be determined using the same method as above with the minimum depth of 0.002 meters (0.00656 feet). Additionally, the inventory of the confined spill area can be used to determine the applicability of a pool fire in that region. By using the minimum depth and either the API suggested area for an unconfined spill or a known area for a confined spill, the minimum volume to sustain a pool fire can be calculated. If the inventory in the defined area does not meet this minimum volume then the flames are not expected to spread as the depth of the pool would be less than 0.002 meters (0.00656 feet) and therefore the external pool fire is not expected to result in overpressure.

If sufficient flammable liquid inventory is present in the fire area, then further investigation of the burn time of the flammable liquid can determined if the pool fire can be considered a non-applicable overpressure scenario. While the minimum volume of liquid to sustain a pool fire may be present, the length of the fire may prove to be too short to result in overpressure. The time needed for a pool fire to raise a piece of equipment from operating conditions to relief pressure is another factor in the determining the possibility of an external pool fire resulting in overpressure. A transient fire calculation can be done for a vessel in the fire circle to obtain the time needed to reach overpressure. Combining the required time for an external fire exposure to cause overpressure with the burn rate of the liquid fuel, provides another condition the flammable liquid inventory of a system's surroundings that impacts the applicability of external fire. In order to evaluate the duration of a fire, the burn rate of the fuel is needed. There have been multiple studies performed on the burn rate of different materials that have resulted in various equations for the burn rate. But empirical data for a wide range of fuels have been found to correspond closely with a ratio of the heat of combustion to the heat of vaporization as given in equations 8 and 9 and this relationship is used for this analysis [2]:

$$\dot{m}''_{max}\left(\frac{kg}{m^2 s}\right) = 1 \times 10^{-3}\frac{\Delta h_c}{\Delta h_{v,sen}} \tag{8}$$

$$\dot{m}''_{max}\left(\frac{lb}{ft^2 s}\right) = 2.05 \times 10^{-4}\frac{\Delta h_c}{\Delta h_{v,sen}} \tag{9}$$

Where $\dot{m}''_{max}$ is the maximum mass burning rate per unit area, Δh_c is the heat of combustion of the material, and $\Delta h_{v,sen}$ is the heat of vaporization of the material. Note that the mass burn rate for uncontrolled spills have been shown to be significantly lower than the mass burn rate of controlled fires. Therefore, the NFPA suggests the burning rates of unconfined spills be estimated as 20% of the maximum mass burning rate [2]. Once the maximum mass burning rate for a liquid is known, the burn time of the fire can be determined. The units of the burn rate are given as mass per area-time. In order to isolate the time factor, the mass of the material must be divided by the mass burn rate that is multiplied by the area of the spill. Since the liquid inventory is typically recorded as a volume and not as a mass this numerator mass factor can be converted to a volume multiplied by the density of the fluid in order to obtain equation 10:

$$t_{bcontrolled} = \frac{V \cdot \rho}{\dot{m}''_{max} \cdot A} \tag{10}$$

$$t_{buncontrolled} = \frac{V \cdot \rho}{0.2 \cdot \dot{m}''_{max} \cdot A} \tag{11}$$

where $t_{bcontrolled}$ is the burn time of a controlled fire, $t_{buncontrolled}$ is the burn time of an uncontrolled fire, V is the volume of the flammable liquid, ρ is the fuel density, $\dot{m}''_{max}$ is the maximum mass burning rate per unit area, and A is the area of the fire. Overpressure due to an external pool fire may occur if the burn time of fire exposure exceeds the time it takes for a vessel to reach overpressure. Therefore, in addition to having the depth of the spill at least 2.0 millimeters, the volume of liquid inventory must also be able to sustain a fire long enough to result in overpressure for a vessel.

An example of an uncontrolled spill of m-xylene pool fire and a vessel containing water is used for analysis. A simplified transient fire calculation on the water vessel is done to obtain a time for the water vessel to reach overpressure. Then the maximum burning rate will be calculated along with the required volume of m-xylene to fuel pool fire to last long enough to cause overpressure in the water vessel. A simplified transient fire approach was taken in Appendix B to determine the time needed for a vessel containing pure water to raise from operating conditions to operating pressure and dew point and then how much time is needed to fill the vapor space in the vessel due to an external fire. Additional analysis can be done to determine the time needed to raise the vessel to relief pressure but for this analysis the time for the first two steps is conservatively taken as the time to overpressure the vessel, 43.6 minutes.

Based on equations 1-7 the minimal volume for a 2,500 ft^2 fire of m-xylene is 122.6 US gal, which translates to 889.5 lb_{mass} of m-xylene at atmospheric pressure and 60°F. Equation 9 combined with the heat of combustion to heat of vaporization ratio, obtained from the graph in appendix A, yields the maximum burn rate of m-xylene for an uncontrolled spill:

$$\dot{m}''_{max}\left(\frac{lb}{ft^2 s}\right) = \left(2.05 \times 10^{-4} \frac{\Delta h_c}{\Delta h_{v,sen}}\right) * 0.20 \tag{12}$$

$$\dot{m}''_{max}\left(\frac{lb}{ft^2 s}\right) = 2.05 \times 10^{-4}(70) * 0.20 \tag{13}$$

$$\dot{m}''_{max}\left(\frac{lb}{ft^2 s}\right) = 0.002867 \frac{lb}{ft^2 s} \tag{14}$$

Once the maximum burn rate is acquired, the time of a fire can be calculated. First, the minimum volume required to sustain a 2,500 ft^2 is evaluated to see if that minimum volume (122.6 US gallons) can create a fire that will last long enough to result in overpressure. The time of this minimum volume m-xylene fire is calculated below,

$$t_b = \frac{V \cdot \rho}{\dot{m}''_{max} \cdot A}$$

$$t_b = \frac{\left(\frac{122.6\ gal}{7.48\ gal/ft^3}\right) \cdot 54.23 \frac{lb}{ft^3}}{0.002867 \frac{lb}{ft^2 s} \cdot 2500\ ft^2}$$

$$t_b = 124.1\ seconds$$

This fire is expected to burn out in less than 3 minutes. This calculation is assuming the entire pool is burning at its maximum rate for the whole duration of the fire and does not take into account the initial time it takes for the flames to spread to the whole pool from the ignition source. However, the time it takes for the flame to spread is substantially less than the burn time and therefore does not need to be considered in the duration calculation [2]. This minimum volume of m-xylene to have a pool fire that covers an area of 2,500 ft^2 results in a fire that is predicted to burn out well before overpressure in the water vessel is expected to occur due to the external fire. The volume required for an m-xylene pool fire of 2,500 ft^2 to burn long enough (43.6 minutes) to result in overpressure is calculated below:

$$V_{req} = \frac{t_b \cdot \dot{m}"_{max} \cdot A}{\rho} \tag{15}$$

$$V_{req} = \frac{(43.6 * 60)s \cdot 0.002867 \frac{lb}{ft^2 s} \cdot 2500\ ft^2}{7.177\ \frac{lb}{ft^3}} \tag{16}$$

$$V_{req} = 2613\ gal \tag{17}$$

$$Volume\ of\ m - xylene = 2{,}613\ gal \tag{18}$$

This volume of m-xylene is the minimum amount to have a sustained pool fire of 2,500 ft^2 for at least 43 minutes. The volume needed for a 5000 ft^2 fire is double the 2500 ft^2 volume, 5,226 gallons. The flammable liquid within a plant or fire circle can be inventoried and the possibility of a lasting pool fire determined. If a pool fire is determined to be an applicable overpressure scenario given the available inventory, the effects of site specific layout and features can be considered on the severity of the possible fire scenario.

4. Factors Affecting Severity of a Pool Fire

Pools of liquid fuel also are strongly influenced by the layout and features of a facility, such as grading, trenches, and diversion walls. In the aforementioned Third Coast disaster, the investigative boards determined that the destruction caused could have been drastically reduced by adequate drainage and diking at the plant [4]. Pools will naturally tend toward the low points of drainage ditches in a plot, and often these features are designed to carry the flammable liquid away from equipment and processes. The Occupational Health and Safety standards for flammable liquids state "the route of [a] drainage system shall be so located that, if the flammable liquids in the drainage system are ignited, the fire will not seriously expose tanks or adjoining property" [6]. If a pool fire were to occur in this case, the pool may be isolated away from processes and the risk of overpressure due to the impingement of flames upon equipment would be significantly reduced. Similarly, grading, drainage, and sloping in a plot may be enough to guide pools of liquid toward a specific area of the plant resulting in a lowered pool fire risk in other areas of the facility. Layer of Protection Analysis (LOPA) refers to this type of protection as "Postrelease Protection" and states that it "provide[s] a high level of protection if maintained properly" [7]. Dikes and drainage systems are specifically listed as an effective means to reduce the frequency of large consequences resulting from an exterior pool fire [7]; however, these measures require an extra layer of scrutiny when they are being considered to reduce the risk of exterior fire. Although walls are useful in diverting flammable materials away from particular vessels and processes, they also tend to increase the effects of fire on the side the pool fire is isolated. According to overpressure relieving standards set forth by API, a pool fire confined by walls or embankments can result in higher heat fluxes than an open pool fire [3].

Finally, the incident history of a facility serves as strong indicator of potential sources of leaks and other problematic issues that can lead to a pool fire. Conversely, plant history can also be used in conjunction with the aforementioned criterion to identify areas that have a low risk of exterior fire relative to other areas. If certain procedures have been identified multiple times to result in a release of flammable liquids, then the factors mentioned previously, such as grading and vessel inventory, can be considered in an attempt to predict the behavior of the hypothetical flammable pool and the relative risk of overpressure in the event that the pool is ignited. LOPA literature suggests that sufficient historical data is very effective at adjusting failure rates to reflect local conditions and situations [7].

5. Conclusion

Fires at refineries, chemical plants, or gas plants are serious and dangerous events. There have been numerous fires that have resulted in extreme damages, injuries, and even deaths. The devastating fires at the PEPCON and Third Coast Industries facilities are prime examples of just how dangerous an external fire within a facility can be. These fires resulted in extensive damages and highlight that external fire should not be overlooked and that proper precautions should be taken to prevent similar incidents from occurring. The risk associated with an external pool fire should never be taken lightly; however, in certain cases overpressure protection in case of an external fire may not be needed as an external pool fire is not feasible for the protected equipment. Certain facilities may have overly conservative relief loads for the external fire scenario that could result in a large, calculated flare relief that may not be plausible for the facility. These relief loads from completing external fire analysis on a facility could cause a facility to design for large disposals systems that are not needed or make unnecessary modifications to existing facilities, incurring unnecessary expense. To ensure overpressure protection is being implemented correctly in the case of an external fire, there are multiple conditions that can be evaluated which, if met, could cause the external pool fire scenario to be not applicable as a source of overpressure. The surrounding chemical inventory, site grading and drainage, and facility history are all factors that dictate whether or not external fire will be a potential source of overpressure. Additionally, the volume and burn rate of the surrounding chemical inventory can be quantified to determine the size and burn time of a potential fire from a loss of containment. The size and burn duration of a fire based on a volume of a flammable liquid can determine if a pool fire could be sustained in a plant. By evaluating each one of these circumstances carefully, it may become acceptable to exclude external fire as a source of overpressure.

6. Appendix A

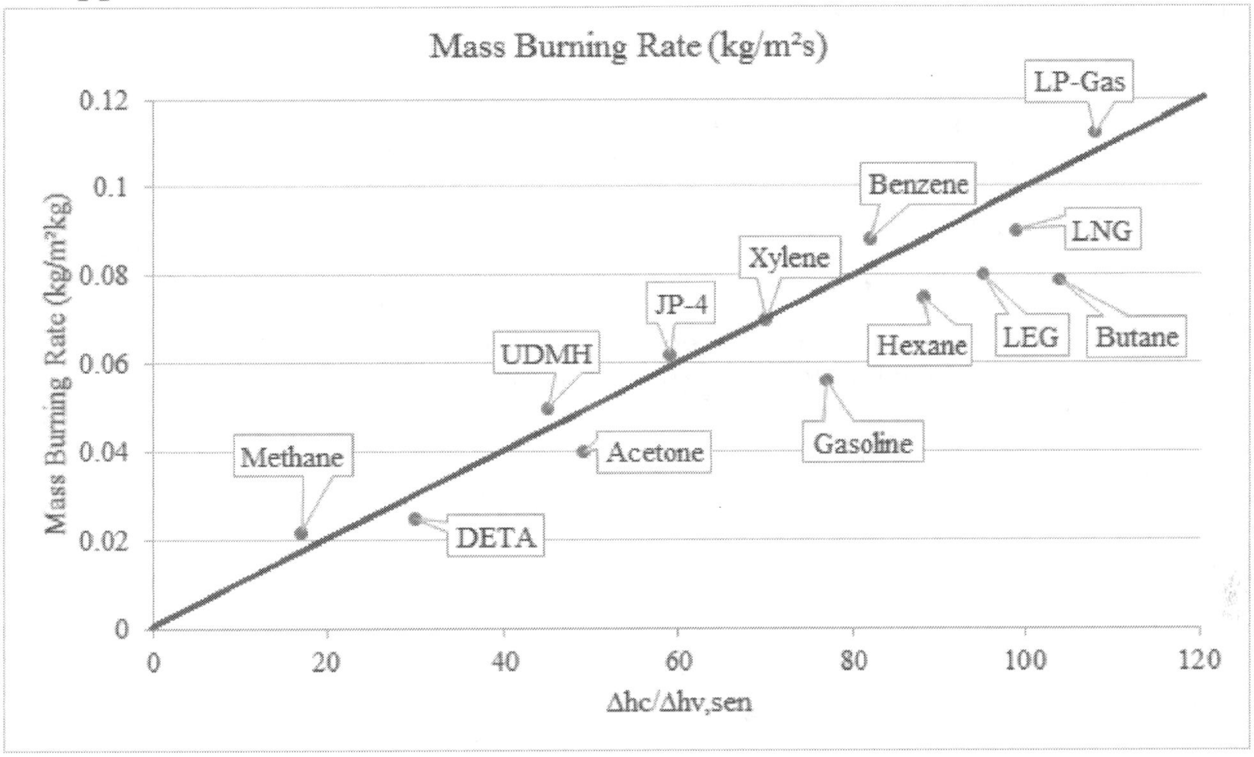

Figure A.1 Mass burning rate per unit area versus the thermochemical property of files burning as pan fires as a function of line 1.A [7]. The xylene value of 0.07 kg/m²s was used in the quantitative analysis.

7. Appendix B

This is the simplified calculation to determine the time taken to overpressure a water vessel for the example analysis.

Table B.1. Geometry and operating conditions of water vessel used in analysis

Description	MAWP (psig)	MAWT (°F)	L -T/T (ft)	Dia (ft)	%LL
Water Vessel	100	450	15	8	25%
	Composition	**Operating P (psig)**	**Operating T (°F)**	**Relief Pressure (psig)**	
	100% Water	80	275	121	
	Vessel Type	**Btm Head**	**Top Head**		
	Vertical	Elliptical	Elliptical		
Volumes	**Total (ft^3)**	**Liquid (ft^3)**	**Vapor (ft^3)**		
	888.0235	318.3481	569.6722		
Phase Density (lb/ft^3)		57.2873			
Mass of phase (lb)	18237.30	18237.30	0.00		
Molecular Weight (lbmol)	18.02	18.02	18.02		
Mass Heat of Vaporization (Btu/lb)		901.94			
Surface Areas	**Total (ft^2)**	**Liquid (ft^2)**	**Vapor (ft^2)**		
	515.7	165.0	320.7		
		Aws= 1.1 Aliq (ft^2)			
		181.5			

Using the API 521 4.4.13.2.4.2 eq (6) [3] the total heat input due to an external fire can be determined using the wetted surface area calculated above, 181.5 ft^2.

Table B.2. API 521 heat input due to external fire table

Heat Input Due to External Fire $Q = C * F * A_{wst}^{0.82}$	Adequate Firefighting/ Drainage?	API Fire Constant	Environment Factor	Q (Btu/hr)
API 521 4.4.13.2.4.2 eq (6)	Yes	21000	1.0	1,494,798.75

The first time step was determined as the time needed to heat the residing liquid to bubble point at operating pressure.

Table B.3. Step 1-Time to Heat Residing Liquid to Bubble Point

	Liquid (ft^3)
Volume (ft^3)	318.3481
Pressure (psig)	80
Temperature (°F)	75.00
Phase Density (lb/ft^3)	57.29
Mass of phase (lb)	18237.30
Molecular Weight (lbmol)	18.02
Mass Heat of Vaporization (Btu/lb)	901.94
Heat to Raise to Bubble Point (Btu, HYSYS)	974195.8
Time to Raise to Bubble Point (hr)	0.65
Time to Raise to Bubble Point (min)	**39.10**

The next time step is the time it takes to vaporize enough of the liquid to fill the vapor space of the vessel, after this point the pressure in the vessel will begin to rise but in this analysis the time taken from these first two steps are used as the time to overpressure, this is a simplified and conservative method as there are many factors that impact a transient fire calculation.

Table B.4. Step 2-Time to Vaporize Liquid to Fill Vapor Space @ Op Pressure

Volumes	**Vapor (ft^3)**
	593.8
Phase Density (lb/ft^3) @ OP Pressure and dew point	0.21
Mass of Vapor (lb)	125.16
Molecular Weight (lbmol)	18.02
Mass Heat of Vaporization (Btu/lb)	901.94
Heat to Fill Vapor Space (BTU)	112,885.47
Time to Fill Vapor Space (hr)	0.08
Time to Fill Vapor Space (min)	**4.53**

Total time needed to fill vapor space: 39.10 mins + 4.53 mins = **43.63 minutes.**

8. References

[1] Routley, J. Gordon, *Fire and Explosions at Rocket Fuel Plant – Henderson, Nevada,* United States Fire Administration Technical Report Series, May 4, 1988

[2] P.J. Dinenno, D. Drysdale Ph.D., C.L. Beyler Ph.D, and W.D. Walton, *SFPE Handbook of Fire Protection Engineering 4th Edition*, National Fire Protection Association, 2008.

[3] American Petroleum Institute, *Pressure-relieving and Depressuring Systems – API Standard 521 – 6th ed.,* January 2014

[4] *Investigation Report – Petroleum Products Facility Incident – Third Coast Industries*, U.S. Chemical Safety and Hazard Investigation Board, Printed Mar. 2003

[5] National Fire Protection Association, *NFPA 30 – Flammable and Combustible Liquids Code, 2015 Edition,* May 19, 2014

[6] Occupational Health and Safety Administration, *Occupational Health and Safety Standards 1910.106 – Flammable Liquids,* United States Department of Labor, Accessed 2/11/2016, Web.

[7] Center for Chemical Process Safety, *Layer of Protection Analysis: Simplified Process Risk Assessment,* New York, NY: Wiley, 2001. E-book.

441525 *Quantification of Double Contingencies in Overpressure Design Philosophy By Fault Tree Analysis*

Abstract:

The design of relief systems, specifically relief valve sizing, remains a crucial factor in mitigating the consequences of overpressure accidents in chemical plants. General criteria from engineering standards (API and ASME) and established engineering practices of operating companies and engineering contractors have provided guidelines in sizing relief valves. Sizing relief valves using standard and independent scenarios such as external fire, blocked outlet, and cooling water failure is a straight-forward matter however in instances where double contingencies or double jeopardy would apply, it is increasingly more difficult to assess the scenarios.

The paper will provide the general pathway to determine quantitatively if a double jeopardy scenario should become credible by using a deductive failure analysis method called Fault Tree Analysis. Fault Tree Analysis can analyze the overpressure state of the system using Boolean logic to combine lower level events such as two separate failures in a double contingencies scenario. A case study for seven permutations of control valve failure and its bypass valve positions will be compared and discussed with reference to Fault Tree Analysis results. Three of the permutations are based on engineering practices from several of operating companies and an assessment is made as to which approach is more realistic for design.

Fault Tree Analysis has been selected for study since it is considered an appropriate mathematical tool for analysis. By taking into account operation and design experience, Fault Tree Analysis can assist design engineers to incorporate or dismiss certain double jeopardy scenarios.

Transportation Risk Management - One Company's Approach

Elizabeth Lutostansky
Air Products and Chemicals, Inc.
Allentown, PA
lutostem@airproducts.com

Gary Miller
Air Products and Chemicals, Inc.
Allentown, PA
millergl@airproducts.com

Martin Dennehy
Air Products PLC
Hersham, United Kingdom
dennehma@airproducts.com

Prepared for Presentation at
American Institute of Chemical Engineers
2016 Spring Meeting
12th Global Congress on Process Safety
Houston, Texas
April 10-14, 2016

Transportation Risk Management - One Company's Approach

Elizabeth Lutostansky
Air Products and Chemicals, Inc.
Allentown, PA
lutostem@airproducts.com

Gary Miller
Air Products and Chemicals, Inc.

Martin Dennehy
Air Products PLC

Keywords: transportation risk assessment, TRA, quantitative risk assessment, QRA, societal risk

Abstract

Air Products has a long history in process safety risk management. The company evaluates the risk of transporting hazardous substances in addition to its risk assessments for fixed sites and pipelines. The Center for Chemical Process Safety (CCPS) book, *Guidelines for Chemical Transportation Risk Analysis*, has been used as the starting point for the Air Products approach to this subject. Over the past few years, the company has refined its transportation assessment methodology and calculation tools and has established a work process where the businesses and supply chains are integrated into the transportation risk analysis (TRA) process.

The purpose of this paper is to summarize Air Products' TRA process, including criteria for deciding if TRA is required, selection of release scenarios considered in a transportation risk analysis, assumptions made regarding populations and weather conditions and the overall risk review process. An example risk analysis will be presented. Comments on the comparison of TRA with pipeline transportation risk analysis will also be provided.

1 Introduction and Background

In 1985, Air Products embarked on a formal program of quantifying risk to the public from its operating facilities, focusing on sites storing or handling toxic materials or significant quantities of flammable materials. Having started the program with fixed sites, a logical extension was to include pipelines and transportation. Air Products has done quantitative risk analyses (QRAs) for transportation since the late 1980s. A procedure for the company's transportation risk management process was formalized in 2007.

In the regulatory field, in 1991, the UK Health and Safety Commission Advisory Committee on Dangerous Substances published a report [1] which quantified the risk of transportation activities in the UK involving hazardous substances. In 1995/6, the Netherlands introduced societal risk criteria for dangerous goods transportation, on a per kilometer basis. Hong Kong carried out transportation QRAs for the territory in 1997 and proposed societal risk criteria. In the Netherlands, the Risk Based Management II hazardous goods transport QRA software has been developed and is used to evaluate pubic risk from hazardous goods transportation to assist land use planning [2].

The Center for Chemical Process Safety published its *Guidelines for Chemical Transportation Risk Analysis* in 1995 [3], and this forms the initial basis for the quantitative risk assessment for transportation of hazard materials by Air Products. The CCPS book covers hazardous material transportation by road, rail, pipeline and waterway. Air Products transportation is mostly by road or pipeline, with some raw materials supplied by railcar.

2 Air Products Transportation Risk Assessment Process

An environmental health and safety risk to the public occurs when a hazardous chemical or gas is released from its packaging during transportation. For this reason, a Transportation Risk Analysis (TRA) is conducted by Air Products for highly hazardous products and raw materials that Air Products has ownership of during transportation. TRAs are carried out by multi-disciplined teams. The evaluation addresses:

- package integrity
- product security
- emergency response preparedness

for the material transportation activity. These are essential elements of the Transportation Management System, as described in reference [4]. As part of the TRA, potential consequences and likelihood of release scenarios are analyzed via a screening level quantitative risk assessment.

3 Thresholds for TRA

The chemicals which would be considered for TRA are those on the Air Products' transportation Chemicals of Concern (COC) list which is based on the US Department of Homeland Security COC list [5]. Only materials which Air Products owns during transportation are assessed in a TRA risk review.

The relevant Air Products business area and supply chain have a responsibility to identify new products and raw materials in need of assessment and to notify the TRA Single Point of Contact (SPOC) so that a TRA can be triggered. Changes to the product supply chain (for example, supply to new regions or larger package sizes) also require notification to the TRA SPOC. The TRA SPOC triggers the risk review process if the threshold for the particular chemical, as shown in Table 1, is exceeded. If the threshold is not exceeded, no further action is required and risk reduction provided from the broader transportation management systems (see reference [4]) is deemed adequate.

Table 1. Thresholds per Package that Require TRA

Material Classification	Threshold Mass per Package
United Nations Globally Harmonized System (UN GHS) Acute Toxicity Category 1 and 2:	$\geq$ 20 kg
UN GHS Acute Toxicity Category 3:	$\geq$ 500 kg
Flammable and Pyrophoric materials with: auto-ignition temperature < 55°C flash point < 60°C boiling point < 60°C	$\geq$ 1500 kg

4 TRA Review Meeting

The relevant Air Products business area and supply chain representatives provide the transportation information to the TRA SPOC who distributes it to the TRA team members. TRA team members include a product safety transportation SPOC, process safety experts, senior representatives of operations, engineering, supply chain and/or technology as appropriate and other subject matter experts as needed. The product safety transportation SPOC facilitates and organizes the team. The process safety experts perform the quantitative risk analysis. Other subject matter experts could provide expertise on emergency response, transportation security and contracts with customers. Table 2 lists the type of information needed for the transportation risk review.

Table 2. Transportation Data for TRA Review Meeting

1	Data on material and package, codes and specifications, inventory, ownership, filling arrangements, shelf life, arrangements for residuals
2	Route details, transportation mode, packages per shipment, shipments per year, specific restrictions
3	Emergency response arrangements
4	Security arrangements

This information is reviewed in the TRA meeting along with the results of the quantitative risk assessment. The QRA is carried by an experienced process safety engineer following an approved calculation procedure. This procedure uses a spreadsheet tool to collect the data on release scenarios and frequencies, along with weather and wind assumptions, ignition probability and population data, so the data can be input into the DNV GL SAFETI software. The following sections summarize the assumptions used by Air Products for a screening TRA study.

5 Typical Product Release Scenarios

A representative set of foreseeable release scenarios which could result from a hazardous material truck accident or external fire, or a railcar accident, is selected by an Air Products process safety engineer. A typical list is shown in Table 3.

Table 3. Typical Product Release Scenarios

Scenario Number	General Description	Detailed Description
1	Catastrophic rupture of container as a result of damage incurred in a vehicle accident.	Release of total contents of container in a short period, typically 3 minutes.
2	Large continuous leak due to a valve torn from container vessel or the largest liquid line breaking, as a result of damage incurred in a vehicle accident.	Entire contents released, existing safeguards fail to operate.
3	If the material is toxic, a smaller continuous leak is assumed to occur, as a result of damage incurred in a vehicle accident.	Leak size selected by iteration such that one to two fatalities are predicted for the release, and the leak is assumed to come from damage to a valve or cracking of a pipe. The leak frequency is selected on the basis of leak size.
4	Where the container is fitted with a pressure relief device, and the material is toxic, an external fire demand is assumed to cause the device to open.	Release of toxic material at the relief device sizing rate. Vessel and contents are heated (heat provided by a tire fire or diesel fire).
5	Hazardous overpressure (boiling liquid, expanding vapor explosion, BLEVE), if the properties of the material are such that this can occur.	Pressure relief devices fail to operate if fitted. Vessel and contents heated until failure (heat provided by a tire fire or diesel fire). Vessel fails at high temperature and pressure.

If the chemical is toxic only, the hazard from releases 1 to 4 to the public are from toxic vapor inhalation. If the chemical is flammable only, the hazards are from fireball or jet fire radiation, flash fire or delayed vapor cloud explosion. Where the material is both toxic and flammable, any of these consequences could occur. The hazardous overpressure (boiling liquid, expanding vapor explosion, BLEVE) case creates a hazard from blast overpressure and impulse and from fragments.

6 Population Types Potentially Affected

There are two types of populations which could be impacted by the hazardous releases noted in Table 3.

6.1 Off-road or off-rail populations

The off-road or off-rail populations are the people in buildings or outdoors sufficiently close to the road or railway line to be impacted by the hazards. It is assumed that 7% of building occupants are outdoors during the day [6]. Buildings surrounding the transportation route will have a range of applicable ventilation rates. As a simplification, an average building ventilation rate of 4 air changes per hour is assumed. The following off-road population densities were assumed for the segments of the route [3]:

- Rural – 20 people/mile^2
- Built up Rural – 100 people/mile^2
- Suburban – 1000 people/mile^2
- Urban – 3000 people/mile^2
- Super-urban – 10,000 people/mile^2

6.2 On-road populations

It is assumed that the road accident which results in a chemical release also causes a traffic jam or tailback of cars, buses, trucks, etc. behind the vehicle accident, and that the vapor cloud from the release may also cause a traffic jam or tailback on the opposite road lane. This traffic jam or tailback assumption is not considered appropriate for a railcar accident. The population density in the traffic is assumed to be 0.05 people per m^2 (129,500 people/mile^2), derived from reference [1].

7 Frequency and Probability

There are several sources of data for transportation accident frequency rates and product release probabilities. Some of these are listed in Table 4.

Table 4. Data Sources for Transportation Accident Frequency Rates

Reference number (Section 16)	Comments
2	The Risk Based Management II hazardous goods transport QRA software, RBM II, contains transportation accident frequency rates and product release probabilities for use in the Netherlands [2].
3	The CCPS book, *Guidelines for Chemical Transportation Risk Analysis,* contains values for transportation accident frequency rates and product release probabilities based on data from the United States [3].
7	The authors suggest that truck accident rates and probabilities of release given an accident provided in Table 6 of reference [7] are appropriate for use as default values in hazardous materials routing studies in the United States.
8	Tables 2.5, 2.6 and 2.7 in reference [8] provide release frequency values for road tankers, derived from reference [1] (United Kingdom estimates). 10% of releases are assumed to be catastrophic ruptures and the remainder are assumed to be to be holes.

Air Products has adopted its accident frequencies from *Guidelines for Chemical Transportation Risk Analysis* [3]. For release probabilities, the values in Table 5 have been adopted for the types of vessels the company uses in transportation. The probabilities are based on a review of published data for similar equipment and engineering judgment.

Table 5. Release Probability Values Adopted by Air Products

Type of container	Probability of leak	Probability of rupture
Flammable tank truck/Unpressurized drum	11%	1.0%
Flammable liquefied gas tanker/Low pressure cylinder	3.0%	0.40%
Toxic liquefied gas tanker or drum	1.4%	0.20%
Gas tube trailer/High pressure seamless cylinders	1.24%	0.087%
ISO module	0.12%	0.0090%
Cryogenic, double walled tanker	0.30%	0.040%

20% of the leak probability value is assigned to the large leak release and 80% of this value is assigned to the small leak, in scenarios where a small leak is considered. Adjustment factors are assigned to the probability values where additional protective measures are employed, such as containers inside protective cages or in rigid-sided vehicles.

For the external fire case, an external fire frequency value of $2.5x10^{-8}$ per kilometer was adopted by Air Products based on values reported in reference [9].

The accident frequency and release probability values are used in the following equations to provide the release case frequencies.

For ruptures and leaks, the release frequency can be expressed as:

$$\left(\frac{Distance}{trip}\right) \times \left(\frac{Trips}{year}\right) \times (Probability\ of\ release) \times (Distance\ weighted\ accident\ frequency)$$

Equation 1

For events caused by external fires, the frequency is:

$$\left(\frac{Distance}{trip}\right) \times \left(\frac{Trips}{year}\right) \times (freq.of\ fire\ per\ distance) \times (prob.of\ event)$$

Equation 2

8 Risk Screening Weather for Consequence Modeling

For QRA screening, the weather conditions noted in Table 6 are used with 8 wind directions, each of equal probability. The weather conditions were selected to capture the range of potential wind conditions. The low wind speed, stable atmospheres are expected to result in the large hazard distances for toxic releases. Whereas the highest wind speed case could result in horizontal plumes or higher pool evaporation rates. The weather probabilities were obtained by averaging weather data from locations throughout the United States.

Table 6. Weather Conditions for Consequence Modeling

Atmospheric Stability	Wind Speed	Probability
F	1.5 m/s	35%
D	5 m/s	54%
D	9 m/s	11%

9 Ignition Probabilities for Flammable Materials

For materials which are flammable, immediate and delayed ignition probabilities must be specified. Equations based on statistical data are proposed to predict the ignition probability of transportation accident hydrocarbon spills as a function of the release size and the substance spilled in reference [10]. Air Products has taken this data and applied adjustment factors for other materials to account for material ignitability (as indicated by auto-ignition temperature and minimum ignition energy). The values are shown in Table 7.

Table 7. Ignition Probability Values

Ignitability		Container rupture		Large leak		Small leak		Relief device operates	
Class	**Example material**	**Imm. Ign. Prob.**	**Delay Ign. Prob.**	**Imm. Ign. Prob.**	**Delay Ign. Prob.**	**Imm. Ign. Prob.**	**Delay Ign. Prob.**	**Imm. Ign. Prob.**	**Delay Ign. Prob.**
Very high	Acetylene Hydrogen	0.9	0.1	0.5	0.09	0.26	0.04	Depends on size of relief orifice – estimate values based on ignition probabilities of large and small leaks	
High	Ethylene Oxide Ethylene Hydrogen sulfide	0.55	0.22	0.22	0.09	0.11	0.04		
Aver-age	Ethane Propane Butane	0.25	0.17	0.1	0.07	0.05	0.03		
Low	Propylene Methane	0.225	0.16	0.09	0.06	0.04	0.03		
Very Low	Carbon monoxide Ammonia	0.04	0.01	0.02	0	0	0		

10 Risk Screening

Societal risk screening results in the form of an FN curve are evaluated. Experience has shown that it is important to have a representative range of low frequency/high consequence and high frequency/low consequence release cases, as discussed in section 5, to produce realistic FN curves for comparison with risk criteria such as those mentioned in section 1. Risk integral values, representing the areas under FN curves, are useful for comparing the risk of different transportation activities and assessing them relative to risks from fixed facilities and pipelines.

11 Reducing the Risk from an Accident

The results of the risk screening assist the review team in deciding which risk reduction measures should be implemented. Options for reducing the probability of an accident include the following:

- reducing the total distance the product is transported (for example, the number of trips per year)
- selection of the mode of transport (rail, road, barge)
- selection of the route of transport (for example, to maximize the distance travelled in areas with low accident frequency statistics)

Other options may also reduce risk, but the effectiveness may be more difficult to judge. These options may include:

- selection of the carrier (assessed for performance attributes)
- training of people involved in the transportation process (drivers, loaders, etc.)
- maintenance and inspection of the transport equipment
- systems increasing the stability of the vehicle
- taking measures to improve security

Options for reducing the consequences of an accident include the following:

- reducing the inventory of hazardous material in a container or reducing the size of the container
- selection of the route of transport (for example, reducing the distance travelled through more densely populated areas)
- taking into account weather/atmospheric stability conditions (for example, to avoid travelling when atmospheric stability conditions would cause hazardous vapor clouds to disperse over larger distances or areas)

Options for reducing the probability of release caused by a vehicular accident include the following:

- improving the quality of the containment (for example, shell thickness of tanks)
- installing crash-buffers (on rail tank cars or road trailers)
- reducing the speed of the vehicle

12 Risk Elevation and Revalidation

Where risk reduction measures are agreed upon by the team, the risk screening is updated to assess the degree of risk reduction achieved. If the resultant risk still exceeds the company's risk elevation criteria, the issue is elevated to a Functional/Business Risk

Review and possibly an Environmental, Health and Safety (EHS) Risk Council Review (senior executive review). When appropriate, the TRA outcome will include a maximum allowable global shipment threshold under which the business can operate before TRA revalidation. If the actual number of shipments approach the threshold, the business area manager must contact the TRA SPOC for a TRA re-assessment.

13 Example Risk Analysis

As an example, the transportation of anhydrous ammonia iso-containers can be evaluated. The shipped iso-container contains 11,500 kg of anhydrous ammonia. The projected total transport is 373,500 km/year, with 450 trips per year going a distance of 830 km/trip. The transportation route goes through some highly populated areas and is estimated to be 15% urban and 85% rural. The resulting FN curve is shown in Figure 1.

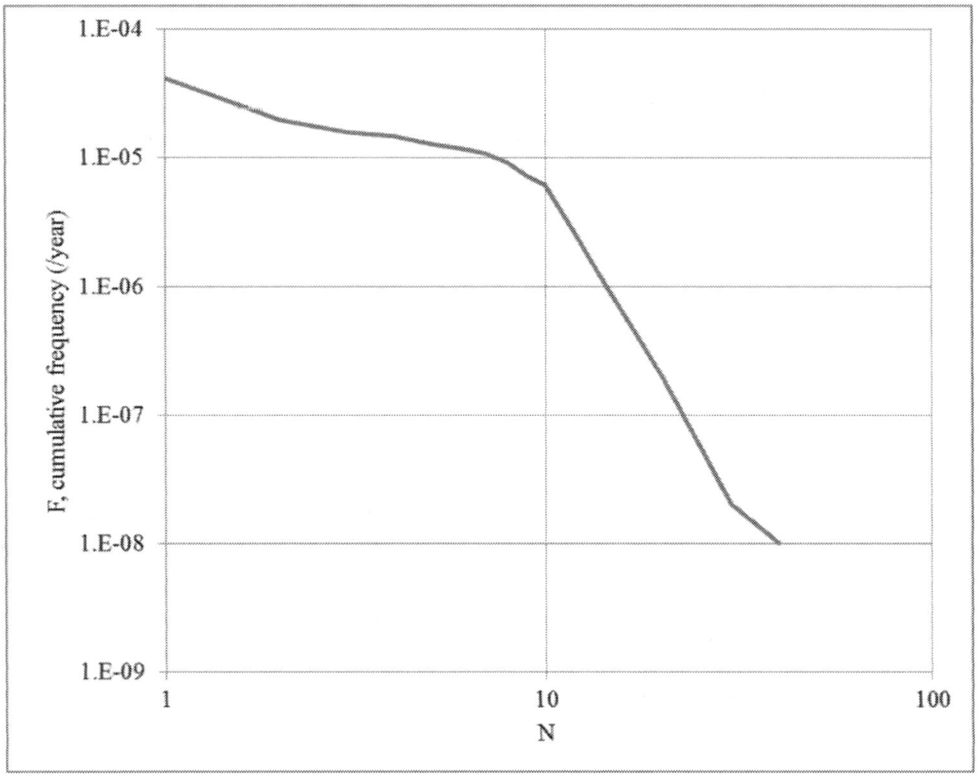

Figure 1 – FN Curve for Initial Route

In order to reduce the risk, rerouting the shipments away from urban population centers can be considered. An alternative route of the same length is surrounded by 3.3% urban population and 96.7% rural population. The impact of rerouting through the area with lower population density is shown in Figure 2.

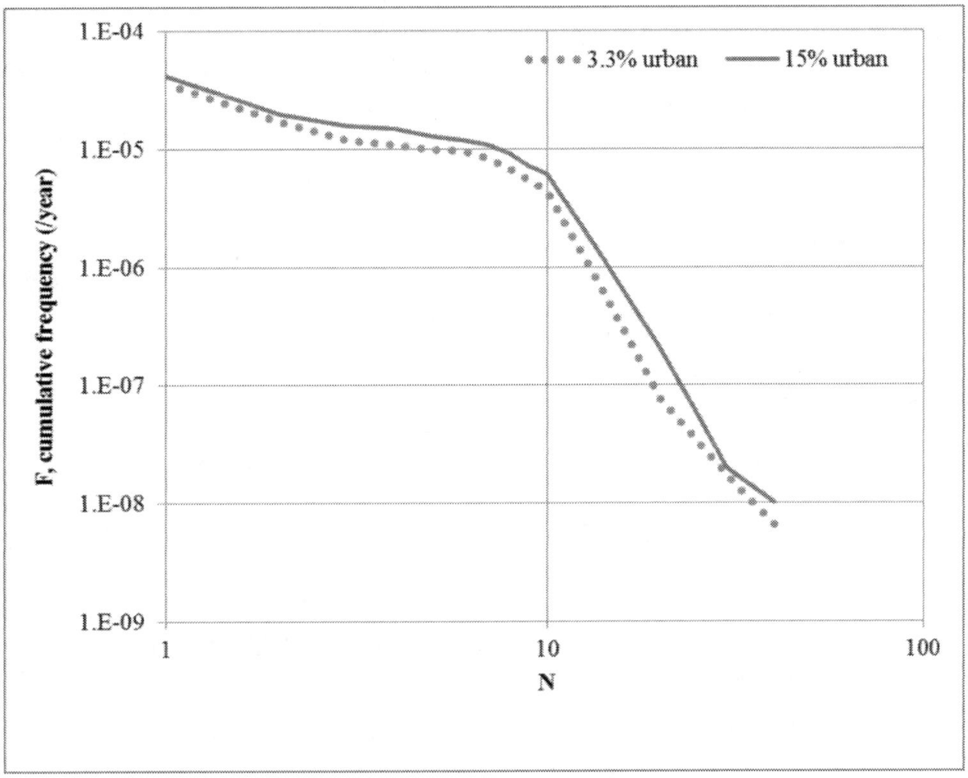

Figure 2 – Comparison FN Curve for Alternative Routes

While the reduction in risk does not appear particularly significant on the logarithmic FN plot, the reduction of the risk integral from rerouting is 22.6%.

14 Comparison of Transportation Risk with Pipeline Risk

There are some key differences in the risk from transportation vehicles and from pipelines. These differences affect the way in which pipeline risk is modeled and evaluated. These differences include:

- Pipeline route

Once established, the pipeline route is unlikely to be changed – particularly for underground pipelines. Therefore, individual point populations around the pipeline are typically considered rather than using generic population densities.

- Release scenarios

Pipeline risk scenarios can include above ground and below ground releases. Pipeline releases are impacted by upstream and downstream inventories and locations of automated shut-off valves, where applicable.

Transportation risk analyses at Air Products are focused on releases attributed to vehicle accidents, whereas pipeline risk analyses consider many causes, such as external interference, corrosion and ground movement.

- Risk mitigation methods

With pipeline risk, there are fewer opportunities to reduce risk once the pipeline is in place. Therefore, population encroachment and customer contract specifications should be considered in the initial risk assessment.

As with transportation risk assessment, business area leaders are involved in the risk assessment process. Pipeline operations representatives are also involved to insure compliance with local regulations and emergency preparedness.

15 Conclusion

Air Products believes that transportation risk analysis (TRA) is an important tool to allow the company to understand its transportation risks and that TRA complements its risk assessment activities for fixed sites and pipelines. The use of risk integrals, in addition to FN curves, gives a broader picture of a company's total portfolio of risk, so that resources for risk reduction can be prioritized.

16 References

[1] Health and Safety Commission, Advisory Committee on Dangerous Substances. "Major Hazard Aspects of the Transport of Dangerous Substances, Her Majesty's Stationary Office, London, 1991

[2] Background document RBM II Version 2.0, Project: 091552, Advisory BV AVIV, Langestraat 11 - 7511 HA Enschede, November 2011.

[3] CCPS. *Guidelines for Chemical Transportation Risk Analysis,* Center for Chemical Process Safety, American Institute of Chemical Engineers, New York, NY, 1995.

[4] CCPS. *Guidelines for Chemical Transportation, Safety, Security and Risk*, Center for Chemical Process Safety, American Institute of Chemical Engineers, New York, NY, 2008.

[5] United States Department of Homeland Security, 6 CFR Part 27, Appendix A, Chemical Facility Anti-Terrorism Standards.

[6] Guidelines for Quantitative Risk Assessment, CFR 18E, Publication Series on Dangerous Substances (PGS 3). RVIM, 2005.

[7] D. Harwood, J. Viner and E. Russell. "Procedure for Developing Truck Accident and Release Rates for HAZMAT Routing," Journal of Transportation Engineering, Vol. 119, No.2, (1993), 189-199.

[8] Land Transport Accident Statistics, OGP Report No. 434 – 9, March 2010, International Association of Oil and Gas Producers, Available at http://www.ogp.org.uk/pubs/434-09.pdf Accessed on February 24, 2016.

[9] D.F. Dunlap. "Large Truck Accidents Involving Tire Failure", Final Report UM-HSRI-SA-74-7, prepared for The Rubber Manufacturers Association, June 1974.

[10] A. Ronza, J.A. Vilchez and J. Casal, 'Using Transportation Accident Databases to Investigate Ignition and Explosion Probabilities of Flammable Spills, Journal of Hazardous Materials 146 (2007) 106–123.

An "Intolerable" Risk: Applying the UK Environmental Risk Assessment Methodology to Freedom Industries, WV

Mark Manton
ABS Group
The Beacons, Birchwood, Warrington, WA3 3WJ, UK
mmanton@abs-group.com

John Farquharson
ABS Group
Knoxville, TN
jfarquharson@abs-group.com

Mike Nicholas
Environment Agency
Surry, UK
mike.nicholas@environment-agency.gov.uk

Prepared for Presentation at
American Institute of Chemical Engineers
2016 Spring Meeting
12th Global Congress on Process Safety
Houston, Texas
April 11-23, 2016

An "Intolerable" Risk: Applying the UK Environmental Risk Assessment Methodology to Freedom Industries, WV

Mark Manton
ABS Group, Warrington, United Kingdom
mmanton@abs-group.com

John Farquharson
ABS Group, Knoxville, TN
jfarquharson@abs-group.com

Mike Nicholas
Environment Agency
Surry, UK
mike.nicholas@environment-agency.gov.uk

Keywords: environmental risk assessment MATTE Intolerable ALARP

Abstract

The first safety legislation in the UK was the Factory Act of 1802. Further acts were added to define how safety was to be managed but still left many workers uncovered. It all changed 40 years ago when the Health and Safety at Work Act replaced the compliance-driven approach with a goal-setting approach. This goal-setting approach puts the "duty" on employers to protect their workers but doesn't not tell them how to do this. The same approach was also used when the EU wrote the Seveso Directive for process safety management. The Seveso-III Directive[1] demands that:

> "*the operator of an upper-tier establishment produce a safety report for the purposes of: demonstrating that ... possible major-accident scenarios have been identified and that the necessary measures have been taken to prevent such accidents and to limit their consequences for human health and the environment.*"

In other words, the onus is on the sites to define what can go wrong and what they are doing to prevent it happening. Also of note is that human health and the environment both require protection in their own right.

After the Buncefield incident the resulting report proposed a methodology using Layer of Protection Analysis (LOPA) and Cost-Benefit Analysis (CBA) to demonstrate that the risks from tank overfills had been reduced to As Low As Reasonably Practicable (ALARP). The HSE's guidance on ALARP demonstration for harm to human health were clear but the definitions of harm to the environment were qualitative and open to different interpretations between operators and the regulators. For humans outside a facility, the HSE's guidance defines a risk as being "intolerable" if the frequency of a fatality exceeds 10^{-4}/yr. (i.e. > one fatality per 10,000 years). What then is the environmental equivalent of such harm to people?

The Chemicals Downstream Oil Industry Forum (CDOIF) is a collaboration of industry and the regulators. A working group with members from UK refineries, chemical and petrochemical plants, tank storage facilities, the Health and Safety Executive (HSE), the Environment Agency (EA) and the Scottish Environment Protection Agency (SEPA) followed up on the Buncefield Report to develop an approach to define whether environmental risks are tolerable and/or reduced to ALARP thereby enabling the environment to be managed within the existing UK risk-based framework. For the environment, the CDOIF methodology defines the tolerability criteria in terms of extent of damage and duration of harm to each receptor[2]. This provides a straightforward approach, using LOPA, to determine whether the risks from a site are tolerable. Cost-benefit analysis is needed to determine whether the outcome is ALARP or if additional barriers are needed.

The Freedom Industries incident provides a real test for the approach. In 2014, a leak through a tank floor of 7,500 US gallons of 4-methylcyclohexanemethanol (MCHM) led to contamination of the Elk River in West Virginia. The Elk is a tributary of the Kanawha River and is used for the drinking water supply of Charleston. As a result of the contamination, 369 people were treated, 13 people were hospitalized and 300,000 people were without potable water for ~10 days. Had an environmental risk assessment, following the CDOIF approach, been performed of the risks from Freedom Industries site (before the incident occurred), the assessment would have identified the risk as "intolerable," i.e. demanding improvements in the facilities or shutting the facility down (if it were in the UK).

1 Introduction

Broadly speaking, the world's safety management systems are divided between compliance-based approaches and risk-based approaches. These divisions in large part reflect the historical legal systems within each country.

1.1 Brief History of UK's Risk-Based Approach

Some of the UK's historical process safety legislation includes: Factory Act 1802, Mines Act 1842, Metalliferous Mines Regulation Act 1872, Quarries Act 1894, Factories Act 1937, Agriculture (Safety, Health and Welfare Provisions) Act 1956 and the Nuclear Installations Act 1959 [1]. This led to a plethora of government departments, including: Factory Inspectorate, Explosives Inspectorate, Employment Medical Advisory Service, Nuclear Installations Inspectorate, Safety and Health Division from the Department of Energy, Mines Inspectorate, Safety in Mines Research Establishment, British Approvals Service for Electrical Equipment in Flammable Atmospheres and the Alkali and Clean Air Inspectorate. Despite all of the inspections done by these regulators, there were over 1,000 work-related deaths per year in the late 1960s and over 5 million workers were not covered by any health and safety legislation [2]. Lord Robens was appointed by the Labour government to review options. He concluded that, if anything, there was already too much legislation. He proposed a radical change - moving from specific targets to setting goals. His report, which was translated in to the Health and Safety at Work Act of 1974 (HSWA) [3], swept away all previous legislation and inspectorates, replacing them with the Health and Safety Executive (HSE). The detailed requirements of earlier legislation were replaced with this key phrase:

*"It shall be the **duty** of every employer to ensure, so far as is reasonably practicable, the health, safety and welfare at work of all his employees ..."*

In other words…

- "We" are not going to tell you what you must do.
- If anyone gets hurt, then, by definition, you have not done your **duty**.
- So, effectively, "you" are **guilty until proven innocent**, and
- You can "prove" your innocence by having implemented "All Measures Necessary" so far as is reasonably practicable.

This change was not universally accepted, and the Labour peer Lady Summerskill argued that the changes "would bring rejoicing to the hearts of every irresponsible employer of labour who used the voluntary approach as an escape from his obligations" [4]. Nonetheless, the Act radically changed the way of working and the approach of the regulators from prescriptive compliance to a goal-setting, risk-based approach to health and safety.

1.2 Snapshot of Risk-Based versus Compliance-Based Safety Management

The HSWA is now 40 years old. Over this time, there has been a significant reduction in the number of work-place deaths, from 651 deaths in 1974 to 142 in 2014 (i.e. 2.9 deaths per 100,000 to 0.48 deaths per 100,000 workers).

The precise cause and effect are unclear but, within the European Union (EU) (where there is a consistent method of reporting), the UK had the lowest rate of workplace fatalities at 0.58 per 100,000 workers in 2012 (Figure 1) [5]. If the UK had performed at the same level as, for example, France (2.64 fatalities per 100,000 workers), then instead of the 150 workplace fatalities, there would have been 682 fatalities (or close to the same rate as in the UK in 1974). The majority of these deaths were due to behavioral safety issues (e.g. falling through weak roofs) rather than process safety, but it is likely that the goal-setting drive of the HSWA has led to this difference.

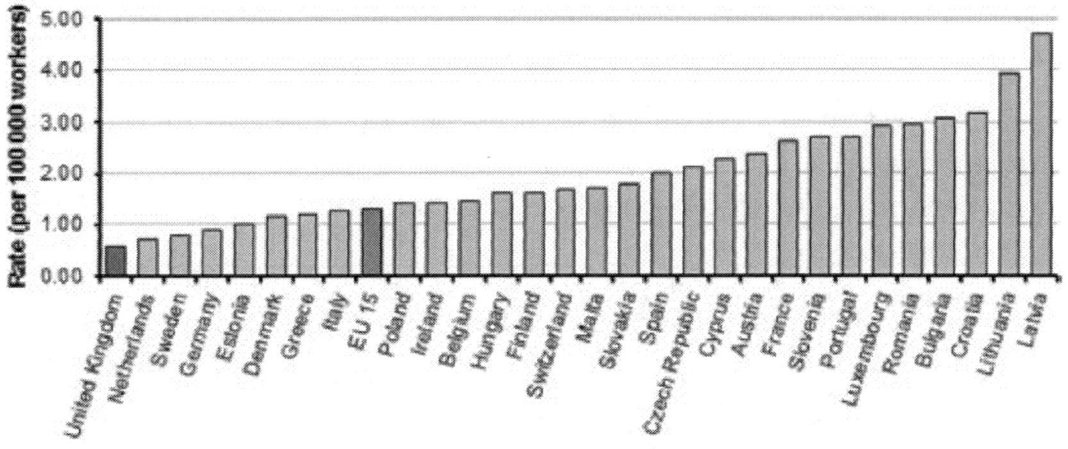

Figure 1: Standardised incidence rates (per 100,000 workers) of fatal accidents at work for 2012

Note: Figures exclude road traffic accidents and accidents on board transport in the course of work.

It cannot be explicitly concluded that the better performance in the UK is only due to the application of a risk-based versus compliance-based approach, because there are too many other cultural factors at work. It is nonetheless tempting to draw this conclusion.

1.3 Risk-Based Process Safety in the UK

Since the implementation of the HSWA the UK became part of the EU, bringing a slightly different approach to safety. Safety legislation in other European countries tends to be more compliance-driven than the UK's risk-based approach, which sets goals rather than prescribing rules. The combination of the two cultures led to the EU's Seveso directive, or the Seveso-III Directive [6] in its latest update. The key clause in the Directive demands compliance in the shape of companies submitting a Safety Report every five years to their local Competent Authority. The Safety Report requirement, however, is definitely goal-setting and risk-based (Table 1).

Table 1: Safety Report Content (as per Article 9 of the Seveso-III Directive)

Member States shall require the operator to produce a safety report for the purposes of: (a) demonstrating that a major-accident prevention policy and a safety management system for implementing it have been put into effect ... (b) demonstrating that **major-accident hazards have been identified** and that the necessary measures have been taken to prevent such accidents and to limit their consequences for human health and the environment; (c) demonstrating that adequate safety and reliability have been incorporated into the ... **operation** and **maintenance** of any ... equipment ... connected with its operation which are linked to major-accident hazards ... (d) demonstrating that internal emergency plans have been drawn up and supplying information to enable the external plan to be drawn up in order to take the necessary measures in the event of a major accident;

So, the HSWA simply tells you not to hurt anyone and Seveso-III (or the Control of Major Accidents and Hazards Act, COMAH, as it is implemented in the UK) tells you to "demonstrate" that you have:

- really thought about how you could hurt someone or the environment,
- done something about it in terms of [safety critical] equipment and [safety critical] activities/operations, and
- a plan in place, and tested, for the possibility that the preventative measures don't work and things goes wrong.

1.4 Reducing Risks to As Low As Reasonably Practicable (ALARP)

In 2001, the HSE documented the approach to risk targets in Reducing Risks, Protecting People (R2P2) [7]. This document explicitly addressed tolerable vs. intolerable risks by describing society's tolerance of risk with fatal consequences. The model presented a three-tiered scheme that classifies risk into the following regions (Figure 2):

- **Unacceptable Region** - Any activity within this region would not be allowed; risk must be reduced.

- **Tolerable Region** – The risk associated with activities within this region is deemed to be tolerable but further risk reduction measures should be considered for implementation.
- **Broadly Acceptable Region** – The risk associated with activities within this region is considered adequately controlled and further risk reduction is not required.

The ALARP regulatory approach enables the HSE to avoid the development of precise standards relating to technology that is continually changing. It is up to each operator to define their own risk thresholds. However, justification is required if these differ from the HSE's guidance:

- If a worker is exposed to a risk of death greater than 1 in 1,000 per year (1×10^{-3} annual exposure), the risk is intolerable.
- If a member of the public is exposed to risk of death greater than 1 in 10,000 per year (1×10^{-4} annual exposure), the risk is intolerable.
- If a worker or individual in the public is exposed to risk of death less than 1 in a million per year (1×10^{-6} annual exposure), the risk is broadly acceptable.

Regarding intolerable risks, the HSE advises that "if the risk is in this region, then ALARP cannot be demonstrated and action must be taken to reduce the risk almost irrespective of cost" [8].

Most companies in the UK define their tolerability thresholds at 1 in 10,000 per year (1×10^{-4} annual exposure) for both workers and members of the public.

Figure 2 summarizes these standards visually in a Tolerability of Risk (TOR) diagram. The values shown are the individual risk values (fatalities per year) that are typically used in the UK. In the region between the thresholds, industry must demonstrate that each employee's risk of fatality is ALARP, which demands that the cost involved in further risk reduction is grossly disproportionate to the benefit (in terms of a reduced potentials for loss of life) to be gained.

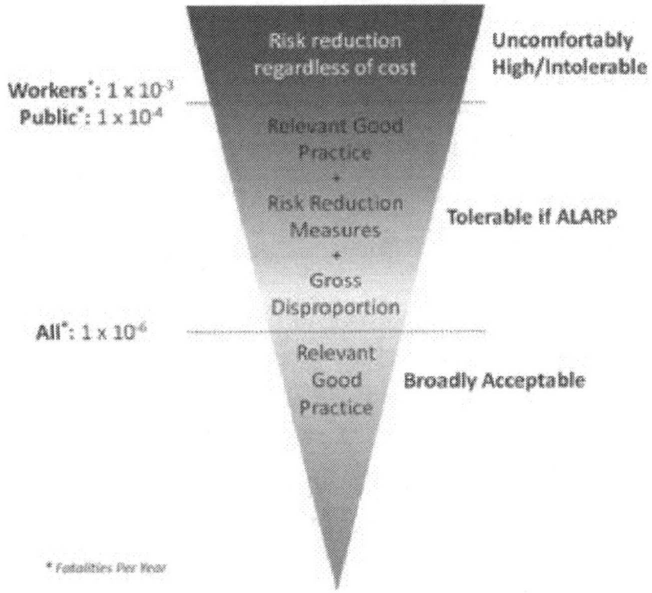

Figure 2. HSE Framework for Risk Tolerability

The HSE is responsible for the safety of workers, and their scope does not extend to the environment. While R2P2 provided clear guidance for the tolerability threshold for fatalities, it did not define the tolerability threshold for harm to the environment.

2 What is a Major Accident to The Environment (MATTE)?

Process safety issues within the European Union are managed by member countries by their local applications of the Seveso-III Directive. Under the Seveso-III Directive, all member states are obligated to report all Major Accidents to the EU Commission. This is in itself only a reporting requirement to enable the commission to compile data on such incidents. The definition of a major accident is included in the Directive as "an occurrence … leading to serious danger to human health or the environment" [6]. This leaves an element of subjective judgment around the level of danger that would be considered "serious." However, thresholds for reporting Major Accidents to the EC give an indication of serious harm thresholds, as included in Annex VI of the Directive, see Table 2.

For unintended releases of dangerous substances, harm to people and damage to property, the reporting thresholds are more or less the same as for a Tier 1 process safety event as per American Petroleum Institute (API) recommended practice API-754, supported by the international offshore oil industry via the Oil and Gas Producers (OGP) Report OGP-456 [9]. While both API-754 and OGP-456 make mention of the environment, they just defer to "local legislative requirements." Such local definitions exist in Europe in the Seveso-III Directive (Table 2).

Table 2: Major Accident to The Environment (as per Seveso-III)

2. Injury to persons and damage to real estate:
. . .
(f) the interruption of drinking water, electricity, gas or telephone services for more than 2 hours (persons × hours): the value is at least 1,000.
3. Immediate damage to the environment:
(a) permanent or long-term damage to terrestrial habitats:
(i) 0.5 hectare (ha) or more of a habitat of environmental or conservation importance protected by legislation;
(ii) 10 or more ha of more widespread habitat, including agricultural land;
(b) significant or long-term damage to freshwater and marine habitats:
(i) 10 kilometers or more of river or canal;
(ii) 1 ha or more of a lake or pond;
(iii) 2 ha or more of delta;
(iv) 2 ha or more of a coastline or open sea;
(c) significant damage to an aquifer or underground water:
1 ha or more.

Unfortunately there is no clear definition of "permanent or long-term damage." Note also, these criteria only provide reporting requirements; they do not define the risk criteria.

3 Buncefield

This was the regulatory environment at the time of the Buncefield explosion on the 11th of December, 2005. The incident cost the UK almost £1 billion ($1.5 billion), a significant proportion from disruption in supply of jet fuel to Heathrow airport [10]. After the explosion, there were two phases of investigations and improvements: the Buncefield Standards Task Group, completed mid 2007 with 48 recommendations [11], and the Process Safety Leadership Group (PSLG), completed December 2009 [12]. Both phases had extensive working groups with representatives from industry (refining, chemicals production and fuel storage terminals), regulators (the HSE, Environment Agency (EA) and the Scottish Environment Protection Agency (SEPA)) and the union (UNITE). During the PSLG phase, there were ~90 people, the majority from industry, in the six separate working groups covering Human Factors, Scope, Control Instrumentation (LOPA/Safety Integrity Level [SIL]), Bunding & Tertiary Containment, Emergency Response and Mechanical Integrity (Tank Design). The outcome was the "Safety and environmental standards for fuel storage sites. Process Safety Leadership Group Final report" in December 2009 [12].

COMAH sites, or "establishments"[1], in the UK similar to Buncefield were obliged to gap assess their operations against the 267 page PSLG report and to deliver LOPAs, or equivalent methodologies. They were required to demonstrate that the risks to people and the environment from a tank overflow of gasoline were tolerable and that they had reduced the risks to ALARP. With the clear definition of the tolerable limit being 10^{-4} off-site fatalities per year and guidance on how to demonstrate ALARP via a cost-benefit analysis, most sites found it straightforward to reach an agreement with the HSE on whether improvements were needed or not. This was not the case for agreements with the EA regarding the environmental risks, however. The PSLG Report defined the tolerability threshold as a function of the degree of impact of the event, varying from 10^{-4} per year for "major" or "catastrophic" events to 10^{-2} per year for "severe" events to 10^{-1} per year for "noticeable" events. Problems arose from disagreements regarding the classifications of these terms in the PSLG Report (Table 3).

Given the qualitative nature of these definitions, the operators and the regulators could arrive at different classifications for the same event, resulting in an unsatisfactory situation.

The implementation and further management of process safety in the UK exists through collaboration between the regulators and industry. There was a clear need for a new definition of risk assessment for the environment. This led to another Working Group under the auspices of the follow-on overarching body, the Chemicals and Downstream Oil Industry Forum (CDOIF) [13].

[1] 'establishment' means the whole location under the control of an operator where dangerous substances are present in one or more installations, including common or related infrastructures or activities; establishments are either lower-tier establishments or upper-tier establishments [6].

Table 3: Environmental Categories as per PSLG (Buncefield) Report

Category	Definitions	
6	Catastrophic	- Major airborne release with serious off-site effects - Site shutdown - Serious contamination of groundwater or watercourse with extensive loss of aquatic life
5	Major	- Evacuation of local populace - Temporary disabling and hospitalisation - Serious toxic effect on beneficial or protected species - Widespread but not persistent damage to land - Significant fish kill over 5 mile range
4	Severe	- Hospital treatment required - Public warning and off-site emergency plan invoked - Hazardous substance releases into water course with ½ mile effect
3	Significant	- Severe and sustained nuisance, eg strong offensive odours or noise disturbance - Major breach of permitted emissions limits with possibility of prosecution - Numerous public complaints
2	Noticeable	- Noticeable nuisance off site, eg discernible odours - Minor breach of permitted emission limits, but no environmental harm - One or two complaints from the public
1	Minor	- Nuisance on site only (no off-site effects) - No outside complaint

4 CDOIF – Environmental Risk Tolerability for COMAH Establishments

A Working Group was set up in 2010 to resolve the issues arising from the PSLG report definitions of the categories of environmental events. Specifically, the drive was to define the tolerability thresholds for different environmental events and to define the environmental risk assessment framework to be used by COMAH sites in the UK, for inclusion in their Safety Reports or risk assessments. The guideline was published in 2013 on the HSE's web site. More recently, worked examples have been added for both a complex site and a simple tank storage site [14].

4.1 Source, Pathway, Receptor Concept

Key to the approach is the concept of Source → Pathway → Receptor. The source is the hazard that could be released into the environment and the pathway is the route the pollutant takes to the receptor. The types of environmental receptors that need to be considered are: terrestrial habitats, freshwater habitats, marine habitats and groundwater bodies.

Each habitat is further divided in to sub-categories with each have differing relative "value." Examples of these sub-categories include areas of national importance, such as National Parks and areas of international environmental importance, including Ramsar sites [15] (typically wetlands which are stop-off sites for migrating sea birds), rivers and aquifers where water is extracted for human consumption.

4.2 Screening Methodology

The methodology to be followed in this approach is as follows:

1. Identify receptors.
2. Identify the major accident scenarios.
3. Assess the unmitigated consequence and assign a severity level.
4. Determine the duration of harm and assign a duration/recovery category.
5. Identify tolerability thresholds.
6. Assess mitigated establishment risk.
7. Compare establishment risk to the thresholds.
8. Determine if more detailed assessment is required.

These steps are now described in more detail.

4.2.1 Identify Receptors

Review the site and its surroundings to identify all receptors that could be affected by any spill. For example: is the site sitting on top of impermeable clay in a low sensitivity environmental setting or is it sitting by a river where the water is extracted for human consumption?

4.2.2 Identify the Major Accident Scenarios

Define all credible scenarios which could have a MATTE potential on the identified receptor, i.e. the accident scenarios from hazard and operability studies (HAZOPs) or process hazard analyses (PHAs) with high environmental consequence.

4.2.3 Assess Unmitigated Consequence and Assign Severity Level

Assume that **none** of the barriers to prevent an event function as desired (e.g. catastrophic failure of a tank and simultaneous complete failure of all bunding and diking). How large an area of the receptor will be affected? What is the severity/harm category (Table 4)?

4.2.4 Determine Duration of Harm and Assign Duration/Recovery Category

How long will the pollutant remain in the environment? Assign a duration/recovery category that would be associated with the unmitigated consequences from each accident scenario. Longer-term harm will be less tolerable than an impact of short duration. Often there will be opportunities for clean-up and remediation to reduce environmental harm; however, these should be disregarded at this stage. Assign the duration/recovery category using Table 5.

Table 4: Categories for Severity for <u>Selective</u> Receptors (this is a sub-set of entire list of receptors)

Receptor Type		Significant	Severe	Major	Catastrophic	Receptors
Harm/Severity Level->		1	2	3	4	Include
Land or Surface Water	Designated Land/Water Sites (internationally important)	<0.5ha or <5% (<5% LF/Pop)	>0.5ha or >5% of site area	25-50% of site area	>50% of site area	Ramsar
Land	Widespread Habitat	<10ha	Contamination of >10ha of land	100-1,000 ha	>1,000ha	Land/water used for agriculture, forestry, fishing or aquaculture
Water	Widespread Habitat	<10ha	Contamination which prevents fishing or renders it inaccessible to the public.			Land/water used for agriculture, forestry, fishing or aquaculture
Water	Fresh and estuarine water habitats	Damage less than "severe"	Reduction in water quality by one class for 2-10km of watercourse or 2-20ha or 10 50% area of	10 times "severe"	100 times "severe"	
Water	Source of Public or Private Drinking Water	Damage less than "severe"	Interruption of drinking water supply (where persons affected x duration in hours > 1,000)	>1 x 10^7 person-hours interruption (a town of ~100,000 people losing supply for month)	>1 x 10^9 person-hours interruption (~1 million people losing supply for 1 month)	
Land	Built environment	Damage less than "severe"	Damage sufficient for designation of importance to be withdrawn.	Feature of built environment entirely destroyed.		"Listed" buildings, ancient monuments, conservation area, etc.
Land	Particular species (National %)	Damage less than "severe"	Loss of 1-10% of animal or 5-50% of plant ground cover.	Loss of 10-90% of animal or 50-90% of plant ground cover.	Total loss (>90%) of animal or plant ground cover.	
Surface Water	Marine	Damage less than "severe"	2-20 ha littoral or sub-littoral zone, 100-1,000 ha of open sea 100-1,000 dead sea	10 times "severe"	100 times "severe"	

Table 5: Duration / Recovery Category

Description	Short term	Medium term	Long term	Very long term
Harm Duration Category ->	1	2	3	4
LAND	≤3 years	> 3 years or > 2 growing seasons for agricultural land	> 20 years	> 50 years
SURFACE WATER (except drinking water)	≤1 year	> 1 year	>10 years	>20 years
GROUNDWATER BODY		Harm affecting non-drinking water source	Harm affecting public drinking water source	
BUILT ENVIRONMENT Listed buildings		Can be repaired in >3 years	Feature destroyed, cannot be rebuilt	World heritage site feature destroyed

Note: defining "harm to public drinking water sources" as a "long-term" duration is a work-around in order to provide a consistent approach for cases where the actual duration of the harm to humans cannot be quantitatively predicted ahead of the event.

4.2.5 Identify Tolerability Thresholds

Combine the harm/severity level with the duration/recovery category to define the tolerability threshold, using the data in Table 6.

Table 6: Matrix Defining the Tolerability and Broadly Acceptable Thresholds

Severity of Harm		1 Short term	2 Med. Term	3 Long Term	4 Very Long Term
Catastrophic	4	Sub-MATTE duration	**10^{-4}** 10^{-6}	10^{-5} 10^{-7}	10^{-5} 10^{-7}
Major	3	Sub-MATTE duration	**10^{-3}** 10^{-5}	**10^{-4}** 10^{-6}	10^{-5} 10^{-7}
Severe	2	Sub-MATTE duration	**10^{-2}** 10^{-4}	**10^{-3}** 10^{-5}	**10^{-4}** 10^{-6}
Significant	1	Sub-MATTE duration	Sub-MATTE harm	Sub-MATTE harm	Sub-MATTE harm

Harm Duration Category

Tolerability threshold in **bold** and broadly acceptable threshold in grey.
Cell color-coding only for visual effect.

4.2.6 Assess Mitigated Establishment Risk

Using a risk assessment method, such as LOPA, and applying the initiating event frequency for the accident scenario and the probability of failure on demand (PFD) of the preventative and mitigation barriers, determine the residual risks to the receptor from the threat.

Repeat the exercise for all of the accident scenarios on site that could affect the receptor. Then sum the frequencies for all accidents scenarios to arrive at the total risk to the receptor from the establishment.

4.2.7 Compare Establishment Risk to the Thresholds

Compare the calculated outcome to the tolerability and the broadly acceptable thresholds (Table 6). If the residual risks are in the intolerable region, then improvements are required, irrespective of the costs of these improvements. If these cannot be economically implemented then the facility should shut down.

4.2.8 Determine if More Detailed Assessment is Required

If the residual risk is between the intolerable and the broadly acceptable levels, the site should determine what additional barriers could be added to reduce the risks still further. The site may perform a cost-benefit analysis, comparing the costs for the additional barriers against the benefits from avoiding the worst consequences of the possible accident. The benefits should include the fines that will not need to be paid, the costs of cleaning-up and remediating the harm and increases in insurance premiums. The benefits typically exclude the lost operating margin because are the responsibility of the business itself.

If the ratio of the costs to the benefits exceeds a defined proportionality factor (e.g. 10x), then the costs can be said to be gross disproportionate and the current situation is ALARP [8].

4.3 Comparing Environmental and People Risks

The widely accepted individual risk threshold for an intolerable risk to a person outside the establishment is 10^{-4} fatalities per year [7].

Based on the CDOIF classification, there is an intolerable risk to the environment of 10^{-4} per year if an event will be a:

- Catastrophic MATTE with a medium-term duration (e.g. 1,000 hectares [~2,500 acres] for between 3 and 20 years), or
- Major MATTE with a long-term duration (e.g. 100 hectares [~250 acres] for between 20 and 50 years, or
- Severe MATTE with a very long-term duration (e.g. 10 hectares [~25 acres] for more than 50 years).

In other words, these are *de facto* considered to be the environmental equivalent of a fatality. Note: these levels and cut-offs are still subject to modification, but they do give a starting point that feels proportionate.

5 Applying the CDOIF Environmental Risk Assessment Methodology to Freedom Industries

5.1 Freedom Industries Event Background

The test for the approach is to apply it to a specific example. The example chosen is that of Freedom Industries near Charleston, West Virginia (Figure 3).

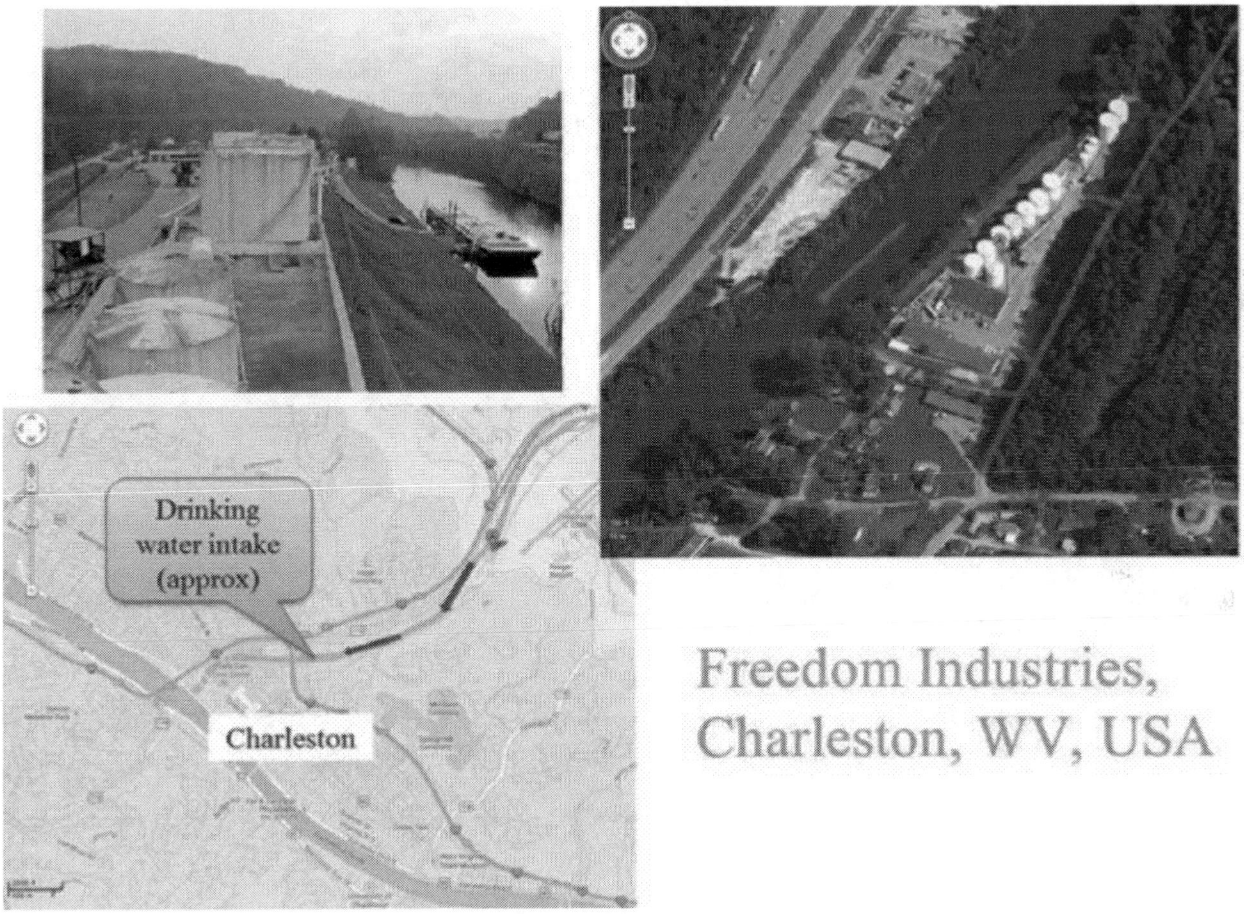

Figure 3: Overview of Freedom Industries site

The site stored various chemicals, including crude 4-Methylcyclohexanemethanol (MCHM). On January 9th, 2014, a total of 7,500 gallons of MCHM leaked through the floor of the tank through two small holes of ~0.4 inch and ~0.75 inch diameters and into the Elk River [16]. The Elk is a tributary of the Kanawha River and is the source of drinking water for Charlestown, West Virginia. After the alarm had been raised, a "Do Not Use" order was issued and 300,000 people were effectively without a water supply for 10 days [17]. 369 patients were treated for possible exposure following the release into the drinking water and 13 were hospitalized; of these, 38% exhibited symptoms of nausea, 29% had skin rashes, and 28% experienced vomiting [16].

5.2 Applying CDOIF Methodology to Freedom Industries

Assuming the CDOIF Methodology had been applied to the site before the event occurred, some of the relevant threats, consequences and potential barriers are shown in this bowtie diagram (Figure 4). Note: this is a simplified overview, considering only the known risks from mechanical tank failures. It is incomplete and ignores other potential risks such as from tank overfill, which would need to be added to get the complete picture.

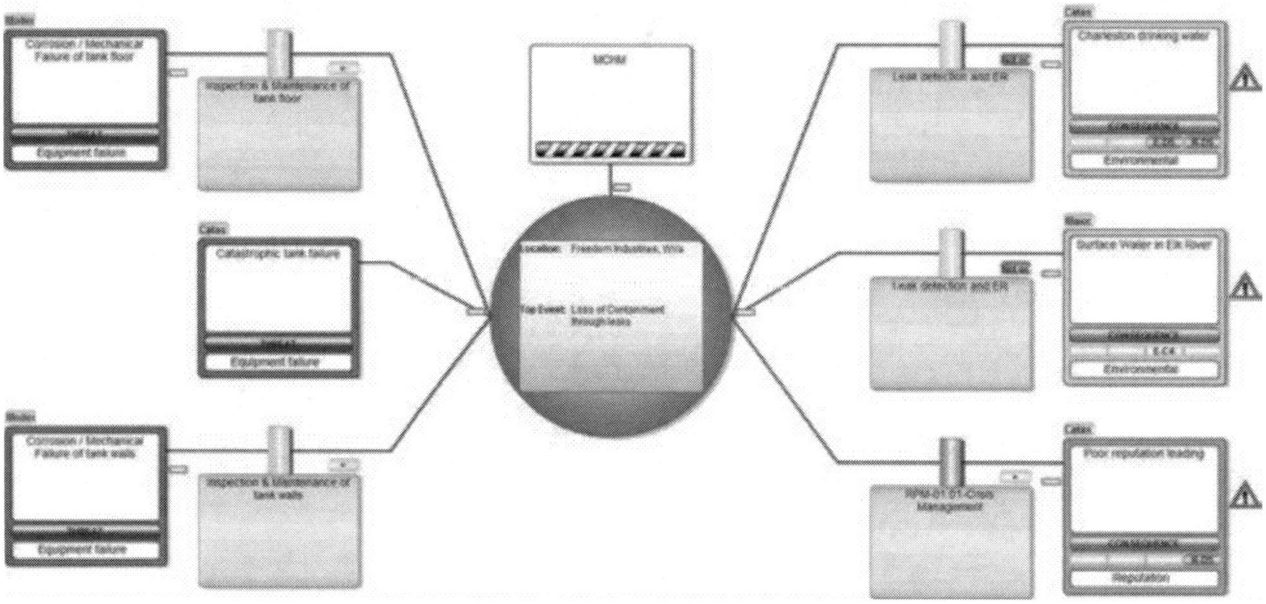

Figure 4: THESIS™ BowTie for Freedom Industries Environmental Risk (single storage tank – mechanical failure modes)

Applying the steps of the methodology (as per Section 4.2):

5.2.1 <u>Identify Receptors</u>

The receptors surrounding the Freedom Industries site include:

- Soil on the site
- Groundwater on the site
- Freshwater in the Elk River:
    - The linear parameters for the river
    - Animal and plant life in the river
    - The drinking water intake downstream

5.2.2 Identify the Major Accident Scenarios

The specific scenario here (because it happened) is leakage through the floor of the tank.

Had the site performed a recent PHA, this would probably have identified other scenarios leading to significant environmental harm. The PHA would, for example, have identified the threats of (see Figure 4):

- catastrophic failure of the tank
- leaks from tank walls and
- tank overfill

For the sake of this exercise, we will restrict ourselves initially to just the threat of a leak through the tank floor.

5.2.3 Assess the Unmitigated Consequence and Assign a Severity Level

The unmitigated consequence would be complete emptying of the tank through the tank floor. The severity level needs to be assessed for each receptor:

Soil on site: contamination probably not greater than 10hectares [~25 acres], so a sub-MATTE. No further consideration needed.

Groundwater: unlikely to be a receptor in itself, but acts as a pathway to the river.

Surface water: an estimate would have to be made on the contamination, i.e. the volume of contaminants that would be reached over some time frame versus normal flow of water in the river and whether this would reduce the water quality by one or more classes. Making a guess, it is likely that the severity level, as per Table 4, would have been between a Severe and Major category (severity level 2 or 3).

Water as drinking water: the Elk River supplies drinking water to Charleston. Drinking water was unavailable for > 10^7 person-hours (300,000 people x 10 days x 24 hrs). The severity level would therefore have been major (severity level 3).

5.2.4 Determine the Duration of Harm and Assign a Duration/Recovery Category

Surface water: harm unlikely to continue for more than 1 year, so harm duration is sub-MATTE.

Drinking water: harm affects public drinking water so a harm/duration of "Long Term" (Table 5), or severity level of 3, would be appropriate.

5.2.5 Identify tolerability thresholds

Combining the severity level of major (3) and the harm/duration category of long term (3) leads to a tolerability threshold (Table 6) of 10^{-4} per year and a broadly acceptable threshold of 10^{-6} per year.

5.2.6 Assess Mitigated Establishment Risk

According to Table 2.1 of the OGP's Risk Assessment Data Directory [18], the probability of a liquid spill outside a single tank is 2.8×10^{-3} per year.

The most common barrier to prevent a loss of containment through a tank floor is scheduled regular internal inspection of the tank and corrective maintenance if problems are found. The PFD for this barrier is 0.1 (considering it to be a "human response to an abnormal condition" [19]), resulting in a residual risk of 2.8×10^{-4} (Table 7).

Table 7: LOPA for Freedom Industries accident scenario of tank floor leakage

Factor	PFD*	Frequency	Comment
Initiating Event			
IEF (for "Liquid spill outside tanks"):		2.8 x 10^{-3}	Table 2.1 in OGP-434-03 Report [17]
Preventative Barriers			
Inspection and repair	0.1		Scheduled regular internal inspection and non-destructive testing
Mitigation Barriers			
Secondary containment	1		Not applicable for tank floor leaks
Tertiary containment	1		Not applicable. No on-site, but outside the bund, storage.
Residual Risk per tank		2.8x10^{-4}	INTOLERABLE, improvements needed
Residual risk for establishment//facility		3x10^{-3}	~10 tanks; definitely Intolerable!

*PFD = Probability of Failure on Demand

5.2.7 Compare the Establishment Risk to the Thresholds

The risk from a single tank is intolerable. For the complete risk assessment, the other accident scenarios from this tank should also be considered, as well as the risk of a leak from the other tanks. The residual risk will get worse. This outcome means that either the site should shut down or make improvements such that the risk is no longer intolerable.

5.2.8 Determine if More Detailed Assessment is Required

The further assessment would be to investigate what options are available to reduce the residual risk. As shown in Table 8, it may be possible to implement some of the following improvements:

- Installing a liner under the tank would mean that any leaks from the tank floor would appear in the bund and would be seen during operator walk-arounds, hence a valid barrier for "leak detection and emergency response" and a PFD of 0.1.
- If the bund itself were lined and the rain-water drain valve removed then this would be an additional valid barrier with a PFD of 0.1. (The bund would then have to be pumped out after rain, but there would be no risk of the rainwater drain valve being left open.)

Together these improvements would reduce the residual risk to 2.8x10^{-5}/yr, i.e. into the tolerable-if-ALARP region.

Further options would then be needed to be evaluated via a cost-benefit analysis, such as:

- Enhanced emergency response to shut off the drinking water intake for Charlestown. This would prevent the contamination of drinking water but would add a new consequence of harm to the reputation of the site.
- Enhanced, more frequent, internal inspections.

Would these additional barriers be cost-effective? If not then the site would have reduced the risks to ALARP for this scenario.

A complete risk assessment would have to consider all threats that could lead to environmental harm.

Table 8: Improvement Options for Freedom Industries

Factor	PFD*	Frequency	Comment
Residual risk for establishment//facility		$2.8x10^{-3}$	~10 tanks; definitely Intolerable!
Improvement Options			
Leak detection & Emergency Response	0.1		Liner under floor and auditable Standard Operating Procedure to drain tank.
Improved secondary containment	0.1		Liner in dike and rainwater drain removed
Residual Risk		$2.8x10^{-5}$	Tolerable (needs ALARP determination)
Enhanced Inspection frequency	0.1		Possibly up to 1 per year (?)
Enhanced external emergency response	0.1		Documented emergency response to shut-off of Charlestown drinking water inlet
Residual Risk		$3x10^{-7}$	Broadly acceptable

*PFD = Probability of Failure on Demand

6 Conclusions

In the UK, a working group composed of participants from industry (oil refiners, tank and terminal operators, chemical and petrochemical sites), the regulators (HSE, EA and SEPA), and the union (UNITE) worked together to define the tolerability criteria for environmental harm.

The tolerability threshold is a function of both the size and effect of the contamination to the environmental receptors and the duration that the harm continues. The resulting guidance has been published on the HSE's website and all COMAH sites in the UK (major chemical, pharmaceutical, manufacturing, tankage storage, petrochemical, whiskey distilleries, refining, etc. sites) are being asked to include environmental risk assessments using the CDOIF methodology in their next submission of their 5-yearly Safety Report.

The CDOIF Environmental Risk Assessment approach provides a somewhat complicated, but very logical and quantifiable, approach to environmental risk assessment and so enables sites to demonstrate that they have reduced the risks to the environment to ALARP.

Had this approach been applied to Freedom Industries of West Virginia, then the conclusion would have been that the environmental risks resulting from the site were intolerable. An equivalent COMAH site in the UK would either be shut down or obligated to implement significant improvements.

7 References

[1] HSE, "The History of HSE [in the UK]," [Online]. Available: http://www.hse.gov.uk/aboutus/timeline/#a1900.

[2] R. C. Browne, "Safety and health at work: The Robens Report," *Brit. J. industr. Med.,* vol. 30, pp. 87-94, 1973.

[3] UK Government, "Health and Safety at Work Act (1974)," 1974. [Online]. Available: http://www.legislation.gov.uk/ukpga/1974/37/contents.

[4] M. Esbester, "The Health and Safety at Work Act, 40 years on," August 01 2014. [Online]. Available: http://www.historyandpolicy.org/opinion-articles/articles/the-health-and-safety-at-work-act-40-years-on. [Accessed 11 January 2015].

[5] D. Leigh, "European Comparisons. Summary of UK Performance," Health and Safety Executive, 2015.

[6] EU, "The Seveso-III Directive," European Commission, 2012. [Online]. Available: http://ec.europa.eu/environment/seveso/legislation.htm.

[7] Health and Safety Executive, "Reducing risks, Protecting People: HSE's decision-making process," 2001. [Online]. Available: http://www.hse.gov.uk/risk/theory/r2p2.htm. [Accessed February 2016].

[8] HSE, "Guidance on ALARP Decisions in COMAH. SPC/Permissioning/37, Version 3," [Online]. Available: http://www.hse.gov.uk/foi/internalops/hid_circs/permissioning/spc_perm_37/. [Accessed Feb 2016].

[9] OGP-456, "Process Safety Recommended Practise for Key Performance Indicators," International Oil and Gas Producers, 2011.

[10] The Major Incident Investigation Board, "The Buncefield Incident - 11 December 2005. The final report of the Major Incident Investigation Board.," Crown Publcation Service, 2008.

[11] Buncefield Standards Task Group, "http://www.hse.gov.uk/comah/buncefield/bstgfinalreport.pdf," [Online].

[12] PSLG, Safety and environmental standards for fuel storage sites, The (UK) Office of Public Sector Information, 2009.

[13] HSE, "Chemical and Downstream Oil Industries Forum," [Online]. Available: http://www.hse.gov.uk/aboutus/meetings/committees/cif/index.htm. [Accessed 2016].

[14] Chemical Downstream Oil Industry Forum (CDOIF), "CDOIF Resources," [Online]. Available: http://www.hse.gov.uk/aboutus/meetings/committees/cif/resources.htm. [Accessed Feb 2016].

[15] The Ramsar Convention Secretariat, Gland, Switzerland, "About the Ramsar Convention," [Online]. Available: http://www.ramsar.org/. [Accessed 2016].

[16] CSB, "Freedom Industries Chemical Release," Chemical Safety Board, Sept 2014. [Online]. Available: http://www.csb.gov/freedom-industries-chemical-release-/. [Accessed 14 January 2016].

[17] A. J. Whelton, L. McMillan, M. Connell, K. M. Kelley, J. P. Gill, K. D. White, R. Gupta, R. Dey and C. Novy, "Residential Tap Water Contamination Following the Freedom Industries Chemical Spill: Perceptions, Water Quality, and Health Impacts," *Environ. Sci. Technol.,* vol. 49, no. 2, pp. 813-823, 2015.

[18] OGP-434-03, "Risk Assessment Data Directory," International Oil and Gas Producers, 2010.

[19] CCPS, Initiating Events and Independent Protection Layers In Layer of Protection Analysis, Whiley, 2015.

[20] EUGRIS, "Glossary: Receptor," EUGRIS (European groundwater and contaminated land remediation information, [Online]. Available: http://www.eugris.info/FurtherDescription.asp?Ca=2&Cy=0&T=Receptor:%20Ecological&e=34.

[21] M. Nicholas, "CDOIF Environmental Risk Tolerability for COMAH establishments," in *New guidance from The Chemical and Downstream Oil Industries Forum (CDOIF) on establishment risk*, Manchester, 2014.

11TH GLOBAL CONGRESS ON PROCESS SAFETY

50th LOSS PREVENTION SYMPOSIUM

The Eye Opener – Preparing for and Performing Facility Siting to Best Communicate with Upper Management

Michael James
Senior Chemical Engineer
Eastman Chemical Company
P.O. Box 511, Building 18
Kingsport, TN 37662
mjames@eastman.com

Prepared for Presentation at
12th Global Congress on Process Safety
Houston, TX
April 10 – 14, 2016

UNPUBLISHED

The Eye Opener – Preparing for and Performing Facility Siting to Best Communicate with Upper Management

Michael James
Senior Chemical Engineer
Eastman Chemical Company
P.O. Box 511, Building18
Kingsport, TN 37662
mjames@eastman.com

Keywords: Facility Siting, dispersion modeling, consequence analysis, risk assessment, dosage, shelter-in-place, temporary haven, evacuation, toxicity, toxic chemical release, toxic load, downwind concentration

Abstract

Upon completion of a facility siting study, a major step in the critical path for implementation is acceptance from upper management on the remediation plan. Communicating with upper management occurs at the end of several steps which can be strategically mapped in order to maximize upper management's understanding of the needs of a site's remediation requirements.

Using an example chemical plant, this paper will detail the process of initiating a facility siting study, provide a novel approach to prioritizing building remediation requirements, and discuss how the resulting information can best be framed and communicated to upper management. In addition to discussing strategies for effective communication with upper management on proposed facility siting strategies, the paper also details an online system that could be used for keeping facility siting data evergreen after the study is completed.

1. Introduction

Facility siting is a requirement based on Occupational Safety and Health Administration (OSHA) Regulation 29 CFR 1910.119 [1]. While this paper will not detail the requirements of a facility siting study, it will provide recommendations for background to include when discussing facility siting projects with upper management.

For the purposes of this paper, the term upper management represents the group of individuals assigned to manage risk at a site or enterprise level. This is the group that will fund remediation efforts based on the siting project findings.

The main focus in communicating facility siting findings is to achieve the following goals:

1. Ensure that the information is specific to each building at risk.
    - Provide a rationale for prioritizing high risk versus lower risk buildings.
    - Provide a prioritized action list, by building, based on this rationale.

2. Ensure that the study accounts for the ability of personnel to safely shelter for a specified event duration as determined by the site.
    - Per the site emergency plan, ensure that individuals can shelter safely for a specified period of time.
    - This will require evaluation of air turnover rates to determine indoor exposure levels.
3. Use a single visual aid to clearly represent the highest risks of the site.
    - Using aerial image of the site overlaid with occupied building locations and a color coding for impact level is recommended.

Background and implementation of these goals will be discussed in detail in this paper.

For reference, a generic plant site image will be overlaid with specific data. Any similarity of the data to actual conditions is purely coincidental. The plant site is shown in Figure 1.

Figure 1. Example Plant Site

2. Before Starting a Facility Siting Study

Keep in mind that the end goal is to have the right data to communicate to upper management. To this end, there are some steps that should be considered prior to starting the study. In chartering a facility siting study, the primary steps should be as follows:

2.1 Define a loose scope

As with any project, having a clear, concise scope is critical. However, at this early stage, flexibility in scope will also be important. The scope will be refined in subsequent steps.

Within a process safety management group, define the following requirements for the study:

- Chemicals to be included.
    - Based on site data of highly hazardous chemical quantities and locations, determine which chemicals are to be included.
    - For toxic chemicals, the "Substance Hazard Index" [2] may aid in determining appropriate chemicals.
    - For flammables, consider using NFPA Hazard Rating System [3] as a tool.
    - Ultimately, process safety management subject matter experts on the site will make the determination of the chemical list.
- Risk-based / consequence-based choice of method.
    - If the results of this facility siting study will be compared against results of studies at other sites, consider using a risk-based approach.
    - Risk-based methods can facilitate the comparison of high-priority building improvements across sites.
    - Consequence-based approaches may be more appropriate for ranking projects within a single geographic location.
- Basis of risk levels.
    - Establish the tolerable levels of risk, in terms of acceptable frequencies of events that could result in increasing numbers of fatalities.
    - The risk criteria for the study are typically based on site or corporate risk tolerance specification.
    - Figure 2 is an example of specified guidelines for allowable levels of risk.

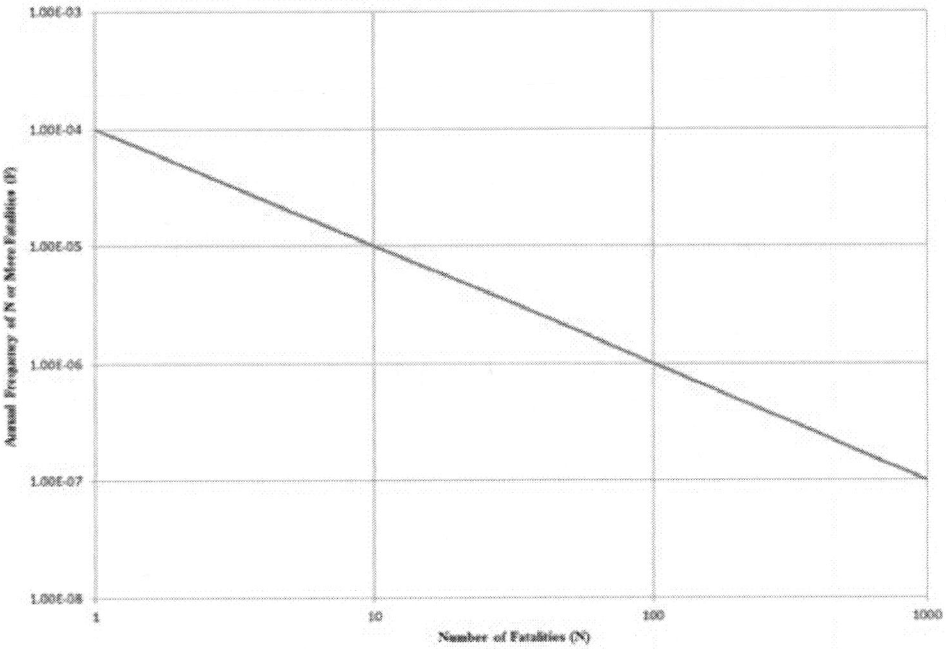

Figure 2. Example Acceptable Societal Fatal Risk Levels

- Specification of safe sheltering duration.
    - For sheltering during a toxic event, document the specified time for which a shelter should be able to safely house occupants.
- Determination of the normally achievable air turnover rates and identification of air turnover rates requiring special evaluation.
    - Understanding the air turnover rate is critical to determine how toxic vapors will infiltrate a shelter [4].
    - The methodology of this study will assess required air turnover rates to provide for safe sheltering during an event.
    - Air turnover rates below a threshold value may require special designs, such as a "shelter-in-a-shelter."
    - Buildings requiring very low air turnover rates for safe sheltering will be considered at high risk.
    - Engineering Toolbox lists air turnover rates for several buildings within the range of 4 - 10 air changes per hour (ACH) [5]. An ACH value of 10 would equate to 10 times the volume of the building flowing through the structure in one hour.

Table 1. Risk Levels Based on Air Turnover Requirements

Risk Level	Air Turnover Rate Minimum (ACH)	Air Turnover Rate Maximum (ACH)
High	N/A	0.1
Medium	0.11	9.99
Low	10	N/A

If, however, a building requires lower air handling rates, it will be considered higher risk. For this study, medium ACH risk will be assigned a value of 1 ACH. High ACH risk will be assigned a value of 0.1 ACH. This does not reflect a real-world analysis, and is only used as an example calculation for this paper.

- High, Medium and Low Levels of Occupancy
    - As another tool for prioritizing results, it is recommended to define occupancy levels as high, medium and low.

Table 2. Sample Occupancy Level Criteria

Risk Level	Occupancy Minimum	Occupancy Maximum
High	31	N/A
Medium	11	30
Low	1	10

- Define the combined Occupancy Level/Air Turnover Rate criteria that will determine how buildings are prioritized.
    - High-Occupancy/High-Risk buildings may be the highest priority for a site.
    - Documenting the rationale for remediation timing based on building risk levels will be critical during the implementation phase.

2.2 Discuss Individual Requirements of Site Divisions

While the process safety management group can define 80 to 90 percent of the scope, certain divisions on-site may have specific needs, including:

- Transient-occupied locations not considered.
    - Are there satellite labs or rooms that were not covered according to the occupancy definition above?
    - Are there locations where contractor groups will amass that were not defined previously?
- Occupancy levels not determined.
    - Division representatives should provide you with a current estimate of personnel for the occupied buildings in the area.

2.3 Formalize scope

Develop a charter document with a scope detailing the information gathered in the previous steps. Additionally, including the following will aid in study refinement and in keeping the study up to date in an evergreen system:

- Sufficient data on the release scenario cases used for each chemical in the study. (Including the scenario information will allow for quick re-creation of scenarios as needed.)
- Basis for determining the frequency of each scenario.
- Weather conditions assumed in the study.
- Aerial image of the site with occupied building locations.
- Table of current building occupancy levels.

3. Conducting the Facility Siting Study - Toxics

This section details a novel approach for assessing risk from toxics on site. This method will allow for prioritization of site risks based on the ability for individuals to safely shelter from a toxic release. It will also allow for grouping of buildings by air turnover risk and occupancy.

In a typical facility siting study, the outdoor chemical concentration is used to determine the severity of release. The consequence of the event will be measured assuming that an individual will remain in the area for the duration of the release.

For a risk-based study, a typical assessment would include risk contours drawn around the release location [6], as shown in Figure 3.

Figure 3. Risk Contours Based on Probit Analysis

The risk contours in Figure 3 represent the frequency of a specific number of fatalities per year, based on distance from the hazard. While this is an agreed-upon approach that is standard in facility siting, the following factors may not considered in some analyses:

- Duration of exposure.
  - Individuals exposed to this release, if able, will flee soon after the start of the incident.
  - This is more pertinent to large, continuous releases.
- Exposure of individuals in occupied buildings.
  - As facility siting is concerned with the safety of personnel inside occupied structures, the ventilation rates and consequence to personnel indoors should be considered.
- Duration of safe sheltering.
  - The site should have a documented duration for each building detailing the time that the building will be habitable as a shelter.
  - This data should be compared to the results of a facility siting analysis to determine if personnel can safely shelter per the documented habitable guidelines.

To account for the above factors, probit analysis can be used in a building-specific analysis. The concentration infiltrating the structure can be used to determine the total toxic dose over time; the toxic dose can then be used to assess occupant safety [2].

Understanding the concentration profile from the studied scenarios for each building is critical. The concentration profile is the outdoor concentration versus time at the distance from the leak source to the building. While this will be a large set of data when considering all scenarios, it will allow for specific analysis of building occupant exposure to toxics. Consider including this as a requirement for deliverables if the project is outsourced. The analysis of this data can be automated for large datasets.

Table 2. Example Concentration Profile

Time (min)	Concentration at Building Exterior (ppm)
0	0
1	130
2	190
3	163
4	283
...	...
28	193
29	180
30	64
31	0

To analyze impact to each building, this paper recommends the following method:

1. Determine the frequency of the scenario impacting the specific building.
2. Based on the frequency of the scenario, determine the maximum tolerable consequence. This can be expressed as a maximum tolerable number of fatalities for this scenario.
3. Determine the ventilation rate into the building being analyzed.
4. Using the methods discussed in "Simplified Methods of Probit Analysis in Consequence Analysis" [2], develop a dose-response profile of impact over time, expressed as the probability of fatality.

Table 3. Dose-Response Analysis for Building Occupants

Time (min)	Outdoor Conc (ppm)	Indoor Conc (ppm)	Dosage	% Probability
-	-	-		#NUM!
1.00	53.00	8.83	39.01	0.0%
2.00	134.00	29.69	518.91	0.0%
3.00	241.00	64.91	3,066.57	0.0%
4.00	202.00	87.76	9,024.27	0.0%
10.00	511.00	318.05	277,644.36	3.9%
11.00	628.00	369.71	396,563.79	7.6%
12.00	656.00	417.42	552,026.58	13.0%
13.00	658.00	457.52	743,809.42	19.7%
14.00	780.00	511.27	979,167.72	27.46%
15.00	841.00	566.22	1,270,167.61	36.0%
16.00	848.00	613.18	1,618,468.80	44.6%
17.00	706.00	628.65	2,004,069.39	52.4%
18.00	705.00	641.38	2,407,355.28	59.0%
19.00	688.00	649.15	2,823,735.22	64.6%
20.00	554.00	633.29	3,234,960.53	69.2%
21.00	501.00	611.24	3,622,297.35	72.7%
22.00	452.00	584.70	3,980,043.81	75.5%
23.00	476.00	566.58	4,311,490.99	77.8%
24.00	355.00	531.32	4,613,150.93	79.6%
25.00	319.00	495.93	4,877,276.93	81.0%
26.00	277.00	459.44	5,105,796.92	82.1%
27.00	241.00	423.04	5,300,822.04	83.0%
28.00	223.00	389.70	5,466,234.56	83.7%
29.00	156.00	350.75	5,603,678.91	84.3%
30.00	118.00	311.96	5,713,849.59	84.7%
31.00	-	259.96	5,796,298.77	85.0%

Chemical	Chlorine
Exponent	2.00
Probit_ConstA	(8.29)
Probit_ConstB	0.92
Initial_ACH	10.00
Long_Term_ACH	10.00

- Table 3 shows the outdoor and indoor concentrations as well as an estimate of the probability of fatality to occupants.
  - The concentration is integrated over the period of exposure to calculate dosage.
  - The probability of fatal exposure is calculated using a probit transformation.
- Per the facility siting guidelines of our example study, this shelter has been specified to allow for 30 minutes of habitable occupancy.
- The "Initial_ACH" and "Long_Term_ACH" will be modified to ensure that the probability of fatality over 30 minutes of exposure is within tolerable site guidelines.
  - The initial ACH value represents turnover when the air handlers are active.
  - The long-term value represents turnover when the air handlers are turned off.
  - Older buildings and those with loose seals may have higher long-term air turnover rates.

5. Determine tolerable probability based on occupancy and risk level.

   - The tolerable probability of fatality is calculated as follows:

$$Probability = \frac{Tolerable\ Exposures\ per\ Risk\ Guidelines}{Building\ Occupancy}\ x\ 100\%$$

6. Vary "Initial ACH" and "Long Term ACH" to match the target probability at the duration of habitable occupancy.
   - A conservative approach is to assume that the air handlers remain active throughout the incident. Initial and long-term ACH values will then be the same.
   - Lower ACH values will result in lower indoor concentrations, which will in turn result in lower probability values.
   - MS Excel "Goal Seek" is a reasonable tool for performing this iterative analysis.

Table 4. Optimized ACH Value to Meet Target Probability

Time (min)	Outdoor Conc (ppm)	Indoor Conc (ppm)	Dosage	% Probability
-	-	-		#NUM!
1.00	123.00	1.03	0.53	0.0%
2.00	132.00	2.12	3.29	0.0%
3.00	234.00	4.05	13.73	0.0%
4.00	211.00	5.77	38.59	0.0%
10.00	525.00	25.76	1,515.72	0.0%
11.00	566.00	30.26	2,305.28	0.0%
12.00	677.00	35.65	3,398.55	0.0%
13.00	660.00	40.85	4,868.45	0.0%
14.00	772.00	46.95	6,804.83	0.00%
15.00	795.00	53.18	9,320.76	0.0%
16.00	825.00	59.61	12,511.48	0.0%
17.00	775.00	65.57	16,438.08	0.0%
18.00	679.00	70.68	21,086.08	0.0%
19.00	615.00	75.22	26,413.26	0.0%
20.00	631.00	79.85	32,430.46	0.0%
21.00	517.00	83.49	39,104.29	0.0%
22.00	528.00	87.20	46,391.79	0.0%
23.00	445.00	90.18	54,259.88	0.1%
24.00	407.00	92.82	62,633.99	0.1%
25.00	390.00	95.30	71,482.61	0.1%
26.00	314.00	97.12	80,739.51	0.2%
27.00	264.00	98.51	90,307.78	0.3%
28.00	246.00	99.74	100,133.91	0.3%
29.00	101.00	99.75	110,082.92	0.5%
30.00	148.00	100.15	120,073.17	0.6%
31.00	-	99.32	130,020.37	0.7%

Chemical	Chlorine
Exponent	2.00
Probit_ConstA	(8.29)
Probit_ConstB	0.92
Initial_ACH	0.50
Long_Term_ACH	0.50

   - Setting an example target of 1% probability of fatality, an ACH of 0.5 will be required per Table 4. A value greater than 0.5 ACH will result in probability of fatality above 1%.

- Per Table 4, the concentration outside of the building is negligible until one minute after the incident starts. Therefore, the duration of safe sheltering for this study will be from minute 1 to minute 31. This will fit the study's standard sheltering time of 30 minutes.

7. Based on the required ACH value and the guidelines in the charter document, determine the air turnover risk level of the building. (Refer to Table 1.) ACH values in the high-risk category may require more immediate attention than others.

Table 1. Risk Levels Based on Air Turnover Requirements (Repeated)

Risk Level	Air Turnover Rate Minimum (ACH)	Air Turnover Rate Maximum (ACH)
High	N/A	0.1
Medium	0.11	9.99
Low	10	N/A

- In the Table 4 example, the ACH of 0.5 would fall into the medium risk category.

8. Determine the occupancy category per the occupancy level guidelines in the charter document.

Table 2. Sample Occupancy Level Criteria (Repeated)

Risk Level	Occupancy Minimum	Occupancy Maximum
High	31	N/A
Medium	11	30
Low	1	10

- In this example, the personnel count of this building is 15. This would be classified as medium occupancy.

Repeat the process for all analyzed buildings and scenarios.

4. After the Study – Communicating Results and Plan to Upper Management

Once the study has been completed and the results digested by the process safety group, there are some key points that should be highlighted when presenting to upper management.

4.1 Visual aid for summarizing site issues

One of the initial goals defined in this paper was to develop a single visual tool that can be used to show the level of risk at a given plant site.

While risk contours are typically used, consider using an aerial image of the plant with color coded identifiers for risk levels of each building. In order to develop these, first assess the grouping of buildings by occupancy level and air turnover risk.

Table 5. Breakdown of Building Counts in Different Risk Categories

Air Turnover Risk	Occupancy		
	Low	Medium	High
Low	9	6	4
Medium	4	4	2
High	2	2	1

In this example, one building was in both the high air turnover risk and high occupancy categories. This would be considered to be the highest risk building on-site. Prioritization of other buildings would proceed as per the charter document.

When similar studies are performed at various plant sites within a corporation, prioritization of planned remediation projects across sites is greatly simplified by the use of this common methodology.

For a concise visual aid, the color scheme in Table 6 will be applied to the building location pins on the site aerial.

Table 6. Color Scheme for Building Pin Overlay

Air Turnover Risk	Occupancy	Building Pin Color
High	High	Pink
High	Medium	Red
High	Low	Red
Medium	High	Red
Medium	Medium	Yellow
Medium	Medium	Yellow
Medium	Low	Yellow
Low	Any	Not Shown

The high risk building is colored pink to highlight its severity. Other high risk buildings are marked with a red pin.

Buildings in the low risk category for air turnover will not be included on the aerial overlay. These buildings have been shown to be within the tolerable risk level at a reasonable air turnover estimate.

Figure 4. Aerial View of Risk Categorization

This aerial view makes immediately evident the highest risk building. It is a very accessible graphic and straightforward to create based on this method of analysis.

4.2 Historical Facility Siting Fines

Upper management should be given a description of the regulations on facility siting, and a review of previously levied fines for facility siting deficiencies will help in driving the importance of completing remediation work in a timely manner. To make the presentation as relevant as possible, the presented fines should be related to chemicals that are also at the site being studied.

While the fines may not need to be directly from a facility siting, they should be along the lines of potential or historical occupant exposures.

Table 7. Historical OSHA Actions

Company	Location	Violation	Fine Amount
BP	Texas City, TX	409 citations resulting from investigation after 2005 explosion.	\$21 Million
Lawter International	Moundville, AL	"...employees were being exposed to BF3 releases..."	\$263,500
Loveland Products	Fairbury, NE	"...failed to address facility siting. This would include a review of the locations of each building and how they could be potentially affected by the ammonia process..."	\$7,700
Arctic Glacier	Fairport, NY	"The Process Hazard Analysis did not address Facility Siting"	\$5,500

These fines will convey to management the severe financial consequence of not protecting workers from chemical hazards.

4.3 Providing Upper Management Recommendations for Implementation

If the recommended facility siting method discussed in this paper is employed, it will be straightforward to detail the recommended actions. A timeline of proposed activities and the leveling of priority across multiple sites can be presented.

5. Maintenance of Facility Siting Data

A vast amount of data will be generated in the study. It will be critical to have a tool for data storage and analysis.

Facility siting studies should be performed on a set frequency so that, as the site changes, personnel safety is maintained to a consistent level. In between facility siting revalidations, there is a need to determine the impact on personnel risk as a result of building occupancy changes, addition of new chemicals, removal of highly hazardous chemicals, or a change in source concentration.

Consider development of a controlled access, web-based tool for the process management group. This can be performed in-house by a data analytics group or outsourced. This tool should maintain all scenario data, chemicals, building occupancy, frequency guidelines, and risk level thresholds. "Scenario data" is referencing the concentration profiles specific to all buildings as impacted from each scenario.

This tool should have the following features:

Inputs

- Allow users to update any databases with current information.
- A standard upload of scenario data that will parse and store the data for analysis.
- Data editing tool for on-the-fly comparison of occupancy and chemical addition/removal scenarios. This data editing should not overwrite the existing data.

Analysis

- An optimizing routine that will assess required air turnover rates and maintain this data.
- For blast and thermal concerns – a system that will track the number of potential fatalities versus the study's tolerable risk.

Output

- Reporting of buildings by risk and air turnover category.
- Visual overlay of buildings on aerial, color coded per Figure 4.
- "Gap" calculation for blast and thermal exposures.

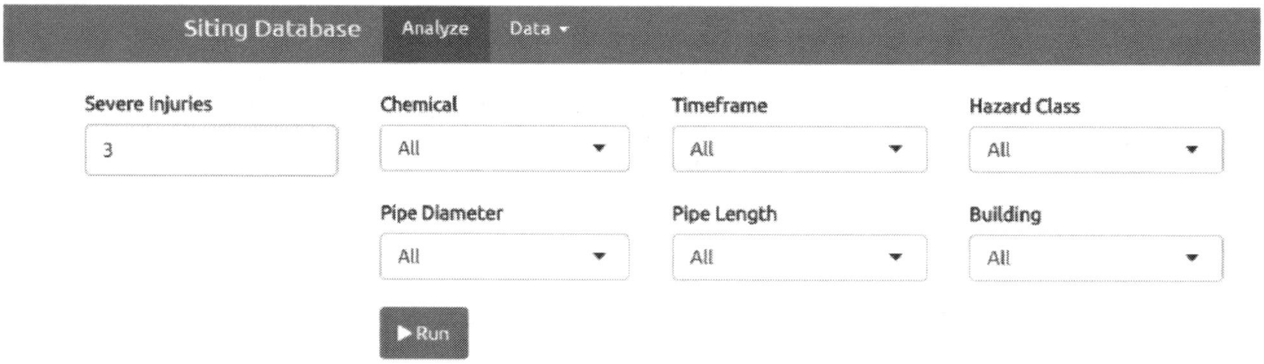

Figure 5. Evergreen Facility Siting Tool

Use of such a tool will allow for concise revalidation of facility siting studies without the investment of commissioning an entire new study. In addition, it will allow for on-the-fly analysis that would otherwise take a significant investment in time.

6. Conclusion

This paper details a method for planning and conducting a facility siting evaluation based on an example site. The output data from the siting analysis was used to visually represent to upper management the relative risks to various buildings on site. This paper then provided guidance on educating management on facility siting regulations and reviewing historical fines levied against companies for not properly managing facility siting. Following these guidelines will allow for straightforward acceptance of facility siting results from upper management.

Finally, the paper describes potential implementation of an evergreen system for data storage and analysis. This system allows for easy revision of site risk analysis in response to changes in building occupancy, and process changes.

References

[1] "Process safety management of highly hazardous chemicals," Feb. 8, 2013, https://www.osha.gov/pls/oshaweb/owadisp.show_document?p_table=STANDARDS&p_id=9760 (Accessed Mar 3, 2016).

[2] "Substance Hazard Index," 2016. http://www.iadclexicon.org/substance-hazard-index (Accessed Mar 9, 2016).

[3] "NFPA HAZARD RATING SYSTEM," 2016. http://www.ehs.neu.edu/laboratory_safety/general_information/nfpa_hazard_rating (Accessed Mar 9, 2016).

[4] James, M., "Simplified Methods of Using Probit Analysis in Consequence Analysis," *Process Safety Progress*. First Published Online June 3, 2014, DOI: 10.1002/prs.11686.

[5] "Air Change Rates for typical Rooms and Buildings," http://www.engineeringtoolbox.com/air-change-rate-room-d_867.html (Accessed Mar 9, 2016).

[6] "The Significance of Hazard Endpoints in Quantitative Risk Analysis," 2016, http://www.questconsult.com/papers/hazard-endpoints-qra (Accessed Feb 11, 2016).

441807 Update on OSHA PSM RAGAGEP Guidance

Abstract:

Abstract

RAGAGEP in PSM Enforcement (RAGAGEP 201)

James R. Lay, PE

Lisa A. Long

Mike L. Marshall, PE

Jeffrey J. Wanko, PE

In 2015 OSHA issued important guidance to field personnel on RAGAGEP enforcement of 29 CFR 1910.119, Process Safety Management of Highly Hazardous Chemicals. In this session, OSHA National Office engineers will address key aspects of these guidance documents with practical examples.

441463 PSM/RMP Modernization Programs in California - New Developments and Correlation to Evolution at the Federal Level

Abstract:

Paper 396953, presented at the 2015 Global Congress on Process Safety, provided a snapshot of "Changes in the PSM/RMP Regulatory Framework (California and National)". Post-publication, additional significant changes have been proposed by the agencies involved in the previous initiatives. On May 26, 2015, Cal/OSHA issued a comprehensive update to its proposed regulation for "Process Safety Management for Refineries," and CalEPA/CalOES issued an update to the California Accidental Release Prevention (CalARP) Program, creating a new program level "Program 4" that applies only to petroleum refineries. Both of these regulations coalesced as part of an overall initiative resulting from the efforts of the California Interagency Refinery Task Force. Subsequently, on September 24, Cal/OSHA made some minor changes and released its updated "PSM for Refineries" for public comment.

The final language for both of these new California petroleum refinery requirements is anticipated by December 2015, and promulgation is anticipated for Summer 2016. In parallel, Federal OSHA and U.S. EPA are creating updates to Process Safety Management (PSM) and Risk Management Plan (RMP) requirements as part of the PSM/RMP Modernization/Expansion Programs designed to address significant accidents that occurred in the past few years.

The changes being proposed for these programs are extensive, currently being formulated, and are not synchronized between the various agencies. The focus of this paper will be to provide relevant background information, a summary of the proposed changes, and tips for addressing key technical and regulatory challenges, as well as providing the reader with a "roadmap" for effective implementation.

The timing of GCPS-2016 is ideal for unveiling the final regulatory requirements in California and for identifying strategies for addressing these desynchronized requirements. The content of this paper will be important to communicate to GCPS attendees in the April 2016 time frame, and key topics include:

- Background – Pre-2012 PSM/RMP Universe
- Overview of Key Safety Management System Program Modernization Initiatives
- Key Elements of the Various SMS Regulatory Initiatives
- SMS Modernization Status and Potential SMS Program Impact
- What Should I Be Doing Now?

441338 PSM: Keeping Abreast of Legal Developments in Order to Stay Compliant

Abstract:

Compliance with PSM is an ongoing challenge. To stay ahead of the curve, industry needs to be aware of how the PSM Standard is being enforced by OSHA and interpreted by the Courts. Attendees will learn about the most recent legal developments involving the PSM Standard and where the next compliance challenges will be.

442269 Combustible Dust Hazard Assessment: A Risk Based Approach

Abstract:

A CCPS Concept Book on performing Combustible Dust Hazard Assessments (CDHA) will be published later this year. This book is not meant to duplicate existing resources, such as NFPA 68, 69, or 654, but rather is designed to be a practical guide on how to do CDHAs, with an emphasis on the combustion hazards of dusts. Two methods are described for doing a CDHA: 1) a traditional, standards compliance based approach, and 2) a risk based approach. The heart of the book is the application of both hazard assessment approaches to three example processes. The book also provides an overview of the basics of combustible dust fires and explosions, although it is not meant to be a thorough treatment of the science of combustible dust hazards. A list of standards and books for further study are provided. The hazards inside selected process equipment items and inside of buildings are described. Methods to prevent and mitigate these hazards are also described. There are special considerations that occur in identifying and assessing hazards in existing facilities, such as existing housekeeping and ignition control, process changes since the initial startup of the facility, and changes in what is considered to be good engineering practice since the facility was built. One chapter of the book is dedicated to the impact of these items in a new facility.

The target audience for this book is primarily technical personnel involved in process research and development, and in designing and running solids handling plants and processes. Combustible dust and hazard analysis subject matter experts will find the book useful as a guideline to develop a consistent methodology for risk assessment of processes handling combustible dusts.

12TH GLOBAL CONGRESS ON PROCESS SAFETY

LPS 50th LOSS PREVENTION SYMPOSIUM

Who's Afraid of Combustible Dust? – An Update on Hazardous Area Classification

Samuel A. Rodgers
Honeywell
15801 Woods Edge Road, Colonial Heights, VA 23834
Samuel.Rodgers@honeywell.com
Tel 804-520-3706

Prepared for Presentation at
American Institute of Chemical Engineers
2016 Spring Meeting
12th Global Congress on Process Safety
Houston, Texas
April 11-23, 2016

Who's Afraid of Combustible Dust? – An Update on Hazardous Area Classification

Samuel A. Rodgers
Honeywell
15801 Woods Edge Road, Colonial Heights, VA 23834
Samuel.Rodgers@honeywell.com

Keywords: hazardous area classification, electrical classification, combustible dust, fibers, flyings

Abstract

NFPA 70 (NEC) has included the Zone system for hazardous area classification since 2005, however the major NFPA dust standards have purposefully discouraged its use due to perceived inconsistencies with the traditional Division system. ANSI/ISA-61241-10 has been the US standard for implementation of the Zone system, superseded by ANSI/ISA 60079-10-2 which is consistent with the IEC 60079-10-2. Recently NFPA 499 has included guidance for hazardous area classification using the Zone system, as well as clarifications necessary to unify the application of the two approaches. The traditional Division system is compared to the Zone system, including the various U.S. and IEC definitions of combustible dust and dust groups, and implications for users. NFPA 652 and 654 bring another variation to the combustible dust definition to address hazards inside process equipment.

1 Introduction

NFPA 70 *National Electrical Code* [1] specified hazardous area classification for combustible dusts, as well as ignitable fibers and flyings, according to the Division system until 2005. Combustible dusts are referred to as Class II materials. Ignitible fibers and flyings, such as nylon or cotton filaments, are referred to as Class III materials.

NFPA-499 *Recommended Practice for the Classification of Combustible Dusts and of Hazardous (Classified) Locations for Electrical Installations in Chemical Process Areas* [2] has provided guidance for Class II materials under the Division system since 1991, but no separate guidance has existed for Class III materials that present a similar hazard.

With the 2005 edition of NFPA 70 an alternative Zone system was introduced for dusts. Either system, Divisions or Zones, is permitted to be used to determine whether a location is considered hazardous (classified) for the installation of electrical equipment. The 2005 and subsequent editions of NFPA 70 reference ANSI/ISA-61241-10 (12.10.05)-2004, *Electrical Apparatus for Use in Zone 20, Zone 21, and Zone 22 Hazardous (Classified) Locations —Classification of Zone 20, Zone 21, and Zone 22 Hazardous (Classified) Locations* [3]. This document was superseded by ANSI/ISA-60079-10-2 (12.10.05)-2013 *Explosive Atmospheres – Part 10-2: Classification of areas - Combustible dust atmospheres* [4], which is the U.S deviation document for IEC-60079-10-2 *Explosive atmospheres – Part 10-2: Classification of areas - Explosive dust atmospheres* [5] and is anticipated to be updated consistent with the 2015 edition of the same.

Inconsistencies between the Division and ISA/IEC Zone systems, specifically with regard to Group placement of metal particulate, fibers and flyings as well as the treatment of dust layer accumulations on electrical equipment, led a number of NFPA dust-related occupancy standards to discourage use of the Zone system by referencing compliance with NEC Article 500.5(C) and (D).

Until recently NFPA-499 did not include guidance for classification according to the alternative Zone system. The committee proposed and balloted second draft of NFPA 499, to issue as the Annual 2016 edition, includes guidance for both Division Class III materials presenting an explosion hazard and Zone Group III materials in general. The similarities and differences between IEC 60079-10-2:2015 and NFPA-499 and the related definitions of combustible dust are the main topics of this paper.

2 Hazards of Combustible Particulate Solids

2.1 Combustible Dust Definitions

What exactly is a combustible dust and what potential hazards does it pose? Much attention has been given to whether or not a maximum particle size should be included in the definition of a combustible dust. Another variation in the combustible dust definitions is the oxidizing medium, with the electrical documents specifying only air and the occupancy standards including other process-specific oxidants. As late as the 2000 edition of NFPA 654 *Standard for the Prevention of Fire and Dust Explosions from the Manufacturing, Processing, and Handling of Combustible Particulate Solids* [6] a nominal size of 420 microns or 40 mesh was specified as an upper bound to combustible dust. This was in addition to terminology reflecting the potential hazards of fire and explosion presented by the materials.

> **Combustible dust:** Any finely divided solid material that is 420 microns or smaller in diameter (material passing a U.S. No. 40 Standard Sieve) and presents a fire or explosion hazard when dispersed and ignited in air. [NFPA 654:2000]

As late as the 2004 edition, NFPA 499 employed essentially the same definition with the exception that oxidizing medium is not mentioned.

> **Combustible dust:** Any finely divided solid material 420 microns or less in diameter (i.e., material passing a U.S. No. 40 Standard Sieve) that presents a fire or explosion hazard when dispersed. [NFPA 499:2004]

Beginning with the 2006 edition, NFPA 654 removed the particle size boundary from the definition, leaving the phenomenological description of a material presenting fire and explosion hazards. This was due to a recognition of test results showing particles exhibiting the explosion hazard from rapid propagation of a deflagration when suspended in a cloud – even though the particles were significantly larger than the arbitrary size limit in major dimension. Additionally NFPA 654 broadened the oxidizing medium to include not only air but also an oxidizing medium that is relevant to the process. While the dust test methods discussed below apply specifically to air as the oxidizing medium, process equipment can operate at elevated oxygen concentrations or higher temperatures or pressures. A material determined to be "not explosible" in air could be "explosible" inside the process equipment. Although standard test methods and electrical equipment identification are based on near ambient air as the oxidizing medium, the occupancy standards warn the user that deflagration hazards can exist within process equipment and electrical equipment design should consider this potential.

> **Combustible dust:** A combustible particulate solid that presents a fire or deflagration hazard when suspended in air or some other oxidizing medium over a range of concentration, regardless of particle size or shape. [NFPA 654:2006]

The upper bound particle size remained in a non-mandatory annex to the 2006 NFPA 654 definition indicating a typical value, as opposed to an absolute limit. This typical upper bound was modified to coincide with the European value of 500 microns. Later the 2013 edition of NFPA 654 modified the definition to replace the broad fire hazard with the more specific hazard of flash-fire, taken from the 2011 edition of NFPA 921 *Guide for Fire and Explosion Investigations*.[7] This NFPA 921 flash-fire definition specifies that the fuel is diffuse or, in other words, suspended in a cloud.

> **Flash-fire:** A fire that spreads by means of a flame front rapidly through a diffuse fuel, such as dust, gas, or the vapors of an ignitible liquid, without the production of damaging pressure. [NFPA 921:2011]

> **Combustible dust:** A finely divided combustible particulate solid that presents a flash fire hazard or explosion hazard when suspended in air or the process-specific oxidizing medium over a range of concentrations. [NFPA 654:2013]

Historically, NFPA 70 has addressed combustible dusts and defined dust groups. However it had not defined combustible dust, per se; or even extracted a definition from NFPA 499; instead NFPA 499 informally provided the definition. Then the 2011 edition of NFPA 70 formally included the 2008 NFPA 499 definition in both Article 500 and Article 506. As a result the electrical and occupancy standards formally disagreed on the definition of combustible dust in both particle size and oxidizing medium.

Combustible Dust. Any finely divided solid material that is 420 microns (0.017 in.) or smaller in diameter (material passing a U.S. No. 40 Standard Sieve) and presents a fire or explosion hazard when dispersed and ignited in air. [NFPA 499:2008]

NFPA 499 in the 2013 edition moved toward the NFPA 654 definition, removing the upper bound on particle size and relying solely on the potential hazards. This would have unified the definitions of combustible dust for electrical and occupancy standards with regard to particle size.

Combustible Dust. Finely divided solid particles that present a dust fire or dust explosion hazard when dispersed and ignited in air. [NFPA 499:2013]

However, the 2014 edition of NFPA 70 did not continue to extract the definition from NFPA 499, instead maintaining an upper particle size and increasing it to 500 micron.

Combustible Dust. Dust particles that are 500 microns or smaller (material passing a U.S. No. 35 Standard Sieve as defined in ASTM E 11-09, *Standard Specification for Wire Cloth and Sieves for Testing Purposes*) and present a fire or explosion hazard when dispersed and ignited in air. [NFPA 70:2014]

2.2 Particle Size Effect on Hazards

Both the flash-fire and explosion hazards are the result of rapid propagation of combustion through a suspended dust cloud. If combustion propagates through a small volume suspended cloud in a relatively larger volume enclosure, a rapidly expanding flame front will envelope equipment and personnel in its path. As the dust burns, the original volume of the cloud will expand on the order of 10 times. If the original suspended cloud instead fills more than a small fraction of the surrounding enclosure before ignition, pressure will concurrently increase and potentially rupture the enclosure. The difference between these two hazards is the degree of containment. The explosion pentagon in Figure 1 illustrates the five criteria for a combustible dust explosion: fuel, oxidant, suspension, ignition and containment. Remove only the containment and the result is a flash-fire.

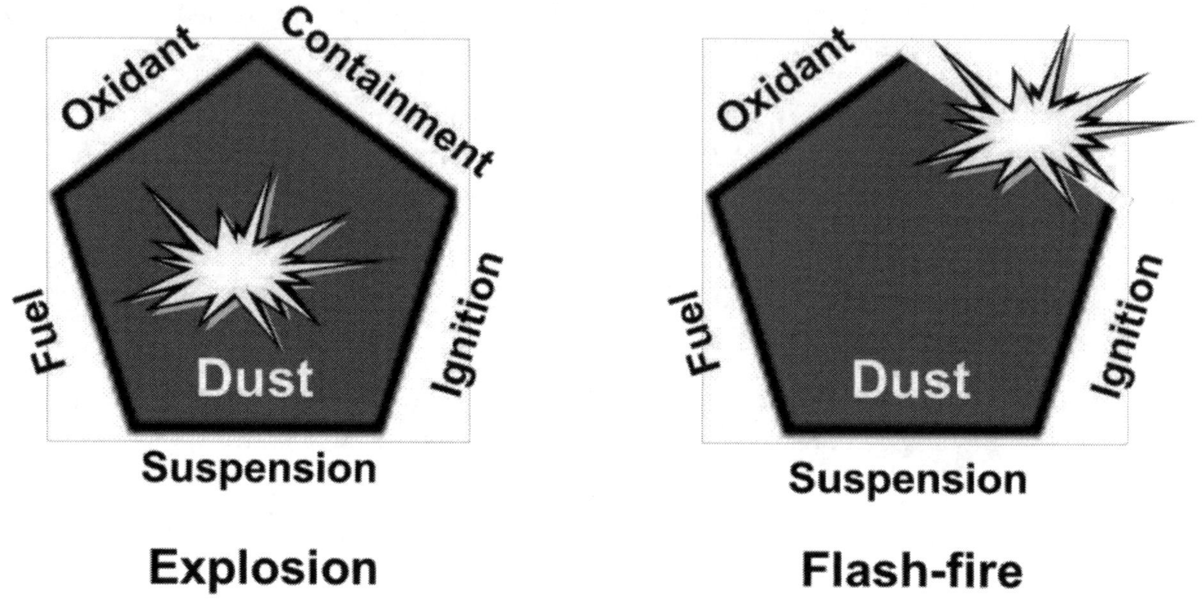

Figure 1. Explosion vs Flash-fire

Not all combustible particulate solids, fibers or flyings will burn fast enough to propagate combustion in a suspended cloud, but some will. The 2016 edition of NFPA 652 *Standard on the Fundamentals of Combustible Dust* [8] and related occupancy standards generally call these materials combustible dust, regardless of particle size.

For the entire range of definitions, notwithstanding the particle size, classification as a combustible dust relies on the results of tests such as ASTM E-1226 [9] and ISO 6184-1 [10] that measure the maximum pressure and rate of pressure rise in an unvented enclosure generated by ignition of a cloud of the particulate suspended in air. Both tests use an essentially spherical test chamber with a dust injection nozzle and ignitor. While the basis test volume is 1 m^3, testing is often done in a smaller 20 L chamber that is calibrated to give approximately the same results. The physical design of the dust dispersion nozzle sets an experimental limit on the nominal particle size that can be dispersed, but this is significantly higher than the 420 or 500 micron limits. These tests have set an arbitrary lower limit of achieved maximum pressure as the threshold for determining whether the dust presents a dust explosion hazard. If the ratio of maximum to initial pressure in absolute units exceeds 2, when corrected for the effect of the ignitor, then the sample is determined to be “explosible”. This is intended to ensure that the test reflects the ability of the suspended material to propagate combustion beyond the point of ignition. Neither of these test standards prescribe an a-priori particle size limit on the ability to propagate combustion.

Focusing on the potential hazards, it is clear that smaller particle size increases the likelihood that a solid material could be suspended in a cloud. It is also intuitively obvious that the higher surface area per unit mass of smaller particles increases the likelihood that a solid material will burn quickly enough to propagate combustion when suspended in a cloud. The maximum rate of pressure rise for a solid particulate at its optimum concentration in an unvented enclosure, normalized to a 1 m^3 volume, is expressed as K_{St} in bar-m/sec.

$$K_{St} = \left(\frac{dP}{dt}\right)_{max} \cdot V^{1/3}$$

A larger value of K_{St} means the particulate is burning faster. Figure 2 from NFPA 68 *Standard on Explosion Protection by Deflagration Venting* [11] shows the typical behavior of K_{St} and P_{max} increasing as particle size is decreased.

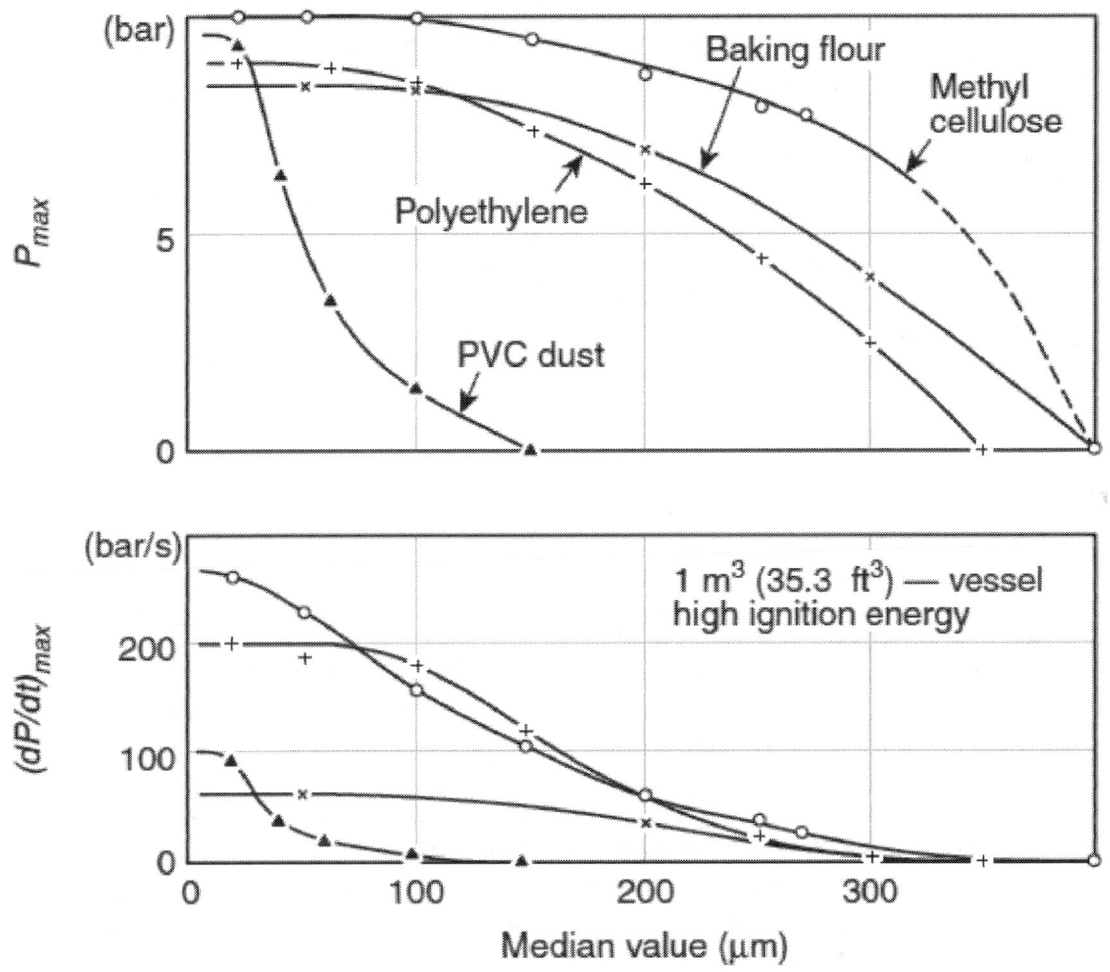

Figure 2. Effect of Particle Size of Dusts on the Maximum Pressure and Maximum Rate of Pressure Rise [NFPA 68 Figure B.2.4.1]

What is not clear is an intrinsic limit, no matter the chemical composition or shape of the material, where the particle is too large to be suspended in a cloud or too large to burn quickly enough to propagate combustion when suspended in a cloud.

Figure 2 shows that rate of combustion, as measured by K_{St}, generally increases as particle size decreases. Table 1 compares the dimensions of spherical, cylindrical and circular flake particulates with the same surface area per unit volume (or per unit mass). The ratio of the largest to smallest dimension of a non-spherical particulate is its aspect ratio. Cashdollar provided a benchmark that a cylindrical shape with aspect ratio larger than 3:1 could be considered a fiber. A 69 micron diameter filament, 1000 microns long, has a surface to volume equivalent to a 400 micron spherical particle, as does a 1000 micron circular flake about 180 microns thick. The 69 micron diameter particulate has an aspect ratio of 14.5:1. 69 microns is not an unusually small diameter for a single polymeric filament in a fiber bundle. In fact, nylon 6 and 6-6 are sold in the range of 10-30 micron filament diameter, while Spectra® polyethylene is sold in the range of 20-40 micron filament diameter. While median diameters for metal flake are generally less than 500 micron, some larger particles are used. Thicker metal flake used in paint formulations can have aspect ratios of 10 or 20:1 and thus thickness of 50 to 100 microns for a 1000 micron diameter flake, but higher aspect ratios of 30 to 100:1 are also typical.

Table 1. Surface/Volume Ratio of Particulates

Surface/Volume, cm^2/cm^3	**Sphere Diameter, μm**	**1000 μm long Cylinder Diameter, μm**	**1000 μm diameter Circular Flake Thickness, μm**
1200	50	8	17
600	100	17	36
300	200	34	77
150	400	69	182

Table 1 shows that non-spherical solid particulate with larger than 500 micron major dimension can have a surface to volume (surface to mass) ratio greater than that for a 200 micron diameter sphere – so it is reasonable to expect that such non-spherical materials, even though they exceed the arbitrary 500 micron limit, could present a similar hazard.

The single particle size criterion, be it 420 or 500 microns, fails to address not only particle shape, but also particle size distribution and surface morphology. Longer fibrous materials, termed fibers/flyings, are known to propagate combustion in a suspended cloud. Ural, et al [11] and Dastidar [12] have presented data verifying explosibility of mixtures of what they term "wood flour" and "wood dust" as well as mixtures of agricultural dusts. Their wood flour was material with mean particle size of approximately 60 micron and their much coarser wood dust was screened to between 500 and 850 micron particle size. Figure 3 from reference [12] shows a typical coarse wood dust particle that could pass through the screens, since its smaller dimension is less than 850 micron. Figures 4 and 5, redrawn from reference [12], show their data for 100% larger particle size wood dust as well as mixtures, verifying that it clearly should be viewed as presenting a flash-fire or explosion hazard. In both figures, data are shown for tests in the 20 L sphere (blue) and 1 m^3 sphere (pink). It is known that marginally combustible carbonaceous dusts can test positive in the 20 L sphere and fail to reach the pressure threshold in the 1 m^3

sphere [13]. The positive test in the 1 m^3 sphere verifies the sample will propagate combustion when suspended in a cloud even though it clearly exceeds the arbitrary 500 micron limit. Further, Figure 5 shows that the common situation, where the particle size distribution includes fractions of smaller particulate, can dramatically change explosibility as evidenced by the rapid increase in K_{St} as the fines fraction increases. To hold that materials with median particle size larger than 500 microns cannot present a flash-fire or explosion hazard, simply by virtue of their size, is not sustainable.

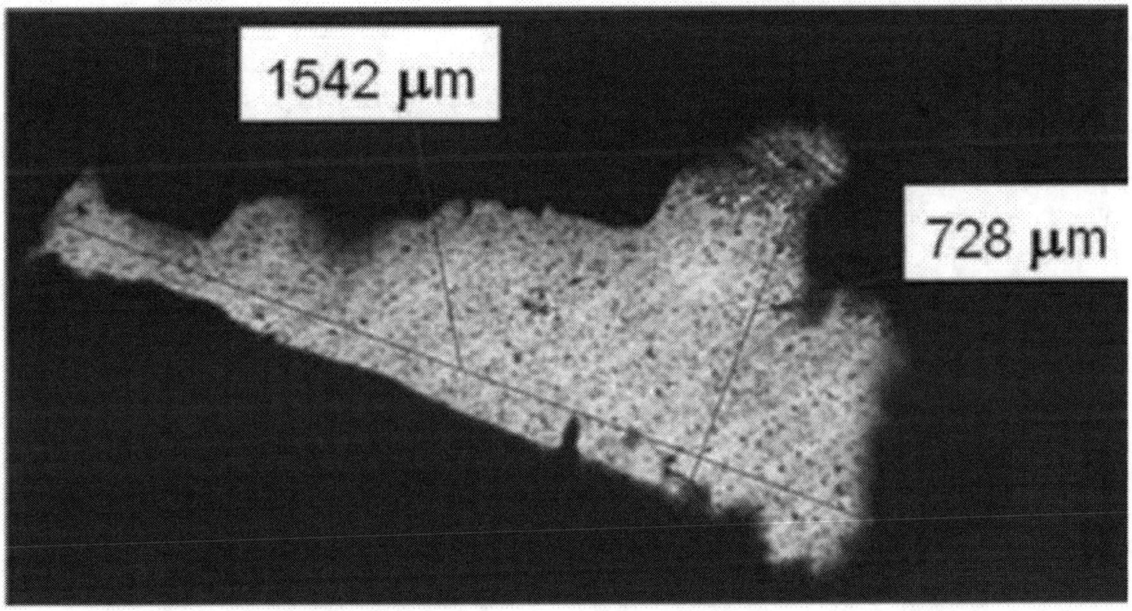

Figure 3. Micrograph of a Wood Dust Particle [12]

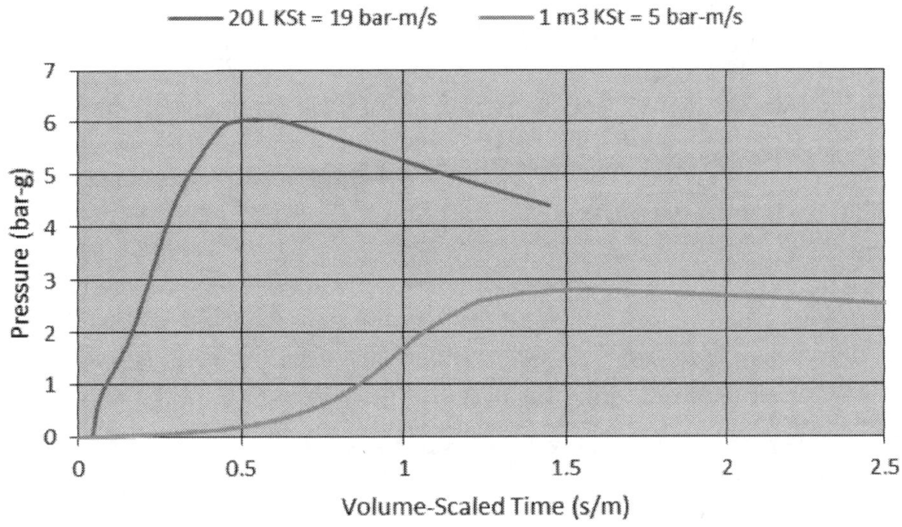

Figure 4. Pressure-Time Traces of 100% Wood Dust Deflagration Tests [12]

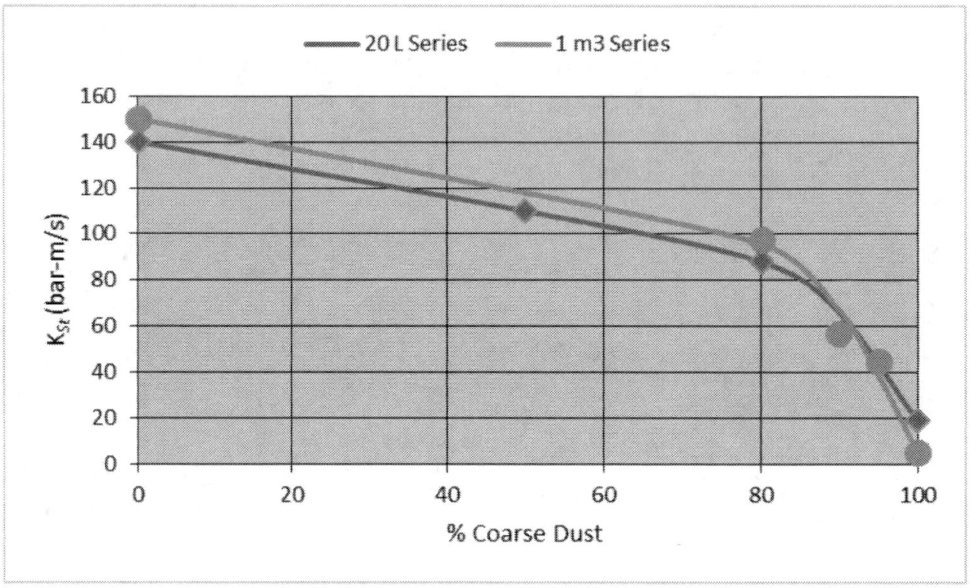

Figure 5. Effect of Added Fines on K_{St} [12]

On the other hand, the argument is often made that we do not want to be advocating the testing of golf balls or lumber. If a material can be ignited and smolder or burn in a pile, then if a handful of that material thrown into the air also creates a lingering cloud, it is likely that the material would present a flash-fire or explosion hazard. The aforementioned is not a recommended test method, but a thoughtful way to determine whether testing is warranted. It is important to remember another rule of thumb that the smallest 5-10% of the particle size distribution can dominate the explosibility parameters of a material.

2.3 Division Locations

Under the Division System, NFPA 70 historically defines two Classes of hazardous locations for combustible particulate, Class II and Class III.

> **Class II Locations.** Class II locations are those that are hazardous because of the presence of combustible dust. [70:500.5(C)]
>
> **Class III Locations.** Class III locations are those that are hazardous because of the presence of easily ignitible fibers or flyings, but in which such fibers or flyings are not likely to be in suspension in the air in quantities sufficient to produce ignitible mixtures. [70:500.5(D)]

Class III locations are clearly those where ignitible fibers or flyings are present, but because they are not likely to be in suspension they do not pose a potential flash-fire or explosion hazard, rather only a very rapidly burning surface fire hazard. Where do suspensions of ignitible fibers or flyings that pose an explosion hazard fit in the Division scheme? Prior to the 2011 edition of

NFPA 70, if a combustible particulate presented an explosion hazard it could be considered a Class II location. However, with the formal adoption of a specific particle size boundary for combustible dust, these hazardous fibers or flyings that could be expected to be in suspension could no longer be considered Class II. This same formal definition now sets a de-facto lower limit on fibers/flyings such that combustible particulate larger than 500 micron would be a Class III material.

2.4 Zone Locations

According to NFPA 70, Article 506 covers "the requirements for the zone classification system as an alternative to the division classification system covered in Article 500, Article 502, and Article 503 for electrical and electronic equipment and wiring for all voltages in Zone 20, Zone 21, and Zone 22 hazardous (classified) locations where fire and explosion hazards may exist due to combustible dusts or ignitible fibers/flyings." While NFPA 70 Article 503 for Division classification appears to eliminate coverage for locations where "explosion hazards may exist due to … ignitible fibers/flyings", Article 506 clearly includes these locations. Article 506 for Zone classification references ANSI/ISA-61241-10 (12.10.05)-2004, which is the U.S. National Deviation of the basis IEC document. The 2004 ISA document defines combustible dust broadly and does not include a particle size boundary.

> **Combustible dust:** dust, fibres or flyings that can burn or glow in air and could form explosive mixtures with air at atmospheric pressure and normal temperatures.

Since 2004, the ISA document has been superceded by ANSI/ISA-60079-10-2 (12.10.05)-2013 *Explosive Atmospheres – Part 10-2: Classification of areas - Combustible dust atmospheres.* The recent 2015 update of the same splits out combustible dust and the subgroup of fibers/flyings with potential for explosion hazards as follows.

> **Combustible dust:** finely divided solid particles, 500 μm or less in nominal size, which may form an explosive mixture with air at atmospheric pressure and normal temperatures
>
> **Combustible flyings:** solid particles including fibers, greater than 500 μm in nominal size, which may form an explosive mixture with air at atmospheric pressure and normal temperatures
>
> **Dust:** generic term including both combustible dust and combustible flyings
>
> **Explosive dust atmosphere:** mixture with air, under atmospheric conditions, of flammable substances in the form of dust, which, after ignition, permits self-sustaining propagation.

2.5 Explosible Particulate

While the current Division locations appear to eliminate the possibility of materials larger than 500 micron dimension presenting a flash-fire or explosion hazard, the Zone locations clearly accept such a potential and call it an explosive dust atmosphere. Instead of "explosive", which conjures up condensed phase explosive materials like dynamite, another term often mentioned in relation to combustible dust is "explosible". Dust testing results such as K_{St} and P_{max} are termed explosibility parameters. Therefore an inclusive term, explosible particulate, is proposed for generally spherical particulate through higher aspect ratio fibrous and flake materials that present similar hazards. Figure 6 illustrates this overlap. The parenthetical wording, "process specific oxidizing medium", is an addition suitable for occupancy standards.

> **Explosible particulate [proposed]:** Combustible dust, fibers or flyings that present a flash-fire or explosion hazard when suspended in air (or the process specific oxidizing medium) over a range of concentrations

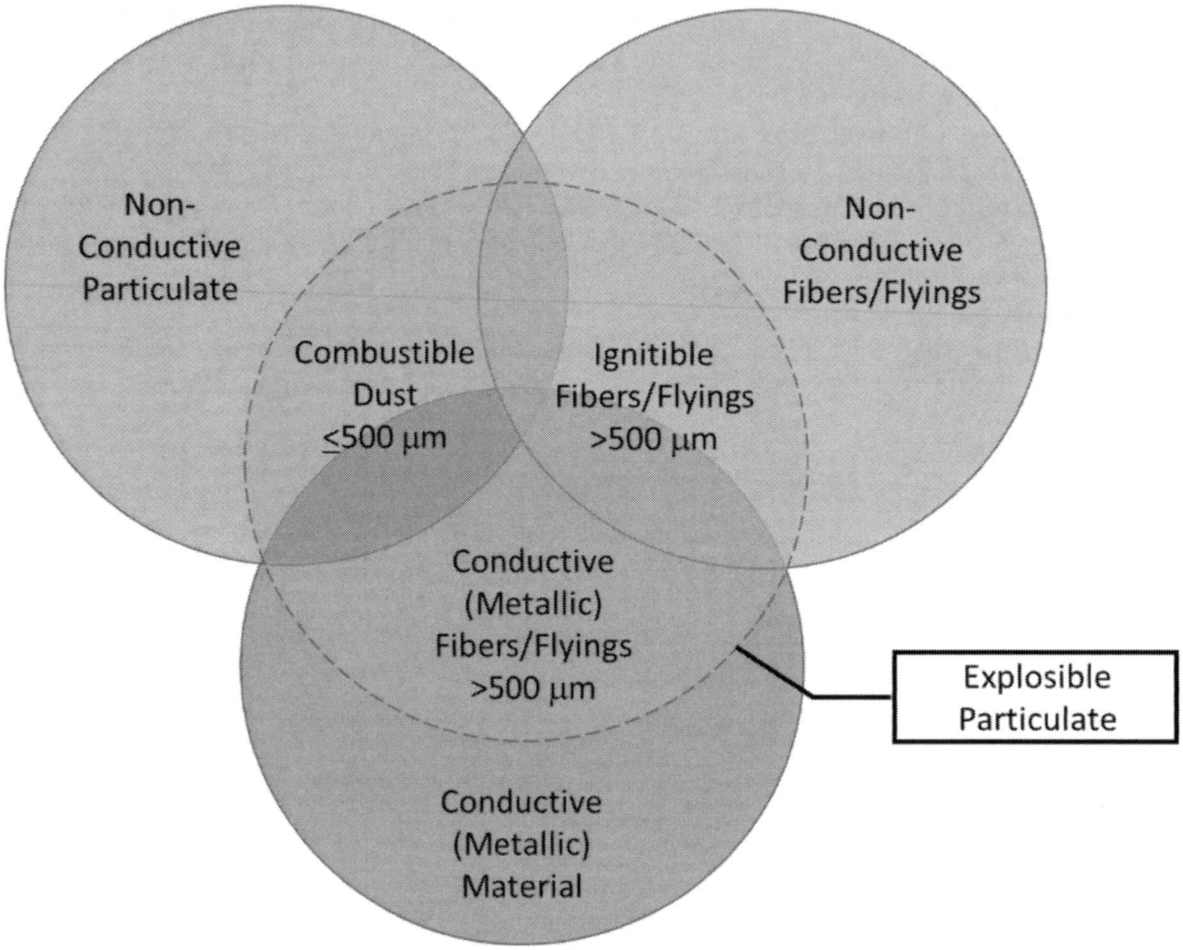

Figure 6. Explosible Particulate – Combustible Dust, Fibers or Flyings

3 Material Groups

The Division system includes three Class II dust material Groups, E, F and G. It also includes Class III fibers/flyings. The Zone system includes only three explosible material Groups, IIIA,

IIIB and IIIC to cover the same range of materials. As is further explained, Zone Group IIIA cannot encompass all of Class III. Table 2 shows how these Groups are equated.

Table 2 Comparison of Division and Zone Dust Groups

Division System Class – Group	Zone System Group
Class II – Group E	IIIC
Class II – Group F, G	IIIB
Class III (sub-set)	IIIA

3.1 Determination of NFPA Division Material Groups

NFPA 70 addresses Class II combustible dusts in Article 500.6(B), dividing them into Groups E, F, and G.

> ***Group E.*** Atmospheres containing combustible metal dusts, including aluminum, magnesium, and their commercial alloys, or other combustible dusts whose particle size, abrasiveness, and conductivity present similar hazards in the use of electrical equipment.
>
> ***Group F.*** Atmospheres containing combustible carbonaceous dusts that have more than 8 percent total entrapped volatiles (seeASTM D 3175, *Standard Test Method for Volatile Matter in the Analysis Sample of Coal and Coke*, for coal and coke dusts) or that have been sensitized by other materials so that they present an explosion hazard.
>
> ***Group G.*** Atmospheres containing combustible dusts not included in Group E or Group F, including flour, grain, wood, plastic, and chemicals.

Class III fibers/flyings are described in NFPA 70 informational notes to Article 500.5(D)(1):

> **Informational Note No. 1**: Such locations usually include some parts of rayon, cotton, and other textile mills; combustible fibers/flyings manufacturing and processing plants; cotton gins and cotton-seed mills; flax-processing plants; clothing manufacturing plants; woodworking plants; and establishments and industries involving similar hazardous processes or conditions.
>
> **Informational Note No. 2**: Easily ignitible fibers/flyings include rayon, cotton (including cotton linters and cotton waste), sisal or henequen, istle, jute, hemp, tow, cocoa fiber, oakum, baled waste kapok, Spanish moss, excelsior, and other materials of similar nature.

Based on the examples given in NFPA 70, metallic particulate or metallic fibers, such as metal flake or steel wool, are not consistent with Class III fibers/flyings. Rather, such metallic fibers/flyings are best considered as Group E materials.

3.2 Determination of NFPA Zone Material Groups

Article 506.6 establishes the equality of Division and Zone Material Groups

> **Group IIIC.** Combustible metal dust.
>
> > Informational Note: Group IIIC is equivalent to Class II, Group E....
>
> **(B) Group IIIB.** Combustible dust other than combustible metal dust.
>
> > Informational Note: Group IIIB is equivalent to Class II, Groups F and G....
>
> **(C) Group IIIA.** Solid particles, including fibers, greater than 500 μm in nominal size, which may be suspended in air and could settle out of the atmosphere under their own weight.
>
> > Informational Note No. 1: Group IIIA is equivalent to Class III.
> >
> > Informational Note No. 2: Examples of flyings include rayon, cotton (including cotton linters and cotton waste), sisal, jute, hemp, cocoa fiber, oakum, and baled waste kapok.

IEC defines the materials and arrives at a similar situation as follows.

> **Conductive Dust:** combustible dust with electrical resistivity equal to or less than 10^3 ohm-m [classified as Group IIIC]
>
> **Non-Conductive Dust:** combustible dust with electrical resistivity greater than 10^3 ohm-m [classified as Group IIIB]
>
> **Combustible flyings:** solid particles including fibers, greater than 500 μm in nominal size, which may form an explosive mixture with air at atmospheric pressure and normal temperatures [classified as Group IIIA]
>
> > Note 1 to Entry: Examples of flyings include rayon, cotton (including cotton linters and cotton waste) sisal, jute, hemp, cocoa fiber, okum and waste kapok

Based on the examples given for combustible flyings, metallic particulate or metallic fibers, such as metal flake or steel wool, are not consistent with Group IIIA. Rather, such metallic fibers/flyings are best considered as Group IIIC materials.

3.3 Lack of Clarity for Metal Fibers/Flyings

A direct reading of the IEC material group definitions indicates that particle size is the main factor between Zone Group IIIA and Groups IIB/IIIC, with fibers/flyings larger than 500 micron in Group IIIA and smaller particulate next subdivided into non-conductive and conductive dust groups IIIB and IIIC, respectively. While the 2015 edition of the IEC 60079-10-2 says that the appropriate combustible dust group must be identified, it can only be inferred from the examples of combustible flyings that conductive materials are not included in Group IIIA. The almost identical examples of Division Class III materials also suggest that conductive, metallic flyings are not included.

NFPA 70 uses the term "combustible metal dust" to define Group IIIC. Interestingly, the 2016 NFPA glossary of terms lists the only NFPA definition for "combustible metal dust" as that found in NFPA 484 *Standard for Combustible Metals* [14]. This definition clearly does not include a particle size limit.

> ***Combustible Metal Dust.*** A combustible particulate metal that presents a fire or explosion hazard when suspended in air or the process specific oxidizing medium over a range of concentrations, regardless of particle size or shape. [NFPA 484:2015]

While the upcoming 2017 edition of NFPA 499 does not address the classification of metallic dust and flyings larger than 500 micron in a definition, the annex to the definition of Zone Group IIIA eliminates metallic materials.

> **3.3.4.6 Group IIIA:** Solid particles, including fibers, greater than 500 μm in nominal size that might be suspended in air and could settle out of the atmosphere under their own weight; Group IIIA is equivalent to Class III.
>
> **A3.3.4.6** Examples of flyings include rayon, cotton (including cotton linters and cotton waste), sisal, jute, hemp, cocoa fiber, oakum, and baled waste kapok. Group IIIA materials are larger particle size Group IIIB materials and specifically do not include metal dust, or metal fibers/flyings.

With these two pieces of support, notwithstanding a specific interpretation, it is more conservative at this time to consider any metallic dust or fiber/flying that presents a flash-fire or explosion hazard when suspended in air as a Zone Group IIIC and Division Group E.

3.4 Equating Class III with Group IIIA

While both Class III and Group IIIA address natural or man-made non-conductive organic fibers/flyings, the Zone Group IIIA definition is specific to fibers/flyings that may form an explosive mixture in air, or equivalently a flash-fire or explosion hazard. The Class III definition says that the fibers/flyings are not likely to be in suspension in air, thus only a very rapidly burning surface fire hazard. These two are not really equal and Zone Group IIIA is only a subset of Division Class III.

The subjective difference is the very low likelihood of fibers/flyings being in suspension and this can be the result of a number of mitigating factors. While size itself might be assumed to be a factor, it is rather the tendency for longer fibers to co-mingle and form larger aggregate structures that are more difficult to dislodge. These same aggregate structures also resist breaking apart while in suspension, with the result that there are large empty spaces between aggregate structures and the flame front is less likely to propagate.

Even though NFPA 70 formally equates Group IIIA materials to Class III materials in the informational note above, it recognizes that Division Class III locations present a different hazard than Zone Group IIIA locations. As further explained below, Class III electrical equipment is not suitable for Group IIIA; instead Class II equipment is permitted.

Hazardous area classification for Zone Group IIIA materials is by the same procedure as Group IIIB and IIIC. NPFA 499, in the upcoming 2017 edition, recommends that similarly hazardous Division Class III materials use its guidance for classification where the fibers/flyings present an explosion hazard

> **1.2.5** This recommended practice provides guidance for classification of Class III or Group IIIA combustible fibers/flyings and of hazardous (classified) locations where these present a dust fire or dust explosion hazard when dispersed and ignited in air.

A step-wise combined classification approach embodying both the combustible metal dust and Class III interpretations would be as follows:

1. If the material is a combustible metal dust (conductive), regardless of particle size, and simultaneously presents an explosion hazard, it is Zone Group IIIC (or Division Group E).

2. Next, if the material is not metallic (non-conductive), equal to or less than 500 micron nominal size and simultaneously presents an explosion hazard, it is Zone Group IIIB (or Division Group F or G).

3. Otherwise, if the material is not metallic (non-conductive), greater than 500 micron nominal size and

    a. is not likely to be in suspension in the air in quantities sufficient to produce ignitible mixtures, it is Division Class III

    b. simultaneously presents an explosion hazard, it is Zone Group IIIA (or Division Group F or G).

4 Division and Zone Equipment Equivalency

Without going into the definitions of Divisions and Zones for dust, there are Division 1 and 2 locations for both Class II and Class III material. There are similarly Zone 20, 21 and 22 locations for Zone Group III materials. NFPA 70 identifies suitably equivalent equipment through permissions and exceptions. Class II, Division 1 equipment is suitable for Zone 20 or 21 while Zone 21 equipment is not necessarily good for Class II, Division 1. Class II, Division 2 equipment is good for Zone 22 and vice-versa.

502.6 Zone Equipment. Equipment listed and marked in accordance with 506.9(C)(2) for Zone 20 locations shall be permitted in Class II, Division 1 locations for the same dust atmosphere; and with a suitable temperature class.

Equipment listed and marked in accordance with 506.9(C)(2) for Zone 20, 21, or 22 locations shall be permitted in Class II, Division 2 locations for the same dust atmosphere and with a suitable temperature class.

503.6 Zone Equipment. Equipment listed and marked in accordance with 506.9(C)(2) for Zone 20 locations and with a temperature class of not greater than T120°C (for equipment that may be overloaded) or not greater than T165°C (for equipment not subject to overloading) shall be permitted in Class III, Division 1 locations.

Equipment listed and marked in accordance with 506.9(C)(2) for Zone 20, 21, or 22 locations and with a temperature class of not greater than T120°C (for equipment that may be overloaded) or not greater than T165°C (for equipment not subject to overloading) shall be permitted in Class III, Division 2 locations.

506.20 Equipment Installation.

(A) Zone 20. In Zone 20 locations, only equipment listed and marked as suitable for the location shall be permitted.

Exception: Equipment listed for use in Class II, Division 1 locations with a suitable temperature class shall be permitted.

(B) Zone 21. In Zone 21 locations, only equipment listed and marked as suitable for the location shall be permitted.

Exception No. 1: Apparatus listed for use in Class II, Division 1 locations with a suitable temperature class shall be permitted.

Exception No. 2: Pressurized equipment identified for Class II, Division 1 shall be permitted.

(C) Zone 22. In Zone 22 locations, only equipment listed and marked as suitable for the location shall be permitted.

Exception No. 1: Apparatus listed for use in Class II, Division 1 or Class II, Division 2 locations with a suitable temperature class shall be permitted.

Exception No. 2: Pressurized equipment identified for Class II, Division 1 or Division 2 shall be permitted.

The equivalent installations permitted show that Zone equipment is suitable for combustible dust and fibers/flying and that Class III equipment is not suitable for combustible dust. A significant difference between Class III and Class II equipment is the omission of seals to ensure the particulate-air mixture is kept out. This highlights the fact that the ignitible fibers/flyings that do present an explosion hazard require equipment identified for either Zones or for Class II materials.

5 Comparison of Division and Zone Dust Accumulation Criteria

NFPA 499 has historically included visual dust accumulation on surfaces as a criteria for establishing Class II Division locations. Where the dust accumulation obscures the underlying surface color a Division 2 location is warranted. Where the dust accumulation exceeds 1/8 inch layer thickness a Division 1 location is warranted. While the NFPA 70 definitions of Division 1 and Division 2 focus on the likelihood of a suspended dust cloud being present, NFPA 499 has provided guidance that surface accumulations directly affect the likelihood that a suspended dust cloud will be present. In other words, if an accumulated dust layer is occasionally or continuously present on electrical equipment, as a result of operations and routine housekeeping, a Division 1 or 2 classification is warranted. No other considerations are required – the potential presence of a dust cloud is inferred.

On the other hand the ISA Zone approach does not make a direct connection between dust accumulation layer thickness and likelihood that a dust cloud will be present. While an accumulated dust layer is to be considered as a potential source of a dust cloud, the ISA approach makes a point that the determination of Zone 20, 21, or 22 is not directly connected to layer thickness. Evaluation of the dust layer hazard is, in fact, a separate clause. There is no assumption of hazard posed by dust layer accumulation. Annex examples of ISA application require an affirmative evaluation of the likelihood that any given accumulation, no matter the thickness, could be suspended as a cloud by identified normal, maintenance or abnormal operations. It is a risk-based evaluation. This can be location dependent as well, further complicating the determination of the appropriate Zone.

NFPA 499 similarly recognizes a risk-based decision concerning dust layers, but it is in the opposite direction – determining that mitigating factors reduce the likelihood of dust cloud formation sufficiently to omit classification, as opposed to identifying factors that support dust cloud formation. The default is that accumulated dust layers result in classified locations. It is not the intention that this comparison suggest the ISA approach, properly applied, will result in a different result. The ISA document properly references personnel certification schemes to ensure sufficient knowledge to make such a determination. Rather, it is to emphasize that the NFPA 499 presumption of hazard from dust accumulation with defined thresholds can be significantly easier to apply.

The 2017 edition of NFPA 499 associates the same three Division layer thickness descriptions to Zones as in Table 3, maintaining the current direct approach for evaluating the potential for dust cloud formation.

Table 3. Division and Zone Determination Guidelines based on Dust Layer Thickness

Layer Condition	Division Classification	Zone Classification

>3.0 mm (1/8 in.)	Division 1	Zone 20 or 21
≤3.0 mm (1/8 in.), and surface color not discernible	Division 2	Zone 22
Surface color discernible under dust layer (or combustible metal dust is not in a hazardous concentration)	Unclassified	Unclassified

6 Conclusions

There has been and continues to be confusion and disagreement between a physical size definition and a broader phenomenological definition of a combustible dust. The confusion arises in part because of the use of the definition, establishing equipment design characteristics versus acknowledging potential hazards to the operating area, respectively. The confusion has advanced from informal to formal and resolution is necessary. Progress has been made with the inclusion of the Zone classification system in NFPA 70 and NFPA 499 where it is now recognized that fibers/flyings can also present a flash-fire or explosion hazard and therefore require an electrical equipment design/installation that prevents ingress of the material. NFPA 70 has clearly done this by omitting any use of Division III electrical equipment for Group IIIA combustible flyings that "may form an explosive mixture with air…".

The option of an inclusive term, explosible particulate, provides a path forward for both the electrical and occupancy standards. Focusing the occupancy standards on explosible particulate, inclusive of the electrical particle size-limited combustible dust as well as ignitable fibers/flyings, maintains the goals of both.

Based on the NFPA 70 informational note describing Division Class III materials, the similar IEC note describing Zone Group IIIA materials, the NFPA 484 definition of combustible metal dust and 2017 NFPA 499 annex description of Group IIIA, it is appropriate at this time to consider any metallic dust or fiber/flying, that presents a flash-fire or explosion hazard when suspended in air, as a Zone Group IIIC and Division Group E material.

The recent NFPA 499 update to include the Zone system provides the opportunity for application in a more easily implemented and similar fashion to the Division system. Specifically the direct connection between observed dust accumulation on electrical equipment and necessary Zone classification is analogous to the Division classification approach.

Armed with knowledge, there is no need to fear as long as the hazards are understood and the equipment is chosen correctly. Who's afraid of Combustible Dust anyway?

7 References

[1] NFPA 70 *National Electrical Code,* Batterymarch Park, Quincy, MA: NFPA Publications.

[2] NFPA-499 *Recommended Practice for the Classification of Combustible Dusts and of Hazardous (Classified) Locations for Electrical Installations in Chemical Process Areas,* Batterymarch Park, Quincy, MA: NFPA Publications.

[3] ANSI/ISA-61241-10 (12.10.05)-2004, *Electrical Apparatus for Use in Zone 20, Zone 21, and Zone 22 Hazardous (Classified) Locations —Classification of Zone 20, Zone 21, and Zone 22 Hazardous (Classified) Locations,* International Society of Automation, Research Triangle Park, NC.

[4] ANSI/ISA-60079-10-2 (12.10.05)-2013 *Explosive Atmospheres – Part 10-2: Classification of areas - Combustible dust atmospheres,* International Society of Automation, Research Triangle Park, NC.

[5] IEC-60079-10-2 (Edition 2.0 – 2015-01) *Explosive atmospheres – Part 10-2: Classification of areas - Explosive dust atmospheres,* International Electrotechnical Commission

[6] NFPA 654 *Standard for the Prevention of Fire and Dust Explosions from the Manufacturing, Processing, and Handling of Combustible Particulate Solids*, Batterymarch Park, Quincy, MA: NFPA Publications.

[7] NFPA 921 *Guide for Fire and Explosion Investigations,* Batterymarch Park, Quincy, MA: NFPA Publications.

[8] NFPA 652 *Standard on the Fundamentals of Combustible Dust,* Batterymarch Park, Quincy, MA: NFPA Publications.

[9] ASTM Standard E1226, 2012a, "*Standard Test Method for Explosibility of Dust Clouds,*" ASTM International, West Conshohocken, PA, 2010, DOI: 10.1520/E1226-12A, www.astm.org.

[10] ISO 6184–1, *Explosion protection systems — Part 1: Determination, of explosion indices of combustible dust in air*, International Organization for Standardization, 1985

[11] E. Ural, S. Johnson and A.G. Dastidar, "Effect of Particle Shape and Size Distribution on Dust Explosion Risks" 8th Global Congress on Process Safety, April, 2012, Houston, TX.

[12] A.G. Dastidar. Is This Combustible "Dust"? , Available at http://www.solidsonline.com/articles/1009-Is-This-Combustible--Dust--1, Accessed on February 6, 2016.

[13] S.A. Rodgers and E.A. Ural, Practical issues with marginally explosible dusts—Evaluating the real hazard. Process Saf Prog 30 (2011), 266–279

[14] NFPA 484 *Standard for Combustible Metals,* Batterymarch Park, Quincy, MA: NFPA Publications.

NFPA Combustible Dust Standards – 2016 Update

Susan Bershad
National Fire Protection Association
1 Batterymarch Park
Quincy, MA 02169
sbershad@nfpa.org

Prepared for Presentation at
American Institute of Chemical Engineers
2016 Spring Meeting
12th Global Congress on Process Safety
Houston, Texas
April 11-23, 2016

NFPA Combustible Dust Standards – 2016 Update

Susan Bershad
National Fire Protection Association
1 Batterymarch Park
Quincy, MA 02169
sbershad@nfpa.org

Keywords: NFPA, Combustible Dust, NFPA 652

Abstract

Every year, destructive and deadly dust-related fires and explosions affect a wide range of industries around the globe. In the United States alone, 50 combustible dust accidents occurred between 2008 and 2012. To manage the dust-related fire, flash-fire, and explosion hazards in industries that use dust collection and handling equipment, or have processes that may generate combustible dust, the NFPA has recently published NFPA 652, Standard on the Fundamentals of Combustible Dust. This paper will provide an update of the NFPA combustible dust standards. Now that NFPA 652 has been published, this paper will review how the industry and commodity-specific dust standards (NFPA 664: Prevention of Fires and Explosions in Wood Processing and Woodworking Facilities, NFPA 654: Prevention of Fire and Explosion from the Manufacturing, Processing, and Handling of Combustible Particulate Solids, NFPA 61: Prevention of Fires and Dust Explosions in Agricultural and Food Processing Facilities, NFPA 484: Combustible Metals, and NFPA 655: Standard for Prevention of Sulfur Fires and Explosion) interact with NFPA 652 to provide requirements for the protection of facilities with combustible dusts. An update will be provided on the status of the commodity specific standards and how they are evolving to correlate with NFPA 652.

1.0 Introduction

Significant incidents involving combustible dust have brought increased attention to industries that generate, manufacture, handle, store, and transport solids. The Chemical Safety Board dust study in 2006 highlighted an apparent lack of awareness of the hazards posed by combustible dusts [1]. NFPA 652, the Standard on the Fundamentals of Combustible Dust, is a new NFPA standard that provides the fundamental requirements for combustible dusts and interacts with the commodity-specific dust standards [2]. An important provision in NFPA 652 is the requirement to perform a Dust Hazards Analysis

(DHA). This analysis is used to identify and evaluate combustible dust hazards, to establish whether existing safeguards are adequate, and to determine whether or not new safeguards are needed.

As a result of several high-profile incidents involving combustible dusts in the 2000's, OSHA established a National Emphasis Program (NEP) on combustible dusts in 2008 and began a series of targeted inspections for industries involved in combustible solids or powders storage, handling, or use [3]. It should be noted that there is currently no federal standard for general industry that addresses comprehensive safeguards to protect workers from combustible dust fire and explosion hazards. In 2009, OSHA announced its intent to prepare a comprehensive standard for combustible dust and published an Advanced Notice of Proposed Rulemaking (ANPRM) [4]. This effort is currently delayed, with the next step in the process, the SBREFA review (Small Business Regulatory Enforcement Fairness Act) scheduled for August of 2016.

2.0 NFPA Technical Committee Structure

After the announcement by OSHA that they planned to initiate rulemaking to develop a comprehensive combustible dust regulation for general industry, NFPA examined its group of combustible dust standards and decided to change the structure of the dust committee projects. Many of the industry and commodity-specific standards trace their origins back to the 1920's and 1930's. In 2010, two changes in committee structure were made. The first was the creation of an oversight, or correlating committee, that was tasked with ensuring consistency and coordination of requirements across all of the varied industrial processes where solids handling produces dust. The second was the creation of a committee that was tasked with defining the fundamental requirements that should be applied to "to all facilities and operations that manufacture, process, blend, convey, repackage, generate, or handle combustible dusts or combustible particulate solids" [2]. Setting the fundamental requirements for combustible dusts allows the technical committees responsible for the industry and commodity-specific standards to then establish additional safeguards that are uniquely applicable to their processes, equipment, or type of dust. The commodity-specific standards would be the source for any requirements that are not deemed fundamental.

The current technical committee structure is presented in Figure 1. There are five technical committees, responsible for the five industry and commodity-specific standards, as well as a related standard, NFPA 91, the *Standard for Exhaust Systems for Air Conveying of Vapors, Gases, Mists, and Particulate Solids.*

Figure 1: Structure of NFPA Combustible Dust Committees

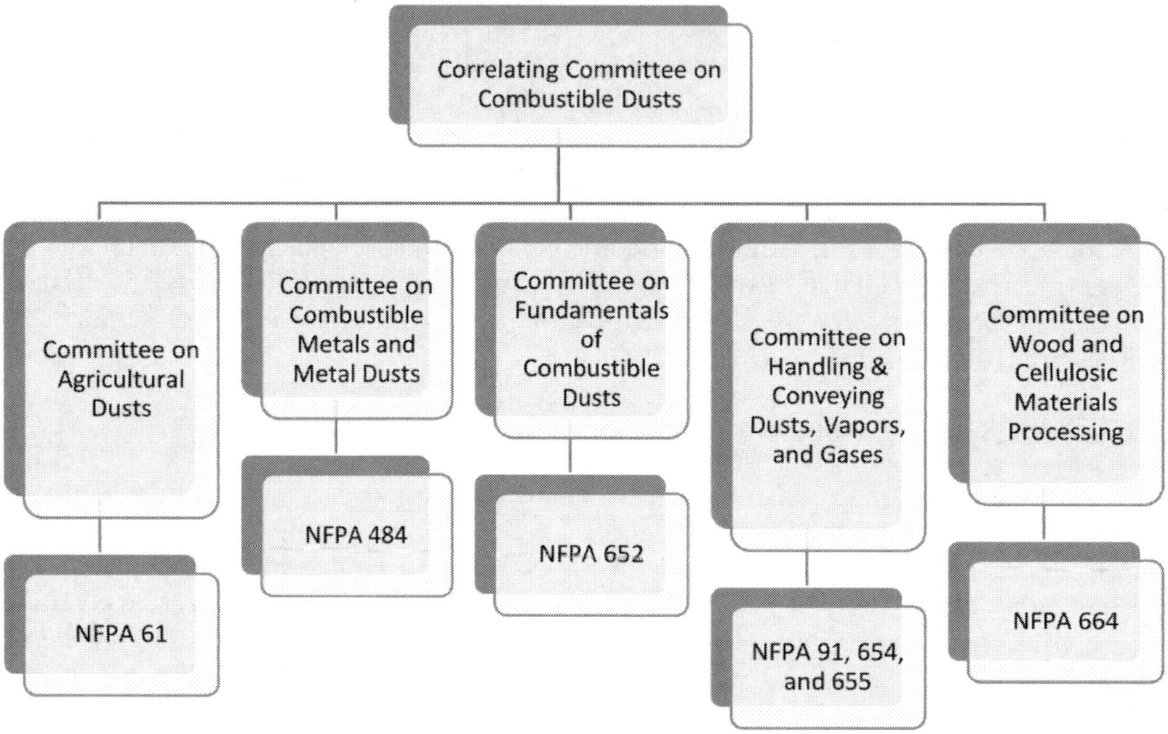

NFPA processes its standards in one of two revision cycles, annual and fall. For annual documents, the approximately two year cycle concludes in June. For fall documents, the cycle concludes in November. There are two main steps in the process, first draft and second draft. Public input is submitted for the first draft and public comment for the second draft. Any member of the public can make a public input or public comment. Public comment is submitted on the changes proposed for the first draft. During both of the steps in the process, the committees consider, evaluate, and respond to proposed changes to the document and then vote on the resulting draft. The last step in the process is the NITMAM (Notice of Intent to Make an Amending Motion). The NITMAM process allows interested parties to contest the content of the second draft that was approved by the technical committee. Documents that do not receive NITMAMs are approved by the NFPA Standards Council as consent documents. Those NITMAMs that are approved become Certified Amending Motions that are then heard by NFPA membership at the NFPA annual meeting.

The Correlating Committee directs the activities of the Technical Committees. It is responsible for resolving conflicts, achieving correlation among the recommendations of the technical committees, correcting errors and omissions, and assuring that the committee activities have been conducted in accordance with NFPA regulations. The Correlating Committee has the authority to choose between alternatives presented by the Technical Committee, but only as necessary for correlation, consistency, and the correction of errors and omissions. The Correlating Committee meets to review the first and second draft report for each standard.

Table 1 shows the status of the five combustible dust standards as well as the related explosion protection standards as of January 2016. Note that the explosion protection standards do not fall under the combustible dust committees, but are included here as they are related to the combustible dust standards.

Table 1 – NFPA Dust and Explosion Protection Standards - Status and Revision Cycles

NFPA Document and current edition	Revision Cycle	Status
NFPA 652 – Standard on the Fundamentals of Combustible Dust – 2016 edition	A2018	The 2016 edition was issued in September of 2015 Public input for the next edition will close June 29, 2016.
NFPA 67 – Guide on Explosion Protection for Gaseous Mixtures in Piping Systems – 2016 edition	F2018	The 2016 edition was issued in the fall of 2015. Public input for the next edition will close January 5th, 2017
NFPA 69 – Standard on Explosion Prevention Systems – 2014 edition	F2018	Public input for the next edition will close January 5th, 2017
NFPA 61 – Standard for the Prevention of Fires and Dust Explosions in Agricultural and Food Processing Facilities – 2013 edition	A2016	Second draft is posted for review. Deadline for NITMAM is February 19, 2016.
NFPA 664 – Standard for the Prevention of Fires and Explosions in Wood Processing and Woodworking Facilities – 2012 edition	A2016	Second draft is posted for review. Deadline for NITMAM is February 19, 2016
NFPA 654 – Standard for the Prevention of Fire and Dust Explosions from the Manufacturing, Processing, and	A2016	Second draft is posted for review. Deadline for NITMAM is February 19, 2016.

Handling of Combustible Particulate Solids – 2013 edition		
NFPA 655 – Standard for Prevention of Sulfur Fires and Explosions – 2012 edition	F2016	Second draft meeting was held in January 2016.
NFPA 68 – Standard on Explosion Protection by Deflagration Venting - 2013 edition	F2017	Public input closed January 7, 2016. First draft meeting was held in February 2016.
NFPA 484 – Standard for Combustible Metals –2015 edition	A2018	First of two first draft meetings was held in August of 2015. The second is scheduled for March of 2016.

The first edition of NFPA 652 was published in September of 2015. Three of the industry and commodity-specific standards, NFPA 61 (agricultural dust), 664 (wood dust), and 654 (chemicals, plastics, and general dust) have gone through the second draft process, including correlating committee review. As of January 2016, the second drafts for all three are posted for further review and possible NITMAM. NFPA 655 is currently in the second draft phase. NFPA 484, the Combustible Metals document, is currently in first draft. The document cycle was recently changed from A2017 to A2018 to give the technical committee more time for task group work to reorganize the document. NFPA 652 is on a three-year revision cycle and is currently open for public input. The NFPA Standards Council recently approved new revision schedules for 61, 664, and 654 in order to allow the documents to align with 652 on a slightly accelerated pace. NFPA 61 and 664, which were previously on a five year revision cycle, have been changed to a 3 ½ year cycle.

3.0 Overview of NFPA 652 – Standard on the Fundamentals of Combustible Dust

NFPA 652 provides the basic principles of and requirements for identifying and managing the fire and explosion hazards of combustible dusts. The purpose of the standard is to provide the minimum general requirements necessary to manage the fire, flash fire, and explosion hazards posed by combustible dusts. It directs the user to other NFPA standards for industry and commodity-specific requirements. An outline of the contents of NFPA 652 is presented in Table 2.

Table 2 –Outline of NFPA 652

Chapter	Description
Chapter 1	Administration
Chapter 2	Referenced Publications
Chapter 3	Definitions
Chapter 4	General Requirements
Chapter 5	Hazard Identification
Chapter 6	Performance-Based Design Option
Chapter 7	Dust Hazards Analysis (DHA)
Chapter 8	Hazard Management: Mitigation and Prevention
Chapter 9	Management Systems
Annex A	Explanatory Material
Annex B	Dust Hazards Analysis (DHA) – Example
Annex C	Accumulated Fugitive Dust
Annex D	Informational References

Chapters 1, 2, and 3 are similar in content to all NFPA documents. Chapter 1, Administration, contains information on the scope, purpose, and application of the document. As with any NFPA document, the user should start with these sections to determine if the document applies to their operations. NFPA 652 applies to facilities and operations that manufacture, process, blend, convey, repackage, generate, or handle combustible dusts or combustible particulate solids [2]. It is equally important to review what the document does not apply to, in order to check if your operation is excluded and/or falls under another standard. NFPA 652 does not apply to:

- Storage or use of consumer quantities
- Storage or use of commercially packaged materials at retail facilities
- Display of materials in original packaging in mercantile occupancies
- Warehousing of sealed containers
- Storage or use of materials used in farm buildings.

Chapter 1 also contains information on conflicts, retroactivity, and equivalency. The conflict provisions in NFPA 652 provide important information on how NFPA 652 interacts with the industry and commodity-specific standards. Basically, if a requirement in an industry or commodity-specific standard differs from NFPA 652, the industry or commodify-specific standard shall be permitted to be used. If the requirement in an industry or commodity-specific standard prohibits the requirement, the prohibition shall be applied. If the requirement in the industry or commodity-specific standard neither prohibits nor provides a requirement, the requirement in 652 applies.

To illustrate this provision, let's use an example of painting a dust collector. Note that there are no requirements in any of the dust standards to paint dust collectors, nor are there any requirements as to what color paint must be used. This is only an example.

Theoretical example of conflict provision in 652:

- NFPA 652 requires that all dust collectors be painted, but does not specify the color
- NFPA 61 requires that all dust collectors be painted pink – Collectors must be painted pink.
- NFPA 664 is silent on paint for a dust collector – Collectors must be painted but can be any color.
- NFPA 654 prohibits painting dust collectors – Collectors must remain unpainted.

Chapter 1 also contains important information on retroactivity. Generally, NFPA standards are not meant to be retroactive except where it is explicitly stated. NFPA Standards do not normally require a facility to be redesigned retroactively. Typically it is the management system requirements, such as housekeeping, employee training, management of change, etc., that are to be applied retroactively. NFPA 652 is explicit in terms of which sections are to be applied retroactively to a facility.

Chapter 4 of 652 contains general requirements on owner/operator responsibilities as well as objectives for the standard. Objectives and compliance options are provided so that there is a consistent basis for choosing either the prescriptive or the performance-based option for application of the standard per Chapter 6. Objectives are provided for Life Safety, Mission Continuity, and the Mitigation of Fire Spread or Explosion.

Chapter 5 contains testing requirements for determining whether a dust is combustible or explosible. It is the responsibility of the owner/operator to determine the properties of their dust. Historical or published data can be used if it is representative of current operations. It is up to the owner/operator of the facility to determine if the data is accurate. Chapter 5 provides a considerable amount of supplemental explanatory material in the Annex to aid the user on testing, test methods, and sampling. The annex material for Chapter 5 also contains tables of representative data on dust properties extracted from several other standards including NFPA 61, 484, and 499 (Recommended Practice for the Classification of Combustible Dust and of Hazardous (Classified) Locations for Electrical Installations in Chemical Process Areas).

Chapter 6 outlines the requirements for performance-based design. This is an alternative to the prescriptive design requirements in Chapter 8. A performance-based design must be prepared by a qualified individual and documented in a format and content acceptable to the authority having jurisdiction (AHJ). This documentation must demonstrate how the performance-based design provided meets the objectives outlined in Chapter 4 (i.e., provides protection equivalent to the prescriptive design requirements of the standard). A performance-based design must still meet the housekeeping, personnel protective equipment, and management system requirements in Chapters 8 and 10.

Chapter 7 contains requirements for the performance of a Dust Hazards Analysis (DHA). The requirement to perform a DHA is to be applied retroactively to existing processes and facilities. The DHA requirements for new and existing facilities are summarized as follows:

- For new processes/facility compartments and those that are undergoing material modification, a DHA must be completed as part of the project
- For existing processes/facility compartments that are not undergoing material modification, the owner/operator must schedule and complete a dust hazard analysis within a three-year period of the effective date of NFPA 652.

A material modification is defined as modifications or repairs/maintenance that exceed 25% of the original cost.

There is no requirement or deadline in the standard as to how quickly recommendations or action items that result from the DHA need to be addressed. The timeframe for closure should be based on the risk posed by the hazard, and the time needed to implement the recommendation, but the recommendation should be implemented as soon as practicable.

NFPA 652 provides general information and criteria, but does not specify how to conduct a DHA. Any one of several methods can be used. Annex B of 652 presents an example of how to conduct a DHA. It should be noted that a DHA is not intended to meet the requirements for a PHA under the OSHA PSM standard. Additional information on Dust Hazards Analysis is provided in this paper.

Chapter 8 contains requirements for mitigation and prevention measures. An outline of Chapter 8 is presented in Table 3.

Table 3 - Outline of Chapter 8 of NFPA 652 – Hazard Management

8.1	Inherently Safe Design (reserved)
8.2	Building Design
8.3	Equipment Design

8.4	Housekeeping
8.5	Ignition Source Control
8.6	Personnel Protective Equipment
8.7	Pyrophoric Dusts (reserved)
8.8	Dust Control
8.9	Explosion Prevention/Protection
8.10	Fire Protection

The standard requires that buildings be designed to minimize areas where dust can accumulate. If a dust deflagration hazard exists in a building or building compartment, then deflagration venting per NFPA 68 must be provided. The standard allows the use of segregation, separation, or detachment as hazard management strategies in building design.

Equipment must be designed to contain dust. If equipment cannot be designed to contain dust, then dust collection must be provided. Pneumatic conveying systems, dust collection systems, and centralized vacuum systems must be designed with sufficient air flow and velocity to capture, collect and control dust, as appropriate. Air material separators are only allowed indoors under specific conditions outlined in the standard. Recycling of air material separator exhaust to buildings is also only allowed under specific conditions.

Section 8.4 contains requirements for housekeeping and is one of the provisions applied retroactively. Dust accumulations are required to be kept below the threshold quantities listed in each of the industry or commodity-specific dust standards. NFPA 654 specifics a threshold quantity of 1/32" at 75 lb/ft3, adjusted for lower bulk densities. NFPA 664 (wood) specifies 1/8" at 20 lb/ft3 bulk density. NFPA 484 (metals) does not allow accumulation, but infers a daily cleaning schedule. NFPA 61 (agricultural dust) specifies that dust is to be removed concurrent with operations and refers to NFPA 654.

Section 8.5 contains requirements for the control of ignition sources. Grinding, chipping, and other operations that produce sparks and open-flame ignition sources must be controlled by a hot work permit system in accordance with NFPA 51B. Other ignition sources include hot surfaces, bearings, electrical equipment and wiring, electrostatic discharges, open flames, industrial trucks, and friction and impact sparks. All of the requirements of Section 8.5 are applied retroactively with the exception of the requirements for hot surfaces (8.5.4), bearings (8.5.5) and flexible connectors (8.5.7.1.4).

Chapter 8 also contains requirements for personnel protection equipment (to be applied retroactively), dust control, and fire protection. It contains requirements on when equipment must be provided with explosion protection using one or more of the methods contained in Section 8.9.3.2, including deflagration venting in accordance with NFPA 68 or deflagration suppression in accordance with NFPA 69. Keep in mind that NFPA 652 as well as the industry and commodity-specific dust standards will point the user towards NFPA 68 an NFPA 69 for specific requirements when explosion protection is needed.

All of Chapter 9 on Management Systems is retroactive, and therefore applies to new and existing facilities. It includes the following elements:

- Operating Procedures and Practices
- Inspection, testing, and maintenance
- Training and hazard awareness
- Contractors
- Emergency planning and response
- Incident Investigation
- Management of Change
- Document Retention
- Management Systems Review

4.0 Dust Hazards Analysis

A Dust Hazards Analysis (DHA) is defined in NFPA 652 as a systematic review to identify and evaluate the potential fire, flash fire, or explosion hazards associated with the presence of one or more combustible particulate solids in a process or facility [2]. As stated previously, it is not intended for the DHA in 652 to comply with all the elements of the process hazard analysis (PHA) in OSHA 1910.119. While the DHA can comply with the PHA requirements as outlined in OSHA's process safety management standard, other methods can be used to meet the performance based approach in NFPA 652. If the process that contains the combustible dust falls under OSHA 1910.119, however, there is a legal requirement to comply with that regulation. It is up to the owner/operation to make the determination as to whether or not their process falls under the PSM standard.

The purpose of a Dust Hazards Analysis is to identify hazards in a process and to document how those hazards are being managed. Where the hazard is not being managed, the DHA identifies possible means to manage the hazard. The Dust Hazards Analysis is used to answer the follow questions:

1. What does "normal" look like for the process?
2. What can go wrong (and how)?
3. How bad, and how likely, can it be?
4. What protections currently exist?
5. What is the risk of continuing to operate as now?
6. Is additional protection warranted (and, if so, what)?

Chapter 7 of NFPA 652 outlines the methodology for conducting a DHA. A DHA must be conducted for each part of the process where combustible dust is or could be present and for each building or building compartment where combustible dust is present. If there are multiple processes or multiple buildings or building compartments that are identical and pose the same hazard, a single DHA can be conducted. Data on combustible dust properties obtained in accordance with Chapter 5 of NFPA 652 is used to conduct the DHA.

The DHA is used to evaluate where fire, flash fire, or explosion hazards exist. If these hazards exist, specific scenarios must be identified and evaluated to determine whether or not adequate safeguards are in place to manage these hazards.

Annex B of NFPA 652 presents an example of a Dust Hazards Analysis for a size reduction operator for wood chips. In this example, the facility receives woods chips via rail car and truck. The chips are unloaded and conveyed pneumatically to a storage silo. The silo is equipped with a cyclone that receives air from the silo. Cyclones are designed to use particulate inertia near the perimeter of the cyclone to separate solids from the air stream. Annex B of 652 walks through a DHA for the entire size reduction process. This paper is just going to illustrate the DHA process for the cyclone. Figure 2 presents an illustration of the cyclone.

Figure 2 – Example of Cyclone for Dust Hazard Analysis

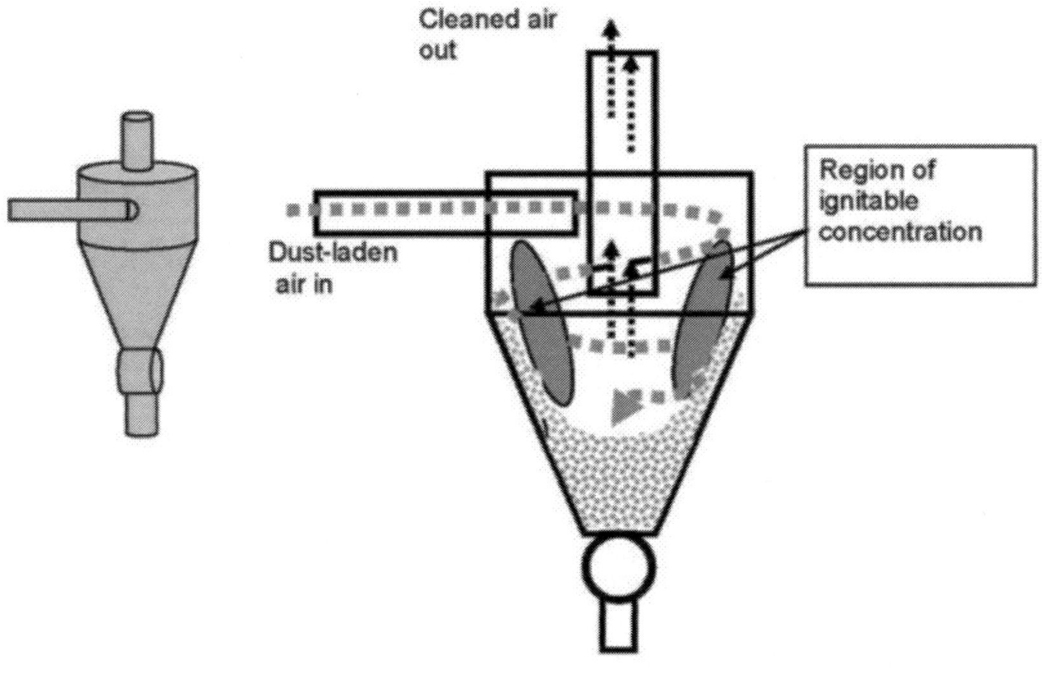

It is assumed that the material being unloaded (wood chips) is a combustible particulate solid, the material has been tested, and we know the properties of the dust. The fine particulates resulting from the attrition of the wood chips during the unloading process are a combustible dust.

The DHA asks and answers the following questions:

Is the particulate deflagrable (explosible)?

If there are any fines in the process particulate they will be separated, at least partially, from the larger particulates and concentrated by the cyclone. Since the fan creates fines and there is particle attrition in the duct work, the answer would be yes.

Is the particulate suspended in air? Yes

Is there sufficient concentration to propagate a flame front?

That depends on the quantity of fine, combustible particulate (dust) per unit of mass of total particulate moved and the volume of air to move it. Calculations should be performed to determine if there is sufficient dust per unit of air volume to exceed the minimum explosible concentration (MEC). The likely answer is yes.

Are there competent igniters available?

The material could have been ignited as it was offloaded into the silo. Tramp metal or other foreign objects may be present that can create sparks as they hit the walls of the duct. Fans in the system could also introduce possible ignition sources. The likely answer is yes.

Based on the answers to the questions above, the cyclone would be considered an explosion hazard. The next step would be to determine what hazard management measures are currently in place, what risks are posed by the equipment and process, and what additional hazard control measures are needed. An assessment of what additional hazard control measures are needed typically results in recommendations from the DHA.

Note that there is nothing in NFPA 652 that mandates a timeframe for implementing recommendations from a DHA. Each facility needs to develop their own time frame based on risk and ease of implementation of the recommendation. Operations that present the highest risk should be addressed first. Recommendations that are complex and require significant time and capital may take longer to implement. Interim mitigations should be considered where risks are intolerable, and appreciable time is needed to implement associated recommendations.

5.0 Next Steps and Conclusion

NFPA 652, the dust fundamentals document, was recently issued in September of 2015. As of January 2016, all three of the A2016 industry and commodity-specific documents, NFPA 61, 654, and 664 have been posted as second draft. Pending any NITMAMs to the documents, they will be issued in June of 2016. During the most recent revision cycle, the correlating committee worked to align the requirements in these standards, both in content and in format, with NFPA 652. NFPA 484, the Combustible Metals standard, which is currently in first draft, is working towards this same goal.

The intended outcome of the correlating committee is starting to take effect. The second draft of NFPA 61 contains requirements for conducting a Dust Hazards Analysis, and a new chapter on performance-based design. Both 654 and 664 are planning to align their outlines (table of contents) with NFPA 652. All of the documents are working towards having common definitions for terms used in the standards.

One question we frequently get is whether or not the NFPA envisions a single combustible dust standard. The simple answer is yes. The correlating committee structure and direction would enable the possibility of a single document with general requirements followed by commodity-specific chapters. The NFPA Standards Council has just moved to shorten the revision cycles of several of the documents in order to facilitate the coordination of the documents. It is up to the users of the documents and the individual technical committees, however, as to whether or not this happens and, if so, how fast it takes place.

References

[1] U.S. Chemical Safety and Hazard Identification Board, Combustible Dust Hazard Study, U.S. Chemical Safety and Hazard Investigation Board, Washington, DC, 2006.

[2] National Fire Protection Association, NFPA 652, Standard on the Fundamentals of Combustible Dust, 2016, Quincy, MA

[3] OSHA, Combustible Dust National Emphasis Program, Available at https://www.osha.gov/pls/oshaweb/owadisp.show_document?p_table=DIRECTIVES&p_id=3830, accessed on January, 12th, 2016

[4] OSHA, Combustible Dust; Advance Notice of Proposed Rulemaking, Docket No. OSHA-2009-0023, Available at https://www.osha.gov/pls/oshaweb/owadisp.show_document?p_table=FEDERAL_REGISTER&p_id=21152, accessed on January 12th, 2016

Evaluating the Effect of Obstruction Separation Distance on Flammable Gas Dispersion using Fire Dynamics Simulator

Haavard R. Boehmer, P.E.
Jason A. Sutula, P.E., PhD
JENSEN HUGHES
3610 Commerce Dr. Suite 817
Baltimore, MD, 21046
hboehmer@jensenhughes.com
jsutula@jensenhughes.com

Prepared for Presentation at
American Institute of Chemical Engineers
2016 Spring Meeting
12th Global Congress on Process Safety
Houston, Texas
April 11-23, 2016

Evaluating the Effect of Obstruction Separation Distance on Flammable Gas Dispersion using Fire Dynamics Simulator

Haavard R. Boehmer,
Jason A. Sutula
JENSEN HUGHES
3610 Commerce Dr. Suite 817
Baltimore, MD, 21046
hboehmer@jensenhughes.com
jsutula@jensenhughes.com

Keywords: LNG hazard; Vapor dispersion; CFD; FDS; Obstructions; Siting Analysis

Abstract

When Computational Fluid Dynamics (CFD) models are used for field siting studies on gas-handling facilities, the separation distance between structures is an important consideration. Previous studies have evaluated the CFD model Fire Dynamics Simulator (FDS) against experimental data for flammable gas release and dispersion in obstructed environments. These studies found that FDS gave results with an acceptable accuracy when compared to the experimental data. However, in these studies only a limited set of configurations of obstructions in the flow path were evaluated, as these experiments typically included only a single configuration. Additionally, the prior studies did not use FDS 6, which included significant changes to species transport compared to earlier versions.

The current study consisted of two elements; (1) evaluate FDS version 6.2 against full-scale gas release test data, and (2) determine the effects of different obstruction configurations on gas dispersion and concentration measurements, as predicted by FDS.

Full-scale test data using a tracer gas, propylene, as a corollary to liquefied natural gas (LNG) and liquefied petroleum gas (LPG) releases was modeling using FDS version 6.2. The gas was released over land, rather than water, and included multiple large obstructions, comparable to buildings, in the dispersion path. The analysis determined that FDS provided a good match for the horizontal plume shape and predicted the highest median gas concentration values within 40%. This is considered well below the 300% uncertainty recommended based on similar studies using previous versions of the FDS code. Building on this result, a systematic study of different obstruction separation distances was performed. As the separation distance was both increased and decreased, the channeling effects, eddy formations, and the resulting flammable gas concentrations were evaluated.

1 Introduction

Facility siting analyses are necessary to determine the overall hazards relative to the built environment within a particular site. Of particular interest is the hazard assessment of flammable gas releases, such as liquefied natural gas (LNG). Analyses of these hazards are guided by federal regulations such as 49 CFR 193, Liquefied Natural Gas Facilities: Federal Safety Standards [1]. When computational fluid dynamics (CFD) models are used for field siting studies on gas-handling facilities, the separation distance between structures is an important consideration that can affect the overall development and extension of a flammable vapor cloud during an accidental release.

Early studies on the release of flammable vapor clouds focused on the collection of data for large releases over unobstructed land and open water [2] [3]. Models with simplifying assumptions, such as DEGADIS [4] and HGSYSTEM [5], were developed and validated against this data to provide the industry with a means to assess the hazards [6] [7]. More recently, studies have been performed on vapor dispersion using commercial CFD code such as FLUENT [6] [8] [9]. These studies have demonstrated good modeling agreement when tested against large-scale data series. While a powerful tool, the use of commercial CFD codes such as FLUENT requires a large, recurring financial investment and restrict the user to proprietary software.

In an effort to appeal to a broader user base at a lower cost, the Fire Dynamics Simulator (FDS), which was developed by the National Institute of Standards and Technology (NIST), was examined for its applicability to gas and vapor dispersion scenarios. Studies utilizing FDS prior to version 6 by Ryder *et al.* [10], Pape and Kim [11], Molina-Torres *et al.* [12], Mouilleau and Champassith [13], and Kohout [14], have demonstrated reasonable agreement with experimental data for flammable gas release and dispersion in obstructed environments when appropriate safety factors were considered. However, in these studies, only a limited set of configurations of obstructions in the flow path were evaluated, as these experiments typically included only a single geometric configuration. Additionally, the prior studies did not use FDS 6 which included significant changes to species transport compared to earlier versions.

The current study consisted of two elements; (1) evaluate FDS version 6.2 against full-scale gas release test data, and (2) determine the effects of different obstruction configurations on gas dispersion and concentration measurements, as predicted by FDS.

2 Fire Dynamics Simulator

The FDS program has been under development at the National Institute of Standards and Technology (NIST) Building Fire Research Laboratory (BFRL) for over a decade [15] [16]. The software and source code is freely available. The model is specifically designed with fire scenarios in mind and continues to be improved and to have its results verified. The FDS model is a large eddy simulation model and is appropriate for use in evaluating pressure conditions created by turbulent flows. Various versions of the model have been applied to a number of outdoor and indoor fire scenarios, heat transfer problems and gas transport analysis. The FDS program is increasingly being used for non-fire applications such as dispersion modeling (see [10], [11], [13], [14],) which can be used in applications such as evaluating human exposure to vehicle exhaust [17]). Validation studies for FDS are described in the fire protection literature [18] .

The model requires that the space under consideration be discretized into smaller volumes or grid cells in order to resolve the smoke and transport equations. The exact size of these volumes is a compromise between computational speed and accuracy. A paper by Zhang *et al.* [19] examined the effect of grid size on mean temperature and velocity quantities. The paper concluded that scenarios without high turbulence and/or buoyancy could be accurately calculated with larger grid cell sizes. The gas transport scenario that was modeled in this analysis has lower levels of turbulence and buoyancy, especially remote from the gas release area; thus, coarser grid cells were appropriate for this model. For the wind field far from the building, a very coarse grid on the order of several meters is sufficient to establish the rough flow pattern and is necessary to avoid tens of millions of cells, which would be computationally impractical.

3 Full Scale Test Series

To evaluate how well the improved species transport model in FDS would be able to model the gas dispersion in a full scale configuration, select scenarios from a previously performed Mock Urban Setting Test (MUST) series were replicated using FDS version 6.2. The test series is described by Yee and Biltoft [20] [21]. The tests were performed in September of 2001 at the U.S. Army Dugway Proving Ground (DPG) Horizontal Grid test site in the Great Basin desert in Utah.

3.1 MUST Description

The focus of the test series was to acquire dispersion data for intentional or accidental gas release in an urban setting of near full-scale configuration. The urban mockup consisted of a 12 x 10 grid of shipping containers measuring 12.2 m long, 2.42 m wide, and 2.54 m high. The obstructions were spaced 7.9 m in the long direction, and 12.9 m in the short direction. The layout of the obstruction grid as presented by Yee and Biltoft is shown in Figure 3-1.

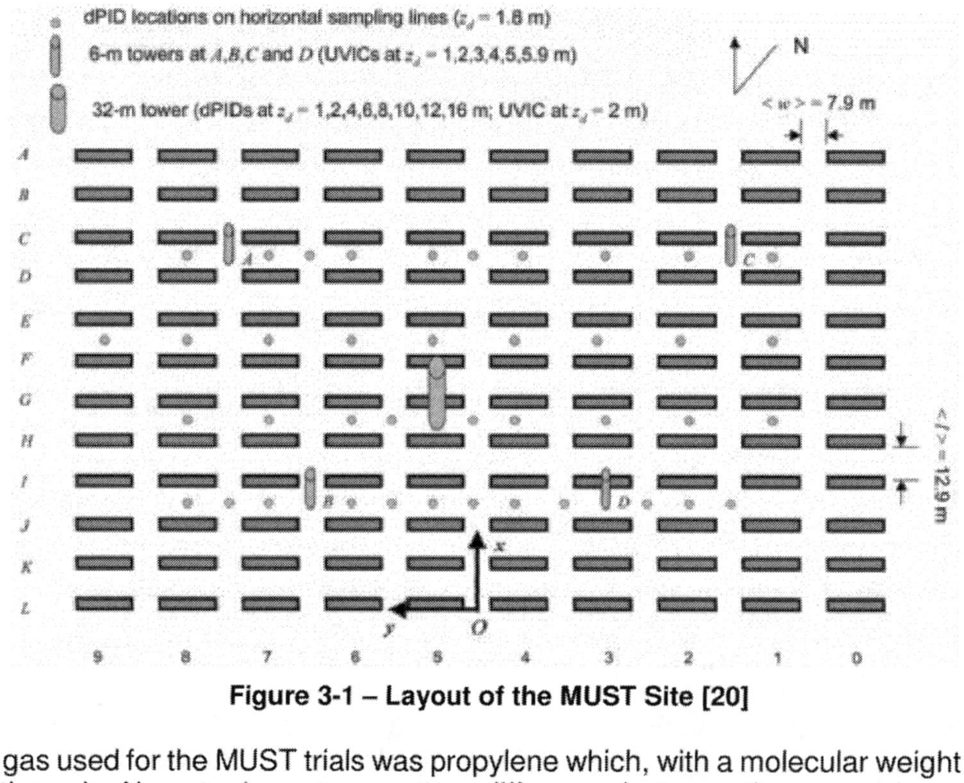

Figure 3-1 – Layout of the MUST Site [20]

The tracer gas used for the MUST trials was propylene which, with a molecular weight of 42 g/mol, is heavier than air. Absent a large temperature difference between the gas and surrounding air, such gases will tend to spread along the ground rather than immediately rise upward as a less dense gas would. Based on its heavier molecular weight, propylene can be considered a reasonable approximation for LNG and LP releases, which exhibit similar behavior after release due to the low gas temperature [2] [14]. A series of four sampling lines measured the concentration of propylene as it dispersed through the obstacle grid. The wind vector for each of the trials was reported, primarily from the south at a velocity of 1.6 – 7.9 m/s. The tracer gas was released between the second and third rows of obstacles and was carried into the obstacle grid towards the measurement devices by the prevailing wind. The test used a 15 minute continuous, steady-state gas release.

Detailed concentration results were reported by Yee and Biltoft [20] for trial number 5, which was modeled in FDS and the results compared. Trial 6 was also modeled. The wind velocity, wind angle, and propylene release rate from trials 5 and 6 are shown in Table 3-1.

Table 3-1 – Properties of MUST Trials Modeled in FDS

Num.	Wind Velocity [m/s]	Wind Angle	Release Rate [lpm]
5	2.69	-26	200
6	1.89	-10	200

4 FDS Model Setup

The details of the most important parameters and inputs to the FDS model are discussed below. The model domain used for the FDS simulations was 210 m by 260 m by 20 m and consisted of approximately 1,444,740 cells. Surrounding the gas release point and including the first row of measurement devices, a cell size of 0.50 m was used, and, in general, 1.0 m cells were used upwind. In the areas at the edge of the domain downwind and to the west, 2.0 m cells were used. In an analysis of the ability of FDS to accurately model LNG dispersion, Kohout found that a 2 m x 2 m x 1m grid size was sufficient for large-scale domains [14]. One model simulation was run using half size, 0.25 m cells around the release point and compared to the equivalent large cell run. The results demonstrated that the peak gas concentration measurements were in the same range seen for the different configurations, and the lateral plume shape was broadly similar. The conclusion from this comparison was that the 0.50 m cell size yielded resolution-independent results.

4.1 Modeling Wind

The wind characteristics measured during the MUST trial were used to develop a wind field in the FDS model, based on the average velocity and wind angle. The numerical grid in the model was extended 40 m beyond the south edge of the obstruction grid to allow the wind to develop before it encountered the obstructions.

The wind was defined in the FDS model using a vent on the south edge of the domain. The vent supplied air with a normal component equal to $S \times \sin\alpha$, where S is the wind velocity reported by Yee and Biltoft [20] for the relevant MUST trial (m/s), and α is the wind angle, measured from the y-axis. Similarly, the vent had a tangential component of velocity equal to $S \times \cos\alpha$. In the trials reported here (i.e., 5 and 6), the tangential component pointed in the negative y-direction. The development of the wind vector field in the FDS model is shown below. Figure 4-1 is a visualization of the velocities early in the simulation before the wind front has reached the end of the obstruction grid. Figure 4-2 shows conditions later in the simulation when the wind field is fully developed throughout the obstruction grid. The propylene gas was released once the wind field was fully developed past the release location at approximately 3 minutes.

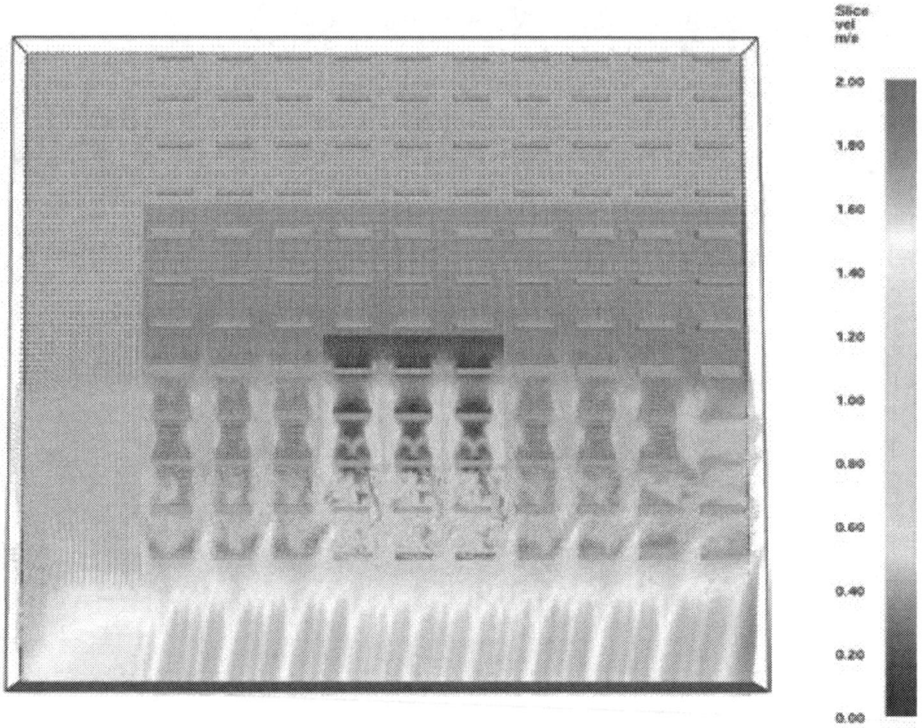

Figure 4-1 – Wind Field Early in Simulation

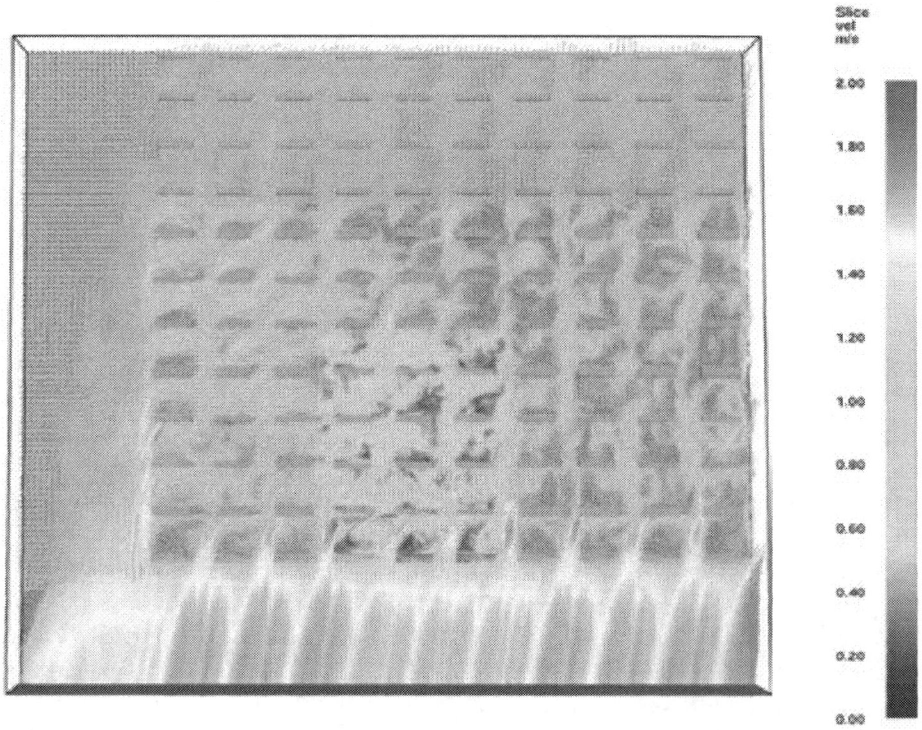

Figure 4-2 – Fully Developed Wind Field

4.2 Measurement Devices

The measurement devices were placed as reported by Yee and Biltoft [20] for the MUST trials, shown in Figure 3-1, with point devices measuring gas concentration as volume fraction in three rows parallel to the y-axis (i.e., running east-west).

In addition to the point devices, the gas concentration was also evaluated via the animated planar slices at predefined locations in the x, y, and z planes. Iso-surface output, which depicts an animated gas-front at a given concentration, was also included.

4.3 Obstruction Separation Analysis

To simplify the evaluation of the effects of obstruction separation distance, the other parameters such as wind and release rate were kept constant. The 5th MUST trial was chosen, as it featured a modest wind velocity at a slight diagonal angle, allowing for analysis of the gas dispersion both along the x and y-axis. A steeper wind angle would require the model domain to be extended further in the y-direction to include the entire wind field, greatly increasing the required computation time.

Trial 5 was used as the baseline. The distance between the obstructions in the y-direction was changed to examine the effect of obstruction spacing on gas dispersion. The range of separation distances modeled in FDS is shown in Table 4-1.

Table 4-1 – Obstruction Separation Distances Modeled in FDS

Run #	Obstruction Separation Distance
1	2 m
2	4 m
3	6 m
4	7 m
5	8 m*
6	9 m
7	10 m
8	11 m
9	12 m
10	13 m
11	14 m
12	15 m
13	16 m
14	17 m
15	18 m
16	20 m
17	22 m

** Original MUST trial separation distance.*

The obstructions were not moved in the x-direction, where the separation was kept constant at 12.9 m. The measurement devices remained in the same locations as in the original MUST trial, as shown in Figure 3-1.

5 Results and Discussion

The results from the FDS model simulations are discussed first in the context of comparing to a set of the MUST data as presented by Yee and Biltoft [20], and then evaluating the effect of increasing or decreasing the separation distance between the obstructions. An exhaustive set of measurement parameters and locations can be examined with a FDS model, so only a limited subset of the data generated is evaluated. With time and computational resources, further examination of other outputs could be undertaken.

5.1 Comparing Gas Concentration Results from FDS and MUST Trial

The data from MUST trial 6 is presented showing the mean propylene concentration at each of the measurement devices placed along the y-axis in the test field.

The MUST data was post-processed and presented as the mean concentration over the 200-second time exhibiting the lowest wind vector fluctuations [20]. As discussed by Mouilleau and Champassith [13], this is not directly translatable to FDS results, as there are no fluctuations in the prescribed wind velocity vector. Instead, concentration results from multiple 300 s moving average periods from FDS were found to give results with lower variation. Using 200 s moving average periods gave large variations in concentration at a given measurement location.

Similar to the work by Mouilleau and Champassith [13] using FDS version 4, the results from a series of eleven 300 s moving average periods obtained from FDS were compared to the results of MUST trial 6 as reported by Yee and Biltoft [20] and is shown in Figure 5-1. The results from a full 15 minute median concentration from FDS are also shown as a red dotted line.

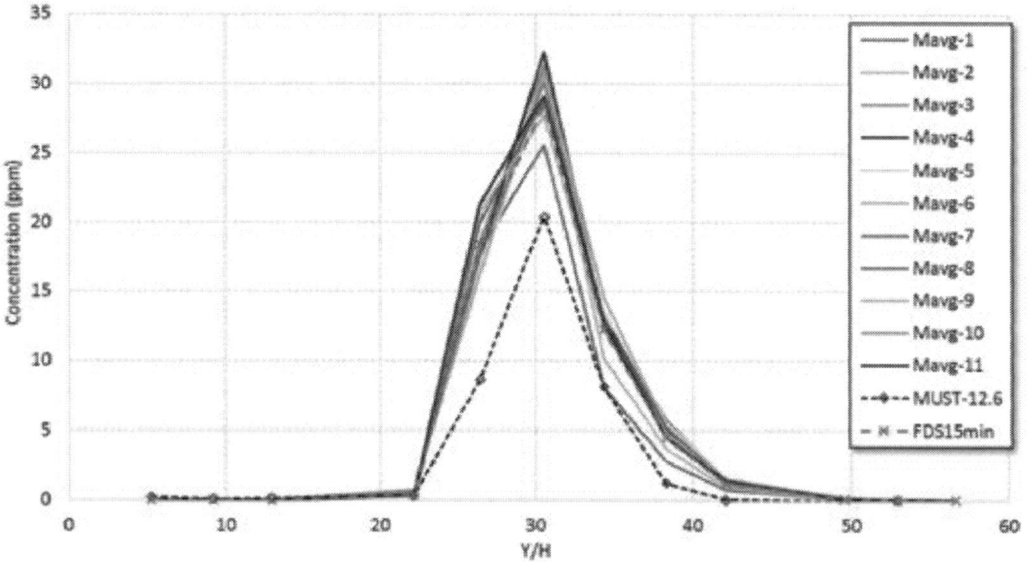

Figure 5-1 – 300 s Moving Average Gas Concentration in FDS and the MUST Trial 6 Results

From the figure it is clear that the 15 minute median concentration results from FDS fall within the results of the 11 moving average periods, and was chosen as a single representative value for the average value from the FDS model. Comparing only the 15 minute median value with the MUST trial results for the two rows of measurement devices, at X/H = 12.6 and X/H=25.1, is shown in Figure 5-2.

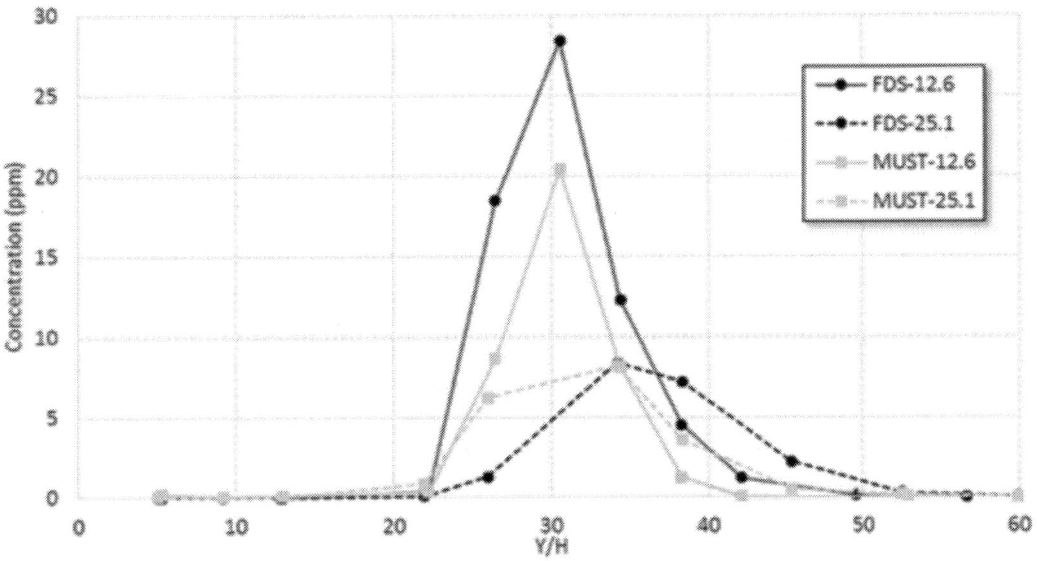

Figure 5-2 – Comparison of Gas Concentration in FDS and the MUST Trial 6

The FDS model results show higher concentrations for the three devices closest to the plume centerline, indicating that FDS predicts less plume spread, gas mixing, and air entrainment than seen in the tests. This could be due to the relatively coarse grid resolution used, which reduced the number of cells available for dissipation and entrainment, keeping the gas plume more compact. However, the model with twice the resolution resulted in a plume shape almost identical to the above. This finding suggests that additional study should consider the lateral gas mixing of downwind gas plumes in FDS when compared to the large-scale test data.

Evaluating the median gas concentration at the first and second measurement lines, FDS demonstrated good agreement with the reported test results, both in terms of the horizontal plume shape and the individual measurement values. The peak value at the plume centerline was 40% higher in FDS than in the experiment. Considering the multiple variables, wind fluctuations, and uncertainties, this difference is within the expected margin of error, which for gas transport problems in FDS was found to be 15% in a room-scale test [18]. In his work with FDS 5.5., Kohout [14] determined that an uncertainty of 300% should be applied to the results when using FDS to model LNG dispersion in obstructions and sloped terrain at the scale considered in these simulations. The 40% uncertainty found in this study (which given the over prediction in FDS will generally be conservative for hazard analysis) is an improvement on Kohout's findings.

5.2 Effect of Obstruction Separation Distance on Gas Concentrations

The gas concentration measured along the first row of devices with obstruction separation distance increased in increments of 1 m, from 2 m to a total 22 m separation is shown in Figure 5-3.

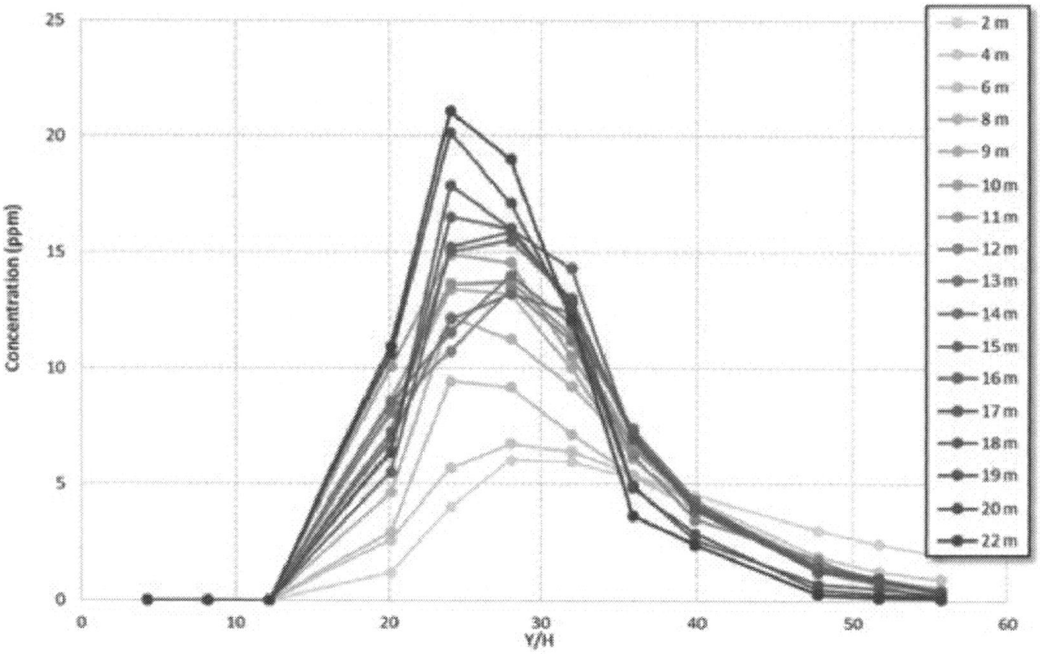

Figure 5-3 – Gas Concentrations at the First Row of Devices with Increasing Obstruction Separation Distance

From Figure 5-3, it is clear that there is little change in the near-Gaussian plume shape. The tails do not change significantly at the right side, i.e., diagonally downwind. On the left side of the plume centerline, the concentration measured at the 4th device (Y/H = 20) show a tendency to increase with increasing separation. The two simulations with the largest separation distance (Y = 20 m and 22 m) exhibit a significant jump in concentrations up through the initial measurement locations, and drop to the lowest values downwind of the peak (Y/H > 35). This concentration development is expected, as a tighter obstruction grid serves to restrict the lateral movement of the propylene gas and force it toward a path normal to the y-axis, while a looser grid allows the gas to more easily travel in the x-direction.

The 15 minute median concentration at the device with the highest concentration (generally device number 5 or 6) is plotted for each of the model scenarios in Figure 5-4. Using the maximum value of all the devices on the measurement line, rather than the same device fixed in place for all of the runs, will avoid large changes in value between runs as the device is either behind or between the obstructions in relation to the plume as the obstructions are moved. Alternatively, the location of the device in relation to the obstructions could be kept constant, e.g., always centered between two.

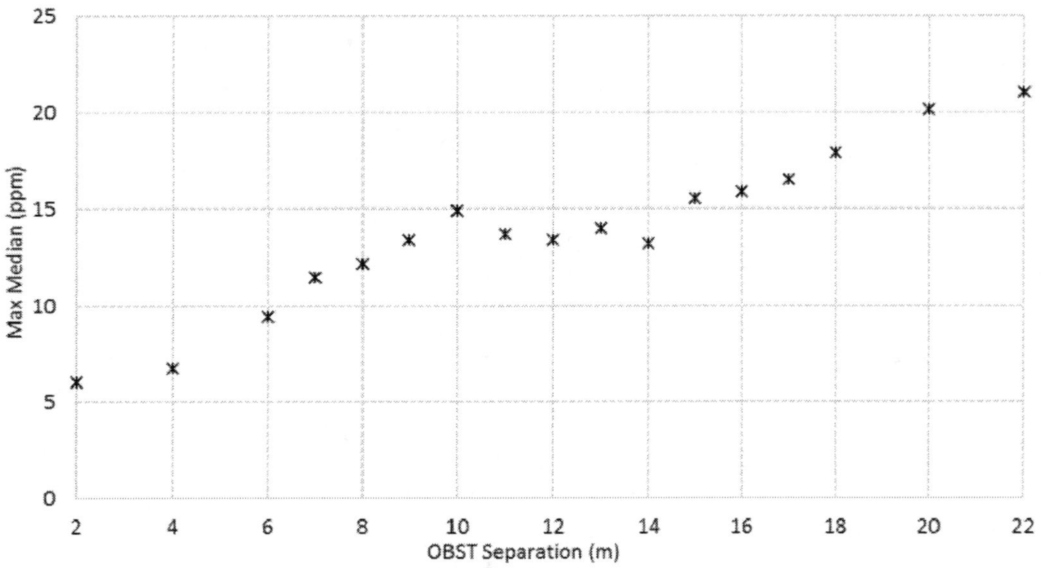

Figure 5-4 – Highest 15 Minute Median Concentration Measured on the First Row of Devices with Varying Obstruction Separation

The median concentration tends to increase as obstruction separation increases. At the smallest separation distances, very small amounts of propylene gas move between obstructions to the measurement line; the majority of the gas flows over the top of the obstruction. This can be seen in Figure 5-5, which shows a snapshot of the animated ISO surface of 10 ppm gas concentration for the simulations with a 2 m separation. The tracer gas is channeled to the right, between the obstructions, in the direction of the prevailing wind.

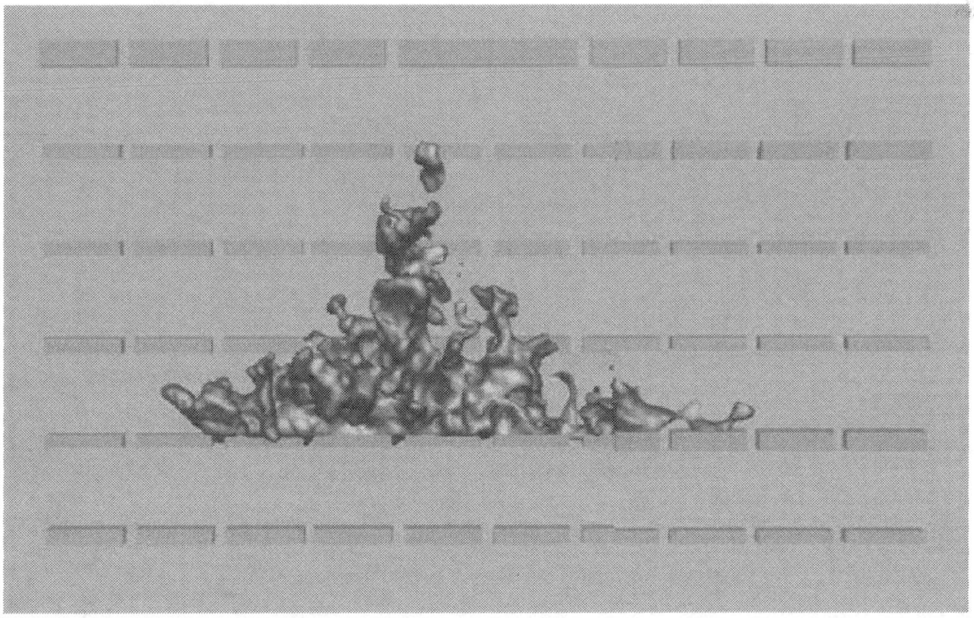

Figure 5-5 – ISO Surface Showing the Gas Front at 10 ppm with 2 m Separation

For comparison Figure 5-6 depicts the same gas concentration with a 17 m separation, where the majority of the gas cloud moves diagonally through the obstruction grid.

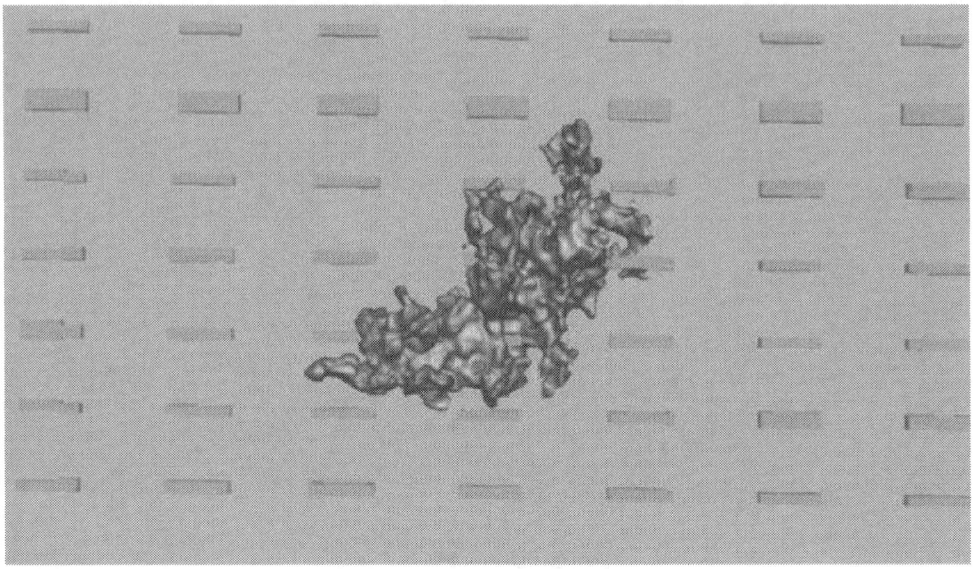

Figure 5-6 – ISO Surface Showing the Gas Front at 10 ppm with 17 m Separation

The highest measured 15 minute median concentration at the second row of devices is plotted alongside the 1st row measurements in Figure 5-7.

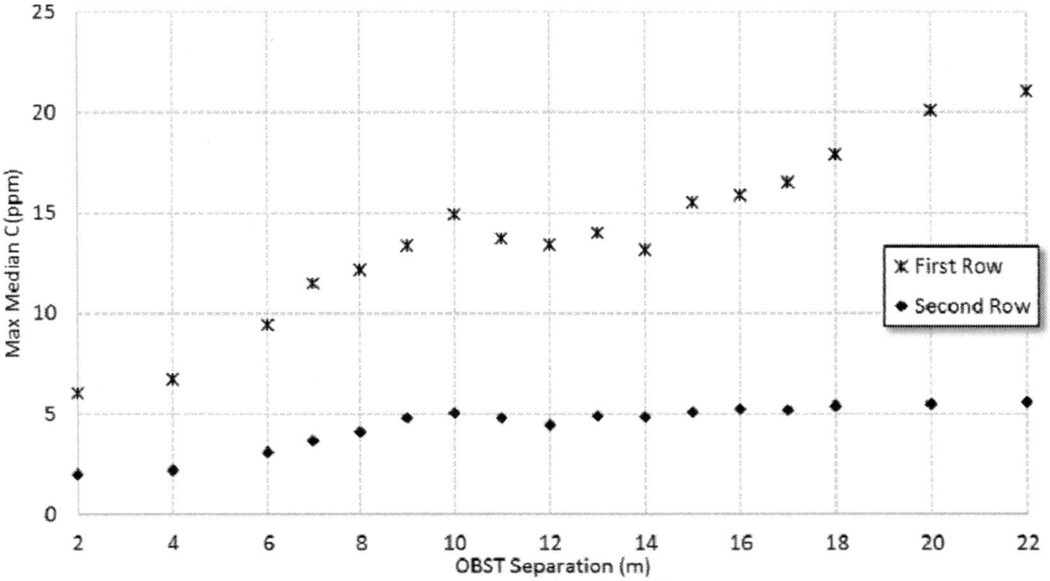

Figure 5-7 - Highest Measured 15 Minute Median Concentration on Both Rows of Devices with Increasing Obstruction Separation

In all modeled scenarios with varying obstruction separation distance, the highest median gas concentration at the 2nd row of measurement devices was approximately 1/3 of the concentration at the 1st row. The 2nd row measurement was 32-37% of the 1st row up to a separation distance of 17 m, after which it dropped to 27% for the 20 m and 22 m separation distance simulations. The wider separation between the obstructions allowed the plume to widen more downstream, and, therefore, increased the gas diffusion through mixing. This led to a greater difference between the 1st and 2nd row concentration measurements at the wider separation distances.

For the full range of separation distances modeled in FDS, the results demonstrated reasonably accurate predictions and captured the changes in gas flow patterns associated with input changes as small as 1 m, or two grid cells. No extreme values or unexpected results were observed in either the pointwise measurements or the visual and animated results.

The MUST trial and FDS simulations performed here used were based on release of heavier than air propylene in gaseous form. In many hazard analysis applications the release will be of a liquefied gas into ambient air. Modeling the pooling, vaporization and boiling of a liquefied gas is highly complex and there is currently no algorithm in FDS to perform these calculations. Separate calculations would have to be made and the results used as source term inputs to the FDS model, introducing additional uncertainties.

6 Conclusions

Version 6.2 of the Fire Dynamics Simulator was used to model the dispersion of a propylene tracer gas through a grid of obstructions under different wind conditions given a steady-state release rate. The model simulations replicated a full-scale gas release test, the MUST series performed in the Utah desert in 2001 and reported by Yee and Biltoft [20]. Two trials, 5 and 6, were modeled in FDS and used for this analysis.

Trial 6 of the MUST series was reproduced in FDS based on the reported parameters. The test results for the median gas concentration as described were compared to the values found in the FDS model. The analysis determined that FDS provided a good match for the horizontal plume shape and predicted the highest median gas concentration values within 40%. This is considered well below the 300% uncertainty recommended based on similar studies using previous versions of the FDS code reported by Kohout [14].

The effects of obstruction separation distance on the gas concentration were evaluated using Trial 5 from the MUST series as the baseline. Obstruction separation distances between 2 m and 22 m were modeled in FDS. The FDS model demonstrated good ability to predict changes in gas flow at all separation distance values modeled, and changes in configurations as small as two grid cells were well captured by the FDS model. The results showed a general trend of increased peak gas concentrations as the separation distance increased, and predicted a shift of the plume towards the upwind direction. Future work should be performed to evaluate the lateral downwind plume mixing predicted by FDS. The gas plume in the FDS model was found to be more compact with higher gas concentration in the center and lower at the edges compared to the test results. The dispersion og gas given a batch or unsteady release rate should also be studied.

Significant upgrades to the species transport submodel in version 6 of FDS make it well suited to predicting flammable vapor dispersion and appear to be a marked improvement over previous versions. Based on these results, the use of FDS for field siting analyses should be strongly considered, for its increased accuracy of results as well as broad support, ready availability and low financial cost required to produce useful analyses.

7 References

[1] Code of Federal Regulations, "49 CFR 193, Liquefied Natural Gas Facilities: Federal Safety Standards," 2016.

[2] A. Luketa-Hanlin, "A review of large-scale LNG spills: Experiments and modeling," *Journal of Hazardous Materials,* vol. A132, p. 119–140, 2006.

[3] R. Koopman et al., "Burro Series Data Report," LLNL/NWC 1980 LNG Spill Tests, December 1982.

[4] J. Havens, "A Dispersion Model for Elevated Dense Gas Jet Chemical Releases," EPA-450/4-88-006a, 1998.

[5] L. Post, "HGSYSTEM 3.0, Technical Reference Manual," TNER.94.059, Shell Research Ltd., 1994.

[6] F. Gavelli, E. Bullister and H. Kytomaa, "Application of CFD (Fluent) to LNG spills into geometrically complex environments," *Journal of Hazardous Materials,* vol. 159, p. 158–168, 2008.

[7] D. Hissong, "Keys to modeling LNG spills on water," *Journal of Hazardous Materials,* vol. 140, p. 465–477, 2007.

[8] R. Qi, Liquefied Natural Gas (LNG) Vapor Dispersion Modeling with Computational Fluid Dynamics Codes, Submitted to the Office of Graduate Studies of Texas A&M University: A Dissertation, 2011.

[9] F. Gavelli, M. Chernovsky, E. Bullister and H. Kytomaa, "Quantification of source-level turbulence during LNG spills onto a water pond," *Journal of Loss Prevention in the Process Industries,* vol. 22, pp. 809-819, 2009.

[10] N. R. Ryder, R. A. Sutula, C. F. Schemel, A. J. Hamer and V. Van Brunt, "Consequence modeling using the fire dynamics simulator," *Journal of Hazardous Materials,* vol. 115, pp. 149-154, 2004.

[11] R. Pape and M. R. Kim , "Computational Fluid Dynamics Methods of Calculating Gas Dispersion in a Building or Outside as a Result of Gas Leaks," in *Natural Gas & LP Coordination Conference & Exhibition*, 2005.

[12] W. Molina-Torres, B. Kim, R. Mentzer and M. Mannan, "Forced Dispersion Analysis of LNG Vapor Mitigation Using Fire Dynamic Simulator," *2013 AIChE Spring Meeting & 9th Global Congress on Process Safety,* 2013.

[13] Y. Mouilleau and A. Champassith, "CFD simulations of atmospheric gas dispersion using the Fire Dynamics Simulator (FDS)," *Journal of Loss Prevention in the Process Industries,* vol. 22, pp. 316-323, 2009.

[14] A. J. Kohout, "Evaluation of Fire Dynamics Simulator for Liquefied Natural Gas Vapor Dispersion Hazard," University of Maryland, College Park, College Park, MD, 2011.

[15] K. B. McGrattan, H. R. Baum and R. G. Rehm, "Large Eddy Simualtion of Smoke Movement in Three Dimensions," *Interflam '96,* pp. 189-198, 1996.

[16] K. McGrattan, R. McDermott, S. Hostikka, J. Floyd, K. Overholt, C. Weinschenk and K. Overholt, "NIST SP 1019, 6th edition, Fire Dynamics Simulator, User's Guide," National Institute of Standards and Technology, Gaithersburg, MD, 2015.

[17] H. R. Boehmer, J. E. Floyd and M. J. Ferreira, "Modeling of Carbon Monoxide Dispersion from Vehicle Exhaust in a Partially Enclosed Roadway Using FDS," in *Fire and Evacuation Modeling Technical Conference*, Gaithersburg, MD, 2014.

[18] K. McGrattan, S. Hostikka, R. McDermott, J. Floyd, C. Weinschenk and K. Overholt, "NIST SP 1018-3 Fire Dynamics Simulator Technical Reference Guide, Volume 3: Validation," Institute of Standards and Technology, Gaithersburg, MD, 2015.

[19] W. Zhang, A. Hamer, M. Klassen, D. Carpenter and J. R. Richard, "erification of the Turbulence Statistics for Fire Dynamic Simulator in a Room Fire Model," in *Proceedings of the Third Technical Symposium on Computer Applications in Fire Protection Engineering,* Baltimore, MD, 2001.

[20] E. Yee and C. A. Biltoft, "Concentration Fluctuation Measurements in a Plume Dispersing Through a Regular Array of Obstacles," *Boundary-Layer Meteorology,* pp. 363-415, 2004.

[21] C. A. Biltoft, "Customer Report for Mock Urban Setting Test," Defense Threat Reduction Agency, Alexandria, VA, 2001.

[22] J. Floyd, R. McDermott, S. Hostikka and K. G. McGrattan, "NIST SP 1019-6 Fire Dynamics Simulator (Version 6) User's Guide," Institute of Standards and Technology, Gaithersburg, MD, 2013.

Transient Large-Scale Chlorine Releases in the Jack Rabbit II Field Tests: Near Source Release Data and Preliminary Analysis

Tom Spicer, Ph.D., P.E.
Martin Dept. of Chemical Engineering, University of Arkansas, Fayetteville, AR
tos@uark.edu

Scott Wallace
Olin Corporation, Niagara Falls, NY
BSWallace@Olin.Com

Cristian E. Tabara
Sipex Sun
Dugway Proving Ground, Dugway, UT
cristian.e.tabara.civ@mail.mil
sipex.sun.civ@mail.mil

Prepared for Presentation at
American Institute of Chemical Engineers
2016 Spring Meeting
12th Global Congress on Process Safety
Houston, Texas
April 11-23, 2016

Transient Large-Scale Chlorine Releases in the Jack Rabbit II Field Tests: Near Source Release Data and Preliminary Analysis

Tom Spicer, Ph.D., P.E. [(1)]
Martin Dept. of Chemical Engineering, University of Arkansas, Fayetteville, AR
tos@uark.edu

Scott Wallace
Olin Corporation, Niagara Falls, NY

Cristian E. Tabara
Sipex Sun
Dugway Proving Ground, Dugway, UT

Keywords: Consequence Modeling; Chemical Security; Hazard Identification and Risk Analysis

Abstract

Sponsored by the Chemical Security Analysis Center (CSAC) of the U.S. Department of Homeland Security, the Defense Threat Reduction Agency (DTRA) of the U.S. Department of Defense, and Transport Canada, the Jack Rabbit II tests were designed to release liquid chlorine at ambient temperature in quantities of 5 to 20 T for the purpose of quantifying the behavior and hazards of catastrophic chlorine releases at scales represented by rail and truck transport vessels. Phase 1 of the two-year testing campaign was conducted at Dugway Proving Ground, Utah, in August and September of 2015. Five successful field trials were conducted in which chlorine was released in quantities of 5 to 10 tons through a 6-inch circular breach in the tank and directed vertically downward at 1 m elevation over a concrete pad. Data from the test program is being made available. This paper summarizes the design of the tank, release mechanism, and the instrumentation used to quantify the internal dynamics of the tank during the release, including the mass release rate. In common with other toxic industrial chemicals (TICs) such as ammonia, liquid chlorine stored at ambient temperature will flash when depressurized to atmospheric pressure forming an aerosol. This paper also discusses a preliminary analysis of the available data to model the release rate as a function of time under such circumstances.

(1) Corresponding author: +1.479.575.6516

1. Introduction

Initiated by the U.S. Department of Homeland Security (DHS) Transportation Security Administration (TSA) Freight Rail HAZMAT division, the Department of Homeland Security (DHS) Chemical Security Analysis Center (CSAC) planned and conducted the Jack Rabbit (I) test series, a series of mid-sized field tests on the two highest priority toxic inhalation hazard (TIH) materials, chlorine and anhydrous ammonia [1]. The one and two (short) ton releases were conducted at the Dugway Proving Grounds (DPG) in April and May of 2010 for the overall goal of improving the understanding of the consequences of large scale TIH releases. Key findings of the tests were the significant persistence of the denser-than-air clouds formed during the release (more prevalent with increased release size) [2] and the extent to which chlorine removal could mitigate the impact of a release [3].

As a continuation of the previous tests, the Jack Rabbit II (JRII) test program was started in 2014 to focus on large scale, controlled chlorine field experiments representative of releases from railcars and tank-trucks (5 to 20 tons). The broad scientific goals of the test program are to collect data on the release source, cloud transport and dispersion, chemical reactions with the environment (including potential mitigation effects), exposure effects on equipment and infrastructure, and effect of releases within a mock urban test environment including indoor infiltration. In addition, the program will provide data, training materials, and guidance for improved emergency response efforts and support the Homeland Security Enterprise by providing quality-assured data and scientifically-based guidance to partners and stakeholders to address assessment of TIH hazards.

Preparations for the tests were organized using a series of working groups to handle specific aspects of the program including:

- Data Quality Working Group
- Instrumentation Working Group
- Mock Urban and Indoor Environment Working Group
- Modeling Working Group
- Dissemination and Near Source Working Group

Oversight and coordination of activities of the working groups is through the program's Scientific Advisory Group (SAG) with final approval by CSAC.

This paper focuses on the preparations for the test program made through the Dissemination and Near Source Working Group (DNSWG) which included representatives from government, academe, and industry to provide support for safely conducting chlorine releases in the JR II program in a reliable and repeatable fashion. In addition to the issues surrounding the dissemination vessel (disseminator), the DNSWG considered issues in the vicinity of the source including aerosol and rainout characterization. Planning and conduct of the tests were done in close coordination with The Chlorine Institute and its representatives and member organizations, particularly CHLOREP® (the Chlorine Emergency Plan). This summary also considers data taken in the disseminator during the 2015 tests and how that data can be used to determine the mass release rate (and release thrust) as a function of time. Additional data was taken during the test program including extensive downwind concentration measurements and indoor concentration and temperature measurements, but these aspects of the test program are not considered here.

2. Dissemination Vessel and Support System

The dissemination system consists of a custom-built vessel (disseminator) and support system including load cells for continuous force (mass) measurement during the release (Figure 1). The system was centered on a 25 m diameter concrete pad designed to collect any liquid rainout and to protect the gravel pad constructed on the playa at DPG from the chlorine aerosol jet anticipated during the tests. A 1 in lip was installed at the edge of the pad to keep liquid from flowing onto the gravel pad.

Figure 1. Dissemination vessel on the 25 m concrete pad

As indicated in Figure 2, the disseminator was manufactured with two vent/fill ports in the vessel headspace (items 1 and 9), an emergency drain (item 10), and a standard pressure relief valve for chlorine service (item 7). In addition, the disseminator was equipped with ports to allow for measurement of vessel pressure, temperature, differential pressure, and liquid depth. The disseminator was designed with a nominal capacity of 10 T of chlorine (2000 gal) with a length-to-diameter ratio comparable to that of a chlorine rail car. Plans for a separate dissemination vessel with a capacity of 1000 gal (for 5 T releases) were quickly abandoned due to cost and logistical considerations, and 5 T releases were simply made with the disseminator partially filled. The disseminator has an external tangent-to-tangent length of 5.64 m (222 in) and outside diameter of 1.37 m (54 in) with a uniform vessel wall thickness of 1.27 cm (0.5 in). The vessel was built with four dissemination ports oriented as follows: 90 degrees upward, horizontal, 45 degrees downward, and 90 degrees downward (Figure 3).

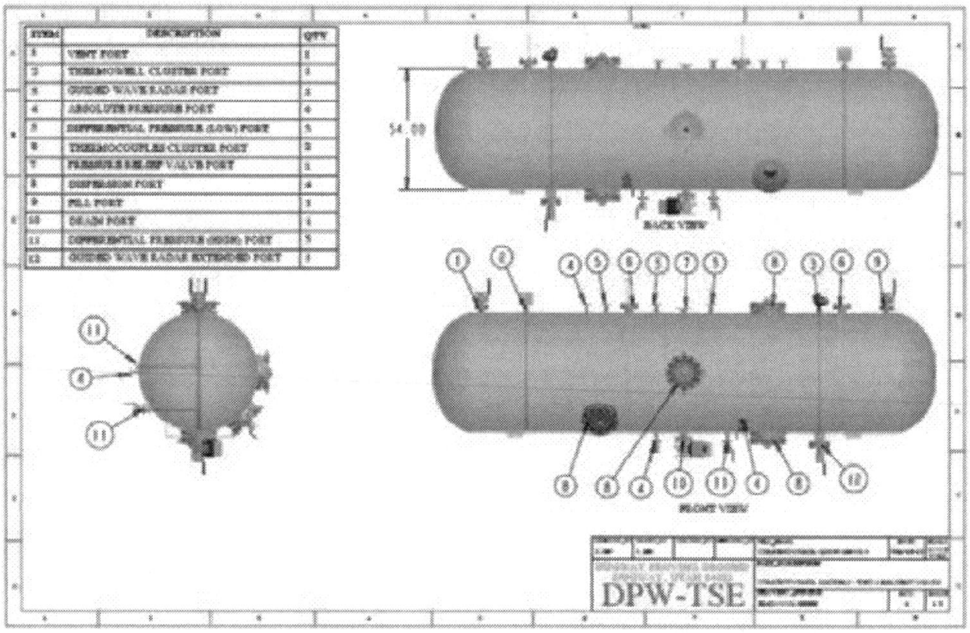

Figure 2. Disseminator with instrumentation and dissemination ports

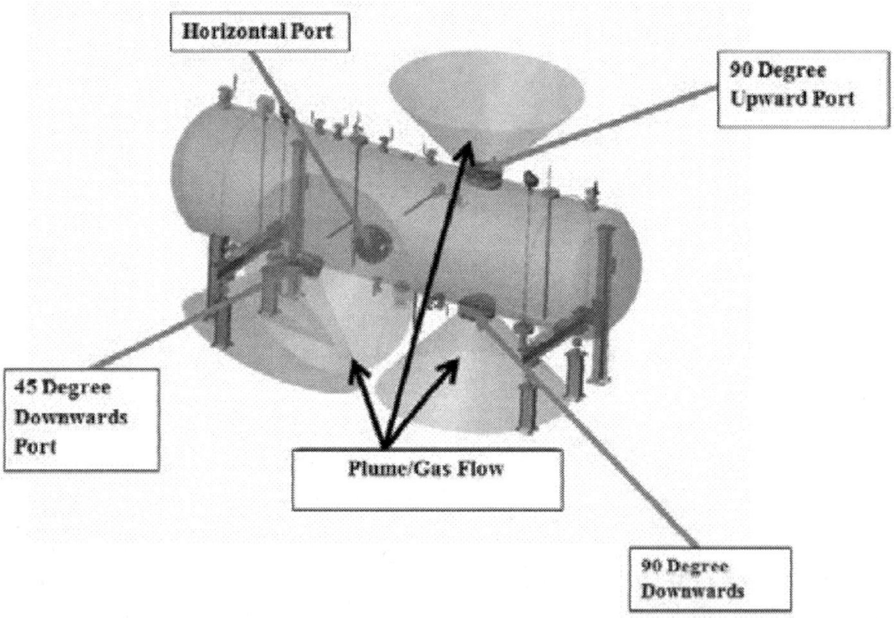

Figure 3. Disseminator vessel showing dissemination ports and release directions

The disseminator support system (Figure 4) consists of an I-beam structure instrumented with seven load cell assemblies for dynamic force and mass measurement. The system pillars were attached to a steel plate using welded bolts, and the plate was attached to the concrete pad using concrete anchors sealed with a chlorine-resistant epoxy. The disseminator was bolted to the horizontal crossbeams and also strapped to the cross beam using pre-tensioned steel cables. Four load cell assemblies measured the vertical forces, and three assemblies measured horizontal forces during the release. The load cell assemblies (consisting of flexures on either side of a load cell) were wrapped in a chlorine-resistant material to avoid damage from corrosion. The support system was designed to ensure that the down-facing dissemination port was located 1 m (3.28 ft) above ground level (AGL) and that the tank remained level during the release. Note that the ports oriented upward and downward were off the vessel midplane so that the load cells measuring vertical forces could be used to determine the mass of the disseminator contents separately from the release thrust.

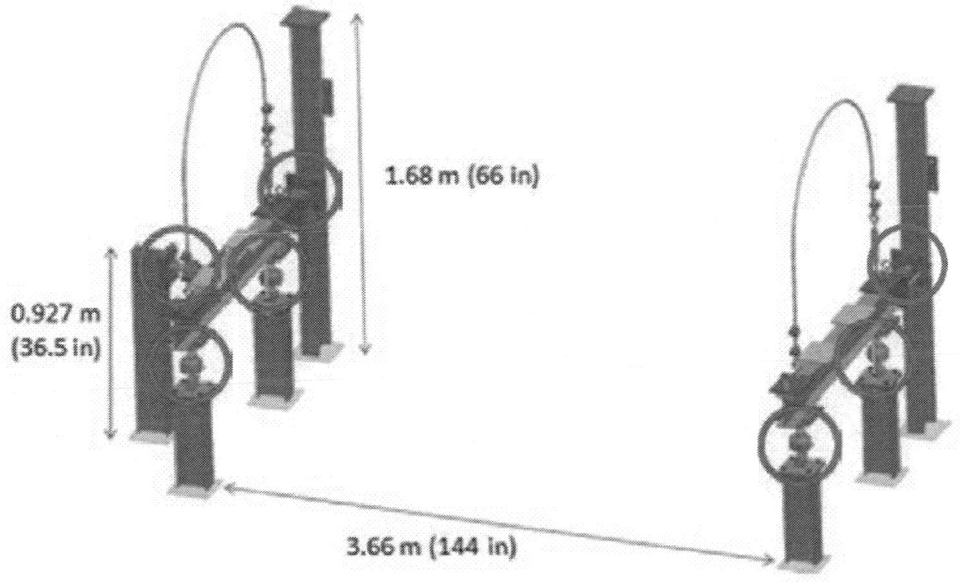

Figure 4. Disseminator support system showing load cell assemblies

Figure 5 illustrates the features of a dissemination port which was designed to provide an instantaneous opening on the disseminator in such a way as to replicate a vessel rupture. The dissemination port was designed to minimize the effect on the release of any valves or external piping as had been used in the Jack Rabbit I tests. Each port can be used for releasing chlorine via a 15.2 cm (6 in) diameter penetration or a 6.4 cm (2.5 in) diameter penetration using a threaded insert (yellow insert indicated in the Figure 5). During 2015 trials, only the 15.2 cm (6 in) diameter port was used, so the insert was not present during any of the tests reported here. The plug illustrated in Figure 5 sealed the chlorine inside the vessel until the time of release. The bottom surface of the plug was covered with a Viton® gasket, and the top threaded portion of the plug extended through a modified blind flange (labeled Flyaway Flange). During filling operations, the Flyaway Flange was secured with 12 standard bolts. Immediately before a release, the standard bolts were replaced with 6 explosive bolts in alternating holes, and at the designated time, the explosive bolts were energized, and the Flyaway Flange assembly (Flyaway Flange and plug) was pushed off the vessel by its internal pressure. Because of the significant risk to the program objectives as well as to ensure safe operations, preliminary tests were made with pressurized air and water on a smaller prototype vessel, and modifications to the disseminator and Flyaway Flange were made to allow the explosive bolts to operate correctly. Torque requirements for the plug were also determined to keep chlorine from leaking past the Viton® gasket. To avoid damage to the disseminator or its instruments during the release, the Flyaway Flange assembly was secured to the concrete pad by steel cables to control the direction of its flight after the explosive bolts were activated (visible in Figure 1). The vessel was filled with chlorine in such a way as to minimize any inerts in the vessel head space using a caustic scrubber on site.

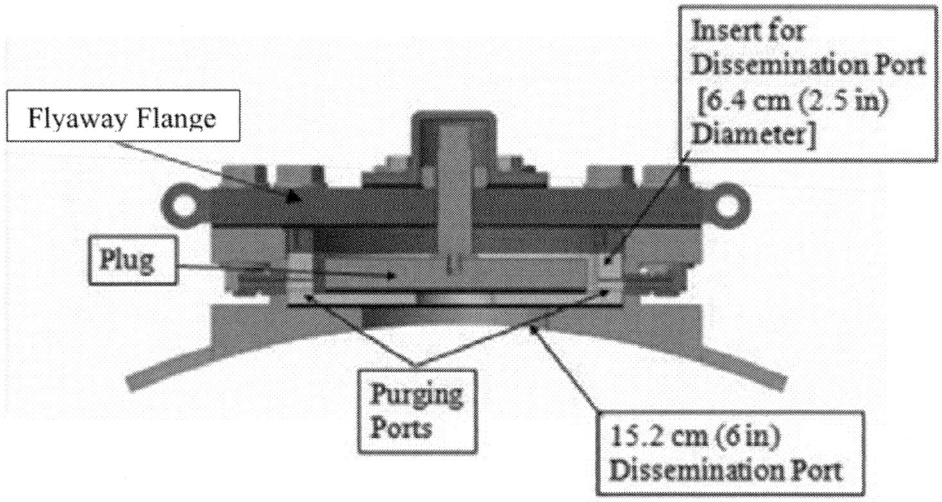

Figure 5. Dissemination port assembly

3. Dissemination Vessel Instrumentation

As discussed above, the disseminator was extensively instrumented to monitor the vessel internal dynamics during a release. Because the releases were expected to be very rapid, instruments were selected for the desired response time. However, there was also concern that any chlorine remaining in the disseminator after a release could absorb water from the ambient (humid) air creating a corrosive environment, so the instrumentation selected was intended to provide redundant measurements so that if a faster response instrument failed, a slower response instrument would be available to provide information about the tank contents before a release and during filling.

- Load Cells. As discussed above, seven load cell assemblies consisting of flexures (Force Measurement Systems FMS-UF-PF4-10K/FMS-UF-PF4-6K) mounted on both sides of a load cell (Force Measurement System 2000-S-10K-4-U/2000-S-5K-4-U) were placed on the support system to measure the chlorine mass in the vessel as a function of time as well as release thrust forces.

- Bare-Wire Thermocouples. Temperature profiles were measured vertically at increments of 10 percent of the tank diameter at two axial locations in the tank using bare wire thermocouples (Type K 20AWG Teflon expected to have a response time of 0.25 s). A sketch of the thermocouple assembly is shown in Figure 6.
- Absolute Pressure. Absolute pressure (Rosemount 3051T) was measured at four locations each of which were aligned with the top of one of the 6 in port openings (90 degrees upward, horizontal, 45 degrees downward, and 90 degrees downward, see Figure 3).
- Differential Pressure. Differential pressure (Rosemount 3051C) was measured vertically between the top of the tank and at elevations that correspond to the top of each of the 6 in port openings (horizontal, 45 degree downward, and 90 degree downward dissemination ports, see Figure 2). The differential pressure ports were outfitted with a well-line to measure the differential pressure in the center of the tank.
- Guided-Wave Radar (GWR). One GWR (Rosemount 5300) measured the liquid chlorine depth. An extended port was added to the disseminator to accommodate the additional probe length required to measure chlorine liquid level to the vessel bottom.
- Thermowell Thermocouples. Temperature profiles were planned to be made in a thermowell, but the thermowell-to-tank seal was damaged during installation, so the thermowell was not installed.

Vessel instrumentation was selected and implemented so as to be fit for the purpose of modeling depressurization and release processes.

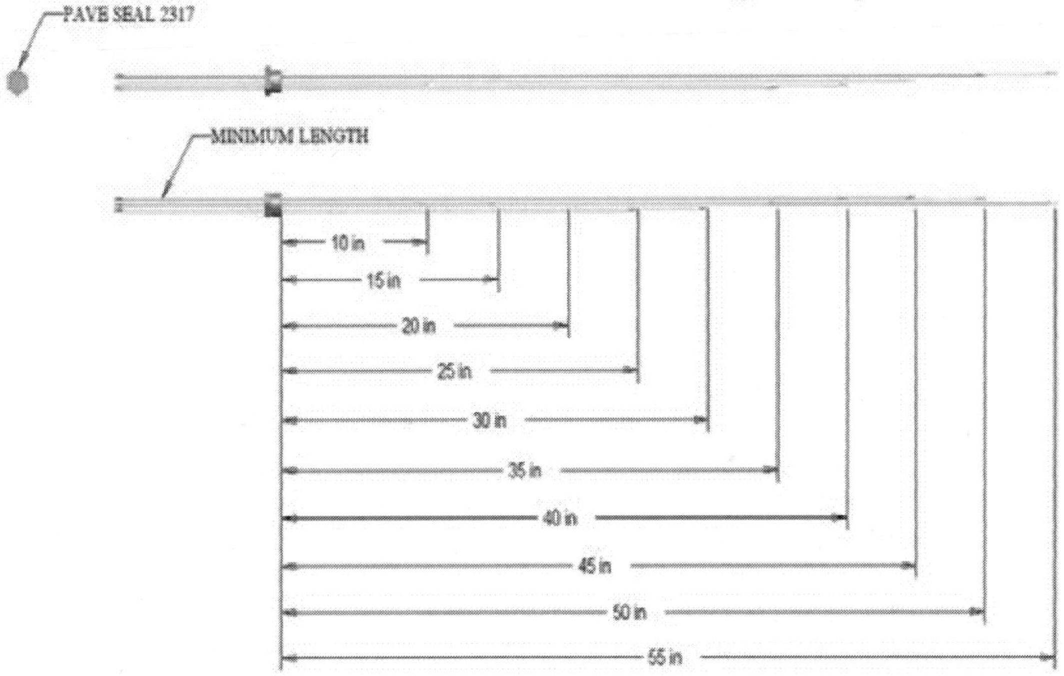

Figure 6. Thermocouple assembly with Pave Seal NPT connector

5. Test Data and Analysis

Table 1 summarizes the tests conducted during the 2015 season. All times are local times (MST). The release time is taken to be the time in the data acquisition system when the load cells reflect a significant change. (The data acquisition system was specified to take data at 100 Hz.) In Table 1, the chlorine mass released was calculated from the difference between the four vertical load cells before and after the release. In Trial 5, the data acquisition system failed during the release, so the final (tare) mass was taken to be the average of the final mass in the previous four tests.

Table 1. Jack Rabbit II Test Summary for 2015 Trials

Trial	Date	Release Time	Mass Released (T)	Wind Direction	Wind Speed (m/s)	Air Temperature (°C)	Pressure (Pa)	Relative Humidity (%)
1	8/24	7:35:45.50	5.01	147°	2.0	17.7	87354	39.2
2	8/28	9:24:20.00	9.03	158°	4.2	22.7	87512	32.8
3	8/29	7:56:55.31	5.03	169°	3.9	22.7	87100	30.2
4	9/1	8:38:47.05	7.73	183°	2.3	22.6	86928	26.8
5	9/3	7:28:17.26	9.20	182°	2.7	22.2	86655	26.5

5.1 Consistency of static chlorine mass measurements

To compare chlorine mass measurements made using the load cells with the other measurements available, it is necessary to know the volume of liquid (V_L) as a function of liquid depth in the vessel. Based on formulas given in [4],

$$V_h = \frac{\pi}{24} D^3$$

$$V_L = (L_t - 2H_h)\left[(h - R)(Dh - h^2)^{1/2} + R^2 \cos^{-1}\left(\frac{R - h}{R}\right)\right] + 2\,V_h \left(\frac{h}{D}\right)^2 \left(3 - \frac{h}{R}\right)$$

where h is the liquid depth and R and D are the inner radius and diameter of the cylindrical vessel, respectively. The volume of one head is V_h, and the depth of each head is H_h (both quantities excluding the straight flange), so the length of the cylindrical middle is $L_t - 2\ H_h$ where L_t is the tangent-to-tangent internal tank length. For a 2:1 semi-elliptical head as specified for the dissemination vessel used here, the volume V_h of a single head is given by the formula above, and the head depth H_h is D/4. As written above, the formula for V_L is suitable for all values of h including the full volume (h = D). Using a tangent-to-tangent internal length of 221 in and inside diameter of 53 in (uniform shell thickness of 0.5 in), the calculated vessel capacity is 2026 gal. DPG measured the volume of the disseminator by filling it with a metered quantity of water and found the volume to be 2020 gal. In the calculations that follow, a volume of 2026 gal will be used.

Instruments were installed on the disseminator up to the start of the testing period leaving no time for field adjustments (such as zeroing the differential pressure or DP cells so that they measured zero when no liquid was present in the tank). The data acquisition system configuration could not record differential pressures that were negative (DP was negative with no liquid present), so the leads to the DP cell leads were switched after the first test to record instrument readings. Consequently, the readings from the DP cells should be interpreted as:

$$\Delta P = \rho_f g(H_t - H_R) - \rho g h$$

where ΔP is the DP cell reading (with switched legs), ρ_f is the fluid density of the DP cell leg (specific gravity of 1.85 taken to be 1850 kg/m^3), H_t is the elevation of the top of the vapor space DP cell connection, and H_R is the elevation of the high pressure DP cell leg. As measured from the supporting steel plate, H_t is 99.35 in AGL (41+53+0.5+4.85 in), and the three elevations of the high pressure legs (H_R) are bottom-to-top: 35.65 in (41-0.5-4.85 in), 48.5 in, and 70.5 in. The chlorine liquid density is evaluated at the initial liquid temperature using the DIPPR database. For simplicity here, only the DP cell at the tank bottom is used for comparison.

Table 2 compares initial liquid mass measurements made with load cells, GWR, and (single) DP cell. In Trials 3-5, the GWR consistently measured liquid levels lower than indicated by the load cells (by roughly 10%). In Trials 2-5, the DP cells consistently measured liquid levels significantly higher than indicated by the load cells (from 10 to 37% too high). A detailed analysis of the mass of chlorine delivered to DPG and the caustic/chlorine consumed during the tests in filling/purging operations was consistent with the chlorine mass measured by the load cells.

Table 2. Comparison of Initial Chlorine Mass Measured By Load Cells, GWR, and DP Cells

Trial	Average Initial Liquid Temperature (°C)	Initial Mass by Load Cells (kg)	Initial Mass by GWR (kg)	Initial Mass by DP Cell (kg)
1	-	4543	-	-
2	18.7	8193	(out of range)	10420
3	18.0	4565	4125	6278
4	17.7	7017	6550	9066
5	20.6	8345	7992	9125

5.2 Dynamic mass and thrust measurements

As indicated above, the load cell measurements provide the most consistent measurement of mass in the vessel under static conditions, and the load cell measurements can also be analyzed as a function of time. The vertical release ports were positioned with a moment arm away from the center of mass of the tank and its contents to determine dynamic mass separately from the force due to the release. From a position facing into the (historic) mean wind direction and also facing the tank (consistent with the orientation in Figures 1-3), the load cells were designated as front (north side), back (south side), right (west side), and left (east side). The vertical release ports were located on the right (west side) centered 37 in from the axial tank center (center of mass). Because the disseminator did not move during the releases, the sum of the vertical forces and torques (about the center of mass) must be zero:

$$\Sigma F_z = 0 = M - T_z - (\Sigma F_E + \Sigma F_W)$$

$$\Sigma \tau = 0 = T_z R_T + \Sigma F_W R_L - \Sigma F_E R_L$$

where M is the mass of chlorine in the vessel, T_z is the vertical thrust due to the jet release with moment arm R_T (37 in), and F_E and F_W are the load cell forces on east and west ends, respectively, with moment arm R_L. The torque equation does not account for the vertical change of the center of mass as the vessel empties. Also, the downward force due to atmospheric pressure on the top of the vessel opposite the jet is ignored. The load cell forces were tared with measurements after the release was complete. The vertical thrust is found from the second equation, and the chlorine mass from the first. Figure 7 shows the resulting chlorine mass and thrust as a function of time for Trial 1. As indicated in the figure (and characteristic of all of the trials), there was some scatter associated with the processed data as was anticipated (particularly the thrust), and this can be largely removed by using a simple time averaging of the data to 1 s resolution; Figure 8 shows the time averaged (calculated) data for Trials 1-4. As discussed above, the data acquisition system failed on Trial 5, and values between (roughly) 38:59 and 39:02 during Trial 4 did not change indicating some additional data acquisition problem.

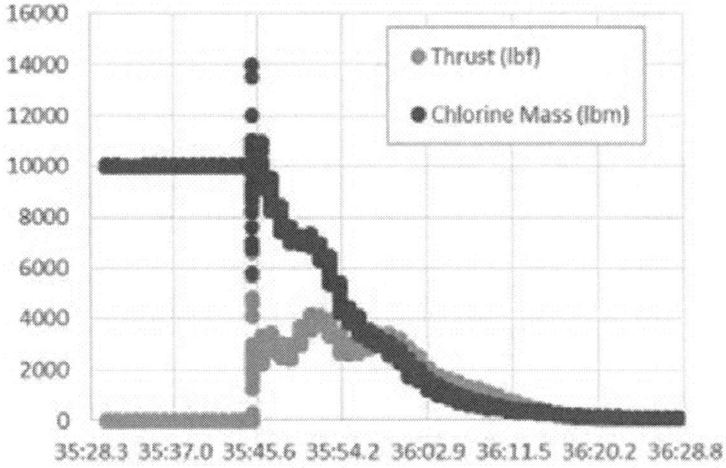

Figure 7. Chlorine mass and release thrust as a function of time for Trial 1

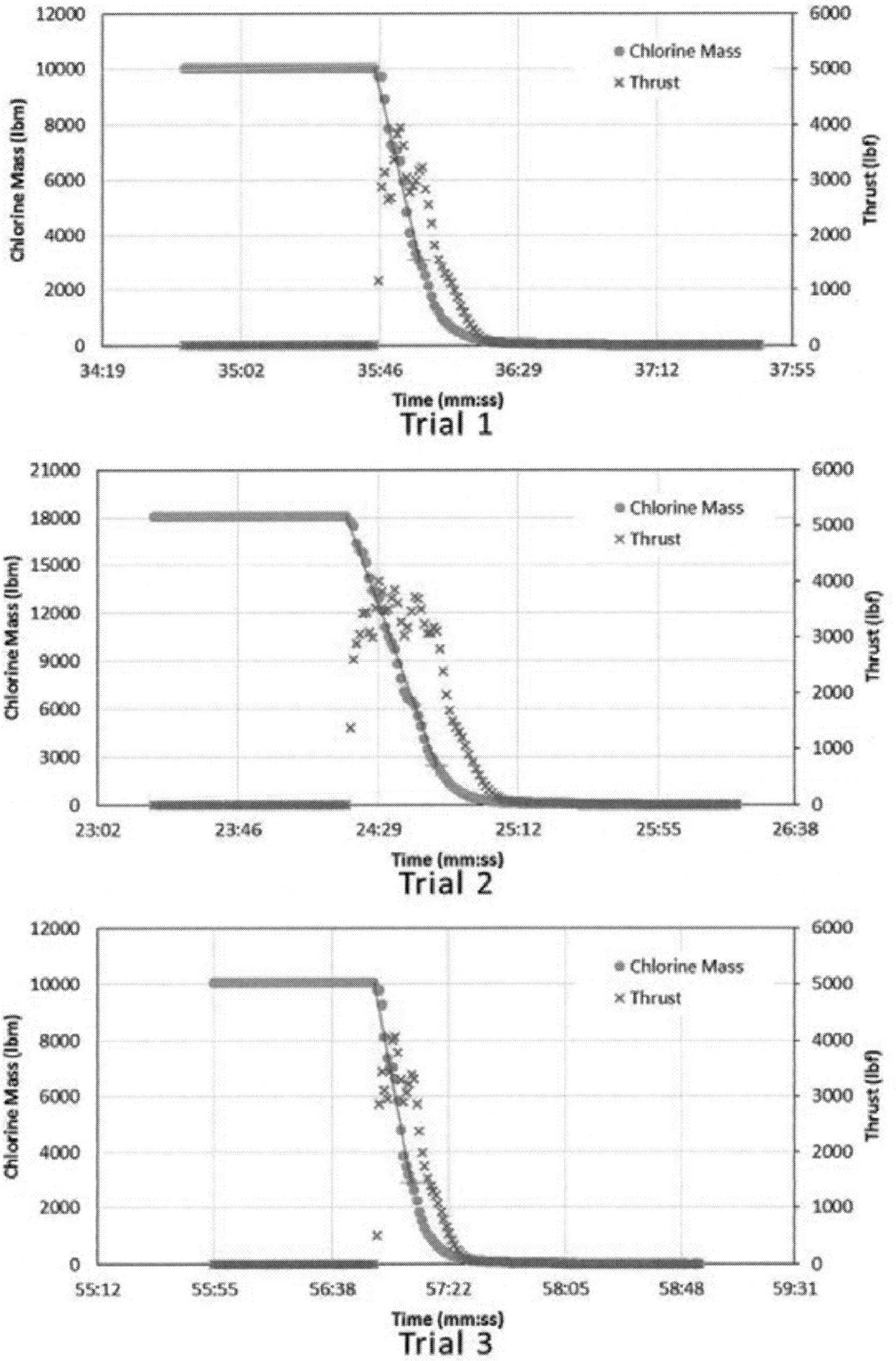

Figure 8. Time Averaged Chlorine Mass and Release Thrust as a Function of Time for Trials 1-4

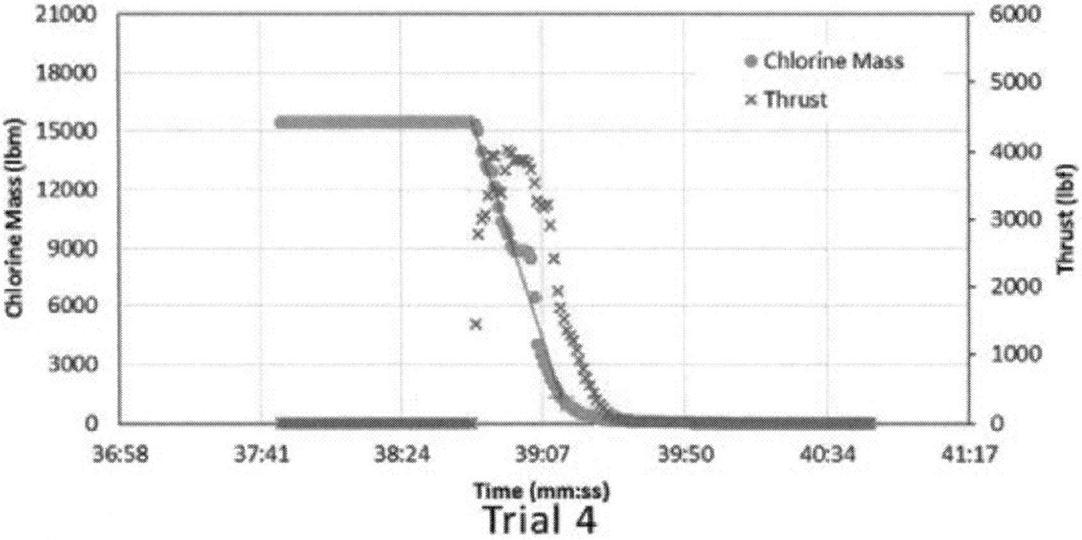

Figure 8. Time averaged chlorine mass and release thrust as a function of time for Trials 1-4 (continued)

The time averaged data was used to determine the initial mass release rate (E) as indicated by the red lines in Figure 8 (calculated as the difference between the initial chlorine mass and the mass at a given time divided by the corresponding time interval). To determine the average initial release rate, an averaging time period was used. The start of the time period was taken to be the last set of values before the release (left end of the red line), and the end of the time period was taken to be when the next (average) mass rate would drop below 90% of the maximum (average) mass rate up to that time (right end of the red line). As indicated in the figure, this approach matched the cumulative mass as a function of time reasonably well with the possible exception of Trial 4 (likely an artifact of the data acquisition problems but probably sufficient for engineering estimates). Table 3 summarizes these initial mass rates as well as the maximum average initial release rate over the same time interval. At later times, mass release rates can be calculated directly from the data using standard finite difference approximations because the data are sufficiently smooth.

Table 3. Average and Maximum Estimated Initial Mass Release Rate and Thrust

Trial	(Average) Initial Mass Release Rate (lbm/s)	Maximum (Average) Initial Mass Release Rate (lbm/s)	Average (Initial) Thrust (lbf)	Maximum Thrust (lbf)
1	494	546	3010	3940
2	557	609	3300	4000
3	511	562	3150	4050
4	524	573	3460	4020

As indicated in Figure 8, there was still considerable variation in the estimated thrust, but average values seem reasonably stable. To determine the average initial thrust, a time period was chosen starting at the time before significant changes in the load cell measurements were observed, and the end of the time period was taken to be when the next thrust would drop below 90% of the average thrust up to that time. (The averaging time periods for the thrust and the initial mass release rate were within 2 to 4 s on every test.) The average thrust so estimated is summarized in Table 3. (All values are shown to three significant figures.) Table 3 also includes the maximum thrust during the averaging period (roughly 15 to 30% larger than the average). (Note that the 1 s averaged thrust information was used to calculate the mass release rate in Figure 8, not the average thrust in Table 3.)

There were no attempts to measure the degree of flashing at the discharge point for several reasons, but an estimate can be made using the available data. The standard approach for estimating the (incompressible) release rate (E_{inc}) is

$$E_{inc} = C_D A\left(2\left((P_d + \rho_L g h) - P_a\right)\rho_L\right)^{1/2}$$

using $C_D = 0.61$ and taking the disseminator pressure P_d to be the vapor pressure at the measured liquid temperature. However, the mass release rate predicted by E_{inc} (using ρ_L) is roughly twice what was observed indicating some flashing was present at the release point (or internally). One way to estimate the degree of flashing would be to use an aerosol density (based on flash fraction f_v) in place of the liquid density in the equation for E_{inc}. Table 4 summarizes calculations of E_{inc} indicating the flash fraction is around 1%, which is significantly lower than the flash fraction at ambient pressure. (Trial 1 is excluded because there is no liquid temperature available to estimate P_d.) Consequently, the release is an underexpanded jet.

Table 4. Average and Estimated Initial Mass Release Rate

Trial	(Average) Initial Mass Release Rate (kg/s)	E_{inc} with $f_v = 0$ (kg/s)	E_{inc} with $f_v = 0.6\%$ (kg/s)	E_{inc} with $f_v = 1.3\%$ (kg/s)	E_{inc} with $f_v = 2\%$ (kg/s)
2	253	446	386	282	233
3	232	441	381	279	230
4	238	439	380	277	229

For an underexpanded jet, the thrust is made up of two components: momentum flow in the z-direction and pressure forces. Consequently,

$$T_z = EU_j + \left(P_j - P_a\right)A$$

where U_j and P_j represent the jet velocity and pressure, respectively. The momentum flow (EU_j) must be evaluated in the z-direction only. For incompressible, single phase flow, the relevant velocity in the z-direction would be the velocity in the vena contracta, so $U_j = E/(C_D A \rho_L)$.

Recognizing that flashing is present in these releases, U_j can be approximated as $E/(C_D A \rho_j)$ where ρ_j is the two-phase density at conditions represented by subscript j.

There are several possibilities for evaluating P_j:

- The flow depressurizes to ambient pressure at the jet conditions (vena contracta). This assumption would result in the largest estimate of U_j ($U_j = T_z/E$). The flow would not be in equilibrium at the jet conditions. Velocities estimated in this manner ranged from 58 to 62 m/s, and flash fractions are between 0.45 and 0.6%.
- The flow depressurizes to P_j which is given by the vapor pressure of chlorine at the jet temperature T_j. T_j can be estimated from the adiabatic flash fraction f_v ($T_j = T_L - f_v \Delta H_{vap}/C_{pL}$ where ΔH_{vap} is the heat of vaporization and C_{pL} is the liquid heat capacity), and the adiabatic flash fraction can be used to estimate ρ_j. With these assumptions (flow is in homogeneous equilibrium), the only undetermined value in the above equation is the flash fraction f_v. Flash fractions estimated in this way are between 0.4 and 0.9% (fluid properties were evaluated at the provisional jet temperature T_j). For these test conditions, the pressure forces are roughly twice as large as the momentum flow forces. Velocities (U_j) estimated in this manner are between 20 and 25 m/s.
- The flow depressurizes to P_j, and the change in specific volume can be estimated using

$$\frac{\upsilon}{\upsilon_o} = \omega\left(\frac{P_o}{P} - 1\right) + 1$$

where υ is the (two-phase) specific volume, ω is the correlating parameter given in [5] as

$$\omega = \frac{C_{pL} T_o P_o}{\upsilon_o}\left(\frac{\Delta V_{vap}}{\Delta H_{vap}}\right)^2$$

where subscript o denotes stagnation conditions. For the disseminator conditions, ω = 10.2, and $\rho_j = 1/\upsilon = \rho_L P_j/(\omega(P_d - P_j) + P_j)$ neglecting the liquid head. Velocities estimated in this manner ranged from 17 to 26 m/s, and flash fractions are between 0.1 and 1.0%

All of the approaches considered here estimate the flash fraction to be around 1% at or near the release exit, but the exit velocity estimate still has some uncertainty which may be able to be resolved using high speed video recordings of the release.

Regardless of confirming evidence, the flow in these experiments is initially subsonic, and consequently, the depressurization process would be expected to be sufficiently fast so that P_j would rapidly approach P_a with thermodynamic equilibrium to be restored on longer time scales. (Note that the flow is likely sonic or choked at later times in the release when high pressure gas flow begins, which is likely related to the time when the thrust is larger than the force due to the remaining mass in the disseminator as clearly shown in Figure 7 and characteristic of all of the tests.) Consequently, the release velocities are significantly larger than would be predicted assuming single phase incompressible flow. If the aerosol formation process is governed by aerodynamic processes (e.g., a critical Weber number), the maximum droplet size would be inversely proportional to the square of the velocity. As the maximum aerosol droplet size decreases, the rainout also decreases. For these experiments, there was clear evidence of liquid chlorine on the concrete pad during and after the releases, but the quantity that rained out was generally considered small in comparison with the amount of chlorine released. (The quantity of rainout has been questioned, and further attempts to quantify the rainout are planned for 2016.)

5.3 Test program data availability

The test program data set is currently going through a verification and validation process. As discussed above, instruments were installed on the disseminator up to the start of testing leaving no time for field adjustments. During the final test, the data acquisition system failed shortly after the release started, so there is only data available for the first few seconds of the release. (The cause of this failure is unknown but could have been a power failure since the system started working again after the test was complete.) At this point, all instruments have been subjected to a post-test calibration process, and all instruments were within calibration limits.

There are plans to make this data (and all data from the test program) available to the public via the Homeland Security Information Network (HSIN) web site. HSIN access is based on a nomination/acceptance process into a Community of Interest (COI). To begin the process, email HSIN.outreach@hq.dhs.gov with your full name and email address including the requested COI (Chemical) and the reason for access: "To become a member of the Jack Rabbit II community to have access to the information and data from the test program."

6. Conclusions

Phase I of the Jack Rabbit II tests have been completed with controlled releases of liquid chlorine at ambient temperature for the purpose of quantifying the behavior and hazards of catastrophic chlorine releases at scales represented by rail and truck transport vessels. Five successful field trials were conducted in which chlorine was released in quantities of 5 to 10 tons through a 6-inch circular breach in the disseminator vessel and directed vertically downward at 1 m elevation over a concrete pad in a mock urban environment.

This paper summarized the design of the tank, release mechanism, and the instrumentation used to quantify the internal dynamics of the tank during the release, including the release rate and thrust generated by the release. The purpose-built release mechanism worked successfully in all tests allowing for safe operations around the chlorine-filled disseminator before the tests and an instantaneous, reproducible opening in the vessel at the time of release.

Data from the dissemination vessel instrumentation have been analyzed, and dynamic chlorine mass and jet thrust measurements were determined from load cell measurements. Despite variation inherent in the chlorine jet release, the mass release rate and jet thrust can be reasonably estimated from the data. From this data, it can be concluded that the release was conducted so that a minimal amount of flashing occurred at the release point (around 1%). Finally, the observed mass release rate is consistent with that predicted using incompressible flow if the fluid density reflects the flash fraction, but if the flash fraction is ignored, the mass release rate predicted using incompressible flow will be roughly twice the observed value.

Acknowledgements

This field test program was sponsored by the Chemical Security Analysis Center (CSAC) of the U.S. Department of Homeland Security (DHS), the Defense Threat Reduction Agency (DTRA) of the U.S. Department of Defense, and Transport Canada. In addition to the program's Dissemination and Near Source Working Group, we acknowledge the contribution of many people who planned and conducted the first season of tests: DHS CSAC including Adolfo Negron, Shannon Fox, and Mark Whitmire; DTRA including Rick Fry; DPG including Damon Nicholson, Petr Serguievski, and Allison Hedrick; Chlorine Institute including Therese Cirone and Henry Ward; and the JR II Scientific Advisory Group (SAG) including Leo Stockham, Joe Chang, Steve Hanna, Allison Hedrick, Mike Sohn, and Thomas O. Spicer, III. Special thanks to Michael Donovan for discussions regarding the analysis of the data.

References

[1] Fox, S.B., and D.P. Storwold, Jr., "Project Jack Rabbit: Field Tests," Chemical Security and Analysis Center, Science and Technology Directorate, US Department of Homeland Security, CSAC 11-006, 2011.

[2] Hanna, S., R. Britter, E. Argenta, and J. Chang, "The Jack Rabbit chlorine release experiments: Implications of dense gas removal from a depression and downwind concentrations," Journal of Hazardous Materials, vol. 213-214, 406-412, 2012.

[3] Hearn, J.D., R. Weber, R. Nichols, M.V. Henley, and S. Fox, "Deposition of Cl_2 on soils during outdoor releases," Journal of Hazardous Materials, vol. 237-238, 307-314, 2012.

[4] Couper, J.R., W.R. Penney, J.R. Fair, and S.M. Walas, "Chemical Process Equipment: Selection and Design," Elsevier, 2nd, 2005.

[5] Epstein, M., R.E. Henry, W. Midvidy, and R. Pauls, "One-Dimensional Modeling of Two-Phase Jet Expansion and Impingement," Therma-Hydraulics of Nuclear Reactors II (2nd Int. Topical Meeting Nuclear Reactor Thermal-Hydraulics, Santa Barbara, CA, Jan 1983.

12TH GLOBAL CONGRESS ON PROCESS SAFETY

Modelling and validation of dispersion following an instantaneous vapor or two-phase release from a pressurized vessel

Henk W.M. Witlox
DNV GL Software
3 Cathedral Street, London SE1 9DE, UK
henk.witlox@dnvgl.com

Mike Harper
DNV GL Software
Exchange Street, Stockport, SK0 3ET
michael.harper@dnvgl.com

David Webber
Integral Science and Software Ltd.
484 Warrington Road, Culcheth, Warrington, WA35RA
dave@musical.demon.co.uk

Prepared for Presentation at
American Institute of Chemical Engineers
2016 Spring Meeting
12th Global Congress on Process Safety
Houston, Texas
April 11-23, 2016

Modelling and validation of dispersion following an instantaneous vapor or two-phase release from a pressurized vessel

Henk W.M. Witlox
DNV GL Software
3 Cathedral Street, London SE1 9DE, UK
henk.witlox@dnvgl.com

Mike Harper
DNV GL Software

David Webber
Integral Science and Software Ltd.

Keywords: dispersion, consequence modelling, hazard identification and risk analysis

Abstract

This paper presents a new model for the initial dispersion phase of energetic instantaneous expansion following a catastrophic vessel rupture. This model has been implemented as a new sub-model in the Phast dispersion model and is based on more sound physical principles than the old sub-model included in the current release of Phast. It allows for both vapor and two-phase releases. In the case of two-phase releases, droplets are assumed to expand radially, potentially leading to time-varying droplet rainout and the formation of a spreading evaporating liquid pool.

For ground-level vapor or two-phase releases the correctness of the numerical predictions are confirmed against an analytical solution. The model has been validated against previously published experimental data for pressurized vapor releases and pressurized two-phase releases with and without rainout. This includes ground-level releases for nitrogen vapor and flashing liquid propylene, and elevated flashing liquid releases for Freon 11, Freon 12, propane and butane. Overall the old model tends to under-predict the cloud radius and cloud speed versus time, while the new model more closely agrees with experimental data. Therefore the new model produces smaller concentrations and doses, and is less conservative. For two-phase releases the new model predicts an increased amount of rainout, which again is more in line with the experimental data.

1 Introduction

Hazardous flammable and/or toxic materials are often stored in pressurized vessels. This paper deals with the modelling of the dispersion following a catastrophic rupture of the vessel leading to an instantaneous loss of containment. The Unified Dispersion Model (UDM) in the hazard assessment software package Phast can account for two-phase jet, heavy and passive dispersion including droplet rainout and pool spreading/evaporation. The model can deal with a wide range of scenarios including both pressurized jet releases and unpressurized releases. As part of the current work the model has been improved for pressurized instantaneous releases. The new model includes an initial phase of energetic rapid expansion (modeled by a new UDM sub-model), and a subsequent phase of dispersion where equations are adopted applicable for unpressurized releases.

Witlox [1] provides an overview of the UDM model in Phast including model verification and validation. A new version of the UDM has been developed for future inclusion in Phast (scheduled to be released within the next year), which includes improved modelling of time-varying dispersion including potential rainout using the so-called observer concept (Witlox and Harper [2]).

The current UDM model in Phast includes a simplistic sub-model for the calculation of the initial dispersion phase of energetic expansion for pressurized instantaneous releases. The main limitations of this model are that this sub-model does not account for gravity effects, and that it presumes a single droplet size moving along a fixed upward angle resulting in too little rainout. Moreover droplets currently start at the edge of the cloud and therefore may erroneously rainout beyond the bund wall, if present.

Following the above a literature review has been carried out regarding both previous theoretical and experimental work. Amongst others this accounts for experimental work for ground-level pressurized releases for nitrogen vapor (Landis et al. [3]) and flashing liquid propylene (Maurer et al. [4]), and elevated flashing liquid releases for Freon 11 (Pettitt [5]), Freon 12, propane and butane (Schmidli [6]).

This paper first outlines the mathematical equations governing the new UDM sub-model called INEX. This model allows for both vapor and two-phase releases. In case of two-phase releases, the model accounts for droplet dispersion and potential time-varying droplet rainout to form a spreading evaporating liquid pool. Subsequently the paper describes the analytical verification of INEX for a number of specific cases. Finally it describes the validation of the model against experimental data for both cases of a pressurized vapor release and a pressurized two-phase release.

2 New UDM sub-model (INEX)

The Phast discharge model presumes isentropic expansion from the stagnation conditions to the atmospheric pressure without air entrainment. The initial UDM dispersion state is based on the post-expansion discharge results. During the initial UDM INEX stage of energetic expansion the air entrainment is dominated by a large radial expansion velocity. Figure 1 depicts the subsequent dispersion phases during the INEX stage. The cloud is modelled as a sphere while elevated, as a truncated sphere during touching down and as a hemisphere after full touchdown. In case of a 2-phase release, the liquid droplets are assumed to be uniformly distributed throughout the cloud volume during the INEX expansion and thus travelling radially at a speed proportional to their distance from the cloud center. Therefore rainout starts when the lower edge of the cloud hits the ground while it ends at full touchdown (see Figure 1). A transition from INEX to the standard UDM model is applied, when the INEX air entrainment reduces to the

UDM air entrainment, or if the INEX spread rate reduces to the UDM gravity spreading rate. Any remaining liquid will rainout at this transition if the cloud is grounded, or possibly at a later stage (using standard UDM droplet modelling) if elevated.

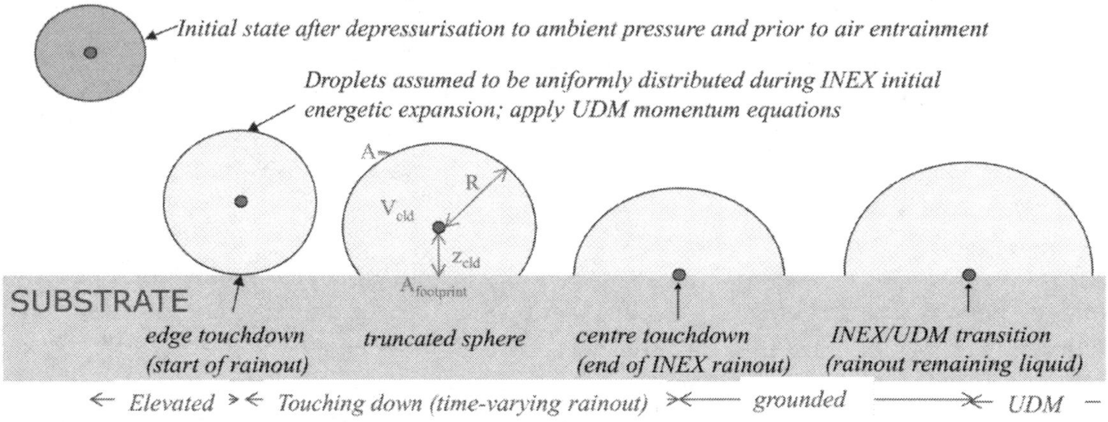

Figure 1. INEX dispersion phases for two-phase instantaneous release

From the above cloud geometry, the cloud volume V_{cld} can easily be expressed as a function of the cloud radius R and the cloud center-line height z_{cld}. The INEX radial momentum (kg m/s) is defined as $I_r = m_{cld} U$. Here the radial cloud expansion speed $U=dR/dt$, and the total cloud mass $m_{cld} = m_{cL} + m_{cv} + m_{wa}$, where m_{cL} is the chemical liquid mass, m_{cv} the chemical vapor mass and m_{wa} is the mass of wet ambient air added to the cloud. The key INEX assumption is that the radial momentum is constant, apart from the loss of momentum due to rainout. Thus the following differential equations are applied in INEX for the radial momentum I_r, the entrained mass of wet air m_{wa}, and the rained out liquid mass $m_{cL}{}^{ro}$,

$$\frac{dI_r}{dt} = -\frac{dm_{cl}{}^{ro}}{dt}U \tag{1}$$

$$\frac{dm_{wa}}{dt} = \rho_{wa}\frac{dV_{cld}}{dt} \tag{2}$$

$$\begin{aligned} \frac{dm_{cL}{}^{ro}(t)}{dt} &= 0\ , \quad if\ z_{cld} \geq R(cloud\ elevated)\ or\ if\ z_{cld} = 0\ (cloud\ grounded) \\ &= \max\left\{K_D \frac{m_{cL}}{V_{cld}} A_{footprint}\left[\frac{z_{cld}}{R}U - \frac{dz_{cld}}{dt}\right], 0\right\},\quad if\ 0 < z_{cld} < R\ (cloud\ touching\ down) \end{aligned} \tag{3}$$

Here t is the time (s), and the change in cloud volume is set as

$$\begin{aligned} \frac{dV_{cld}}{dt} &= AU\ , \quad if\ z_{cld} \geq R(cloud\ elevated)\ or\ if\ z_{cld} = 0\ (cloud\ grounded) \\ &= AU + A_{footprint}\frac{dz_{cld}}{dt}\ , \quad if\ 0 < z_{cld} < R\ (cloud\ touching\ down) \end{aligned} \tag{4}$$

where A is the cloud surface area above the ground, and $A_{footprint}$ is the cloud footprint area (see Figure 1).

The first term in the right-hand side of rainout equation (3) represents the rainout due to cloud expansion and the second term represents the rainout due to the cloud center-line height z_{cld} reducing. In the derivation of first term it is presumed that the radial cloud velocity linearly increases, and thus it can be derived that the vertical downward component at the footprint equals $(z_{cld}/R)\ U$. The maximum value of the parameter $K_D=1$, which presumes that all liquid hitting the ground will rainout.

The initial cloud speed $U(t=0)=U_o$ is set as $U_o=f_{kinetic}\ E_{exp}^{\ 0.5}$. Here the specific expansion energy (J/kg) is set as $E_{exp} = h_{st} - h_f$, where h_{st} is the specific stagnation enthalpy and h_f is the specific final enthalpy after expansion to ambient pressure. Furthermore the fraction of total energy converted to kinetic energy is set to $f_{kinetic}=0.04$ following the recommendation of Pattison [7] based on a best fit to data from Schmidli [6].

In addition to the above equations, the standard UDM equations are applied during the INEX stage to evaluate the cloud center location (from momentum equations including gravity effects), the cloud temperature, the cloud volume V_{cld} and the cloud phase distribution (from UDM two-phase thermodynamics equations). Following rainout, the model carries out pool spreading/evaporation calculations and the model accounts for pickup of vapor from the time-varying pool by the instantaneous cloud. For this purpose the so-called observer concept is applied to evaluate the time-varying dispersion in line with the paper by Witlox and Harper [2].

3 Analytical solutions

Ground level release

For a ground-level release, the INEX cloud is a hemisphere of radius R, and integration of Eq. (2) yields

$$m_{wa} = \rho_{wa}\left(V_{cld} - V_{0}\right) = \rho_{wa}\,\frac{2\pi}{3}\left(R^3 - R_o^{\ 3}\right) \tag{5}$$

where R_o is the initial cloud radius (following depressurization to ambient and prior to air entrainment). In case of a two-phase ground-level release, it is presumed that 50% of the liquid rains out immediately, while the remaining liquid moves out radially in upward directions. Therefore according to Eq. (1) radial momentum $I_r = m_{cld}U$ is constant and therefore [m_o = released vessel mass after immediate rainout]:

$$m_o U_o \approx m_{cld} U \approx \left[m_0 + m_{wa}\right]\frac{dR}{dt} = \left[m_0 + \rho_{wa}\,\frac{2\pi}{3}\left(R^3 - R_o^{\ 3}\right)\right]\frac{dR}{dt} \tag{6}$$

By means of separation of variables R and t, the above equation can be integrated to yield,

$$t = \frac{R}{U_o}\left\{1+\left[\frac{\rho_{wa}}{\rho_o}\right]\left[\frac{1}{4}\left(\frac{R}{R_o}\right)^3-1\right]\right\} \quad -\frac{R_0}{U_o}\left\{1-\frac{3}{4}\left[\frac{\rho_{wa}}{\rho_o}\right]\right\}, \quad U = \frac{U_o}{1+\left[\frac{\rho_{wa}}{\rho_o}\right]\left[\left(\frac{R}{R_o}\right)^3-1\right]} \qquad (7)$$

where $\rho_o = m_o/(2\pi R_o^3/3)$ is the initial chemical density after the initial immediate rainout. Thus the above expression is used to analytically evaluate cloud radius R and cloud expansion speed U versus time t.

Elevated release (non-evaporating liquid)

A pressurized instantaneous elevated release is considered of a sub-cooled liquid at ambient temperature, which does not evaporate during the dispersion stage ($m_{cv}=0$; $m_{cld} = m_{cL}+m_{wa}$). In this case isothermal mixing occurs between the released liquid and the ambient moist air, and the cloud volume V_{cld} and the rained out liquid mass m_{cL}^{ro} can be evaluated analytically as (ρ_{cld} = cloud mixture density, ρ_{cL} = density of released liquid):

$$V_{cld} = \frac{m_{cld}}{\rho_{cld}} = \frac{m_{wa}}{\rho_{wa}} + \frac{m_{cL}}{\rho_{cL}} \ , \ m_{cL}^{ro} = \left\{1-\frac{V_{cld}}{(4/3)\pi R^3}\right\} m_{cL}(t=0) \qquad (1)$$

It was verified that the numerical INEX time-varying predictions for cloud volume and liquid rainout indeed satisfied the above equations during elevated dispersion and cloud touching down.

4 Model verification and validation

4.1 Ground-level nitrogen vapor releases (Landis et al)

Landis et al. [3] carried out experiments involving an instantaneous release of nitrogen vapor from a cylindrical tank (length 12", radius 2") with a range of stagnation temperatures (273 or 303K) and pressures (4.2, 8.2, 21.5 and 71.7 bara). These experiments have been modelled as hemispherical grounded releases.

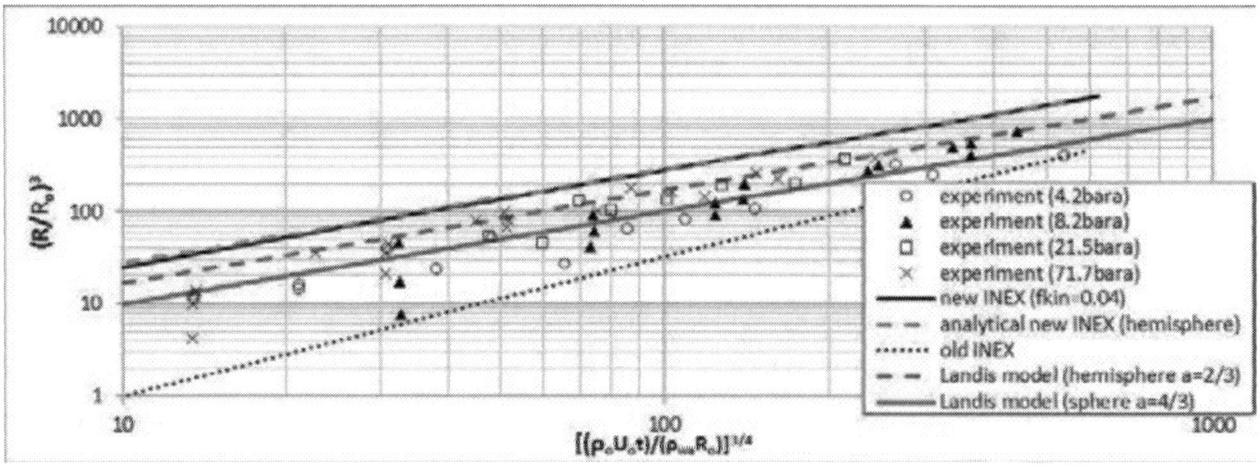

Figure 2. Validation of cloud volume versus time against Landis nitrogen experiments

Figure 2 plots the dimensionless cloud volume versus a dimensionless time, where the INEX numerical predictions are chosen to correspond to the experiment with 273 K and 71.7 bara. The plot confirms perfect matching of the above analytical solution against the new INEX numerical model. The new INEX model slightly over-predicts the data, while the old INEX model significantly under-predicts. In this context note that the old INEX radius R is always set presuming a sphere (including grounded releases), and therefore its initial radius R_o is a factor $2^{-1/3}$ smaller than the new INEX hemispherical radius.

The plot also includes results from the model by Landis et al. [3], who presume frontal capture of the cylindrical tank (for setting air entrainment) in conjunction with a hemispherical assumption (the corresponding figure 5 in their paper appears to show an incorrect curve). This differs from our assumption of air entrainment across the entire hemispherical cloud surface area A. Neither do they report initial values ρ_o, R_o or U_o and this could also partially account for difference of our model with the data. Since the model by Landis has been customized specifically to the cylindrical release geometry for their experiments, this model would be expected to be more accurate than our model which presumes instantaneous spherical releases.

4.2 Ground-level flashing propylene liquid releases (Maurer et al)

Maurer et al. [4] carried out experiments involving an instantaneous release of heated flashing propylene liquid from a cylindrical tank with stagnation pressure 60 barg, stagnation temperature 323 or 353 K, and considering a range of tank sizes (release masses 0.125, 0.42, 1.95, 6.55, 15.6 or 452 kg). These experiments have been modelled as hemispherical grounded releases.

Maurer adopts a 'normalized' time t/L_G (s/m) where L_G is the cube root of the volume (V_G) of twice the released mass ($2M$) of propylene in the gaseous phase at 0 Celsius and 1bar. For the case of $M=452kg$ (tank length 2.8m, tank diameter 0.7m), Figure 3a plots the cloud speed U versus t/L_G, where $L_G=7.83m$. Figure 3b plots the radius R versus time t. The INEX model starts from time t=0 after the initial isentropic expansion to atmospheric pressure, and (by comparing experimental radii with initial INEX radius) this is ~ 0.04s after the start of the release. Therefore we have shown in both plots experimental data obtained from Maurer's paper, both with and without a time shift of 0.04s. The plot again confirms perfect matching of the analytical solution with the new INEX numerical predictions. The new INEX model closely agrees with the data, although appears to slightly over-predict. The old INEX model significantly under-predicts for smaller times and more closely agrees with the data for the visible cloud at larger times.

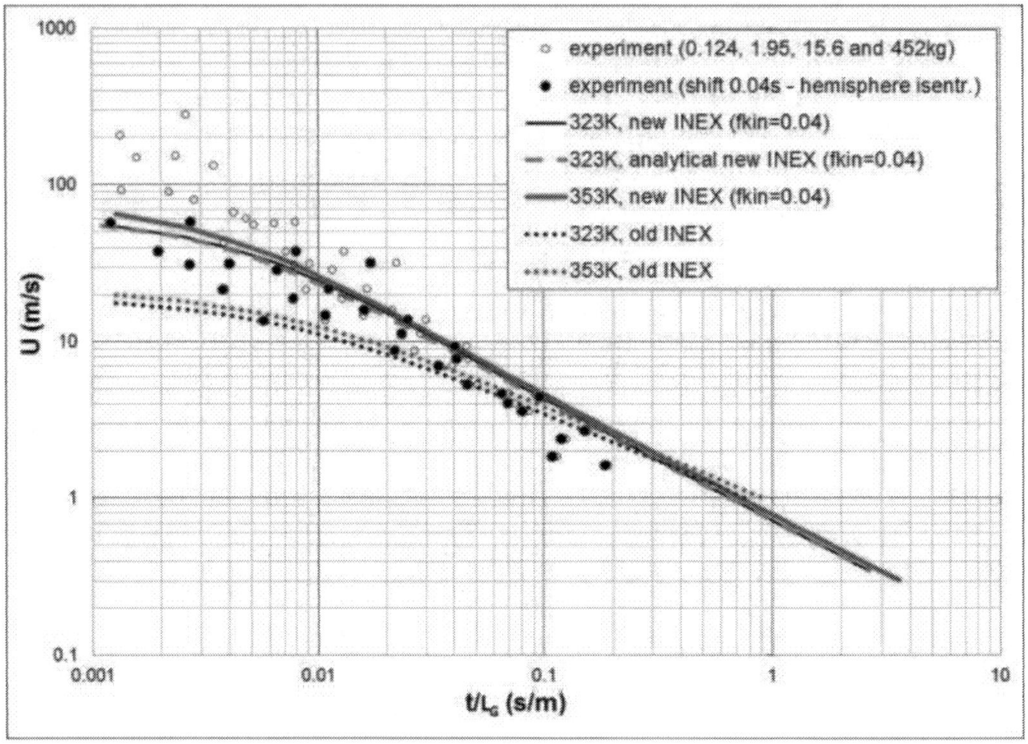

(a) speed versus scaled time

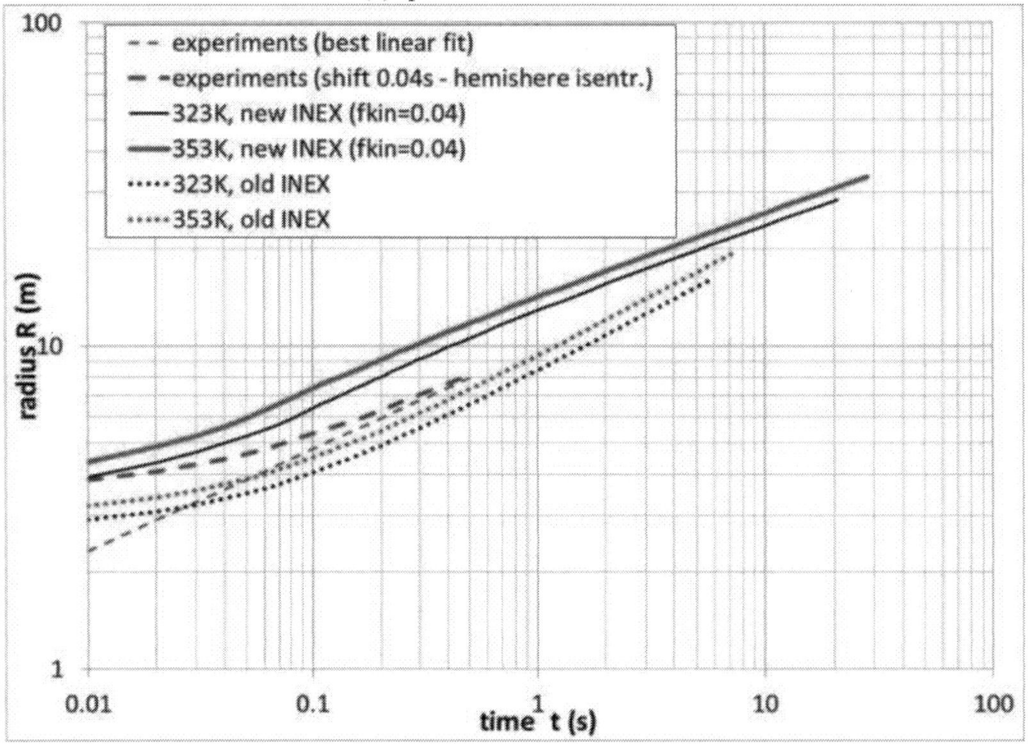

(b) radius versus time

Figure 3. Validation against Maurer experiment (propylene, 452kg, 60barg, 323K or 353K)

4.3 Elevated flashing Freon, propane and butane liquid releases (Pettitt, Schmidli)

Following earlier PhD's at South Bank University, Pettitt [5] carried out experiments involving a catastrophic failure of a 1 litre elevated glass sphere by hammer impact. The glass sphere was elevated at 0.25m height and filled with a saturated liquid. A range of experiments was carried out including variation of material (Freon 11, 113 or 114), % fill, and stagnation pressure. The conditions were chosen such as to avoid rainout. In line with our model assumptions, he confirmed the cloud increases as a sphere, with expansion speed U decreasing with time and with droplet speeds independent of size and moving in radial directions.

As part of a PhD at the Swiss Federal Institute of Technology, Schmidli [6] carried out similar experiments to Pettitt. He considered a range of flask sizes (0.1, 1 or 1 litre), materials (R-114, R-12, propane and butane) and superheats. His experiments included both grounded and elevated releases, and unlike Pettitt he also investigated effects of rainout.

Table 1 includes the measured initial cloud expansion speed U(t=0), alongside a theoretical speed based on isentropic expansion. From this the fraction of energy $f_{kinetic}$ converted to kinetic energy is given, and the value of $f_{kinetic}$ is seen to increase with the stagnation pressure P_{st}. For the Schmidli experiments the value of $f_{kinetic}$ is in line with the adopted value by the new INEX of 0.04 as recommended by Pattison (1994), while for the Pettitt experiments it is significantly higher.

Table 1. Observed initial cloud speed U(t=0) versus theoretical speed

Experiment	Material	T_{st} (K)	P_{st} (barg)	Flask (ltr)	Fill %	Mass (kg)	Exp.speed (m/s)	Theor. Speed(m/s)	$f_{kinetic}$
Pettitt	Freon11	324.0	1.40	1	50	0.707	11	47.4	0.054
		328.3	1.72	1	50	0.701	19	54.8	0.120
		332.6	2.06	1	50	0.696	26	61.9	0.176
		339.0	2.64	1	50	0.687	44	72.8	0.365
		342.3	2.96	1	50	0.682	51	78.3	0.425
		346.9	3.45	1	50	0.676	60	86.0	0.487
		352.4	4.10	1	50	0.668	66	95.3	0.479
		357.2	4.71	1	50	0.661	72	103.3	0.486
		364.1	5.70	1	50	0.651	78	115.0	0.460
Schmidli	R12	295.1	4.98	1	100	1.320	14.4	99.5	0.021
	Propane	278.1	4.50	2	100	1.043	18.4	149.6	0.015
		285.1	5.73	2	100	1.022	28.5	171.7	0.028
		291.1	6.93	2	100	1.004	39.6	190.8	0.043
	Butane	315.1	2.99	1	100	0.552	21.6	126.4	0.029
		325.1	4.22	1	100	0.539	33	155.8	0.045

Figure 4 includes INEX validation results of cloud speed U versus cloud radius R against experimental data by Pettitt for a 1 litre vessel with a 100% fill of Freon 11 and stagnation pressure of 4.10barg. Experimental data include minimum, mean and maximum values. It is seen that the old INEX model significantly under-predicts, while the new INEX model slightly under-predicts. The figure also includes the result from an additional new INEX run, where a value of the fraction of total energy converted to kinetic energy, $f_{kinetic}=0.15$, is applied such that the INEX speed matches the observed data at the initial INEX radius (after isentropic expansion to ambient pressure). It is seen that this results in improved agreement against the experimental data.

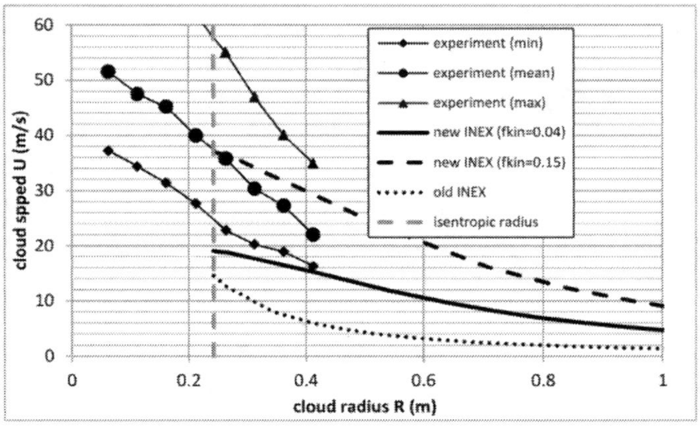

Figure 4. Cloud speed validation against Pettitt experiments (Freon 11, 100% fill, 4.1 barg, 1 litre)

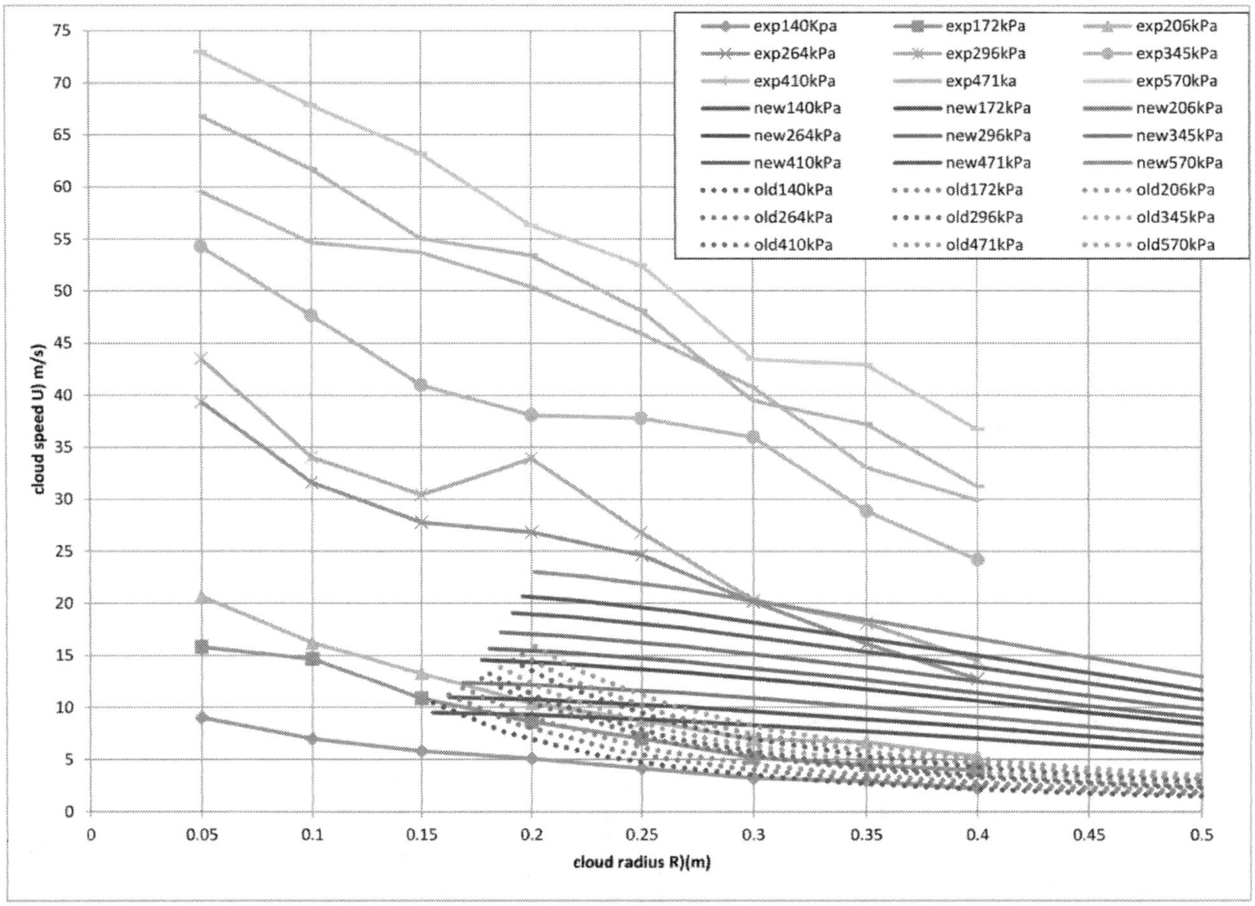

Figure 5. Cloud speed validation against Pettitt experiments (Freon 11, 50% fill, 1 litre)

Validation of cloud speed U was also carried out against Freon 11 experiments by Petitit with 50% fill and pressures ranging between 1.4 and 5.7 barg and similar conclusions were derived as above. Figure 5 shows that the experimental cloud speeds are larger than those predicted by both the old and new INEX (except for the lower pressures) with the new INEX results considerably closer to the experimental data. Again more close agreement with the experimental data would be obtained by increasing $f_{kinetic}$ such that the initial radius $R(t=0)$ equals the experimental observed speed.

Figure 6 includes INEX predictions against experimental data by Schmidli for R-12 (stagnation temperature 22C corresponding to 52.1C superheat and saturated stagnation pressure 6 bara, ambient humidity 38 or 83%) and propane (5C or 18C corresponding to 47.4C and 60.4C superheat and pressures of 5.52 and 7.94bara). Again experimental data are shifted such that at time $t=0$ the INEX isentropic radius is reached.

For R-12 the visible aerosol cloud radius slightly increased with humidity, while the effect of humidity on INEX predictions is insignificant. New INEX predictions of cloud radius are larger than those of the old INEX and closer to the experimental data. For propane the effect of superheat appears to be larger for the experiments than for the new model, but this may result from increased condensation of the water in the moist air, and hence a more visible cloud.

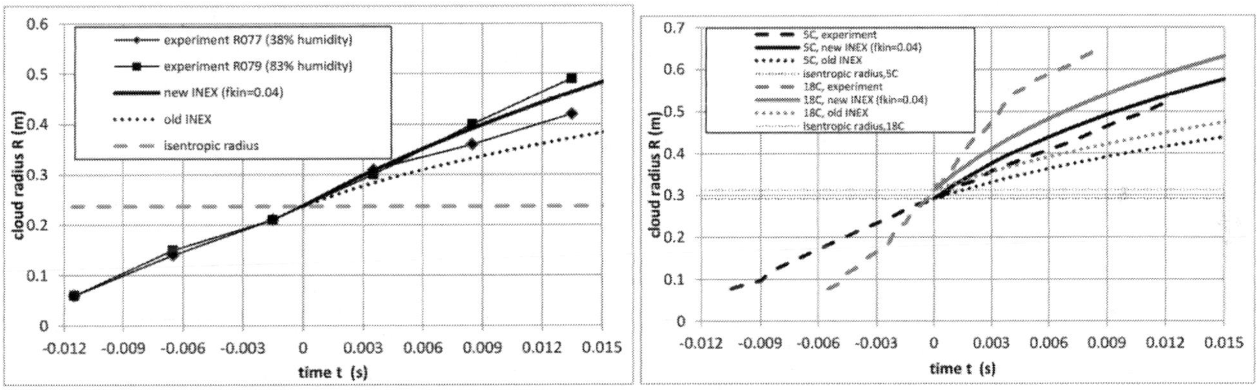

(a) R-12, 1 litre, 22C, 0.62m elevation (b) propane, 2 litre, 5 or 18C, 1m elevation

Figure 6. Cloud radius versus time for Schmidli experiments (100% fill)

INEX validation of rainout

In Phast 7.2 the initial droplet size (Sauter Mean Diameter, SMD) for instantaneous releases is always based on the CCPS flashing correlation (Witlox et al. [8]). On the other hand Schmidli suggests to use a 'Weber' criterion based on the reduced post-expansion final velocity u_f ($f_{kinetic}=0.04$). Table 2 includes results of SMD predictions. It is seen that the flashing SMD is about twice as small as the mechanical SMD for R-12, and very similar for propane.

Figure 7 plots liquid rainout (kg) versus time (s) using the new INEX model for the R-12 and propane experiments using both droplet size correlations. It demonstrates the subsequence stages of (a) elevated INEX cloud (no rainout), (b) INEX cloud touching down (time-varying rainout), and (c) INEX/UDM grounded cloud (no rainout). For the case of R-12 (mechanical breakup) it also shows the final instantaneous rainout due to an INEX/UDM transition for a grounded plume.

Table 2. UDM input data for selected Schmidli experiments (elevated, 100% fill)

Fluid	temperature /SH (C)	elevation (m)	observed % rainout			FLASHING CORRELATION			MECHANICAL CORRELATION		
						SMD (μm)	predicted rainout		SMD (μm)	predicted rainout	
			100ml	1 ltr	2 ltr		1 ltr	2 ltr		1 ltr	2 ltr
R-12	22/52.1	5D	24±4			214	20.7%		428	54.0%	
Propane	5/47.4	1		26±6	30±9	155		5.1%	180		6.5%
	12/54.4	1		13±2	26±17	134		4.3%	136		4.4%
	18/60.4	1		16±3	25±5	119		3.6%	110		3.1%
Butane	42/42.6	1				179			244		
	52/52.6	1				148			160		

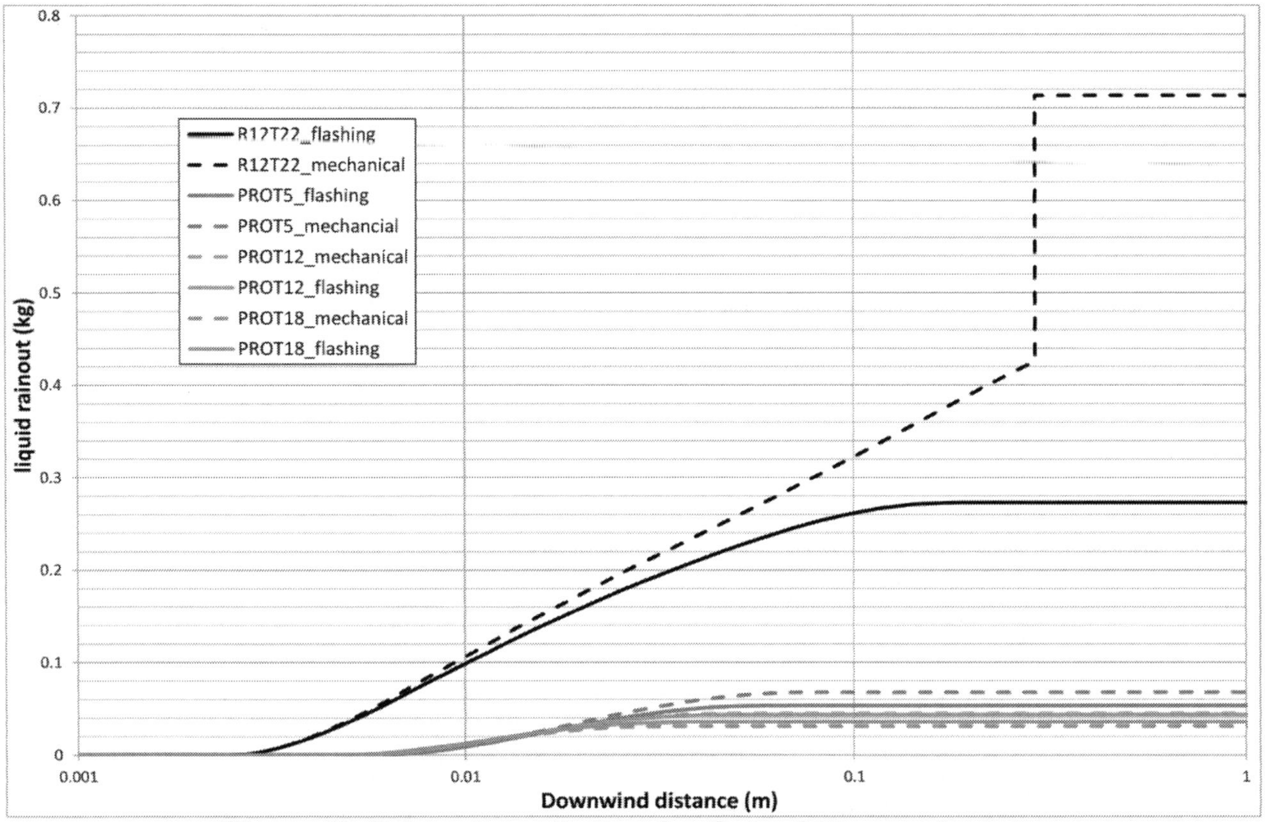

Figure 7. Rainout prediction for Schmidli experiments

It is seen that the significant smaller SMD in the case of flashing breakup for R-12 results in significant less rainout. Table 2 compares results of the thus predicted overall % liquid rainout. It is seen that for R-12 accurate results are obtained in case of flashing SMD, while results are over-predicted in case of mechanical SMD. For propane rainout is under-predicted. No rainout was predicted using the old INEX model, and therefore improved rainout predictions are obtained using the new model.

Note that so far research projects have very much focused on development and validation of droplet size correlations associated with orifice releases and much less work has been carried out on droplet size for instantaneous releases. The above Weber mechanical-breakup correlation may not be applicable for the above superheated release, while the above CCPS flashing correlation is based on experimental data for superheating orifice releases.

5 Discussion and conclusions

A new model (INEX) has been formulated for the initial dispersion phase of initial energetic expansion following a pressurized instantaneous release. The correctness of the numerical predictions has been confirmed against analytical solutions for ground-level vapor or two-phase releases and elevated non-evaporating liquid releases.

INEX is based on more sound physical principles than the previous model. Overall the old model tends to under-predict the cloud radius and cloud speed versus time. The new model more closely agrees with the experimental data, and provides larger predictions of the cloud radius and cloud speed versus time. Therefore the new model produces smaller concentrations and doses, and is less conservative. For two-phase releases the new model predicts an increased amount of rainout, which again is more in line with the experimental data.

Experiments identified so far derive the cloud radius and cloud speed from the visible cloud front. This results in added uncertainty since the visible cloud expansion velocity depends amongst others on humidity, and this may e.g. explain some of the discrepancy in results shown in this paper between the experiments by Schmidli and Pettitt for measured | fraction $f_{kinetic}$ of total energy converted to kinetic energy. Therefore additional experimental work to measure concentrations is strongly recommended to provide a sounder basis for model validation. In addition experimental data including additional measurements of droplet sizes and rainout would be useful. This may assist in developing improved droplet size correlations applicable for instantaneous releases.

Acknowledgements

Financial support of the work reported in this paper was provided by DNV GL Software and RIVM (Dutch Government). The contents of this paper including any opinions and/or conclusions expressed, are those of the authors alone and do not necessarily reflect the policy of these organizations.

References

[1] H.W.M. Witlox, "Overview of consequence modelling in the hazard assessment package Phast", Paper 7.2, 6th AMS conference on applications of air pollution meteorology, Atlanta, USA, 17-21 January 2010

[2] H.W.M Witlox and M. Harper, "Modelling of time-varying dispersion for releases including potential rainout", Process Safety Progress 33 (3), pp. 265-273 (2014)

[3] J.G. Landis, R.E Linney and B.F. Hanley, "Dispersion of instantaneous jets", Process Safety Progress 13 (1), pp. 35-40 (1994)

[4] B. Maurer, H. Schneider, K. Hess and W. Leuckel, "Modelling of vapour cloud dispersion and deflagration after bursting of tanks filled with liquefied Gas", Int. Loss Prevention Symposium, Heidelberg (1977)

[5] G.N. Pettitt, "Characterisation of two-phase releases", PhD Thesis, South Bank Polytechnic, London (1990)

[6] J. Schmidli, The initial phase of sudden releases of superheated liquid, Ph.D. Thesis, ETH, Zurich (1993)

[7] M.J. Pattison, E.C. "Cloud Project – Initial stages of release", Draft note, Department of Chemical Engineering and Chemical Technology, Imperial College, London (1994)

[8] H.W.M. Witlox, M. Harper, A. Oke, P. Kay and P.J. Bowen, "Sub-cooled and flashing liquid jets and droplet dispersion, I. Overview and model implementation/validation", Journal of Loss Prevention in the Process Industries 23, pp. 831-842 (2010)

442487 Feasibility Study for Hydrogen Sulfide (H_2S) Optical Gas Imaging

Abstract:

Infrared (IR) cameras have been used extensively to detect hydrocarbon leaks from process equipment for both process safety and environmental applications. Recent advancements in the field have allowed for continuous and autonomous detection as well as quantification of hydrocarbon emissions. While these advances have led to new applications for hydrocarbon leak detection, optical gas imaging to date has not be utilized for detection of H_2S leaks. H_2S is a very toxic chemical commonly found in oil and gas streams and some process streams. Due to the toxic nature of the gas, it is highly desirable to remotely image and detect the presence of H_2S. Unfortunately, due to its IR spectral characteristics there has been no demonstrated method to achieve IR imaging of H_2S plumes.

A series of tests were conducted to evaluate the feasibility of imaging H_2S and to determine the sensitivity of an infrared imager in terms of ppm-m. Various spectral regions were examined for suitability of H_2S imaging. The results of the test show that it is feasible to image H_2S in the SWIR region. The results of these tests and the corresponding sensitivity of an H_2S imager are presented, along with potential applications for this technology.

Design Considerations for Oily Water Lift Stations (OWLS)

William J Celenza
KBR Inc
Houston, TX
William.celenza@kbr.com

Janani Sankar, AIChE Member

Prepared for Presentation at
American Institute of Chemical Engineers
2016 Spring Meeting
12th Global Congress on Process Safety
Houston, Texas
April 11-23, 2016

Design Considerations for Oily Water Lift Stations (OWLS)

William J Celenza
KBR Inc
Houston, TX
William.celenza@kbr.com

Janani Sankar
AIChE Member

Keywords: Oily Water Sewer, Pump Station, Sump, Gravity Drain

Abstract

Oily water, a mixture of hydrocarbon residues and water, is contaminated water released during normal operations at an oil refinery. Process wastewater flows are collected by an oily water or industrial sewer system that can be combined with area drains to collect potentially contaminated surface runoff from the on plot process paved areas as well as areas where spillages and wash water may occur. Surface runoff includes precipitation and fire water which is drained to catch basins, or area drains, through a network of gravity drainage piping and manholes. A clean storm water system collects precipitation from unpaved areas, roofs, roads and other areas not likely to be subject to contamination for direct offsite discharge. An oily water system needs to safely transfer collected flows to treatment or storage facilities to prevent contamination from impacting human health or the environment.

Refineries that segregate the oily water collection system from storm or fire water flows may drain runoff by gravity directly to one influent lift station at the entrance to the treatment plant, without using any intermediate lift stations. However, with combined systems the reliance on gravity conveyance can lead to excessive excavation depths and increased sewer construction complexity. The use of lift stations strategically placed throughout a combined collection network can simplify underground sewer piping networks, accommodate high water tables, and provide functional isolation for complexes or units within a refinery.

The design of an Oily Water Lift Station (OWLS) is distinct from that of municipal or highway related storm water lift stations due mainly to the industrial nature of the inflows, which typically are non-flammable but can include limited amounts of flammable and combustible hydrocarbons. The potential for explosive atmospheres in oily water lift stations, typically sealed and vented for VOC emission compliance, presents what is to some an unacceptable risk and has led to a design philosophy of minimizing or even eliminating the use of lift stations in this service, regardless of the increase in installation complexity. Also system flexibility can be increased and environmental risk associated with operating an Oily Water Drainage System (OWDS) can be better managed by including OWLS. Even though relying on lift stations presumes lower system reliability, appropriate pump selection, use of an emergency power supply, and automatic control and alarm systems have been successfully used to reduce this risk.

This paper describes methods for managing risks associated with operating OWLS through design considerations and related operational and maintenance procedures. Compliance with current US codes and standards for fire, electrical, and flammable vapor emission issues are discussed for gravity drain systems with lift stations, as well as the use of complementary closed drain pressurized collection systems, chemical sewers, and dedicated collection systems for constituents not compatible with atmospheric gravity drain systems. However, specialized systems such as Benzene Waste Operations NESHAPs compliant collection networks are not included in this discussion.

1 Background and Purpose

Modern refineries segregate direct contact process wastewater, such as desalting washes, process condensates and sour water, as well as solvent discharges in dedicated collection systems. Segregation allows for pre-treatment near the point of generation and potential cascading or reuse onsite with the aim of reducing end-of-pipe treatment loads. Wastewater is then typically pumped directly to the Wastewater Treatment Plant (WWTP) headworks. Also, flammable, toxic, corrosive, reactive or high pour-point liquid materials are segregated into closed drain systems (CDS), consisting of dedicated piping and vessels connected to selected drains that collect these liquid types for recovery or safe disposal. Hydrocarbons from tank drains, solvents or sour water from sulfur recovery or tail gas treatment units, and amines from acid gas removal units are examples of process streams which should use CDS. Typically only residuals or cleaning washes from these operations would be drained to a gravity Oily Water Drainage Systems (OWDS).

Oil-free wastewater such as boiler blowdown, cooling system blowdown, and demineralizer waste if not needed or suitable for reuse is typically collected then pumped around the headworks and primary treatment section of the WWTP as shown in Figure 1.

Not many refineries have segregated OWDS and process area storm water drainage systems possibly due to local regulations. But if segregated systems exist, the water is directed to separate treatment systems and the storm water system is likely to follow first flush principles subject to local regulation.

The combined OWDS will collect potentially contaminated storm water runoff from the on plot process paved areas and any areas where spillages and wash water may occur. This system will also collect and drain all fire water. Collecting these flows ensures that any contaminated runoff will not be discharged to the environment, or infiltrate into the groundwater. The areas being drained to the OWDS system are usually concrete paved.

Sanitary sewer systems should be entirely segregated from an OWDS. Sanitary wastewater may eventually be commingled with the oily water if the mixing may prove beneficial for the treatment of the combined wastewater at the WWTP.

As schematically shown in Figure 1, only some wastewater generated from refining operations is conveyed to the site WWTP via an underground gravity OWDS. The wastewater streams presented in Table 1 when only nonvolatile, nonflammable, and non-toxic components are present are typically managed in a refinery gravity OWDS. Since these streams are either continuously or potentially oil contaminated, treatment through the primary treatment section of a refinery WWTP is always recommended. Also as indicated in Table 1, the process area storm water is collected through the OWDS as a combined system. When compared to the typical influent for a refinery API Separator treatment system, the OWDS pollutant levels would generally be lower in chemical oxygen demand (COD), phenols, sulfides and ammonia but higher in oil & grease and total suspended solids (TSS). The API separator influent includes the OWDS combined with other wastewater streams at the WWTP headworks.

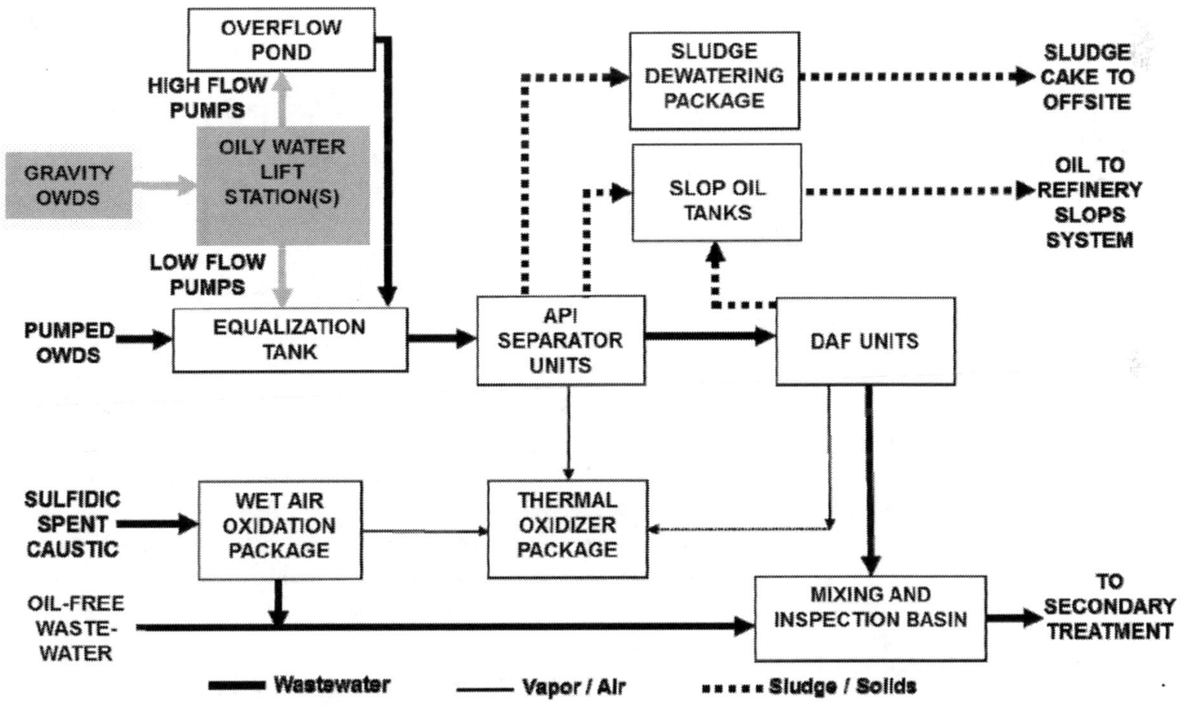

Figure 1 Refinery Wastewater Collection and Primary Treatment System

An OWLS, as part of the OWDS, receives flows that are incidental in nature and thus highly variable in flowrate as well as pollutant loading. In addition to storm water and fire water events, incidents such as hydrocarbon spills, hose station wash-downs, testing of fire water sprays, process vessel washes, pump baseplate working loses, etc. need to be anticipated.

Table 1 Wastewater typically managed in a refinery gravity OWDS (derived from US EPA 1980)

<table>
<tr><th rowspan="2">Wastewater Source</th><th colspan="7">Typical Conventional Pollutant Levels (mg/l)</th></tr>
<tr><th>Oil & Grease (1)</th><th>COD</th><th>TSS (2)</th><th>Phenolic Cmpnds</th><th>Ammonia Nitrogen (3)</th><th>Sulfides (3)</th><th>pH</th></tr>
<tr><td>Hydrocarbon Equipment Drains (tanks, vessels, etc.) for dewatering, cleaning, hydrotesting, etc.</td><td rowspan="6">< 1,000</td><td rowspan="6">< 300</td><td rowspan="6">< 1,500</td><td rowspan="6">< 5</td><td rowspan="6">Trace</td><td rowspan="6"><10</td><td rowspan="6">5 to 9</td></tr>
<tr><td>Pump Station Drains (casing, base plate, etc.)</td></tr>
<tr><td>Utility Drains (sample points, level gauges, instruments, etc.)</td></tr>
<tr><td>(Un)loading Station Drains</td></tr>
<tr><td>Pipe Trench Drains</td></tr>
<tr><td>Area Drains (Wash, Storm, Fire Water)</td></tr>
</table>

Notes:

(1) Oily Water is a mixture of non-flammable hydrocarbon residues and water, released during normal operations and is not expected to create hazardous conditions if discharged directly to a gravity OWDS.

(2) Can include fine dust/carbonate type solids that have been observed to produce a sticky paste when mixed with water and oils at some sites.

(3) For instance, Sour Water which contains more than 10 mg/l of hydrogen sulfide with other components such as ammonia, phenols, and benzene would not be permitted as discharge to an OWDS.

2 OWDS SYSTEM DESCRIPTION

The basic types of gravity OWDS are "dry" or "flooded". Dry systems consist of free-draining sewer networks interconnecting catch basins and manholes. A typical catch basin or manhole has a sand trap and a liquid seal as shown in Figure 2. The sewers free drain to an OWLS that transfers the wastewater to a site WWTP for treatment or temporary storage.

Flooded systems maintain continuous wastewater gravity flow within the sewer piping and drain directly over a weir located at the entrance to an OWLS or to an API Separator at a WWTP. The system is flooded by the weir at these locations; and the catch basin and manhole components are the "dry-box" type with sewer inlet and outlet lines located flush with the flat bottom (Mason 1984). Water entering the OWDS will cause the level to rise and water will eventually flow into the OWLS by overflowing the weir. The weir needs to be designed to manage the peak instantaneous flow. Flooded systems may be impractical in locations sufficiently dusty where excessive solids may cause clogging (Lees 2012a, Mason 1984).

Compared to a flooded OWDS, a dry OWDS requires deeper excavation to accommodate free draining sewer lines and operation of the OWLS nearly empty. Thus a dry OWDS typically requires more intermediate OWLS for the same topographic conditions, such as available terrain slope, and water table elevation.

Per the OWDS in Figure 3, raw wastewater streams normally are introduced into the collection system through individual or area drains, which can be open to the atmosphere or sealed to prevent wastewater contact with the atmosphere. Individual drains may be dedicated to a single source or piece of equipment.

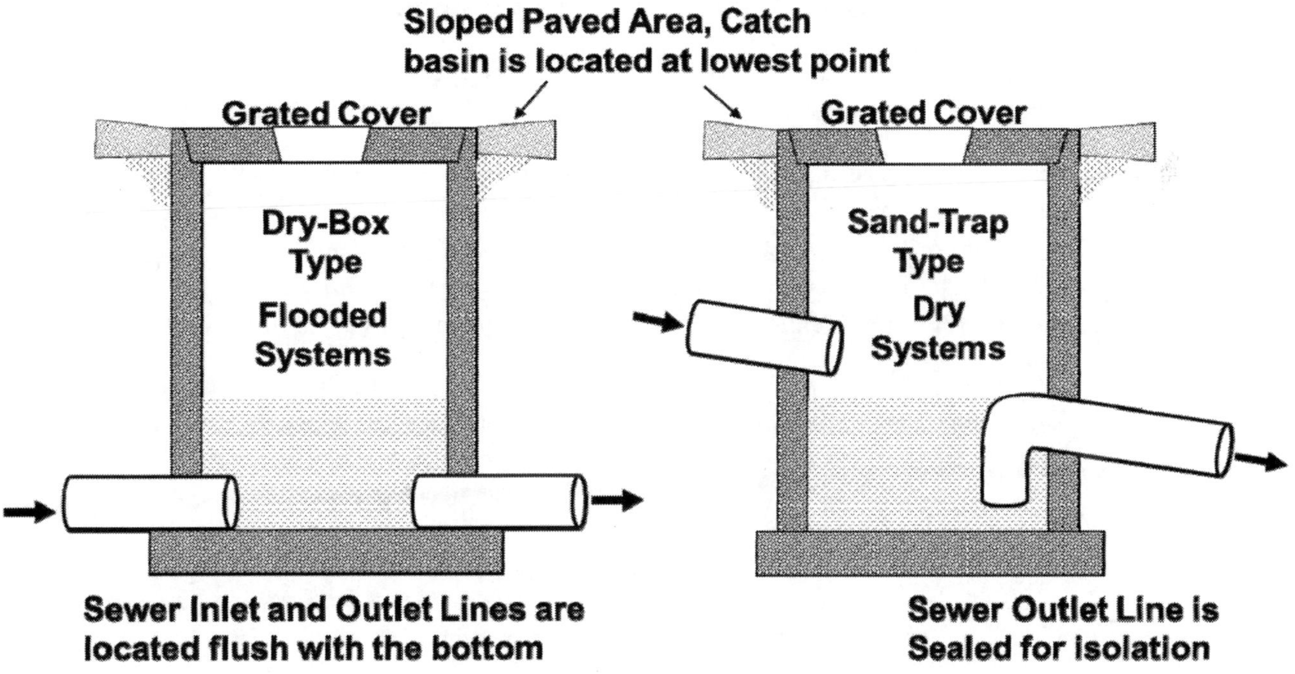

Figure 2 Dry-box "Flooded" vs. Sand-Trap "Dry" Type Catch Basins

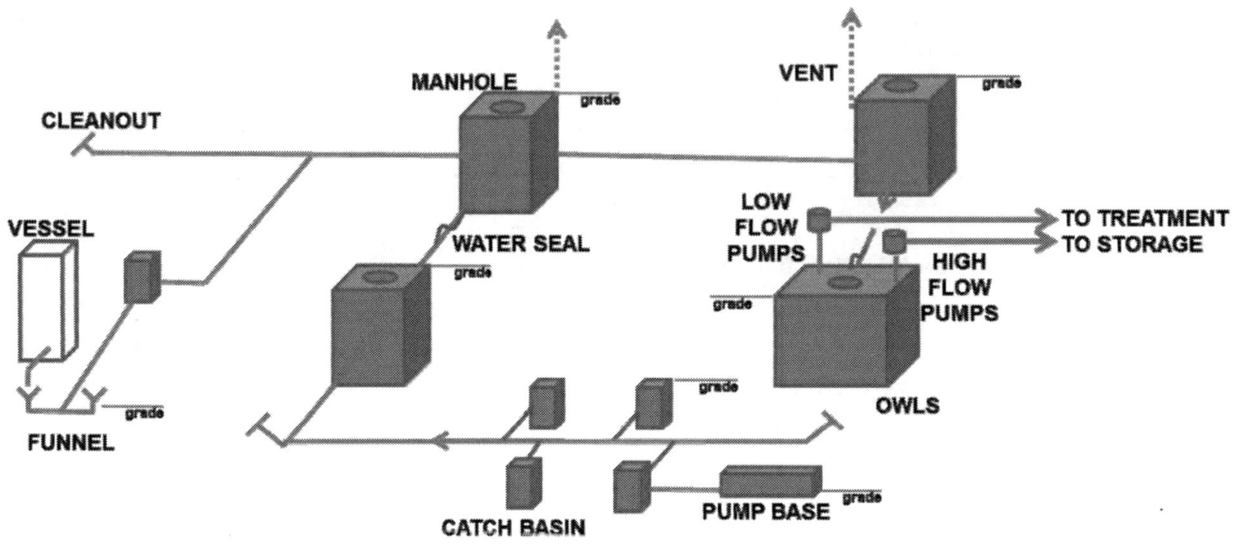

Figure 3 OWDS, Local Oily Wastewater Lift Station (OWLS) System

Drain Pipe / Funnels are outlet pipes that collect oily water from points such as pump bases, drain points, or vessel drains and discharge through a catch basin or manhole into the OWDS.

Area drains, or catch basins, serve several sources and are located centrally among the sources or pieces of equipment that they serve. Catch basins have a grate on top, an integral sediment trap and outlet liquid seal for dry systems and also serve as a collection drain for surface flows. Individual catch basin drainage areas are typically limited to approximately 5,000 square feet to ensure rapid surface runoff and to isolate groups of equipment for fire protection

Manholes in sewer lines permit service, inspection, and cleaning of a line. They may be located where sewer lines intersect or where there is a significant change in direction, grade, or sewer line diameter. Manholes include an outlet liquid seal for dry systems.

An OWLS, as part of the OWDS, is usually the last collection unit before transfer for treatment or storage, accepting wastewater from one or several sewer lines. An OWLS is a retaining structure, usually under atmospheric pressure, located below the normal grade of the gravity OWDS network. The collected wastewater is normally pumped to the WWTP.

An OWLS is sized to manage the larger of the following design flow rate cases:

- Normal process operation flows plus the peak fire water flow from the local risk area; or

- Cumulative storm water flow, typically based on a 5-year or 10-year return frequency, to prevent flooding that would disrupt operations or prevent a safe shutdown.

In some regions, the allowance for fire water may be more than the volume allowed for a 5-year or even 10-year storm water event; while in other regions the storm water event sizes the system. In those cases where fire water dominates, the normal process flows would be added to the storm water for the sizing analysis. This analysis also depends on the size of the collection area, since storm water flows are proportional to the ground surface area while fire water flows are not.

An OWDS with a network of intermediate OWLS is shown in Figure 4. The OWLS pumps typically would discharge into common headers, with one header dedicated to convey the wastewater to the WWTP for treatment and another header to storage.

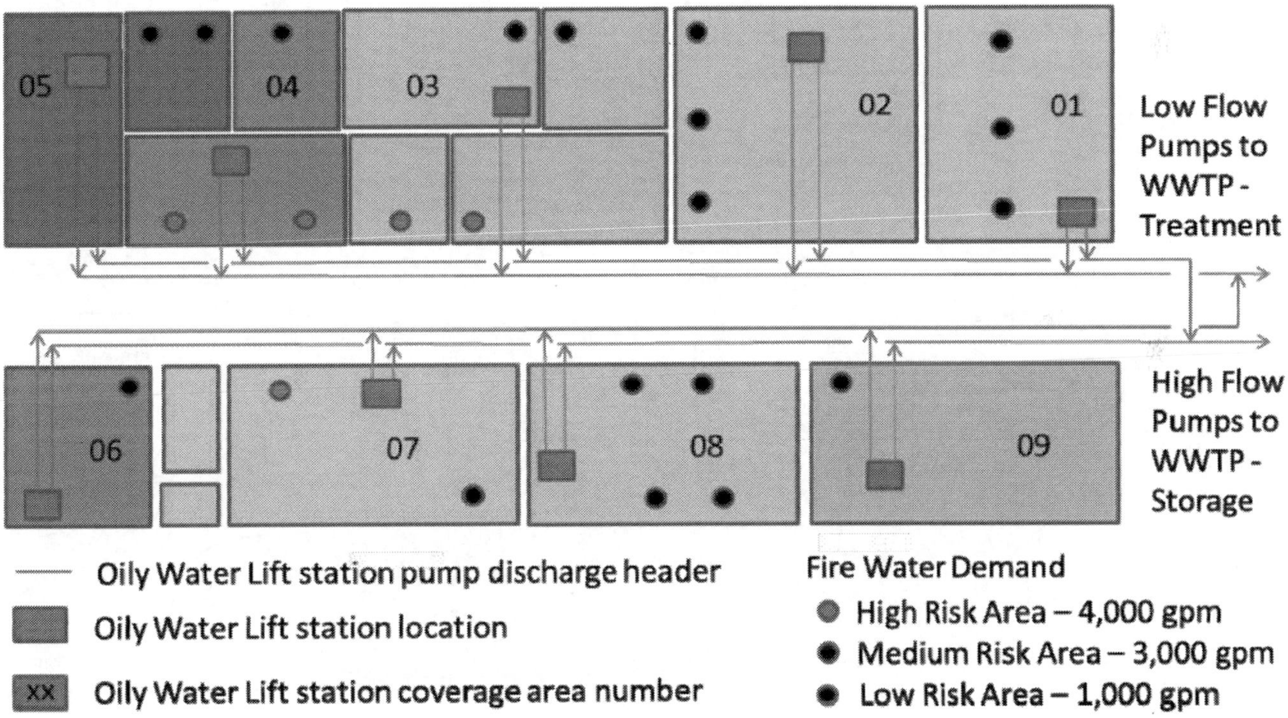

Figure 4 OWDS With A Network of Intermediate OWLS

Depending on the difference between normal and peak or high inflows, an OWLS may have one or two sets of pumps. If the high flow is less than 10% more than normal flow, a single set of pumps sized for the high flow condition can be used efficiently. If these flows vary by more than 10%, two sets of pumps, one for the normal and one for the high flow condition may be more efficient. For a two pump set system, a single or dual pumping chamber OWLS may be considered. Figures 5 and 6 show plan and section views of a dual chambered OWLS. Normal flows are managed in a smaller chamber with the larger chamber of the same OWLS reserved for the high flows.

All OWDS wastewater initially enters the low flow or process flow chamber and the high flow or wet weather chamber receives overflow from the low flow chamber. The minimum required working volumes of the chambers are related to their pump capacities, with the low flow pumps typically around 50 gpm or less and the high flow pumps 500 gpm or more each. However as indicated in Figure 6, the low flow chamber must also accommodate an underflow baffle and overflow weir rated at the design inlet flow. These weir dimensions often dictate the footprint shape, and often result in a working volume larger than the required minimum.

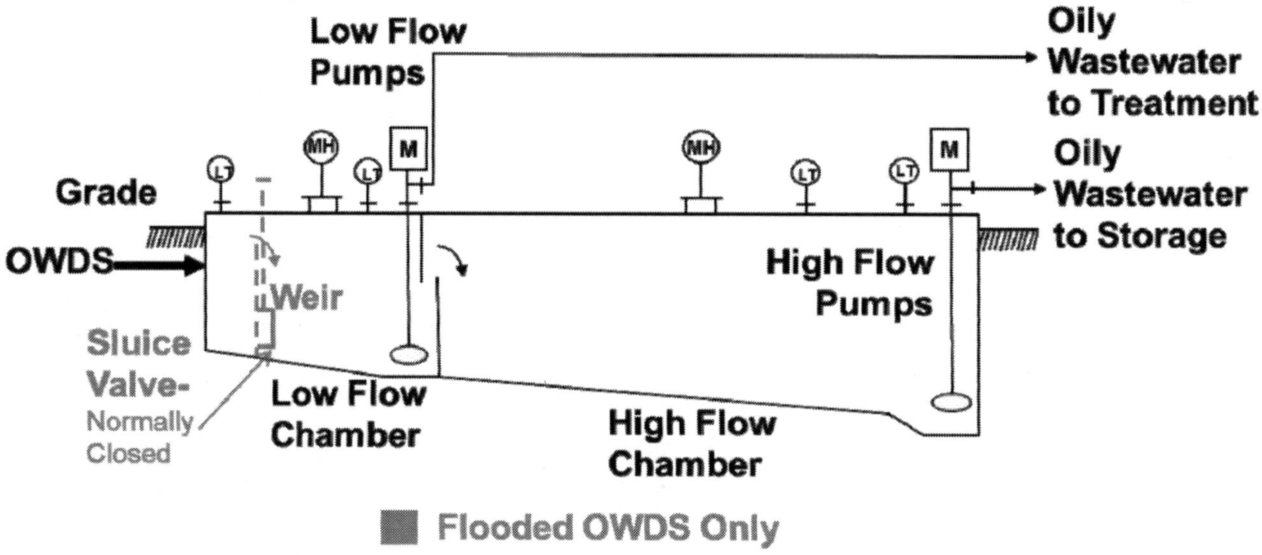

Figure 5 Dual Chamber Oily Water Lift Station (OWLS) – Section A-A

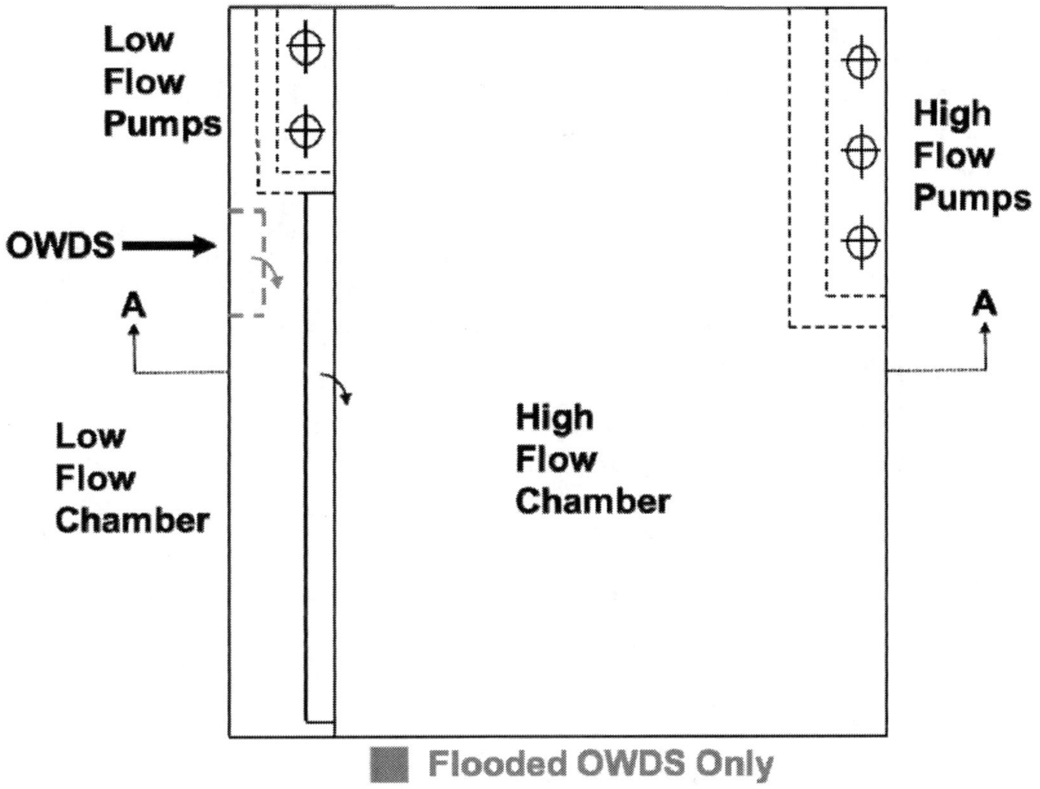

Figure 6 Dual Chamber Oily Water Lift Station (OWLS) – Plan View

3 SAFETY CONSIDERATIONS

A majority of safety incidents occur during startup or in auxiliary areas such as transport, storage, waste management, and recovery of raw materials or products (Lamberti 2011). Flammable liquids in wastewater discharged to an OWDS can cause fires / explosions in collection systems-particularly near the point of discharge, where high temperatures promote evaporation into a relatively fixed volume of air and form vapors that are trapped within the collection system. As the heated gases expand within the confined space, they create underground pressure seeking an opportunity to vent. Due to the vapor space confinement within OWDS and OWLS, when a flammable vapor cloud is ignited, the combustion probably will give rise to an overpressure resulting in a vapor cloud explosion (VCE) rather than a fire (Lees 2012b). This pressure will blow off at the weakest point, such as a manhole cover, a liquid seal trap, and / or an OWLS cover. Once dislodged, more oxygen may be introduced triggering additional explosions. A VCE can be ignited by electric sparks, friction, surfaces such as manhole covers heated by the sun, or heat generated by chemical reactions. Often workers use metal tools that can accidentally strike against a concrete surface or manhole cover creating sparks. (US EPA, 1992). Also, potential pump-related ignition sources include defective pumps, dry running pumps and faulty cables on submersible pumps (Bloch 2015).

Twenty years of historical data from a refinery and terminal facilities operator recorded a total of three dozen OWLS incidents in dry type systems (confidential 2008). Forty percent of the incidents involved a fire / explosion, while the rest were overflows. The main contributors to the incidents were:

- Presence of flammable liquids that formed an explosive vapor mixture – fire incident,
- Malfunction of level instrumentation - overflow,
- Failure of pumps - overflow.

Table 2 summarizes methods for managing risks associated with operating the components of an OWDS, preventing system failures, and increasing reliability.

Table 2 Methods for Managing Risks Associated with oily water lift stations

Design Considerations	
Electrical Area Classification	OWDS are classified as Class I, Zone 0 locations for manholes, catch basins, OWLS, underground piping, and vents. Zone 0 is defined as an area where an explosive gas mixture is continuously present or expected to be present 1,000 or more hours/year or 10% of the time (API RP 505, reaffirmed 2013).
Seals / Isolation	All flows into a dry OWDS enter through sealed catch basins or through a sealed connection to a manhole. All sealed connections, including sections of a drainage line located between liquid seals, from upstream of the first sealed manhole in an OWDS to downstream of the farthest catch basin, need to be vented to direct the release of any flammable or hazardous gases and to prevent a vacuum which could drain out the water seal connection. A water seal typically consists of an elbow or a tee with an outlet extending downward to provide a minimum of a 6 inch seal.
	In exceptionally hot and dry locations the seals in dry OWDS may require water addition to avoid evaporation; and a flooded OWDS may require a water purge to maintain a level of working liquid (Mason 1984).
	The use of a network of independent OWLS permits the isolation of individual process areas, units or complexes during site maintenance shutdowns. Whereas a site-wide sewer network increases the risk of a fire spreading and may lead to effluent stagnation during emergency situations.
Managing	Closed Drain Systems (CDS) are needed to manage flammable liquids and toxic materials from process water draw-offs, hydrocarbon contaminated

Flammable Liquids and Corrosive or Toxic Materials	cooling water, de-inventorying equipment drainage, and streams which could otherwise create hazardous conditions if they were discharged directly to OWDS. Even though no continuous flammable liquid source connections should exist, flammable liquid hydrocarbon content less than 50 mg/l may be considered for drainage. A combustible liquid at a temperature within 15 F (8 C) of its flashpoint would be considered flammable.
	Examples of toxic materials include: spent caustic, benzene, phenol, sour water that contains H2S (>10 mg/l) and ammonia (> 0 mg/l). Corrosive materials have a pH < 3 (acid) or pH > 12 (base).
Venting	Since flammable gases may accumulate in an OWDS, manhole covers and OWLS covers should be sealed or gasketed and vented to a safe location away from any potential ignition sources.
	End-of-Line Deflagration Flame Arresters may be used on vent lines to prevent flames from entering an OWLS, such as if a flammable gas mixture flowing from the vents were to encounter an ignition source.
Piping line depths	The OWDS piping in onplot unit areas is located under reinforced concrete slabs. Thus the configuration of the standard catch basin or funnel / drain pipe assembly vs. Minimum Cover Over Top of Pipe typically sets the minimum depths. Since a connection at the extremity of an OWDS sets the network depth, drain lines at these locations may be embedded in concrete to limit depth if the required soil-bearing loads are high.
	In a dry OWDS the slope for a free draining line is usually 0.6% to 0.8%. A flooded OWDS line slope would only require a nominal 0.1% slope (Mason 1984).
	For a dry OWDS, all gravity flow lines entering an OWLS must be above the maximum water level that will be maintained during operation. This design requirement ensures that the line will not become overfilled and not work properly. A 6-inch elevation difference is typically required to not impede incoming flow and cause a backflow and accumulation of sediment (confidential report 2008).
	Typically, when gravity OWDS lines are installed in trenches deeper than 20 feet, the installation cost increases significantly due to more complex excavation and trench shoring techniques (US EPA 2000a)
Siting	When locating OWLS, consideration for the distance to major process equipment should be considered. Offset, such as 150 to 250 ft. from process equipment, 30 to 70 ft. from occupied buildings, and 10 ft. from pipeways; may limit options and even lead to consideration of off unit plot locations.
	If a high water table or large distances are an impediment to the design of an OWLS, even after considering converting from a dry OWDS to a flooded and / or segregation of the OWDS from the storm / fire water, then additional OWLS may need to be installed.
	The number of local OWLS is determined by the following:

	• Type of OWDS, dry vs. flooded. • Site topography, available elevation change. • Minimum gravity flow line velocity, • Maximum invert depth to prevent the OWLS bottom elevation from being located below the water table, • Prevention of gravity flow lines from crossing large underground site utility lines, • Prohibition of locating near process equipment, inhabited buildings, drainage swales, diked areas, or designated pipeline corridors. • Avoidance of the pump lift station head from exceeding 20 ft. to allow the use of self-priming pumps.
	For OWLS vent locations, a safe location should be considered as no closer than 50 ft. horizontally from a furnace or other ignition source; and if within 100 ft. consider ending the vent line at least 50 ft. above grade.
Downstream impact	As the number of OWLS pumping to an API Separator increases, the turbulence impact from on-off pumping on the oil / water separation process should be considered. Either installation of a break tank with equalization or ensuring that the API Separator design is robust enough to manage the potential fluctuations and turbulence is suggested.
	Due to potential disruption of the pump(s), such as by mechanical or power supply reasons, the WWTP design may need to include quick operational turn down.
Pump Selection	Due to a Class 1 Zone 0 Electrical Area Classification, pumps with submersible motors would not be permitted. Bloch (2015) further discusses pump types for Zone 0 service.
	Several types of centrifugal pumps are frequently used in OWLS applications. Straight-flow centrifugal pumps are suitable for low-flow / high head conditions, angle-flow are appropriate for low to moderate heads, and mixed flow for large flows at low head (US EPA 2000c). Concerns for causing emulsification that would impact downstream treatment may also guide pump selection.
	Self-priming pumps have been found to be more reliable in OWLS service (confidential report 2008). If the required pump flow / discharge pressure is within the operating range of a self-priming pump, and the OWLS depth does not exceed a 20 ft. suction lift then the use of a self-priming pump may be an option. Use of a piping foot valve or anti-siphon valve would be recommended in this application. Or an option would be a vertically suspended line shaft type pump.
	US EPA (2000b and c) states that the number of pumps should be chosen to correspond as nearly as possible to flow fluctuations. In OWLS with peak inflows less than 700 gpm, two pumps are customarily installed with each

	able to meet the peak influent rate, subject to flow variability concerns. For instance, if the peak flow is less than 10% more than normal flow, a single set of pumps sized for the peak flow condition can be used efficiently. For larger OWLS or OWLS with greater variability, strategies include: installing pumps of different sizes, using parallel pumps, trimming impellers, installing variable speed drives or two-speed motors in order to prevent frequent starting / stopping and to not require excessive OWLS storage capacity. However for OWLS with rapid flow increase or high head losses, a single peak pump may be favored because two pumps operating together yield only slightly higher flows than one pump. Parallel peak pumping is favored for large OWLS with relatively flat system head curves, where multiple pumps deliver substantially more flow and provide more flexibility.
	Ursino (2014) states that adding a redundant pumping element in standby is less effective than increasing the number of pumping units that may function in parallel. Thus a larger number of smaller online pumps may prove more reliable than a less number of larger pumps, despite provisions for standby operation.
	Overall pump efficiency depends on the type of pump, the control system, solids loading, and the fluctuation of influent. Well-designed pump systems may operate at up to 75 to 85 percent efficiency most of the time (US EPA 2000c); however, discharge header conditions may limit efficiencies to closer to 50 percent at times (US EPA 2000a). Also, the mechanical efficiency of the pump impeller would normally decrease with increased solids handling ability, in relation to the solid size to be passed.
	Installing an inlet OWLS screen with means to clean it, will provide for filtering out incoming debris to protect the pumps from blockage.
Power Supply	Including separate transformer feeds to all pump systems can help to ensure a continuous source of power is provided. Emergency generators whether onsite standby or portable with quick connection can also be used as a backup power source. However, in the case of a power failure, options to de-inventory an OWLS would be: a) Use of Vacuum Trucks with suction hose access through a sump cover manway and, b) Use of portable diesel pumps with suction hose access through a sump cover manway or by way of dedicated connections on pump piping lines.
Instrumentation and Controls	Instrumentation to ensure overflows or running pumps dry does not occur is recommended; however, faulty or complex instrumentation should not hold up pumps from automatically operating when needed for instances such as removing fire water. The confidential report (2008) states: 1. Provide two radar level transmitters (LTs) for redundant level measurement in the OWLS. Level indication shall be in the control room. LTs should be configured to fail low such that pumps will not start on

	transmitter failures. LTs are preferred over switches because switches are blind devices that can fail in an undetected manner. Typically one LT should be designed to control the sequenced start-stop of pumps. Each LT should indicate its level on a DCS display in the control room and be provided with low and high level alarms. The two associated LTs should be compared and alarm on deviation, to provide more confidence that the level signal is true. 2. Maintain the water seal within an OWLS by having the incoming flow line termination point at 6 inches below the pump shut off level. Also the OWLS high-high level shall be reset below the incoming flow line. 3. A flow or pressure transmitter should be provided on each pump to trip on low flow. An orifice based flowmeter is recommended, but a pressure transmitter may be considered if the pump discharge pressure will reliably drop to a known pressure on loss of flow. Flow transmitters (FTs) are preferred over switches because switches are blind devices that can fail in an undetected manner. FTs should trip its associated pump on low flow and be configured to fail low. Low flow tripping should be active when the pump is on automatic or manual operation. However, consider low flow alarming only for pumps likely to be used for fire water service, to allow for enhanced availability. 4. Local controls should be located at a safe distance from the edge of the OWLS – 10 ft. away or more - to keep operators away from potential explosion exposure. Consider a local analog flow indicator from each pump FT and a local analog level indicator for each sump. Each pump should also have a local hand-auto-off (HOA) switch and push button start. However, since operating the pumps in manual mode increases the risk of pumps dry running, this switch should be locked or car-sealed in auto mode. Also consider providing the control room DCS with an indication of HOA switch position. 5 The control room DCS should be provided with a running and stopped indicator for each pump. Also consider DCS logic that will monitor both the OWLS level and pump flow / running conditions, so to provide an alarm if a pump is running with acceptable flow but the level is not moving.
<u>Operational and Maintenance Procedures</u>	
<u>Spills</u>	Operating procedures will need to stress the importance of detecting and isolating oil spills.
	The dual chamber version of the OWLS contains an overflow / underflow weir system to segregate free oil in the inlet portion of the system. However, an automated oil removal system is usually not installed. As a result, oil will collect in this first chamber and needs to be removed

	manually. The main function of the underflow / overflow weir system is to avoid oil carryover into the second or high flow chamber since only the low flow or dry chamber discharges directly to the WWTP's oil removal system.
	During vessel turn-around after depressurization, equipment inventory is drained into a CDS. Only residual inventories of flammable and toxic liquids, less than 26 gallons from equipment should be drained to an OWDS under special procedures. Flushing the sewer with water for dilution after draining until the percent lower explosive limit (%LEL), determined by lowering an LEL tester sample hose near the liquid level, is reduced to < 10% is suggested. Before draining any small amounts of flammable liquids or hazardous materials to the OWDs, also consider: (a) wind direction, (b) positioning of any ignition source beyond 50 ft., (c) H2S or ammonia content, and (d) use of proper personal protective equipment (PPE) including breathing apparatus.
Removal of Surface Oil	For lighter hydrocarbons, the insoluble oil tends to float on the surface. When the OWLS pumps down, the oils often are retained when the low level is reached. Special pump designs are available to allow for removal of much of this oil, and / or vacuum truck access can be provided to remove surface oil.
By-pass for maintenance	Facilities for bypassing an OWLS should be considered, such as emergency pump connections on the underground gravity inflow line and on the outlet pump discharge line. In addition, provisions for placement of an isolation sluice gate on underground inflow lines to assist in a temporary pump-around may be practical.
Pump Failure	Confidential report (2008) states that OWLS pump failures are mainly due to: • Dry running: from dirt plugging the pump strainers or faulty level instruments. • Lack of lubrication: poor dry packing design or poor maintenance practices. • Vertical pump application: more prone to failure than horizontal self-priming pumps because of their suspended long shaft, which is more likely to wear against pump bushings and packing if dirt or vibration increases. Also more difficult to detect problems as components are below the surface and not easily accessible. • Lack of effective preventative maintenance (PM): even if an OWLS pump PM program exists; confirm that all pumps are covered.
	Maintenance goal should be to have Mean Time Between Failures (MTBF) for pumps above two years.
Temperature	From a materials of construction consideration, the temperature of any drainage should not be higher than the temperature rating of the drainage piping. Hot steam condensate is not recommended to be discharged to the

	OWDS.
	In the case that Potentially Contaminated Condensate (PCC) has been detected from a leaking exchanger, cooling would be required before any PCC would be approved for draining to an OWDS subject to flammability concerns. Collecting PCC in a CDS for pumping directly to a WWTP based dedicated cooler should be considered. Cooling to at least 140F is recommended in either case. For operational input: 1. In the event of any contamination of steam condensate with hydrocarbons, identify the source of contamination and isolate for correction. 2. If for any reason, hydrocarbon-contaminated hot steam condensate is directed to OWDS, then it must be cooled by dilution with utility or fire water.
Solids	Recognize that EPA HAZARDOUS WASTE CODE F037 pertains to Petroleum refinery primary oil/water/solids separation sludge –including any conveyances; sumps; and storm water units receiving dry weather flow. Also, in some regions fine dust/carbonate which produces a sticky paste when mixed with water and oils can be expected in an OWDS. Any sand or grit that collects in process areas and gets washed into an OWDS may mix with hydrocarbons and create sticky oily solids. One pound of dry solids can create ten pounds of oily sludge (IPIECA 2010). Solids issues favor the use of non-clog pumps, typically open impeller trash pumps, since they have been demonstrated to be reliable for handling oily wastewater and provide protection against oily waste causing pump failure.
Firewater	The OWDS function to remove fire water is essential to firefighting activities, so OWLS pumps must be set-up to operate when required. The primary concern is to contain firewater runoff onsite. Firewater within unit battery limits should gravity drain through the OWDS and collect in a local OWLS. Both the OWDS and OWLS in a combined system should be designed for firewater loads. For dual chamber OWLS, the excess inflow will overflow to the high flow chamber. When the level in the high flow chamber reaches the high flow pump start level, the high flow pumps will start and the low flow pumps stop. However, an operator can shut off the low flow pump and eventually rely on only the high flow pumps without loss of containment. The high flow pumps will discharge to storage, thus isolating the collected firewater from the WWTP treatment system. Note, the low flow pumps discharge directly to the WWTP treatment system but their relatively small flow rate will limit the impact of firewater on the WWTP. Once collected in storage, firewater can be analyzed and even pre-treated before discharging through the WWTP for treatment at a controlled rate to limit potential impact. Any significant free oil layer should be manually skimmed off before discharging back to the WWTP.

	The first flush approach while applicable to storm water, may not necessarily apply to firewater runoff which can remain contaminated throughout a fire fighting event. Therefore, the disposition of any collected firewater runoff will have to be determined on a case-by-case basis, and should be expected to depend on the specific firefighting scenario. Firewater can carry environmental contaminants, with a scenario including flammable and/or combustible materials, such as petroleum products, foam solution containing perfluorooctane sulphonates, dissolved and particulate materials from combustion processes, materials generated through quenching, as well as heavily contaminated solids and grit. Since the hazards and associated risks, such as unknown health effects and the impact on a biological treatment system are difficult to estimate in practice, both the liquids and solids should be considered as potentially hazardous wastes. Also, if the analyses indicate that the detained solids / sludge in an OWDS / OWLS after a fire event may not be recommended for treatment at the onsite WWTP then arrangements for offsite treatment and/or disposal via licensed waste contractors may be necessary. Note, even if the firewater runoff liquid is determined acceptable for the onsite WWTP, the solids may require a different disposal route.

4 CONCLUDING REMARKS

The management of oily water involves managing process safety hazards. Since all hydrocarbons can form an explosive atmosphere under the right temperature and pressure conditions; methods for mitigating and controlling risks associated with operating OWLS should be firmly established.

Safety hazards associated with OWLS can be mitigated through proper design, permits and procedures. The "safety culture" must extend to control auxiliary areas such as wastewater management systems; and should consider the OWLS as a hydrocarbon process facility in regards to HAZOPs and process safety analysis. The most labor-intensive task for OWLS is routine preventive maintenance (US EPA 2000c); however regardless of the excellence of a maintenance program, equipment use causes wear and ultimate failure or breakage of parts should be accounted for in planning.

5 REFERENCES

API RP 505, (Reaffirmed 2013), Recommended Practice for Classification of Locations for Electrical Installations at Petroleum Facilities Classified as Class I, Zone 0, Zone I, and Zone 2.

Bloch, H.P. (August 2015). Consider Zone 0 vertical pumps for slop pits. Hydrocarbon Processing. Vol. 94 No. 8.

Confidential Report, (2008).

IPIECA (2010), Petroleum refining water / wastewater use and management. Operations Best Practice Series. Refinery Water Management Task Force. Storm water and sewerage section.

Lamberti, P.G., JL Woodward. (March 2011). Learning from Events: Major Process Safety Incidents from Wastewater in Process Chemical Plants, Or Take Care of Your Wastewater or It Will Take Care of You. Process Safety Progress Vol. 30, No. 1.

Lees, F.P. (2012). Lees' Loss Prevention in the Process Industries (4th Ed) Elsevier. (a)Chapter 10 – Plant Siting and Layout, Section 10.17 Drain Systems.

(b)Chapter 17—Explosion, Section 17.28 Vapor Cloud Explosions

Mason, G.S., C. Arnold. (1984) Contain Liquid Spills and Improve Safety with a Flooded Storm Water Sewer. Chemical Engineering V 91 (Sept 17th), pp 105-109.

Ursino, N. (2014), P. Salandin. Combined effects of mechanical availability and storage capacity in pumping station risk assessment. Procedia Engineering V 70, pp 1679-1687.

US EPA, (July 1980) Treatability Manual Volume II Industrial Descriptions, Petroleum Refining Category Section14, EPA-600/8-80-042b

US EPA, (June 1992) Guidance to Protect POTW Workers From Toxic And Reactive Gases and Vapors, Section 3.1. EPA 812-B-92-001.

US EPA, (September 2000a) Wastewater Technology Fact Sheet – Sewers, Force Main, EPA 832-F-00-071.

US EPA, (September 2000b) Collections Systems Technology Fact Sheet – Sewers, Lift Station, EPA 832-F-00-073.

US EPA, (September 2000c) Wastewater Technology Fact Sheet – In-Plant Pump Stations, EPA 832-F-00-069.

12TH GLOBAL CONGRESS ON PROCESS SAFETY

Critical Considerations in the Pressure Relief Analysis of Refinery Sour Water Systems

James (Jim) R. Bridgett
james.bridgett@siemens.com

Robert (Bob) Siml
robert.siml@siemens.com

Gabriel Gandhi Martiniano Ribeiro de Andrade
andrade.gabriel@siemens.com

Siemens Energy, Inc.
4615 Southwest Freeway, Suite 900,
Houston TX 77027

Prepared for Presentation at
American Institute of Chemical Engineers
2016 Spring Meeting
12th Global Congress on Process Safety
Houston, Texas
April 10-14, 2016

UNPUBLISHED

Critical Considerations in the Pressure Relief Analysis of Refinery Sour Water Systems

James (Jim) R. Bridgett
james.bridgett@siemens.com

Robert (Bob) Siml
robert.siml@siemens.com

Gabriel Gandhi Martiniano Ribeiro de Andrade
andrade.gabriel@siemens.com

Siemens Energy, Inc.
4615 Southwest Freeway, Suite 900,
Houston TX 77027

Keywords: safety, pressure relief, two-phase flow, inherently safer design, sour water

Abstract:

Sour water systems are an integral part of all refineries. They are designed to collect sour water (i.e., water containing hydrogen sulfide, other sulfur compounds, and trace amounts of hydrocarbons) from numerous sources for treatment prior to disposal. Due to the complexity of these systems, the pressure relief analysis is inherently difficult. This paper discusses critical considerations in the pressure relief analysis, including the impact of light hydrocarbons entering the sour water system, design flaws that set traps for operators, deficiencies in operating procedures, and some critical considerations for the pressure relief analysis. Also presented will be potential mitigations for existing systems and recommended best practices for the design of new systems.

1. Introduction

Sour water collection and treatment systems are an integral part of all refineries. Although not a core production unit, the proper function of the system is important to the safe, reliable economical operation of a refinery. While it might be tempting to assume that non-core production units such as sour water systems are low risk and therefore don't require the same level of overpressure protection analysis and review as other parts of the refinery, the work recently done by Siemens to support the study of a process upset in a sour water system strongly indicates that is not the case. Since sour water systems inherently contain numerous high to low pressure transitions, the overpressure protection analysis is much more complex and difficult than might first appear. This paper will discuss in detail the lessons learned from our recent work and how those lessons can be used to strengthen the overpressure protection of sour water and similar systems. Improving the overpressure protection of these systems is a key part of reducing risks associated with operating these types of systems to as low as reasonably practicable (ALARP).

2. Initiating Event

A typical refinery sour water system consists of multiple sources, including streams that are pumped or pressured in on level control, as well as streams that are manually drained.

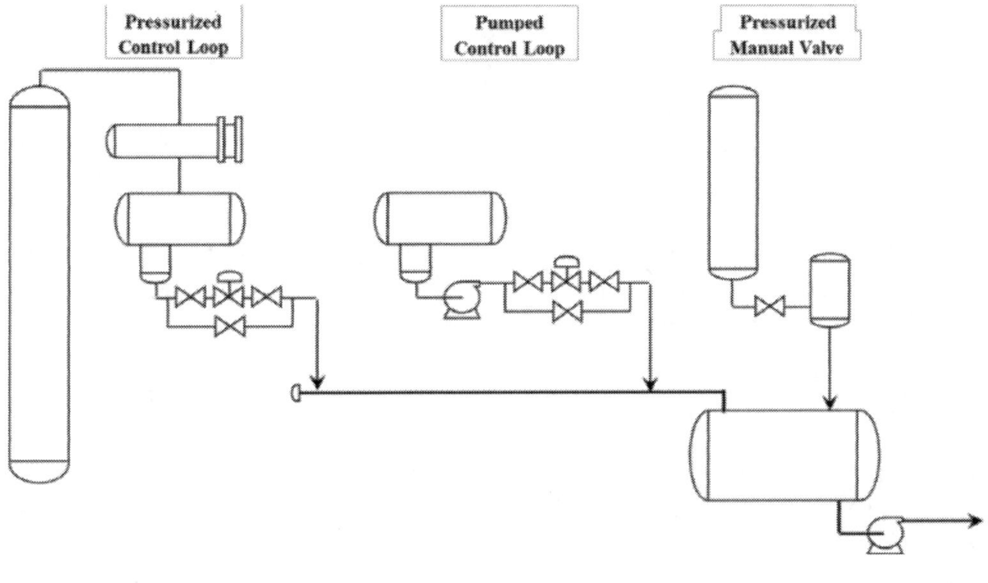

Due to a process upset, the feed to the column contained much more water than normal. Water was entering the overhead accumulator and separating out in the water boot faster than the interface level control could remove the water to the sour water system. Therefore the outside operator was asked to open the bypass valve around the interface level control valve, to reduce the water level.

It appears that the operator opened the bypass valve gradually, in steps, over a few minutes. The water level in the boot decreased in corresponding steps, and then sharply dropped to zero. At the same time the pressure in the sour water drum increased rapidly. Over about 30 seconds, the pressure increased from the normal operating pressure of approximately 10 psig to at least 55 psig. The actual pressure the vessel reached is not known because 55 psig was top of scale for the pressure transmitter on the sour water drum. As the pressure in the drum spiked, both the pressure control vent valve and the relief valve on the sour water drum opened as designed, releasing vapors and/or gases to the flare.

The pressure transmitter remained "pegged" at the maximum reading of 55 psig for another 30 seconds. The pressure decreased to approximately 15 psig over the next 4 minutes. As the pressure in the system dropped, the relief valve closed. Over the next 30 minutes or so, the pressure gradually dropped from 15 psig and stabilized at the normal system pressure of 10 psig. As the pressure dropped, the vent valve returned to normal operating position.

In response to the loss of interface level and the pressure increase at the sour water drum, the operator immediately closed the level control bypass valve. The interface level was re-established in 10 to 15 minutes.

3. Preliminary Review

The above information is based on both process data as captured by the data historian and information from operations. Siemens first conducted a preliminary evaluation to validate the process data and time line as described above. This evaluation was intended to answer the following questions:

a. What was the root cause of the pressure rise in the sour water system?
b. How much liquid hydrocarbon, if any, entered the sour water system?
c. What maximum pressure did the sour water system reach?
d. Did the time line make sense based on estimated flow rates through the bypass valve, etc.?

Since these questions can be answered using standard and well understood calculation techniques, detailed calculations will not be presented in this paper. Instead, the focus will be on the results and conclusions.

3.1 Root Cause of the Sour Water System Pressure Rise

Three potential root causes of the pressure spike in the sour water drum were considered:

a. *Release of trace amounts of hydrocarbons, hydrogen sulfide and other dissolved gases from the additional sour water flow that occurred when the level control bypass valve was opened.*

   Based on the dimensions of the water boot, the maximum amount of sour water that could have been drained to the system was about 15 gallons (approximately 120 pounds). This volume of water is simply too small, and would have contained too little dissolved hydrocarbons, hydrogen sulfide, etc. to contribute significantly to the pressure rise in the sour water drum.

   If the additional sour water was the only source of light gases and vapors into the system, this analysis indicates that the process vent valve would have had more than sufficient capacity to handle the event; as a result, the pressure would not have reached the relief valve set point.

b. *High pressure hydrocarbon vapors from the overhead accumulator (i.e. "gas blow-by" scenario).*

   For a gas blow-by type scenario to occur, all of the liquid in the overhead accumulator would have to have been (drained) transferred to the sour water drum. Since the hydrocarbon level in the overhead accumulator did not drop to zero at any time during this upset, this ruled out the gas blow-by scenario as the root cause of the pressure spike in the sour water drum.

In addition, even though there was more than enough hydrocarbon in the overhead accumulator, the high level alarm in the sour water drum never activated, no frost line occurred, and the interface level was quickly reestablished.

c. *Liquid hydrocarbon from the overhead accumulator flashing as it entered sour water drum.*

Primarily by a process of elimination, the root cause of the pressure spike in the sour water system was determined to be a moderate amount of liquid hydrocarbon entering the sour water drum (see 3.2 for details). Due to the difference in pressure between the overhead accumulator and the sour water drum, the hydrocarbon flashes as it enters the sour water system. The process vent valve could not respond fast enough and/or did not have enough capacity by itself to vent the vapor volume generated to the flare, and as a result, the pressure in the system increased until the relief valve opened.

Additional evidence that supports this conclusion includes:

- Little to no change in the hydrocarbon level in the overhead accumulator during this process upset.
- The relatively short duration of the maximum pressure reading and the relatively quick return of the system to normal pressure.
- Interface water level quickly being reestablished.

3.2 How much, if any, liquid hydrocarbon entered the sour water system

Based on the available empty volume of the sour water drum and the documented pressure changes, this analysis indicates that about 100 pounds of hydrocarbon vapor was required to increase the pressure in the sour water drum from the normal 10 psig to 55 psig. This sets a lower limit on the amount of hydrocarbons that could have entered the sour water drum.

The total amount of hydrocarbon that entered the sour water drum, in addition to the amount required to pressure up the system, includes the following:

a. *Amount released to the flare through the process vent valve.*

   Based on the flow coefficient (C_V) of the control valve and estimated vessel pressures plus the established time line, the best estimate of the total amount of material released to the flare through the vent valve is about 300 pounds. This estimate includes approximately 200 pounds over the 5 minutes as the pressure spiked and then decreased to 15 psig and at least 100 pounds over the next 30 minutes as the pressure dropped from approximately 15 psig back to the normal 10 psig.

b. *Amount released to the flare through the relief valve*

   Since there was no indication that release was two phase or liquid, the flow through the relief valve was based on standard relief device vapor sizing equations using ASME certified parameters for the relief valve orifice area and coefficient of discharge. Based on the time line, the pressure relief device was open for less than 2 minutes. During this time, approximately 90 to 110 pounds of material was released through the relief valve to the flare.

Overall, it is estimated that approximately 500 pounds (110 gallons) of hydrocarbon drained from the overhead accumulator into the sour water drum during this upset.

3.3 What maximum pressure did the sour water drum reach?

Based on the recorded process data, it was known that the pressure in the sour water drum reached at least 55 psig. Since 55 psig was the maximum reading of the pressure transmitter, the actual pressure reached was not known. To address this question, a process simulator was used to plot the vapor pressure of the overhead accumulator hydrocarbon vs. temperature. Unfortunately, the temperature of the sour water drum was not available. Therefore, the temperature of the overhead accumulator, 85 °F, was used as the best available estimate of the temperature in the sour water drum. This temperature is consistent with cooling water temperatures expected during the time of year the upset occurred.

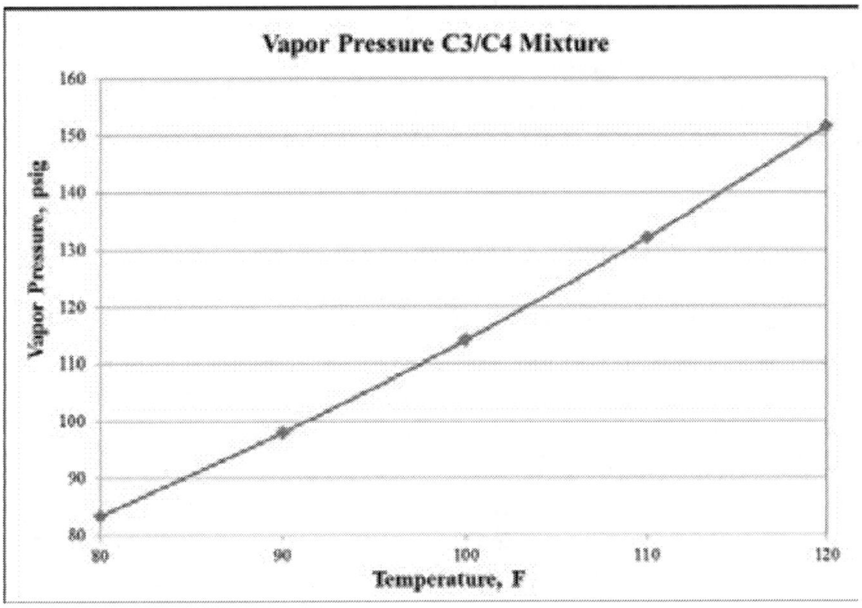

Using the estimated temperature of the sour water system, the maximum potential pressure in the sour water drum is approximately 90 psig when liquid hydrocarbon is present. Based on the capacities of the relief valve and process vent valve, the best estimate for the pressure in the sour water drum during this upset is 60 to 70 psig.

This pressure is just above the top of scale pressure for the pressure transmitter on the sour water drum and is an indication that the pressure transmitter was reasonably accurate.

3.4 Does the time line make sense based on estimated flow rates through the bypass valve, etc.?

As previously indicated, the estimated volume of sour water in the overhead accumulator boot was approximately 15 gallons. The interface control station consists of a 1 inch control valve and a 2 inch gate valve as the bypass. Although it is not unusual for gate valves to be used as a control valve bypass it is generally recognized that gate valves are not suitable for flow control. It is unusual to see an oversized gate valve as a bypass. Obviously, due to both size and flow characteristics, at the same pressure drop, the capacity of the gate valve is much greater than the control valve.

For sour water, the wide open capacity of the bypass was as high as 500 gpm based on the normal operating pressure in the overhead accumulator (185 psig) and the normal pressure in the sour water drum (10 psig). This estimate assumes little or no flashing occurs across the control valve.

For the hydrocarbon, the wide open capacity of the bypass was approximately 150 gpm based on the same pressure differential before the pressure in the sour water drum started to rise. This flow rate is much lower than the flow rate on water due to the lower density of the hydrocarbon and the flashing of the hydrocarbon liquid across the bypass.

In general, the estimated flow rates through the bypass valve are consistent with the rapid short duration of the pressure spike in the sour water drum.

Some time and effort was spent to investigate whether the flow characteristics of the gate valve also contributed to this process upset.

The number of turns from fully closed to fully open for a 2 inch gate valve varies between 11 and 13, depending on manufacturer. Definitive opening characteristics for gate valves were not readily available in the open literature, but it was generally noted that due to the geometry of the valve, the flow characteristics of gate valves are highly non-linear. There is some anecdotal evidence that gate valves may flow as much as 80% of capacity at only 20% open, but this is not well documented and should only be used with appropriate care and judgement as a relative indicator of the opening characteristics of gate valves.

While the evidence is not as definitive as desired it is doubtful that the bypass valve was ever fully open and was closed quickly once the interface level indication in the water boot went to zero (0) and the pressure spiked in the sour water drum. All indications are the 2 inch gate valve was grossly oversized and not suitable for control of flow out of the water boot.

4. Overpressure Protection Analysis

The following scenarios would typically be considered during the overpressure protection analysis of the sour water drum:

a. External fire
b. Blocked liquid outlet (overfilling)
c. Process vent control valve failing closed
d. Failure of automatic controls on each feed stream

Of these scenarios, a, b and c did not play a role in this process upset, and the analysis of these scenarios is fairly straightforward using standard overpressure protection assumptions and calculation methods appropriate to each type of scenario.

The failure of automatic controls cases were the primary focus for two reasons:

a. A failure of automatic controls and/or associated bypass played a central role in the process upset being studied.
b. Sour water systems inherently contain numerous high to low pressure interfaces at control valves or manual valves.

The failure of automatic controls scenarios typically include:

1. The control valve failing or being driven 100% open
2. The control valve bypass valve being 100% open
3. Some combination of control valve and bypass valve open, per client preference.

For all of these cases, the required relief rate calculation is based on:

a. Control valve flow coefficient (C_v) provided by the valve manufacturer
b. The pressure differential across the control valve. The typical assumption used in the calculation is normal upstream pressure minus downstream relief pressure.

When evaluating control valve and bypass scenarios the first step in the analysis is to determine if the liquid in the upstream system will fit in the downstream system.

If the liquid will fit, then the relief device on the downstream vessel should be sized for the vapor breakthrough or gas blow-by case.

If the liquid does not fit, then the relief device on the downstream vessel should be sized for the "liquid displacement" case. Liquid displacement is a phenomenon where vapor or two-phase material entering a liquid full system displaces the liquid. The relief device must be sized to handle the displaced liquid at the volumetric flow rate entering the system. Typically, sizing for liquid displacement requires a considerably larger relief device.

Liquid Displacement

Liquid Full **Volumetric Flow In = Liquid Volume Flow Out**

In this particular case, the hydrocarbon is flashing at the lower pressure in the sour water system. Initially, the liquid will tend to separate out and accumulate in the drum. The flow through the relief device would only be vapor. Note that any liquid hydrocarbon that accumulates in the drum would continue to flash and generate additional vapor due to transfer of heat from the sour water. In general assuming all the liquid will vaporize is conservative. Evaluating the potential for liquid to collect and instantaneously generate a higher vapor rate due to mixing is very difficult to predict and is beyond the scope of this paper.

One of realizations of this analysis is as liquid separates out and begins to fill the drum, the space available for liquid-vapor disengagement is reduced. At some point, the vapor velocity in the drum will increase to the point that the liquid will not separate out and instead will be re-entrained out the relief device. This two phase release case can significantly impact the size of the relief valve. Therefore any process vessel with a source of two-phase flow must be evaluated for vapor-liquid disengagement and the relief system must be sized for two-phase flow if re-entrainment will tend to occur from the level rising.

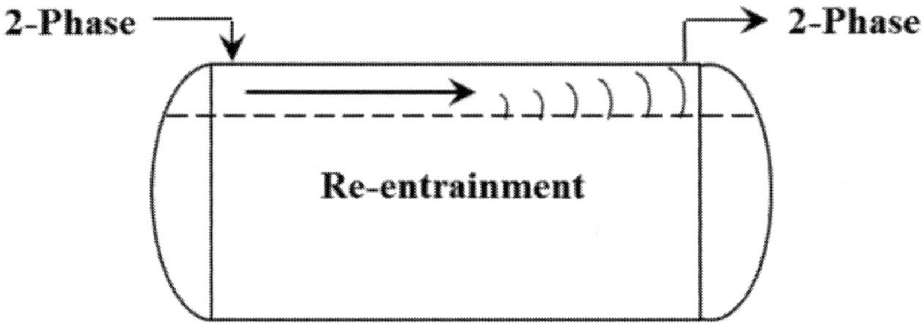

One of the most surprising realizations is liquid displacement will not occur due to the entry of two-phase flow when downstream vessel is not initially liquid full, even when there is enough liquid in the upstream system, because re-entrainment will occur first.

5. Conclusions

- Overpressure protection analysis of sour water systems in particular, and other similar collection systems in general, is not simple and straightforward. Particular attention must be paid to all high pressure to low pressure interfaces and ensure all credible scenarios have been identified.
- Liquid displacement will not occur when the flow entering is two phase if the vessel is not initially liquid full because re-entrainment will occur first.
- An oversized gate valve is not suitable for use as the bypass valve and can set a trap for operations.
- Any process upset is an opportunity for continuous improvement.

6. Recommendations

Corrective Actions for Existing Systems

- Check for oversized bypass valves:
  - Consider downsizing, installing a mechanical stop, or restricting orifice.
  - Consider replacing gate valves with globe valves.
- Review the overpressure protection analysis.
- Consider increasing the span of pressure transmitter.
- Consider recording process temperatures in the data historian

Inherently Safer Design for New Systems

- It's generally not practical to design the sour water system for max upstream pressure of every source. To the degree practical, design for the vapor pressure of the hydrocarbon at upper end of the sour water system operating temperature.
- In the design phase, balance the size of the water boot versus design of control and bypass valves to ensure the operator has adequate time to respond.
- Bypass valves for use in normal operation should generally be globe valves of similar size to the control valve.
- If oversized bypass valves are required for start-up, shut down, or other specific operating conditions, consider installation of a separate bypass valve that is car-sealed closed during normal operations.
- When reasonably practicable, the process vent valve should be sized to keep the relief valve from opening.

Disclaimer:

The information contained in this document represents the current view of the authors at the time of publication. Process safety management is complex and this document cannot embody all possible scenarios or solutions related to compliance. This document is for informational purposes only. Siemens makes no warranties, express or implied, in this paper.

For more information, please contact:

Siemens Energy, Inc.
Process Safety Consulting (SAF)
4615 SW Freeway, Suite 900
Houston, TX 77027
1 (800) 658-8809
+1 (713) 570-2940
Website: www.siemens.com/energy/oilandgasconsulting
Email: info.ogconsulting.energy@siemens.com

442429 Adequate Pressure Relief

Abstract:

Pressure relief valves are reliable and effective means of overpressure protection that protect personnel and equipment from the dangers of overpressure. Relief systems design validation and ensuring appropriate relief protection should take place not only during initial design but on a regular interval as changes are made to equipment and/or processes. Changes and tie-ins can drastically influence the capability of existing systems to safely handle overpressure situations. This presenstation reviews the requirements and illustrates how seemingly minor process modifications result in severly underrated pressure relief scenarios.

Additionally a case study is presented for a boiler plant which shows distribution pattern of equipment not meeting code requirements as well as distribution pattern of various types of deficiencies/concerns.

The presentation also describes benefits of rigorously simulating column systems while estimating relief requirements, steady state set up for simple systems, simulation set up for complex systems with convergence issues and benefits of dynamic simulations.

PREVENTION OF COMBUSTIBLE DUST EXPLOSIONS

12TH GLOBAL CONGRESS ON PROCESS SAFETY

Steven D. Emerson
Emerson Technical Analysis, LLC

Tutorials in Process Safety – LPS
2:30 PM Tuesday 12 April 2016

Speaker:
Steven D. Emerson

President – ***Emerson Technical Analysis, LLC***
Corpus Christi, TX & Las Vegas, NV

The firm specializes in the safe management of hazardous, explosive, energetic or flammable materials

Active in GCPS (especially PPSS, PSM^2)

PE (ChE), BSc, MEA, PhD

2

PREVENTION OF COMBUSTIBLE DUST EXPLOSIONS

Steven D. Emerson
Emerson Technical Analysis, LLC

ETA 3

OUTLINE OF TALK

1. Fundamentals
2. Unique Dust Explosion Features
3. Types of Combustible Dusts
4. Equipment Susceptible to Dust Explosions
5. Prevention & Mitigation
6. Standards & Regulation

ETA 4

FUNDAMENTALS

Imperial Sugar Company - Port Wentworth, GA
February 7, 2008
13 Fatalities and 42 injuries

ETA 5

What Is a Dust Explosion ?

Many solid materials become highly flammable when reduced to fine powders.

ETA 6

Fire Triangle

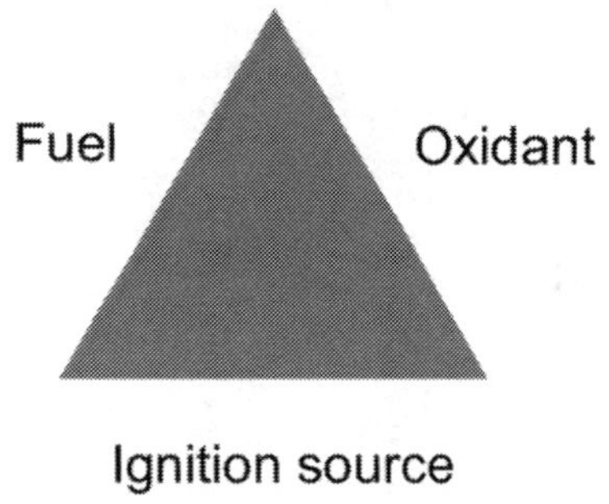

The fire triangle

Like all fires, a **combustible dust fire** occurs when:

fuel (the combustible dust) is exposed to
heat (an ignition source)
in the presence of
oxygen (air).

Removing *any one* of these elements eliminates the possibility of a fire --- or an explosion.

ETA 7

Dust Explosion Pentagon

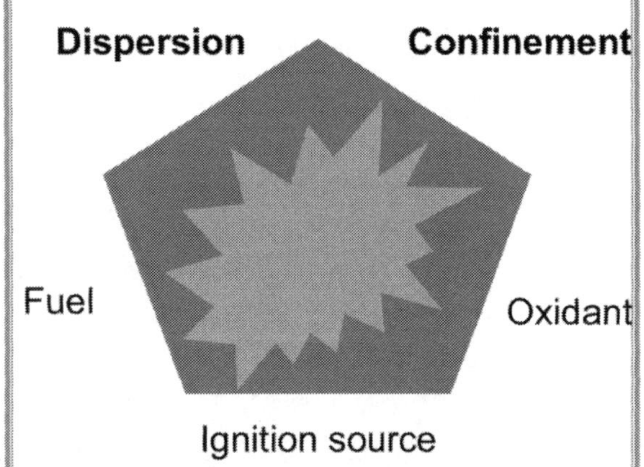

The dust explosion pentagon

A **combustible dust explosion** requires two additional elements:

Dust suspension and
Confinement.

Removal of *either* the suspension or the confinement *prevents* an explosion.

However, fire may still occur.

ETA 8

UNIQUE EXPLOSION FEATURES

West Pharmaceutical Services – Kinston, North Carolina
January 29, 2003
Polyethylene powder
6 Fatalities and 38 injuries

ETA 9

Primary and Secondary Explosions

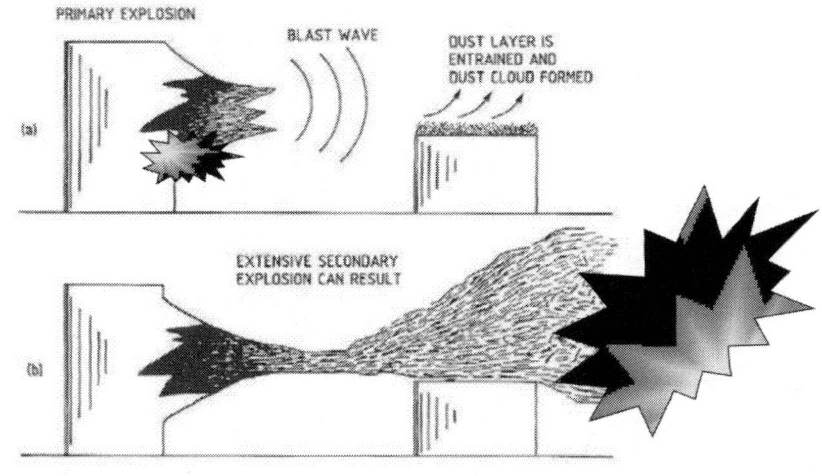

Figure 1.7 *Illustration of how the blast wave from a primary explosion entrains and disperses a dust layer, which is subsequently ignited by the primary dust flame*

ETA 10

Important Distinction Combustible Dust Explosion vs. Vapor Cloud Explosion

- Primary Explosions of Combustible Dust Clouds usually ***inside*** Process Equipment
  - (e.g., mills, baghouses, cyclones, etc.)
- VCEs usually ***outside*** Process Equipment, because equipment is supposed to contain *only* fuel.

11

Explosible Range

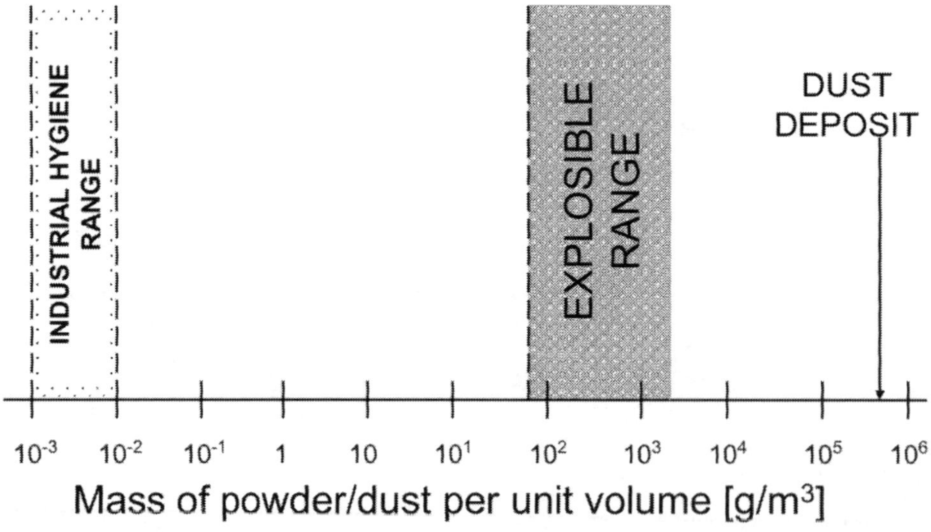

Eckhoff, Rolf, Dust Explosions in the Process Industries, 3rd ed, Gulf Publishing (2003)

12

TYPES OF COMBUSTIBLE DUSTS

Hayes Lemmerz International – Huntington, IN
October 29, 2003
aluminum dust
1 Fatality and 1 severely burned

ETA 13

Combustible Dusts

Dusts that can cause explosions include:

- **Metals** (e.g., Al, Mg, Ti)
- **Chemicals** (e.g., drugs, dyestuff)
- **Plastics** (e.g., urea formaldehyde resin, polyethylene, polystyrene)
- **Food** (e.g., starch, flour, sugar, cocoa, feedstuffs)
- **Wood**
- **Coal**

ETA 14

Dust Chemistry

Heats of combustion per mole O_2 consumed

SUBSTANCE	OXIDATION PRODUCT	KJ/mole O_2
Mg	MgO	1240
Al	Al_2O_3	1100
Fe	Fe_2O_3	530
Cu	CuO	300
Sucrose	CO_2+H_2O	470
Starch	CO_2+H_2O	470
Polyethylene	CO_2+H_2O	390
Carbon	CO_2	400
Coal	CO_2+H_2O	400
Sulphur	SO_2	300

ETA 15

Equipment for explosibility measurements

1m^3 Vessel

1000 L

ETA 16

Equipment for explosibility measurements

Useful to measure

P_{max},
$(dP/dt)_{max}$, K_{st},
LOC,
MIE,
MEC

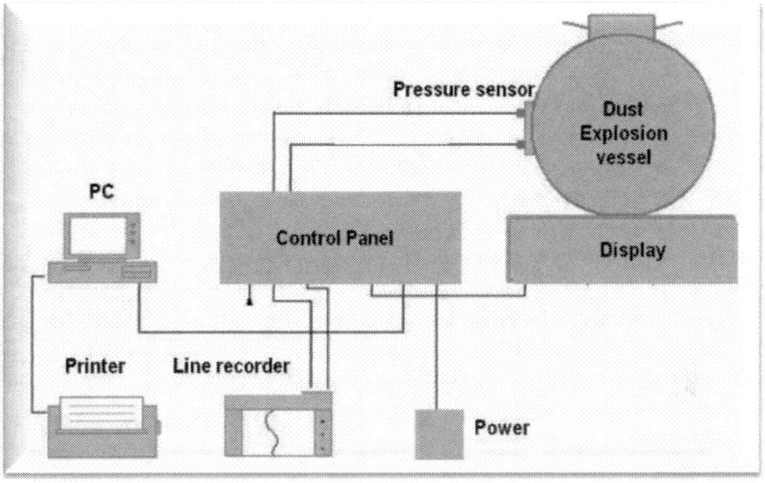

ETA 17

Pressure/time-diagram of a dust explosion

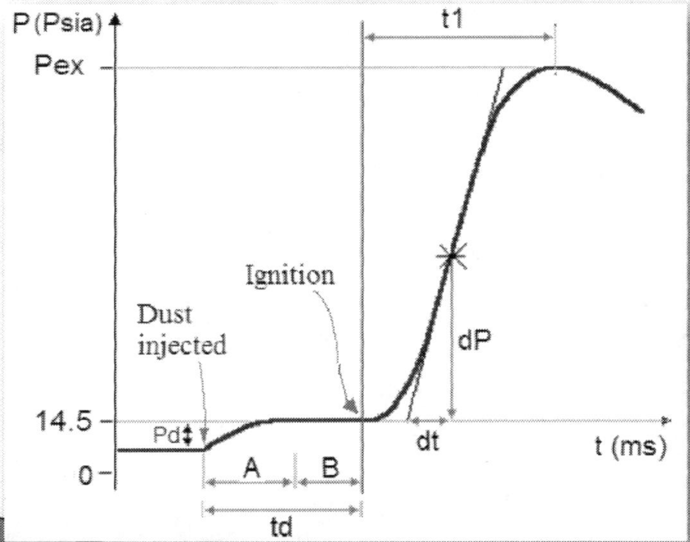

t_d: total delay time

P_d: Expansion pressure

t1: Duration of combustion

P_{ex}: Explosion overpressure

dP/dt: Rate of pressure rise with time

ETA 18

Explosibility characteristics

Combustible dusts are classified by their K_{st} values in increasing order of **explosion violence** as follows:

$\mathbf{K_{st}} > 300$	Group St3: very strong	↑ *Increasing Explosibility*
$200 < \mathbf{K_{st}} =< 300$	Group St2: strong	
$0 < \mathbf{K_{st}} =< 200$	Group St1: weak	
$\mathbf{K_{st}} = 0$	Group St0: non-explosive	

Dust Classification

Kst	**St3**	**e.g. fine metal dusts**
	St2	**e.g. organic pigments**
	St1	**e.g. grain dust**
	St0	**non-explosive**

EQUIPMENT SUSCEPTIBLE TO PRIMARY EXPLOSIONS

Port Grain Silo – Corpus Christi, TX
1981
9 Fatalities and 30 Injuries

Eckhoff, Rolf, Dust Explosions in the Process Industries, 3rd ed, Gulf Publishing (2003)

21

Equipment Involved in Dust Explosions

EQUIPMENT	US (1985 – 1995)		UK (1979 – 1988)		Germany (1965 – 1980)	
	No.	%	No.	%	No.	%
Dust Collectors	156	42%	55	18%	73	17%
Grinders	35	9%	51	17%	56	13%
Silos/ Bunkers	27	7%	19	6%	86	13%
Conveyors	32	9%	33	11%	43	10%
Dryers/ Ovens	22	6%	43	14%	34	8%
Mixer/ Blender	>12	>3%	7	2%	20	5%
Other/Unknown	84	23%	95	31%	114	27%
Total	372	100%	303	100%	426	100%

Guidelines for Safe Handling of Powders and Bulk Solids CCPS, AIChE.

22

Silos and Hoppers

A smoldering nest in a dust deposit can be discharged into an combustible dust cloud and initiate a dust explosion.

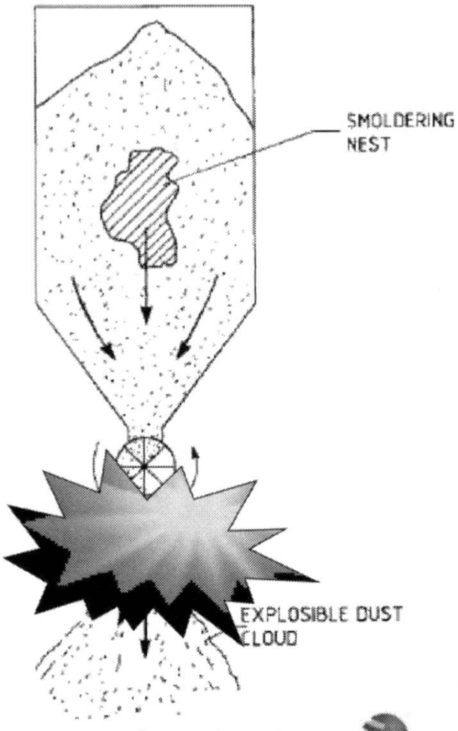

Eckhoff, Rolf, Dust Explosions in the Process Industries, 3rd ed, Gulf Publishing (2003)

ETI 23

PREVENTION and MITIGATION

ETI 24

How to Prevent

PREVENTION		MITIGATION
Preventing ignition	**Preventing dust clouds**	
Smoldering	Inert gases	Isolation
Open flames	Inert dust	Venting
Hot surfaces	Intrinsic inerting	Pressure-resistant containment
Electric sparks. Electrostatics	Dilution of dust concentration	Automatic suppression
Mechanical strikes		Housekeeping

Prevention - Ignition

- Electrical equipment
  - Class II, Division 1 and 2
    - Combustible Dusts
  - Class I and Class III may also be an issue depending on other materials on site
  - Consider equipment both inside dust handling equipment and outside

Mitigation - Isolation

- **Screw conveyors**

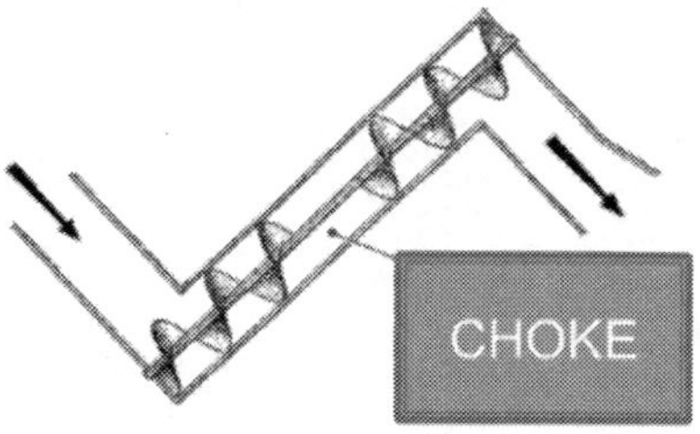

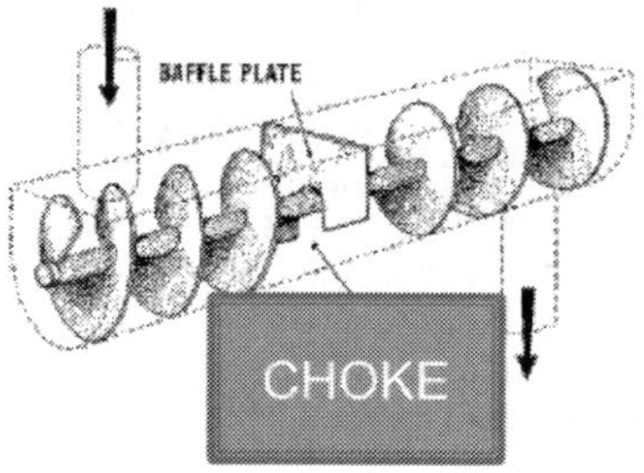

Screw conveyors - prevent transmission of dust explosion

ETA 27

Mitigation - Isolation

- **Rotary locks**

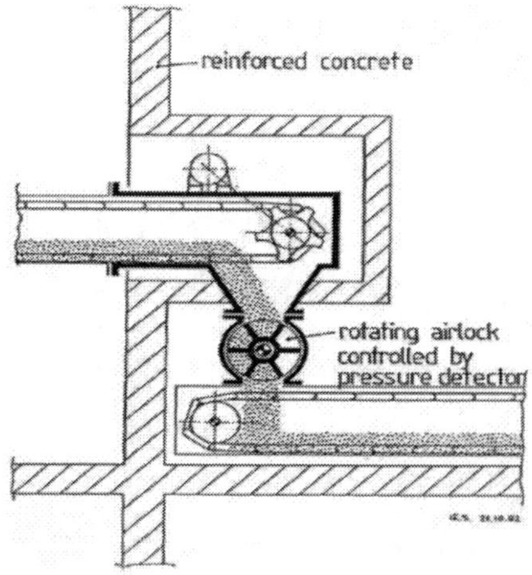

ETA 28

Mitigation – Isolation & Venting

- **Passive devices in ducts**

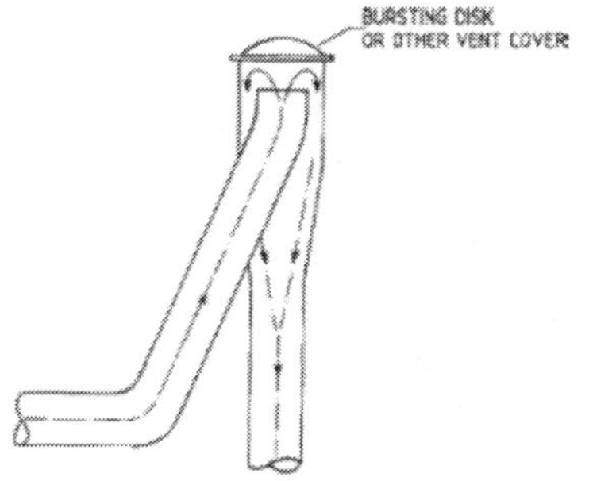

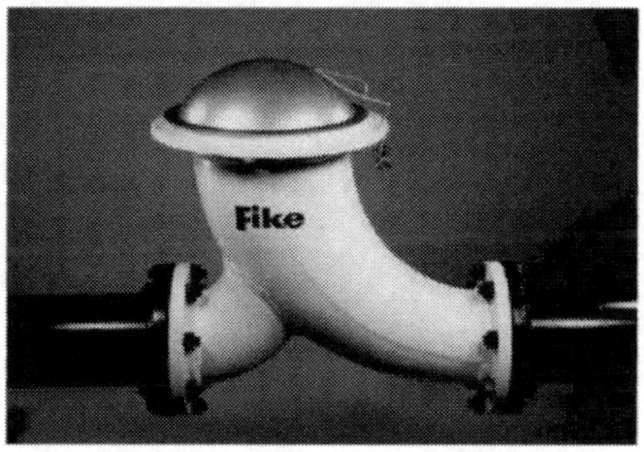

To interrupt dust explosions in ducts
Both vents and changes flow direction

ETA 29

Mitigation - Isolation

- **Active devices in ducts**

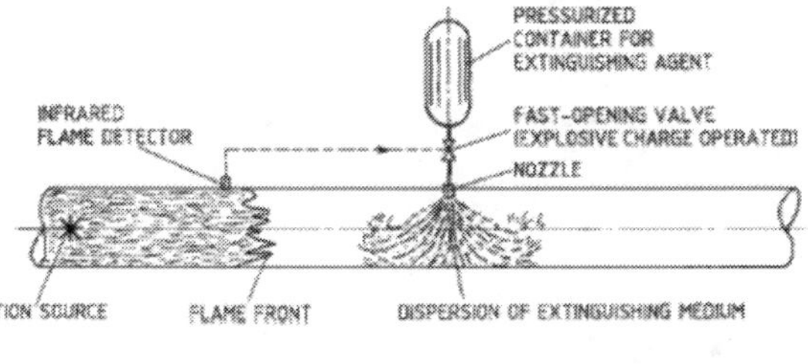

Extinguishing agent ahead of flame

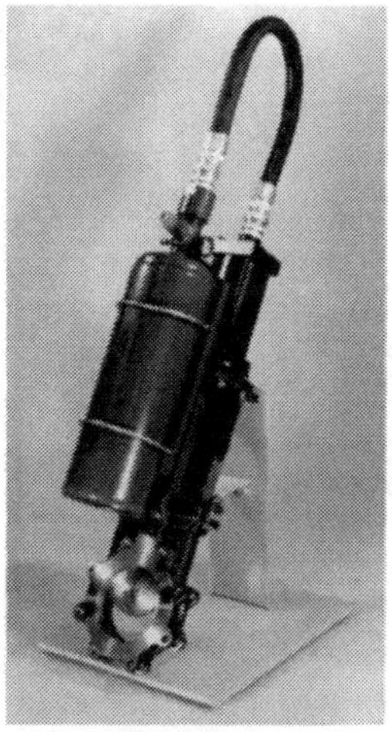

Automatic valve closing ahead of flame

ETA 30

Prevention & Mitigation HOUSEKEEPING

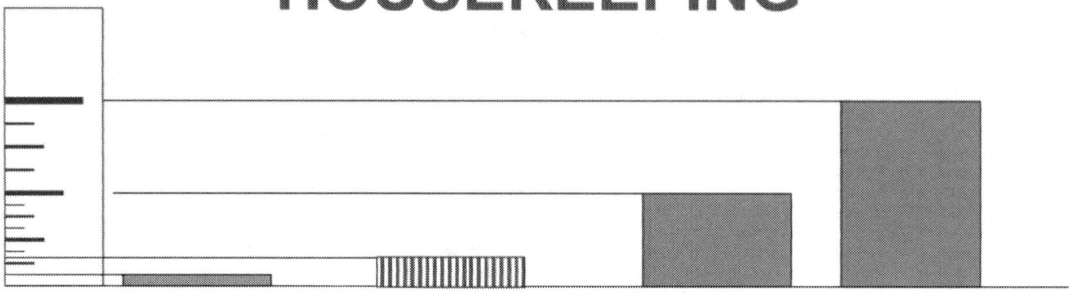

1/32"	3/64"	1/4"	1/2"
0.031"	0.047"	Typical accumulations above ceiling at West Pharmaceuticals	
NFPA 654	U.S. Dime Coin		

31

HOUSEKEEPING Dust Layer Thickness

- Assume a 10' x 10' x 10' room
- Theoretical thickness to yield Optimum Concentration (OC)

Dust Material	OC oz/ft^3	Depth
Cornstarch	0.5	0.15"
Cork	0.2	0.125"
Sugar	0.5	0.075"
Wood	1.0	0.47"
Polyethylene	0.5	0.180"

NFPA 499 Classification of Combustible Dusts and Hazardous Locations for Electrical Installations, Appendix A

32

STANDARDS & REGULATIONS

NFPA Standards for Prevention of Fires and Explosions in:

61 - Ag Products and Food
484 - Combustible Metals
652 - Fundamentals – NEW
654 - Combustible Particulate Solids
655 - Sulfur
664 - Wood processing and Woodworking

---

68 – Deflagration Venting
69 – Explosion Prevention Systems

ETA 33

OSHA Timeline

•1999-2003	High-Profile Catastrophic Dust Explosions (e.g., Ford, West)
•November 2006	CSB issues report on *"Combustible Dust Hazards"*
•October 2007	OSHA NEP on Combustible Dust (reissued March 2008)
•February 2008	Imperial Sugar Savannah, GA
•October 2009	OSHA Status Report on NEP
•October 2009	OSHA ANPR issued
•January 2010	Comments due on ANPR
•Present	Internal OSHA review.

ETA 34

OSHA DUST NEP

4900 violations at 1100 inspected facilities

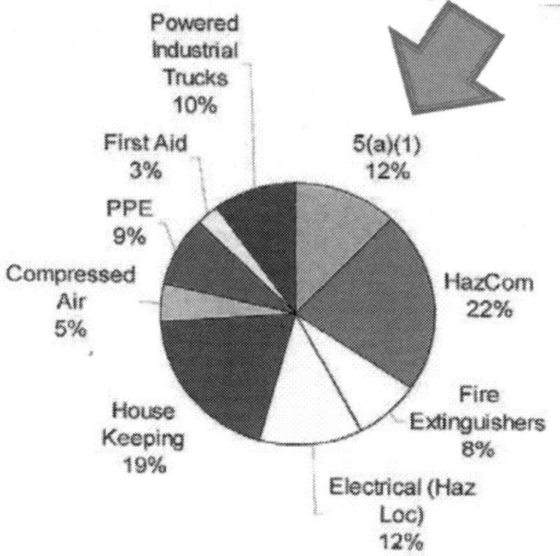

ETA 35

THANK YOU

QUESTIONS?

ETA 36

Practical Application of Advancing Process Safety Tools for the Petroleum Refining and Petrochemical Industries

Rodney Reibold
Marathon Petroleum Company
539 S. Main, Findlay, OH 45840
rrreibold@marathonpetroleum.com

Prepared for Presentation at
American Institute of Chemical Engineers
2016 Spring Meeting
12th Global Congress on Process Safety
Houston, Texas
April 11-13, 2016

UNPUBLISHED

Practical Application of Advancing Process Safety Tools for the Petroleum Refining and Petrochemical Industries

Rodney Reibold
Marathon Petroleum Company
539 S. Main, Findlay, OH 45840
rrreibold@marathonpetroleum.com

Keywords: AFPM, API, Advancing Process Safety, APS, Hazard Identification, Practices Sharing, Event Sharing, API 754, Human Reliability, Process Safety Site Assessments

Abstract

Over the past 5 years, the American Fuel & Petrochemical Manufacturers (AFPM) and American Petroleum Institute (API) have developed process safety tools for use by the refining and petrochemical industries to assist in improving process safety across the entire industry. The suite of programs are called Advancing Process Safety. This paper describes what industry is seeing in the way of process safety performance, the tools and programs available to these industries to help them improve performance, and practical ways companies are using the tools to improve performance.

1. Background and Development of Advancing Process Safety Programs

The petroleum refining and petrochemical manufacturing industry is committed to protecting the health and safety of its workers, contractors, neighbors, and the environment in which it operates. For decades, industry has worked together to develop safe work practices and standards aimed at reducing risks of major industry hazards and preventing process safety incidents. Despite these efforts, process safety incidents occur. As a result, industry, government, non-governmental organizations (NGO's) and the media have increased focus and attention on process safety.

In 2010, a group of refining and petrochemical industry executives belonging to the then NPRA (National Petrochemical and Refiners Association) now AFPM, tasked process safety professionals within their companies to develop a working group to explore ways to collectively improve process safety performance across the refining and petrochemical industries. The goal was to develop programs that would create a significant improvement in process safety performance. Shortly after work began, API was asked to join the effort to leverage the strengths and resources of both organizations to successfully develop and implement the programs. Over a two year period, the group developed a suite of six programs currently being utilized by the petroleum refining and petrochemical industries. These programs are Process Safety Metrics, Process Safety Event Sharing, Hazard Identification and Practices Sharing Program, Process Safety Regional Networks, Process Safety Site Assessment, and Focused Improvement. Together these programs are called the Advancing Process Safety (APS) program.

A key objective of the APS program is to develop tools and information, aimed at improving the industry's process safety performance, and to make these tools and information freely available to member companies. The APS program is able to use data streams unique to the program and specific to our industry to target areas for improvement.

This paper introduces and illustrates some of the tools and information available to member companies and how these tools and information can be used to raise process safety awareness and improve process safety performance.

2. Advancing Process Safety Data Collection and Analysis

AFPM and API members provide the following data on an annual basis to APS to be analyzed by industry experts:

- Facts regarding Process Safety Events (PSEs) per *ANSI/API RP 754 Process Safety Indicators for the Refining and Petrochemical Industries,*
- Tier 1 and 2 PSE summary reports to the Event Sharing database, and
- Results and observations from the API Process Safety Site Assessments.[1][2]

The following sections describe the data collected, the analysis process, and the resulting observations for industry improvement. This data analysis occurs on an annual basis. At the time of this paper, the most recent full year of data was 2014 process safety data, which was reported and analyzed in 2015. For the purposes of this paper, this will be referred to as the 2015 data analysis.

2.1 Data Analysis

Each group, Process Safety Metrics, Event Sharing, and Site Assessment individually analyze their data and present their observations during the annual Deep Dive Analysis meeting. The meeting participants consist of over forty process safety experts from member companies.

The results of the analyses are combined into one Deep Dive analysis to provide insight into opportunities for improvement that will be discussed later. It should be noted that data from each program (Process Safety Metrics, Event Sharing Database, and Site Assessments) are discrete data sets, and therefore some observations might vary from data set to data set. Trends observed in the 2015 data analysis are discussed in the following sections.

2.1.1 API 754 Process Safety Metric Data

The experts analyzed the 2014 Process Safety Metric data submitted by 59 companies representing 199 sites and were able to make several observations.[3] The data consisted of 921 events: 268 Tier 1 PSEs and 653 Tier 2 PSEs.

Figure 1 represents the four Tiers of performance indicators, as explained in API RP 754. For more information, please refer to Appendix A for API RP 754 Tier 1 and 2 definitions.

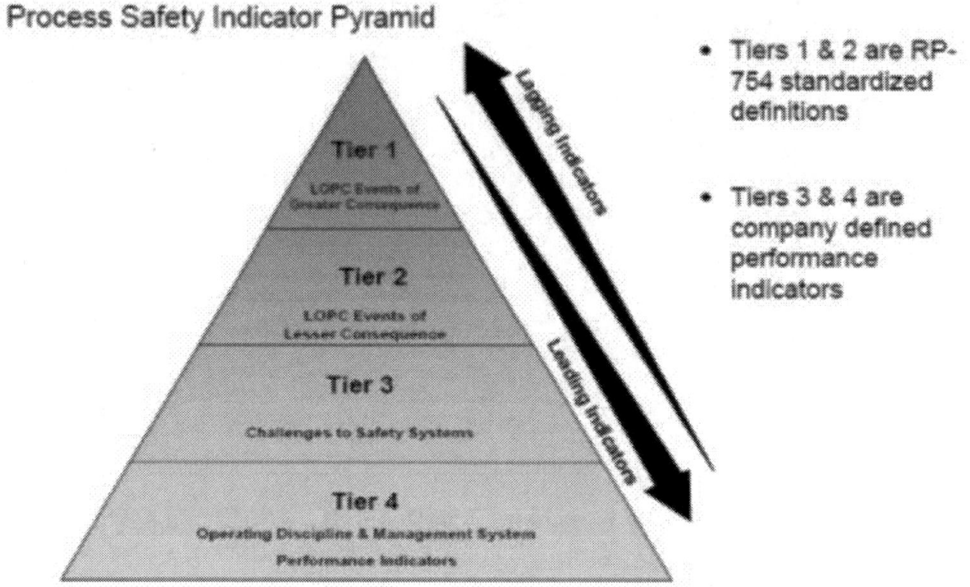

Figure 1. API RP 754 Process Safety Indicator Pyramid

Basic Observations:

- Mode of Operations – 70% of reported incidents occurred during Normal Operation with a majority of those events occurring during Steady State (72%). The next largest category of releases occuring during Normal Operation occurred with Loading and Unloading acvtivities (10%).
- Point of Release – Loss of Primary Containment (LOPC) from Piping Systems and Tanks accounted for 63% of the incidents reported (Piping Systems 49% and Tanks 14%). Figure 2 illustrates that the largest subset of releases from Piping Systems were caused by Piping and valve leaks.

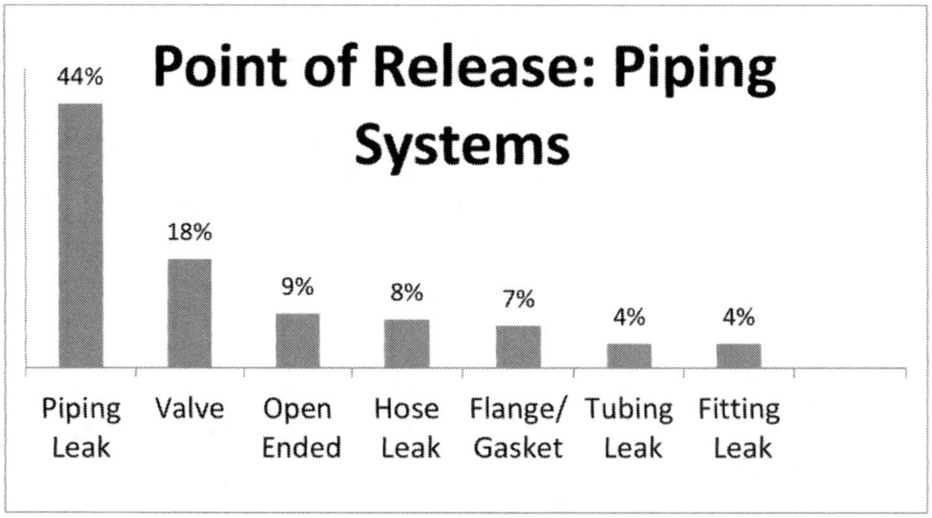

Figure 2. Points of Release from Piping System PSEs

- Piping Leaks – The top contributing factors to piping leaks were fixed equipment inspection, equipment reliability and design. Fixed equipment inspection issues are related to programs for detecting and mitigating metallurgical issues such as corrosion or cracking, before piping failure.
- Valves – Observations from incidents that involved releases from valves indicated that a majority of events were related to "Valves Left Open" and a majority of those events occurred during "normal operation". See Figure 3.

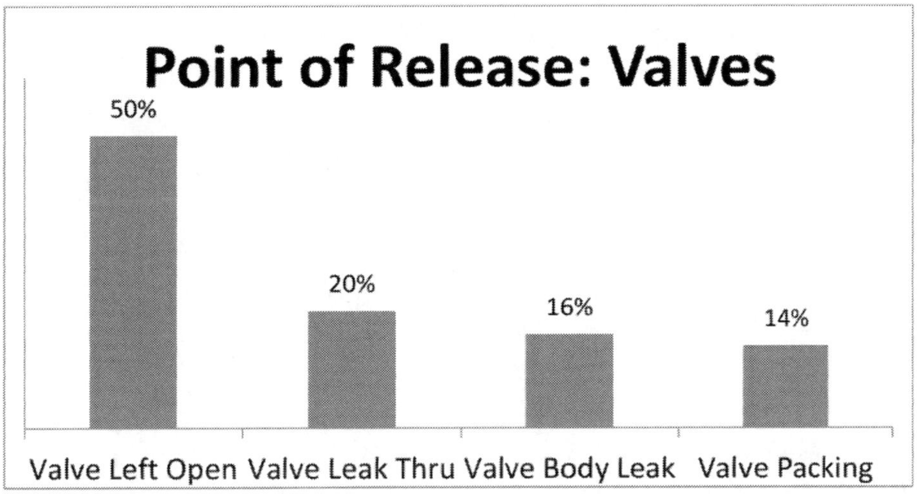

Figure 3. Valve Related PSE Subcategories

- Release from Atmospheric Tanks – Atmospheric tanks (including floating roof tanks) were the second largest point of release locations for PSE Tier 1 and 2 events after Piping Systems. A majority of the events occurred in the Tank Farm followed by the Sulfur Recovery unit. The largest group of those events were considered "overfill" events. Human factors and equipment reliability were the top causes for tank related events and human factors and procedures were the main causes contributing to the overfilling of atmospheric tanks. Figure 4 illustrates the types of releases from tanks.

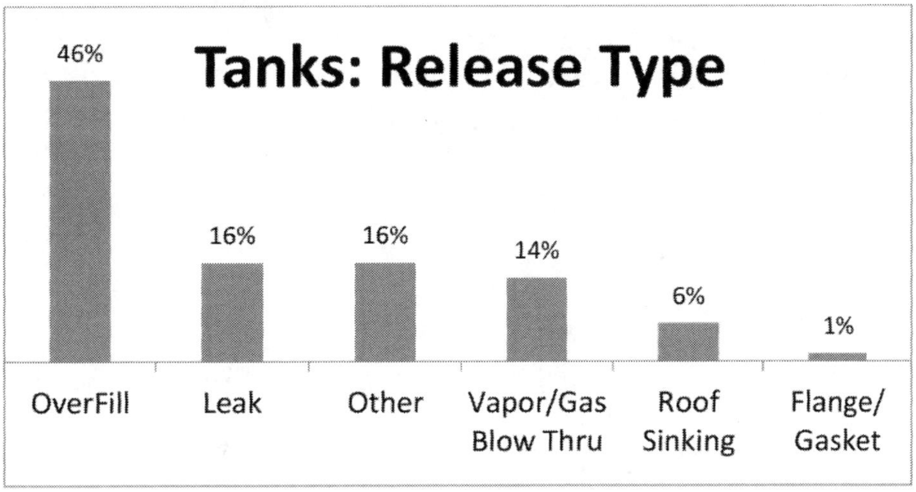

Figure 4. Release Type for Tank PSEs

- Causes – One of the limitations of API RP 754 statistics is that the data does not contain the cause of the incident as the member company determined during their incident investigations. The information provided under API RP 754 includes a synopsis of each process safety event that occurred. The team analyzing this data was able to identify some apparent causes for about 50% of the data, by reading the event description that was included. Although the analysis was based on the limited sample size, the team concluded that human factors, equipment reliability and inadequate fixed equipment inspection were the primary apparent causes of the events. Figure 5 shows the apparent causes of the PSEs, derived from the 2015 API RP 754 data analysis.

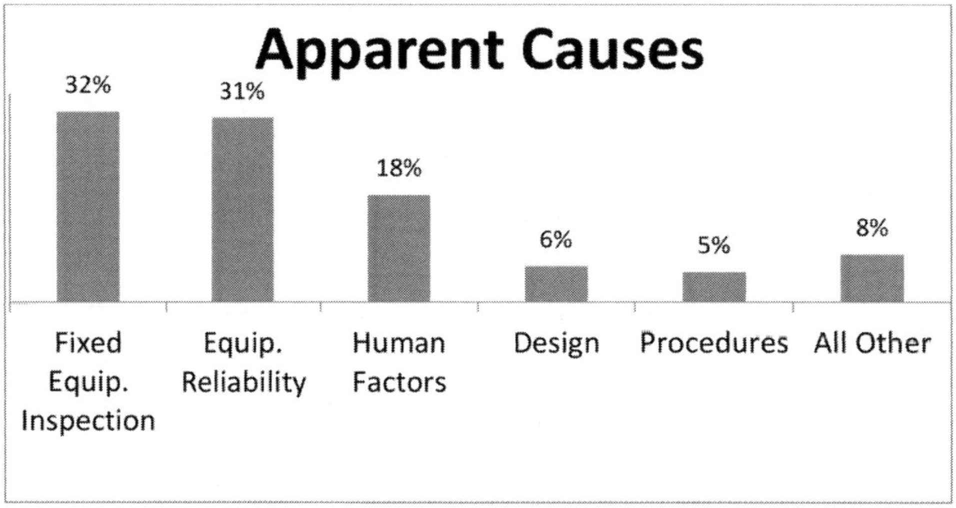

Figure 5. Overall Apparent Causes of PSEs

2.1.2 Event Sharing Database

The Event Sharing Database is a smaller subset of the API RP 754 data and provides more descriptors of the event, causes as determined by the member company incident investigations, and lessons learned.[4] At the time of the 2015 data analysis, approximately 428 process safety events were reviewed from the AFPM Event Sharing Database. Eighty percent of those events were API Tier 1 events that occurred from 1989 to 2014. Some general themes observed from the deep dive analysis include LOPC as a result of valves being left open or inadvertently opened, loss of primary containment originating from small bore piping and/or tubing (2" or less in diameter), and corrosion related LOPC events. Figure 6, illustrates the proportion of these Human Reliability events and how they break down into sub-categories.

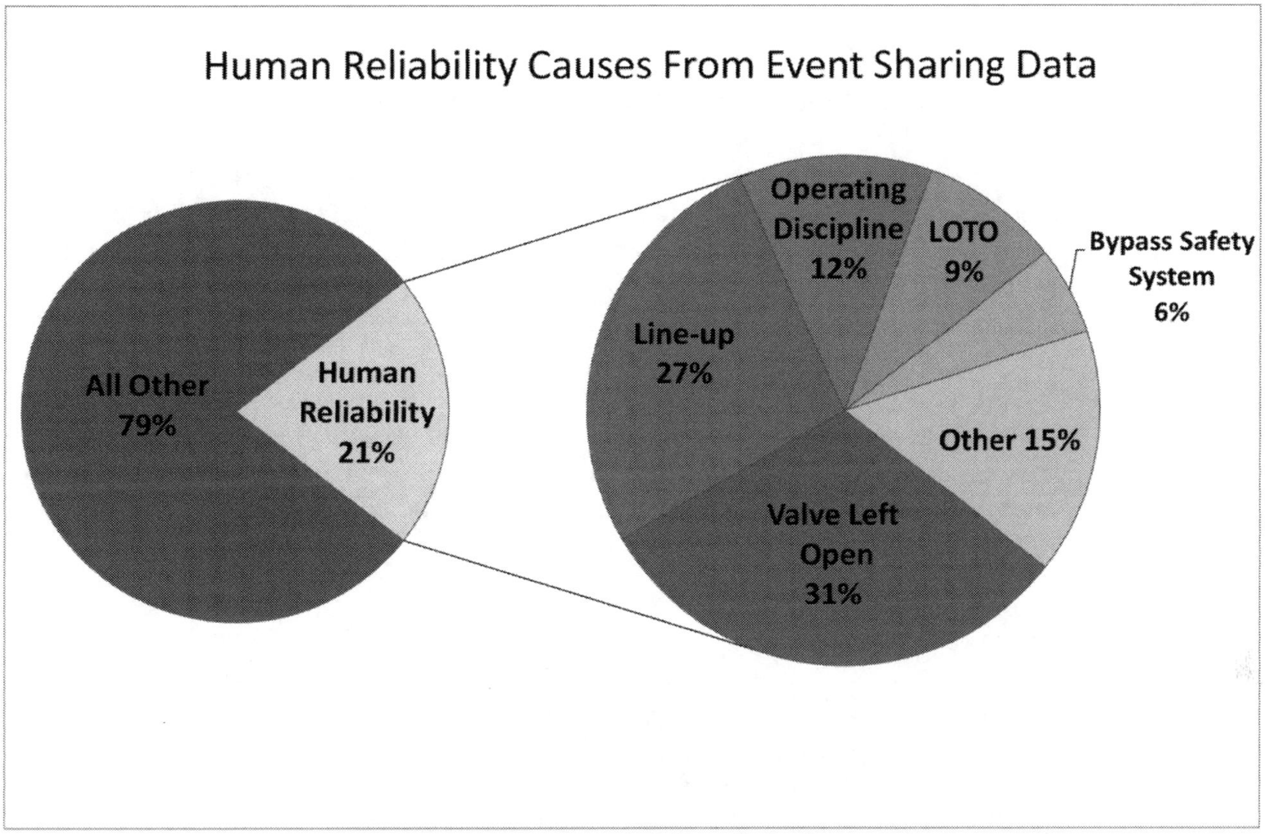

Figure 6. Human Reliability Sub-Categories

Open Valves – Twenty-one percent of the events from the database were identified as being caused by human reliability problems, including valves that were inadvertently left open to atmosphere or closed valves that were inadvertently opened. Nearly 1/3 of this subset involved bleeders, vents, and drains with 40% of the "open valve" events occurring during the startup of equipment.

Operating procedures and human factors (i.e., operating discipline) were identified in the database as the two most prevalent causes for these types of events. "Procedures" were often named as a cause because they did not mention the valve(s) in question or the specific task(s) being performed. In only a few cases were the procedures found to be unclear or just not followed.

Small Bore Piping – LOPC events involving small diameter piping continue to be prevalent in the 2015 data analysis with at least 19% of the reported events being related to piping or tubing 2" in diameter or less.

Piping Corrosion – Events associated with corrosion represented about 19% of events in the database. Corrosion continues to relate to a significant number of PSEs.

Winterization – Freeze-related events trended much higher in 2014 than historical rates. In 2014, 15% of events entered were associated with freeze-ups compared to 6% in the entire database.

2.1.3 Site Assessment Program

Observations from the Site Assessment program are shown in Figures 7 and 8.[2]

Basic observations:

- The lowest scores came from Mechanical Integrity (MI) and Process Hazards Analysis (PHA).
- When looking at the protocol categories, there appeared to be a wide distribution of scores for topics like shift turnover, operator rounds, and facility siting which indicates that some sites are doing those things very well and that others have an opportunity to learn from those sites.

A note about assessment scores: Higher is better, but the scoring isn't quite the same as a college exam. In the Site Assessment world, a score of 75% represents strong performance that meets the requirements of the assessment protocol. To achieve a 100% score, a program must show it meets or exceeds the requirements of the protocol, show a cycle of continuous improvement, must demonstrate sustainability, and show evidence of being embraced by the workforce.

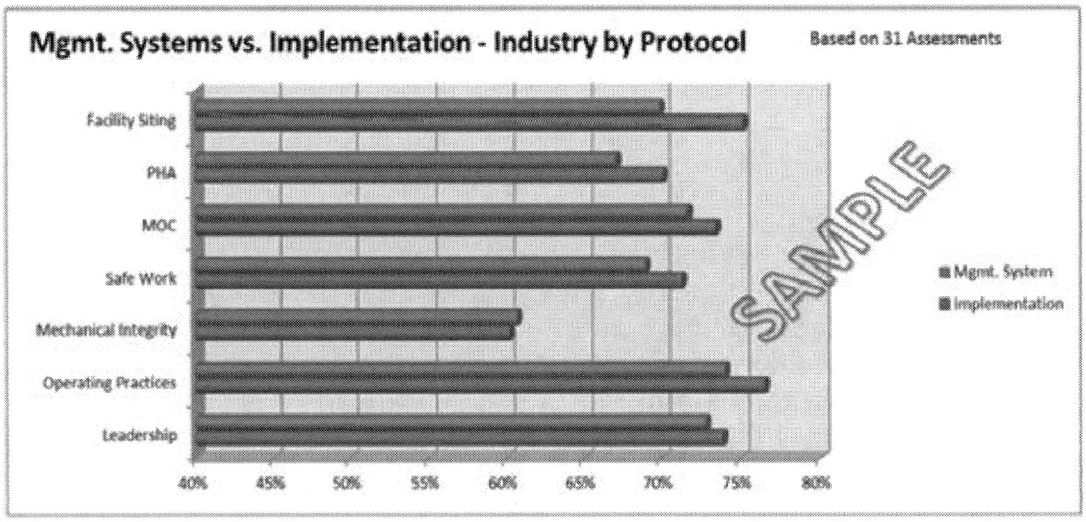

Figure 7. Management System versus Implementation by Protocol

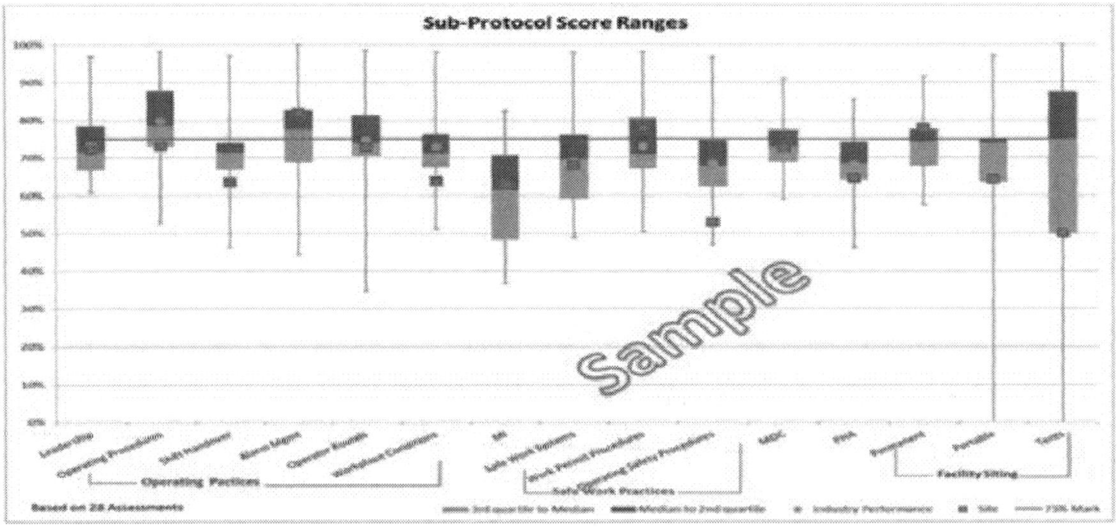

Figure 8. Site Assessment Sub-Protocol Scores

2.2 Using the Data to Improve

Does the data above look familiar? Are you experiencing similar types of events at your facility? It is likely that similar types of incidents ARE occurring at your processing facilities. The APS data indicate that these common themes are occurring over a broad cross-section of the industry; small plants and large plants, refineries and chemical plants, domestic and international. It is incumbent upon each company to collect, analyze, and make improvements based on their own data. The foundation of APS is to collect, analyze, and develop tools for improvement based on the industry's data.

The Advancing Process Safety program has used the data for industry improvement initiatives and to develop many tools for member company sites to use to improve their own performance. The following are a few actions that have been pursued based on the annual deep dive analysis:

- Developed and proposed a number of revisions to strengthen API 570 (Piping inspection code) and API RP 574 (piping inspection practices) in the areas of small bore piping.[5][6] The main idea is to increase awareness of small bore piping issues as part of a company's routine piping inspections procedures.
- Developed a number of Hazard Identification documents to address hazards associated with equipment, or work practices on equipment, uncovered in the data analysis. Examples include topics such as Piping Dead-legs, Winterization, and Small Bore Piping.[7]
- Developed a number of Hazard Identification documents and Shared Practices to address common personnel issues uncovered in the data analysis. Examples include topics such as Shift Change issues, Operator Communications, Operator Line-ups, and Operator Evaluation Rounds.[7]
- Developed a program called "Walk the Line" to address common Human Reliability issues. This includes prepackaged training materials, prepared tool box meeting outlines (scripts), informational newsletters, and videos.[8]

All of these materials and much more are available to member companies for use in their process safety improvement journey. The following section describes how information and tools developed by the APS Program can be used to improve your process safety performance.

3. Advancing Process Safety Tools and Ways to Use Them

The following sections outline the individual APS programs, tools, and information as well as how some companies are using them to improve their process safety performance. They are organized according the APS program where the information or tools are developed, and are freely available to member companies. It should be noted that this paper only discusses a few of the ways these tools can be used. In practice, there are many more ways that each of the tools discussed below can be used. The products described in these sections can be found on the AFPM Safety Portal website.[9]

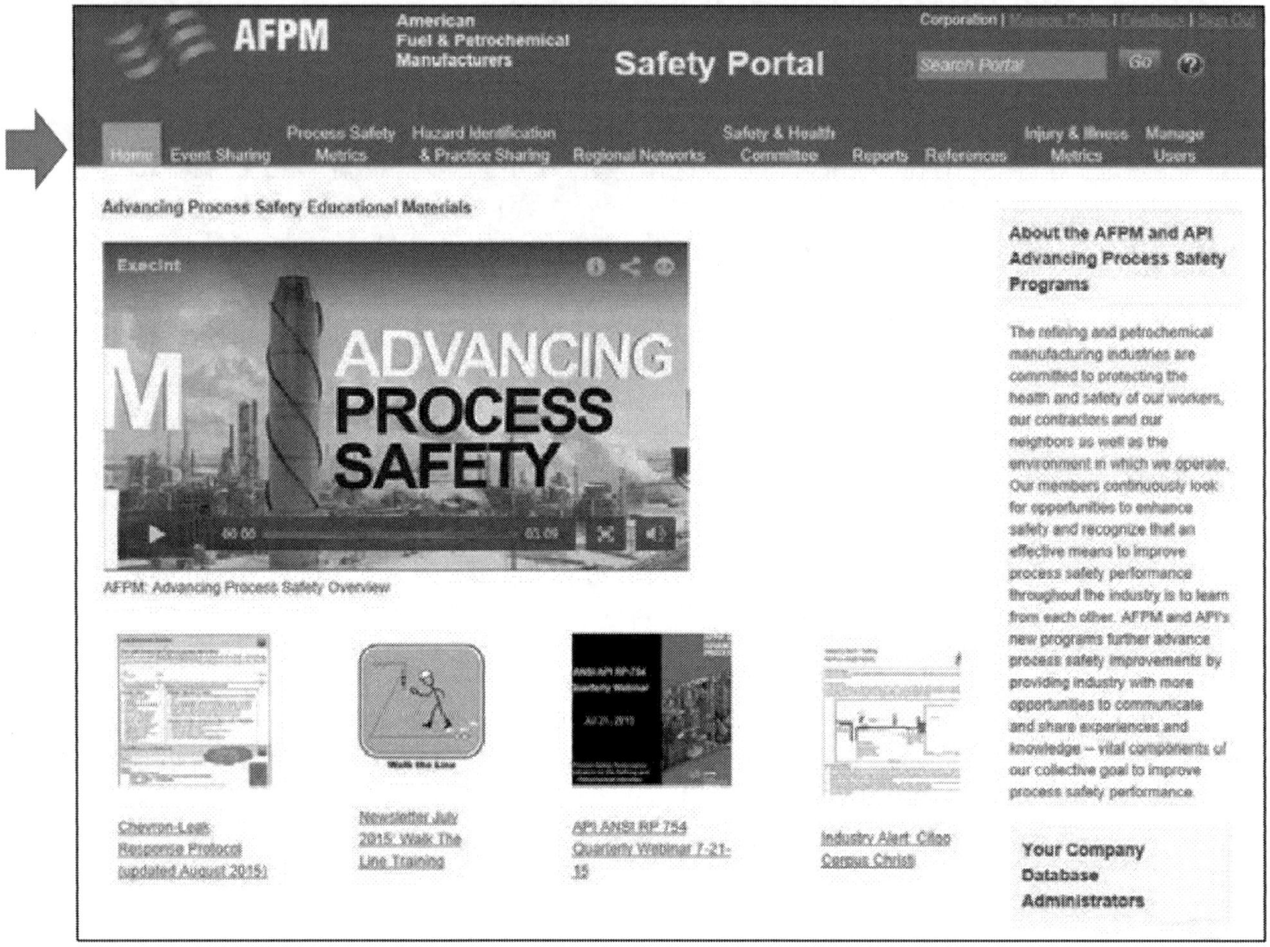

Figure 9. AFPM Safety Portal

3.1 API 754 Process Safety Metrics

The purpose of API RP 754 is to identify and track leading and lagging process safety indicators.[1] Knowledge of these indicators and taking appropriate action based on those indicators reduces the overall risk of major hazards in industry. On an annual basis AFPM and API publish a Process Safety Event Report and data analysis tool for both Tier 1and Tier 2 events.

The following illustrates a few practical examples of how APS member companies use Process Safety Metrics data.[3]

3.1.1 Benchmarking Against Industry

Figure 10 shows how benchmarking can be used by a company to compare its process safety performance to industry performance. Smaller companies, without a central corporate Process Safety Management (PSM) organization, may also use this tool to compare their sites to each other without need for a centralized compilation of data. Performance can be compared based on multiple criteria including the type of facility, size of facility, exposure hours, type of incident, incident year, or for specific process units. Comparisons can be done by both PSE count and PSE rate. In this example, a fictional company has four sites and is comparing its number of Tier 1 Process Safety Events over a three year period to other refining facilities of any size. The metric database is an extremely flexible benchmarking tool. By making simple choices from drop down menus, similar comparisons can be quickly created based on a wide variety of criteria. Figure 11 shows a comparison of Site D's performance on its fluid catalytic cracking (FCC) unit compared to other small facility FCC units. Figure 11 also shows that 44 of the FCC units at small facilities went the entire period without a Tier 1 or Tier 2 process safety event.

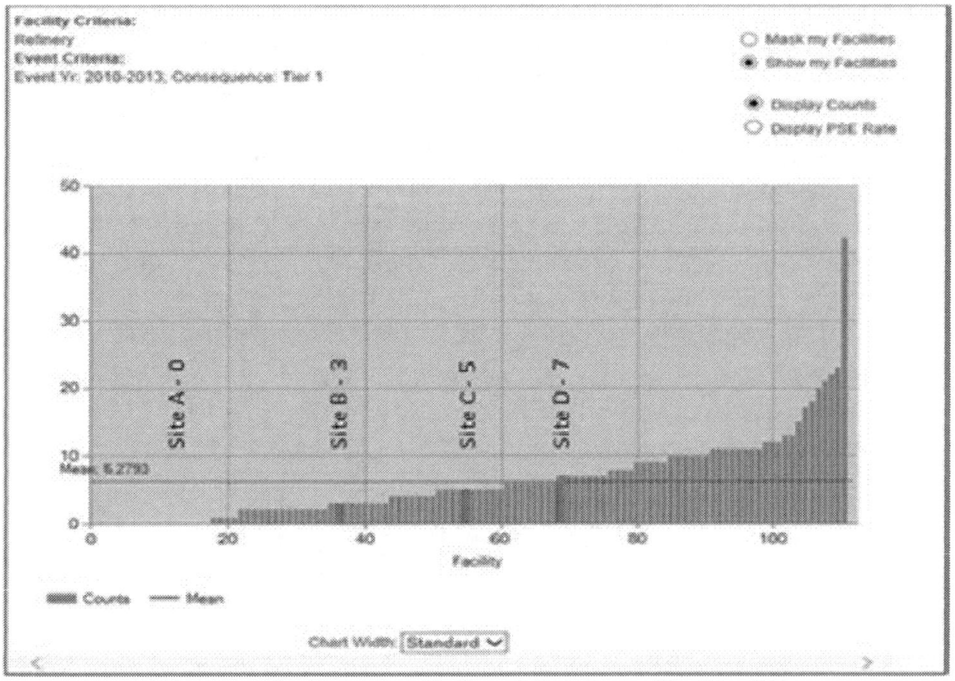

Figure 10. Process Safety Metrics – By Facility

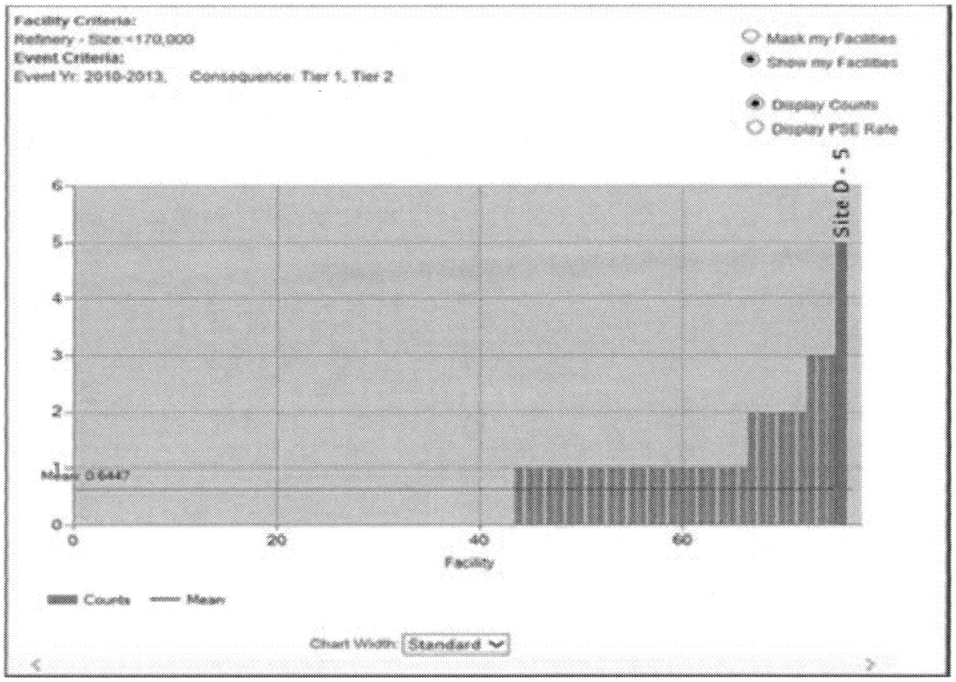

Figure 11. Process Safety Metrics – By Unit

- Goal Setting – Companies set process safety performance goals for their sites. This tool allows a company to identify what level of performance they will need to target to be an industry leader.

- Learning – The benchmarking tool can be used to identify successful practices within a company so that they can be shared across their sites. For instance in the example in Figure 10, Site D can use the tool to identify that Site A has implemented successful PSM programs and practices. Site D can then engage Site A to understand these programs and practices to implement at their own site.

- Prioritization of Resources – The benchmarking tool can be used to identify where resources need to be focused to improve PSM performance. For instance in the example in Figure 11, Site D would be able to identify that the large majority of their PSEs are occurring in their FCC unit, and that they are the worst performer in FCC process safety among similar sized sites. With this information, the site can target improvement efforts at the FCC. The site may also use this tool to identify which of their sister sites have strong FCC process safety performance, and direct resources to learning from the other sites and implementing existing successful practices to improve performance.

3.2 Event Sharing

The Event Sharing Database is a smaller subset of the API RP 754 Process Safety Metrics data and provides additional descriptors of the event, causes, and lessons learned. Where the main intent of API RP 754 data collection is **benchmarking**, the main intent of the APS Event Sharing program is **learning**.[4] Data submitted to APS includes a description of the event and equipment involved, as well as the cause(s) and learning(s) identified by root cause analysis performed by the member company. Individuals have the opportunity to search the database for specific information or export the database in a variety of forms. Key information about the event, cause(s), and learning(s) are clearly laid out, as is other information not shown here. See Figure 12. Other tools derived from the database are Safety Bulletins and the Annual Learnings Report.

AFPM
American
Fuel & Petrochemical
Manufacturers

Event Summary Report

Process Safety Event ID Number: 531

Event Year: 2014

Facility Type: Refinery

Process Type: Tank Farm/Storage Facility/Offsite

Equipment Involved at Point of Release: Atmospheric Tank | Process Liquids Storage | large tank (>1320 gal / 5000 L) | Release from vents , overflows, or around seals on floating roofs | Vapor/gas blow thru to tank

Material Type: Flammable

Mode of Operation: Emergency shutdown

Equipment involved in initiation Event: Instrumentation | Temperature Instrument

Event Description:

The Hydrocracker Unit experienced an emergency. The SIS trip was on a high bed temperature reading (1-out-of-1) which was later determined to be false. In response to the HCU trip, Operations began executing the emergency procedure associated with this trip activation. Part of this procedure was lining diesel product up to slop. The diesel stripper was pumped to slop to 30% level and the pump stopped. The level continued to decrease to … the manual valve to slop was closed … was signed off as complete. Howev… was at 0% output, liquid and then va… padding system on the fractionator. … natural gas makeup. The natural ga… escape around the tank roof seals r… identified, the natural gas make up v… the high temperature reading, which… leak which allowed process fluid to … be analyzed at the next opportunity.

Cause and Learning Information:

Initiating Event: Blow through from hydrocracker unit during unit shutdown resulting in crude oil on roof of floating roof tank. Manual block valve was left open.

Event Cause 1: Design | Specifications, codes, or practices less than adequate

Event Description - EC1: False temperature indication caused by the instrumentation on a one-out-of-one voting trip. Allowed a single point of failure to trip the HCU.

Event Cause 2: Procedures | Operations | Procedure followed incorrectly (e.g. steps out of order)

Event Description - EC2: The manual block valve from the diesel stripper to slop header within the unit remained open. The Boardman does not recall positive affirmation that the valve was closed but procedure step was signed off

Learning 1: A closed control valve may not be adequate to prevent blow thru to an atmospheric tank when the upstream vessel loses liquid level.

Learning 2: In certain circumstances, gas padding systems can provide enough pressure and flow to cause hazards at storage tanks in the event of a loss of liquid level.

Learning 3: Emergency shutdowns with one-out-of-one voting are susceptible to spurious shutdowns due to instrument malfunction.

Figure 12. Event Sharing – Sample Event

There are a number of ways companies use information gathered through Event Sharing submissions including:

- PHA Preparation and Conduct – One tenet of conducting a process hazard analysis is to identify and evaluate incident scenarios which could have catastrophic consequences. The Event Sharing Database is a convenient tool for supplementing a site's own incidents to ensure hazard scenarios from the broader industry are identified and evaluated.[4] The database can be queried to extract events related to the process unit in question. The query can be further refined to pinpoint a specific type of equipment (e.g. Fired Heater) or a mode of operation (e.g. Start-up). The events can be prescreened by the PHA leader and appropriate events selected for review by the PHA team. The team can consider the database events as well as the causes and learnings as they evaluate hazard scenarios, causes, and safeguards.

- Procedure and Safe Work Practice Development – A company or facility can use the database when developing procedures or practices. The database can be queried to find events that relate to a specific unit and further refined to find events where procedures or safe work practices were listed as a cause. The company or site can then use the information in these events to ensure they have adequate measures incorporated into their procedures or practices to prevent a similar incident from happening.

- Mode of Operation Scenarios – In preparation for starting up a unit after turnaround, an operations foreman can query the database to find events on the same type of unit which occurred during start-up. The foreman can use the learnings from the event(s) to ensure the same incident does not happen during the pending start-up. For example: Many events during start-up are related to bleeder valves left open when feed is started to the unit. The Foreman could use this information to verify that a unit walk down occurs prior to start-up which looks for open bleeder valves.

- Design Reviews – When designing a new unit or making modifications to an existing unit, a design engineer can use the database to find events on the same type of unit where Design was listed as a cause. The engineer can then take appropriate measures to ensure their design will not be susceptible to a similar event.

- Tool Box Topics – The Event Sharing database is an excellent source of events to use as tool box topics or "What-if" drills. An event for a pertinent unit or task can be selected for discussion of what happened, how to respond to a similar event, and how to prevent a similar event from occurring in the first place.

3.2.1 Process Safety Bulletins

There are times when it is important to highlight key and/or urgent information to companies when a high learning value event or trend has been identified. Each Bulletin contains detailed information about events, causes, and lessons. Process Safety Bulletins are located in the Safety Portal under the Reference tab. Figure 13 is a sampling of recent bulletins which were issued to member companies.

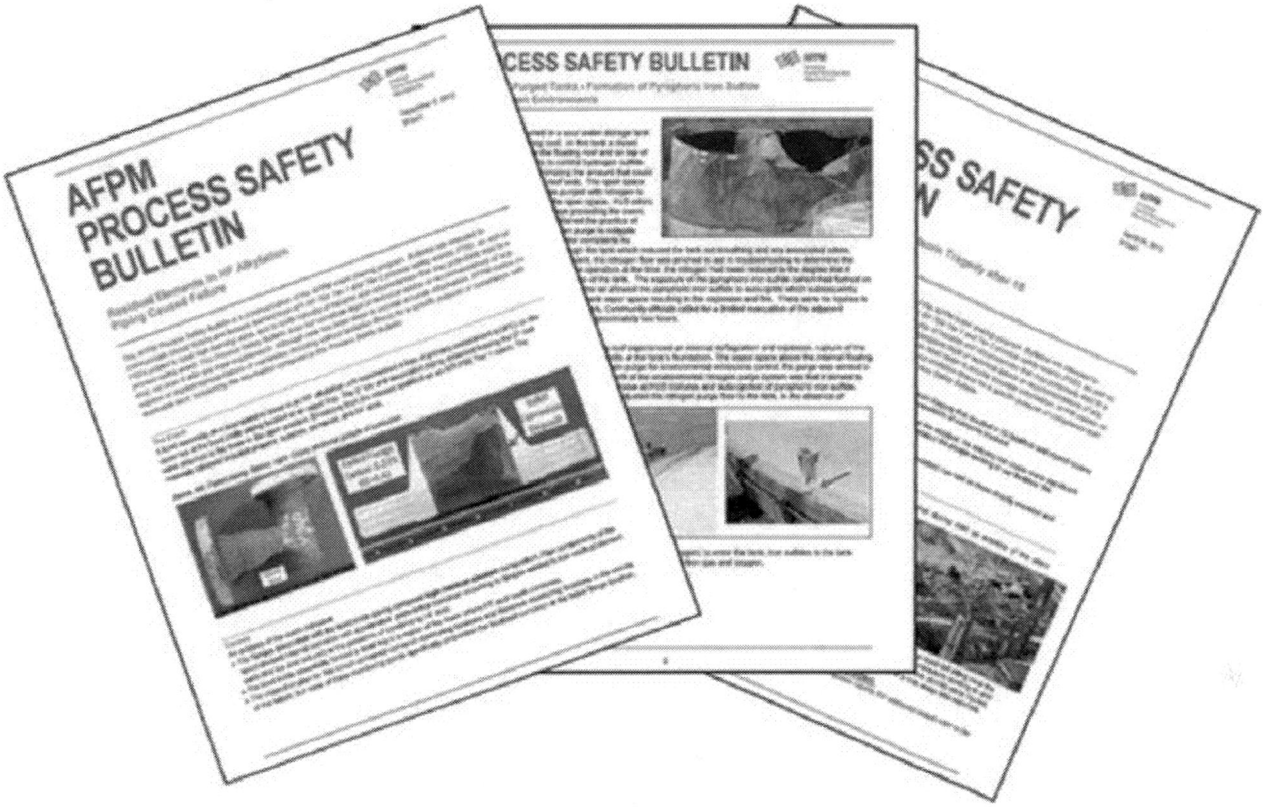

Figure 13. Process Safety Bulletin - Samples

Companies can use Process Safety Bulletins in a number of ways, including:

- Awareness – A company may use the bulletins to inform personnel of a serious incident that has occurred or a risk that exists, which could affect them.
- Standards, Specifications, and Practices – Subject matter experts can use the bulletins to evaluate the Standards, Specifications, and Practices for their company or facility to make sure that they have adequate requirements, features, and/or safeguards to prevent a similar event from happening at their company.
- PHA Preparation and Conduct – The process safety bulletins are appropriate for use by PHA teams to identify hazards and ensure adequate safeguards are in place.
- Leadership/Management Review – A facility leadership team can review the content of the bulletins to increase awareness of hazards and to determine if action at the facility is necessary to prevent future incidents. Any actions can be assigned and tracked to completion by the facility.
- Tool Box Topics – Like the Event Sharing database, the Process Safety Bulletins are an excellent resource to use as tool box topics or "What-if" drills. A pertinent bulletin can be selected for discussion of what happened, what the hazard is, how to respond to a similar event, or how to prevent a similar event from occurring at your site.

3.2.2 Annual Learnings Report

The Annual Learnings Report contains the results of an annual data analysis that identifies event trends in the industry. Figure 14 is a sampling of a recent Learnings Report.

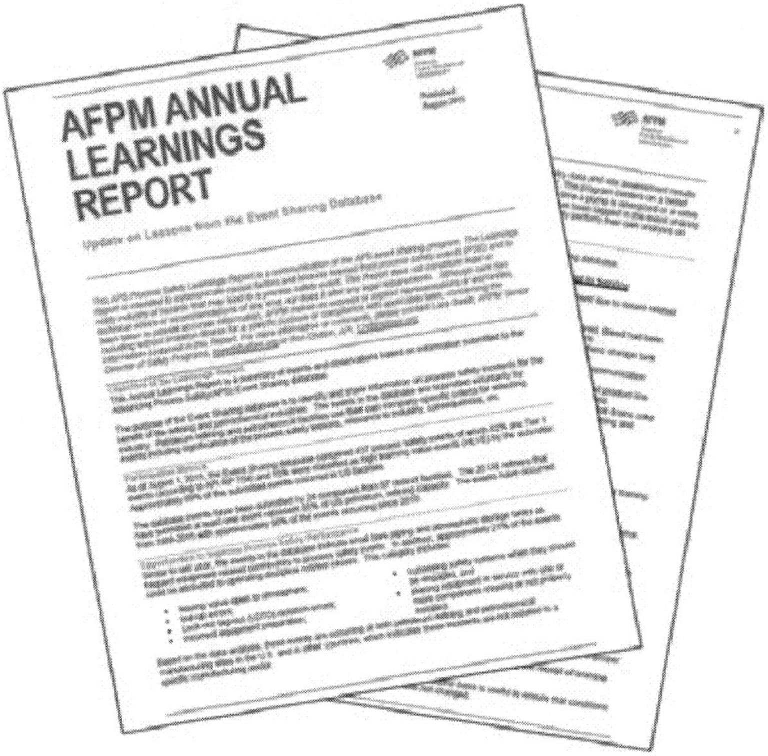
AFPM ANNUAL LEARNINGS REPORT

Update on Lessons from the Event Sharing Database

Figure 14. Event Sharing Annual Learnings Report

In addition to the same uses as outlined for the Process Safety Bulletins, the Learnings Report is also an opportunity for introspection and a chance to recognize "blind spots." A company can consider that "If this is an industry trend, is it also happening at our facilities?" Where appropriate, the company may choose to take measures or actions intended to prevent the events from happening at their facilities.

3.3 Hazard Identification

The AFPM Process Hazard Identification program supports the sharing of process safety hazards.[7] The program provides information and ready guides for potentially overlooked and not widely known process safety hazards in order to improve process safety awareness. The goal is to create tools that capture and share knowledge that could be used by other companies or sites when developing new process safety practices or improving existing ones. A sample Hazard ID document is shown in Figure 15. The Hazard ID documents are located on the AFPM Safety Portal.[7] Current Hazard ID documents are shown in Figure 16.

Each Hazard ID document contains sections with information on:

- Purpose and Use of the document,
- Scope,

- Examples of Inherent Concerns or Hazards,
- Detailed breakdown of potential causes for the hazards,
- A listing of available references and resources, and
- Links to known industry events related to the hazard.

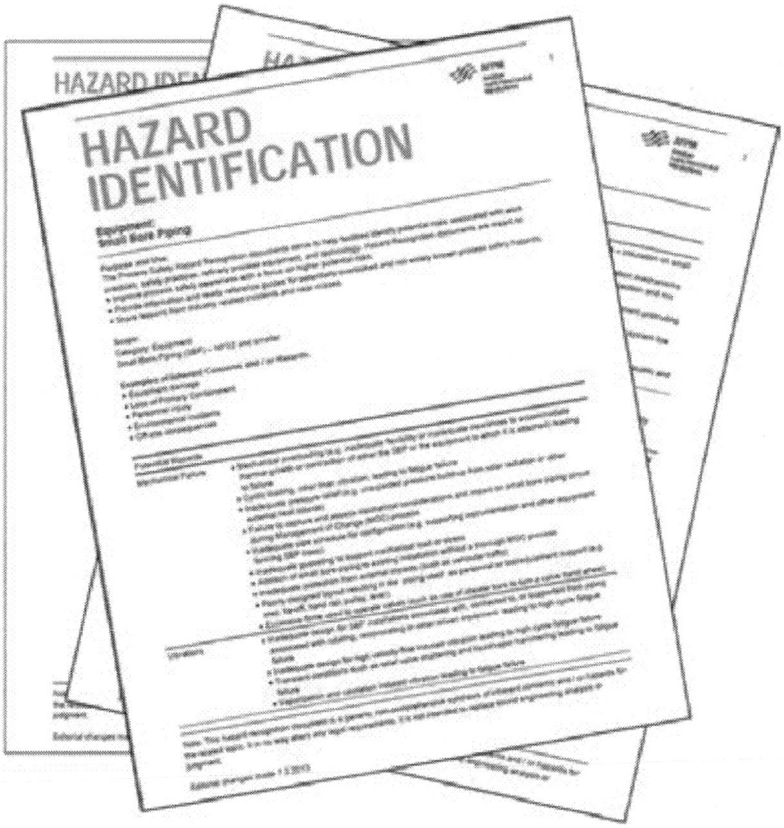

Figure 15. Sample Hazard Identification Tool

Hazard ID documents can be used in a number of ways, including:

- PHA – The Hazard ID documents are useful during hazard reviews as a check to make sure that no credible hazards were missed or omitted by the review team. The Hazard ID documents can be prescreened by the PHA leader and appropriate documents selected for review by the PHA team. The team can consider these hazard identification tools as they evaluate hazard scenarios, to ensure no credible hazards are missed, and to verify they have adequate safeguards or controls to address the identified hazards. The References and Resources section can be used to help the team determine what actions to take if they feel the existing safeguards are not adequate.
- Pre-Job Hazard Assessments – A number of the Hazard ID documents are associated with recurring tasks in a Refining environment such as vacuum truck operation, sampling, lighting or tuning a fired heater, live flare work, etc. These

documents are great tools for a pre-job hazard review to facilitate discussion and calibrate a technician before they go do a task. They are brief, to the point, and include a robust listing of potential hazards. During the meeting the Hazard ID document can be used to discuss how each potential hazard will be mitigated during the upcoming task.

- Procedure and Safe Work Practice Development – A company or facility can use the Hazard ID tools when they are developing procedures or practices. The Hazard ID portal can be reviewed to identify documents that are applicable to procedures or practices that are being developed, or are up for their periodic review. The company or site can then use the information in these documents to ensure they have adequate measures incorporated into their procedures or practices to mitigate the identified hazards. Another way to use these tools would be to stay abreast of new Hazard ID documents as they are created, and proactively complete a gap assessment with your existing management systems to ensure all identified hazards are adequately addressed.
- Design Reviews – When designing a new unit or making modifications to an existing unit, a design engineer can review the Hazard ID tools during the design to ensure hazards are identified and mitigated.
- Training Development – The Hazard ID documents can be useful tools when preparing training materials. One key aspect of refinery training is to include an overview of the hazards of the process. These tools are a great way to make sure all of the bases are covered in this portion of the training.

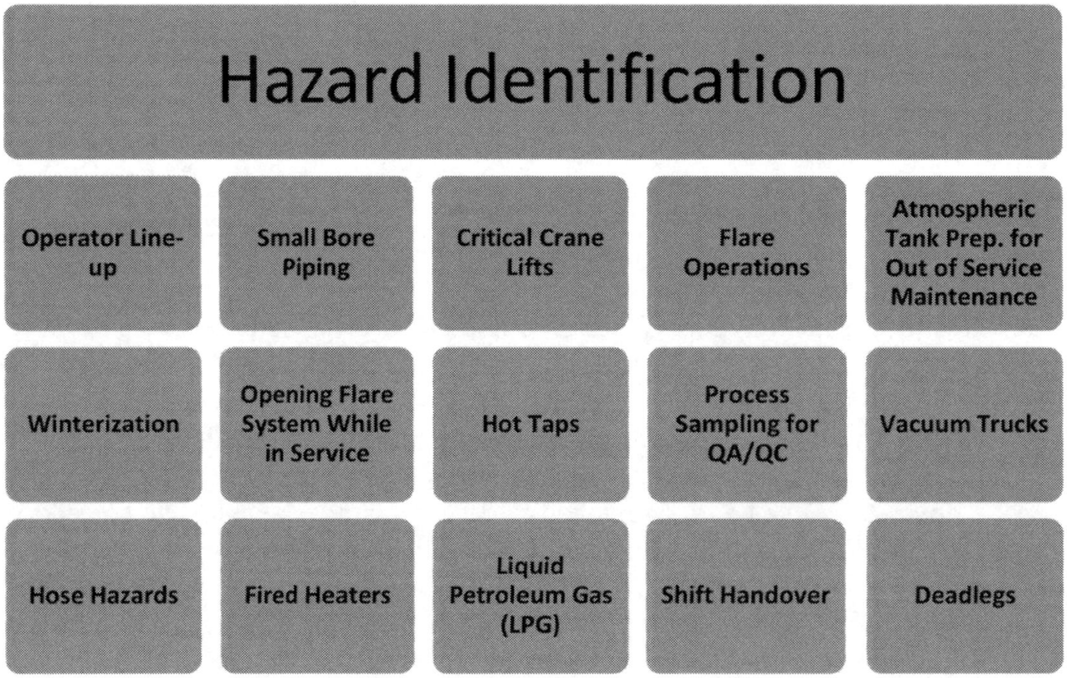

Figure 16. Currently Available Hazard ID Documents

3.4 *Practice Sharing Documents*

Practice Sharing Documents are designed to share information on process safety practices in order to help improve process safety performance and awareness throughout industry.[7] The goal is to capture and share knowledge that could be used by other companies or sites when developing new process safety practices or improving existing ones. The Practices being shared have been successfully used by an industry member, but this does not mean they are right for everyone, or that they will produce similar results at your site. Rather, they are an option to consider when implementing or adjusting programs and practices at a site. Figure 17 shows a sample Practice Sharing document.

Each Practice Sharing document contains sections with information on:

- Purpose and Use of the document,
- Scope,
- A description of the practice,
- Implementation suggestions, and
- References and Resources.

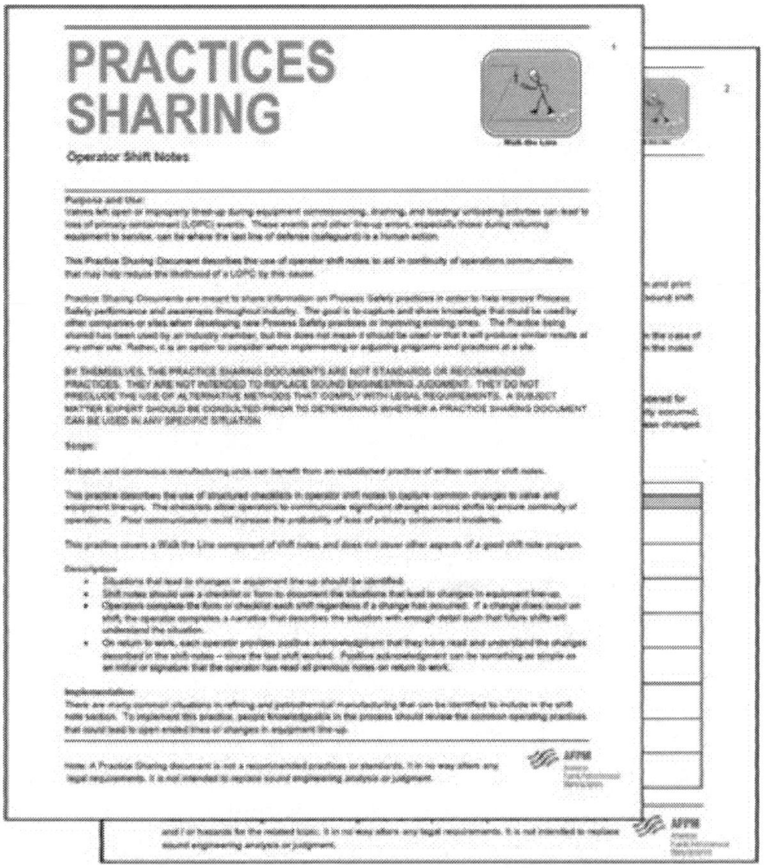

Figure 17. Sample Practices Sharing Tool

Practice Sharing Documents can be used in the following ways:

- Procedure and Safe Work Practice Development – A company or facility can use the Practice Sharing tools when they are developing or updating procedures or practices. A shared practice can be helpful when a facility is developing a new practice, by providing related concepts so that the facility is not starting from scratch. Similarly, the practice shared could be consulted by a site developing a related practice, to see if there are any aspects in the shared practice that the site might want to implement. Where a company or facility has an existing practice of their own, the shared practice can be useful during the periodic reviews of the existing practice to determine if there are any useful concepts to incorporate from the shared practice.
- Incident Investigation – When a site has completed a root cause investigation and is looking for ways to address a gap in site practices, a relevant shared practice could be a useful tool for developing a new site practice or modifying an existing one.

Practices Sharing is a new program developed in 2015. Currently there are nine Practices Sharing Documents available:

- Critical Bleeder Valves,
- Threaded Lanyard Plug,
- Operating Instructions,
- Operator Shift Notes,
- Shift Change Meeting,
- Shift Change Communications,
- Process Safety Field Audit,
- Operator Evaluation Rounds, and
- Informal Unit Walk-Through.

More practices will be identified and shared as time goes on. These practices can be located on the AFPM Safety Portal.

3.5 *Process Safety Site Assessments*

The Process Safety Site Assessment (PSSA) Program is a collaborative industry effort managed by API, focused on higher risk process safety activities in petroleum refining and petrochemical facilities.[2] This is **NOT** a run-of-the-mill compliance audit! This program primarily involves the assessment of a site's Process Safety systems by a small team of independent and credible third-party industry-qualified process safety assessors. Each assessor is an industry veteran subject matter expert with decades of experience. The expertise of the assessors is the strength of this program and what makes it so successful. Through the use of industry-developed protocols that span over 1,000 questions in total, the site assessments evaluate both the quality of the written programs and the effectiveness of field implementation in eight focus areas: Leadership, Management of Change, Mechanical Integrity, Safe Work Practices, Operating Practices, Facility Siting, Process Hazards Analysis, and HF Alkylation/API RP 751. More information about the Process Safety Site Assessment program is available HERE.

There are two primary ways to use the Site Assessment Program

Actual Assessment

When you have a Site Assessment performed at your site, you will get the benefit of seasoned veterans, looking for soft spots in your critical programs, using an industry developed protocol. The site assessment protocols are not just a check-the-box verification of compliance with the PSM regulations, but far more. The site assessment team evaluates your written programs and how your facilities implement them, using over 1000 very specific protocol questions. The resulting report provides a scoring which shows your performance in overall programs, sub-programs, and also drills down to very specific protocol provisions. The team also provides the facility management with a benchmarking report that compares your performance against the industry or against sister sites within your company. Figure 18 shows a sample page from an assessment protocol. One of the many benefits of a site assessment is having access to industry experts in each area; it's "like going to school." While they are on-site, they are also available to answer questions and engage in discussions on topics pertinent to the protocols.

Benefits of a Site Assessment Include:

- Identification of soft spots in your programs or their implementation.
- Suggestions for how to improve.
- Access to hundreds of years of combined expertise from the assessors. It's almost a training opportunity for your people.
- Obtaining a scored report and benchmarking against the industry.
- Having multiple sites assessed and use the scores to benchmark them against each other, to identify strong programs that could be shared from one site to another.
- Satisfies API 751 audit requirements with an additional audit.

Self-Assessment

Conduct a self-assessment, using subject matter professionals from your company. The same protocols used by the expert assessors are available, to member companies to use in self-assessments. These free tools are an excellent way to evaluate some of your site's critical programs. You won't get benchmarking, but you can score your own assessment to find areas for improvement. It's as simple as this: (1) APS member companies request the protocols from the API PSSA webpage, (2) put a team of your own employees together to apply the protocols to your site, (3) identify areas where improvements could be made in either your management systems or their implementation, and (4) go make the selected improvements.

Benefits of a Self-Assessment Include:

- The protocols alone can serve as roadmap for a successful program.
- Assessing your programs against the items suggested in the protocols can identify weak areas in your programs or their implementation.

- The protocols can also be used when developing new management systems as a check to ensure key aspects are covered in the program or practice being developed.
- Provide a standardized way to compare multiple sites within a company.

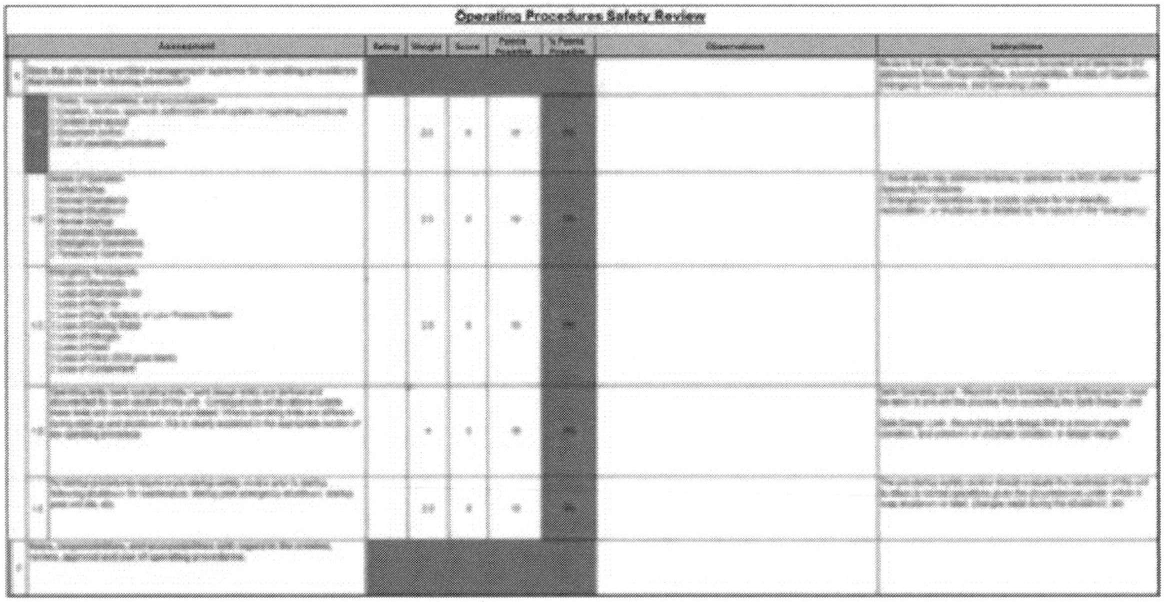

Figure 18. Sample Site Assessment Protocol

3.6 *Human Reliability – "Walk the Line"*

Analysis completed by the APS Program has determined that approximately 20% of Tier 1 and Tier 2 process safety events are caused by human reliability issues. Types of events in this group include:

- Leaving a valve open to atmosphere,
- Valve line-up errors,
- Lock-out tag-out (LOTO) isolation errors,
- Incorrect equipment preparation,
- Bypassing safety systems when they should be engaged, and
- Placing equipment in service with one or more components missing or not properly installed.

"Walk the Line" is a program developed by the APS members to create a set of tools, free to member companies, to address the issues and cultures that lead to human reliability related events.[8] Walk the Line addresses the human reliability associated with conduct of operations and specifically, operating discipline by providing guidance for operations leadership and support to set the expectation to "walk the line" (to know exactly where a material will flow in a process). It reinforces the fundamental responsibility of operations to know with 100% certainty where a material will flow between any two points in a process. Walk the Line addresses the culture of operator line-up and provides tools that help the operator and front-line operations supervision become successful in meeting this responsibility.

The Walk the Line program is meant to generically address loss of primary containment in the refining and petrochemical industry caused by operator error when making a valve line-up to transfer process materials or operate a process. While there are many common causes and solutions, each company and site should have an understanding of their specific causes. Here is how you can use Walk the Line at your site or company:

- Set a Goal of Zero Walk the Line Incidents and commit to it.
- Analyze your LOPC Data.
- Identify Walk the Line Causes.
- Review and customize Walk the Line tools.
- Train employees using your site's customized Walk the Line tools.
- Repeat the discussions often in various meetings, tool box talks, and informal settings with process operations personnel.

The suite of Walk the Line tools are available to APS members HERE.

- Newsletters – The Walk the Line Newsletters are useful for increasing awareness about these Human Reliability issues, and for planning your sites effort to confront these causes and shift culture. The newsletters are suitable for top to bottom awareness and planning, from management level review to "boots on the ground" front line personnel.
- Tool Box Topics – The tool box topics offer short, scripted tool box meeting topics aimed at identifying common incident scenarios, engaging front line personnel in discussions, and providing suggested solutions to prevent incidents. There are 10 tool boxes related to Walk the Line topics, that are available for immediate use. While the tool box topics are fully complete, it is expected that each company or site might modify them to better align with their work practices or culture.
- Hazard ID Documents and Practices Sharing – A number of the tools being developed as part of the Walk the Line project are Hazard Identification Documents and Shared Practices. These have been previously discussed in this paper as has their suggested uses. Preparation of these documents for the Walk the Line project is on-going and these tools will be made available to APS member companies as they are completed. A development schedule is included on the Walk the Line page of the Safety Portal.
- Training Materials – Packages of "ready to go" training materials have been prepared and are available for off the shelf use. There are packages intended for both front line personnel and front line supervision.

4.0 Conclusion

The AFPM and API have been working together on an industry led initiative, Advancing Process Safety, with the overall objective of improving our process safety performance. Data obtained through the APS programs of Metric Reporting, Event Sharing, and Site Assessment identify high priority areas for improvement. Each area for improvement is being addressed through a APS work group, where tools and materials are developed and made available for AFPM and API member companies.

References

[1] API. "Process Safety Performance Indicators for the Refining and Petrochemical Industries," ANSI/API Recommended Practice 754, First Edition, April 2010.

[2] AFPM. AFPM Safety Portal Process Safety Site Assessment Programs, Available at http://safetyportal.afpm.org/SiteAssessment.aspx – Login credentials required. Accessed on February 10, 2016.

[3] AFPM. The AFPM Safety Portal Process Safety Metrics Database, Available at http://safetyportal.afpm.org/ProcessSafetymetrics-access.aspx – Login credentials required. Accessed on February 10, 2016.

[4] AFPM. The AFPM Safety Portal Event Sharing Database, Available at http://safetyportal.afpm.org/EventSharing-access2.aspx – Login credentials required. Accessed on February 10, 2016.

[5] API. "Piping Inspection Code: In-service Inspection, Rating, Repair, and Alteration of Piping Systems," API 570, Fourth Edition, February 2016.

[6] API. "Inspection Practices for Piping System Components," API Recommended Practice 574, Third Edition, November 2009.

[7] AFPM. AFPM Safety Portal Hazard Identification & Practice Sharing, Available at http://safetyportal.afpm.org/HazardIdentification-access.aspx – Login credentials required. Accessed on February 10, 2016.

[8] AFPM. AFPM Safety Portal – Walk the Line, Available at http://safetyportal.afpm.org/walktheline.aspx - Login credentials required. Accessed on February 10, 2016.

[9] AFPM. AFPM Safety Portal, Available at http://safetyportal.afpm.org/default.aspx – Login credentials required. Accessed on February 10, 2016.

[10] AFPM. AFPM Safety Portal – References, Available at http://safetyportal.afpm.org/References.aspx - Login credentials required. Accessed on February 10, 2016.

Appendix A: API RP 754 Tier 1 and Tier 2 Definitions

A Tier 1 or Tier 2 Process Safety Event begins with an unplanned or uncontrolled release of any material, including non-toxic and non-flammable materials from a process that results in one or more consequences described in the RP. The consequences that define a Tier 1 and Tier 2 process safety events are tabulated below.

Tier 1 Consequences	Tier 2 Consequences
An employee, contractor or subcontractor "days away from work" injury and/or fatality	An employee, contractor or subcontractor recordable injury
A hospital admission and/or fatality of a third-party	
An officially declared community evacuation or community shelter-in-place	
A fire or explosion resulting in greater than or equal to \$25,000 of direct cost to the Company	A fire or explosion resulting in greater than or equal to \$2,500 of direct cost to the Company
A pressure relief device (PRD) discharge to atmosphere whether directly or via a downstream destructive device that results in one or more of the following four consequences: ▪ liquid carryover; ▪ discharge to a potentially unsafe location; ▪ an on-site shelter-in-place; ▪ public protective measures (e.g. road closure); and, a PRD discharge quantity greater than the threshold quantities in Table 1 in any one-hour period.	A pressure relief device (PRD) discharge to atmosphere whether directly or via a downstream destructive device that results in one or more of the following four consequences: ▪ liquid carryover; ▪ discharge to a potentially unsafe location; ▪ an on-site shelter-in-place; ▪ public protective measures (e.g. road closure); and, a PRD discharge quantity greater than the threshold quantities in Table 2 in any one-hour period.
A release of material greater than the threshold quantities described in Table 1 of the standard in any one-hour period	A release of material greater than the threshold quantities described in Table 2 of the standard in any one-hour period

442500 Forge Bonding: A Safer Metal Joining Process

Abstract:

When planning in-service repairs or enhancements that involve the joining of metals; sparks, flames and even elevated temperatures are a concern for refinery and storage terminal owners and operators. Processes that generate uncontrolled elevated temperatures that exceed Auto Ignition Temperatures (AITs) published by American Petroleum Institute (API) or published Hot Surface Ignition Temperatures (HSITs) are often scrutinized and classified as "hot work" within these facilities. A metal bonding process has been developed that in inherently safe in these environments where flammable hydrocarbons are present.

Friction welding is a solid-state welding process that generates heat through mechanical friction between work pieces in relative motion to one another, with the addition of a lateral force to plastically displace and fuse the materials. Technically, because no melt occurs, friction welding is not actually a welding process in the traditional sense, but a forging technique. An advanced form of friction welding called forge bonding has been developed for portable entry into refineries, improving worksite safety by reducing the time at temperature and the hot surface area at the bonding site. In addition, forge bonding has a distinct advantage over welding in that it can join dissimilar metals that have largely different melting points.

An introduction to the forge bonding process followed by results from extensive ignition testing to determine the safety margin of multiple chemicals commonly found in refineries when in the presence of forge bonding, will be presented. The dependency of the ignition temperature of hydrocarbons on the hot surface area and the ignition lag time are discussed along with support from data from API 2216 and other well known publications. A presentation of the problem followed by an investigation of the forge bonding solution will prove that ignition of common hydrocarbons is not possible in the presence of the low energy levels created by the small hot surface area, short ignition lag time and moderate temperatures.

Real life implementations add further proof that forge bonding is an alternative to welding that is inherently safe in refineries, storage terminals and anywhere flammable hydrocarbons are present. This technology has far reaching implications for improved safety in other industries, such as pharmaceutical, pipelines, mining and marine to name just a few.

16AIChE
Spring Meeting, Houston, TX

Best Safety Practice for Refinery and Petrochemical Plants During the Major T&I (Turnaround & Inspection)

Ayman Natto
Safety Department, Technical Safety Section, Rabigh Refining and Petrochemical Company (Petro Rabigh),
Rabigh, Saudi Arabia
ayman.natto@petrorabigh.com

Prepared for Presentation at
American Institute of Chemical Engineers
2016 Spring Meeting
12th Global Congress on Process Safety
Houston, Texas
April 10-13, 2016

UNPUBLISHED

Safety Practice for Refinery and Petrochemical Plants During the Major T&I (Turnaround & Inspection)

Ayman Natto, Process Safety Engineer
Safety Department, Technical Safety Section, Rabigh Refining and Petrochemical Company (Petro Rabigh),
Rabigh, Saudi Arabia
ayman.natto@petrorabigh.com

Keywords: Safety Practice during T&I, Safety in Major shutdown, Process Safety roles in shutdown, Pre-Start up Safety Review, Management of Change, Mechanical Completion Certificate.

Abstract

Planned shutdowns or Turnaround and Inspections (T&I) are normal activities for gas and oil companies the world over, but Petro Rabigh carried out a major shutdown at its huge refinery and petrochemical complex consisting of Refinery, Monomer, Polymer and Utility plants. This paper is aimed at the best Safety Practice implemented during the T&I which was completed at the end of December 2015. It explains the Safety Program developed and implemented by Safety Department to execute the T&I safely. Planning was a key element of the T&I preparation. A dedicated taskforce was established for the T&I planning with representatives from Manufacturing, Maintenance, Engineering, Safety, and other involved departments. The taskforce addressed various safety programs including Work Permit System, Critical Lift Plan, Confined Space Entry, Process Hazard Analysis (PHA) etc. Moreover, critical jobs were identified and Job Safety Analysis (JSA), Management of Change (MOC), Mechanical Completion Certificate (MCC) and Pre-Startup Safety Review (PSSR) were conducted for each of these jobs. Special training and presentations were given to both Petro Rabigh and Contractor Employees on the lessons learned from T&I incidents experience worldwide. A huge contractor workforce and the massive scale of activities were managed safely with the above innovative implementation of the practice, Petro Rabigh would like to improve in the 2019 T&I with many of its new Phase II plants.

1. Introduction

Most oil and gas companies and organizations worldwide have periodic planned plant shutdowns known as Turnaround and Inspections (T&I). They covers activities such as inspection, overhaul, modification and the installation of new parts or equipment. Activities that cannot be performed when the plant is in normal operation are also carried out during the shutdown period.

Rabigh Refining and Petrochemical Company (Petro Rabigh) is located on a 3,000 acre site in Rabigh, on the Red Sea coast of western Saudi Arabia. Petro Rabigh - a joint venture between Saudi Aramco and Sumitomo Chemical Company, began operations in April 2009, with the capacity to produce 18.4 million tons of petroleum based products and 2.4 million tons of ethylene and propylene-based derivatives per annum.

This integrated Refinery and Petrochemical complex has a 400,000 barrels per day Refinery with Distillation, Hydrotreating, Hydrogen Production and Recovery plants, and petrochemical plants consisting of the world's largest High Olefin Fluidized Catalytic Cracking plant, Ethane Cracker plant, Alkylation & Isomerization Unit, Propylene Oxide, Mono Ethylene Glycol, Polyethylene, Polypropylene, High Density Polyethylene plants, etc. These are supported by a vast network of utilities, such as a storage tank farm and marine berthing facility for the receipt of crude oil/raw materials and dispatch of finished products. Moreover, Petro Rabigh is currently constructing many new production plants - tied-in with existing plants during the 2015 T&I - for its Phase II project, known as Petro Rabigh II.

The first T&I at Petro Rabigh was successfully completed in 2011 with about 7 million man-hours logged, over seven weeks. With proper planning, preparation and perfect execution, Petro Rabigh was able to exhibit an enviable safety record, viz., two lost-time injury incidents of contractor employees, three medical treatment cases and five first-aid injury incidents. A huge contractor workforce and the massive scale of activities were managed safely with the above innovative practices and by team work, cooperation and dedication from all. Petro Rabigh would like to continue these practices in the 2019 T&I with improvement in the field of Process Safety, in addition to the hundreds of tie-in points for the Petro Rabigh II.

2. T&I Preparation

This was a major complex-wide turnaround, as it involved all the plants and utilities in the Petro Rabigh Complex. More than one hundred CAPEX works and Warranty Claim works were also included in the T&I scope of turnaround, adding to its daunting magnitude.

Based on the work scope and schedule, the peak contractor manpower in the Petro Rabigh Complex during the turnaround was over 13,000. This was in addition to the regular company employees, nearly 4000. On average the daily man-hours logged was 165,000 per day during the twelve weeks long turnaround period. At the end of the turnaround, the total man-hours logged were more than 15 million.

Petro Rabigh developed a T&I link in the company main web page which showed the day to day status of all Petro Rabigh T&I activities. Safety was one the main activities and included data such as daily safety observations, daily safety performance dashboards, safety flyers, incident flyers, lessons learned, near misses and safety suggestions. This link was updated daily by the Petro Rabigh Safety Department below is the screen print from the web page of that time:

2.1. Contractor Safety Management

Planning is Good Business. A Good Plan ensures safe execution.

It was essential that all the contractor organizations were made sufficiently aware of Petro Rabigh' s minimum safety requirements. The following actions were implemented during the T&I:

No. of Contractor Companies	57
Approved Hazard Identification Plan (HIP)	66
Approved CVs Of the Safety Representative	181

2.1.1. Safety Orientation sessions

Petro Rabigh provided about 13,000 attendees for the Safety Orientation to all the contractor employees through a specially developed Safety Orientation video film. This film was produced in five different languages for the different nationalities who would be working in the Turnaround.

2.1.2. Turnaround Safety Forum Sessions

Turnaround Safety Forums for Contractors and Petro Rabigh staff were conducted before the Turnaround. The Forum included short presentations and discussions on the safe execution of work activities related to turnaround, lessons learned from the 2011 T&I and best industrial safety practices etc. to further enhance safety awareness and improve safety communication during the execution.

In addition, several sessions of Petro Rabigh Work Permit Receiver Training were provided to the selected, qualified Contractor Supervisors to understand the work permits system requirements and standards. This ensured alignment of safe work systems of the contractors with the Petro Rabigh Work Permit System.

2.1.3 Confined Space Standby Man Training

Thousands of confined spaces were opened during the T&I, with high risk especially in such a large refinery and petrochemical plant. Petro Rabigh trained over 1100 standby personnel to understand their roles and responsibilities upon confined space entry. Petro Rabigh issued a standby man certificate for the trained contractor employees to ensure the confined spaces were monitored and managed by a competent person.

2.2. Critical jobs

From among the listed jobs for the Turnaround, critical jobs were identified by a team of Petro Rabigh Operations, Maintenance, Engineering and Safety representatives and a JSA was conducted by the team for each such job. Similar review was done for all scaffold erection plans and critical lift plans.

Awareness Training was also given to Operations and Maintenance staff of Petro Rabigh covering JSA and Scaffold Safety.

2.2.1 Tools Inspection and Certifications

A dedicated team of Petro Rabigh Maintenance Department examined all tools and equipment of the contractors and issued approval sticker on only the properly functioning and safe tool / equipment. This prevented the contractors from using damaged tools and equipment in Petro Rabigh work site. Such inspection also covered Heavy Equipment – such as cranes, loaders, forklifts, low-bed trailers etc. through approved Third Party agencies.

Petro Rabigh established the certification requirement for the Scaffold Inspectors, Scaffold Supervisors and Crane Riggers.

All such measures in the planning stage collectively helped in managing the Turnaround safely.

3. T&I Safety Implementation and Monitoring

All the planning and preparation were converted into implementation, which ensured an excellent safety record during the Turnaround.

The Safety Department hired 30 contractor safety advisors to monitor the T&I in addition to dedicated Petro Rabigh safety Inspectors and Engineers who went around the contractor work areas to evaluate contractor safety performance, observing the activities, identifying unsafe actions, initiating talks with the workers and supervisors, establishing rapport, correcting unsafe practices at site and investigating incidents. Checklists were developed to record field observations.

The data provided by the Safety Inspectors was analyzed on a daily basis and transformed into a daily dashboard called the Turnaround Daily Safety Performance Dashboard. The dashboard indicated the Top Risk Category Items, the Top Three Unsafe Behaviors, the percentage of unsafe activities – areas-wise, etc.

Petro Rabigh Safety personnel conducted a total of 61,022 daily formal interactions with contractor employees and Petro Rabigh staff, in each area. Meetings were established with contractor management directly, to highlight important concerns to improve safety. Such meetings were held twice every week.

Daily Safety interaction meetings were held by Safety Department with contractor safety representatives on day shifts and night shifts.

The daily Tool Box Talks became very popular and served as a means of safety communication. Topics of the tool box talk sessions were selected based on the Top Risk Category of the dashboard for the previous day. This session was conducted in the beginning of each shift by each contractor.

Field safety observations and analysis were listed together and released in the form of Safety Flyers by Petro Rabigh. The lessons learned from the near miss and injury incidents were also released in the form of Safety Flyers.

3.1. Safety Observation Tours

Based on the statistics of observations shown in Table 1, "Sub-standard use of PPE" and "Unsafe Tools and Equipment" have appeared as our Top Risk Categories in turns. Top Three Unsafe Observations range from various categories related to "Housekeeping, Scaffold Inspection, Confined Space, Basic Safety Rules, and Barricading" etc.

The dashboard along with the Top Risk Category and Unsafe Behaviors generated constructive discussions in T&I meetings thereby providing focus on areas of concern for necessary improvement. Also, it included the percentage of observed people working safely, indicating the status of improvement in safety performance and reflecting the joint efforts of the whole organization.

The daily dashboard will continue to highlight the priority areas where everyone's efforts must be concentrated.

3.1.1 Safety Observation Statistics

Table 1. Safety Observation Statistics

Counselling meetings were conducted with low performing contractors top management to communicate safety concern and observations and action plan was demanded from contractors to reduce safety observations at site. Visible improvement was observed in contractor performance after these meetings.

Area	Total No. of Observations
Refinery	1439
Monomer	2781
Polymer	621
Utilities & Terminal	646
Shops and Laydown Yards	1560
Total	7047

3.1.2 Safety Flyers and Incident Alerts

In order to address Top Risk Categories and Unsafe Behaviors, safety flyers were developed and disseminated among all T&I concerned parties, including contractors.

The flyer information contents were discussed during Day and Night T&I meetings, and also during contractors' Toolbox Talks.

3.2. Near Miss & Safety Suggestions Reports

3.2.1 Lessons learned

Petro Rabigh encouraged contractors and employees to enter lessons learned whenever on the T&I web page visible to all Petro Rabigh employees to avoid recurrences and learn from it. About 550 lessons learned were received during the T&I which was already communicated and closed in the site.

3.2.2 Near Miss & Safety Suggestions

Petro Rabigh requested employees to submit near miss incidents and safety suggestions through the Petro Rabigh web page throughout the year, and this practice continued during the T&I as shown in Table 2.

Table 2. Near Miss & Safety Suggestion during the T&I

Total number of PRC submitted Safety Suggestions	01
Total number of PRC submitted Near Misses	10
Total number of Contractors submitted Safety Suggestions	06
Total number of Contractors submitted Near Misses	44

4. Process Safety through the T&I

Petro Rabigh had about 25 major modifications that were considered as a project which included designing, reviewing and constructing that had completed during the T&I in addition to the hundreds of tied-in points for the Phase II Project. The Safety Department had to ensure that all these projects complied with the process Safety regulation and standards. By implementing the Engineering and Process Safety elements smooth and timely transition from construction to start-up phase was achieved by implementing the Engineering and Process Safety elements smooth and timely transition from construction to start-up phase was achieved.

4.1 Management of Change (MOC)

Management of Change (MOC) is a proactive process with a two-fold objective: first, to challenge the need for change and to analyze its wider impacts on the business goals; and second to evaluate, authorize changes before they are made and ensure proper closure after the changes are completed. Its purpose is to establish the procedures necessary to ensure that the all the risks arising from proposed changes are decreased to an acceptable level. It is also an excellent communication tool to inform impacted groups of areas that a change is being considered and/or about to occur.

The core elements of the MOC process that are implemented in all modifications during the T&I:

- Agreement on the justification and reasons for the change by appropriate management,
- Risk analysis of the proposed change by a multi-disciplinary team of competent people who have required knowledge and experience, including vendors and specialists,
- Approval of complex changes or design deviations by engineering managers or technical authorities of sufficient experience and knowledge,
- Develop operating procedures to supplement or implement the change,
- Communicate the change to all concerned and train the workforce, those who are directly affected by the change,
- Conduct a pre-startup safety review to ensure system is ready for safe start, i.e. that all actions from the risk analysis are implemented, operating procedures are developed and staff are trained,
- Monitor the change during implementation period and capture any lesson learned for future reference, and
- Close out the change after specified time period otherwise initiate a permanent change process.

The MOC process is also critical in day to day operations in case of safe operating envelope excursion during process upset; it requires immediate action such as shut down or reporting of incident for further investigation if conditions were normalized quickly. Overall a robust MOC process and its effective implementation is an integral element of chemical process safety in order to prevent and control major incidents risks.

4.2 Mechanical Completion Certificate (MCC)

An important initiative during this T&I was the implementation of MCC through a Committee chaired by Proponent representative and comprised of Construction Agency, Inspection Organizations, Safety and other representatives as appropriate, depending upon the nature of the facility. The committee had to complete the following MCC steps: (1) accepting new facilities as mechanically complete, (2) commissioning and start-up, and (3) final performance acceptance and financial closeout. The Construction Agency was responsible for delivering a completed facility as defined in the project scope of work. After the facility was accepted as Mechanically Complete, the Commissioning, Start-Up and Performance Acceptance became the responsibility of the Proponent Organization with the Construction Agency providing necessary design and construction personnel, technicians and vendor representatives to assist in these activities.

For all facilities and especially where an existing facility was undergoing an upgrade or retrofit, both the Construction Agency and the Proponent were to assign personnel during the design phase to establish a Pre- Commissioning, Commissioning and Start-Up Plan.

4.3 Pre Startup Safety Review (PSSR)

Pre Startup Safety Review (PSSR) aids in the transfer of projects from the construction stage to the operations stage. It provides assurance that commissioning and start-ups can be carried out in a safe manner following the installation of new facilities. In particular, PSSR aims to confirm that what was built and installed conforms to the original design and standards, and that the hazards during start-up of the project have been properly addressed and controlled. Petro Rabigh dedicated a team to conduct about 25 PSSR for the projects taken up during the T&I, in all four Operation departments.

Nominated leaders from each Operation department led their respective PSSR teams (comprising of members from Operation, Engineering, Maintenance, Project Management Team, Safety and Environment) and started conducting PSSR for such modifications.

5. Conclusion

T&I execution is a major challenge that needs team work, cooperation, dedication, and flexibility in order to be successfully accomplished. With proper planning, implementation of safety programs, innovative practices and active involvement of contractors, Petro Rabigh completed the T&I scope safely, timely and reliably.

The Application of a High Temperature Chemical Equilibrium Program for Assessing Combustion and Other Chemical Reactivity Hazards

E. Winter
DuPont Engineering Research and Technology
Bldg. 1282 Chambers Works Plant
Deepwater, NJ 08023
eric.r.winter@dupont.com

S. Dharmavaram
DuPont Safety, Health, & Environment and Operating Excellence CoE
974 Centre Road
Wilmington, DE 19805
seshu.dharmavaram@dupont.com

Prepared for Presentation at
American Institute of Chemical Engineers
2016 Spring Meeting
12th Global Congress on Process Safety
Houston, Texas
April 11-23, 2016

UNPUBLISHED

The Application of a High Temperature Thermal Equilibrium Program for Assessing Combustion and Other Chemical Reactivity Hazards

E. Winter
DuPont Engineering Research and Technology
Bldg. 1282 Chambers Works Plant
Deepwater, NJ 08023
eric.r.winter@dupont.com

S. Dharmavaram
DuPont Safety, Health, & Environment and Operating Excellence CoE
974 Centre Road
Wilmington, DE 19805
seshu.dharmavaram@dupont.com

Keywords: high temperature chemical equilibrium, decomposition, chemical reactivity, Pmax, maximum deflagration pressure

Abstract

Unexpected energetic chemical reaction events in chemical processing plants can pose chemical reactivity hazards. DuPont uses a high temperature thermodynamics program to predict the equilibrium state of chemical reactions. The program can be used to calculate the equilibrium pressure of a closed system, equilibrium temperature, and number of moles generated for a deflagration event. The worst case ingredient composition can be determined (e.g. determining the optimal fuel/oxidant composition that will give the worst case deflagration pressure in a closed vessel). The program will help determine if a vessel could exceed its burst pressure. Results from a combustion reaction of toluene in air will be presented. Highly energetic chemical reactions which progress at detonation conditions should be evaluated differently. If vessel rupture from the detonation event will occur, then blast waves can be estimated based on the calculated energy released. A decomposition reaction of dinitrobenzene will be presented. The heats of reaction between chemicals are important values to understand when looking at chemical reactivity hazards. The reaction between ethylene and chlorine will be discussed. Adiabatic Flame Temperature calculations will also be presented for the ethylene chlorine reaction.

1 Introduction

Unexpected energetic chemical reaction events in chemical processing plants can pose chemical reactivity hazards. A chemical reactivity hazard is a situation with the potential for an uncontrolled chemical reaction that can result directly or indirectly in serious harm to people, property or the environment [1]. Understanding the effects of potential chemical reactions in a process can help to determine if a process hazard exists. DuPont uses an in-house high temperature chemical equilibrium program called "HT65" to predict the equilibrium state of chemical reactions. A high temperature chemical equilibrium program allows the user to understand several important "worst case" equilibrium properties such as the maximum pressure that can be developed from a deflagration inside a closed vessel and the maximum heat release or work that can occur from the decomposition or reaction of substances. The purpose of this paper is to discuss computational methods in which chemical reactivity hazards can be examined.

2 Chemical Equilibrium Programs

"HT65" is E. I. DuPont de Nemours Co.'s high temperature chemical equilibrium software program first written by Arthur Squyres in 1965. The program has had several minor updates over the years. The program calculates chemical equilibria at high temperatures using equations that represent the thermodynamic properties of the various products of combustion over a wide range of temperatures. The software program uses a procedure which fits the specific heat data of the JANAF thermochemical tables. The program can be used to understand how much heat and pressure can be released in an energetic chemical reaction. For the specified initial loading density and amounts of various materials and their energies of formation, the program finds the ideal product composition that maintains the atoms balance and minimizes the total free energy for the formation of product species.

HT65 is similar to other free-energy minimization programs that are available publicly. NASA has a program publicly available called Chemical Equilibrium and Applications (CEA) which is available for use online [2]. There are also programs available through universities such as Colorado State's Chemical Equilibrium Calculator [3]. ASTM's CHETAH program can provide some energy release evaluations and prediction of chemical thermodynamic properties as well as aid in providing hazard classification evaluations [4].

High temperature chemical equilibria computational tools can be used to predict when explosion hazards may exist. These tools can also help reduce the number of experimental tests that are performed in order to understand safe operating limits for chemical processes.

Chemical equilibrium programs can be used to predict pressure and temperature equilibrium states for worst case adiabatic (no heat loss to surroundings) conditions. Ingredient composition can be changed to determine worst case conditions (for example determining the optimal air/fuel composition that will give the worst case deflagration pressure in a closed vessel). The equilibrium calculator is a very useful tool for estimating the maximum combustion pressure near the stoichiometric point, especially since these pressures are frequently beyond the pressure rating of the experimental apparatus [5].

3 Calculation to Determine Maximum Deflagration Pressure

An important step for process hazards analysis is determining the maximum pressure that might occur in a vessel. The Maximum Pressure, P_{MAX}, is defined as the maximum pressure developed in a contained deflagration of an optimum mixture [6]. In an enclosed vessel containing a fuel-air mixture, the deflagration pressure rise is typically six to ten times the initial [absolute] pressure [7]. High temperature chemical equilibrium calculations tend to provide a higher equilibrium pressure than what is obtained in actual testing. Calculations for equilibrium are idealized, minimum free energy state calculations that calculate idealized products and do not account for heat losses. Calculations for equilibrium do not account for the kinetics and inefficiencies of the reactions. Therefore empirical tests ultimately result in the formation of a less than "ideal" product distribution and lower maximum pressures obtained. A test method for P_{MAX} in gases can be performed in an approximately spherical calibrated test vessel of at least 5 liter capacity with an initially quiescent mixture and a low energy ignition source (<100 Joule) [7]. It should be noted that empirical tests performed in labs in small scale vessels have heat losses due to nearby boundary walls of the vessel. These heat losses may not be as great in large vessels of a production facility so a production scale vessel may have a greater deflagration pressure rise than what is seen at the small scale. High temperature chemical equilibrium calculations help to provide a worst case P_{MAX} that helps to provide an upper limit which can be used for inherently safer process design.

HT65 calculations for the combustion reaction between toluene and air are performed in this paper and compared to experimental results. An unexpected combustion of Toluene and air within a process vessel may lead to over pressurization of the vessel leading to vessel bursting. For this example we will assume the combustion of toluene and air occurs within a quiescent process vessel such that only a deflagration occurs and not a detonation which will be discussed later. The concentration of toluene and air may vary in the vessel over time, and the design of the vessel needs to contain the worst case deflagration pressure that could occur.

HT65 can be used to calculate the high temperature equilibrium pressure based on the input composition and their internal energies of formation and the density of the mixture. HT65 sorts through a list of potential products and their enthalpies from the available atoms of the initial ingredients to determine which products and compositions minimize the total free energy. HT65 Calculations in "Constant Volume Deflagration" mode correspond to a minimum value of the Helmholtz free energy. Heat is prescribed as internal energy in this type of calculation. The equilibrium temperature and equilibrium pressure are determined by the energy available, starting density and product composition.

The HT65 program requires the user to specify the composition and density of the mixture. The maximum deflagration pressure calculated by HT65 is usually found near the stoichiometric composition. In order to cut down on the number of calculations performed, it is usually a good idea to determine the stoichiometric concentration, if known. First, using a balanced chemical reaction, we can find the stoichiometric conditions for combustion of toluene. Typical combustion products of carbon dioxide and water are used as the basis for the equation.

Balanced Stoichiometric Chemical Reaction Equation:

$$1 mol\, C_7H_8 + 9\, mol\, O_2 + 33.93\, mol\, N_2 \rightarrow 7\, mol\, CO_2 + 4\, mol\, H_2O + 33.93\, mol\, N_2$$

The stoichiometric fuel concentration for this reaction is calculated as 2.27 mol% toluene.

The stoichiometric fuel/air composition is a good starting point for finding the worst case equilibrium pressure; however it is not necessarily the worst composition. By having the HT65 program search for the minimum free energy, many iterations of potential product species combinations are performed in order to determine the one combination that gives the lowest free energy.

Using HT65 and varying the input composition of toluene and air allows us to find the composition of toluene and air giving the highest equilibrium pressure. HT65 calculations were performed at initial conditions of 1 atmosphere (14.7 psia) and 25°C. In the chart below, eleven different HT65 calculations were made around stoichiometric condition by varying the toluene concentrations in air between 0.015 and 0.051 mole fractions.

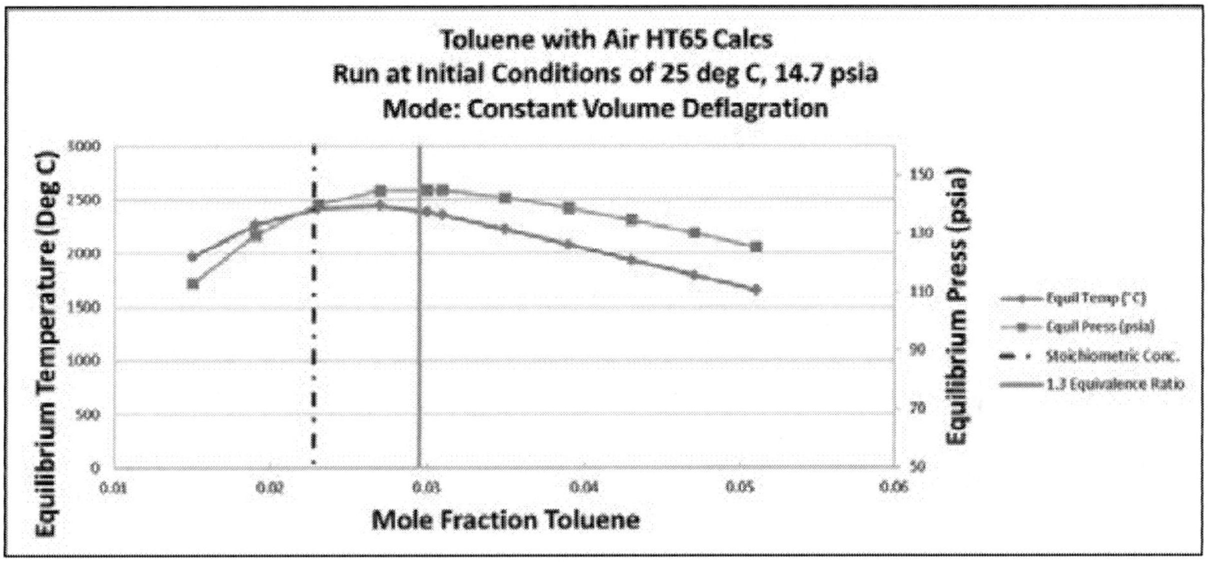

Figure 1: HT65 Calculation Results for Toluene in Air Deflagrations at Constant Volume

The worst case equilibrium pressure is found to be 145 psia which occurs at an initial loading of 2.95 mol% toluene in air. The equilibrium pressure at stoichiometric concentration (2.27 mol% toluene) is slightly less at 140 psia. P_{MAX} for toluene in air was found experimentally to be 7.8 bar (113 psia) using a 5 liter test vessel [7]. The experimental pressure rise is 7.7 times the initial absolute pressure. The worst case HT65 calculated pressure rise is 9.9 times the initial absolute pressure.

The HT65 calculated worst case mixture occurs at a slightly more fuel rich condition than the stoichiometric mixture. In this case, slightly more fuel allows an optimal product balance and free energy minimization. A higher density condition (such as at a lower temperature) may give higher equilibrium pressure results so understanding the range of operating conditions is important in evaluating a process. Species present in the products list include carbon dioxide and water like the balanced stoichiometric equation but also include reduced species such as carbon monoxide and hydrogen. See the table below.

Table 1: HT65 Reaction Products from 2.95 mol% Toluene in Air Deflagration

HT65 Products	Mole Percent
N2	69.9178
CO	10.1685
H2O	8.9262
CO2	8.6715
H2	1.5920
HO	0.2908
H	0.2042
NO	0.1517
O2	0.0537
O	0.0235
HO2	0.0001
N	3.42E-05
HNO	2.69E-05
CHO	2.27E-05
N2O	2.24E-05
NO2	1.62E-05
H3N	1.51E-05
H2N	6.44E-06
H2O2	6.44E-06
HN	5.04E-06
CHN	3.36E-06
CHNO	2.80E-06
CH2O2	1.96E-06
HNO2	1.12E-06
HNO2	1.12E-06
CH2O	8.40E-07

The product species lists calculated by HT65 represents the chosen combination of available species in the program that satisfy the material balance and minimize free energy. Complex products are not available and are not used within the program. This can be an issue if one wants to try to use the program to predict whether or not a particular substance could be produced in a reaction. Theoretical programs such as HT65 are coded for idealized high temperature chemical equilibrium calculations and should not be used to predict whether or not toxic, pollutant or otherwise hazardous material may actually be present in a reaction. The lack of the species in the HT65 product list does not mean it is not present in the real life event. Kinetic or other factors may cause a real deflagration event to produce different species than what are predicted by a theoretical equilibrium program.

Optimal fuel rich mixtures are sometimes identified by their equivalance ratio. The equivalence ratio is the ratio of the worst case fuel/air molar ratio divided by the stoichiomteric fuel/air molar ratio. For toluene and air at ambient temperature and pressure the optimal 2.95 mol% toluene in air mixture is calculated to have a 1.3 equivalence ratio.

So far the equilibrium calculations we have been talking about assume a deflagration occurs and not a more powerful detonation reaction. Transition to detonation can occur when flame propagation is able to accelerate. Turbulence causes flames to stretch, which increases the net flame surface area that is exposed to unburned material, which in turn leads to increased flame speed [6]. If flame acceleration is possible, then deflagration calculations are not sufficient in understanding the potential peak pressures that can be achieved in a system. Within a vessel, flame acceleration can occur if there are internal obstacles and/ or if there is a large length to diameter ratio of the vessel's shape. Flame acceleration can also occur in piping and vent systems connecting to and from vessels. Flames entering a vessel from another vessel or from connecting piping can reach overdriven detonation conditions which can create peak pressures that are many times the maximum deflagration pressure. Expert guidance should be sought during design and hazards review to determine the potential for flame acceleration and pressure generation that is higher than deflagration pressure.

If a deflagration inside a vessel filled with gas causes the vessel to rupture, pressure waves can be created by the rupture. The pressure waves can be estimated based on the rupture pressure and volume of the vessel. If a vessel ruptures which contains liquid near its critical point then additional work energy needs to be included for the liquid as it expands to a gas.

See the attached HT65 output file for details of the calculated results in the appendix section.

4 Calculation to Determine Detonation Energy

Highly explosive chemical reactions can generate significant energy which if not able to be controlled can cause significant damage and injury. Having a way to estimate the power of an explosive chemical reaction will help in performing consequence analysis and understanding the severity of an uncontrolled chemical reaction. Highly energetic chemical reactions which progress at detonation conditions need further evaluation than deflagration reactions.

HT65 can be used to evaluate Chapman Jouguet (CJ) Detonation conditions for gases as well as condensed phases. The Chapman Jouguet condition is where the detonation propagates at a velocity at which the reacting gases just reach sonic velocity. The Chapman Jouget condition holds approximately in detonation waves in high explosives [8, 9]. HT65 calculates the properties of a Chapman Jouguet detonation assuming the detonation front does mechanical work on the explosive by compressing it to a reduced volume, after which chemical reaction to equilibrium takes place at the decreased volume. This reduced volume and the resulting pressure is used to compute the detonation velocity. Detonation in gases can get more complicated due to Deflagration to Detonation (DDT) transition, pressure piling and reflection/ retonation effects in pipelines and long vessels and the CJ condition may not be the worst peak pressure that should be used for evaluating the safety of a system.

If vessel rupture from the detonation event will occur, then blast waves can be estimated based on the calculated energy released. HT65 will compute the changes in product composition that occur as the gaseous combustion products from a CJ detonation expand to atmospheric pressure. The work these gases can do can then be computed. The work from the expansion table is used as the estimate of explosion energy. Work performed by the detonation of pure 2,4,6-Trinitrotoluene (TNT) at 25 deg C and pure molten 1,3-Dinitrobenzene (m-DNB) at 100 deg C can be calculated using HT65 and compared to other methods of determining energy release such as New Jersey's Department of Environmental Protection.

The New Jersey Department of Environmental Protection Bureau of Release Prevention Toxic Catastrophe Prevention Act Program has released a guidance document for calculating the blast wave radius for reactive hazard substances based partly on the US Environmental Protection Agency's TNT Equivalency [10, 11]. The EPA TNT Equivalency method is used to determine the endpoint distance for which a blast wave reaches 1 psi overpressure. TNT equivalency methods can also be used to calculate blast overpressures and reflected pressures at various distances from the blast center [12]. The equation is based off of experimental work done with TNT. In order to use the method for your material of interest, the TNT equivalent explosive power of your material must be determined. Several ways to obtain the TNT equivalency can be used to meet the New Jersey requirements. The heat of reaction of the substance should be related to the equivalent amount of TNT. The heat of reaction may be obtained through literature review, calculation or calorimetric testing. The heat of reaction for many substances is not available through literature so New Jersey also allows using a value of 28% of the heat of combustion because the ratio of the energy of explosion to the heat of combustion for many substances such as TNT has found to be 28% [10].

The heat of combustion for m-DNB is given in the design Institute for Physical Properties (DIPPR) as -4125.2 cal/g. 28% of the heat of combustion gives -1155.1 cal/g. The heat of combustion for TNT in DIPPR is -3584.0 cal/g. 28% of this is -1003.5 cal/g.

A detonation with expansion calculation for m-DNB in HT65 was calculated with a load density of 1.354 g/cc and a heat of formation of 1.70 kcal/mol. The detonation with expansion work is calculated to be -922.2 cal/g.

TNT was calculated with a load density of 1.35 g/cc and a heat of formation of -16 kcal/mol. The detonation with expansion work is calculated with HT65 to be -962.5 cal/g. This is close to the EPAs heat of explosion value for TNT at -1,024 cal/g. In order to determine the TNT equivalency of DNB using HT65, the work result must be corrected for the discrepancy in the TNT value. The TNT equivalent corrected HT65 value for m-DNB is -981.1 cal/g.

$$-922.2\frac{cal}{g}\times\frac{-1024\frac{cal}{g}TNT_{literature}}{-962.5\frac{cal}{g}TNT_{HT65}}=-981.1\frac{cal}{g}$$

The HT65 calculated and corrected value for m-DNB has a lower heat release than the value calculated by taking 28% of the heat of combustion.

Table 2: HT65 Calculations and NJ Heat of Reaction Results for TNT and m-DNB

Material	HT65 Load Density	HT65 Heat of Formation	HT65 Calculated Expansion Work	Corrected Expansion Work	DIPPR Gross Heat of Combustion	28% of Heat of Combustion
	g/cc	kcal/mol	(-) cal/g	(-) cal/g	(-) cal/g	(-) cal/g
TNT	1.35	-16.00	**962.5**	1024	3584.0	**1003.5**
m-DNB	1.354	1.70	**922.2**	981.1	4125.2	**1155.1**

See the attached HT65 output files for details of the calculated results in the appendix section.

Heat release can be calculated for the explosive decomposition of mixtures. Dilution of explosives with inert materials can reduce the overall heat release and can make some materials insensitive to detonation. For example, although water and DNB do not readily mix, keeping the mixture in suspension can help reduce the overall potential heat release and help to make a process safer [13, 14]. As additional water is added, the oxygen balance approaches zero, however water is inert and helps to reduce the overall heat release as it can absorb heat upon vaporization. Below is a figure presenting heat release, oxygen balance and equilibrium temperatures for DNB/ Water mixtures calculated using HT65 [13, 14].

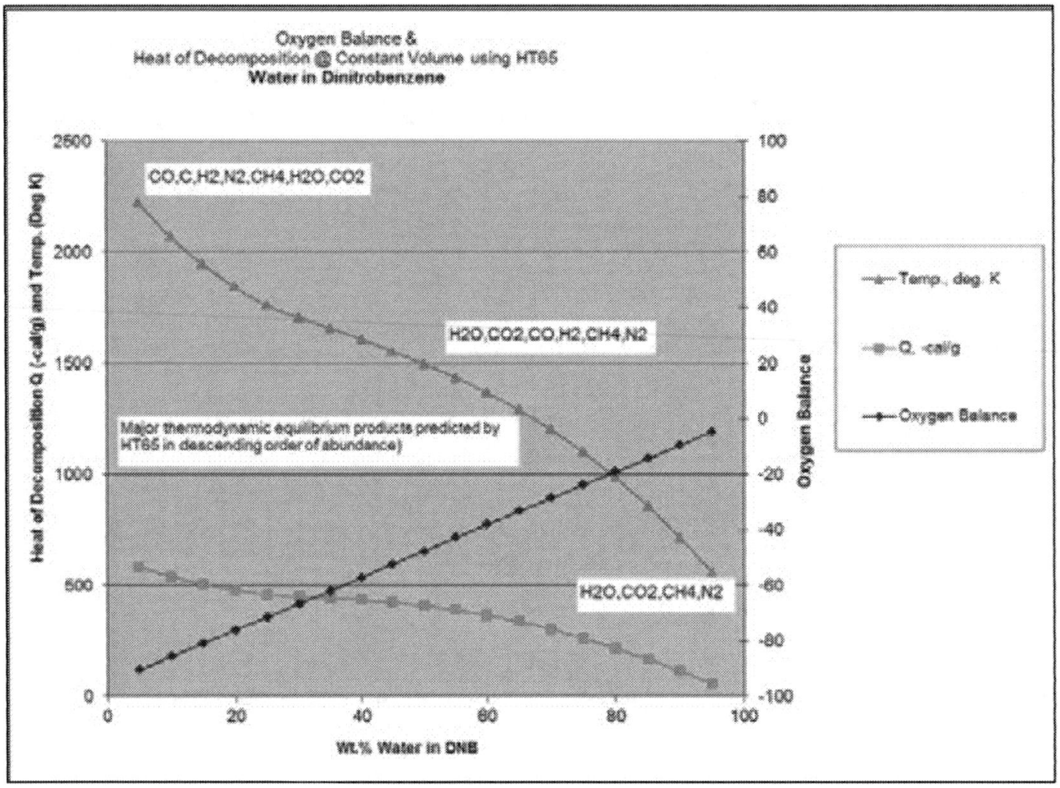

Figure 2: HT65 Calculations for Mixtures of DNB and Water

5 Calculations for Heat of Chemical Reactions

An indicator of a potential chemical reactivity hazard is the amount heat released per unit mass of the reactants [1]. Programs like the ASTM Chetah [4] can calculate the standard enthalpy of a balanced reaction if both the reactants and products involved in the reaction are inputted into the program. Proper selection of the assumed products and their state are needed in order to perform the calculation and interpret the results properly.

For the Chetah method, the calculation is based on the sum of the heats of formation of the products minus the sum of the heats of formation of the reactants. The products and reactants are calculated at the same temperature.

As an example, take the reaction of ethylene with chlorine in the vapor phase. Assuming that the reaction just makes carbon and hydrogen chloride, a balanced reaction can be generated and inputted into Chetah.

$$1\ mol\ C_2H_4(g) + \ 2\ mol\ Cl_2(g) \rightarrow 4\ mol\ HCl(g) + 2\ mol\ C(s)$$

Chetah gives the standard enthalpy of reaction as -100.79 kcal/mol

Using the enthalpy of reaction, a calculation can be performed on a stoichiometric reaction of 1 mol ethylene and 2 mols of chlorine which gives a heat of reaction per unit reaction mass of -592.6 cal/g.

An HT65 explosion with expansion type program is run for comparison. This calculation is performed at constant volume, prescribing heat as internal energy, and minimizes Helmholtz free energy. HT65 calculates product species with primarily hydrogen chloride and carbon; however additional species are included such as hydrogen, chlorine, monoatomic chlorine and monoatomic hydrogen. Some formation of these minor constituents is favorable at the high temperature equilibrium conditions.

For an HT65 explosion type calculation, the heat of reaction is calculated using the sum of the standard enthalpies of formation of the products minus the sum of the internal energies of formation of the reactants. HT65 uses the product composition that it finds that minimizes the free energy. With the HT65 program, you only need to enter the reactant mixture, and it will calculate the heat of reaction per unit reaction mass based on the high temperature equilibrium product composition.

HT65 calculates a heat of reaction value of -489.7 cal/g. The HT65 product composition results in a higher (lower absolute value) total standard enthalpy of formation at 25°C for the products chosen by HT65 compared to those inputted into Chetah.

The heat of reaction value from HT65 should be used with caution when comparing values from other methods. The HT65 heat release represents the available heat release at the high temperature condition and product composition. If the actual real world process system is not adiabatic, additional reactions and heat release may occur as the system loses heat to surroundings. In a process with a vessel that can withstand the high pressure of the reaction, the heat from the high temperature products species will start to be absorbed by the surroundings over time. This heat loss will shift the chemical equilibrium towards more favorable product species at lower temperatures which could release additional heat.

For a process vessel that cannot withstand the pressure generated by the reaction (if a constant volume condition exists), a vessel burst may occur. Work will be performed by the product species as they expand to atmospheric pressure. The work of expansion and the heat of reaction can be calculated by HT65 using the explosion with expansion type calculation.

HT65 calculates the expansion work to be 250.3 cal/g. The heat of reaction is calculated to be -565.2 cal/g. This heat of reaction is based on the sum of the standard enthalpy of formation of the products determined by HT65 to be in equilibrium at the high temperature condition minus the sum of the internal energies of formation of the original reactants. See the attached HT65 output file for details of the calculated results in the appendix section.

6 Calculations for Adiabatic Flame Temperature

Another indicator of a potential chemical reactivity hazard is the adiabatic flame temperature. The determination of the adiabatic flame temperature can be of interest in looking at the thermochemical potential of a chemical reaction. For a chemical reaction, the adiabatic flame temperature corresponds to the temperature that results when the ingredients are reacted. An adiabatic flame temperature calculation can be at either constant pressure or constant volume. The temperature is calculated and compared to a baseline reference temperature to get an understanding of the potential for the mixture to be flammable/ reactive. The comparison is better when made to similarly structured chemicals that have already had flammability testing performed to understand the relevant temperature limits. The adiabatic flame temperature result does not necessarily mean that a mixture is flammable or that reacts quickly. The adiabatic flame temperature comparison assumes that the flammability/ reactivity of a mixture is based only on thermochemistry and not on kinetic effects which may not be appropriate. Empirical tests should always be conducted to understand the true nature of the reaction. Calculations for adiabatic flame temperature can be used as an aid in reducing the number of tests required to obtain flammability information associated with a chemical reaction.

HT65 and NASA's CEA program both can be used to calculate the adiabatic flame temperature of a chemical reaction. For Dupont's HT65 program, this is a type 3 Adiabatic Flame Temperature calculation at constant pressure. For NASA's CEA program, this is a constant pressure "hp" type. These calculations use the enthalpy of formation of the reactants to determine the equilibrium adiabatic flame temperature by minimizing the Gibbs Free energy. Pressure Volume work is included in the calculation in order to keep the pressure constant.

Both programs calculate an adiabatic flame temperature of 2384 Kelvin for the ethylene and chlorine reaction.

The table below shows the significant product species in mass fractions from the HT65 program and NASA's CEA program. The product species are in close agreement but not exact.

Program:	DuPont HT65	NASA CEA
Species	mass frac	mass frac
C (s)	0.141077	0.14111
HCl	0.806638	0.8069
Cl	0.049785	0.04953
H2	0.001336	0.0013306
H	5.96E-05	5.9102E-05
C2H2	0.000126	0.00011175
C2HCl	7.78E-05	0.00004105
Cl2	0.000901	0.00090262

See the attached NASA and HT65 output files for details of the calculated results in the appendix section.

It is important to remember that the kinetics or the rate of the reaction is an important factor for determining a chemical reactivity hazard. There are no predictive tools available to assess the kinetics of different reactions. The only means of determining the rates of temperature increase, reaction completion, pressure rise, etc. is to conduct detailed tests [1].

7 Conclusion

It is important to understand the effects of potential chemical reactions in a process to determine if a process hazard exists. Chemical equilibrium calculators can be used to understand the worst case pressure rise and energy release of chemical reactions. These calculations can help engineers understand whether vessels can burst and help them determine the information needed to conduct consequence analysis modelling for blast wave impacts. Understanding the potential of chemical reactivity hazards will help ensure that proper design and safeguards are in place.

8 References

[1] S. Dharmavaram. "Chemical Reactivity Hazards Assessment and Management Workshop", E. I. du Pont de Nemours and Company. October 2013.

[2] S. Gordon and B. McBride. NASA Reference Publication 1311. National Aeronautics and Space Administration. October 1994. NASA Chemical Equilibrium and Applications program available online at http://www.grc.nasa.gov/WWW/CEAWeb/ceaHome.htm.

[3] Dandy, D. S., Chemical Equilibrium Calculation, Available at: http://navier.engr.colostate.edu/~dandy/code/code-4/index.html.

[4] ASTM Computer Program for Chemical Thermodynamic and Energy Release Evaluation, CHETAH 9.0, ASTM, West Conshohocken, PA, 2009.

[5] D. A. Crowl and G. Tao. "Equations for Flammability Parameters, P_{MAX} and K_G", AIChE 10th Global Congress on Process Safety, 2014.

[6] NFPA 67. Guide on Explosion Protection for Gaseous Mixtures in Pipe Systems, National Fire Protection Association, Quincy, MA, 2013.

[7] NFPA 68. Standard on Explosion Protection by Deflagration Venting, National Fire Protection Association, Quincy, MA, 2013.

[8] Cooper, Paul W. Explosives Engineering, Wiley-VCH.New York, 1996.

[9] Fickett, Wildon; Davis, William C. Detonation, Berkeley: U. Calif. Press, 1979.

[10] NJ Department of Environmental Protection, "Guidance on Hazard Assessment for Reactive Hazard Substances (RHS) and RHS Mixtures", Revised July 26, 2012.

[11] EPA, Chemical Emergency Preparedness and Prevention Office, "Risk Management Program Guidance for Offsite Consequence Analysis", March 2009, Appendix C.

[12] Process Safety Institute, Consequence Analysis Methods, Knoxville, TN, 1992.

[13] R. W. Trebilock and S. Dharmavaram. "Intrinsic Hazards Assessment: Chemical Reactivity Hazards Assessment for Nitration Processes and Nitrochemicals Handling", AIChE 10th Global Congress on Process Safety, 2014.

[14] R. W. Trebilock and S. Dharmavaram. "Assessment of Chemical Reactivity Hazards for Nitration Reactions and Decomposition of Nitro-Compounds", American Chemical Society Symposium Series Vol. 1155 Chapter 10, pp 121-139. November, 2013.

9 Appendix

HT65 Output File for Toluene in Air Explosion (with Expansion). 2.95 mol% Toluene in Air at 25 deg C and 14.7 psia starting conditions.

```
         TDATA file is hitm05

         DENING file is dening.
1 EXPLOSION CALCULATION WITH EXPANSION
    PROBLEM NO.      0
0
                                                                                          MOLE PC.
0 INGREDIENTS              NO.     WT. PC.    MOL. WT.   N,MOL.       HF
                                                                    KCAL/MOL

  TOLUENE (Expl at 25 C)   504     8.8537     92.138 0.0960916      11.34     2.9500      0.029
  AIR (Expl at 25C)        502    91.1463     28.832 3.1612520      -0.59    97.0500      0.971
0 ATOMIC COMPOSITION          G. ATOMS/100 G.
          C                      0.672641
          H                      0.768733
          N                      4.997939
          O                      1.321403
0 HF, KCAL./100 G.                -0.78
0 LOADING DENSITY, G./CC.      0.126300E-02
0 PERFORMANCE PARAMETERS
                                 INITIAL              EXPANDED
    T, DEG. K.                     2670.8               1697.7
    VOLUME, CC./G.               791.7656             4956.0846
    P, BAR.                      10.0                  1.01
    MOLS. GAS/100 G.               3.57025               3.55602
    M, G./MOL. GAS                28.009                28.121
    CV, CAL./MOL./DEG.             7.695                 7.296
    K, SP. HT. RATIO             1.258243              1.272380
    H, KCAL.                       -0.78                -56.48
    Q, CAL./G.                    -557.0                -579.0
    Q, CAL./CC.                     -0.7                  -0.7
0 VIRIAL COEFFICIENTS      0.427     0.204     0.041      9.819
0EXPANSION TABLE
     T           V           P          CV         K          WK          Q           Q
  DEG.K.       CC/G.       KBAR.                             CAL/G.     CAL/G.      CAL./CC

  2670.8    791.766       0.010      7.695      1.258        0.0      -557.0        -0.7
  2364.6   1372.895       0.005      7.626      1.261       98.3      -572.4        -0.7
  2060.5   2378.599       0.003      7.504      1.265      184.3      -576.8        -0.7
  1784.2   4113.929       0.001      7.353      1.270      258.6      -578.5        -0.7
  1697.7   4956.085       0.001      7.296      1.272      281.5      -579.0        -0.7


0 ELAPSED TIME FOR THIS PROBLEM WAS  0.00 SECONDS
1 PRODUCTS, MOLS./100 G.
    PROBLEM NO.      0            18-MAR-16
                                    INITIAL              EXPANDED
       CHN                        0.00000012           0.00000002
       CHNO                       0.00000010           0.00000000
       CHO                        0.00000081           0.00000000
       CH2O                       0.00000003           0.00000000
       CO                         0.36304190           0.32520144
       CO2                        0.30959318           0.34744002
       H                          0.00728872           0.00007848
       HN                         0.00000018           0.00000000
       HO                         0.01038138           0.00001498
       H2                         0.05683832           0.08301396
       H2N                        0.00000023           0.00000000
       H2O                        0.31868719           0.30130526
       H2O2                       0.00000023           0.00000000
       H3N                        0.00000054           0.00000040
       N                          0.00000122           0.00000000
       NO                         0.00541682           0.00000161
       NO2                        0.00000058           0.00000000
       N2                         2.49623947           2.49896871
       N2O                        0.00000080           0.00000000
       O                          0.00083828           0.00000001
       O2                         0.00191613           0.00000002
       HNO                        0.00000096           0.00000000
       HNO2                       0.00000004           0.00000000
       HNO2                       0.00000004           0.00000000
       HO2                        0.00000238           0.00000000
       CH2O2                      0.00000007           0.00000000
```

HT65 Output File for liquid m-Dinitrobenzene detonation with expansion calculation. M-DNB in molten liquid state at 100 deg C.

```
HT65 Case Details
Case Name:     1                                   Changed: 1100 03/04/04
Description:  ONE COMPONENT
Prob. Type:  )DETONATION EXPANSION             Initial Temp Guess (°K):  3000.00
Entry Units: )Weight %       Density From: )Entry
Density (g/cc):    1.35400

 P,,,,,,,,,,,,,,,,,,,,,,,,,,,,,,,,,,,,,,,,,T P,,,,,,,,,,,,,T P,,,,,,,,,T
 .Order   Use            Compound Name            . .  Weight %      . .       .
 5,,,,,,,,,,,,,,,,,,,,,,,,,,,,,,,,,,,,,,,,,6 5,,,,,,,,,,,,,6 5,,,,,,,,,6
 .  1      Y  )DINITROBENZENE, M-, LIQ, 100C     . .   100.0000    . .  ________ .
 Φ,,,,,,,,,,,,,,,,,,,,,,,,,,,,,,,,,,,,,,,,,Γ Φ,,,,,,,,,,,,,Γ Φ,,,,,,,,,Γ
USERNAME =          JOHNSTJE
         TDATA file is  user5:[keanet.ht65.xqt]hitemptdata.
        DENING file is  user5:[keanet.ht65.xqt]hitempdening.
 C - J DETONATION CALCULATION WITH EXPANSION
   PROBLEM NO.      1           03/04/04

ONE COMPONENT

 INGREDIENTS                  NO.     WT. PC.    MOL. WT.  N,MOL.       HF
                                                                     KCAL/MOL

 DINITROBENZENE, M-, LIQ, 701  100.0000   168.114 0.5948345      1.70

 ATOMIC COMPOSITION           G. ATOMS/100 G.
         C                        3.569007
         H                        2.379338
         N                        1.189669
         O                        2.379338

 HF, KCAL./100 G.                   1.01

 LOADING DENSITY, G./CC.          1.35400

 PERFORMANCE PARAMETERS
                                  INITIAL (C-J)       EXPANDED
   T, DEG. K.                       3697.2              428.2
   VOLUME, CC./G.                   0.5535          1143.7702
   P, KBAR.                        108.019              0.001
   MOLS. GAS/100 G.                2.60378            3.25152
   M, G./MOL. GAS                   38.406             30.755
   CV, CAL./MOL./DEG.               16.489              7.329
   K, SP. HT. RATIO               1.279943           1.271159
   E, KCAL./100 G.                   25.17             -91.21
   Q, CAL./G.                      -1024.1             -931.9
   Q, CAL./CC.                     -1386.6            -1261.7
   W, M./SEC.                       1421.7
   D, M./SEC.                       5611.6
```

VIRIAL COEFFICIENTS 0.427 0.204 0.041 9.819

EXPANSION TABLE

T DEG.K.,	V CC/G.	P KBAR.	CV	K	WK CAL/G.,	Q CAL/G.,	Q CAL./CC
3697.2	0.554	108.019	16.489	1.280	-241.5	-1024.1	-1386.6
2944.2	0.739	47.031	13.778	1.251	66.7	-978.2	-1324.5
2526.9	0.919	25.439	12.613	1.230	214.1	-945.7	-1280.4
2190.9	1.183	13.151	11.861	1.209	328.5	-924.5	-1251.7
1933.6	1.593	6.631	11.420	1.193	419.1	-917.0	-1241.6
1739.1	2.267	3.323	11.192	1.184	493.9	-921.0	-1247.1
1600.0	3.306	1.772	11.104	1.181	553.6	-931.9	-1261.7
1437.7	5.272	0.884	10.856	1.183	611.5	-931.9	-1261.7
1289.6	8.779	0.441	10.581	1.188	663.2	-931.9	-1261.7
1153.9	15.041	0.220	10.276	1.193	709.2	-931.9	-1261.7
1029.4	26.195	0.110	9.937	1.200	750.1	-931.9	-1261.7
915.2	45.990	0.055	9.582	1.207	786.4	-931.9	-1261.7
810.6	80.978	0.027	9.220	1.216	818.4	-931.9	-1261.7
715.0	142.554	0.014	8.848	1.225	846.5	-931.9	-1261.7
627.9	250.406	0.007	8.466	1.235	871.0	-931.9	-1261.7
548.7	438.326	0.003	8.071	1.246	892.3	-931.9	-1261.7
476.7	763.806	0.002	7.655	1.260	910.7	-931.9	-1261.7
411.5	1323.612	0.001	7.203	1.276	926.5	-931.9	-1261.7
428.2	1143.770	0.001	7.329	1.271	922.2	-931.9	-1261.7

ELAPSED TIME FOR THIS PROBLEM WAS 1.23 SECONDS

HT65 Output File for liquid Trinitrotoluene detonation with expansion calculation. Note density of TNT used is 1.35 g/cc and heat of formation is -16 kcal/mol.

```
            HT65 Case Details
Case Name:      3                                   Changed: 1419 03/01/04
Description:    3-COMPONENT
Prob. Type:   )DETONATION EXPANSION           Initial Temp Guess (°K): 3000.00
Entry Units: )Weight %        Density From: )Entry
Density (g/cc):     1.35000

P,,,,,,,,,,,,,,,,,,,,,,,,,,,,,,,,,,,,,,,,,,,T P,,,,,,,,,,,,,,T P,,,,,,,,,,T
.Order   Use            Compound Name              . .  Weight %     . .        .
5,,,,,,,,,,,,,,,,,,,,,,,,,,,,,,,,,,,,,,,,,,,6 5,,,,,,,,,,,,,,6 5,,,,,,,,,,6
. 1      Y   )TNT (2,4,6-TRINITROTOLUENE),X    . .   100.0000    . .  ________ .
Φ,,,,,,,,,,,,,,,,,,,,,,,,,,,,,,,,,,,,,,,,,,,Γ Φ,,,,,,,,,,,,,,Γ Φ,,,,,,,,,,Γ

TDATA file is  user5:[keanet.ht65.xqt]hitemptdata.

       DENING file is  user5:[keanet.ht65.xqt]hitempdening.
 C - J DETONATION CALCULATION WITH EXPANSION
   PROBLEM NO.     1          03/01/04

3-COMPONENT

 INGREDIENTS                NO.    WT. PC.    MOL. WT.   N,MOL.       HF
                                                                   KCAL/MOL

 TNT (2,4,6-TRINITROTOLUE 701  100.0000   227.141 0.4402552    -16.00

 ATOMIC COMPOSITION           G. ATOMS/100 G.
         C                        3.081786
         H                        2.201276
         N                        1.320766
         O                        2.641531

 HF, KCAL./100 G.                   -7.04

 LOADING DENSITY, G./CC.          1.35000

 PERFORMANCE PARAMETERS
                                   INITIAL (C-J)       EXPANDED
   T, DEG. K.                          3938.7              358.5
   VOLUME, CC./G.                      0.5490          1006.8913
   P, KBAR.                           119.956              0.001
   MOLS. GAS/100 G.                   2.78069            3.41919
   M, G./MOL. GAS                      35.962             29.247
   CV, CAL./MOL./DEG.                  14.592              6.324
   K, SP. HT. RATIO                  1.325621           1.314220
   E, KCAL./100 G.                      20.56            -103.29
   Q, CAL./G.                         -1013.0             -954.8
   Q, CAL./CC.                        -1367.5            -1289.0
   W, M./SEC.                          1519.7
   D, M./SEC.                          5847.1
```

VIRIAL COEFFICIENTS 0.427 0.204 0.041 9.819

EXPANSION TABLE

T DEG.K.,	V CC/G.	P KBAR.	CV	K	WK CAL/G.,	Q CAL/G.,	Q CAL./CC
3938.7	0.549	119.956	14.592	1.326	-276.0	-1013.0	-1367.5
3065.5	0.741	51.496	12.065	1.294	80.9	-979.1	-1321.7
2594.1	0.922	27.794	11.041	1.269	242.9	-950.1	-1282.7
2220.8	1.182	14.314	10.414	1.243	366.6	-933.3	-1260.0
1942.7	1.581	7.210	10.076	1.222	462.8	-931.0	-1256.9
1738.0	2.229	3.621	9.926	1.210	541.0	-940.3	-1269.4
1600.0	3.160	1.990	9.892	1.204	600.3	-954.8	-1289.0
1421.8	4.961	0.992	9.646	1.207	659.9	-954.8	-1289.0
1260.6	8.130	0.495	9.372	1.212	712.3	-954.8	-1289.0
1114.4	13.717	0.247	9.065	1.219	758.4	-954.8	-1289.0
981.5	23.538	0.123	8.726	1.228	798.8	-954.8	-1289.0
860.9	40.722	0.061	8.384	1.237	834.1	-954.8	-1289.0
751.7	70.646	0.031	8.037	1.247	864.8	-954.8	-1289.0
653.2	122.492	0.015	7.690	1.258	891.3	-954.8	-1289.0
564.7	211.829	0.008	7.344	1.271	914.1	-954.8	-1289.0
485.4	364.874	0.004	6.999	1.284	933.5	-954.8	-1289.0
414.8	625.404	0.002	6.649	1.299	950.0	-954.8	-1289.0
352.1	1065.777	0.001	6.284	1.316	963.9	-954.8	-1289.0
358.5	1006.891	0.001	6.324	1.314	962.5	-954.8	-1289.0

ELAPSED TIME FOR THIS PROBLEM WAS 1.16 SECONDS

PRODUCTS, MOLS./100 G.
PROBLEM NO. 1 03/01/04

	INITIAL (C-J)	EXPANDED
C S	1.19573855	1.08585558
C	0.00000002	0.00000000
C H	0.00000075	0.00000000
C N H	0.01252809	0.00126023
C N H O	0.13131047	0.00027315
C H O	0.04988024	0.00008355
C H2	0.00001254	0.00000000
C H2O	0.00843962	0.00027026
C H3	0.00872937	0.00002732
C H4	0.08949470	0.16396778
C N	0.00007295	0.00000000
C O	0.59551311	1.39058791
C O2	0.65599478	0.43870678
C2	0.00000003	0.00000000
C2H2	0.01931087	0.00001167
C2H4	0.05359197	0.00032168
C2H4O	0.00025772	0.00000000
C2N2	0.00629501	0.00000000
C3	0.00000008	0.00000000
C3O2	0.03413060	0.00000008
C4N2	0.00154904	0.00000000
H	0.00019065	0.00000130
N H	0.00006056	0.00000000
H O	0.00048334	0.00000005
H2	0.03119106	0.38778991
N H2	0.00195060	0.00000028
H2O	0.40061912	0.37275564
H2O2	0.00000122	0.00000000
N H3	0.09886229	0.00684294
N2H4	0.00086146	0.00000000
N	0.00000091	0.00000000
N O	0.00129009	0.00000000
N O2	0.00000085	0.00000000
N2	0.52848334	0.65619463
N2O	0.00003185	0.00000000
O	0.00000145	0.00000000
O2	0.00000311	0.00000000
N H O	0.00005232	0.00000000
H N O2 C	0.00000549	0.00000000
H N O2 T	0.00000601	0.00000000
H O2	0.00000085	0.00000000
N2H2	0.00009619	0.00000000
C H4O	0.00817938	0.00002626
C H2O2	0.02400122	0.00006391
C2H6O	0.01720188	0.00000058

Chetah Reaction Thermochemistry Results of Ethylene and Chlorine Reaction.

ASTM CHETAH 9.0

COMMITTEE E-27 CHEMICAL THERMODYNAMIC AND ENERGY APPRAISAL PROGRAM

REACTION THERMOCHEMISTRY

Reaction:

C2H4 (g) + 2Cl2 (ref)aq)	---->	4ClH (g) + 2C (ref)aq)

Temp	delHrxn	delGrxn	logK	delCp	delSrxn
K	kcal/mol	kcal/mol		kcal/(mol-K)	kcal/(mol-K)
298.15	-100.790	-107.464	78.772	0.005	0.022
348.15	-100.534	-108.604	68.175	0.005	0.023
400.00	-100.306	-109.822	60.003	0.004	0.024

Reactant(s):

	Compound Name:	Molecule 1		
	Formula:	C2H4	Amount:	1 Mole(s)
	Molecular Weight:	28.054	State:	Gas
	Symmetry Number:	4	Optical Isomers:	0
	Group(s)			

	Description		Count
	C2H4	Ethylene	1

	Compound Name:	Molecule 2		
	Formula:	Cl2	Amount:	2 Mole(s)
	Molecular Weight:	70.905	State:	Ref
State:	Aqueous			
State:	Liquid			
State:	Solid			
	Symmetry Number:	N/A	Optical Isomers:	N/A
	Group(s)			

	Description		Count
	Cl2	Chlorine	1

Product(s):

	Compound Name:	Molecule 3		
	Formula:	ClH	Amount:	4 Mole(s)

	Molecular Weight:	36.461	State:	Gas
	Symmetry Number:	1	Optical Isomers:	0
	Group(s)			

	Description	Count
	ClH Hydrogenchloride	1

	Compound Name:	Molecule 4		
	Formula:	C	Amount:	2 Mole(s)
	Molecular Weight:	12.011	State:	Ref
State:	Aqueous			
State:	Liquid			
State:	Solid			
	Symmetry Number:	N/A	Optical Isomers:	N/A
	Group(s)			

	Description	Count
	C Carbon	1

Literature References

1	Chao,J., and Zwolinski,B.J., J.Phys.Chem.Ref.Data, V4, 251 (1975).
2	Hall,K.R., et al, "Selected Values Of Properties Of Hydrocarbons and Related Compounds," TRC Hydrocarbon Project, Thermodynamics Research Center, Texas A&M Univ. College Station Texas Loose Leaf Data Sheets, Extant April(1983) Supplement A-85.
3	Robie,R.A., Hemingway,B.S., and Fisher,J.R., Thermodynamic Properties Of Minerals and Related Substances at 298.15 K and 1 Bar Pressure and At Higher Temperatures, U.S Geol.Survey Bull. 1452, (1979).
4	(a). Stull,D.R., et al, JANAF Thermochemical Data, 1963-Date. (b). Stull,D.R., and Sinke,G.C., "Thermodynamic Properties Of The Elements," ACS Advances In Chemistry Series No. 18, American Chemical Society, Washington, D.C., 1956.
5	Robie,R.A., Hemingway,B.S., and Fisher,J.R., Thermodynamic Properties Of Minerals and Related Substances at 298.15 K and 1 Bar Pressure and At Higher Temperatures, U.S Geol.Survey Bull. 1452, (1979).

HT65 Output File for ethylene and chlorine explosion with expansion calculation.

```
         TDATA file is hitm05

         DENING file is dening.
1 EXPLOSION CALCULATION WITH EXPANSION
    PROBLEM NO.      0           04-MAR-16
0
                                                                                          MOLE PC.
0 INGREDIENTS                     NO.    WT. PC.    MOL. WT.   N,MOL.        HF
                                                                          KCAL/MOL

  CL2 EXPL 25 C                   501    83.5058     70.906 1.1776972      -0.59    66.7000      0.667
  ETHYLENE EXPL 25C               502    16.4942     28.053 0.5879658      11.96    33.3000      0.333
0 ATOMIC COMPOSITION                  G. ATOMS/100 G.
         C                              1.175932
         Cl                             2.355394
         H                              2.351863
0 HF, KCAL./100 G.                        6.33
0 LOADING DENSITY, G./CC.             0.233940E-02
0 PERFORMANCE PARAMETERS
                                         INITIAL             EXPANDED
    T, DEG. K.                            2758.5               1997.4
    VOLUME, CC./G.                       427.4600            3887.2569
    P, BAR.                              13.1                 1.01
    MOLS. GAS/100 G.                      2.44083              2.37065
    M, G./MOL. GAS                        40.970               42.182
    CV, CAL./MOL./DEG.                     9.596                9.443
    K, SP. HT. RATIO                     1.207087             1.210439
    H, KCAL.                                6.33               -42.63
    Q, CAL./G.                            -489.7               -565.2
    Q, CAL./CC.                             -1.1                 -1.3
0 VIRIAL COEFFICIENTS        0.427     0.204      0.041      9.819
0EXPANSION TABLE
     T          V            P           CV          K          WK          Q           Q
  DEG.K.      CC/G.        KBAR.                               CAL/G.     CAL/G.      CAL./CC

  2758.5    427.460      0.013       9.596      1.207          0.0      -489.7        -1.1
  2554.5    758.549      0.007       9.598      1.207         73.6      -515.8        -1.2
  2355.6   1346.515      0.003       9.575      1.208        141.0      -537.3        -1.3
  2160.7   2390.054      0.002       9.520      1.209        202.5      -554.3        -1.3
  1968.3   4240.308      0.001       9.426      1.211        258.4      -566.8        -1.3
  1997.4   3887.257      0.001       9.443      1.210        250.3      -565.2        -1.3
0 ELAPSED TIME FOR THIS PROBLEM WAS  0.00 SECONDS


1 PRODUCTS, MOLS./100 G.
    PROBLEM NO.      0           04-MAR-16
                                         INITIAL             EXPANDED
       C          S                   1.16773264          1.17588877
       C                              0.00000078          0.00000000
       CCl                            0.00000322          0.00000000
       CH                             0.00000041          0.00000000
       CHCl3                          0.00000005          0.00000000
       CH2                            0.00000153          0.00000000
       CH2Cl2                         0.00000058          0.00000001
       CH3                            0.00000683          0.00000002
       CH3Cl                          0.00000205          0.00000004
       CH4                            0.00000268          0.00000004
       C2                             0.00000026          0.00000000
       C2H2                           0.00253212          0.00001355
       C2H4                           0.00000030          0.00000000
       C3                             0.00000322          0.00000000
       Cl                             0.16535183          0.03373973
       ClH                            2.18101457          2.31984179
       Cl2                            0.00391821          0.00090234
       H                              0.00855992          0.00031014
       H2                             0.07787105          0.01583825
       CCl2                           0.00001263          0.00000005
       CCl3                           0.00000016          0.00000000
       C2H                            0.00044473          0.00000032
       C2Cl2                          0.00008269          0.00000058
       C2HCl                          0.00101758          0.00000685
       CHCl                           0.00000303          0.00000000
```

NASA CEA Output File for ethylene and chlorine “hp” calculation.

```
         NASA-GLENN CHEMICAL EQUILIBRIUM PROGRAM CEA2, FEBRUARY 5, 2004
                   BY  BONNIE MCBRIDE AND SANFORD GORDON
      REFS: NASA RP-1311, PART I, 1994 AND NASA RP-1311, PART II, 1996

***********************************************************************
********

 prob case=22227594  hp p(psia)=14.7
 o/f=5.0627
 reac
  fuel  C2H4            wt%= 100.0 t,k= 298.15
  oxid  CL2             wt%= 100.0 t,k=  298.15
   output massf
 output short
 output trace= 1e-8
 end

         THERMODYNAMIC EQUILIBRIUM COMBUSTION PROPERTIES AT ASSIGNED

                                             PRESSURES

 CASE = 22227594

             REACTANT                            WT FRACTION      ENERGY
TEMP
                                                  (SEE NOTE)      KJ/KG-MOL
K
 FUEL        C2H4                                 1.0000000     52500.000
298.150
 OXIDANT     CL2                                  1.0000000         0.000
298.150

 O/F=    5.06270  %FUEL= 16.494301  R,EQ.RATIO= 2.995506  PHI,EQ.RATIO=
2.995506

 THERMODYNAMIC PROPERTIES

 P, BAR             1.0135
 T, K              2383.91
 RHO, KG/CU M    2.1074-1
 H, KJ/KG           308.68
 U, KJ/KG          -172.26
 G, KJ/KG         -15500.7
 S, KJ/(KG)(K)      6.6317

 M, (1/n)           41.213
 MW, MOL WT         27.768
 (dLV/dLP)t       -1.01028
 (dLV/dLT)p         1.2312
 Cp, KJ/(KG)(K)     2.2310
 GAMMAs             1.1452
```

SON VEL,M/SEC 742.1

MASS FRACTIONS

CCL	3.5618-6
CCL2	1.0995-6
CCL3	1.3991-8
CHCL	7.4984-7
CHCL2	3.8114-7
CH2	1.2428-8
CH2CL	3.1091-7
CH2CL2	2.0851-8
CH3	8.3631-8
CH3CL	6.4270-8
CH4	5.5148-8
C2CL	1.1661-7
C2CL2	2.5107-6
C2H	5.3353-7
C2HCL	4.1050-5
C2H2,acetylene	1.1175-4
C2H2,vinylidene	5.7666-8
C2H3CL	1.1627-8
*C3	6.4389-8
C3H3,2-propynl	1.1431-8
C4H2,butadiyne	5.7058-7
*CL	4.9530-2
CL2	9.0262-4
*H	5.9102-5
HCL	8.0690-1
*H2	1.3306-3
C(gr)	1.4111-1

* THERMODYNAMIC PROPERTIES FITTED TO 20000.K

NOTE. WEIGHT FRACTION OF FUEL IN TOTAL FUELS AND OF OXIDANT IN TOTAL OXIDANTS

HT65 Output File for ethylene and chlorine adiabatic flame temperature calculation.

```
          TDATA file is hitm05

         DENING file is dening.
1 HIGH TEMPERATURE EQUILIBRIUM CALCULATION
    PROBLEM NO.      0            07-MAR-16
0
                                                                                                  MOLE PC.
0 INGREDIENTS                 NO.     WT. PC.     MOL. WT.   N,MOL.        HF
                                                                        KCAL/MOL

  CL2 Type3 25 C              501    83.5058     70.906 1.1776972       0.00    66.7000        0.667
  ETHYLENE Type3 25C          502    16.4942     28.053 0.5879658      12.55    33.3000        0.333
0 ATOMIC COMPOSITION             G. ATOMS/100 G.
          C                          1.175932
          Cl                         2.355394
          H                          2.351863
0 HF, KCAL./100 G.                     7.38
0 PERFORMANCE PARAMETERS
    T, DEG. K.                         2383.8
    P, PSIA.                             14.7
    MOLS. GAS/100 G.                  2.42683
    M, G./MOL. GAS                     41.206
    CP, CAL./MOL./DEG.                 11.453
    K, SP. HT. RATIO                 1.209947
    H, KCAL.                             7.38
0 ELAPSED TIME FOR THIS PROBLEM WAS  0.00 SECONDS
1 PRODUCTS, MOLS./100 G.
    PROBLEM NO.      0            07-MAR-16
      C        S              1.17456956          0.48399421          0.66522864
      C                       0.00000008          0.00000003          0.00000004
      CCl                     0.00000022          0.00000009          0.00000013
      CH                      0.00000002          0.00000001          0.00000001
      CH2                     0.00000010          0.00000004          0.00000006
      CH2Cl2                  0.00000002          0.00000001          0.00000001
      CH3                     0.00000061          0.00000025          0.00000034
      CH3Cl                   0.00000014          0.00000006          0.00000008
      CH4                     0.00000028          0.00000012          0.00000016
      C2                      0.00000001          0.00000000          0.00000001
      C2H2                    0.00048239          0.00019877          0.00027321
      C2H4                    0.00000001          0.00000001          0.00000001
      C3                      0.00000020          0.00000008          0.00000011
      Cl                      0.14042368          0.05786311          0.07953029
      ClH                     2.21228636          0.91159674          1.25294942
      Cl2                     0.00127038          0.00052348          0.00071949
      H                       0.00590892          0.00243484          0.00334658
      H2                      0.06625430          0.02730081          0.03752376
      CCl2                    0.00000078          0.00000032          0.00000044
      C2H                     0.00006200          0.00002555          0.00003511
      C2Cl2                   0.00000639          0.00000263          0.00000362
      C2HCl                   0.00012867          0.00005302          0.00007287
      CHCl                    0.00000016          0.00000007          0.00000009
```

12TH GLOBAL CONGRESS ON PROCESS SAFETY

LPS 50th LOSS PREVENTION SYMPOSIUM

Guidelines for Chemical Reactivity Evaluation and Application to Process Design Update After Twenty Years

Marabeth S. Holsinger
The Dow Chemical Company
Midland, MI 48667
msholsinger@dow.com

Dow Co-Authors:
Steven Horsch (Freeport, TX)
Scott Tiper (Midland, MI
Kevin Shaughnessy (Midland, MI)

Prepared for Presentation at
American Institute of Chemical Engineers
2016 Spring Meeting
12th Global Congress on Process Safety
Houston, Texas
April 10 – 14, 2016

UNPUBLISHED

ABSTRACT

Originally published by the Center for Chemical Process Safety/AIChE in 1995, this book provided state-of-the-art technology in theory and test methods for the evaluation of hazardous chemical reactions, and guidance for applying this information to process design and operation. Over the past twenty years, there have been technological advances in the test methods for evaluating and quantifying scenario-based reactive chemicals hazards, as well as the methodologies for scale-up to manufacturing plants. Building on this foundational book and drawing from related publications, this paper describes current approaches at Dow in the areas of reactive chemicals testing and thermokinetic modeling. This paper also illustrates the types of results which can be expected and how they are applied in scenario-based hazard evaluation efforts, with examples. It also describes factors which may influence the selection of appropriate test methods. Our intent is that by sharing enhanced practices for identifying and managing uncontrolled chemical reactions, the chemical process industries can realize benefits from improved reliability of operations, fewer incidents over the life of a process, and reduced consequences of incidents that may occur.

INTRODUCTION

The Guidelines for Chemical Reactivity Evaluation and Application to Process Design book (CCPS, 1995) is focused on hazardous chemical reactivity and the prevention of runaway chemical reactions which have the potential to result in injury, property damage, or environmental harm. The target audience is anyone involved with chemicals or processes at any scale. The intent is to raise awareness of their shared responsibility to prevent potential reactive chemicals accidents by:

- having knowledge of the inherent potential energy in every process step or unit operation,
- understanding the conditions whereby the energy may be released, and
- ensuring adequate controls exist to prevent process accidents, releases of material or energy, fires, and explosions.

These fundamental principles, along with the concept of "Owner Responsibility", are key elements of Dow's Reactive Chemicals Program which has evolved over the past fifty years into a recognized industry model (CCPS, 2008; Frurip, 2012). One important aspect of this program involves formalized and documented periodic Reactive Chemicals/Process Hazard Analysis (RC/PHA) reviews conducted by multi-functional review teams for new and existing Dow facilities, new leaders, and capital projects. Each team is commonly comprised of representatives from the production facility, Technology Center, and Process Safety and Reactive Chemicals functions, along with "outside-in" expertise from R&D, engineering, and manufacturing. The review team provides a company-wide, consistent, and as accurate as currently available perspective on hazard identification and the evaluation of adequacy of hazard controls, reaching consensus on whether additional controls are needed or if additional hazard or risk evaluation is warranted. The most effective and thorough RC/PHA reviews occur when experts from Process Safety and Reactive Chemicals gather with plant personnel to prepare well in advance of the review. Using Dow's RC/PHA Questionnaire as a guide, they evaluate the chemistry and supporting reactive chemicals test data to identify and document potential hazards for every unit operation, including conditions leading to such hazards (Whiting et. al., 1997).

Over the past twenty years, there have been many technological advances in estimation and prediction methods, test methods, instrumentation, and modeling for reactive chemical hazard evaluation (CCPS, 1995, 2008, 2010; Duh, et. al., 1996; Frurip, et. al., 1997, 2004, 2007, 2008, 2009, 2012; Grossel, 2008;

Johnson, et. al., 2003; Kossoy, et. al, 2001; Kossoy and Benin, 2003; Leggett, 2001; Mannan, et. al., 2001; Melhem, et. al., 1995) Whereas the fundamental thermodynamic, kinetic, and engineering principles haven't changed, there are many exciting new opportunities for exploiting these technologies to improve the safety, productivity, and reliability of operations.

Building on this foundational book, and drawing from related publications, this review is an attempt to present some of the more important tools, calorimetric instrumentation and methods, and modeling utilized within Dow for identifying and managing chemical stability and reactivity hazards. It also provides current approaches for over-pressure protection, with examples. Our intent is that by sharing enhanced practices for identifying and managing uncontrolled chemical reactions, the chemical process industries can realize benefits from improved reliability of operations, fewer incidents over the life of a process, and reduced consequences of any incidents that occur.

This review is not intended to be comprehensive, but informative regarding reactive chemical hazard identification. In keeping with the original book's scope, this review precludes hazards associated with flammability and transportation. In each section of this paper, advances in technologies are discussed along with guidance on use of the methods and instrumentation, interpretation of the results, accuracy and precision estimates, and common pitfalls. Detailed description of instrumentation is avoided. The reader is urged to refer to other literature sources for additional information.

ADVANCEMENTS IN CALCULATIONAL AND PREDICTIVE METHODS FOR THERMAL AND REACTIVITY HAZARD EVALUATIONS

CHETAH

CHETAH® (CHEmical Thermodynamic And Hazard evaluation) computer program for Energy Release Evaluation and Prediction of Chemical Thermodynamic Properties was introduced in 1974 (Seaton, Freedman, and Treweek, 1974) and commercialized by ASTM (Harrison, et. al., 2005; Harrison, 2009; Britton and Harrison, 2014). It is a unique tool for predicting both thermochemical and flammability properties, and certain reactive chemicals hazards. These hazards are associated with a pure chemical, a mixture of chemicals, or a chemical reaction. The calculations use stored information, user-entered information, or information from a chemical's molecular structure using a library of molecular fragments or groups.

Through the years, there have been many key technical enhancements to CHETAH. Its database has been greatly expanded to include thermodynamic properties for a wide variety of common chemicals and Benson groups including:

- Largest Database for Benson Groups (~1000)
- Large whole molecule database from DIPPR (>1200)
- Database for Liquid and Solid Benson Groups (>400) (Domalski, et. al., 1993)
- Database for Species in Aqueous Solution (>875) (Wagman, et. al., 1982)

Once a molecule is described, CHETAH provides its thermodynamic properties and makes the molecule available for estimation of reaction heats; estimation of thermodynamic properties of individual substances; prediction of the tendency of a compound or a mixture to propagate a deflagration or a detonation; and flammability property estimations.

For the latter, flammability properties may be estimated at 298 K and at higher temperatures for a wide variety of chemicals using methods outlined by Britton (Britton, 2002; Britton and Frurip, 2003; Britton and Harrison, 2014). These properties include: Limiting Oxygen Concentration (LOC), Lower Limit Flame Temperature (LLFT), Maximum Flame Temperature (T-Max), Fundamental Burning Velocity (Su) of Single and Mixed Fuels, Quenching Distance (qd), Lowest Minimum Ignition Energy (LMIE), Lower Flammable Limit (LFLT) at Temperatures Other than 298 K. Flammability calculations are limited to compounds composed of C, O, H, N, and Cl. Mixtures of compounds are also allowed. Values for pure components are calculated for all the above. Mixture calculations are available for LFL and LOC. For Chlorine containing compounds, calculations are made for LFL, LLFT, and LOC but not for the other parameters, as appropriate correlations are not available.

CHEMICAL REACTIVITY WORSHEET (CRW)

The Chemical Reactivity Worksheet (CRW) predicts potential hazards from the unintentional mixing of chemicals with other chemicals, with absorbents, and with materials of construction. As free software, it is one of the most comprehensive tools available to emergency responders, the chemical industry, and academia for identifying and managing chemical reactions with the potential to release heat or toxic gaseous byproducts. Within Dow, the CRW is promoted as the Most Effective Technology (MET) for predicting chemical compatibility and used in conjunction with reactivity testing as appropriate.

Four versions of CRW have been developed to date by volunteer scientists from the US government, chemical industry, and academia including but not limited to: The Center for Chemical Process Safety (CCPS), The National Oceanic and Atmospheric Administration (NOAA), The Dow Chemical Company, The Environmental Protection Agency (EPA) Office of Emergency Management, DuPont, Koch Industries, Chevron Phillips, and The Materials Technology Institute (MTI). Since the release of version 3.0 (CRW3)(Gorman et. al, 2014) there have been over 200,000 downloads further building on this tool's reputation as the gold standard for determining and displaying chemical compatibility. Key features of CRW include:

- Compatibility determinations based on Reactive Groups and Special Hazard Codes
- Extensive database of specific chemicals (~5200)
- Ability to input data and add custom chemicals (expandable database)
- Common absorbent guidance
- Ability to display/print compatibility charts (up to 100x100)

The newest version (CRW4) has just been released through (CCPS, 2016). The expanded capabilities of CRW4 include:

- Training videos
- Materials of Construction section
- Refinement of Reactive Groups (68 total)
- Improved file sharing and chart display

Limitations

The CRW only accounts for reactions between two chemicals at a time (i.e. pair-wise reactions). If you add more than two substances into the reactivity prediction worksheet, the chemicals may react together in ways that CRW's pair-wise reactions can't predict. For example, glycerin and nitric acid react very little when they're mixed together. However, if sulfuric acid is added to the mix, the three chemicals react together to produce the powerful explosive, nitroglycerin. CRW doesn't predict this reaction, because it looks at the pairs (glycerin + nitric acid; glycerin + sulfuric acid; and nitric acid + sulfuric acid), rather than the three chemicals together. In other case, one of the chemicals may act as a catalyst which accelerates the rate of reaction between other chemicals in the mixture. For example, nickel carbonyl catalyzes many polymerization reactions and other kinds of desired synthetic organic reactions. In both of these cases, the reactivity hazard may be greater than predicted by the tool. Therefore, it is prudent to perform reactive chemicals testing to analyze for potential synergistic reactivity effects.

YOSHIDA'S CORRELATION

Basis of the Correlation

Yoshida's correlation is an empirical method derived from a series of approximately thirty-five sealed-cell DSC (Differential Scanning Calorimetry) tests on known high energy materials in the powder form, mainly nitro compounds (Yoshida, 1987). The original test data set in Yoshida's correlation were all classified for their ability to propagate an explosion under high energy stimulus. Some of the materials were known not to propagate an explosion, as shown by test data from BAM (German Republic Materials Testing Laboratory), even though they contained high energy groups (Appendix A).

Although the method is not new, we have implemented it within Dow in a spread sheet tool as a convenient means to screen for potential impact sensitivity hazards. It visually depicts a material's hazard potential when subjected to low-to-moderate energy forces (e.g. crushing, scraping with a spatula, dropping containers, hitting with a hammer, puncturing with a fork truck tine)

For this correlation, DSC test data are plotted as the log10 of the integrated energy of the exothermic decomposition versus the log10 (extrapolated onset T in °C minus 25). Typically, materials having a DSC decomposition energy less exothermic than about -250 cal/g (-1046 J/g) will not be an explosion hazard when subjected to low-to-moderate energy forces.

Example

A DSC in a sealed-N2 capillary is run on an unknown sample. A large exotherm is detected at 205°C with a reported enthalpy of -450 cal/g (-1883 J/g). Within Dow, we enter the DSC detected onset temperature instead of the extrapolated onset temperature because the software reports it, and it is typically lower use in this correlation resulting in more conservative prediction. For this example, the data are plotted in **Figure 1**.

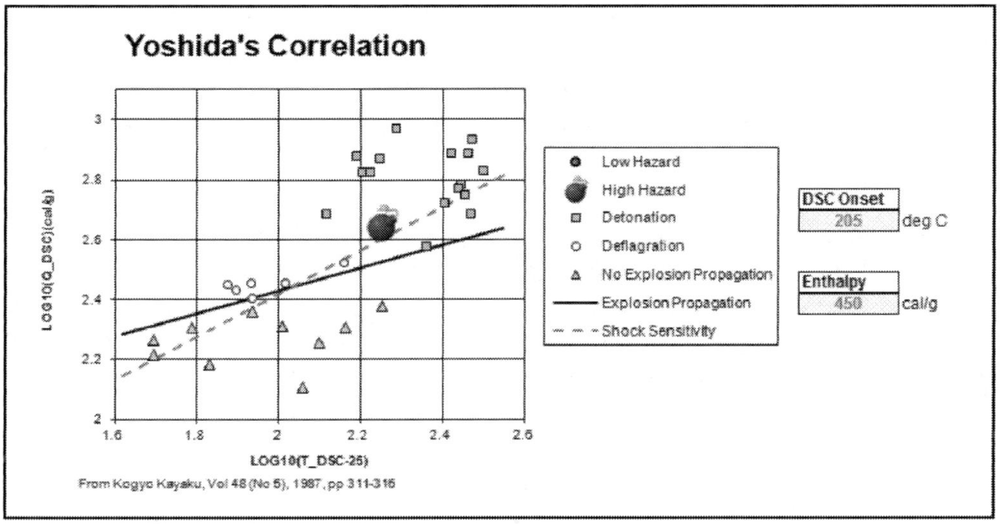

Figure 1: Yoshida's Correlation for an unknown material based on DSC in N2.

The predicted result (large black icon) puts the material very clearly in the region of known detonable materials. Thus, further investigation of the material is warranted and prudent. This may include drop-weight or BAM Fallhammer testing (Gregory and Salven, 1991), and further calorimetry testing such as Accelerating Rate Calorimetry (ARC) (Townsend and Tou, 1980). Ultimately, depending on the scale of the process and the results of the testing, more sophisticated testing such as the United Nations Card Gap or other high energy testing may be needed, depending upon the storage, handling, or transportation scenario.

Finally, if a molecule has one or more high energy groups (such as nitro, azo or azide, or peroxy) shown in Appendix 1 and the Yoshida result is negative, it is prudent to perform testing to confirm this result. This is especially true if the quantities of the material are expected to be significant in the subject process (kg vs. mg).

Limitations

The original author cautioned against applying the correlation to liquids (Yoshida, 1987). There has been work to improve this correlation (Oxley et al., 1999, 2010; Chervin and Bodman, 2003; and Bodman and Chervin, 2004). In general, this method should be used to flag the need for more data if the correlation indicates a propensity for the material to be an explosive hazard (e.g. by drop weight, external card gap testing, UN-United Nations Manual of Tests). On the other hand, if the test indicates a point on the graph "well away" from the boundary in the "safe" direction, then depending on circumstances of scale, process conditions, and other factors, no further testing may be warranted and the risk accepted. This may be even more justified if the material contains no known energetic groups (Appendix 1).

ADVANCEMENTS IN THERMAL HAZARD EVALUATION TESTING AND METHODS

Measuring Process Heats

The heat of reaction is an important parameter in the safe, successful scale-up of chemical processes. Generally, heats of reaction are used to calculate or simulate temperature rises for potential hazard scenarios and to calculate heat loads for the design of heat exchange systems. Within Dow, we measure the heat experimentally using any of a number of calorimetric techniques including Accelerating Rate

Calorimetry (ARC), Simple Mixing Calorimetry, Heat Flow Microcalorimetry, Differential Scanning Calorimetry (DSC), VSP, and Reaction Calorimetry (RC-1). On occasion, specialized calorimetric techniques are developed and applied to address specific problems. The extent of the reaction (i.e. conversion of reactants to intermediates or products) is determined to report the heats per mol of a specific reactant such as the limiting reagent.

Experimental determination of reaction heats is not always practical, feasible or required. Predictive techniques are therefore, necessary and advantageous. Within Dow, we use the so called "Analog Reaction Method" to estimate heats of reaction for model reactants and products from existing heats of formation in literature (Frurip, et. al., 1995) If no published or proprietary thermochemical data exist, estimations of heats of formation can be performed using CHETAH or *ab initio* quantum mechanical calculations.

Heat of Mixing

The "two drop" mixing calorimeter was originally developed at Dow to measure the heat and pressure associated with the intentional or inadvertent mixing of chemicals (Hofelich et al., 1994). For hazard evaluations, heats of mixing experiments are often followed by a DSC or other temperature-ramped screening tests. This is done to determine if it is operationally feasible to reach secondary decomposition reactions by loss of control of the desired reaction in a system with limited heat loss.

Dow has evaluated numerous replacement options for heat of mixing/two-drop calorimetry for basic screening. We have selected the Thermal Hazard Technology (THT) micro reaction calorimeter (µRC) on the basis of our internal evaluation against our desired functionality and performance criteria (e.g. accuracy, sensitivity, reproducibility, temperature and pressure operational ranges). This benchtop instrument has a rapid equilibration time. It is easy to use, portable, and accommodates small sample sizes. It has multiple operating modes including titration calorimetry, isothermal calorimetry and differential scanning calorimetry for applications in hazard analysis, thermal stability, heat capacity (Cp) determination, and reaction kinetic evaluations It also offers a pressure cell withstanding 160 psi vs. the 2-drop cell that withstands 30 psi making it less prone to leaks.

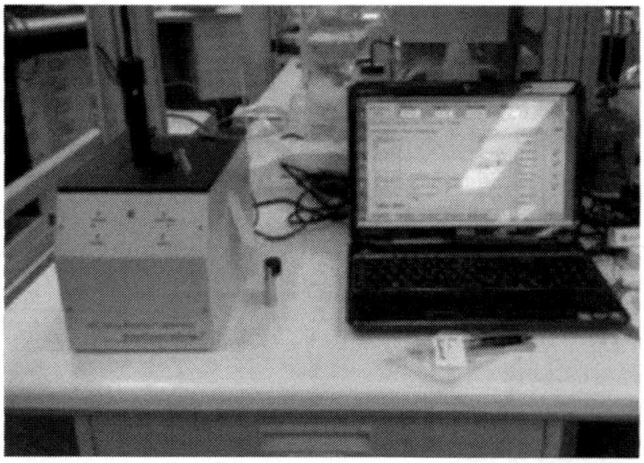

Figure 2. Thermal Hazard Technology (THT) micro reaction calorimeter (µRC)

Differential Scanning Calorimetry (DSC)

When screening for potential thermal hazards, screening calorimeters and simple reaction calorimeters should be applied first to identify candidates for further advanced tests (Frurip, 2004, 2008). Differential Scanning Calorimetry (DSC) is used to study the thermal and oxidative stability of materials or mixtures (Whiting, 1988). A typical DSC test uses ~1 mg material, sealed in a glass capillary or ampoule which is heated from room temperature to an elevated temperature (400°C). The DSC results indicate whether a material undergoes an exothermic or endothermic reaction and a general temperature range in which the reaction occurs. The DSC is often used to determine if further testing is required. The main pros and cons are:

- This test is often run before an adiabatic calorimetry test, a drop weight test, or when preliminary quantitative data is desired,
- Heats of reaction and decomposition can be determined for systems generating significant pressure (up to 100 atm),
- Small changes in the composition of a material can have a significant effect on its thermal behavior, and
- No pressure information is available from this technique.

Interpretation and application of the results to process scenarios requires extensive expertise to avoid mis-interpretation or mis-application of the data (Hofelich and LaBarge, 2002).

Adiabatic calorimetry

Adiabatic calorimetry is used primarily as an aid in thermal hazard evaluation to measure rates of energy output, temperature, and pressure rises of desired or undesired chemical reactions. Under certain conditions, the data may be scaled to larger vessels. Two principal types of adiabatic calorimetry are routinely used in the hazard evaluation context, see below.

ARC - High thermal inertia adiabatic calorimetry

The ARC was developed at Dow (Townsend and Tou, 1980), and is still extensively used within Dow. This type of calorimetric testing provides an opportunity to examine the thermal stability of materials under adiabatic conditions. Tests are conducted using up to 4g of material, which is not well mixed. The sample is heated until there is evidence of self-heating (wait and search phase). If no self-heating is detected, the sample temperature is rapidly raised by 5 or 10°C (heat mode) and the search for self-heating is repeated. Once an exotherm is detected, the calorimeter switches to a tracking mode, also under adiabatic conditions, until the exotherm is complete. Pressure is also measured throughout the experiment. The data are analyzed to provide heats of reaction and other relevant information, and exported for Arrhenius-type kinetic analyses. The effective use of "open-cup" ARC purged with air for evaluating the potential for runaway reactions in carbon beds has been previously described (Hofelich et. al., 2001). This principal may be extended to other types of solids, organics on insulation, and other similar systems.

VSP - Low thermal inertia adiabatic calorimetry

A test similar in overall operation to the heat-wait-search mode may be performed to obtain thermodynamic and kinetic parameters for materials and reactions. Typical sample sizes are 40–70 g and can be well-stirred. The data are used in relief design for fire cases or reactive systems (Fisher, 1992; Chakrabarti, 1995; FAI, 2002). Several modes of testing can be used to study different aspects of reactive venting. Some of the conditions that can be simulated are as follows:

- exothermic reactions, with or without external fire, with or without gas generation
- endothermic reactions with external fire, with or without gas generation
- degree of disengagement of vapor from the liquid reaction mass relieving through the emergency relief nozzle
- turbulent or laminar flow
- area-to-charge venting test (blow-down) or simulation of an existing relief

TS^{u} – Thermal Screening Unit

The thermal screening unit is an instrument specifically designed for the fast and efficient thermal hazard screening of liquids, solids and heterogeneous systems (Singh and Simms, 2001). The TS^{u} is an alternative device to DSC and DTA for use in hazard assessments and provides information regarding "detected onset temperatures" for exothermic events and rates and magnitudes of pressure rise in chemical systems. TS^{u} uses the same sample container as the ARC, so it offers universal consumable parts.

Within Dow, we utilize the TS^{U} :

- For screening worst-case pressure scenarios for relief design. Later, the systems can be run in the VSP.
- For quantifying non-condensable gases under isothermal conditions
- In combination with DSC to obtain both thermokinetics (from DSC) and pressure data from TSu
- To generate results fast and economically. If there are no significant exothermic and pressure generation events, then no need to get more sophisticated tests. A test in the ARC may take up to three days longer to complete. DSC and TSU are cheaper and less costly to maintain. Using these instruments to generate thermokinetic and pressure data save wear and tear on the ARC.

Figure 3: Hazard Evaluation Lab (HEL) Thermal Screening Unit (Ts^{u})

Advantages over the ARC

- Provides a reasonably accurate measurement of detected onset temperature for materials having either low or high activation energy
- Provides an accurate measurement of pressure
- Easy to set up and maintain: less costly to maintain than an ARC
- Easier to recover after a ruptured sample sphere in TS^U than ARC. Returning an ARC to service can take up to two weeks longer than returning a TS^U to service (considering repair, calibration, and performance validation.)
- Can be set up for injection runs

Limitations

The TS^U is not a true calorimeter, so it cannot quantify heat generation. Also, the instrument is not adiabatic, so the kinetics cannot be modeled. They are qualitative at best.

Micro-calorimetry

Many process scenarios require more sophisticated measurements of total heat or heat flow. Examples of these include packed-bed, plug-flow reactors, systems where mass transfer may be rate limited and extreme high or low temperature scenarios. Specialized calorimetric experiments can be designed to quantitatively measure the heats and heat rates under these conditions. Further, systems with extremely limited heat transfer, such as large storage vessels (>10,000 gallons), powdery solids (e.g. from large drying operations), materials spilled on insulation, may undergo thermal runaways at very low rates of heat generation. These systems require isothermal measurement of heat rates in the milliwatt to microwatt range. Within Dow, we utilize many types of micro-calorimeters including the TAMIII and Setaram C-80 in hazard evaluation for validation of extrapolated kinetic models from ARC data to lower temperatures.

ADVANCES IN THERMOCHEMICAL MODELING FROM CALORIMETRIC DATA

CRISIS and RUNAWAY Spread Sheets for Rapid Screening

Within Dow, we developed spread sheet processes for the quick, flexible evaluation of potential thermal runaway reaction hazards. The CRISIS spread sheet is extremely useful for helping reactive chemicals experts within the company provide effective assistance for unplanned events, of which ~20% originate from Dow customers and CHEMTREC (Chemical Transportation Emergency Center). From calorimetric data, CRISIS can predict Time-to-Maximum Rate (TMR) and Temperature of No Return (TNR). See the following sections for details.

TSS CISP for Advanced Thermochemical Modeling

The assessment, control, and mitigation of thermal reaction hazards rely on the use of kinetic models. Within Dow, we utilize the Thermal Safety Software (TSS) developed by ChemInform St. Petersburg (CISP) Ltd. for complex thermal hazard evaluation scenarios to perform advanced thermokinetic modeling and to simulate reaction behavior under defined conditions (Kossoy and Koludarova, 1995; Kossoy and Benin, 1997, 2003; Kossoy et al., 2001; Kossoy and Hofelich (2003); Kossoy and Sheinman, 2004; Kossoy and Akhmetshin, 2007). The following is a description of the type of information obtained from modeling, with examples of how the information may be used to address specific scenarios.

Time to Maximum Rate (TMR)

The "TMR" or "Time to Maximum Rate" may be considered the time frame for a runaway event to reach the point where the event has the highest potential to become problematic. This value is useful for Emergency Responders as it may be used to decide on critical next steps. For example, if the estimated TMR is 2 hours vs. 2 weeks, entirely different paths forward may be decided to mitigate the event. In the former case, emergency cooling and evacuation may be necessary whereas in the latter case, addition of diluents, transfer of contents, or other less urgent mitigation measures may be considered.

WARNING: TMR from calorimetric self-heat rate data does not consider pressure or vessel integrity and therefore may or may not indicate the time to actual physical destruction.

The calculation of TMR is based on a truncated version of the approximate general analytical solution for TMR as proposed by Townsend and Tou, 1980. The general analytical solution for TMR is:

$$\theta_m = \frac{RT^2}{\phi m_T E_A} - \frac{RT_m{}^2}{\phi m_m E_A}$$

Where θ_m is the time to maximum rate, R is the gas constant, ϕ is the thermal inertia (so called PHI factor), T is the current temperature, m_T is the heat rate at the current temperature, E_A is activation energy, T_m is the temperature at maximum rate, and m_m is the maximum rate. In practice, m_m is taken as the maximum heat-rate observed in the calorimetric experiment; however, this is not technically correct because simply multiplying the maximum rate observed in the calorimeter by the PHI factor will not provide the actual maximum rate. One of the most common approaches, for estimating the "true" maximum rate is the equation proposed by Fisher, 1992.

$$m_m = \left[\frac{dT}{dt}\right]_{\phi=1} = \phi * exp\left[\frac{E_A}{R}\left(\frac{1}{T_{mes}} - \frac{1}{T_a}\right)\right]\left[\frac{dT}{dt}\right]_{\phi>1}$$

Where T_{mes} is the measured maximum temperature in the calorimeter and T_a is the temperature at the "true" maximum rate and is calculated by:

$$T_a = T_o + \phi * (T_{mes} - T_o)$$

where T_o is the detected onset temperature of the exotherm.

Townsend and Tou demonstrated the above equation is accurate and conservative for high activation energies (E_A > 30 kcal/mol) and low thermal inertia (ϕ < 2) values and remains conservative but less accurate at low activation energies (E_A < 20 kcal/mol) and high thermal inertia (ϕ > 2) values.

Often, in emergency situations, the maximum heat-rate and the temperature the maximum heat-rate occurs at are not known. However, for activation energies greater than 20 kcal/mol and a low thermal inertia the above equation can be reduced to:

$$\theta_m \sim \frac{RT^2}{\phi m_T E_A}$$

This truncated version is used for the calculation in the TMR tab of the crisis tool. It is important to note that for both equations if the temperature and heat-rate data come directly from process scale equipment and not from a calorimeter, the thermal inertia factor should be assumed to be 1. And the m_T is that observed and the equation of Fisher need not be used. Conversely, if the temperatures and heat rate data are taken from a calorimetric experiment the thermal inertia associated with the experiment must be used to accurately determine the TMR.

Note – TMR should not be evaluated for E_A > 200 kJ/mol since this may be indicative of an autocatalytic system. Such systems require more testing to confirm this mechanism, and advanced modeling using TSS CISP described in another section.

Example

A plant has a contamination event where two materials were inadvertently mixed in a 5000 L storage tank. The tank started at 25°C and seems to be rising by 3 deg. per hour (0.05 deg C/min) for the past 5 hours and it is now at 40°C. No test data exist for the mixture but the contents are clearly reacting and generating heat. Using only this minimal data (current T and observed self heat rate) and making conservative assumptions about the activation energy (20 kcal/mol), TMR is predicted to be 195 minutes or approximately 3 hours. Using this conservative result, the plant emergency responders may choose to immediately transfer the contents to another tank containing cold inert diluents and internal cooling coils.

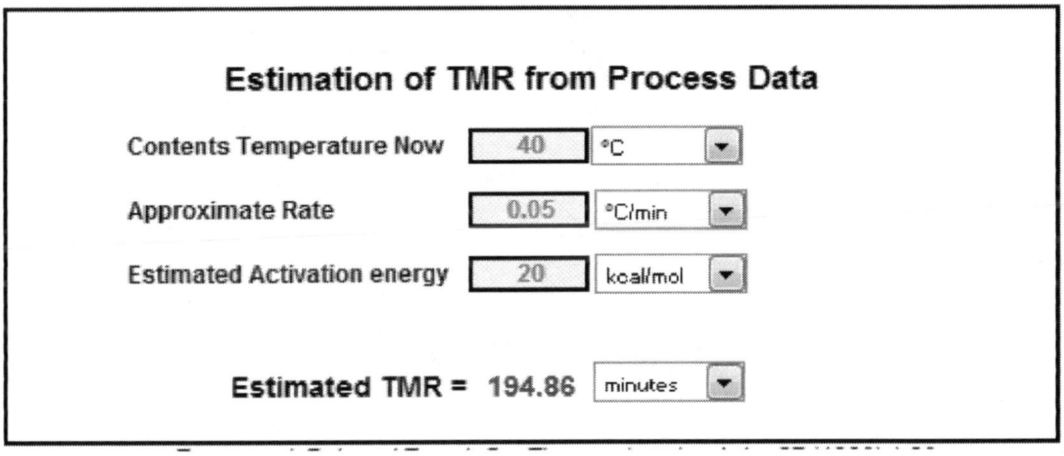

Adiabatic TMR from DSC data

The basis for this calculation comes from Keller, et. al., 1997. This method uses DSC detected onset T (NOT extrapolated onset T but first deviation from baseline, sometimes referred to as lower integration limit) for an exothermic event and using some simple assumptions, predicts the so-called Time to Maximum Rate, TMR under adiabatic conditions. This time, while quite conservative, can give the emergency response team an idea of the timeframe for any emergency mitigation actions that might be planned. Some times in the literature, TMR is (perhaps erroneously) referred to as Time to Explosion. In simpler terms TMR gives an estimate of how much time do we have to do something about the runaway? For example, a TMR value of hours and days is much more desirable than seconds and minutes.

For most applications, one needs to estimate TMR from a given temperature. Typically the most conservative (shortest TMR) will come from choosing a lower Ea but that will depend on the relative positions of the T_{onset} with respect to the desired starting temperature for the TMR calculation.

In other cases, it is useful to predict temperature for a given TMR to address the question *"At what temperature will we have X amount time before "bad" things potentially happen?"* This is particularly useful in a true emergency situation such that an Emergency Response Team may want to know at what temperature the TMR becomes some critical value such that some action needs to occur, such as evacuation, etc. For example, the Emergency Response Team might want to know at what temperature will the TMR be 8 hours; a typical batch time.

Details of the Calculation

The basic principle for this estimation is based on the simple fact that since the calorimetric detection limit of a DSC is known ($\dot{q}_{onset}$ = 45mW/g for our current instrumentation), then the first deviation of the DSC signal above the baseline threshold and the temperature at which that occurs, T_o, defines a single heat rate and temperature point for the system. Then, by assuming an activation energy (Ea), or a range of reasonable activation energies from ~20-40 kcal/mol, the entire kinetic profile is determined for a given reaction order (zeroth order in this case) and may be used, as in this module, to determine Adiabatic Time to Maximum Rate (TMR_{ad}) from any temperature T_o by the following expression:

$$\mathrm{TMR_{ad}} = \frac{c_p * R * T_0^2}{\dot{q}(T_0) * E_a}$$

Where $\dot{q}\ (T_o)$ or more simply $\dot{q}_o$ is determined by a simple and conservative zeroth-order Arrhenius extrapolation:

$$\dot{q}_0 = \dot{q}_{onset} * \exp\left(\frac{E_a}{R} * \left(\frac{1}{T_{onset}} - \frac{1}{T_0}\right)\right)$$

Above, Cp is the heat capacity of the material and may be estimated also and R is the gas constant. The above treatment ignores the effect of the concentration decline and thus the calculated TMR is shorter than real (more conservative).

Limitations

As a caution, the DSC data required for this is for samples under N_2 atmosphere. DSC data in air indicate potential lower temperature oxidative chemistry and is typically not applicable to this calculation which is for the homolytic degradation of a material with energy release.

Note also that it is important to have a good value (or estimate) of the heat capacity (Cp) since the TMR calculation scales **directly** with Cp. In general a lower Cp will be the most conservative. The same amount of released energy will cause a smaller rise in the temperature of the reactants.

Finally, although this estimate can be useful, estimates of TMR from ARC data are typically more accurate since ARC data allow for a more quantitative determination of Ea.

Note – TMR should not be evaluated for E_A > 200 kJ/mol since this may be indicative of an autocatalytic system. Such systems require more testing to confirm this mechanism, ad advanced modeling using TSS CISP described in another section.

Example

During an unexpected event in a plant, a high temperature alarm in a tank of a reactive material is tripped at 40^{0}C (normal storage is at 25^{0}C) and the tank is now at 45^{0}C. Only DSC data exist on the material and it shows a detected onset (lower integration limit) of 163.8^{0}C. The plant personnel ask:

1. At what T will the TMR be 4 hours when extreme emergency actions (e.g. evacuations) may need to be implemented?

2. What is the TMR estimate from the current T of 45^{0}C?

As seen below, using a conservative Ea of 50 kJ/mol and a Cp from the product literature of 0.5 cal/g/deg C the T would need to be ca. 94^{0}C to result in a 4 hour TMR. Also, as can been seen in the screen shot below, from 45^{0}C the TMR is predicted to be 37 hours (1.5 days). In this case the plant now knows that they appear to have a reasonable amount of time (many hours as opposed to many minutes!) to mitigate the situation by using reasonable intervention methods say, cooling the material, adding a shortstopping agent, adding an inert diluent, etc. before the extreme measures to stop the reaction, say adding a shortstopping agent that causes the reaction to be wasted or requires a significant plant outage to recover before a really dangerous situation occurs.

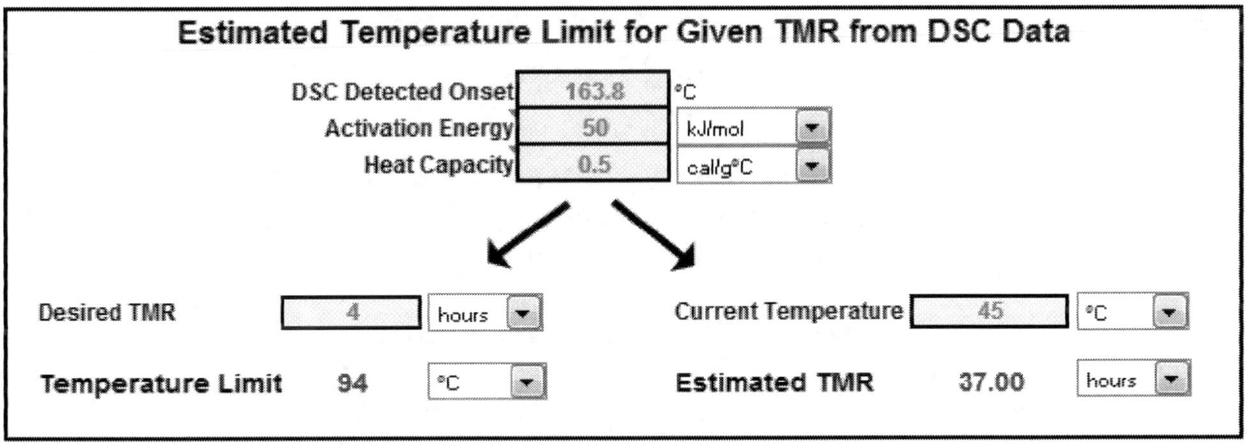

Both these data points give the plant personnel and emergency responders some comfort that there may be time to mitigate the event before the worst effects of the runaway are manifested. Note that this calculation assumes that the chemistry occurring in the material is exactly the same as observed in the DSC test. If the observed temperature rise is due to, for example some contamination issue, especially if the added material was a catalyst for the decomposition of the contents, then this calculation would be erroneous and thus not directly applicable. Also, hazardous gas-producing reactions which may not be very energetic, but which may rupture containers, must be considered.

Temperature of No Return (TNR)

Introduction

The Temperature of No Return (TNR) is defined as the temperature where heat gains from the chemical reaction (and any other heat sources) outpace the heat losses in a system (**Figure 4**). TNR can be a critical point in an emergency situation. By determining an accurate TNR, one can determine whether a process is already in a "runaway" situation or is currently under control. Further, it is important to have a sense of how long it will take to reach the maximum reaction rate (Time to Maximum Rate or TMR) if and when the TNR is exceeded so the TNR can be used as a key parameter in evaluating results from the other TMR tabs in CRISIS as well. In fact the TMR from TNR is often used to evaluate the relative hazards of a process and how critical it is to ensure that TNR is never exceeded. For example, if TMR from TNR is weeks to months, this situation may well be of lesser concern than if the TMR from TNR is minutes to hours.

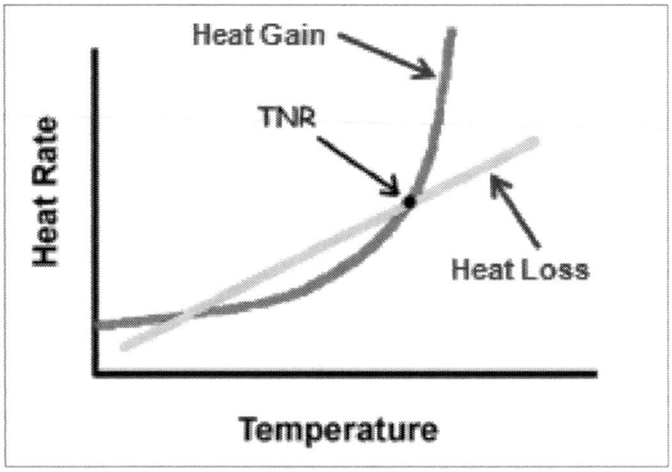

Figure 4. Heat Gains (exponential with temperature) vs. Heat Losses (linear with temperature). The intersection is the Temperature of No Return or TNR.

Determination of the TNR is accomplished by understanding the heat generation rate of the reaction and the heat removal capabilities of the system. Above, TNR, the system always exhibits a higher heat generation rate than capability to lose heat. The TNR is a system dependent parameter and unless two systems are identical a TNR calculated for one system cannot be leveraged to another. Smaller reaction vessels lose heat at a far faster rate than large reactors or storage tanks, since the heat loss is proportional to the surface area, related to tank diameter squared, while the heat generation is related to the volume, or the tank diameter cubed – a similarly dimensioned tank twice the size loses 4 times as much heat, but generates 8 times as much heat – so there is a relative doubling of the potential heat generation divided by heat loss.

Within Dow, the CRISIS spread sheet tool uses a zeroth-order kinetic approximation and allows the user to input the proper kinetic parameters from calorimetric data if known. The user then selects from a list of typical container geometries which define the known heat losses from these vessels. Furthermore, the spread sheet assumes Semenov theory for heat transfer. Specifically, that the thermal conductivity of the material in the vessel is high and the main resistance to heat transfer is the surrounding air (i.e. no temperature gradient exists within the contents of the vessel). Therefore, the TNR estimation are NOT conservative for poorly conducting systems like solid particulates and should not be used! For poorly conducting systems, ones where there is a large temperature gradient within the vessel. A different approach based on Frank-Kamenetskii theory is required.

Basis for the Calculation

The reaction rate can be described by the following differential equation (key assumption – Arrhenius kinetics, batch system):

$$C_o \frac{d\alpha}{dt} = C_o^n (1-\alpha)^n k_o e^{-\frac{E_a}{RT}}$$

Where:

$\alpha = Extent\ of\ Conversion$
$C_o = Concentration\ at\ t = 0$
$n = reaction\ order$
$k_o = Frequency\ Factor\ or\ pre - exponential\ factor$
$E_a = Activation\ Energy$
$R = Gas\ Constant$
$T = Absolute\ Temperature$

Heat Generation rate

Assuming Arrhenius kinetics and a batch system the generic heat generation rate Ġ(T) can be estimated by the equation below.

$$\dot{G}(T) = \Delta H_{rx} * \left(\frac{-r}{C_0}\right) = \frac{d\alpha}{dt} = \Delta H_{rx} * C_o^{n-1} * (1-\alpha)^n * k$$

Where:

$$-r = C_o^n * (1-\alpha)^n * k$$

and

$$k = k_o e^{-\frac{E_a}{RT}}$$

$$\Delta H_{rx} = Heat\ of\ reaction$$
$$k = Rate\ constant$$
$$k^* = Pseudo\ rate\ constant$$

The "Pseudo Rate Constant" can then be defined as:

$$k^* = C_o^{n-1} * k = k_o^* e^{-\frac{E_a}{RT}}$$

And Ġ(T) can be written as:

$$\dot{G}(T) = \Delta H_{rx} * \left(\frac{-r}{C_0}\right) = \frac{d\alpha}{dt} = \Delta H_{rx} * (1-\alpha)^n * k^*$$

or

$$\dot{G}(T) = \Delta H_{rx} * \left(\frac{-r}{C_0}\right) = \frac{d\alpha}{dt} = \Delta H_{rx} * (1-\alpha)^n * k_o^* e^{-\frac{E_a}{RT}}$$

Therefore, the heat generation rate for a zeroeth order Arrhenius reaction in a batch system is:

$$\dot{G}(T) = \Delta H_{rx} * \left(\frac{-r}{C_0}\right) = \frac{d\alpha}{dt} = \Delta H_{rx} * k_o^* e^{-\frac{E_a}{RT}}$$

Heat Removal Rate

The heat removal rate is governed by the following algebraic equation:

$$\dot{R}(T) = \dot{Q} = UA\Delta T$$

Where:

$\dot{Q} = Heat\ Removel\ Rate$
$U = Overall\ Heat\ Transfer\ Coefficient$
$A = Wetted\ Surface\ Area\ Avaliable\ for\ Heat\ Transfer$
$\Delta T = Temperature\ Difference\ between\ Reaction\ Mass\ and\ Ambient$

The TNR is the point where:

$$\dot{R}(T) = \dot{G}(T)$$

or

$$UA\Delta T = \Delta H_{rx} * k_o^* e^{-\frac{E_a}{RT}}$$

This is the basis for the TNR calculation in the CRISIS spread sheet tool. The user has the ability to select overall heat transfer coefficient, reaction mass and heat transfer area for a specific type of vessel from a drop down menu or enter the parameters manually.

To calculate the TNR using the CRISIS spread sheet tool the key inputs are activation energy, pre-exponential factor, heat capacity, heat transfer coefficient, surface area, reaction mass and ambient temperature. The reaction mass is important because the inputs for the TNR tab typically come from the other tabs in the spreadsheet that report temperature rates rather than actual heat rates. Therefore, the reaction mass and heat capacity are used to convert the heat removal rate to a temperature rate as well. For common vessels and containers the heat removal parameters can be selected from the drop down tap next to the cell or they can be defined by the user by selecting "other" in the heat loss scenario cell. The kinetic information can be directly input if known or can be estimated from ARC or DSC data. In emergency situations where the heat capacity is not explicitly known a good approximation for non-halogenated organics is 0.5 cal/(g °C), for halogenated organics 0.34 cal/(g°C) and for aqueous systems 1.0 cal/(g °C).

ADVANCEMENTS IN OVER-PRESSSURE PROTECTION STRATEGIES

Conventional pressure relief devices provide highly reliable and cost-effective protection for most pressure containing equipment. However there are scenarios where conventional relief devices are not practical, not reliable, or not cost effective. For these scenarios, the American Society of Mechanical Engineers (ASME) Boiler and Pressure Vessel Code, Section VIII, Division 1, UG-140(b) can be used to mitigate overpressure scenarios by system design or by a combination of system design and pressure relief devices.

Within Dow, High Integrity Protection Systems (HIPS) or Alternate Overpressure Protection (AOP) are used to mitigate scenarios or reduce the size of the relief devices by system design (Tipler et al. 2013) (Stack, 2010)(ASME, 2009). HIPS is a collection of protection layers that is used in lieu of, or in combination with, a relief device to mitigate a single overpressure scenario. In Dow, the HIPS solution must include as a minimum a Safety Instrumented System (SIS) of Safety Integrity Level (SIL) 1. LOPA is used to define the required protection layers and the SIL requirements for the SIS protection layers. An AOP system is also a collection of protection layers that is used in lieu of, or in combination with, a relief device to mitigate a single overpressure scenario. As with HIPS, LOPA (Layers of Protection Analysis) is used to identify all of the required protection layers. In contrast to HIPS, AOP does not include fully automated protection layers such as Safety Instrument Systems (SIS) and Basic Process Control System (BPCS) Protective Systems because a fully automated loop is neither practical nor effective. For AOP, the layers of protection include operator response to alarm and other Safety Related Protection Systems (SRPS). SRPS includes operating discipline (e.g. sampling and checklist to test for positive identification), fire protection systems (e.g. drainage to a remote containment area), and safety design features (e.g. unique connections).

Implementing a HIPS or an AOP system is not an insignificant task. The analysis, specification, and documentation of a HIPS or an AOP system is often more complex, more time consuming, and more costly than designing and installing a conventional relief device. It is imperative that there are management systems in place to maintain the integrity of the protection layers for both HIPS and AOP. This can be accomplished by predictive preventative maintenance and testing, management of change, periodic audits, periodic reviews of the Process Hazard analysis, and periodic revalidation of the LOPA.

The following are examples of on how to mitigate scenarios with HIPS and AOP that are difficult to mitigate with conventional relief devices.

HIPS Example: Runaway reaction of tower bottoms – Loss of cooling or power

The heating media supply temperature (e.g. steam) for distillation towers may be greater than the temperature of no return (TNR) where heat gains outpace the heat losses. In addition, significant amounts of gas may be produced at elevated temperatures without heat detection (i.e. from thermoneutral, near-thremoneutral or endothermic reactions.). Loss of power or cooling can result in a runaway reaction (e.g. decomposition) if the heating media supply valve should leak through, when process controls are reducing or closing the heating media, causing the tower bottoms to exceed the TNR. Use of a relief device to protect against overpressure from such a runaway reaction may be impractical if the relieving temperatures exceed the maximum allowable working temperature (MAWT) or if the decomposition gas generation rate is excessive. (See Figure 5.)

Distillation Tower and Reboiler
Scenarios:

1. Loss of Power
2. Loss of cooling (e.g. cooling pump failure)

Problem: Both scenarios will exceed the MAWT of the equipment and lead to additional decomposition reactions which are difficult to relieve.

Solution: Use HIPS to mitigate both scenarios

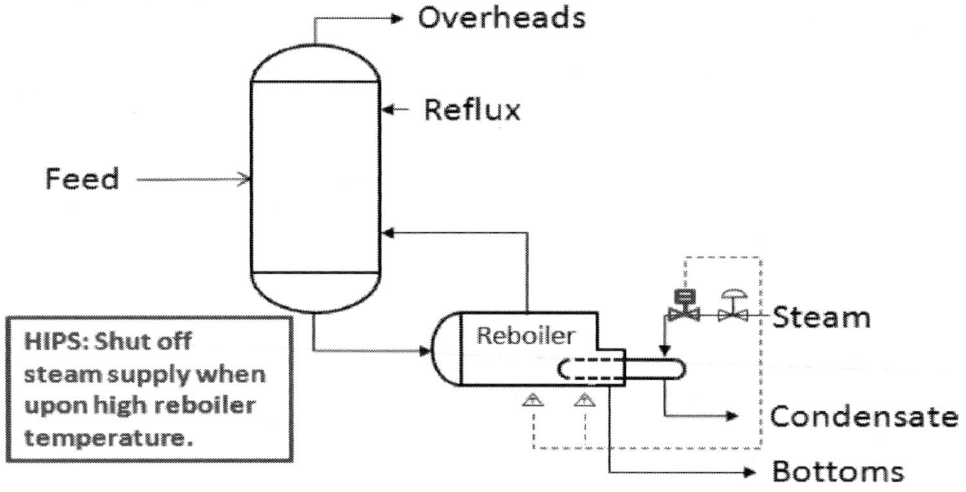

Figure 5: Distillation Tower Bottoms Runaway

In this example, a high temperature interlock is used to close an EBV in the steam supply line. In addition, the bottom recirculation pump may need to be shut off, especially if the system is insulated. In addition, a LOPA credit can also be taken for the basic process control action to close the steam control valve since the initiating event is loss of power or cooling and the valve is designed to fail closed. The required safety integrity level (SIL) for the high temperature EBV interlock is determined by LOPA.

Hydrogen Peroxide Storage Tank – Contamination

Contamination of hydrogen peroxide storage tank by unloading an incompatible material into the hydrogen peroxide tank can result in an exothermic, decomposition reaction. Hydrogen peroxide decomposition may occur if contaminated with iron, caustic, or an organic material (CEFIC, 2012). Contamination may occur if the wrong material is added to storage tank for a variety of reasons including: 1) supplier sent the wrong material, 2) supplier mislabeled the container, or 3) logistics group located the wrong material at the unloading station.

AOP Example: Iron contamination of hydrogen peroxide storage tank
Scenarios:

1. Fire with decomposition
2. Contamination with reaction

Problem: Required relief size for the contamination and reaction scenario is impractical

Solution:
• Size the relief for the fire with decomposition scenario
• Use AOP to mitigate the contamination with reaction scenario

In this example, AOP is provided by Safety Related Protection Systems (SRPS) which mitigate the contamination with reaction scenario. SRPS may include confirmation of identity from a certificate of analysis, sampling with positive identification, dedicated unloading stations, unique unloading connections, dedicated or unique containers, and unloading procedures with associated checklist. A human reliability analysis (HRA) may be required to confirm the reliability of the unloading procedures. An alternative approach for providing overpressure protection may be to use a frangible roof. However, the advantage to using AOP is that it mitigates both the overpressure and the hazards associated with the effluent for the contamination scenario (e.g. toxic exposure).

Use of technology may also afford protection against these types of scenarios. For example, plant personnel can verify the identity and enter the railcar or truck number and dome seal numbers into the process control system. The computer system ensures that the information matches the correct material before allowing the unloading EBV to open (permissive function). Likewise, bar coding can be used to confirm the identity of the material before the unloading valve would be allowed to open. This is especially useful for batch processes that have a number of manual additions. LOPA, with supporting analyses such as HRA, is then used to confirm that there are adequate lines of defense to mitigate the contamination scenario (close the risk gap).

SUMMARY AND CONCLUSIONS

Building on the fundamental engineering principles and guidelines described in CCPS, 1995, Dow leaders and subject matter experts in reactive chemicals, process safety, and engineering solutions continue to advance the science and technology for identifying and managing uncontrolled chemical reactions. This is demonstrated by the diversity of innovative test methods and approaches which Dow employs to gather critical information and solve real-world problems. As part of our commitment to Responsible Care, we have also continued to leverage key aspects of Dow's Reactive Chemicals Program, including RC/PHA protocols. By sharing our enhanced practices, our intent is for the chemical industry to realize benefits from improved reliability of operations, fewer incidents over the life of a process, and reduced consequences of any incidents that may occur.

ACKNOWLEDGEMENTS

The authors wish to thank the following for their contributions to this work: Rob Bellair, Dave Gorman, Paul Morton, Min Sheng, and Klaas van Gelder of the Reactive Chemicals Capability within The Dow Chemical Company. We also thank Dave Frurip who retired from Dow in Sept. 2014.

REFERENCES

ASME, 2009

ASME Code UG-140, **Pressure Vessels with Overpressure Protection By System Design**, *2007 ASME Boiler and Pressure Vessel Code, Section VIII Division 1*. 2009b Addenda (June 2009).

Bodman and Chervin, 2004 G.T. Bodman, S. Chervin

Use of ARC in Screening for Explosive Properties

Journal of Hazardous Materials, **115(1-3)** (2004), pp. 101-105.

Britton, 2002 L.G. Britton

Using Heats of Oxidation to Evaluate Flammability Hazards

Process Saf. Prog., **21** (2002), pp. 31–54.

Britton and Frurip, 2003 L.G. Britton, D.J. Frurip

Further Uses of the Heat of Oxidation in Chemical Hazard Assessment

Proc. Saf. Prog. **22** (2003), pp. 1–19

Britton and Harrison, 2014 L.G. Britton, B.K. Harrison

Using CHETAH to Estimate Lower Flammable Limit, Minimum Ignition Energy, and other Flammability Parameters

Proc. Safety Prog., **33** (2014), pp. 314–328

Chakrabarti, 1995 A. Chakrabarti & Conference

International symposium on runaway reactions and pressure relief design, Boston, MA (United States), 2-4 Aug 1995. (1995). *Experimental approaches to testing in the VSP for different types of reactions for emergency relief design by simplified equations*. United States.

CCPS, 1995

Guidelines for Chemical Reactivity Evaluation and Application to Process Design, Center for Chemical Process Safety. New York, NY: American Institution of Chemical Engineers (1995).

CCPS, 2008

Guidelines for Hazard Evaluation Procedures, 3rd ed., Center for Chemical Process Safety, New Jersey: John Wiley & Sons, Inc. (2008). The public version of Dow's Reactive Chemicals Program is provided on CDROM. See also sections 3 – Hazard Identification Methods (utilizing existing material properties and developing compatibility charts).

CCPS, 2010

Guidelines for Safe Handling of Powders and Bulk Solids, Center for Chemical Process Safety, New Jersey: John Wiley & Sons (2010). See section 4.3.3 – Laboratory Test Methods for Chemical Incompatibility Hazards.

CCPS, 2016 D. Gorman, J. Farr, A. Hielscher, T. Nguyen, G. Baran, B. Drake, E. Ford, D. Frurip, K. Mulligan, J. Ryan, D. Sliva, D. Viveros

Chemical Reactivity Worksheet Version 4 (CRW4) was published by The Center for Chemical Process Safety on 3-March-2016 at http://www.aiche.org/ccps/resources/chemical-reactivity-worksheet-40

CEFIC, 2012
Bulk Storage Guideline of Hydrogen Peroxide, CEFIC, March 2012

Chervin and Bodman, 2003 S. Chervin and G.T. Bodman
Method for Estimating Decomposition Characteristics of Energetic Chemicals
Proc. Saf. Prog., **22** (2003), pp. 241–243. doi: 10.1002/prs.680220412.

Domalski et al., 1993 E.S. Domalski and E.D. Hearing
Liquid Benson groups and their corrections are an implementation of this publication.
J. Phys. Chem. Ref. Data, **22(4)** (1993), pp. 816-829

Duh et al., 1996 Yih-Shing Duh, Chang-Chia Hsu, Chen-Shan Kao, Shuh Woei Yu
Applications of reaction calorimetry in reaction kinetics and thermal hazard evaluation
Thermochimica Acta, **285(1)** (1996) pp. 67-79
FAI, 2002 FAI
VSP2 Manual and Methodology
Fauske & Associates, Inc., Burr Ridge, Illinois, USA (2002)

Fisher et al., 1995 H.G. Fisher
(1992). *Emergency relief system design using DIERS technology: The Design Institute for Emergency Relief Systems (DIERS) project manual*. New York, NY: Design Inst. for Emergency Relief Systems.

Frurip et al., 1995 D.J. Frurip, A. Chakrabarti, J.R. Dowey, H.D. Ferguson, S.K. Gupta, T.C. Hofelich, M.S. LaBarge, A.J. Pasztor Jr., L.M. Peerey, S.M. Eveland, R.A. Suckow (1995).
Determination of Chemical Process Heats by Experiment and Prediction. United States: American Institute of Chemical Engineers, New York, NY United States).

Frurip et al, 1997 D.J. Frurip, T.C. Hofelich, D. Leggett, J. Kurland, J. Niemeir
A Review of Chemical Compatibility Issues
Proceedings AIChE Prevention Symposium (1997)

Frurip, 2004 D.J. Frurip
The role of ASTEM E27 methods in hazard assessment: Part I
Process Saf. Prog., **23** (2004), p. 266

Frurip, 2008 D.J. Frurip
Selection of the Proper Calorimetric Test Strategy in Reactive Chemicals Hazard Evaluation
Organic Process Research & Development, **12(6)** (2008) pp. 1287-1292.

Frurip et al., 2009 D.J. Frurip, K. Bakkum, H. Johnstone, A. Tesolin-Gee, A.; B. Vallieu, K. van Gelder
Reactive Chemicals Emergency Response and Post-event Calorimetric Testing
Proc. Saf. Prog., **29(1)** (2009), pp. 2-10

Frurip, 2012 D.J. Frurip
The Essential Elements of a Successful Reactive Chemicals Program,
Chemical Engineering Transactions, **26**, (2012), pp. 1-6. DOI: 10.3303/CET1226001

Frurip and Elwell, 2007 D.J. Frurip and T.I. Elwell
Effective use of Differential Scanning Calorimetry in Reactive Chemicals Hazard Evaluation, *Proc. Saf. Prog*, **26(1),** (2007) pp. 51.

Gorman et al., 2014 D. Gorman, J. Farr, R. Bellair, W. Freeman, D. Frurip, A. Hielscher, H. Johnstone, M. Linke, P. Murphy, M. Sheng, K. van Gelder, D. Viveros
Enhanced NOAA Chemical Reactivity Worksheet for Determining Chemical Compatibility
Proc. Saf. Prog., **33(1)** (2014) pp. 4-18.

Grossel, 2008 S.S. Grossel
Guidelines for Hazard Evaluation Procedures, 3rd ed., AIChE Center for Chemical Process Safety
Hoboken, NJ: John Wiley & Sons, Inc. (2008)

Harrison, 2009 B.K. Harrison
CHETAH 9.0 Users Manual,
The ASTM Computer Program for Chemical Thermodynamics and Energy Release Evaluation.
West Conshohocken, PA: ASTM

Hofelich et al., 1994 T.C. Hofelich, J.B. Powers, D.J. Frurip
The Determination of Compatibility via Thermal Analysis and Mathematical Modeling
Proc. Saf. Prog., **13(4)** (1994), pp. 227-233

Hofelich et al., 2001 T.C. Hofelich, M.S. LaBarge, D.A. Drott
Prevention of thermal runaways in carbon beds
J. Loss Prev. in the Proc. Ind., **12(6)** (1999), pp. 517-523

Hofelich and LaBarge, 2002 T.C. Hofelich and M.S. LaBarge
On the Use and Misuse of Detected Onsets Temperature of Calorimetric Experiments for Reactive Chemicals
J. Loss. Prev. Proc. Ind., **15(3)** (2002) pp 163-168.

Johnson et al., 2003 R. W. Johnson, S. W. Rudy and S. D. Unwin,
Essential Practices for Managing Chemical Reactivity Hazards. New York, NY: American Institute of Chemical Engineers (2003). See section 4.4 and table 4.8.

Keller et al., 1997 Keller, A.; Stark, D.; Fierz, H.; Heinzle, E.; Hungerbuhler, K.
Estimation of the Time To Maximum Rate Using Dynamic DSC Experiments
J. Loss Prev. Proc. Ind., **10(1)** (1997), pp. 3141.

Kossoy and Koludarova, 1995 A. Kossoy, E. Koludarova
Specific Features of Kinetics Evaluation in Calorimetric Studies of Runaway Reactions
J. Loss Prev. Proc. Ind., **8(4)** (1995), pp. 229–235

Kossoy and Benin, 1997. A. Kossoy, A. Benin
Developments in the field of methodology and software for thermal hazards assessment,
Proceedings of 21st Meeting of DIERS Users Group of the AIChE, Pittsburgh, Pennsylvania, USA (1997).

Kossoy et al., 2001 A. Kossoy, A. Benin, Yu Akhmetshin,
From Experimental Data via Kinetic Model to Predicting Reactivity and Assessing Reaction Hazards (methodology and software), Proceedings of Mary Kay O'Connor Process Safety Center 2001 Annual Symposium, *Beyond Regulatory Compliance, Making Safety Second Nature*, College Station, Texas, USA, October 2001, pp. 321–335.

Kossoy and Benin, 2003 A. Kossoy, A. Benin
Further Development of the CISP Thermal Safety Software: New Features
Proceedings of the Annual Meeting of European DIERS User's Group (EDUG), April 7-8, Sent Etienne (2003)

Kossoy and Hofelich, 2003 A. Kossoy, T. Hofelich
Methodology and software for reactivity rating
Process Safety Prog., **22 (4)** (2003), pp. 235-240

Kossoy and Sheinman, 2004 A.A. Kossoy, I.Y. Sheinman
Evaluating Thermal Explosion Hazard by Using Kinetics-Based Simulation Approach
Process Safety and Environmental Protection, **82(6)** (2004), pp. 421-430

Kossoy and Akhmetshin, 2007 A. Kossoy, Yu Akhmetshin
Identification of Kinetic Models for the Assessment of Reaction Hazards
Proc. Saf. Prog., **26 (3)** (2007), pp. 209–220

Leggett, 2001 David J Leggett
Safe Process Development from Reaction Hazards Testing
Thermochimica Acta, **367-368**, (2001), pp. 351-365

Mannan et al., 2001 M.S. Mannan, W.J. Rogers, A. Aldeeb
A Systematic Approach to Reactive Chemicals Analysis In Symposium Series No. 148, *Hazards XVI: Analysing the Past, Planning the Future, Symposium Proceeding* (pp. 41-58). Melbourne, Australia: Institution of Chemical Engineers (2001).

McIntosh and Nolan, 2001 R.D. McIntosh, P.F. Nolan
Review of the selection and design of mitigation systems for runaway chemical reactions
J. Loss Prev. Proc. Ind. **14** (2001), pp. 27– 42

Melhem et al., 1995 G.A. Melhem, H.G. Fisher, D.A. Shaw
An Advanced Method for the Estimation of Reaction Kinetics, Scaleup, and Pressure Relief Design
Proc. Saf. Prog., **14(1)** (1995), pp. 1-21

Oxley et al., 1999 J. C. Oxley, J.L. Smith, E. Rogers, M.A. Hamad, E. Resende
Small Scale Explosivity Testing
Journal of Energetic Materials, **17(4)** (1999), pp. 331-343.

Oxley et al., 2010 J.C. Oxley, J.L. Smith, K. Marimaganti
Developing Small-Scale Tests to Predict Explosivity
J Therm. Analysis and Calorimetry, **102(2)** *(*2010) pp. 597-603.

Singh and Simms, 2001 J. Singh, C. Simms
The Thermal Screening Unit (Ts^u) - A Tool for Reactive Chemical Screening. In Symposium Series No. 148, *Hazards XVI: Analysing the Past, Planning the Future, Symposium Proceeding* (pp. 67-79). Melbourne, Australia: Institution of Chemical Engineers.

Stack, 2010 Robert J. Stack
High Integrity Protection System Design Using a Risk Base Approach
2010 Spring Meeting and 6th Global Congress on Process Safety Conference Proceedings, San Antonio, TX: AIChE (2010)

Tipler et al., 2013 S. Tipler, D. Eure, K. First, R.J. Stack, T.J. Wagner, J.W. Champion
Practical Examples of Alternate Overpressure Protection Systems: An Owner's Perspective
2013 Spring Meeting and 9th Global Congress on Process Safety Conference Proceedings, San Antonio, TX: AIChE (2013)

Townsend and Tou, 1980 D.I. Townsend, J.C. Tou
Thermal Hazard Evaluation by an Accelerating Rate Calorimeter
Thermochim. Acta, **37** (1980), pp. 1–30.

Wagman et al., 1982 D.D. Wagman, W.H. Evans, V.B. Parker, R.H. Schumm, I. Halow, S.M. Bailey, K.L. Churney, and R. L. Nuttall
The NBS Tables of Chemical and Thermodynamic Properties
Journal o f Physical Chemical Reference Data, Supplement No. 2, **11** (1982)

Whiting, 1988 L.F. Whiting
Evaluation of a Capillary Tube Sample Container for Differential Scanning Calorimetry
Thermochim. Acta, **136** (1988), p. 231

Whiting, 1997 L. Whiting, M. LaBarge, J. Steinmetzer
Global Technical Information Management in thermal Hazard Evaluation
J. Thermal Analysis, **49** (1997), pp.1617-1624

Yoshida et al., 1987 T. Yoshida, F. Yoshizawa, M. Itoh, T. Matsunaga, M. Watanabe, M. Tamura,
Prediction of Fire and Explosion Hazards of Reactive Chemicals (I). Estimation of Explosive Properties of Self-Reactive Chemicals from SC-DSC Data
Kogyo Kayaku, **48(5)** (1987), pp. 311-316.

APPENDIX A: HIGHLY ENERGETIC MATERIALS BY FUNCTIONAL GROUP

The following list of functional groups may be useful to determine if a chemical has high potential for reactivity, explosivity, shock sensitivity and/or thermal instability hazards.

Reactive Substance Class	Functional Groups
Acetylenic compounds, conjugated	-C≡C-C≡C- -C≡C-C=C-
Acetylides, metal derivatives of (e.g., Copper acetylide, Mercury acetylide, Silver acetylide, Aluminum acetylide)	-C≡C-Metal (Metal=Cu, Hg, Ag, Al)
Acetylene, halogenated derivatives of	-C≡C-X (X=F, Cl, Br, I)
Aminachromium peroxocomplexes	$H_3N \rightarrow Cr\text{-}OO\text{-}$
Aminemetal oxosalts	$[N \rightarrow Metal]^+ Z^-$
Arenediazoates	Ar-N=N-O-R
Arenediazo aryl sulfides	Ar-N=N-S-Ar
Arenediazoniumolates	$C\text{-}N_2^+\text{-}O^-$
Azides, metal Azides, organic (acyl, halogen, nonmetal)	$M\text{-}N_3$ $R\text{-}N_3$
Aziridine	
Azo compounds	-C-N=N-C
N-Azolium nitroimidates	$=N^+\text{-}N\text{-}NO_2$
Bis(arenediazo) oxides	Ar-N=N-O-N=N-Ar
Bis(arenediazo) sulfides	Ar-N=N-S-N=N-Ar
Constrained Rings (≤ 4 atoms) - C_1 - C_4 epoxides (i.e. ethylene oxide, propylene oxide, butylene oxide) - Azetidine - Methylcyclopropene	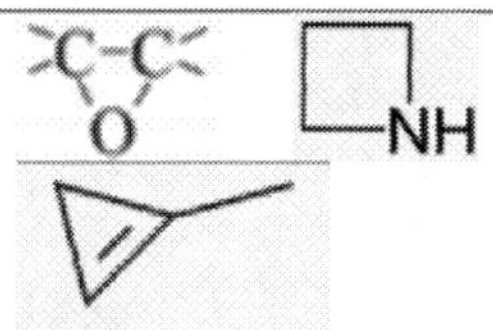
Diazeno compounds	-C-N=NH
Diazirines	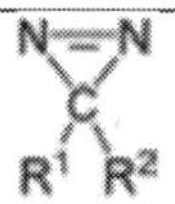

Diazonium Salts, Diazonium Carboxylates	$-C-N_2^+Z^-$ (Z = F, Cl, Br, I, CO_3)
Diazonium sulfides and derivatives	$-C-N_2^+S^-$
Difluoroamino compounds	$-N-F_2$
Fluorinated Aryl-LI, Mg, and Al compounds	F_n-Ar-Li(MgX) $(CF_3)_n$Ar-Li(MgX) for n = 1-4
Fulminates (metal compounds) or aci-nitro salts, oximates(e.g., Mercury (II) fulminate, Potassium fulminate, Silver fulminate)	Metal-C≡N→O >C=N-O-Metal
N-halogen compounds (excluding: NBS, N-Bromosuccinimide) and NCS (N-Chlorosuccinimide)	>N-X
High-nitrogen compounds	-N=N-N=N-
Hydrazides, Metal	M-N=N-R
Hydrazinium salts	N_2H_4 * Z (e.g., Z = HNO_3)
Hydroxylamine	NH_2OH
Hydroxyl ammonium salts	$-N^+-OH\ Z^-$
Nitrides	M_3N
Nitrogen Trihalides	NX_3 (X = F, Cl, Br, I)
N-Nitro compounds	$>N-NO_2$
N-Nitroso compounds	>N-N=O
Oxosalts of nitrogenous bases	$NH_4^+EO_n$ E = N, I, Cl
Oxygen-Halogen Groups - Chlorite salts (ClO_2^-) - Halogen oxides (e.g., Cl_2O – dichlorine monoxide) - Hypohalites (except bleach) (e.g., ClO^-) - Perchlorate compounds (e.g., ClO_4^-, $NH_4^+ClO_4^-$, alkyl perchlorates, etc.) - Perchloryl compounds (e.g., ClO_3^-)	-O-X
Ozonide	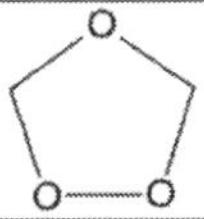
Peroxides - cyclic, diacyl, dialkyl, etc) - metal peroxides - Hydrogen Peroxide ("Yes" if ≥30%) - Alkylhydroperoxides	R-O-O-R (R = H, C, Metal)

Peroxoacids, Peroxo salts, Superoxides	R-C(=O)-O-OH Metal-OO^- MO_2
Picric Acid, Metal Picrates	$C_6H_2(OH)(NO_2)_3$ Metal-$OC_6H_2(NO_2)_3$
Poly Nitro alky compounds, Poly Nitro aryl compounds	R-$(NO_2)_n$ for $n \geq 2$ Ar-$(NO_2)_n$ for $n \geq 2$
Poly(dimercuryimmonium salts)	-N=Hg^+=N-
Tetrazoles	N-N=N-N=C-
Triazenes	>C-N=N-N(R)-C< (R=H, CN, OH, NO)
N,N,N-trifluoroalkylamidines	-C(NF)NF_2
STRONG Oxidizers or Reducing Agents (e.g., conc. nitric acid, some peroxides, diazo compounds, nitro compounds, nitrites, nitrates, perchlorates, bromates...)	**R9, H271 – in contact with combustibles may cause explosion**

Abbreviations: **Ar** = aromatic (benzene); **M** = metal; **R** = organic chain; **X** = halogen; **E** = nonmetal; **Z** = anion; **n** = integer variable; all other abbreviations are for the element symbols from the periodic table of elements

Inherently Safer Quenching of HCN Polymerization Emergencies

David E. Herrick
Universidad de los Andes
Carrera 1E No 19A-40, Bogotá, Colombia
de.herrick@uniandes.edu.co

Felipe Muñoz Giraldo & Nicolas Rios Ratkovich
Universidad de los Andes
Carrera 1E No 19A-40, Bogotá, Colombia
fmunoz@uniandes.edu.co
n.rios262@uniandes.edu.co

S. Dharmavaram
DuPont Safety, Health, & Environment and Operating Excellence CoE
974 Centre Road
Wilmington, DE 19805
seshu.dharmavaram@dupont.com

Prepared for Presentation at
American Institute of Chemical Engineers
2016 Spring Meeting
12th Global Congress on Process Safety
Houston, Texas
April 11-23, 2016

Inherently Safer Quenching of HCN Polymerization Emergencies

David E. Herrick
Universidad de los Andes
Carrera 1E No 19A-40, Bogotá, Colombia
de.herrick@uniandes.edu.co

Felipe Muñoz Giraldo & Nicolas Rios Ratkovich
Universidad de los Andes
Carrera 1E No 19A-40, Bogotá, Colombia
fmunoz@uniandes.edu.co
n.rios262@uniandes.edu.co

S. Dharmavaram
DuPont Safety, Health, & Environment and Operating Excellence CoE
974 Centre Road
Wilmington, DE 19805
seshu.dharmavaram@dupont.com

Keywords: Hydrogen Cyanide, Polymerization, Hydrolysis, Mixing, Phosphoric Acid, Sulfuric Acid

Abstract

Storage of hydrogen cyanide (HCN) is common in the industry. One of its hazards is runaway polymerization resulting in explosion of the storage tank. A key layer of protection against such an event is addition of acid to control or "quench" the polymerization reaction and prevent explosion. Although mixing of the acid with HCN is critical to this safety layer, very little information is available on this subject. This paper provides some rudimentary data to guide HCN industry members to evaluate their existing quenching systems and design new systems.

Introduction

Hydrogen cyanide (HCN) is produced in large quantities (3 billion lbs in 2005) [1] for a variety of purposes, including production of methyl methacrylate monomer for the production of acrylic plastics, sodium or potassium cyanide as a lixiviant for gold mining or electroplating metals, or the production of organic chemicals such as methionine. Most HCN is produced either by reacting ammonia with methane (in the Andrussow process - which also uses air- or in the BMA process) or as a byproduct of Acrylonitrile production [2].

The toxicity hazards of HCN are well-known, but its reactivity hazards also produce substantial risk for those storing HCN in tanks. HCN is capable of a variety of reactions in storage, the most important of which is polymerization. The reaction is both auto-catalytic and exothermic, and if not stopped it can result in thermal runaway. The mechanism involves reaction between HCN molecules and CN^- ions created from dissociation of this weak acid. The reaction produces polymer and releases ammonia (NH_3). Both the polymer and the ammonia act as catalysts by creating more CN^- ions [3].

Most HCN product is "inhibited" or "stabilized" by the addition of small amounts of acid or sulfur dioxide. Sulfur dioxide inhibits the polymerization reaction by forming a complex with the CN^- ion and preventing it from reacting. Acids inhibit the polymerization reaction by neutralizing any bases (including the ammonia and polymer produced by the reaction). This reduces the concentration of CN^- ion by shifting the acid dissociation equilibrium toward the HCN molecule [4].

Uncontrolled polymerization of an HCN storage tank can be a serious process safety issue. In 1962 a storage tank containing approximately 5000 kg of HCN exploded at the ICI Cassel Works in Billingham, UK [5]. The explosion was equivalent to approximately 5 kg of TNT and generated pressures in excess of 50 bar. The event caused extensive damage to the facility (Figure 1). The industry currently has storage tanks with capacity as high as 150,000 kg, which is an order of magnitude more than the tank in the 1962 incident. The explosion of such a tank could cause fatalities from the explosion and toxic release, major equipment damage, and even threaten the industry's right to operate.

An HCN storage tank can undergo a polymerization event at any time. In order to prevent the polymerization event from starting, inhibitor concentration is maintained at a minimum level (approximately 0.1% wt), the tank is kept refrigerated (approximately 5-10°C), the tank is monitored to assure that is has not been contaminated, and the tank material is circulated to prevent regions of high temperature or low inhibitor concentration. Tank temperature and pressure are monitored to ensure that a polymerization reaction has not initiated. Product color, which usually resembles water with a slight blue tint, will turn yellow, brown, or completely dark when the polymerization is advanced [6].

Figure 1: Explosion of HCN Storage Tank at ICI Cassel Works, 1962 [5].

When polymerization starts to become a problem in a storage tank, the heat generation rate exceeds the capacity of the refrigeration system. Since most HCN tanks are operated near atmospheric pressure, the temperature will rise to the HCN boiling point of 26°C and the heat of reaction will be absorbed by vaporization of HCN. This is called the first exotherm. The reaction will continue at constant temperature, vaporizing HCN product that will be leave the tank through its normal vent until the vaporization rate exceeds its capacity. Then HCN will start to vent from the tank's emergency venting system (e.g. relief valve or emergency vent) [6].

The tank operator will be quite busy managing the tank temperature and pressure and the impact of the HCN leaving containment. But the relative stability of operating conditions during this period can give a false sense of security that the situation is under control. The heat sink provided by vaporization of the liquid HCN in the tank can only last until it has all been either consumed in the reaction or vaporized. When there is no more HCN to vaporize, the heat of reaction can no longer be absorbed by vaporization and is converted into sensible heat. The temperature rises exponentially and the reaction culminates in an explosion. This is called the second exotherm [6].

While the first exotherm is in progress, the tank operator has an opportunity to add inhibitor to try to prevent the explosion. This is generally referred to as "emergency quenching." The effectiveness of emergency quenching depends on the ability to get sufficient acid to the location where the polymerization is occurring within the tank. Thus, the ability to mix acid with HCN is of primary importance.

There are several obstacles to good mixing in an HCN emergency quenching system. The first is that the tank can be large compared with the amount of acid added (e.g. 20 liters of acid in 200,000 liters of HCN). Another issue is that the specific gravity of the acid is usually much larger than the HCN (e.g. 1.8 vs 0.7). The viscosity of strong acid is another barrier to mixing. For these reasons, mixing is a critical factor in the design of a quenching system. However, there are very little data currently available for the design of an HCN polymerization quenching system.

Industry practice for emergency quenching is quite varied in terms of the inhibitor used, the equipment used to drive the flow, and the addition point. A survey emergency quenching practices at HCN-handling sites performed by the HCN Industry Conference in 2012 showed that ten sites used SO_2, five used sulfuric acid, three used acetic acid, and two used phosphoric acid [7]. Note that some sites have systems for adding both SO_2 and an acid, so the total number of sites is less than 20.

A similar survey presented at the 2007 HCN Industry Safety Conference [8] showed variety in the methods of adding inhibitor. Of the six sites that added acid for emergency quenching, four injected into a "pumped loop" and two injected via dip pipe. Of these same sites, three used gravity to drive flow and three sites used nitrogen pressure. Nitrogen pressure is sometimes favored because, in addition to providing higher driving force for flow, when the acid flow stops the nitrogen enters behind it to create turbulence which helps mix the acid in the HCN.

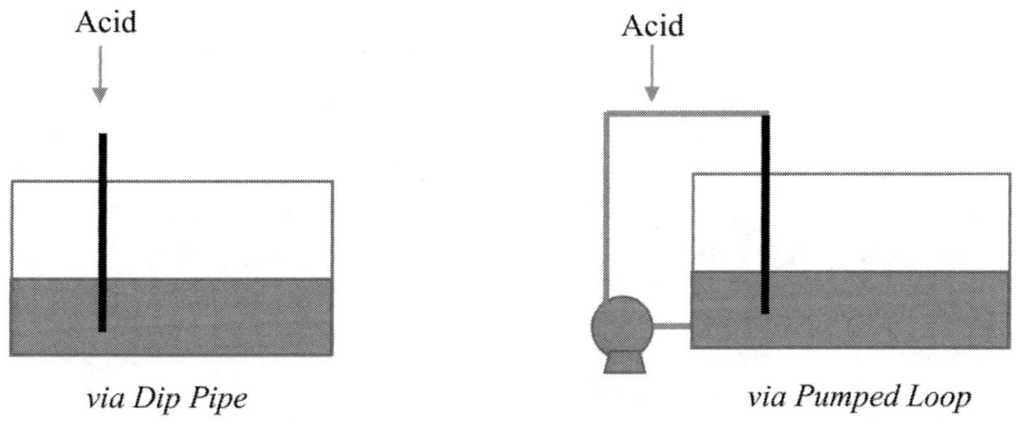

Figure 2: Acid injection locations for Emergency Quenching

Gravity addition of acid using a dip pipe is sometimes preferred because it is very reliable during an emergency since there are no moving parts to fail. But this method is unlikely to provide good mixing. The acid will tend to collect at the bottom of the vessel and mix very slowly with the HCN. The velocity of the circulating liquid inside the tank is usually very low (perhaps 5 cm/min) so there will be no turbulence to assist in mixing.

Adding acid to the circulation line of the tank allows for much more effective mixing of the acid in the HCN and delivery to the points where it is needed. Most HCN tanks have a circulation line in order to move the liquid through an external heat exchanger and provide cooling. A simple static mixer can ensure that the entire circulating flow contained a uniform concentration of diluted acid.

Hydrolysis Reaction

Sulfuric acid is a very popular inhibitor used for emergency quenching. However, over-addition (e.g. 1:1 mole ratio of acid to HCN) can will produce a hydrolysis reaction which could also result in an explosion. This reaction was described in early literature on HCN [9][10].

The large amount of acid needed to produce explosive hydrolysis may seem like an unrealistic scenario, but the HCN industry is keenly aware of its potential. An HCN tank exploded at an American Cyanamid plant in Fortier, Louisiana in June 1963 [11]. Concentrated sulfuric acid was used as an inhibitor added through a rotameter. A bypass allowed a small flow to continue into the tank unnoticed. Eventually the concentration rose to dangerous levels and the tank exploded, causing extensive damage within a 9 meter radius. Debris from the explosion pierced a nearby Stripper Feed Tank and caused a deflagration which blew the tank off its foundation and moved it 50 meters away. For this reason, HCN tank operators are hesitant to provide any direct connection with sulfuric acid storage. This can be problematic as the acid then must come from a smaller reservoir and have its own motive force for driving it into the HCN storage tank.

Considering the hydrolysis hazard of sulfuric acid, phosphoric must be considered inherently safer since their use cannot result in explosive hydrolysis at any ratio with HCN [12]. Although there is a possibility of some hydrolysis of HCN when strong phosphoric acid is added, its impact would be limited to product quality. Product quality of HCN in a tank in danger of explosive polymerization is a very minor concern compared with the assurance that over-addition of acid will not cause an explosive hydrolysis reaction.

The five sites who used sulfuric acid for emergency quenching in 2012 likely continue to do so today. For them to substitute an inherently safer acid such as phosphoric, they require data to justify that the change will be both effective and safe. In the 2014 HCN Safety Conference, one of the objections to using phosphoric acid was that it did not mix as well as sulfuric. Providing an answer to this comment was incentive for starting this work.

Methods

Mixing of HCN with sulfuric acid and phosphoric acid was studied by both computer modeling and experimentation.

Modeling was carried out using the StarCCM+ software. The basis was a horizontal cylindrical storage tank with the following parameters.

Table 1: Modeling input parameters

Storage Tank size	14 ft diam. x 60 ft length
Mass of HCN contained	350,000 lb
Circulation rate	100 gpm
Circulation inlet location	2m from end of tank
Acid addition location	3m from end of tank
Circulation return location	20m from end of tank
Amount of acid added	1 gal
Acid addition rate	1 gal/minute
Acid composition	93% H_2SO_4… or… 85% H_3PO_4
Tank composition	>99% HCN at 5°C

Experimental work was conducted using acetonitrile as a substitute for HCN. Acids (93% H2SO4 and 85% H3PO4) were used with dye added in order to visually show mixing. Acetonitrile was added to a round-bottom flask. Then acid was added near the bottom of the flask to simulate addition through a dip pipe. A video recording was made for approximately 30 seconds to show the degree of mixing. Then acid was added dropwise from a point well above the liquid level. A video was made to show the degree of mixing of the plunging drops.

Table 2: Experimental Conditions

Acid composition	98% H_2SO_4… or… 85% H_3PO_4
"HCN" composition	Acetonitrile (reagent grade)
Dye added to acid	Food coloring (H_3PO_4) Dye (H_2SO_4)
Temperature	Ambient (approx. 15°C)
Elevation	2600 m above sea level

Results and Discussion

Modeling shows that injecting acid through a dip tube into a circulating tank is not very effective to even establish a minimum acid concentration as an inhibitor (Figure 3). After 10 minutes there are still very large parts of the tank that have not been affected by the emergency quenching acid. In the actual polymerization emergency the situation is even more serious. The model only shows the acid concentration in a tank full of pure HCN. In the actual polymerization event NH_3 and polymer are being generated continuously. So even where the acid reaches the polymerization event, it will be quickly consumed by the accumulated basic components.

The choice of acid seems to have little impact on the mixing. Figure 4 shows the concentration profile for phosphoric acid at 10 minutes after injection. Although sulfuric acid appears to be slightly more well-mixed (Figure 3), both acids are qualitatively identical in their inability to reach the majority of the tank contents.

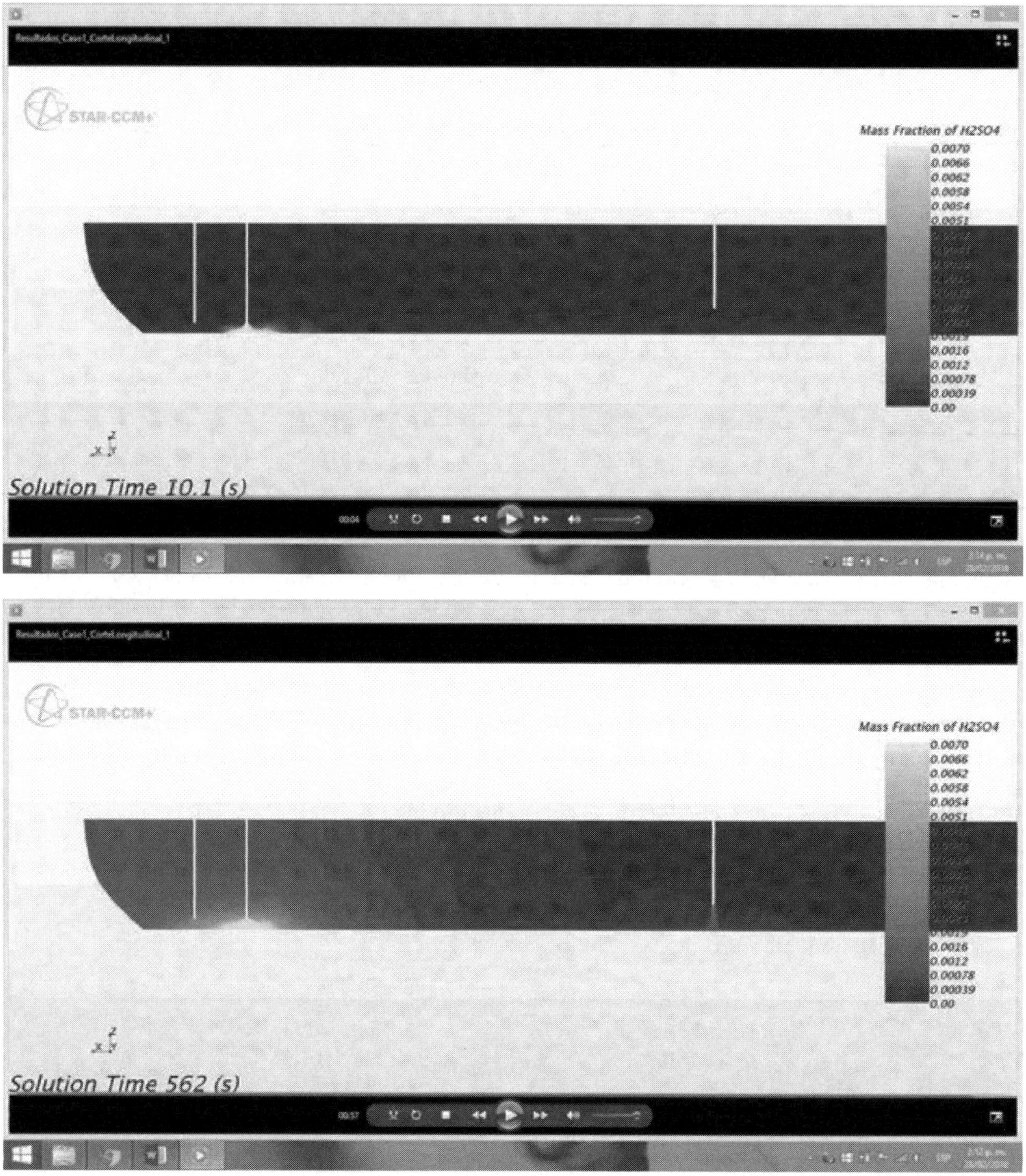

Figure 3: Dip Pipe Addition of 1 gal of H_2SO_4 at 10 s and 562 s after injection

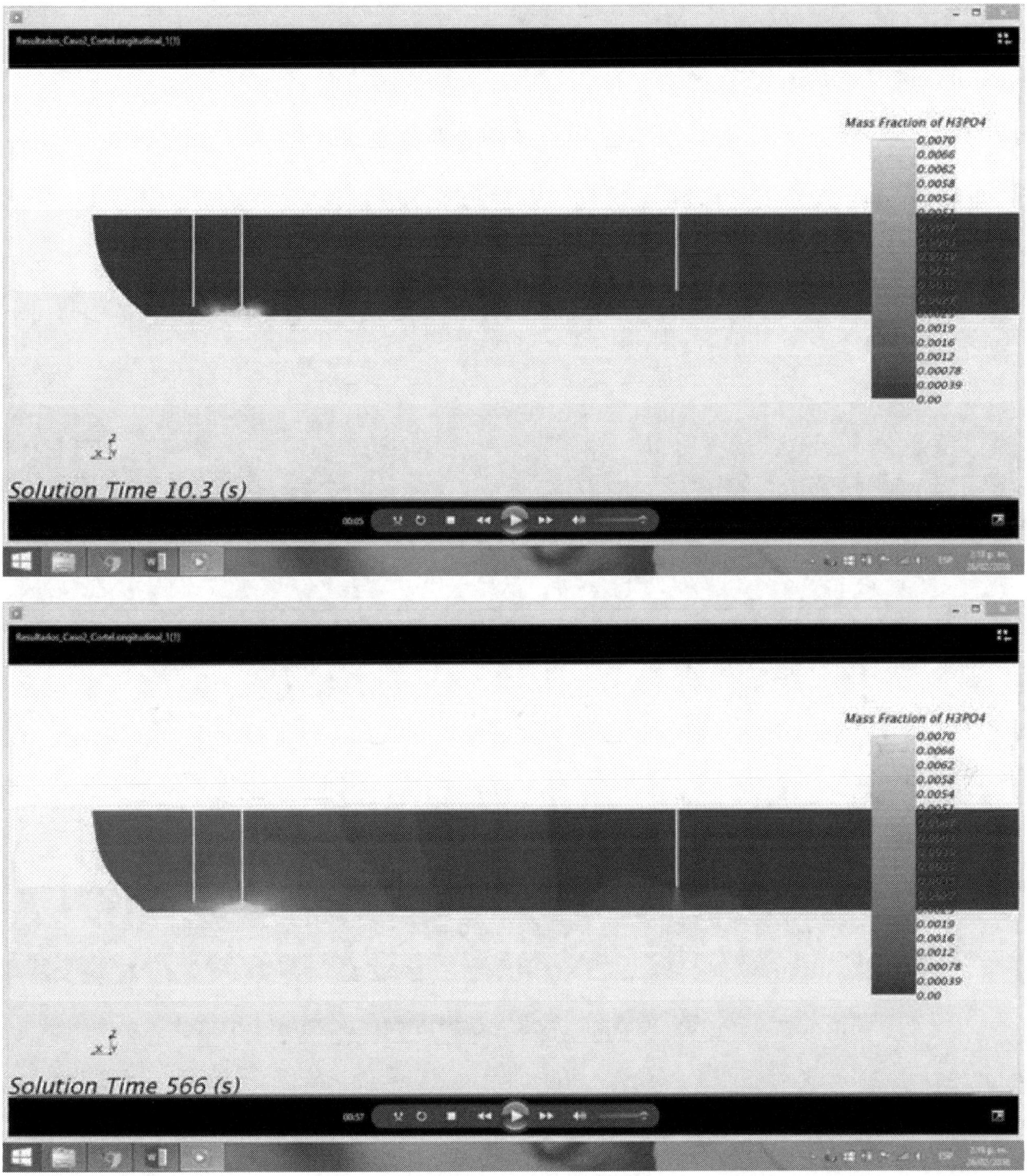

Figure 4: Dip Pipe Addition of 1 gal of H_3PO_4 at 10 s and 566s after injection

Experimental Results

Sulfuric acid behaved as expected when added to the bottom of a flask filled with acetonitrile. The acid fell to the bottom of the flask and remained there for several minutes without any apparent change (Figures 5-8). The size and color of the purple area marked by the dye did not change during the time it was observed (several minutes). The stagnant nature of the flask simulates what would happen if acid were added to an HCN storage tank with no circulation in the vessel. While lack of circulation would be one reason for poor mixing, it is not the only reason – the modeling results above confirm that even with circulation, the majority of the acid remains at the bottom of the tank where it was injected even after 10 minutes.

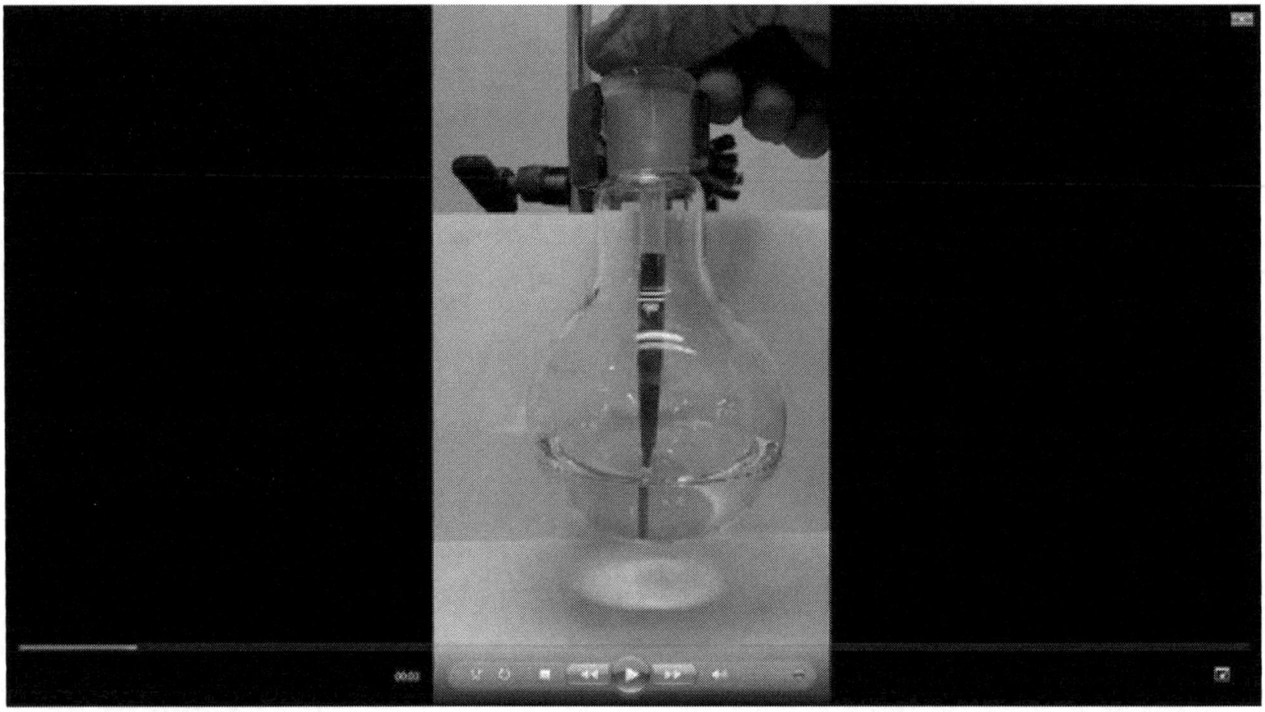

Figure 5: Begin Sulfuric Acid addition to bottom of flask

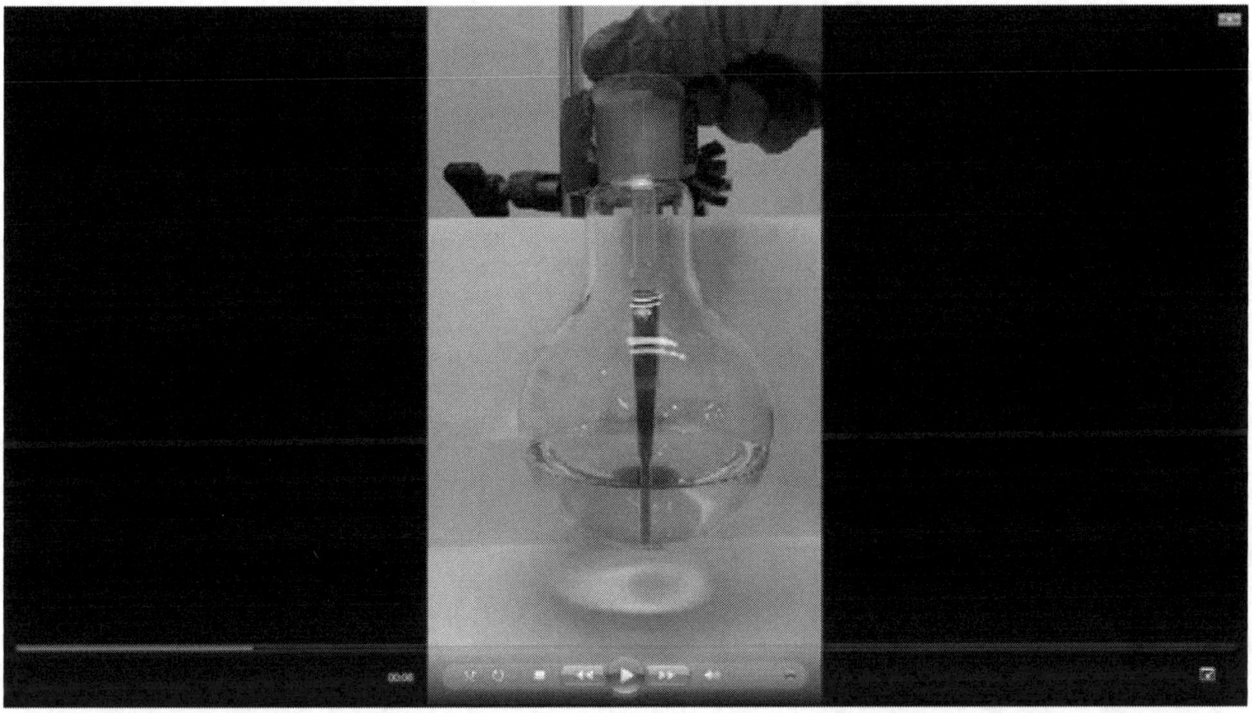

Figure 6: End Sulfuric Acid addition to bottom of flask

Figure 7: Sulfuric Acid – 10 seconds after addition

Figure 8: Sulfuric Acid – 30 seconds after addition

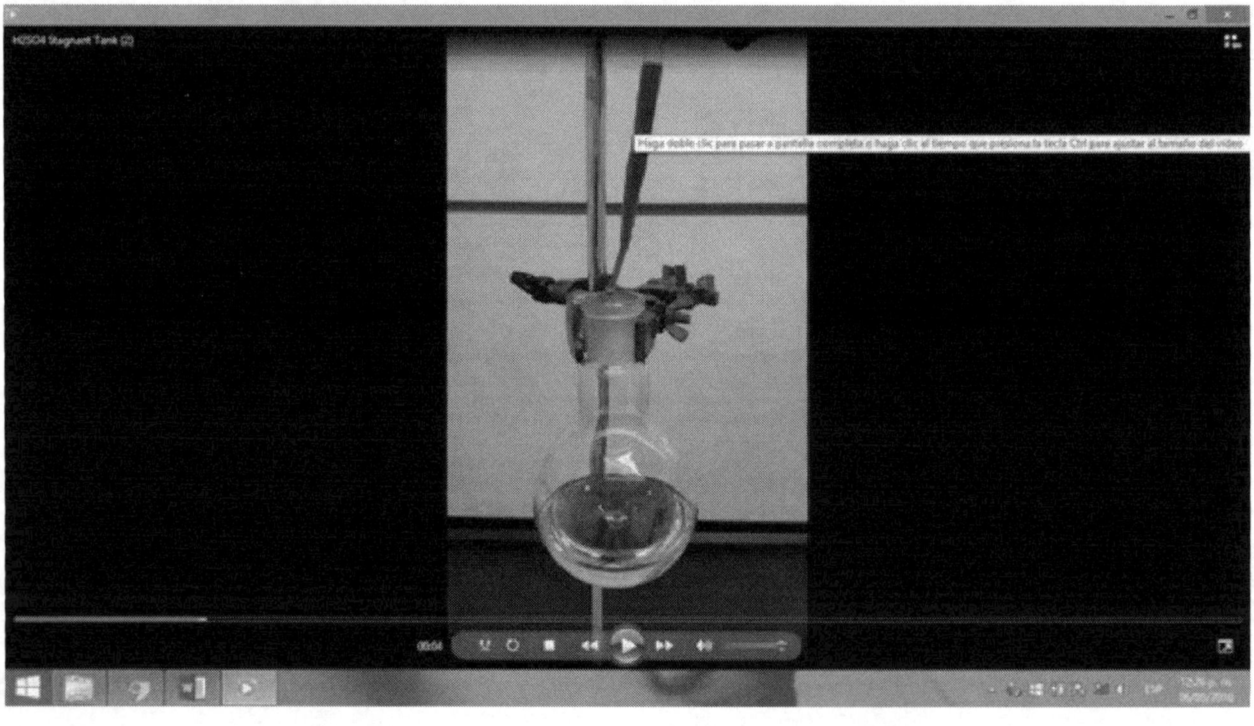

Figure 9: Sulfuric Acid – preparing to add drops from high above liquid level

Drops of acid were added to the flask from a height of approximately 10 cm above the liquid surface. When the drops entered the liquid acetonitrile, the dye disappeared so it was assumed that the acid was mixing well with the acetonitrile. However, closer inspection revealed that the drops held together fairly well even after impact with the surface, then they dropped quickly to the bottom (Figure 10). The process of entry certainly provided a certain amount of mixing which could not be quantified by this experiment. This mixing appeared to allow the dye to be extracted from the acid-rich phase to the acetonitrile-rich phase where it was apparently more soluble. But despite the absence of dye in the droplets, it was apparent from the index of refraction that the drop of acid immediately settled to the bottom of the flask.

While the still photographs capture the effect, the video of the experiment provides much more convincing evidence.

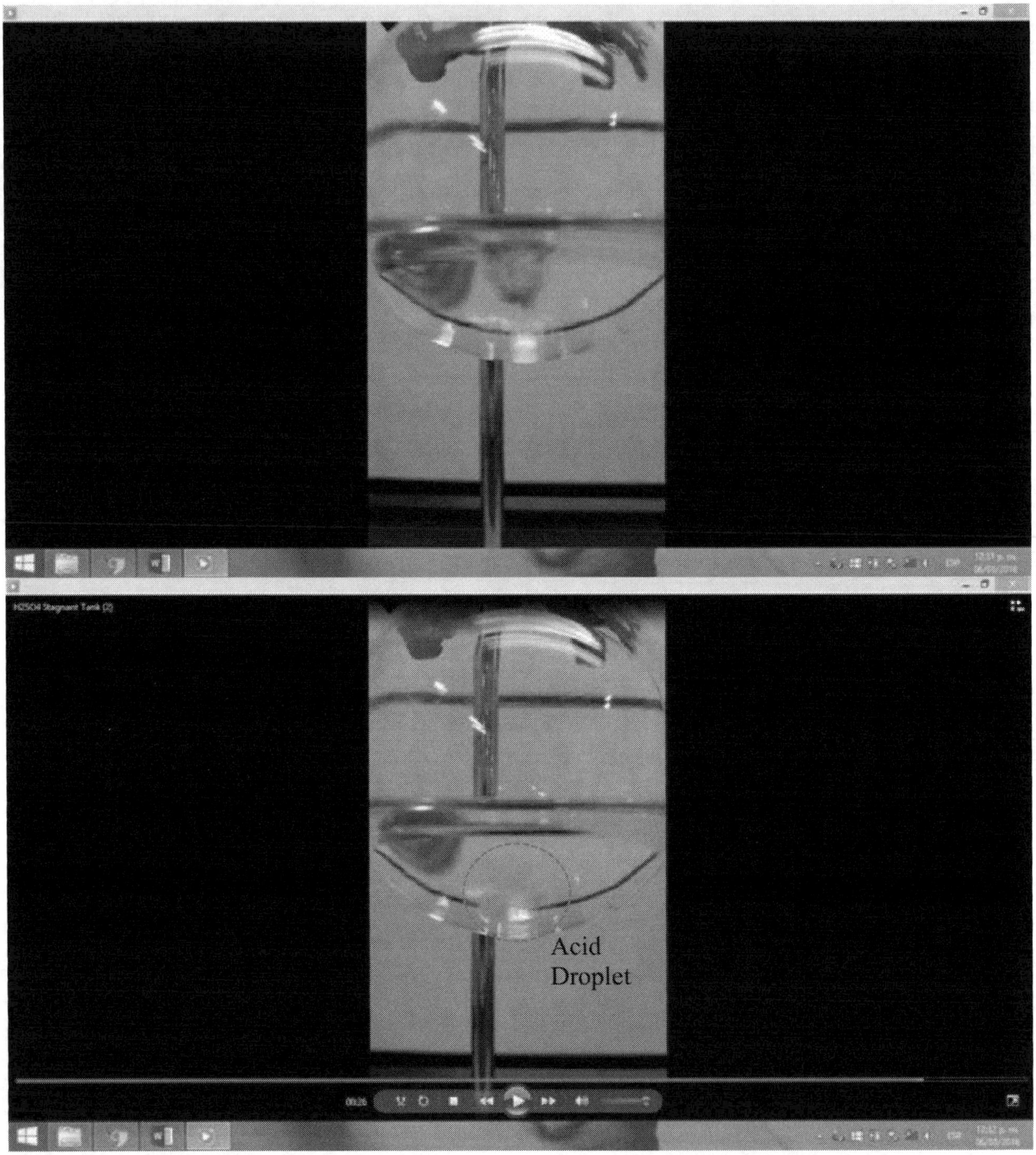

Figure 10: Sulfuric Acid – drop entering and falling to bottom

Phosphoric acid was added to acetonitrile in the same manner, but with a different color dye to prevent confusion in the videos and photographs. It behaved in qualitatively the same manner as the sulfuric acid. The colored area remained unchanged for the duration of the experiment, indicating that there was no significant mixing between the acetonitrile and the acid (Figure 11).

Just as in the case of sulfuric acid, phosphoric acid drops entered the liquid and quickly lost their color from the dye. Because the dye used for sulfuric acid was not as strong, it was slightly more difficult to see the acid drops fall to the bottom. But the photos show this effect (Figure 12) and the video show it even more convincingly.

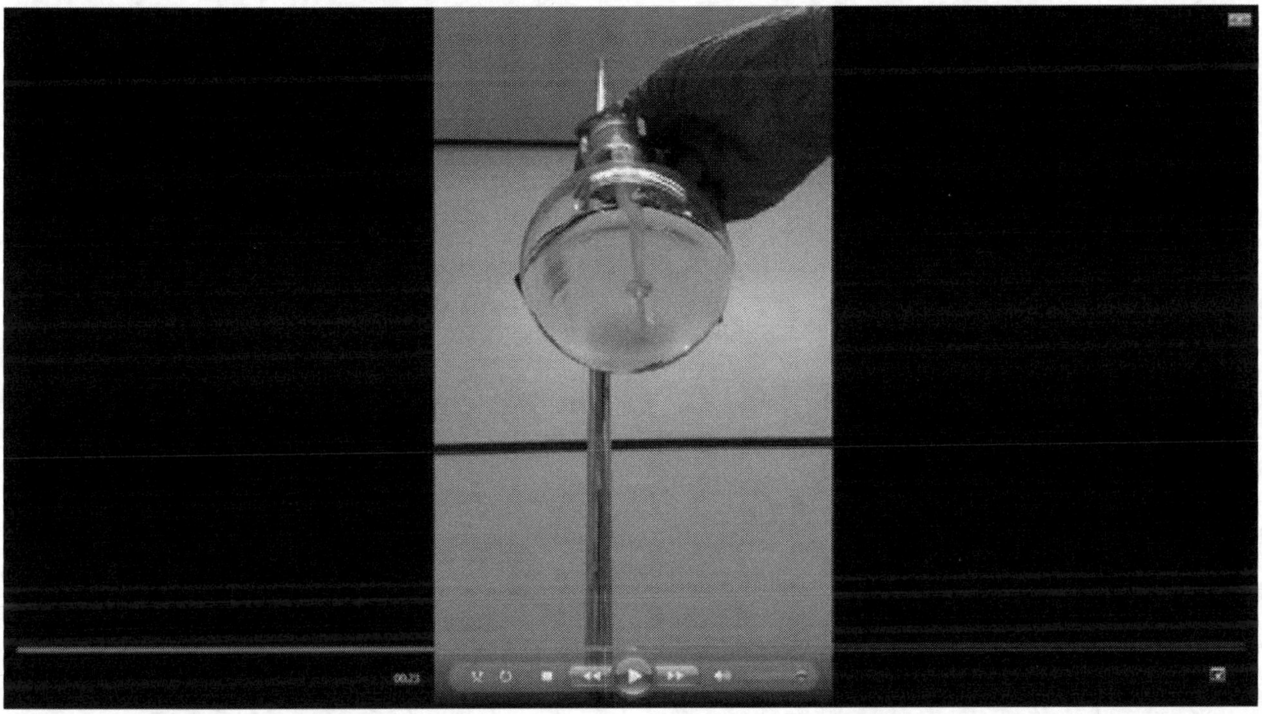

Figure 11: Phosphoric Acid – addition to bottom of flask

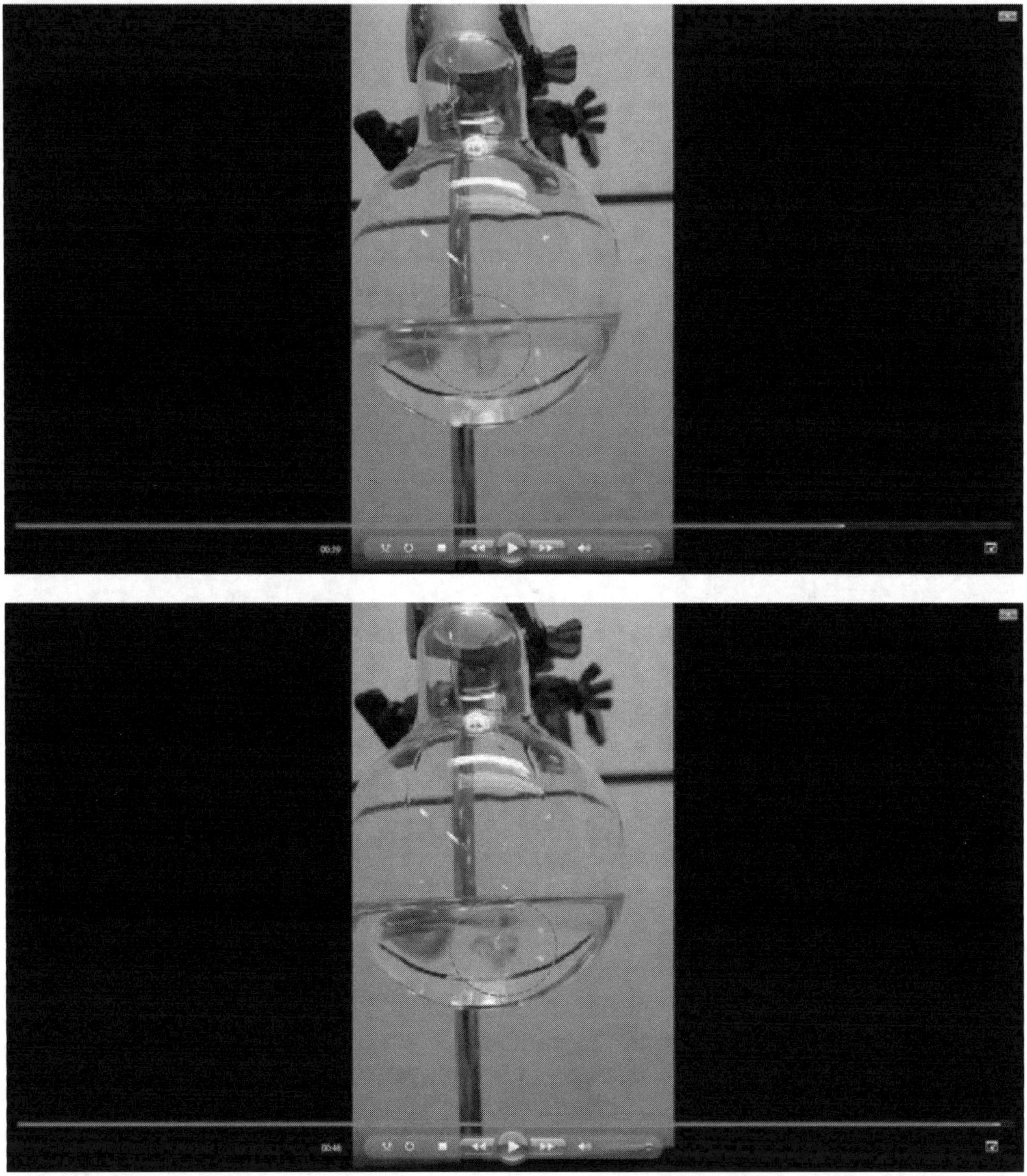

Figure 12: Phosphoric Acid – drops entering and falling to bottom

Conclusions

When strong sulfuric and phosphoric acid are added through a dip pipe to an HCN storage tank for emergency quenching, unless some means of mixing is provided the majority of the acid will remain at the bottom of the tank for a considerable period of time. The use of phosphoric acid does not negatively impact the situation, since the type of acid is not a strong factor in the amount of mixing. Other means of mixing, such as injection into a circulation line, should be strongly considered for emergency quenching systems.

Phosphoric acid should be considered superior to sulfuric acid for emergency quenching for the following reasons:

- Phosphoric acid is inherently safer. Over-addition will not cause explosive hydrolysis like sulfuric acid.
- Because it will not cause explosive hydrolysis, phosphoric acid can be added in large quantities immediately when a polymerization event is detected. The ability to add massive quantities of acid without fear of hydrolysis simplifies the emergency quenching procedure. Operators can add as much acid as the tank will hold without needing to wait for management approval.
- Because it will not cause explosive hydrolysis, phosphoric acid can be directly connected by pipeline from an acid storage tank to an HCN storage tank. This simplifies the addition of acid during an emergency. Sulfuric acid cannot be connected directly to the HCN storage tank, so manually-intensive means of transporting and adding the acid must be used.

References

[1] R. McKinney and A. Clarke, "Introduction to Cyanides," Presentation to the 2011 HCN Industry Safety Conference.

[2] "Hydrogen Cyanide" Wikipedia, accessed 2/28/2016, https://en.wikipedia.org/wiki/Hydrogen_cyanide

[3] A. Clarke and R. McKinney, "Hazardous Reactions Involving HCN," Presentation to the 2011 HCN Industry Safety Conference.

[4] Discussion with Steven Grise (DuPont), 2013.

[5] J.H. Openshaw, "1962 HCN Storage Tank Polymerisation," Presentation to the 2002 HCN Industry Safety Conference, Breckenridge, Colorado, USA

[6] Fauske & Associates, "Thermal Data and Vent Sizing Tests on Hydrogen Cyanide," Presented to American Cyanamid Co. and E.I. duPont de Nemours Co., December 1985.

[7] J. Armitage and D. Penn, "Panel Session –Storage Part 3: Emergency Stabilisation," Presentation to the 2012 HCN Industry Safety Conference

[8] M. Jelpke, "Quench Survey," Presentation to the 2007 HCN Industry Safety Conference

[9] A. Cobb, J. Walton, "The Reaction of Hydrogen Cyanide with Sulfuric and Phosphoric Acids," Journal of Physical Chemistry, Vol 41, Issue 3, pp 351-363, March 1937.

[10] V. Krieble, A. Peiker, "The Hydrolysis of Hydrogen Cyanide by Acids II," Journal of the American Chemical Society, Vol 55, Issue 6, pp 2326-2331, June 1933.

[11] J. Harvey Openshaw. "Hazardous Reactions in the HCN/Acid/Water System," Presentation to the 2003 HCN Industry Safety Conference

[12] R. Banavali, M. Chang, S. Fitzwater, R. Mukkamala. "Thermal Hazards Screening Study of the Reactions between Hydrogen Cyanide and Sulfuric Acid and Investigations of their Chemistry," Ind. Eng. Chem. Res., 2002, 41(2), pp 145-152.

Testing Flameless Vents

Bill Stevenson
CV Technology, Inc.
Jupiter, Florida
bstevenson@cvtechnology.com

Prepared for Presentation at
American Institute of Chemical Engineers
2016 Spring Meeting
12th Global Congress on Process Safety
Houston, Texas
April 11-23, 2016

Testing Flameless Vents

Bill Stevenson
CV Technology, Inc.
Jupiter, Florida
bstevenson@cvtechnology.com

Keywords: Vents, Flameless, Dust, Dust Loading

Abstract

Flameless vents were developed and marketed for the first time over 25 years ago. The technology has evolved and been applied effectively in a wide range of applications involving many different types of dust. As the application and installation of flameless vents has expanded, there have been both successes along with some failures pointing to a need to develop more knowledge about the safe application of the technology. Questions include learning the effective limits for dust loading, vessel size, combustion characteristics such as K_{st} and P_{max}, melting and fibrous dust effects, and pressure variations. How would flameless vents hold up when exposed to environmental factors such as vibration, elevated temperature, and corrosive environments? Perhaps the most significant testing involved full-scale tests with optimal explosion concentrations. Although there has been some skepticism concerning the need for such rigorous testing, the reasons for conducting tests in this manner will be explained and their significance made clear. This new test methodology was developed and conducted by FM Approvals at their new test facility in Rhode Island. Our company, CV Technology, Inc., introduced flameless venting to North America and having assumed that mantle of leadership, also is the first to undergo the full battery of these new tests. The purpose of this paper is to outline what was learned and why the results are vitally important to the future safe use of flameless venting devices.

1. A Short History

Explosion venting has been proven to be a very effective method for deflagration protection. While venting does nothing to prevent ignition, it does a great deal to avoid a damaging overpressure from occurring that could result in unacceptable damage. Explosion venting is a damage-limiting technique. Flameless vents are a subset of explosion vents and are characterized by not allowing flame or dust to escape through the device.

Quench tubes, the first style of flameless vents, were first conceived in the 1980s and have been available commercially since the early 1990s. They were the first in a new category of protection device now known as flameless vents. Over the course of time they have proven to be remarkably effective, both in terms of the protection they afford and their reliability, which is unmatched. They are at once simple in concept and complex in application. The purpose of this paper is to present arguments that they offer the best option, where applicable, for dust explosion protection among a plethora of alternatives currently available.

A new technology was born out of an idea for situations where the vented vessel was inside a building and the vent discharge was directed outdoors via a duct. The idea started with a question: Could a vent duct's length be safely extended? The problem that is reflected by this question is the fact that vent ducts create back pressure: The longer the duct, the greater the back pressure. This can be compensated for by increasing vent area, but of course there is a limit. When the limit is exceeded there is not room on the vessel for the required vent area. So the expanded question was: Could vent duct length be extended while simultaneously minimizing the rate at which back pressure is developed?

The inquiry started with an investigation into the cause(s) of the back pressure. It turns out that as a vent opens to allow burned and unburned dust, pressure, and flame to escape into the duct, there is an increase in turbulence at the boundary between the duct wall and the flame front. The turbulence in turn increases the rate of oxidation with a commensurate increase in the rate of rise of pressure, *dp/dt*. Obviously, the longer the duct the more turbulence is exacerbated and the greater the *dp/dt*. The next step was to investigate the possibility that lining the duct with some sort of barrier material could ameliorate the turbulence, which in turn would allow the duct length to be extended. Ducts were lined with a variety of materials to observe the effects, and it was discovered that certain configurations of materials, notably stainless steel mesh, resulted in the flame collapsing in the duct. It remained to determine what the optimal lining would be and how long the tube needed to be to extinguish the flame and to eliminate the duct wall.

So with that as a backdrop let's look at the anatomy of a quench tube to better understand how it works. Figure 1 shows a cutaway of a quench tube.

Figure 1: Quench Tube

At the base is the explosion vent with the flame arresting part superimposed over the outlet side. It has two important jobs: To protect the mesh so that it will be free and clear of contamination until such time as there is a deflagration; and to relieve the developing pressure at the beginning of a deflagration. Superimposed over the vent, the mesh is held in place by an external frame. The mesh also has two important jobs, to trap burning and unburned dust during a deflagration; and to act as a heat sink to quench the flame. In a properly applied quench tube most dust is trapped in the mesh, and no flame escapes into the surrounding environment. At the same time the maximum reduced pressure P_{red} achieved at the pinnacle of the event will be low enough to avoid vessel damage beyond acceptable limits. It should be clear that for this outcome to be reliably achieved both the vent disc and the mesh must remain intact and unencumbered until activated by a deflagration. Otherwise the quench tube could be compromised to such an extent that it may not provide the required level of protection.

2. Testing

Even though quench tubes have proven to be quite reliable in most installations, their increasing usage has resulted in some losses and has driven a need to understand more about them. Test protocols have been developed and improved upon in both Europe and in the USA to ascertain their capabilities and limits. The most recent protocol, and by far the most stringent, has been developed by FM Approvals[1]. There are many aspects to the FM Approval protocol including both performance and administrative matters such as quality assurance and documentation that are not necessary to outline here. The principal thing that needs to be considered for now is related to the performance testing methodology itself. First, testing is comprised of three parts: unvented vessel, vented vessel using a rupture disc only, and vented vessel using a rupture disc plus flame arresting device. All tests are accomplished full scale in a test vessel of 8 m^3. Moreover, testing is conducted for examples, at 750 g/m^3 using corn starch or at 1000 g/m^3 for sugar (a melting dust). These are nominal values that can be changed for other dusts such as fibrous dust or metal dust, more on these shortly. The reader is encouraged to read the referenced Approval Standard for the full story.

The idea is to test the flameless vent at or near the optimal concentration for a given dust to define a near worst-case condition. There are many differences worthy of note in this approach. By testing using actual rupture discs from the same lot and comparing P_{red} for disc alone with P_{red} for disc plus flame arrestor, the calculated efficiencies are much more realistic than is the case when comparing an actual result against a theoretical one. One very important difference is in the dust loading methodology. In the EN standardized tests[2] that our flameless vents have undergone the dust concentrations varied from 250 g/m^3 to 500 g/m^3. Even lower dust concentrations can be utilized if necessary to achieve a certain result. The difference between the lower values for dust concentration in the European tests and the FM approach of testing at or near c_{st} are significant. This is true for two reasons: First, testing at or near an optimal concentration, $c_{st.}$ can always be expected to result in higher pressures; and second, at higher dust loadings the mesh in the flame arrestor elements becomes more fully loaded. This latter factor affects both the capacity of the mesh to pass the hot gases being generated per unit time during a deflagration and the effectiveness of the mesh to absorb heat. In plain English, the higher the dust loading on the mesh, the more plugged it becomes. Plugged mesh is not desirable as it will affect efficiency ratings and the limits of applicability for flameless vents.

3. Practical Implications

Some have asserted that while the FM test protocol is definitely more rigorous than the European one, it is overkill and not at all necessary. The principle concern is that the dust concentrations are too high and entirely unnecessary because rarely, so the argument goes, do vessels handle dust concentrations as high as the optimal concentration for any given dust. In order to test this argument, I decided to take a look at dust collectors to learn what might be a realistic expectation for dust concentrations inside a typical dust collector. This choice was made for some very good reasons: Dust collectors seem to be the favored locus for dust explosions based on loss history, second they are perhaps the most protected of all vessel types by application, and third there is quite a bit of data available about them.

Bob Zalosh wrote a very comprehensive article on dust explosions in dust collectors[3]. He looked at various dusts, various filter configurations, and assumed very low dust concentrations (i.e. well below Lower Explosible Limit) at the inlet to collectors. Moreover, he even assumed only the row of bags being pulsed would provide fuel for a partial volume deflagration. He then calculated dust concentrations near c_{st} anyway. My only criticism of Bob's article is that I think he was too conservative in calculating the dust loading factor, for if a partial volume deflagration is the incipient event, clearly a full volume event can be expected in immediate follow up. In private correspondence Todd Baker, Chief Engineer at Horizon Systems, ran some experiments in his lab and found that for grain dust the thickness of the caked layer of dust on the filter cartridges quickly built up to ~1/16"-1/8" or more. He then removed that layer of dust and weighed it and determined, if that amount of dust were entrained, the concentration would be approximately 1800 g/m^3-1900 g/m^3. As a first order approximation it would be reasonable to predict that 50% of the caked layer of dust would be entrained and subsequently involved in a deflagration. 900 g/m^3-950g/m^3 would approximate c_{st} for a typical grain dust. In private correspondence Nathan Egbert, Global Director of R&D at Schenck Process, answered the same question via his lab director, Mike Althouse "I would say 1/8" on bags and 1/16" on vertical cartridges. Horizontal cartridges would probably average 1/4-3/8". Clearly, for all practical purposes the answer to the question concerning typical dust loading on filter elements is quite similar no matter how the estimate is made. I have looked inside a lot of dust collectors over the years and can't recall any that had less dust on the filters than the numbers cited here. Arguments that the amount of dust entering or exiting a dust collector can be used to evaluate the amount of dust that is inside the enclosure are just not taking into account what cakes on and sticks to the filters, waiting to be dislodged. And it is what is inside the dust collector that needs to be considered. I conclude from this admittedly ad hoc evaluation that FM's decision to test at or near c_{st} is reasonable and prudent considering both the near worst-case reactivity of any given dust and expected dust loading in typical operating conditions.

Of course there is more to evaluating the nature of the risk posed by combustible dust than the reactivity and concentration of an entrained sample. For a flameless vent in addition to all of the other variables that would need consideration for all explosion vent types, there is the added variable of the flame arresting apparatus. As mentioned earlier, mesh which is plugged even partially has a negative effect on the overall performance of any flameless vent. Plugging can be increased by two further characteristics, is the dust a fibrous or a melting dust? If the dust to be handled possesses either of these characteristics, then testing should be conducted to determine how these variables affect the performance of the device. It should be clear as well that the need to test fibrous or melting dusts must be at or near c_{st} to have any validity. This is so because a test-run at a low concentration of these types of dusts is unlikely to load the mesh enough to be helpful in knowing how to size a flameless vent.

Next I would like to address the special case of metal dusts. Metal dusts have the unfortunate habit of burning to completion when ignited. For this reason the flameless vent must be designed in a special way using more and various materials in the flame arresting element to withstand the high heat and relatively long duration of the flameball contained inside the device. The required more elaborate construction could itself restrict air flow and thus cause a loss of efficiency. There is a very real time-plus-temperature related aspect to venting a metal dust deflagration through a flameless vent. Accordingly, when undergoing tests using metal dusts it is imperative that the dust concentrations be controlled at or near c_{st}. A minimal dust loading in a test for metal dust would not be appropriate. In this special case, the FM Approval test methodology is particularly important.

Finally I would like to comment on a trend in the market toward lower-cost alternatives to the quench tube-type flameless venting device that has been the focus of this presentation. There are box-type units and at least one self-seating relief valve type device. At best these devices have no more than approximately 1/3 of the surface area in their flame arresting element as compared to a quench tube with nominally similar vent area. None of these devices has yet undergone the FM test protocol. Based on our experience it is doubtful that any would pass, or at the very least the limits of their applicability would be severely restricted. Just in case someone should think I am being unfair to others, my company manufactures a box-type flameless vent of this type.

Since flameless venting technology has been an approved protection under the auspices of both OSHA and the NFPA, these devices have found widespread acceptance and application. In NFPA 68, the criteria for use and suitability rests upon independent third party testing at a suitable lab. As a member of the Technical Committee responsible for 68, I expect any such testing to be conducted in a manner that reflects a reasonable approximation of the conditions to which the device would be expected to encounter in practice. Having gone through both the European and more recently the FM Approvals test methodologies, I can assure you that there is simply no comparison.

4. Conclusion

Flameless vents have been on the market for a long while now and have proven to be a mostly reliable protection device. They solve a lot of problems and afford the user with an excellent alternative where they can be safely used. With experience it became obvious that the parameters for testing and safe application needed refinement. The most recent and most rigorous testing methodology has been developed by FM Approvals. Using the FM test methodology stresses the devices under more realistic conditions than had been developed heretofore. Efficiencies are determined that can be believed, and using full scale volume and dust loadings that approach optimal concentrations gives comfort that a device that can pass this test is going to provide the expected safety and reliability. At the same time, the very rigor of this test protocol raises questions about suitability for difficult to handle dusts or less robust devices that on the surface seem to offer the same or similar protection at a much lower cost. The prudent person would be wise to carefully consider the options and the test methodology that was used to verify performance limits.

5. References.

[1] Approval Standard for Explosion Venting Devices, Class Number 7730, FM Approvals LLC, April 2014

[2] EN 14797 Explosion Venting devices

[3] Robert Zalosh, "Dust Collector Explosions: A Quantitative hazard evaluation method," Journal of Loss Prevention in the Process Industries, 36, Elsevier 2015

Using a burning rate model for deflagration vent sizing

Michelle R. Murphy
ioKinetic, LLC
95 Stiles Road, Salem, New Hampshire 03079, USA
murphy@ioKinetic.com

Dr. Georges Melhem
ioMosaic Corporation
95 Stiles Road, Salem, New Hampshire 03079, USA
melhem@ioMosaic.com

Prepared for Presentation at
American Institute of Chemical Engineers
2016 Spring Meeting
12th Global Congress on Process Safety
Houston, Texas
April 11-23, 2016

Using a burning rate model for deflagration vent sizing

Michelle R. Murphy
ioKinetic, LLC
95 Stiles Road, Salem, New Hampshire 03079, USA
murphy@ioKinetic.com

Dr. Georges Melhem
ioMosaic Corporation

Keywords: enter a few keywords here to aid searching

Abstract

NFPA 68 provides simplified calculation methods for sizing deflagration vents for combustible dust explosions. These simplified methods apply to a limited set of process parameters and equipment. For applications outside those limitations, full-scale testing or a performance-based approach is needed. This paper presents a methodology for using burning rate models to predict explosion behavior in enclosed vessels. Measured deflagration index (K_{St}) values are used to determine an estimate of the burning rate. Once the burning rate of a material is known, application to various process parameters and equipment can be made using dynamic simulation. Predicted burning rates for one material will be presented.

1 Introduction

Being able to accurately size deflagration vents is critical to safe, efficient operation of combustible dust handling equipment. NFPA 68 [1] provides simple calculation methods for sizing deflagration vents for combustible dust deflagrations. These calculation methods have two limitations, they apply only to a limited set of operating parameters and they result in conservative vent sizes.

The simple calculation methods are limited to specific operating parameters of vessel geometry, initial pressure, maximum pressure, deflagration index and vent opening pressure. Many combustible dust unit operations fall outside the narrow limitations of the NFPA vent sizing calculations.

The NFPA 68 calculation methods for low L/D are limited to the following applications:

- $L/D \leq 6$; ≤ 8 for some top-fed bins, hoppers, and silos
- $P_i \sim 1$ bara $\pm$ 0.2 bara (~ between -3 and 3 psig)
- 5 barg $\leq P_{max} \leq$ 12 barg (72.5 psig $\leq P_{max} \leq$ 174 psig)

- 10 bar-m/sec $\leq K_{St} \leq 800$ bar-m/sec
- $m^3 \leq V \leq 10{,}000\ m^3$
- $P_{stat} \leq 0.75$ barg (~ 11 psig)

Operation outside these limits requires a performance-based design or a full-scale test. (See Figure 1)

The calculations understandably err on the conservative side, sometimes making the installation of such oversized vents impossible. In this paper, an alternative approach is presented, applying the fundamental parameter of laminar burning velocity.

Laminar burning velocity is a more fundamental parameter of the dust. According to Eckhoff [2], the laminar rate at which a laminar combustion wave or reaction zone propagates relative to the unburnt gas is called the fundamental or laminar burning velocity, S_u.

Implementation of this parameter will allow for more accurate simulation of real-life conditions. It may be used to address the performance-based option within NFPA 68 for operation outside the applicability range of the provided equations.

Figure 1: Decision tree for applicability of NFPA 68 vent calculation methods

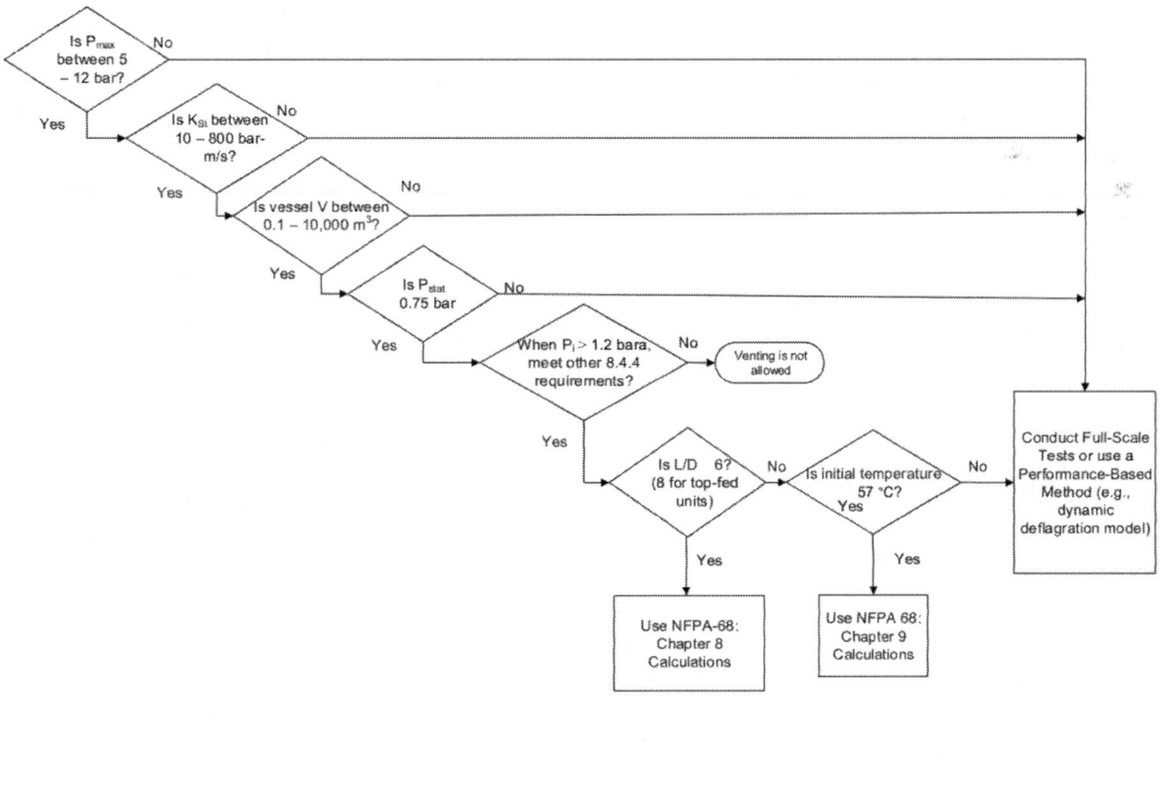

2 Methodology

Traditional vent sizing applications rely on K_{St} and P_{max} values from explosion severity testing. In explosion severity testing, P_{max} and dP/dt_{max} values are measured directly. The K_{St} value is calculated using the normalization of measured maximum pressure rise rate, according to:

$$K_{St} = \frac{dP}{dt}_{max} \times \sqrt[3]{V}$$

This normalization is intended to correlate measured K_{St} values from different volume test vessels.

Dunjo, et. al. [3] explained the importance of defining laminar flame speed in characterizing explosive reaction behavior. In that work, Dunjo, et. al. applied the methodology of Metghalchi and Keck [4] to estimate laminar burning velocities of flammable gases at atmospheric conditions.

It is proposed that the same relation can be used to determine laminar burning velocities of combustible dusts at atmospheric conditions from measured explosion severity parameters. Once a reference laminar burning velocity is known, it can be used to simulate burning behavior in full-scale vessels. Better understanding of the burning behavior in deflagration events allows for better design of mitigation systems, including deflagration vents.

3 Calculation Method

Metghalchi and Keck fit gas burning velocities from several fuels to the following relation:

$$S_u = S_{u0}\left(\frac{T_u}{T_{u0}}\right)^{\alpha}\left(\frac{P}{P_0}\right)^{\beta}$$

where:

$S_{u0} = reference\ laminar\ flame\ speed\ (^{cm}/_{sec})$
$T_{u0} = 298K$
$T_u = unburned\ gas\ temperature, in\ K$
$P_0 = reference\ pressure, 1\ atm$
$P = pressure, in\ atm$
α, and β are fitted constants (See Dunjo, et.al.)

Using a dynamic simulation algorithm, calculations are performed at each time step to determine the combustion reaction products, flame temperature, reactant temperature, and pressure. Determination of combustion products is accomplished by direct minimization of the Gibbs free energy. The algorithm employed is found in the SuperChems™ component of ioMosaic's Process Safety Office™. More information about the algorithm employed can be found in Melhem [5].

To complete the simulations, the following information must be defined:

- Volume of the vessel
- Initial temperature and pressure
- Initial or reference flame speed
- Initial flame speed reference temperature and pressure
- Temperature and pressure exponents
- Reactants and potential products
- Explosion parameters

Using an iterative approach, the reference laminar flame speed can be determined by matching measured deflagration parameters. This reference laminar flame speed could be used to predict behavior at different starting conditions and in different vessel geometries.

Standard explosion severity testing measures/calculates the following burning parameters:

- P_{max} – maximum pressure reached during the deflagration
- dP/dt_{max} – maximum pressure rise rate reached during deflagration
- K_{St} –deflagration index

The volume of the vessel, initial temperature and pressure, reactants and explosion parameters are all defined by the test set-up and results of an explosion severity test. An initial guess at the reference flame speed is made and simulations are completed until the simulated pressure rise rate matches the experimental value.

The temperature and pressure exponents are a function of the initial fuel to oxidizer ratio vs. the stoichiometric fuel to oxidizer ratio as shown in Dunjo, et.al.

4 Example

Burning behavior of Niacin was measured in a 20L sphere apparatus. In this test, the sphere was evacuated to -0.6 barg. The dust was placed in a dispersion cup and blown into the sphere to reach atmospheric pressure. The dust was ignited with 2-5 kJ Simex igniters. The test was repeated, at different concentrations, until maximum values for both pressure and pressure rise rate are found. Two more test series around the measured maximums are run to confirm the results. Turbulence is a function of ignition delay time. To remove variability, the test is conducted with a specified ignition delay time (tv) of 60 ms.

The results are presented in Table 1. Using the standard approach, P_{max}, dP/dt_{max}, K_{St} are found as an average of three series of values. The values in bold show the maximum measured values for a particular series. Averaging these three sets gives the values below:

P_{max} = 7.9 bar

dP/dt_{max} = 857 bar/sec

K_{St} = 233 bar-m/sec

Table 1: Explosion severity results of Niacin

Test Series	Series	Concentration (g/m^3)	P_{max} (bar)	dP/dt (bar/sec)	tv eff (ms)
2	1	60	.0	0	60
3	1	125	4.7	443	60
4	1	250	7.0	663	60
5	1	500	**7.8**	**822**	60
6	1	750	6.6	657	60
7	1	1000	6.9	757	60
8	1	1250	6.6	716	60
9	1	1500	5.9	521	60
11	2	250	**7.8**	789	60
12	2	500	7.8	806	60
13	2	750	7.6	**860**	60
15	3	250	7.8	809	60
16	3	500	**8.0**	**889**	60
17	3	750	7.5	874	60

Using the algorithm to complete the simulations, multiple iterations were run until a value for S_{u0} was found that allowed the simulation to match the measured dP/dt_{max} of 857 bar/sec. A value of 0.4309 cm/sec was found for S_{u0}.

Figure 2 shows the pressure history for the matching simulation. Figures 3 and 4 show how the pressure rise rate and, similarly, K_{St} change with time.

Figure 2: Pressure history of simulation of 500 g/cm3 Niacin in 20L sphere apparatus

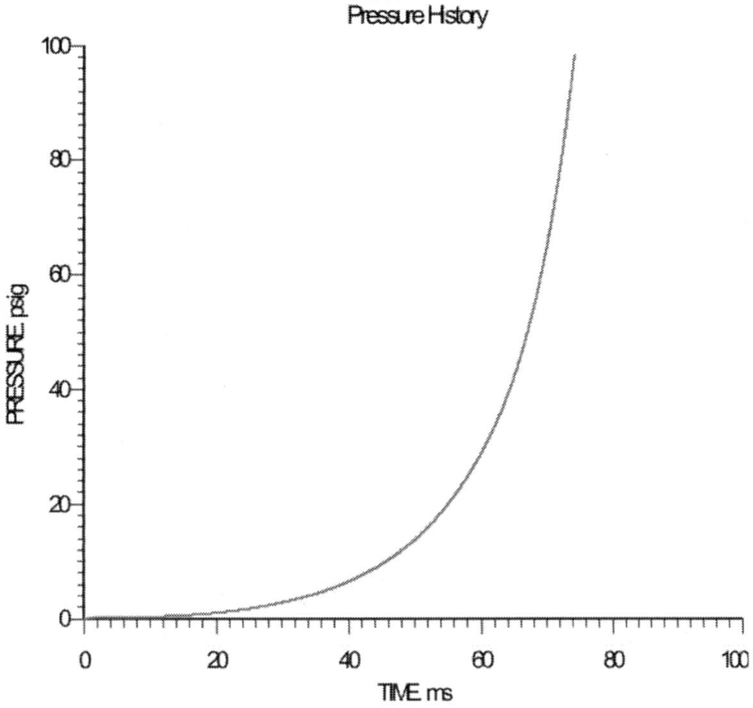

Figure 3: Pressure rise rate history of 500 g/cm3 Niacin in 20L sphere apparatus

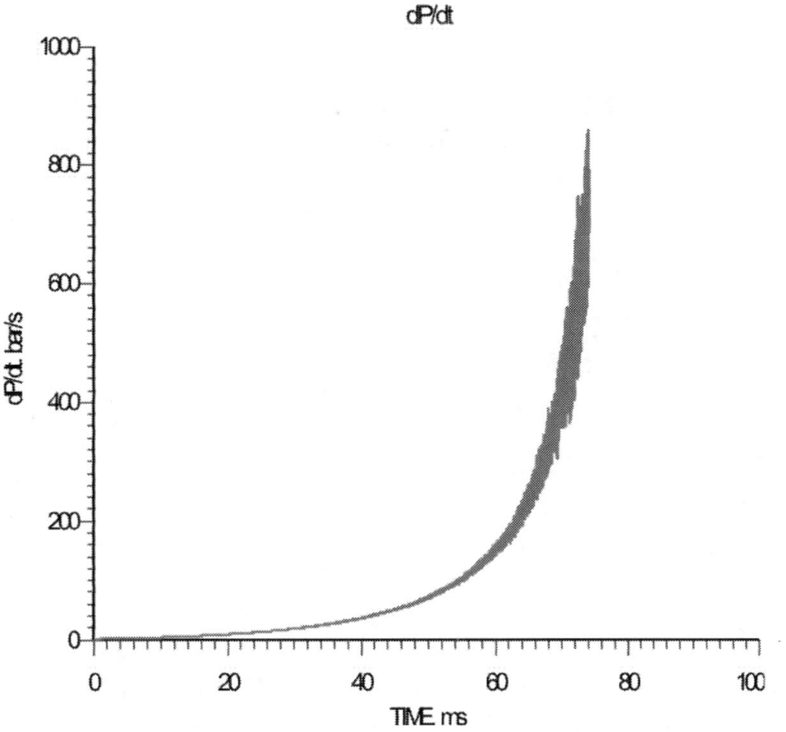

Figure 4: Explosion index (K_{St}) history of 500 g/cm3 Niacin in 20L sphere apparatus

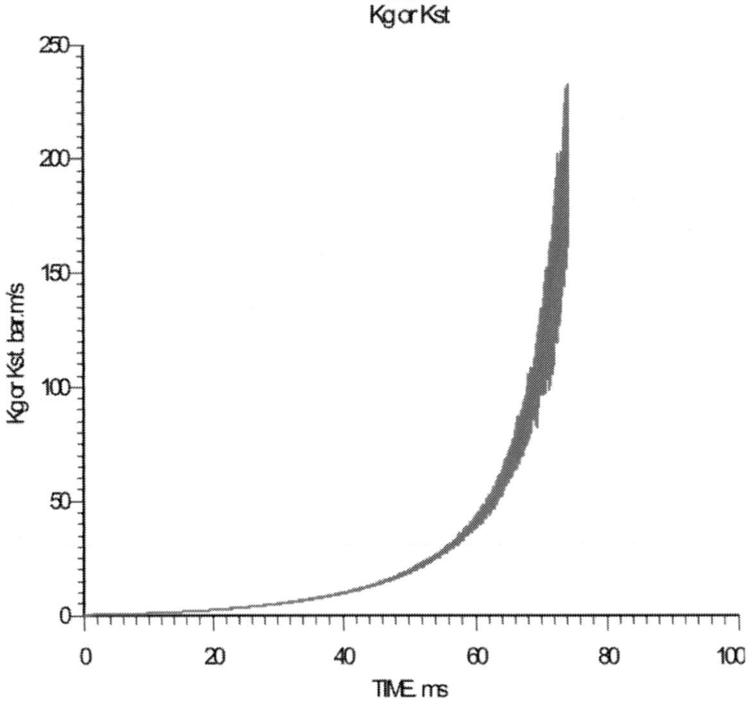

With a reference laminar flame speed determined, the simulator should be able to predict the behavior at different conditions. This will allow extrapolation to conditions of higher temperature and/or pressure.

It is important to note that standard 20L sphere testing in this range is expected to have an accuracy of ±10% for each measured value. The variability in results is evident in the experimental data shown in Table 1. In this particular data set, the deviation is +3.7% and -4.1% for dP/dt_{max}. With this in mind, it would be expected that the simulation values would be within ±4% of the experimental.

5 Deflagration Vent Sizing

Using the same dynamic simulator, it is possible to input the full-scale operating parameters of volume, temperature, pressure as well as installed vents to simulate full-scale behavior of systems under vented deflagration conditions. In so doing, acceptable vent sizes can be determined.

6 Conclusions

A methodology for estimating reference laminar flame speeds from experimental data is presented. It was shown that application of the methodology can be used to size deflagration vents. This approach may be used for meeting the performance-based option in NFPA 68 for operation outside of the applicability limits of the supplied equations. Next steps could include further validation with additional materials and comparison with actual vented systems data.

7 References

[1] NFPA 68, Standard on Explosion Protection by Deflagration Venting, 2013, National Fire Protection Association, NFPA, Quincy, Massachusetts, December 2012.

[2] Eckhoff, Rolf K., Dust Explosions in the Process Industries, Second Edition, Butterworth-Heinemann, Oxford, Great Britain, 1997, pp 256 – 297.

[3] Dunjo, Jordi, Daniel Nguyen and Michelle Murphy, "Laminar Flame Speeds Data Collection. Ensuring reliable data for explosions characterization." An ioMosaic Corporation Whitepaper, January 23, 2014.

[4] Metghalchi, M. and J.C. Keck, "Laminar Burning Velocity of Propane-Air Mixtures at High Temperature and Pressure," *Combustion and Flame* 38 (1980) 143-15.

[5] Melhem, G. A., "A Detailed Method for Estimating Mixture Flammability Limits Using Chemical Equilibrium," Process Safety Progress, Vol 16, No.4 pp. 203-218.J.A. Klein and B.K. Vaughen. "Implement an Operational Discipline Program to Improve Process Plant Safety," *Chemical Engineering Progress* (2011), pp. 49-52.

Nomenclature

L/D – length to diameter ratio

dP/dt – pressure rise rate

K_{St} – deflagration index for dusts

P_i – initial pressure

P – pressure

P_0 – reference temperature, 1 atm

P_{max} – maximum pressure developed in an unvented vessel

P_{stat} – static activation pressure

S_u – laminar burning velocity

S_{u0} – reference laminar flame speed

tv – ignition delay time in explosion severity testing

tv eff – measured ignition delay time in explosion severity testing

T_u = unburned gas temperature

T_{u0} – reference temperature, 298K

V – volume of the enclosure

Dust Explosions - *Testing Requirements of the New NFPA 652*

Vahid Ebadat, Ph.D.
Chief Technical Officer - Process Safety, DEKRA Insight

113 Campus Drive, Princeton, New Jersey 08540, USA
Email: vahid.ebadat@dekra.com
www.dekrainsight.com

Prepared for Presentation at
American Institute of Chemical Engineers
2016 Spring Meeting
12th Global Congress on Process Safety
Houston, Texas
April 11-23, 2016

Dust Explosions - *Testing Requirements of the New NFPA 652*

Vahid Ebadat, Ph.D.

Chief Technical Officer - Process Safety, DEKRA Insight

113 Campus Drive, Princeton, New Jersey 08540, USA

Email: vahid.ebadat@dekra.com

www.dekrainsight.com

Keywords: NFPA 652, Combustible Dust, Explosible Dust, Laboratory Testing, Dust Hazard Analysis, DHA, Dust Explosion Hazard Assessment, Ignition Sensitivity, Explosion Severity

1. INTRODUCTION

A major concern in many process industries, including the chemical, pharmaceutical, food, metals, wood, and pulp/paper is the risk of flash fires, explosions, or otherwise unplanned loss of containment and release of energy due to the use or presence of flammable and/or combustible materials. An explosible atmosphere is typically created when the right amount of a fuel (flammable gas or vapor, or combustible dust) is mixed with the right amount of an oxidant, usually the oxygen in air. It should be noted that all combustible materials could create an explosible mixture with air. Ignition of an explosible atmosphere occurs if the ignition-source energy exceeds the minimum energy that is required to ignite the mixture at the given process conditions. As one can consequently imagine, a flash-fire or explosion hazard can exist during the transfer, handling, processing, and packaging of many flammable/explosible materials. These events can lead to catastrophic events involving fatalities, injuries, environmental impact, facility damage, and economic losses. It is therefore necessary that precautions be taken to prevent their inadvertent ignition and/or protect employees, contractors, visitors, the public, and the facility, against the effects of fires and explosions. Table 1 lists some of the recent dust-related explosions in the past 12 years or so.

Table 1. Some Recent Dust-Related Explosions

Date	Company / Location	Dust	Fatalities / Injuries
5/16/2002	Rouse Polymerics Int., Vicksburg, MS	*Rubber dust*	*5 killed,7 injured*
1/29/2003	West Pharmaceutical Services, Kinston, NC	Polyethylene dust	6 killed, 38 injured
2/20/2003	CTA Acoustics, Corbin, KY	Phenol-Formaldehyde Resin dust	7 killed, 42 injured
2/07/2008	Imperial Sugar, Savannah, GA	Sugar dust	14 killed, 44 injured
2/24/2010	China starch factory	Starch dust	19 killed, 50 injured
4/23/2012	Prince George Wood Mill, Canada	Wood Dust	2 killed, 20 injured
8/02/2014	Zhongrong Plating (Hubcap polishing) Company, Jiangsu Province, China	Aluminum Dust	169 killed, >200 injured
6/27/2015	Water Park, Taiwan	Starch based colored powder	1 killed, 498 injured
7/17/2015	Wood Flour Mill, Cheshire, UK	Wood Dust	4 killed, 35 injured

In response to these continuing incidents, and since OSHA has yet to issue a consolidated dust standard, the National Fire Protection Association (NFPA) has recognized the common threads through all its existing dust-hazard-related standards and sought to consolidate the best general practices for all combustible solids, irrespective of industry and powder/dust type, in the new NFPA 652: Standard on the Fundamentals of Combustible Dust (Ref. 1).

According to the new NFPA 652, safety of the occupants, facility, processes, and equipment shall be achieved by considering the following the (prescriptive) requirements:

- Determine the combustibility/explosibility characteristics of process dusts and powders. Absence of previous incidents shall not be used as the basis for considering a dust as non-combustible or non-explosible. It shall also be assumed that routine powder handling will, through attrition, create small particles (dust) that could more readily give rise to flash fire and explosion hazards.
- Perform a Dust Hazard Analysis (DHA) if materials handled and processed have been identified as combustible and/or explosible. DHA is a systematic review to identify and evaluate potential dust fire, flash fire, and explosion hazards and must be conducted or led by a qualified person with demonstrated ability to deal with hazards related to the processing and managing of combustible particulate solids. Additionally, DHA must be based upon laboratory test data, historical data, or published data of representative combustible particulate solids that could be present in the facility. It is noteworthy that a DHA does not need to comply with the Process Hazards Analysis (PHA) requirements of the OSHA regulation, 29 CFR 1910.119 (e) (Ref.2).

A documented risk assessment acceptable to the AHJ is permitted to be conducted to determine the level of design and features to be provided for the building design, equipment design, ignition source control, explosion prevention/protection, and certain aspects of fire protection. The risk assessment is permitted to be conducted when the user seeks to demonstrate that some alternative to the prescriptive requirements of the standard would provide equivalent protection.

- Establish written management systems and procedures for operating the new and existing facility and equipment to prevent or mitigate fires, deflagrations, and explosions from combustible particulate solids, including:
    - Operating Procedures and Practices
    - Inspection, Testing, and Maintenance
    - Training and Hazard Awareness
    - Contractors Qualifications and Training
    - Emergency Planning and Response
    - Incident Investigation
    - Management of Change
    - Documentation Retention
    - Management Systems Review
    - Employee Participation

It should be noted that if any one component of the prescriptive requirements found in Chapter 8 proves to be impractical, NFPA 652 permits the use of a performance-based approach, provided that it meets the life-safety requirements of this standard. Performance-based design will achieve the life-safety (fire and explosion) objectives if ignition has been prevented, or no person, other than those in the immediate proximity of the ignition, is exposed to untenable conditions due to the fire and/or explosion. Additionally, no critical structural element of the building is damaged to the extent that it can no longer support its design load during the period of time necessary to affect complete evacuation. It should be noted that most US companies would not accept injury or fatality of even the most exposed individual.

This paper focuses on those requirements of the NFPA 652 that are related to process safety data. As it has been recognized universally, process safety data is essential for the effective understanding and management of dust-cloud flash fires and explosions. Different strategies for the collection of process safety data are discussed in a White Paper by DEKRA Insight (Ref 3).

2. PROCESS SAFETY DATA

NFPA 652 requires that the Dust Hazard Assessment (DHA) shall be based upon laboratory test data of material that is representative of combustible particulate solids present in the process or the facility. The "owner/operator" (YOU) shall start by establishing whether or not a material is combustible/explosible. One could begin this process by referring to the properties tables in Appendix A of NFPA 652 for any information that it might contain on the combustibility/explosibility of the material. If a raw material is being processed/handled, one could also seek information from the manufacturer or supplier. For an intermediate or final product or a dust deposit, it is recommended to either look for any appropriate/applicable test results that might already be available. Otherwise, it is recommended that a sampling plan be created, have appropriate samples collected from the plant, and tested by an accredited laboratory. Note: Test results are strongly influenced by particle size and shape, moisture content, age, and presence of contaminants. Be sure that the test results are appropriate for your material!

2.1 The First Step - Formation of a Flammable/Explosible Atmosphere

A flammable atmosphere is formed when a sufficient amount of a dust is mixed (suspended) with an oxidant. As is highlighted above, in order to assess flash-fire and explosion hazards in a facility and to select the most appropriate basis of safety, the fire and explosion characteristics of the dust(s) that could be present must be determined for the handling and processing conditions. Handling and processing conditions would include temperatures, conveying rates, transfer rates to and from containers, and concentrations in air in equipment and in work areas.

The first step in the assessment of flash-fire and explosion hazards is to establish whether a dust cloud or a hybrid atmosphere [a mixture of dust and vapor], if formed, is flammable/explosible under the process/operating conditions. For some common materials, this information may be available in the Safety Data Sheet (SDS) from vendors, manufacturers' technical bulletins, or other publically-available sources. Otherwise, some well-established tests such as Explosion Classification (Go/No Go Screening), Fire Train, and Minimum Explosible Concentration tests for powders/dusts can be considered as outlined in Table 2. NFPA 652 requires that explosibility, ignition sensitivity, and explosion severity of dusts shall be determined in accordance with applicable American Society for Testing and Materials (ASTM) test standards or other equivalent test methodologies. The methods used for conducting the tests listed in Table 2 can be found in the Annual Book of ASTM Standards, Volume 14.05 (Ref. 4), but also in publications of the International Electrotechnical Commission (IEC) (Ref. 5), and the European Committee for Electrotechnical Standardization (CENELEC) (Ref. 6).

Table 2. Explosibility Tests

Flammable Atmosphere	Examples	Initial Test(s)	Follow up Tests
Explosible Dust Cloud	• Headspaces of storage vessels and equipment during powder transfer, blending, mixing, fluidization, etc. • Buildings where combustible powders are handled/processed in open vessels, containers • Poor housekeeping resulting in accumulation of unsafe levels of dust on various building and equipment surfaces	One or a number of the following tests: • Explosibility Screening (Go/No Go) in modified Hartman, 20-Liter, or 1-m^3 spherical vessel • Fire Train • Minimum Explosible Concentration (MEC) • Limiting Oxygen Concentration (LOC)*	• Simulation of actual process / operating conditions
Hybrid (gas/vapor coexisting with dust cloud even when individually below their respective LFL and MEC levels)	• At open manways during transfer of powder to vessels containing flammable liquid • Fluidized bed drying of solvent-wet powders	One or both tests: • Explosibility Screening (Go/No Go) in modified Hartman, 20-Liter, or 1m^3 spherical vessel • Minimum Explosible Concentration • Limiting Oxygen Concentration (LOC)*	• Simulation of actual process / operating conditions

*LOC testing is usually only considered when the basis of safety is decided to be inert gas blanketing

The data obtained from the test(s) in Table 2 will only provide the information that is necessary for determining whether or not an explosible atmosphere exists in a vessel or a building under optimum or actual operating/process conditions. These data do not provide any information on the ease of ignition nor on the severity of the explosion. A very small percentage [about 5.5%] of the powders or dusts that have been submitted to Chilworth for testing were found to be "non-explosible" when dispersed in air and submitted to high-energy ignition sources.

The data obtained from the Limiting Oxidant Concentration (LOC) test are used for evaluating the options for the avoidance of oxidant (usually air) as a basis for ensuring safety from flash fires and explosions.

2.2 Determining Ignition Sensitivity

Once the flammability/explosibility of a material or an atmosphere is confirmed, then testing normally falls into two groups, "Ignition Sensitivity" and "Explosion Severity".

Ignition Sensitivity - There are a number of tests that are designed to determine the sensitivity of a flammable atmosphere to ignition by the type of ignition source(s) that could be present during normal and also foreseeable abnormal operating conditions. So, the selection of appropriate "ignition sensitivity" test(s) requires a good understanding of the type of ignition source(s) that could be associated with specific unit operations and activities in a facility. For example, simple gravity transfer of a powder from a container into another vessel may only give rise to electrostatic discharges if conductive components are not grounded or if non-conductive components are utilized. In this example, the only "Ignition Sensitivity" test to consider would be Minimum Ignition Energy (MIE). The MIE test is used primarily to assess the susceptibility of flammable atmospheres to ignition by electrostatic discharges, and it is defined as the lowest-energy electrostatic spark that is capable of igniting a flammable atmosphere at its optimum concentration for ignition. Depending on the MIE of the dust cloud, some additional tests may need to be considered, such as the powder's Volume Resistivity and Electrostatic Chargeability.

Table 3 below presents the descriptions of the initial "ignition sensitivity" test and any follow-up tests that are normally considered for the common ignition sources that might exist in a facility.

Table 3. Ignition Sensitivity Testing Guide

Ignition Source / Hazard	Examples	Initial Test(s)	Follow up Test(s)
Electrostatic Discharges	Discharges from ungrounded conductive (e.g. metal) objects and people, insulating objects such as plastic containers, liners, hoses, etc., non-conductive liquids, insulating powders	• (Capacitive) Minimum Ignition Energy	If MIE < 25 mJ: • Volume Resistivity • Electrostatic Chargeability
Electrical Arcs and Sparks	Arcs/sparks from electric appliances (motors, switches, etc.) during normal operation or by accidentally cutting the connecting circuit	• Minimum Ignition Current (Ratio)	• Simulation of actual process / operating conditions
Self-Heating/Auto Ignition	Unintentional heating (e.g. during mechanical milling) of powders, various powder drying operations, bulk storage of powders after heating	One or a number of the following tests: • Differential Scanning Calorimeter • Accelerating Rate Calorimeter (ARC) • Bulk Powder "Screening" • Aerated Powder "Screening" • Air-Over-Layer "Screening" • Minimum Ignition Temperature - Dust Cloud	One or a number of the following tests: • Isothermal Bulk Powder • Isothermal Aerated Powder • Isothermal Air-Over-Layer • Basket • Adiabatic Dewar
Hot Surfaces	Dust clouds, and/or dust deposits in contact or in close very close proximity to hot surfaces; for example, uninsulated steam pipes, lighting fixtures, and hot process equipment (exterior surfaces of process equipment, electric motors, etc.)	• Minimum Ignition Temperature - Powder Layer • Minimum Ignition Temperature - Dust Cloud	• Simulation of actual process / operating conditions
Friction / Grinding Sparks and heating	Breakdown of bearings, misalignment of screw conveyers and belt conveyors, entrainment of foreign objects causing sparks and hot surfaces (hot particles, hot spots)	One or a number of the following tests: • Friction Sensitivity • Minimum Ignition Temperature - Dust Cloud • Inductive Minimum Ignition Energy	• Simulation of actual process / operating conditions
Impact Energy	Energy release resulting from the (accidental) impact between two hard objects in the presence of a flammable vapor or a combustible solid/powder	• Impact Sensitivity	• Simulation of actual process / operating conditions

The methods used for conducting the tests listed in Table 2 can be found in the Annual Book of ASTM Standards (Ref. 4), International Electrotechnical Commission (Ref. 5), and the European Committee for Electrotechnical Standardization (Ref. 6).

The results of laboratory tests provide important information concerning the sensitivity of a flammable atmosphere to ignition by sources that might be present within a dust cloud. This information would be a powerful tool in the hands of process safety practitioners for determining the applicability and practicality of "elimination of ignition sources" as the primary basis for safe operation.

It should be noted that avoidance of ignition sources as a primary basis of safety not only requires a proper understanding of the dust-cloud's ignition sensitivity, but it also requires rigid management of all credible ignition sources under normal and foreseeable abnormal operating conditions.

Finally, even if all foreseeable ignition sources cannot be reliably controlled at all times, avoidance of ignition sources should still be considered as a secondary basis of safety (that is, to reduce the frequency of undesirable ignitions).

2.3 Determining Explosion Severity

The severity of an explosion that is caused by ignition of a dense dust cloud in a confined space or enclosure is determined by the maximum rate of pressure rise resulting from the expanding fireball and the maximum explosion pressure. These explosion properties are typically measured in a 20-liter or 1-cubic-meter spherical vessel. The pressure-rise rate is used to calculate the Deflagration Index for combustion of a dust cloud (K_{st}). These data can be used in the design of explosion protection measures, as summarized in Table 4.

Table 4. Application of Maximum Explosion Pressure and Explosion Severity Data

Basis of Safe Operation	Purpose	Required Test	Discussion
Explosion Resistant Equipment	Enclosure is designed to withstand the maximum pressure resulting from the ignition and deflagration of the flammable atmosphere	• Maximum Explosion Pressure (P_{max}) (Ref. 7, 8)	Although possible, explosion containment is generally not feasible for large vessels/equipment due to the high costs associated with ensuring that ~~they~~equipment is strong enough to withstand the maximum explosion pressure (typically 8- 10 bar). Smaller items such as transfer lines, screw conveyors and rotary valves can often be more easily designed for containment. Operating enclosures/equipment built for containment can be cumbersome, as all openings and directly- connected equipment also have to be resistant to the maximum explosion pressure
Explosion Relief Venting	One section/area of the enclosure is designed to be weaker than the rest of the enclosure so that in case of an explosion the weaker section would rupture/open and relieve the explosion products (pressure, flame, combustion gases, and unburnt fuel) to a safe location	• Maximum Explosion Pressure (P_{max}) (Ref 7, 8) • Maximum rate of pressure rise $(dp/dt)_{max}$ (Ref 7, 8), or K_{st} value.	Explosion venting is the most commonly used protection measure, particularly for dust handling/processing equipment such as dust collectors, large bins, bucket elevators, and dryers. The explosion relief vent must be located to direct any products of an explosion (i.e., fireball, pressure, combustion gases and particulates) to a safe area outside of the building.
Explosion Suppression	Explosion is detected (by infrared or pressure sensors) at the early stages of the explosion, and a suitable suppressant is then injected into the enclosure to quench the developing fireball	• Maximum rate of pressure rise $(dp/dt)_{max}$ (Ref 7, 8), or K_{st} value.	Explosion suppression is commonly used when the location of the vessel / equipment prevents the use of relief venting, or if the material is hazardous to health and/or the environment. Explosion suppression is generally more

			expensive than relief venting and requires regular inspection and testing.
Explosion Isolation	Propagation of an explosion from a protected enclosure to any interconnected equipment or to the work area is prevented by providing mechanical or chemical isolation systems	• Maximum Explosion Pressure (P_{max}) (Ref 7, 8) • Maximum rate of pressure rise $(dp/dt)_{max}$ (Ref 7, 8), or K_{st} value.	Explosion isolation is usually considered in conjunction with explosion protection measures (containment, relief venting, and suppression).

It is important to note that providing any type of explosion protection presumes that a deflagration could occur within the protected enclosure. However, if the explosion protection and isolation systems are correctly specified, designed, manufactured, installed, and maintained, there would be no harm to people in the vicinity and no unforeseen damage to the equipment and the facility.

3. CONCLUDING REMARKS

Effective management of flash fires and explosions resulting from the use, handling, and processing of combustible powders and dusts often involves:

- Establishing a sound process safety-management system and safety-management competency within the organization, including:
    - Having appropriate data for the understanding of the combustibility, self-heating, electrostatic properties, and explosion characteristics of the fuel(s)
    - A thorough understanding of all operations and processes, and review of all pertinent information (P&ID's and flow sheets; equipment specifications, including materials of construction and operating limitations; process descriptions; and operating and maintenance procedures.
    - Identifying locations where explosible dust cloud atmospheres are, or could be, present during normal and foreseeable abnormal operating conditions, including within equipment and accidental or deliberate spillages.
    - Identifying potential ignition sources (including potential electrostatic sources) that could be present under normal and foreseeable abnormal conditions
    - Defining one or more "Basis of Safety" – including proper process and facility design – to prevent and/or minimize the occurrence of flash fires and explosions and protect people and facilities against their consequences.
- Establishing a safety culture that starts at the top but pervades consistently through the whole organization.
- Regular review and maintenance of process data and information, process-control systems, processing equipment, and the facility to ensure the continued safety of people, the community, and the business.

"Process Safety Data" is the starting point for assuring process safety.

REFERENCES

1. National Fire Protection Association, "Standard on the Fundamentals of Combustible Dust", NFPA 652 (2016).
2. U. S. Department of Labor, Occupational Safety and Health Administration, "Process Safety Management of Highly Hazardous Chemicals", Title 29 of the Code of Federal Regulation, Part 1910, Paragraph 1910.119 (February 24, 1992).
3. Rowe, S. [DEKRA Insight], "Process Safety Data: A Critical Ingredient in Process Safety Excellence" www.dekrainsight.com (2014).
4. American Society for Testing and Materials, "Annual Book of ASTM Standards", Volume 14.05 – "Hazard Potential of Chemicals; Thermal Measurements", http://www.astm.org
5. International Electrotechnical Commission (IEC), http://www.iec.ch
6. European Committee for Electrotechnical Standardization (CENELEC), http://www.cenelec.eu
7. American Society for Testing and Materials, – "Standard Test Method for Explosibility of Dust Clouds", ASTM E1226, http://www.astm.org
8. International Organization for Standardization, Explosion Protection Systems", ISO 6184, http://www.iso.org (1993).

12TH GLOBAL CONGRESS ON PROCESS SAFETY
50th LOSS PREVENTION SYMPOSIUM

Transition Path Sampling for Process Safety Events

Ian H. Moskowitz, Warren D. Seider, and Amish J. Patel
Department of Chemical and Biomolecular Engineering
University of Pennsylvania
Philadelphia, PA 19104-6393

Jeffrey E. Arbogast
American Air Liquide
Newark, DE 19702

Prepared for Presentation at
American Institute of Chemical Engineers
2016 Spring Meeting
12th Global Congress on Process Safety
Houston, Texas
April 11-23, 2016

Abstract:

Chemical manufacturing processes have the potential for catastrophic accidents – those involving loss of human life or major environmental impacts. The numerous ways through which a complex chemical process can fail is limited neither by our experience, nor by our imagination; consequently, so-called *un-postulated* safety-event scenarios pose a major concern. By combining extensive chemical process data, process modeling and state-of-the-art path sampling algorithms, this work aims to uncover rare and un-postulated process safety events, to identify secondary process variables that can portend such events (so they can be incorporated in alarm systems), and to characterize their rates of occurrence.

Transition path sampling (TPS) has been a widely used by the nano-scale modeling community. Its goal is to identify rare-event trajectories that are worthy of investigation; such as the dissociation of a weak acid in aqueous solution or the flipping of phospholipids in a lipid bilayer. The types of rare-events that are best handled by transition path sampling are those in which the time-scale of the event is far shorter than the time-scale over which the event may take place. This ratio of time-scales is similar to a safety-event in a chemical manufacturing process; the event may occur once every five years of operation, while the event takes place over just a five-hour period. To simulate hundreds of such events, a typical process simulation with expected disturbances would have to be simulated for 500 years! It would be computationally beneficial to use a sampling technique that focuses on the safety trajectories of interest, and excludes the vast number of safe operating trajectories.

The details of this Monte-Carlo approach are outlined, and the ability to discover un-postulated rare events demonstrated. The safety trajectories investigated with TPS are used to build event-specific alarms, whose simulated false-positive and false-negative rates are explored. Extensive process, control, and alarm data are essential and are highlighted throughout the presentation. A dynamic model of an air separation unit (ASU) is the test bed for the TPS technique.

Introduction

In response to typical disturbances in manufacturing processes, the basic process control system normally maintains operation within safe thresholds. Occasionally, large disturbances violate operation and safety thresholds (as special-cause events) and alarm and safety interlock systems are required to return the process to its typical operating conditions. The latter events occur rarely, but can be expensive (due to unplanned shutdowns and/or lost product) and even rarer they can be catastrophic (with human health losses, major environmental impacts, …). There is clear motivation to study these events and to identify trends amongst them, but the available process and alarm data are statistically insufficient.

To gain a fuller understanding of how a process moves from typical operation into a safety-event, the process could be simulated over long time periods, during which occasional safety-event trajectories would be encountered, assuming appropriate disturbances and fluctuations are included. To obtain a statistically sufficient set of trajectories (typically on the order of just a few hours), the simulation time would have to be on the order of decades! Clearly this would be computationally infeasible, and a technique is needed to target safety-event trajectories while sharply reducing the time spent simulating typical operations.

A similar challenge, identified in the molecular dynamics community, involves the molecular dynamics of reacting molecules. The reaction events occur very quickly – on the order of nanoseconds, wheras the time between reactions is on the order of milliseconds. Said differently, a simulation millisecond would be needed to capture a nanosecond long reaction trajectory. While the time (and length) scales of the rare events in molecular dynamics are far smaller than those studied in process simulations, the *ratio of rare event times and the waiting period between them* is similar.

This work uses a dynamic simulation model of a simplified, non-commercial air separation unit (ASU) as a test bed for demonstrate application for the identification of rare event trajectories. The rare event that this presentation will focus on is the impurity of the LAR product due to excess nitrogen. While this particular rare event has limited likely safety consequences, it provides a test bed example on which to propose, develop, and refine a technique for generating many trajectories of a rare event. For this particular hypothetical example, the rare-event of concern occurs when the mole fraction of nitrogen rises above 0.01. As the application of the method expands, safety-events with significant potential safety consequences can be efficiently and effectively studied. The methods presented in this case study are generally applicable for many manufacturing processes and rare safety-events.

Overview of Transition-Path Sampling (TPS)

In TPS, which has evolved over the past two decades, the probabilities of locating rare reactive paths are increased significantly [1,2]. First, just one rare trajectory leading to the event of interest is identified. Then, a new "trial" trajectory is generated by perturbing the initial trajectory, and when the trial trajectory is also reactive, it is accepted with a finite probability. A new trial trajectory is then generated, and the procedure is repeated. By focusing only on the reactive paths, the waiting time between the trajectories of interest is eliminated, and consequently, computation times are sharply reduced.

To begin a TPS algorithm, using a dynamic (molecular or process) model, it is necessary to locate one rare trajectory from an initial state to a desired rare target state. One is illustrated schematically in Figure 1a. Here, it proceeds from an initial state (e.g., acceptable nitrogen in the LAR – green) and ends at a rare state (e.g., impure LAR product – red). Then, any time is randomly selected (t') and the independent variables (e.g. the liquid molar holdup and composition of each tray, see the *TPS Demonstration* section) are perturbed (Δx). The model is integrated forward from t' to final time t_f, and in reverse from t' to 0. When another desired rare trajectory is found with its end points lying within the red and green zones, as in Figure 2b, the new trajectory is accepted with a finite probability that depends on the relative likelihood of observing the two trajectories. The trajectories that are a result of more typical process initial conditions and fluctuations are selected at a higher rate. The TPS algorithm is presented in Table 1.

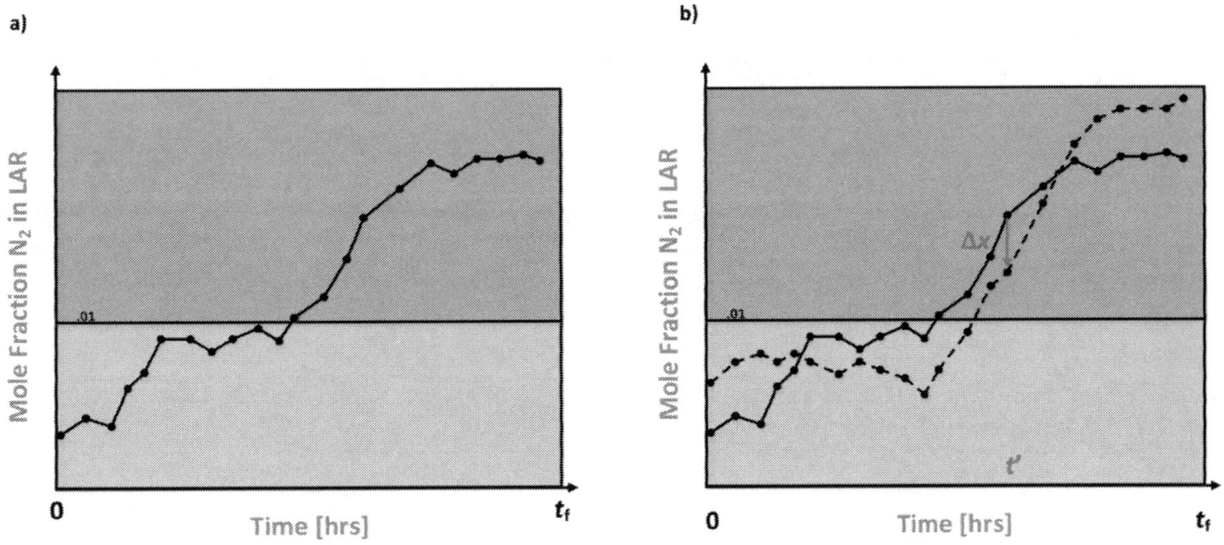

Figure 2. Typical trajectories between initial desired states

Table 1. TPS Algorithm

1.	Identify an initial safety event trajectory
2.	Choose a random time, t', along the trajectory
3.	At t', perturb state variables, x_i, to x_i'
4.	Integrate forward from t' to t_f, and backwards from t' to 0
5.	Determine if this trial trajectory identifies a safety-event
6.	If yes, accept the trial safety event trajectory as the new trajectory with a probability, min $(1, p'/p)$. Here, p'/p is the relative likelihood of the occurrence of the trial perturbed trajectory with respect to the unperturbed trajectory. If the trial trajectory is not accepted, set the previous trajectory to be the new trajectory.
7.	Return to step 2.

Example ASU Model

The proposed approach is demonstrated using a dynamic model of a simplified, non-commercial ASU process, motivated by the process flow diagram shown by NETL[3], and further modeling considerations were motivated by Huang et al. [4]. Saturated vapor air is separated into a nitrogen,oxygen, and argon product in this process, shown in Figure 2. Three distillation columns are used – a high pressure column (HPC) at 5.5 bar beneath a low pressure column (LPC) at 1.25 bar, plus a crude argon column (CAR) for separating argon. As connecting streams expand from 5.5 bar to 1.25 bar, saturated liquids are maintained at a constant temperature. Three components are included in this model – nitrogen, argon, and oxygen, listed in order of their volatilities.

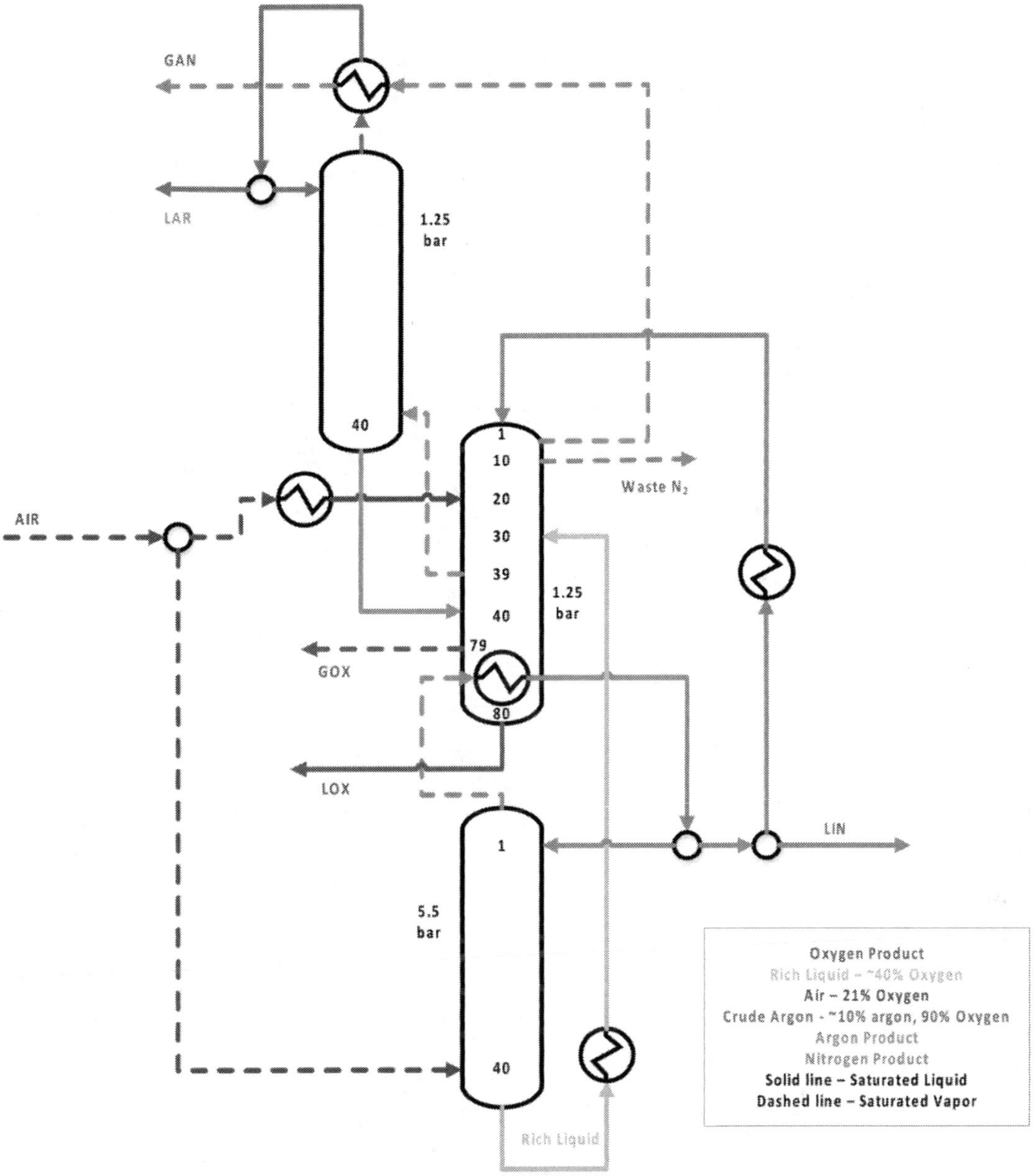

Figure 2. Process flow diagram of the ASU.

Saturated vapor air is fed to the sump of the HPC. Off the top of this column, nitrogen product is condensed while boiling liquid in the sump of the low-pressure column (LPC), with a fraction of the liquid nitrogen (LIN) drawn as product, a fraction returned to the HPC as reflux, and the remainder sent to the top of the LPC as reflux. The oxygen-rich liquid from the sump of the HPC is fed to the middle of the LPC, along with saturated liquid air. Gaseous nitrogen product (GAN) is taken off the top of the LPC, with liquid oxygen (LOX) and gaseous oxygen (GOX) product is removed from the bottom of this column. A vapor side-stream of approximately 10% argon, 90% oxygen, and trace nitrogen is sent to the sump of the CAR. Here, liquid argon (LAR) is produced using the GAN stream for cooling in the condenser. Some of this LAR is taken as product, with the remainder refluxed to the CAR column.

In the simulated model, nearly all of the nitrogen entering the CAR escapes in the LAR product. Because nitrogen is the lightest component, it rises up the column – making the LAR difficult to condense if nitrogen content is too high. While nitrogen is removed in the LAR product stream, some is returned in the column reflux. When the LAR purity cannot be met, the CAR column undergoes abnormal operation – not ideal for maintaining safety. For this reason, the excursion of nitrogen content in the LAR product is considered a safety-event for the context of this presentation. In a similar commercial ASU process, the impact upon plant operation would be especially pronounced due to the significant heat integration of the plant, the use of rich liquid from the HPC (instead of GAN) in the condenser of the CAR column, and the return of vaporized rich liquid to the LPC [5].

The trays of each column are modeled at equilibrium using the MESH (mass balance, phase equilibrium, summation of mole fractions, and heat balance) equations, an empirical equation relating the liquid holdup to the liquid flow rate, and the Peng-Robinson equation of state. This model is modified from that presented by Huang et al. [4] for model-predictive control. The overall material balance of tray i is represented by Eqn. (1):

$$\frac{d}{dt}(M_i) = L_{i-1} + V_{i+1} - L_i - V_i + F_i \tag{1}$$

where M_i is the liquid molar holdup, L is the liquid molar flow rate, V is the vapor molar flow rate, and F_i is the molar feed flow rate (zero for most trays). It should be noted that only the liquid holdup is considered here – the vapor, being far less dense than the liquid, has a negligible holdup. The component material balances describe the composition of each tray, and is shown in Eqn. (2):

$$\frac{d}{dt}(x_{i,j}M_i) = x_{i-1,j}L_{i-1} + y_{i+1,j}V_{i+1} - x_{i,j}L_i - y_{i,j}V_i + z_{i,j}F_i \;\forall\, j \in \{N_2, Ar, O_2\} \tag{2}$$

where x is the liquid molar fraction, y is the vapor molar fraction, and z is the feed molar fraction (phase unspecified). The heat balance is:

$$\frac{d}{dt}(M_i h_i^L) = h_{i-1}^L L_{i-1} + h_{i+1}^V V_{i+1} - h_i^L L_i - h_i^V V_i + h_i^F F_i \tag{3}$$

where h^L is the liquid molar enthalpy, h^V is the vapor molar enthalpy, and h^F is the feed molar enthalpy (phase unspecified).

An empirical relationship for the liquid flow rate as a function of the liquid molar holdup is assumed:

$$L_i = k_d M_i \tag{4}$$

where k_d was determined by Huang et al. [4] to be 0.5 min^{-1} .

The vapor-liquid equilibrium in each tray is calculated by equating the mixture fugacities of the species in the vapor and liquid phases:

$$\phi_{i,j}^L x_{i,j} = \phi_{i,j}^V y_{i,j} \tag{5}$$

where $\phi_{i,j}^L$ is the liquid fugacity coefficient of species j on tray i and $\phi_{i,j}^V$ is the vapor fugacity coefficient of species j on tray i, each calculated using the Peng-Robinson equation of state, at the temperature of each tray, T_i, the column pressure, and the mole fractions of the associated phases. The enthalpies are also calculated using the equation of state as described by Peng et al. [6]. Lastly, the mole fractions sum to unity in the vapor on each tray:

$$1 = \sum_j y_{i,j} \tag{6}$$

TPS Demonstration

The first step of the TPS algorithm, shown in Table 1, is the identification of a rare safety-event trajectory. This trajectory begins with little nitrogen in the CAR column – thus, there is little nitrogen entering the column from the LPC. The vapor mole fraction profile along the LPC is shown in Figure 3a at the start of this trajectory – notice that the argon composition peaks close to tray 39, the tray whose vapor sidedraw is fed to the CAR column, while the nitrogen fraction is sufficiently low. At the beginning of this trajectory (which spans the total of 24 hours) overly aggressive setpoints for the production rate of GOX and LAR are simulated. These setpoints are probably unlikely in typical operation, as discussed shortly. With large amounts of vapor from the bottom of the LPC column, liquid higher in the column vaporizes to replace it, and the argon bubble begins dropping along the column. The mole fraction profiles at 1-, 4-, and 12-hr are shown in Figures 3b, 3c, and 3d, respectively. As the nitrogen enters the CAR, it rises into the LAR product, whose nitrogen mole fraction along the trajectory is shown in Figure 3e. This trajectory, with the mole fraction of nitrogen exceeding 0.01, predicts a safety-event – it begins with less than 0.001 mole fraction nitrogen in the LAR product and finishes with almost 0.02.

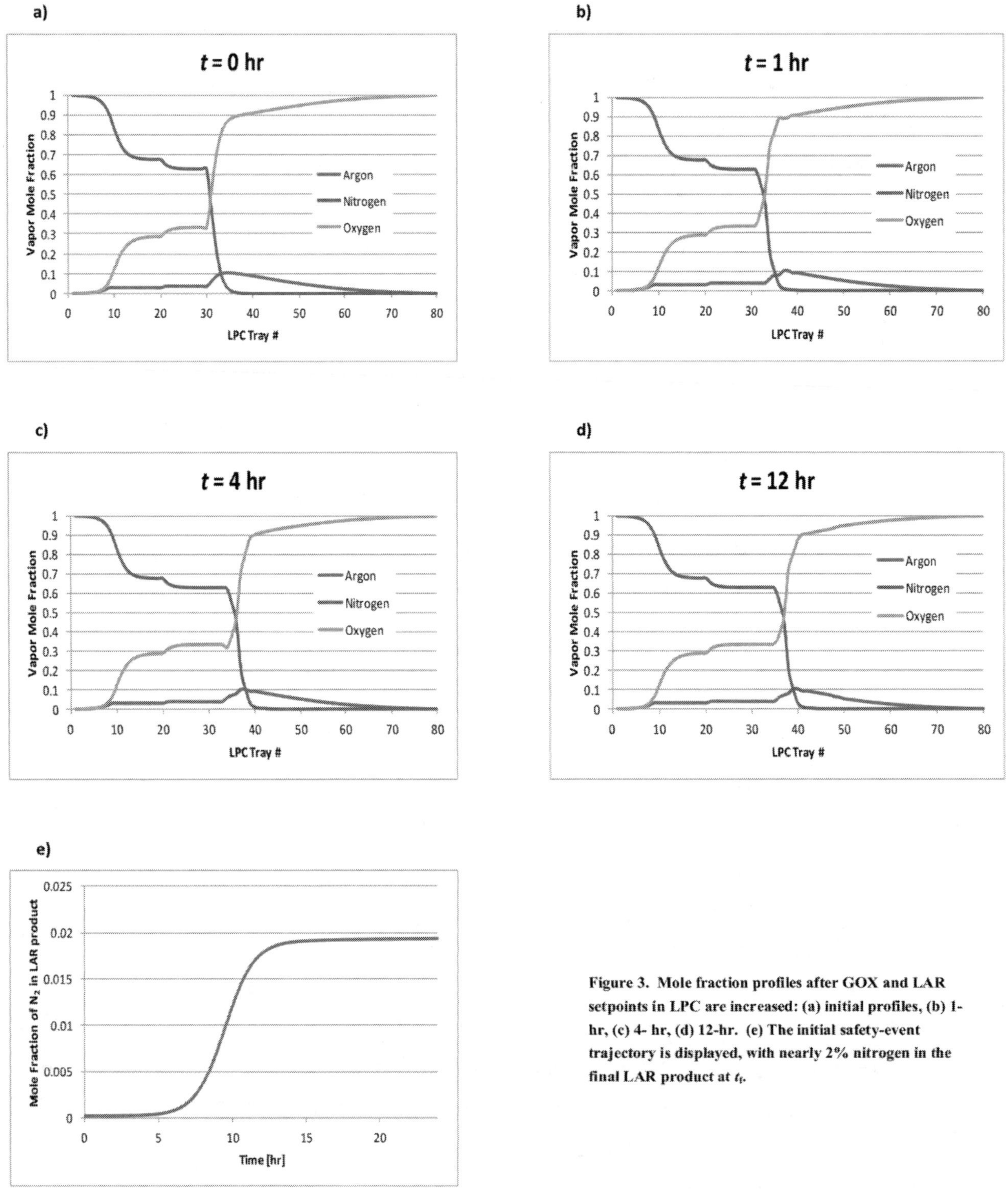

Figure 3. Mole fraction profiles after GOX and LAR setpoints in LPC are increased: (a) initial profiles, (b) 1-hr, (c) 4- hr, (d) 12-hr. (e) The initial safety-event trajectory is displayed, with nearly 2% nitrogen in the final LAR product at t_f.

The second and third steps involve choosing a random time t' (t' = 18 hours) and perturbing all the independent variables, respectively. The independent variables on each tray are $\left\{\frac{dM_i}{dt}, \frac{dx_{i,j}}{dt}, P_{SP}\right\}$, where P_{SP} is the product flow rate setpoint of each of the three liquid products. Each independent variable is sampled from a normal distribution centered at $\left\{\frac{dM_i}{dt}|_{t'}, \frac{dx_{i,j}}{dt}|_{t'}, P_{SP}|_{t'}\right\}$, with a user-defined variance. These distributions play a critical role in determining the types of trajectories uncovered using TPS, to be discussed in the oral presentation. The fourth step of the TPS algorithm involves the forward integration of the perturbed system to $t = t_f$, and the backward integration to $t = 0$. The trial trajectory is shown in Figure 4 – although there is less nitrogen in the LAR at t_f, it is also a rare event of concern. Step six involves determining the relative probability of the trial trajectories compared to the initial trajectory. Likelihood distributions are built using long simulations, generating simulated data by which the model initial conditions and P_{SP} perturbations are evaluated. The 'unrealistic' P_{SP} values used to generate the initial trajectory are relaxed into more realistic values through iterations of TPS. Said differently, as more realistic demand profiles are chosen using random perturbations, the trajectories become more realistic, while continuing to generate safety events. In this way, new and un-postulated safety events are generated, and postulated safety events are better understood. The likelihood distributions play an important role in determining the safety event trajectories generated by TPS – and will be discussed in the oral presentation. As this algorithm is repeated, new safety event trajectories are generated, one of which is displayed in Figure 4.

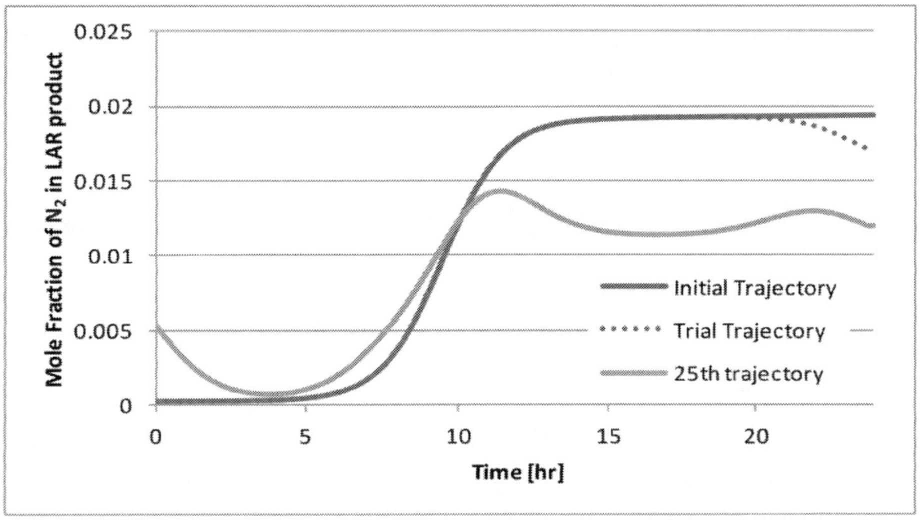

Figure 4: Various rare safety-event trajectories generated using TPS.

Conclusions and Future Work

TPS is an efficient method for generating rare safety-event trajectories. As highly unlikely initial trajectories are relaxed into more realistic trajectories, un-postulated safety events may be discovered. The approaches for generating and accepting these trajectories leave room for user flexibility – these must be tuned appropriately. As trajectories are generated, common trends among them are identified. The long term goal is to influence operation and design enabling previously un-postulated, not adequately appreciated, and/or poorly understood, safety-events to be anticipated and handled appropriately. This novel technique may provide an efficient, stochastic approach to studying rare safety-events in manufacturing processes and should be applicable across all manufacturing industries.

Work Cited

1. P.G. Bolhuis, D. Chandler, and P.L. Geissler. "Transition path sampling: Throwing ropes over rough mountain passes, in the dark," Annual Review of Physical Chemistry, pp. 291-318, 2002.
2. C. Dellago, P. Bolhuis, and P. L. Geissler. "Transition path sampling," Advances in Chemical Physics, pp. 1-84, 2002.
3. Commercial Technologies for Oxygen Production. National Energy Technology Laboratory, available at http://www.netl.doe.gov/research/coal/energy-systems/gasification/gasifipedia/commercial-oxygen, Accessed on March 15, 2016.
4. R. Huang, V. M. Zavala, and L. T. Biegler. " Advanced step nonlinear model predictive control for air separation units," Journal of Process Control, pp. 678-685, 2009.
5. History and Technological Process – Cryogenic Air Separation, The Linde Group, available at http://www.linde-engineering.com/internet.global.lindeengineering.global/en/images/AS.B1EN%201113%20-%20%26AA_History_.layout19_4353.pdf, accessed on March 17, 2016.
6. D. Y. Peng and D. B. Robinson. "A new two-constant equation of state," Industrial & Engineering Chemistry Fundamentals, pp. 59-64, 1976.

12TH GLOBAL CONGRESS ON PROCESS SAFETY
LPS 50th LOSS PREVENTION SYMPOSIUM

Analysis of Overflow Incidents: Searching for the Demons of Situation Awareness

Nicholas P. Sands, CAP, PE
Manufacturing Technology Fellow, DuPont Protection Technology

Palaniappan Chidambaram
Global Solutions Architect, DuPont Sustainable Solutions
DuPont Company (Singapore) Pte Ltd
1 Maritime Square, #10-54, HarbourFront Centre, Singapore 099253

Abstract

Alarms are intended to provide indications to operators to take the necessary actions to prevent undesired consequences. Thus, alarms are often considered as a layer of protection during Process Hazard Analysis (PHA) or Safety/Hazard Review. Without due consideration for human factors that are necessary to ensure its effectiveness, alarms only give a false sense of security.

In several process industry incidents, the designed layers of protection using the operator response to alarm failed. There are several mechanisms for this failure relating to systems or human factors. Among the human factors, there are several situation awareness demons such as attentional tunnelling, data overload and misplaced salience as identified by Endsley (2011).

This paper will examine several overflow incidents for the causes specifically related to the failure of alarms, the human factors component and situation awareness demon, and how it can be addressed during the facility life cycle. The incidents presented in this paper are from a major chemical manufacturing company and from the chemical industry.

Summary

Alarms are commonly used as layers of protection to prevent vessel overflow events in the process industries. A review of industry incidents shows that human error is often a factor in overflow incidents. However in several incidents, the designed layers of protection using the operator response to alarm failed to prevent the consequence. There are several mechanisms for this failure, relating to systems or human factors. Among the human factors, there are several situation awareness demons such as attentional tunnelling, data overload, and misplaced salience as identified by Endsley (2011).

Background

A review of articles on incidents shows that overflow incidents continue to occur, and that human error is a significant factor in overflow events. Further, human error as a factor in overflow incidents may be increasing. The studies indicate that an increased focus on human factors in design may reduce the overflow incidents.

Overfilling of vessels can occur in any process that handles liquid and is of primary concern when the liquids are hazardous in nature, with the potential to cause personnel, environmental, asset and business impact in case of loss of containment. A study of 242 storage tank accidents in North America, Asia and Europe that took place between 1960 – 2003, shows that 30 percent of the incidents were due to operational and maintenance error. Overfilling was identified as the most frequent type of incident in operational error category. Among the 15 overfilling incidents reviewed, 13 cases led to fire and explosion (Chang, 2006).

A further survey of 330 accidents shows human error as the cause in 72 incidents and among them, 12 were overfilling incidents. A review of incidents data set in 21st century showed that the human factor contribution increased to 35 percent for the period 2000-2013 as compared to their contribution share in 1961-2013 (Hemmatian, 2014).

Waite (2013) highlighted 10 tank overfilling incidents that happened between 1972-2009; with eight out of 10 incidents resulting in fatalities and one incident resulting in 43 injuries. These incidents are similar in nature to that of refinery incidents in which columns have been flooded, resulting in loss of containment through the relief system. Waite highlights the need to challenge and review the operator mental models such as reliance on pressure or liquid head without understanding their relation with density, and on out-flow information from a vessel to confirm liquid is not accumulating and to address them adequately during Process Hazard Analysis with special focus on human factors (Waite, 2013).

While these analyses point toward human error, a more focused review of the human and organizational factors that influence such errors will enable prevention of such incidents. A detailed review of factors that led to failure of the operator response to alarms, including failures to alarm, may lead to changes in the design of protection layers. A framework was developed in a recent paper, using concepts from the standard ANSI/ISA-18.2-2009 Management of Alarm Systems for the Process Industries and human factors concepts from the airline industry.

Alarms and Alarm Management

The definition of an alarm, as stated in standard IEC62682 Management of Alarm Systems for the Process Industries, is an audible and/or visible indication to the operator of an equipment malfunction, process deviation or abnormal condition that requires a timely response. The operator response to alarms can correct conditions such as product quality problems, equipment malfunctions, and operating issues that could lead to potentially catastrophic process hazards. Because of the range of consequences related to the alarms directed to a single operator, the alarm system has to be managed for them to be effective. In an unmanaged alarm system, lower priority alarms can obscure the operator's response to critical alarms. The identification of alarms as a protection for process hazards such as a vessel overflow, the design of the alarms and the design of the Human Machine Interface (HMI) are important steps in management of the alarm system as shown in the alarm management lifecycle, described in ISA-18.2 and IEC62682 (Figure 1).

The lifecycle puts activities into stages, and the stages can be mapped to other lifecycles or processes. The identification stage of the lifecycle would include Hazard Review and Layer of Protection of Analysis (LOPA) in a project process or the Hazard and Risk Assessment and Allocation of Protection Layers stages of the Safety Instrumented System lifecycle. Identification activities create potential alarms for the rationalization stage. If the identified potential alarm satisfies the criteria for rationalization, it becomes an alarm. Often alarms identified as safeguards in a PHA, do not pass rationalization for reasons such as the operator not having an action to prevent the consequence or the time to take that action.

The detailed design stage of the lifecycle includes the activities of basic alarm design, advanced alarm design, and HMI design. Alarm dead bands and time delays are part of the basic alarm design. Suppressing the alarm based on process or other conditions is part of the advance alarm design. The way alarms are presented, including the alarm priorities, is part of the HMI design. Mistakes in the basic design of an alarm can cause an alarm to become a nuisance.

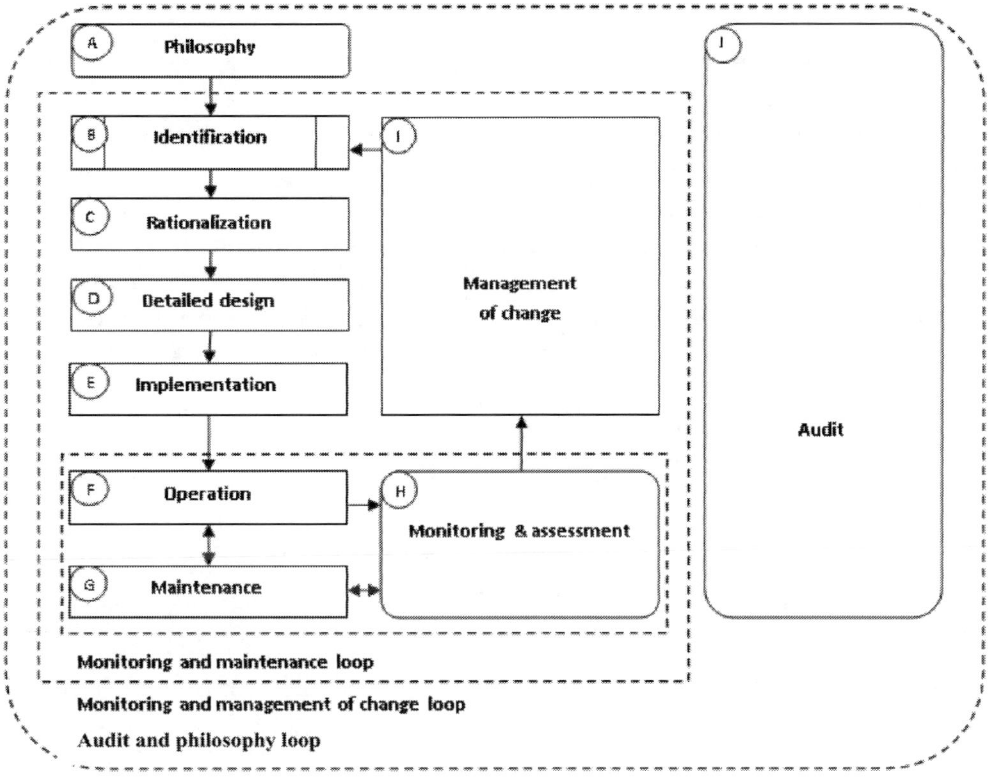

Figure 1 The Alarm Management Lifecycle

Identification of Alarms – Hazard Review

Operator responses to indications and alarms are often considered as effective safeguards without giving much consideration as to their effectiveness and independence in HAZard and OPerability Study (HAZOP) or "What If" analysis. Often it is due to insufficient understanding of what is an alarm and the requirements of an alarm to be an effective safeguard.

It is essential to educate the team that alarms should not be identified as safeguards if the operator:

1) does not have a corrective action than can break the chain of events and bring the process to a safe state;
2) does not have time to take the corrective action between the time the alarm occurs and the time the consequence occurs (process safety time); and
3) training is not provided for operators and documentation of corrective action is not available.

By doing so, a number of configured alarms can be reduced, hence minimizing alarm flood potential and encouraging a culture of responding to alarms if it is activated, instead of ignoring or suppressing them. It is essential for the HAZOP/"What If" team to resist the temptation to add alarms that are ineffective and management systems to ensure those added are managed effectively.

Operator Response to Alarms

ISA-18.2 provides a model for the operator response to an alarm, as shown in figure 2. The model is useful as a starting point for evaluating human error. If the alarm is annunciated, but the operator does not see the alarm, the failure is in detection. If the operator detects the alarm, but does not choose the correct response to the alarm, the failure is in diagnosis. If the operator determines the correct action, but cannot complete the action or complete the action in time, the failure is in the response to the alarm (Dunn, 2013).

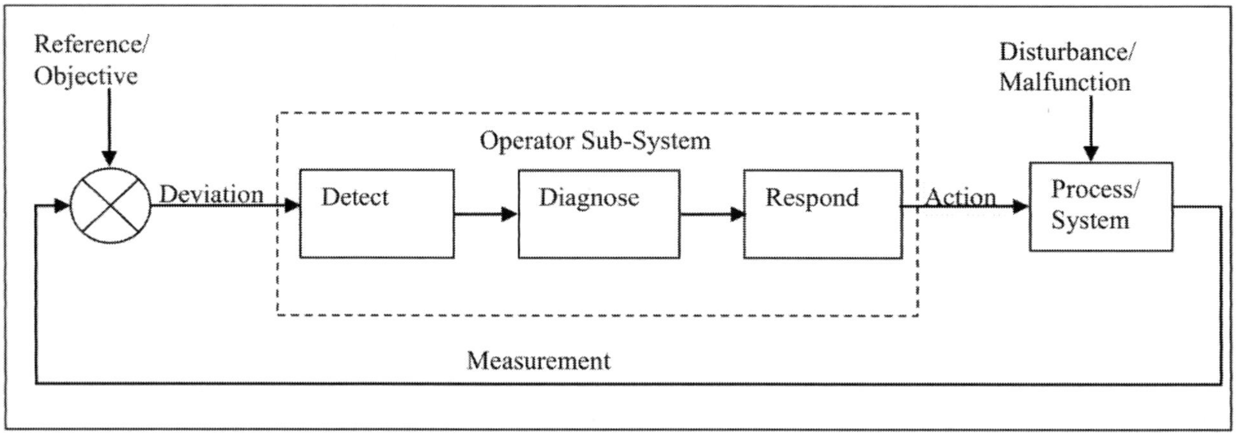

Figure 2. Feedback model of operator process interaction

Alarm Failure Mechanisms

Dunn et al combined a set of alarm failures in a fishbone diagram, shown in figure 3. The top half of the figure are failure mechanisms related to operator response and the bottom half of the figure are failure mechanisms related to the design and maintenance activities of the alarm management lifecycle (Dunn, 2013).

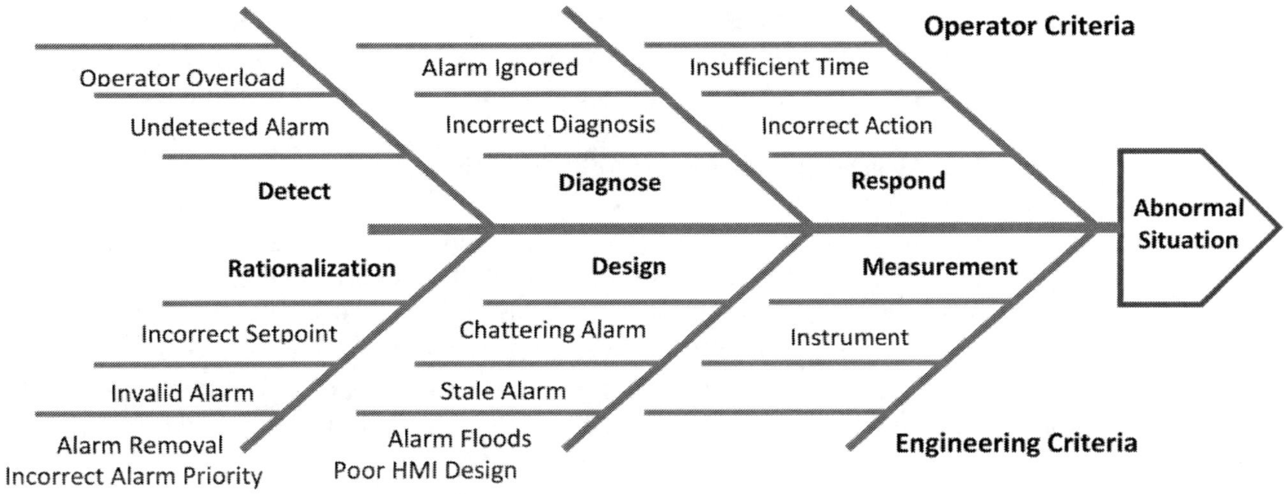

Figure 3. Failure Mechanisms (Fishbone)

Situation Awareness Demons

To further evaluate the human factors that contribute to human error in response to an alarm, the concepts of situation awareness and detrimental factors or situation awareness demons, can be borrowed from the work of Endsley (2011). "Situation awareness is being aware of what is happening around you and understanding what that information means to you now and in the future" (Endsley, 2011).

The situation awareness demons were restated specific to alarm management (Dunn, 2013):

- Attentional Tunnelling – Focusing on one area or issue to the extent that alarms from another area or issue are excluded;
- Requisite Memory Trap – Over dependence on operator memory for effective response to alarms, perhaps under different operating conditions;
- Workload, Anxiety, Fatigue and Other Stressors – (WAFOS) physical or psychological stress that can impact performance;
- Data Overload – Over dependence on the operator to make sense of excessive amounts of alarm information;
- Misplaced Salience – Incorrect alarm priority or HMI representation of alarm importance and other status information;
- Complexity Creep – complex response or complex HMI features that increase response complexity;
- Errant Mental Models – Thought process that incorrectly interprets alarms or mistakenly discounts relevant alarms; and
- Out-of-the-Loop Syndrome – Process or response automation that causes the operator to not know what to do when automation fails.

Neither the failure mechanisms nor the situation awareness demons are mutually exclusive, therefore more than one can be related to a specific incident. It is helpful to identify only the minimum number of failures or demons related to an incident.

Alarms – Are we crediting appropriately in LOPA?

Operator response to alarms is often given more credit without a detailed evaluation and such practices result in a higher residual risk. In 2007, as part of HSE initiative, Bellamy conducted a review of seven LOPA studies from different companies related to overfill scenarios in tank farm. He concluded that the demonstration of independence of protection layers and justification for credit assigned for operator response to high-level alarms was lacking in general (Bellamy, 2007). It is critical to ensure that operator response to alarms are effective, especially for batch processing facilities, unloading/loading operations and in non-routine operations, where reliance on human intervention is high. Myers (2013) emphasizes the need to consider the site specific culture while estimating human error probabilities. Myers also highlighted typical issues in the LOPA when considering human error as initiating events and human intervention as independent protection layers.

Kletz (2002) highlights that incident reports are superficial, and often stop at the first evident cause or immediate causes without further investigation into ways to avoid hazards, organizational factors that contributed to the incidents and identification of the factors that influenced the human error instead of just listing human error as cause. This results in increased potential for incident recurrence. Many incident investigations do not contain the details necessary to identify the failure mechanisms and the situation awareness factors.

Questions that capture some of the information during an investigation specific to alarms include:

- Was there another process upset occurring at the same time?
- What was the alarm rate at the time of the incident?
- Was the alarm or alarms of interest frequent alarms or nuisance alarms?
- Was the alarm or alarms of interest easy to distinguish in the HMI?
- Was consequence of deviation and associated response documented and training provided?

Overflow Incidents Analysis

The incidents described below are reduced to the facts needed for the failure analysis:

Incident 10 Tank Overflow

During a start-up of a process unit, a tank of organic solvent, above its flash point, overflowed causing a vapor cloud, which did not ignite. The level rise generated at least two different alarms prior to the overflow. Two other units in the span of control of the same control room operator were conducting maintenance tests that generated a significant number of alarms. The operator did not respond to the tank-level alarms. A field operator reported the vapor cloud.

The facts related to the alarm system failures can be stated as follows:

- The operator was distracted by the maintenance testing in the two units.
- The operator did not respond to the tank high-level alarms.
- The tank high-level alarms were not nuisance alarms.

The failures can be mapped to the following failure mechanisms:

- Diagnose Failure
    - Operator did not respond to the tank high-level alarms.
        - Operator was distracted by the maintenance testing.
        - Operator may have believed the alarms were triggered by maintenance testing.

The failures can be mapped to the following SA conditions:

- Attention Tunnelling
    - Operator was distracted by the maintenance testing.
- Errant Mental Model
    - Operator did not connect tank high level alarm to unit that was in start-up.

Incident 11 Sump Overflow

Hazardous waste sump overflowed when redundant sump pumps lost prime and cavitated due to failure of common seal flush system. High level and high-high level alarms sounded before overflow. A process upset was occurring simultaneously and the control room operator was addressing the upset and did not respond to the alarms. A field operator identified the overflow. ~500 gallons of material was spilled.

The facts related to the alarm system failures can be stated as follows:

- The operator was focused on the upset in the operating area.
- The operator did not respond to the sump high-level alarms.
- The sump high-level alarms were not nuisance alarms.

The failures can be mapped to the following failure mechanisms:

- Diagnose Failure
    - Operator did not respond to the sump high-level alarms.

        - Operator was focused on the upset in the operating area.

The failures can be mapped to the following SA conditions:

- Attention Tunnelling
    - Operator was focused on the upset in the operating area.

Incident 12 Sump Overflow

During a plant shutdown, a check valve failed, allowing backflow of a hazardous material from a tank to a sump. An alarm indicated the high sump level condition, and the control room operator contacted the field operator to empty the sump. Later a second high sump level alarm occurred and the control operator did not contact the field operator. The sump overflowed, spilling ~4,500 lbs of material until the spill was discovered.

The facts related to the alarm system failures can be stated as follows:

- The operator was less attentive to alarms during the plant shutdown.
- The operator did not respond to the second high level sump alarm.
- The sump high level alarms were nuisance alarms due to rain.

The failures can be mapped to the following failure mechanisms:

- Diagnose Failure
    - Operator did not respond to the sump high level alarm.
        - Operator was not as vigilant due to the plant shutdown.

The failures can be mapped to the following SA conditions:

- Errant Mental Model
    - Operator was not as vigilant due to the plant shutdown.

Incident Analysis summary

The results of above discussed overflow incidents 10-12 have been included in the previous set of analysis by Dunn (2013) to demonstrate the importance of understanding situational awareness and alarm failure mechanisms. As evident from table 1 below – errant mental models and inability to diagnose are key contributors to incidents where alarms failures were identified as contributors:

	Failure Mechanisms (FM)								Situation Awareness (SA)								
Incident	Rationalization	Design	Measurement	Detect	Diagnose	Respond	FM Totals		Attention Tunneling	Requisite Memory Trap	Workload, Anxiety, Fatigue...	Data Overload	Misplaced Salience	Complexity Creep	Errant Mental Models	Out-of-the-Loop	SA Totals
1		X		X	X		3				X	X			X		3
2		X	X	X	X		4								X		1
3		X	X	X			3										0
4	X				X		2		X						X		1
5	X				X		2								X		1
6					X		1								X	X	2
7					X		1		X			X					1
8					X		1								X		1
9					X		1					X		X	X		3
10					X		1		X						X		1
11					X		1		X								1
12					X		1								X		1
	2	3	2	3	11	0	18		4	0	1	3	0	1	9	1	14
% of FM/SA Total	11%	17%	11%	17%	44%	0%			14%	0%	7%	21%	0%	7%	57%	7%	
FM/SA % of Incidents	17%	25%	17%	25%	92%	0%			33%	0%	8%	25%	0%	8%	75%	8%	

Table 1. Overflow Incidents – Alarm Failure Analysis

Conclusions

Overflow incidents are common, and the contribution of human error is a significant factor. In this paper, we have highlighted the need to ask the right questions during incident investigations in order to learn and understand the human factors related to alarm failures. We have demonstrated the importance of understanding the human factors related to alarms and addressing them adequately during the alarm life cycle using alarm failure analysis framework. Improvements can be made by ensuring situation awareness demons and potential failure mechanisms are addressed adequately in the project life cycle when alarms are identified as safeguards and credited as protection layers.

References

Bellamy, L.J., Geyer, T.A.W., 2007, *Development of a working model of how human factors, safety management systems and wider organizational issues fit together,* Health and Safety Executive 2007

Chang, J. I., Lin, C-C., 2006, *A study of storage tank accidents*, Journal of Loss Prevention in the Process Industries 19 (2006) 51–59

Dunn, D. G., Sands, N. P., Stauffer. T., 2013, *When Good Alarms Go Bad: Learnings from Incidents*, 70th Annual Instrumentation and Automation Symposium – Texas A&M University

Endsley, M. 2011, *Designing for Situation Awareness: An Approach to User-Centered,* 2nd Design, Second Edition

Hemmatian, B., Abdolhamidzadeh, B., Darbra, R.M., Casal, J., 2014, *The significance of domino effect in chemical accidents,* Journal of Loss Prevention in the Process Industries 29 (2014) 30-38

Kletz, T.A, 2002 *Accident Investigations - Missed Opportunities,* Institution of Chemical Engineers Trans IChemE, Vol 80, Part B, January 2002

Myers, M.P, 2013, *Layer of Protection Analysis e Quantifying human performance in initiating events and independent protection layers*, Journal of Loss Prevention in the Process Industries 26 (2013) 534-546

Waite, P., 2013, *The Chemical Engineer Magazine,* iChemE (March 2013)40-44

INHERENTLY SAFER PROCESS DESIGN FOR VERTICAL KNOCK-OUT DRUMS AND VAPOR/LIQUID SEPARATORS IN GAS PROCESSING

By
George Maliakkal
Bechtel Corporation
Oil, Gas and Chemicals
Houston, Texas
gjmaliak@bechtel.com

David Rodriguez
Bechtel Corporation
Oil, Gas and Chemicals
Houston, Texas
drodrigu@bechtel.com

Prepared for Presentation at
American Institute of Chemical Engineers
2016 Spring Meeting
12th Global Congress on Process Safety
Houston, Texas
April 11-13, 2016

INHERENTLY SAFER PROCESS DESIGN FOR VERTICAL KNOCK-OUT DRUMS AND VAPOR/LIQUID SEPARATORS IN GAS PROCESSING

George Maliakkal
Bechtel Corporation
Oil, Gas and Chemicals
Houston, Texas
gjmaliak@bechtel.com

David Rodriguez
Bechtel Corporation
Oil, Gas and Chemicals

Keywords: Inherently Safer, Vapor Liquid Separators, Pool Fire, Depressuring, Fireproofing

ABSTRACT

In gas processing and LNG production, large diameter (> 2 meter) vertical vapor/liquid separator vessels and compressor suction knock-out drums are common. These separate small (<0.1wt%) fraction of flammable liquid from large volumes of gas. The liquid level in these vessels is typically controlled above the bottom tangent line of the vessel permitting storage of large quantity of flammable liquid at the bottom of the vessel, often in excess of the threshold volume for pool fire. To safeguard against hazards of pool fire, such vessels must be designed for fire-depressuring. Piping, adjacent structures and equipment associated with the vessels must also be fireproofed. Inclusion of a small diameter (< 0.5 meter) boot below the bottom of the vessel enables relocation of the level control from the main vessel to the boot. The large liquid inventory in the bottom of the main vessel is removed. The liquid inventory in the boot will be reduced, at least by one order of magnitude to well below the threshold for pool fire, eliminating pool fire potential. This makes the system inherently safer while eliminating costs associated with fire-depressuring and fireproofing.

1.0 INTRODUCTION

In gas processing plants and LNG plants, high pressure-large diameter vertical vessels handling hydrocarbon vapor and liquid are common. The hydrocarbon liquids in these drums are generally highly volatile and flammable. These vessels may be compressor suction drums or vapor/liquid separators. These vessels separate small fractions of liquid from large volumes of gas. The liquid level in these vessels are normally controlled in the bottom section of the vessel, above the bottom tangent line. For large diameter vessels, this type of level control will result in large inventory of flammable liquid to be present in the vessel all the time. The inventory can be in excess of the threshold volume for pool fire, which may occur in case of a leakage or spill. In that case, to safeguard against pool fire, equipment, structures and piping in the vicinity of the vessel must be fireproofed. Also, the process system associated with the vessel must be designed for fire-depressuring. These requirements make the system more expensive to design and build, even as the hazard of fire associated with carrying large inventory of flammable liquid continues to be present. The method proposed here, of controlling the liquid level in a boot at the bottom of the vessel with inventory less than the threshold volume for pool fire, will make the vessel and associated system inherently safer while reducing capital expenditure to design and build.

2.0 TRADITIONAL LEVEL CONTROL IN VERTICAL DRUMS

2.1 VESSELS WITH HEMISPHERICAL HEADS

Figure 2.1 shows typical level control for a vertical vapor/liquid separator (in this case a compressor suction drum) in a gas processing plant, where the feed gas to the drum generally has only a small fraction of entrained liquid. The volume below the high liquid level is considered as inventory for pool fire assessment. This includes the volume in the cylindrical section as well as the hemispherical section. The generally accepted industry threshold for pool fire is that the flammable liquid volume be greater than 5 cubic meters or ≈180 cubic feet. Generally, to facilitate fabrication, large diameter vessels with high design pressures are built with hemispherical heads. The wall thickness for hemi-head is one half of that for ellipsoidal head for vessel with the same diameter and design pressure.

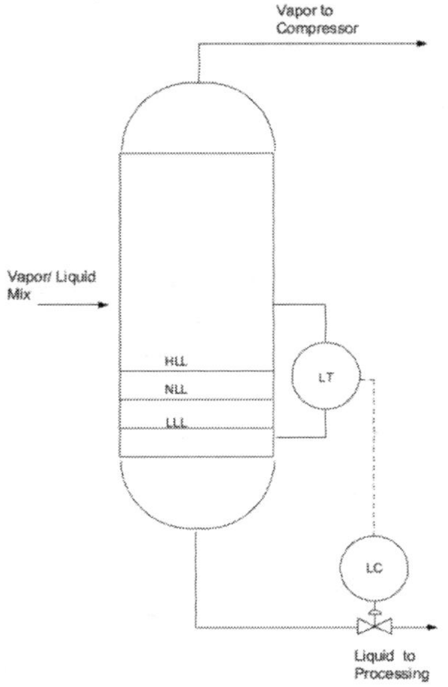

Fig. 2.1: Vessel with Hemispherical Head

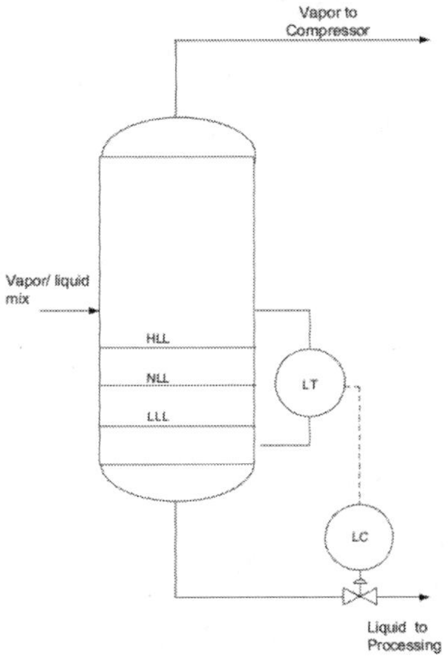

Fig. 2.2: Vessel with 2:1 Ellipsoidal Head

2.2 VESSELS WITH ELLIPSOIDAL HEADS

Vessels with lower design pressures will have 2:1 ellipsoidal heads. Clearly, the vessels with ellipsoidal heads carry less inventory in the head. Figure 2.2 illustrates vertical vessel with ellipsoidal head. Whether hemispherical or ellipsoidal head, the total liquid inventory below high operating liquid level should not exceed the guideline (5 m^3) if one has to avoid pool fire.

3.0 APPLICATIONS OF LEVEL CONTROL IN A BOOT

3.1 HORIZONTAL FILTER/COALESCERS IN GAS PROCESSING

It is common practice in hydrocarbon industry to collect the separated liquids in a "boot" and remove the liquids under level control from the boot, when the vapor/liquid separator is horizontal. Such is the case with filter/coalescers. Typical illustration is shown in Figure 3.1.

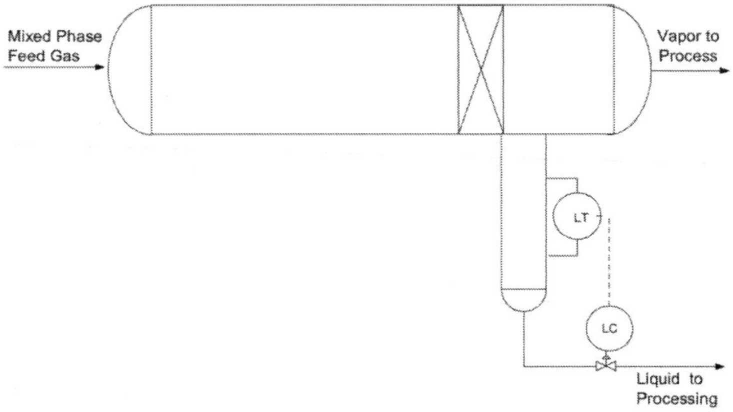

Fig. 3.1: Filter/Coalescer in Gas Processing

3.2 VACUUM COLUMN IN CRUDE OIL PROCESSING

In petroleum refining industry level control in a boot is used at the bottom of vacuum tower in crude processing, with the intent to minimize residence time of vacuum resid in order to prevent coking at the prevailing high temperature. This is illustrated in Figure 3.2.

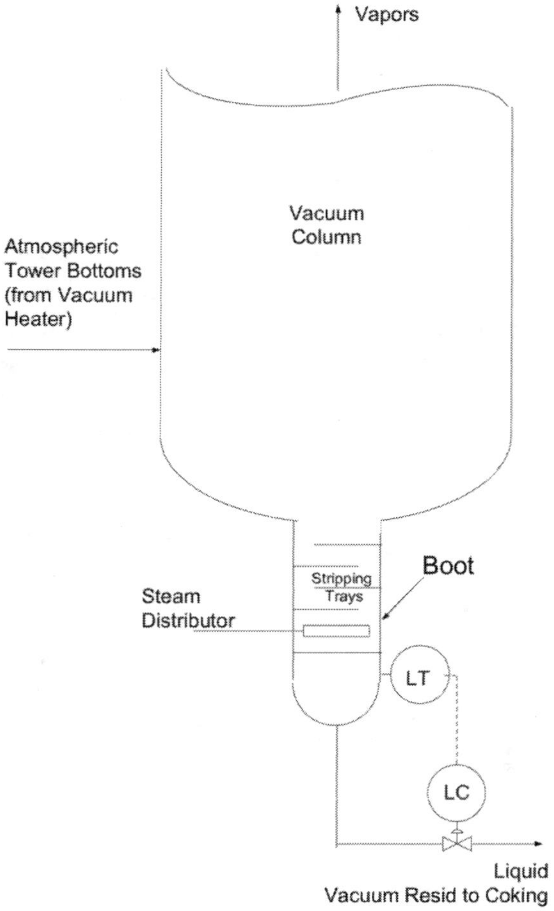

Fig. 3.2: Vacuum Column with Boot in Crude Oil Processing

4.0 BASIS FOR LIQUIDS REMOVAL BY LEVEL CONTROL IN A BOOT

As described in Section 3.0 a "boot" design is used for applications of liquids removal both in horizontal and vertical vessels. The liquid removal operation is independent of the vapor/liquid separation process that occurs in the body of the main vessel. Therefore, liquid removal by level control in a boot can be incorporated into any vessel for vapor/liquid separation, regardless of vessel geometry. Only question is: are there tangible benefits for doing so and under what circumstances?

4.1 APPLICABILITY OF LEVEL CONTROL IN A BOOT FOR VERTICAL VAPOR/LIQUID SEPARATORS

In LNG and gas processing, large volume flow rates of gas, sometimes exceeding a billion standard cubic feet of gas per day, are flowed through vapor/liquid separators and compressor suction drums in order to remove traces of liquid. The prevailing pressures range from low to high, from 30 barg to 120 barg. The presence of liquid can be because of entrainment from previous processing step, condensation due to temperature drop or due to retrograde condensation due to pressure drop. These liquids are volatile hydrocarbons, typically molecular weight that of heptane. The proportion of liquid present in these gas streams is rather small, generally less than 0.1 weight %. Large diameter vessels are needed to process the large volumes of gas. High design pressures leads to high wall thickness for the vessel. Vertical vessels 5 meters in diameter with 125 millimeter wall thickness are not uncommon.

The traditional method of removing liquid from these vertical vessels by controlling the level of liquid in the main diameter results in carrying large inventory of liquid in the bottom of the vessel. For a 5 meter diameter vessel with hemispherical bottom, this inventory below high operating liquid level is estimated to be 60 cubic meters, which is far in excess of the 5 cubic meter limit for pool fire. With the addition of a small diameter "boot" at the bottom of the vessel, the level control for the small amount of liquid being removed in the separator can be shifted to the boot. Because only very small amounts of liquid is collected in the vessel, the boot can be as small as a tenth of the diameter of the main vessel. The boot can be designed to carry less than 1 cubic meter of liquid inventory, with no liquid inventory in the main vessel. The inventory is much lower than the threshold 5 m^3 for pool fire. The avoidance of pool fire makes the design inherently safer with significant benefits related to safety and cost. The traditional level control and level control in a boot are compared for a typical vapor/liquid separator in Figures 4.1a and 4.1b.

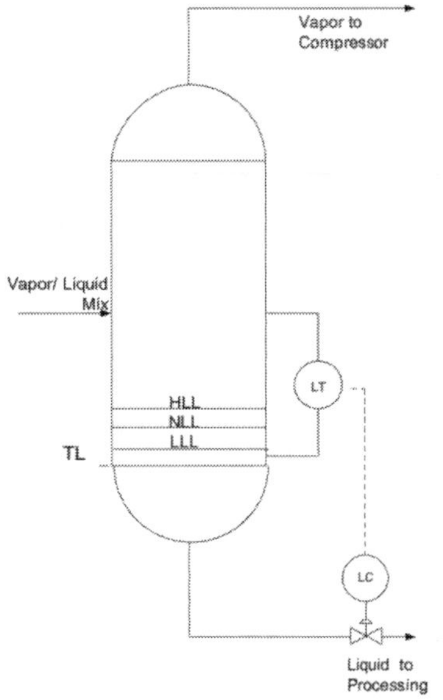

Fig. 4.1a Traditional Level Control

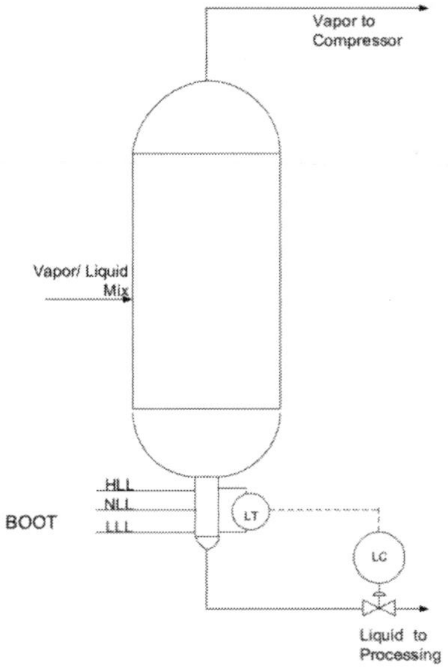

Fig. 4.1b Level Control in a Boot

5.0 SAFETY BENEFITS OF LEVEL CONTROL IN A BOOT

5.1 AVOIDANCE OF POOL FIRE

The liquid level control in a boot design is applicable to practically all vapor/liquid separators and compressor suction drums encountered in gas processing, being that these gas streams have only very small liquid content. The inventory of flammable hydrocarbons in the boot will be well below the threshold for pool fire, thus avoiding hazards associated with pool fire in the event of a spill. In comparison, control of level in the body of the vessel comes with large inventory of flammable hydrocarbons with the associated hazard of pool fire.

5.2 *AVOIDANCE OF FIRE DEPRESSURING*

API 521 requires that vessels subject to pool fire be designed for fire depressuring. For vessels thicker than 1", the mandatory depressuring criterion is from design pressure to 50% of design pressure in 15 minutes in the case of a fire emergency. If there is no potential for a pool fire, as is the case with boot level control, mandatory depressuring rate is not applicable. Depressuring can be spread over longer time to suit the convenience of plant operations and maintenance. Accordingly, the depressuring load and depressuring header size can be much smaller with the boot option. The relief load to the flare is also reduced.

6.0 COST SAVINGS OF LEVEL CONTROL IN THE BOOT

6.1 *AVOIDANCE OF FIRE PROOFING*

With pool fire comes the requirement for fireproofing. The fireproofing is required for all equipment, structures and piping within several meters radius around the vessel and several meters height above grade. Material such as *pyrocrete 241* or equal is used for fireproofing. The cost associated with such fire proofing can run into the thousands of dollars, depending on the size of the vessel. With no fire proofing needed with a boot design, the cost savings can be substantial.

6.2 *SAVINGS RELATED TO VESSEL SIZE/COST RELATED TO BOOT*

The boot allows the high liquid level for the vessel to be set in the boot, whereas with level control on the main vessel diameter, the high liquid level will be set well above vessel tangent line. Once the high liquid level is moved to the boot, the full depth of the head is free of liquid. The chance of liquid re-entrainment due to inlet vapor distribution is eliminated. The T/T height of the vessel can be at least less by length equal to one radius in the case of hemi-head vessel and half radius for vessel with ellipsoidal head. In comparison, the design with level control in the boot will require the addition of a small diameter cylindrical section, below the bottom of the vessel.

7.0 EXAMPLE VERTICAL VESSEL DESIGN WITH BOOT

Figure 7.0a and 7.0b shows process details of a vertical vessel (compressor suction drum) design with and without boot in gas processing application.

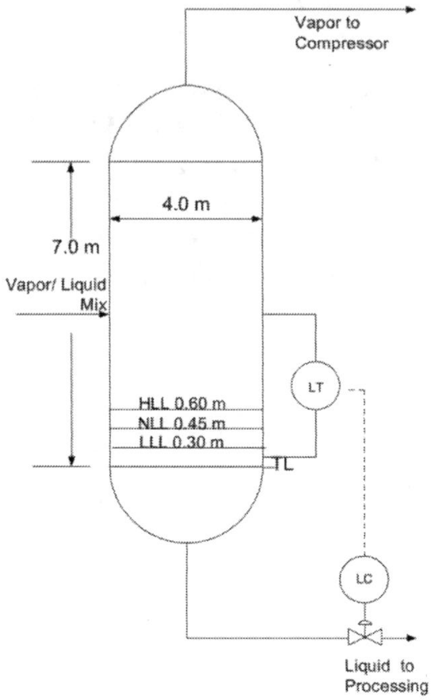

Fig. 7.0a Traditional Level Control

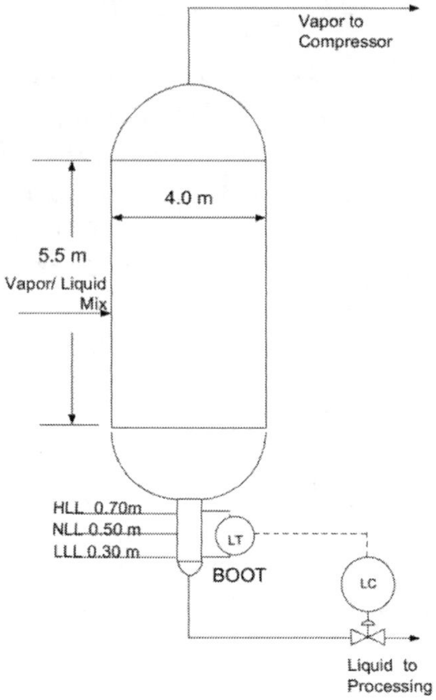

Fig. 7.0b Level Control in a Boot

Feed gas is natural gas with maximum of 0.1 wt% liquid dispersed in the gas. For residence time purposes, the assumption is that the liquid is completely separated in the vessel. The two designs are compared in Table 7.0. With the boot, the inventory of flammable liquid is reduced from 24.5 m3 to 0.2 m3, a hundred-fold reduction, which is far below the threshold for pool fire of 5 m3. Accordingly no fireproofing or no fire depressuring are needed for design with a boot.

TABLE 7.0

EXAMPLE VERTICAL VESSEL DESIGN W/ AND W/O BOOT

COMPRESSOR SUCTION DRUM

Row No.	CONFIGURATION		VERTICAL VESSEL	VERTICAL VESSEL WITH BOOT
1	LEVEL CONTROL LOCATION		ON MAIN DIAMETER OF VESSEL	ON THE BOOT
2	GAS FLOW RATE	kg/hr	880,000	880,000
3	GAS FLOW RATE	mmscfd	1,050	1,050
4	LIQUID CONTENT	wt%	0.1	0.1
5	LIQUID (TYPICAL)		HEPTANE	HEPTANE
6	VOLUME RATE OF LIQUID	m^3/hr	1.35	1.35
7	VESSEL DIAMETER	m	4	4
8	VESSEL T/T	m	7	5.5
9	BOOT DIAMETER	m	-	0.5
10	BOOT HEIGHT	m	-	1
11	HLL (above bot tan)	m	0.6	-
12	LLL (above bot tan)	m	0.3	-
13	HLL (above BOOT bot tan)		-	0.7
14	LLL (above BOOT bot tan)		-	0.3
15	VOL BETWEEN HLL AND LLL	m^3	3.8	0.08
16	LIQUID RESIDENCE TIME (LLL TO HLL)	minutes	165	4
17	VESSEL FLAMMABLE LIQUID INVENTORY (BELOW HLL)	m^3	25	0.2
18	DESIGN FOR POOL FIRE		YES	NO
19	FIREPROOFING REQUIRED		YES	NO
20	FIRE DEPRESSURING REQUIRED?		YES	NO
21	FIRE DEPRESSURING RATE	kg/hr	310,000	84,000
22	DEPRESSURING HEADER SIZE	in.	16	10

Note on Residence Time:

While controlling the liquid level in the main diameter gives a residence time of 165 minutes between high and low liquid levels, the level control on the boot gives a residence time of 4 minutes. A residence time of 4 minutes is consistent with process engineering practice for liquids pressured out of vessels, as is the case with most of these separators that operate at relatively high pressures in gas processing applications. The potential for vapor breakthrough is likewise no different than other similar process systems. With 165 minutes of residence time for the main diameter level control, the level control valve is exercised very infrequently, whereas with the boot design, it is exercised more frequently, which may be in favor of boot design.

8.0 CRITERIA FOR SELECTING A BOOT FOR LEVEL CONTROL

Vessels of various diameters with ellipsoidal head and hemispherical head were evaluated to determine the breakpoint diameter at which boot design is suggested to avoid pool fire. Since the subject analysis applies only to separators with small fraction of liquid compared to the vapor flow, the operating levels would be set by minimum gaps between various level settings for purposes of level measurement rather than by residence time considerations. This led to a minimum high liquid level (HLL) of 0.625 meter above bottom tangent line. The volume of liquid inventory below this level was calculated for various diameters for vessels with hemispherical and ellipsoidal heads. The table below indicates that below diameter of 2.48 meters for ellipsoidal heads and below diameter of 2.18 meters for hemispherical heads, the flammable inventory is less than the threshold of 5 cubic meter for pool fire. Boot level control is suggested for vessels above these diameters to avoid pool fire by the "5 m^3 inventory" criterion and the consequent mandatory mitigation requirements.

TABLE 8.0

Vertical Vessel with Ellipsoidal (2:1 SE) Head					
Diameter	meter	2.00	2.25	2.48	2.50
	feet			8	>8
Volme, head	m3	1.05	1.49	2.00	2.05
HLL (above BTL)	m	0.625	0.625	0.625	0.625
Volume, cylinder (to HLL)	m3	1.96	2.49	3.02	3.07
Liquid Inventory	m3	3.01	3.98	5.02	5.12
Meets Pool Fire Criterion OF 5 m3?		No	No	Yes	Yes
Candidate for Vessel with Boot?				Yes	Yes

Vertical Vessel Hemi Head						
Diameter	meter	2.00	2.18	2.20	2.25	2.50
	feet		7		>7	
Volme, head	m3	2.09	2.71	2.79	2.98	4.09
HLL (above BTL)	m	0.625	0.625	0.625	0.625	0.625
Volume, cylinder (to HLL)	m3	1.96	2.33	2.38	2.49	3.07
Liquid Inventory	m3	4.06	5.05	5.16	5.47	7.16
Meets Pool Fire Criterion OF 5 m3?		No	Yes	Yes	Yes	Yes
Candidate for Vessel with Boot?			Yes	Yes	Yes	Yes

As was indicated in the example in Section 7.0, level control in the boot reduces flammable inventory to very low levels, to a fraction of a cubic meter. The hazard associated with carrying flammable inventory is nominally proportional to the inventory, and any means by which this inventory could be reduced should be considered for safer design. Thus, even vessels with diameter as low as 1 meter (with inventory ≈1 cubic meter) are candidates for level control in a boot.

9.0 CONCLUSION

Exercising liquid level control in a boot below the bottom rather than in the main vessel section is a viable and preferred method for removing separated liquid from vertical vapor/liquid separators and compressor suction drums that are encountered in gas processing and LNG plants, where the liquid fraction in the mixed phase feed stream is small. For vessels over 2 meter diameter, the inventory of flammable liquid below high liquid level is likely to be greater than 5 cubic meters, the generally accepted threshold spill volume for pool fire. Fire-depressuring and fireproofing of equipment, structures and piping in the vicinity of the vessel become mandatory. For such vessels, having liquid level control in a boot reduces the inventory to a fraction of a cubic meter, making the vessel inherently safer without the need for fire depressuring and fireproofing. Even for smaller diameter (<2 meters) vertical separator vessels, the significant reduction in flammable inventory when considering level control in a boot as compared to level control in the main vessel may warrant a boot design because of being inherently safer.

10.0 REFERENCES

[1] Pressure-relieving and Depressuring Systems, API Standard 521, Sixth Edition, January 2014

[2] Fireproofing Practices in Petroleum and Petrochemical Processing Plants, API Recommended Practice 2218, Third Edition, July 2013

In-breathing requirement for a tempered tank exposed to prolonged sudden rainfall

Presenter/Dr.-Ing. Davide Moncalvo
Dr.-Ing. Michael Davies
Braunschweiger Flammenfilter GmbH (PROTEGO Germany)
Industriestraße 11, 38122 Braunschweig, Germany
davide.moncalvo@protego.com

Prof. Dr.-Ing. Roman Weber
Prof. Dr.-Ing. Reinhard Scholz
Institute of Energy Processes Engineering and Fuel Technology, TU Clausthal
Agricolastraße 4; 38678 Clausthal-Zellerfeld, Germany
weber@ievb.tu-clausthal.de

Prepared for Presentation at
American Institute of Chemical Engineers
2016 Spring Meeting
12th Global Congress on Process Safety
Houston, Texas
April 11-23, 2016

In-breathing requirement for a tempered tank exposed to prolonged sudden rainfall

Presenter/Dr.-Ing. Davide Moncalvo
Dr.-Ing. Michael Davies
Braunschweiger Flammenfilter GmbH (PROTEGO Germany)
Industriestrasse 11, 38122 Braunschweig, Germany
Davide.moncalvo@protego.com

Prof. Dr.-Ing. Roman Weber
Prof. Dr.-Ing. Reinhard Scholz
Institute of Energy Processes Engineering and Fuel Technology, TU Clausthal

Keywords: enter a few keywords here to aid searching

Abstract

In this paper a model is presented aimed to determine the air in-breathing requirements to compensate for collapse of a tempered tank when vapors condense inside due to the strong cooling brought by a prolonged intense rainfall. The effect of liquid product evaporation on tank cooling dynamics is also discussed. The presented model shows that vapor condensation strongly affects the in-breathing requirement and that liquid evaporation affects the cooling dynamics.

In order to verify this calculation procedure with measurements, a test program with a cylindrical API 12F tank will be initiated.

1 Introduction

Tempered tanks are an equivalent way to address liquid storage tanks. Protection of these tanks against corrosion of the bare metal surface due to rainfall is accomplished through either painting or applying an organic coating. Painting, though cheaper, does not protect the stored liquid from noticeable cooling in case of prolonged rainfall. Since liquid storage tanks have design pressures near atmospheric levels, a prolonged rainfall may result in vapor condensation, which may result in tank deformation and eventually collapse. In order to prevent this occurrence the general practice is the installation of vacuum breathers with a cumulative capacity larger than the so-called in-breathing requirement of the tank.

The principal standards governing breathing of low-pressure liquid storage tanks are ISO 28300 [1], API 2000 7th Edition [2], and EN 14015 [3]. These standards, as far as in-breathing requirements are concerned, vaguely refer to an earlier work of Förster et al. [4], where the rainfall was modeled as a uniform rain coat on the tank. Fullarton [5,6] evinced the effect of condensation on the in-breathing requirement of a tank filled with humid air saturated with steam and methanol vapors. On one hand, condensation releases latent heat to the walls thus dampening their cooling and therefore the in-breathing requirement. On the opposite hand, it is responsible for additional in-breathing to fill in the volume previously occupied by condensed vapor. A later work by Salatino et al. [7] tried to quantify the effect of condensation but he lacked proper modeling for the dampening of the wall temperature.

In this paper the authors are developing a model for an uninsulated liquid-filled tank exposed to a severe rain shower after a prolonged, uninterrupted solar exposure. Both vapor condensation and liquid evaporation will be addressed in the mass transfer dynamics. For simplicity the vapor space would be assumed to contain initially only liquid vapors.

2 Thermodynamical model

A sketch of a tank exposed to a rain shower of given intensity and impingement angle is shown in Fig. 1. In general, rain and ambient air temperatures differ. For a first analysis the rain film is assumed to be of constant thickness along the lateral walls and the roof. In the following model the bulk product inside the tank is assumed to have a uniform temperature and that the liquid level height is assumed to remain constant.

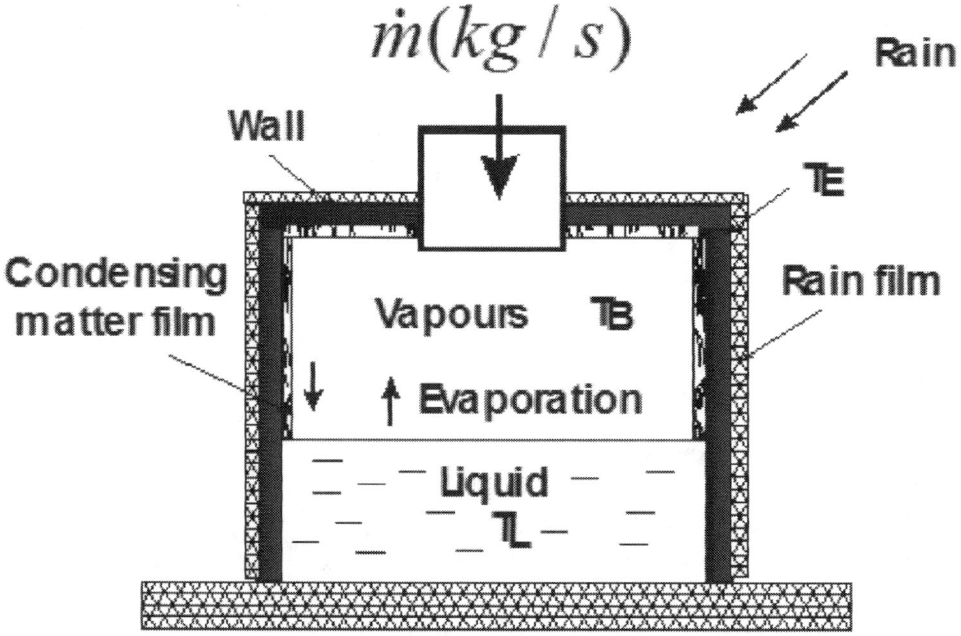

Figure 1: Model for the tank exposed to a rainfall

In this model the condensed vapors are assumed to form a film that is at the same temperature of the bulk product and condensation is expected to occur mostly at the cold tank walls, while evaporation mostly at the liquid free surface. Additionally, the eventuality of a non - uniform wall temperature distribution is neglected. Under these assumptions, it is possible to write a separate energy balance for the tank product and the tank walls. The energy balance for the tank product in Eq. (1) is derived by the authors. According to this equation, the rate of cooling of the gas bulk temperature is affected by both the internal heat transfer to the walls and the warming-up of the in-breathed colder air.

$$\left(M_{B\,liq}c_{B\,liq} + M_{B\,vap}c_{B\,vap}\right) dT_B/dt = -h_{int}A_{lat}(T_B - T_{walls}) + \dot{M}_{air}c_{air}(T_{amb} - T_B) \quad (1)$$

The energy equation for the tank walls in Eq. (2) relates the temperature change in the walls to the internal convective heat inflow from the tank product, the external heat transfer to the cold rain film and the heat release to the walls from the condensing liquid film. Its structure is similar to the one used by Fullarton [6], differing from it in the estimation of the rate of condensation in the tank.

$$M_{walls}c_{walls}\frac{dT_{walls}}{dt} = h_{int}A_{lat}(T_B - T_{walls}) - U_{rain}A_{lat}(T_{walls} - T_{rain}) + \dot{M}_{cond}\Delta h_{cond} \quad (2)$$

As long as no product breaths out of the tank, there is no loss in the mass of stored product but merely a transfer from one phase to the other. However, some air inflow is still needed to compensate for the volume loss of the bulk gas vapors due to rain cooling. In Eq. (3a) it will be assumed that the instantaneous change in the mass of the product vapors must be compensated by an ambient air flow of same magnitude. In Eq. (3b) the temporary change in the mass of the vapors is split into a term for the volume loss over time and the phase change mechanisms.

$$-\,dM_{B\,vap}/dt = \dot{M}_{air} \quad (3a)$$

$$dM_{B\,vap}/dt = d\left[p_B\,V_{B\,vap}/\hat{R}\,T_B\right]/dt + \dot{M}_{evap} - \dot{M}_{cond} \quad (3b)$$

The rate of phase change, either via condensation or evaporation, is directly related to the change in saturation concentration within the gaseous phase; see Eq. (3c) and (3d)

$$\dot{M}_{cond} = \rho_{mix}\beta[y(T_B) - y^*]A_{liq\,film} \quad (3c)$$

$$\dot{M}_{evap} = \rho_{mix}\beta[y(T_B) - y^*]A_{free} \quad (3d)$$

The overall heat transfer coefficient to the rain film is calculated as shown in Eq. (4a), considering that the rain mass flux wetting the tank $\dot{m}_{film}$ is less than the rain intensity on ground $\dot{m}_{rain}$ in Eq. (4b), following the formula of Förster et al. [4]

$$1/U_{rain} = 1/h_{rain} + 1/h_{free} = 1/\left(\dot{m}_{film}c_{rain}\right) + 1/h_{free} \quad (4a)$$

$$\dot{m}_{film} = \dot{m}_{rain}\;[\pi D^2/4 + DH\,cot\theta]/[\pi D^2/4 + \pi DH] \quad (4b)$$

For reference, in Continental Europe a rain of an intensity of 75 kg/(m2 h) occurs once a year, a rain of 150 kg/(m2 h) intensity occurs once every twenty years, while a pouring rain with an intensity of 225 kg/(m2 h) happens once in a century.

3 Results

For a given tank size, wall thickness and specified weather conditions, i. e. temperature and pressure, the above set of balance equations Eq. (1) to (3) can be integrated to calculate the temperature profiles of the bulk product and the walls, as well as the in-breathing flow rate as a function of time.

Fig. 2 and 3 show the in-breathing flow rates for a 2651 m³ tank at an initial temperature of 49 °C containing water and heptane, for rain intensity levels from 75 to 225 kg/m²h. Ambient and rain temperature are both equal to 15°C. Liquid level is below 1 %; at these low filling levels evaporation is negligible. These figures clearly show that when the rain intensity increases from 75 to 225 kg/(m² h), the in-breathing flow rate requirement rises by a factor of 1.7 for steam and 1.45 for heptane vapors. In each graph the maximum value of the cooling rate is added in parentheses and reported for completeness in Table 1. The maximum cooling rate coincides with the moment when the maximum air flow rate is breathing into the tank, which usually occurs shortly after the start of the rain shower (3 – 10 minutes). Both the maximum in-breathing flow rate and the cooling rate increase with the rain intensity, though not linearly. In a previous publication [8], the authors found that accounting for vapor condensation augments the calculated maximum in-breathing flow rates by approximately 25 % for the tank filled with steam and roughly 65 to 80 % for the tank filled with heptane vapors. As already observed by Fullarton [5], air in-breathing in order to refill the volume occupied by condensed vapors clearly dominates over any dampening effect due to latent heat release. This explanation should also work for partially filled tanks.

Furthermore in the latest editions of ISO 28300 [1], API 2000 [2] and EN 14015 [3] a constant cooling rate of 28 K/h is assumed for a tank volume of 2651 m³. Table 1 clearly shows that even with a rain shower of 75 kg/(m² h) intensity this value is not at all conservative.

Table 1: In-breathing and cooling rates for the tank of 2651 m³ volume filled with water and heptane vapors in function of rain intensity. Filling level is below 1 %.

Rain intensity (kg/m²/h)	Maximum in-breathing (Nm³/h)		Maximum cooling rate (K/h)	
	Water tank	Heptane tank	Water tank	Heptane tank
75	640	496	32	29
150	937	640	52	45
225	1097	721	67	50

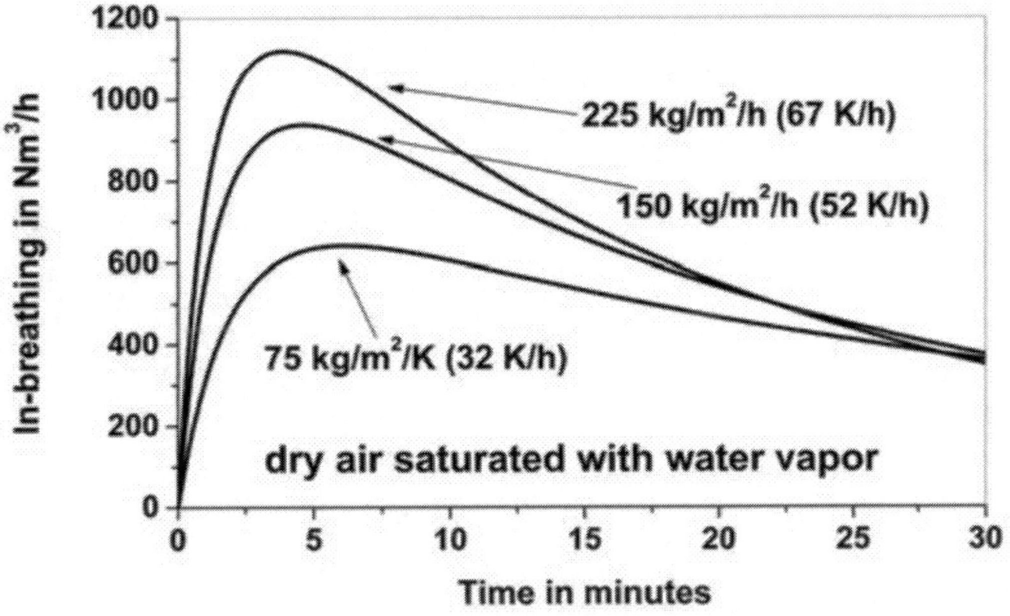

Figure 2: In-breathing as a function of time for a water tank of 2651 m³ volume under a rainstorm of 75, 150 and 225 kg/m2/h intensity. Filling level is below 1 %. Source [8]

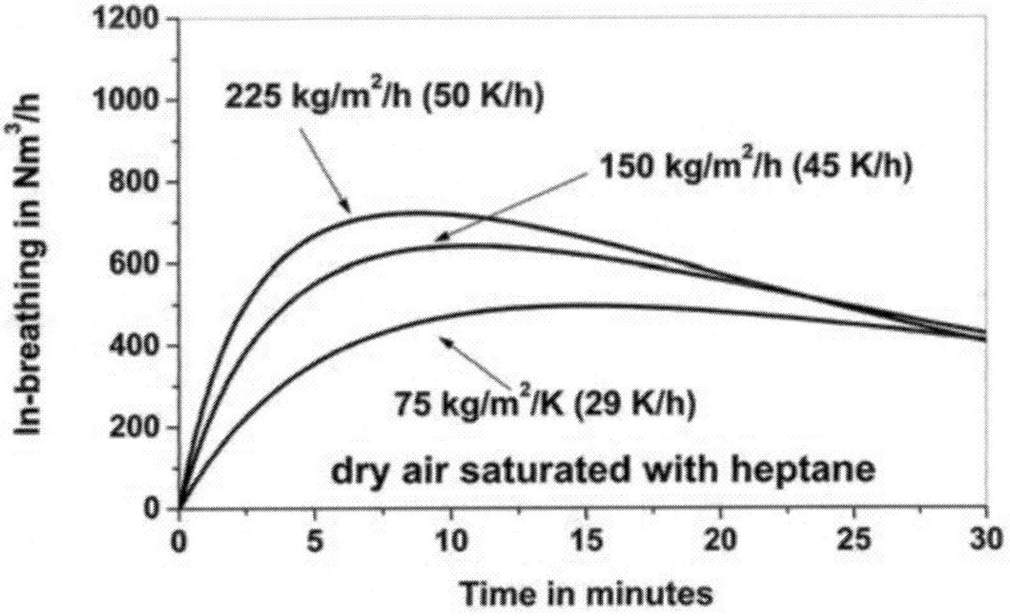

Figure 3: In-breathing as a function of time for a heptane tank of 2651 m³ volume under a rainstorm of 75, 150 and 225 kg/m2/h intensity. Filling level is below 1 %. Source [8].

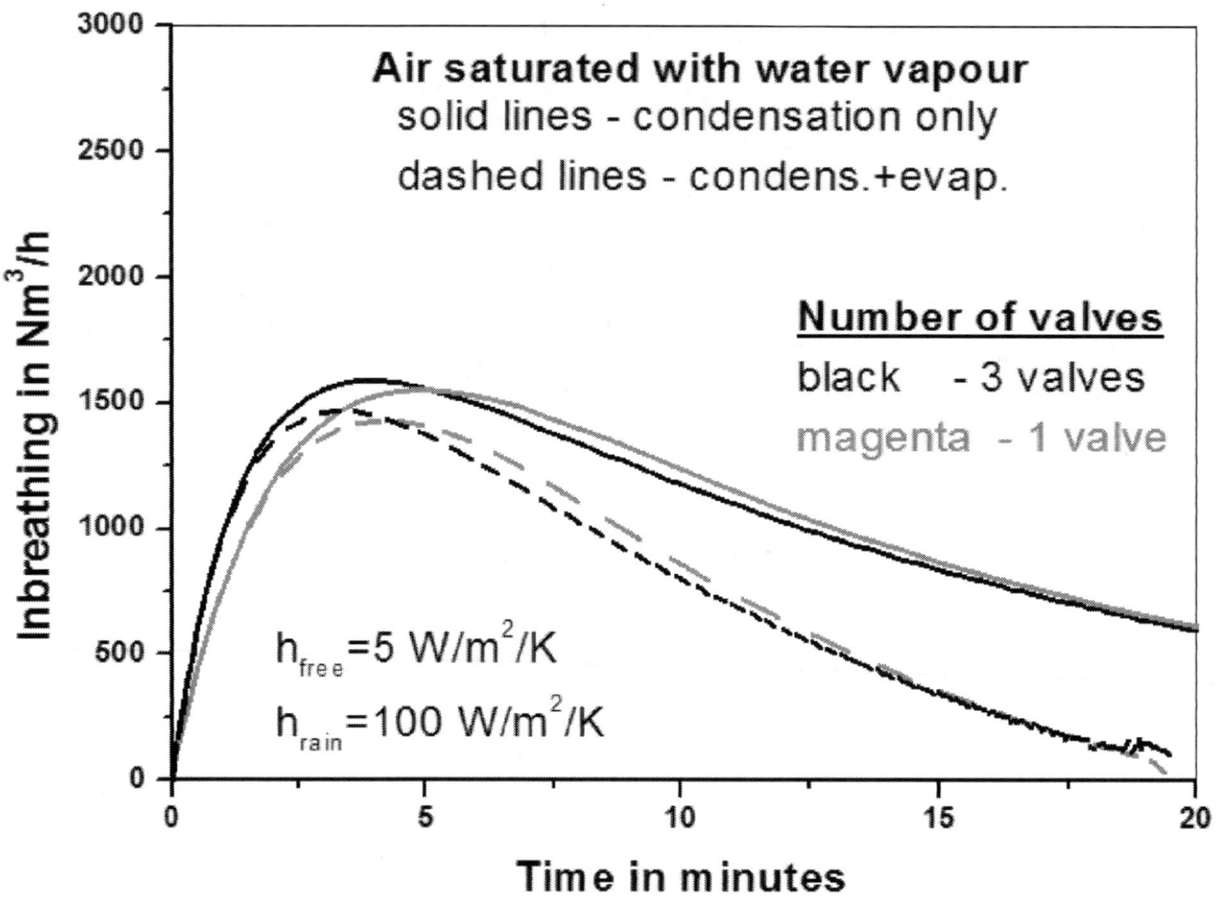

Figure 4: In-breathing as a function of time for a water tank of 4003 m³ volume with a free convection and rain equivalent heat transfer of 5 and 100 W/m²K. Filling level is 50 %-vol.

In Fig. 4 the in-breathing flow rate as a function of time is shown for a water tank with a filling level of 50 %-vol. From this picture it can be clearly seen that taking into consideration the evaporation of liquid strongly reduces the flow of in-breathed air over time, as expected. Additionally, it has been investigated how strongly the simultaneous in-breathing through several vacuum valves of given type and size is impacting the breathing requirements. For a near atmospheric tank the capacity of a breather valve can be assumed proportional to the instantaneous incompressible flow capacity though an ideal nozzle, as shown in Eq. (5). In this equation K_{valve} has a similar meaning to the discharge coefficient for safety relief valves as described in API 520 [9] and ISO 4126-1 [10] and is dependent on type and size of the device.

$$\dot{M}_{valve} = K_{valve} \cdot A_{throat} \cdot \sqrt{\rho_{air}[p_{amb} - p_{tank}(t)]} \quad (5)$$

On behalf of Fig. 4 it may be concluded that the valve throat area is large enough so that one device suffices, since the peak of demand for ambient air is uninfluenced by the quantity of vacuum breathers.

4 Measurements and future work

The above discussion clearly points out the importance of cooling rates on in-breathing. In this context, proper calculations of the heat transfer, particularly concerning the heat transfer coefficient, are essential. More precisely for a given rain intensity and impingement angle, the rate of cooling and the in-breathing flow rate have to be measured as functions of time in order to verify the proposed model.

The first set of measurements should involve an API 12F tank filled with air, that is saturated with steam and heptane vapors, in order to model the effect of condensation. In a second set of measurements the tanks should be filled with saturated water at several filling levels. The effect of precipitation may be recreated by ad-hoc showers in order to avoid random changes of rain intensity and impingement angle. Temperature and pressure inside the tank and at the walls as well as air flow rate shall be recorded.

Future extension of the test program, for instance to tanks of larger or smaller size or using other storage media, like methanol or ethanol, will be discussed on the base of the results obtained from the first round of tests.

5 Conclusions

A physical model for the calculation of the in-breathing requirement and the cooling rate of low-pressure liquid filled storage tanks exposed to a prolonged rainfall has been presented in this paper. The model relies in part on a previously published model, which has demonstrated a substantial effect of vapor condensation on ventilation requirements, which is not accounted for in the most recent editions of the standards ISO 28300, API 2000 and EN 14015. The model presented here includes in comparison with the previous one the effect of liquid evaporation. Our calculations showed that evaporation has a severe impact on the thermodynamics of the tank cooling. The quantity of top mounted vacuum breathers showed a rather negligible impact in the determination of the in-breathing requirements: nevertheless, this conclusion should not be generalized based on one calculation for only one scenario. Realizing the short-comings of the current venting models in the standards, among the other aspects an arbitrary specification of the medium cooling rates, the need for a comprehensive set of measurements dedicated to this important issue proves necessary. A series of experiments involving a cylindrical API 12F tank at several liquid filling levels will be started soon.

6 Nomenclature

A	Surface area	[m²]
c	Specific heat	[J/(kg K)]
D	Tank diameter	[m]
h	Heat transfer coefficient	[W/(m² K)]
Δh_{cond}	Specific latent heat of evaporation	[J/kg]
H	Tank height	[m]
$\dot{m}$	Mass flux	[kg/(m² s)]
K	Proportionality constant	[--]
$\dot{M}$	Mass flow rate	[kg/s]
M	Mass	[kg]
p	Pressure	[Pa]
t	Time	[s]
$\hat{R}$	Specific gas constant	[J/(kg K)]
T	Temperature	[K]
U	Overall heat transfer coefficient	[W/(m² K)]
V	Volume	[m³]
y	Mass fraction of the condensing vapor in the gaseous phase	[--]
y*	Mass fraction of the condensing vapor at saturation temperature	[--]
β	Vapor mass transfer coefficient	[m/s]
ρ	Density	[kg/m³]
Θ	Rain impingement angle	[--]

amb	Ambient	
B	Bulk product	
cond	condensation	
evap	evaporation	
film	liquid film (condenses or rain)	
free	Free or natural convection; free liquid surface	
int	Internal, i. e. between tank walls and bulk product	
lat	Lateral surface incl. roof	
liq	Liquid phase	
mix	Mixture of product vapors and in-breathed air	
throat	Throat, i. e. minimum surface area of a venting device	
vap	Vapor or gaseous phase	

7 References.

[1] ISO 28300, Petroleum, petrochemical and natural gas industries – Venting of atmospheric and low-pressure storage tanks, First Edition, 2008

[2] API 2000, Venting atmospheric and low-pressure storage tanks, Seventh Edition, 2014

[3] EN 14015. Specification for the design and manufacture of site built, vertical, cylindrical, flat-bottomed, above-ground, welded, steel tanks for the storage of liquids at ambient temperature and above, 2005

[4] H. Förster, K. Schampel, H. Steen, Atmungsvorgänge infolge von Witterungseinflüssen an Lagertanks für brennbare Flüssigkeiten, PTB Report W-22, ISSN 0341-6739, 1984

[5] D. Fullarton, Sicherheitstechnische Überlegungen bei Atmungsvorgängen in Lagertanks – Einfluss der Kondensations von Produktdämpfen, Chem. Ing. Tech. 58 (1), 1986, 38 – 40

[6] D Fullarton, E.U Schlünder, Näherungsweise Bestimmung der Austauschfläche bei der Kondensation von Gas—Dampfgemischen, Chem. Eng. Proc., 18 (5), 1984, 283- 292

[7] P. Salatino, G. Volpicelli, P. Volpe, On the design of thermal breathing devices for liquid storage tanks, Proc. Saf. Env. Prot. 77 (B), 1999, 354-359

[8] D. Moncalvo, M. Davies, R. Weber, R. Scholz, Breathing losses from low pressure storage tanks, J. Loss Prev. Proc. Ind., accepted

[9] API 520 Sizing, selection and installation of pressure-relief devices, Ninth Edition, 2014

[10] ISO 4126-1, Safety devices for protection against excessive pressure. Part 1: safety valves, Third edition, 2013

12TH GLOBAL CONGRESS ON PROCESS SAFETY

Methodology to Determine Relief Load for Trans-Critical Pool Fire

Craig Powers
Aspen Technology, Inc.
20 Crosby Dr, Bedford MA 01730
craig.powers@aspentech.com

Prepared for Presentation at
American Institute of Chemical Engineers
2016 Spring Meeting
12th Global Congress on Process Safety
Houston, Texas
April 11-23, 2016

Methodology to Determine Relief Load for Trans-Critical Pool Fire

Craig Powers
Aspen Technology, Inc.
20 Crosby Dr, Bedford MA 01730
craig.powers@aspentech.com

Keywords: supercritical pool fire pressure relief

Abstract

Adequate pressure relief design is of high priority in a variety of industries in order to maintain compliance with governmental regulations including OSHA 1910.119 and Pressure Equipment Directive 97/23/EC. Accurate relief sizing reduces the potential risk of damages to personnel, equipment, and production schedule. One of the most common overpressure scenarios for relief sizing is external pool fire. Among the most challenging types of fire analysis is a trans-critical pool fire, where the process fluid is subcritical at the onset of relief then progresses to become fully supercritical during the fire. Often, process simulation is required to get accurate fluid properties at different times during the fire, then a relief load is estimated using the 'alternative method' outlined in the 6th edition of API Std 521, Section 4.4.13.2.4.4, which examines the change in specific volume over time. This paper will examine, through process simulation, the effect of using several alternative methods to determine the required relief load for trans-critical fires.

1 Introduction

Pressure relief systems are an essential component of any process unit design. Relief systems must be designed to meet regulatory requirements, including 29 CFR 1910.119 and Pressure Equipment Directive 97/23/EC. Accurate relief sizing reduces the potential risk of damages to personnel, equipment, and production schedule. One of the most common overpressure scenarios for relief sizing is external pool fire. External pool fire scenarios are often calculated using the wetted vessel method outlined in the sixth edition of API Standard 521, section 4.4.13.2.4. [1] As an alternative, the Guidelines for Pressure Relief and Effluent Handling Systems suggests performing a batch distillation calculation. [2] For systems without phase change, API also offers a stepwise expansion calculation. Similar methods are also available in the literature. [3] [4] [5]

Upstream oil and gas production facilities present significant challenges in performing fire calculations. Because these facilities handle fluids at near-wellhead conditions, they operate at high pressures. The hydrocarbon liquids in the high pressure separators in these facilities may have relatively high fractions of methane, leading to critical pressures that are close to operating conditions. When performing a fire calculation, it may prove impractical to obtain a heat of vaporization to calculate the required relief based on a boiling calculation. Stepwise flash calculations are complicated by the potential for the liquid to disappear rapidly as the pressure passes the cricondenbar (the maximum pressure at which two phases are present, which may be higher than the critical pressure). Moreover, in three-phase systems, a water phase may remain, which will also influence relief calculations.

In this paper, we will explore the use of process simulation to calculate fire relief requirements for fluids operating near their critical pressure. The method that is presented involves a combination of spreadsheet calculations and process simulator unit operations. Limitations and potential extensions will also be discussed.

2 Calculation Method

2.1 Required Relief Load

Fundamentally, a vessel exposed to fire goes through a batch distillation process: the fire heats the vessel; the contents heat up, increasing the pressure until the opening pressure of the pressure relief device is reached. Then, as heat input continues, the density in the vessel will decrease as material is removed—in a typical vessel containing liquid, the material relieved is principally vapor evolved from the boiling liquid, but in vessels containing only vapor or supercritical fluid, the relief is from the changing density of the vessel contents.

To model this process, a single-stage batch distillation is performed in discrete steps. The system is maintained at the relief pressure (design pressure plus 21% allowable overpressure), and the required relief volume is the excess material due to vaporizing liquid and density changes in all phases. The calculation may be performed using a process simulator combined with a spreadsheet calculation.

The process simulator portion of the calculation involves two parts. The first part takes the vessel contents to relief conditions. First, the relative quantities of liquid and vapor are adjusted to agree with the specified liquid level in the system. This may be accomplished using a separator to split the liquid and vapor, a tee to remove the excess material (typically in the vapor phase), a mixer to combine the phases back together, and an adjustment calculator to achieve the volume ratio target. A conceptual diagram of this process is shown in Figure 1.

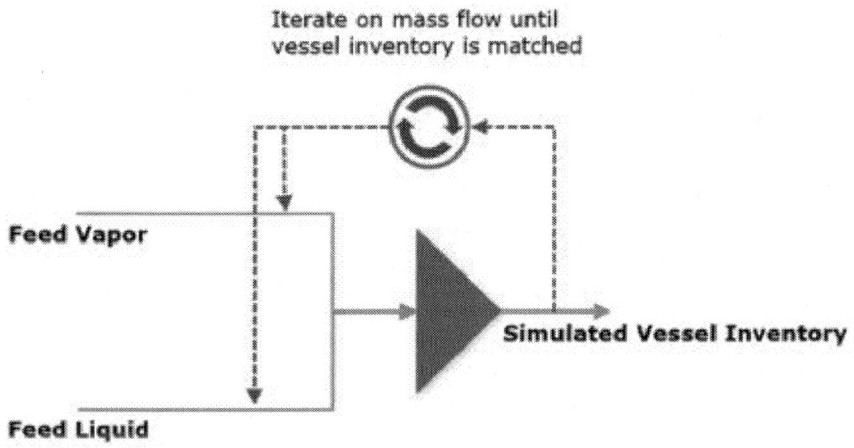

Figure 1. Feed Rebalance Operation

Then, the resulting stream is compressed to the relief pressure and heated until the bulk density matches the bulk density at operating conditions. This may be accomplished using a compressor or pump unit operation to increase the pressure of the material, a heater to increase the temperature, and another adjustment calculator to vary the temperature to attain the bulk density target.

The stream from the previous steps is then used as the input into the stepped fire calculation. Each calculation stage involves a heater to perform the step fire heat input, a separator to split the resulting liquid and vapor phases, a tee to pull the relief off of the vapor phase, and a mixer to produce the new vessel contents. A sample flowsheet of this process is shown in Figure 2. The output of the heater may be based on a temperature step, a target vapor fraction, or a duty step. The amount of vapor removed (the relief quantity) is obtained from the spreadsheet calculation detailed below. The following properties are used as inputs to the spreadsheet calculation: bulk specific enthalpy entering the step, bulk specific enthalpy leaving the heater, fraction of each phase leaving the heater, density of each phase leaving the heater, and molecular weight of each phase leaving the heater. These are then used in the spreadsheet to obtain the required relief on each step.

It is assumed that there is enough freeboard in the vessel so that the relief is entirely vapor. In cases with low freeboard or materials that tend to foam, a disengagement calculation is recommended. Methods for doing this calculation are available in the literature. [6]

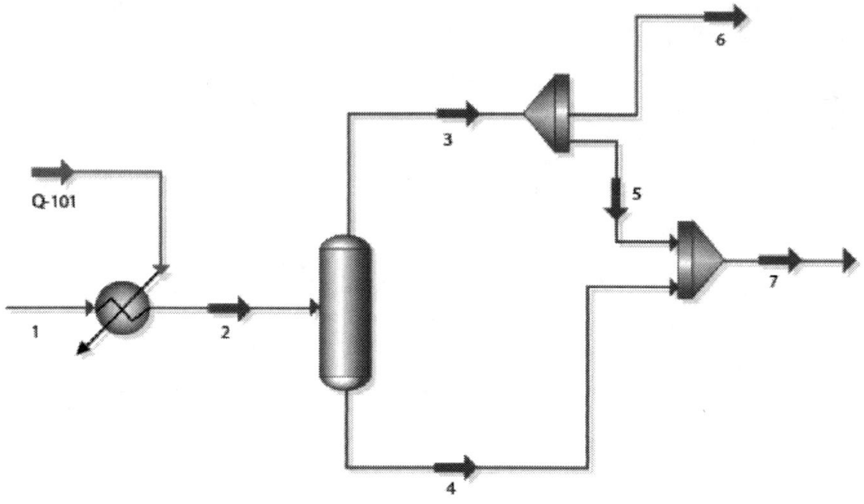

Figure 2. Flowsheet For Each Analysis Step

The calculations in the spreadsheet are as follows.

The volume taken up by the liquid after heat input in step (n) is:

$$V_\ell = \frac{m_\ell}{\rho_\ell}$$

Where: V_ℓ is the volume of liquid, m_ℓ is the mass of liquid, and ρ_ℓ is the mass density of liquid after heat input

The volume of vapor relieved in step (n) is:

$$V_{rel} = \frac{m_{total} f_v}{\rho_v} - (V - V_\ell)$$

Where: V_{rel} is the volume of vapor relieved, m_{total} is the total mass at the beginning of the step (both liquid and vapor), f_v is the mass fraction vapor after heat input, and V is the total vessel volume

The mass of vapor relieved in step (n) is:

$$m_{rel} = V_{rel} \rho_v$$

Where: m_{rel} is the mass of vapor relieved and ρ_v is the vapor density after heat input

The liquid and vapor properties may be taken from stream "2" in Figure 2.

The mass from this step is used to divide the vapor from this step into the part that is relieved (stream "6") and the part that is retained (stream "5"). The accounting in the process simulator may be easier if the initial flowrate is set to be the same as the initial total mass of the vessel contents (e.g. if the mass is calculated to be 5,000 lb, then the initial flowrate is set to 5,000 lb/h)—then, the amount of vapor removed in the simulator follows directly from the value of m_{rel} calculated above (e.g. if the mass relieved in step (n) is 250 lb, then the vapor relief stream in the simulator would have a flowrate set to 250 lb/h).

At this point, the process simulator is ready for the next calculation step, but the results obtained thus far are not sufficient for a relief calculation. To evaluate a pressure relief valve, a flowrate is required. As noted before, steps are typically taken by temperature, vapor fraction, or enthalpy, so an additional calculation must be taken to determine how much time passed.

Based on the volume of liquid, the wetted area of the vessel may be calculated. In this paper, the simple API heat flux equation is used to calculate the fire heat input into the vessel: [1]

$$Q = CFA_{ws}^{\ 0.82}$$

Where: Q is the heat input into the vessel, C is a unit conversion constant, F is a factor to account for the presence of fireproof insulation (the value of F is 1.0 for an uninsulated vessel), and A_{ws} is the wetted surface area of the vessel

The amount of time that passed in the calculation step is:

$$\Delta t = (\widehat{H}_{final} - \widehat{H}_{initial})m_{total}/Q$$

Where: Δt is the amount of time passed in the calculation step, $\widehat{H}_{final}$ is the specific enthalpy on step (n) after heating (but before material is removed for relief, stream "2"), $\widehat{H}_{initial}$ is the specific enthalpy on step (n) before heating (stream "1"), and m_{total} is the total amount of material present

Then, the required relief flow rate is simply:

$$w = m_{rel}/\Delta t$$

Depending on the behavior of fluid, the maximum relief requirement may be on the first step of the procedure, or it may occur later. The maximum relief is best characterized based on the required pressure relief device area, as the literature as shown that this maximum may not correspond to the peak of either mass or volumetric flow. [3] Strictly, the system should be analyzed up to a projected vessel failure condition (or other standard end point, e.g. a time limit), although there may be an earlier step where it is reasonable to conclude that the peak relief has already passed.

As an alternative to the simple API heat flux equation, a more detailed treatment of heat absorption could be used. A combination radiation/convection Stefan-Boltzmann model is presented in Annex A of API 521 and in Appendix A of Scandpower report 27.207.291/R1. [7]

2.2 Required PRD Area and PRD Capacity

There are two methods offered for calculating relief device required area and capacity in API Standard 520 Part I that are applicable for vapor or supercritical systems. [8] One is for vapor sizing, recommended for use when the compressibility Z is within a range of $0.8 < Z < 1.1$. Kim and coworkers have studied the extension of this method to cases involving more extreme conditions, as might be seen in relief of a supercritical system. [9] The other method offered in API 520 Part I is the general direct integration method.

The unmodified API vapor method is unlikely to give satisfactory results for relief of a supercritical fluid as in this case. Both Kim et al.'s modification of the API Vapor and the direct integration method have been observed to give reasonable results. Kim et al.'s method is easier to use as part of a table calculation for screening purposes, while direct integration may be easily used in a process simulator that supports calculating a property table for a stream, which makes it suitable for a more exact calculation of required relief on a specifically-identified stream. In the cases that follow, the method of Kim et al. is used to screen for the step with the largest required PRD area, then the required PRD area for that specific relief stream is calculated using the direct integration method.

3 Applications

3.1 High Pressure Separator, Case (a)

The equipment to be analyzed is a high pressure separator operating at 90 °F and 1200 psig, with a design pressure of 1440 psig. The separator normally handles a three-phase fluid. For the purposes of this analysis, the separator is assumed to be a horizontal vessel with 2:1 ellipsoidal heads, a diameter of 5 ft, a tan/tan length of 10 ft, and an initial liquid level of 2 ft. To account for associated process piping, an additional 15% is added to the calculated wetted area. No credit is taken for adequate drainage and firefighting. The overall composition of the stream entering the separator is given in Table 1. Calculations were performed using Aspen HYSYS V8.8. Initial estimates for required orifice area were calculated using the method of Kim et al., while the final required orifice area was calculated using direct integration. A phase envelope of the system prior to relief is shown in Figure 3.

Table 1. Case (a) Composition

Comp.	Mole Fraction
N2	0.001
CO2	0.02
C1	0.73
C2	0.135
C3	0.057
iC4	0.013
nC4	0.02
iC5	0.007
nC5	0.006
nC6	0.003
nC7	0.002
nC8	0.001
nC9	0.0003
nC10	0.0001
H2O	0.0046

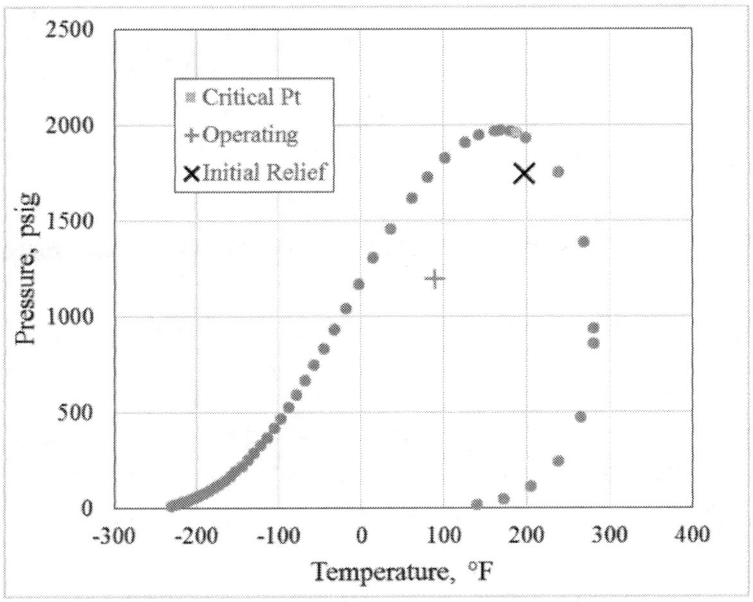

Figure 3. Phase Envelope for Case (a)

3.1.1 API Wetted Vessel / Boiling Liquid Calculation

The simplest form of the API wetted vessel / boiling liquid fire calculation requires application of the fire heat flux to boil the liquid contents of the vessel. The liquid is typically taken at operating conditions, then brought to its bubble point at the required relief pressure. In this case, with 21% allowable overpressure for a fire scenario, the relief pressure is 1742 psig.

This simplified calculation fails in this case, because the cricondenbar of the light liquid phase is below the relief pressure.

3.1.2 Stepwise Boiling Liquid Calculation

As an alternative to the simplified problem setup above, consider a more rigorous calculation of the vessel contents at relief conditions. The quantities of the liquid and vapor phases are adjusted to match the operating volume fraction, then an isochoric flash is performed to bring the fluid to the relief pressure, as discussed in the Calculation Method section above.

When this process is performed, at relief pressure, there is still a substantial amount of hydrocarbon liquid present. This may be surprising given the failure of the simplified boiling liquid calculation; however, the inclusion of the vapor in the constant density flash is very important. In this case, a significant amount of methane enters the liquid phase as the mixture is compressed; in fact, comparing the light liquid at relief to the light liquid at operating conditions, approximately 50% more methane is present on a mole fraction basis (0.47 vs. 0.29).

However, this calculation is still problematic, as there are several dramatic phase transitions that complicate the use of a boiling liquid calculation. First, around 225 °F, the hydrocarbon liquid phase ceases to vaporize, becoming a supercritical phase that the simulator identifies as vapor. Then, around 265 °F, the phase identification changes from liquid to vapor. Finally, around 620 °F, the water phase is vaporized. Stepping by temperature will give unrealistically large requirements in the steps where the large phase transitions occur. The two most likely alternatives both fail in this case: stepping by vapor fraction cannot work with the supercritical phase transition in the hydrocarbon phase, and stepping by enthalpy results in a flash failure. Moreover, the relatively small amount of liquid that vaporizes before the phase transition suggests that this method (based solely on liquid vaporization) is a poor choice.

3.1.3 Supercritical / Vapor Expansion Calculation

Given that the hydrocarbon liquid portion is a supercritical fluid at the relief pressure of 1742 psig, an alternative method of calculation would be to separately consider fluid expansion in the vapor/hydrocarbon portion of the system and boiling of the water. This method gives a required relief rate of 4,380 lb/h and a required orifice area of 0.03 in^2. The required relief rate for boiling the water portion of the feed is significantly smaller, 830 lb/h with a required area of 0.011 in^2.

3.1.4 Detailed Step Calculation

The method presented earlier in this paper is applied to the system, with heat input taken based on the simple API heat flux. To account for reduced heat transfer to supercritical / gas space, a correction factor of 0.3 was applied to the full-vessel heat transfer. [2] For any step, the larger of the wetted heat transfer (without correction factor) or total vessel heat transfer (with correction factor) was used. Maximum relief occurs on the initial step, with a required relief load of 4,370 lb/h and a required orifice area of 0.03 in^2 was obtained by a direct integration calculation. Although the water phase might be expected to flash suddenly, this was not observed in the simulator, as it disappeared gradually over a broad temperature range, with the vessel contents becoming entirely vapor around 375 °F.

Selected results from this calculation are presented in Table 2.

Table 2. Selected Results from Case (a)

T, °F	$\hat{H}_{initial}$, BTU/lb	$\hat{H}_{final}$, BTU/lb	VF, molar	HC frac, molar	ρ_v, lb/ft^3	ρ_ℓ, lb/ft^3	m_{rel}, lb	Δt, s	w, lb/h	A, in^2
202.8	-1567.6	-1562.9	0.664	0.256	11.85	24.47	54.69	45.07	4369	0.0321
207.8	-1565.6	-1560.8	0.690	0.230	11.92	24.49	54.49	46.59	4210	0.0307
212.8	-1563.7	-1558.9	0.718	0.201	11.99	24.56	54.33	48.30	4049	0.0294
247.8	-1555.9	-1551.0	0.918	0	12.10	58.48	63.40	54.87	4160	0.0298
252.8	-1556.3	-1551.5	0.918	0	11.84	58.34	61.01	53.97	4070	0.0294
257.8	-1556.7	-1551.8	0.918	0	11.59	58.19	58.73	53.15	3978	0.0289

3.2 High Pressure Separator, Case (b)

The equipment to be analyzed is a high pressure separator operating at 5 °C and 64.5 bar, with a design pressure of 80 bar(g). The separator normally handles a two-phase fluid with no water present. For the purposes of this analysis, the separator is assumed to be a horizontal vessel with 2:1 ellipsoidal heads, a diameter of 2 m, a tan/tan length of 5 m, and an initial liquid level of 0.75 m. To account for associated process piping, an additional 15% is added to the calculated wetted area. No credit is taken for adequate drainage and firefighting. The overall composition of the stream entering the separator is given in Table 3. Calculations were performed using Aspen HYSYS V8.8. Initial estimates for required orifice area were calculated using the method of Kim et al., while the final required orifice area was calculated using direct integration. A phase envelope of the system prior to relief is shown in Figure 4.

Table 3. Case (b) Composition

Component	Mole Fraction
N2	0.0299
CO2	0.2183
H2S	0.0479
C1	0.5974
C2	0.0461
C3	0.0235
nC4	0.0148
nC5	0.0065
nC6	0.0048
nC7	0.0030
nC8	0.0032
nC9	0.0032
E-Mercaptan	0.0014

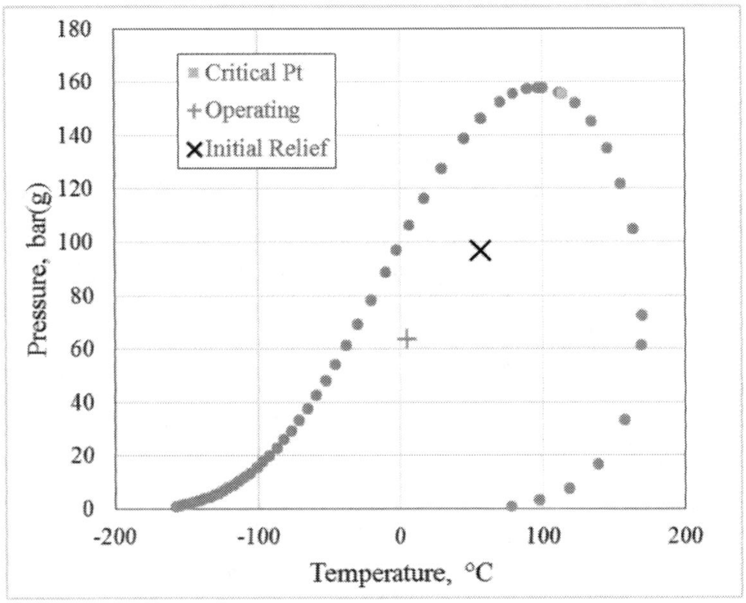

Figure 4. Phase Envelope for Case (b)

3.2.1 API Wetted Vessel / Boiling Liquid Calculation

Because the liquid portion of the separator feed is below its critical pressure at the relief pressure of 96.8 bar(g), a simplified boiling liquid calculation may be performed. Based on the liquid initially at its bubble point and finally at 10 mole percent vaporized, with sensible heat included, the heat of vaporization is 730 kJ/kg. The required relief rate is 3,840 kg/h with a required orifice area of 0.45 cm^2.

Note that using increasing spans of 10 mole percent vaporized (e.g. initially 10% vaporized, finally 20% vaporized) will decrease the heat of vaporization, increase the required relief rate, and increase the required orifice area, assuming that no changes in liquid level are considered. However, at higher fractions vaporized, the liquid level would be reduced, and less heat input to the fluid would be expected. This effect is part of the motivation for the step calculations.

3.2.2 Stepwise Boiling Liquid Calculation

The stepwise boiling liquid calculation was conducted as described under Case (a). As would be expected given the success of the simple boiling liquid calculation in this case, the stepwise boiling liquid calculation begins without issue. Similarly to the previous case, this calculation fails due to the phase behavior of the liquid as vapor is removed and its temperature increases. Around 170 °C, the liquid exceeds its cricondenbar. Phase identification in supercritical fluids is a difficult problem; in this case, the simulator continues to identify the material as liquid phase until it reaches 250 °C, at which point the phase identification changes to all vapor. For a boiling liquid calculation, this is a major concern, as all of the liquid turning into vapor over a short step requires an excessive amount of area for adequate relief. Alternatively, the initial step could be considered, but the required relief rate then is unreasonably low at 160 kg/h with a required device area of 0.02 cm^2.

3.2.3 Supercritical / Vapor Expansion Calculation

Because the liquid contents of the vessel are not supercritical at initial relief conditions, a supercritical / vapor expansion calculation cannot be conducted except by changing the initial state of the material in the vessel.

3.2.4 Detailed Step Calculation

The method presented earlier in this paper is applied to the system, with heat input taken based on the simple API heat flux. To account for reduced heat transfer to supercritical / gas space, a correction factor of 0.3 was applied to the full-vessel heat transfer. [2] For any step, the larger of the wetted heat transfer (without correction factor) or total vessel heat transfer (with correction factor) was used. Maximum relief occurs on the initial step, with a required relief load of 4,310 kg/h with a required orifice area of 0.52 cm^2. Note that the required area calculated here is 18% higher than the area requirement from the simple API wetted vessel calculation.

Selected results from this calculation are presented in Table 4.

Table 4. Selected Results from Case (b)

T, °C	$\hat{H}_{initial}$, kJ/kg	$\hat{H}_{final}$, kJ/kg	VF, molar	ρ_v, kg/m³	ρ_ℓ, kg/m³	m_{rel}, kg	Δt, s	w, kg/h	A, mm²
76.3	-4007.2	-3949.0	0.582	137.83	519.2	437.4	374	4207	51.9
96.3	-3843.1	-3784.7	0.614	135.16	491.0	365.9	356	3696	46.5
116.3	-3698.7	-3639.3	0.654	135.17	456.6	325.6	348	3367	42.8
196.3	-3290.9	-3230.2	1	153.16		294.9	373	2849	35.8
216.3	-3230.2	-3177.1	1	139.29		246.8	294	3017	39.3
236.3	-3177.1	-3124.6	1	128.36		194.7	265	2650	35.6

4 Conclusion

The detailed stepwise fire calculation method presented in this paper offers significantly more robust performance for fluids near their critical point. In particular, simple boiling liquid calculations often fail in these cases, and may be less conservative than the detailed method when they do not fail. The single-phase expansion method may give comparable results for some cases, but is not suitable for cases where two phases are present in the vessel at the initial relief conditions. As an additional benefit, the detailed method is suitable for incorporating more advanced heat input methods, such as the detailed convection-radiation Stefan-Boltzmann model.

5 References

[1] American Petroleum Institute, *API Standard 521 6th Ed.: Pressure-relieving and Depressuring Systems,* Washington, DC: API Publishing Services, 2014.

[2] Center for Chemical Process Safety of the American Institute of Chemical Engineers, Guidelines for Pressure Relief and Effluent Handling Systems, New York: American Institute of Chemical Engineers, 1998.

[3] R. Ouderkirk, "Rigorously Size Relief Valves for Supercritical Fluids," *Chemical Engineering Progress,* pp. 34-43, August 2002.

[4] R. C. Doane, "Designing for Pressure Safety Valves in Supercritical Service," *Hydrocarbon Processing,* pp. 63-67, January 2010.

[5] M. A. Grolmes, "Vent Sizing for Pressure Relief from the Super-Critical State - Revisited," in *DIERS User Group Meeting*, Las Vegas, April 2013.

[6] H. Fisher, H. Forrest, S. Grossel, J. Huff, A. Muller, J. Noronha, D. Shaw and B. Tilley, Emergency Relief System Design Using DIERS Technology, New York: American Institute of Chemical Engineers, 1992.

[7] Scandpower Risk Management AS, *Guidelines for the Protection of Pressurised Systems Exposed to Fire,* 2004.

[8] American Petroleum Institute, *API Standard 520 Part I 9th Ed.: Sizing, Selection, and Installation of Pressure-relieving Devices,* Washington, DC: API Publishing Services, 2014.

[9] J. S. Kim, H. J. Dunsheath and N. R. Singh, "Proper Relief-Valve Sizing Requires Equation Mastery," *Hydrocarbon Processing,* December 2011.

12TH GLOBAL CONGRESS ON PROCESS SAFETY

Dynamic Simulation Modeling to Protect Flare Headers from Air Ingress

Jesse Mumm
Bechtel Oil, Gas and Chemicals
Houston, Texas
jlmumm@bechtel.com

Satyajit Verma
Bechtel Oil, Gas and Chemicals
Houston, Texas
sverma4@bechtel.com

Freeman Self
Bechtel Oil, Gas and Chemicals
Houston, Texas
feself@bechtel.com

Prepared for Presentation at
American Institute of Chemical Engineers
2016 Spring Meeting
12th Global Congress on Process Safety
Houston, Texas
April 11-23, 2016

Dynamic Simulation Modeling to Protect Flare Headers from Air Ingress

Jesse Mumm
Bechtel Oil, Gas and Chemicals
Houston, Texas
jlmumm@bechtel.com

Satyajit Verma
Bechtel Oil, Gas and Chemicals
Houston, Texas
sverma4@bechtel.com

Freeman Self
Bechtel Oil, Gas and Chemicals
Houston, Texas
feself@bechtel.com

Keywords: flare, air ingress, dynamic simulation, flare header, make-up gas, vacuum protection

Abstract

A flare header system is subject to vacuum formation inside when a hot relief stops and the internal gases cool and/or condense. If make-up gas is not provided at an adequate rate, air may ingress from outside and potentially create a flammable air-hydrocarbon mixture inside the header. With the presence of a pilot as a ready source of ignition, explosions inside the header system could occur with potentially catastrophic consequences. This paper addresses six major considerations for selecting, sizing and specifying make-up gas requirements. The pros and cons of the locations and combinations of various types of sensors—such as differential temperature, actual temperature, header pressure, gas flow, and oxygen concentration—are discussed. Finally, recommendations are provided for instrumented controls to activate make-up gas addition, in addition to methods for determining the rate and duration of addition. Graphs and charts showing temperature and pressure profiles as well as vacuum formation as a function of time are presented to illustrate the concepts.

1. Overview of Vacuum Formation Mechanism

The pressure inside a flare header fluctuates depending upon the nature of relief events and ambient temperature conditions. On one hand, back pressure generated during a relief event may exceed the design pressure of the header, while on the other hand, severe vacuum may form inside the header after the relief has stopped and potentially exceed the design vacuum. In a header designed to mechanically withstand the extremes of high and low pressures, still another condition may exist. Because the flare stacks, when not equipped with a liquid seal drum or other means of preventing air ingress, are essentially open to the atmosphere, the flare headers are prone to air ingress from outside, even at small vacuums. This can potentially result in explosive air-hydrocarbon mixtures inside the header.

Air ingress into a flare header system is a distinct possibility when cooling or condensation of hot vapors happens after a release event is over and the header is stagnant. The resulting rate of reduction in volume can easily be larger than the constant seal purge flow to the flare stack. Under these circumstances, outside air will be drawn into the header and the resulting mixture of air and hydrocarbons in some sections of the header may fall within the flammability envelope. Because sources of ignitions are ubiquitous—the pilot being a guaranteed source—explosions are likely. The consequences of an explosion can range from damage to the flare header system to significant injury to plant personnel, lost production, and harm to company reputation.

In order to avoid the ingress of atmospheric air into the flare header, make-up gas must be injected into the flare header to compensate for the reduction in volume of vapors in the header.

2. Dynamic Simulation Model

Dynamic simulation enables one to rigorously model the process, equipment, and controls using differential and algebraic equations resulting from mass and energy balances, phase equilibria, heat transfer, and hydraulics. These equations are integrated over time to predict how the process and its equipment will respond to various disturbances.

A detailed computer model of a flare header system was prepared and divided into small segments. The model included pressure relief valves situated along the header, a flare knock-out drum, and a flare stack with a velocity seal that was open to atmosphere. The system did not contain a liquid seal drum at the base of the stack. Various reliefs, one at time, were introduced into the header system and the resulting flow conditions and heat transfer inside the header followed as a function of time and length of the header using appropriate heat transfer equations and thermodynamic properties of the released streams. Commonly available dynamic simulation software was used for this purpose.

For simplicity and speed of convergence, each header segment was considered to be a well-mixed region in terms of thermal and physical properties. The models were integrated long after relief was stopped to capture the cooling of relief gases, as well as the cooling of the header itself that took place due to the header being at a temperature higher than the ambient temperature. Included in the study were cases where the cooling vapor will partially or completely condense. The flow of gases through the flare stack was monitored as well as the temperature and pressure profiles along the length of the header. As a result of vacuum formation mentioned above, flow reversal is noted in many cases within a few seconds after the hot relief stops and cooling begins. The methods of initiating make-up gas injections, required make-up gas flow rates, and duration of the injections were investigated to determine the resulting header conditions and develop a control scheme that would prevent flow reversal at the stack that could otherwise result in air ingress into the header system. Cases studied included warm releases, cryogenic releases, and combination releases into the flare header system.

3. Establishing the Design Parameters

In order to provide a robust but economical design for instrumentation, piping, and make-up gas usage, the following six major design parameters should be considered:

1. Cooling/condensation mechanisms
2. Sensors to detect relief
3. Location of temperature sensors
4. Make-up gas flow rates
5. Location of make-up gas addition point
6. Starting and stopping of make-up gas

For existing systems with instrumentation and piping constraints, additional variations and simplifications will also be discussed.

3.1 Cooling/Condensation Mechanisms

There are two cooling mechanisms to consider when designing a make-up gas system:

- Cooling from ambient temperature or rain – The gas in the header is cooled by heat transfer through the pipe wall to the ambient air. If only cooling of the gas is involved, meaning no condensation occurs, the heat transfer is typically limited by the inside heat transfer coefficient. The occurrence of rain has minimal effect in these instances

  If condensation of the gas also occurs, the outside heat transfer coefficient becomes more important. Rain will increase the outside heat transfer coefficient, but the impact is small compared to other variables. Additionally, while a layer of ice on the header may provide insulation from cold ambient temperatures, it can also absorb heat from the gas. However, as with the effect of rain, the impact is not generally significant within the modeling capabilities.

- Cooling/Condensation from a cold header – Cooling and/or condensation of the flare gas may occur due to a cold header, which has been cooled below ambient temperature. In these instances, the header is cooled by a previous cold release. These cases are most common in cryogenic plants such as ethylene or LNG. The duration of the initial cold release has a large impact on the conditions of the header; the longer the initial cold release, the colder the initial temperature of the header.

3.2 Sensors to Detect Relief

When designing or configuring the control scheme for make-up gas injection, two important aspects to consider are how to detect the hot release and where to locate the sensors. This section will discuss possible methods of detecting relief as well as the importance of properly locating the sensors.

There are several methods to sense when relief occurs that, if cooled or condensed, will cause volume contraction in the header. This section will discuss advantages of each of the following detection methods:

- Temperature difference (gas minus ambient) – The temperature difference measures the difference between the gas temperature and the ambient temperature, under the assumption that the ambient temperature drives the cooling
- Actual temperature of gas being flared
- Pressure in header
- Oxygen composition in header

3.2.1 Temperature Difference (Gas Minus Ambient)

For cases where contraction is expected to occur through cooling, the rate of volume contraction is dependent on the rate of heat loss from the header to the ambient surroundings. Because the rate of cooling is driven by the temperature difference between the hot gases inside the header and the ambient temperature, the trigger to start make-up injection should be based on a temperature differential measurement. When this temperature difference exceeds a pre-determined criterion, make-up gas should be introduced. Once the temperature in the header drops back below this criterion, the addition of make-up gas can safely be stopped.

Measuring temperature along the entire header to ensure that all points are below the specified threshold before stopping the make-up gas flow would be ideal for reliefs where condensation or cooling are possible, however, installing many such measurements points may be infeasible. Specifying an additional time to keep injecting make-up gas after the threshold has been reached may be necessary to ensure the temperature downstream of the sensor has sufficiently cooled.

Identifying the minimum temperature difference criteria to inject make-up gas is crucial to ensuring the system can prevent ingress under a variety of conditions. Even at low temperature differentials a vacuum can form, however any seal purge gas injected upstream of the velocity seal of the flare stack may feed the vacuum up to a certain threshold ΔT before the addition of make-up gas is required.

This ΔT threshold that the seal gas can maintain will generally become the criterion to inject make-up gas, however to avoid any potential constant injection of make-up gas, additional consideration should be made to ensure the ΔT between the make-up gas and ambient temperature does not exceed this value.

3.2.2 Actual Temperature of Gas Being Flared

Cooling cases: Using actual temperature instead of differential temperature to detect hot vapor reliefs is generally not a recommended solution due to the difficulty of establishing an effective criterion separate from ambient temperature. The severity of contraction is dependent on the temperature differential and not the actual temperature of the gas, and because the ambient conditions can vary significantly due to changes in weather, a single temperature criterion may not be suitable for all conditions and relief cases. For example, the criterion should be chosen so that it will effectively cover low ambient temperature conditions, but on a hot day the ambient temperature may approach the criterion and unnecessarily trigger make-up gas injection. Conversely, if the criterion is set too high, on a cold day the temperature differential between relieving gas and ambient might exceed the minimum threshold of the seal gas while the actual temperature remains below the criterion and therefore fail to trigger make-up gas.

Condensation cases: For cases where condensation can occur, actual temperature readings may be used because the dew point is independent of the ambient temperature and a condensable gas must necessarily be above its dew point initially in order for contraction from condensation to occur. Oftentimes the relieving temperature of a condensable gas is much higher than the minimum threshold for air ingress due to cooling only, so a higher temperature criterion may be used. Because contraction due to cooling may be still observed in addition to condensation, however, it may be advantageous to use a single temperature differential criterion that accounts for both cooling and condensation instead of actual temperature for purposes of simplification.

For cases when condensation can be expected to occur only if the header has been pre-cooled to below the dew point of a relief, the actual temperature of the header itself should be measured. For these cases, condensation can be expected at temperatures well below ambient, so the actual temperature criteria to begin make-up injection should be based on the dew point of the condensable gas. Because the header itself is the source of cooling in these cases and not the ambient temperature, the skin temperature of the header wall should be measured and gas should be injected for as long as the header is below the dew point of any condensable reliefs. Ideally the temperature along the entire header would be measured to ensure condensation is no longer possible, however it may be necessary to calculate the required time to warm up the header using make-up gas and use this data to set a minimum duration for make-up gas injection.

3.2.3 Pressure in Header

Pressure sensors are generally not sufficiently sensitive or fast enough to detect ingress and trigger make-up flow. If located near the flare stack, the pressure reading may be less than atmospheric pressure because of the buoyancy effect. Additionally, if the control action is not fast enough, the ingress of air may have already occurred and the header will pressurize as air fills the vacuum, thus fooling the control action.

3.2.4 Oxygen Composition in Header

Because the purpose of make-up gas injection is to prevent the formation of flammable air-hydrocarbon mixtures, the amount of oxygen in the header is of significant concern. However, using measurements of oxygen to trigger make-up gas injection is difficult for several reasons. First, the sensor will not sense the presence of oxygen until air has already started to ingress. Additionally, because such a sensor would need to be located near the flare stack in order to sufficiently detect the ingress of air, the potential distance between the relief and the measurement may introduce a significant time delay and even more air may ingress before the make-up gas can mitigate. Finally, the reliability and speed of such measurements is a concern, especially when compared to temperature measurements.

3.3 Location of Temperature Sensors

Improper choice of location for temperature sensors could introduce potential delays in taking preventive action. The preferred location to detect relief is on the main header immediately downstream of where the relief lateral ties into the header.

Sensing the temperature further downstream, such as at the knockout drum, can be ineffective for many reasons. First, the relieving gas will cool as it travels down the header and the temperature difference between the gas and ambient will begin to narrow and will not be representative of the temperature upstream. Secondly, as a result of this cooling and the holdup of the upstream portion of the header, a time lag in sensing the temperature may be introduced.

For the example of a cooling vapor relief case, Figure 1 compares sensing the temperature in the header downstream of the relief lateral tee to sensing it at the knockout drum. The set point to initiate make-up flow was 5°C above ambient ($\Delta T = T_{gas} - T_{ambient}$) and a hot vapor relief started at t=10 minutes. As shown in this figure, when the sensor to trigger make-up was located near the lateral, the temperature reached the set point seconds after relief began. When the sensor was located at the knockout drum, the temperature reached the set point after nearly 10 minutes. Because of this delay, if the duration of the relief was less than 10 minutes, the make-up gas would not be triggered and air ingress could occur.

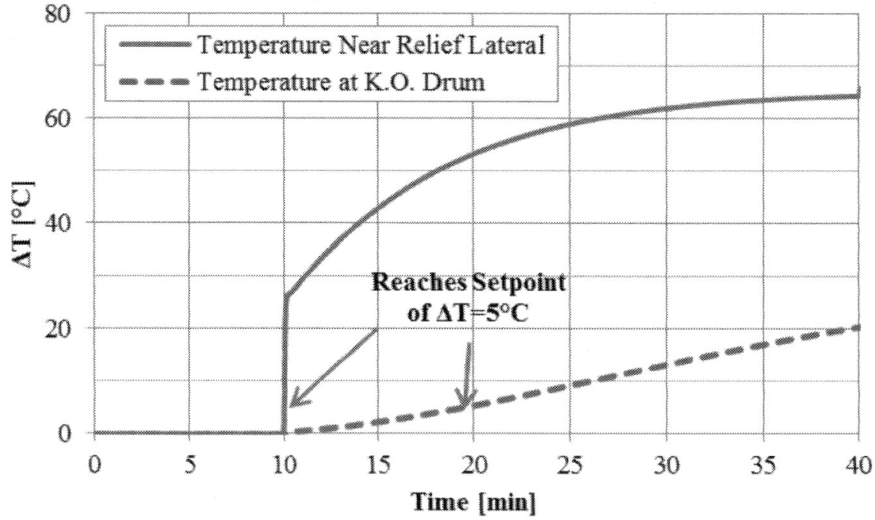

Figure 1: Sensing temperature at the relief lateral and at the knockout drum for a hot vapor relief.

Sensing the temperature inside the header is suitable for most cases. Sensing directly in the relief tailpipe or relief lateral may provide a faster response, but may also be infeasible if multiple reliefs must be considered. Additionally, locating the sensor in the lateral will not adequately measure the cooling of the header by the make-up gas, which is injected into the header rather than the lateral. This may unnecessarily extend the duration of make-up injection. On the other hand, sensing in the lateral could be used in conjunction with other sensors for special cases, such as individual cases that require separate higher make-up flows or are determined to require significantly faster response times.

3.4 Make-up Gas Flow Rates

Although the required make-up gas rate is a function of many factors, generalizations may be made to facilitate the establishment of the rate. These factors include the rate of cooling, location of the make-up gas injection point, potential for condensation, relief rate, seal purge rate, and duration of the relief. The following sections describe how these factors can impact the required rate of make-up gas.

3.4.1 Seal Purge Rate vs. Ambient Temperature

Seals are provided for flares open to the atmosphere, regardless of whether a pipe flare or a flare utilizing burners similar to industrial furnace burners (sometimes called sonic burners) are used. These seals are located on the open line upstream of the burner. The two most common types are velocity and kinetic seals. Seal purge gas is provided to increase the effectiveness. Husa (1) and API (2) offer recommendations for flow rates.

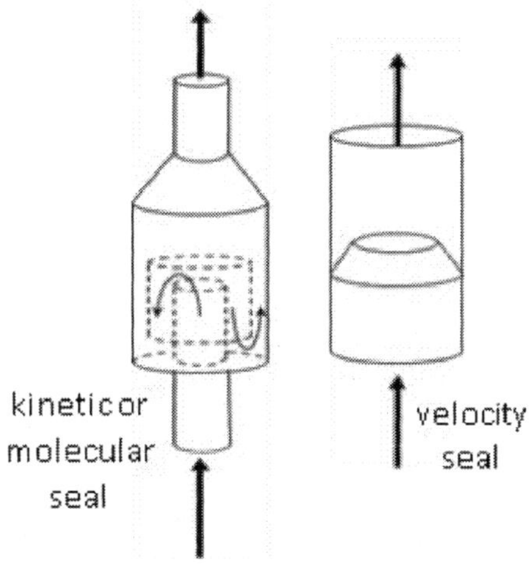

Figure 2: Velocity and kinetic seals.

However, even though these rates are not sufficient in most instances to protect against cooling and condensing of the flare gas, the seal purge does offer some minimal protection. This is dependent upon the header volume, the seal purge flow rate, and the rate of heat loss to the ambient (and therefore the differential temperature between the ambient and hot gases inside the header). The higher the seal purge rate, the higher the differential temperature range up to which the header is protected.

3.4.2 The Impact of Relief Duration on Make-up Gas Rates

The duration of relief has a significant impact on the formation of a vacuum. The temperature differential between the relieving gas and pipe wall is highest immediately after the relief begins, so relieving for short duration would yield the highest instantaneous rate of contraction. As relief continues, the pipe wall will begin to warm up with the gas in the header. This will decrease the heat transfer rate between the gas and the pipe wall but increase the overall total volume of gas that will contract.

The highest potential instantaneous rate of air ingress occurs simultaneously with the highest instantaneous rate of contraction. In order to eliminate any potential for air ingress, the make-up gas rate should be sufficient to prevent air ingress at this point. In order to determine the maximum temperature differential, and therefore highest rate of contraction, simulations should be conducted at the minimum ambient temperature.

The dynamic simulation example illustrates when the maximum rate of air ingress can occur. Figure 3 compares the volumetric flow rate measured at the base of the flare stack for a hot vapor release of methane at 130,000 Sm^3/h and 100°C at durations of 1, 5, and 30 minutes. A negative volumetric flow rate would indicate that a vacuum has formed and air has begun ingressing through the flare stack. As shown in the figure, the maximum rate of air ingress for this case occurred at a very short relief duration of 1 minute, but over time, longer relief durations exhibited a higher sustained rate of ingress because of the higher overall volume of relief gas present in the system.

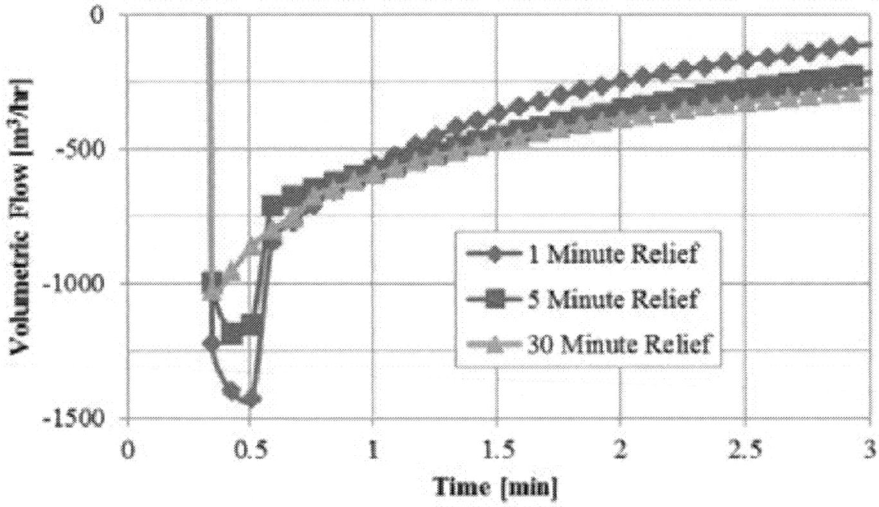

Figure 3: Volumetric flow rate at base of flare stack for hot vapor cooling case at varying relief durations.

<u>3.4.3 Gas Composition and Cooling vs. Condensation</u>

Hot releases of vapor that cool but do not contain condensable gases will exhibit slower and lower rates of volume contraction, with the heat transfer rate usually limited by the inside heat transfer coefficient. On the other hand, vapor containing condensable gases exhibit quick and high rates of volume contraction, with the rate of contraction strongly affected by the composition. In these instances the heat transfer rate will be limited by the outside heat transfer coefficient.

This difference is demonstrated by comparing Figure 3 for the cooling only case to Figure 4, which compares the volumetric flow rate measured at the base of the flare stack for a condensable vapor release of light naphtha at 30,000 Sm^3/h and 150°C at durations of 1, 2, and 5 minutes. As for the cooling only case, the maximum rate of air ingress occurred at a very short relief duration of 1 minute because the difference between ambient and the relieving gas remained the highest. At longer relief durations, the header may even warm above the condensing temperature of the gas, resulting in a slower rate of contraction because the header would need to cool before the gas could be condensed. This is demonstrated by the delayed negative flow rate peak for the 5 minute relief duration.

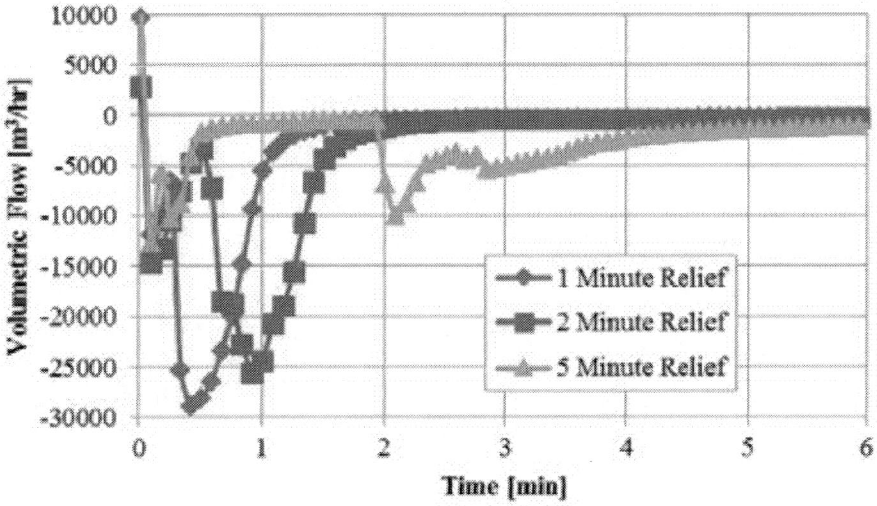

Figure 4: Volumetric flow rate at base of flare stack for hot vapor condensation case at varying relief durations.

3.4.4 Establishment of Make-up Gas Rate

Once the potential for vacuum formation and air ingress is understood, additional dynamic simulations can be performed to determine the flow rate of make-up gas necessary to prevent air ingress. The actual required rate is not necessarily equal to the maximum reverse flow observed, depending on the location of the make-up gas injection point and temperature and composition of the make-up gas. The make-up gas may compensate for volume contractions and displace the hot gas in the header and therefore reduce the overall rate of contraction. For cases with condensation, the make-up gas may also dilute the condensing gas and favorably affect the dew point of the mixture.

Figure 5 shows the volumetric flow rate at the base of the flare stack for a condensable relief with and without make-up flow. For this case, a hot vapor stream of light naphtha consisting of mostly C5-C6 hydrocarbons was released into the header for five minutes. At an ambient temperature of 20°C, this mixture will start to condense after an initial period of cooling. Without make-up, significant rates of contraction were observed immediately after relief subsided. A make-up flow rate of 6500 Sm^3/hr was required to prevent air ingress.

The condensing cases required a significantly higher make-up gas rate than the cooling only cases in Figure 3. For the cooling only cases shown in Figure 3, a make-up gas rate of only 975 Sm^3/hr was deemed sufficient to prevent air ingress.

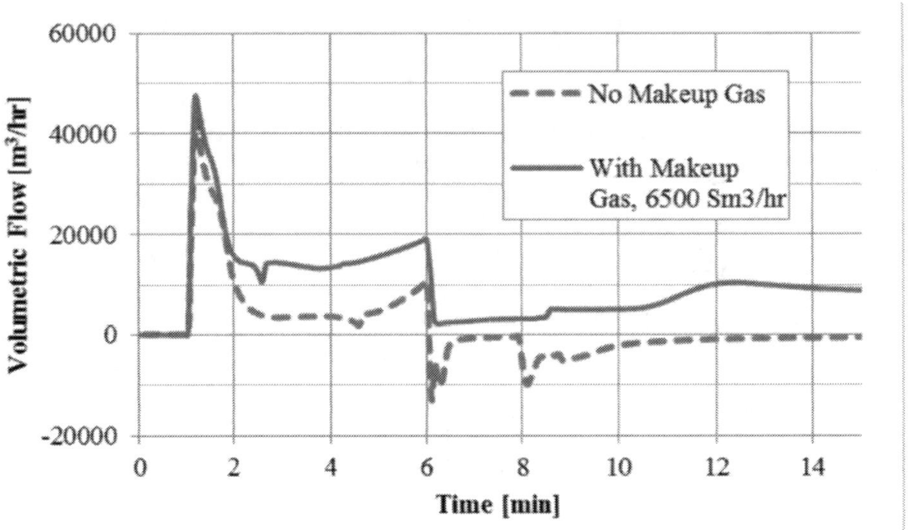

Figure 5: Volumetric flow rate at the base of the flare stack for a condensable release (light naphtha) with and without make-up gas flow.

3.4.5 Reliability of Make-up Gas

It is important to note that—considering the consequences of failure to provide adequate make-up gas to counter the potential vacuum formation—the source of the make-up gas must be highly reliable. Usually the source for the pilot gas, seal purge gas and the make-up gas are the same.

3.5 Location of Make-up Gas Addition Point

The preferred location of the make-up gas addition point is at the farthest end of the header, upstream of all reliefs, so that the flowing gas fills the hot spot first, compensates for the volume contractions, and displaces the remaining hot gases in the header. This method has an implicit advantage of reducing the total volume contraction requirement by pushing the hot gases outside of the header and potentially favorably affecting the dew point of a condensable mixture by dilution.

This location is the optimum regardless of how the other parameters are established. If the make-up gas addition point is located far downstream of the relief tie-in, such as near the flare stack or the knockout drum, the make-up gas would have to flow in the reverse direction to the relief point where the volume contraction is occurring.

Figure 6 demonstrates the impact that the location of the addition point has on the required make-up gas rates. The volumetric flow rates at the base of the flare stack when make-up gas is injected upstream of the relief and when it is injected downstream at the knockout drum for a condensable relief case are compared. For this case, a hot vapor stream of light naphtha consisting of mostly C5-C6 hydrocarbons was released into the header for five minutes. At an ambient temperature of 20°C, this mixture will start to condense after an initial period of cooling. For both cases, make-up gas was injected at a rate of 6500 Sm^3/hr. As shown in the figure, when make-up gas was injected near the knockout drum, negative flow was observed through the flare stack, whereas positive flow was maintained when the make-up gas was injected upstream of the relief.

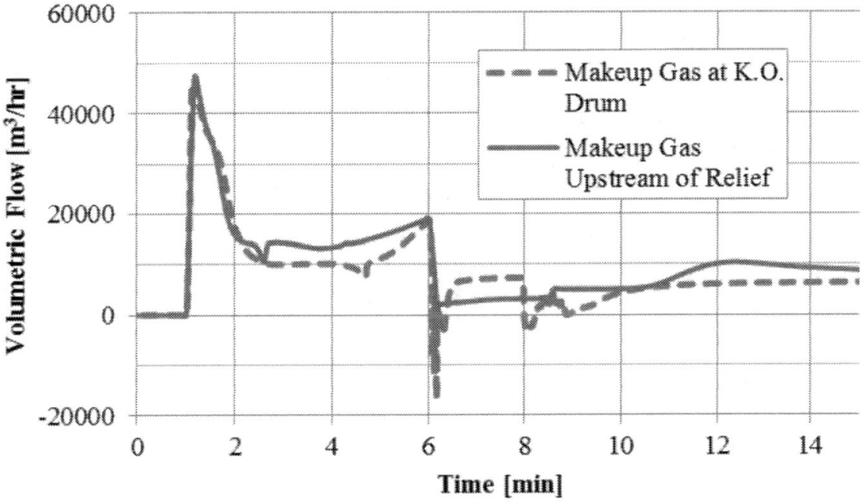

Figure 6: Volumetric flow rate at the base of the flare stack for a condensable release (light naphtha) when make-up gas is injected upstream of the relief and near the knockout drum.

3.6 Starting and Stopping of Make-up Gas

In the recommended control scheme, make-up gas is injected once relief is detected in the header via the appropriate temperature criteria. Valves and control schemes are provided to inject the make-up gas upstream of where the relief enters the header. Once a pre-set temperature condition in the header is satisfied, make-up addition can be safely stopped. Note that because the addition of make-up gas is initiated while the relief is occurring, the total backpressure at the discharge of the relief valve will be higher than the estimated backpressure when only the relief valve is discharging. Appropriate backpressure corrections should be made for the increased header flow rates.

The criteria for starting and stopping the injection of make-up gas depend on whether the expected reliefs will result in contraction due to cooling only, due to condensation and cooling by ambient conditions, or due to condensation and cooling by a pre-cooled header. The appropriate criteria for each condition are discussed in the following sections.

3.6.1 The Case of Cooling Only of Relief Gases – Starting and Stopping Make-up Gas

Starting make-up gas flow: As described in section 3.2.1, the difference between the temperature of gas in the header and the ambient temperature should be used to detect relief flow and to trigger make-up gas for cases where only cooling is expected.

To prevent excess injection of make-up gas, a flow override may be incorporated into the control scheme where make-up flow would not be injected until the flow rate close to the relief tie-in point drops below a specified minimum threshold. Both the temperature and the flow measurements should be located close to the relief point to minimize any potential delays that could result in air ingress.

Stopping make-up gas flow: After relief subsides, make-up gas should continue to be injected for as long as the temperature difference between the gas in the header and ambient remains above the minimum threshold value. Because the header will cool fastest near to the make-up gas injection point, once the temperature at the sensor drops below the threshold value, the header may still contain hot gas further downstream of the sensor location. To account for this, make-up gas should be programmed to continue to be injected for a defined duration to allow the gas temperature to stabilize.

Dynamic simulation can be used to determine how long it will take for the entire header to cool below the minimum ΔT at a given make-up flow rate. To determine the maximum time required, the relief simulation should initially be run long enough to establish the maximum steady temperature profile along the header before stopping the relief and allowing the header to cool.

Figure 7 shows the temperature profile along the example header for the hot vapor relieving case at various times after 30 minutes of relief ends. As shown in the figure, it took 90 minutes for the temperature along the entire header to fall below the threshold of 25°C (5°C above the ambient temperature of 20°C) assuming a make-up gas rate of 950 Sm^3/hr. This information can be used to determine how much additional time the make-up gas should be injected, depending on where the sensor is located relative to the make-up gas injection point and the relative rate of cooling at that location.

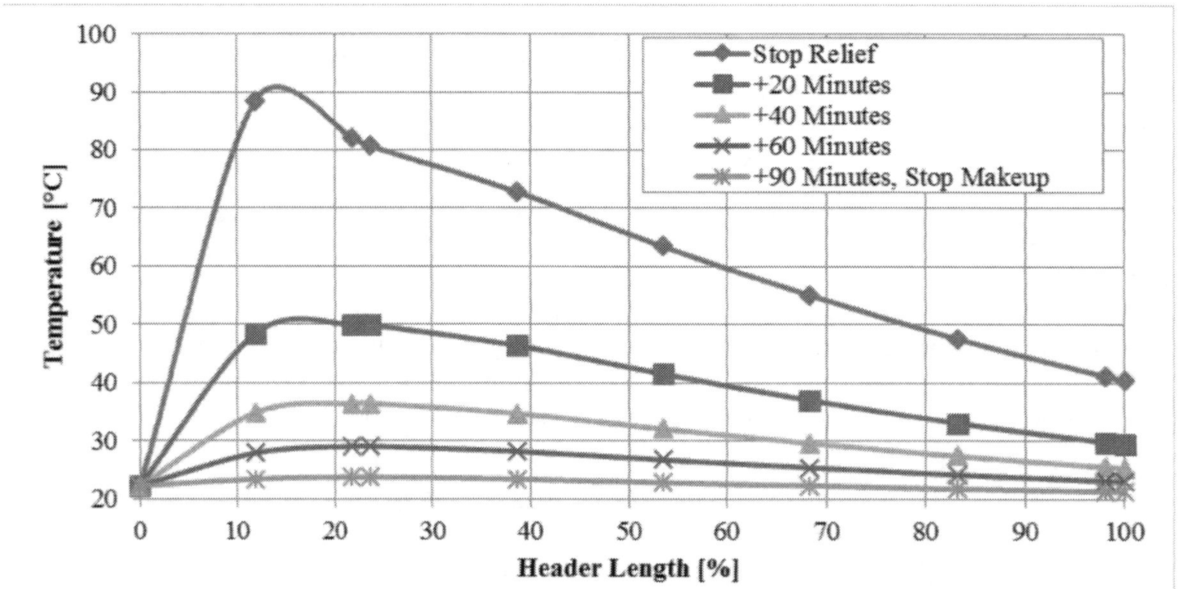

Figure 7: Temperature profile along header at various time intervals after hot vapor relief subsides.

3.6.2 Condensation of Relief Gases due to Ambient Conditions – Starting and Stopping Make-up Gas

Starting make-up gas flow: For cases where condensation can occur at or above ambient temperature, or as a result of rain or other weather conditions outside of the header, starting make-up gas should similarly be triggered by a minimum temperature difference criterion because the dew point of the gas will typically be well above the expected ΔT threshold for air ingress across all ambient temperatures as well as above the temperature of the make-up gas source. Otherwise, actual temperature may be used.

Unlike for cooling only cases, flow control may not be feasible when condensation is expected. Because the rate of contraction due to condensation is significantly faster than for cooling, a minimum flow override for injecting make-up gas only at high temperature and low flow may not provide a fast enough response to prevent vacuum formation and subsequent ingress of air.

Stopping make-up gas flow: As for cooling only cases, make-up flow should be stopped based on a minimum temperature difference threshold because the hot release may warm the header wall, resulting in a slower sustained contraction due to cooling after condensation begins to wane. Similarly, since it may be infeasible to measure temperature at all points along the header to ensure none exceed the minimum ΔT threshold, make-up gas should be injected for a pre-determined minimum duration.

3.6.3 Condensation of Relief Gases in a Cold Header – Starting and Stopping Make-up Gas

Starting make-up gas flow: A cold release in a warm header typically does not require injection of make-up gas because the header will warm and cause the gas to expand, which will result in a net forward flow out of the header. However, certain cases may exist where either a hot or cold release that is not normally condensable at ambient temperature occurs after the header has been cooled to below the dew point of the release.

Starting and stopping make-up gas for such a case requires a different approach than for the standard case of condensing that occurs at or above ambient temperature. Instead of injecting make-up gas once the temperature climbs above a specified threshold, it becomes necessary to inject make-up gas when the temperature of the header wall itself drops below the dew point of any potential subsequent reliefs.

Take for example a flare system that would experience releases of both cryogenic methane (boiling point of -162°C) and propane (boiling point of -42°C). There may be an instance in which a cryogenic release of methane may cool the header to temperatures to well below -42°C. If propane were to be released into the cooled header, it would condense and air ingress could occur. To prevent this, make-up gas would be injected once the temperature of the header drops below -42°C in order to warm the header. After the temperature along the entire header increases above this point, make-up gas would no longer be needed.

Stopping make-up gas flow: Ideally, after a cold release, the temperature at multiple points along the header would be measured to ensure that the entire header length has been warmed above the minimum condensing criteria. The make-up gas should continue for a specified duration to ensure that the temperature in the interim space between sensors has had adequate time to warm. This time can be established using dynamic simulation by calculating the time it takes to warm the entire header from the coldest state to above the threshold using only make-up gas.

Figure 8 shows the initial cooling by a cryogenic release of methane at three points along the header, followed by the subsequent warming of the header by make-up gas. A cold release of cryogenic methane at -130°C was released into the header for 3 hours, causing the entire header to cool to well below the boiling point of propane. Make-up gas was injected after the methane release subsided, and the header began to warm back up. As shown in the figure, it took nearly 40 minutes for the make-up gas to warm the entire header to above -42°C. This duration establishes how long the make-up gas should be injected once the cold release subsides.

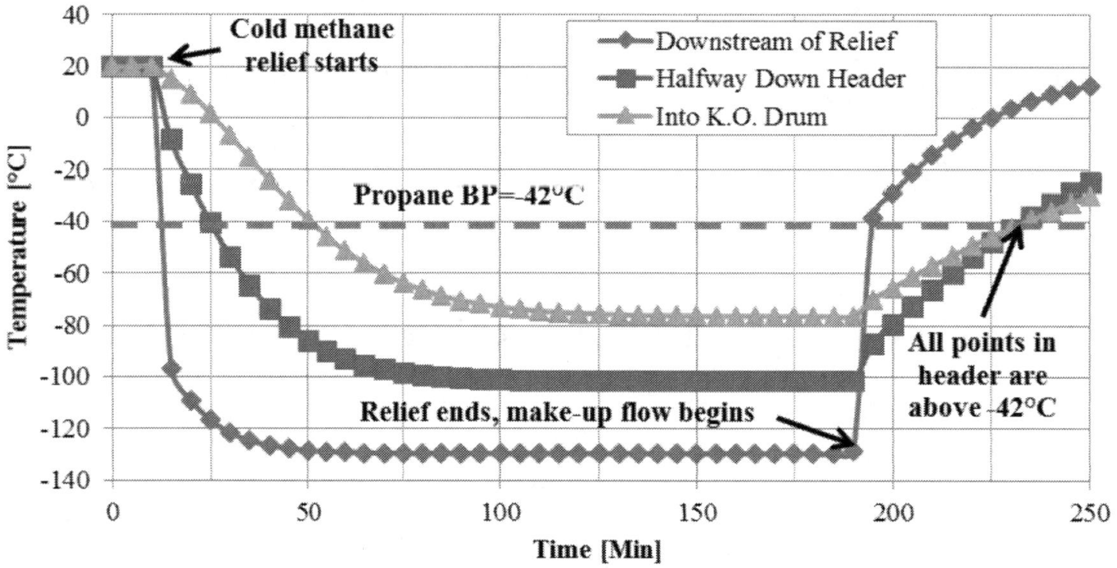

Figure 8: Temperature at three different locations along the header over time as the header is initially cooled by cryogenic methane for 3 hours, then subsequently warmed by make-up gas.

3.7 Potential Simplification of the Make-up Gas Addition Control Scheme

All the above scenarios were simulated with one pressure relief valve (PRV) relieving and a dedicated make-up gas scheme. An ideal make-up addition schematic is shown in Figure 9. In an ideal design, each PRV on the flare header system will have a dedicated sensor for relief and a make-up gas addition scheme that will add make-up gas only when a relief is sensed, with the addition rate and starting and stopping times tuned to that particular PRV. In this scheme, little or no make-up gas will be wasted.

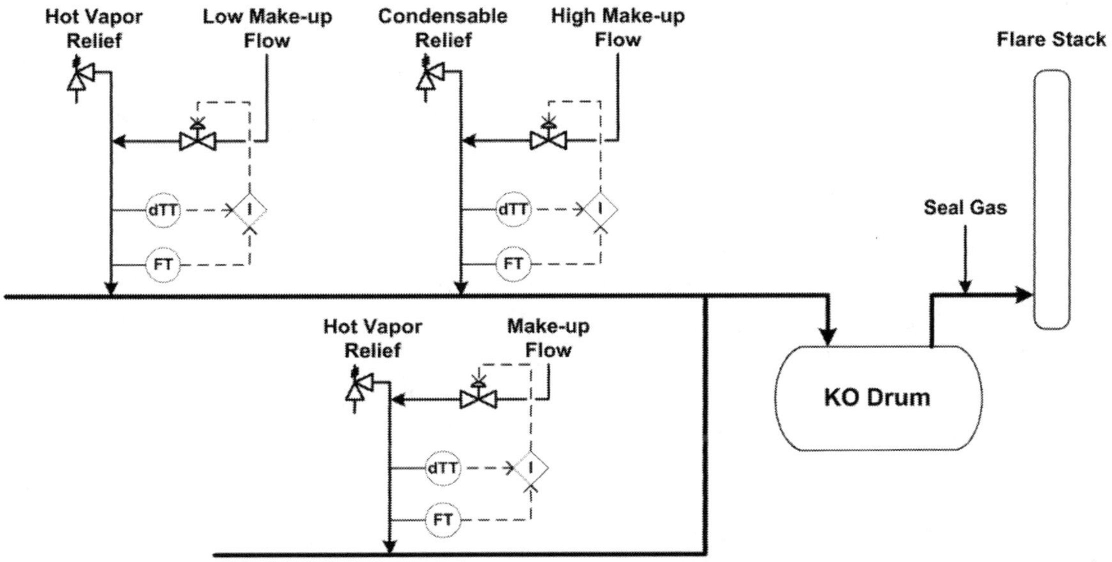

Figure 9: Ideal make-up gas addition scheme; dedicated temperature and flow sensors and separate make-up flows are provided for each PRV.

Obviously, such a scheme for each PRV on the flare header is generally not implemented because of economic and maintenance considerations. To reduce instrumentation, consideration could be made to group the PRVs that require low rates of make-up gas (essentially those where the relief would only be cooling) and the PRVs that require large rates of make-up gas addition (essentially those reliefs that may condense in the header). The location of the PRV laterals on the header and its branches would also be a consideration for such grouping. The temperature/differential temperature sensors and the flow meters should then be appropriately located to capture the start of relief to trigger make-up gas flow, as well as to measure the temperature subsidence in the header in order to stop the make-up gas flow at a safe point in time. Finally, if the flow meters are considered unreliable within the flow range required for reliable control purposes, they could be taken out of the control scheme altogether. In these circumstances the make-up gas flow duration would need to be increased in order to ensure early start and late stop.

A more typical schematic is shown in Figure 10. Note that in all schemes, the make-up gas addition point is as far upstream of all PRVs as possible so that it not only cools at the temperature sensor location, but effectively displaces the hot/condensable gases out of the header. Additionally, the differential temperature sensor is located downstream of all the PRVs, but as close as possible to the sources of relief in order to minimize any potential delays.

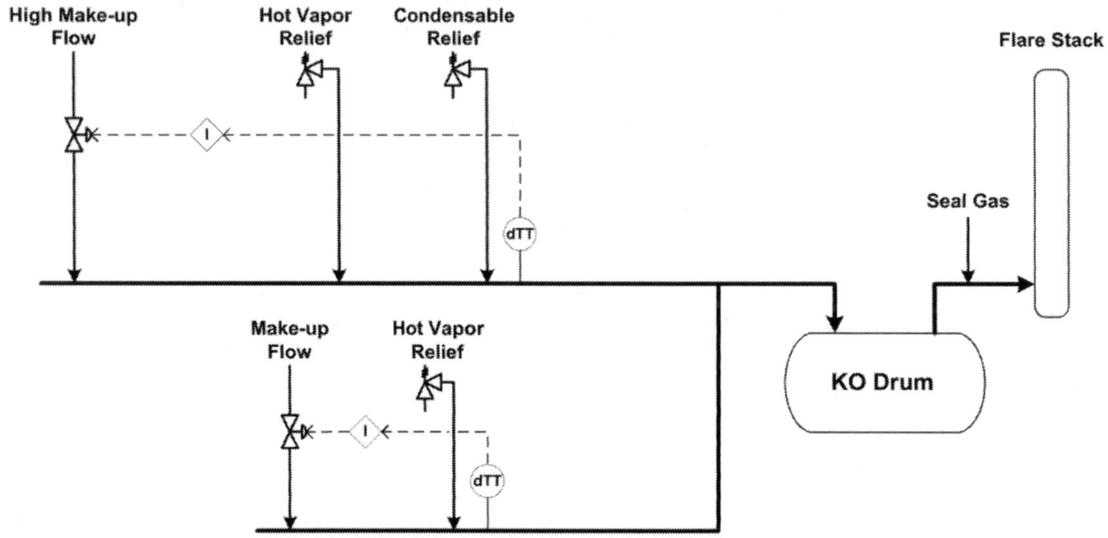

Figure 10: Simplified make-up gas addition scheme; a common temperature sensor and a single make-up flow rate cover all PRVs.

4. Conclusion

Designing an adequate and safe make-up gas injection system to protect a flare header system from various types of reliefs requires a thorough consideration of the physical properties of the relief gases, the locations of temperature and flow instrumentations, the seal purge rate, and the size of the flare header as well as prudent selection of when to start and stop make-up gas flow. The source of the make-up gas supply should be highly reliable. Simplification from a fully instrumented system is possible, albeit at the expense of potentially wasting make-up gas.

5. References

[1] Husa, H.W., "How to Compute Safe Purge Rates", Hydrocarbon and Petroleum Refiner, Vol. 43, No. 5, May 1961.

[2] Pressure-relieving and Depressuring Systems, API Standard 521, Sections 5.7.6 and 5.7.8, Sixth Ed., January 2014.

Additional references not cited:

[3] Reed, Robert, "What is Flare's Proper Purge Rate?" The Oil and Gas Journal, February 14, 1972.

[4] Bader, Adam; Baukal, Charles; Bussman, E., "Selecting the Proper Flare Systems", CEP, July 2011.

Abstract ID# 445580 – Accepted by *12th GLOBAL CONGRESS ON PROCESS SAFETY*

Title: Basic Concepts in Risk Analysis

Author: Vic Edwards, VHE Technical Analysis, Houston, TX

vhe@alumni.rice.edu

The chemical process industries often work with large inventories of materials that are hazardous due to flammability, explosivity, reactivity, corrosivity, and/or toxicity. The resulting hazards of these process systems must be identified, eliminated when possible, minimized, and/or mitigated.

Risk from a hazardous process event is a function of both the consequences of the event (what could happen) and the frequency with which it could occur (how often). It is essential to minimize risks in the chemical process industries. This is done by analysis of the components of the risk, coupled with management of the risks by reducing consequences and/or frequencies of potentially hazardous events.

Minimization of risk is the primary goal of process safety management (PSM). Presented here are the basic concepts of risk analysis that are at the core of PSM.

VICTOR H. (VIC) EDWARDS, PHD, PE

CHEMICAL ENGINEERING CONSULTANT

- Director Process Safety (retired), IHI E&C, 30 years
- 2015 Norton H. Walton/Russell L. Miller Award for Process Safety Engineering
- Chair, 2013 Global Congress on Process Safety
- Chair, 1992 Process Plant Safety Symposium
- Five DuPont Awards for HSE Excellence
- 2007 HSE Excellence in Design Award
- 2003 Service Award-Texas A&M University Process Safety Center
- 1998 Employee of the Year
- Projects include biochemical and chemical processes, biomass energy, coal gasification, methanol-to-gasoline, LNG, chemical weapons demilitarization, hazardous waste disposal, offshore hydrocarbons facilities
- >60 publications, including editor-Section 10 of Perry's Chemical Engineers' Handbook
- AIChE Fellow & Life Member; Member ACS, AAAS, NFPA, NSPE, and New York Academy of Sciences

BASIC CONCEPTS IN RISK - OUTLINE

- Many Process Chemicals are Hazardous, Particularly in Large Quantities
- The Chemical Industry is One of the Safest Places to Work
- What is Risk?
- Hazard Identification and Risk Management
- Example - What-If Analysis
- ALARP
- Risk Matrix
- Layers of Protection and LOPA Example
- Recommended Reading
- Questions?

PUBLIC PERCEPTION – CHEMICALS AND THE INDUSTRY ARE *RISKY!*

- Everything is made of chemicals
- The chemical process industries give us many valuable products-from agriculture to potable water
- Yet the public perceives all things *"Chemical"* as *a Risk* to health and safety
- This perception is partly due to the public's limited knowledge about chemicals and the industry
- This perception is also partly due to the media coverage whenever there is a significant industrial incident
- This perception is also due partly to the occasional large process incidents (Bhopal-thousands killed; tens of thousands injured; BP Gulf of Mexico)

THE CHEMICAL PROCESS INDUSTRIES - AMONG THE SAFEST OCCUPATIONS

US Fatal Occupational Injury Rates for 2012 - for a Few Occupations	
Fishers and Related Workers	120.8
Logging Workers	129.9
Structural Iron and Steel Construction	37
Roofers	42.2
Agriculture - Crop Production	32.5
Taxi and Chauffeurs	16.2
Protective Services (police, fire fighters)	15
Average Fatality Rate	3.4
Grocery Store	2.5
Chemical Manufacturing	1.8
Finance and Insurance	0.3

Fatal Injuries per 100,000 full-time workers per year

- Roy Sanders, Chemical Process Safety, 4th Edition, 2015

WHAT IS RISK?

- *RISK is the combination of three attributes:*

1. ***Hazard*** - What can go wrong?
2. ***Consequence*** - How bad can it be?
3. ***Likelihood*** - How often might it happen?

- *Guidelines for Risk-Based Process Safety*, CCPS, 2007

RISK IN THE PROCESS INDUSTRIES

- The chemical process industries often have large inventories of hazardous materials, often at high temperatures and pressures.
- Loss of containment can lead to major fires, explosions, and/or toxic gas releases.
- Because of careful and skillful design, construction, operation, maintenance and management of process facilities, the frequency of major process incidents is low.
- The risk of large consequence process incidents is comparatively low (but still too high) because the frequency is kept low.

HAZARD IDENTIFICATION AND RISK ANALYSIS (HIRA)

- *Hazard – Caged Tiger:* Safe Condition

  A Hazard is a Source of Danger
- *Initiating Event – Gate Left Open - Escaped Tiger:* Risk Condition
- *Consequences – Dangerous Situation!!! – About to Suffer the Consequences of Hazard's Damaging Capability*

- Roberto Blanco, Process Safety Progress, 2014

RISK MANAGEMENT IS.....

- *WHAT*.....the systematic application of management policy, procedures, and practices to the tasks of analyzing, assessing, and controlling risk.....
- *WHY*.....in order to protect employees, the general public, and the environment as well as company assets while avoiding business interruption.

- Guidelines for Chemical Process Quantitative Risk Analysis, CCPS, 2nd Edition, 2000.

HAZARDS IDENTIFICATION AND RISK ANALYSIS (HIRA): NON-SCENARIO BASED

- Preliminary Hazard Analysis
- Safety Review
- Relative Ranking
- Checklist Analysis

- Guidelines for Hazard Evaluation Procedures, 3rd Edition, CCPS, 2008.

HAZARDS IDENTIFICATION AND RISK ANALYSIS (HIRA): SCENARIO-BASED

- What-If/Checklist Analysis
- Hazard and Operability Review (HAZOP)
- Failure Modes, Effects and Criticality Analysis (FMECA)
- Event Tree Analysis (ETA)
- Fault Tree Analysis (FTA) – *Detailed - Quantitative Risk Analysis*
- Cause-Consequence Analysis (CCA) and Bow-Tie Analysis
- Layer of Protection Analysis (LOPA)
- *- Guidelines for Hazard Evaluation Procedures, 3rd Edition, CCPS, 2008.*

HIRA TEAMS

- Most hazard identification and risk analysis activities are conducted by a team of experts
- Team members typically represent process technology, process engineering, process safety, operations, maintenance, mechanical, instrumentation and control
- Some team members attend all meetings, but specialists may be brought in only when appropriate
- For small facilities, meetings may last a day or two, but large facilities may require weeks or months to complete
- Some HIRA studies, such as Fault Tree Analysis, may be conducted by one analyst.

TYPICAL WHAT-IF ANALYSIS WORKSHEET

Area:			Meeting Date:	
Drawing Number:			Team Members:	
What If	***Hazard***	***Consequence***	***Safeguards***	***Recommendations***

WHAT-IF EXAMPLE: DAP REACTOR TYPICAL WHAT-IF QUESTIONS

1. The wrong feed material is delivered instead of phosphoric acid?
2. The phosphoric acid concentration is too low?
3. The phosphoric acid is contaminated?
4. The valve controlling phosphoric acid feed is closed or plugged?
5. Too much ammonia is supplied to the reactor?
6. Agitation stops in DAP Reactor?
7. Valve on outlet of DAP Reactor is closed or plugged?

- CCPS, 2008

EXAMPLE WHAT-IF ANALYSIS

- **What If** – *Wrong feed material is delivered instead of phosphoric acid?*
- **Hazard** – *Contaminant incompatibility*
- **Consequences** – *Potentially hazardous phosphoric acid or ammonia reactions with contaminants, or production of off-specification product*
- **Safeguards** – *Reliable vendor; plant material handling procedures*
- **Recommendation** – *Ensure adequate material handling and receiving procedures and labeling exist*

WHAT RISK IS ACCEPTABLE?

- ***ALARP*** – As Low As Reasonably Practicable
- The concept that efforts to reduce risk should be continued until the incremental sacrifice (in terms of cost, time, effort, or other expenditure of resources) is grossly disproportional to the incremental risk reduction achieved.

  - *Guidelines for Risk-Based Process Safety*, CCPS, 2007

RISK MATRIX - EXAMPLE

Risk	Serious Danger in Immediate Area	Serious Danger Inside Battery Limits	Serious Danger Site Wide	Serious Danger Offsite
More Than Once Per Year	Action Required Unless Risk ALARP	Action Required at First Opportunity	Immediate Action Required	Immediate Action Required
Once Every Few Years	Action Required Unless Risk ALARP	Action Required Unless Risk ALARP	Action Required at First Opportunity	Immediate Action Required
Once in the Facility's Lifetime	No Action Required	Action Required Unless Risk ALARP	Action Required Unless Risk ALARP	Action Required at First Opportunity
Not Expected in the Facility's Lifetime	No Action Required	No Action Required	Action Required Unless Risk ALARP	Action Required Unless Risk ALARP

PROTECTIVE LAYERS IN PROCESS SYSTEMS

- At the core – Basic Process Design
- BPCS and Operator Response
- Safety-Instrumented Systems
- Pressure Relief Systems
- Procedures and Training
- Emergency Response
- Community Response

LAYER OF PROTECTION ANALYSIS (LOPA)

- Semi-Quantitative Tool for Analyzing and Assessing Risk
- Uses Order-of-Magnitude Categories for
  - Initiating Event Frequency
  - Consequence Severity
  - Probability of Failure-Independent Protective Layers(IPLs)

LOPA EXAMPLE-COOLING WATER FAILURE- RUNAWAY REACTION WITH LOSS OF CONTAINMENT

- ***Risk Tolerance–One in 1,000,000 years_Frequency = 10^{-6}***
- Frequency of Loss of Cooling Water-Once in 10 years = 10^{-1}
- Probability of Reactor Runaway on Cooling Water Loss = 0.5
- ***Frequency of Unmitigated Consequence = 5×10^{-2} per year***

Thus Unmitigated Risk is Unacceptably High

LOPA EXAMPLE – MITIGATIONS & RISK

- BCPS Alarm and Human Action – 10^{-1}
- Pressure Relief Valves – 10^{-2}
- Safety Instrumented System (SIS) – 10^{-3}
- Total of All Independent Protection Layers – 10^{-6}

- *Frequency of Mitigated Consequences – 5 x 10^{-8} per year*
- *Risk Tolerance – 10^{-6} per year > 5 x 10^{-8} per year*

  \- ***Therefore, Risk Tolerance Met with SIS***

RECOMMENDED READING

- Guidelines for Risk Based Process Safety, AIChE CCPS and John Wiley, Hoboken, NJ (2007).
- Guidelines for Hazard Evaluation Procedures, 3rd Edition, AIChE CCPS and John Wiley, Hoboken, NJ (2008).
- Guidelines for Engineering Design for Process Safety, 2nd Edition, AIChE CCPS and John Wiley, Hoboken, NJ (2012).
- Roy E. Sanders, Chemical Process Safety – Learning from Case Histories, 4th Edition, Elsevier, Inc., Oxford UK (2015).

OUR NEXT SPEAKER, DR. VINICIUS SIMOES, WILL ADDRESS QUANTITATIVE RISK ANALYSIS

THANKS FOR YOUR ATTENTION!

QUESTIONS?

SPARE SLIDES

RISK ANALYSIS IS.....

- *WHAT*.....The development of a quantitative estimate of risk.....
- *HOW.....based on engineering evaluations and mathematical techniques for combining estimates of incident consequences and frequencies.*

- Guidelines for Chemical Process Quantitative Risk Analysis, CCPS, 2nd Edition, 2000.

RISK ASSESSMENT IS…..

- ***WHAT…..*The process by which the results of a risk analysis are used to make decisions…..**
- ***HOW…..*<u>either</u> through relative ranking of risk reduction strategies…..**
- **…..<u>or</u> through comparison with risk targets (risk criteria).**

  - Guidelines for Chemical Process Quantitative Risk Analysis, CCPS, 2nd Edition, 2000.

12TH GLOBAL CONGRESS ON PROCESS SAFETY

50th LOSS PREVENTION SYMPOSIUM

Assessing Potential Damage Mechanisms and Selecting Effective Inspection & Testing Methodology for Fixed Equipment

F. Russ Davis, CSP
Mistras Group
4000 Underwood, LaPorte, TX 77571
russ.davis@mistrasgroup.com

Prepared for Presentation at
American Institute of Chemical Engineers
2016 Spring Meeting
12th Global Congress on Process Safety
Houston, Texas
April 11-13, 2016

Assessing Potential Damage Mechanisms and Selecting Effective Inspection & Testing Methodology for Fixed Equipment

F. Russ Davis, CSP
Mistras Group
4000 Underwood, LaPorte, TX 77571
russ.davis@mistrasgroup.com

Keywords: Asset Integrity, Damage Mechanism, Non-destructive Examination (NDE)

Abstract

Ensuring that process fixed equipment is safe to operate is essential in today's chemical, refining, upstream, midstream and all other oil & gas and chemical operations. Process equipment must be effectively inspected and tested correctly in order to lower the risk of operations to an acceptable level. The current regulatory environment and public opinion will no longer tolerate accidents that were preventable had effective inspections and testing been performed correctly. This paper will discuss potential damage mechanisms that degrade process equipment and evaluate the effectiveness of the various non-destructive examinations (NDE) and alternative inspection methods necessary for proactive condition-based monitoring. Internal and external corrosion mechanisms will be discussed. Damage mechanisms such as stress corrosion cracking (SSC), thermal fatigue, caustic embrittlement and hydrogen-induced stress corrosion cracking (HIC), along with several others, will be discussed. Various NDE methods will be evaluated for effectiveness at detecting the different potential damage mechanisms.

A safe and reliable plant is the desired end result. The implementation of a proactive inspection and testing program will detect damage to fixed equipment and mitigate the damage before an incident occurs.

1. Background

1.1 On August 6, 2012, the Chevron U.S.A. Inc. Refinery in Richmond, California experienced a catastrophic pipe rupture in the #4 Crude Unit. The incident occurred in piping referred to as the "4-sidecut" stream, one of several process streams exiting in the refinery's C-1100 Crude Unit Atmospheric Column. The pipe rupture occurred on a 52-inch long component of the 4-sidecut 8-inch line (the 52-inch component). At the time of the incident, light gas oil was flowing through the 8-inch line at a rate of approximately 10,800 barrels per day (bpd).

The ruptured pipe released flammable, high temperature light gas oil, which then partially vaporized into a large, opaque vapor cloud that engulfed nineteen Chevron U.S.A. Inc. (Chevron) employees. At 6:33 p.m., approximately two minutes following the release, the released process fluid ignited. Eighteen of the employees safely escaped from the vapor cloud just before ignition; one employee, a Chevron refinery firefighter, was inside a fire engine that was caught within the fireball when the process fluid ignited. Because he was wearing full-body fire-fighting protective equipment, he was able to make his way through the flames to safety. Six Chevron employees suffered minor injuries during the incident and subsequent emergency response efforts.

In the weeks following the incident, approximately 15,000 people from the surrounding communities sought medical treatment at nearby medical facilities for ailments including breathing problems, chest pain, shortness of breath, sore throat, and headaches. Approximately 20 of these people were admitted to local hospitals as inpatients for treatment.

1.2 In October 2009 a leaking pipeline in a storage facility in Jaipur, India led to a fire and explosion which spread throughout the facility. Over 500,000 people were evacuated. There was an estimated loss of $65 million in property and product. The facility will be out of service for 12 to 18 months to be rebuilt. The environmental impact is devastating.

1.3 In November 2009, Silver Eagle Refinery in Woods Cross, Utah suffered a devastating explosion and fire when a 10 inch pipe attached to a reactor ruptured. The explosion and fire caused onsite as well as offsite impact. Surrounding houses were dislodged from their foundation. The U.S. Chemical Safety Board recommended a complete shutdown of the refinery while failure of the mechanical integrity program was evaluated. Inspection was being performed on this line but failed to recognize significant wall loss.

1.4 On July 28, 2005 a hydrogen fire in the resid hydrotreater unit (RHU) at the BP refinery in Texas City, Texas resulted in approximately $30 million in damage to the refinery. The Chemical Safety Board (CSB) report stated that "BP should have required positive materials verification of these pipe elbows using an x-ray fluorescence test device. This would have identified the mistake in the reassembly of the identically appearing elbows before the unit was returned to service. The accident would not have occurred." [1]

Process systems are continuing to fail causing unacceptable consequences that can and have affected safety, the environment, production and the company profitability. These failures often occur not because we do not inspect the equipment but because we do not *correctly* inspect the fixed equipment systems.

Mechanical integrity (asset integrity) programs have been in place in refineries and chemical plants for several years but equipment failures continue to severely impact safety and the environment and result in millions of dollars lost each year. Non-destructive examination (NDE) is being used as the primary method for detecting damage to fixed equipment.

There is no single non-destructive examination method that will properly detect all damage mechanisms. Many process systems require the use of multiple NDE methodologies to detect the various damage mechanisms. This paper discusses some of the current NDE methods available for use in the process plants, several frequently encountered damage mechanisms and suggests various NDE methods for the detection of these damage mechanisms. I conclude with a discussion of the benefits of selecting the most effective testing protocols for fixed equipment based on potential damage that may affect the equipment and piping.

2. Damage Mechanisms

Process units should have a full damage mechanism review so that fixed equipment inspection plans can be developed that test for all potential damage that may occur. API 571 "*Damage Mechanisms Affecting Fixed Equipment in the Refining Industry*"[3] supplies very good recommended practices for identifying potential damage mechanism. While API 571 is directed to the refining industry most of the damage mechanisms identified are applicable to the chemical industry as well. The results of the various damage mechanisms are the same for both refining and chemical industries. The metal is either lost due to wastage/corrosion or the integrity is lost due to cracking. The CSB has issued recommendations noting that the process industry should include a damage mechanism review and likelihood of failure analysis as part of the process hazard analysis. The PHA should identify the risk of operating fixed equipment in the given environment of each process.

Figure 1 shows documentation associated with a unit damage mechanism review. Equipment that may be affected by damage has been identified and the damage mechanism is noted.

Figure 1

Figure 5-71 – Hydroprocessing - Hydrotreating, Hydrocracking

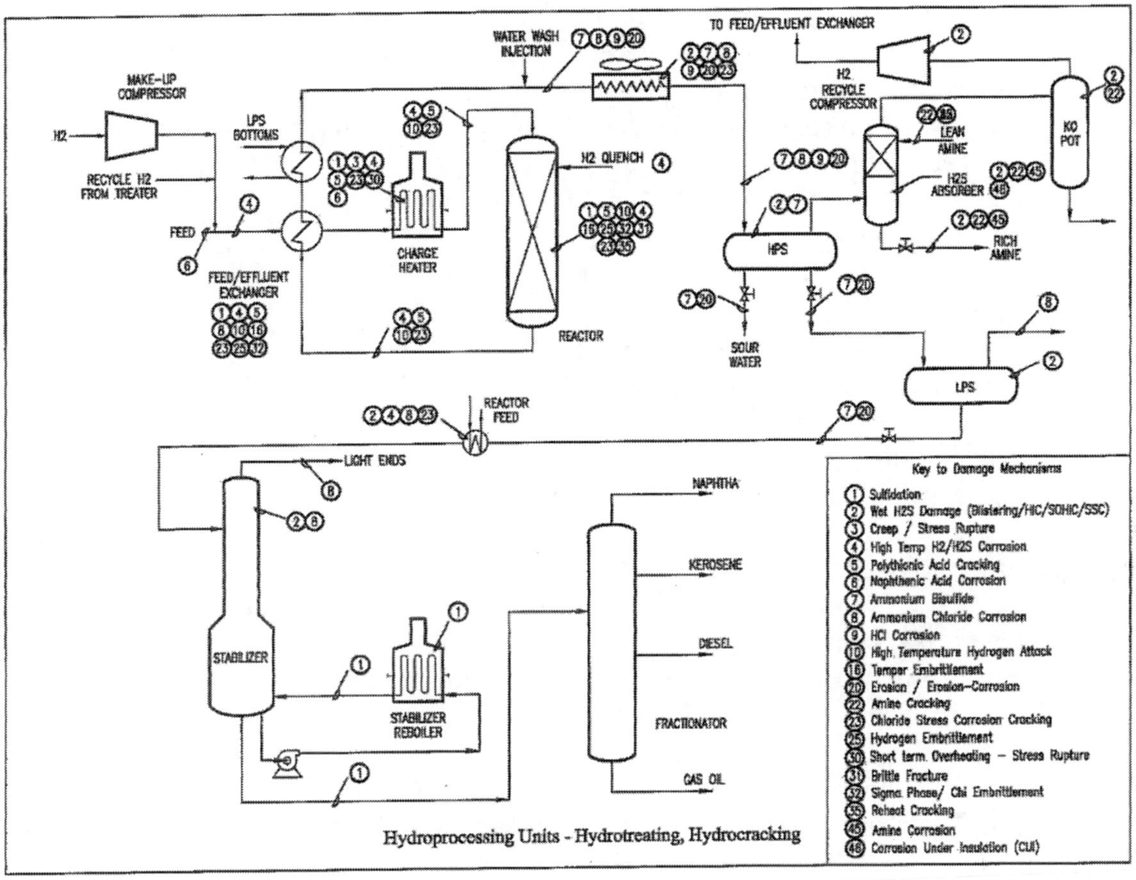

Hydroprocessing Unit Damage Mechanism Map (Figure 5-71, API RP 571)[3]

The following is a discuss some of the most common damage mechanisms that can adversely affect fixed equipment in refineries and chemical plants alike.

2.1 Erosion/Erosion Corrosion

- The removal of surface material as a result of impingement by solids, liquids, or vapors.
- Erosion Corrosion the more common damage mechanism.
- Damage and damage rates are dependent upon:
    - Corrosivity of the process
    - pH
    - Temperature
    - Flow rate
    - Material of construction

2.2 Corrosion Fatigue

- Cracking that develops in a corrosive environment when cyclic loading is present.
- Cracking dependent on number of cycles.

Operational startups and shutdowns in batch operations result in an increase in operational cycles which can lead to corrosion fatigue.

2.3 Chloride Stress Corrosion Cracking (Cl SCC)
- Caused by the presence of chlorides in a system constructed of stainless steel materials.
- Dependent on level of chlorides, temperature, and the presence of stress.
- Chloride concentration in the process, run temperatures, and pH level are parameters that significantly affect Cl SCC.

2.4 Caustic Corrosion
- Caused by caustic in the process system.
- Produces general and localized corrosion.
- Problems arise when:
  - Caustic is added to process streams where it was not previously used.
  - Caustic-containing cleaning materials are used.

2.5 Caustic Stress Corrosion Cracking
- Caused by embrittlement of susceptible materials in the presence of caustic, stress and elevated temperature.
- Temperature is a critical factor. Higher operating temperatures affect/accelerate caustic stress corrosion.

2.6 Microbial Corrosion (MIC)
- Caused by living organisms found in aqueous environments, especially where flow is stagnant.
- Stagnant flow in idle heat exchangers creates the possibility of MIC

2.7 Wet H2S Damage
- Types include:
  - Hydrogen blistering
  - Hydrogen induced cracking
  - Stress-oriented hydrogen induced cracking (SOHIC)
  - Sulfide stress corrosion cracking (SSC)
- Affects carbon steel and low alloy steels
- Critical factors: pH, H_2S levels, contaminants and temperature
- Operational changes in the critical parameters effect likelihood of damage.

2.8 Amine Stress Corrosion Cracking
- Carbon steel and low-alloy steels are susceptible.
- Amine concentration, stress and temperature are the critical factors.
- Operational changes such as higher amine concentration or higher temperatures can make a system more susceptible to amine cracking.

2.9 Corrosion under Insulation (CUI)
- Affects primarily carbon steel and low alloy steels.

- Susceptible temperature ranges are from 10°F to 350°F
- Can affect stainless steels especially in the coastal regions where chlorides may be present in the environment.

2.10 Sulfidation/Naphthenic Acid Corrosion

- Affects carbon steel and other alloy steels due to the reaction with sulfur compound0s in high temperature process streams.
- Major factors for sulfidation corrosion are temperature and sulfur content of the process stream.
- Primarily caused by H_2S and other reactive sulfur species.

3. Non-Destructive Testing Methodologies

Selecting the most effective non-destructive examination (NDE) methodology is critical to detecting damage. In many cases multiple NDE methods are necessary to fully evaluate the equipment or piping system. This section identifies the advantages and disadvantages of many NDE methods. Figure 2 below identifies the NDE method and briefly discusses the pros and cons of the method.

Figure 2 Advantages & Disadvantages of Select NDE Methods

Method	Advantages	Disadvantages
Manual Ultrasonic thickness testing	Quantitative inspection, thickness measurements are easily identified, verification of thickness readings is relatively fast. Can be performed quickly without advanced safety measures.	UT Thickness methodology is effective in identifying general corrosion since wall loss is more wide spread, but may not be sufficiently effective for localized corrosion. Temperature limitations. Spot thickness data collection. Limited area coverage. Requires direct access to inspection area. Cost control is based on access and production rate.
Automated Ultrasonics	Quantitative inspection, thickness measurements. More detailed coverage than typical manual UT thickness scanning. Provides adequate data for Fitness for Service analysis.	Temperature limitations. Covers limited area, requires direct access to inspection area, cost control is based on access and production rate, Close proximity piping runs and pipe geometry can restrict inspection area.

Radiography Film	Qualitative & quantitative inspection, thickness measurements are possible, identifies OD & ID conditions, no insulation removal required	Longer exposure times, accuracy ±10%, film processing required, results delayed, environmental issues w/chemicals, image adjustment not possible, archival restrictions
Radiography Digital Radiography (DR)	Qualitative & quantitative inspection, accurate thickness measurements, identifies OD & ID conditions, no insulation removal required, short exposure times, smaller exclusion zone, instant image review, quicker processing	Covers limited area, requires direct access to inspection area, cost control is based on access and production rate, radiation hazards, detector panel is fragile, rigid & heavy, cumbersome in tight spaces, 110 v power required, connection to laptop, one system/crew
Radiography Computed Radiography (CR)	Qualitative & quantitative inspection, accurate thickness measurements, identifies OD & ID conditions, no insulation removal required, short exposure times, smaller exclusion zone, software image enhancement, fewer re-shots, electronic media, multiple crews/processing unit	Covers limited area, requires direct access to inspection area, cost control is based on access and production rate, radiation hazards, imaging plate care required
Real-time Radiographic Testing	Quantitative inspection, identifies OD conditions, fast-immediate results for identification of welds, no insulation removal required, no radiation hazards, production good on open piping runs	Close proximity piping runs and pipe geometry can restrict inspection, cannot identify ID conditions, production slower in close spaced piping, inspection unit is cumbersome in tight spaces
Guided Wave Ultrasonics (GUL)	Rapid inspection of long lengths of pipe (1500 – 3500 feet/day), 100% volumetric coverage, limited insulation removal for inspection, accurate location of pipe features and corrosion, semi-qualitative analysis of wall loss, temperature range to 650° F, most cost effective for very long lengths of pipe	Qualitative screening, requires secondary method for verification of anomalies, geometry can restrict inspection

Pulsed Eddy Current (PEC)	Fast, no insulation removal, easily deployed, high temperature capability, non-contact, qualitative data	Close proximity piping can limit access, no inspection of piping < 3 inches
Electro-Magnetic Acoustic Transducer, (EMAT) High Temperature Ultrasonic Thickness	Temperature range up to 1000°C. Continuous thickness scanning can detect gradual change and random isolated/localized corrosion. Rapid scanning of large components, elbows, reducers, tees, headers. Can inspect up to 1000 feet per day, Quantitative thickness measurements are easily identified, verification of thickness reading is relatively fast. Can be performed quickly without advanced safety measures.	Covers limited area, requires direct access to inspection area, cost control is based on access and production rate.
Alternating Current Field Measurement (ACFM)	Identifies and sizes surface breaking cracks. May be used on coated materials and surfaces containing minor scaling. Temperature up to 900°F.	Limited to non-ferrous metals. Requires 110 volt access.
Acoustic Emission	Detects small-scale damage, applicable to cracking and active corrosion. External application does not require equipment to be cleaned and cleared, can be used on all metallurgies.	Requires follow-up with ultrasonic shearwave to size and characterize indications.
Eddy Current	Primarily used to detect cracking in tubing. Fast and requires only minimal cleaning of tubes. Sensitive to small cracks and indications. Surface Eddy Current testing effective in detecting surface breaking or slightly subsurface cracking.	Most effective on stainless steel materials. Does not fully characterize indications or corrosion. Only conductive materials can be inspected, surface finish and roughness may affect the test.

4. Selecting Effective Methodologies

Understanding the potential damage that may adversely affect process equipment and piping and the NDE tools that can effectively detect this damage is critical to lowering the risk of operating hazardous process facilities. Detecting damage and making necessary repairs extends the service life of process fixed equipment leading to less downtime for the units. This section will discuss two major damage mechanisms results, corrosion and cracking, and the effective inspection and testing methodology that can detect and characterize the damage before failure.

4.1 Corrosion/erosion corrosion

4.1.1 Visual inspection by a competent inspector has been part of NDE inspections for as long as there has been equipment to inspect. Visual inspection with photo documentation is a very good tool for recording damage mechanisms for future evaluations. There are however many draw backs to visual inspections. Inspectors cannot see through insulation and must make entry into possibly hazardous confined spaces in order to perform visual inspection. Internal visual inspection can be very expensive. Cleaning, clearing and preparing a vessel or storage tank for a visual inspection can cost as much as US $10,000 per vessel. Visual inspection augmented with additional NDE is still the tried and proven inspection method of choice.

4.1.2 Automated ultrasonics which includes methods such as Large Structure Inspection (LSI) or long range guided wave (GUL) is valuable in locating corroded areas. The tried and proven manual UT method is also valuable. LSI corrosion mapping is a very good tool for identifying and quantifying of internal corrosion. LSI is performed externally eliminating the need to clean and clear a vessel. Below is an example of LSI mapping:

Item scanned

LSI Scan Results

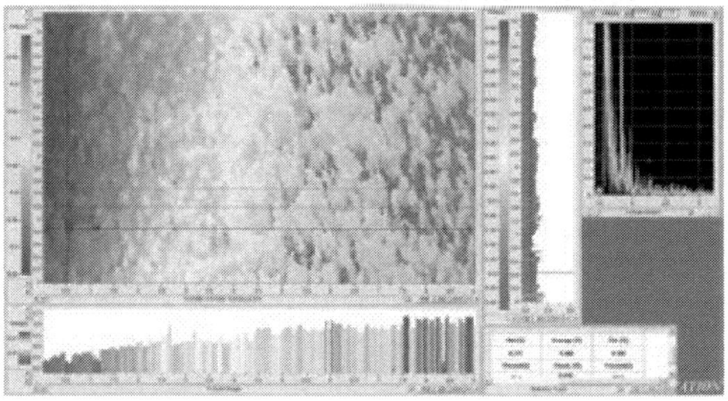

4.1.3 Guided wave ultrasonics is a method for testing long runs of pipe for corrosion. GUL can be effective in locating corrosion in underground piping and insulated piping minimizing the need to dig up a pipeline or to remove all the insulation. GUL is especially suited to finding corrosion under insulation (CUI).

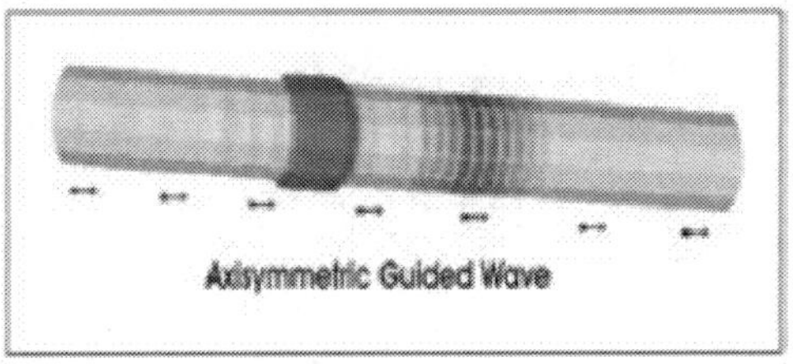

(Field guided wave)

4.1.4 Phased array ultrasonics can detect flange face corrosion without disassembling the flanges. This is proved to be a very effective tool for hydrofluoric acid (HF) service.

4.1.5 Corrosion under fireproofing can be extremely difficult to detect. The pulsed eddy current (PEC) methodology is effective in locating the corrosion without removing the fireproofing.

4.1.6 X-ray inspection can readily detect corrosion. Computed radiography allows for a permanent record and accurate measurements of corrosion and remaining wall thickness.

4.2 Cracking

4.2.1 Ultrasonic shear wave testing and phased array testing are effective for detecting and characterizing cracking.

4.2.2 ACFM detects cracks in non-ferrous metals that are open to the surface

4.2.3 Surface Eddy Current testing is very effective for the detection of surface breaking and slightly subsurface cracking.

4.3 NDE/Damage Mechanism Matrix

4.3.1 Corrosion/damage mechanism mapping followed up with selection of effective testing and inspection methodology must be performed in order for each facility to safely operate. Table 1 below is a quick and easy matrix for selecting a basic methodology for inspection. Specific applications should always be determined by a subject matter expert in the testing methodology that is selected.

Table 1	RECOMMENDED NDE METHODS FOR DETECTING DAMAGE									
DAMAGE MECHANISM										
	RT	**UT-THICKNESS**	**AUT**	**LONG RANGE UT**	**UT SHEAR WAVE**	**AE**	**ET**	**ROBOTICS**	**Visual**	**MT-PT**
Corrosion - vessels		X	X			X		X	X	
Erosion-vessels		X	X					X	X	
Corrosion - Storage tanks		X				X		X	X	
Corrosion-piping	X	X	X	X						
Erosion-piping	X	X	X	X						
Cracking-vessels			X		X	X				X
Cracking-piping	X		X		X	X				X
Corrosion under Insulation	X	X		X						
Exchanger tube corrosion							X	X	X	
Exchanger tube erosion							X		X	
Exchanger tube cracking							X			

RT – Radiographic testing
UT – Ultrasonic testing
AUT – Automated ultrasonic testing
PT- Dye penetrant testing
AE – Acoustic emission testing
ET – Eddy-current testing
MT – Magnetic flux testing

Summary

Understanding all of the potential damage mechanisms is a priority in the selection of effective inspection and testing methodology. The use of non-effective inspecting and testing is wasted money and resources. Where failure is not an acceptable alternative a facility that handles hazardous chemicals must be proactive in the selection of inspection and testing methods. The process hazard analysis (PHA) must evaluate the risk of operating process equipment and piping. Risk must be reduced to an tolerable level. Effective inspection and testing of fixed equipment needs to be performed in order to properly manage facility risk.

References

1. U.S Chemical Safety Board, *CSB FINAL INVESTIGATION REPORT, REPORT NO. 2012-03-I-CA, JANUARY 2015, CHEVRON RICHMOND REFINERY PIPE RUPTURE AND FIRE*

2. U.S Chemical Safety Board, *CSB issues safety bulletin on BP Texas City major fire: Better material identification needed, errors during systems maintenance cited; fire caused $30 million in property damage,* Oct. 15, 2006

3. API Recommended Practice 571, *Damage mechanisms Affecting Fixed Equipment in the Refining Industry,* 2nd edition, April 2011

4. API RP 581 *Risk Based Inspection Technology*, 2nd edition September 2008

Pressure Relief Valve Inspection and Maintenance

Thiago Trotta
Siemens Energy, Inc.
4615 Southwest Fwy, Suite 900
Houston, TX 77027
thiago.trotta@siemens.com

Charles Kashou
Nancy Faulk
Siemens Energy, Inc.
4615 Southwest Fwy, Suite 900
Houston, TX 77027
charles.kashou@siemens.com, nancy.faulk@siemens.com

Prepared for Presentation at
American Institute of Chemical Engineers
2016 Spring Meeting
12th Global Congress on Process Safety
Houston, Texas
April 11-13, 2016

Pressure Relief Valve Inspection and Maintenance

Thiago Trotta
Siemens Energy, Inc.
4615 Southwest Fwy, Suite 900
Houston, TX 77027
thiago.trotta@siemens.com

Charles Kashou, Nancy Faulk
Siemens Energy, Inc.

Keywords: PRV inspection interval, PRV maintenance, PRV risk-based inspection

Abstract

Determining the correct interval for pressure relief valve (PRV) inspection, testing, and maintenance remains a major challenge for facilities covered by the U.S. Occupational Safety and Health Administration (OSHA) Process Safety Management (PSM) Standard. To this end, guidance is provided by API Standard 510, *Pressure Vessel Inspection Code: In-Service Inspection, Rating, Repair, and Alteration*, by API Recommended Practice 576, *Inspection of Pressure-relieving Devices*, and by API Standard 520, *Sizing, Selection, and Installation of Pressure-Relieving Devices, Part II-Installation.* Furthermore, ASME BPVC, Sections I and VIII, provide general guidelines for the repair of pressure relief valves.

However, the testing and inspection interval listed, ten years, is the *maximum* time span between shop inspections and overhaul. Further direction is often requested for determining the proper interval for valves in typical process services, especially in cases of PRV chattering. Recent API STD 520 Part II guidance on performing engineering analyses for PRV installations, based on service and specific installation, is included here.

In this paper, a decision-making approach to determining these intervals based on a combined understanding of risk-based inspection, quality assurance, engineering analyses, and facility experience is presented. The approach provides process operators and managers with additional assistance in making this determination.

1 Introduction

Pressure-relieving devices (PRD) are installed on pressure vessels, boilers and low pressure tanks in order to relieve excess pressure which may result from the overpressure causes described in the American Petroleum Institute Standard 521 (API STD 521) [1]. These include process upsets, operator error, external fires, and other scenarios. Failure to function properly on demand could result in vessel overpressure, and possibly loss of containment with subsequent explosions, fires, or toxicity impacts. There are also consequences associated with leakage of

PRDs. To ensure personnel safety as well as protection of equipment, facilities and the environment, it is essential that the PRD be properly designed, installed, inspected regularly, and maintained in good operating condition.

Inspections are used to determine the general physical and operating conditions of the PRD, and assess whether it meets the requirements for a given installation and service. These inspections are of two types: visual on-stream inspections or shop inspection and overhaul. How often should these inspections take place?

Determining the correct interval for a pressure relief valve (PRV, which is a type of PRD) inspection and testing, as well as any indicated maintenance, remains a major challenge for facilities covered by the U.S. Occupational Safety and Health Administration (OSHA) Process Safety Management (PSM) Standard [2]. To this end, guidance is provided by API Standards 510 and 520 PT II, as well as API RP 576 [3] [4] [5]. Furthermore, ASME BPVC, Sections VIII and I, provide general guidelines for the repair of pressure relief valves [6] [7]. Note that a *definite* time interval between inspections or tests should be established for *every* PRV on operating equipment.

Guidance is often requested for determining the proper interval for valves in typical process services, especially in cases of pressure relief valve instability. This paper first describes the instabilities which may affect PRV performance and their causes, then describes how the inspection interval may be determined based on risk-based inspections as well as additional engineering analysis.

2 Instability Phenomena and Causes

The principal reason for inspecting and maintaining PRVs is to ensure that they can provide this protection when needed. Inspections determine the condition of the valve and look for signs of damage or other concerns. In addition to corrosion, deposits, and plugging, there are instability phenomena which can cause damage to the seating surface of a PRV and prevent it from performing properly. These are described in the following sections.

2.1 Cycling, Fluttering and Chattering

The total non-recoverable pressure loss between the protected equipment and the PRV should not exceed 3% of the PRV set pressure, to prevent valve instability; with the following exceptions to the aforementioned criteria [8]:

- Thermal relief valves: usually relief devices designed solely to protect against the overpressure caused by liquid hydraulic expansion, due to ambient or process heating, are largely oversized for the relief requirement, which could already lead to chattering. More than that, it also may result in exceeding the 3% criteria. However, upon closely looking into the liquid hydraulic expansion phenomenon, this would only be a concern for isolatable equipment, and should it happen, the amount of trapped liquid would be limited. In that case, as the relief device opens and discharges a certain flow of liquid, it would reclose due to loss of built up pressure on the inlet of the valve. It would not immediately reopen because it would take some time for the pressure to build up again. In that situation, chattering would not be expected for this case.

- Remotely-sensed pilot operated relief valves: this arrangement for pilot operated relief valves permits the relief device to sense the pressure directly from the protected equipment, and by doing that, the relief device would not try to reclose due to high non-recoverable pressure losses.

The concern in exceeding the 3% rule is that it may result in one of the following instability phenomena which may be observed on pressure relief devices [9].

2.1.1 Cycling

Cycling refers to a low frequency movement of opening and reclosing of a relief device, and it tends to occur when the relief requirement is small when compared to its capacity. The pressure in the system is decreased, and then built back up again periodically. As it consists in a low frequency movement, it does not usually result in damage to the PRV; however, it does impact the ability of the valve to reseat and may lead to wear over time.

2.1.2 Fluttering

Flutter occurs when the device is open, but moving parts of the valve are rapidly reciprocating; e.g., the disk does not contact the seat, instead it reciprocates around a point, creating pressure pulsations. Over time it may lead to the valve becoming stuck with full or partial opening. This phenomenon wears out the components of the valve.

2.1.3 Chattering

Chattering is a very high frequency and high amplitude oscillation; in the worst case, the disc may move between its two extremes of completely closed and maximum lift. The frequency of the oscillations may exceed several hundred hertz [10]. Different from the cycling, the main consequences of this event are possible loss of containment, damage to moving parts of the device, and damage to the equipment to which the valve is connected. Chattering can be very destructive to the valve seat.

A recent method has been proposed in API STD 520 PT II to investigate if a relief device would really chatter upon its relief [11]. Experience has shown that a relief device may start chattering when the unrecoverable pressure losses are below this recommended limit; conversely, it may not even chatter if the losses are above 3% or result in failures due to relieving events. Because the relationship between inlet pressure drop and chattering is not well understood, detailed requirements for an engineering analysis are the responsibility of the user. The standard states that the user's engineering analysis may be qualitative or quantitative and shall be documented. Readers are cautioned that an engineering analysis *shall not be applied* if the relief device has a history of chattering.

2.2 Causes of PRV Instability

The main causes for PRV instability are discussed in API STD 520 PT II and are listed below [12]. PRV instability is a complex occurrence and often cannot be attributed to a single cause.

2.2.1 Excessive inlet pressure losses

As the relief devices opens, the pressure on the inlet nozzle will decrease due to the pressure losses resulting from friction on the inlet piping. If the pressure losses are significant, the inlet pressure may fall below the reseating value for the PRV, which will result in the device reclosing, and as the pressure on the protected equipment builds up again the valve will reopen. This effect may lead to chattering on the relief device.

2.2.2 Excessive built-up backpressure

Excessive built-up backpressures: built up backpressure due to flow on the discharge piping of the PRV results in a force on the valve disc, which may lead to reclosing the valve. If the backpressure is excessively large, the valve will close and reopen again as the resisting force will cease. This is also a cause for chattering on the relief device.

2.2.3 Acoustic interaction

The rapid opening of a relief device will result in an also rapid drop on the pressure upstream of the valve disc. As a result, a pressure reduction wave will travel through the upstream fluid and as it comes in contact with an upstream reservoir it will reflect and travel back as a compression wave towards the valve disc. If this wave reaches the disc before it closes, the valve may discharge at a stable manner or it may flutter. If the disc is closed, however, it may lead to a cycling or a chattering effect.

2.2.4 Retrograde condensation

If the fluid during the relief scenario is at a supercritical state (e.g., pressure relief devices protecting high pressure system on hydrocracker units), the drop in pressure as the valve opens may lead to partial condensation of the fluid. When this condensation occurs, there will be an increased pressure drop effect due to the contraction of the fluid from supercritical to liquid.

2.2.5 Improper valve selection

Vapor certified valves will open in a more rapid manner than liquid certified ones. In this case, for services with no possibility of vapor relief, only liquid, this device would open much faster than what is required, as the pressure buildup on a liquid service is also slower, which may lead to reclosing the valve right after it starts relieving.

2.2.6 Oversized relief devices

Where the device's capacity is excessively larger than the relief requirement, when the valve is fully open, the pressure on the protected system will lower which may result in reclosing the PRV (if the pressure falls below the reseating pressure), and reopening as the pressure builds up again.

2.2.7 Body bowl choking

One additional cause for PRV instability is given by the Center for Chemical Process Safety (CCPS) [13]. The velocity at the outlet flange of a conventional PRV reaches sonic velocity in cases of larger PRV sizes and higher set pressures. This phenomenon is known as body bowl choking. When this occurs, the pressure in the valve body rises regardless of the back pressure at the PRV outlet. This pressure rise in the PRV body may in turn cause reduced lift and/or unstable motion of the valve.

3 Determining Inspection Interval

API RP 576 states that the normal interval between shop inspection/overhaul is determined by operating experience and the environment the relief device is subjected to; therefore, it is expected that a valve with one or more of the following would require a shorter interval than a valve in clean, non-fouling or non-corrosive service [14]:

- corrosive or fouling service;
- subject to vibration;
- low differential between set and operating pressures;
- frequent operational upsets; or
- leakage problems.

Establishing shop inspection/test history for a valve is therefore vital to the process of establishing its test interval. Where several sequential tests of the "as received" valve reflect test results consistent with the cold differential test pressure (CDTP), and no change in service is expected for the valve, an increase in the test interval may be considered, if allowed by local regulations. On the other hand, if the tests show a history of erratic results, evidence of PRV instability, or significant deviation from CDTP, the test interval should decrease and/or the valve installation should be modified to improve performance.

4 Risk-Based Inspection of PRV's

Note that the maximum interval between relief device inspections or tests is ten years, per API RP 576; this is the *maximum* time span allowed between shop inspections and overhaul, unless qualified by a risk-based inspection (RBI) assessment [15]. This approach is accomplished through RBI method of utilizing "Risk" to manage and prioritize an inspection program based on API RP 581 [16].

The following steps are based on a recent RBI case study, in which it is assumed that all PRV's were inspected in accordance with API RP 576, and includes the assumption that all PRV's have been sized, selected and installed per API STD 520 PT II.

The details of Consequence of Failure (COF) and Probability of Failure (POF) regarding PRV's will be demonstrated below.

- COF can be analyzed by looking at the types of overpressure scenarios, the discharge location of the relief valve, and if there were multiple relief valves in parallel.

- POF can be analyzed by looking at the number of overpressure scenarios, fluid service severity, relief valve type, if a rupture disk existed upstream of the relief valve, and inspection history.

The resulting risk, recommended inspection interval, and next inspection date can then be determined.

4.1 PRV RBI decision-making

The following decision-making occurred in two steps: (1) Reviewing applicable sections from API RP 581 [17]; and (2) Reviewing equipment and piping applicable data to determine what items that addressed COF and POF could actually be analyzed.

4.2 Risk Matrix

The case study is performed using API RP 581 5x5 risk matrix [18].

4.3 Consequence of Failure (COF)

Consequence is set up as A – E (X-axis of the Risk Matrix), equivalent to Very Low, Low, Moderate, High, and Very High (respectively). The COF is evaluated by looking at an individual PRV and noting what equipment (and/or piping) or pieces of equipment are protected by that particular valve. Look at the RBI analysis for the protected equipment and start with the COF for that equipment. If there is more than one piece of equipment protected by the valve, use the highest COF to be conservative. If a pump is the piece of equipment being protected then defer to the piping associated with the pump and use the COF for the piping. If the equipment or piping has not been part of the equipment RBI study then start with a COF of “C” (Moderate).

After determining the starting COF for each PRV, look at the types of overpressure scenarios, the discharge location of the relief valve, and if there are multiple relief valves in parallel with the valve that is being reviewed.

4.4 Overpressure Scenarios

Note: In this example consider the overpressure scenarios as have already been identified. Utilize API RP 581 Table 7.2 (Default Initiating Event Frequencies) and consider the overpressure scenarios that have an Event Frequency of 1 per 10 years or less, per the case study. Therefore the following are considered.

The following scenarios have an Event Frequency of 1 per year.

- Runaway Chemical Reaction

The following scenarios have an Event Frequency of 1 per 5 years.

- Tower Pump Around Failure or Reflux Pump Failure

The following scenarios have an Event frequency of 1 per 10 years.

- Blocked Discharge, without Administrative Controls in Place

- Loss of Cooling Water Utility
- Thermal / Hydraulic Expansion Relief, without Administrative Controls in Place
- Control Valve (CV) Failure, Initiating event is same direction as CV normal fail position (i.e. Fail safe)
- Liquid Overfilling without Administrative Controls

4.5 Discharge location

Adjustments can be made to the PRV consequence based on the discharge location for each PRV, For example:

- Atmosphere – Increase the (COF value)
- Flare system – Decrease the (COF value)
- Closed Process - Increase the (COF value)
- Closed Drain (underground piping, or liquid portion of the flare system) - Decrease the (COF value)

4.6 Multiple Valves in Parallel

In case of multiple valves in parallel the consequence level should be adjusted based on the number of valves in parallel. For example if there were 2 valves in parallel then adjust down one level. If there were 3 or more valves in parallel then adjust down two levels.

4.7 Probability of Failure (POF)

Probability (Likelihood) is set up as 1 – 5 (Y-axis of the Risk Matrix) equating to Highly Unlikely, Unlikely, Possible, Somewhat likely, and Very Likely (respectively). All PRVs started with a likelihood category of "3" otherwise known as "Possible." Look at the number of overpressure scenarios, fluid service severity, relief valve type, and if a rupture disk existed upstream of the relief valve, and the inspection history.

4.8 Number of Overpressure Scenarios

Look at the Estimated Demand Rate (from applicable scenarios in Table 7.2 – Default Initiating Event Frequencies) total number of scenarios (Total for all equipment protected). The POF was adjusted based on the number of overpressure scenarios.

4.9 Fluid Service Severity

There are two different types of PRV Service Severity as described in API RP 581 [19]. One is based on "Fail to Open" cases and the other is based on "Leak" only cases. Determine the severity level for each PRV based on the descriptions in API RP 581 Categories of PRD Service Severity (Fail Open Case) Table 7.4 and (Leak Case) Table 7.11. The probability will be adjusted relative to fluid service based on the determined severity level for the fluid service. If the device protected multiple pieces of equipment, note the service for each piece of equipment and note the severity for each one. Use the worst severity to make the adjustments, For example:

Fail to Open

- MILD – reduce by one level
- MODERATE – keep same number
- SEVERE – increase by one number

Leak

- MILD – reduce by one number
- MODERATE – keep same number
- SEVERE – increase by one number

4.10 Valve Type

Because the Probability of Failure on Demand vs. Time in Service is based on Weibull parameters which were determined using industry failure rate data, using the valve type as a variable for probability of failure was deemed unreasonable in this example (See API RP 581 Fail to Open Table 7.5 / Figure 7.2 and Leakage Table 7.12 / Figure 7.3).

4.11 Rupture Disk Upstream

If there was a rupture disk upstream of the PRV then the probability should be slightly lowered.

4.12 Historical Inspection Data

Review the case study inspection data and determine which types of relief device inspections have been performed. In this example we can determine the actual inspection interval for the last two inspections by comparing the date for the last shop inspection/overhaul and the second to last shop inspection/overhaul. If there was no second to last shop inspection/overhaul, then the other option is to use the date three years prior to the last shop inspection/overhaul.

Evaluate if a visual On-stream inspection had been performed.

Evaluate if the last shop pre-test "Passed" or "Failed". Further consideration should be made to see if the "Failed" shop test was when the PRV popped higher than 30% of the set pressure. A "Failed" pre-test should also determine if the pre-test never had a pop and/or the valve leaked from the beginning of the pre-test. These were the only ones that were considered failures.

If the PRV popped lower than the set pressure then the PRV shop test was designated as "Passed," per the case study.

If the "Failed" pre-test did pop or leak above or below the set pressure, record "Above" if the valve failed to pop or if the PRV popped higher than 30% of the set pressure. Record "Below" if the valve leaked below the set pressure. If the PRV popped lower than the set pressure then the PRV shop test was considered and recorded as "Passed."

Grade the last inspection data based on API RP 581 Table 7.7 – Inspection and Testing Effectiveness. A = Highly Effective to D = Ineffective. The probability was adjusted based on the following: If last inspection was graded Highly Effective and Passed adjusted probability down one letter. If last inspection was graded:

- Highly Effective and Failed keep probability the same value.
- Usually Effective and Passed keep probability the same value.
- Usually Effective and Failed adjust probability up one number.
- Fairly Effective and Failed adjust probability up one number.
- Ineffective or no data then adjust probability up two numbers.

Record the next inspection due date based on the case study data.

Determine if the next inspection was overdue relative to the date of the current analysis. If the next inspection is overdue, then the probability of failure should be increased by one number.

Resulting Risk

By now the PRV has a defined COF (shown as Impact on the 5 x 5 Risk Matrix) A – E, Major, Significant, Minor, and Low (respectively). It also has a defined POF (shown as Likelihood on the 5 x 5 Risk Matrix) 1 – 5, Highly Unlikely, Unlikely, Possible, Somewhat likely, and Very Likely (respectively). These are based on the operating company's risk-ranking acceptance criteria.

With these defined Consequence and Probability Numbers and Letters, there are resulting risks of High Risk, Medium High Risk, Medium Risk and Low Risk.

Note: The distribution of the number of PRV's are placed in the Risk Matrix proper designations.

As a result, the summary of the consequent risk distribution shall be as follows:

- High Risk = (Total numbers of PRV's assigned in matrix)
- Medium High Risk = (Total numbers of PRV's assigned in matrix)
- Medium Risk = (Total numbers of PRV's assigned in matrix)
- Low Risk =(Total numbers of PRV's assigned in matrix)

Based on the resulting risk information, the next inspection intervals should be focusing on high risk PRV's while the inspection frequency regarding the low risk PRV's can be reduced or even greatly reduced compared to the usual inspection program. This results in considerable inspection and general cost reduction. Therefore, implementing RBI provides a dependable methodology of determining the best combination of "inspection frequencies and inspection scopes".

5 Role of Pressure Relief Analysis

Pressure relief analysis (PRA) studies and periodic revalidations provide a comprehensive database of potential sources of overpressure and can be used to develop a concern list summarizing any deviations from industry and operating company standards. During a pressure relief analysis on a unit, it is possible to identify and quantify:

- Non-recoverable pressure losses on the inlet piping to the relief devices;

- Built-up backpressures on the outlet piping;
- Undersized overpressure scenarios;
- Oversized overpressure scenarios;
- Acoustic interaction on the piping;
- Improper valve selection;
- Improperly installed relief valves;
- Liquid static head on piping that could prevent the relief devices to open prior to the protected equipment exceeding its MAWP plus allowable accumulation;

Obtaining all the above mentioned data, which are outputs of PRA, is beneficial to identify relief devices that could present instability (even if they never presented before) and to judge in a more assertive approach which actions can be taken to prevent instability for that devices.

6 Conclusion

As previously mentioned, the 10-year interval should be considered as a maximum time between inspections. A risk-based approach, as illustrated on this paper, is able to improve on the inspection management of PRVs by providing a dependable methodology of determining the best combination of "inspection frequencies and inspection scopes".

7 Acronyms and abbreviations used

API – American Petroleum Institute
ASME – American Society of Mechanical Engineers
BPVC – Boiler & Pressure Vessel Code
CCPS – Center for Chemical Process Safety
CDTP – Cold Differential Test Pressure
COF – Consequence of Failure
CV – Control Valve
MAWP – Maximum Allowable Working Pressure
OSHA – Occupational Safety and Health Administration
POF – Probability of Failure
PRA – Pressure Relief Analysis
PRD – Pressure Relief Device
PRV – Pressure Relief Valve
PSM – Pressure Safety Management
RBI – Risk Based Inspection
RP – Recommended Practice
STD – Standard

8 References

[1] American Petroleum Institute, *API Standard 521, Pressure-relieving and Depressuring Systems,* 6th Edition, 2014, Table 1.

[2] U.S. Department of Labor Occupational Safety and Health Administration, *29 CFR 1910.119 – Process Safety Management of Highly Hazardous Chemicals*, 1992.

[3] American Petroleum Institute, *API Standard 510, Pressure Vessel Inspection Code: In-Service Inspection, Rating, Repair, and Alteration,* 10th Edition, 2014.

[4] American Petroleum Institute, *API Standard* 520, *Sizing, Selection, and Installation of Pressure-Relieving Devices, Part II-Installation*, 6th Edition, 2015.

[5] American Petroleum Institute, *API Recommended Practice* 576, *Inspection of Pressure-relieving Devices*, 3rd Edition, 2009.

[6] American Society of Mechanical Engineers, *ASME Boiler and Pressure Vessel Code, Section VIII – Division 1,* 2015 Edition.

[7] American Society of Mechanical Engineers, *ASME Boiler and Pressure Vessel Code, Section I,* 2015 Edition.

[8] API STD 520 PT II §7.3.4

[9] API STD 520 PT II §7.1 – 7.2

[10] Borg, A. and Jakobsson, S., On the Stability of Pressure Relief Valves: A numerical study using CFD, Master's Thesis, Chalmers University of Technology, 2014, http://publications.lib.chalmers.se/records/fulltext/211055/211055.pdf, Accessed 27 January 2016.

[11] API STD 520 PT II §7.3.6

[12] API STD 520 PT II §7.2

[13] Center for Chemical Process Safety (CCPS), Effect of Body Bowl Choking on Pressure Relief Valve Stability, AIChE Academy Webinar, available at http://www.aiche.org/academy/videos/conference-presentations/effect-body-bowl-choking-on-pressure-relief-valve-stability, Accessed 25 January 2016

[14] API RP 576 §6.4.2

[15] API RP 576 §6.4.1.3

[16] American Petroleum Institute, *API Recommended Practice 581,Risk-Based Inspection Technology* , 2nd Edition, 2008

[17] API RP 581 Part I §7

[18] API RP 581 Part I Figure 4.2

[19] API RP 581 Part I §7.6

Disclaimer:

The information contained in this document represents the current view of the authors at the time of publication. Process safety management is complex and this document cannot embody all possible scenarios or solutions related to compliance. This document is for informational purposes only. Siemens makes no warranties, express or implied, in this paper.

Verification of in-Service LP Tanks Pressure Ratings

Archie Buchman and Jose Ortega
Smith and Burgess, LLC
7600 W. Tidwell Rd. Ste 600 Houston, Texas 77040
Archie.Buchman@smithburgess.com **Jose.Ortega@smithburgess.com**

Bill Frenk
Frenk Water Technologies, LLC
2798 Appleridge Ln Byron, IL 61010
bfrenk@fwtwater.net

Prepared for Presentation at
American Institute of Chemical Engineers
2016 Spring Meeting
12th Global Congress on Process Safety
Houston, Texas
April 11-23, 2016

Verification of in-Service LP Tanks Pressure Ratings

Archie Buchman and Jose Ortega
Smith and Burgess, LLC
7600 W. Tidwell Rd. Ste 600 Houston, Texas 77040
Archie.Buchman@smithburgess.com **Jose.Ortega@smithburgess.com**

Bill Frenk
Frenk Water Technologies, LLC
2798 Appleridge Ln Byron, IL 61010
bfrenk@fwtwater.net

Keywords: Low-Pressure, Tanks, Relief Device, Inbreathing, Outbreathing

Abstract

The scope of the paper is to discuss the importance of knowing the pressure and vacuum ratings of low-pressure storage tanks in facilities while ensuring API 2000 venting requirements are met. A back-calculation method is used to estimate tank ratings (pressure and vacuum) with retirement thicknesses for a low-pressure tank found using general industry software and original tank drawings. This method would require the venting requirement to be known based on equations in API 2000. With the requirements of low-pressure tank vents being fulfilled at an estimated rating, facility's maintenance and inspection personnel can track the corrosion of the tank walls until estimated retirement thicknesses are reached. If this study is implemented, findings would result in improved lifetime of tanks, verification of pressure/vacuum ratings, and venting compliance with API 2000 for the lifetime of the low-pressure tanks. The findings in this paper will be based on API 650 Standard, as this code will govern the sizing for most tanks found in the oil and gas industry.

1 Introduction

Smith and Burgess has come across many in-service Low-Pressure (LP) tanks that have a pressure rating of atmospheric. A majority of these facilities are under Process Safety Management (PSM) with their Process and Instrumentation Diagrams (P&ID) and vessel nameplate information being vague on tank pressure and vacuum ratings. Most of the tanks state their pressure rating is atmospheric, which directly deems any tank constructed following the American Petroleum Institute (API) guidelines unfit for service if the Pressure/Vacuum Relief Vent (PVRV) has any set pressure other than atmospheric. A vessel's integrity is limited to the amount of information and knowledge possessed by the engineer. Site engineers must determine if their tank farm is currently fit for service. This decision cannot be made without knowing the pressure and vacuum rating of each tank in its current or future state. Engineering analysis can be performed on collected data to determine if the vessel is fit for service at the time of evaluation. This time of evaluation could range from the current time and thickness to an extended retirement date and thickness. By performing a detailed study on the expected thicknesses, weather conditions, and live loads, an estimated pressure and vacuum rating can be determined and applied to the LP tank for engineering studies and overall review of the asset's integrity. This would include a detailed analysis of the tank's relief devices against overpressure and over vacuum along with expected corrosion based on frequently testing the wall thicknesses around the tank. A summary on the deficiencies found in the ventilation requirements will be presented in this case study. This case study will discuss the inputs, calculations, and findings that assist an engineer with the necessary information to determine if their tank farms are "adequate" at a future date and thickness, while ensuring that the asset will remain serviceable during the time of evaluation.

2 Methodology

Different methodologies can be applied to a facility, which is based on the information requested/required to make common engineering decisions about tank farms. These can range from determination of the current pressure rating at the previous thickness ratings to a future pressure rating at the facilities' normal retirement thickness (limited to company guidelines or recognized and generally accepted good engineering practices (RAGAGEP). The following paragraphs will describe a few approaches to fit the need of each facility.

Current Rating

Current tank rating can be found by using thickness readings at the current time and conditions. This results in the highest pressure and vacuum ratings for the low pressure tank. All process safety engineers would promote applying this study for the relief device documentation, however it is not possible when contrasting the mechanical integrity of low-pressure tanks to the benefits. The mechanical team at the site argue this idea, as the new pressure rating would only last until the next series of thickness readings. At the time of this testing, the "calculated retirement" thicknesses would be breached and therefore deem the study performed as old and outdated.

Based on this approach, the current rating methodology may seem unreasonable for an asset integrity review even though relief device studies would significantly benefit from the high working pressure.

HAZOP/PHA Perspective

The risk associated of not knowing the true rating of tanks along with the methodology may hinder any risk study performed by site facilitator. The hazard analysis is to evaluate the amount of risk under normal operation of the tank. The re-rating study to be applied would cover the asset's current rating at the time of evaluation. By applying the current thickness reading and additional inputs, a current pressure and vacuum rating could be calculated. With this new rating, the relief system documentation could be constructed or updated to determine if any deficiencies are apparent. If only the design pressure and vacuum values are available at the time of study, the risk studies would suffer from "inaccurate" maximum working pressure and vacuum ratings during evaluation in the relief system documentation. A relief system study based on the design pressure and vacuum would produce overly conservative ventilation capacity and not capture the actual risk of operating the low pressure tank.

Longevity of asset

The mechanical engineering mindset would be geared to maximize the amount of corrosion allowable while ensuring that proper ventilation is met. This would be done through a back-calculation method of performing the relief device documentation to calculate the required relief rates for each tank to be evaluated. Once the required relief rates are calculated, the tank re-rate can be tailored to the lowest corrosion allowance while ensuring the ventilation requirements are met. However, if this method is selected and applied to low-pressure tanks in service, the relief device adequacy may suffer as the pressure and vacuum rating will be minimal and would produce more inadequacies in the pressure/vacuum relief vents.

Method for Case Study

This case study will be based on the future retirement thickness being fixed at an in-between value for the bottom shell course to give a new pressure rating until the tank's fixed thickness rating is breached. This is the best methodology as it will produce longer tank life as well as decent pressure and vacuum ratings when determining the adequacy of low-pressure vents.

3 Inputs Required by Code

A general industry software was selected that utilizes design formulas from API 650, API 620, ASME Sec VIII Div 1 (bottom plate vacuum), as well as Brownwell & Young – Process Equipment Design (roof structural support). The required input to arrive at the tank current maximum allowable working pressure (MAWP) and maximum allowable working vacuum (MAWV) was determined by entering the original tank's details. These details include information from the tank's thickness readings, materials of construction, and other design criteria per API guidelines. As most tanks vary in amount of corrosion, an updated thickness reading is required to adequately determine the current pressure rating of the in-service low-pressure tank. A walkdown of each tank can be performed to gather the additional input to ensure the tank is currently maintained, while also recording any changes from the previous inspection. This walkdown includes inspection of the tank's nozzles, dimension, anchor bolts, and rafters. Most of the tanks included in the study were fixed roof. Information on the welding of the roof to the shell was determined either by the field walkdown or the original tank drawing. This is the information of a basic tank configuration: diameter, shell height, type of roof, specific gravity of contents, maximum temperature, material, roof support structure (when applicable), and maximum liquid level. Corrosion allowance was not entered as the tank was being re-rated in the corroded condition.

4 Assumptions

Over the course of the study some assumptions were required to complete the pressure ratings. These include adjusting the thicknesses to ensure the vacuum rating was at least 1"WC (API 650 5.2.1.b) to meet the code requirement. These assumptions would follow code suggestions: 20 lb/ft2 roof live load (API 650 5.2.1 f), 140 mph wind speed, and maximum liquid level taken as 2'-0" less than the shell height. If there was no information on X-ray radiography on the tank, an assumed joint efficiency of 0.70 was used. Note that seismic loading was not considered given the geographical location where the study occurred. All connections were assumed to be reinforced per API standards.

5 Calculations

Equations used in the re-rate of the tanks were based on API 650, ASME Sec VIII Div 1, UG 34, and Brownwell and Young – Section 4.3 Roof Design. A sample re-rate calculation for a low pressure tank can be sent upon request.

Shell

Throughout each of the shell course calculations, the applicable Code paragraph is referenced. Shell calculations were based the "One foot method" as all of the tanks were less than 200 ft in diameter. The "One foot method" calculates the thickness required at points 1 ft above the bottom of each shell course. The "pass-fail" criteria used for each tank was to determine a retirement thickness such that the resulting MAWV was at least 1" WC and the MAWP was sufficient for the "Maximum Liquid Operating Level". Occasionally, it was necessary to reduce the "MLOL" for the tank to "pass" at the desired minimum retirement thickness.

Roof

Roof calculations can be performed on a structurally supported conical roof with a single center column or without. The calculations follow both API 650 and Brownwell and Young for rafter design.

Bottom

For a non-annular flat bottom without roof support columns and a vacuum condition at near empty, calculations from ASME Sec VIII Div 1, UG-34 (Unstayed Flat heads and covers) were used to determine the appropriate thickness to ensure the bottom plate would not flex under vacuum.

6 Findings

Smith and Burgess evaluated 131 low-pressure tank based on the methods described above during the relief system documentation. The findings included deficiencies in the vacuum and pressure scenarios identified in API 2000. The findings and overall percentage are presented in Table 1 below.

Table 1: Deficiencies found during Relief System Evaluation

Scenario	# of Tanks with Issues	Overall Percentage
Inbreathing	36	27%
Outbreathing	30	23%
Inadvertent Valve Opening (Pressure)	34	26%
Fire Exposure	106	81%
Inert Pad Failure	31	24%

Based on the findings during the tank evaluation at this facility, the low-pressure tank farm needed to be evaluated and updated to ensure proper ventilation was installed and maintained. The study performed for this site resulted in a higher deficiency percentage as the focus was on improving the longevity of the low-pressure tank. Because of this focus of improving tank longevity, the results came back with lower than expected pressure and vacuum ratings.

7 Conclusion

The results ranged from relief device inadequacy during normal operation (inbreathing and outbreathing) to relief device inadequacy during emergency venting (external fire), possibility reissuing currently out of service low-pressure tanks for prolonged lifetime, and design flaws during construction for the study on the 131 tanks. The majority of the tanks had inbreathing concerns. It should be noted that for a number of tanks analyzed, the number of rafters were often less than the amount required. Some original tank drawings were outdated, as they stated the tank was to be built to the latest API specifications. However the tanks in the study were built in the early 1960's. Because of this phrasing, it eluded to the tanks being built to today's API specification standards. In actuality, they were built to the specifications of the year they were built. Thicker roof retirement thicknesses were needed to achieve the desired MAWV due to the wider spacing for the rafters. In conclusion, the study produced results to help any engineer determine if a low-pressure tank is fit for service regardless if it is excluded from PSM guideline and requirements. It also ensures the pressure relief vents will be adequately sized for the normal operating conditions of each tank within the tank farm. This study has produce stunning results that low pressure tank farm if not reviewed by engineers can hold many deficiencies. This is due to the exclude and minimalist requirement from PSM standards.

8 References

[1] API 2000 7th Edition. Venting Atmospheric and Low-pressure Storage Tanks. American Petroleum Institute, March 2014

[2] API 650 12th Edition. Welded Tanks for Oil Storage. American Petroleum Institute, March 2013

[3] API 620 12th Edition. Design and Construction of Large, Welded, Low-pressure Storage Tanks. American Petroleum Institute, October 2013

[4] ASME Section VIII Division 1. Boiler and Pressure Vessel Code. American Society of Mechanical Engineers, March 2013

[5] L.E. Brownell and E.H. Young, Process Equipment Design, January 15, 1959, Copyright 1959

[6] ANSI/AISC 360-10, Specification for Structural Steel Buildings. American National Standard Institute, June 22, 2010

CSB INVESTIGATION OF THE WEST (TEXAS) FERTILIZER FIRE AND EXPLOSION

Presenter:
Johnnie A. Banks, CFEI, Supervisory Investigator,
U.S. Chemical Safety and Hazard Investigation Board, Washington, DC
johnnie.banks@csb.gov

Prepared for Presentation at
American Institute of Chemical Engineers
2016 Spring Meeting
12th Global Congress on Process Safety
Houston, Texas
April 10-14, 2016

CSB INVESTIGATION OF THE WEST (TEXAS) FERTILIZER FIRE AND EXPLOSION

Presenter:
Johnnie A. Banks, CFEI, Supervisory Investigator,
U.S. Chemical Safety and Hazard Investigation Board, Washington, DC
johnnie.banks@csb.gov

Keywords: *Fertilizer Grade Ammonium Nitrate (FGAN), Hazard Awareness, Emergency Planning and Response, FGAN Storage Practices, Land Use Planning*

Abstract

On April 17, 2013, a fire and explosion occurred at the West Fertilizer Company (WFC), a fertilizer blending, retail, and distribution facility in West, Texas. The violent detonation fatally injured 12 emergency responders and three members of the public. Local hospitals treated more than 260 injured victims, many of whom required hospital admission. The blast completely destroyed the WFC facility and caused widespread damage to more than 150 offsite buildings. The WFC explosion is one of the most destructive incidents ever investigated by the U.S. Chemical Safety and Hazard Investigation Board (CSB) as measured by the loss of life among emergency responders and civilians; the many injuries sustained by people both inside and outside the facility fenceline; and the extensive damage to residences, schools, and other structures.

The CSB conducted an exhaustive investigation that lasted more than two years and which yielded 19 recommendations directed to ten recipients. The tragic loss of life to the civilians and emergency responders, in addition to issues related to community-wide property damage were central in the development of these recommendations, and cover a range of elements which include:

- Regulatory oversight
- Insurance
- Emergency response
- Emergency planning
- Land use planning
- Technical conditions

This presentation will provide an overview of the incident, consider the key findings from the investigation, and illustrate the relationship of these findings to the recommendations. Click here for the final West Fertilizer Investigation Report and related videos and information.

Process Safety Incidents: Big Picture Revelations

Paul Baybutt
Primatech Inc.
Columbus, Ohio, USA
paulb@primatech.com

Prepared for Presentation at
American Institute of Chemical Engineers
2016 Spring Meeting
12th Global Congress on Process Safety
Houston, Texas
April 11-23, 2016

Process Safety Incidents: Big Picture Revelations

Paul Baybutt
Primatech Inc.
Columbus, Ohio, USA
paulb@primatech.com

Keywords: process safety instruments; process safety management; risk management programs

Abstract

Many incidents have helped to define and develop process safety. Each has provided valuable learning opportunities. However, it is even more important to identify insights that can be obtained from an analysis of a large set of incidents that represents those that typically occur. This larger picture illuminates trends and commonalities and provides learning opportunities that are even more important than the causes of any one individual incident.

The Chemical Safety Board has published the results of numerous investigations of process safety incidents and the Center for Chemical Process Safety has published information on incidents that have helped to define process safety. These data have been analyzed to identify commonalities and trends so that measures to help protect against future incidents can be developed. Recommendations are made to address key issues identified.

1.0 Introduction

Process safety regulations have been promulgated in various parts of the world to address protecting process plants against catastrophic incidents in response to the occurrence of a number of major incidents such as Flixborough, Seveso, and Bhopal [1]. However, despite the widespread adoption of process safety practices, major incidents continue to occur. The continued occurrence of such incidents is unacceptable and the issue of what can be done to protect against them must be addressed. This paper addresses the issue of why such incidents continue to occur.

Process safety regulations contain requirements for periodic audits for programs to identify any omissions or deficiencies and some process safety programs require periodic management reviews. However, while omissions and deficiencies must be identified and corrected, knowledge of what omissions and deficiencies have actually resulted in process safety incidents is essential to help prevent further incidents.

Compilations of incident descriptions have been published by various authors, for example, [1 - 7]. Usually, incidents are organized in various ways, such as by root cause, release type, equipment involved, operation being conducted, and process safety element involved. Descriptions of individual incidents also can be found, for example, in reports from the U.S. Chemical Safety Board (CSB) [8], and in journals such as the Loss Prevention Bulletin [9]. Such publications usually focus on providing a description of the incident, identifying its causes, and providing recommendations for corrective actions. Certainly, lessons can be learned from each incident. However, it is also important to take an overall view to identify commonalities and trends across incidents so that measures can be identified that will help to protect against future incidents. In particular, the identification of issues that have contributed to multiple incidents is especially important so that future incidents may be prevented. For example, the CSB has flagged reactive hazards [10], combustible dusts [11], and hot work [12] as significant issues across multiple incidents. This paper identifies additional issues that were important contributors in numerous incidents.

CSB investigation reports have been analyzed concerning management system failures [13], process hazard analysis (PHA) [14], and the hierarchy of controls [15]. The evolution of process safety standards and legislation following nine key process safety incidents has been addressed in a recent article [16].

Section 2 identifies the incidents reviewed and the analysis approach followed. Section 3 identifies and discusses the common contributors that were found for the incidents studied and provides recommendations to address them. Section 4 provides a discussion of the results. Section 5 provides some additional noteworthy observations that were made during the review of incident descriptions. Section 6 identifies three issues that underlie all of the other issues. Each underlying issue is discussed and recommendations are made for addressing them. Conclusions are presented in section 7.

2.0 Incidents Addressed and Analysis Approach

Incidents investigated by the Chemical Safety Board (CSB) [8] and incidents described by the Center For Chemical Process Safety (CCPS) [1] were reviewed. CCPS provides descriptions of over 100 incidents. Some date back to the early 20th century but most occurred within the past 40 years. They cover a variety of industries including chemicals, oil and gas production and refining, aviation, shipping, and nuclear power generation. The CSB has issued over 60 incident investigation reports, case studies, and safety bulletins dating back to 1998 when the CSB began investigating incidents. The CSB incidents cover various sectors of the process industries. The CSB currently is investigating one incident from 2010, three from 2013, three from 2014 and one from 2015. Sixty-eight CSB incidents and 12 CCPS incidents making a total of 80 incidents were studied (see Tables 1 and 2, respectively). About 40% of the CSB incidents occurred in processes covered by the Occupational Safety and Health Administration's (OSHA's) process safety management (PSM) regulations. The descriptions of incidents and their causes were taken at face value.

Table 1. CSB Incidents Reviewed.

No.	Incident	Incident Date	Report Date	PSM[1]	PHA[2]
1	Sonat Exploration Co. Catastrophic Vessel Overpressurization	03/04/1998	09/21/2000	N	N
2	Morton International Inc. Runaway Chemical Reaction, Explosion and Fire	04/08/1998	08/16/2000	N	Y
3	Concept Sciences Hydroxylamine Explosion	02/19/1999	02/01/2002	Y	Y
4	Tosco Avon Refinery Petroleum Naphtha Fire	02/23/1999	03/21/2001	Y	N
5	Bethlehem Steel Corporation Gas Condensate Fire	02/02/2001	12/06/2001	N	N
6	BP Amoco Thermal Decomposition Incident	03/13/2001	05/20/2002	N	Y
7	Motiva Enterprises Sulfuric Acid Tank Explosion	07/17/2001	08/28/2002	N	Y
8	Georgia-Pacific Corp. Hydrogen Sulfide Leak	01/16/2002	11/20/2002	N	N
9	Third Coast Industries Petroleum Products Facility Fire	05/01/2002	03/06/2003	N	U
10	DPC Enterprises Festus Chlorine Release	08/14/2002	05/01/2003	Y	U
11	First Chemical Corp. Reactive Chemical Explosion and Fire	10/13/2002	10/15/2003	N	N
12	Environmental Enterprises Hydrogen Sulfide Release	12/11/2002	09/17/2003	N	U
13	Catalyst Systems Inc. Reactive Chemical Explosion and Fire	01/02/2003	10/29/2003	N	N
14	BLSR Operating Ltd. Vapor Cloud Deflagration and Fire	01/13/2003	09/17/2003	N	U
15	West Pharmaceutical Services Dust Explosion and Fire	01/29/2003	09/23/2004	N	N
16	Technic Inc. Ventilation System Explosion and Fire	02/07/2003	08/20/2004	N	N
17	CTA Acoustics Dust Explosion and Fire	02/20/2003	02/15/2005	N	N
18	D.D. Williamson & Co. Catastrophic Vessel Failure	04/11/2003	03/12/2004	N	N
19	Honeywell Chemical Releases (3 incidents)	a. Chlorine 07/20/2003 b. Antimony pentachloride 07/29/2003 c. Hydrogen fluoride 08/13/2003	08/08/2005	Y N Y	Y N N
20	Isotec / Sigma Aldrich Nitric Oxide Explosion	09/21/2003	08/24/2004	Y	Y
21	Hayes Lemmerz Dust Explosions and Fire	10/29/2003	09/27/2005	N	N
22	DPC Enterprises Glendale Chlorine Release	11/17/2003	02/28/2007	Y	Y
23	Giant Industries Oil Refinery Explosion and Fire	04/8/2004	10/26/2005	Y	Y
24	MFG Chemical Inc. Toxic Gas Release	04/12/2004	04/11/2006	Y	N
25	Formosa Plastics Vinyl Chloride Explosion	04/23/2004	03/06/2007	Y	Y
26	Sterigenics Ethylene Oxide Explosion	08/19/2004	03/30/2006	Y	Y
27	Marcus Oil and Chemical Tank Explosion and Fire	12/03/2004	06/06/2006	N	N

No.	Incident	Incident Date	Report Date	PSM[1]	PHA[2]
28	Acetylene Service Company Gas Explosion	01/25/2005	01/26/2006	Y	Y
29	BP Texas City Refinery Explosion and Fire	03/23/2005	03/20/2007	Y	Y
30	Formosa Plastics Propylene Fire and Explosions	10/06/2005	07/20/2006	Y	Y
31	Bethune Point Wastewater Plant Explosion and Fire	01/11/2006	03/13/2007	N	N
32	Synthron Chemical Reaction and Vapor Cloud Explosion	01/31/2006	07/31/2007	Y	N
33	Partridge Raleigh Oilfield Explosion and Fire	06/05/2006	06/12/2007	N	N
34	Universal Form Clamp Co. Explosion and Fire	06/14/2006	04/10/2007	Y	N
35	EQ Hazardous Waste Plant Explosions and Fire	10/05/2006	04/16/2008	N	U
36	CAI / Arnel Chemical Plant Explosion	11/22/2006	05/13/2008	Y	N
37	Valero Refinery Propane Fire	02/16/2007	07/09/2008	Y	Y
38	Barton Solvents Explosions and Fire	07/17/2007	06/26/2008	N	U
39	Xcel Energy Company Hydroelectric Tunnel Fire	10/02/2007	08/25/2010	N	Y
40	Barton Solvents Flammable Liquid Explosion and Fire	10/29/2007	09/18/2008	N	U
41	T2 Laboratories Inc. Reactive Chemical Explosion	12/19/2007	09/15/2009	N	N
42	Imperial Sugar Company Dust Explosion and Fire	02/07/2008	09/24/2009	N	U
43	Goodyear Heat Exchanger Rupture and Ammonia Release	06/11/2008	01/27/2011	Y	U
44	Packaging Corporation Storage Tank Explosion	07/29/2008	03/04/2010	N	N
45	Bayer CropScience Pesticide Waste Tank Explosion	08/28/2008	01/20/2011	Y	Y
46	INDSPEC Chemical Corporation Oleum Release	10/11/2008	09/30/2009	Y	Y
47	Allied Terminals Fertilizer Tank Collapse	11/11/2008	05/26/2009	N	N
48	Veolia Environmental Services Flammable Vapor Explosion and Fire	05/04/2009	07/21/2010	Y	N
49	ConAgra Natural Gas Explosion	06/09/2009	02/04/2010	N	U
50	CITGO Refinery Fire and Hydrofluoric Acid Release	07/19/2009	12/09/2009	Y	Y
51	Caribbean Petroleum Refining Tank Explosion and Fire	10/23/2009	10/21/2015	N	N
52	Silver Eagle Refinery Explosion	11/04/2009	04/14/2014	Y	U
53	NDK Crystal Inc. Explosion	12/07/2009	11/14/2013	N	U
54	DuPont Corporation Toxic Chemical Releases (3 incidents)	a. Methyl chloride 01/22/2010 b. Oleum 01/23/2010 c. Phosgene 01/23/2010	09/20/2011	U U Y	U U Y
55	Kleen Energy Natural Gas Explosion	02/07/2010	06/28/2010	N	U
56	Tesoro Refinery Explosion and Fire	04/02/2010	05/01/2014	Y	Y
57	Horsehead Holding Company Explosion and Fire	07/22/2010	03/11/2015	N	U

No.	Incident	Incident Date	Report Date	PSM[1]	PHA[2]
58	Millard Refrigerated Services Ammonia Release	08/23/2010	01/15/2015	U	U
59	E. I. DuPont De Nemours Co. Hotwork Explosion	11/09/2010	04/19/2012	Y	Y
60	AL Solutions Dust Explosion	12/09/2010	07/16/2014	N	U
61	Carbide Industries Fire and Explosion	03/21/2011	02/07/2013	N	U
62	Hoeganaes Corporation Flash Fires and Explosion	05/27/2011	01/05/2012	U	U
63	Chevron Refinery Fire	08/06/2012	01/28/2015	Y	Y
64	US Ink Fire	10/09/2012	01/15/2015	U	U

[1] Covered by OSHA's PSM standard.
[2] PHA was performed
Y – Yes
N – No
U – Unknown

Table 2. CCPS Incidents Reviewed.

No.	Incident	Incident Date
1	ELF refinery, LPG tank farm, Feyzin, France	01/04/1966
2	DSM Nypro plant, Flixborough, UK	06/01/1974
3	ICMESA, chemical plant, Seveso, Italy	07/10/1976
4	BP hydrocracker unit, Grangemouth, UK	03/21/1987
5	Shell refinery, FCC unit, Norco, LA	05/05/1988
6	Occidental Petroleum, Piper Alpha Platform, North Sea, UK	07/06/1988
7	Phillips Chemical Company, polyethylene reactor, Pasadena, TX	10/23/1989
8	Total refinery, FCC unit, La Mede, France	11/09/1992
9	Texaco Oil Refinery, FCC Unit, Milford Haven	07/24/1994
10	Napp Technologies, Lodi, NJ	04/21/1995
11	Tosco Avon Refinery, hydrocracker unit, Martinez, CA	07/21/1997
12	Sonatrach LNG plant, Skikda, Algeria	01/19/2004

All incident descriptions were first reviewed to identify commonalities. A second reading of the incident descriptions was performed to confirm the initial findings and document characteristics of the incidents. A third reading of the incident descriptions was conducted to corroborate the findings and provide further detail.

3.0 Incidents Addressed and Analysis Approach

This section identifies common contributors to the incidents. Of course, each incident has its own individual characteristics but there is a remarkable degree of similarity in some deficiencies and omissions across the incidents. The focus in this paper is on issues common to many of the incidents.

Incidents where each issue was a contributor are identified, the percentage of CSB incidents where the issue was present is given, aspects of each issue are discussed, and recommendations are made to address the issues.

3.1 Process Design

Many incidents occurred in processes that arguably had deficient designs {1, 3, 6, 15, 16, 17, 18, 21, 25, 29, 31, 34, 37, 40, 41, 42, 57, 58, 64} and {*10, 11*}[3]. This issue was present in 28% of CSB incidents and 18% of incidents covered by OSHA's PSM standard. Deficiencies arose in the initial design and as a result of modifications.

Hazards were not recognized, designs created hazards, safety systems were inadequate, process instrumentation was insufficient, human factors issues were not addressed, hazards of scale-up were not recognized, and throughput was increased without review. In some cases deficient equipment design contributed to incidents {54a, 54c}. Possibly designers were unaware of the problems and/or design reviews were inadequate but these reasons are not acceptable.

Design deficiencies can be addressed by making process design and engineering an element in process safety programs [17]. Examples of topics that should be addressed by requirements for process design and engineering include design practices, design reviews, inherent safety, hazardous area classification, pressure relief, alarm management, damage mechanisms, human factors, facility siting, competency, and training for designers. Also, process design and engineering should be subject to the same management system requirements as other elements in a process safety program such as written policies and procedures, document management, assigning roles and responsibilities, auditing, management review, and continuous improvement. Process design and engineering should be an element of process safety management systems in order to provide a specific focus on the activities involved and to ensure the application of management system practices to it. Also, the competency of process designers must be addressed.

For new designs and process modifications, proof of safety should be required rather than proof of danger. This will require a change in culture.

---

[3] Braces or curly brackets, that is, { }, are used to enclose the number designations of incidents from Table 1, using normal font, and Table 2, using italic font.

3.2 Safeguards

Many incidents involved processes that were not adequately protected by safeguards {2, 7, 8, 9, 11, 12, 13, 17,18, 20, 21, 22, 24, 25, 26, 28, 29, 30, 32, 34, 35, 36, 37, 38, 40, 41, 42, 46, 48, 49, 50, 51, 54a, 54b, 54c, 60, 62, 64} and *{1, 3, 4, 5, 6, 9, 11}*. This issue was present in 56% of CSB incidents and 57% of incidents covered by OSHA's PSM standard. Needed safeguards were not present or those that were present were insufficient, unreliable, disabled, or lacked independence. Commonly, the hierarchy of controls was not applied (see Figure 1). Often reliance was placed on procedural safeguards which are at the lower end of the hierarchy (19a, 22, 25, 29, 32, 46, 54c}. Process risks should be matched with safeguards in the hierarchy of controls. Higher consequence scenarios, such as a chlorine release in a densely populated area, require safeguards higher in the hierarchy and likely multiple, independent safeguards. Procedural safeguards typically are viewed as less reliable, and, therefore, inadequate for protection against high-consequence events.

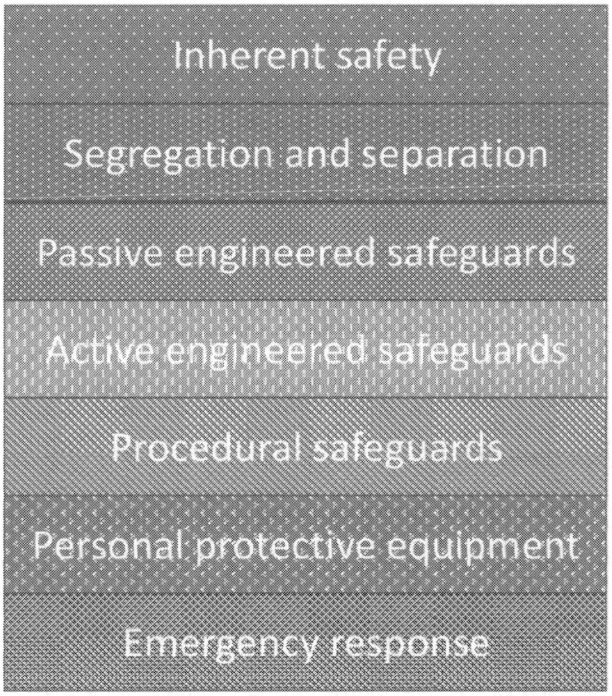

Figure 1. Hierarchy of Controls.

Inherently safer designs or practices could have prevented or mitigated some incidents {29, 39, 55, 56, 54b, 63}. Currently, there are no requirements to address inherently safer systems or the hierarchy of controls in Federal process safety regulations. Process safety regulations in New Jersey require an inherently safer technology review [18] and in Contra Costa County, California require that inherently safer systems be addressed [19]. Proposed amendments to California's process safety regulations for refineries contain a requirement to perform a hierarchy of controls analysis [20].

The operability of safeguards was an issue during several incidents. Explosions and fires disabled safeguards {15, 30, 42} and {*7*}, and toxic releases and fires prevented access to controls {19a, 30, 37, 50} and {*6*}.

Currently, there is no requirement in US process safety regulations to address the design basis for safeguards, evaluate or document the effectiveness of process safeguards, or show that safeguards adequately reduce risks. Proposed amendments to California's process safety regulations for refineries contain a requirement to perform a hierarchy of controls analysis and a safeguards protection analysis (SPA) [20]. The SPA should be performed before a PHA study and updated after the PHA study has been completed. The SPA should include the preparation of a Safety Requirements Specification (SRS) for each individual safeguard [21]. An SRS documents the functional and integrity requirements for a safeguard. Functional requirements describe what the safeguard does to prevent a hazardous state of the process and integrity requirements specify how well the safeguard should perform and what is needed to support that performance level. The SPA should validate that safeguards adequately protect against the hazards of the process, have a specified integrity level, have been installed correctly, are functional, are being maintained appropriately, etc.

Once the SPA has been performed, the PHA team can focus on the qualification of safeguards for specific hazard scenarios [21, 22]. Qualification of safeguards addresses whether safeguards are appropriate for each hazard scenario in a PHA study. Information developed in a PHA study may necessitate revisions to the SPA and SRSs. For example, it may be determined that a redundant power supply is needed to support the integrity required for a scrubber system. SPAs should be updated during management of change (MOC) reviews and when developing recommendations from incident investigations.

3.3 Operations and Maintenance

Problems with procedures were common {1, 2, 3, 5, 6, 11, 12, 13, 14, 18, 19b, 19c, 21, 22, 25, 28, 29, 32, 33, 36, 41, 42, 45, 46, 47, 51, 54c, 59, 61, 62} and {*4, 9*}. . This issue was present in 44% of CSB incidents and 43% of incidents covered by OSHA's PSM standard. In numerous incidents, procedures had not been developed, were not followed, were incomplete, or were otherwise deficient. Such situations are unacceptable. Properly written procedures must be developed following best practices for all process operations and activities before they occur and for all stages and modes of operation. Procedures are needed to deal with any situation where there is the potential for a process safety incident. Procedures must be workable in all circumstances where they apply and personnel need to be trained in the procedures and practice them regularly in drills. Facility personnel must be competent.

In six CSB investigations that identified problems with standard operating procedures (SOPs), the CSB noted that each incident involved SOP deviations that became "necessary violations" to get the job done {45}. Such situations cannot be tolerated.

Existing process safety regulations require written operating procedures. However, the regulations should be clarified to encompass all process tasks and activities and a mechanism is needed to ensure the regulatory requirements are followed.

Deficiencies in maintenance were present in a number of incidents {5, 12, 21, 29, 31, 32, 51, 54c}. Issues regarding the adequacy of inspections were common {4, 7, 47, 53, 56, 62} and {*2, 8*}. For some incidents, there was no or an inadequate preventive maintenance (PM) program {11, 13, 16, 41} and no or inadequate MI program for some incidents {10, 23, 29, 52, 61, 63}. Such issues were present in 34% of CSB incidents and 32% of incidents covered by OSHA's PSM standard. Existing process safety regulations address the integrity of equipment but a mechanism is needed to ensure the requirements are followed for all safety-critical equipment. Current US process safety regulations require that covered equipment comply with recognized and generally accepted good engineering practices (RAGAGEP) but the term is not defined. OSHA issued a Request for Information (RFI) in advance of rulemaking to revise the PSM standard [23]. Amongst other topics, the RFI sought comment on whether an explicit definition of RAGAGEP should be provided, and, if so, what definition should be used. The current OSHA PSM regulation provides a limited list of covered equipment [24] and the RFI sought comments on whether mechanical integrity requirements should apply to all equipment designated as safety-critical equipment.

The CCPS Risk Based Process Safety (RBPS) element, Conduct of Operations, provides requirements for operational discipline that tolerate no deviation from approved procedures, even if the outcome is inconsequential or desirable [26, 27]. Appropriate work procedures must be determined, documented, and enforced. The importance of following procedures must be emphasized to facility personnel. The adoption of an element on operational discipline would help in this regard coupled with an element that addresses process safety culture. The CCPS RBPS program contains both these elements but they are not included in existing US regulations. Additionally, deviations from approved procedures should be identified and investigated as process safety incidents.

An operational discipline element would encourage and enforce strict adherence to established procedures and imbue facility personnel with the belief that no operation or activity should be performed without a governing procedure. When people's lives are at risk it should not be acceptable to perform activities without proper preparation or by improvising.

Another deficiency present in several incidents was the acceptance of practices as normal and acceptable by facility personnel that outside observers would view as deviant and unacceptable {17, 19c, 21, 29, 42, 57, 62, 63}. This phenomenon has been called "normalization of deviance" [28]. It involves allowing operation of a process that deviates from established limits by relaxing performance standards rather than identifying and correcting the underlying causes of deviations so the deviations become the new, normal operating conditions. Hazards have been normalized in some cases {12, 27, 46, 56}. Normalization of deviance was present in 18% of CSB incidents and 18% of incidents covered by OSHA's PSM standard.

While the acceptance of a lesser performance standard in such cases may not immediately result in an incident, continued successful operation instills a sense of false security. It will be only a matter of time before uncorrected hazardous conditions eventually result in an incident.

Unfortunately, facility personnel can become so accustomed to exposure to hazardous situations that the risks are not recognized even when their lives are at stake. Just because a facility has not experienced a catastrophic incident, perhaps ever in its lifetime, does not mean one could not happen tomorrow.

Normalization of deviance can be addressed by ensuring that any deviations from established performance standards are identified and resolved as soon as they occur. Operational discipline will help in this regard.

3.4 Abnormal and Non-Routine Operations

Many incidents involved non-routine operations and situations {1, 3, 4, 5, 6, 10, 11, 16, 17, 19c, 20, 21, 23, 24, 26, 27, 29, 32, 33, 39, 41, 43, 45, 46, 47, 49, 52, 54a, 54c, 55, 56, 58, 59, 60, 61, 62, 63, 64} and {*1, 7, 9*}. In some cases, they were not recognized as non-routine operations and were not reviewed to ensure that work could proceed safely. Flaws in hot work {7, 31, 33, 44, 59} and other safe work practices {4, 5, 6, 11, 14, 23, 39, 43} and {*6*} also contributed to incidents. Lack of equipment isolation was a contributor to some incidents {11, 33, 59}. Abnormal and non-routine operations were involved in 62% of CSB incidents and 64% of incidents covered by OSHA's PSM standard.

People play an intimate role in non-routine operations. This both increases the likelihood of human error being a factor and places people closer to process hazards. Furthermore, a number of incidents occurred in processes that involved manual operations or were manually controlled {2, 12, 13, 16, 18}.

Current US process safety regulations contain requirements to develop and implement safe work practices and utilize hot work permits. However, the regulations do not specifically address non-routine operations. There are requirements for written procedures under the Operating Procedures element for specific phases of operation listed in the regulations (initial startup, normal operations, temporary operations, emergency shutdown, emergency operations, normal shutdown, and startup following a turnaround or after an emergency shutdown). However, other phases of operation are not specified including commissioning, standby, bypass, unattended operation, maintenance on live equipment, regeneration, troubleshooting, operation with disabled safety systems, cleaning, decommissioning, etc. Process safety programs must address all phases of operation that are or may be experienced by a process. Emphasis should be placed on non-steady state operations including abnormal operations because they are the process states in which most incidents have occurred.

Other elements of process safety programs besides operating procedures should include requirements to address non-routine operations, for example, mechanical integrity and PHA [29]. A separate element on non-routine operations may be merited to provide requirements in a single place. Absent specific requirements to address non-routine operations, many non-routine operations are likely not to be viewed as such and their hazards not recognized or addressed.

Procedures are needed for identifying and planning for non-routine work and abnormal situations, e.g. restart after emergency shutdown or an extended unplanned shutdown. Processes should be operated defensively and conservatively. Procedures should specify alternative courses of action if safety preconditions cannot be met. Furthermore, job hazard analysis [29] should be performed immediately prior to any non-routine activity.

3.5 Human and Organizational Factors

Human and organizational factors contributed to many incidents, for example, valve positions were not clear {23, 51}, valve operation was difficult {56}, lighting was insufficient {51}, operators were distracted {18}, operators were fatigued {29, 45}, operators had insufficient information {57}, procedures were complex and lengthy {45}, communications were problematic {25, 29, 43, 45}, layout was error prone {25}, staffing was not sufficient {25, 29, 45, 51}, operators became desensitized to an alarm due to its history of unreliability {54a}, the work environment encouraged operations personnel to deviate from procedures {29}, a computerized control system was poorly designed {29}, a new control system posed challenges {45}, and there were difficulties with organizational structure {25, 29}. Undoubtedly, problems with safety culture played an important role in many incidents, with BP Texas City being the best known {29}. Most CSB investigations have not explicitly identified contributions to incidents from human and organizational factors so it is not possible to provide meaningful statistics for their contributions to incidents. Undoubtedly, human and organizational factors played a role in most accidents.

Fundamentally, all of the incidents reviewed resulted from human failures ranging from design to operation to maintenance. Even when equipment failures such as corrosion were the basic cause of an incident, the underlying cause usually was human failure {63}.

Some incidents involved conflicts and tradeoffs between process safety and other imperatives {4, 7, 29, 39, 45, 51, 54c, 58, 63}, often financial but sometimes other issues such as noise attenuation {36}.

OSHA's PSM standard requires that "the process hazard analysis shall address human factors" [24]. However, it is not sufficient simply to require PHA studies to address human factors. Certainly, PHA teams must address the role of human factors in influencing human failures but this approach leads to insufficient consideration of human factors issues for processes. Commonly, human factors are addressed by using a checklist with questions that are asked about the entire process so that analysts can try and think of any problems. A much more detailed approach is needed. Otherwise, many aspects of human factors will be missed because of practical limits on the level of detail that can be addressed in PHA studies, limitations on the ability of PHA methods to address human factors, and time pressures on PHA teams. Also, human factors issues affect each of the elements of a process safety program including PHA itself [31]. It is not reasonable to expect a PHA team to address human factors issues that impact its own performance. The Contra Costa industrial safety ordinance requires the development and implementation of a written human factors program [19]. Proposed amendments to California's process safety regulations for refineries contain a requirement to develop, implement and maintain an effective written human factors program that goes beyond the requirements of Contra Costa's industrial safety ordinance [20].

All process safety management programs should have a distinct element that addresses human factors for processes with specific requirements to be addressed. Many aspects of human factors for process safety have been addressed by CCPS [32] and they should be covered by process safety programs.

For many incidents, it is difficult to believe that the conditions that resulted in the incident were allowed to occur. An independent observer easily can conclude that surely someone could have intervened at some stage in the sequence of events that occurred and taken action to prevent the incident. However, such a view does not take account of the psychological factors that influence decision making by people, particularly in groups and when under pressure to achieve competing goals. Factors such as group think, peer pressure, and other cognitive biases interfere with objective decision making [33]. Questionable decision making played a role in numerous incidents, for example {58, 63}. Facility personnel need to apply critical thinking in their daily work activities [34]. Critical thinking can be taught and key facility personnel should be so trained.

3.6 Process Hazard Analysis

Failures in PHA are a common finding in CSB incident investigation reports. Indeed, issues with PHA were present for all CSB incidents where a PHA study was performed. Of particular note is that CSB found that incidents were not identified by the PHA studies performed. However, many of the incidents involved deliberate operation of processes in ways that a PHA team would not consider credible. PHA teams look for ways in which a process may deviate from its design intent but usually the specified design intent identifies what should be done, not what should *not* be done. It is the latter that occurs in many incidents and, while PHA teams try to identify ways that a process may be misoperated, it is unreasonable to expect a PHA team to identify the ways in which a process may be maloperated. Moreover, there is essentially an infinite number of ways in which a process may be maloperated and it would be impractical to ask a PHA team to try and identify all of them. Maloperation must be addressed through operating discipline, proper change reviews, and designing robust processes.

In some cases, the CSB found that PHA studies did not adequately address safeguards because safeguards listed in PHA studies were non-specific or general {19a, 56, 63} and, in many cases, relied too heavily on administrative procedures. For example, safeguards were recorded as "design", "standard operating procedures", "testing and inspection", and "proper emergency response procedures". PHA studies must identify specific safety-critical equipment, for example, a specific relief valve, and specific safety-critical actions, for example, activation of an emergency quench system by an operator. PHA teams should consider the hierarchy of controls when evaluating existing safeguards and deciding on recommendations for further risk reduction. PHA team leaders and team members must be competent [35].

Some CSB incident investigation reports noted the failure to validate critical assumptions used in PHAs {45, 56, 29}. However, such validations should not be part of PHA. They should be performed before a PHA study is performed [21]. Such validations require a different skill set than that required for PHA team members. Moreover, if they were included in PHA studies they would add complexity to the studies and detract from their primary goal of scenario identification. Of course, the adequacy of safeguards for individual hazard scenarios must be addressed by the PHA team.

Some incidents have occurred when more than one activity was being performed simultaneously or sequentially and the activities conflicted {4, 7, 8, 25, 33, 43, 47, 48, 59}. Simultaneous and sequential operations (SIMOPS) reviews [36] should be performed as adjuncts to PHA studies to address such issues. However, they are not yet common and there are no regulatory requirements to perform them.

PHA studies were known to be performed for 22 of the 68 CSB incidents that were reviewed, although incident descriptions did not indicate whether or not a PHA study was performed for 21 of the incidents. However, it is likely that a PHA study was not performed for most of these incidents. Of the 22 PHA studies that were performed, only 4 studies were conducted for processes not covered by OSHA's PSM standard (see Table 1). This finding suggests that PHA studies are only performed when required by regulations.

3.7 Process Changes

Many incidents involved changes that occurred but were not reviewed for their impact on process safety {2, 4, 6, 7, 8, 16, 17, 18, 21, 25, 27, 29, 37, 45, 46, 54c, 59, 63} and {*2, 9, 11*}, and inadequate reviews were performed in some cases {7, 15, 19a, 29, 32, 45, 54a, 56, 63}. Issues with process change reviews were present in 34% of CSB incidents and 39% of incidents covered by OSHA's PSM standard. Omissions and deficiencies can arise because changes are not recognized as requiring MOC reviews, personnel lack awareness of MOC requirements, or MOC reviews are bypassed for expediency. These problems relate to the competency of facility personnel and operational discipline. If personnel are involved in unauthorized and unreviewed changes, they cannot be viewed as competent to work in the facility in the same way as if they maloperated a process or performed maintenance work incorrectly. The establishment of a competency management program should help in this regard. It should be reinforced by operational discipline.

Also, MOC reviews required by the OSHA PSM standard have a narrow definition of covered changes. They cover only changes to PSI, procedures, and facilities that affect a covered process [24]. These requirements are neither global nor sufficiently specific and they are open to interpretation. Organizational changes have played a role in several incidents {29, 45, 56, 64} but they are not currently covered by the regulations, although OSHA's RFI addresses this omission [23]. However, incidents have involved other types of changes not explicitly defined in US process safety regulations including changes in feedstock {4, 63}, addition or removal of equipment {16, 22, 37, 42, 59}, degraded equipment {59}, scale-up of recipes {32}, and prolonged operation {6}. Furthermore, MOC programs should encompass operational deviations and variances, and abnormal situations as well as preplanned changes [37]. Regulations and best practices should specify that any change that may impact on or affect a release of a highly hazardous chemical should be covered by MOC requirements.

Additionally, activities involved in making process changes can be hazardous and are not covered by current US process safety regulations. MOC reviews of changes should include the performance of job hazard analysis for the activities involved.

3.8 Previous Incidents

Information from previous incidents that could have helped to prevent incidents was not used in many cases {2, 5, 6, 11, 17, 19a, 20, 21, 25, 26, 29, 41, 42, 45, 51, 53, 54a, 54c, 60, 61, 63, 64} and {*4, 11*}. Issues with previous incidents were present for 32% of CSB incidents and 29% of incidents covered by OSHA's PSM standard. Previous incidents were not investigated properly, or at all, and near misses were ignored. The issue of retaining information from previous incidents in the corporate memory of companies is well known [38]. However, information cannot be retained if it has not first been developed.

Facility personnel may not have the time or inclination to study incident descriptions and they may not see or understand their relevance to their own facilities. Reliance cannot be placed on people to address lessons learned from previous incidents unless process safety programs contain specific requirements to do so. OSHA's PSM standard requires that previous incidents be addressed [24]. However, the regulatory requirement is part of the PHA element which simply states that the PHA shall address previous incidents. This weak requirement is likely to result in the inadequate consideration of how information from previous incidents can be used to prevent repeat or similar incidents. Process safety programs should have a separate element on previous incidents, including near misses, that requires companies to identify relevant previous incidents from within the company and from other companies in the process industries, compile pertinent information on them, analyze the information in the context of reducing risks in existing processes, and develop recommendations to do so and track them to implementation. The previous incidents review should be performed periodically, e.g. annually, and documented. It should address both the issue of ensuring that incidents are identified, reviewed, and lessons learned are implemented, and that information is retained.

3.9 Facility and Stationary Source Siting

The proximity of facilities to members of the public was an issue in multiple incidents {2, 3, 9, 10, 15, 16, 17, 18, 19a, 20, 22, 23, 24, 25, 26, 27, 29, 30, 32, 35, 36, 41, 45, 46, 48, 51, 52, 53, 63}. This issue was present for 43% of CSB incidents and 64% of incidents covered by OSHA's PSM standard. It raises concerns regarding land use planning in the US. Proximity of facility personnel to hazards in plants also was an issue in numerous incidents {11, 19a, 20, 21, 25, 26, 29, 32, 41, 48, 61} and {*2, 5, 8, 9, 12*]. This issue was present for 16% of CSB incidents and 25% of incidents covered by OSHA's PSM standard. The consequences of some incidents have been exacerbated by the presence of non-essential personnel {29, 49, 56}.

OSHA's PSM standard requires that "the process hazard analysis shall address facility siting" [24] and EPA's RMP rule requires that "the process hazard analysis shall address stationary source siting [39]. However, it is not sufficient simply to require PHA studies to address facility and stationary source siting. Certainly, PHA teams must address the impact of facility and stationary source siting on hazard scenarios but this approach likely leads to insufficient consideration of facility and stationary source siting issues for processes. Aspects of facility and stationary source siting will be missed because of practical limits on the level of detail that can be addressed in PHA studies, limitations on the ability of PHA methods to address facility and stationary source siting, and time pressures on PHA teams.

Process safety management programs should have a distinct element that addresses facility and stationary source siting for processes with specific requirements to be addressed. These requirements should include topics that often are overlooked including using designs that facilitate minimal operator presence; limiting the number of people exposed to hazards, particularly at more hazardous times, by moving unnecessary people away from non-routine operations; training and requiring people to leave when hazardous conditions or situations arise; and assessing the potential for domino events.

The problem of the proximity of people to hazards is compounded by poor emergency planning and response which was present for about 40% of these incidents (see next section).

3.10 Emergency Preparedness and Response

Deficiencies in emergency preparedness and response existed in many cases {5, 8, 10, 11, 16, 19a, 20, 22, 24, 25, 32, 34, 35, 39, 42, 43, 45, 51, 54a, 54b, 63, 64} and {*1, 3, 6, 10, 12*}. This issue was present for 32% of CSB incidents and 39% of incidents covered by OSHA's PSM standard. Deficiencies occurred in hazard communication, evacuation procedures and emergency egress, community notification, emergency preparedness, knowledge of responders, and coordination between responders. OSHA's RFI addresses requiring facilities to coordinate emergency planning with local emergency response authorities [23].

An effective way to help ensure that emergency response plans are in place and will work is to hold regular, periodic drills to test them. It is important to ensure that emergency preparedness and response correlates emergency response with actual hazard scenarios and addresses access to fire water monitors and HAZMAT equipment during incidents and provisions for containment of fire fighting water to prevent environmental pollution.

A hazard analysis should be performed for emergency response activities to identify issues before emergency response is needed. Such a hazard analysis should incorporate task analysis, job safety analysis, and a SIMOP study.

3.11 Audits and Inspections

CSB investigations found that audits by companies, regulators, insurance providers, and others often did not identify process safety omissions and deficiencies and that audits and inspections missed issues that contributed to incidents in several cases {17, 18, 22, 42, 56, 59, 64}. Inadequate audits were identified in several cases {4, 22, 29}. Issues with audits and inspections were present for 13% of CSB incidents and 18% of incidents covered by OSHA's PSM standard.

The CSB noted that in the case of a refinery incident {56} a refinery National Emphasis Program (NEP) inspection was carried out about 18 months before the incident occurred but the mechanical integrity issues that were at the heart of the incident were not identified. NEP audits are the most comprehensive ones that have been performed by regulators. This audit was conducted by a team of people that spent several weeks at the refinery and it focused on the unit that subsequently experienced the catastrophic incident investigated by the CSB.

OSHA's PSM standard requires that a company conduct a compliance audit every three years to "verify that the procedures and practices developed under the standard are adequate and are being followed" [24]. This terse requirement can make for perfunctory audits. Furthermore, process safety compliance audits do not focus on fundamental inadequacies in the design of a process safety program; rather they evaluate the design against regulatory requirements and their implementation. Audits may not properly address non-routine operations because usually they do not occur when audits are performed. Additionally, audits are conducted under time limitations and use sampling techniques.

More comprehensive audits should be performed. They should cover all aspects of a process safety program including safety culture, roles and responsibilities, Auditors must be competent and should be certified [40]. OSHA's RFI addresses requiring third-party audits, increasing the required frequency of compliance audits, and requiring specific time frames for responding to deficiencies found by audits. There is an international standard for auditing management systems [41]. Process safety regulations should require audits be conducted in compliance with the standard.

3.12 Equipment Damage Mechanisms

Equipment damage mechanisms played an important role in several incidents {4, 10, 19a, 20, 53, 56, 63}. Interestingly, all but one of these incidents occurred in a facility covered by OSHA's PSM standard. The CSB has noted the need to perform damage mechanism reviews (DMRs) {63}. Proposed amendments to California process safety regulations include requirements for conducting DMRs [20].

3.13 Process Safety Regulations

Incidents occurred in facilities both covered and not covered by US process safety regulations. Many facilities that are not covered by process safety regulations in the US pose serious process safety risks, for example, facilities handling reactive chemicals and dusts. These issues have been recognized by the CSB [10, 11].

In some cases, facilities that are covered by process safety regulations had not implemented a process safety program {24, 32, 34, 36, 48}.

3.14 Compliance with Industry and Company Standards and Practices and Regulations

Non-compliance with industry and company standards and practices were common in numerous incidents {1, 4, 8, 9, 13, 15, 16, 17, 21, 24, 27, 28, 29, 31, 33, 40, 42, 43, 47, 50, 51, 56, 54b, 54c, 62, 64} and {2, 7}. This issue was present for 38% of CSB incidents and 29% of incidents covered by OSHA's PSM standard. Also, there were failures to follow regulations for multiple incidents {8, 12, 16, 24, 25, 33, 34, 36, 39, 40, 41, 64}. This issue was present for 18% of CSB incidents and 14% of incidents covered by OSHA's PSM standard.

Non-compliance with standards, practices and regulations is a fundamental issue for process safety management. If standards, practices and regulations are ignored, the comprehensiveness of their content does not matter and incidents will continue to occur. Certainly, management review and continuous improvement may go some way to address this problem and the inclusion of this CCPS RBPS element [25] was addressed in OSHA's RFI [23]. The RFI also addressed the inclusion of the CCPS RBPS element on measurement and metrics [25] to track the effectiveness of process safety management programs. Regulators in California have proposed amendments to the state's process safety regulations for refineries that will require the use of process safety performance indicators [20]. Public disclosure of process safety performance would encourage management to pay close attention to ensuring those measures reflect favorably on the company. Thus, for example, it can be expected that there would be fewer action items that remain unclosed past their due dates. Of course, if such an approach were to be applied, it may still be ignored by some companies. Changes in regulatory enforcement may be needed.

A recent case in the UK contrasts markedly with US practices. Phosphine was released from a plant requiring 4,500 people to shelter in place for 2-3 hours and causing disruption to a motorway [42]. The U.K. Health and Safety Executive (HSE) investigated the incident and found that an equipment failure caused loss of containment. The company was fined a total of £333,000 and ordered to pay costs of £110,000 after pleading guilty to an offence under the U.K. Health and Safety at Work Act of 1974 having failed to properly assess and act upon the risk of the equipment failing. Such a penalty for an incident of relatively low consequences contrasts with smaller penalties for greater consequence incidents in the United States.

After the prosecution, an HSE inspector said, "This case should serve as a warning to other companies dealing with harmful substances that they need to get their processes absolutely right, in order to ensure the safety of the public, if they don't they will face the consequences." In virtually, every incident analyzed in this paper, the company did not get it right. There are very few, if any process safety accidents; they are preventable incidents.

3.15 Industry Standards and Guidelines

Numerous CSB incident investigation reports identified flaws and omissions in industry standards and recommended practices {9, 14, 16, 25, 26, 29, 36, 37, 48, 51, 53, 54c, 55, 56, 63} from organizations such as the American Petroleum Institute (API), the American Society of Mechanical Engineers (ASME), the Compressed Gas Association (CGA), and the National Fire Protection Association (NFPA). Issues with industry standards and guidelines were present for 22% of CSB incidents and 32% of incidents covered by OSHA's PSM standard.

Such standards and practices usually are subject to periodic review and updating as part of a process of continuous improvement so such omissions and deficiencies ultimately can be addressed. However, the CSB has criticized the permissive nature of some of these standards and recommended practices {56} which leave options open to companies not to follow some of the guidance. Also, the CSB has criticized the process by which standards are set by some organizations, for example API {56}.

This finding suggests that CSB should investigate a diversity of incidents to identify as many deficiencies and omissions as possible in industry standards and guidelines.

3.16 Product Stewardship

Some incidents involved the use of highly hazardous chemicals obtained by one company from another which did not recognize their hazards {14, 17}. In another case, a company was not informed by a supplier that process safety regulations would need to be implemented before receiving shipments {24}. While chemical supplier companies will not necessarily know the details of how purchasers intend to use the chemicals, they should at least communicate their known hazards to the purchaser. Process safety considerations should be a key part of product stewardship initiatives such as that in the ACC's Responsible Care Code of Management Practices [43]. Suppliers of highly hazardous chemicals have responsibilities in supplying such chemicals to other companies.

Some additional issues were identified for a few CSB incidents (see Table 3).

Table 3. Additional Issues Noted for CSB Incidents.

Issue	Examples
Process boundaries defined too narrowly	59
Inadequate workforce participation	56 / *24*
Inadequate process safety information	3, 6
Inaccurate or incomplete safety data sheets	2, 5, 15, 38
Loss of process design and safety information due to poor management of records	11
Lack of hazard recognition and understanding	14, 15, 21, 26, 31
Inadequate communication of plant hazards	5, 17
Need for positive materials identification during commissioning, modifications, and maintenance	10
Inadequate or no pre-startup safety review	3, 13, 29, 45
No or inadequate near miss reporting	54c
Inadequate or no incident investigation	12, 20
Inadequate contractor management	39, 59, 64
Outstanding action items	2, 4, 6, 7, 16, 25, 45, 54b, 54c, 63 / *27*
Inadequate oversight by management	4, 7, 10, 12, 29, 32

4.0 Discussion of the Analysis

A summary of the results of the analysis is provided in Table 4. Clearly the top issues are abnormal / non-routine work and safeguards which are about equally important for all incidents and for PSM incidents and stationary source siting for PSM incidents. These are followed by procedures which are about equally important for all incidents and for PSM incidents and stationary source for all incidents. Compliance with standards and practices, process changes, emergency preparedness and response, maintenance, and previous incidents come next.

Table 4. Summary of Analysis for CSB Incidents.

Issue	% Incidents	% PSM Incidents
Design	28	18
Safeguards	56	57
Procedures	44	43
Maintenance	34	32
Normalization of deviance	18	18
Abnormal and non-routine operations	62	64
Process changes	34	39
Previous incidents	32	29
Stationary source siting	43	64
Facility siting	16	25
Emergency preparedness and response	32	39
Audits and inspections	13	18
Compliance with industry and company standards and practices	38	29
Failure to follow regulations	18	14
Industry standards and guidelines	22	32

Notable differences between the issues for all incidents and PSM incidents occur in some cases. There are fewer issues for both design and compliance with standards and practices for PSM incidents than for all incidents, but there are more issues for PSM incidents than for all incidents for each of stationary source siting, facility siting, and industry standards and guidelines. The difference for stationary source siting is particularly significant.

The lowest percentage of incidents with any of the issues is 13% and the highest is 64%. This is poor process safety performance.

Time trends for the CSB incidents are provided in Table 5. The numbers of incidents displayed per year in the table largely reflects resources available and decisions made by the CSB. However, it appears that the trend for virtually all issues is for them to continue.

Table 5. Time Trends for CSB Incidents.

Issue	1998		1999		2000		2001		2002		2003		2004		2005		2006		2007		2008		2009		2010		2011		2012	
	N	P	N	P	N	P	N	P	N	P	N	P	N	P	N	P	N	P	N	P	N	P	N	P	N	P	N	P	N	P
Design	1	-	-	1	-	-	1	-	-	-	5	-	-	1	-	1	1	1	2	1	1	-	-	-	2	-	-	-	1	-
Safeguards	1	-	-	-	-	-	1	-	4	-	4	2	-	3	-	3	1	3	2	1	1	1	2	2	3	1	1	-	1	-
Procedures	2	-	-	1	-	-	1	-	3	-	5	2	-	1	-	2	1	1	1	-	2	2	1	-	-	2	2	-	-	-
Maintenance	-	-	-	1	-	-	2	-	2	1	3	-	-	1	-	1	1	1	1	-	1	-	2	1	-	2	2	-	-	1
Normalization of deviance	-	-	-	-	-	-	-	-	1	-	2	1	1	-	-	1	-	-	-	-	1	1	-	-	1	-	1	-	-	1
Abnormal and non-routine operations	1	-	-	2	-	-	3	-	1	1	4	2	1	3	-	1	2	1	2	-	2	3	1	1	4	3	2	-	1	1
Process changes	1	-	-	1	-	-	2	-	1	-	5	-	1	1	-	2	-	1	-	1	-	2	-	-	1	3	-	-	-	1
Previous incidents	1	-	-	-	-	-	2	-	1	-	2	2	-	2	-	1	-	-	1	-	1	1	2	-	2	1	1	-	1	1
Stationary source siting	1	-	-	1	-	-	-	-	1	1	4	3	1	4	-	2	1	2	1	-	-	2	2	2	-	-	-	-	-	1
Facility siting	-	-	-	-	-	-	-	-	1	-	1	2	-	2	-	1	-	1	1	-	-	-	-	1	-	-	1	-	-	-
Emergency preparedness and response	-	-	-	-	-	-	1	-	2	1	1	3	-	2	-	-	1	2	1	-	1	2	1	-	2	-	-	-	1	1
Audits and inspections	-	-	-	1	-	-	-	-	-	-	2	1	-	-	-	1	-	-	-	-	1	-	-	-	-	2	-	-	1	-
Compliance with industry and company standards and practices	1	-	-	1	-	-	-	-	2	-	5	-	1	1	-	2	2	-	1	-	2	1	1	1	1	2	1	-	1	-
Failure to follow regulations	-	-	-	-	-	-	-	-	2	-	1	-	-	2	-	-	1	2	3	-	-	-	-	-	-	-	-	-	1	-
Industry standards and guidelines	-	-	-	-	-	-	-	-	1	-	2	-	-	2	-	1	-	1	-	1	-	-	2	1	1	2	-	-	-	1

N – Non PSM Covered
P – Covered by PSM

There is an average of 5 issues per incident and as many as 13 issues. Some CSB investigations appear to be more thorough than others so the number of issues identified in the investigation reports does not necessarily reflect all of the issues that were present. Consequently, these numbers may be low. However, it is clear that most incidents involved multiple issues and this conclusion would not change. It is likely that the underlying reason for multiple issues being present for the majority of incidents is a systemic problem with the process safety management system and process safety culture.

5.0 Further Observations

This section provides some additional noteworthy observations that were made during the review of incident descriptions.

5.1 Age of Plants

Many process facilities in the US were constructed some time ago. Of course the longer a process operates the more exposure there is to risk. Plant aging factors need to be addressed by process safety programs.

Equipment deteriorates over time and this has played a role in some incidents {63}. Damage mechanism reviews can help to address this issue. Old equipment may be used, for example, a tank constructed in 1929 failed catastrophically in 2008 {47}. Use of old equipment must be subject to close review.

As time passes, industry standards and best practices evolve but the designs of processes are frozen in time. Thus, gaps between current industry standards and best practices and existing company standards and practices may exist. Use of dated standards and practices has played a role in several incidents {30}. Outdated technology also has played a role in some incidents {29}. Process safety programs should require periodic formal reviews to identify and evaluate the feasibility of adopting current standards, best practices, and technology retrospectively. Current standards, best practices, and technology should be addressed when developing recommendations to reduce risk, for example, in PHA studies, MOC reviews, and incident investigations. OSHA's PSM standard does not require employers to evaluate updates to applicable RAGAGEP or to examine new RAGAGEP after evaluating and documenting compliance with mechanical integrity requirements. OSHA's RFI sought comments on the best approach for requiring the evaluation of updates to applicable RAGAGEP. However, RAGAGEP is defined narrowly by CCPS which is the definition being advanced by OSHA. It covers generally approved ways to perform specific engineering, inspection, or mechanical integrity activities, such as fabricating a vessel, inspecting a storage tank, or servicing a relief valve. There are many more types of standards and practices that could be included in the definition of RAGAGEP.

Informal procedures may become institutionalized with the passage of time with operators passing them on verbally without others being aware {46}. Key tribal knowledge may be lost through departures of personnel including retirements {11, 21, 54}. An aging workforce may display diminishing physical and mental faculties. These are human factors issues that should be addressed by a human factors program.

Often process facilities have been built in rural areas away from the public but over time housing developments are built near the facility. This issue was present for several incidents. It must be addressed through land use planning.

5.2 Magnitude of Consequences

The consequences of some incidents could have been much worse. For example, releases occurred under favorable weather conditions {22}, with the wind blowing away from populated areas {10}, when projectiles created by explosions missed hitting adjacent processes {11}, and at a time when the number of people potentially exposed was at a minimum {3, 36}. In some cases, people exposed to serious hazards miraculously escaped harm {63} and avoided being caught in the collapse of structures {49}.

These situations were serendipitous. However, it does suggest a hazard control strategy in which riskier activities, such as non-routine work, should be timed for when the risk to receptors is at a minimum. Also, facility personnel in the vicinity of a riskier activity who are not part of the activity should be relocated for the duration of the activity. Such considerations should be an integral part of planning for non-routine work.

Some injuries occurred when people investigated process abnormalities {32, 45, 64} or tried to fix leaks {62, 63}. Such situations must be taken seriously and not considered part of normal operations. Appropriate precautions must be taken.

6.0 Underlying Issues

This section identifies three issues that underlie all of the other issues. Each underlying issue is discussed and recommendations are made for addressing it.

6.1 Competency of Facility Personnel

The CSB has not explicitly addressed the issue of competency in many incident investigations. However, it was mentioned as a contributor in at least one CSB incident {29} and two CCPS incidents {2, 4}. In reading incident descriptions it is hard not to reach the conclusion that the people responsible for the processes were not competent in one or more ways in many cases. This issue likely underlies most of the process safety incidents that have occurred.

The decisions made by personnel ranging from managers to operators can be questioned for many incidents. While facility personnel were not obviously incompetent, in numerous cases clearly they were not competent to perform the work required of them safely. In particular, facility personnel often lacked an awareness and/or understanding of one or more critical aspects of a process, such as its hazards {6, 8, 11, 24, 32, 36, 39, 41, 44, 54a, 54c, 62}. Inadequate operator and mechanic training was an issue in a number of incidents {1, 2, 3, 10, 11, 12, 13, 14, 18, 21, 22, 25, 28, 29, 32, 39, 42, 45, 64}.

OSHA's RFI sought comments on the addition of an element on process safety competency [23] as described by CCPS [25]. Certainly, facilities should implement a competency management program for process safety but the CCPS RBPS competency approach should be modified [44]. Critical personnel should be certified and periodically re-certified. For example, operators and mechanics should not be allowed to operate or maintain a process without demonstrating their competence, including such topics as knowledge and understanding of process hazards, the higher risks of non-steady state operations such as startup, previous incidents, the safe operating envelope for the process, handling abnormal and emergency situations, conditions under which the process must be shut down, use of permit-to-work systems, and the identification and management of changes. Responsibility for supervision of the operation of a process should be formally assigned to a single person and the responsibility formally transferred to another individual as necessary. There must be supervisory oversight of processes at all times.

In some incidents, the availability of technical expertise has been an issue {29, 32, 45}. When critical decisions are faced regarding process safety, someone must be present who understands mass and energy balances, physical and chemical phenomena that may occur, process controls and safeguards, hazards present, and the safe operating envelope, that is tolerable ranges of values of safety-critical parameters. Individuals should be designated as engineering and technical authorities who ensure the proper application of engineering standards and technical practices and adjudicate when deviations are proposed. They must be available for consultation at all times and they must have the authority to shut down processes without the need for consultation with managers and without recrimination. They must be free of production or commercial pressures and should report directly to upper management. A stop work authority that authorizes personnel to stop work in the face of imminent risk or dangerous activity and an ultimate work authority that requires designation of who has the authority for operational safety and decision making are addressed by OSHA's RFI [23].

6.2 Standards for Process Safety Elements

Process safety programs comprise multiple elements such as employee participation, PSI, PHA, etc. For example, OSHA's PSM standard has 14 elements [24] and CCPS's RBPS program has 20 elements [25]. However, it is possible for companies to address each of these elements in widely varying ways because their requirements are general in nature and do not address the effectiveness of their contribution to risk reduction. For example, PHA studies can be conducted with widely varying levels of quality. A solution to this problem is to establish minimum performance standards for each element. These standards should be established by an international organization such as ISO so they apply around the world. There is precedent for such standards in the IEC 61511 standard for safety instrumented systems [45 - 47] and the Hazard and Operability Study (HAZOP) standard [48], although the latter is outdated. These standards are still performance-based in nature, as should be new standards for other process safety elements.

One benefit of standards for process safety elements is that they can be updated more easily than regulations. Current US process safety regulations are static and unable to adapt to innovation and advances in the management of major hazard risks.

This problem could be overcome if the US process safety regulations simply established overall requirements for each process safety element and required companies to follow RAGAGEP for each element. There is already a precedent for doing so with the PSI and mechanical integrity PSM elements although RAGAGEP is interpreted more narrowly in that context. The proposal is to broaden the meaning of RAGAGEP, establish RAGAGEPs for all process safety elements, and require that RAGAGEPs be followed. The process safety regulations would effectively update automatically as RAGAGEPs evolve. Continuous improvement is an essential element of a management system for any surety such as quality, environmental protection, and occupational health and safety, and is part of CCPS's RBPS program [25].

6.3 Safety Goals and Risk Tolerance Criteria

In an least one incident, the CSB concluded that one company had knowledge of and acceptance of high risk {29}. Most companies do not really know what risk their process facilities pose because they do not evaluate risk as there are no requirements to do so under the current US process safety regulatory regime. Indeed, it is possible that similar facilities may pose quite different risks because the regulations do not contain any safety goals or risk tolerance criteria.

The goal of process safety programs is to ensure safety and control risks. Consequently, some form of safety goals or risk tolerance criteria are needed [49]. Absent some metric to judge achievement of process safety goals, resources may be allocated inappropriately. There is a current precedent for the use of numerical risk tolerance criteria provided by the New Jersey Toxic Catastrophe Prevention Act (TCPA) process safety regulations [18]. Although the particular approach used is erroneous [50], companies in the state have accepted the use of numerical criteria.

Many jurisdictions around the world employ the tolerability of risk framework developed by the UK HSE including the As Low As Reasonably Practicable (ALARP) principle [51]. Its use should be incorporated into US regulations. Although the ALARP principle usually incorporates the use of numerical criteria, it can be used without specifying such criteria but using its essential element that risk should be reduced until the incremental expenditures of resources is grossly disproportionate to the value of the incremental risk reduction achieved. This is a performance-based requirement that is consistent with the current philosophy of US process safety regulations. Use of the ALARP principle would help to accommodate improved process safety practices without the need for periodic modification of regulations.

One of the difficulties faced by PHA teams is the use of risk ranking in studies to determine if recommendations for risk reduction, that is, additional safeguards, are needed. The CSB has been critical of the subjectivity involved and the criticism is warranted [31]. The CSB has suggested that Layers of Protection Analysis (LOPA) be used for this purpose. Of course, LOPA requires the use of risk tolerance criteria that are allocated from facility criteria [52]. Proposed amendments to the California process safety regulations include requirements for such an analysis [20]. Simply put, the adequacy of safeguards cannot be determined without risk tolerance criteria.

Use of risk tolerance criteria with LOPA raises the issue of what type of risk is being addressed and should be addressed. Certainly, no loss of life is acceptable but in US society single fatalities are more tolerable than multiple fatalities as evidenced by the relative attention paid by the media to single versus multiple fatality incidents. Perhaps the focus in process safety should be on protecting groups of people, particularly members of the public.

All process safety elements would benefit from the use of appropriate performance indicators [25, 53, 54]. Taken together, their use helps to control overall facility and process risks.

7.0 Conclusions

This survey of incidents covered a range of company and facility sizes, and types of processes. Also, it included facilities covered and not covered by process safety regulations. The study results can be used by companies to focus on improving areas where other companies have fallen short.

Of the 80 incidents studied, only the explosion resulting from corrosion failure of piping at the Shell refinery in Norco, LA, that occurred in 1988 {*5*} was truly unexpected owing to a unique corrosion phenomenon that occurred under conditions that were expected to be less corrosively onerous and with pH within the tolerable range. However, the failure occurred as a result of process changes and it is possible that the incident could have been prevented. All other incidents were predictable and therefore preventable.

Process safety programs aim to identify possible incidents before they occur and implement practices to prevent them. Remarkably, many of the incidents reviewed were attributable to deficiencies or omissions in the design, operation, or maintenance of processes. Therefore, it should have come as no surprise that these processes experienced incidents.

The quote, "It is said that there is nothing new in safety - it has all happened before" was published by CCPS [1]. Virtually every incident or something like it had happened somewhere before. All the incidents involved some type of deficiency or omission in adhering to established process safety practices. In many cases there were multiple deficiencies and omissions. All of these incidents were predictable and preventable. Just because an event may be described as unthinkable does not necessarily mean it is unpredictable.

Several recommendations can be made to strengthen process safety programs and regulations.

7.1 Recommendations for Operating Companies

- Add process safety elements on:
    - Design and engineering

  - Human factors review
  - Facility and stationary source siting review
  - Safeguards protection analysis
  - Operational discipline
  - Competency management
  - Previous incidents review
  - Best practices review
- Address non-routine operations throughout all process safety elements.
- Require the use of JSA for non-routine work.
- Expand the types of changes addressed under the MOC element to include any change that may impact on or affect a release of a highly hazardous chemical.
- Make compliance audits more comprehensive.
- Require annual emergency response drills for facilities.
- Require that process risks be managed using the ALARP principle.
- Require facilities to set safety goals and/or risk tolerance criteria and demonstrate compliance with them.

7.2 Recommendations for Standards Organizations

- Develop international standards for key process safety elements that can be referenced by process safety regulations.

7.3 Recommendations for the Chemical Safety Board

- Expand the incident attributes that are addressed in investigations to gain more insights into process safety performance, for example, explicitly address competency, human and organizational factors, safety culture, plant aging factors, and operational discipline.
- Provide an appendix to each investigation report that characterizes each attribute for the incident. Not only is information on deficiencies and omissions valuable but also knowledge of what was done correctly or that there were no problems in certain areas is valuable.
- Compile a data base of incidents and their attributes that is searchable.

- Provide periodic analyses of the full data base of incidents looking for commonalities and trends and make recommendations for improvements in process safety practices.
- Publish the criteria used for the selection of incidents that are investigated by the CSB and update as appropriate.
- Expand on the current documentation in incident investigation reports to explain how the criteria were applied in selecting the incident for investigation.

The CSB adopted a motto in its early days. It still applies:

Salus populi suprema lex esto.

(People's safety should be the highest law.)

8.0 References

[1] J. Atherton and F. Gil, Incidents That Define Process Safety, Wiley-AIChE/CCPS, New York, NY, 2008.

[2] K. Gugan, Unconfined Vapor Cloud Explosions, Gulf Publishing, Houston, TX, 1979.

[3] T. A. Kletz, Learning from Accidents, 3rd edition, Routledge, Abingdon, UK, 2001.

[4] T. A. Kletz, What Went Wrong?: Case Histories of Process Plant Disasters and How They Could Have Been Avoided, 5th edition, Elsevier, Burlington, MA, 2009.

[5] T. A. Kletz, Still Going Wrong!: Case Histories of Process Plant Disasters and How They Could Have Been Avoided, Butterworth-Heinemann, Burlington, MA, 2003.

[6] Roy E. Sanders, *Chemical Process Safety: Learning from Case Histories*, Butterworth-Heinemann; 4th edition, Butterworth-Heinemann, 5th edition, 2015.

[7] The 100 Largest Losses 1974-2013, Large Property Damage Losses in the Hydrocarbon Industry, 23rd Edition, Marsh and McLennan.

[8] http://www.csb.gov.

[9] http://www.icheme.org/lpb.aspx

[10] Hazard Investigation: Improving Reactive Hazard Management, U.S. Chemical Safety Board, Washington, DC, 2002.

[11] Combustible Dust Hazard Study, U.S. Chemical Safety Board, Washington, DC, 2006.

[12] Seven Key Lessons to Prevent Worker Deaths During Hot Work In and Around Tanks, Safety Bulletin, U.S. Chemical Safety Board, Washington, DC, 2010.

[13] A. S. Blair, Management system failures identified in incidents investigated by the U.S. Chemical Safety and Hazard Investigation Board, Process Safety Progress, Volume 23, Issue 4, pages 232–236, December 2004.

[14] M. Kaszniak, Oversights and omissions in process hazard analyses: Lessons learned from CSB investigations, Process Safety Progress, Volume 29, Issue 3, pages 264–269, September 2010.

[15] Amyotte, P. R., MacDonald, D. K. and Khan, F. I. (2011), An Analysis of CSB Investigation Reports Concerning the Hierarchy of Controls. Process Safety Progress, Volume 30, Issue 3, pages 261–265, September 2011.

[16] T. Kerin, The evolution of process safety standards and legislation following landmark events—what have we learnt?, Process Safety Progress, DOI: 10.1002/prs.11762.

[17] P. Baybutt, Process Design - A Missing Process Safety Element (letter to the editor), Process Safety Progress, Volume 35, Issue 1, March, 2016.

[18] New Jersey Department of Environmental Protection, Title 7, Chapter 31, Toxic Catastrophe Prevention Act Program, Consolidated Rule Document.

[19] Contra Costa County Ordinance, Chapter 450-8-Risk Management.

[20] Proposed General Industry Safety Order (GISO) §5189.1, Process Safety Management for Petroleum Refineries, State of California, Department of Industrial Relations, Division of Occupational Safety and Health, Version 5.0 - September 14, 2015.

[21] P. Baybutt, On the Validation of Safeguards for Process Hazard Analysis, Process Safety Progress, Volume 32, Issue 2, pps. 165–169, June 2013.

[22] P. Baybutt, "What risk reduction measures should be credited in process hazard analysis?", Process Safety Progress, Vol. 31, No. 4, pps 359–362, December, 2012.

[23] Process Safety Management and Prevention of Major Chemical Accidents, Request for Information, Federal Register Volume 78, Number 236, Monday, December 9, 2013.

[24] Process Safety Management (PSM) of Highly Hazardous Chemicals, 29 CFR 1910.119.

[25] Guidelines for Risk Based Process Safety, Wiley-AIChE/CCPS, New York, NY, 2007.

[26] Risk Management Program for Chemical Accident Prevention, 40 CFR Part 68.

[27] Conduct of Operations and Operational Discipline, Wiley-AIChE, New York, NY, 2011

[28] D. Vaughn, Risky Technology, Culture, and Deviance at NASA, 1st edition, University of Chicago Press, Chicago, IL, 1996.

[29] P. Baybutt, Process hazard analysis for phases of operation in the process life cycle, Process Safety Progress, Vol. 31, No. 3, pps 279–281, September, 2012.

[30] G. Swartz, Job Hazard Analysis: A Guide to Identifying Risks in the Workplace, Government Institutes, Rockville, MD, 2001.

[31] P. Baybutt, The Role Of People And Human Factors In Performing Process Hazard Analysis And Layers Of Protection Analysis, J. of Loss Prevention in the Process Industries, Vol. 26 pps.1352-1365, 2013.

[32] Human Factors Methods for Improving Performance in the Process Industries, Wiley-AIChE/CCPS, New York, NY, 2007.

[33] P. Baybutt, The Psychology of Decision Making in Process Hazard Analysis, 11th Global Congress on Process Safety, Austin, TX, April 26 - 30, 2015.

[34] B. N. Moore and R. Parker, Critical Thinking, Eleventh Edition, McGraw- Hill Education, New York, NY, 2015.

[35] P. Baybutt, Competency requirements for process hazard analysis (PHA) teams, Journal of Loss Prevention in the Process Industries, Volume 33, January 2015, Pages 151-158.

[36] Guidance on SIMOPS, IMCA M 203, International Marine Contractors Association, March, 2010.

[37] Management of Change, Safety Bulletin No. 2001-04-SB, U.S. Chemical Safety Board, 2001.

[38] T. A. Kletz, Lessons from Disaster: How organisations have no memory and accidents recur, Institute of Chemical Engineers, Rugby, UK, 1993.

[39] Accidental Release Prevention Requirements: Risk Management Programs Under Clean Air Act Section 112(r)(7), Federal Register / Vol. 61, No. 120 / Thursday, June 20, 1996.

[40] P. Baybutt, Competency requirements for process safety auditors, Process Safety Progress, Volume 34, Issue 2, pages 179–182, June 2015.

[41] Guidelines for auditing management systems, ISO 19011:2011, International Organization for Standardization, Geneva, Switzerland, 2011.

[42] http://press.hse.gov.uk/2016/chemical-company-fined-for-dangerous-gas-release/?ebul=hsegen&cr=8/08-feb-16, last accessed 2/8/16.

[43] Responsible Care® Process Safety Code, American Chemistry Council.

[44] P. Baybutt, The meaning and importance of process safety competency, Process Safety Progress, DOI: 10.1002/prs.11767.

[45] ANSI/ISA—84.00.01—2004 Part 1 (IEC 61511-1 Mod), Functional Safety: Safety Instrumented Systems for the Process Industry Sector — Part 1: Framework, Definitions, System, Hardware and Software Requirements.

[46] ANSI/ISA—84.00.01—2004 Part 2 (IEC 61511-2 Mod), Functional Safety: Safety Instrumented Systems for the Process Industry Sector — Part 2: Guidelines for the Application of ANSI/ISA—84.00.01—2004 Part 1 (IEC 61511-1 Mod).

[47] ANSI/ISA—84.00.01—2004 Part 3 (IEC 61511-3 Mod), Functional Safety: Safety Instrumented Systems for the Process Industry Sector — Part 3: Guidance for the Determination of the Required Safety Integrity Levels — Informative.

[48] Hazard and Operability Studies (HAZOP Studies) – Application Guide, IEC 61882:2001, International Electrotechnical Commission, Geneva, Switzerland, 2001.

[49] P. Baybutt, Letter to the Editor, Managing process risks, Process Safety Progress, Volume 34, Issue 2, page 106, June 2015.

[50] P. Baybutt, Numerical Risk Tolerance Criteria in the United States: a Critique of the Risk Criterion Used for the New Jersey Toxic Catastrophe Prevention Act, Journal of Loss Prevention in the Process Industries 32 (2014), Volume 32, November 2014, Pages 428-435.

[51] P. Baybutt, The ALARP Principle in Process Safety, Process Safety Progress, Vol. 33, Issue 1, Pages: 36–40, March 2014.

[52] P. Baybutt, Allocation of risk tolerance criteria, Process Safety Progress, Volume 33, Issue 3, pages 227–230, September 2014.

[53] Guidelines for Process Safety Metrics, Wiley-AIChE, New York, NY, 2009.

[54] Guidelines for Integrating Management Systems and Metrics to Improve Process Safety Performance, Wiley-AIChE, New York, NY, 2016.

Accidents during Turnarounds, Cleanings, and other Infrequent Operations

Michael C. Stern
Exponent, Inc.
Natick, Massachusetts
mstern@exponent.com

Vijay Somandepalli,[1] Noah Budiansky,[1] Ali Reza,[2] Timothy J. Myers[1]
Exponent, Inc.
[1]Natick, Massachusetts
[2]Los Angeles, California

Prepared for Presentation at
American Institute of Chemical Engineers
2016 Spring Meeting
12th Global Congress on Process Safety
Houston, Texas
April 11-13, 2016

Accidents during Turnarounds, Cleanings, and other Infrequent Operations

Michael C. Stern
Exponent, Inc.
Natick, Massachusetts
mstern@exponent.com

Vijay Somandepalli, Noah Budiansky, Ali Reza, Timothy J. Myers
Exponent, Inc.

Keywords: Process Safety, Turnarounds, Startup, Shutdown, Cleaning

Abstract

Infrequent operations, such as turnaround services and cleaning operations, pose unique hazards due to their lesser familiarity to operators and plant personnel. For example, cleaning chemicals and methods can have unforeseen reactions with scale material, the chemistry of which may not be well-understood. Turnaround services may create a false sense of security because the plant is not in operation, which eliminates many of the commonly encountered hazards. Turnaround services, however, can be hazardous, especially if proper precautions have not been taken when shutting down the plant. Avoidance of accidents during these infrequent operations requires similar hazard analyses to those for standard operations, but these may not be thoroughly performed if the operation is assumed safe. Multiple case studies will be presented on accidents that have occurred during infrequent operations and the lessons learned from those incidents will be discussed.

1 Introduction

Operations including startups, shutdowns, turnarounds, and cleanings can pose additional risk for facilities due to their infrequency and uniqueness. The infrequency of these types of operations provides less opportunities for operators to gain experience and understanding of the procedures. The uniqueness, such as replacing or modifying a part that has been relatively untouched for years, can lead to unintended changes in the process when operation resumes.

Variability is a common risk factor for cleaning and turnaround operations. The same cleaning procedure or chemicals that might have worked for a certain type of scale could be hazardous if used for to clean out a different contaminant. Turnarounds and cleanings may also involve the use of contractors who are less familiar with the facility and may require site specific training to safely operate.

In this paper, the heighted risks of different infrequent operations are discussed and relevant case studies are presented. The case studies, some previously published and some presented for the first time, span a wide range of industries from manufacturing facilities to chemical plants, but all experienced or observed failures during an infrequent and transient operations.

2 Cleaning Operations

Both scheduled and unscheduled cleaning operations are often required at chemical and manufacturing facilities to maintain the equipment functionality and minimize contaminants in the products. Cleaning with chemicals can pose hazards due to undesired chemical reactions between the scale material and the cleaner. In certain cases when multiple cleaners are used, the cleaners could react to form hazardous substances or gases. Table 1, shown below, provides a few examples of chemical mixtures that could form explosive, toxic, or flammable substances leading to hazardous conditions. As will be described in the Condea Vista case study below, even steam cleaning can lead to an explosion in a process vessel when contacted with an aluminum powder containing sludge.

Table 1. Potential reaction products and hazards for certain chemical mixtures

Chemical Type One	Chemical Type Two	Reaction Products	Hazards
Acetone	Peroxides	Acetone-Peroxides	Explosives
Salts with (Cl^-, F^-, CN^-) ions	Acids	Acid gases	Corrosive and toxic
Metal Powders	Water or Steam	Hydrogen	Flammable

When cleaning areas that are already hazardous locations, such as areas near equipment containing flammable liquids or combustible dusts, the cleaning process can be a potential ignition source if proper precautions are not taken. Routine maintenance or cleaning actions can also release combustible dust accumulations creating a hazardous environment in an area where ignition controls are not present, leading to a fire or explosion. The case studies of the Jahn Foundry and CTA Acoustics combustible dust explosions provide examples of how ignition of disturbed accumulations can lead to catastrophic explosions.

2.1 Condea Vista Chemical Plant Explosion[1]

On October 13, 1998, an explosion and fire occurred inside a 3000-gallon Hastelloy reactor (R-2) that was part of the linear alkyl benzene (LAB) process at the Condea Vista plant in Baltimore, Maryland. Process materials released from the various vessels and process piping by the explosion fueled a fire that took approximately 2 hours to extinguish.

The investigation revealed that prior to the explosion, the bottom of R-2 was plugged with approximately 180 gallons of a sludge-like mixture that likely consisted of a mixture of coarse aluminum powder, aluminum chloride and various hydrocarbons. The explosion occurred approximately 10 minutes after steam was introduced at the bottom of the reactor in an effort to break up the sludge accumulation.

Figure 1. Photographs showing the overview of the explosion (left) and top of the exploded R-2 reactor (right).

A series of small-scale chemical reactivity tests were performed to investigate the potential reactions between steam and manufactured sludge that was similar to the material accumulated inside the R-2 reactor. Rapid runaway chemical reactions were observed after steam was injected into this mixture, while pressures in excess of 700 psig and temperatures in excess of 400 F were recorded before the test vessels ruptured.

A review of possible chemical reactions between steam, aluminum and aluminum chloride confirms that the predominant reaction mechanism is the exothermic hydrolysis of aluminum chloride to produce HCl vapor. Significant amounts of hydrogen gas can also be generated as the aluminum reacts with the acid/steam mixture after it is stripped of its protective oxide coating. Exponent concluded that the root cause of this accident was direct and sustained contact between the steam and accumulated sludge at the bottom of the reactor.

2.2 Dust Explosion at CTA Acoustics[2,3]

At approximately 7:30 am on February 20, 2003, a fire in at the CTA Acoustics facility in Corbin, Kentucky, lead to multiple secondary explosions and fires resulting in the death of seven workers and 37 injuries.

CTA Acoustics manufactured thermal acoustic insulation. The insulation material consisted of fiberglass bound together by phenolic resin powder. The product was used in automotive applications including the insulation on the hood or floor board of a car. CTA had four fiberglass forming lines, 401, 402, 403 and 405. The CSB report on the incident stated that the explosion originated near line 405 where dust generated during cleaning was ignited by the oven on line 405.

The CSB investigation concluded that workers started a routine cleaning of the dust collection system on line 405 thirty minutes before the incident occurred. The cleaning was expected to take approximately one hour. The night shift had reported significant clogging on the line. During the cleaning, the operators observed that the transition leading to the baghouse was the source of the clog and proceeded to use compressed air to break up the clog. When the baghouse was restarted at approximately 7:30 am, an initial explosion occurred that spread to the rest of the facility through secondary explosions fueled by clouds of combustible dust that were dislodged during the initial explosion.

2.3 Dust Explosion at the Jahn Foundry[3]

Figure 2. Photographs showing dust deposits on exhaust hoods in Shell Mold department (left) and the structural damage to the foundry (right).

At approximately 3:04 PM, on February 25, 1999, a series of explosions and a subsequent fire caused extensive damage within the Shell Mold building at the Jahn Foundry Corp. in Springfield, MA. Three employees suffered fatal injuries and nine others were injured. The foundry produced gray iron cast products for various customers.

The casting process included the production of shell molds, into which the molten material was ultimately poured. The molds were produced from a 95/5 percent mixture of sand and phenolic resin, added as a thermosetting bonding agent to hold the mold together. Even though the resin was supplied in the form of a fine powder, a dust collection system was not installed in the Shell Mold building. However, an odor control system designed to remove volatile organic compounds emitted from the resin during heating of the shell molds was sufficiently powerful to entrain fine particulates. The odor control system consisted of hoods or canopies above each shell mold station that vented into duct manifolds that connected to a scrubber tower where volatile organic compounds were removed.

An engineering investigation of the incident concluded the explosions were initiated due to the ignition of resin dust dislodged from an overhead exhaust duct or inadvertently dispersed during routine maintenance. A subsequent deflagration (dust explosion) within the odor control system was fueled by excessive, resin rich deposits of dust within the exhaust ducts. A second, larger dust explosion in the upper half of the Shell Mold building caused the majority of the damage to the structure. This second explosion was fueled by resin rich dust deposited on various horizontal surfaces on the upper level of the Shell Mold building.

Personnel at Jahn Foundry were generally unaware of the specific dust explosion hazard associated with this product and the foundry did not comply with the requirements of NFPA 654, *Standard for the Prevention of Fire and Dust Explosions from the Manufacturing, Processing and Handling of Combustible Particulate Solids*

3 Startup and Shutdown Operations

Operator inexperience and unidentified operational differences can lead to additional hazards when starting up or shutting down a plant. For plants that only shut down for maintenance annually or less, the operators will have only limited experience with the tasks required for startup and shutdown due to the infrequency of the operation. As a result, lessons learned from previous experiences may be less likely to be remembered, recorded, and passed on to future operators.

Differences that occur in the process during startup can be difficult to predict and assess. In systems with dust generation, for example, the properties of the collected dust during startup can vary from the dust generated during steady-state operations. It can be difficult to quantify these differences because process sampling during startup is complicated by the inherent transience of the operation.

In processes that include recycle loops or process material recycled downstream, the feedstock during startup can vary in considerable but unidentified ways. Small process changes, equipment maintenance and replacement, and other alterations to the system can introduce small but significant changes into the performance of the system that could lead to equipment failures if the changes are not anticipated. The case of the ammoxidation reactor incident described below provides an example of a chemical reactor that was damaged during startup due to the changes that occurred while the reactor was offline.

3.1 Ammoxidation reactor failure during startup

Several startup attempts of an ammoxidation reactor led to unplanned shutdowns within hours of starting due to over-temperature alarms. The ammoxidation reactor was a packed-bed counter-current tube-in-shell reactor where the tubes were packed with catalyst and an organic heat transfer fluid was pumped through the shell to cool the tubes and remove heat from the exothermic reaction. After a two-day period involving several attempted startups that were followed with subsequent shutdowns due to over-temperature conditions, the reactor was taken offline. Inspection of the reactor revealed a significant leak of heat transfer fluid and severe localized damage to the top tubesheet. The reactor was repaired by replacement of tubesheet sections and a number of tubes.

The reactor was brought back online and was used to produce a different isomer and performed without issue. After two months, the reactor was switched back to producing the original isomer. After two unsuccessful startup attempts that led to shutdowns due to over-temperature conditions, the reactor was inspected and thermal damage was observed.

Investigation of the reactor incidents revealed that prior to the startup, new catalyst was installed to replace the old catalyst that had been in the reactor. The catalyst packing gradient had also been adjusted such that higher concentrations of catalyst were present near the beginning of the reactor tubes. The new catalyst packing gradient and the increased activity of the fresh catalyst both contributed to increased reaction rates within the reactor during operation. It was also discovered that the heat transfer fluid reacted exothermically when exposed to conditions inside the reactor tubes.

It was determined that the failure of the reactor was a direct result of thermal expansion of tubes that occurred as feed rates were increased in the reactor and over-temperature conditions occurred. The thermal expansion led to the mechanical failure of a tube-to-tubesheet connection, which allowed heat transfer fluid to enter the tube side of the reactor. Exothermic reaction of the heat transfer fluid in the tube side of the reactor during repeated startups of the reactor led to the melting and heat damage of the tubesheet.

4 Plant Turnarounds

Plant turnarounds that involve significant maintenance, cleaning, replacement of equipment, and plant modifications can pose additional hazards due to the uniqueness and unfamiliarity of the activities, the involvement of contractors who are not as intimately aware of the hazards at the plant, and the difficulty of entirely purging the process lines and equipment of hazardous chemicals. Corrosion or contaminant buildup that occurred during the operation of the process can alter the condition of the process equipment in ways unbeknownst to the operators until it is inspected during the turnaround. The case study presented below provides an example of equipment that became corroded during operation but was detected before it created hazardous conditions.

4.1 Corrosion in a Flue Gas Desulfurization Absorber

All equipment is subject to both periodic planned shutdowns to conduct routine maintenance and inspections and unplanned shutdowns to replace components or make repairs. During these shutdowns careful inspections of the equipment are important to prevent accumulating degradation of a system or fix other unforeseen issues.

During a routine shutdown to perform periodic cleanout of a flue gas desulfurization (FGD) absorber associated with an incineration plant, corrosion damage was observed in the corrosion resistant liner of the absorber. FGD systems are designed to remove sulfur dioxide and other aggressive gaseous species from power plant and incineration emissions by spraying the flue gas with a limestone slurry, producing a calcium sulfite (gypsum) or other reaction species solid that falls to the bottom of the tower. The FGD slurry environments can have pH values as low as 1 and chloride concentrations as high as 200,000 ppm. The resulting slurry can be very corrosive and highly alloyed metals (or non-metallic coatings) are required to prevent corrosion of the liner to prevent exposure of the underlying carbon steel support structure.

In this case, 2205 duplex stainless was selected as the absorber liner was selected to prevent corrosion based on the nominal operating conditions, but corrosion of the duplex stainless steel liner was observed around a cleanout door. The corrosion damage had penetrated through the stainless steel liner in the time frame since the last inspection and caused minor damage to the underlying structure. A corrosion failure analysis was conducted and it was determined that the isolated environment in the gap between the cleanout access door and the door sill was more aggressive than the nominal conditions in the absorber. Laboratory corrosion testing indicated that corrosion resistance of the stainless steel liner, that was highly resistant to corrosion under the nominal design conditions, was significantly reduced in this area due to accumulated ions resulting in a reduced pH. In this case, periodic inspection of the equipment resulting in detection of corrosion degradation that was ultimately repaired prevented more serious failures from occurring.

5 Discussion and Conclusions

A review many of the industrial failures that have been investigated by Exponent shows a significant number that occurred during infrequent and transient operations including startups, shutdowns, cleanings (scheduled and unscheduled), and chemical transfers. While the specific incidents vary substantially in their magnitude and cause, an unknown change is the system can be traced to the cause. In three cases it was accumulation (combustible dust or aluminum sludge) and in another cases it was a change in a process variable (catalyst packing) that occurred during a turnaround that caused a failure upon startup.

Preparation for these upsets require thorough process hazard analyses (PHAs) and an effective management of change (MoC) program to identify these key variables before they become an issue and develop strategies for operators to employ when these events occur. The case of the FGD unit highlights the importance of careful inspection during turnarounds to identify corrosion or other changes in equipment. In some cases, such as combustible dust incidents, significant industry and governmental focus has raised awareness of the hazards of combustible dust. An OSHA National Emphasis Program and updated NFPA standards have been developed to address these hazards. In other cases, the lessons learned from the incidents should be used in future PHAs to mitigate the likelihood of similar incidents in the future.

6 References

[1] A. Reza, A. Kemal and P. Markey. "Runaway Reactions in the Aluminum, Aluminum Chloride, HCl and Steam System. An Investigation of the 1998 CONDEA Vista Explosion in Maryland," 36th Annual Loss Prevention Symposium. AICHE Spring National Meeting, March 2002.

[2] CSB. *Combustible Dust Fire and Explosions: CTA Acoustics*. United States Chemical Safety and Hazard Investigation Board, Report No. 2003-09-I-KY, February, 2005.

[3] T. Myers and A. Ibarreta. "Investigation of the Jahn Foundry and CTA Acoustics dust explosions: similarities and differences," *Journal of Loss Prevention*, pp. 740-745, 2009.

12TH GLOBAL CONGRESS ON PROCESS SAFETY

Biomass Industry Explosion Case Study

Amy E. Theis, P.E.
Fauske & Associates, LLC
16W070 83rd Street, Burr Ridge, IL USA 60524-5802
Theis@Fauske.com

Timothy L. Cullina, M.S., P.E. and Kenneth N. Kurko
Fauske & Associates, LLC
16W070 83rd Street, Burr Ridge, IL USA 60524-5802
Cullina@Fauske.com and Kurko@Fauske.com

Prepared for Presentation at
American Institute of Chemical Engineers
2016 Spring Meeting
12th Global Congress on Process Safety
Houston, Texas
April 11-23, 2016

Biomass Industry Explosion Case Study

Amy E. Theis, P.E.
Fauske & Associates, LLC
16W070 83rd Street, Burr Ridge, IL USA 60524-5802
Theis@Fauske.com

Timothy L. Cullina, M.S., P.E. and Kenneth N. Kurko
Fauske & Associates, LLC

Keywords: Explosion, thermal stability, combustible dust, biomass, incident, dust hazard analysis (DHA), management of change (MOC)

Abstract

The focus of this case study is an explosion that occurred in a biomass processing facility. There were three pieces of equipment involved including a dryer, conveyor and a cyclone. Thankfully, there were no injuries or fatalities due to this incident. This event occurred at a biomass company, resulting in damaged equipment with one piece needing to be completely replaced. Other equipment was able to be repaired and returned to service.

Material handled in the process had been observed to smolder on prior occasions, but the cause of these near miss incidents was not investigated or identified. The material was a potential combustible dust and there was also the possibility for flammable gas to be generated due to decomposition. Through material characterization, several hazards were identified. Combustible dust and calorimetry testing were employed to evaluate the thermal stability, reactivity and combustibility of the material. In addition, off-gases were evaluated to determine if flammable gas was generated. This case study will outline the test design strategy to obtain key process safety information, identify the multiple hazards based on the test results, review existing explosion safeguards and determine a postulated series of events that could have led to the incident. Lessons learned from the incident will be reviewed including recommended risk reduction strategies that could be used to prevent future events. One example is the importance of obtaining appropriate process safety information to facilitate data driven decisions to be made regarding safe operating conditions.

1 Introduction

An explosion occurred at a biomass processing facility. Fortunately no one was injured or killed in this incident. It is believed that a fire, smoldering or other ignition source began in the conveyor. Due to a lack of explosion protection, this led to a subsequent explosion in a connected piece of equipment. Findings indicate that the source of the explosion was consistent with a combustible dust deflagration that appeared to have ignited in or near the transfer equipment and accelerated through the duct work to a cyclone.

Lessons learned presented in this paper are based on experimental test results provided by the customer as well as testing performed by Fauske & Associates, LLC after the incident. Deficiencies that yield opportunities for improvement at this facility include a lack of hazards identification for the process, lack of combustible dust management systems and lack of a risk assessment.

2 Background

2.1 Process Description

In the process, the biomass is fed to a dryer that heats the material to elevated temperatures. The biomass has some residence time in the dryer before it is transferred through the bottom to a conveyor. Forced air is fed to the dryer and then goes to a cyclone where the entrained biomass is collected downstream. See Figure 1 for a process flow diagram.

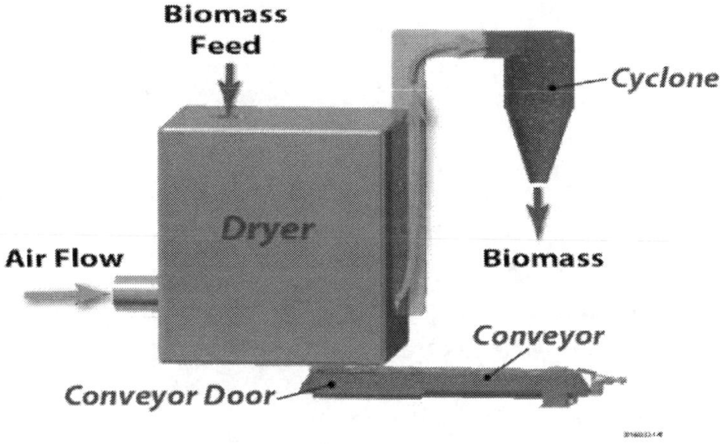

Figure 1 Biomass Process Flow Diagram

2.2 Explosion Event Description

On the day of the explosion event, a fire occurred earlier during the process. It is postulated that heated material accumulated in the conveyor, which led to self-decomposition at elevated temperatures, ultimately resulting in a fire. Employees used a water spray to suppress the flames, but perhaps this did not fully extinguish embers.

When sufficient oxygen was available, hot embers from the transfer equipment could have been drawn into the duct work and provided an ignition source for airborne biomass particulate solids. After ignition of dust, gases, or both, the rapidly expanding pressure wave and flame front moved into all of the available space in the first few milliseconds of the event.

The pressure wave blew the conveyor access door open and ejected burnt material. The flame front propagated to the dryer, fueled by airborne biomass then proceeded to the cyclone, ultimately leading to a rupture along the top of the cyclone. No personnel observed either explosive discharge. All witnesses reported hearing only one explosion. The access door opening and the cyclone failure could have occurred within less than 100 milliseconds of each other, producing a noise identifiable to witnesses as a single explosion. Pressure effects were observed downstream of the cyclone as well.

2.3 Previous Incidents

Based on history from processing similar biomass, personnel believed that smoking and fires, while not desirable, were typical and expected during the process. Fires had not only occurred during processing, but also during transportation of a by-product generated during the process.

In addition, biomass color and odor were not consistent during process runs. Operators believed there was a correlation between initial biomass color, intermediate color/odor, smoking events and fires. A lack of familiarity with combustible dust and gas hazards, especially the potential for deflagrations and explosions by both the management and operators contributed to the incident.

3 Hazard Evaluation

3.1 Material Properties

3.1.1 Biomass #1

Before the explosion, limited combustible dust and calorimetric data were available for this type of biomass. Previous combustible dust test results of a related biomass (Biomass #1) include a minimum explosive concentration (MEC) well above concentrations expected during the normal process conditions and a minimum ignition energy (MIE) value greater than 500 mJ. These results indicate that Biomass #1 was a combustible dust, but was not sensitive to ignition by electrostatic discharge.

Differential Scanning Calorimeter (DSC) testing of Biomass #1 identified one significant exotherm. This exotherm was detected above the normal operating temperature of the process and was confirmed by Carius tube testing. The Carius tube testing also indicated that non-condensable gas was formed during the event.

3.1.2 Biomass #2

After the incident, testing was performed on a sample obtained from biomass being processed at the time of the event (Biomass #2). First, DSC testing was conducted on Biomass #2 per ASTM E 537-12, *Standard Test Method for the Thermal Stability of Chemicals by Differential Scanning Calorimetry* [1] to determine the endothermic and exothermic activity of the material. A Differential Scanning Calorimeter was utilized for this test in conjunction with a 20 μl gold-plated stainless steel high pressure crucible. The DSC cell was heated at a preset rate while the differential heat flow, as a function of temperature, was measured and recorded. Five (5) DSC experiments were run on the sample at different heating rates: 1.0, 2.0, 4.0, 8.0, and 16.0°C/min.

The lowest heating rate DSC test showed that exothermic activity was detected at temperatures just less than the maximum temperature of the normal operating range for the process. Test results indicated that Biomass #2 yielded exothermic activity at lower temperatures compared to Biomass #1. The DSC results are summarized in Figure 2.

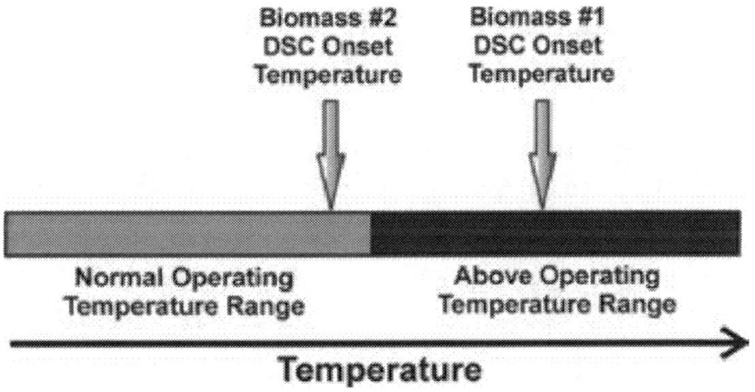

Figure 2 Thermal stability summary for biomass materials

The DSC data were evaluated to determine kinetic parameters for Biomass #2. These results showed that the material exhibited a self-accelerating decomposition temperature (SADT) value according to the United Nations Department of Transportation (UN DOT) H-1 test definition [2] that was within the normal operating temperature range. Similar evaluations were performed for smaller quantities of material (10 kg compared to 50 kg) and were also within the normal operating temperature range of the process.

Kinetic parameters were also used to determine adiabatic times to maximum rate for Biomass #2 as a function of temperature. These results indicated that the temperature that would yield a 24 hour adiabatic time to maximum rate would be in the middle of the normal operating temperature range of the process.

The next test performed on Biomass #2 was conducted using an Accelerating Rate Calorimeter (ARC), which is a high thermal inertia adiabatic calorimeter that is used to obtain data regarding the relationships between time, temperature, and pressure for exothermic reactions. The ARC utilized a 10 ml spherical test cell constructed of stainless steel with ¼" neck. In this experiment, the ARC was run in isothermal mode in order to monitor pressure generation. The instrument was programmed to track a reaction if exothermic activity was detected or maintain the desired temperature. ARC testing was conducted per ASTM E 1981-98, *"Standard Guide for Assessing Thermal Stability of Materials by Methods of Accelerating Rate Calorimetry"* [3].

The ARC isothermal test was performed on Biomass #2 at the maximum operating temperature range of the process over a time period of approximately 6 days. The temperature of the ARC test remained constant for the duration of the experiment. The evidence of non-condensable gas generated indicates that a reaction did occur. After the sample cooled, the non-condensable gas was subjected to a flame. No changes in the flame were observed, indicating that flammable gases were not present.

The last test performed on Biomass #2 was to determine the Minimum Explosible Concentration (MEC) using a 20-L Siwek chamber. The MEC is the minimum concentration of a combustible dust that is capable of propagating a deflagration through a uniform mixture of the dust in air, under the specified conditions of the test. Data obtained from this MEC test method can provide a relative measure of the concentration of a dust in the form of a uniform cloud necessary for an explosion. The test was performed per ASTM E1515, *Standard Test Method for Minimum Explosible Concentration of Combustible Dusts* [4].

Results from the MEC test on Biomass #2 showed that the MEC measured was within the range of what could be expected to occur within the process. If this concentration was present in the process, the material could be readily ignited by a burning/smoldering ember. A comparison of MEC values are shown in Figure 3.

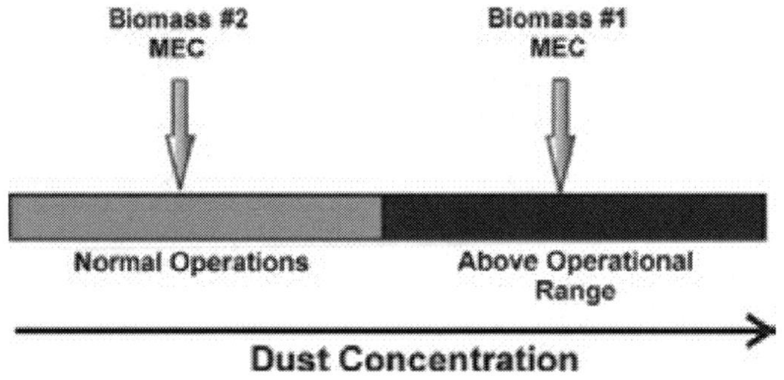

Figure 3 Minimum explosible concentrations summary for biomass materials

3.2 Incident Evaluation

It is postulated that heated material accumulated in the conveyor, which led to self-decomposition at elevated temperatures, ultimately resulting in a fire. Though fire-fighting measured were taken, it is possible that remaining material in the conveyor continued to heat or smolder. This could provide sufficient heat or an ember to serve as an ignition source for airborne particulate solids.

The combustible dust hazards of Biomass #2 handled in an existing process were not adequately identified or understood. It followed that safeguards were not in place to manage the risks associated with each piece of equipment in the process because the hazards and consequences were not characterized or evaluated. Since the hazards were not characterized at the time of the incident, employees were not made aware of the hazards associated with processing the material.

There was a concern whether flammable gas was present that could have created a hybrid mixture in addition to the combustible dust in the system. The presence of a small quantity of volatiles well below the lower flammability limit (LFL) of the gas can create a hybrid mixture with combustible dust that effectively reduces the minimum ignition energy and the minimum explosive concentration required to initiate a deflagration. Testing indicated that non-condensable gas formed during the isothermal experiment was not flammable. The gas tested may not be representative of gases formed during the deflagration event or even after the material was burned, so it is not possible to conclude whether flammable gases were present during the deflagration event. Additional testing is recommended to further evaluate this concern.

The lack of explosion prevention or mitigation devices on the equipment and absence of appropriate training are both a result of this inadequate hazard recognition. Installation of a properly designed explosion vent and installation of isolation technology could have mitigated damage to the cyclone. Figure 4 illustrates the pressure increase in a vessel both with and without the application of explosion mitigation by venting.

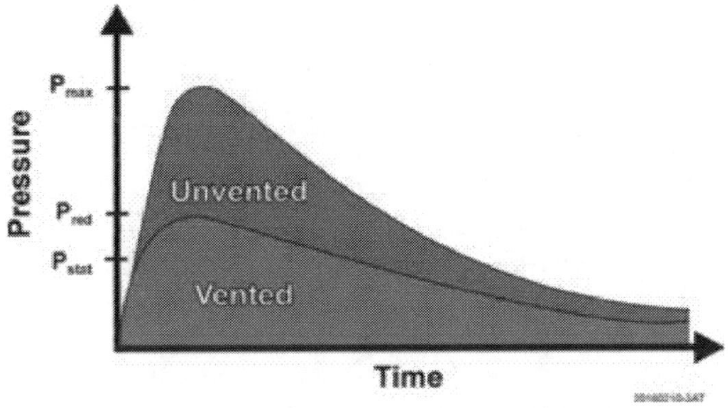

Figure 4 Progress of Explosion With and Without Explosion Protection

There was no record of a dust hazard analysis being performed for this process. Many of the required elements of a comprehensive combustible dust management program as outlined in National Fire Protection Association (NFPA) standards were not utilized to reduce the risk of dust fires and explosions.

3.3 Lessons Learned

This process included equipment covered by NFPA 654, *Standard for the Prevention of Fire and Dust Explosions from the Manufacturing, Processing, and Handling of Combustible Particulate Solids*. While processing Biomass #1, testing was performed to characterize combustible dust hazards and thermal stability for the material. Results indicated that the material was thermally stable in the normal operating temperature range and MEC values were above what would be expected during normal operations. At that time, no further protective measures were implemented. Data was gathered, but a risk basis of safety was not formally established.

When the raw material for the process changed from Biomass #1 to Biomass #2, this modification did not trigger additional testing to be performed. A management of change process, as required in NFPA 654, would have triggered the need to characterize Biomass #2 since it was different from Biomass #1. Even though the process equipment and technology remained the same, the facility changed raw materials which should have triggered Biomass #2 to be evaluated for combustible dust explosibility characteristics. This evaluation should have included thermal stability testing which were also performed previously for Biomass #1.

Fires were not uncommon in this process when processing Biomass #2 and the cause of these fires was not evaluated. It did not appear that an incident investigation process was in place at this facility, as required per NFPA 654 or the most recent NFPA 652, *Standard on the Fundamentals of Combustible Dust*. It appeared that management had allowed an acceptance of deviation to become the norm since fires were expected in the process. The response was for the operator to extinguish the fire and continue processing.

It was not evident if an inspection, cleaning and maintenance schedule was identified for pieces of equipment covered by NFPA 654 and 652. Material was allowed to accumulate in the conveyor which could have self-heated and led to fires observed in this equipment. It is not clear that inspection and access panels were in place to allow adequate cleaning of ductwork and the conveyor. It would be appropriate to establish and review cleaning and maintenance schedules per NFPA 654 and 652 requirements.

A dust hazard analysis (DHA) was not performed on this process. Given the operating temperatures for the dryer, the material should additionally be evaluated for thermal stability. This process should have triggered the need for testing to characterize the material for combustible dust hazards as well as for thermal stability. Once the hazards were identified, the equipment could then be evaluated to determine if additional protective strategies such as explosion protection and isolation were necessary.

After hazards have been identified, employees should be trained and provided guidance on required work practices in order to ensure their safety. Insufficient training contributed to the near miss of exposing employees to the deflagration. Without knowledge of the deflagration potential and the likely effects, employees were preparing to position themselves in high risk areas during the unintended explosion vent.

Data reviews after the incident revealed that employees did not understand the full intent of information recorded during the process. Design of the data forms likely contributed to the uncertainty regarding data collected. Employees did not have an awareness regarding the critical safety parameters and were not adequately trained on how to respond to deviations in the process.

Lessons learned from this case are summarized as follows:

- There was a lack of formal risk based safety documentation for the process.
- A management of change (MOC) process did not trigger data to be obtained when a change in the raw materials for the process were made.
- Sufficient data for the material/process were not obtained to establish safe operating conditions and identify potential upset scenarios.
- Performing a DHA could have identified the hazards, potential ignition sources, potential consequences and provided a framework to evaluate the adequacy of safeguards or mitigation strategies to determine if additional protection was needed.
- Proper hazard identification for the process is critical to ensure that adequate safeguards are in place and that employees receive adequate and appropriate training regarding normal operations and emergency response procedures.

- Routine fires during processing and transportation should serve as warning signs that trigger an investigation to identify preventive and corrective actions.
- The inspection, maintenance and cleaning procedures for dust handling equipment should be consistent with NFPA 654 and 652 requirements.

4 Conclusion

Testing performed after the incident indicated that the new material (Biomass #2) handled in the process was thermally unstable during normal operating conditions as well as being a combustible dust hazard.

In summarizing the lessons learned, it appears that several combustible dust management systems required by NFPA 654 and 652 did not exist, or if they existed, were not effective to prevent this incident. Key process safety information was not obtained to identify the hazards of the process. The company did not understand the warning signs for a potential combustible dust explosion in this process. Performing a DHA on this process could have identified the need to obtain relevant combustible dust and thermal stability data and understand that normal operating conditions could lead to an upset scenario. This information could then be used to propose changes to the process and determine if existing systems and equipment were adequately designed to protect against the consequences of the identified upset scenarios.

The following steps are recommended for identifying hazards associated with a process that contains equipment that could potentially handle combustible dust. These steps are consistent with the practices recommended in NFPA 652.

- Identify equipment in the process that is intended to handle particulate solids and the desired processing conditions (temperature and pressure operating range, equipment design requirements)
- Identify material to be handled in the particulate handling equipment
- Identify the hazards (safety data sheets, combustible dust hazard data, thermal stability data as warranted by process conditions, identify potential consequences due to upset conditions, perform a DHA)
- Understand the consequences in terms of severity and likelihood (determine if flash fire and/or explosion scenarios exist, determine if normal operating conditions or process deviations could lead to thermal decomposition, fire and/or explosion scenarios, identify if flammable gas could be generated)
- Determine process safety controls, safeguards and mitigation strategies (protect against overpressure situations, control and alarm set points, change in process design or procedures, explosion protection systems)
- Implement and maintain the safety measures (document safe operating ranges, perform preventive maintenance and inspections, train appropriate personnel, implement emergency response plans, employ other procedures as identified in NFPA 652)

5 References

[1] ASTM Standard E 537-12. *Standard Test Method for the Thermal Stability of Chemicals by Differential Scanning Calorimetry*, ASTM International, West Conshohocken, PA, 2012.

[2] United Nations. Recommendations on the Transport of Dangerous Goods: Manual of Tests and Criteria (6th ed.), New York and Geneva, 2015.

[3] ASTM Standard E 1981-98 e2. *Standard Guide for Assessing Thermal Stability of Materials by Methods of Accelerating Rate Calorimetry*, ASTM International, West Conshohocken, PA, 2012.

[4] ASTM E1515-14. *Standard Test Method for Minimum Explosible Concentration of Combustible Dusts*, ASTM International, West Conshohocken, PA, 2014.

[5] NFPA. NFPA 652, *Standard on the Fundamentals of Combustible Dust, 2016 Edition.* NFPA, Quincy, MA, September 2015.

[6] NFPA. NFPA 654, *Standard for the Prevention of Fire and Dust Explosions from the Manufacturing, Processing, and Handling of Combustible Particulate Solids, 2013 Edition*, Quincy, MA, June 2012.

[7] Ebadat, V. Testing to Assess Explosion Characteristics of Dust Clouds, NFPA Symposium on Dust Explosion Hazard Recognition and Control, 2009.

[8] U.S. Chemical Safety and Hazard Investigation Board. Investigation Report: Combustible Dust Hazard Study. Rep. No. 2006-H-1, 2009.

[9] Grossel, S.S. and R.G. Zalosh. Guidelines for Safe Handling of Powders and Bulk Solids. Center for Chemical Process Safety, AIChE, New York, 2005.

Atmospheric Tank Explosions: Mechanisms and an Unexpected Case Study

Michael S. Schmidt, P.E.
Principal, Bluefield Process Safety LLC
3101 Cherokee Street
Saint Louis, Missouri 63118, USA
bluefieldsafety@gmail.com

Prepared for Presentation at
American Institute of Chemical Engineers
2016 Spring Meeting
12th Global Congress on Process Safety
Houston, Texas
April 11-13, 2016

Atmospheric Tank Explosions: Mechanisms and an Unexpected Case Study

Michael S. Schmidt, P.E.
Principal, Bluefield Process Safety LLC
3101 Cherokee Street
Saint Louis, Missouri 63118, USA
bluefieldsafety@gmail.com

Keywords: storage tank, atmospheric tank, tank explosion, caustic

Abstract

Most of the literature on catastrophic failures of low pressure tanks focuses on the ignition of explosive mixtures in the vapor space of those tanks, and rightly so, because most catastrophic failures of atmospheric tanks involve explosions of such mixtures. Moreover, there is a general sense in the industry that other than explosions of flammable mixtures, the only other hazard associated with low pressure tanks is the occasional dramatic implosion of a vessel that is inadequately protected against vacuum.

In fact, there are a number of mechanisms that can lead to the catastrophic failure of a low pressure tank that have nothing to do with combustion or unprotected vacuum. Under certain circumstances, even a tank equipped with an atmospheric vent and containing nothing other than salt water can explode, with disastrous impacts.

This paper reviews the mechanisms for catastrophic failure of low pressure tanks, both implosion and explosion, and serves as a reminder for experienced process safety practitioners and as a tutorial for new process safety practitioners of what to look for during a hazard review. It also includes a case study of an atmospheric caustic tank explosion that resulted from a previously unreported mechanism that involved neither combustion nor vacuum.

1 Introduction

It is probably fair to say that atmospheric tanks are not very high on most people's list of hazards to worry about. Usually, they just sit there, storing liquids, doing what they are designed to do. Issues are typically the result of overfilling the tank and, on occasion, leaking components. Nothing sudden and catastrophic. When the words "sudden and catastrophic" are associated with atmospheric tanks, it is usually in relation to the flammable hazards of the liquids being stored. Chang and Lin [1] conducted a study of 242 storage tank incidents and found that 145 of them involved fire and another 61 involved explosion, mostly of flammable vapors from the liquids being stored. Outside of flammability issues, sudden and catastrophic incidents associated with atmospheric tanks are not what we typically worry about. Setting flammability issues aside, which have been addressed thoroughly elsewhere [2], [3], [4], [5] there are three types of atmospheric tank failures that can be sudden and catastrophic: collapses, implosions, and explosions.

Before continuing, however, it is important to first establish what is meant by the term, "atmospheric tank." The ASME boiler and pressure vessel code [6] divides vessels into two categories: pressure vessels, which operate at pressures greater than 15 psig and low pressure vessels, which operate at 15 psig or less. The BPVC applies only to pressure vessels. There are two API standards that apply to low pressure vessels. API 620 [7] applies to tanks that operate between 2½ psig and 15 psig, and at temperatures not greater than 250°F. API 650 [8] applies to tanks that operate at pressures up to 2½ psig, and at temperatures not greater than 200°F. OSHA defines an atmospheric tank as "a storage tank which has been designed to operate at pressures from atmospheric through 0.5 psig" [9].

All atmospheric tanks are low pressure tanks, but not all low pressure tanks are atmospheric tanks. There is not a consensus as to the limit—API 620 uses 2 ½ psig, OSHA uses 0.5 psig, others [10], [11] use a limit somewhere in between. For the purposes of this paper, an atmospheric tank is not intended to operate under vacuum and is not intended to operate at greater than 1 psig, or 30 inch water column. Human lungs have a greater operating pressure range.

2 Catastrophic Atmospheric Tank Failure

2.1 Collapse

One of the most famous catastrophic tank failures of the industrial era is the Great Molasses Flood of 1919 in Boston, Massachusetts [12]. The 50-foot tall, 90-foot diameter cast iron tank, built in a rush during World War I, ruptured and the wave of molasses killed 21 people. There was no fire. There was no explosion. The tank collapsed because it was under-designed for its application.

A more recent example is the failure of a liquid fertilizer storage tank at the Allied Terminals facility in Chesapeake, Virginia in 2008 [13]. The tank, originally designed for service with petroleum products, ruptured while being filled with a 32% solution of urea-ammonium nitrate (UAN). Two welders who were working on the tank as it was being filled for the first time after weld repairs were made were seriously injured when the tank collapsed.

2.1.1 Mechanisms for catastrophic failure

Atmospheric tanks fail for three reasons:

- The design or construction of the tank was inadequate.
- The tank is converted to a service that exceeds the capabilities of the initial design.
- The tank deteriorates over time.

Inadequate design or construction. While an atmospheric tank does not need to withstand high pressures, it does need to withstand the hydrostatic head of the liquid being stored. Atmospheric tanks collapse because the material of construction is not strong enough for the service, as was the case in the Great Molasses Flood, where the tank walls were 50% too thin, even for the relaxed standards of the 1910s [12]. Atmospheric tanks also collapse because the welds holding the tank together are inadequate. The Chemical Safety Board's report on their investigation into the Allied Terminals tank collapse includes an appendix of similar incidents, in which the listed cause of the incidents is "faulty welds" in more than 90% of the incidents where a cause is known [13]. Because of these issues, the CSB report notes that "tanks are more likely to fail when being filled to the maximum level for the first time."

Converted to a new service. Atmospheric storage tanks are designed for a specific service. Many tanks are designed to store petroleum products, for which a typical specific gravity is around 0.8. Simply converting the use from hydrocarbons to water increases the hydrostatic pressure on a tank by 25% when filled to the same level. Converting the use from hydrocarbons to 32% UAN, which has a specific gravity of 1.33, increases the hydrostatic pressure on a tank by two thirds. Converting the use from hydrocarbons to 50% caustic, which has a specific gravity of more than 1.5, almost doubles the hydrostatic pressure on a tank. Converting the service of atmospheric storage tanks to higher density liquids can lead to catastrophic tank failure by exceeding the design strength of the tank. Likewise, increasing the "safe fill level" has a similar impact on hydrostatic pressure, which can also lead to catastrophic tank failure.

Deterioration with time. Lastly, atmospheric storage tanks, like all equipment, deteriorate with time. Walls thin due to corrosion and erosion. Ideally, the thinning is uniform, steady, and slow. Catching this kind of deterioration during a tank inspection that includes thickness testing is straightforward. A tank with an adequate corrosion allowance will give lots of warning before the condition of the tank warrants a change in the operating conditions.

But not all deterioration is uniform. Tanks where nozzles, dip pipes, or eductors are aimed at the wall or floor of the tank are subject to localized thinning. Even a stream running down the interior wall of a tank can cause localized thinning. This weak spot then become the nucleus of a catastrophic failure. Stresses concentrate at a weak spot, the weak spot fails, and the stresses are immediately propagated through the rest of the material. The tank unzips and can collapse suddenly and catastrophically, often leading bystanders to believe there has been an explosion.

2.1.2 Protections against catastrophic failure

The causes of catastrophic tank failures also give a strong hint as to the protections that might be used against them.

Inadequate design or construction. This is best addressed with good equipment specifications according to recognized and generally accepted good engineer practice (RAGAGEP) and appropriate pre-startup safety reviews (PSSRs) to assure that the tank, as designed, constructed, and installed, meets those specifications. During commissioning, when many vulnerabilities in the tank are revealed, personnel should be kept out of harm's way, based on the assumption that the tank will fail.

Converted to a new service. Converting to a new service is a clear case for management of change, and the MOC review should consider changes in level, density, and corrosion rates, both general and local.

Deterioration with time. All equipment deteriorates, which is why a mechanical integrity program that adheres to RAGAGEP is essential. Tank inspections should test for thickness in enough spots to assure that general thinning is detected. Interior inspections should also look for specific evidence of localized thinning, especially at points where nozzles, dip pipes, or eductors are directed at the walls or floor of the tank.

2.2 Implosion

Atmospheric tanks are not designed to withstand vacuum, and they do not have to be exposed to very much vacuum to implode suddenly and catastrophically. The emptier the tank is when exposed to vacuum, the more dramatic the implosion. Tank implosions make such dramatic film that there are countless YouTube videos of tank implosions.

The lore that has been told and retold is of the maintenance team that was steaming out an atmospheric tank to prepare it for vessel entry, but did not finish before the end of the day. Rather than conduct the vessel entry at night, they closed up the tank and left. During the night, a storm front moved in, showering the tank with cold rain. The steam condensed, creating a vacuum, and when the crew returned in the morning they found the tank crumpled like an empty beer can. The crew from *Mythbusters* set out to prove or disprove this scenario using a DOT 111 tank car and decided that this story was "busted" [14]. *Mythbusters* notwithstanding, atmospheric storage tanks are not DOT 111 tank cars and the scenario should be considered credible.

Unlike explosions, implosions do not produce shock waves or flying shrapnel. Breaches may occur at the creases that form, but generally any loss of containment is not significant. The primary hazard associated with implosions are the potential injuries to personnel in the immediate area. The Center for Disease Control and Prevention reports of a worker killed while repairing a tank when it imploded [15]. Similarly, a worker in Spokane, Washington was killed in 2011 when the sewage digester he has checking imploded [16].

2.2.1 Mechanisms for catastrophic failure

Atmospheric tanks implode when they experience an internal pressure lower than they are capable of withstanding, i.e. a vacuum. Source of vacuum include the deliberate vacuum created by vacuum pumps and steam vacuum jets, as well as the incidental vacuum created by the pump suction of other pumps during pump out. Even in the absence of a pump, a gravity drain to empty a tank and other forms of emptying a tank of its liquid contents will create a vacuum. Finally, as described above, phase changes, particularly condensing vapors will create a vacuum.

2.2.2 Protections against catastrophic failure

In the absence of a source of vacuum, protections are unnecessary. However, all atmospheric tanks have a source of vacuum, if for no other reason than they are emptied. Protection include vacuum breakers, pressure/vacuum conservation vents, nitrogen pads, and most commonly, atmospheric vents. As long as these are properly sized and designed and remain fully operational, they are sufficient to protect against catastrophic implosion.

2.3 Explosion

Most atmospheric tank explosions involve the storage of flammable liquids, where it is the explosion of the flammable material that is the concern and which is outside the scope of this paper. There are also physical explosions that can occur, resulting from overpressure in the tank for reasons other than sudden combustion.

2.3.1 Mechanisms for catastrophic failure

Atmospheric tanks explode when they experience an internal pressure higher than they are capable of withstanding, which is not much. Sources of this high pressure include the pressure of inlets to the tank and heating within the tank.

Inlet line pressure can be of either gases or liquids. When the tank is open to atmosphere, the pressure drop at the inlet is complete, and the pressure in the inlet is not transferred to the interior of the tank. However, when the tank is blocked off from the atmosphere, the pressure in the tank will approach the pressure of the inlet line. Gases will flow into the tank until the pressure in the tank matches the supply pressure. Liquids will flow into the tank, compressing the headspace, until the headspace pressure matches the deadhead pressure of the liquid line. A tank does not need to be liquid full to reach the deadhead pressure of the liquid line.

Heating a vessel can result from either external heating or the heat of reaction, planned or unplanned, within the vessel. In either case, the liquid will expand and the vapor pressure will increase. Liquid expansion is only a concern when the tank is blocked off from atmosphere and the tank is completely full of liquid. In this case, since liquids are incompressible fluids, any additional expansion will sharply increase the pressure. However, atmospheric tanks are not absolutely rigid, so the effect will be to cause the tank to bulge, rather than to cause it to burst.

Increased vapor pressure is the real concern. Vapor pressure increases as temperature increases, and is independent of head space. An atmospheric tank blocked in while containing any liquid will experience an increase in pressure when heated, and atmospheric tanks are not designed to contain much increase. The vapor pressure of water, for instance, will increase by 1 psi when heated from 72°F to 113°F; it will increase by 5 psi when heated from 80°F to 165°F.

2.3.2 Protections against catastrophic failure

The key to protecting against physical explosion of atmospheric tanks is to assure that any source of pressure is released from the vessel before the pressure becomes too high. The obvious means is an atmospheric vent. Other protections include pressure/vacuum conservation vents and venting entryways or "trashcan lid."

To be useful, however, it is imperative that these protections be properly sized and designed and then remain fully operational. A vent that is too small or blocked is no protection. An atmospheric vent may be directly to atmosphere, or pass through a seal pot. Many creatures—

birds, wasps, rodents—find the outlet of an atmospheric vent a perfect place to build a nest. Bird screens, when in place, will keep out birds and rodents, but not nest-building insects. Likewise, a seal pot that has frozen solid will not vent. Vent lines that contain monomer vapor, whether directly to atmosphere or through a device can become plugged with polymer or simply glued shut.

Pressure vessels can be protected with rupture discs and pressure relief valves, but these require set pressures that are too high to be of use in atmospheric tanks. Likewise, pressure vessels are frequently protected from inlet pressure with regulators on the inlet. Unfortunately, atmospheric tanks operate at such low pressures that to regulate inlet pressures down to the operating pressure of the tank would reduce flow to unacceptably low rates.

Weak-seam roof tanks are frequently used to protect against the failure of atmospheric tank walls in the event of overpressure. They are designed so that the roof-to-shell joint will fail before any of the seams in the shell or the floor of the tank fail, thus preventing a release of the contents. However, this approach must be used with care. Accessories, such as railing or stairs, can make the roof-to-shell joint stronger than intended, as can the workmanship of an enthusiastic welder. (Welders are rarely scolded for making their welds too strong.) In general, no credit should be given for a weak-seam roof on a tank smaller than 10 meters (30 feet) in diameter because of the concern that the joint will be too strong [17]. When a weak-seam roof is used, the roof itself should be restrained with some kind of tether or hinge. Otherwise, the roof, ripped away from its weak seam, will be launched as a projectile.

3 A Case Study

While process air was blowing through the transfer line from a caustic storage tank to a caustic day tank, the caustic day tank failed catastrophically, launching the top head of the tank approximately 125 feet, where it came down through the roof of a nearby process building and triggered an impromptu evacuation in which one employee fell and had to receive medical treatment.

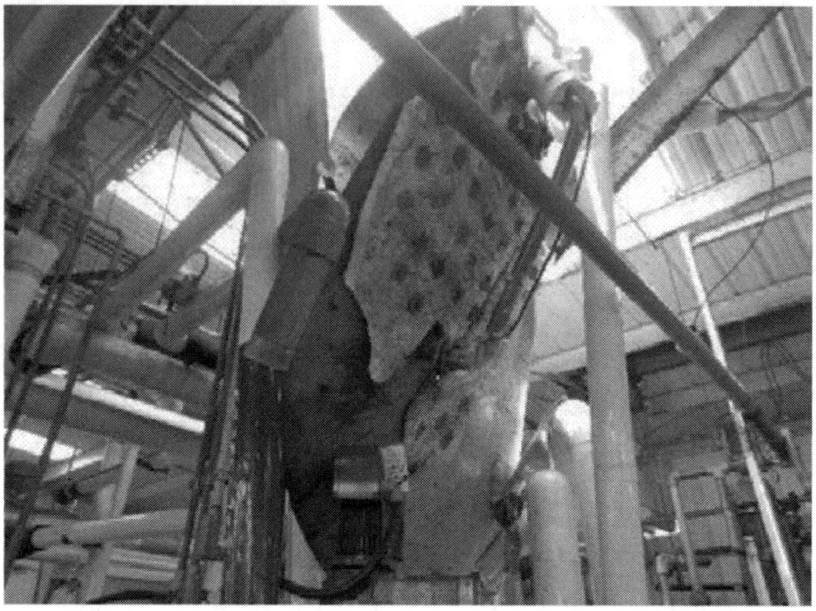

Figure 1. Caustic Day Tank head, after coming down through the roof of a nearby process building.

3.1 Process Background

The process consisted of a 3-inch transfer line constructed of 304 stainless steel that connected the discharge of a caustic transfer pump at the caustic storage tank to the top of a caustic day tank. A 50% solution of NaOH in water entered the caustic day tank through a dip pipe to the bottom of the tank. The transfer line was purged after each transfer of 50% caustic solution from the caustic storage tank to the caustic day tank by blowing out the line from the caustic storage tank to the caustic day tank for about 10 to 15 minutes with process air.

When the transfer line was being prepared for maintenance, there was a more extensive purge procedure. The normal procedure for purging the transfer line was to blow out the line with process air, blowing from the caustic storage tank to the caustic day tank for about 10 to 15 minutes. The next step was to flush the line with city water for 5 to 10 minutes, again flowing from the caustic storage tank to the caustic day tank. After the water flush, any residual water in the transfer line was blown out with process air, again blowing from the caustic storage tank to the caustic day tank for about 10 to 15 minutes.

The transfer line was routinely purged following transfers and was purged successfully for maintenance a month before the incident.

3.2 Incident Description

On a balmy, spring Friday morning, preparation began for maintenance on a leaking caustic transfer line from the caustic storage tank to the caustic day tank at about 9:30 am. The first step – purging the caustic from the transfer line with process air – was completed without incident.

The second step – purging residual caustic from the transfer line with city water – started at around 9:50 am. This step was also completed without incident, and at around 10 am the operator disconnected the water hose and left it to drain while he turned his attention to other matters in the unit.

At around 10:30 am, the operator returned to the caustic storage tank, disconnected the water hose, and reconnected the air line. He then opened the air supply valve in order to blow the water out of the transfer line.

At 11:10 am (the time shown on the time-stamp from the video monitoring the area), while the process air was still blowing through the transfer line from the caustic storage tank to the caustic day tank, the caustic day tank failed catastrophically. The pressure was sufficient to separate the conical head from the top of the tank and launch it approximately 125 feet, where it came down through the roof of a nearby process building to the east.

While the sides and bottom of the caustic day tank remained intact after the incident, the damage was enough to consider the caustic day tank a total loss. The piping attached to the top head of the caustic day tank was also damaged beyond repair. There were no fatalities or serious injuries associated with the incident, but one employee did require medical treatment for a sprained ankle following a fall that occurred during his evacuation from the immediate area of the incident.

3.3 Possible Issues

An explosion of a caustic tank can result from:

- Sources of pressure considerably higher than the design of the tank was capable of handling,
- Blockage in the equipment that lowered the pressure the tank was capable of handling, or
- Material failures in the tank that lowered the pressure the tank was capable of handling.

Guidance documents from suppliers of 50% caustic solution suggest the following as sources of high pressure, causes of blockage in caustic equipment, and causes of equipment material failure at pressures lower than design pressure [18], [19].

3.3.1 Possible sources of high pressure in a generic caustic tank

- Reaction of caustic solution with reactive metals (aluminum, magnesium, zinc, tin, chromium, brass, and bronzes made with zinc or tin), releasing hydrogen that forms a flammable mixture with air and subsequently ignites.
- Plant air or nitrogen used to transfer caustic solution by pressure.
- Steam used to clean out transfer lines.
- Direct steam heating of frozen caustic, exposing equipment to steam pressure.
- Pump deadhead pressure in unvented tank or line.
- Diluting strong caustic with water, rather than diluting strong caustic by adding it to water.
- Heating caustic solution while a crust of frozen caustic is formed over the top of the liquid

3.3.2 Possible causes of blockage in generic caustic equipment

- Block valves inadvertently closed.
- Lines plugged with frozen caustic solution.

3.3.3 Possible causes of generic equipment failure at lower than design pressure

- "Caustic embrittlement" – stress-corrosion failure which occurs at areas of mechanical stress or where fabricating operations produce high residual stress such as at bends in piping and at welds.
- Using iron and steel as materials of construction for equipment handling 50% caustic solutions at temperatures above 140°F. While iron and steel are the usual materials for equipment handling 50% caustic solutions below 140°F, nickel, nickel alloys, and stainless steel are required at higher temperatures and for higher concentrations.
- Using unsupported plastic pipe, which can fail because of the high density of 50% caustic solution, and using fiberglass reinforced plastic pipe except for specific conditions for which it is suited.
- Continued heating after caustic soda is melted, which may damage the special lining on the inside of railcars. The temperature of 50% caustic solution should never exceed 150°F (66°C) in lined cars.
- Wet or saturated steam above a maximum of 15 psig and 250°F. Steam at higher temperatures will damage rail car linings.

3.4 Conditions at Incident Scene

3.4.1 Noted from video

There were several dozen closed circuit video cameras positioned around the plant, three of which captured evidence of the catastrophic failure of the caustic day tank:

- The truck entrance camera, located about 600 feet southeast of the caustic day tank and aimed northwest.
- The tram gate camera, located about 275 feet southwest of the caustic day tank and aimed northeast.
- The main crosswalk camera, located about 175 feet west of the caustic day tank and aimed east.

The time stamps on the truck entrance camera and the tram gate camera coincide closely, and both showed the initial cloud release from the tank rupture at 11:10:19. The main crosswalk camera did not show the cloud release from the tank rupture, but shuddered at the time the cloud was released. Both the truck entrance camera and the main crosswalk camera captured an employee falling down and then getting up and walking away, hardhat in hand, at 14 seconds after the tank rupture.

Figure 2. Detail of screen capture from truck entrance camera at 11:10:19.647, showing launch of the caustic day tank top and associated cloud.

The main crosswalk camera, which was closest to the caustic day tank, shuddered for 0.4 sec, consistent with the passing of a shockwave. Neither of the more distant cameras showed the shudder. The more distant cameras, however, showed the initial cloud released. The cloud formed and was completely dispersed within 2 seconds. There was no evidence at all in those videos of combustion or of a secondary BLEVE. The shockwave confirms that there was an explosion. The absence of combustion or BLEVE and the quick dispersion of the cloud that was released confirm that the explosion resulted from simple overpressure, not from the ignition of a flammable mixture of hydrogen and air in the headspace or from the boil-up of the contents of the vessel.

The truck entrance camera was the only view that showed the launch of the top head. In the first 0.2 seconds, a cloud appeared above the west side of the tank. Then, 0.2 seconds after the initial release of the cloud, the conical head fully separated from the tank and began its 125-foot trajectory eastward, where it came down through the roof of a nearby process building. The video suggests that the separation began on the west side of the conical head and the break proceeded almost instantly around the tank head, propelling the conical head upward before the pressure inside the tank could be relieved.

3.4.2 Noted from field observations

The most relevant observation was that of the condition of the vent from the caustic day tank. A 3-inch gooseneck constructed from glued heavy-wall PVC pipe and elbows, the vent was attached to the top of the caustic day tank at a flanged nozzle 40 inches from the center peak of the conical head. The vent sheared off during the incident just above the flanged connection.

The internal surface of the vent (the surface that would have been visible from the outside while it was still attached to the functioning tank) was open and free of any obstruction. However, from the point the nozzle was attached to the top head, through the flange and up to the first elbow, the vent was encrusted with a white solid that tested positive for NaOH. The morphology of the solid was consistent with a slow accumulation over an extended period of time, rather than the result of an incident of overfilling the tank and then freezing in the vent. Moreover, there was no evidence of accumulation or discoloration on the head under the discharge of the vent, which would have been consistent with discharge through the vent onto the surface of the conical head.

The accumulation of solids within the vent choked it down to a diameter of about a half inch, and the rough, convoluted surface of the choked vent would have rendered it less useful than a clean half-inch pipe as a vent.

Figure 3. Vent nozzle, showing build-up of caustic.

Likewise, there were no solids built up on the interior walls of the vessel or on the interior surface of the head. This suggests that the buildup inside the vent line was the result of an entrained mist of 50% caustic solution being carried to the vent, where it collected and evaporated, leaving behind solid NaOH. The monthly vent inspections conducted on the tank since it was put into service in late 2003 would not have detected the gradual accumulation of solids.

Figure 4. View of interior of the caustic day tank after the incident. The thin white solid is caustic that dried there at the time of the incident. The interior shows some staining, but no build-up of the NaOH encrustation seen in the vent nozzle.

A thorough inspection of the vents of the two caustic tanks in the area showed no similar build-up of solids. However, one of the caustic storage tanks was put into service about 5 years before the incident and the other caustic storage tank was put into service about a year before the incident, so there would not have been the same opportunity for buildup.

The caustic day tank was specified to be constructed of 304 stainless steel. A field inspection of the vessel was consistent with that specification. The metal looked like stainless steel and was not magnetic; short of testing the metallurgy, this was considered enough to conclude that the tank was built as specified. Suppliers of 50% caustic recommend that carbon steel is a suitable material of construction for up to 140°F; over 140°F, they recommend 304 stainless steel to provide good corrosion resistance. This tank was operated at ambient temperatures. There was no reason, then, to believe that temperature or corrosion played any role in this incident.

The conical head separated completely. An inspection of the head showed that the separation was at the weld of the head to the 3"x3" angle that formed the top flange of the vertical cylinder of the tank. While the failure of an atmospheric tank is typically expected to be in the form of a split, where pressure is released without accompanying shockwave or shrapnel, an explosive separation of the head from the body of the vessel is not unheard of, e.g., Barton Solvents, Valley Center, Kansas, on July 17, 2007 [20].

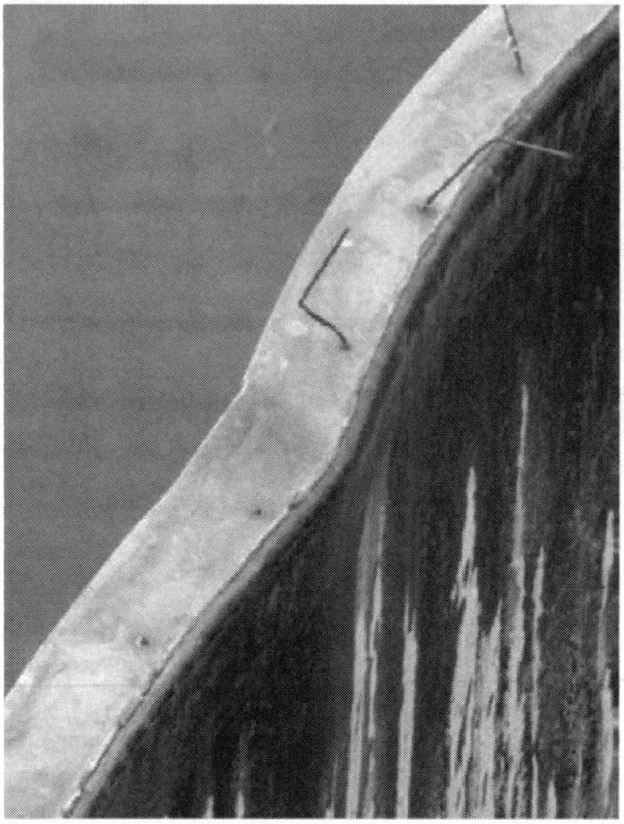

Figure 5. The separation of the head from the body of the caustic day tank was at the weld of the head to the 3"x3" angle that formed the top flange of the vertical cylinder of the tank.

3.4.3 Noted from interviews

The pressure gauge on the process air was not read at the time of the incident. The production supervisor noted that the pressure should be set at around 35 to 40 psig, while the maintenance engineer stated that the process air pressure is normally around 80 psig, and that the regulator was normally wide open, with little motivation to crank it back closed. This is consistent with the pressure reading observed in the field, which indicated that the process air pressure was at 90 psig. The caustic day tank itself has a design pressure of atmospheric (0 psig) and was pneumatically tested at 5 psig at the time it was manufactured in 2003.

The supply pressure for the city water used to flush the transfer line is around 40 to 60 psig. This means that the line could have been successfully flushed into the caustic day tank, even if the pressure in the tank had risen to 35 psig during the initial purge. During the second process air purge, which lasted for around 45 minutes before the incident, there would have been more time for the pressure to build.

3.5 Contributing Factors

3.5.1 Non-contributing factors

The investigation was able to eliminate many of the generic issues identified by caustic solution suppliers.

In regard to possible sources of high pressure, the materials of construction – 304 stainless steel and PVC – make hydrogen formation as a result of reaction with reactive metals an extremely unlikely factor. This is further supported by the absence of any evidence of combustion during the incident. Likewise, neither steam nor a pump were in use, so steam pressure and pump deadhead pressure did not factor in. Temperatures were well above 65°F, the temperature at which frozen 50% caustic solution become a concern.

The purging procedure did include a step where water was added to strong caustic, a step that is normally discouraged. However, this did not create the high pressure that resulted in the incident, and overall, is a reasonable way to deal with the flush water.

In regard to possible causes of blockage in caustic equipment, there were no block valves to be inadvertently closed, and ambient temperatures were high enough to prevent lines from being plugged with frozen caustic solution.

In regard to possible causes of equipment failure at lower than design pressure, the materials of construction – 304 stainless steel and PVC – also eliminate the concerns with caustic embrittlement and with temperatures above 140°F. There was no unsupported plastic pipe to fail, and steam was not in use.

3.5.2 Direct causes

A "direct cause" is the uncontrolled release of energy or hazardous material that cannot be absorbed safely. In the case of this incident, the direct cause was high pressure in the caustic day tank, the source of which was the approximately 80 to 90 psig process air used to blow down the transfer line from the caustic storage tank to the caustic day tank.

3.5.3 Indirect causes

An "indirect cause" is the action or condition that results in direct causes. In the case of this incident, the indirect cause was the plugged tank vent on the caustic day tank. A contributing factor was probably the extended time that process air was allowed to blow into the tank. However, the vent was gradually becoming clogged and eventually, even shorter periods would have been sufficient to overpressure the vessel.

3.5.4 Basic causes

A "basic cause" is the policy, procedure, or decision that leads to indirect or direct causes. There were three basic causes. The first basic cause was the procedure that called for blowing fluids into the caustic day tank at pressures higher than the rated pressure of the tank. While pressures higher than atmospheric pressure, either as a result of pressurized fluids or pumping, are absolutely necessary to effect transfer into the tank, the higher pressure also made a functioning atmospheric vent an imperative element of normal operation.

The second basic cause was a tank design and operation that resulted in carrying entrained 50% caustic solution mist to the vent, where it could deposit and plug the vent.

The third basic cause was the vent inspection procedure. In hindsight, it is now clear that monthly visual inspection did not detect that the vent was becoming plugged. In the absence of this hindsight, there is nothing to fault the vent inspection procedure. Caustic solution is not volatile, and there is no general expectation in the industry that vents on caustic solution tanks are subject to plugging. However, now that the problem has been identified, it should be addressed.

3.6 Recommendations

3.6.1 Vent design to assure pressure relief is adequate

Dow Chemical has the following recommendation in regard to vents on caustic solution storage tanks: "In addition to a solution high-level alarm, the tank should have a liquid-level gauge and a vent. The vent should have a cross-sectional area at least four times that of the fill line (which is normally 2"), extending from the top of the tank to within about 3' of ground level. The vent, which also acts as an overflow, should be visible from the tank truck or tank car loading area so that personnel will see and/or hear caustic discharging from this overflow pipe in case the tank is overfilled. Valves should never be installed in the vent line. It must be open to the atmosphere at all times" [18].

In this particular case, the vent and fill are both 3-inch lines. The vent pipe is a gooseneck, rather than a pipe that extends to within 3 feet of ground level. Modifications to the vent piping may be helpful.

3.6.2 Vent inspection to catch any buildup

It was unexpected to discover that the vent was plugged with caustic. The likely cause of the plugging was the deposit of entrained mist in the vent. The mist, composed as it was of 50% caustic solution, would be the source of solids. The practice of blowing process air into the tank through the dip pipe assures that there for mist to be entrained after every transfer into the tank. The evidence is that process air carries the mist all the way to the vent. As a result, this is an issue that should be expected in any tank containing solutions of dissolved solids that can be entrained as mist.

In this particular case, the purpose of the vent inspection was to look for evidence of external blockage, primarily as the result of natural activity, e.g., bird nests, wasp nests, etc. In the case of tanks storing solutions of dissolved solids, a more thorough inspection of the vents is in order. This may involve the use of boroscopes and cameras, or as an alternative, the installation of inspection and clean-out ports in the vent line. The frequency of inspection should be adjusted as experience dictates, beginning with quarterly inspection, but probably settling on annual inspection.

3.6.3 Limit use of dip pipes for inlet piping

The practice of blowing process air or nitrogen into a tank through a sub-surface dip pipe assures that mist will be entrained. While the density of aqueous solutions such as 50% caustic is high enough to lead to the expectation that mist droplets will settle out before reaching a vent, especially if the level is low enough, droplets that are small enough will be carried to the vent, regardless of the level. Operating at lower liquid levels will slow down the rate of blockage, but not stop it, and the improvement would come at the cost of lost capacity.

A more effective solution is to eliminate the dip pipe so that process air is not blown through the caustic solution to create a mist in the first place. Neither Dow Chemical nor PPG Industries show dip pipes in their example configurations of storage tanks [18], [19]. Because aqueous solutions of solids are neither flammable nor non-conducting, the usual concerns regarding static discharge from free-falling liquid are simply not an issue.

In this particular case, the dip pipe also serves to avoid erosion of the head and vessel walls that result from feeding directly through a tank nozzle. Cutting the dip pipe off a few inches below the entrance to the head and fitting it with a slight elbow to direct the free-falling liquid away from the wall still avoids erosion problems but eliminates the mist that forms by sparging process air through the caustic solution.

3.6.4 Pressure regulation to reduce potential for overpressure

Atmospheric tanks are frequently connected to high pressure sources. Examples include fill lines, where the pressure can be as high as the deadhead pressure of the transfer pump; water supplies, which typically run around 40 to 60 psig; and plant air, which is often in the range of 80 to 120 psig. Any one of these can be a source of overpressure, but to regulate them down to atmospheric pressure would render them useless. While the source pressure must be above atmospheric pressure in order for these connections to work, they should not be any higher than necessary.

In this particular case, the plant air supply used to blow out the lines would have been just as effective at 20 psig as it was at 90 psig. Regulating the air supply pressure down creates no burden. The water pressure and the transfer pump pressure need to be where they are.

4 Conclusions

Atmospheric tanks are relatively fragile, especially when compared to vacuum vessels, low pressure vessels, and high pressure vessels. They are also widely used and considered relatively benign. However, because they are considered relatively benign, they are often overlooked as a hazard and neglected as a maintenance priority.

When atmospheric tanks contain flammable materials, the consequence of concern is most often a tank fire. Beyond that, the typical consequence of concern associated with atmospheric tanks is a leak. The impacts of these events are usually limited to injuries and low-consequence impacts to the environment, although there have been some notable exceptions [21], [22]. Rarely is the concern one of sudden, catastrophic failure. However, an atmospheric tank can fail suddenly and catastrophically, with life-threatening consequences. The catastrophic failure can result in a collapse, an implosion, or an explosion. Collapses result when a tank is badly installed or when a tank is misused. The best way to avoid collapses is to first make sure the tank is properly designed and installed, and then to use it within the limits of that design.

Explosions can result from the ignition of flammable vapors, but surprising to many, can also occur for purely physical reasons. Implosions result primarily when sources of vacuum are overlooked or neglected. Atmospheric tanks are called atmospheric tanks because they vent to atmosphere. When that vent becomes inoperable, it becomes all too easy for the pressure in an atmospheric tank to become lower or higher than the thin walls of the tank can withstand. So it is imperative for operators to understand what causes a vent to become inoperable, and to prevent that from happening. The response in a HazOp to "pressure-too high" or to "pressure-too low" cannot be, "That can't happen, we're open to atmosphere."

5 Recommendations

5.1 Regarding catastrophic tank collapse

To avoid catastrophic vessel collapse, apply rigorous management of change and pre-startup safety review when putting an atmospheric tank into new or different service. In particular, pay attention to the impact of using a tank for higher density liquids. When a tank is first put into service is the time when the tank is most vulnerable. Resist the temptation to fast-track atmospheric tank commissioning when it is being filled for the first time; act as though you know it will fail and deploy personnel accordingly.

Once an atmospheric tank has been successfully put into service, it should provide good service. The greatest concern will be wall thinning that compromises the strength of the vessel walls. Ideally, the thinning will be uniform and caught with a minimum of testing. However, uniform thinning is rarely the actual experience, so special attention needs to be paid to localized thinning due to erosion, which can only be detected by looking for it. Pay attention to where nozzles, dip pipes, and eductors are aimed and confirm that weak spots are not being created.

5.2 Regarding catastrophic tank implosion

Avoiding catastrophic tank implosion requires identifying and controlling sources of vacuum and assuring that measures to protect tanks from implosion remain operable. Obvious vacuum sources include vacuum pumps, pump out and other forms of emptying a tank of its liquid contents, and condensing vapors (typically, but not limited to, steam). When the presence of any of these is unavoidable, they should only be present with full awareness of their hazard and the total reliance on the protection device.

All of the devices to protect atmospheric tanks against implosion have mechanisms by which they can fail. Facilities that take credit for these devices – atmospheric vents, conservation vents, inerting pads, and vacuum breakers – need to periodically validate that they are working as designed. The means to validate the functionality of protective devices needs to be designed with the same care with which the device itself is designed.

5.3 Regarding catastrophic tank explosion

In addition to the obvious concerns of flammable vapor combustion, avoiding catastrophic tank explosion requires identifying and controlling sources of pressure. Obvious pressure sources include high inlet gas pressure, high inlet liquid pressure, and heating. Again, the presence of any of these may be unavoidable, so they should only be present with full awareness of their hazard and the importance of the protective devices.

As with any device, the devices to protect atmospheric tanks against overpressure and explosion have mechanisms by which they can fail. The term "fail-safe" does not mean they cannot fail; it means that in the absence of energy sources, they are supposed to go to the safe state. It does not mean that they cannot fail to a dangerous state. Facilities that take credit for these devices – typically atmospheric vents and conservation vents – need to periodically validate that they are working as designed. As with all protective devices, the means to validate the functionality of protective devices needs to be designed with the same care with which the device itself is designed.

For atmospheric vents in particular, the design needs to allow for easy inspection of the entire vent. While this may involve the use of boroscopes and cameras, the installation of inspection and clean-out ports in the vent line may be more cost effective and easier to implement. It is not enough to have inspection measures, however. They must be used and at a sufficient frequency to detect a developing problem in time to address it. The frequency of inspection may be adjusted as experience dictates, but two failed inspections in a row indicates that the test interval is too long.

5.4 Regarding use of dip pipes in atmospheric storage tanks

Dip pipes have two important features when used in atmospheric storage tanks. The most important is that when the liquid being stored is both non-polar and flammable, dip pipes prevent free-falling liquid, which is known to create static sparks, a source of ignition of the flammable vapors. The other feature is that dip pipes direct liquid away from areas of the tank vulnerable to erosion. When these features are not required, dip pipes should not be used.

This is of particular concern when the liquid being stored is a solution of dissolved solids. When the line and its dip pipe are blown clear, which is a fairly common practice, the gas used to blow the line clear will sparge through the liquid and carry some quantity of mist to the vent. At the vent, the droplets will collect, evaporate, and leave a deposit of solids. Unless inspections explicitly look for this deposit, it is likely to go unnoticed. When dip pipes are installed and their use with solutions of dissolved solids involves blowing them out, vent inspections for blockage should not be limited to looking for nests or other biological activity, but should also include inspection for solids build-up.

6 References

[1] Chang, J.I. and C-C Lin. "A study of storage tank accidents," *Journal of Loss Prevention*, 19 (2006) 51-59.

[2] Atherton, W. and J.W.Ash. *Review of Failures, Causes & Consequences in the Bulk Storage Industry*, National Lightning Safety Institute. 2006. http://lightningsafety.com/nlsi_lls/Causes-of-Failures-in-Bulk-Storage.pdf (accessed 02-Sep-2015).

[3] EPA. "Catastrophic Failure of Storage Tanks," *Chemical Safety Alert*, EPA 550-F-97-002b, Chemical Emergency Preparedness and Prevention Office, USEPA, Washington, DC. May 1997. www.nrt.org/production/nrt/nrtweb.nsf/AllAttachmentsByTitle/A-107tanks/$File/tanks.pdf?OpenElement (accessed 02-Sep-2015).

[4] Marsh. "Atmospheric Storage Tanks," *Risk Engineering Position Paper-01*, Rev 1.0, Marsh & McLennan Companies, February 2011. uk.marsh.com/Portals/18/Documents/Atmospheric%20Storage%20Tanks_lowres.pdf (accessed 02-Sep-2015).

[5] OGP. "Storage incident frequencies," Risk Assessment Data Directory, Report No. 434-3, International Association of Oil & Gas Producers, March 2010. http://www.ogp.org.uk/pubs/434-03.pdf (accessed 02-Sep-2015).

[6] ASME. *Boiler and Pressure Vessel Code, Section VIII-Rules for Construction of Pressure Vessels*, 2015 Ed. American Society of Mechanical Engineers, New York, NY. 2015.

[7] API. *API 620: Design and Construction of Large, Welded, Low-pressure Storage Tanks*, 12th Edition. American Petroleum Institute, Washington, DC. 2013.

[8] API. *API 650: Welded Tanks for Oil Storage*, 12th Edition. American Petroleum Institute, Washington, DC. 2013.

[9] OSHA. *29 CFR 1910.119, Process safety management of highly hazardous chemicals.*, Occupational Safety and Health Administration, U.S. Department of Labor, Washington, DC. Last revised 08-Feb-2013.

[10] API. *API 2000: Venting Atmospheric and Low-Pressure Storage Tanks*, 7th Edition. American Petroleum Institute, Washington, DC. 2014.

[11] Ennis, T. "Pressure Relief Considerations for Low-Pressure (Atmospheric) Storage Tanks," *IChemE Symposium Series No. 151*, Institution of Chemical Engineers, Rugby, UK. https://www.icheme.org/communities/subject_groups/safety%20and%20loss%20prevention/resources/hazards%20archive/~/media/Documents/Subject%20Groups/Safety_Loss_Prevention/Hazards%20Archive/XIX/XIX-Paper-63.pdf (accessed 02-Sep-2015).

[12] Schworm, P. "Nearly a century later, structural flaw in molasses tank revealed," The Boston Globe, January 14, 2015. www.bostonglobe.com/metro/2015/01/14/nearly-century-later-new-insight-into-cause-great-molasses-flood/CNqLYc0T58kNo3MxP872iM/story.html (accessed 15-Feb-2016).

[13] CSB. *Investigation Report: Allied Terminals, Inc.-Catastrophic Tank Collapse*. No. 2009-03-I-VA, U.S. Chemical Safety and Hazard Investigation Board, Washington, DC, May 2009. 49 pp.

[14] Savage, A. "On MythBusters' FIRST-EVER Implosion," *TESTED*. January 22, 2016. www.tested.com/art/makers/561346-mythbusters-first-ever-implosion/ (accessed 16-Feb-2016).

[15] FACE. "Worker Dies While Repairing a Vacuum Evaporator Tank in Virginia," FACE 8725. Fatal Accident Circumstances and Epidemiology Project, NIOSH, CDC, Atlanta, GA. October 15, 2014. www.cdc.gov/niosh/face/In-house/full8725.html (accessed 16-Feb-2016).

[16] Clouse, T. "Ruling upheld in sewage death: Supreme Court says CH2M Hill liable for $6.5 million," The Spokesman-Review, Spokane, WA, May 27, 2011. www.spokesman.com/stories/2011/may/27/ruling-upheld-in-sewage-death/ (accessed 16-Feb-2016).

[17] CCPS. *Guidelines for Engineering Design for Process Safety*, Center for Chemical Process Safety, American Institute of Chemical Engineers, New York, NY. 2010. p. 110.

[18] Dow. *Caustic Soda Solution Handbook*, Form Number 102-00011-0810XBBI, The Dow Chemical Company, Midland, Michigan. August 2010. 23 pp.

[19] PPG. *NaOH-Caustic Soda*, #0100 Rev 0406 pdf, PPG Industries, Inc., Monroeville, Pennsylvania. 2008. 63 pp.

[20] CSB. *Case Study: Barton Solvents*. No. 2007-06-I-KS, U.S. Chemical Safety and Hazard Investigation Board, Washington, D.C., June 26, 2008. 14 pp.

[21] DoJ. *Former Freedom Industries official sentenced for role in chemical spill.* U.S. Attorney's Office, Southern District of West Virginia, February 1, 2016. www.justice.gov/usao-sdwv/pr/former-freedom-industries-official-sentenced-role-chemical-spill (accessed 15-Feb-2016).

[22] Asahi. Police push for charges against TEPCO, execs over pollution problems after nuclear disaster. The Asahi Shimbun, Asia & Japan Watch. October 3, 2015. http://ajw.asahi.com/article/0311disaster/fukushima/AJ201510030052 (accessed 15-Feb-2016).

Explosions in SMPO Hydrogenation Unit

Presenter/Principal Author: Nick Gonzales
Affiliation: Shell Global Solutions
Address: 3333 Hwy 6 South, Houston TX 77082
Email address:nick.gonzales@shell.com

Prepared for Presentation at
American Institute of Chemical Engineers
2016 Spring Meeting
12th Global Congress on Process Safety
Houston, Texas
April 11-23, 2016

LEARNING FROM INCIDENTS
AWARENESS ALERT

For Industry-Wide Sharing — September 2015

Explosions in SMPO Hydrogenation Unit

Figure 1 – Hydrogenator structure after the explosion.

Target audience for this alert

- Process Safety and Process Engineers
- Production staff in Refining/Chemical and hydrogenation processes

What happened

On June 3rd 2014, a Styrene Monomer and Propylene Oxide (SMPO) unit was in the pre start-up phase after the scheduled shutdown and replacement of catalyst in the hydrogenation reactors. The reaction system contained two fixed-bed adiabatic, trickle-flow reactors. Operations were heating up the hydrogenation unit in preparation for the catalyst reduction step. The heat-up was performed by the circulation of Ethyl Benzene (EB) through the reaction system under a nitrogen blanket and continuous nitrogen purge. During this heat up, one hydrogenation reactor and an additional upstream separation vessel suddenly exploded. Two contractor workers, who were in the immediate process area, were injured with burns from hot catalyst particles. The explosion caused extensive damage requiring over a year to repair.

Why it happened

The incident investigation concluded that both the reactor and vessel had failed because of overpressure, resulting from unexpected exothermic reactions between Ethyl Benzene (EB) and copper chromite oxide catalyst which produced an excessive amount of gas.

Prior to the incident, it was generally believed that under inert nitrogen conditions EB would not react with the passivated catalyst. However, it was subsequently determined and proven in the laboratory that two exothermic reactions actually occurred during the heat up:

- EB reacted with active oxygen present in the catalyst (mainly but not limited to copper chromate $CuCr^{VI}O_4$). The catalyst contained more copper chromate than the previous catalyst used at the site. The redox reaction started when the temperature exceeded approximately 90°C. The target temperature at the end of the heat up phase was 130°C, but due to the exothermic reaction the temperature increased further in local dry zones of the trickle bed reactors.
- Copper oxide ($Cu^{II}O$) was reduced by EB when localized temperatures reached 180°C. This redox reaction generated large amounts of heat and gaseous molecules (such as CO_2 and H_2O) leading to the overpressure.

AWARENESS ALERT

The early heat generation by the EB-chromate reaction, in combination with reduced cooling capacity in localized dry zones created localized temperatures in the reactor bed, high enough to initiate the EB-copper oxide reaction. Contributing factors include:

Low nitrogen flow: In a trickle flow reactor, sufficient nitrogen flow is required to achieve a proper liquid distribution from the distribution trays to wet the catalyst in the catalyst bed. The actual nitrogen flow was lower than needed, which led to non optimum liquid distribution in both trickle bed reactors causing dry zones in the top part of the catalyst beds.

Low EB flow: The EB flow to the hydrogenator was, at times, lower than needed for proper wetting of the catalyst and the flow variations led to preferential or rivulet flow and stagnant wet zones in the trickle flow reactor catalyst bed. These conditions led to non optimal heat removal and as a consequence high localized temperatures

Increasing Pressure: About thirty minutes before the explosion, the vent gas valve on the second separator tripped closed on high level and stayed closed. The system pressure increased up to the nitrogen supply pressure and caused the nitrogen flow to the reactor to decrease. The increasing pressure reduced the evaporative cooling of the EB from the catalyst.

Reactive hazard understanding: A Reactive Hazard Assessment (RHA) had been conducted in 2011 for this section of the SMPO unit. The RHA was based on process development work, design knowledge, literature search and application of a tool that was developed and used for identifying unintended reactions. The RHA only identified the reactions with hydrogen during the reduction phase. The reactions of EB with the catalyst in the absence of hydrogen were not identified in the RHA. This, together with the previous operational reduction experiences, reinforced the belief that EB was an inert medium in combination with the catalyst.

Lessons learned

- The passivated copper chromite hydrogenation catalyst contains bonded oxygen in the form of copper chromate and copper oxide prior to reduction. Where a hydrocarbon and oxidizing substance (i.e. both a "fuel" and reactive oxygen) are combined and in sufficient quantities, an exothermic reaction may occur and the only mitigation available may be the total amount of cooling provided.
- Heating up in EB, prior to the reduction step with hydrogen, was seen as normal and safe from a chemical reaction point of view. This influenced the choices made during the process design, process development work, the design of the trickle bed reactor, qualification of the catalyst and operational procedures.
- The purpose of a Reactive Hazard Assessment (RHA) is to identify consequences of the potential reactions, intended and unintended, between the components of the reaction streams. At the time of the Assessment, The composition of the reactants and the potential initiation temperatures were not understood adequately. Post incident investigation suggests that suitable analytical tools such as Thermogravimetric Analysis (TGA) may help to identify such exothermic reactions.
- A series of incremental and seemingly innocuous changes can allow unknown reactions to proceed and run out of control, resulting in a severe incident. Introduction of a new catalyst or catalyst supplier can introduce new components or changes even in small quantities of components that could lead to unidentified and unintended chemical reactions. The catalyst selection and subsequent change processes should address these with the focus on reactive hazards in both transient and steady state operations. Note that this may also apply to changes in process chemicals in other services (e.g. polymerization inhibitors, cleaning solutions, etc.).

AUTHOR INDEX

AUTHOR INDEX

AUTHOR INDEX